Y0-AFY-564

The chemistry of the

**metal–carbon bond**

Volume 2

# THE CHEMISTRY OF FUNCTIONAL GROUPS

*A series of advanced treatises under the general editorship of Professor Saul Patai*

The chemistry of alkenes (2 volumes)
The chemistry of the carbonyl group (2 volumes)
The chemistry of the ether linkage
The chemistry of the amino group
The chemistry of the nitro and nitroso groups (2 parts)
The chemistry of carboxylic acids and esters
The chemistry of the carbon–nitrogen double bond
The chemistry of amides
The chemistry of the cyano group
The chemistry of the hydroxyl group (2 parts)
The chemistry of the azido group
The chemistry of acyl halides
The chemistry of the carbon–halogen bond (2 parts)
The chemistry of the quinonoid compounds (2 parts)
The chemistry of the thiol group (2 parts)
The chemistry of the hydrazo, azo and azoxy groups (2 parts)
The chemistry of amidines and imidates
The chemistry of cyanates and their thio derivatives (2 parts)
The chemistry of diazonium and diazo groups (2 parts)
The chemistry of the carbon–carbon triple bond (2 parts)
The chemistry of ketenes, allenes and related compounds (2 parts)
The chemistry of the sulphonium group (2 parts)
Supplement A: The chemistry of double-bonded functional groups (2 parts)
Supplement B: The chemistry of acid derivatives (2 parts)
Supplement C: The chemistry of triple-bonded functional groups (2 parts)
Supplement D: The chemistry of halides, pseudo-halides and azides (2 parts)
Supplement E: The chemistry of ethers, crown ethers, hydroxyl groups and their sulphur analogues (2 parts)
Supplement F: The chemistry of amino, nitroso and nitro compounds and their derivatives (2 parts)
The chemistry of the metal—carbon bond Volume 1
The chemistry of peroxides

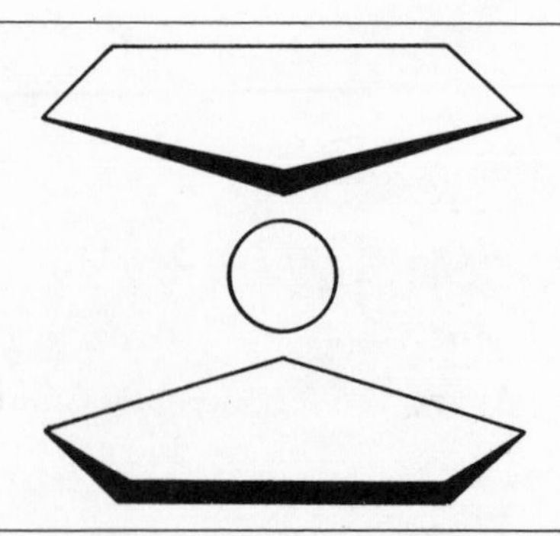

# The chemistry of the
# **metal–carbon bond**
## Volume 2
## The nature and cleavage of metal—carbon bonds

*Edited by*

FRANK R. HARTLEY
*The Royal Military College of Science*
*Shrivenham, England*

and

SAUL PATAI
*The Hebrew University*
*Jerusalem, Israel*



**1985**
JOHN WILEY & SONS
CHICHESTER–NEW YORK–BRISBANE–TORONTO–SINGAPORE
*An Interscience ® Publication*

***Library of Congress Cataloging in Publication Data***

Main entry under title:

The Nature and cleavage of metal—carbon bonds.

(The Chemistry of the metal—carbon bond; v. 2)
(The Chemistry of functional groups)
'An Interscience publication.'
Includes indexes.
1. Organometallic compounds. 2. Chemical bonds.
I. Hartley, F. R. II. Patai, Saul. III. Series.
IV. Series: Chemistry of functional groups.
QD410.C43 1984 vol. 2 547′.05s [547′.05] 83-14504
[QD411]

***British Library Cataloguing in Publication Data***

The Chemistry of the metal—carbon bond.—
(The chemistry of functional groups)
Vol. 2: The nature and cleavage of metal–carbon bonds
1. Organometallic compound
I. Hartley, Frank II. Patai, Saul
III. Series
547′.05 QD411

ISBN 0 471 90282 9

Filmset and printed in Northern Ireland by The Universities Press (Belfast) Ltd., and bound by the Pitman Press, Bath, Avon.

# Volume 2—Contributing authors


| | |
|---|---|
| Michael H. Abraham | Chemistry Department, University of Surrey, Guildford, Surrey GU2 5XH, UK |
| John J. Alexander | Department of Chemistry, University of Cincinnati, Cincinnati, Ohio 45221, USA |
| Philip J. Barker | Research School of Chemistry, Australian National University, P.O. Box 4, Canberra, A.C.T. 2600, Australia |
| Ron J. Cross | Chemistry Department, The University, Glasgow G12 8QQ, Scotland, UK |
| Priscilla L. Grellier | Chemistry Department, University of Surrey, Guildford, Surrey GU2 5XH, UK |
| Louis S. Hegedus | Department of Chemistry, Colorado State University, Fort Collins, Colorado 80523, USA |
| Mike D. Johnson | Department of Chemistry, University College London, 20 Gordon Street, London WC1H 0AJ, UK |
| John P. Oliver | Department of Chemistry, Wayne State University, Detroit, Michigan 48202, USA |
| Eric S. Paterson | Department of Chemistry, University of Aberdeen, Meston Walk, Old Aberdeen AB9 2UE, Scotland, UK |
| Chris J. Pickett | ARC Unit of Nitrogen Fixation, University of Sussex, Brighton, East Sussex BN1 9RQ, UK |
| John K. Stille | Department of Chemistry, Colorado State University, Fort Collins, Colorado 80523, USA |
| Jim L. Wardell | Department of Chemistry, University of Aberdeen, Meston Walk, Old Aberdeen AB9 2UE, Scotland, UK |
| Jeremy N. Winter | Department of Chemistry, University of York, Heslington, York Y01 5DD, UK |

# Foreword

*The Chemistry of the Metal–Carbon Bond* is a multi-volume work within the well established series of books covering *The Chemistry of Functional Groups*. It aims to cover the chemistry of the metal—carbon bond as a whole, but lays emphasis on the carbon end. It should therefore be of particular interest to the organic chemist. The general plan of the material will be the same as in previous books in the series with the exception that, because of the large amount of material involved, this will be a multi-volume work. The first volume was concerned with:

(a) Structure and thermochemistry of organometallic compounds.
(b) The preparation of organometallic compounds.
(c) The analysis and spectroscopic characterization of organometallic compounds.

The present volume is concerned with cleavage of the metal—carbon bond. It also includes the chapter on the structure and bonding of main group organometallic compounds deferred from the first volume. The chapter on the photochemical cleavage of metal—carbon bonds has not yet been completed and will be included in a later volume. Subsequent volumes will be concerned with the use of organometallic compounds for the formation of new carbon—carbon, carbon—hydrogen, and other carbon—element bonds. In classifying organometallic compounds we have used Cotton's hapto-nomenclature ($\eta -$) to indicate the number of carbon atoms directly linked to a single metal atom.

In common with other volumes in *The Chemistry of the Functional Groups* series, the emphasis is laid on the functional group treated and on the effects which it exerts on the chemical and physical properties, primarily in the immediate vicinity of the group in question, and secondarily on the behaviour of the whole molecule. The coverage is restricted in that material included in easily and generally available secondary or tertiary sources, such as *Chemical Reviews* and various 'Advances' and 'Progress' series as well as textbooks (i.e. in books which are usually found in the chemical libraries of universities and research institutes) is not, as a rule, repeated in detail, unless it is necessary for the balanced treatment of the subject. Therefore, each of the authors has been asked *not* to give an encyclopaedic coverage of his subject, but to concentrate on the most important recent developments and mainly on material that has not been adequately covered by reviews or other secondary sources by the time of writing of the chapter, and to address himself to a reader who is assumed to be at a fairly advanced postgraduate level. With these restrictions, it is realised that no plan can be devised for a vôlume that would give a *complete* coverage of the subject with *no* overlap between the chapters, while at the same time preserving the readability of the text. The Editors set themselves the goal of attaining *reasonable* coverage with *moderate* overlap, with a minimum of cross-references between the chapters of each volume. In this manner sufficient freedom is given to each author to produce readable quasi-monographic chapters. Such a plan necessarily means that the breadth, depth and thought-provoking nature of each chapter will differ with the views and inclinations of the author.

The publication of the Functional Group Series would never have started without the support of many people. Foremost among these is Dr Arnold Weissberger, whose reassurance and trust encouraged the start of the task and who continues to help and advise. This volume would never have reached fruition without Mrs Trembath's and Mrs Baylis's help with typing and the efficient and patient cooperation of several staff members of the Publisher, whose code of ethics does not allow us to thank them by name. Many of our colleagues in England, Israel and elsewhere gave help in solving many problems, especially Professor Z. Rappoport. Finally, that the project ever reached completion is due to the essential support and partnership of our wives and families.

FRANK HARTLEY
Shrivenham, England

SAUL PATAI
Jerusalem, Israel

# Contents

# List of Abbreviations Used

| | |
|---|---|
| A | appearance potential |
| acac | acetylacetone |
| ac | acrylonitrile |
| acacen | bis(acetylacetonate)ethylenediamine |
| aibn | azobisisobutyronitrile |
| all | allyl |
| An | actinide metal |
| ap | antiplanar |
| appe | $Ph_2AsCH_2CH_2PPh_2$ |
| bae | bis(acetylacetonate)ethylenediamine |
| bipy | 2,2′-bipyridyl |
| cd | circular dichrosim |
| cdt | *E,E,E*-cyclododeca-1,5,9-triene |
| cht | cycloheptatriene |
| CI | chemical ionization |
| CIDNP | chemically induced dynamic nuclear polarization |
| CNDO | complete neglect of differential overlap |
| coct | cyclooctene |
| cod | cycloocta-1,5-diene |
| cot | cyclooctatetraene |
| Cp | $\eta^5$-cyclopentadienyl |
| Cp* | $\eta^5$-pentamethylcyclopentadienyl |
| C.P. | cross-polarization |
| Cy | cyclohexyl |
| dba | dibenzylideneacetone |
| dbn | 1,5-diazabicyclo[5.4.0]non-5-ene |
| dbu | 1,8-diazabicyclo[5.4.0]undec-7-ene |
| dccd | dicylohexylcarbodiimide |
| def | diethyl fumarate |
| diars | *o*-bis(dimethylarsino)benzene |
| dibah | diisobutylaluminium hydride |
| diop | 2,3-*o*-isopropylidene-2,3-dihydroxy-1,4-bis(diphenylphosphino)butane |
| dme | 1,2-dimethoxyethane |
| dmf | dimethylformamide |
| dmfm | dimethyl fumarate |
| dmm | dimethyl maleate |
| dmpe | bis(1,2-dimethylphosphino)ethane |
| dotnH | bis(diacetylmonoxime)propylene-1,3-diamine |
| dpm | dipivaloylmethanato |
| dppb | bis(1,4-diphenylphosphino)butane |
| dppe | bis(1,2-diphenylphosphino)ethane |
| dppm | bis(1,1-diphenylphosphino)methane |

| | |
|---|---|
| dppp | bis(1,3-diphenylphosphino)propane |
| dmso | dimethyl sulphoxide |
| $E_{1/2}$ | half-wave potential |
| ece | electron transfer–chemical step–further electron transfer |
| ee | enantiomeric excess |
| EI | electron impact |
| $E_p$ | peak potential |
| ESCA | electron spectroscopy for chemical analysis |
| eV | electronvolt |
| Fc | ferrocene |
| FD | field desorption |
| FI | field ionization |
| fmn | fumaronitrile |
| fod | $F_3C(CF_2)_2COCH{=}C(O)CMe_3$ |
| Fp | $Fe(\eta^5\text{-}C_5H_5)(CO)_2$ |
| Fp* | $Fe(\eta^5\text{-}C_5H_5)(CO)(PPh_3)$ |
| FT | Fourier transform |
| hfac | hexafluoroacetone |
| hfacac | hexafluoroacetylacetonato |
| hmdb | hexamethyl(Dewar)benzene |
| hmpa | hexamethylphosphoramide |
| hmpt | hexamethylphosphorotriamide |
| HOMO | highest occupied molecular orbital |
| $I$ | ionization potential |
| ICR | ion cyclotron resonance |
| $I_D$ | ionization potential |
| INDOR | inter-nuclear double resonance |
| LCAO | linear combination of atomic orbitals |
| lda | lithium diisopropylamide |
| Ln | lanthanide metal |
| LUMO | lowest unoccupied molecular orbital |
| M | metal |
| $M$ | parent molecule |
| ma | maleic anhydride |
| MAS | magic angle spinning |
| *m*-cpba | *m*-chloroperbenzoic acid |
| MNDO | modified neglect of diatomic overlap |
| map | 2-methyl-2-nitrosopropane |
| ms | millisecond |
| nbd | norborna-1,5-diene |
| nbs | *N*-bromosuccinimide |
| nmp | *N*-methylpyrrolidone |
| oA | *o*-allylphenyldimethylarsine |
| Pe | pentenyl |
| phen | *o*-phenanthroline |
| pmdeta | Pentamethyldiethylenetriamine |

| | |
|---|---|
| ppm | parts per million |
| PRDDO | partial retention of diatomic differential overlap |
| psi | pounds per square inch |
| py | pyridyl |
| pz | pyrazolyl |
| RT | room temperature |
| salen | bis(salicylaldehyde)ethylenediamine |
| salophen | bis(salicylaldehyde)-*o*-phenylenediamine |
| SCE | Saturated calomel electrode |
| set | single electron transfer |
| SOMO | singly occupied molecular orbital |
| sp | synplanar |
| SPT | selective population transfer |
| tba | tribenzylideneacetylacetone |
| tcne | tetracyanoethylene |
| thf | tetrahydrofuran |
| tmeda | tetramethylethylenediamine |
| tms | tetramethylsilane (only used in this context as a free standing symbol) |
| tms | trimethylsilyl (only used in this context either with a dash after it or adjacent to a chemical symbol) |
| tond | 1,3,5,7-tetramethyl-2,6,9-trioxobicyclo[3.3.1]nona-3,7-diene |
| tos | tosyl |
| ttfa | thallium tristrifluoroacetate |
| un | olefin or acetylene |
| X | halide |

The Chemistry of the Metal—Carbon Bond, Volume 2
Edited by F. R. Hartley and S. Patai


CHAPTER 1

# Electrochemical cleavage of metal—carbon bonds

C. J. PICKETT
*ARC Unit of Nitrogen Fixation, University of Sussex, Brighton, East Sussex BN1 9RQ, UK*



## I. INTRODUCTION

Electron-transfer reactions of organometallic compounds can be studied in great detail by the application of modern electrochemical methods. Thermodynamic data can be obtained, products prepared by electrosynthesis, unstable intermediates generated, and their decay rates or spectroscopic properties measured. In addition, rates of heterogeneous electron transfer from or to a substrate in solution can be estimated and adsorption phenomena studied.

There are several excellent reviews and texts on the fundamentals and application of electrochemical methods to the study of the electrode reactions of molecules[1–6]; in particular the recent book by Bard and Faulkner is recommended[7]. The electrochemical reactions of coordination and organometallic compounds have been the subject of several reviews, although metal—carbon bond cleavage reactions have not been discussed specifically[8–11].

## II. THE METAL—CARBON SIGMA BOND

### A. Introduction

Transition metal and main group metal—carbon $\sigma$-bonds are commonly cleaved on electrochemical oxidation or reduction, and the organic products of such cleavage reactions depend on the nature and fate of the organic moiety produced. Anodic oxidation can lead to the formation of an organic free radical, to a carbonium ion, or to nucleophilic attack on the activated carbon centre. Similarly, cathodic reduction can give a radical or a carbonium ion or result in electrophilic attack. Generally, the anodic or cathodic reactions involve electrochemically irreversible one- or two-electron transfer as the primary oxidation or reduction step. There are some examples of metal—carbon bond cleavage preceded by electrochemically reversible charge transfer in cases where an electron is removed from, or added to, an orbital with little metal—carbon bond character.

Aryl and alkyl radicals, generated by anodic or cathodic cleavage reactions of organometallic compounds, often undergo independent oxidation or reduction chemistry to give carbonium ions or carbanions. Whether such electron-transfer reactions takes place is determined by the redox potentials of the $R^{\bullet}/R^{+}$ and $R^{\bullet}/R^{-}$ couples relative to that imposed at the anode or cathode, respectively. Data for simply alkyl and aryl radicals are limited, although there have been several attempts to estimate oxidation and reduction potentials of such radicals: $E_{1/2}$ for the $CH_3^{\bullet}/CH_3^{-}$ couple is reported[12] to be $-1.47$ V *vs.* SCE and that for the $C_6H_5CH_2^{\bullet}/C_6H_5CH_2^{-}$ couple to be $-1.35$ V.

### B. Main Group

#### 1. Oxidation

a. *Mechanisms.* Electrochemical oxidation of a range of homoleptic main group metal alkyls has been studied by Klinger and Kochi[13]. The metal alkyls listed in Table 1 undergo anodic oxidation at a Pt electrode in acetonitrile containing 0.1 M $[NEt_4][ClO_4]$ in an *overall* two-electron process which leads to the formation of a carbonium ion (equation 1). Analysis of cyclic voltammetric data led to the authors to conclude that the primary electron transfer step was a totally irreversible single electron transfer (equation 2).

$$M—R \xrightarrow{2e} M^{+} + R^{+} \tag{1}$$

$$M—R \xrightarrow[E_p]{e} M^{+} + R^{\bullet} \tag{2}$$

The heterogeneous rate constant, $k_E$, determined for the rate-limiting step (equation 3) at various electrode potentials, $E$, was found to correlate with the homogeneous

TABLE 1. Cyclic voltammetric oxidation peak potentials ($E_p$) and ionization potentials ($I_D$) for metal alkyls[13]

| Compound | $E_p$ (V) | $I_D$ (eV) |
|---|---|---|
| $Et_4Si$ | 2.56 | 9.78 |
| $Me_4Ge$ | 2.64 | 10.02 |
| $Et_4Ge$ | 2.24 | 9.41 |
| $Me_4Sn$ | 2.48 | 9.69 |
| $Et_4Sn$ | 1.76 | 8.93 |
| $EtSnMe_3$ | 1.96 | 9.10 |
| $Bu^s{}_4Sn$ | 1.45 | 8.45 |
| $Bu^i{}_4Sn$ | 1.77 | 8.76 |
| $Pr^i{}_4Sn$ | 1.51 | 8.46 |
| neopentyl$_4$Sn | 1.80 | 8.67 |
| $Bu^t{}_2SnMe_2$ | 1.25 | 8.22 |
| $Me_4Pb$ | 1.80 | 8.90 |
| $Et_4Pb$ | 1.26 | 8.13 |
| $Et_2PbMe_2$ | 1.56 | 8.45 |
| $Pr^n{}_2Hg$ | 1.39 | 8.29 |
| EtHgMe | 1.70 | 8.84 |

$$\text{M—R} \xrightarrow[e]{k_E} \text{M—R}^{+\bullet} \tag{3}$$

rate constants measured using various $Fe^{3+}$ oxidants of known formal potentials. It had been established earlier that the homogeneous oxidation for these alkyls was an *outer-sphere* process, independent of steric effects, and it was therefore concluded that the electrochemical oxidation at the electrode also involved simple outer-sphere electron transfer[13]. This conclusion was reinforced by the observation that the cyclic voltammetric oxidation peak potential, $E_p$, correlated linearly with ionization potential data, $I_D$ (Table 1), and showed no apparent deviation attributable to steric effects.

The overall two-electron oxidation for the metal alkyls according to equation 1, listed in Table 1, is a consequence of the rapid one-electron anodic oxidation of the organic radical (reaction 4), generated in the irreversible step shown in reaction 2.

$$\text{R}^\bullet \xrightarrow{-e} \text{R}^+ \longrightarrow \text{products} \tag{4}$$

From Table 1, it can be seen that the neutral metal alkyls oxidize at relatively positive potentials. Thus one-electron oxidation of the alkyl radical readily takes place at the potentials necessary to achieve the homolytic cleavage.

The primary oxidation of anionic homoleptic alkyls and aryls is more facile than the oxidation of neutral homoleptic alkyl species. At the potentials necessary to achieve anodic metal—carbon bond cleavage, the alkyl or aryl radicals generated are generally electro-*in*active. Thus, electrochemical oxidation of tetra-alkyl or-aryl aluminates is an overall one-electron process which leads to the formation of organic radicals which are not oxidized at the anode (equations 5 and 6)[14–16].

$$[\text{AlEt}_4]^- \xrightarrow{-e} [\text{AlEt}_3] + \text{Et}^\bullet \tag{5}$$

$$[\text{AlPh}_4]^- \xrightarrow{-e} [\text{AlPh}_3] + \text{Ph}^\bullet \tag{6}$$

The tetraphenylborate anion shows interesting behaviour. At a Pt electrode in an aqueous electrolyte it undergoes an irreversible two-electron oxidation with *intra*molecular carbon—carbon bond formation (equation 7).

$$MR_2 \xrightarrow[\text{Pt}]{-2e} M^{2+} + R—R \tag{7}$$

Electrolysis of a mixture of $[B(C_6H_5)_4]^-$ and $[B(C_6D_5)_4]^-$ produces only $C_6H_5—C_6H_5$ and $C_6D_5—C_6D_5$ with no cross-coupled $C_6H_5—C_6D_5$. In the presence of water the two-electron oxidation proceeds according to reactions 8 and 9[17].

$$[BPh_4]^- \xrightarrow{-2e} [BPh_2]^+ + Ph—Ph \tag{8}$$

$$[BPh_2]^+ \xrightarrow{H_2O} [BPh_2OH] + H^+ \tag{9}$$

b. *Organic products.* One-electron oxidation of main group alkyls and aryls according to reaction 2 generates a radical which can have a variety of fates (i–v).

(i) Hydrogen abstraction from a solvent molecule to produce an alkane (reaction 10):

$$R^{\cdot} \xrightarrow{H^{\cdot}} RH \tag{10}$$

(ii) Dimerization (reaction 11):

$$2R^{\cdot} \longrightarrow R—R \tag{11}$$

(iii) Disproportionation (reaction 12):

$$2RCH_2CH_2^{\cdot} \longrightarrow RCH{=}CH_2 + RCH_2CH_3 \tag{12}$$

(iv) Organometallic formation *via* attack on the electrode (reaction 13):

$$nR^{\cdot} \xrightarrow{M} [MR_n] \tag{13}$$

(v) Attack on a solution substrate (e.g. reaction 14):

$$R^{\cdot} + CH_2{=}CH_2 \longrightarrow RCH_2CH_2^{\cdot} \longrightarrow \text{products} \tag{14}$$

Here it can be noted that a radical generated by *cathodic* reduction will have the same independent solution chemistry.

The fate of radicals generated by one-electron oxidation of Grignard reagents {MgRX} (reaction 15) has been well studied[18–24].

$$\{MgRX\} \xrightarrow{-1e} R^{\cdot} \rightsquigarrow \text{products} \tag{15}$$

Although the earlier investigations of product distributions were not carried out potentiostatically, and the electroactive species present in {MgRX} solutions are not well defined, they do illustrate some of the factors which determine which of the pathway(s), (i–v), predominate. Whether or not the electroactive species present in Grignards are $[MgR_2]$, $[MgR_2X]^-$, $[MgRX_2]^-$, $[MgR_3]^-$, and/or [MgRX], all of which have been suggested as Schlenk equilibria species[21], there is little doubt that anodic oxidation of Grignard solutions generates radicals rather than carbonium ions.

Oxidation of {MgBrMe} in an ethereal electrolyte at a Pt anode gives $CH_4$ and $C_2H_6$ (reaction 16). The yields of methane and ethane formed by pathways (i) and (ii) are sensitive to the nature of the ether and to the concentration of the Grignard. Di-

$$\{MgBrMe\} \xrightarrow[Pt]{-e} CH_3^{\bullet} \begin{cases} \xrightarrow{(i)} CH_4 \\ \xrightarrow{(ii)} C_2H_6 \end{cases} \quad (16)$$

*n*-butyl ether gives a higher yield of methane than does diethyl ether, presumably because H$^{\bullet}$ abstraction from the former is more facile[18]. Oxidation of *ca.* 1.0 M solutions of {MgBrMe} in $Et_2O$ gives predominantly $CH_4$ whereas *ca.* 3.0 M solutions give predominately $C_2H_6$[19]. Here, higher current densities and hence concentrations of $CH_3^{\bullet}$ at the anode favour the second-order dimerization reaction.
In contrast, electrooxidation of ethyl- or *n*-propyl-magnesium bromides at a Pt electrode in $Et_2O$ gives mainly disproportionation products according to pathway (iii) (reactions 17 and 18)[20].

$$\{MgBrEt\} \xrightarrow[Pt]{-e} Et^{\bullet} \xrightarrow{(iii)} C_2H_6 + CH_2{=}CH_2 \quad (17)$$

$$\{MgBrPr^n\} \xrightarrow[Pt]{-e} Pr^{n\bullet} \xrightarrow{(iii)} CH_3CH{=}CH_2 + CH_3CH_2CH_3 \quad (18)$$

Longer chain and branched alkylmagnesium Grignards were found to oxidize in $Et_2O$ at a Pt anode to give predominantly dimerization products according to (ii) (reactions 19 and 20)[12].

$$\{MgBr(n\text{-hexyl})\} \xrightarrow[Pt]{-e} n\text{-}C_{12}H_{26} \quad (19)$$

$$\{MgBrBu^s\} \xrightarrow[Pt]{-e} \text{3,4-dimethylhexane} \quad (20)$$

Anodic oxidations of {MgBrR} on anodes such as Au, Ag, Cu, Fe, Ni, or Co proceed similarly to those on Pt. They may be considered as inert, although not necessarily indifferent anodes[9]. The formation of chemisorbed intermediates such as alkyl or hydrogen radicals and alkenes on the electrode surface may influence the kinetics of the dimerization (ii) or disproportionation (iii) pathways and hence the product distribution[9]. Anodes such as Hg, Sn, Pb, Cd, Zn, Al, and Mg are not inert and are attacked by the anodically generated radicals to give organometallic products. Thus oxidation of ethylmagnesium chloride at a sacrificial Pb anode yields tetraethyllead (reaction 21)[22].

$$\{MgClEt\} \xrightarrow[-e \ (iv)]{Pb} [PbEt_4] \quad (21)$$

The anodic discharge of organo-magnesium and -aluminium complexes at sacrificial anodes has received considerable attention as such processes have been of commercial interest in terms of lead tetraalkyl production and, in combination with cathodic deposition, metal refining[22].

The anodic oxidation of Grignard reagents in the presence of olefins has been studied by Schäfer and Küntzel[23]. Electrolysis of {MgBrBu$^n$} in $Et_2O$ containing $LiClO_4$ as a supporting electrolyte in the presence of styrene produces diphenyldodecanes at Pt or Cu anodes (reactions 22 and 23). Employing a diaphragm-less flow cell and working at current densities of *ca.* 10 mA cm$^{-2}$, current yields of 10 and

$$\{MgBrBu^n\} \xrightarrow[+\text{styrene}]{\text{Pt},-e} C_6H_5\dot{C}H(n\text{-pentyl}) + C_6H_5CH(Bu^n)CH_2^{\bullet} \quad (22)$$

$$CH_3(CH_2)_4CH(Ph)CH_2CH(Ph)(CH_2)_2CH_3 + CH_3(CH_2)_4CH(Ph)CH(Ph)(CH_2)_2CH_3 \quad (23)$$

5% for the 6,8- and 6,7-diphenyldodecane isomers, respectively, were obtained at Pt anodes. The current yield for the 6,7-isomer was found to be considerably higher on a Cu anode, although the reasons for this are not understood.

Oxidation potentials of the electroactive species present in solutions of Grignard reagents have not been adequately assessed by potentiostatic techniques. Empirically it has been found that decomposition potentials decrease as the branching of the alkyl group increases and that methyl magnesium Grignards are harder to oxidize than their ethyl analogues[24]. The latter observation is paralleled by $E_p$ data listed for the various homoleptic methyl and ethyl metal alkyls listed in Table 1.

Electrochemical oxidation of main group alkyls to generate carbonium ions has been poorly studied in terms of the identification of the organic products of electrolysis (reaction 4). It has been shown that the electrochemical oxidation of alkane boronates, $[B(OH)_3R]^-$, proceeds according to reactions 2 and 4 at a Pt electrode in an aqueous electrolyte (reaction 24)[17].

$$[B(OH)_3R]^- \xrightarrow[\text{Pt}]{-e} [B(OH)_3R]^{\bullet} \longrightarrow [B(OH)_3] + R^{\bullet} \xrightarrow{-e} R^+ \longrightarrow \text{products} \quad (24)$$

In this cleavage reaction the organic product distribution is similar to that observed in the Kolbe reaction which is known to proceed via carbonium ion formation (reactions 25 and 26).

$$RCOO^- \xrightarrow{-e} R^{\bullet} + CO_2 \quad (25)$$

$$R^{\bullet} \xrightarrow{-e} R^+ \longrightarrow \text{products} \quad (26)$$

The organic products derived from carbonium ions, generated by electrochemical oxidation of various *organic* molecules, have been thoroughly investigated[25] and these studies should well serve to illustrate the types of product to be expected from electrooxidation of metal alkyls and aryls according to reaction 1. In this, we assume that the fate of the free carbonium ion is independent of the nature of its precursor. The 3,3-dimethylbutyl cation[26] undergoes quenching reactions typical of a carbonium ion. These are as follows:

(vi) Nucleophilic attack (e.g. reaction 27):

$$Me_3CCH_2\overset{+}{C}H_2 \xrightarrow{OH^-} Me_3CCH_2CH_2OH \quad (27)$$

(vii) Base attack (e.g. reaction 28):

$$Me_3CCH_2\overset{+}{C}H_2 \xrightarrow[-H_2O]{OH^-} Me_3CCH{=}CH_2 \quad (28)$$

(viii) Fragmentation (e.g. reactions 29 and 30):

$$Me_3CCH_2CH_2^+ \longrightarrow Me_2C{=}CMe_2 + Me_3C^+ \quad (29)$$

$$Me_3C^+ \xrightarrow{H_2O} Me_2C{=}CH_2 + Me_3COH \quad (30)$$

(ix) Rearrangement (e.g. reaction 31):

$$Me_3CCH_2\overset{+}{C}H_2 \longrightarrow Me_3\overset{+}{C}CHME \xrightarrow{\text{(vi) or (vii)}} \text{products}$$
$$\downarrow$$
$$Me_2\overset{+}{C}CHMe_2 \xrightarrow{\text{(vi) or (vii)}} \text{products} \quad (31)$$

In $CH_3CN$ electrolytes, carbonium ion attack on the solvent gives *N*-alkylacetamides (reaction 32). This is presumably the fate of the carbonium ions produced by electrooxidation of the homoleptic alkyls listed in Table 1.

$$R^+ + MeCN \longrightarrow RN{=}\overset{+}{C}Me \xrightarrow{H_2O} RNHCOMe \quad (32)$$

The reaction of alkenes with aqueous $Hg^{II}$ salts ($SO_4^{2-}$, $ClO_4^-$ or $NO_3^-$) leads to Hg—C $\sigma$-bond formation (reaction 33). Anodic oxidation of hydroxyalkyls of this

$$Hg^{2+} + H_2O + R^1CH{=}CHR^2 \underset{}{\overset{-H^+}{\rightleftharpoons}} R^1CH(OH)CH(Hg^+)R^2 \quad (33)$$

type has been investigated by *in situ* generation of the organometallic in an electrochemical cell. The reaction of propene with $Hg^{2+}$ in 1 M $HClO_4$ generates an alkyl complex which is oxidized at a Pt anode to a range of carboxylic acids (reaction 34)[27,28].

$$[MeCH(OH)CH_2Hg^+] \xrightarrow[-ne]{Pt} \underset{33\%}{MeCOOH} + \underset{30\%}{HCOOH} + \underset{24\%}{EtCOOH} \quad (34)$$

The primary step in this conversion is believed to involve a two-electron oxidation of the metal alkyl with the formation of a carbonium ion (reaction 35).

$$[MeCH(OH)CH_2Hg^+] \xrightarrow{-2e} MeCH(OH)\overset{+}{C}H_2 + Hg^{2+} \quad (35)$$

The carboxylic acid products of the anodic oxidation were explained by invoking hydride or alkyl shifts followed by further oxidation of the carbonyl intermediates to the acid products (reactions 36 and 37).

$$[MeCH(OH)\overset{+}{C}H_2] \xrightarrow{-H^+} MeCH_2CHO \longrightarrow EtCOOH \quad (36)$$

$$[MeCH(OH)\overset{+}{C}H_2] \xrightarrow{-H^+} MeCOMe \longrightarrow \longrightarrow MeCOOH + HCOOH \quad (37)$$

### 2. Reduction

a. *Mechanisms and products.* The electrochemical reduction of organo-$Hg^{II}$ compounds has been investigated more thoroughly than any other main group organometallic species[29–31]. Their cathodic reactions therefore provide a useful starting point for a discussion of the factors pertinent to metal—carbon bond cleavage. In addition, organomercury intermediates are often formed in the reduction of other metal alkyls and aryls at a Hg *cathode* (see below).

Dialkyl and diaryl $Hg^{II}$ compounds of the type $[HgR_2]$ generally reduce in an irreversible two-electron process at a dropping mercury electrode with concomitant metal—carbon bond cleavage (reaction 38)[29,30].

$$[HgR_2] \xrightarrow[\text{dmf},H_2O \atop 2H^+]{2e} Hg + 2RH \qquad (38)$$

In protic media, the carbanions are probably not produced in the free state according to reaction 39. It is more likely that protonation of carbon occurs in the

$$[HgR_2] \xrightarrow{2e} Hg + 2R^- \qquad (39)$$

activated complex and is concerted with Hg—C bond cleavage. The polarographic $E_{1/2}$ values for the reduction of a wide range of $[HgR_2]$ compounds are listed in Table 2. Butin *et al.*[29] and Denisovich and Gubin[31] have analysed $E_{1/2}$ values in terms of the nature of R. Butin *et al.* suggest that there is a linear correlation between $\alpha E_{1/2}$ and the p$K_a$ values of RH, where $\alpha$ is the electron-transfer coefficient for the heterogeneous electron-transfer step. Denisovich and Gubin point out, however, that $E_{1/2}$ and the p$K_a$ of RH directly correlate since $\alpha$ shows little variation in a range of $[HgR_2]$ species. Both groups used their relationships to estimate unknown p$K_a$s of RH acids[29,31]. A linear correlation between Hammett's $\sigma$-constant and $E_{1/2}$ has been observed for compounds of the type $[Hg(CH_2C_6H_4X\text{—}p)_2]$; as is usually the case, the more withdrawing the substituent the easier the complex is to reduce. In addition, $E_{1/2}$ values for various $[HgR_2]$ have been shown to correlate with the electron affinity of the $R^\bullet$ radical[31].

Organomercury compounds of the type [HgRX], where X = Cl, Br, I, and their solvated cations $[HgR]^+$ are generally reduced in two *single*-electron steps which usually have well separated $E_{1/2}$ values (reactions 40–42)[20,29–31]. The $Hg^I$ radicals

$$[HgRX] \xrightarrow{e} [HgR]^\bullet + X^- \qquad (40)$$

$$[HgR]^+ \underset{-e}{\overset{e}{\rightleftharpoons}} [HgR]^\bullet \qquad (41)$$

$$[HgR]^\bullet \xrightarrow[H^+]{e} Hg + RH \qquad (42)$$

formed in reaction 41 are unstable and have been shown to undergo disproportionation reactions via dimeric intermediates (reaction 43).

$$2[HgR]^\bullet \longrightarrow \{Hg_2R_2\} \longrightarrow [HgR_2] + Hg \qquad (43)$$

TABLE 2. Polarographic half-wave potentials ($E_{1/2}$)[a]

| R | $-E_{1/2}$ ($HgR_2$) | $-E_{1/2}$ ($HgR^\bullet$) |
|---|---|---|
| $CH_3$ | 2.88 | 1.96 |
| Ferrocenyl | 2.68 | 1.84 |
| $C_6H_4OMe$ | 2.69 | 1.81 |
| $C_6H_4Me$ | 2.67 | 1.83 |
| $C_6H_5$ | 2.58 | 1.78 |
| $C_6H_4F$ | 2.51 | 1.65 |
| $CH_2{=}CHCH_2$ | 2.13 | 1.62 |
| $\sigma$-$C_5H_5$ | 0.72 | 0.75 |

[a] Conditions: 90% dioxane, 0.1 M $[NEt_4][ClO_4]$.

Dimerization is, however, a relatively slow process and reduction at the potential associated with the second electron transfer (reaction 42) produces RH. It should be noted that $[HgR_2]$ species formed according to reaction 43 are harder to reduce than the corresponding $[HgR]^{\bullet}$ intermediates, hence the irreversible two-electron reduction process associated with the former (Table 2). $E_{1/2}$ for the second-electron transfer of the compounds $[HgCl(p\text{—}XC_6H_4)]$ correlates with Hammett's $\sigma_p$ constant; again, the more withdrawing the substituent X, the easier is the reduction of $[HgR]^{\bullet}$[31].

The conclusions which have been drawn from the various relationships between $E_{1/2}$, the electron affinity of $R^{\bullet}$, and $pK_a$ values of RH are that for the $[Hg_2R_2]$ species, an electron is added to an orbital which has predominantly Hg—C antibonding character and, for $Hg^IR$ species, an electron is added to a half-filled Hg—C orbital. Following the analysis of Klinger and Kochi[13] (see above), it is reasonable to suggest that on a mercury cathode the intimate mechanism of reduction of the $[HgR_2]$ and $[HgR]^{\bullet}$ species is also an outer-sphere process.

Certain organo-mercury compounds show a modified cathodic behaviour whereby reduction in a single one-electron process to give a free-radical (reaction 44) or in a

$$[Hg(CCl_3)Cl] \xrightarrow[-Cl^-]{1e} Hg^I(CCl_3) \longrightarrow Hg + CCl_3^{\bullet} \qquad (44)$$

single two-electron process to give Hg and a stable organic species (reaction 45) takes place[24].

$$[Hg(CH{=}CHCl)Cl] \xrightarrow[-Cl^-]{+1e} \{Hg^ICH{=}CHCl\} \xrightarrow[-Cl^-]{+1e} Hg + CH{\equiv}CH \qquad (45)$$

In the former case, we can suppose that if the $Hg^I$—alkyl is formed, it decays at a rate considerably faster than the dimerization–disproportionation pathway (reaction 43). Similarly, if $\{Hg^I(CH{=}CHCl)\}$ is formed it is either at least as facile to reduce as the parent compound or rapidly loses $CH{\equiv}CH$ to generate an electroactive $Hg^I$ species (reaction 46).

$$\begin{array}{ccc} Hg^I(CH{=}CHCl) & \longrightarrow & Hg^I + CH{\equiv}CH + Cl^- \\ & & \downarrow {\scriptstyle +e \text{ fast}} \\ & & Hg \end{array} \qquad (46)$$

The reduction of $[MgR_2]$ species has been investigated at a mercury electrode in dimethoxyethane–0.1 M $[NBu_4][BF_4]$. Complexes in which $R = Me$, Et, $Pr^i$, or Ph were reduction electro-inactive but for $R = C_5H_5$, $PhCH_2$, or $CH_2{=}CHCH_2$ reduction was accessible in this medium (Table 3)[32]. It was proposed that these organomagnesium species underwent irreversible one-electron reduction according to reactions 47 and 48, although the fates of the carbanion and radical were not determined.

$$[MgR_2] \xrightarrow{1e} \{MgR\} + R^- \qquad (47)$$

$$\{MgR\} \longrightarrow Mg + R^{\bullet} \qquad (48)$$

Solutions of Grignard reagents, {MgRX}, where $R = Me$, Et, $Pr^i$, $Bu^i$, or Ph, were found to show the same polarographic behaviour as a mixture of $[MgR_2]$ and $[MgBr_2]$. Thus both the natural *and* the synthetic Grignard showed two reduction waves, the first attributable to $[MgBr_2]$ and the second to [MgBrR]. It was suggested

TABLE 3. Half-wave potentials ($E_{1/2}$) for the reaction $MgR_2 \xrightarrow[1e]{-E_{1/2}}$ products in dimethoxyethane containing $[NBu_4][ClO_4]$

| R | $E_{1/2}{}^a$ |
|---|---|
| $C_6H_5CH_2$ | 2.74 |
| $CH_2{=}CHCH_2$ | 2.65 |
| $C_5H_5$ | 2.50 |

$^a$ $E_{1/2}$ *vs.* $10^{-3}$ M $AgClO_4$/Ag.

that reduction of [MgBrR] took place according to reactions 49 and 50, although

$$\{MgBrR\} \xrightarrow{e} \{MgR\} + Br^- \quad (49)$$

$$\{MgR\} \longrightarrow Mg + R^{\bullet} \quad (50)$$

again the organic products of controlled potential electrolyses were not identified[33]. Evidently, at the potentials necessary for the reduction of [MgRX], $R^{\bullet}$ attack on Hg did not take place although further reduction of $R^{\bullet}$ to a carbanion was not established.

Participation of a mercury cathode in the electrochemical reduction reactions of metal alkyls and aryls is, however common, and is well illustrated by the cathodic reduction of $[SbPh_4]^+$[34]. At −0.9 V *vs.* SCE, reduction take splace according to reactions 51–53 (cf. reaction 43).

$$[SbPh_4]^+ \xrightarrow[-0.9\,V]{1e} [SbPh_3] + Ph^{\bullet} \quad (51)$$

$$Ph^{\bullet} + Hg \longrightarrow HgPh^{\bullet} \quad (52)$$

$$HgPh^{\bullet} \longrightarrow \tfrac{1}{2}Hg + \tfrac{1}{2}HgPh_2 \quad (53)$$

At −1.4 V, however, benzene was produced via further reduction of the $Hg^{I}Ph$ intermediate (reaction 54) (cf. reaction 42).

$$[HgPh]^{\bullet} \xrightarrow[H^+]{e} Hg + PhH \quad (54)$$

The complex $[Pb(OAc)_2Ph_2]$ shows a similar potential-dependent behaviour[35]. Thus one-electron reduction affords the $Pb^{III}$ organometallic according to reaction 55:

$$[Pb(OAc)_2Ph_2] \xrightarrow[-1.1\,V,\ Hg]{1e} [Pb(OAc)Ph_2] + AcO^- \quad (55)$$

whereas two-electron reduction on Hg affords the *trans*-metallated aryl (reaction 56).

$$[Pb(OAc)_2Ph_2] \xrightarrow[-1.6\,V,\ Hg]{2e} Pb + [HgPh_2] + 2AcO^- \quad (56)$$

$[PbCl_2Ph_2]$ and $[BiClPh_2]$ both undergo overall two-electron reduction on Hg to give the *trans*-metallated product $[HgPh_2]$ and the metal[35].

## C. Transition Metal Alkyls, Aryls, and Acyls

### 1. Introduction

Electrochemical cleavage of transition metal—carbon $\sigma$-bonds parallels that of main group electrochemistry in so far as the products depend on the fate of radicals, carbonium ions, or carbanions generated free in solution or incipiently bound to the metal centre. An additional feature of the electrochemistry of certain transition metal complexes containing $\sigma$-bonded carbon is that single electron transfer can lead to the formation of relatively stable intermediates with slow homogeneous decay kinetics to the organic and metal-based products; however, irreversible one- or two-electron transfer processes tend to dominate the electrode reaction of transition metal alkyls, aryls, and acyls, as they do main group electrochemistry. Considerable attention has been paid to the generation of nucleophilic transition metal complex anions in the presence of organic substrates, e.g. alkyl halides, as this can result in the formation of more or less reactive M—R intermediates which decompose to give radical, carbanion, or carbonium ion derived products. Much of the work in this area is based upon the studies relevant to the cobalamin coenzymes.

### 2. Reduction

Table 4 shows reduction potential data obtained by Denisovitch and Gubin for a range of complexes of the general formula $[FeR(\eta^5—C_5H_5)(CO)_2]$ which undergo irreversible two-electron reduction according to reaction 57 in a $CH_3CN$ electrolyte at a dropping mercury electrode[36].

$$[FeR(\eta^5—C_5H_5)(CO)_2] \xrightarrow[H^+]{2e} [Fe(\eta^5—C_5H_5)(CO)_2]^- + RH \quad (57)$$

It is not clear whether or not a carbanion is released and scavenges protons from the solvent or whether protonation is concerted with the formation of an incipient carbanion. As with reduction potentials for the irreversible reduction of Hg—C compounds, $E_{1/2}$ for the Fe series appears to correlate reasonably well with the p$K_a$ values of the respective hydrocarbons.

Irreversible two-electron reduction with hydrocarbon formation and the formation

TABLE 4. Potentials ($E_{1/2}$) for the reduction of $[FeR(\eta^5—C_5H_5)(CO_2]$ according to reaction 57

| R | $-E_{1/2}{}^a$ |
|---|---|
| iso—$C_3H_7$ | 2.27 |
| n-$C_3H_7$ | 2.22 |
| $C_2H_5$ | 2.16 |
| $CH_3$ | 2.13 |
| $C_6H_5$ | 1.96 |
| $CH_2{=}CHCH_2$ | 1.96 |
| $C_6H_5CH_2$ | 1.93 |
| $C_5H_5$ | 1.32 |

[a] $E_{1/2}$ at a mercury electrode *vs.* SCE; $CH_3CN$–0.1 M $[NEt_4][ClO_4]$; 25 °C.

of a generally stable metal carbonyl anion has been observed in many cathodic reactions of transition metal alkyl and aryl carbonyl complexes (e.g. reaction 58)[37].

$$[Mn(CO)_5CH_3] \xrightarrow[H^+]{2e} [Mn(CO)_5]^- + CH_4 \quad (58)$$

Varying the ligands around a given transition metal centre can change the mechanism of cathodic cleavage. Thus, whereas complexes $[Fe(C_6H_4X-p)(\eta^5—C_5H_5)(CO)_2]$ reduce in an irreversible two-electron step according to reaction 57, substitution of one carbonyl ligand by a tertiary phosphine results in irreversible one-electron transfer and homolytic cleavage as the primary step (reaction 59)[38,39].

$$[Fe(C_6H_4X)(\eta^5—C_5H_5)(CO)(PR_3)] \xrightarrow{1e} [Fe(\eta^5—C_5H_5)(CO)(PR_3)]^- + [C_6H_4X]^{\bullet} \quad (59)$$

(X = ring substituent)

The electrode reactions of several alkyl and aryl complexes of the early transition metals of the type $[M(\eta^5—C_5H_5)_2R_2]$, where M = Ti, Zr, Hf, have been investigated at Pt, vitreous carbon and Hg electrodes in non-aqueous electrolytes. One-electron irreversible reductions are observed and, perhaps surprisingly, a cyclopentadienide ion is released (rather than $R^{\bullet}$) and relatively stable $M^{III}$ radicals result (reaction 60)[40].

$$[Ti(\eta^5—C_5H_5)_2Me_2] \xrightarrow[\text{Pt, thf electrolyte}]{1e} [Ti(\eta^5—C_5H_5)Me_2]^{\bullet} + [C_5H_5] \quad (60)$$

Electrochemical formation and reduction of complexes containing a Co—C $\sigma$-bond have been investigated extensively, principally because such reactions may bear upon those of the $B_{12}$ coenzymes. The mechanisms of Co—C bond scission, the role of co-ligands and supporting electrolyte, and the products of cleavage have been studied in detail for complexes of the type **1** and many interesting conclusions have been drawn.

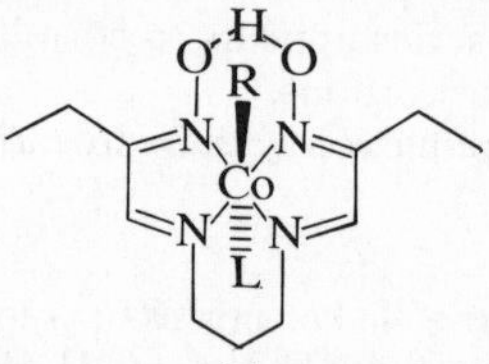

(**1**)
[CoR(salen)(L)]

Complexes **1**, in which L = solvent or other base and R = alkyl or aryl, and related species with tetradentate Schiff's base ligands in the equatorial plane, [CoR(chel)(L)], generally undergo *reversible* one-electron reduction to generate more or less stable $Co^{II}$—R species (reaction 61)[41,42]

$$[CoR(chel)(L)] \underset{-e}{\overset{e}{\rightleftharpoons}} [CoR(chel)(L)]^- \quad (61)$$
$$E_{1/2}$$

The reversible reduction potentials for this process are sensitive to the nature of R, L, and chel, and in addition, the cation of the supporting electrolyte can have a small influence on $E_{1/2}$. It is found that $E_{1/2}$ for the $Co^{III}$—R/$Co^{II}$—R (and $Co^{II}$—R/$Co^{I}$—

R) couples depend on the p$K_a$ of the corresponding carbon acid, RH (cf. Fe—C, Hg—C discussed above)[43]. The role of axial ligands L in determining the redox potentials and Co—C bond stabilities in these compounds has received attention in the context of the base-on, base-off chemistry of the cobalamin coenzymes. For example, it has been suggested that for L = 1,5,6-trimethylbenzimidazole, in a weakly coordinating solvent medium, the base-on, base-off equilibrium (reaction 62) exists.

$$L—Co^{II}—R \rightleftharpoons L + Co^{II}—R \quad (62)$$

At high concentration of the base, L, the decomposition of $[Co^{II}Me(salen)]^-$ species via a *trans*-methylation pathway is inhibited[43,44] The $Co^{II}$—C bond cleavage and products depend to a large extent on the nature of R. For example, when R = Ph an effective carbanion is developed, whereas when R = Me homolytic scission takes place (reactions 63 and 64)[43].

$$[Co^{II}Ph(salen)(dmf)]^- \xrightarrow{H^+} [Co^{II}(salen)(dmf)] + PhH \quad (63)$$

$$[CoMe(salen)(dmf)]^- \longrightarrow [Co^{I}(salen)(dmf)] + Me^{\bullet} \quad (64)$$

The $[Co^{I}(salen)(L)]^-$ species are powerful nucleophiles and can react with alkyl halides, aryl halides, sulphonium compounds, and even quaternary ammonium cations to give stable $Co^{III}$—R complexes (reaction 65)[43,44]. Electrocatalytic reduction of RX to $R^{\bullet}$ products can be achieved (reactions 64 and 65), the overall rate of

$$[Co^{I}(salen)(L)]^- + RX \longrightarrow [Co^{III}R(salen)(L)] + X^- \quad (65)$$

which is determined by the homogeneous Co—R formation and cleavage steps. Two competing pathways appear to operate in the reaction of $[Co^{I}(salen)(L)]^-$ with $NR_4^+$. In addition to nucleophilic attack by the $Co^{I}$ on the quaternary ammonium cation, electron transfer can also take place (reaction 66).

$$[Co^{I}(salen)(L)]^- + NR_4^+ \begin{cases} \nearrow [Co^{III}R(salen)(L)] \\ \searrow [Co^{II}(salen)(L)] + NR_3 + R^{\bullet} \end{cases} \quad (66)$$

Electrochemical reduction of certain $Ni^{II}$ complexes to the $Ni^{I}$ oxidation state generates powerful nucleophiles which react with alkyl and aryl halides to form $Ni^{III}$—C bonds. The $Ni^{III}$—C bond is generally thermally labile and homolytically cleaves to give a $Ni^{II}$ product and a radical. For example, the $Ni^{II}$ square-planar complexes **2** and **3** reduce in a reversible one-electron step in a $CH_3CN$ electrolyte to

Me Me Me H H N 2+ N Ni N N H H Me Me Me

**(2)**

N O Ni N O

**(3)**

give stable $Ni^{I}$ species; however, if such reductions are carried out in the presence of an alkyl bromide or iodide, reactive $Ni^{III}$—alkyls are formed via an oxidative addition (reaction 67). Homolytic cleavage can regenerate the $Ni^{II}$ precursor complex

$$Ni^{II} \underset{-e}{\overset{+e}{\rightleftharpoons}} Ni^{I} \xrightarrow{RX} RNi^{III}X \quad \rightarrow R^{\bullet} + X^{-} \text{ (regenerating } Ni^{II}) \tag{67}$$

and an alkyl radical, thus effecting the catalytic reduction of the alkyl halide. However, in certain cases (**3**, $Bu^{n}I$) the $Ni^{III}$—R bond is sufficiently stable for further electron transfer to take place with the generation of a carbanion precursor (reaction 68).

$$Ni^{II} \underset{-e}{\overset{e}{\rightleftharpoons}} Ni^{I} \xrightarrow{RX} RNi^{III}X \xrightarrow{e} RNi^{II}X \longrightarrow R^{-} + \{Ni^{II}\} \tag{68}$$

Apparently, in this case the $Ni^{II}$ species is electroinactive and is not capable of partaking in further reduction cycles of RX, and the reduction therefore terminates with an overall two-electron stoichiometry via the electron transfer–chemical step–further electron transfer (ece) mechanism (reaction 68). Products arising from the electrolysis of RX in the presence of **2** or **3** have been identified as those normally arising from $R^{\bullet}$ and $R^{-}$ intermediates. Interestingly, $RCH_2CN$ is formed during the electrolysis of **2** or **3** in the presence of RX, presumably via reactions 69–72[45,46].

$$R^{\bullet} + CH_3CN \longrightarrow RH + {}^{\bullet}CH_2CN \tag{69}$$

$${}^{\bullet}CH_2CN + R^{\bullet} \longrightarrow RCH_2CN \tag{70}$$

$$R^{-} + CH_3CN \longrightarrow RH + {}^{-}CH_2CN \tag{71}$$

$${}^{-}CH_2CN + RX \longrightarrow RCH_2CN + X^{-} \tag{72}$$

$[NiX_2(PR_3)_2]$, where X = halide and R = alkyl or aryl, similarly electrocatalyse the reduction of alkyl and aryl halides in non-aqueous electrolytes[47–49].

### 3. *Oxidation of transition metal alkyls, aryls, and acyls.*

Electrochemical oxidation of transition metal—carbon $\sigma$-bonded species is generally less well studied than electrochemical reduction, although this is being redressed. As a part of a more general study of the kinetics of the heterogeneous oxidation of organometallic compounds, the irreversible oxidations of $Co^{III}$—alkyls and $Pt^{II}$—alkyls at a Pt electrode have been investigated and kinetic and activation energy parameters determined[50].

$Co^{III}$—alkyls and —aryls with tetradentate Schiffs base ligands, e.g. **1** and **2**, undergo reversible one-electron oxidation to generate labile $Co^{IV}$ compounds, which readily transfer the $\sigma$-bonded organic group to nucleophiles or undergo homolytic cleavage. One-electron oxidations of $[Co^{III}Et(salen)]$ and a range of related Shiffs base and glyoxime complexes were shown by cyclic voltammetry to be electrochemically reversible in a $CH_3CN$ electrolyte at a Pt electrode. It was found that pyridine reacts rapidly with the $Co^{IV}$—alkyl complexes to displace the alkyl group via nucleophilic substitution at the $\alpha$-carbon atom. Kinetic results suggested that an $S_N2$ mechanism was operative for the glyoxime complexes whereas the salen complexes reacted via an internal nucleophilic displacement $S_Ni$ (reaction 73). In some cases the

$$Co^{III} \xrightarrow{-e} Co^{IV} \xrightarrow{Pyr} RCo^{IV}Pyr \longrightarrow Co^{II} + R\overset{+}{N}\text{(C}_5\text{H}_{10}) \qquad (73)$$

$Co^{IV}$—C bond undergoes rapid unimolecular scission, thus $[Co^{III}(Me)_2(chel)]$ (chel = macrocyclic *N*-donor ligand), undergo irreversible one-electron oxidation in a $CH_3CN$ electrolyte with the concomitant homolytic cleavage of *one* Co—C bond. In the absence of H or Cl donors, the predominant organic product is ethane, formed via radical–radical coupling (reaction 74)[51,52].

$$MeCo^{III}Me \xrightarrow[\text{CH}_3\text{CN electrolyte}]{-e} MeCo^{IV}Me \xrightarrow{\text{fast}} MeCo^{III} + Me^{\bullet} \longrightarrow EtH \qquad (74)$$

Electrochemical oxidation of $[Fe(\eta^5—C_5H_5)(CH_3)(CO)_2]$ in $CH_3CN$ leads to CO insertion rather than to radical or carbonium ion products (reaction 75). In contrast,

$$[Fe(\eta^5\text{-}C_5H_5)(CH_3)(CO)_2] \xrightarrow[\text{CH}_3\text{CN electrolyte, 0.88 V }vs.\text{SCE, Pt}]{-le} [Fe(\eta^5\text{-}C_5H_5)(COCH_3)(CO)(CH_3CN)]^+ \qquad (75)$$

oxidation of $[Fe(\eta^5—C_5H_5)(COCH_3)(CO)_2]$ in the presence of EtOH as a nucleophile affords ethyl acetate and unidentified metal products (reaction 76)[53,54].

$$[Fe(\eta^5—C_5H_5)(COCH_3)(CO)_2] \xrightarrow[\text{1.3 V }vs.\text{ SCE, CH}_3\text{CN electrolyte +10\% EtOH, Pt}]{-3e} EtOCOCH_3 \qquad (76)$$

The anion $[Fe(\eta^5—C_5H_5)(CN)(COCH_3)(CO)]^-$ oxidizes to the neutral $Fe^{III}$ species in an electrochemically reversible one-electron step. The paramagnetic product is stable below 0 °C but at 20 °C under an atmosphere of CO, Fe—C cleavage takes place liberating acetone and acetaldehyde (reaction 77)[53].

$$[Fe(\eta^5—C_5H_5)(CN)(COCH_3)(CO)]^- \underset{+e,\ 0\,^\circ\text{C}}{\overset{-e}{\rightleftharpoons}} [Fe(\eta^5\text{-}C_5H_5)(CN)(COCH_3)(CO)] \xrightarrow{20\,^\circ\text{C, CO}} \underset{40\%}{CH_3COCH_3} + \underset{6\%}{CH_3CHO} + \text{unidentified metal products} \qquad (77)$$

Oxidation of $[Cr(COPh)(CO)_5]^-$ gives a stable hydroxycarbene product without Cr—C bond scission (reaction 78)[55].

$$[Cr(COPh)(CO)_5]^- \underset{\text{0.48 V }vs.\text{ SCE, CH}_3\text{CN electrolyte, Pt}}{\overset{-e}{\rightleftharpoons}} [Cr(COPh)(CO)_5] \xrightarrow{Et_4N^+} [Cr\{C(OH)Ph\}(CO)_5] \qquad (78)$$

## III. π-BONDED HYDROCARBON LIGANDS

The electrochemical behaviour of a vast range of closed-shell transition metal complexes containing π-bonded hydrocarbon ligands has been studied in more or less detail; in particular, those complexes containing $\eta^5$-cyclopentadienyl or $\eta^6$-arene ligands are amongst the most well studied. Cyclopentadienyl and arene complexes of the transition metals are generally inert with respect to anodic metal—carbon bond cleavage and one-electron *oxidations* of their closed-shell complexes tend to give stable 17-electron products (reactions 79–81).

$$[Fe(\eta^5\text{—}C_5H_5)_2] \underset{+e}{\overset{-e}{\rightleftharpoons}} [Fe(\eta^5\text{—}C_5H_5)_2]^{+\cdot} \tag{79}$$

$$[Cr(\eta^6\text{—}C_6H_6)_2] \underset{+e}{\overset{-e}{\rightleftharpoons}} [Cr(\eta^6\text{—}C_6H_6)_2]^{+\cdot} \tag{80}$$

$$[Mn(\eta^5\text{—}C_5H_5)(CO)_3] \underset{+e}{\overset{-e}{\rightleftharpoons}} [Mn(\eta^5\text{—}C_5H_5)(CO)_3]^{+\cdot} \tag{81}$$

Such oxidations are metal centred rather than ligand based. Although complexes with these $\pi$-bonded hydrocarbon ligands are robust with respect to one-electron oxidation, at more extreme potentials multi-electron anodic reactions are possible which can lead to metal—carbon bond cleavage (reaction 82)[56].

$$[Cr(\eta^6\text{—}C_6H_6)_2] \xrightarrow{-3e} Cr^{3+} + 2C_6H_6 \tag{82}$$

Closed-shell tertiary-phosphine, -phosphite, and carbonyl complexes of Groups VI, VII, and VIII possessing $\eta^6$-arene, $\eta^5$-cyclopentadienyl, $\eta^4$-diene, or $\eta^2$-acetylene ligands generally undergo electrochemically reversible one-electron oxidations at Pt electrodes in non-aqueous solvents to give more or less stable 17-electron metallo-radicals. Whereas the $\eta^4$-compounds show interesting chemical reactions subsequent to the electron-transfer step, metal—carbon bond cleavage has not been observed[57,58].

Electrochemical *reductions* of closed-shell complexes containing cyclopentadienyl, arene, and other $\pi$-bonded hydrocarbon ligands give rise to several interesting types of reaction involving 'partial' or full metal—hydrocarbon ligand bond cleavage. One-electron reduction of $\pi$-bonded hydrocarbon ligand compounds generally involves the addition of the electron to a $\pi^*$-antibonding orbital which in certain cases gives rise to fairly stable radical anions which can be studied by epr, ir, and, where appropriate, Mössbauer spectroscopy (reactions 83(a) and (b)[59–61].

$$[Fe(CO)_4(olefin)] \underset{-e}{\overset{e}{\rightleftharpoons}} [Fe(CO)_4(olefin)]^- \tag{83a}$$

$$[Fe(CO)_3(diene)] \underset{-e}{\overset{e}{\rightleftharpoons}} [Fe(CO)_3(diene)]^- \tag{83b}$$

The cobaltocenium cation undergoes two successive reversible one-electron reductions in aprotic media to give the neutral and anionic species (reaction 84)[63,64]; $[Ni(\eta^5\text{—}C_5H_5)_2]^{2+}$ behaves analogously[65].

$$[Co(\eta^5\text{—}C_5H_5)_2]^+ \underset{-e}{\overset{e}{\rightleftharpoons}} [Co(\eta^5\text{—}C_5H_5)_2]^0 \underset{-e}{\overset{e}{\rightleftharpoons}} [Co(\eta^5\text{—}C_5H_5)_2]^- \tag{84}$$

Reversible one-electron reduction of arene and cycloheptatriene closed-shell complexes to give reactive metallo-radicals in aprotic media is also well established (reactions 85 and 86)[66–68].

$$[Fe(\eta^5\text{—}C_5H_5)(\eta^6\text{—}C_6H_6)]^+ \underset{-e}{\overset{e}{\rightleftharpoons}} [Fe(\eta^5\text{—}C_5H_5)(\eta^6\text{—}C_6H_6)]^0 \tag{85}$$

$$[Cr(\eta^5\text{—}C_5H_5)(\eta^7\text{—}C_7H_7)]^+ \underset{-e}{\overset{e}{\rightleftharpoons}} [Cr(\eta^5\text{—}C_5H_5)(\eta^7\text{—}C_7H_7)]^0 \tag{86}$$

Although one- and two-electron electrochemical reductions of $\pi$-bonded hydrocarbon ligated species give rise to more or less stable neutral and anionic species, the addition of electrons makes the ligand susceptible to electrophilic attack at a carbon centre and consequent 'partial' metal—hydrocarbon ligand bond cleavage, e.g. $\eta^5 \rightarrow \eta^4$ conversions. Thus, reduction of the cobaltocenium cation in the presence of $CO_2$ as an electrophile proceeds according to reaction 87[69].

$$[Co(\eta^5\text{—}C_5H_5)_2]^+ \xrightarrow[CO_2,\ \text{dmf},\ -1.90\ \text{V vs. SCE}]{2e} [Co(\eta^5\text{—}C_5H_5)(\eta^4\text{—}C_5H_5CO_2^-)] \quad (87)$$

Similar electrophilic attack has also been observed in the electrochemical reductions of $[Ni(\eta^5\text{—}C_5H_5)_2]$ and $[Fe(\eta^5\text{—}C_5H_5)(\eta^6\text{—}C_6H_6)]^+$ (reaction 88)[70,71].

$$[Fe(\eta^5\text{—}C_5H_5)(\eta^6\text{—}C_6H_6)]^+ \xrightarrow[+H^+]{2e} [Fe(\eta^5\text{—}C_5H_5)(\eta^5\text{—}C_6H_7)] \quad (88)$$

Partial metal—hydrocarbon ligand cleavage can also occur via intermolecular carbon—carbon bond formation as in the $\eta^7 \rightarrow \eta^6$ conversion in reaction 89[72].

$$2[Cr(\eta^7\text{-}C_7H_7)(CO)_3]^+ \xrightarrow{2e} [(CO)_3Cr(\eta^6\text{-}C_7H_7)\text{—}(\eta^6\text{-}C_7H_7)Cr(CO)_3] \quad (89)$$

Full reductive cleavage of cyclic hydrocarbon ligands from the metal is rare, although the electrochemical reduction of certain Group IV metallocenes is reported to lead to the release of the cyclopentadienide anion (reaction 90)[73].

$$[Ti^{IV}(\eta^5\text{—}C_5H_5)_2Me_2] \xrightarrow[\text{thf, electrolyte}]{e} [Ti^{III}(\eta^5\text{—}C_5H_5)Me_2] + C_5H_5^- \quad (90)$$

This is surprising as reduction of $[TiCl_2(\eta^5\text{—}C_5H_5)_2]$ results in expulsion of $Cl^-$ (reaction 91).

$$[TiCl_2(\eta^5\text{—}C_5H_5)_2] \underset{-e,\ \text{thf, electrolyte}}{\overset{e}{\rightleftharpoons}} [TiCl_2(\eta^5\text{—}C_5H_5)_2]^- \xrightarrow{-Cl^-} [TiCl(\eta^5\text{—}C_5H_5)_2] \quad (91)$$

Electrochemical reduction of the $\pi$-allyl complexes $[Fe(\eta^3\text{—}C_3H_5)(CO)_2(NO)]$ and $[Co(\eta^3\text{—}C_3H_5)(CO)_3]$ is an overall two-electron process with cleavage of the hydrocarbon ligand from the metal (reaction 92). In contrast, reduction of

$$[Fe(\eta^3\text{—}C_3H_5)(CO)_2(NO)] \xrightarrow[H^+]{2e} [Fe(CO)_2(NO)]^- + C_3H_6 \quad (92)$$

$[Fe(\eta^3\text{—}C_3H_5)(CO)_3X]$, where X = Cl, Br or I, results in halide expulsion rather than the loss of the allyl ligand; the Ru analogues behave similarly (reaction 93)[74,75]

$$[Fe(\eta^3\text{—}C_3H_5)(CO)_3X] \xrightarrow{e} [Fe(\eta^3\text{—}C_3H_5)(CO)_3]^{\bullet} + Cl^- \quad (93)$$

There are numerous studies concerned with the relationships between structure and oxidation or reduction potentials of cyclopentadienyl and arene ligand complexes. The ease of reduction of 18-electron metallocenes is determined largely by the net charge on the complex, for example reduction becomes more difficult by *ca.* 1.5–2.0 V in the following order ($-E_{1/2}^{RED}$): $[Ni(\eta^5—C_5H_5)_2]^{2+} < [Co(\eta^5—C_5H_5)_2]^+ < [Fe(\eta^5—C_5H_5)_2]$[76]. Oxidation potentials for the 18e/17e couples, not unexpectedly, show the opposite trend.

## IV. TRANSITION METAL CARBONYL, CARBENE, CARBYNE, AND ISOCYANIDE COMPLEXES

### A. Introduction

The electrode reactions of metal carbonyl complexes are amongst the most thoroughly studied of organometallic electrochemical processes. Anodic and cathodic M—CO bond cleavage reactions are well established and factors influencing products, pathways, and mechanisms moderately well understood. Little attention has been given to the electrochemical behaviour of isocyanide or carbene complexes and detailed electrochemical studies are few. The following sections therefore deal almost exclusively with the behaviour of transition metal carbonyl complexes.

### B. Oxidation

The anodic oxidation of 18-electron mononuclear carbonyl complexes can usually be achieved at Pt or vitreous carbon anodes in non-aqueous electrolytes at potentials up to *ca.* +2.5 V *vs.* SCE. The primary electrochemical event is usually the removal of a single electron from a metal-based orbital to give a more or less stable 17-electron metallo-radical. The stability of this species is determined by the nature of the metal centre, the co-ligand environment, and the electrolyte media. These are discussed in more detail below.

The removal of an electron from a metal-based orbital can result in dissociation of the M—CO bond as a consequence of diminished $Md\pi—CO\pi^*$ back-bonding: qualitatively the increased effective nuclear charge at the metal centre lowers the energy of the $d\pi$ orbitals thereby decreasing $d\pi–\pi^*$ overlap. This weakening of $d\pi–\pi^*$ back-bonding is not sufficiently compensated by a corresponding increase in M—CO $\sigma$-bond strength, hence the CO ligand is invariably more labile in its 17-electron complexes than in its 18-electron precursors. For example, $[Mo(CO)(S_2CNEt_2)_2(dppe)]$ is indefinitely stable in a thf electrolyte and shows $\nu$(CO) at 1790 $cm^{-1}$. Controlled potential oxidation of this species generates the unstable radical cation, $\nu$(CO) 1930 $cm^{-1}$, which decays via CO loss with a first-order rate constant of 0.22 $s^{-1}$ at 25 °C (reaction 94)[77]. M—CO bond cleavage following single

$$\underset{18e}{[Mo(CO)(S_2CNEt_2)_2(dppe)]} \underset{+e}{\overset{-e}{\rightleftharpoons}} [Mo(CO)(S_2CNEt_2)_2(dppe)]^+ \longrightarrow [Mo(S_2CNEt_2)_2(dppe)]^+ + CO \quad (94)$$

electron transfer also takes place via a disproportionation pathway, thus $[Cr(CO)_6]$ oxidizes reversibly in a $CH_3CN$ electrolyte to give the unstable $[Cr(CO)_6]^{+\cdot}$ radical cation[78]. Decomposition of this species takes place via disproportionation (reaction 95). Further irreversible one-electron oxidation of $[Cr(CO)_6]^{+\cdot}$ is observed at more

$$[Cr(CO)_6] \underset{+e}{\overset{-e}{\rightleftharpoons}} [Cr(CO)_6]^{+\cdot} \overset{-e}{\longrightarrow} \{Cr(CO)_6\}^{2+} \underset{fast}{\longrightarrow} Cr^{2+} + 6CO \quad (95)$$

positive potentials[79]. $[Cr(CO)_6]^{+\cdot}$ is also susceptible to attack by nucleophiles and this species and other radical cations are generally more stable when electrolytically generated in trifluoroacetic acid media which has low nucleophilicity (reaction 96)[80,81]

$$[Fe(CO)_5] \underset{+e}{\overset{-e}{\rightleftharpoons}} [Fe(CO)_5]^{+\cdot} \quad \text{CF}_3\text{COOH} \tag{96}$$

First-row mononuclear binary carbonyls give 17-electron products upon electrochemical oxidation with stabilities which decrease in the order $[V(CO)_6] > [Cr(CO)_6]^{+\cdot} > [Fe(CO)_5]^{+\cdot} > [Ni(CO)_4]^{+\cdot}$. Electrochemical oxidation of $[Mo(CO)_6]$ and $[W(CO)_6]$ are totally irreversible processes which lead to M—CO bond cleavage, in a multielectron step (reaction 97)[79]. However, substitution

$$[Mo(CO)_6] \xrightarrow[-5e]{CH_3CN} Mo^{V} + 6CO \tag{97}$$

of the carbonyls by more basic ligands such as dppe or 2,2′-bipyridyl can stabilize 17-electron Mo and W carbonyls with respect to M—CO bond cleavage. For example, *cis*-$[Mo(CO)_2(dppe)_2]$ undergoes single electron transfer to give the *cis* radical cation, which rapidly isomerizes to the *trans*-isomer via an intramolecular process (reaction 98)[82]. This radical cation is isolable and undergoes further electron transfer in the

$$cis\text{-}[Mo(CO)_2(dppe)_2] \xrightarrow[thf]{-e} trans\text{-}[Mo(CO)_2(dppe)_2]^{+\cdot} \tag{98}$$

presence of a nucleophile such as $F^-$ to give a stable $Mo^{II}$ dicarbonyl, an 18-electron closed-shell species (reaction 99)

$$trans\text{-}[Mo(CO)_2(dppe)_2]^{+\cdot} \xrightarrow[-e]{F^-} [MoF(CO)_2(dppe)_2]^{+} \tag{99}$$

The stability of *trans*-$[Mo(CO)_2(dppe)_2]^{+\cdot}$ *vis à-vis* $[Mo(CO)_6]^{+\cdot}$ must be attributed largely to the relatively good electron donating diphosphine ligands preserving $Mod\pi \rightarrow CO\pi^*$ back-bonding capacity and to the lower electrophilicity of the metal centre. It has also been suggested that charge is delocalized over the diphosphine ligands in such compounds, i.e. the oxidation is substantially ligand based[83]. It is possible that the bulky dppe ligands substantially protect the metal centre from nucleophilic attack except by the smallest of nucleophiles (such as $F^-$) and that this contributes to their relative stabilities; this is consistent with the relatively greater stability of $[Cr(CO)_6]^{+\cdot}$ compared with $[Mo(CO)_6]^{+\cdot}$ (metal atom size) and with $[Fe(CO)_5]^{+\cdot}$ (potential coordination site). Electrochemical oxidation of complexes $[Cr(CO)_5(L)]$ at Pt or vitreous carbon electrodes in non-aqueous electrolytes give radical cations the stability of which depends upon the nature of L[84]. Decomposition pathways may involve M—CO bond cleavage such as those discussed for $[Cr(CO)_6]^{+\cdot}$ etc. or may proceed via dissociation of L or other pathways. What is fairly well established is that the stability of $[Cr(CO)_5(L)]^{+\cdot}$ increases as the donicity of L increases, for example, cyclic voltammetric studies indicate the following stability order $L = I^- > CH_3CN > CO$. This trend parallels the formal oxidation potentials for the $Cr^I/Cr^0$ couples, the more negative the oxidation potential of the 18-electron complex, the more stable is the radical cation[85]. Again this can be rationalized in terms of $d\pi - \pi^*$ back-bonding and metal centre electrophilicity. Oxidation potentials of various octahedral complexes of the type $[M(CO)_{6-n}(L)_n]^{y+}$ have been discussed in terms of the Lewis basicity of L, the degree of substitution $n$

and the net charge y; however, kinetic and thermodynamic data relating M—CO bond cleavage to these parameters are generally lacking[86–88]. The carbonyl anions such as $[Mn(CO)_5]^-$, $[Re(CO)_5]^-$, and $[Mo(\eta^5\text{—}C_5H_5)(CO)_3]^-$ undergo single electron irreversible oxidations to give dimeric products, e.g. reaction 100[79].

$$[Re(CO)_5]^- \xrightarrow[\text{thf}]{-e} \tfrac{1}{2}[Re_2(CO)_{10}] \qquad (100)$$

Dinuclear metal carbonyls tend to undergo irreversible two-electron oxidation at Pt or vitreous carbon electrodes in non-aqueous electrolytes which leads to M—M bond cleavage rather than M—CO bond cleavage, e.g. reaction 101[79].

$$[Mn_2(CO)_{10}] \xrightarrow[CH_3CN]{-2e} 2[Mn(CO)_5(CH_3CN)]^+ \qquad (101)$$

Although there have been several studies of the electrochemical oxidation reactions of various carbene, carbyne, and isocyanide complexes generally in non-aqueous electrolytes, metal—carbon ligand bond cleavage reactions of such species have not been recognized. Investigations have primarily been concerned with relationships between redox potentials for the 18-electron/17-electron couples and structure rather than with product identification after controlled potential electrolysis in the cases where the 17-electron species are known to be unstable[86,87,89–91].

### C. Reduction

Electrochemical reduction of certain transition metal carbonyls in non-aqueous electrolytes can involve M—CO bond cleavage if an electron is added to a primarily M—CO anti-bonding orbital (cf. photolytic cleavage of M—CO species). The addition of a single electron to a closed-shell, 18-electron complex must result in a heterolytic metal—ligand cleavage reaction of some sort unless the charge can be delocalised over an accessible ligand $\pi^*$ system (cf. reduction of arene and cyclopentadienyl complexes discussed above; phthalocyanines, porphyrins, and 2,2′-bipyridyl also 'stabilize' 19- and 20-electron mononuclear systems). Binary mononuclear carbonyls undergo irreversible one-electron reductions in tegrahydrofuran electrolytes to give dinuclear, and in some cases polynuclear, metal carbonyl anions via M—CO bond cleavage[79,92]. For example, $[M(CO)_6]$ complexes, where M = Cr, Mo, or W, reduce at potentials of *ca.* −2.1 V *vs.* SCE according to reaction 102.

$$[M(CO)_6] \xrightarrow{1e} \tfrac{1}{2}[M_2(CO)_{10}]^{2-} + CO \qquad (102)$$

Similarly, $[Fe(CO)_5]$ and $[Ni(CO)_4]$ show analogous reduction processes (reaction 103)[79,92,93–95] Cyclic voltammetry of $[M(CO)_6]$ in thf and $CH_3CN$ electrolytes at a Pt

$$[Fe(CO)_5] \xrightarrow{1e} \tfrac{1}{2}[Fe_2(CO)_8]^{2-} + CO \qquad (103)$$

electrode reveals the formation of an intermediate in the reduction process which is not the final product of electrolysis, i.e. $[M_2(CO)_{10}]^{2-}$, and which has a lifetime of several milliseconds. It has been suggested that this intermediate is the 17-electron pentacarbonyl (reaction 104), although this has not been confirmed by independent

$$[M(CO)_6] \xrightarrow{1e} [M(CO)_6]^{-\cdot} \xrightarrow{\text{fast}} [M(CO)_5]^{-\cdot} + CO \longrightarrow [M_2(CO)_{10}]^{2} \qquad (104)$$

spectroscopic detection, e.g. by epr. Nevertheless, reduction of $[Mo(CO)_6]$ in the presence of an alkyl halide, RX, results in the formation of radical-derived hydrocarbon products and $[Mo(CO)_5X]^-$ in an overall two-electron process consistent with reactions 104 and 105[96].

$$[M(CO)_5]^{-\cdot} + RX \xrightarrow{e} [M(CO)_5X]^- + R^\cdot \qquad (105)$$

Electrochemical reduction of substituted metal carbonyls need not necessarily involve M—CO bond cleavage, indeed the addition of an electron to an orbital other than one with M—CO antibonding character might be expected to lead to an increase in $d\pi$–$\pi^*$ back donation and hence a strengthening of the M—CO bond. For example, electrochemical reduction of $[Mn(CO)_5X]$, where X = halide or $[Mn(CO)_5R]$ and R = alkyl or aryl, proceeds via Mn—X or Mn—R bond cleavage rather than Mn—CO scission (reaction 106)[97].

$$[Mn(CO)_5Cl] \xrightarrow[\text{thf}]{2e} [Mn(CO)_5]^- + Cl^- \qquad (106)$$

An interesting example of a complex with a multiplc choice of metal—ligand cleavage reactions is provided by the electrochemistry of $[FeH(CO)(dppe)_2]^+$. Two-electron reduction of this species in a thf electrolyte at a Pt electrode leads to the cleavage of an Fe—P bond rather than to Fe—CO or Fe—H scission (reaction 107)[98].

$$[FeH(CO)(dppe)_2]^+ \underset{\substack{-2e \\ \text{thf}}}{\overset{2e}{\rightleftharpoons}} \left[ \text{Fe(H)(CO)}(\text{Ph}_2\text{PCH}_2\text{CH}_2\text{PPh}_2)(\text{Ph}_2\text{PCH}_2\text{CH}_2\text{PPh}_2\text{-}\kappa P) \right]^- \qquad (107)$$

Ligand-centred electrochemistry of carbene, carbyne or isocyanide complexes remains largely unexplored.

## V. CONCLUSIONS

The electrochemistry of organometallic compounds has long promised to be an area in which interesting reactions might be achieved, complementary to those involving thermal or photochemical activation of the metal—carbon bond. In so far as anodic or cathodic reactions of certain organometallic alkyls and aryls lead to the generation of incipient or free carbonium ions, carbanions, or radicals as reactive intermediates, and hence give rise to metal—carbon bond cleavage products derived from these, this promise has in part been realised. However, synthetically useful electrochemical cleavage reactions of metal alkyls and aryls remain elusive. This can be attributed to the preponderance of radical formation reactions over those of carbonium ion or carbanion generating pathways. Also, electrochemical studies of organometallic compounds have tended to be physico-chemically orientated rather than preparative, although in the last few years this imbalance has been somewhat redressed. In particular, electrode reactions of cyclic $\pi$-bonded hydrocarbon ligands such as cyclopentadienyls and arenes have recently been shown to give interesting ligand-based chemistry which will no doubt be extended to other hydrocarbon ligands. Stepwise electrochemical reduction of coordinated CO to formaldehyde, methanol, etc., e.g. reaction 108, also remains to be explored. Interestingly, electrochemical

$$M(CO) \xrightarrow{4e,\,4H^+} M + CH_3OH \quad (\text{M} \xrightarrow{CO} \text{M(CO)}) \qquad (108)$$

reduction of $CO_2$ to CO mediated by certain $Co^{II}$ and $Ni^{II}$ macrocyclic ligand complexes has been demonstrated; whether or not such reactions involve M—C bond formation and cleavage remains unclear[99].

In conclusion, electrochemical metal—carbon ligand bond cleavage reactions and related ligand centred electrochemistry still hold considerable promise for the future.

## VI. ACKNOWLEDGEMENTS

I thank Dr D. R. Stanley for reading the manuscript and for helpful comments.

## VII. REFERENCES


1. R. N. Adams, *Electrochemistry at Solid Electrodes*, Marcel Dekker, New York, 1969.
2. D. T. Sawyer and J. L. Roberts, *Experimental Electrochemistry for Chemists*, Wiley, New York, 1974.
3. D. Pletcher, *Chem. Soc. Rev.*, **4,** 471 (1975).
4. Z. Galus, *Fundamentals of Electrochemical Analysis*, Ellis Horwood, Chichester, 1976.
5. D. D. Macdonald, *Transient Techniques in Electrochemistry*, Plenum, New York, 1977.
6. A. M. Bond, *Modern Polarographic Methods in Analytical Chemistry*, Marcel Dekker, New York, 1980.
7. A. J. Bard and L. R. Faulkner, *Electrochemical Methods*, Wiley, New York, 1980.
8. J. A. McCleverty, in *Reactions of Molecules at Electrodes* (Ed. N. S. Hush), Wiley–Interscience, New York, 1971, p. 363.
9. H. Lehmkuhl, in *Organic Electrochemistry* (Ed. M. M. Baizer), Marcel Dekker, New York, 1973, p. 621.
10. L. I. Denisovich and S. P. Gubin, *Russ. Chem. Rev.*, **46,** 27 (1977).
11. C. J. Pickett, in *Specialist Periodical Reports, Electrochemistry*, Vol. 8 (Ed. D. Pletcher), Royal Society of Chemistry, London, 1983, p. 81.
12. B. Jann, J. Schwarz, and R. Breslow, *J. Am. Chem. Soc.*, **102,** 5741 (1980). and references cited therein.
13. R. J. Klinger and J. K. Kochi, *J. Am. Chem. Soc.*, **102,** 4791 (1980).
14. K. Ziegler, H. Lehmkuhl, and E. Lindner, *Chem. Ber.*, **92,** 2320 (1959).
15. H. Lehmkuhl, *Chem.-Ing.-Tech.*, **36,** 612 (1964).
16. R. Schaefer, *Dissertation*, Technische Hochschule Aachen, 1961.
17. D. H. Geske, *J. Phys. Chem.*, **63,** 1062 (1959); **66,** 1743 (1962).
18. W. V. Evans and E. Field, *J. Am. Chem. Soc.*, **58,** 2284 (1936).
19. W. V. Evans and F. H. Lee, *J. Am. Chem. Soc.*, **56,** 654 (1934).
20. L. W. Gaddum and H. E. French, *J. Am. Chem. Soc.*, **49,** 1295 (1927).
21. W. V. Evans and D. Braithwaite, *J. Am. Chem. Soc.*, **61,** 898 (1939).
22. D. G. Braithwaite, *Chem. Abstr.*, **66,** 76155 (1967). Ger. 1,231,700 (Cl.C.07f).
23. H. Schäfer and H. Küntzel, *Tetrahedron Lett.*, 3333 (1970).
24. W. V. Evans, F. H. Lee, and C. Lee, *J. Am. Chem. Soc.*, **57,** 489 (1935).
25. M. M. Baizer (Ed.), *Organic Electrochemistry*, Marcel Dekker, New York, 1973.
26. P. S. Skell and P. H. Reichenbacher, *J. Am. Chem. Soc.*, **90,** 2309 (1968).
27. N. L. Weinberg, *Tetrahedron Lett.*, 4823 (1970).
28. M. Fleischmann, D. Pletcher, and G. M. Race, *J. Chem. Soc.* (*B*), 1746 (1970).
29. K. P. Butin, I. P. Beletskaya, A. N. Kashin, and O. A. Reutov, *J. Organomet. Chem.*, **10,** 197 (1967).
30. K. P. Butin, M. T. Ismail, and O. A. Reutov, *J. Organomet. Chem.*, **175,** 157 (1979).
31. L. I. Denisovich and S. P. Gubin, *J. Organomet. Chem.*, **57,** 87 and 99 (1973).
32. T. Psarrs and R. E. Dessy, *J. Am. Chem. Soc.*, **88,** 5132 (1966).
33. R. E. Dessy, W. Kitching, T. Psarrs, R. Salinger, A. Chen, and T. Chivers, *J. Am. Chem. Soc.*, **88,** 460 (1966).
34. M. D. Morris, P. S. McKinney, and E. C. Woodbury, *J. Electroanal. Chem.*, **10,** 85 (1965).
35. R. E. Dessy, W. Kitching, and T. Chivers, *J. Amer. Chem. Soc.*, **88,** 453 (1966).
36. L. I. Denisovich, S. P. Gubin, Yu. A. Chapovskii, and N. A. Ustynyuk, *Izv. Akad. Nauk SSSR, Ser. Khim.*, 924 (1968).
37. L. I. Denisovich, S. P. Gubin, A. A. Ioganson, N. E. Kolobova, and K. N. Anismov, *Izv. Akad. Nauk SSSR, Ser. Khim.*, 258 (1969).
38. L. I. Denisovich and S. P. Gubin, *J. Organomet. Chem.*, **57,** 109 (1973).

39. L. I. Denisovich, I. V. Polovyanuk, B. V. Lokshin, and S. P. Gubin, *Izv. Akad. Nauk SSSR, Ser. Khim.*, 1964 (1971).
40. A. Chaloyard, A. Dormand, J. Tirouflet, and N. El Murr, *J. Chem. Soc., Chem. Commun.*, 214 (1980).
41. M. F. Lappert, C. J. Pickett, P. I. Riley, and P. I. W. Yarrow, *J. Chem. Soc., Dalton Trans.*, 805 (1981).
42. G. Costa, A. Puxeddu, and E. Reisenhofer, *J. Chem. Soc., Chem. Commun.*, 993 (1971).
43. G. Costa, A. Puxeddu, and E. Reisenhofer, *J. Chem. Soc., Dalton Trans.*, 1519 (1972).
44. A. Puxeddu, G. Costa, and N. Marisch, *J. Chem. Soc., Dalton Trans.*, 1489 (1980).
45. C. Gosden and D. Pletcher, *J. Organomet. Chem.*, **186,** 401 (1980).
46. C. Gosden, J. B. Kern, D. Pletcher, and R. Rosas, *J. Electroanal. Chem. Interfacial Electrochem.*, **117,** 101 (1981).
47. S. Sibille, M. Trouple, J. F. Fauvarque, and J. Perichon, *J. Chem. Res.* (*S*) 147 (1980).
48. M. Trouple, Y. Rollin, S. Sibille, J. Perichon, and J. F. Fauvarque, *J. Organomet. Chem.*, **202,** 435 (1980).
49. J. Perichon, S. Sibille, J.-C. Folest, J. Coulombeix, M. Troupel, and J. F. Fauvarque, *J. Chem. Res.* (*S*), 268; also 24, 26 (1980).
50. R. J. Klinger and J. K. Kochi, *J. Am. Chem. Soc.*, **104,** 4186 (1982).
51. M. E. Vol'pin, I. Y. Levitin, A. L. Sigaw, J. Halpern, and G. M. Tom, *Inorg. Chim. Acta*, **41,** 271 (1980).
52. W. H. Tamblyn, R. J. Klinger, W. S. Hwang, and J. Kochi, *J. Am. Chem. Soc.*, **103,** 3161 (1981).
53. R. J. Klinger and J. Kochi, *J. Organomet. Chem.*, **202,** 49 (1980).
54. R. H. Magnusson, S. Zulu, W.-M. T'sai, and W. P. Bering, *J. Am. Chem. Soc.*, **102,** 6887 (1980).
55. R. J. Klinger, J. C. Huffman, and H. K. Kochi, *Inorg. Chem.*, **20,** 34 (1981).
56. S. Valcher, *Ric. Sci.*, **8,** 1007 (1965).
57. P. K. Baker, K. Broadley, N. G. Connelly, B. A. Kelly, M. D. Kitchen, and P. Woodward, *J. Chem. Soc., Dalton Trans.*, 1710 (1980).
58. N. G. Connelly, R. L. Kelly, and M. W. Whiteley, *J. Chem. Soc., Dalton Trans.*, 34 (1981).
59. R. E. Dessey and R. L. Pohl, *J. Am. Chem. Soc.*, **90,** 1995 (1968).
60. R. E. Dessy, F. E. Stary, R. B. King, and M. Waldrop, *J. Am. Chem. Soc.*, **88,** 471 (1966).
61. R. E. Dessy, R. B. King, and M. Waldrop, *J. Am. Chem. Soc.*, **88,** 5112 (1966).
62. R. E. Dessy, J. C. Charkoudian, T. P. Abelo, and A. L. Rheingold, *J. Am. Chem. Soc.*, **92,** 3947 (1970).
63. W. E. Geiger, *J. Am. Chem. Soc.*, **96,** 2632 (1974).
64. N. El Murr, R. Dabard, and E. Laviron, *J. Organomet. Chem.*, **47,** C13 (1973).
65. S. P. Gubin, S. A. Smirnova, and L. I. Denisovich, *J. Organomet. Chem.*, **30,** 257 (1971).
66. A. N. Nesmeyanov, L. I. Denisovich, S. P. Gubin, N. A. Vol'kenan, E. I. Sirotkina, and I. Bolesova, *J. Organomet. Chem.*, **20,** 169 (1969).
67. D. Astruc, R. Dabard, and E. Laviron, *C. R. Acad. Sci.*, **608,** 1269 (1969).
68. C. Furlani, A. Furlani, and L. Sestili, *J. Electroanal. Chem.*, **9,** 140 (1965).
69. J. Tirouflet, E. Laviron, R. Dabard, and J. Komenda, *Bull. Soc. Chim. Fr.*, 857 (1963).
70. A. N. Nesmeyanov, N. A. Vol'kenav, P. V. Petrovskii, L. S. Kotova, V. A. Petrakova, and L. T. Denisovich, *J. Organomet. Chem.*, **210,** 103 (1981).
71. J. R. Hamon, D. Astruc, and P. Michand, *J. Am. Chem. Soc.*, **103,** 758 (1981).
72. A. N. Romanin, A. Venzo, and A. Cecconi, *J. Electroanal. Chem. Interfacial Electrochem.*, **112,** 147 (1980).
73. A. Chaloyard, A. Dormand, J. Tirouflet, and N. El Murr, *J. Chem. Soc., Chem. Commun.*, 214 (1980).
74. G. Paliani, S. M. Murgia, and G. Cardaci, *J. Organomet. Chem.*, **30,** 221 (1971).
75. G. Cardacci, S. M. Murgia, and G. Paliani, *J. Organomet. Chem.*, **77,** 253 (1974).
76. L. J. Denisovich and S. P. Gubin, *Russ. Chem. Rev.*, **46,** 27 (1977).
77. B. A. L. Crichton, J. R. Dilworth, C. J. Pickett, and J. Chatt, *J. Chem. Soc., Dalton Trans.*, 419 (1981).
78. C. J. Pickett and D. Pletcher, *J. Chem. Soc., Chem. Commun.*, 660 (1974).
79. C. J. Pickett and D. Pletcher, *J. Chem. Soc., Dalton Trans.*, 879 (1975).
80. C. J. Pickett and D. Pletcher, *J. Chem. Soc., Dalton Trans.*, 636 (1976).

81. R. J. Klinger and J. K. Kochi, *J. Am. Chem. Soc.*, **102,** 4791 (1980).
82. A. M. Bond, R. Colton, and J. J. Jackowski, *Inorg. Chem.*, **17,** 105 (1978).
83. A. M. Bond, J. W. Bixler, E. Mocellin, S. Datta, E. J. Janes, and S. S. Wreford, *Inorg. Chem.*, **19,** 1760 (1980).
84. J. A. Connor, E. M. Jones, G. R. McEven, M. K. Lloyd, and J. A. McCleverty, *J. Chem. Soc., Dalton Trans.*, 1743 (1973).
85. J. Chatt, C. T. Kan, G. J. Leigh, C. J. Pickett, and D. R. Stanley, *J. Chem. Soc., Dalton Trans.*, 2032 (1980).
86. P. M. Triechel and G. E. Dureen, *J. Organomet. Chem.*, **39,** C20 (1972).
87. P. M. Triechel, G. E. Dureen, and H. J. Mueh, *J. Organomet. Chem.*, **44,** 339 (1972).
88. C. J. Pickett and D. Pletcher, *J. Organomet. Chem.*, **102,** 327 (1975).
89. E. O. Fischer, F. J. Gamnel, J. O. Besenhard, A. Frank and D. Neugebauer, *J. Organomet. Chem.*, **191,** 261 (1980).
90. R. Riedce, H. Kofima, and K. Oelfele, *Angew. Chem.*, **92,** 550 (1980).
91. A. Bell, D. D. Klendworth, R. E. Wild, and R. A. Walton, *Inorg. Chem.*, **20,** 4457 (1981).
92. P. Lemoine and M. Gross, *C. R. Acad. Sci., Ser. C*, **280,** 797 (1975).
93. P. Lemoine and M. Gross, *Electrochim. Acta*, **1,** 121 (1976).
94. N. El Murr and A. Chaloyard, *Inorg. Chem.*, **21,** 2206 (1982).
95. A. M. Bond, R. Colton, and J. J. Jackowski, *Inorg. Chem.*, **14,** 174 (1975).
96. C. J. Pickett and D. Pletcher, *J. Chem. Soc., Dalton Trans.*, 749 (1976).
97. L. I. Denisovich, A. A. Ioganson, S. P. Gubin, N. E. Kolobova, and K. N. Anisimov, *Izv. Akad. Nauk SSSR, Ser. Khim.*, 258 (1969).
98. C. J. Pickett and D. R. Stanley, unpublished results.
99. B. Fisher and R. Eisenberg, *J. Am. Chem. Soc.*, **102,** 7361 (1980).

The Chemistry of the Metal—Carbon Bond, Volume 2
Edited by F. R. Hartley and S. Patai


CHAPTER **2**

# Heterolytic cleavage of main group metal—carbon bonds

MICHAEL H. ABRAHAM and PRISCILLA L. GRELLIER

*Chemistry Department, University of Surrey, Guildford, Surrey GU2 5XH, UK*

# I. INTRODUCTION

## A. General Introduction

The non-transition metal—carbon bond is polarized in the sense $M^{\delta+}—C^{\delta-}$ and thus heterolytic cleavage of such a bond can formally take place by electrophilic attack at the carbon centre, although an electrophile will actually first interact with the filled metal—carbon $\sigma$-bond. The main types of electrophilic reagents that have been studied in some detail are the halogens, various acids, and metal salts and esters; the latter reagents lead to the well known metal-for-metal exchange reactions. A number of books[1–5] and important reviews[6–9] have been devoted to accounts of the various types of electrophilic substitution and to the mechanism of those substitutions in organometallic compounds of the non-transition metals.

Throughout this work, the organometallic substrate is denoted as $RMX_n$, where R is the group cleaved from the non-transition metal atom, M. The entity $MX_n$ is the leaving group where X may or may not be equal to R. In many cases it is useful to consider the electrophilic reagent as E—N, where E and N are electrophilic and nucleophilic centres, and to represent the substitution as $N—E\ R—MX_n$.

If R is an alkyl group, a fully developed carbanion, $R^-$, may be formed, but more generally the alkyl group may simply display 'carbanion character' during the substitution. Aryl groups, of course, are very unlikely to exist as simple carbanions under normal circumstances. A carbanion derived from an alkyl group, if fully developed and free in solution, will be pyramidal in shape ($sp^3$), and, like the isoelectronic nitrogen (amine) derivatives, will invert rapidly. The tendency of a C—H bond to undergo carbanion formation, and hence to be susceptible to electrophilic attack, will very largely be a function of the nature of the metal concerned. Approximate enthalpies of reaction for a number of methyl—metal compounds are given[4] in Table 1 and show, as expected, that formation of the methyl anion is favoured when the metal is electropositive such as lithium, and when the reaction is carried out in a polar solvent. However, the alkyls of these electropositive metals are not easy to handle, and are not simple in terms of structure, being polymeric in most solvents in which they are reasonably stable. Thus considerable

TABLE 1. Standard enthalpy changes[4], $\Delta H^0$, for the formation of the methyl anion from various organometallic compounds at 25 °C

| Reaction | $\Delta H^0$ (kcal mol$^{-1}$) |
|---|---|
| $MeH(g) \longrightarrow Me^-(g) + H^+(g)$ | 392 |
| $\frac{1}{2}Me_2Hg(g) \longrightarrow Me^-(g) + \frac{1}{2}Hg^{2+}(g)$ | 340 |
| $\frac{1}{2}Me_2Zn(g) \longrightarrow Me^-(g) + \frac{1}{2}Zn^{2+}(g)$ | 332 |
| $MeLi(g) \longrightarrow Me^-(g) + Li^+(g)$ | 158 |
| $MeH(aq.) \longrightarrow Me^-(aq.) + H^+(aq.)$ | 54 |
| $\frac{1}{2}Me_2Hg(aq.) \longrightarrow Me^-(aq.) + \frac{1}{2}Hg^{2+}(aq.)$ | 45 |
| $\frac{1}{2}Me_2Zn(aq.) \longrightarrow Me^-(aq.) + \frac{1}{2}Zn^{2+}(aq.)$ | 10 |
| $MeLi(aq.) \longrightarrow Me^-(aq.) + Li^+(aq.)$ | −17 |

effort has gone into the design of organometallic compounds of the less electropositive metals, for instance mercury, in which the potential carbanion is stabilized through suitable substituents in the alkyl group. In this way, kinetics of carbanion formation from compounds such as [PhCH($CO_2$Et)HgBr] may be studied without the complications that arise in the study of lithium or magnesium compounds.

We briefly survey the various mechanisms that have been postulated for electrophilic substitution at saturated carbon, and then at aromatic carbon. Until recently, relatively few mechanisms had been put forward for electrophilic substitution of alkyls of non-transition metals, in contrast to alkyls of transition metals for which a number of mechanisms based on one- or two-electron oxidation of the organometallic substrate have been suggested. There seems now, however, to be not so much distinction between non-transition metals and transition metals, with respect to organometallic reaction mechanisms, and it seems likely that in the future a number of mechanisms originally devised to account for observations on alkyls of transition metals will be applied to compounds of the non-transition metals.

Throughout this work, rate constants are given in units of $s^{-1}$ or $l\,mol^{-1}\,s^{-1}$, and activation parameters calculated from second-order rate constants are expressed on the molar concentration scale.

## B. Mechanisms of Electrophilic Substitution at Saturated Carbon

### 1. *Unimolecular electrophilic substitution ($S_E1$)*

This mechanism is characterized by an initial slow, reversible ionization to form a carbanion, which can then react rapidly with the electrophile. When the back-reaction of the ionization step can be neglected, and when the substrate and electrophile are present in reasonably low concentrations, then the kinetic form is expected to be first order with respect to $RMX_n$.

$$RMX_n \underset{\text{fast}}{\overset{\text{slow}}{\rightleftharpoons}} R^-\overset{+}{M}X_n \quad (1)$$

$$R^- + EN \xrightarrow{\text{fast}} RE + N^- \quad (2)$$

If the carbanion, $R^-$, is a relatively long-lived species in solution, then the stereochemical outcome will be predominant racemization if the organometallic substrate is optically active due to an asymmetric $\alpha$-carbon atom. However, if the carbanion is shielded by the leaving group, or held in some preferred configuration by substituents on the alkyl group, then the product RE may be formed with inversion or retention of configuration.

### 2. *Bimolecular electrophilic substitution* ($S_E2$)

This is a general mechanism which can further be subdivided. Various notations have been suggeted for the subdivisions; we shall use a systematic description in which the subdivisions are all regarded as variants of the general $S_E2$ mechanism.

a. $S_E2$ (*open*). Dominant electrophilic attack, formally at the $\alpha$-carbon atom, but actually at the metal—carbon $\sigma$-bond, may proceed via an open transition state in which four pairs of electrons are involved with the $\alpha$-carbon atom. Inversion and retention of configuration at the $\alpha$-carbon atom are both energetically possible (equations 3 and 4), and both stereochemical outcomes have been observed.[10,11]

$$RMX_n + EN \longrightarrow |N{-}E\ R{-}MX_n|^{\neq} \longrightarrow N^- + ER + {}^+MX_n \qquad (3)$$

$$RMX_n + EN \longrightarrow \left| R\ \begin{matrix} MX_n \\ E{-}N \end{matrix} \right|^{\neq} \longrightarrow N^- + RE + {}^+MX_n \qquad (4)$$

Regardless of stereochemistry, the kinetic form of the $S_E2$ (open) mechanism will be that of second-order kinetics, first order in $RMX_n$ and first order in EN, when the two reactants are present in dilute solution at the same order of concentration. Because of the charge separation involved in formation of the transition states, reactions proceeding by mechanism $S_E2$ (open) should be accelerated by polar solvents and by the presence of inert salts.

b. $S_E2$ (*cyclic*). More or less synchronous electrophilic and nucleophilic attack results in a cyclic transition state (equation 5), where retention of configuration at the

$$RMX_n + EN \longrightarrow \left| R\ \begin{matrix} MX_n \\ N \\ E \end{matrix} \right|^{\neq} \longrightarrow RE + NMX_n \qquad (5)$$

$\alpha$-carbon atom is the only possible stereochemical outcome. Other notations that have been used to describe this mechanism include $S_Ei$, $S_F2$ and $S_E2$. Under the same conditions as mentioned for the $S_E2$ (open) mechanism, the kinetic form second order in each reactant will be observed, so that on simple kinetic and stereochemical grounds it is not easy to distinguish between mechanisms $S_E2$ (open) and $S_E2$ (cyclic). However, the latter would be expected to respond rather less to polar solvents and to inert added salts.

It has been suggested[12] that four-centre concerted reactions of type 5 are forbidden on grounds of orbital symmetry. However, the nucleophilic part of the reagent EN invariably possesses a non-bonding pair of electrons and the transition state is more correctly written as in equation 6.

$$RMX_n + EN \longrightarrow \left| \begin{matrix} R{-}MX_n \\ E{-}N\colon \end{matrix} \right|^{\neq} \longrightarrow RE + NMX_n \qquad (6)$$

These types of processes are termed pseudopericyclic and are not orbital symmetry forbidden[13].

The cyclic mechanism is not restricted to four-centre transition states, but may take place through six-centre transition states, as in the acidolysis of organometallic compounds (equation 7)[14].

$$RMX_n + MeCO_2H \longrightarrow \left| \text{R} \cdots \overset{X_n}{M} \cdots :O \!=\! C\text{—Me} \cdots \text{H—O} \right|^{\neq} \longrightarrow RH + MeCO_2MX_n \quad (7)$$

c. $S_E2$ (*coord*). A strongly nucleophilic centre in the reagent may first coordinate to the substrate $RMX_n$. Electrophilic cleavage then follows, with stereochemical retention of configuration at the $\alpha$-carbon atom (equation 8).

$$RMX_n + EN \rightleftharpoons \underset{\substack{\uparrow \\ EN}}{RMX_n} \longrightarrow RE + NMX_n \quad (8)$$

The kinetic situation is not simple, and depends on the various rate constants for the three elementary reactions involved. If the complex is present in but very small concentration, application of the steady-state approximation reveals that second-order kinetics will be observed[15]. Should the complex be formed rapidly and irreversibly, the kinetic form will be first order in whichever of $RMX_n$ or EN is present at the lower initial concentration. In general, however, the kinetics are complicated; Chipperfield[16] has discussed in some detail the situation in which an irreversible biomolecular elementary reaction is followed by a unimolecular decomposition of the intermediate complex (equation 9).

$$RMX_n + EN \longrightarrow C \longrightarrow RE + NMX_n \quad (9)$$

### 3. Mechanisms involving rearrangement

If an allylic substrate undergoes electrophilic substitution, the possibility of rearrangement of the allyl entity has to be considered. When an allylic anion is formed, the electron rich centre will not solely be the $\alpha$-carbon atom, but also the $\gamma$-carbon atom, for example, as in equation 10.

$$CH_3CH\!=\!CHCH_2MX_n \longrightarrow CH_3CH\!=\!CH\bar{C}H_2 \longleftrightarrow CH_3\bar{C}HCH\!=\!CH_2 \quad (10)$$

Attack of the electrophile at the $\alpha$-carbon atom will lead to mechanism $S_E1$, or if attack takes place on the $RMX_n$ substrate to one of the $S_E2$ variants described above. Electrophilic attack at the $\gamma$-carbon atom will lead to substitution with rearrangement, denoted as $S_E1'$, $S_E2'$, etc. There is a fundamental difference between these substitutions with rearrangement and those without, namely that attack at the $\gamma$-carbon atom involves an unsaturated carbon atom, whereas attack at $\alpha$-carbon remains a substitution at saturated carbon. Mechanisms $S_E2'$ (open) and $S_E2'$ (cyclic) are illustrated by reactions of a crotyl compound (equations 11 and 12).

$$\underset{\text{E—N}}{CH_3CH\!=\!CH\!-\!CH_2\!-\!MX_n} \longrightarrow CH_3\underset{\substack{| \\ E}}{C}H\!-\!CH\!=\!CH_2 + N^- + {}^+MX_n \quad (11)$$

$$CH_3CH{=}CH \cdots E \leftarrow \cdots CH_2 \quad N{:}\ \ MX_n \longrightarrow CH_3CH(E){-}CH{=}CH_2 + NMX_n \qquad (12)$$

### 4. Mechanisms involving catalysis

Nucleophiles may intercede in electrophilic substitutions either by coordinating to the substrate or to the electrophile itself. This 'external' nucleophilic effect must be distinguished from the effect of the nucleophilic part of the electrophilic reagent in mechanisms such as those shown in equations 5, 6, and 8.

The nucleophile may complex reversibly with the substrate (equation 13) and the new complex may then react by an $S_E1$ type of mechanism (equation 14), where B denotes the nucleophilic catalyst.

$$RMX_n + B \underset{\text{fast}}{\overset{\text{fast}}{\rightleftharpoons}} RMBX_n \qquad (13)$$

$$RMBX_n \xrightarrow{\text{slow}} R^- + MBX_n \qquad (14)$$

$$R^+ + EN \xrightarrow{\text{fast}} RE + N^- \qquad (15)$$

In a sequence such as equations 13–15, the rate of formation of RE will be proportional to $[RMX_n][B]$.

If a reaction taking place by mechanism $S_E2$ (open) is catalysed by a nucleophile, and if it is concluded that one molecule of the nucleophile is involved in the transition state, then in principle the nucleophile may be carried into the transition state either by the substrate, as $RMBX_n$, or by the electrophilic reagent, as EBN. Generally, it is expected that the former will be the case,

$$RMX_n + B \rightleftharpoons RMBX_n \qquad (16)$$

$$RMBX_n + EN \longrightarrow RE + NMX_n + B \qquad (17)$$

because $RMBX_n$ will be more reactive than $RMX_n$ towards an electrophile, whereas EBN will be less reactive than EN towards the substrate.

$$EN + B \rightleftharpoons EBN \qquad (18)$$

$$RMX_n + EBN \longrightarrow RE + NMX_n + B \qquad (19)$$

If the transition state is cyclic, with 'internal' nucleophilic assistance, the above argument will be reversed because $RMBX_n$ will be less susceptible than $RMX_n$ to internal nucleophilic attack, whilst EBN will be more powerful than EN in completing the cyclic transition state. Thus the expected mode of catalysis will now be via equations 18 and 19. This will also be true for nucleophilic catalysis of a reaction proceeding by mechanism $S_E2$ (coord).

It has been suggested by Gielen *et al.*[17] that a solvent molecule may function as a nucleophilic catalyst and might formally be incorporated in the transition state (equation 20).

$$R_4Sn + Br_2 \xrightarrow{\text{MeOH}} \left| \begin{matrix} & SnR_3 \longleftarrow OHMe \\ R & \\ & Br{-}Br \end{matrix} \right|^{\neq} \longrightarrow RBr + BrSnR_3 + MeOH \qquad (20)$$

They suggested that the solvent plays a major role in influencing the mechanism of $S_E2$ reactions: in polar, nucleophilic, solvents, external nucleophilic catalysis by the solvent leads to $S_E2$ (open) mechanisms, but in solvents with little nucleophilic power, internal nucleophilic assistance by EN now competes effectively with the solvent and results in the $S_E2$ (cyclic) mechanism. Unfortunately, there seem to be few direct methods of testing these suggestions; although mechanism 20 results in inversion of configuration at the tin atom, whereas a cyclic mechanism would lead to retention of configuration at the tin atom, no experiments on these lines have been reported.

### 5. *Electron transfer mechanisms*

In recent years, Kochi and co-workers have published numerous papers concerning the electrophilic cleavage of metal—carbon bonds via electron transfer mechanisms. They dealt initially[18] with alkyl transfers from a series of organolead(IV) compounds to hexachloroiridate(IV), a compound known to participate in outer-sphere oxidations involving one-electron changes. More recently, Fukuzumi and Kochi have described charge-transfer complexes betwen alkylmetal compounds and typical electrophilic reagents such as halogens[19] and mercury(II) halides[20,21], and have suggested that the cleavage of the organometallic substrate involves the formation of a charge-transfer complex. Their general mechanism[20] for electrophilic cleavage is given in equations 21–23, where $K_{DA}$ is the formation constant for the

$$\mathrm{RMX}_n + \mathrm{EN} \overset{K_{DA}}{\rightleftharpoons} [\mathrm{RMX}_n\mathrm{EN}] \tag{21}$$

$$[\mathrm{RMX}_n\mathrm{EN}] \xrightarrow{k_E} [\mathrm{R\overset{+}{M}X}_n\mathrm{\overset{-}{E}N}] \tag{22}$$

$$[\mathrm{R\overset{+}{M}X}_n\mathrm{\overset{-}{E}N}] \xrightarrow{\text{fast}} \mathrm{RE} + \mathrm{NMX}_n \tag{23}$$

charge-transfer complex, and $k_E$ represents the rate constant for conversion of the charge-transfer species to the ion pair $[\mathrm{R\overset{+}{M}X}_n\mathrm{\overset{-}{E}N}]$ by the electron-transfer step 22. The final products in equation 23 arise through fragmentation of the cation $\mathrm{RMX}_n$ and recombination within a solvent cage.

Fukuzumi and Kochi[10,20] correlated the effects of various alkyl groups, R, the effects of solvent, and the effects of the electrophile in a series of substitutions of tetraalkyltins, $R_4Sn$, by the electrophiles bromine, iodine and mercury(II) chloride, through the general equations 24 and 25.

$$\Delta G_r^{\neq} = \Delta h\nu_{CT} + \Delta G_r^s \tag{24}$$

$$\Delta G_r^{\neq} = \Delta E + I_D + \Delta G_r^s \tag{25}$$

In equation 24, $\Delta G_r^{\neq}$ is obtained through $-RT\ln(k/k_0)$, where $k$ and $k_0$ are the rate constants for substitution of $R_4Sn$ and $Me_4Sn$ by a given electrophile in a given solvent; $\Delta h\nu_{CT}$ is the relative charge-transfer transition energy of $R_4Sn$ with the electrophile in the solvent concerned; $\Delta G_r^s$ in the relative solvation energy of $R_4Sn$ on oxidation to $R_4Sn^+$, again with respect to the standard $Me_4Sn$. Fukuzumi and Kochi were able to obtain values of $\Delta G_r^s$ in solvent acetonitrile[22], and then estimated those in other solvents through a modified Born equation[20]. Equation 25 may then be used, since $\Delta E = \Delta h\nu_{CT} - I_D$, in order to express the steric effect of the tetraalkyltins, $\Delta E$, provided that $I_D$, the corresponding ionization potential, is known.

### C. Mechanisms of Electrophilic Substitution at Aromatic Carbon

No example of a unimolecular electrophilic substitution of an aryl organometallic compound has yet been reported, and the electrophilic cleavage of metal—aryl bonds usually proceeds through mechanisms analogous to the familiar bimolecular electrophilic aromatic substituted mechanism. Demetallations of metal—aryl compounds are thought to involve species such as $[1—E—1—(MX_n)\text{-}C_6H_5]^+$ (**1**), where *E* is the electrophile, similar to the Wheland intermediates ($\sigma$-complexes) postulated in the common aromatic substitutions.

The rate-determining step can be either the formation of the intermediate **1** or the subsequent cleavage of the metal—aryl bond. When the electrophile is an acid or is a halogen, formation of species **1** is often rate determining. For example, the acidolysis of various aryl—metal compounds by HCl is slower in $D_2O$/dioxane than in $H_2O$/dioxane[23], and other isotope studies also indicate the rate-determining formation of **1**[24,25].

It is possible for aromatic electrophilic substitutions to proceed via a $\pi$-complex instead of, or as well as, the $\sigma$-complex. Hashimoto and Morimoto[26] have suggested that in the substitution of $ArSnEt_3$ compounds by mercury(II) acetate in tetrahydrofuran a $\pi$-complex is involved. It is well known that potential electrophiles such as $Ag^+$ and bromine form $\pi$-complexes with aromatic compounds, and Fukuzumi and Kochi[21] have described charge-transfer complexes between mercury(II) salts and hexamethylbenzene. If the transition state for electrophilic substitution does resemble a $\pi$-complex, then it would be expected that substituent effects in the aryl system would be appreciably less marked than in cases where the transition state resembles a $\sigma$-complex.

## II. ELECTROPHILIC CLEAVAGE OF NON-TRANSITION METAL—CARBON BONDS BY METAL COMPOUNDS

### A. General Introduction

The electrophilic substitution of an organometallic substrate by a metal salt or ester, that is the metal-for-metal exchange reaction, is one of the most common general reactions in organometallic chemistry. Many of these reactions are rapid under normal laboratory conditions and are of considerable preparative value. Thus aryl or alkyl compounds of the more electropositive metals such as lithium or magnesium are used extensively to prepare the corresponding aryl or alkyl compounds of less electropositive metals such as zinc, cadmium, mercury, boron, silicon, germanium, tin, and lead. Detailed procedures for the preparation of organometallic compounds by the metal-for-metal exchange reaction are well documented (see, for example, refs.[27–29]) and so will not be listed here.

### B. Cleavage of Group IA Metal—Carbon Bonds

#### 1. *Cleavage of lithium—carbon bonds by silicon*(IV) *compounds*

Seyferth and Vaughan[30] have reported that reaction 26 takes place with retention of configuration at the site of substitution.

$$trans\text{-}MeCH{=}CHLi + Me_3SiCl \longrightarrow trans\text{-}MeCH{=}CHSiMe_3 + LiCl \quad (26)$$

## C. Cleavage of Group IIA Metal—Carbon Bonds

### 1. Cleavage of magnesium—carbon bonds by mercury(II) compounds

It was reported by Jensen and Nakamaye[31] that reaction 27 in ether occurs with retention of configuration at the carbon atom undergoing substitution.

$$\text{(2-MgBr)bicyclo[2.2.1]heptane} + \text{HgBr}_2 \longrightarrow \text{(2-HgBr)bicyclo[2.2.1]heptane} + \text{MgBr}_2 \quad (27)$$

### 2. Cleavage of magnesium—carbon bonds by boron(III) compounds

Breuer[32] has studied the stereochemistry of the reaction of the steroidal Grignard reagent **2** with diborane in tetrahydrofuran, to yield the corresponding organoborane. The reaction took place with retention of configuration at the carbon atom undergoing substitution.

BrMg $C_8H_{17}$ H H

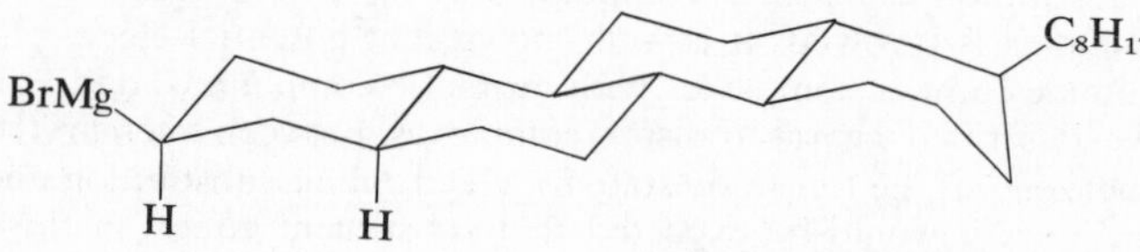

**(2)**

### 3. Cleavage of magnesium—carbon bonds by silicon(IV) compounds

Reaction 28 was found by Reid and Wilkins[33] to follow second-order kinetics in ether solvent; activation parameters are given in Table 2. A cyclic, four-centred transition state was suggested[33].

$$\text{MeMgX} + \text{R}_3\text{SnX}' \longrightarrow \text{MeSiR}_3 + \text{MgXX}' \quad (28)$$

TABLE 2. Activation parameters[33] at 0 °C for the reaction

$$\text{MeMgX} + \text{R}_3\text{SiX}' \longrightarrow \text{MeSiR}_3 + \text{MgXX}'$$

| $X$ | $R_3SiX'$ | $\Delta G^{\neq}$ [kcal mol$^{-1}$ (±0.05)] | $\Delta H^{\neq}$ (kcal mol$^{-1}$ (±0.6)] | $\Delta S^{\neq}$ [cal K$^{-1}$ mol$^{-1}$ (±2.5)] |
|---|---|---|---|---|
| Cl | $Me_3SiCl$ | 21.2 | 8.0 | −48 |
| Cl | $ClCH_2SiMe_2Cl$ | 19.4 | 8.3 | −42 |
| Br | $Me_3SiF$ | 20.86 | 8.8 | −44 |
| Br | $Me_3SiCl$ | 20.47 | 8.1 | −45 |
| Br | $Me_3SiBr$ | 20.21 | 10.6 | −35 |
| Br | $ClCH_2SiMe_2Cl$ | 19.72 | 11.5 | −30 |
| I | $Me_3SiF$ | 21.04 | 13.5 | −27 |
| I | $Me_3SiCl$ | 20.34 | 10.6 | −34 |
| I | $Me_3SiBr$ | 19.32 | 11.3 | −29 |
| I | $Me_3SiI$ | 18.46 | 9.2 | −34 |
| I | $ClCH_2SiMe_2Cl$ | 20.52 | 9.8 | −39 |
| I | $PhSiMe_2Cl$ | 20.12 | 11.2 | −33 |
| I | $p$-tolyl$SiMe_2Cl$ | 20.19 | 10.1 | −37 |

## D. Cleavage of Group IIB Metal—Carbon Bonds

### 1. Cleavage of zinc—carbon bonds by mercury(II) compounds

Abraham and Rolfe[34] investigated reaction 29 in ether and tetrahydrofuran. They showed that the kinetic form was first order in each reactant, and recorded the second-order rate constants for reaction of $Et_2Zn$ at 0 °C as 6.0 l $mol^{-1}$ $s^{-1}$ in ether and 0.64 l $mol^{-1}$ $s^{-1}$ in tetrahydrofuran. At 35 °C relative rate constants in ether were $Me_2Zn$ 100, $Et_2Zn$ 450, *n*-$Pr_2Zn$ 1700 and *i*-$Pr_2Zn$ 2200; it was suggested that reaction took place via a cyclic four-centred transition state. In tetrahydrofuran at 35 °C, all four dialkylzincs reacted at about the same rate, however.

$$R_2Zn + PhHgCl \longrightarrow RHgPh + RZnCl \quad (29)$$

### 2. Cleavage of mercury—carbon bonds by palladium(II) and iron(III) compounds

The action of $[PdCl_2(PhCN)_2]$ on *threo*- or *erythro*-$MeCH(NMe_2)CH(HgCl)Me$ in tetrahydrofuran was reported by Bäckvall and Åkermark[35] to yield the corresponding *threo*- or *erythro*-$MeCH(NMe_2)CH(PdCl)Me$ compound, that is with retention of configuration at the carbon atom undergoing substitution.

Wong and Kochi[36] have studied the reaction of dialkylmercurys with various iron(III) complexes in acetonitrile; however, these reactions seem to be outer-sphere electron transfer processes and are not simple electrophilic substitutions.

### 3. Cleavage of mercury—carbon bonds by mercury(II) compounds

These reactions are amongst the best known electrophilic substitutions, and have been discussed in detail a number of times[1,2,4,7,8]. The earlier work will therefore be dealt with briefly, and attention will be focused on the more recent investigations and reports.

The one-alkyl mercury-for-mercury exchange reaction has been the subject of extensive kinetic and mechanistic studies, in particular by Reutov and coworkers and by Hughes and Ingold and coworkers. Since the products and reactants are chemically identical, the exchange is usually followed by the transfer of radioactive mercury ($^{203}Hg$, denoted by $\overset{*}{Hg}$) from the mercury(II) salt to the organomercury(II) salt:

$$RHgX + \overset{*}{Hg}X_2 \longrightarrow R\overset{*}{Hg}X + HgX_2 \quad (30)$$

Following earlier studies by Nefedov and Sintova[37–40], a re-investigation by Hughes *et al.*[41] established that reaction 30 [R = (−)-*sec*-butyl; X = OAc) took place with retention of configuration at the carbon atom undergoing substitution, and that reaction 30 (R = Me; X = Br) followed a kinetic form first order in each reactant in ethanol at 100 °C. Added lithium nitrate and added water accelerated the reaction, and the rate constants increased along the sequence X = Br < I < OAc ≪ $NO_3$. It was concluded that in ethanol, these one-alkyl exchange reactions took place by a simple bimolecular $S_E2$ (open) mechanism with retention of configuration. Hughes and Volger[42] obtained rate constants for substitution of a number of alkylmercury(II) salts by $HgBr_2$ and explained the order of decreasing reactivity R = Me > Et ≈ $Pe^{neo}$ > $Bu^s$ as due to steric nonbonded compressions in the transition state. The various recorded rate constants and activation parameters are collected in Table 3.

Abraham *et al.*[43] later calculated the steric effect of various alkyl groups in RHgBr

TABLE 3. Second-order rate constants, $10^5 k_2$ and activation energies in ethanol for the reaction

$$RHgX + \overset{*}{H}gX_2 \longrightarrow R\overset{*}{H}gX + HgX_2$$

| R | X | $10^5 k_2$ ($l\ mol^{-1}\ s^{-1}$) | $\Delta H^{\neq}$ ($kcal\ mol^{-1}$) | Ref. |
|---|---|---|---|---|
| Me | Br | 35[a] | 17.7 | 38 |
| Et | Br | — | 10.9 | 37 |
| $Pr^n$ | Br | 9.7[a] | 19.4 | 40 |
| Me | Br | 12.8[b]; 0.5[c]; 0.0007[d] | 19.2 | 41, 42 |
| Me | I | 101[b] | | 41, 42 |
| Me | OAc | 500[c] | | 41, 42 |
| Me | $NO_3$ | 169[d] | | 41, 42 |
| Et | Br | 5.4[b] | | 41, 42 |
| $Pe^{neo}$ | Br | 4.2[b] | | 41, 42 |
| $Bu^s$ | OAc | 30.9[c] | | 41, 42 |

[a] At 100 °C.
[b] At 100.2 °C.
[c] At 59.8 °C.
[d] At 0.0 °C.

on the rate constants for substitution by $HgBr_2$, using both an open transition state model and a cyclic four-centred model; only in the former case could agreement be obtained between calculated and observed rate constants (Table 4). Interestingly, the steric effect of $\alpha$-methyl groups (as in R=Et, $Pr^i$, $Bu^t$) is much more pronounced than the effect of $\beta$-methyl groups (as in $Pr^n$, $Bu^i$, $Pe^{neo}$) in the open transition state **3**

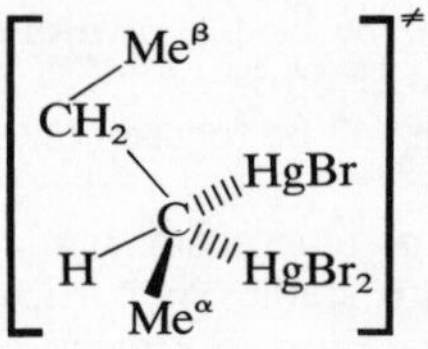

$Me^{\beta}$ $CH_2$ HgBr C H $HgBr_2$ $Me^{\alpha}$ ≠

**(3)**

TABLE 4. Calculated and observed relative rate constants for the substitution of RHgBr by $HgBr_2$[43]

| | 60 °C | | 100 °C | |
|---|---|---|---|---|
| R | Calc.[a] | Obs. | Calc.[a] | Obs. |
| Me | 1 | 1 | 1 | 1 |
| Et | 0.515 | | 0.502 | 0.42 |
| $Pr^n$ | 0.461 | | 0.451 | |
| $Bu^i$ | 0.413 | | 0.405 | |
| $Pe^{neo}$ | 0.387 | | 0.377 | 0.33 |
| $Pr^i$ | 0.068 | | 0.076 | |
| $Bu^s$ | 0.049 | 0.062 | 0.056 | |
| $Bu^t$ | $2\times10^{-6}$ | | $0\times10^{-6}$ | |

[a] Based on the transition state (**3**) with each C····Hg bond length 2.288 Å and with the $Hg\hat{C}Hg$ angle 84°.

so that the alkyl sequence of reactivity in this substitution with retention is different to that in typical $S_N2$ (inv) reactions.

The effect of added bromide ions on reaction 30 (R = Me, X = Br) in ethanol was later studied[7,41,44], and the pronounced catalytic effect of bromide ion was suggested to be due to formation of the species $HgBr_3^-$ (when added bromide ion was less than the concentration of $HgBr_2$) which allows the formation of a cyclic transition state with a bridging bromine atom (**4**). When the concentration of added bromide ion is

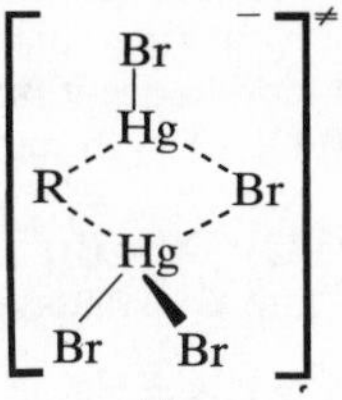

(**4**) 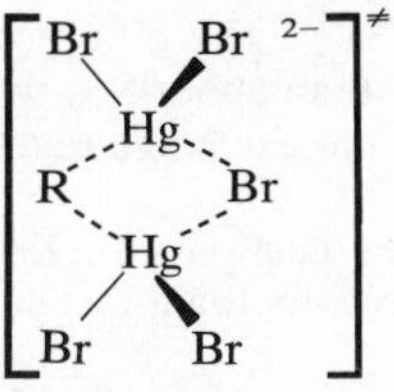 (**5**)

greater than that of $HgBr_2$, formation of the species $RHgBr_2^-$ as well as $HgBr_3^-$ leads to two-anion catalysis via a bridged cyclic transition state such as **5**. Ions other than $Br^-$ also catalyse the one-alkyl exchange, the catalylic power being in the order of increasing coordinating ability towards $HgX_2$ or RHgX, viz. $NO_3^- < OAc^- < Cl^- < Br^- < I^-$[44]. Lucchini and Wells[45] later suggested that in the one-anion catalysed reaction the bromide ion is brought into the transition state, not as $HgBr_3^-$, but as the species $RHgBr_2^-$, and that in the two-anion catalysed reaction the species $RHgBr_3^{2-}$ reacts with $HgBr_2$. Abraham *et al.*[46] then extended their calculations of steric effects to these anion-catalysed exchanges and showed that a model based on transition state **5** for the two-anion catalysis yielded rate factors in good agreement with experiment. However, for the one-anion catalysis, their calculations did not support the unsymmetrical transition state **4**, but were in accord with the symmetrical transition state **6** suggested by Jensen and Rickborn[2].

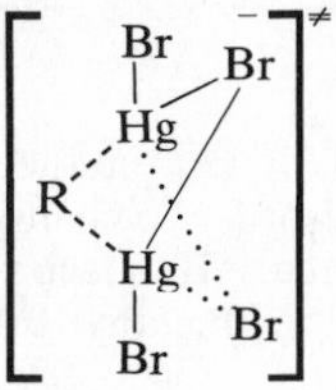

(**6**)

It should be noted that whereas there is little controversy over the possibility of bridged transition states in the anion-catalysed reactions, the assumption of open transition states such as **3** has been criticized[2,5] on the grounds of violation of the principle of microscopic reversibility. It has been shown[47], however, that an unsymmetrical transition state **3** does not necessarily violate the principle, and that reaction paths can be constructed that are of overall symmetry with respect to forward and back reactions and yet still include transition state **3**[47].

The one-aryl exchange reaction 30 (R = Ph) has been investigated by Reutov and coworkers[48]. In anhydrous toluene reaction 30 (R = Ph; X = Cl) was found to be second order overall[48]; the second-order rate constant at 30.1 °C was $0.071\,l\,mol^{-1}\,s^{-1}$ and the activation energy reported as $6.09\,kcal\,mol^{-1}$. A four-centred cyclic transition state was tentatively suggested[48]. Previous workers[49] had found that in benzene saturated with water reaction 30 (R = Ph; X = Cl) was kinetically of first order with respect to $HgCl_2$ and of order 1.7 with respect to PhHgCl; an activation energy of only 3.5 kcal $mol^{-1}$ was reported[49].

Reutov and coworkers[50] showed also that reaction 30 (R = Ph; X = Br) in methanol was kinetically of first order with respect to each reactant. At 22 °C the rate constant was reported to be 0.117 l $mol^{-1}$ $s^{-1}$ with $E_a = 12.8 \pm 1$ kcal $mol^{-1}$ and $\Delta S^{\neq} =$ −22 cal $K^{-1}$ $mol^{-1}$ [50]. From the kinetic results in Table 3 it may be calculated that for the one-alkyl exchange 30 (R = Me; X = Br) in ethanol at 22 °C the rate constant is only $1.1 \times 10^{-7}$ l $mol^{-1}$ $s^{-1}$. The difference in solvent (methanol/ethanol) cannot lead to a rate factor larger than about 5 at the most, and hence it may be deduced that the rate factor Ph/Me in the exchange reaction 30 (X = Br) in methanol or ethanol is at least $2 \times 10^5$.

The aryl exchange reaction 31 (X = Br, $ClO_4$; X′ = Br, OAc, $ClO_4$) has been studied[50] using aqueous methanol as solvent; rate constants are given in Table 5.

$$PhHgX + \overset{*}{Hg}X'_2 \longrightarrow Ph\overset{*}{Hg}X' + HgXX' \qquad (31)$$

Unlike the above aryl exchange reactions, that between $C_6F_5HgBr$ and $\overset{*}{Hg}Br_2$ in dimethyl sulphoxide was reported[51] to be first order overall (first order in $C_6F_5HgBr$ and zero order in $HgBr_2$) with a first-order rate constant of $8.5 \times 10^{-6}$ $s^{-1}$ at 30 °C. It was suggested[51] that coordination of the solvent to the mercury atom in $C_6F_5HgBr$ helps to stabilize the carbanion centre. Thus in benzene, practically no exchange takes place unless bromide ion (in the form of $Bu_4NBr$) is added, when a rapid second-order exchange occurs, possibly via the intermediate **7**[51].

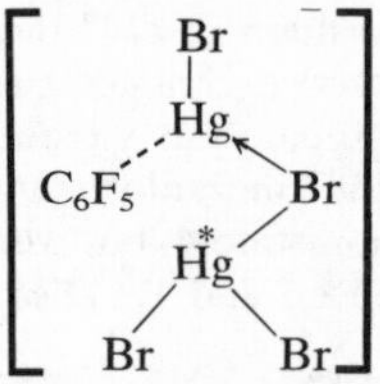

Br
Hg
$C_6F_5$ Br
$\overset{*}{Hg}$
Br Br

(**7**)

In a novel study, Peringer and Winkler[52] studied the exchange between $PhHg^+$ and $Hg^{2+}$ and found that the second-order exchange was faster in methanol than in dimethyl sulphoxide. The two mercury(II) species were introduced as their trifluoromethanesulphonate salts solvated by dimethyl sulphoxide.

TABLE 5. Second-order rate constants[50] in methanol + 10% water for the reaction

$$PhHgX + \overset{*}{Hg}X'_2 \rightleftharpoons Ph\overset{*}{Hg}X' + HgXX'$$

| X | X′ | Temperature (°C) | $k_2$ (l $mol^{-1}$ $s^{-1}$) |
|---|---|---|---|
| Br | Br | 20 | 0.157 |
| Br | OAc | 4 | 2.71 |
| Br | $ClO_4$ | 4 | Very fast |
| $ClO_4$ | Br | 4 | 0.684 |
| $ClO_4$ | $ClO_4$ | 4 | Very fast |
| $ClO_4$ | $ClO_4$ | −20 | Very fast |

TABLE 6. Second-order rate constants[53–55], $10^4k_2$ ($l\,mol^{-1}\,s^{-1}$), at 70 °C for the reaction

$$p\text{-}XC_6H_4CH_2HgBr + \overset{*}{Hg}Br_2 \longrightarrow p\text{-}XC_6H_4CH_2\overset{*}{Hg}Br + HgBr_2$$

| | X | | | | |
|---|---|---|---|---|---|
| Solvent | F | Cl | H | Me | Pr$^i$ |
| Quinoline | 0.34 | 0.36 | 0.411$^a$ | 0.57 | 0.66 |
| dmso | 0.83 | 0.83 | 1.03 | 1.22 | 1.22 |

$^a$ $\Delta H^{\neq} = 18.2 \pm 0.9$ kcal mol$^{-1}$.

There have been a large number of investigations into benzyl exchange reactions. The exchange 30 (R = benzyl; X = Br) has been studied by Reutov *et al.*[53,54] using quinoline as the solvent; they showed that the kinetic form was second order overall and first order in each reactant. In dimethyl sulphoxide the same reaction was again shown to be kinetically of first order in each reactant[55]. Rate constants were obtained for reaction of a number of *p*-substituted benzylmercury(II) bromides, and are given in Table 6[53,55]. As with the related one-alkyl exchanges above, the benzyl exchange reaction was also accelerated by added bromide ion in dimethyl sulphoxide. With the *p*-$NO_2$ substituent, the mechanism of the benzyl exchange changed from $S_E2$ to $S_E1$ in dimethyl sulphoxide[56], the reaction now being overall of first order in *p*-nitrobenzylmercury(II) bromide. The first-order rate constant at 70 °C was reported as $3.78 \times 10^{-5}\,s^{-1}$ with an activation energy of $18 \pm 1$ kcal mol$^{-1}$. This $S_E1$ process is also accelerated by added bromide ion[57].

The exchange between $\alpha$-carbethoxybenzylmercury(II) bromide and mercury(II) bromide (reaction 32) has been studied intensively[58–66]. In 70% aqueous dioxane,

$$PhCH(CO_2Et)HgBr + \overset{*}{Hg}Br_2 \longrightarrow PhCH(CO_2Et)\overset{*}{Hg}Br + HgBr_2 \qquad (32)$$

the exchange takes place with retention of configuration at the carbon atom undergoing substitution[62], and follows second-order kinetics[62], thus indicating an $S_E2$ mechanism. The same kinetic form was observed[60,61,63–65] using a number of other solvents; the results are collected in Table 7. However, in dimethyl sulphoxide, the exchange proceeds with racemization of optically active (−)-PhCH($CO_2$Et)HgBr, the rate of racemization being identical with the rate of exchange[62], and both following a kinetic form first order only in the organomercury substrate[62,66]. The effect of

TABLE 7. Second-order rate constants, $10^4k_2$, at 60 °C and activation parameters for the reaction

$$PhCH(CO_2Et)HgBr + \overset{*}{Hg}Br_2 \longrightarrow Ph(CO_2Et)\overset{*}{Hg}Br + HgBr_2$$

| Solvent | $10^4k_2$ ($l\,mol^{-1}\,s^{-1}$) | $E_a$ (kcal mol$^{-1}$) | $\Delta H^{\neq}$ (kcal mol$^{-1}$) | $\Delta S^{\neq}$ (cal K$^{-1}$ mol$^{-1}$) | Ref. |
|---|---|---|---|---|---|
| 70% aq. dioxane | 28 | | | | 62 |
| Pyridine | 660 | 16.3 | 15.7 | −17 | 60, 63 |
| dmf | 2.35 | 16.1 | 15.5 | −29 | 64 |
| 80% aq. ethanol | 10.3 | 25.1 | 24.5 | +1 | 65 |

introducing *p*-substituents into the benzyl group is to accelerate the rate through electron-withdrawing groups: relative rate constants in dimethyl sulphoxide at 30 °C are[66] *p*-$NO_2$ 4.04, *p*-I 1.31, H 1 and *p*Me 0.71, and replacement of bromide by chloride in equation 32 reduces the rate at 58.9 °C by a factor of 1.8[62]. All these experimental observations are in accord with an $S_E1$ mechanism for the exchange in dimethyl sulphoxide. The $S_E1$ exchange is accelerated by the presence of lithium nitrate but decelerated by potassium or lithium perchlorate. These effects are very mild in comparison with the very large accelerating effect of added bromide ion; detailed experiments indicate that two bromide ions may be incorporated into the transition state[62], no doubt introduced as $PhCH(CO_2Et)HgBr_3^-$.

It seems established that in the one-alkyl exchange of saturated alkyl groups, retention of configuration at the carbon atom undergoing substitution is the rule[41,44,67]. This seems also to be the case for exchange between *cis*-(or *trans*)-β-chlorovinylmercury(II) halides and mercury(II) halides, retention of geometrical configuration being observed[67–69].

Kitching *et al.*[70] have used n.m.r. spectroscopy to study the one-alkyl exchange of allyl- and β-methylallyl-mercury(II) halides; exchange was very rapid, probably via an $S_E2'$ mechanism involving allylic rearrangement.

The two-alkyl mercury-for-mercury exchange reaction, or *syn*-proportionation, may be represented by the general equation 33. As for the one-alkyl exchange, the stereochemistry

$$R_2Hg + HgX_2 \longrightarrow 2RHgX \qquad (33)$$

of reaction 33 indicates retention of configuration at the carbon atom undergoing substitution for the cases R = $Bu^s$ [10,71], $PhCH(CO_2Me)$[72,73], 1,4-dimethylpentyl[74], and *cis*-2-methoxycyclohexyl[14,75], so that this may be regarded as an $S_E2$ rule for reaction 33.

Charman *et al.*[10] showed that reaction 33 (R = $Bu^s$; X = Br) in acetone or ethanol

TABLE 8. Second-order rate constants and activation parameters for the reaction

$$R_2Hg + HgX_2 \longrightarrow 2RHgX$$

| R | X | Solvent | $k_2$ ($l\,mol^{-1}\,s^{-1}$) | Temperature (°C) | $\Delta H^{\neq}$ ($kcal\,mol^{-1}$) | $\Delta S^{\neq}$ ($cal\,K^{-1}\,mol^{-1}$) | Ref. |
|---|---|---|---|---|---|---|---|
| $Bu^s$ | Br | Acetone | 2.4 | 25 | | | 10 |
| $Bu^s$ | Br | Ethanol | 0.39 | 25 | | | 10 |
| $Bu^s$ | OAc | Ethanol | 5.3 | 0 | | | 10 |
| $Bu^s$ | $NO_3$ | Ethanol | 7.6 | −46.6 | | | 10 |
| Me | I | Dioxane | Very slow | 25 | | | 76 |
| Et | I | Dioxane | 0.0163 | 25 | 12.3 | −28 | 76 |
| $Pr^n$ | I | Dioxane | 0.0186 | 25 | 12.0 | −28 | 76 |
| $Pr^i$ | I | Dioxane | 0.0160 | 25 | 12.0 | −29 | 76 |
| cyclo-Pr | I | Dioxane | 0.0767 | 25 | 12.8 | −22 | 76 |
| $Pr^n$ | I | Benzene | 0.112 | 25 | 11.0 | −28 | 77 |
| Me | Cl | Dioxane | 0.00259 | 36 | | | 78 |
| Me | Br | Dioxane | 0.00354 | 36 | | | 78 |
| Me | I | Dioxane | 0.00154 | 36 | | | 78 |
| Me | Cl | Methanol | 1.05 | 36 | | | 78 |
| Me | Br | Methanol | 0.30 | 36 | | | 78 |
| Me | I | Methanol | 0.05 | 36 | | | 78 |

TABLE 9. Second-order rate constants[79] at 20 and 40 °C and activation parameters in methanol for the reaction

$$R_2Hg + HgX_2 \longrightarrow 2RHgX$$

| R | X | $k_2$ ($l\,mol^{-1}\,s^{-1}$) 20 °C | 40 °C | $\Delta H^{\neq}$ ($kcal\,mol^{-1}$) | $\Delta S^{\neq}$ ($cal\,K^{-1}\,mol^{-1}$) |
|---|---|---|---|---|---|
| Me | Cl | 0.65 | 2.7 | 12.0±0.7 | −18±2 |
| Et | Cl | 6.0 | 19.6 | 10.1±0.3 | −20±1 |
| $Pr^n$ | Cl | 4.5 | 15.6 | 10.8±0.3 | −19±1 |
| $Pr^i$ | Cl | 0.6 | 3.3 | 14.4±0.7 | −10±2 |
| Me | Br | 0.30 | 1.1 | 10.5±0.4 | −25±1 |
| Et | Br | 2.9 | 9.9 | 9.5±0.4 | −24±1 |
| $Pr^n$ | Br | 2.3 | 8.5 | 10.9±0.5 | −20±2 |
| $Pr^i$ | Br | 1.5 | 5.7 | 11.2±0.3 | −19±1 |

was first order in each reactant and that for reaction of $Bu^s{}_2Hg$ with mercury(II) salts the rate constant increased in value along the series $HgBr_2 < Hg(OAc)_2 < Hg(NO_3)_2$, thus suggesting[10] an open transition state, mechanism $S_E2$ (open); however, for a criticism of this assignment see refs. 2 and 4.

The two-alkyl exchange has also been studied by Dessy and coworkers[76,77] and by Rausch and van Wazer[78]; their kinetic data are assembled in Table 8. More recently, Reutov and coworkers[79] have studied both the effect of the alkyl group, R, and the halide, X, on the rate of the two-alkyl exchange in solvents methanol and dimethylformamide. They observed that the reactivity sequence in R was temperature dependent in methanol, as was also the relative reactivity of mercury(II) chloride and bromide. The kinetic results are summarized in Tables 9 and 10. Reutov and coworkers[79] suggested that these two-alkyl exchange reactions proceed by mechanism $S_E2$, although the effect of dimethylformamide solvent is complex.

Diarylmercurys react readily with mercury(II) halides; this aryl exchange reaction has been studied by Dessy and coworkers[76,77], who investigated both solvent effects and substitutent effects; the results are summarized in Table 11. A four-centred

TABLE 10. Second-order rate constants[79] at 20 and 40 °C and activation parameters in methanol and dimethylformamide for the reaction

$$Et_2Hg + HgX_2 \longrightarrow 2EtHgX$$

| X | Solvent | $k_2$ ($l\,mol^{-1}\,s^{-1}$) 20 °C | 40 °C | $\Delta H^{\neq}$ ($kcal\,mol^{-1}$) | $\Delta S^{\neq}$ ($cal\,K^{-1}\,mol^{-1}$) |
|---|---|---|---|---|---|
| Cl | MeOH | 6.0 | 19.6 | 10.1±0.3 | −20±1 |
| Br | MeOH | 2.9 | 9.9 | 9.5±0.4 | −24±1 |
| I | MeOH | 0.3 | 1.2 | 12.7±0.5 | −18±2 |
| $MeCO_2$ | MeOH | 110 | | | |
| Cl | dmf | 6.4 | 14.3 | 7.2±0.7 | −30±3 |
| Br | dmf | 6.1 | 15.8 | 8.1±0.4 | −27±1 |
| I | dmf | 0.9 | 2.9 | 9.9±0.2 | −25±1 |

TABLE 11. Second-order rate constants at 25 °C and activation parameters for the reaction

$$(XC_6H_4)_2Hg + HgI_2 \longrightarrow 2XC_6H_4HgI$$

| X | Solvent | $k_2$ ($l\,mol^{-1}\,s^{-1}$) | $\Delta H^{\neq}$ (kcal $mol^{-1}$) | $\Delta S^{\neq}$ (cal $K^{-1}\,mol^{-1}$) | Ref. |
|---|---|---|---|---|---|
| 4-MeO | Dioxane | 71.5 | | | 76 |
| 4-Me | Dioxane | 13.1 | 10.6 | −20 | 76 |
| 4-Ph | Dioxane | 2.25 | 12.0 | −19 | 76 |
| H | Dioxane | 1.97 | 12.8 | −16 | 76 |
| 4-F | Dioxane | 0.42 | 12.9 | −19 | 76 |
| 4-Cl | Dioxane | 0.092 | 14.5 | −17 | 76 |
| H | Dioxane (5% $H_2O$) | 3.5 | | | 77 |
| H | Dioxane (10% $H_2O$) | 6.8 | | | 77 |
| H | Ethanol | 62.8 | 11.7 | −13 | 77 |
| H | Benzene | 29.2 | 7.6 | −28 | 77 |
| H | Cyclohexane | 15.9 | 7.6 | −31 | 77 |

cyclic transition state was suggested for the two-aryl exchange between diphenylmercury and mercury(II) iodide[77], although it was recognized that electrophilic attack at carbon must be the dominant feature of the mechanism. In dioxane, diphenylmercury is about 120 times as reactive as diethylmercury towards mercury(II) iodide (see Tables 8 and 11). Reutov and coworkers[80] have studied the two-alkyl exchange between dibenzylmercury and mercury(II) halides using pyridine as solvent (Table 12), but it is not possible from their results to assess the comparative reactivity of the benzyl group. This can be achieved, however, from the results of a general survey of organomercury reactivity carried out by Reutov and coworkers[81] using dimethylformamide as solvent and using as reactant $HgCl_2$ in both the absence and presence of added chloride ions. The latter considerably retard the rate of reaction [except with $(C_6F_5)_2Hg$], the most probable explanation being that the reactants are now $R_2Hg$ and the ion $HgCl_3^-$. Rate constants for reaction with $HgCl_2$ itself are given in Table 13, it now being clear that dibenzylmercury is not particularly reactive.

The action of mercury(II) chloride on mixed alkylarylmercurys has been studied by various groups, who have shown that when alkylphenylmercurys are allowed to react with labelled mercury(II) chloride, all the radioactive label appears in the phenylmercury(II) chloride (reaction 34, R = Et, n-Bu)[76,82–84]. The reaction is thus an electrophilic aromatic substitution.

$$PhHgR + \overset{*}{Hg}Cl_2 \longrightarrow Ph\overset{*}{Hg}Cl + RHgCl \qquad (34)$$

TABLE 12. Second-order rate constants[80], $10^3k_2$, at 46 °C in pyridine and activation parameters for the reaction

$$(PhCH_2)_2Hg + HgX_2 \longrightarrow 2PhCH_2HgX$$

| X | $10^3k_2$ ($l\,mol^{-1}\,s^{-1}$) | $\Delta H^{\neq}$ (kcal $mol^{-1}$) | $\Delta S^{\neq}$ (cal $K^{-1}\,mol^{-1}$) |
|---|---|---|---|
| Cl | 4.05 | 11.6 | −28 |
| Br | 3.33 | 14.9 | −24 |
| I | 1.19 | 19.5 | −12 |

TABLE 13. Second-order rate constants[81] in dimethylformamide for the reaction

$$R_2Hg + HgCl_2 \longrightarrow 2RHgCl$$

| R | Temperature (°C) | $k_2$ ($l\,mol^{-1}\,s^{-1}$) |
|---|---|---|
| Ph | 20 | Fast |
| $CH_2{=}CH$ | 20 | Fast |
| Et | 25 | 8.2 |
| $PhCH_2$ | 25 | 0.028 |
| $C_6F_5$ | 25 | $2.5\times10^{-4}$ |
| $CH_2CO_2CH_3$ | 25 | $1.7\times10^{-6}$ |
| $(CF_3)_2CH$ | 70 | Slow |

Reaction 33, the *syn*-proportionation reaction, is in principle reversible, and the interaction of two molecules of an organomercury(II) halide is known as a symmetrization reaction. Usually agents such as ammonia are used to bring about the symmetrization reaction, and there have been several studies on these induced symmetrizations. Reutov and coworkers[72,73,85] showed using diastereoisomeric (−)-menthyl esters of $\alpha$-carboxybenzylmercury(II) bromide that reaction proceeded with retention of configuration at the carbon atom underoing substitution, and also showed that the stoichiometry of the symmetrization was that in equation 35, with

$$2RHgBr + 2NH_3 \longrightarrow R_2Hg + HgBr_2{\cdot}2NH_3 \tag{35}$$

chloroform as the solvent[86,87]. Both Reutov and coworkers[86–91] and Jensen and coworkers[92,93] are agreed that the symmetrization of $\alpha$-carbalkoxybenzylmercury(II) bromides in chloroform is a fourth-order process, second order in RHgBr and second order in ammonia. As suggested before[4,92], there are two possible mechanisms in accord with the kinetics and stoichiometry, either reactions 36 and 37, or reactions 38 and 39.

$$RHgBr + NH_3 \overset{\text{fast}}{\rightleftharpoons} RHgBr{\cdot}NH_3 \tag{36}$$

$$2RHgBr{\cdot}NH_3 \overset{\text{slow}}{\rightleftharpoons} R_2Hg + HgBr_2{\cdot}2NH_3 \tag{37}$$

$$RHgBr + 2NH_3 \overset{\text{fast}}{\rightleftharpoons} RHgBr{\cdot}2NH_3 \tag{38}$$

$$RHgBr{\cdot}2NH_3 + RHgBr \overset{\text{slow}}{\rightleftharpoons} R_2Hg + HgBr_2{\cdot}2NH_3 \tag{39}$$

Reutov and coworkers[94–96] have studied the effect of variation of X and R′ in the symmetrization of $XC_6H_4CH(CO_2R')HgBr$ by ammonia in chloroform. Details are given in Table 14, but it should be noted that the observed rate constants given by $k_2^{obs} = k_4[NH_3]^2$ when an excess of ammonia is used, contain an equilibrium constant, either for reaction 36 or reaction 38 as well as the rate constant for reaction 37 or reaction 39. The co-symmetrization of two $\alpha$-carbethoxybenzylmercury(II) bromides carrying different substituents in the aromatic ring has also been studied by Reutov and coworkers[97,98] using ammonia in chloroform. As for the other symmetrizations of this type, a four-centred transition state of the $S_E2$ (cyclic) type was suggested[97,98].

TABLE 14. Second-order rate constants[a] in chloroform at 20 °C for the reaction[94–96]

$$2XC_6H_4CH(CO_2R')HgBr + 2NH_3 \longrightarrow [XC_6H_4CH(CO_2R')]_2Hg + HgBr_2 \cdot 2NH_3$$

| X | R′ | $10^3k_2^{obs}$ ($l\,mol^{-1}\,s^{-1}$) |
|---|---|---|
| H | Me | 181 |
| H | Et | 110 |
| H | $Pr^i$ | 25.9 |
| H | $Bu^t$ | Slow |
| H | Nonyl | 10.2 |
| H | Menthyl | 6.9 |
| $p$-$NO_2$ | Et | 17 730[b] |
| $m$-Br | Et | 1 445 |
| $p$-I | Et | 676 |
| $p$-Br | Et | 540 |
| $p$-Cl | Et | 470 |
| $o$-Br | Et | 426 |
| $p$-F | Et | 148 |
| H | Et | 110 |
| $m$-Me | Et | 71 |
| $p$-$Pr^i$ | Et | 41 |
| $p$-Et | Et | 40 |
| $p$-Me | Et | 34 |
| $o$-Me | Et | 32 |
| $p$-$Bu^t$ | Et | 28 |

[a] Reactions were carried out in the presence of a large excess of ammonia, and under these conditions the fourth-order rate equation (rate = $k_4[RHgBr]^2[NH_3]^2$) reduces to a second-order expression (rate = $k_2^{obs}[RHgBr]^2$).
[b] Approximate value.

Exchange reactions between two different organomercurys (reaction 40) have been studied in a limited number of cases.

$$RHgX + R'\overset{*}{Hg}X \longrightarrow R\overset{*}{Hg}X + R'HgX \quad (40)$$

Reutov and coworkers[99] reported that reaction 40 [R = $Ph(CO_2Et)CH$; R′ = Ph; X = Br) followed second-order kinetics in pyridine and was first order in each reactant. Reaction 40 (R = $p$-$YC_6H_4$; R′ = $p$-$Me_2NC_6H_4$; X = Cl) also had an overall kinetic order of two, but had fractional orders in each reactant, thus suggesting a complex reaction[99]; the rate constant decreased along the series Y = MeO > H > Cl > $CO_2Et$ but increased when the organomercury chlorides were replaced by organomercury acetates. Replacement of the phenyl group in PhHgX by the allyl group resulted in the isotopic equilibrium being established very quickly[100].

The three-alkyl mercury-for-mercury exchange reaction 41 has been studied by Reutov and coworkers[101–104] and by Hughes and Ingold and their coworkers[105], using the technique of double labelling. As before, the radioactive mercury isotope $^{203}Hg$ is denoted as $\overset{*}{Hg}$, whilst an optically active alkyl group is denoted as $\mathring{R}$. Reaction 41 may be considered as an electrophilic substitution of the substrate $R_2Hg$ by the electrophilic reagent $\mathring{R}'\overset{*}{Hg}X$. Paulik and Dessy[106] made use of reaction 41 (R = Ph; R′ = $CO_2Me$) to prepare unsymmetrical RHgR′ compounds. Reaction 41 (R = R′ = 1,4-dimethylpentyl; X = Br) and reaction 41 (R = R′ = $Bu^s$; X = Br) were

TABLE 15. Second-order rate constants, $10^5k_2$, in ethanol for the reaction

$$Bu^s{}_2Hg + R'\overset{*}{Hg}X \rightleftharpoons Bu^s\overset{*}{Hg}R' + Bu^sHgX$$

| R′ | X | Temperature (°C) | $10^5k_2$ (l mol$^{-1}$ s$^{-1}$) | Ref. |
|---|---|---|---|---|
| $Bu^s$ | Br | 35 | 4.7 | 105 |
| $Bu^s$ | OAc | 35 | 27 | 105 |
| $Bu^s$ | $NO_3$ | 0 | 3400 | 105 |
| $Me_2CH(CH_2)_2CHMe$ | Br | 35 | 5.0[a] | 104 |

[a] Value calculated by extrapolation of values at higher temperatures. $\Delta H^{\neq} = 14.7$ kcal mol$^{-1}$.

both shown to proceed

$$R_2Hg + \mathring{R}'\overset{*}{Hg}X \longrightarrow R\overset{*}{Hg}\mathring{R}' + RHgX \qquad (41)$$

with complete retention of configuration at the carbon atom undergoing substitution, in ethanol as solvent[101–103,105]. Rate constants for reaction 41 ($R = R' = Bu^s$; X = Br, OAc, $NO_3$) are given in Table 15; since the order of increasing reactivity is X = Br < OAc < $NO_3$, and since added lithium salts increased the value of the rate constants when X = Br, $NO_3$, Hughes, Ingold and coworkers[105] suggested mechanism $S_E2$ (open) via the transition state **8**. It is probable, however, that the mechanism is

$$\left[\begin{array}{l} \quad HgR \\ R\!\!\begin{array}{l}\cdot^{\cdot}\\ \cdot_{\cdot}\end{array} \\ \quad Hg{-}X \\ \quad \;| \\ \quad R \end{array}\right]^{\neq}$$

(**8**)

more complicated than just a single elementary bimolecular reaction. Reutov and coworkers[107] have studied reaction 41 ($R = PhCH_2$; $R' = CF_3$, $CCl_3$; X = Cl, I, $OCOCF_3$). They showed that the reaction was first order with respect to each reactant and obtained second-order rate constants in a variety of solvents (Table 16). For this set of reactions, a cyclic transition state was proposed[107].

TABLE 16. Second-order rate constants[107], $10^3k_2$, at 31 °C for the reaction

$$(PhCH_2)_2Hg + CF_3HgX \longrightarrow PhCH_2HgCF_3 + PhCH_2HgX$$

| | $10^3k_2$ (l mol$^{-1}$ s$^{-1}$) | | |
|---|---|---|---|
| Solvent | X = $OCOCF_3$ | X = I | X = Cl |
| Tetrachloroethylene | 720 | 2.10 | 1.70 |
| Dichloromethane | 53.7 | 0.062 | 0.035 |
| Dms | 61.2 | 0.082 | 0.032 |
| Dmf | 59.8 | 0.119 | 0.023 |
| Acetone | 39.4 | 0.086 | 0.0086 |
| Benzene | 11.4 | 0.042 | 0.0065 |
| Pyridine | 0.65 | 0.002 | 0.0005 |

Jensen and Miller[108] reported that the three-alkyl exchange 42 occurred almost instantaneously, with precipitation of PhHgBr, but that the subsequent three-alkyl exchange 43 took place much more slowly. Since all the radioactive mercury in

$$Ph_2Hg + Ph(CO_2Et)CHH\overset{*}{g}Br \longrightarrow Ph(CO_2Et)CHH\overset{*}{g}Ph + PhHgBr \quad (42)$$

$$Ph(CO_2Et)CHHgPh + Ph(CO_2Et)CHHgBr \longrightarrow [Ph(CO_2Et)CH]_2Hg + PhHgBr \quad (43)$$

reaction 42 remains in solution as $Ph(CO_2Et)CHH\overset{*}{g}Br$, presumably transition state **9** is involved[109]. However, in reaction 43 the radioactive mercury is found to be

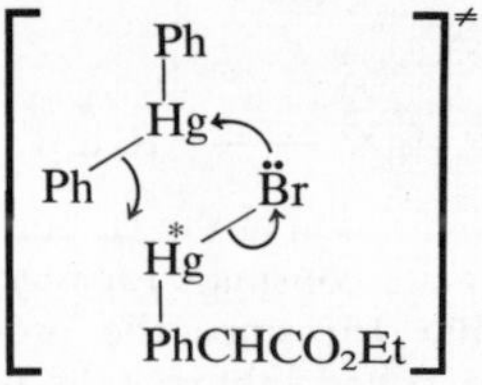

**(9)**

distributed evenly over both products[109]. Transition state **10** would give rise to such an even distribution, but would yield no new products, whereas transition state **11** would give the correct products but with no radioactive scrambling. Clearly, neither **10** nor **11** alone can account for the experimental observations, and it seems as though there is a rapid equilibrium via **10**, followed by a slow reaction through **11**.

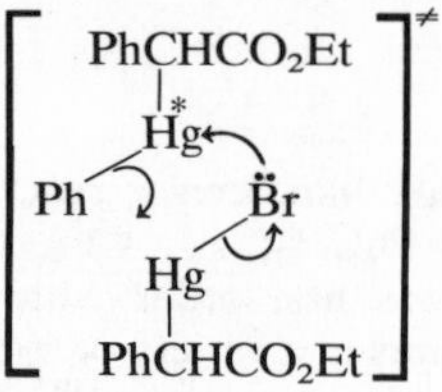

**(10)** 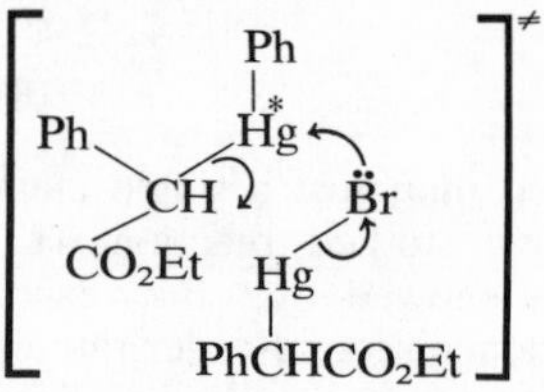 **(11)**

There has been considerable speculation over the four-alkyl (or four-aryl) exchange reaction, written as 44 or 45. Charman *et al.*[110] ruled out the four-alkyl

$$R_2Hg + R'_2\overset{*}{Hg} \longrightarrow R'_2Hg + R_2\overset{*}{Hg} \quad (44)$$

$$R_2Hg + R'_2\overset{*}{Hg} \longrightarrow R'\overset{*}{Hg}R + R'HgR \quad (45)$$

exchange altogether, and suggested that when such a stoichiometric exchange reaction was observed, the reaction sequence actually involved one of the other alkyl-exchange reactions; in their view, the four-alkyl exchange was not an elementary reaction at all[110]. Reutov *et al.*[111], however, have observed a facile exchange between $Hg(CH_2CHO)_2$ and a number of diarylmercurys in acetone under mild conditions, and Pollard and Thompson[112] have reported a similar exchange between diphenylmercury and di(*p*-chlorophenyl) mercury in pyridine. Both sets of workers found only the symmetrical products (reaction 44) and it was suggested that the four-alkyl or aryl exchange 44 took place via a six-centred transition state **12** in which the two mercury atoms were equivalent[2,111,112].

$$\left[ \begin{array}{c} \text{Hg} \\ \text{R} \quad \text{R}' \quad \text{R}' \quad \text{R} \\ \text{Hg} \end{array} \right]^{\neq} \qquad \left[ \begin{array}{c} \text{R}' \\ | \\ \text{Hg} \\ \text{R}' \qquad \text{R} \\ \text{Hg} \\ | \\ \text{R} \end{array} \right]^{\neq}$$

**(12)** **(13)**

On the other hand, unsymmetrical products have been observed by Rausch and van Wazer[78] (reaction 45) ($R = PhCH_2$; $R' = Ph$) and by Dessy *et al.*[113] (reaction 45) ($R = CH_3$; $R' = CD_3$), in both cases under severe conditions. Thorpe and coworkers[114] followed a number of four-alkyl or aryl exchanges, using p.m.r. in dimethyl sulphoxide and pyridine and u.v. spectroscopy in ethanol, under mild conditions (dilute solutions at room temperature). For the cases where $R = CH_2CHO$ and $R' = CH_2CHO$, Ph, $p$-$MeOC_6H_4$, $p$-$Me_2NC_6H_4$, $p$-$MeNHC_6H_4$, $p$-$BrC_6H_4$, $p$-$EtC_6H_4$, and where $R = Ph$, $p$-$ClC_6H_4$, $p$-$MeC_6H_4$ and $R' = p$-$Me_2NC_6H_4$, a facile exchange was observed, shortly after mixing the reactants, that led to the formation of the unsymmetrical products (reaction 45). Thorpe and coworkers[114] suggested that reaction 44 should more properly be represented as two successive unsymmetrical

TABLE 17. Second-order rate constants[116], $k_2$, at 25 °C and activation parameters in dimethylformamide for the reaction

$$Ar_2Hg + R_2Hg \longrightarrow 2ArHgR$$

| Ar | R | $k_2$ ($l\ mol^{-1}\ s^{-1}$) | $\Delta H^{\neq}$ ($kcal\ mol^{-1}$) | $\Delta S^{\neq}$ ($cal\ K^{-1}\ mol^{-1}$) |
|---|---|---|---|---|
| $p$-$MeOC_6H_4$ | PhC≡C | 0.55 | 9.5±0.5 | −28±1 |
| $p$-$MeC_6H_4$ | PhC≡C | 0.25 | | |
| $p$-$PhC_6H_4$ | PhC≡C | 0.20 | 10.1±0.5 | −27±1 |
| $C_6H_5$ | PhC≡C | 0.16 | 9.3±0.5 | −31±1 |
| β-Naphthyl | PhC≡C | 0.09 | 8.5±0.5 | −33±1 |
| 9-Anthryl | PhC≡C | 2.9[a] | | |
| $p$-$ClC_6H_4$ | PhC≡C | 0.024 | | |
| $p$-$BrC_6H_4$ | PhC≡C | 0.016 | | |
| $p$-$EtOCOC_6H_4$ | PhC≡C | 0.0025 | | |
| $p$-$MeOC_6H_4$ | CN | 9.1 | | |
| $p$-$MeC_6H_4$ | CN | 2.0 | 10.9±0.5 | −21±1 |
| $m$-$MeC_6H_4$ | CN | 0.97 | | |
| $p$-$PhC_6H_4$ | CN | 1.2 | | |
| $C_6H_5$ | CN | 0.89 | 9.5±0.5 | −28±1 |
| $p$-$ClC_6H_4$ | CN | 0.039 | 11.8±0.6 | −25±1 |
| $p$-$BrC_6H_4$ | CN | 0.012 | | |
| $p$-$EtOCOC_6H_4$ | CN | 0.0046 | | |
| $C_6H_5$ | $CCl_3$ | 0.61[b] | | |
| $p$-$MeOC_6H_4$ | $CBr_3$ | 10.2[b] | | |
| $p$-$MeOC_6H_4$ | $CF_3CHCOOEt$ | 0.76[b] | | |
| $p$-$MeOC_6H_4$ | $C_6F_5$ | 0.002[b,c] | | |
| $p$-$MeOC_6H_4$ | $(CF_3)_2CH$ | <0.0004[b,c] | | |

[a] At 67 °C.
[b] Ref. 115.
[c] At 60 °C.

exchange reactions:

$$R_2Hg + R'_2\overset{*}{Hg} \rightleftharpoons R'\overset{*}{Hg}R + R'HgR \rightleftharpoons R'_2Hg + R_2\overset{*}{Hg} \quad (46)$$

There is thus no need to postulate transition state **10** because the four-alkyl exchange 45 could take place via a four-centred transition state **13**, according to Thorpe and coworkers[114]. However, such a transition state would certainly be forbidden on grounds of orbital symmetry (see Section 1.A.2.b.) and it is still possible that reactions proceeding with the overall stoichiometry 45 are themselves composed of a series of one-, two-, or three-alkyl (or aryl) exchange reactions.

The exchange reaction 47 has been studied by Reutov and coworkers[115,116] using dimethylformamide solvent. The various exchanges were second order overall, first

$$Ar_2Hg + R_2Hg \rightleftharpoons 2ArHgR \quad (47)$$

order in each reactant, and the recorded rate constants are collected in Table 17. The reactions were described[115] in terms of electrophilic attack of $R_2Hg$ on an aromatic carbon (of $Ar_2Hg$) and it was concluded[116] that the transtiion state for this aromatic electrophilic substitution was intermediate between a $\sigma$- and a $\pi$-complex.

### 4. *Cleavage of mercury—carbon bonds by thallium*(III) *compounds*

Hart and Ingold[7,117] reported that the carbanion, *s*-$Bu^-$, was formed in the reaction between di-*sec*-butylmercury and diethylthallium(III) bromide at 70 °C in dimethylformamide, but Jensen and Heyman[118] later showed that this was not correct.

### 5. *Cleavage of mercury—carbon bonds by lead*(IV) *compounds*

Reaction 48 was reported by Kalman *et al.*[119], who used trifluoroacetic acid as solvent.

$$p\text{-}FC_6H_4HgOCOCF_3 + Pb(OCOCF_3)_4 \longrightarrow p\text{-}FC_6H_4Pb(OCOCF_3)_3 + Hg(OCOCF_3)_2 \quad (48)$$

## E. Cleavage of Group IIIB Metal—Carbon Bonds

### 1. *Cleavage of boron—carbon bonds by mercury*(II) *compounds*

The preparative scope of this set of reactions has been studied by Larock and Brown[120,121], who showed that towards mercury(II) acetate in tetrahydrofuran the order of reactivity of R—B bonds was R = *n*-Butyl > cyclopentyl > cyclohexyl > cyclooctyl > *sec*-butyl > norbornyl. The benzyl—boron bond had about the same order of reactivity as *n*-alkyl—boron bonds. Change in the mercury(II) salt gave the reactivity sequence nitrate ⩾ trifluoroacetate > benzoate > *n*-butyrate ⩾ fluoride > acetate > phthalate. Larock and Brown[121] also showed that reaction of tri-*exo*-norbornylboron with mercury(II) benzoate proceeds with almost complete retention of configuration, apparently in agreement with the earlier report of Matteson and Bowie[122] that the action of mercury(II) chloride on (*R*)-(−)-dibutyl-1-phenylethaneboronate yielded the corresponding organomercury(II) chloride with net retention of configuration. Gielen and Fosty[123], however, showed later that (*erythro*-$Bu^tCHDCHD)_3B$ with mercury(II) acetate in tetrahydrofuran gave the *threo* product, with inversion of configuration, and this has subsequently been confirmed by

Bergbreiter and Rainville[124]. There seems, therefore, to be no general stereochemical rule governing these mercury-for-boron cleavages.

Matteson and Waldbillig[125,126] showed that both the *exo* isomer **14** and the *endo* isomer **15** react with mercury(II) chloride to yield nortricyclylmercury(II) chloride, the boron isotope effect, $^{10}B/^{11}B$, being 1.033 and 1.027, respectively[126], thus suggesting that the boron atom in **14** and **15** is involved in the rate-determining step. Using as the reaction medium a 75% aqueous acetone mixture buffered with phthalate, Matteson and Talbot[127] obtained rate constants for the substitution of **14** and **15** by mercury(II) chloride: at 25 °C the values are $2.08 \times 10^{-3}\,l\,mol^{-1}\,s^{-1}$ for **14** and $4.9 \times 10^{-6}\,l\,mol^{-1}\,s^{-1}$ for **15**. A proposed mechanism for reaction of the more active *exo* isomer **14** involves attack of $HgCl_2$ at the double bond together with skeletal rearrangement with inversion of configuration at the $\alpha$-carbon atom[128].

B(OH)₂ H H B(OH)₂

**(14)** **(15)**

Matteson and coworkers[129,130] also reported relative rate constants for the substitution of a number of boronic acids by mercury(II) chloride in ethanol–water–glycerol (88:8:4) buffered with sodium acetate and acetic acid and using an excess of chloride ion (Table 18). A detailed analysis[4] suggests that electron-donating substituents in the benzylboronic acid accelerate the actual electrophilic substitution of the negatively charged glycerol esters of the boronic acids. In a related study, Matteson and Allies[131] obtained rate constants for substitution of another series of boronic esters by mercury(II) chloride using methanol as the solvent, again buffered with sodium acetate and acetic acid; rate constants are given in Table 19.

An aromatic substitution has been studied by Kuivila and Muller[132], who showed that the substitution of benzeneboronic acid by phenylmercury(II) perchlorate in

TABLE 18. Relative rate constants[129,130] at 40 °C for the reaction

$$RB(OH)_2 + HgCl_2 \longrightarrow RHgCl + ClB(OH)_2$$

in ethanol–water–glycerol (88:8:4)

| R | Relative rate constant |
|---|---|
| $PhCH_2$ | 1 |
| PhCHMe | 0.5 |
| $p$-$ClC_6H_4CHMe$ | 0.04 |
| $p$-$CF_3C_6H_4CHMe$ | 0.0015 |
| $Bu^s$ | 0 |
| $p$-$MeC_6H_4CH_2$ | 0.76 |
| $p$-$ClC_6H_4CH_2$ | 2.15 |
| $m$-$CF_3C_6H_4CH_2$ | 2.43 |

TABLE 19. Second-order rate constants[131], $10^4k_2$, for the reaction between boronic esters and mercury(II) chloride in methanol at 30 °C

| Ester | $10^4k_2$ ($l\ mol^{-1}\ s^{-1}$) |
|---|---|
| $CH_3[B(OMe_2)]$ | 6.7 |
| $CH_2[B(OMe)_2]_2$ | 1370 |
| $CH[B(OMe)_2]_3$ | 2700 |
| $C[B(OMe)_2]_4$ | 1600 |
| $ClHgCH_2B(OMe)_2$ | 20 |
| $PhCH_2B(OMe)_2$ | 2.4 |

TABLE 20. Second-order rate constants[132] for the reaction of $PhHgClO_4$ with $PhB(OH)_2$ in aqueous ethanol at 25 °C

| Water in solvent (%) | $k_2$ ($l\ mol^{-1}\ s^{-1}$) |
|---|---|
| 70 | 13.3 |
| 60 | 8.56 |
| 50 | 5.21 |
| 40 | 3.00 |

aqueous ethanol (reaction 49) was first order in the boronic acid and first order in the

$$PhB(OH)_2 + PhHgClO_4 + H_2O \rightarrow Ph_2Hg + B(OH)_3 + HClO_4 \qquad (49)$$

mercury(II) reactant. The actual substitution was thought to involve the species $PhB(OH)_3^-$ and $PhHg^+$; rate constants are given in Table 20.

### 2. *Cleavage of thallium—carbon bonds by lead (IV) compounds*

Kalman *et al.*[119] reported that reaction 50 (X = F, Cl, Br) took place in trifluoroacetic acid.

$$p\text{-}XC_6H_4Tl(OCOCF_3)_2 + Pb(OCOCF_3)_4 \longrightarrow p\text{-}XC_6H_4Pb(OCOCF_3)_3 + Tl(OCOCF_3)_3 \qquad (50)$$

## F. Cleavage of Group IVB Metal—Carbon Bonds

### 1. *Cleavage of silicon—carbon bonds by lithium(I) compounds*

Organolithium compounds have been found to cleave silicon—carbon bonds, in what amounts to alkyl-exchange reactions. In this way, the exchange of organic groups between silicon and lithium may lead to the formation of preparatively useful reagents. Some typical examples are shown in reactions 51[133], 52 (X = Cl, F)[134,135], and 53[136], and other similar reactions have been reported[137–139].

$$Ph_3SiC{\equiv}CPh + n\text{-}BuLi \longrightarrow n\text{-}BuSiPh_3 + PhC{\equiv}CLi \qquad (51)$$

$$Ph_2HSiC_6X_5 + PhLi \longrightarrow Ph_3SiH + C_6X_5Li \qquad (52)$$

$$Ph_3SiC{\equiv}CSiPh_3 + Ph_3SiLi \longrightarrow Ph_3SiSiPh_3 + Ph_3SiC{\equiv}CLi \qquad (53)$$

Although $n$-BuLi will not easily cleave tetraorganosilicons, MeLi will cleave the Si—Ph bonds in $Ph_4Si$, $Ph_3SiMe$, and $Ph_2SiMe_2$[140]. One of the few cases in which a silicon—carbon bond is broken by a non-organometallic lithium compound is the cleavage of the $SiC{\equiv}CCH{=}CH_2$ bond in $Me_3SiC{\equiv}CCH{=}CH_2$ by lithium piperidide in ether[141].

### 2. *Cleavage of silicon—carbon bonds by beryllium(II) and magnesium(II) compounds*

Beryllium chloride in benzene cleaves a silicon—carbon bond according to equation 54[142].

$$2Me_3SiCH(COMe)CO_2Et + BeCl_2 \longrightarrow \left[ Be \begin{matrix} O{=}C \\ O{-}C(OEt) \end{matrix} CH \right]_2 + 2Me_3SiCl \quad (54)$$

Cleavage of silicon—benzyl bonds takes place when either $(PhCH_2)_3SiH$ or $(PhCH_2)_2SiH_2$ is treated with allylmagnesium chloride in tetrahydrofuran (see reaction 55); tetrabenzylsilicon does not react in this way[143].

$$(PhCH_2)_3SiH + CH_2{=}CHCH_2MgCl \longrightarrow (PhCH_2)_2Si(H)CH_2CH{=}CH_2 + PhCH_2MgCl \quad (55)$$

### 3. *Cleavage of silicon—carbon bonds by iron(II) and iron(III) compounds*

Eaborn and coworkers[144] have reported that the electrophilic reagent $[(C_6H_7)Fe(CO)_3]^+$ readily removes an aryl group from silicon or tin compounds (reaction 56) (M = Si, Sn; Ar = Ph, $p$-$MeOC_6H_4$, $p$-$Me_2NC_6H_4$, 2-furyl, 2-thienyl) in acetone or acetonitrile.

$$ArMMe_3 + [(C_6H_7)Fe(CO)_3]BF_4 \longrightarrow [\text{tricarbonyl}(\zeta^4\text{-5-arylcyclohexa-1,3-diene})\text{iron}] + Me_3MBF_4 \quad (56)$$

Rate constants for reaction 56 (M = Si) in acetonitrile at 45° were reported as Ar = Ph ($<10^{-7}$), $p$-$MeOC_6H_4$ ($1.1\times10^{-6}$), $p$-$Me_2NC_6H_4$ ($5.8\times10^{-3}$), 2-furyl ($1.4\times10^{-4}$), and 2-thienyl ($1.9\times10^{-6}\,l\,mol^{-1}\,s^{-1}$). The corresponding tin compounds were appreciably more reactive than the silicon compounds.

The reaction of tetraethylsilicon with various iron(III) compounds has been reported by Wong and Kochi[36], but these are outer-sphere electron transfer processes and are not simple electrophilic substitutions.

### 4. *Cleavage of silicon—carbon bonds by mercury(II) compounds*

The aromatic substitution 57 was reported by Combes[145] as long ago as 1896. Since then, Manulkin[146] has shown that mercury(II) chloride will cleave an ethyl group

$$(p\text{-}Me_2NC_6H_4)_3SiCl + HgCl_2 \longrightarrow p\text{-}Me_2NC_6H_4HgCl + (p\text{-}Me_2NC_6H_4)_2SiCl_2 \quad (57)$$

from $Et_4Si$ to yield EtHgCl, and will preferentially cleave an aryl group from mixed arylalkylsilicon compounds. The silicon—alkynyl bond in the compounds

$Me_3SiC{\equiv}CPh$, $Et_3SiC{\equiv}CPh$, and $Me_3SiC{\equiv}CSiMe_3$ is broken preferentially by mercury(II) sulphate and a little sulphuric acid in methanol[147].

DeSimone[148] has reported that trimethylsilyl salts commonly used as n.m.r. reference compounds are demethylated very readily by mercury(II) salts in aqueous solution (e.g. see reaction 58). The order of reactivity of mercury(II) salts in aqueous solution was $HgCl_2 \ll Hg(OAc)_2 < Hg(NO_3)_2$; in methanol solution only the nitrate reacted.

$$Me_3Si(CH_2)_3SO_3Na + Hg(OAc)_2 \longrightarrow MeHgOAc + Me_2Si(OAc)(CH_2)_3SO_3Na \quad (58)$$

Mechanistic studies have been carried out on the aromatic substitution reaction 59 using acetic acid[149,150] and 20% aqueous acetic acid[151–153] as solvents; kinetic results are given in Table 21. Reaction 59 (X = *p*-Me) was later re-investigated by Webster

$$XC_6H_4SiMe_3 + Hg(OAc)_2 \longrightarrow XC_6H_4HgOAc + Me_3SiOAc \quad (59)$$

and coworkers[154] in an attempt to uncover the reasons for inconsistences in the kinetic results previously reported. In this careful study, Webster and coworkers showed that both the species $\overset{+}{Hg}OAc$ and $Hg(OAc)_2$ are active in the mercuridesilylation, the former being about 15 times more reactive, and suggested a cyclic transition state for reaction through $Hg(OAc)_2$. Rate constants for a number of $XC_6H_4SiMe_3$ compounds were reported, the order of reactivity being[154] X = H < 2-Me, 4-Me (see Table 22).

TABLE 21. Pseudo first-order rate constants[a], $10^5k_1$, in acetic acid at 25 °C for the reaction

$$ArSiMe_3 + Hg(OAc)_2 \longrightarrow ArHgOAc + Me_3SiOAc$$

| Ar | $10^5k_1$ ($s^{-1}$)[b] | $10^5k_1$ ($s^{-1}$)[c] | $10^5k_1$ ($s^{-1}$)[d] |
|---|---|---|---|
| $C_6H_5$ | 1.83 | 1.03 | |
| $2\text{-}MeC_6H_4$ | 19.8 | 11.6 | |
| $3\text{-}MeC_6H_4$ | 3.75 | 2.67 | 2.51 |
| $4\text{-}MeC_6H_4$ | 21.2 | 11.0 | 10.6 |
| $2,3\text{-}Me_2C_6H_3$ | | 42.5 | |
| $3,4\text{-}Me_2C_6H_3$ | | 28.0 | |
| $2,4\text{-}Me_2C_6H_3$ | | 165.0 | |
| $2,6\text{-}Me_2C_6H_3$ | | v. fast | |
| $3,5\text{-}Me_2C_6H_3$ | | 3.65 | |
| $2,5\text{-}Me_2C_6H_3$ | | 25.0 | |
| $3\text{-}Pr^iC_6H_4$ | | | 3.96 |
| $3\text{-}Bu^tC_6H_4$ | | | 5.50 |
| $4\text{-}EtC_6H_4$ | | | 11.5 |
| $4\text{-}Pr^iC_6H_4$ | | | 12.0 |
| $4\text{-}Bu^tC_6H_4$ | | | 14.0 |

[a] Rates were measured under pseudo first-order conditions, with each reagent in turn in excess. Since the second-order rate constants calculated from both sets of conditions do not agree, the results should not be considered accurate.
[b] Ref. 150; [silane] = 0.017 M, [$Hg(OAc)_2$] = 0.1788 M.
[c] Ref. 150; [silane] = 0.400 M, [$Hg(OAc)_2$] = 0.04 M.
[d] Ref. 152; the electrophile was not identified and it is not clear which reagent is in excess.

TABLE 22. Second-order rate constants[154] at 25 °C in glacial acetic acid for the reaction

$$XC_6H_4SiMe_3 + Hg(OAc)_2 \longrightarrow XC_6H_4HgOAc + Me_3SiOAc$$

| X | [Si][a] | [Hg][b] | $10^3 k_2^{obs}$ (1 mol$^{-1}$ s$^{-1}$) |
|---|---|---|---|
| H | 0.4 | 0.04 | 1.77 |
| H | 0.8 | 0.04 | 1.20 |
| H | $9.2 \times 10^{-3}$ | 0.18 | 6.17 |
| 2-Me | 0.4 | 0.04 | 16.1 |
| 2-Me | 0.55 | 0.057 | 13.1 |
| 2-Me | $9.2 \times 10^{-3}$ | 0.18 | 48.3 |
| 4-Me | 0.2 | 0.02 | 21.8 |
| 4-Me | 0.4 | 0.02 | 14.2 |
| 4-Me | 0.8 | 0.02 | 10.4 |
| 4-Me | $9.2 \times 10^{-3}$ | 0.18 | 46.2 |

[a] Concentration (mol l$^{-1}$) of $XC_6H_4SiMe_3$.
[b] Concentration (mol l$^{-1}$) of $Hg(OAc)_2$.

Roberts[155] has studied the substitution of trimethylallylsilicon by mercury(II) salts in acetonitrile, the reaction being thought to proceed by the slow step 60. With

$$Me_3SiCH_2CH{=}CH_2 + HgX_2 \longrightarrow CH_2{=}CH{-}CH_2HgX + Me_3SiX \quad (60)$$

$HgCl_2$, the observed second-order rate constant at 25 °C was 0.042 l mol$^{-1}$ s$^{-1}$ with $E_a = 12.0$ kcal mol$^{-1}$. Relative rate constants for reaction of a number of mercury(II) salts increased along the series X = I (0.18) < Br (0.9) < Cl (1.0) < OAc (29), and Roberts[155] suggested an open transition state for these reactions.

### 5. *Cleavage of silicon—carbon bonds by gallium*(III) *and thallium*(III) *compounds*

Alkyl groups are cleaved from tetraalkylsilicons and polydimethylsiloxanes by gallium(III) chloride[156,157]. A method of preparing a wide range of arylthallium bis-trifluoroacetates in good yield has been reported by Bell *et al.*[158], who studied reaction 61 using trifluoroacetic acid as solvent, with a wide range of substituents, X.

$$XC_6H_4SiMe_3 + Tl(OCOCF_3)_3 \longrightarrow XC_6H_4Tl(OCOCF_3)_2 + Me_3SiOCOCF_3 \quad (61)$$

### 6. *Cleavage of silicon—carbon bonds by silicon*(IV) *and lead*(IV) *compounds*

Redistribution reactions of organosilicon compounds have been extensively investigated[159–166]. For example, when tetraethylsilicon and tetra-*n*-propylsilicon are heated together in the presence of a little aluminium(III) chloride, an equilibrium mixture of $n\text{-}Pr_xEt_{4-x}Si$ compounds is obtained[159,160] Usually the smaller alkyl groups exchange more easily, but exchange is inhibited by the presence of chlorine atoms on silicon.

A study of the exchange 62 showed the order of increasing reactivity X = *p*-Cl < H < *p*-Me, as expected for an electrophilic aromatic substitution[166].

$$XC_6H_4SiMe_3 + ClMe_2SiSiMe_3 \longrightarrow XC_6H_4SiMe_2SiMe_3 + Me_3SiCl \quad (62)$$

Kalman and coworkers[119] have reported the cleavage of a number of aryltrimethylsilicons by lead(IV) trifluoroacetate in solvent trifluoroacetic acid (compare reaction 50).

### 7. *Cleavage of silicon—carbon bonds by antimony*(III) *and antimony*(V) *compounds*

The formation of tetrafluorosilicon from the cleavage of compounds containing silicon—carbon by antimony(III) fluoride in solvents xylene or ether has been reported by Müller and Müller[167]. Antimony(V) chloride will remove one aryl group from $Ph_2SiCl_2$ and $(p\text{-}ClC_6H_4)_2SiCl_2$ to yield compounds of type $ArSbCl_4$; these then decompose to ArCl and $SbCl_3$[168].

### 8. *Cleavage of germanium—carbon bonds by iron*(III) *compounds*

Kochi and coworkers[36] have reported rate constants for the reaction of tetraethylgermanium with various iron(III) compounds in acetonitrile; these reactions, however, are outer sphere electron transfer processes rather than $S_E$ reactions.

### 9. *Cleavage of germanium—carbon bonds by mercury*(II) *compounds*

Roberts[169] has studied the action of mercury(II) salts on allylgermanium compounds in solvents acetonitrile and ethanol (reaction 63) (R = H, Me; X = Cl, Br); rate constants are given in Table 23.

$$RCH{=}CHCH_2GeEt_3 + HgX_2 \longrightarrow RCH{=}CHCH_2HgX + Et_3GeX \quad (63)$$

When the central metal atom in $RCH{=}CHCH_2MEt_3$ was varied, the order of increasing reactivity towards $HgX_2$ was $M = Si < Ge \ll Sn$[169].

### 10. *Cleavage of germanium—carbon bonds by germanium*(IV) *and tin*(IV) *compounds*

Tetraalkylgermaniums and hexaalkyldigermaniums are dealkylated by germanium(IV) chloride (reaction 64) ($n = 0, 1$); the process is accelerated by polar solvents[170,171].

$$R_3Ge(GeR_2)_nR + GeCl_4 \longrightarrow RGeCl_3 + R_3Ge(GeR_2)_nCl \quad (64)$$

Similar dealkylations take place with tin(IV) chloride (reaction 65) ($n = 0, 1$), the

TABLE 23. Second-order rate constants[169] at 25 °C for the reaction

$$Et_3GeCH_2CH{=}CHR + HgX_2 \longrightarrow Et_3GeX + XHgCH_2CH{=}CHR$$

| R | X | Solvent | $k_2$ (l mol$^{-1}$ s$^{-1}$) |
|---|---|---|---|
| H | Br | Acetonitrile | 39 |
| H | Cl | Acetonitrile | 45 |
| Me | Br | Acetonitrile | 110 |
| H | Br | Ethanol | 0.035 |
| Me | Br | Ethanol | 0.149 |

ease of dealkylation increasing with increase in polarity of the solvent and generally decreasing with increase in the size of the alkyl group[171–173]. Reaction 65 ($R = Et$,

$$R_3Ge(GeR_2)_nR + SnCl_4 \longrightarrow RSnCl_3 + R_3Ge(GeR_2)_nCl \qquad (65)$$

$n = 1$) in acetyl chloride was found to be kinetically first order in each reactant, with a second-order rate constant of *ca.* $2 \times 10^{-6}\,l\,mol^{-1}\,s^{-1}$ [173].

### 11. Cleavage of tin—carbon bonds by lithium(I) compounds

There have been a large number of preparative investigations of transmetallations between tetraorganotins and organolithiums. Vinyl—tin and allyl—tin bonds are readily broken in this way, and preparative details have been published for reactions 66 ($R = Ph$)[174,175] and 67 ($R = Ph$)[176,177].

$$(CH_2{=}CH)_4Sn + 4RLi \longrightarrow 4CH_2{=}CHLi + R_4Sn \qquad (66)$$

$$CH_2{=}CHCH_2SnPh_3 + RLi \longrightarrow CH_2{=}CHCH_2Li + Ph_3SnR \qquad (67)$$

With mixed tetraorganotins, the vinyl and allyl groups are transferred to lithium preferentially over phenyl or alkyl groups[178]. Perfluorovinyllithium has been prepared by the cleavage of phenyl triperfluorovinyltin by phenyl lithium[179], and benzyllithium may similarly be prepared via a transmetallation reaction[180].

A stereochemical investigation showed that both *cis*- and *trans*-propenyltintrimethyl were cleaved by *n*-butyllithium to yield the *cis*- and *trans*-propenyllithium with retention of configuration[30], and *trans*-$\beta$-styryllithium has been obtained from the corresponding tin compound[181].

Seyferth and Cohen[182] have used the transmetallation between tetracyclopropyltin and *n*-butyllithium to prepare cyclopropyllithium, and have described a number of other related reactions.

The mechanism of the transmetallation reaction is not yet fully understood; various equilibria are involved, and the controlling force in these reactions seems to be the relative electronegativity of the organic groups[30].

### 12. Cleavage of tin—carbon bonds by copper(II), silver(I), potassium(I), titanium(IV), vanadium(V), and palladium(II) compounds

Anderson[183] has investigated the cleavage of tetraethyltin brought about by refluxing the liquid with various metallic halides for various lengths of time. In all cases but one, triethyltin halide was formed; the exception was in the reaction with silver(I) chloride, when diethyltin dichloride was formed.

### 13. Cleavage of tin—carbon bonds by iron(II), platinum(II), iron(III), and iridium(III) compounds

The aryl group is cleaved from various aryltrimethyltins by $[C_6H_7Fe(CO)_3]BF_4$ (see reaction 56)[144] and by chloroplatinum(II) complexes[184] to yield [ArPt(cod)Cl] or $[Ar_2Pt(cod)]$ compounds. Analogous reactions were carried out with aryl compounds of the other Group IV metals, the reactivity increasing along the series $Si < Ge \ll Sn < Pb$[184]. Chloroplatinum(II) complexes have also been used to cleave $PhC{\equiv}C$, $CF_2{=}CF$, and $CH_2{=}CH$ groups from organotintrimethyl compounds with formation of the corresponding organo—Pt bonds[185,186]. King and Stone[187] have reported that when ironpentacarbonyl is allowed to react with *n*-$Bu_2Sn(vinyl)_2$, *n*-$Bu_2SnPh_2$, or *n*-$Bu_3SnPh$, the compound $[n\text{-}Bu_2SnFe(CO)_4]_2$ is formed.

Kochi and coworkers[36,188] have reported that iron(III) complexes, $[FeL_3]^{3+}$, where L = 2,2′-bipyridine or various substituted 1,10-phenanthrolines, cleave a variety of organometallic compounds including tetraalkyltins. The rate-determining step is the electron transfer process 68, and the overall stoichiometry of the reaction sequence is given by equation 70.

$$R_4Sn + FeL_3^{3+} \xrightarrow{k_{Fe}} R_4Sn^{+\bullet} + FeL_3^{2+} \tag{68}$$

$$R_4Sn^{+\bullet} \xrightarrow{\text{fast}} R_3Sn^+ + R^\bullet \tag{69}$$

$$R_4Sn + 2[FeL_3](ClO_4)_3 \longrightarrow RClO_4 + R_3SnClO_4 + 2[FeL_3](ClO_4)_2 \tag{70}$$

For a particular alkylmetal, log $k_{Fe}$ varied linearly with the standard reduction potential, $E_0$, of a series of iron(III) complexes as expected for an outer-sphere mechanism, and for a particular iron(III) complex the log $k_{Fe}$ values were linearly correlated with the ionization potentials, $I_D$, of a series of alkylmetal compounds. Rate constants are given in Table 24. With unsymmetrical tetraalkyltins, the selectivity of the group cleaved takes place in step 69, and with methylethyltin compounds the ethyl group is preferentially cleaved.

In the analogous reaction of $R_4Sn$ compounds with hexachloroiridate(IV), the stoichiometry of the overall process was found to be as in equation 71. The proposed

$$R_4Sn + 2[IrCl_6]^{2-} \longrightarrow RCl + R_3Sn^+ + [IrCl_6]^{3-} + [IrCl_5]^{2-} \tag{71}$$

mechanism is similar to that suggested for the iron(III) cleavages reaction 72 now being the rate-determining step, followed by reaction 69. Rate constants for step 72,

TABLE 24. Second-order rate constants[36,188] ($l\,mol^{-1}\,s^{-1}$) for electron transfer from tetralkyltins to iron(III) complexes $[FeL_3]^{3+}$ in acetonitrile at 25 °C

| | L | | | | |
|---|---|---|---|---|---|
| $R_4Sn$ | 4,7-Di-Ph-phen | Bipy | Phen | 5-Cl-phen | 5-$NO_2$-phen |
| $Me_4Sn$ | 0.000166 | 0.000515 | 0.00154 | 0.0172 | 0.256 |
| $Et_4Sn$ | 3.45 | 5.17 | 13.2 | 99.5 | 774 |
| $Pr^n{}_4Sn$ | 10.7 | 19.3 | 46.1 | 290 | 3 200 |
| $Pr^i{}_4Sn$ | | | 6 900 | | |
| $Bu^n{}_4Sn$ | | | 360 | | |
| $Bu^s{}_4Sn$ | 867 | 1150 | 4 500 | 35 200 | 162 000 |
| $Bu^i{}_4Sn$ | 43 | 79.3 | 257 | 1 580 | 9 220 |
| $Pe^{neo}{}_4Sn$ | 52.4 | 111 | 353 | 2 660 | 17 800 |
| $EtSnMe_3$ | | | 1.6[a] | | |
| $Pr^nSnMe_3$ | | | 0.79 | | |
| $Bu^nSnMe_3$ | | | 0.85 | | |
| $Bu^tSnMe_3$ | | | 760 | | |
| $Et_2SnMe_2$ | 1.23 | 2.24 | 5.68[a] | 48.3 | 449 |
| $Pr^n{}_2SnMe_2$ | | | 170 | | |
| $Pr^i{}_2SnMe_2$ | | | 2 300 | | |
| $Bu^n{}_2SnMe_2$ | | | 190 | | |
| $Bu^t{}_2SnMe_2$ | | | 78 000 | | |

[a] Selectivity, $k_{Et}/k_{Me} = 27 \pm 2$.

TABLE 25. Second-order rate constants[36,188], $10^4k_2$, for electron transfer from tetraalkyltins to hexachloroiridate(IV) in acetonitrile at 25 °C

| $R_4Sn$ | $10^4k_2$ ($l\ mol^{-1}\ s^{-1}$) | $R_4Sn$ | $10^4k_2$ ($l\ mol^{-1}\ s^{-1}$) |
|---|---|---|---|
| $Me_4Sn$ | 0.52 | $Pr^nSnMe_3$ | 0.53 |
| $Et_4Sn$ | 6.3 | $Pr^iSnMe_3$ | 65 |
| $Pr^n{}_4Sn$ | 4.6 | $Bu^nSnMe_3$ | 0.3 |
| $Pr^i{}_4Sn$ | 9.9 | $Bu^tSnMe_3$ | 780 |
| $Bu^n{}_4Sn$ | 32 | $Et_2SnMe_2$ | 6.9[a] |
| $Bu^s{}_4Sn$ | 5.9 | $Pr^n{}_2SnMe_2$ | 13 |
| $Bu^i{}_4Sn$ | 1.4 | $Pr^i{}_2SnMe_2$ | 64 |
| $Pe^{neo}{}_4Sn$ | $\sim 10^{-2}$ | $Bu^n{}_2SnMe_2$ | 3.6 |
| $EtSnMe_3$ | 1.4[a] | $Bu^t{}_2SnMe_2$ | 9.4 |

[a] Selectivity, $k_{Et}/k_{Me} = 11 \pm 2$.

given in Table 25,

$$R_4Sn + [IrCl_6]^{2-} \xrightarrow{k_{Ir}} R_4Sn^{+\bullet} + [IrCl_6]^{3-} \tag{72}$$

are no longer correlated with $E_0$ and $I_D$ but are related to the steric properties of the alkyl group R, and it was concluded that an inner sphere contribution to electron transfer was taking place.

### *14. Cleavage of tin—carbon bonds by mercury(II) compounds*

Extensive studies have been carried out by Abraham and coworkers on the electrophilic substitution of tetraalkyltins by mercury(II) salts (reaction 73)[189–210] and

$$R_4Sn + HgX_2 \longrightarrow RHgX + R_3SnX \tag{73}$$

second-order rate constants have been obtained for the cleavage of a series of tetraalkyltins by mercury(II) salts in several solvents. Reported rate constants at 25 °C are given in Table 26 for a number of solvent systems. In addition, Abraham and Johnston[192,193] have determined rate constants using a number of water–methanol mixtures as solvent, and Abraham and Behbahany[197] have similarly used a range of methanol–*tert*-butanol solvents. Activation parameters for many of these systems have been obtained and have been detailed before[4,211]; in general, reaction 73 in the various solvents studied is characterized by negative entropies of activation, ranging from about $-20$ to $-40$ cal $K^{-1}$ $mol^{-1}$, depending on the alkyl group and solvent.

The substitution by mercury(II) iodide is interesting, in that the rate-determining step 73 is followed by a rapid reversible ionization step 74, involving a reactant and a

$$R_3SnI + HgI_2 \rightleftharpoons R_3Sn^+ + HgI_3^- \tag{74}$$

product[189]. This leads to a kinetic situation that had not previously been reported, but a method of solution of the kinetic equations was devised in order to obtain the required rate constant for reaction 73 (X = I)[210].

For substitution by the three mercury(II) salts with X = I, Cl, and OAc, there is a very pronounced effect of the alkyl group in the tetraalkyltin on the rate constant (see Table 26). Studies with unsymmetrical tetraalkyltins of type $Pr^i_3SnR$ in which only the group R is cleaved (Table 27) show conclusively that the major steric effect

TABLE 26. Second-order rate constants, $10^3k_2$, at 25 °C in various solvents for the reaction

$$R_4Sn + HgX_2 \longrightarrow RHgX + R_3SnX$$

| | | $10^3k_2$ (l mol$^{-1}$ s$^{-1}$) | | | |
|---|---|---|---|---|---|
| $R_4Sn$ | X | MeOH | 96% MeOH | Bu$^t$OH | MeCN |
| $Me_4Sn$ | Cl | 1550[a] | 2590[a] | 192[c,e] | 2820[e] |
| $Et_4Sn$ | Cl | 3.33[b] | 6.3[b] | 0.288[c,e] | 18.7[f] |
| $Pr^n{}_4Sn$ | Cl | 0.628[a] | 1.13[a] | 0.0462[c,e] | 4.25[e] |
| $Pr^i{}_4Sn$ | Cl | $<10^{-5}$ [a] | $<10^{-5}$ [a] | $<1.6\times10^{-4}$ [c,e] | $<1.6\times10^{-5}$ [e] |
| $Bu^n{}_4Sn$ | Cl | 0.615[a] | 1.04[a] | 0.0310[c,e] | 4.57[e] |
| $Bu^i{}_4Sn$ | Cl | 0.08[a] | 0.138[a] | 0.0067[c,e] | 1.17[e] |
| $Pe^{neo}{}_4Sn$ | Cl | <0.0025[c,d] | | $<2\times10^{-4}$ [c,d] | |
| $Me_4Sn$ | I | | 710[g] | | |
| $Et_4Sn$ | I | 2.53[i] | 4.75[h] | 0.032[i,j] | 34.3[f] |
| $Pr^n{}_4Sn$ | I | | 0.72[g] | | |
| $Pr^i{}_4Sn$ | I | | $<2\times10^{-4}$ [g] | | |
| $Bu^n{}_4Sn$ | I | | 0.758[g] | | |
| $Bu^i{}_4Sn$ | I | | $5.03\times10^{-2}$ [g] | | |
| $Bu^i{}_3SnMe$ | I | | 213[g] | | |
| $Me_4Sn$ | OAc | $2.33\times10^5$ [d,k] | | 907[d,k] | |
| $Et_4Sn$ | OAc | 1023[d,k] | 1180[i] | 8.38[d,k] | 83.7[f] |
| $Pr^n{}_4Sn$ | OAc | 208[d,k] | | 1.90[d,k] | |
| $Pr^i{}_4Sn$ | OAc | 0.0013[d,k] | | 0.011[d,k] | |
| $Bu^n{}_4Sn$ | OAc | 173[d,k] | | 1.72[d,k] | |
| $Bu^i{}_4Sn$ | OAc | 43.7[d,k] | | 0.633[d,k] | |
| $Pe^{neo}{}_4Sn$ | OAc | 0.183[d,k] | | 0.0157[d,k] | |

[a] Ref. 193.
[b] Ref. 192.
[c] At 40 °C.
[d] Ref. 202.
[e] Ref. 200.
[f] Ref. 198.
[g] Ref. 190.
[h] Ref. 189.
[i] Ref. 197.
[j] Estimated from value at higher temperature.
[k] At 30 °C.

on the rate constant is due to the group being transferred from tin to mercury and that the alkyl groups in the $SnR_3$ entity have little effect[205]. It was suggested[202] that the 'steric' effect of alkyl groups leading to the reactivity sequence R = Me > Et > Pr$^n$ > Bu$^i$ > Pe$^{neo}$ > Pr$^i$ is due to interactions between (a) the moving group and the incoming $HgX_2$ group and (b) the moving group and the leaving $SnC^\alpha$ group.

TABLE 27. Second-order rate constants[205], $10^3k_2$, at 30 °C in methanol for the reaction

$$Pr^i{}_3SnR + Hg(OAc)_2 \longrightarrow RHgOAc + Pr^i{}_3SnOAc$$

| R | $10^3k_2$ (l mol$^{-1}$ s$^{-1}$) | R | $10^3k_2$ (l mol$^{-1}$ s$^{-1}$) |
|---|---|---|---|
| Me | 36 500 | Bu$^i$ | 8.57 |
| Et | 8 280 | Pe$^{neo}$ | 0.0878 |
| Pr$^n$ | 15.9 | Pr$^i$ | 0.00034 |

TABLE 28. Second-order rate constants, $10^3 k_2$ ($l\,mol^{-1}\,s^{-1}$), at 30 °C for the reaction

$$Et_4Sn + Hg(OCOR')_2 \longrightarrow EtHgOCOR' + Et_3SnOCOR'$$

| | R' | | | | | |
|---|---|---|---|---|---|---|
| Solvent | Bu$^t$ | Et | Me | $ClCH_2CH_2$ | $MeOCH_2$ | $ClCH_2$ |
| MeOH[a] | 445 | 870 | 1190 | 1490 | 3880 | 5480 |
| Bu$^t$OH[b] | 3.0 | 11.2 | 19 | 21.2 | 66.0 | 347 |
| MeCN[b] | 175 | 155 | 178 | 550 | 833 | Fast |
| MeCOMe[b] | 19.5 | 20.3 | 24.0 | 128 | 176 | Fast |
| $MeCO_2Et$[b,c] | 3.17 | 4.17 | 5.0 | 37.3 | — | — |

[a] Ref. 204.
[b] Ref. 207.
[c] At 25 °C.

The effect of the mercury(II) salt is pronounced; from the results in Table 26 it can be seen that the order of reactivity is $HgI_2 < HgCl_2 < Hg(OAc)_2$, and further studies showed[204,207] that for a series of mercury(II) carboxylates the effect of electron-attracting substituents is to increase the rate constant (Table 28), as expected for an electrophilic substitution. Earlier work had shown that in methanol or aqueous methanol solvents the species $HgI_3^-$ was unreactive[189], and it was shown also that in methanol there is no catalysis by $Cl^-$ (i.e. that $HgCl_3^-$ is unreactive)[192]. The latter studies showed also that the species $HgCl^+$ was not an active reagent, and that the attacking entity was the simple $HgCl_2$ molecule.

Salt effect studies revealed that in solvents such as methanol, aqueous methanol, *tert*-butanol, and acetonitrile, there were pronounced accelerating effects of added inert salts on reaction 73 (R = Et; X = Cl or I). In some cases[197] the results were analysed through equation 75, in which $k$ and $k_0$ are the rate constants in the

$$\left[\frac{\log(k/k_0)}{I}\right]_{I\to 0} = \frac{4\pi N e^4 Z^2 d}{2303(\varepsilon k T)^2} = JZ^2 d \qquad (75)$$

presence and in the absence of added salt, respectively, $I$ is the ionic strength, and $J$ is a constant for a particular solvent at a given temperature. It is therefore possible to calculate the term $Z^2d$ that refers to a transition state dipole in which charges $+Z$ and $-Z$ are located a distance $d$ apart. Values of $Z^2d$ thus calculated for reaction 73 (R = Et; X = Cl) were very large and if $d$ was taken as 3.1 Å led to values of $Z$ from 0.75 to 0.91 in hydroxylic solvents and 0.47 in acetonitrile, thus leading to the suggestion of mechanism $S_E2$ (open) for these substitutions[197].

Another method of examining transition states in reaction 73 (R = alkyl) has been to dissect solvent effects on rate constants into initial-state and transition-state contributions[211]. For transfers from methanol to aqueous methanol there are very large transition-state effects, and an electrostatic analysis by Abraham and Johnston[199] suggests that the transition state for reaction 73 (R = Et; X = Cl) behaves towards change of solvent as though it contained a charge separation of about ±0.72 unit in the methanol–water solvent system. Abraham and Dorrell[201] later carried out a similar analysis for reaction 73 (R = Et; X = Cl) in the solvent system methanol–*tert*-butanol, and from the Kirkwood equation deduced for the transition state a dipole moment of no less than about 14 D. If a correction for the contribution of the $HgCl_2$ group is applied, then this leads to charge separation of 0.86–0.91 unit, with the dipolar distance again taken as 3.1 Å.

Thus for reaction 73 (R = Et; X = Cl), in hydroxylic solvents, estimates of the charge separation in the transition state are 0.75–0.91 unit from kinetic salt effects, 0.72 unit from an electrostatic analysis of solvent effects, and 0.86–0.91 unit from the Kirkwood equation applied to solvent effects[211], so there seems no doubt that the transition state is highly polar in nature, and that reaction proceeds by mechanism $S_E2$ (open). Abraham and Grellier[202] have discussed the stereochemistry of these reactions, and have pointed out that the steric effects of alkyl groups are more compatible with retention of configuration than with inversion of configuration. Unfortunately, no experimental stereochemical studies have been reported, and since there is no general stereochemical rule for $S_E2$ substitutions it is not possible to assign a stereochemical course unambiguously.

Rate constants for reaction 73 (R = Me, Et; X = Cl, I, OAc) were later reported using a series of alcohols as solvents (Table 29)[206]. The order of reactivity is always $HgI_2 < HgCl_2 < Hg(OAc)_2$, and towards a given electrophile in a given alcohol, the $Me_4Sn$ compound is always more reactive than the $Et_4Sn$ compound.

Roberts[212,213] has also studied the cleavage of $Me_3SnR$ compounds by mercury(II) iodide in methanol, acetonitrile and dimethyl sulphoxide. Except for 6-(trimethylstannyl)-2-norbornene, the methyl group was cleaved in all cases to yield methylmercury(II) iodide. Activation parameters and rate constants for the $Me_4Sn$ reactions are given in Table 30. The effect of the above three solvents on rate constants and activation parameters was later dissected into initial-state and transition-state contributions[209].

An interesting general survey of the cleavage of another series of $Me_3SnR$ compounds by mercury(II) chloride in methanol and dimethyl sulphoxide has been reported;[214,215] rate constants for cleavage of the R or of the methyl group are given in Table 31. As usual, the groups Ph, $CH_2{=}CH$, and $CH_2{=}CHCH_2$ are removed faster than are simple alkyl groups. Activation parameters for a number of the substitutions in methanol were also given (see Table 31). In a later paper[216], it was shown that in presence of an excess of chloride ion in dimethyl sulphoxide or dimethylformamide, the neutral species $HgCl_2$ does not take part in the actual cleavage reaction, the transition state being formed with participation of two chloride ions. It

TABLE 29. Second-order rate constants[206], $10^3k_2$, at 25 °C in alcohol solvents for the reaction

$$R_4Sn + HgX_2 \longrightarrow RHgX + R_3SnX$$

| Reactants | $10^3k_2$ (l mol$^{-1}$ s$^{-1}$) | | | | |
|---|---|---|---|---|---|
| | MeOH | EtOH | Pr$^n$OH | Bu$^n$OH | Bu$^t$OH |
| $Me_4Sn + HgCl_2$ | 1550[a] | 848 | 709 | 529 | 104 |
| $Me_4Sn + HgI_2$ | 310[b] | 154 | 94.2 | 74.8 | 7.0 |
| $Me_4Sn + Hg(OAc)_2$ | $1.1 \times 10^5$ | | | | 828 |
| $Et_4Sn + HgCl_2$ | 3.33[a] | 1.24 | 0.990 | 0.785 | 828 |
| $Et_4Sn + HgI_2$ | 2.53[a] | 0.968 | 0.615 | 0.453 | 0.032 |
| $Et_4Sn + Hg(OAc)_2$ | 470 | 238 | 228 | 202 | 7.7 |

[a] See Table 26.
[b] Ref. 209.

TABLE 30. Activation parameters[212] at 25 °C for the reaction

$$Me_3SnR + HgI_2 \longrightarrow Me_3SnI + RHgI$$

| R | Solvent | $\Delta H^{\neq}$ (kcal mol$^{-1}$) | $\Delta S^{\neq}$ (cal K$^{-1}$ mol$^{-1}$) |
|---|---|---|---|
| Me | MeCN[a] | 9.6±0.1 | −27.8±0.5[d] |
| | MeOH[b] | 10.8±0.5 | −26.9±1.7[d] |
| | dmso[c] | 10.0±0.5 | −23.9±1.7[d] |
| $(CH_2)_2CO_2Me$ | MeCN | 9.3±0.4 | −22.3±1.5 |
| | MeOH | 10.0±0.5 | −29.0±1.8 |
| | dmso | 10.0±0.6 | −23.8±1.9 |
| $(CH_2)_3Me$ | MeCN | 10.0±0.2 | −25.5±0.6 |
| $(CH_2)_2CH{=}CH_2$ | MeCN | 7.8±0.2 | −33.4±0.5 |
| $(CH_2)_3CH{=}CH_2$ | MeCN | 7.5±0.1 | −34.0±0.3 |
| $(CH_2)_4Ph$ | MeCN | 8.4±0.3 | −31.4±1.4 |
| Bicyclo[2.2.1]hept-7-yl | MeCN | 11.1±0.4 | −22.2±1.4 |
| Bicyclo[2.2.1]hept-2-en-7-yl | MeCN | 10.5±0.6 | −24.4±2.1 |
| Bicyclo[2.2.1]hepta-2,5-dien-7-yl | MeCN | 6.3±0.5 | −36.1±1.8 |

[a] $k_2 = 1.91$ l mol$^{-1}$ s$^{-1}$ at 25.2 °C; ref. 213.
[b] $k_2 = 0.47$ l mol$^{-1}$ s$^{-1}$ at 25.2 °C; ref. 213.
[c] $k_2 = 7.5$ l mol$^{-1}$ s$^{-1}$ at 25.2 °C; ref. 213.
[d] Statistically corrected for the four methyl groups.

was suggested[216] that in this case, the complex $HgCl_3^-$ reacts with a complex between $RSnMe_3$ and $Cl^-$.

There is some disagreement over the rate constants and, especially, the activation parameters for the key reaction of tetramethyltin with mercury(II) chloride in methanol. Abraham and coworkers have twice[193,209] examined this reaction and have concluded that $\Delta H^{\neq}$ is about 11 kcal mol$^{-1}$ and $\Delta S^{\neq}$ about −20 cal K$^{-1}$ mol$^{-1}$,

TABLE 31. Second-order rate constants, $10^3 k_2$, at 25 °C in dimethyl sulphoxide[214] and methanol[215] and activation parameters in methanol[215] for the reaction

$$RSnMe_3 + HgCl_2 \longrightarrow RHgCl + Me_3SnCl$$

| R | $k_2^{dmso}$ (l mol$^{-1}$ s$^{-1}$)[a] | $k_2^{dmso}$ (l mol$^{-1}$ s$^{-1}$) | $k_2^{MeOH}$ (l mol$^{-1}$ s$^{-1}$) | $\Delta H^{\neq}$ (kcal mol$^{-1}$) | $\Delta S^{\neq}$ (cal K$^{-1}$ mol$^{-1}$) |
|---|---|---|---|---|---|
| Me | 5 000 | 4 500 | 970 | 16.0 | −5 |
| Bu$^n$ | 3 200 | 3 470[b] | | | |
| CN | | Very fast | | | |
| $CF_3$ | 18 | 18[b] | | | |
| $CH_2{=}CH$ | 112 000 | 188 000 | 660 000[c] | 8.2 | −18 |
| $CH_2{=}CHCH_2$ | Very fast | Very fast | Very fast | | |
| c-$C_5H_5$ | | Very fast | | | |
| Ph | 50 000 | 58 000 | 135 000 | 6.7 | −26 |
| $C_6F_5$ | 960 | 960 | 1 010[d] | 5.5 | −40 |
| $PhCH_2$ | 2 160 | 1 950[b] | Very fast | | |
| PhC≡C | | Very fast | | | |

[a] Rate constants for the reaction in the presence of $Cl^-$ (ref. 216).
[b] Rates for cleavage of a methyl group.
[c] Extrapolated from values at lower temperatures.
[d] Extrapolated from values at higher temperatures.

TABLE 32. Second-order rate constants, $10^3 k_2$ ($l\,mol^{-1}\,s^{-1}$), for the reaction between tetramethyltin and mercury(II) chloride in methanol

| Temperature (°C) | Ref: Ionic strength: | 193 0 | 209 0 | 209 0.1 | 214 0.1 | 217 |
|---|---|---|---|---|---|---|
| 15 | | | | 1260 | | |
| 25 | | 1550 | 1400 | 2540 | 970 | 910 |
| 30 | | | 2000 | 3560 | 1440 | |
| 35 | | | | | | 2550 |
| 40 | | 3840 | 3680 | 6520 | 3690 | |
| 45 | | | | | 5560 | |
| $\Delta H^{\neq}$ ($kcal\,mol^{-1}$) | | 10.6 | 11.3 | 11.2 | 16.0 | 18.2[a] |
| $\Delta S^{\neq}$ ($cal\,K^{-1}\,mol^{-1}$) | | −22 | −20 | −19 | −5 | +2[a] |

[a] Approximate values calculated by the present authors from the two rate constants.

whereas Reutov and coworkers[214] found values of $16\,kcal\,mol^{-1}$ and $-5\,cal\,K^{-1}\,mol^{-1}$. A recent re-investigation by Reutov and coworkers[217] did not resolve the problem (see Table 32); all that can be said is that the value of $-20\,cal\,K^{-1}\,mol^{-1}$ for $\Delta S^{\neq}$ is more compatible with other activation entropies for the same reaction in methanol–water mixtures and for the corresponding reaction between tetramethyltin and mercury(II) iodide in methanol.

Roberts[218] has studied the cleavage of tin—allyl bonds in some detail, using ethanol as solvent and $HgI_2$, $HrBr_2$, and $HgCl_2$ as the mercury(II) salt. Rate constants for the removal of various allyl groups from tin are given in Table 33. The rate constant for reaction of $CH_2{=}CHCH_2SnEt_3$ with $HgI_2$ obtained by Roberts, $2640\,l\,mol^{-1}\,s^{-1}$ at 25 °C, may be compared with the value found by Abraham and Grellier[206] for the corresponding reaction of $Et_4Sn$, viz. $9.68\times10^{-4}\,l\,mol^{-1}\,s^{-1}$ at 25 °C; after correction for the statistical effect, it is found that the allyl—tin bond is cleaved $1.1\times10^{7}$ times as rapidly as is the ethyl—tin bond. Roberts[218] suggested

TABLE 33. Second-order rate constants[218] and $\Delta H^{\neq}$ in ethanol for the reaction

$$R_3SnCH_2CH{=}CHR' + HgX_2 \longrightarrow R'CH{=}CHCH_2HgX + R_3SnX$$

| R | R′ | X | Temperature (°C) | $k_2$ ($l\,mol^{-1}\,s^{-1}$) | $\Delta H^{\neq}$ ($kcal\,mol^{-1}$) |
|---|---|---|---|---|---|
| Et | H | I | 30 | 2900 | 5.2 |
| Et | Me | I | 30 | 15 | |
| Et | Ph | I | 30 | 3.36 | |
| $Bu^n$ | H | I | 30 | 1020 | |
| Et | Ph | Cl | 25 | 178 | |
| Et | Ph | Br | 25 | 13.6 | |
| Et | Ph | I | 25 | 2.8 | 7.8 |
| Ph | H | I | 25 | 25 | 3.4 |
| Ph | Me | I | 25 | 0.54 | |
| Ph | Ph | I | 25 | 0.46 | |
| Ph | H | Cl | 25 | 1500 | |
| Ph | H | Br | 25 | 330 | |

transition state **16** for allylic cleavage reactions, that is via an $S_E2$ process and not an $S_E2'$ with rearrangement.

$$\left[\begin{array}{l} R'CH{=}CH \\ \quad\downarrow \qquad \diagdown \\ \qquad\qquad CH_2 \\ \quad R_3Sn \quad\;\; HgX_2 \end{array}\right]^{\neq}$$

**(16)**

Although Reutov and coworkers[214,215] included the benzyl compound $PhCH_2SnMe_2$ in their reactivity survey (see Table 31), they were unable to deduce the benzyl group reactivity because in dimethyl sulphoxide only the methyl group was cleaved by $HgCl_2$ and in methanol, where the benzyl group was reported to be cleaved, the compound reacted too rapidly for kinetic studies to be carried out. Abraham and Andonian-Haftvan[219] therefore re-investigated this reaction, and also studied a number of other benzyl derivatives, $PhCH_2SnR_3$. When R = Et, $Pr^n$, and $Bu^n$ only the benzyl group was removed by $HgCl_2$ or $Hg(OAc)_2$ in methanol, but when R = Me, only the methyl group was removed from tin. A recent report by Reutov and coworkers[217] confirms that difficulties over the polarographic method of analysis led to the previous incorrect conclusion that in methanol only the benzyl group was removed from $PhCH_2SnMe_3$. Rate constants for reaction of $PhCH_2SnR_3$ compounds are given in Table 34, and lead to the conclusion that towards mercury(II) salts in methanol, the benzyl group is less reactive than the methyl group, but is about 11 times as reactive as is the ethyl group[219].

The aryl group is removed from tin much more rapidly than are simple alkyl groups (see Table 31), so that in unsymmetrical tetraorganotins, $ArSnR_3$, only the aryl group is substituted by mercury(II) salts. Thus Hashimoto and Morimoto[26] were able to study the cleavage of a series of compounds $XC_6H_4SnEt_3$ by mercury(II) acetate in tetrahydrofuran, and by a competitive method establish the influence of the substituent X on the reactivity (Table 35). It was suggested that the transition state did not resemble a $\sigma$-complex, but either resembled a $\pi$-complex or involved

TABLE 34. Second-order rate constants[219], $10^3k_2$, at 25 °C in methanol for the reactions

$$PhCH_2SnR_3 + HgX_2 \begin{cases} \xrightarrow{(a)} PhCH_2SnR_2X + RHgX \\ \xrightarrow{(b)} R_3SnX + PhCH_2HgX \end{cases}$$

| | $10^3k_2$ ($l\,mol^{-1}\,s^{-1}$) | |
|---|---|---|
| R | $HgCl_2$ | $Hg(OAc)_2$ |
| $Me^a$ | 598 | $3.33 \times 10^5$ |
| $Et^b$ | 10.8 | 3670 |
| $Pr^{n\,b}$ | 6.53 | 2250 |
| $Bu^{n\,b}$ | 7.67 | 1900 |

$^a$ Methyl group cleaved.
$^b$ Benzyl group cleaved.

TABLE 35. Relative rates[26] at 20 °C in tetrahydrofuran for the reaction

$$XC_6H_4SnEt_3 + Hg(OAc)_2 \longrightarrow XC_6H_4HgOAc + Et_3SnOAc$$

| X | $k_X/k_H$ | X | $k_X/k_H$ |
|---|---|---|---|
| 4-MeO | 8.30 | 3-Me | 1.21 |
| 4-Me | 1.76 | 3-MeO | 0.72 |
| 4-Cl | 0.25 | 3-Cl | 0.029 |
| H | 1.00 | | |

concerted bond-making and -breaking. A more detailed kinetic study of the reaction of $PhSnEt_3$ with mercury(II) salts in methanol was later carried out by Abraham and Sedaghat-Herati[15], who obtained rate constants and activation parameters for substitution by mercury(II) chloride and mercury(II) iodide (Table 36). Salt effects showed that only the neutral species $HgCl_2$ is active, and that neither $HgCl_3^-$ nor $HgCl^+$ contribute to the overall rate in methanol. The phenyl group is cleaved from tin about $4 \times 10^2$ times as fast as is a methyl group and about $1 \times 10^5$ times as fast as is an ethyl group. The effect of added water is to increase the rate constant, although not to any great extent, whereas added inert salts generally decreased the rate constant, indicating a transition state that is not particularly polar. The activation enthalpies for this aromatic substitution are very low (Table 36), and it was suggested that the substitution took place via a $\pi$-complex intermediate, with the transition state resembling the $\pi$-complex[15].

Although the cleavage of tin—carbon bonds by mercury(II) salts has traditionally been considered to be an electrophilic substitution reaction, Fukuzumi and Kochi[20,21] have recently treated these cleavages in terms of their general charge-transfer mechanism (equations 21–23). These workers also determined rate constants for a considerable number of substitutions by mercury(II) chloride in acetonitrile, methanol, and dichloromethane, and these are given in (Table 37). Values of the rate constants for the symmetrical tetraalkyltins in methanol were taken from previous work, but those for symmetrical tetraalkyltins in acetonitrile were redetermined. In the event, there is reasonable agreement between the values of Fukuzumi and Kochi[20] (Table 37) and those of Abraham and coworkers[198,200] (Table 26), except for the case of the $Pr^i_4Sn/HgCl_2$ reaction, for which the two values are $5 \times 10^{-6}\,l\,mol^{-1}\,s^{-1}$

TABLE 36. Second-order rate constants for the aromatic substitution of $PhSnEt_3$ by mercury(II) salts[15]

| Reactants | Solvent | $k_2$ at 25 °C ($l\,mol^{-1}\,s^{-1}$) | $\Delta H^{\neq}$ ($kcal\,mol^{-1}$)[a] | $\Delta S^{\neq}$ ($cal\,K^{-1}\,mol^{-1}$)[a] |
|---|---|---|---|---|
| $PhSnEt_3 + HgCl_2$ | MeOH | 103.4 | 7.6 | −23 |
| $PhSnEt_3 + HgCl_2$ | 98% MeOH | 130.8 | 8.3 | −20 |
| $PhSnEt_3 + HgCl_2$ | 94% MeOH | 147.8 | 4.4 | −33 |
| $PhSnEt_3 + HgI_2$ | MeOH | 23.9 | 9.4 | −20 |
| $PhSnEt_3 + Hg(OAc)_2$ | MeOH | Very fast | | |
| $PhSnMe_3 + HgCl_2$ | MeOH | 135.0[b] | 6.7[b] | −26[b] |

[a] Activation parameters on the usual molar scale. Note that in ref. 15 they are given on the mol fraction scale.
[b] See Table 31.

TABLE 37. Second-order rate constants[20], $10^3k_2$, in various solvents at 25 °C for the reaction

$$R_4Sn + HgCl_2 \longrightarrow R_3SnCl + RHgCl$$

| | $10^3k_2$ ($l\,mol^{-1}\,s^{-1}$) | | |
|---|---|---|---|
| $R_4Sn$ | $MeCN^a$ | $CH_2Cl_2$ | $MeOH^b$ |
| $Me_4Sn$ | 2400 | 66 | 1550 |
| $Et_4Sn$ | 16 | 1.7 | 3.33 |
| $Pr^n_4Sn$ | 3.7 | 0.9 | 0.628 |
| $Pr^i_4Sn$ | 0.005 | 0.003 | $<10^{-5}$ |
| $Bu^n_4Sn$ | 3.8 | 1.4 | 0.615 |
| $Bu^s_4Sn$ | 0.0007 | 0.0008 | |
| $Bu^i_4Sn$ | 0.56 | 0.37 | 0.08 |
| $Bu^nSnMe_3$ | 1100 | 100 | 1000 |
| $Pr^i_2SnMe_2$ | 2500 | 130 | 230 |
| $Bu^n_2SnMe_2$ | 500 | 150 | 640 |
| $Bu^t_2SnMe_2$ | 46 | 26 | 1.4 |

[a] Compare values in Table 26.
[b] Values for symmetrical $R_4Sn$ from Table 26.

(ref. 20) and $<1.6 \times 10^{-8}\,l\,mol^{-1}\,s^{-1}$ (ref. 200). Rate constants[20] for substitution of tetramethyltin and tetra-*n*-butyltin by various mercury(II) salts in dichloromethane are given in Table 38, together with results from a previous study by Abraham *et al.*[207].

Fukuzumi and Kochi[20] analysed their results in terms of the charge-transfer transition enegies, $h\nu_{CT}$, of the tetraalkyltin–mercury(II) chloride charge-transfer absorption bands. The absorption maxima were found to be independent of solvent, and Fukuzumi and Kochi[21] were able to determine values of $\Delta h\nu_{CT}$, these being defined as $h\nu_{CT}(R_4Sn) - h\nu_{CT}(Me_4Sn)$. Fukuzumi and Kochi[20] also defined relative

TABLE 38. Second-order rate constants[20], $10^3k_2$, in methylene chloride at 25 °C for the reaction

$$R_4Sn + HgX_2 \longrightarrow R_3SnX + RHgX$$

| | $10^3k_2$ ($l\,mol^{-1}\,s^{-1}$) | |
|---|---|---|
| X | $Me_4Sn$ | $Bu^n_4Sn$ |
| Cl | 66 | 1.4 |
| Br | 12 | 0.48 |
| CN | 480 | 2.5 |
| $O_2CMe$ | 330 | 7.4 |
| $O_2CCHMe_2$ | 54 | 9.1 |
| $O_2CCMe_3$ | 35 | 59 |
| $O_2CBu^n$ | 82 | 38 |
| $O_2CCH_2Cl$ | 1 600 | 8.9 |
| $O_2CCHCl_2$ | 48 | 0.83 |
| $O_2CCHF_2$ | 40 000 | 50 000 |
| $O_2CCF_3$ | 260 000 | 160 000 |

solvation energies of the excited ion pair $(R_4Sn^+HgCl_2^-)^*$, formed in the charge-transfer step, going to the thermal ion pair $(R_4Sn^+HgCl_2^-)$ as $\Delta G_r^s = \Delta G^s(R_4Sn) - \Delta G^s(Me_4Sn)$. Thus for the energetics of ion-pair formation relative to that for the $Me_4Sn$ compound, scheme 76 applies.

$$(R_4SnHgCl_2) \xrightarrow{\Delta h\nu_{CT}} (R_4Sn^+HgCl_2^-)^* \xrightarrow{\Delta G_r^s} (R_4Sn^+HgCl_2^-) \qquad (76)$$

Now if it is assumed that the thermal ion pair $(R_4Sn^+HgCl_2^-)$ is a good model for the transition state in the electrophilic substitution of $R_4Sn$ by $HgCl_2$, it follows that the relative value of $\Delta G_r^{\neq} = \Delta G^{\neq}(R_4Sn) - \Delta G^{\neq}(Me_4Sn)$ will be given by equation 77.

$$\Delta G_r^{\neq}(\text{calc}) = \Delta G_r^s + \Delta h\nu_{CT} \qquad (77)$$

Values of $\Delta h\nu_{CT}$ for the $(R_4Sn^+HgCl_2^-)^*$ charge-transfer complexes are given in Table 39, after conversion from electron volts to kcal $mol^{-1}$. Fukuzumi and Kochi determined $\Delta G_r^s$ values in acetonitrile[22], calculated those in other solvents through a simple Born-type electrostatic expression[20], and then compared values of $\Delta G_r^{\neq} - \Delta h\nu_{CT}$ with values of $\Delta G_r^s$[20]. However, from the point of view of solvent effects on the relative reaction rates of tetraalkyltins, it is more pertinent to compare the $\Delta G_r^{\neq}$ values calculated through equation 77 with those determined experimentally using equation 78. In Table 39 are given observed and calculated values of $\Delta G_r^{\neq}$ in

$$\Delta G_r^{\neq}(\text{obs}) = -RT \ln(k^{R_4Sn}/k^{Me_4Sn}) \qquad (78)$$

acetonitrile, methanol, and dichloromethane[20], and also for the more polar solvent

TABLE 39. Comparison of observed values of $\Delta G_r^{\neq}$ with those calculated by the method of Fukuzumi and Kochi[20] for the $R_4Sn/HgCl_2$ reaction[a]

| | | 85% MeOH | | MeOH | | MeCN | | $Bu^tOH$ | | $CH_2Cl_2$ | |
|---|---|---|---|---|---|---|---|---|---|---|---|
| $R_4Sn$ | $\Delta h\nu_{CT}$[b] | Obs.[c] | Calc. | Obs.[d] | Calc. | Obs.[e] | Calc. | Obs.[f] | Calc.[g] | Obs.[h] | Calc. |
| $Me_4Sn$ | 0.0 | 0.0 | 0.0 | 0.0 | 0.0 | 0.0 | 0.0 | 0.0 | 0.0 | 0.0 | 0.0 |
| $Et_4Sn$ | −4.4 | 3.4 | 3.0 | 3.6 | 3.0 | 3.0 | 3.0 | 3.9 | 2.6 | 2.2 | 2.3 |
| $Pr^n_4Sn$ | −3.9 | 4.5 | 4.1 | 4.6 | 4.1 | 3.8 | 4.1 | 4.9 | 3.7 | 2.5 | 3.4 |
| $Bu^n_4Sn$ | −3.5 | 4.6 | 5.1 | 4.6 | 5.1 | 3.8 | 5.1 | 5.2 | 4.7 | 2.3 | 4.5 |
| $Bu^i_4Sn$ | −3.9 | 5.8 | 5.4 | 5.8 | 5.4 | 4.6 | 5.4 | 6.1 | 4.8 | 3.1 | 4.5 |
| $Pr^i_4Sn$ | −3.0 | >11.2 | 8.3 | >10.9 | 8.3 | >11.2[i] | 8.3 | >8.3 | 7.7 | 5.9 | 7.4 |
| $Bu^s_4Sn$ | −3.4 | | | — | 8.0 | 8.9 | 8.0 | | | 6.7 | 7.0 |
| $Bu^nSnMe_3$ | −2.3 | | | 0.3 | 3.2 | 0.5 | 3.2 | | | −2.5 | 2.7 |
| $Bu^n_2SnMe_2$ | −5.8 | | | 0.5 | 2.4 | 0.9 | 2.4 | | | −4.9 | 1.7 |
| $Pr^i_2SnMe_2$ | −6.7 | | | 1.2 | 3.7 | 0.0 | 3.7 | | | −4.0 | 2.7 |
| $Bu^t_2SnMe_2$ | −9.5 | | | 4.2 | 4.2 | 2.3 | 4.2 | | | 0.6 | 3.0 |

[a] All values in kcal $mol^{-1}$ at 25 °C. Observed quantities found through equation 78 and calculated quantities through equation 77.
[b] Refs. 20 and 21.
[c] Ref. 193.
[d] Refs. 20 and 193.
[e] Refs. 20 and 200.
[f] From rate constants at 40 °C given in Ref. 200, using the equation $\Delta G_r^{\neq} = -1.9872 \times 298.15 \ln(k/k_0)$. Were rate constants available at 25 °C, the observed $\Delta G_r^{\neq}$ values would all be slightly greater than those quoted.
[g] Calculated at 25 °C.
[h] Ref. 20.
[i] Ref. 200.

methanol–water (85 : 15)[193], and *tert*-butanol[200]. For the reactions of the symmetrical tetraalkyltins, observed and calculated $\Delta G_r^{\neq}$ values are in reasonable agreement, but the unsymmetrical methylalkyltins nearly always react very much faster than calculated. It is possible that in general the transition state does resemble a thermal ion pair ($R_4Sn^+HgCl_2^-$), but that solvation of the unsymmetrical tetraalkyltin transition states with $Bu^nSnMe_3$ and $R_2SnMe_2$ more resembles the solvation of the $(Me_4SnHgCl_2)^{\neq}$ transition state than a large thermal ion pair. Thus Fukuzumi and Kochi[20] suggest that solvation energy of the ($Bu^n_2SnMe_2{}^+HgCl_2^-$) ion pair lies between those of the ($Pr^n_4Sn^+HgCl_2^-$) and ($Bu^n_4Sn^+HgCl_2^-$) ion pairs, whereas from the results in Table 39 it seems as though the solvation energy of the $(Bu^n_2Sn\text{-}Me_2HgCl_2)^{\neq}$ transition state is somewhere between those of the $(Me_4SnHgCl_2)^{\neq}$ and $(Et_4SnHgCl_2)^{\neq}$ transition states. Nevertheless, the approach of Fukuzumi and Kochi[20] is a most interesting new development in the long-standing area of the effect of solvents on the relative reactivities of alkyl—metal compounds.

Finally, in an interesting paper by Brinckman and coworkers[220] the methylation of $Hg^{2+}$ by $Me_3Sn^+$ in aqueous solution is examined from the point of view of the biomethylation of metals. From the effect of added perchlorate and chloride ion, it was concluded that reaction takes place between $Me_3Sn^+$(aq.) and $HgCl_2$(aq.); at 25 °C the second-order rate constant is $6.02 \times 10^{-3}$ l mol$^{-1}$ s$^{-1}$ with $\Delta H^{\neq} = 13.6$ kcal mol$^{-1}$ and $\Delta S^{\neq} = -23.0$ cal K$^{-1}$ mol$^{-1}$.

### *15. Cleavage of tin—carbon bonds by boron*(III)*, aluminium*(III)*, and thallium*(III) *compounds*

A number of preparative studies have been reported on the alkylation of boron(III) compounds by tetraorganotins. Boron(III) halides will cleave one or two alkyl groups from tetraalkyltins, yielding the corresponding alkylboron dihalide[221,222], but all four phenyl groups may be cleaved from tetraphenyltin[223]. Vinyl boron dihalides have been prepared by the action of boron(III) chloride or boron(III) bromide on tetravinyltin[224].

Similarly, aluminium(III) chloride is reported to cleave one or two tin—alkyl bonds in tetraalkyltins, but all four tin—phenyl bonds in tetraphenyltin[225]. Van Egmond *et al.*[226] have studied the reaction between tetraisobutyltin and aluminium(III) chloride in dichloromethane in some detail and report that the degree of dealkylation depends on the amount of the reagent used.

Good yields of organothallium chlorides are obtained by the action of thallium(III) chloride on tetraorganotins[227], and Okawara and coworkers[228–232] have developed the reaction between tetramethyltin and organothallium dicarboxylates into a synthetic route to unsymmetrical thallium compounds (equation 79) ($X = MeCO_2$, $Me_2CHCO_2$). Reutov and coworkers[233] have investigated polarographically the kinetics of the related reaction 80 ($X = Cl$, $R'CO_2$) in methanol containing $Bu^n_4NClO_4$ as the supporting electrolyte.

$$Me_4Sn + RTlX_2 \longrightarrow MeRTlX + Me_3SnX \qquad (79)$$

The reactions were found to be second order overall, first order in each reactant, and it was concluded that both the free ion $[PhTlX]^+$ and the ion pair $[PhTlX^+, X^-]$ were involved. Rate constants are given in Table 40. In a subsequent study of the common ion effect of $Bu^n_4NOAc$ on reaction 80 ($X = OAc$) in methanol and acetic acid, it was

$$Me_4Sn + PhTlX_2 \longrightarrow MePhTlX + Me_3SnX \qquad (80)$$

shown[234] that in methanol the reaction took place via the free ion and the ion pair (see above, $X = OAc$), the free ion being about 25 times as reactive as the ion pair. In

TABLE 40. Approximate[a] second-order rate constants[233] at 20 °C in methanol for the reaction

$$Me_4Sn + PhTlX_2 \longrightarrow PhMeTlX + Me_3SnX$$

| X | $k_2$ (l mol$^{-1}$ s$^{-1}$) | X | $k_2$ (l mol$^{-1}$ s$^{-1}$) |
|---|---|---|---|
| $CF_3CO_2$ | 1.10 | $MeCO_2$ | 0.20 |
| $ClCH_2CO_2$ | 0.63 | $Pr^iCO_2$ | 0.19 |
| $HCO_2$ | 0.47 | Cl | 0.10 |

[a] The reaction rate decreased as reaction proceeded, so rate constants were extrapolated to zero time. Such values depend on the initial reactant concentrations, so that values in this table must be treated as approximations only.

glacial acetic acid reaction proceeds via a solvent separated ion pair, $[PhTlOAc^+, CH_3CO_2H, OAc^-]$ The complex ion $PhTl(OAc)_3^-$ is unreactive in both solvents[234]. Jewett and Brinkman (see ref. 200) have studied the methylation of $TlCl_3$(aq.) by $Me_3Sn^+$(aq.) in aqueous solution; at 25 °C the rate constant is $4.34 \times 10^{-3}$ l mol$^{-1}$ s$^{-1}$ with $\Delta H^{\neq} = 18.5$ kcal mol$^{-1}$ and $\Delta S^{\neq} = -7.4$ cal K$^{-1}$ mol$^{-1}$.

## 16. *Cleavage of tin—carbon bonds by silicon*(IV), *germanium*(IV), *and tin*(IV) *compounds*

Although not very widely studied, it is known that tetraalkyltins may be dealkylated by various chlorosilicon(IV) and chlorogermanium(IV) compounds to give alkylsilicon(IV) and alkylgermanium(IV) products[172,235,236].

The cleavage of tin—alkyl or tin—aryl bonds by organotin(IV) halides or tin(IV) halides is the well known redistribution reaction, widely used to prepare $R_nSnCl_{4-n}$ compounds where $n = 1$, 2, or 3. A recent example of this reaction is due to Kuivila *et al.*[237], who prepared a series of organochlorotins, $RSnMe_2Cl$, through reaction 81

$$RSnMe_3 + Me_3SnCl \longrightarrow RSnMe_2Cl + Me_4Sn \qquad (81)$$

[R = $MeCO(CH_2)_2$ and $MeCO(CH_2)_3$]. Tin(II) halides have also been used to cleave tetraethyltin, yielding the corresponding triethyltin halide[183].

Tagliavini and coworkers[238] have studied in detail the redistribution reaction 82, and in Table 41 are given the rate constants and activation parameters reported in

$$MeSnR_3 + Me_2SnCl_2 \longrightarrow Me_3SnCl + R_3SnCl \qquad (82)$$

TABLE 41. Second-order rate constants, $10^4k_2$, at 25 °C and activation parameters in methanol for the reaction[238]

$$MeSnR_3 + Me_2SnCl_2 \longrightarrow RSnMe_2Cl + Me_4Sn$$

| $MeSnR_3$ | $10^4k_2$ (l mol$^{-1}$ s$^{-1}$) | $\Delta H^{\neq}$ (kcal mol$^{-1}$) | $\Delta S^{\neq}$ (cal K$^{-1}$ mol$^{-1}$) |
|---|---|---|---|
| $MeSnMe_3$ | 8.1 | 16.0 | −17 |
| $MeSnEt_3$ | 3.5 | 17.6 | −13 |
| $MeSn(Pr^n)_3$ | 2.7 | 14.1 | −26 |
| $MeSn(Bu^n)_3$ | 2.2 | 12.1 | −32 |
| $MeSn(Pr^i)_3$ | 1.3 | 14.2 | −27 |
| $MeSnMe_2Et$ | 6.8 | | |
| $MeSnMeEt_2$ | 7.0 | | |

TABLE 42. Second-order rate constants[a], $10^4k_2$, at 30 °C and activation parameters for the reaction[239]

$$Me_4Sn + Me_2Sn(NO_3)_2 \longrightarrow 2Me_3SnNO_3$$

| Solvent | $10^4k_2^{obs}$ (l mol$^{-1}$ s$^{-1}$) | $\Delta H^{\neq}_{obs}$ (kcal mol$^{-1}$) | $\Delta S^{\neq}_{obs}$ (cal K$^{-1}$ mol$^{-1}$) |
|---|---|---|---|
| MeOH | 20.4 | 12.2 | −30.4 |
| EtOH | 10.6 | 13.0 | −29.2 |
| $Pr^nOH$ | 8.1 | 14.7 | −24.1 |
| $Pr^iOH$ | 2.5 | 13.4 | −30.8 |
| $Me_2CO$ | 21.9 | 13.3 | −26.7 |

[a] All measurements were carried out with excess of $Me_4SN$, i.e. under pseudo first-order conditions. $k_2^{obs} = k_1^{obs}/[Me_4Sn]$.

methanol[238]. Addition of inert salts such as $NaClO_4$ increased the rate constant value, but addition of NaCl or of water decreased the value. It was suggested[238] that reaction actually takes place through the $Me_2Sn^{2+}$ entity, formation of which is inhibited by $Cl^-$. The decelerating effect of water was thought to be due to formation of less reactive species $Me_2Sn(H_2O)_4{}^{2+}$. Tagliavini and coworkers[238] interpreted the reactivity sequence shown in Table 41 as arising from a prior coordination of methanol to the $MeSnR_3$ compound. However, Abraham pointed out that if the observed rate constants are statistically corrected, cleavage of a Me—Sn bond in $Me_4Sn$ takes place less readily than such a cleavage in most of the $MeSnR_3$ compounds; this is not really in accord with a prior coordination step.

In a subsequent paper[239], the related redistribution 83 was studied; rate constants and activation parameters are given in Table 42. Unlike the case of reaction 82, the redistribution 83 takes place readily in acetone, presumably because the nitrate

$$Me_4Sn + Me_2Sn(NO_3)_2 \longrightarrow 2Me_3SnNO_3 \qquad (83)$$

dissociates to some extent even in this solvent. As for the chloride reaction, added water and added common ion ($NO_3^-$) both inhibit reaction 83, again for similar reasons[239].

### 17. *Cleavage of lead—carbon bonds by lithium*(I) *compounds*

Juenge and Seyferth[240] have reported the cleavage of tetravinyllead by phenyllithium in ether to yield tetraphenyllead and vinyllithium.

### 18. *Cleavage of lead—carbon bonds by iridium*(IV) *and iron*(III) *compounds*

Kochi and coworkers[18,36] have obtained rate constants for dealkylation of tetraalkylleads by hexachloroiridate(IV) and by a number of iron(III) complexes. These reactions take place by electron transfer processes, rather than by $S_E$ mechanisms, but rate constants are collected in Table 43.

### 19. *Cleavage of lead—carbon bonds by copper*(II), *silver*(I), *and gold*(III) *compounds*

Copper(II) nitrate reacts readily with tetralkyl- and tetraaryl-leads; one organo group is cleaved from lead, and compounds $R_3PbNO_3$ are produced[241–246]. Numerous workers have reported on the reactions between tetraorganoleads and silver(I)

TABLE 43. Second-order rate constants[36] ($l\,mol^{-1}\,s^{-1}$) in acetonitrile at 25 °C for electron transfer from tetraalkylleads to hexachloroiridate(IV) and iron(III) complexes, $[FeL_3]^{3+}$

| Species[a] | $Me_4Pb$ | $Me_3PbEt$ | $Me_2PbEt_2$ | $MePbEt_3$ | $PbEt_4$ |
|---|---|---|---|---|---|
| $[IrCl_6]^{2-}$ | 0.02 | 0.57 | 3.3 | 11 | 25 |
| $L_1$ | 5.06 | | 566 | | $1.19\times10^4$ |
| $L_2$ | 9.09 | | 1410 | | $3.19\times10^4$ |
| $L_3$ | 25.5 | | 3530 | | $1.07\times10^5$ |
| $L_4$ | 166 | | $2.76\times10^4$ | | $9.3\times10^5$ |
| $L_5$ | 1490 | | $1.96\times10^5$ | | |

[a] $L_1$ = 4,7-di-Ph-phen; $L_2$ = bipy; $L_3$ = phen; $L_4$ = 5-Cl-phen; $L_5$ = 5-$NO_2$-phen.

nitrate, again leading to compounds of type $R_3PbNO_3$ (see, for example, refs. 241–244 and 247–250) and Tagliavini and coworkers[251–254] have shown that gold(III) chloride will dealkylate tetramethyllead to form dimethyllead dichloride.

### 20. *Cleavage of lead—carbon bonds by mercury*(II) *and mercury*(I) *compounds*

Compounds of type $R_4Pb$ are dealkylated by mercury(II) chloride to yield the corresponding RHgCl product; one or two organo—lead bonds may be broken, and in mixed arylalkylleads, the aryl group is preferentially transferred to mercury[255,256]. Mercury(II) acetate will also dealkylate tetraethyllead, although the product seems to be a lead(II) compound[256] (equation 84). Kocheshkov and coworkers, however, reported that mercury(II) acetate in acetic acid reacts with diphenyllead diacetate to yield the expected lead-containing product, phenyllead(IV) triacetate[257,258]. Mercury(I) nitrate is reported to dealkylate tetraalkylleads according to equation 85 (R = Me, Et)[259].

$$Et_4Pb + 3Hg(OAc)_2 \longrightarrow 3EtHgOAc + Pb(OAc)_2 + EtOAc \quad (84)$$

$$R_4Pb + Hg_2(NO_3)_2 \longrightarrow R_2Hg + R_2Pb(NO_3)_2 + Hg \quad (85)$$

### 21. *Cleavage of lead—carbon bonds by thallium*(III) *compounds*

Goddard and Goddard[260,261] have studied the dealkylation and dearylation of $R_4Pb$ compounds by thallium(III) chloride, and have shown that one or two organo—lead bonds may be broken.

### 22. *Cleavage of lead—carbon bonds by germanium*(IV), *tin*(IV), *and lead*(IV) *compounds*

Germanium(IV) chloride is reported to react with tetraethyllead to give ethylgermanium trichloride and triethyllead chloride[262]. Tin(IV) chloride also reacts with $R_4Pb$ compounds and is reported to remove two phenyl groups from $Ph_4Pb$[263].

Redistribution reactions between tetramethyllead and tetraethyllead, catalysed by aluminium(III) chloride, have been studied by Calingaert *et al.*[264] and others[163,225,265], and are thought to proceed by a series of reactions in which alkyl groups are transferred from lead to aluminium, and back again to lead.

### 23. *Cleavage of lead—carbon bonds by antimony*(III), *antimony*(V), *and bismuth*(III) *compounds*

Goddard *et al.*[263] reported that in the cleavage of tetraphenyllead by antimony(III) chloride and by antimony(V) chloride, two phenyl groups are removed to leave diphenyllead dichloride. Antimony(III) chloride cleaves all four groups from tetraethyllead, according to equation 86[266], but with tetravinyllead the products are divinyllead dichloride and vinylantimony dichloride[267].

$$Et_4Pb + 3SbCl_3 \longrightarrow 3EtSbCl_2 + PbCl_2 + EtCl \tag{86}$$

Bismuth(III) trichloride and tribromide both dearylate tetraphenyllead, following equation 87 (X = Cl, Br)[263,268].

$$Ph_4Pb + 2BiX_3 \longrightarrow Ph_2BiX + Ph_2PbX_2 \tag{87}$$

## G. Cleavage of Group VB Metal—Carbon Bonds

### 1. Cleavage of antimony—carbon bonds by antimony(III) compounds

Weingarten and van Wazer[269] have shown that reaction 88 in dimethylformamide is second order, first order in each reactant. The second-order rate constant was $(6.7 \pm 0.5) \times 10^{-5}\,l\,mol^{-1}\,s^{-1}$ at 72 °C and $(5.3 \pm 0.5) \times 10^{-4}\,l\,mol^{-1}\,s^{-1}$ at 100 °C, and the reported activation parameters were $\Delta H^{\neq} = 18$ kcal mol$^{-1}$ and $\Delta S^{\neq} = -25$ cal K$^{-1}$ mol$^{-1}$. A four-centred cyclic transition state was suggested[269].

$$Me_3Sb + SbCl_3 \longrightarrow MeSbCl_2 + Me_2SbCl \tag{88}$$

## H. Inversion of Primary Metal—Alkyl Compounds

The proton magnetic resonance spectra of a number of metal—alkyl compounds have been found to be temperature dependent, and a detailed examination showed that for primary alkyl compounds inversion of configuration at the metal—$CH_2$ centre was taking place[270–275]. Witanowski and Roberts[274] showed that for a series of 3,3-dimethylbutyl (R) compounds, the inversion rate increased along the series

TABLE 44. Activation energies for the inversion of some primary alkyl—metal compounds in various solvents

| Compound | Solvent | $\Delta H^{\neq}$ (kcal mol$^{-1}$) | Ref. |
|---|---|---|---|
| $MeCH_2CHMeCH_2MgBr$ | Anisole | 9.6 | 270, 271 |
| $MeCH_2CHMeCH_2MgBr$ | Ether | 12.0 | 270, 271 |
| $MeCH_2CHMeCH_2MgBr$ | thf | 16.6 | 270, 271 |
| $Me_3CCH_2CH_2MgCl$ | Ether | 10.4 | 272, 275 |
| $Me_2CHCHPhCH_2MgCl$ | Ether | 11.4 | 272, 275 |
| $Me_2CHCHPhCH_2MgCl$ | thf | 17.4 | 272, 275 |
| $(MeCH_2CHMeCH_2)_2Mg$ | Dioxane | 4.9 | 270, 271 |
| $(MeCH_2CHMeCH_2)_2Mg$ | thf | 10.5 | 270, 271 |
| $(MeCH_2CHMeCH_2)_2Mg$ | Ether | 18.2 | 270, 271 |
| $Me_3CCH_2CH_2Li$ | Ether | 14.4 | 274 |
| $(Me_3CCH_2CH_2)_2Mg$ | Ether | 19.4 | 274 |
| $(Me_3CCH_2CH_2)_2Zn$ | Ether | 25.4 | 274 |
| $(Me_3CCH_2CH_2)_2Hg$ | Ether | Very slow | 274 |
| $(Me_3CCH_2CH_2)_3Al$ | Ether | Very slow | 274 |
| $MeCH_2CHMeCH_2Li$ | *n*-Pentane | 4.5 | 275 |
| $MeCH_2CHMeCH_2Li$ | Ether | 8.0 | 275 |
| $MeCH_2CHMeCH_2Li$ | Toluene | 10.5 | 275 |
| (*i*-Hexyl)$_3$Al | Toluene | 28.0 | 275 |
| (*i*-Hexyl)$_3$Al | Ether | Very slow | 275 |

$R_2Hg < R_3Al < R_2Zn < R_2Mg < RLi$, and suggested that the inversion took place via an $S_E1$ process, a carbanion being formed in an initial rate-determining step. Interestingly, the above sequence of reactivity is exactly that predicted on the basis of results given in Table 1, viz. $R_2Hg < R_2Zn < RLi$. Activation enthalpies are listed in Table 44.

## III. HALOGENOLYSIS OF ORGANOMETALLIC COMPOUNDS

### A. General Introduction to Halogenolysis

Halogens will often cleave non-transition metal—carbon bonds quite cleanly according to the general equation

$$RMX_n + Y_2 \longrightarrow RY + YMX_n \tag{89}$$

Although halogenolysis is not preparatively as useful as the metal-for-metal exchanges, for example, there have been a large number of investigations reported on the kinetics of halogenolysis. In such cases, the reactions are mostly carried out in the presence of an excess of halide ion which converts the halogen to the trihalide ion according to equilibrium 90.

$$Y_2 + Y^- \rightleftharpoons Y_3^- \tag{90}$$

In the solvents used for the kinetic studies, the equilibrium constants for equation 90 are large, and since an excess of halide ion is present, the approximation that $[Y_2] = [Y_3^-]$ can be made. Gielen and Nasielski[276] treated the situation in which second-order kinetics are observed as follows, where $k_2^{obs}$ is the observed second-order rate constant and where the concentration of the organometallic substrate is denoted as [R]. Let the velocity of halogenolysis be $v$. Then,

$$v = \frac{-d[R]}{dt} = \frac{d[Y_2]}{dt} = \frac{-d[Y_3^-]}{dt} = k_2^{obs}[R][Y_3^-] \tag{91}$$

Various terms may contribute to the overall kinetic expression so that

$$v = v' + v'' + v'''. \tag{92}$$

Then, following the notation of Abraham[4],

$$v' = k_2^a[R][Y_2] \tag{93}$$

$$v'' = k_2^b[R][Y_3^-] \tag{94}$$

$$v''' = k_3[R][Y_2][Y^-] \tag{95}$$

Combining equations 91–95 leads to

$$k_2^{obs} = k_2^a \frac{[Y_2]}{[Y_3^-]} + k_2^b + k_3 \frac{[Y_2][Y]}{[Y_3^-]} \tag{96}$$

However, the equilibrium constant for equation 90 is given by equation 97, so that the final expression for $k_2^{obs}$ is equation 98.

$$K = [Y_3^-]/[Y_2][Y^-] \tag{97}$$

$$k_2^{obs} = \frac{k_2^a}{K[Y^-]} + k_2^b + \frac{k_3}{K} \tag{98}$$

Thus at constant ionic strength, a plot of $k_2^{obs}$ against $1/[Y^-]$ should yield a straight line of slope $k_2^a/K$ and of intercept $(k_2^b + k_3/K)$. It is therefore possible to dissect the overall rate into contributions corresponding to $v'$ on the one hand and $v'' + v'''$ on the other, but there is no kinetic treatment that will further separate out the $v''$ and $v'''$ terms.

## B. Chlorinolysis of Organomercury(II) Compounds

A stereochemical study[277] showed that 2-X-3-(HgCl)-5,6-(di-$MeCO_2$)-7-oxabicyclo[2.2.1]heptane **17** (X = OH, $CO_2Me$) reacted with chlorine in acetic acid and in chloroform with retention of configuration at the carbon atom undergoing substitution to yield the corresponding chloride together with mercury(II) chloride.

## C. Chlorinolysis of Organothallium(III) Compounds

Bäckvall *et al.*[278] reported a preference for retention of configuration at the carbon atom undergoing substitution when *threo*-RHC(OMe)CDHTl$(OAc)_2$ (R = Ph) was allowed to react with CuCl–KCl in acetonitrile at 80 °C, the ratio of *threo* to *erythro* product being 2:1. With R = *n*-octyl, the *erythro* compound was converted into the product with a preference for inversion, however. It was suggested that the chlorinolysis followed both radical and polar reaction paths.

## D. Chlorinolysis of Organosilicon(IV) Compounds

The aromatic substitution 99 has been found[279] to be kinetically of the second order in 98.5% aqueous acetic acid. However, the observed rate constant decreased during the course of reaction.

$$PhSiMe_3 + Cl_2 \longrightarrow PhCl + Me_3SiCl \qquad (99)$$

Stock and Spector[280] carried out the same reaction in glacial acetic acid and found the system to be kinetically well behaved. The second-order rate constant at 25 °C was reported to be $1.57 \times 10^{-2}\,l\,mol^{-1}\,s^{-1}$.

It has been observed[281] that at fairly low temperatures chlorine will cleave a silicon—carbon bond in $(Me_3Si)_2C{=}CH_2$ to form $Me_3SiCCl{=}CH_2$ and $Me_3SiCl$.

## E. Chlorinolysis of Organotin(IV) Compounds

Reaction 100 was carried out using chloroform as solvent at room temperature, with a slight deficiency of chlorine; a methyl group was cleaved from tin in preference to the trifluoromethyl group[282].

$$Me_3SnCF_3 + Cl_2 \longrightarrow MeCl + Me_2ClSnCF_3 \qquad (100)$$

## F. Brominolysis of Organolithium(I) Compounds

Stereochemical studies have shown that the reaction between optically active 1-methyl-2,2-diphenylcyclopropyllithium and bromine in ether takes place with 95% retention of optical activity and configuration[283]. The brominolysis of *cis*-2-methylcyclopropyllithium by bromine pentane–ether (94:6) is non-stereospecific at −70 °C, although in pure pentane at 30 °C 93% retention of configuration was observed[284]. However, the cleavage of *exo*-2-norbornyllithium by bromine in pentane at 70 °C results in inversion of configuration[285].

## G. Brominolysis of Organomercury(II) Compounds

The stereochemical course of the brominolysis of *cis*- and *trans*-4-methylcyclohexylmercury(II) bromide and of optically active *sec*-butylmercury(II) bromide in various solvents has been extensively studied[286–289]. It was concluded that in non-polar solvents such as $CS_2$, brominolysis takes place by a non-stereospecific free-radical mechanism, whereas in more polar solvents reaction proceeds through some combination of free-radical and polar mechanisms with various degrees of retention of configuration at the carbon atom undergoing substitution. In pyridine, complete retention of configuration was observed.

Reutov and co-workers[290,291] also studied the stereochemistry of the brominolysis of *sec*-butylmercury(II) bromide and reported, in accord with the results of Jensen[286–289], that the stereochemical outcome depends on the solvent used and the experimental conditions. Other workers have also examined the stereochemical course of brominolysis; thus Larock and Brown[121] observed that in reaction 101 the *exo*-mercury(II) salt was converted into the corresponding *exo*-bromide.

$$exo\text{-3-}(HgCO_2Ph)\text{bicyclo[2.2.1]heptane} \xrightarrow{Br_2} exo\text{-3-bromobicyclo[2.2.1]heptane} \quad (101)$$

Gielen and Fosty[123] reported that *threo-tert*-BuCHDCHDHgCl was converted by bromine in pyridine to the *threo*-bromide, again in agreement with results of Jensen and Reutov on the alkylmercury(II) reactions, above.

In all the above cases, the product of the brominolysis reaction is the corresponding alkyl bromide; this is the expected, or 'normal', product. Reutov and co-workers[292], however, have shown that in certain cases other 'abnormal' products are formed. The main brominolysis reaction studied was that between benzylmercury(II) chloride and bromine, using a number of solvents. The expected product, benzyl bromide, was formed in yields from 35–95% depending on the solvent, the other products being various 'anomalous' products including benzyl chloride, benzyl ethyl ether (when ethanol was used as the solvent), and benzyl formate (when dimethylformamide was used as the solvent). Details of the various experiments are given in Table 45. In separate experiments it was shown that the anomalous products are not formed in secondary reactions but are produced as a result of the direct involvement of the solvent in the brominolysis reaction. Reutov and coworkers[292] also observed

TABLE 45. Products formed[292] in the brominolysis of benzylmercury chloride by bromine at 20 °C

| Solvent | Decolorization time (min) | Products (mol-%) | | |
|---|---|---|---|---|
| | | $PhCH_2Br$ | $PhCH_2Cl$ | Other |
| Dimethylformamide | 8 | 59.0 | 2.2 | 38.8 ($PhCH_2OCHO$) |
| Dimethylformamide–water (2:1) | 2.5 | 35.1 | 2.7 | 30.0 ($PhCH_2OH$) 32.1 ($PhCH_2OCHO$) |
| Acetone | 0.66 | 55.1 | 2.3 | 42.2 ($PhCH_2OH$) |
| Acetic acid–benzene (4:1) | 11 | 68.5 | 21.0 | 10.4 ($PhCH_2OAc$) |
| Ethanol–benzene (4:1) | — | 74.6 | 6.0 | 19.4 ($PhCH_2OEt$) |
| Dioxane | 45 | 94.6 | 3.3 | — |
| Dichloromethane | 3 | 68.2 | 31.8 | — |
| Benzene | 27 | 69.3 | 28.7 | 2.0 ($Ph_2CH_2$) |

TABLE 46. Second-order rate constants[293] at 15 °C and activation parameters for the reaction

$$PhCH_2HgCl + Br_2 \xrightarrow{Br^-} PhCH_2Br + HgClBr$$

| Solvent | $k_2^{obs}$ (l mol$^{-1}$ s$^{-1}$) | $E_a$ (kcal mol$^{-1}$) | $\Delta S^{\neq}$ (cal K$^{-1}$ mol$^{-1}$) |
|---|---|---|---|
| Dioxane–water (7:3) | 6.6 | 13.2 | −9 |
| Dimethylformamide | 0.79 | 15.3 | −6 |
| Methanol | 10.3 | 11.4 | −14.5 |

the formation of anomalous products in the brominolysis of cyclohexylmercury(II) bromide (8.7% cyclohexyl formate) and of cyclohexylcarbinylmercury(II) bromide (8.1% cycohexylcarbinol) using dimethylformamide and aqueous dimethylformamide as solvents, respectively.

The kinetics of brominolysis of *sec*-butylmercury(II) bromide were investigated by Reutov and coworkers[290], who used carbon tetrachloride plus a small amount of methanol as the solvent. The kinetic form, in the absence of bromide ion, was second-order and it was concluded that the substitution involved bromine complexed in some way with ether.

Rather more detailed kinetic studies were carried out on benzylmercury(II) chloride, in both the presence and the absence of bromide ion (reaction 102)[293]. Rate

$$PhCH_2HgCl + Br_2 \longrightarrow PhCH_2Br + HgClBr \qquad (102)$$

constants were reported for reaction in the presence of a 20-fold excess of bromide ion (as $NH_4Br$) in dimethylformamide, dioxane–water (7:3), and methanol; these second-order rate constants and associated activation parameters are given in Table 46. Reaction 102 was also studied[294] in the absence of bromide ion, using carbon tetrachloride as the solvent. Under these conditions, the reaction was free-radical in nature with the rate depending on the degree of illumination, but in the presence of small quantities of additives such as water, aliphatic alcohols, or ethers it was found that the reaction rate did not depend on the degree of illumination and that overall second-order kinetics were observed[295,296].

Reutov and coworkers[297] later studied the aromatic substitution reaction 103; even in the presence of bromide ion the reaction was very rapid in dimethylformamide,

$$PhHgBr + Br_2 \longrightarrow PhBr + HgBr_2 \qquad (103)$$

dioxane–water (4:1), and methanol. Although the order with respect to the individual reactants was not established, the reaction was found to be overall second order; rate constants and activation parameters are given in Table 47.

TABLE 47. Second-order rate constants[297] and activation parameters for the reaction

$$PhHgBr + Br_2 \xrightarrow{Br^-} PhBr + HgBr_2$$

| Solvent | $k_2$ (l mol$^{-1}$ s$^{-1}$) | $E_a$(kcal mol$^{-1}$) | $\Delta S^{\neq}$ (cal K$^{-1}$ mol$^{-1}$) |
|---|---|---|---|
| Dimethylformamide | 4.22[a] | 13.4 | −12.0 |
| 80% aqueous dioxane | 89.4[a] | 14.0 | −3.3 |
| Methanol | 129.9[b] | 8.2 | −21.7 |

[a] At 20 °C.
[b] At 19 °C.

TABLE 48. Relative rates[298] for the one-anion catalysed brominolysis of alkylmercury(II) bromide (RHgBr) in methanol at 25 °C

| RHgBr | $k_{rel}$ | RHgBr | $k_{rel}$ |
|---|---|---|---|
| MeHgBr | 1 | $Pe^{neo}HgBr$ | 0.173 |
| EtHgBr | 10.8 | $Pr^{i}HgBr$ | 780 |
| $Pr^{n}HgBr$ | 4.42 | $Bu^{s}HgBr$ | 605 |
| $Bu^{n}HgBr$ | 4.95 | $Bu^{t}HgBr$ | 3370 |
| $Bu^{i}HgBr$ | 1.24 | | |

The one-anion catalysed brominolysis of alkylmercury(II) bromides has been studied in some detail by Sayre and Jensen[298]. The brominolyses of (−)-*sec*-butyl-, (+)-*sec*-butyl-, and *erythro*-1,2-dideuterio-3,3-dimethylbutylmercury(II) bromides by bromine in methanol were shown to take place with 90% retention of configuration, irrespective of the concentration of added bromide ion. Relative rate constants for the one-anion catalysed brominolysis are given in Table 48 and show two opposing trends: successive methyl-for-hydrogen replacement at the $\alpha$-carbon (Me, Et, $Pr^{i}$, and $Bu^{t}$) leads to a marked rate enhancement due to an electronic effect, whereas similar replacement at the $\beta$-carbon (Et, $Pr^{n}$, $Bu^{i}$, and $Pe^{neo}$) leads to a diminution of rate. It was suggested that these results reflect a balance of electronic and steric effects; in the $\alpha$-methylated series electronic effects outweigh steric effects, but in the $\beta$-methylated series the steric effects of the large groups are more powerful than the electronic effects. It is interesting that calculations by Abraham *et al.*[43,46] have shown that for $S_E2$ reactions proceeding by retention of configuration, as in the case under study, the steric effect of $\alpha$-methyl groups is much more pronounced than steric effects of $\beta$-methyl groups. It may therefore be deduced that the electronic effect of $\alpha$-methyl groups, as expected, is very much larger than of $\beta$-methyl groups in the reaction studied by Sayre and Jensen[298], so that in the $\alpha$-methylated series a very large electronic effect overcomes a large steric effect, but in the $\beta$-methylated series, a smaller steric effect is sufficient to dominate a very small or non-existent electronic effect.

## H. Brominolysis of Organoboron(III) Compounds

Brown and coworkers[121,299] have shown that the brominlysis of tri-*exo*-norbornylboron by bromine in tetrahydrofuran proceeds slowly, but with retention of configuration, to yield the corresponding *exo*-norbornylbromide. On the other hand, brominolysis in the presence of methanolic sodium methoxide in tetrahydrofuran takes place more rapidly with inversion of configuration.

Similarly, brominolysis of the acyclic compound (*threo*-$Me_3CCHDCHD)_3B$ by bromine in the presence of methanolic sodium methoxide again takes place with inversion of configuration[124].

Kinetic studies have been carried out on the brominolysis of a number of benzeneboronic acids in aqueous acetic acid in the presence of bromide ion (reaction 104)[300,301].

$$ArB(OH)_2 + Br_2(+H_2O) \xrightarrow{Br^-} ArBr + B(OH)_3 + HBr \qquad (104)$$

The kinetic form was first order in each reactant, i.e. the boronic acid and bromine, and the reported second-order rate constants are given in Table 49.

TABLE 49. Second-order rate constants[300,301], $10^5k_2$, at 25 °C in 80% aqueous acetic acid for the reaction

$$XC_6H_4B(OH)_2 + Br_2(+H_2O) \xrightarrow{Br^-} XC_6H_4Br + HBr + B(OH)_3$$

| X | $10^5k_2$ (l mol$^{-1}$ s$^{-1}$) | X | $10^5k_2$ (l mol$^{-1}$ s$^{-1}$) |
|---|---|---|---|
| 4-MeO | $>7\times10^8$ | 4-Br | 200 |
| 4-Me | 38 000 | 3-I | 34.6 |
| 4-Ph | 10 500 | 3-Br | 21.3 |
| 3-Me | 1 610 | 3-$CO_2$Et | 21.0 |
| 4-F | 1 360 | 3-F | 18.7 |
| H | 484 | 3-Cl | 16.9 |
| 4-Cl | 261 | 4-$CO_2$Et | 5.04 |
| 4-I | 241 | 3-$NO_2$ | 1.46 |

## I. Brominolysis of Organothallium(III) Compounds

Backväll *et al.*[278] have reported that brominolysis of *threo*-PhH(OMe)CCDHTl$(OAc)_2$ by CuBr–KBr in acetonitrile at 80 °C leads to a mixture of *threo*- and *erythro*-bromides in the ratio 2.3:1.

## J. Brominolysis of Organosilicon(IV) Compounds

A number of stereochemical studies have been reported on the brominolysis of organosilicon(IV) compounds, not only with respect to the configuration at the carbon atom undergoing substitution, but also with respect to the configuration at the silicon atom. Kumada and coworkers[302] investigated the cleavage of the silicon—carbon bond in *exo*- and *endo*-norbornylsilicon(IV) pentafluorides by various brominating agents. With bromine in polar solvents such as methanol and tetrahydrofuran, reaction took place with *ca.* 95% inversion of configuration at the *exo*- or *endo*-$\alpha$-carbon atom, but in the non-polar solvents the *endo*-isomer reacted with 99% inversion and the *exo*-isomer with predominant inversion only. *N*-Bromosuccinimide yielded the norbornyl bromide with a high degree of inversion in all the solvents studied, but with copper(II) bromide as the brominolysis reagent, stereochemical scrambling took place. Kumada and coworkers suggested that reaction with bromine took place through an electron-transfer step (equation 105), in which the bromide ion was formed, followed by an $S_N2$ substitution (equation 106) with

$$RSiF_5^{2-} + Br_2 \longrightarrow [RSiF_5^{-\cdot}, Br^\cdot, Br^-] \qquad (105)$$

$$Br^- \overset{\frown}{R}\overset{\frown}{—}SiF_5^{-\cdot} \longrightarrow BrR + SiF_5^{2-\cdot} \qquad (106)$$

$$Br^\cdot + SiF_5^{2-\cdot} \longrightarrow BrSiF_5^{2-} \qquad (107)$$

inversion of configuration at the $\alpha$-carbon atom. The brominolysis of *N*-bromosuccinimide was suggested to occur by an $S_E2$ (inversion) mechanism, whilst the loss of stereospecificity with $CuBr_2$ was attributed to formation of the norbornyl radical as a reaction intermediate[302].

In another type of stereochemical investigation, Eaborn and Steward[303] showed that bromine cleaved the silicon—*p*-methoxyphenyl bond in optically active *p*-methoxyphenylmethyl-1-naphthylphenylsilicon with inversion of configuration at the

TABLE 50. Relative rates[151,152,279,304] at 25 °C in 1.5% aqueous acetic acid for the reaction

$$ArSiMe_3 + Br_2 \longrightarrow ArBr + Me_3SiBr$$

| Ar | $k_{rel}$ | Ar | $k_{rel}$ |
|---|---|---|---|
| Naphth-1-yl | 195 | $o$-$PhC_6H_4$ | 1.8 |
| $o$-$MeC_6H_4$ | 81.5 | $p$-$Me_3SiC_6H_4$ | 1.5[a] |
| $p$-$MeC_6H_4$ | 48.8 | Ph | 1.0 |
| $p$-$EtC_6H_4$ | 45.5 | $p$-$FC_6H_4$ | 0.68 |
| $p$-$Pr^iC_6H_4$ | 32.5 | $m$-$PhC_6H_4$ | 0.41 |
| $p$-$Bu^tC_6H_4$ | 29.2 | $p$-$ClC_6H_4$ | 0.092 |
| $p$-$PhC_6H_4$ | 12.5 | $p$-$IC_6H_4$ | 0.088 |
| Naphth-2-yl | 11.5 | $p$-$BrC_6H_4$ | 0.071 |
| $m$-$Me_3SiCH_2C_6H_4$ | 4.2[a] | $m$-$ClC_6H_4$ | 0.003 |
| $m$-$MeC_6H_4$ | 2.9 | | |

[a] Statistically corrected for the presence of two silicon atoms.

asymmetric silicon atom. Such a result was felt to be inconsistent with the four-centred transition state **18** that had previously been proposed[279], but the six-centred cyclic transition state was thought to be more probable[303].

$$\left[\begin{array}{ccc} Ar & \cdots & SiR_3 \\ \vdots & & \vdots \\ Br & \cdots & Br \end{array}\right]^{\neq} \qquad \left[\begin{array}{ccccc} & SiR_3 & \cdots & Br & \\ Ar & & & & Br \\ & Br & \cdots & Br & \end{array}\right]^{\neq}$$

**(18)** **(19)**

Eaborn and coworkers[151,152,279,304] also studied the aromatic substitution 108 kinetically, in both acetic acid with 1.5% water and glacial acetic acid, both in the presence and in the absence of bromide ion.

$$ArSiMe_3 + Br_2 \longrightarrow ArBr + Me_3SiBr \tag{108}$$

At low bromine concentrations, reaction 108 was first order with respect to each reactant, but at higher bromine concentrations the order with respect to bromine increased. Relative rate constants for substitution of a number of aryl groups were reported[151,152,304] for reaction in the moist acetic acid and are given in Table 50.

Relative rate constants were also reported for the related reaction 109 in the moist acetic acid and are given in Table 51[151,152,305].

$$p\text{-}MeOC_6H_4SiR_3 + Br \longrightarrow p\text{-}MeOC_6H_4Br + R_3SiBr \tag{109}$$

Bromine is reported to cleave silicon—unsaturated carbon bonds as in equations 110[281] and 111[306].

$$(Me_3Si)_2C{=}CH_2 + Br_2 \longrightarrow Me_3SiCBr{=}CH_2 + Me_3SiBr \tag{110}$$

$$PhCH{=}CHSiMe_3 + Br_2 \longrightarrow PhCH{=}CHBr + Me_3SiBr \tag{111}$$

## K. Brominolysis of Organogermanium(IV) Compounds

A variety of reactions are known in which germanium—alkyl bonds are broken by bromine under a variety of conditions. Thus tetramethylgermanium with bromine gives a good yield of trimethylbromogermanium after 1 week at room temperature (reaction 112) (R = Me)[307].

$$R_4Ge + Br_2 \longrightarrow RBr + R_3GeBr \tag{112}$$

TABLE 51 Relative rates[151,152,305] in 1.5% aqueous acetic acid at 25 °C for the reaction

$$p\text{-MeOC}_6\text{H}_4\text{SiR}_3 + \text{Br}_2 \longrightarrow p\text{-MeOC}_6\text{H}_4\text{Br} + \text{R}_3\text{SiBr}$$

| $R_3$ | $k_{rel}$ | $R_3$ | $k_{rel}$ |
|---|---|---|---|
| $Me_3$ | 1100 | $(p\text{-MeOC}_6H_4)_3$ | $8.9^a$ |
| $Me_2Ph$ | 355 | $(p\text{-MeC}_6H_4)_3$ | 3.4 |
| $MePh_2$ | 15 | $(m\text{-MeC}_6H_4)_3$ | 1.5 |
| $(PhCH_2)_3$ | 30 | $(p\text{-ClC}_6H_4)_3$ | 0.12 |
| $ClCH_2Me_2$ | 65 | $(m\text{-ClC}_6H_4)_3$ | 0.052 |
| $(EtO)_3$ | 8.5 | $(o\text{-MeC}_6H_4)_3$ | 0.023 |
| $Ph_3$ | 1.0 | | |

[a] Statistically corrected value.

The corresponding tetraethyl compounds react similarly with bromine in refluxing ethyl bromide[308,309], whereas $Pr^n_4Ge$ with one equivalent of bromine also gave a fairly high yield of the $R_3GeBr$ compound; in the presence of an excess of bromine, the reaction becomes complex. It is reported[310] that reaction 113 takes place with iron powder as a catalyst.

$$Pr^n_3GeF + Br_2 \longrightarrow Pr^n_2GeFBr \qquad (113)$$

Arylgermanium compounds react more easily than do alkylgermanium compounds, and with bromine one or two aryl groups may be cleaved[311–313]. Thus in mixed alkylaryl compounds, it is expected that aryl groups would be cleaved preferentially. The spiro compounds in reactions 114 and 115 are readily broken by

(spiro-Ge bis(cyclopentane)) + $Br_2$ ⟶ (cyclopentane)Ge($(CH_2)_3Br$)(Br) (114)

$Et_2Ge$(cyclobutane) + $Br_2$ ⟶ $Et_2Ge((CH_2)_3Br)(Br)$ (115)

the action of bromine[314–316], but in the mixed compounds (reactions 116–119) only the phenyl group is removed by bromine[314,316,317].

(cyclopentane)$GePh_2$ + $Br_2$ ⟶ (cyclopentane)Ge(Ph)(Br) + PhBr (116)

(cyclopentane)Ge(Ph)(Br) + $Br_2$ ⟶ (cyclopentane)$GeBr_2$ + PhBr (117)

(cyclohexane)$GePh_2$ + $Br_2$ ⟶ (cyclohexane)Ge(Ph)(Br) + PhBr (118)

(cyclohexane)Ge(Et)(Ph) + $Br_2$ ⟶ (cyclohexane)Ge(Et)(Br) + PhBr (119)

Allyl groups are also very readily cleaved by bromine; tri-*n*-butylallylgermanium yields allylbromide[318]; compare also reaction 120[319].

$$Et_3GeR(R=2,4\text{-cyclopentadien-1-yl})+Br_2 \longrightarrow Et_3GeBr+RBr \quad (120)$$

Similarly[320,321], tri-*n*-butylstyrylgermanium forms styryl bromide, PhCH=CHBr.

Substituted allyl groups may be activated, as with $n$-$Bu_3GeCH_2CO_2Me$ where formation of $BrCH_2CO_2Me$ takes place on brominolysis[322], although tetrabenzylgermanium does not seem especially reactive towards bromine[323].

## L. Brominolysis of Organotin(IV) Compounds

Preparative aspects of the brominolysis of organotin compounds have been studied by numerous workers. Grüttner and Krause[324] showed that at −40 °C bromine cleaves one alkyl group from tetraalkyltins, but that at room temperature two alkyl groups could be removed. Usually, these cleavage reactions were carried out in solvents such as chloroform or carbon tetrachloride, or sometimes even in the absence of solvent. At higher temperatures, monoalkyltintribromides may be formed as in reaction 121[325]. Tetraaryltins also react readily with bromine, and in pyridine bromine gives the triaryltin bromide in good yield.

$$Et_2SnBr_2+Br_2 \longrightarrow EtSnBr_3+EtBr \quad (121)$$

With mixed tetraorganotins, bromine removes groups in the following sequence of decreasing preference[326–328]: *o*-tolyl, *m*-tolyl, phenyl, benzyl, methyl, ethyl, *n*-propyl, *n*-hexyl, *n*-heptyl. Other investigators[329–336] have shown that secondary alkyl groups and the *neo*-pentyl group are much less reactive than primary alkyl groups, whereas the perfluorovinyl group is cleaved by bromine even more easily than the phenyl group. Spirotin compounds have also been studied; reaction 122 (R = Me, Et) takes place at 0 °C in acetic acid[337].

$$R_2Sn(CH_2)_5 \text{ (ring)} + Br_2 \longrightarrow R_2Sn(Br)(CH_2)_5Br \quad (122)$$

The 1-adamantyl group seems to be very unreactive; Roberts[338] showed that bromine in $CDCl_3$ cleaves only the methyl group from 1-adamantyltintrimethyl, and that tetra-1-adamantyltin did not react with bromine over a period of weeks.

Gielen and coworkers[339] have used the selective cleavage of alkyl and aryl groups by bromine in methanol or ethanol to prepare various $R_4Sn$ compounds with four different groups linked to tin through a very elegant and rapid synthesis.

Several groups of workers have studied the stereochemistry of brominolysis reactions. Nasielski and coworkers[340] reported that brominolysis of *cis*- and *trans*-2-methylcyclopropyltintrimethyl in acetic acid and in chlorobenzene took place with a very high degree of retention of configuration to yield the corresponding *cis*- and *trans*-2-methylcyclopropyl bromides. The brominolysis of optically active *sec*-butyltin compounds has been studied in considerable detail. Jensen and Davis[11] showed that reaction 123 in methanol at 45 °C proceeded with inversion of configuration at the carbon atom undergoing substitution. Rahm and Pereyre[341] later reported on the brominolysis of a series of *sec*-butyltin compounds, $Bu^sSnR_3$, in a mixture of methanol and cyclohexane at −10 °C. They observed that when the leaving $SnR_3$ group contained smaller alkyl groups such as $Bu^s$ and $Pr^i$, the *sec*-butyl bromide was formed with predominant retention of configuration, but with the $Pe^{neo}{}_3Sn$ leaving group, the stereochemical result was mainly inversion of configuration in the *sec*-butyl group. In a later report, McGahey and Jensen[342] demonstrated that the stereochemical outcome of brominolysis of *sec*-butyltin compounds depended on the choice of solvent and of leaving group, and concluded that there is no preferred stereochemistry for this reaction. Seemingly, there is a delicate balance between retention and inversion pathways. The results of Jensen and McGahey are given in Table 52.

$$Bu^sSn(Pe^{neo})_3 + Br_2 \xrightarrow{Br^-} Bu^sBr + Pe^{neo}{}_3SnBr \qquad (123)$$

Gielen and Fosty[123] examined the stereochemical outcome of the brominolysis reaction 124 ($R = Me$ or $Pr^i$), in chlorobenzene. They showed that at low bromine concentrations the product $Me_3CCHDCHDBr$ was formed with retention of configuration but at higher bromine concentrations inversion of configuration was observed.

$$Me_3CCHDCHDSnR_3 + Br_2 \longrightarrow Me_3CCHDCHDBr + R_3SnBr \qquad (124)$$

Reutov and coworkers[343] have shown that brominolysis of *S*-(+)-(3-methylindenyl)trimethyltin by BrCN in dimethoxyethane proceeds with apparent inversion of configuration to give *R*-(−)-1-bromo-3-methylindene. This result, however, was interpreted as due to two consecutive processes:

$$\text{[3-methylindene, } CH_3\text{]} \longrightarrow \text{[1-methyl-1-}SnMe_3\text{-indene]} \xrightarrow{BrCN} \text{[3-Me-indene]} + Me_3SnCN \qquad (125)$$

The actual reaction with BrCN takes place by an $S_E2'$ process, that is, substitution with rearrangement, involving *anti*-attack by the electrophilic reagent.

The kinetics and mechanism of brominolysis have been extensively studied, especially by Gielen and Nasielski, although only for the brominolysis of *sec*-butyltin compounds in methanol (see later) have both stereochemical and kinetic studies been carried out on the same system. One of the earliest kinetic investigations in this area was that of Gielen and Nasielski[344] on the brominolysis of tetraalkyltins in dimethylformamide (reaction 126), carried out in the presence of bromide ion. If the

$$R_4Sn + Br_2 \longrightarrow RBr + R_3SnBr \qquad (126)$$

TABLE 52. Stereochemical outcome (%) of the bromination of $Bu^sSnR_3$ in various solvents[342]

| | $R_3$ | | | |
|---|---|---|---|---|
| Solvent | $Pr^i{}_3$ | $Pr^i{}_2(Pe^{neo})$ | $Pr^i(Pe^{neo})_2$ | $Pe^{neo}{}_3$ |
| $CCl_4$ | 70% ret. | 74% ret. | 76% ret. | 89% ret. |
| MeOH | 22% ret. | 1% ret. | 35% inv. | 40–65% inv. |
| MeOH[a] | 12% ret. | | | ~100% inv. |
| MeOH[b] | 9% ret. | 4% inv. | | |
| MeOH[c] | 10% ret. | | | |
| MeCN | 9% inv. | 60% inv. | ~100% inv. | ~100% inv. |

[a] With 0.2 M NaBr

[b] With 0.4 M NaBr.

[c] With 0.9 M $NaClO_4$.

equilibrium constant for tribromide ion formation in dimethylformamide at 20 °C is taken as $2.3 \times 10^6$ l mol$^{-1}$, rate constants for reaction 126 at 20 °C are as follows: R = Me, 240; R = Et, 110; and R = Pr$^n$, 14.5 l mol$^{-1}$ s$^{-1}$; the reaction is kinetically first order in $R_4Sn$ and in bromine. Gielen and Nasielski showed also that increase of ionic strength brought about an increase in the second-order rate constants, and suggested that reaction took place through an 'open' transition state.

Reactions 126 and 127 were also studied in acetic acid; the reported second-order rate constants are given in Table 53[344]. For reaction 126 the reactivity sequence observed for R was Me > Et > Pr$^n$ ≈ Bu$^n$ > Pr$^i$; this (steric) sequence differs from that in reaction 127: Et > Me > Pr$^n$ [344]. The same workers later reported[345] rate constants for the brominolyses of various unsymmetrical tetraalkyltins in acetic acid with added bromide ion, and those are given in Table 53.

$$R_3SnBr + Br_2 \longrightarrow RBr + R_2SnBr_2 \tag{127}$$

Gielen and Nasielski[346] also studied the brominolysis of an extended series of tetraalkyltins, in chlorobenzene, in the absence of added bromide ion and in the absence of light. They initially felt that the reaction was a simple second-order electrophilic substitution, but later studies[345] showed that the rate law was made up of a second- and a third-order term; the latter term, however, was important only in the case of tetraisopropyltin. The kinetic results are assembled in Table 54. The same workers also reported similar studies on the brominolysis of a series of $RSnMe_3$ compounds, for which a second-order rate law was found; details are given in Table 54.

Reaction 126 has been shown to follow a second-order law in carbon tetrachloride, and values of the rate constants are given in Table 55. The change in the

TABLE 53. Second-order rate constants[344] at 20 °C in acetic acid for brominolysis of tetraalkyltins[a] and trialkyltin bromides

| $R_4Sn$ | $k_2$ (l mol$^{-1}$ s$^{-1}$)[b] | $k_2$ (l mol$^{-1}$ s$^{-1}$)[c] |
|---|---|---|
| $Me_4Sn$ | 9.6[d] | 1.16 |
| $Et_4Sn$ | 8.1[d] | 0.98 |
| $Pr^n{}_4Sn$ | 1.15 | 0.141 |
| $Bu^n{}_4Sn$ | 1.00 | 0.121 |
| $Pr^i{}_4Sn$ | 0.25 | |
| $Me_3SnBr$ | $7.37 \times 10^{-4}$ | |
| $Et_3SnBr$ | $10.8 \times 10^{-4}$ | |
| $Pr^i{}_3SnBr$ | $1.29 \times 10^{-4}$ | |
| Me—$SnMe_3$ | 2.92 | |
| Et—$SnMe_3$ | 1.21 | |
| $Pr^n$—$SnMe_3$ | 0.36 | |
| $Bu^n$—$SnMe_3$ | 0.55 | |
| $Pr^i$—$SnMe_3$ | 0.03 | |
| Me—$SnMe_2Et$ | 4.33 | |
| Me—$SnMe_2Pr^n$ | 3.46 | |
| Me—$SnMe_2Bu^n$ | 3.40 | |
| Me—$SnMe_2Pr^i$ | 3.20 | |

[a] For unsymmetrical tetraalkyltins values are the statistically corrected second-order rate constants for cleavage of the indicated alkyl group[345].
[b] In the absence of added bromide ion.
[c] In the presence of added bromide ion.
[d] Calculated by a comparison with values for brominolysis by $Br_2/Br^-$.

TABLE 54. Second-order rate constants at 20 °C in chlorobenzene for the brominolysis of symmetrical[346] and unsymmetrical[a] tetraalkyltins

| $R_4Sn$ | $k_2$ ($l\,mol^{-1}\,s^{-1}$) | $R_4Sn$ | $k_2$ ($l\,mol^{-1}\,s^{-1}$) |
|---|---|---|---|
| $Me_4Sn$ | 0.116 (0.112)[b] | $Pr^n$—$SnMe_3$ | 0.0090 |
| $Et_4Sn$ | 1.42 (1.40)[b] | $Bu^n$—$SnMe_3$ | 0.0090 |
| $Pr^n{}_4Sn$ | 0.54 | $Pr^i$—$SnMe_3$ | 0.0206 |
| $Bu^n{}_4Sn$ | 0.51 | $Bu^t$—$SnMe_3$ | ~0.0550 |
| $Octyl^n{}_4Sn$ | 0.62 | Me—$SnMe_2Et$ | 0.0550 |
| $Dodecyl^n{}_4Sn$ | 0.49 | Me—$SnMe_2(Pr^n)$ | 0.060 |
| $Pr^i{}_4Sn$ | 1.61 (0.35)[b] | Me—$SnMe_2(Bu^n)$ | 0.0617 |
| Me—$SnMe_3$ | 0.0285 | Me—$SnMe_2(Pr^i)$ | |
| Et—$SnMe_3$ | 0.0208 | Me—$SnMe_2(Bu^t)$ | ~0.04 |

[a] Values are statistically corrected second-order rate constants for the cleavage of the indicated alkyl group[345].
[b] Values from ref. 345, and include a third-order term in the rate equation.

sequence of reactivity in $R_4Sn$ with the group R on varying the solvent from more polar to the less polar media is very pronounced (see Tables 53–55 and the above results in dimethylformamide). Gielen and Nasielski have interpreted such reactivity changes (see also ref. 4) as due to a change in mechanism from $S_E2$ (open) in the more polar media to $S_E2$ (cyclic) in the less polar media. Fukuzumi and Kochi (see later), however, have interpreted such changes using their general mechanism of substitution throughout.

Faleschini and Tagliavini[347] also studied the brominolysis of unsymmetrical compounds, $RSnR'_3$, in carbon tetrachloride and reported overall rate constants (Table 55). They then calculated rate constants for the cleavage of given alkyl groups, but in these calculations they assumed that in the $RSnR'_3$ compounds only the smaller alkyl group was cleaved; this assumption is not generally valid. However, an analysis of the available rate constants in carbon tetrachloride reveals that the effect of the substituted group is to increase the reaction rate in the order $Me < Et < Pr^n < Pr^i$, whereas the effect of the alkyl group on the leaving $SnR'_3$ entity is to increase the rate in the order $SnMe_3 < Sn(n\text{-alkyl})_3 < Sn(Pr^i)_3$.

Reutov and co-workers[348] have investigated the kinetics of reaction 128 in dimethyl sulphoxide, using several tetraalkyltins. The reaction was found to be

$$RSnMe_3 + Br_2 \xrightarrow{Br^-} RBr + Me_3SnBr \qquad (128)$$

TABLE 55. Second-order rate constants, $10^2k_2$, in carbon tetrachloride for the brominolysis of tetraalkyltins

| $R_4Sn$ | $10^2k_2$ ($l\,mol^{-1}\,s^{-1}$) (20 °C)[346] | $R_4Sn$ | $10^2k_2$ ($l\,mol^{-1}\,s^{-1}$) (35.5 °C)[347] |
|---|---|---|---|
| $Me_4Sn$ | 0.018 | $Me_4Sn$ | 0.05 |
| $Et_4Sn$ | 1.70 | $MeSnEt_3$ | 1.24 |
| $Pr^n{}_4Sn$ | 0.80 | $MeSn(Pr^n)_3$ | 0.70 |
| $Bu^n{}_4Sn$ | 0.97 | $MeSn(Bu^n)_3$ | 0.80 |
| $Octyl^n{}_4Sn$ | 1.25 | $MeSn(Pr^i)_3$ | 9.90 |
| $Dodecyl^n{}_4Sn$ | 1.44 | $EtSn(Bu^n)_3$ | 2.47 |
| $Pr^i{}_4Sn$ | 15.5 | $Pr^nSn(Bu^n)_3$ | 1.83 |

TABLE 56. Rate constants[348] at 25 °C in dimethyl sulphoxide for the brominolysis[a] of $RSnMe_3$ in the presence of bromide ion

| R | $10^4k_2^{exp}/K$ ($s^{-1}$) | R | $10^4k_2^{exp}/K$ ($s^{-1}$) |
|---|---|---|---|
| Me | 1.6 | $CH_2{=}CHCH_2$ | $2.7\times10^5$ [b] |
| Ph | 12.7 | $C_6F_5$ | 1.68 |
| $CH_2{=}CH$ | 20.1 | $PhCH_2$ | 4.10 |

[a] For all groups except $PhCH_2$, cleavage of the R—Sn bond occurs. For R = $PhCH_2$ a methyl group is cleaved.
[b] At 20 °C.

second order overall, but since the equilibrium constant for formation of $Br_3^-$ in dimethyl sulphoxide was not known, the actual second-order rate constants could not be evaluated. In Table 56 are given values of $k_2^{exp}/K$ for a series of $RSnMe_3$ compounds. The related reaction 128 where R = CN was also studied[349] in dimethyl sulphoxide, but in this case the reaction was found to be zero order in bromine and first order in $(CN)SnMe_3$. At 20 °C the first-order rate constant is 0.30 $s^{-1}$. In a similar type of investigation using as the solvent dimethylformamide–carbon tetrachloride (1:1), the effect of a number of electron-withdrawing groups, R, on reaction 128 was studied[350]. Again, only relative rate constants could be obtained (Table 57) for the various $S_E2$ substitutions by molecular bromine. When R = cyclopentadienyl, reaction 128 was found to be a first-order process, the suggested mechanism being $S_E1$; compare also $(CN)SnMe_3$. The results in Table 56 follow a general pattern in halogenolysis of organotin compounds, in that not only the phenyl group but also allyl and vinyl groups are cleaved from tin more rapidly than are the simple alkyl groups.

The only example of a reaction that has been studied both stereochemically and kinetically is the bromination of $Bu^sSn(Pe^{neo})_3$ in methanol (reaction 123). Jensen and Davis[11] showed that in methanol at 45 °C reaction proceeded with inversion of configuration at the $Bu^s$ group. They also reported second-order rate constants for reaction of the $Bu^s$ compound and for a series of $RSn(Pe^{neo})_3$ compounds (Table 58). It was suggested that the combination of kinetic and stereochemical results leads only to the possibility of an $S_E2$ reaction with inversion, through transition state **20**. It

$$[Br{-}{-}Br{-}{-}R{-}{-}Sn(Pe^{neo})_3]^{\neq}$$

**(20)**

TABLE 57. Rate constants at 25 °C for the brominolysis of $RSnMe_3$ in dimethylformamide–tetrachloromethane (1:1) in the presence of bromide ion[350]

| R | $k_2/K$ ($s^{-1}$) | $k_2$ ($l\,mol^{-1}\,s^{-1}$)[a] |
|---|---|---|
| $C_6H_5$ | $6.75\times10^{-3}$ | $1.35\times10^4$ |
| 9-Methylfluorenyl | $2.05\times10^{-2}$ | $4.1\times10^4$ |
| Indenyl[b] | 6.0 | $1.2\times10^8$ |
| Cyclopentadienyl | | $k_1 = 1.7\ s^{-1}$ |

[a] These values were calculated using $K = 2\times10^6\ l\,mol^{-1}$, which is the value for the $Br_3^-$ equilibrium in pure dimethylformamide.
[b] The given values of $k_2/K$ and $k_2$ are not compatible. If $k_2/K = 6.0\ s^{-1}$ then $k_2$ should have the value $1.2\times10^7\ l\,mol^{-1}\,s^{-1}$.

TABLE 58. Second-order rate constants at 45 °C for the brominolysis of $Pe^{neo}{}_3SnR$ compounds in methanol[11]

| R | $k_2$ ($l\ mol^{-1}\ s^{-1}$) | Relative $k_2$[a] |
|---|---|---|
| Me | 16.6 | 1 |
| Et | 2.39 | 0.144 |
| $Pr^n$ | 0.797 | 0.048 |
| $Pr^i$ | 0.125 | $7.53 \times 10^{-3}$ |
| $Pe^{neo}$ | $3.88 \times 10^{-2}$ | $5.84 \times 10^{-4}$ [b] |

[a] Relative values for the cleavage of the $Pe^{neo}{}_3Sn$—R bond.
[b] Statistically corrected.

should be pointed out, though, that since there sems such a fine balance between $S_E2$ reactions with retention and inversion, it cannot be taken for granted that because $Bu^sSn(Pe^{neo})_3$ undergoes substitution with inversion at the $Bu^s$ group, other $RSn(Pe^{neo})_3$ compounds will also undergo substitution with inversion at the alkyl group, R.

The most recent work on brominolysis of tetraalkyltins is due to Fukuzumi and Kochi[351]. These workers observed that on mixing the reactants, transient charge-transfer bands were formed, which then disappeared as the actual brominolysis took place. Rate constants for brominolysis in the absence of added bromide ion were obtained for reactions in solvents hexane and carbon tetrachloride, and are given in Table 59. Reactions were carried out in the dark in order to eliminate light-induced radical chain processes. The results were analysed in terms of the charge transfer mechanism, and will be discussed together with the iodinolysis kinetic results later.

TABLE 59. Second-order rate constants, at 25 °C for the brominolysis of tetraalkyltins in hexane and in carbon tetrachloride[351]

| | $10^4 k_2$ ($l\ mol^{-1}\ s^{-1}$) | |
|---|---|---|
| $R_4Sn$ | Hexane | $CCl_4$ |
| $Me_4Sn$ | 0.43 | 1.9 |
| $EtSnMe_3$ | 2.1 | 9.9 |
| $Pr^nSnMe_3$ | 1.2 | 5.0 |
| $Bu^nSnMe_3$ | 1.2 | 3.8 |
| $Et_2SnMe_2$ | 5.5 | 27 |
| $Pr^n{}_2SnMe_2$ | 4.1 | 18 |
| $Pr^i{}_2SnMe_2$ | 96 | 210 |
| $Bu^n{}_2SnMe_2$ | 3.3 | 17 |
| $Bu^t{}_2SnMe_2$ | 7.1 | 9.5 |
| $Et_3SnMe$ | 28 | 84 |
| $Et_4Sn$ | 51 | 250 |
| $Pr^n{}_4Sn$ | 25 | 110 |
| $Pr^i{}_4Sn$ | 2500 | 2700 |
| $Bu^n{}_4Sn$ | 33 | 130 |
| $Bu^i{}_4Sn$ | 8.4 | 24 |
| $Bu^s{}_4Sn$ | 350 | 650 |

## M. Brominolysis of Organolead(IV) Compounds

Depending on the experimental conditions, one or two organic groups may be cleaved from tetraorganoleads[352]. With tetraalkylleads one group is removed at −70 °C and the second alkyl group at −25 °C, reactions 129 and 130 usually being carried out using an organic solvent. With cleavage of aryl groups, the step-wise stoichiometry of these reactions is less pronounced.

$$R_4Pb + Br_2 \longrightarrow RBr + R_3PbBr \quad (129)$$

$$R_3PbBr + Br_2 \longrightarrow RBr + R_2PbBr_2 \quad (130)$$

Kinetic studies have been carried out on reaction 129 in methanol in the presence of bromide ion[17]. When R = Me or Et the reactions were found to be first order in each reactant. Rate constants at an ionic strength of 0.20 M were found to be $11 \times 10^4\,\mathrm{l\,mol^{-1}\,s^{-1}}$ for brominolysis of tetramethyllead and $9.4 \times 10^4\,\mathrm{l\,mol^{-1}\,s^{-1}}$ for brominolysis of tetraethyllead. These rate constants decreased with decrease in ionic strength, and an $S_E2$ (open) mechanism was postulated.

## N. Iodinolysis of Organolithium(I) Compounds

The optically active compound 1-methyl-2,2-diphenylcyclopropyllithium was found to react with iodine in ether to yield the corresponding cyclopropyl iodide with almost complete retention of optical activity and configuration[283].

## O. Iodinolysis of Organomercury(II) Compounds

Early work by Keller[353] showed that the iodinolysis of phenyl- and alkylmercury(II) iodides by iodine in dioxane was free-radical in nature, with a kinetic form second order in iodine and zero order in organomercury(II) iodide. In the presence of added iodide ion, the free-radical path was suppressed and the kinetic form changed to first order in substrate and first order in $I_3^-$. Winstein and Traylor[354] then studied the iodinolysis of 4-camphylmercury(II) iodide in dioxane–water (95:5) with iodine in the presence of iodide ion and reported a second-order rate constant of $1.27 \times 10^{-3}\,\mathrm{l\,mol^{-1}\,s^{-1}}$ at 55 °C. Reaction 131 was also studied in dioxane–water

$$RHgI + I_2 \xrightarrow{I^-} RI + HgI_2 \quad (131)$$

(90:10) and relative rate constants at 55 °C were found to be as follows: R = Bu$^n$, 66; R = 4-camphyl, 3.1; and R = neophyl, 1; Winstein and Traylor[354] suggested that reaction took place by a four-centred $S_E2$ (cyclic) transition state. In another early study. Lord and Pritchard[355] reported that reaction 132 when carried out with the

TABLE 60. Second-order rate constants, $10^3k_2$, at 28 °C and activation energies for the iodinolysis of $Me_2Hg$ in various solvents[355]

| Solvent | $10^3k_2$ ($\mathrm{l\,mol^{-1}\,s^{-1}}$) | $E_a$ (kcal $\mathrm{mol^{-1}}$) |
|---|---|---|
| Cyclohexane | 1.15 | 7.1 |
| $CCl_4$ | 1.22 | 7.7 |
| Benzene | 6.67 | 8.5 |
| $CHCl_3$ | 24.8 | 8.0 |
| EtOH | 116 | 7.4 |

TABLE 61. Second-order rate constants at 20 °C and activation parameters for the reaction

$$PhCH_2HgCl + I_2 \xrightarrow{I^-} PhCH_2I + HgClI$$

| Solvent | $k_2^{obs}$ ($l\,mol^{-1}\,s^{-1}$) | $E_a$ ($kcal\,mol^{-1}$) | $\Delta S^{\neq}$ ($cal\,K^{-1}\,mol^{-1}$) | Ref. |
|---|---|---|---|---|
| Dioxane–water (70:30) | 0.366 | 10.8 | −23 | 356 |
| Dimethylformamide | 0.215 | 15.3 | −8 | 357 |
| Methanol | 0.807 | 9.6 | −26 | 357 |
| Ethanol | 0.279 | 11.6 | −20 | 357 |
| Butanol | 0.072 | | | 358 |
| Acetonitrile | 0.590 | 10.1 | −27 | 358 |
| Dimethyl sulphoxide | 0.145 | 15.2 | −10.5 | 358 |
| Benzene | 6.6 | 9.7 | −24 | 358 |

rigorous exclusion of light was first order in each reactant. Kinetic results are given in Table 60.

$$Me_2Hg + I_2 \longrightarrow MeI + MeHgI \qquad (132)$$

Reutov and coworkers carried out several investigations on the kinetics of iodinolysis of organomercury(II) compounds, using either benzyl- or aryl-mercury(II) halides as substrates. Reaction 133 was studied using dioxane–water (70:30) as

$$PhCH_2HgCl + I_2 \xrightarrow{I^-} PhCH_2I + HgClI \qquad (133)$$

solvent and cadmium(II) iodide as the source of iodide ion[356]. The kinetic form was first order in $I_3^-$, and the value of $k_2^{obs}$ was found to be independent of iodide ion concentration. It therefore follows from equation 98 that only the terms in $k_2^b$ and $k_3$ contribute to the overall rate. Table 61 shows the quoted rate constants for reaction in aqueous dioxane[356] and various other solvents[357,358]. Reutov *et al.*[359] put forward two possible transition states **21** and **22**, arising from attack of $I_3^-$ on $PhCH_2HgCl$. This corresponds to the $k_2^b$ term in equation 98; the $k_3$ term would involve attack of $I_2$ on $PhCH_2HgClI^-$, possibly through a transition state such as **23**.

**(21)** **(22)** **(23)**

Substituent effects in the benzyl group were also investigated[360] in methanol and dimethylformamide; again cadmium(II) iodide was the source of iodide ion. Rate enhancements were found both for the electron-attracting $p$-$NO_2$ substituent and for the electron-donating $p$-Me group (see Table 62). It was suggested[360] that the mechanism approached $S_E1$ for the $p$-$NO_2$-substituted compound and approached $S_E2$ for the methyl- and halogen-substituted compounds.

Reutov and coworkers[361] also studied substituent effects in reaction 134 in methanol—toluene (Table 63).

$$XC_6H_4CH(CO_2Et)HgBr + I_2 \xrightarrow{I^-} XC_6H_4CH(CO_2Et)I + HgBrI \qquad (134)$$

The mechanism was suggested to be in the boundary region betwen $S_E1$ and $S_E2$.

TABLE 62. Relative second-order rate constants[360] at 20°C in methanol and dimethylformamide for the reaction

$$XC_6H_4CH_2HgCl + I_2 \xrightarrow{I^-} XC_6H_4CH_2I + HgClI$$

| X | MeOH | Dmf | X | MeOH | Dmf |
|---|---|---|---|---|---|
| *p*-$NO_2$ | Very fast | | *p*-Br | | 0.76 |
| *p*-MeO | 11.1 | 11.1 | *p*-Cl | 0.95 | 0.64 |
| *p*-Me | 2.57 | 2.04 | *m*-Br | 0.61 | 0.48 |
| *o*-Me | 2.33 | 1.80 | *m*-F | 0.56 | 0.48 |
| *p*-F | 1.12 | 0.75 | *o*-Cl | 0.31 | 0.32 |
| *m*-Me | 1.11 | 0.92 | *o*-F | 0.29 | 0.32 |
| H | 1.0 | 1.0 | | | |

It should be noted that the rate constants found for reactions 133 and 134 are composite constants, since $k_2^{obs} = k_2^b + k_3/K$ for the case in which the observed rate constant is independent of $[I^-]$. Further, $k_3$ will itself be a composite constant and will include the equilibrium constant for formation of the complex $[RHgXI]^-$. Thus substituent effects on $k_2^{obs}$ may reflect possible variations in the $k_2^b$ to $k_3/K$ ratio and possible substituent effects on the $[RHgXI]^-$ formation constant as well as on the rate constant for reaction of RHgCl with $I_3^-$.

Substitution of arylmercury(II) compounds have been reported; Reutov and coworkers[297] studied the iodinolysis of phenylmercury(II) iodide by $I_2/I^-$ in 80%

TABLE 63. Relative second-order rate constants[361] (probably at 20 °C) in toluene + 1.5% methanol for the reaction

$$XC_6H_4CH(COOEt)HgBr + I_2 \xrightarrow{I^-} XC_6H_4CH(COOEt)I + HgBrI$$

| X | $k_2^{rel}$ | X | $k_2^{rel}$ |
|---|---|---|---|
| *p*-$NO_2$ | 61.5 | H | 1.0 |
| *m*-Br | 7.98 | *m*-Me | 0.7 |
| *p*-I | 4.36 | *p*-$Pr^i$ | 0.457 |
| *p*-Cl | 3.34 | *p*-$Bu^t$ | 0.353 |
| *p*-F | 1.4 | | |

TABLE 64. Second-order rate constants[297] at 20 °C and activation parameters for the reaction

$$PhHgI + I_2 \xrightarrow{I^-} PhI + HgI_2$$

| Solvent | $k_2$ (l mol$^{-1}$ s$^{-1}$) | $E_a$ (kcal mol$^{-1}$) | $\Delta S^{\neq}$ (cal K$^{-1}$ mol$^{-1}$) |
|---|---|---|---|
| dmf | 20.2 | 9.8 | −20.2 |
| 80% aq. dioxane | 78.0 | 11.2 | −13.9 |
| MeOH | 148.1 | 11.1 | −12.9 |

TABLE 65. Second-order rate constants[362,363] in ethanol at 30 °C and activation parameters for the reaction

$$o\text{-}XC_6H_4HgCl + I_2 \longrightarrow o\text{-}XC_6H_4I + HgClI$$

| X | $k_2$ (l mol$^{-1}$ s$^{-1}$) | $E_a$ (kcal mol$^{-1}$) | $\Delta S^{\neq}$ (cal K$^{-1}$ mol$^{-1}$)$^a$ |
|---|---|---|---|
| H | 5.67 | 11.04 | −21.0 |
| Me | 1.50 | 11.15 | −20.0 |
| OH | 11.8 | 8.10 | −29.2 |
| $NO_2$ | 0.014 | 15.96 | −11.0 |

$^a$ At 40 °C.

TABLE 66. Second-order rate constants at 25 °C in carbon tetrachloride for the iodinolysis of dialkylmercurys[19]

| $R_2Hg$ | $k_2$ (l mol$^{-1}$ s$^{-1}$) | $R_2Hg$ | $k_2$ (l mol$^{-1}$ s$^{-1}$) |
|---|---|---|---|
| $Me_2Hg$ | $7.8 \times 10^{-4}$ | $Bu^n{}_2Hg$ | 0.62 |
| EtHgMe | $6.5 \times 10^{-2}$ | $Bu^i{}_2Hg$ | 0.26 |
| $Pr^n{}_2Hg$ | 0.63 | | |

aqueous dioxane, methanol, and dimethylformamide. The reaction was very rapid, and although it was shown to be kinetically of the second order overall, the order with respect to each reactant could not be established; rate constants are given in Table 64. Jha and coworkers[362,363] have studied the substituent effect on a related aromatic substitution, that of arylmercury(II) chlorides with iodine in ethanol. The overall kinetic order was found to be two, and the reported rate constants and activation parameters are given in Table 65.

More recently, Fukuzumi and Kochi[19] have reported second-order rate constants for the iodinolysis of dialkylmercury(II) compounds in carbon tetrachloride (see Table 66). The mechanism of the iodinolysis was described by Fukuzumi and Kochi in terms of their charge-transfer mechanism, as discussed for the halogenolysis of tetraalkyltins (see Section III.W).

## P. Iodinolysis of Organoboron(III) Compounds

Alkylboron(III) compounds are rather inert towards iodine, but aryl—boron bonds are easily broken by the action of iodine. Brown *et al.*[364] have studied the kinetics of iodinolysis of arylboronic acids in aqueous methanol at constant ionic strength and in the presence of an excess of iodide ion (reaction 135). The reaction was first order

$$ArB(OH)_2 + I_2 + H_2O \xrightarrow{I^-} ArI + H_3BO_3 + HI \qquad (135)$$

both in boronic acid and in iodine. Rate constants and activation parameters are given in Table 67. As had previously been shown for the iodinolysis of *p*-methoxybenzene boronic acid[365], the reacting species are molecular iodine and the boronate anion.

TABLE 67. Activation parameters and second-order rate constants at 25 °C in sodium acetate buffer solution for the reaction[364]

$$ArB(OH)_2 + I_2(+H_2O) \xrightarrow{I^-} ArI + B(OH)_3 + HI$$

| Ar | $k_2$ ($l\,mol^{-1}\,s^{-1}$) | $E_a$ ($kcal\,mol^{-1}$) | $\Delta S^{\neq}$ ($cal\,K^{-1}\,mol^{-1}$) |
|---|---|---|---|
| Phenyl | 42.8 | 15.3 | −1.7 |
| Thien-3-yl | 30 000 | 8.6 | −11.0 |
| Thien-2-yl | 416 000 | 7.5 | −9.8 |

## Q. Iodinolysis of Organosilicon(IV) Compounds

A number of preparative studies have been reported on these iodinolyses. Eaborn[366] showed that iodine in the presence of catalytic amounts of aluminium(III) iodide will cleave two ethyl groups from tetraethylsilicon, at the boiling point of the mixture, to form $Et_2SiI_2$ and EtI. Clark *et al.*[367] used the same catalytic system in toluene solvent to form $n$-$C_3F_7I$ from $(n\text{-}C_3F_7)_2SiMe_2$. $CH_2{=}CHCH_2Si(OEt)Me_2$, $CH_2{=}CHCH_2Si(OEt)_2Me$, and $CH_2{=}CHCH_2Si(OEt)_2Ph$ were all reported by Grafstein[368] to react with iodine yielding mixtures of alkyl iodide and ethyl iodide.

The action of iodine chloride on phenyltrimethylsilicon was studied by Stock and Spector[280] (reaction 136) using acetic acid as the solvent. Good second-order kinetics

$$PhSiMe_3 + ICl \longrightarrow PhI + Me_3SiCl \qquad (136)$$

were found, the rate constant at 25 °C being $0.133\,l\,mol^{-1}\,s^{-1}$. This is faster than the analogous chlorodesilylation reaction, and a cyclic four-centred transition state was suggested by Stock and Spector.

## R. Iodinolysis of Organogermanium(IV) Compounds

A number of product analyses have been reported on these reactions[369]; in the presence of $AlI_3$, iodine and tetraalkylgermaniums react according to either reaction 137 when R = Me and Et or to reaction 138 when $R = Bu^n$, $Bu^i$, and $Pe^n$. For $R = Pr^n$ a combination of both reactions takes place.

$$R_4Ge + I_2 \xrightarrow{AlI_3} R_3GeI + RI \qquad (137)$$

$$R_4Ge + I_2 \xrightarrow{AlI_3} R_3GeI + RR \qquad (138)$$

The uncatalysed reaction of $Bu^n{}_4Ge$ and iodine yields tri-$n$-butyliodogermanium[370], and it is reported that in refluxing decahydronaphthalene iodine will cleave one phenyl group from tetraphenylgermanium to give a poor yield of $Ph_3GeI$[371].

## S. Iodinolysis of Organotin(IV) Compounds

There is increasing use of the destannylation reaction as a means of introducing substituents into carbon skeletons; Seitz and coworkers have obtained iodo and 125-iodo derivatives of tamoxifen through iodinolysis of the corresponding tri-butyltin compound[372]. Baekelmans *et al.*[340] have reported that the iodinolysis of *cis*- or *trans*-2-methylcyclopropyltintrimethyl in either methanol or acetic acid takes place

TABLE 68. Second-order rate constants at 20 °C for iodinolysis of tetraalkyltins

| $R_4Sn$ | $10^4k_2$ ($l\,mol^{-1}\,s^{-1}$)[a] | $k_2$ ($l\,mol^{-1}\,s^{-1}$)[b] | $k_2$ ($l\,mol^{-1}\,s^{-1}$)[c] |
|---|---|---|---|
| $Me_4Sn$ | 1.67 | 6.8 | 6.55 |
| $Et_4Sn$ | 10 | 0.8 | 0.833 |
| $Pr^n{}_4Sn$ | 1.30 | 0.1 | 0.136 |
| $Bu^n{}_4Sn$ | 1.0 | 0.04 | |
| $Pr^i{}_4Sn$ | 9.4 | 0.004[d] | |
| $Bu^i{}_4Sn$ | | | 0.071 |
| $Pe^{neo}{}_4Sn$ | | | 0.086 (0.03)[e] |

[a] In chlorobenzene[346].
[b] In the presence of iodide ion in methanol[276].
[c] In the presence of iodide ion in methanol[373].
[d] Ref. 345.
[e] Different values from different sets of experiments.

with complete retention of configuration to yield the corresponding cyclopropyl iodide. As may be seen from stereochemical studies of the brominolysis reaction (see Section L), however, this cannot be taken as a general stereochemical rule.

A very large number of kinetic studies have been reported on the iodinolysis of alkyl- and aryl-tin compounds, starting with the work of Gielen and Nasielski. They showed that in the presence of a large excess of iodide ion, reaction 139, where R = several alkyl groups, followed second-order kinetics in methanol. Analysis as indicated in Section III.A indicated that molecular iodine was the electrophilic reagent and that the rate constant increased with increasing ionic strength of the medium. They concluded that alkyl—tin bonds were broken by an $S_E2$ reaction involving an 'open' transition state[276].

$$R_4Sn + I_2 \longrightarrow RI + R_3SnI \qquad (139)$$

Rate constants for cleavage of symmetrical tetraalkyltins are given in Table 68[276,345,346,373]. Combination of similar kinetic studies, together with relevant product analyses, also enabled rate constants for cleavage of alkyl—tin bonds in unsymmetrical tetraalkyltins to be obtained[345] (Table 69). Also given in Table 69 are rate

TABLE 69. Statistically corrected second-order rate constants[345] for iodine cleavage of the indicated alkyl group from tetraalkyltins in the presence of iodide ion at 20 °C in methanol

| $R_4Sn$ | $k_2$ ($l\,mol^{-1}\,s^{-1}$) | $R_4Sn$ | $k_2$ ($l\,mol^{-1}\,s^{-1}$) |
|---|---|---|---|
| Me—$SnMe_3$ | 1.77 | $Pr^n$—$Sn(Bu^n)_3$ | 0.16[a] |
| Et—$SnMe_3$ | 0.256 | $Bu^n$—$Sn(Bu^n)_3$ | 0.01 |
| $Pr^n$—$SnMe_3$ | 0.056 | Me—$SnMe_2Et$ | 2.34 |
| $Bu^n$—$SnMe_3$ | 0.132 | Me—$SnMe_2(Pr^n)$ | 1.95 |
| $Pr^i$—$SnMe_3$ | 0.01 | Me—$SnMe_2(Bu^n)$ | 1.95 |
| $Bu^t$—$SnMe_3$ | 0.00 | Me—$SnMe_2(Pr^i)$ | 1.51 |
| Me—$SnEt_3$ | 3.58 | Me—$SnMe_2(Bu^t)$ | 0.01 |
| Et—$SnEt_3$ | 0.22 | Me—$Sn(Pr^n)_3$ | 1.68[a] |
| $Pr^n$—$SnEt_3$ | 0.065 | Me—$Sn(Pr^i)_3$ | 0.17[a] |
| $Bu^n$—$SnEt_3$ | 0.060 | Et—$SnEt_2(Pr^n)$ | 0.245 |
| $Pr^i$—$SnEt_3$ | 0.004 | Et—$SnEt_2(Bu^n)$ | 0.247 |
| Me—$Sn(Bu^n)_3$ | 1.42[a] | Et—$SnEt_2(Pr^i)$ | 0.255 |
| Et—$Sn(Bu^n)_3$ | 0.58[a] | | |

[a] Values are taken from ref. 347 and may not be totally reliable (see text).

constants calculated by Abraham from the work of Faleschini and Tagliavini[347], in the light of subsequent criticism[345] of these workers' product analyses. From the kinetic data listed in Tables 68 and 69 it seems that there is a very strong steric effect in the alkyl group that is cleaved, but that alkyl groups on the leaving $SnR_3$ entity exert much less powerful steric effects. If these substitutions in methanol do proceed with retention of configuration at the site of substitution, through a transition state such as **24**, then following the analysis given by Abraham and Spalding[190] for other $S_E2$

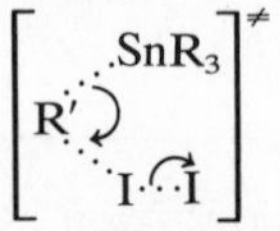

$$\left[ \begin{matrix} & SnR_3 \\ R' & \\ & I \cdots I \end{matrix} \right]^{\neq}$$

**(24)**

reactions, it seems probable that the main steric interactions in **24** are between the moving R′ group and the incoming iodine molecule, and between the moving R′ group and the $SnR_3$ group. However, it should be noted that no experimental determination of the stereochemical course of these iodinolyses of simple aliphatic alkyltin compounds has been reported.

Gielen and Nasielski[374] have also studied the iodinolysis of alkyltin compounds using various other solvents. In dimethyl sulphoxide, with no added iodide ion, a second-order rate equation was followed, and rate constants at 20 °C were given as $0.48\,l\,mol^{-1}\,s^{-1}$ for $Me_4Sn$ and $0.021\,l\,mol^{-1}\,s^{-1}$ for $Et_4Sn$[374]. Studies using acetic acid were carried out again in the absence of iodide ion, and the given rate constants are collected in Table 70[344]. In chlorobenzene, a different sequence of reactivity was observed (see Table 68); this was attributed to formation of a cyclic transition state[6], rather than the open transition state in methanol or acetic acid.

Roberts[212] has studied the iodinolysis of a series of trimethyltin compounds ($RSnMe_3$) in methanol, acetonitrile, and dimethyl sulphoxide and has shown that when R = $Bu^n$, $Pe^n$, —$(CH_2)_4Ph$, —$(CH_2)_2CH{=}CH_2$, —$(CH_2)_3CH{=}CH_2$, —$(CH_2)_2CO_2Me$, and a variety of cyclic groups, only the methyl—tin bond was broken with formation of methyl iodide. The exceptions were when R = 7-norbornadienyl and *anti*-7-norbornadienyl, for which $C_7$—tin bond fission took place. Activation parameters for reaction of $Me_4Sn$ and $Me_3Sn(CH_2)_2CO_2Me$ with iodine in acetonitrile were reported[212].

The iodinolysis of tetra-*n*-butyltin by iodine in a number of non-polar solvents was

TABLE 70. Second-order rate constants in acetic acid at 20°C for the iodinolysis of tetraalkyltins[344] and (statistically corrected) values for the iodine cleavage of the indicated alkyl group from unsymmetrical substrates[345]

| $R_4Sn$ | $k_2$ ($l\,mol^{-1}\,s^{-1}$) | $R_4Sn$ | $k_2$ ($l\,mol^{-1}\,s^{-1}$) |
|---|---|---|---|
| $Me_4Sn$ | 0.22 | $Bu^n$—$SnMe_3$ | $3.17 \times 10^{-3}$ |
| $Et_4Sn$ | 0.082 | $Pr^i$—$SnMe_3$ | $4.6 \times 10^{-4}$ |
| $Pr^n{}_4Sn$ | $9.4 \times 10^{-3}$ | $Bu^t$—$SnMe_3$ | $\sim 5 \times 10^{-5}$ |
| $Bu^n{}_4Sn$ | $7.8 \times 10^{-3}$ | Me—$SnMe_2Et$ | 0.0872 |
| $Pr^i{}_4Sn$ | $\sim 10^{-4}$ | Me—$SnMe_2(Pr^n)$ | 0.0635 |
| Me—$SnMe_3$ | 0.061 | Me—$SnMe_2(Bu^n)$ | 0.0635 |
| Et—$SnMe_3$ | $9.5 \times 10^{-3}$ | Me—$SnMe_2(Pr^i)$ | 0.058 |
| $Pr^n$—$SnMe_3$ | $1.66 \times 10^{-3}$ | Me—$SnMe_2(Bu^t)$ | ~0.005 |

TABLE 71. Fourth-order rate constants[375,376], $k_4$, at 20 °C for the reaction

$$Bu^n{}_4Sn + I_2 \longrightarrow Bu^nI + Bu^n{}_3SnI$$

| Solvent | $k_4$ ($l^3\,mol^{-3}\,s^{-1}$) | Solvent | $k_4$ ($l^3\,mol^{-3}\,s^{-1}$) |
|---|---|---|---|
| *n*-Butyl bromide | 0.625 | Toluene | 0.0333 |
| α-Methylnaphthalene | 0.333 | Ethylbenzene | 0.0208 |
| Carbon tetrachloride | 0.125 | *p*-Xylene | 0.0067 |
| Cyclohexane | 0.050 | *m*-Xylene | 0.0063 |
| Benzene | 0.0417 | Mesitylene | 0.0022 |

found by Jungers and coworkers[375,376] to follow a fourth-order rate law, first-order in $Bu_4Sn$ and third-order in iodine; values of the fourth-order rate constants are given in Table 71. More recently, Isaacs and Javaid[377] have studied the pressure dependence of the rate constant for iodinolysis of tetramethyltin in di-*n*-butyl ether at 29.1 °C. At 1 atm pressure, the second-order rate constant was $9.03\,l\,mol^{-1}\,s^{-1}$ and the volume of activation was calculated to be $-50\,cm^3$. Abraham and Grellier[206] have determined rate constants for iodinolysis of tetramethyl- and tetraethyl-tins in several alcohols in the presence of iodide ion; rate constants are given in Table 72.

Recently, a number of detailed kinetic and mechanistic investigations on the iodinolysis of simple tetraalkyltins have been reported. Abraham *et al.*[373] have reinvestigated the iododemetallation of symmetrical tetraalkyltins in methanol in the presence of iodide ion (see Table 68). Through a combination of product analyses and kinetic salt effect studies, it was shown that the two-step mechanism proposed by Mathieu[378] for the bromodemetallation of tetraalkyltins did not apply to these systems. It should be noted that the Mathieu mechanism, reactions 140 and 141, is indistinguishable on simple kinetic grounds from a single elementary bimolecular reaction between substrate and molecular halogen.

$$R_4Sn + X_2 \rightleftharpoons R_4SnX^+ + X^- \qquad (140)$$

$$X^- + RSnR_3X^+ \longrightarrow XR + R_3SnX \qquad (141)$$

Fukuzumi and Kochi[19] have determined rate constants for the iododemetallation of tetraalkyltins using a wide range of solvents, in the absence of light. The various reported second-order rate constants are collected in Table 73. The results were analysed by Fukuzumi and Kochi in terms of their charge-transfer mechanism,

TABLE 72. Second-order rate constants[206], $k_2$, at 25 °C for the iodinolysis of tetraalkyltins in the presence of iodide ion

| | $k_2$ ($l\,mol^{-1}\,s^{-1}$) | |
|---|---|---|
| Solvent | $Me_4Sn$ | $Et_4Sn$ |
| MeOH | 6.84 | 1.26 |
| EtOH | 1.325 | 0.265 |
| $Pr^nOH$ | 0.438 | 0.095 |
| $Pr^iOH$ | 0.476 | 0.170 |

TABLE 73. Second-order rate constants[22], $10^3k_2$, at 25 °C in various solvents for the iodinolysis of tetraalkyltins

| $R_4Sn$ | $10^3k_2$ (l mol$^{-1}$ s$^{-1}$) | | | | | | | | |
|---|---|---|---|---|---|---|---|---|---|
| | MeCN | $CH_2Cl_2$ | EtOH | $Me_2CO$ | PhCl | $CCl_4$[a] | 1-Cl-2-Br-benzene | dcp[b] | $CHCl_3$ |
| $Me_4Sn$ | 6300 | 5.8 | 1100 | 17 | 0.17 | 0.023 | — | — | — |
| $EtSnMe_3$ | 6300 | 7.9 | 340 | 16 | 0.93 | 0.15 | 46[c] | 14[c] | 35[c] |
| $Pr^nSnMe_3$ | 6600 | 7.8 | 210 | — | 0.68 | 0.14 | 4.2[c] | — | 32[c] |
| $Bu^nSnMe_3$ | 7200 | 8.9 | 320 | 22 | 0.68 | 0.14 | 4.5[c] | 10[c] | 32[c] |
| $Bu^iSnMe_3$ | — | 9.3[c] | — | — | — | — | — | — | — |
| $Et_2SnMe_2$ | 11 000 | 15 | 1300 | — | 0.84 | 0.42 | 8.7[c] | 15[c] | — |
| $Pr^n_2SnMe_2$ | 8900 | 13 | 950 | 26 | 0.74 | 0.38 | 5.3[c] | — | — |
| $Pr^i_2SnMe_2$ | 9200 | 22 | 630 | 17 | — | 1.1 | — | — | — |
| $Bu^n_2SnMe_2$ | 9300 | 14 | 990 | — | 0.72 | 0.32 | 6.1[c] | 9.8[c] | — |
| $Bu^i_2SnMe_2$ | — | 22[c] | — | — | — | — | — | — | — |
| $Bu^t_2SnMe_2$ | 830 | 6.2 | 1.7 | — | 0.19 | 0.40 | — | 1.7[c] | 0.45[c] |
| $Et_3SnMe$ | 10 000 | 20 | 1100 | 27 | 1.1 | 0.80 | 6.5[c] | 16[c] | — |
| $Et_4Sn$ | 1300 | 38 | 280 | — | 1.0 | 0.96 | — | — | — |
| $Pr^n_4Sn$ | 270 | 17 | 44 | 1.1 | 0.13 | 0.61 | — | — | — |
| $Pr^i_4Sn$ | 14 | 2.2 | 12 | — | 0.94 | 1.9 | — | — | — |
| $Bu^n_4Sn$ | 270 | 15 | 34 | 0.57 | — | 0.61 | — | — | — |
| $Bu^i_4Sn$ | 210 | 8.3 | 2.8 | 0.27 | — | 0.26 | — | — | — |
| $Bu^s_4Sn$ | 3.2 | 0.65 | 4.8 | — | — | 1.5 | — | — | — |

[a] At 50 °C.
[b] dcp = 1,2-dichloropropane.
[c] Ref. 19. Additional values of $10^3k_2$: $Et_2SnMe_2$ in $o$-$ClC_6H_4CH_3$, 1.0; $EtSnMe_3$ in $n$-$Bu_2O$, 0.043 (from ref. 19).

TABLE 74. Second-order rate constants[379], $k_2$, and activation parameters at 25 °C in dimethyl sulphoxide for the alternative reactions (A) and (B)

$$RSnMe_3 + I_2 \xrightarrow{I^-} RI + Me_3SnI \qquad (A)$$

$$RSnMe_3 + I_2 \xrightarrow{I^-} MeI + Me_2RSnI \qquad (B)$$

| R[a] | $k_2$ (l mol$^{-1}$ s$^{-1}$) | $\Delta H^{\neq}$ (kcal mol$^{-1}$) | $\Delta S^{\neq}$ (cal K$^{-1}$ mol$^{-1}$) |
|---|---|---|---|
| Me | 52.5 | 11.6 | −37 |
| Ph | 416 | 11.2 | −34 |
| $CH_2{=}CH$ | 628 | 10.5 | −36 |
| $CH_2{=}CHCH_2$ | $2.3\times10^7$ | — | — |
| $C_6F_5$ | 46.7 | 11.9 | −41 |
| $PhC{\equiv}C$ | $3.4\times10^6$ | 7.8 | −28 |
| $c$—$C_5H_5$ | Very fast | — | — |
| CN | Very fast | — | — |
| $C_4H_9$[b] | 48 | 12.9 | −34 |
| $PhCH_2$[b] | 153 | 11.6 | −35 |
| $CF_3$[b] | 1.0 | 10.2 | −50 |

[a] Cleavage of R—Sn bond, reaction (A), except where otherwise indicated.
[b] Cleavage of Me—Sn bond, reaction (B).

considered in detail in Section W. In the case of unsymmetrical tetraalkyltins, they carried out product analyses in order to be able to obtain the selectivity with respect to the alkyl group cleaved, that is for a compound of type $R_2SnMe_2$ or $RSnMe_3$, the ratio of rate constants for cleavage of the group R compared with the group Me. The observed selectivities varied widely with solvent, even showing reversals in some cases.

Reutov and coworkers have published several papers on the iodinolysis of mixed tetraorganotins. They investigated the kinetics of iodinolysis of trimethyltin compounds, $RSnMe_3$, in dimethyl sulphoxide in the presence of iodide ion; rate constants and activation parameters are collected in Table 74. Product analyses were carried out to determine the alkyl—tin bond that was cleaved; details are also given in Table 74[379]. Similar studies were carried out with methanol solvent; again reactions were second order overall, and details are given in Table 75[380]. Different results were found in carbon tetrachloride[381]: firstly for a similar range of $RSnMe_3$ compounds, the R—Sn bond was invariably broken, and secondly the kinetic form was variable. For $RSnMe_3$ compounds containing strongly electron-withdrawing groups, the kinetic form was first order in each reactant, but for other groups it became third order overall and second order in iodine. The latter behaviour was interpreted as being due to the formation of an initial complex, $RSnMe_3{\cdot}I_2$, which then reacts with a second molecule of iodine in the kinetic-determining stage. Results are given in Table 76.

By suitable choice of alkyl group and solvent, Reutov and coworkers were able to observe $S_E1$ processes in the iodinolysis of tetraorganotins. Thus in dimethyl sulphoxide, (CN)$SnMe_3$ undergoes iodinolysis at a rate independent of iodine concentration, with a first-order rate constant of 0.33 $s^{-1}$ at 20 °C[349]. An ion-pair type of $S_N1$ mechanism was suggested:

$$(CN)SnMe_3 \xrightleftharpoons{\text{slow}} CN^-SnMe_3^+ \tag{142}$$

$$CN^-SnMe_3^+ + I_2 \xrightarrow{I^-} ICN + Me_3SnI \tag{143}$$

TABLE 75. Second-order rate constants[380], $k_2$, and activation parameters at 25 °C in methanol for the alternative reactions (A) and (B)

$$RSnMe_3 + I_2 \xrightarrow{I^-} RI + Me_3SnI \quad \text{(A)}$$

$$RSnMe_3 + I_2 \xrightarrow{I^-} MeI + Me_2RSnI \quad \text{(B)}$$

| R[a] | $k_2$ (l mol$^{-1}$ s$^{-1}$) | $\Delta H^{\neq}$(kcal mol$^{-1}$) | $\Delta S^{\neq}$ (cal K$^{-1}$ mol$^{-1}$) |
|---|---|---|---|
| Me | 7.47 | 5.1 | −37 |
| Ph | 425 | 2.7 | −46 |
| $CH_2$=CH | 214 | 2.8 | −40 |
| $CH_2$=$CHCH_2$ | $1.8\times10^{7}$[b] | — | — |
| PhC≡C | $2.5\times10^{6}$ | — | — |
| $C_6F_5$ | 8.0 | — | — |
| CN | Very fast | — | — |
| $C_4H_9$[c] | 7.68 | 1.8 | −48 |
| $PhCH_2$[d] | 24.6 | 2.7 | −43 |

[a] Cleavage of R—Sn bond, reaction (A), except where otherwise indicated.
[b] Rate at 20 °C.
[c] Cleavage of Me—Sn bond, reaction (B).
[d] Cleavage of Me—Sn bond, reaction (B), but see ref. 384 and Table 79.

TABLE 76. Rate constants, $k_2$ and $k_3$ at 20 °C and some activation parameters for the reaction in carbon tetrachloride[381]

$$RSnMe_3 + I_2 \longrightarrow RI + Me_3SnI$$

| R | $k_2$ ($l\,mol^{-1}\,s^{-1}$) | $k_3$ ($l^2\,mol^{-2}\,s^{-1}$) |
|---|---|---|
| Me | — | Very slow |
| $C_6F_5$ | — | 15.0 |
| $PhCH_2$ | — | 27.8 |
| $CH_2{=}CH$[a] | — | 31.2 |
| Ph[b] | — | 232 |
| $PhC{\equiv}C$ | — | 1770 |
| $CH_2{=}CHCH_2$ | — | $1.92 \times 10^6$ |
| $C_9H_7$ | 160 | — |
| $c$-$C_5H_5$ | 500 | — |
| CN | 2300 | — |

[a] $\Delta H^{\neq} = 2.7$ kcal mol$^{-1}$; $\Delta S^{\neq} = -44$ cal K$^{-1}$ mol$^{-1}$.
[b] $\Delta H^{\neq} = 2.2$ kcal mol$^{-1}$; $\Delta S^{\neq} = -42$ cal K$^{-1}$ mol$^{-1}$.

An $S_E1$ process was also observed for the iodinolysis of $(CN)SnMe_3$ and $CH_3COCH_2SnEt_3$ in 1:1 dimethyl sulphoxide–carbon tetrachloride. In the case of cyclopentadienyltintrimethyl, the first-order rate constant varied with iodide ion concentration and the mechanism was thought to be intermediate between $S_E1$ and $S_E2$, whilst the indenyl compound $C_9H_7SnMe_3$ followed the usual second-order rate law for an $S_E2$ process[382] (see Table 77).

Allyltin compounds have been found to react extremely rapidly with iodine, due no doubt to the possibility of substitution with rearrangement. Gielen and Nasielski[276] showed that in the iodinolysis of tetraallyltin by $I_2/I^-$ in acetone the second-order rate constant could be dissected into a term corresponding to $k_2^a$ (attack by molecular iodine) and one corresponding to $(k_2^b + k_3/K)$, that is, direct attack by $I_3^-$ or nucleophilic assistance of $I^-$. Gielen and Nasielski[276] showed also that tetraallyltin was about $4 \times 10^7$ times as reactive as tetraethyltin towards iodine and proposed an $S_E2'$ mechanism (reaction 144).

$$R_3Sn{-}CH_2{-}CH{=}CH_2\;\; I{-}I \longrightarrow R_3Sn^+ + CH_2{=}CH{-}CH_2I + I^- \qquad (144)$$

TABLE 77. Rate constants[382], $k_1$, at 20 °C in 1:1 dimethyl sulphoxide–carbon tetrachloride for the reaction

$$RSnR'_3 + I_2 \xrightarrow{I^-} RI + R'_3SnI$$

| $RSnR'_3$ | $k_1$ ($s^{-1}$) |
|---|---|
| $(CN)SnMe_3$ | 0.20 |
| $CH_3COCH_2SnEt_3$ | 0.25 |
| $(c$-$C_5H_5)SnMe_3$ | 0.8–4.0[a] |
| $C_9H_7SnMe_3$ | 24 000[b] |

[a] The value of $k_1$ varied with $[I^-]$.
[b] This is the value of $k_2^{exp}/K$, where $K$ is the equilibrium constant for the formation of $I_3^-$ (unknown in this solvent).

TABLE 78. Second-order rate constants[383], $k_2$ at 25 °C for the reaction

$$Ph_3SnR + I_2 \xrightarrow{I^-} RI + Ph_3SnI$$

| R | Solvent | $k_2$ ($l\,mol^{-1}\,s^{-1}$) |
|---|---|---|
| $CH_2CH{=}CH_2$ | MeCN | $1.6 \times 10^6$ |
| $CH_2CH{=}CHPh$ | MeCN | $9.5 \times 10^3$ |
| $CH_2CH{=}CH_2$ | MeOH | $9.6 \times 10^5$ |
| $CH_2CH{=}CHPh$ | MeOH | $4.4 \times 10^4$ |
| $CH_2CH{=}CHMe$ | MeOH | $3.5 \times 10^5$ |
| $CH_2CH{=}CH_2$ | dmso | $4.5 \times 10^5$ |
| $CH_2CH{=}CHPh$ | dmso | $5.7 \times 10^4$ |
| $CH_2CH{=}CHMe$ | dmso | $3.6 \times 10^5$ |

Roberts[383] has also studied the iodinolysis of allyltin compounds in a variety of solvents and has dissected the observed rate constants into contributions from $k_2^a$ and $(k_2^b + k_3/K)$. The various second-order rate constants are given in Table 78, and show again the very high reactivity of allyl—tin bonds. The iodinolysis of allyltintrimethyl in dimethyl sulphoxide has been studied by Reutov and coworkers[349], who reported a second-order rate constant of $2.3 \times 10^7\,l\,mol^{-1}\,s^{-1}$. After correction for the statistical effect, this corresponds to a reactivity about $2 \times 10^6$ times that of a methyl—tin bond (see Table 74), comparable to the enhanced reactivity reported by Gielen and Nasielski[276] for the allyl—tin bond.

The benzyl—tin bond is broken by iodine also rather rapidly, although by no means as rapidly as is the allyl—tin bond. Abraham and Andonian-Haftvan[384] showed that in the iodinolysis of benzyltrialkyltins by $I_2/I^-$ in methanol, only the benzyl—tin bond was broken (reaction 145), when R = Et, $Pr^n$ and $Bu^n$, but that both the benzyl—tin and methyl—tin bonds were broken (reactions 145 and 146), when R = Me.

$$PhCH_2SnR_3 + I_2 \xrightarrow{I^-/MeOH} PhCH_2I + R_3SnI \qquad (145)$$

$$PhCH_2SnR_3 + I_2 \xrightarrow{I^-/MeOH} RI + PhCH_2SnR_2I \qquad (146)$$

Kinetic results are given in Table 79, and by comparison with previous studies show that the benzyl group is intermediate in reactivity between phenyl and methyl.

Vinyl—tin bonds are also broken very easily by the electrophilic iodine molecule, as shown by Nasielski and coworkers[385]. Values of the second-order rate constants obtained for a number of compounds in methanol and chlorobenzene are given in Table 80 and show that in methanol at 20 °C the vinyl group from vinyltintrimethyl is cleaved about 170 times as rapidly as the methyl group in tetramethyltin. Compare the results of Reutov and coworkers[380] (Table 75), where the factor may be calculated as 115, again allowing for the statistical effect in tetramethyltin.

The results of Reutov and coworkers[379,380] show also that the phenyl—tin bond is

TABLE 79. Second-order rate constants[384], $k_2$ at 25 °C in methanol for the iododemetallation of $PhCH_2SnR_3$ in the presence of iodide ion

| R | $k_2$ ($l\,mol^{-1}\,s^{-1}$) |
|---|---|
| $Bu^n$ | 6.74[a] |
| $Pr^n$ | 5.90[a] |
| Et | 8.86[a] |
| Me | 10.57[b] |

[a] Only the benzyl group is cleaved in these cases.
[b] Overall total rate; the rate constant for cleavage of the benzyl group is 7.61 $l\,mol^{-1}\,s^{-1}$.

rapidly broken in the iodinolysis reaction (Tables 74 and 75). In methanol the factor Ph—$SnMe_3$/Me—$SnMe_3$ is 230, but falls to 32 in dimethyl sulphoxide.

As well as studies that yield the reactivity of the aryl—tin bond compared with the alkyl—tin bond, there have been numerous investigations on the mechanism of substitution of aryl—tin bonds by iodine. Eaborn and coworkers[386] studied the iodinolysis of a number of aryltin compounds (reaction 147) in carbon tetrachloride,

$$YC_6H_4SnR_3 + I_2 \longrightarrow YC_6H_4I + R_3SnI \quad (147)$$

and observed third-order kinetics, first order in the substrate and second order in iodine. The third-order rate constants are given in Table 81. It was suggested[386] that the transition state resembled a $\pi$-complex, which would account for the small spread of rates with variation of the substituent Y. On the other hand, reaction 147 (R = Me), which was studied by Nasielski and coworkers[387,388] using methanol as solvent in the presence of iodide ion, was thought not to involve $\pi$-complex formation. In methanol, however, the kinetics were second-order overall, first order in substrate and iodine; second-order rate constants are given in Table 82. Further studies were carried out with a series of substrates $ArSnR_3$[389,390]; the rate constants

TABLE 80. Second-order rate constants[385], $k_2$, at 20 °C for the iododemetallation of vinyltrialkyltins in the presence of iodide ion

| Compound | $k_2{}^a$ ($l\,mol^{-1}\,s^{-1}$) | $k_2{}^b$ ($l\,mol^{-1}\,s^{-1}$) |
|---|---|---|
| $H_2C{=}C(Et)SnMe_3$ | 155 | |
| $H_2C{=}CHSnMe_3$ | 306 | 2.8 |
| *cis*-$MeCH{=}C(Me)SnMe_3$ | 14 000 | |
| *trans*-$MeCH{=}C(Me)SnMe_3$ | 53 000 | |
| $Me_2C{=}CHSnMe_3$ | $\sim 10^7$ | |
| $CH_2{=}CHSnEt_3$ | 334 | 23 |
| $CH_2{=}CHSn(Pr^n)_3$ | 164 | 20 |
| $CH_2{=}CHSn(Pr^i)_3$ | 49 | 29 |

[a] In methanol.
[b] In chlorobenzene.

TABLE 81. Third-order rate constants[386], $10^{-2}k_3$, in carbon tetrachloride at 25 °C for the reaction[a]

$$XC_6H_4SnR_3 + I_2 \longrightarrow XC_6H_4I + R_3SnI$$

| X | $10^{-2}k_3$ ($l^2\ mol^{-2}\ s^{-1}$) | X | $10^{-2}k_3$ ($l^2\ mol^{-2}\ s^{-1}$) |
|---|---|---|---|
| 4-MeO | 1145, 1.19[b] | 3-MeO | 37.2 |
| 4-$(C_6H_{11})_3$Sn | 331 | H | 16.6, 0.057[b], 3.1[c], 16[d] |
| 4-Bu$^t$ | 232 | 2-Ph | 5.75 |
| 4-Pr$^i$ | 201 | 4-F | 3.71 |
| 4-Et | 167 | 4-Cl | 1.67, 0.209[c] |
| 4-Me | 125, 23.4[c] | 4-Br | 1.325 |
| 3-Me | 70 | 3-Cl | 0.642, 0.0516[c] |
| 4-Ph | 47.7 | 4-HOOC | 0.242 |

[a] R = cyclohexyl except where otherwise indicated.
[b] R = Ph.
[c] R = Me.
[d] R = Et.

TABLE 82. Second-order rate constants[387,388], $k_2$, at 25 °C and activation parameters for the reaction in methanol

$$ArSnMe_3 + I_2 \longrightarrow ArI + Me_3SnI$$

| Ar | $k_2$ ($l\ mol^{-1}\ s^{-1}$) | $E_a$ (kcal $mol^{-1}$) | Log A |
|---|---|---|---|
| 4-$MeOC_6H_4$ | 32 400 | 3.9 | 7.4 |
| 4-$MeC_6H_4$ | 2510 | 5.0 | 7.0 |
| Naphth-1-yl | 1100 | 6.3 | 7.7 |
| Naphth-2-yl | 1020 | 5.5 | 7.0 |
| 4-$Me_3SnC_6H_4$ | 870[a] | — | — |
| 3-$MeC_6H_4$ | 759 | 5.5 | 6.9 |
| Phenanthr-9-yl | 740 | 6.0 | 7.2 |
| 4-$Me_3SiC_6H_4$ | 513 | — | — |
| $C_6H_5$ | 507 | 6.0 | 7.1 |
| 4-$BrC_6H_4$ | 121 | 6.7 | 7.0 |

[a] Statistically corrected.

TABLE 83. Second-order constants[389], $k_2$, at 25 °C and activation parameters in methanol for the reaction

$$ArSnR_3 + I_2 \longrightarrow ArI + R_3SnI$$

| Ar | R | $k_2$ ($l\ mol^{-1}\ s^{-1}$) | $E_a$ (kcal $mol^{-1}$) | Log A |
|---|---|---|---|---|
| Phenyl | Me | 437 | 5.7 | 6.8 |
| | Bu$^n$ | 107 | 5.9 | 6.3 |
| | Pr$^i$ | 15.9 | 6.8 | 6.2 |
| Naphth-1-yl | Me | 1100 | 6.3 | 7.7 |
| | Bu$^n$ | 398 | 7.5 | 8.1 |
| | Pr$^i$ | 64.6 | 4.6 | 5.2 |
| Naphth-2-yl | Me | 1020 | 5.5 | 7.0 |
| | Bu$^n$ | 257 | 5.4 | 6.4 |
| | Pr$^i$ | 28.8 | 10.3 | 8.6 |
| Pyren-3-yl[a] | Bu$^n$ | 1320 | | |

[a] Ref. 390.

obtained show that as the leaving group $SnR_3$ increases in size, so the rate is decreased (see Table 83). Combination of the rate constant for iododemetallation of $PhSnMe_3$ at 25 °C, 437 l mol$^{-1}$ s$^{-1}$ (Table 82), with that for iododemetallation of $Me_4Sn$ at 25 °C, 6.84 l mol$^{-1}$ s$^{-1}$ (Table 72), yields another value for the Ph—$SnMe_3$/Me—$SnMe_3$ ratio of 256 in methanol.

## T. Halogenolysis of Organotin(IV) Compounds by Interhalogens

McLean and coworkers[391] have investigated the action of ICl and IBr on tetra-*n*-butyl- and tetraphenyl-tin in carbon tetrachloride with a view to establishing a good preparative route to compounds of type $R_3SnX$ and $R_2SnX_2$. In all cases, the more negative halogen atom in ICl or IBr became attached to the tin atom in the organotin halide. Both interhalogens yielded $R_3SnX$ quickly, followed by a slower second step giving $R_2SnX_2$. With IBr, the second step was very slow and the action of IBr on $Bu_4Sn$ to yield $Bu_3SnBr$ was considered to be a successful preparative route.

The action of IBr on tetramethyl- and tetra-*n*-propyl-tin in carbon tetrachloride and cyclohexane yields alkyl iodide and trialkyltin bromide[392]. Complex kinetics were observed, with both a second-order and a third-order term in the total rate equation 148; values of the rate constants obtained are given in Table 84.

$$v = k_2[R_4Sn][IBr] + k_3[R_4Sn][IBr]^2 \tag{148}$$

TABLE 84. Values[392] of $k_2$ and $k_3$ at 25 °C for the reaction

$$R_4Sn + IBr \longrightarrow RI + R_3SnBr$$

| $R_4Sn$ | Solvent | $k_2$ (l mol$^{-1}$ s$^{-1}$) | $k_3$ (l$^2$ mol$^{-2}$ s$^{-1}$) |
|---|---|---|---|
| $Me_4Sn$ | Cyclohexane | $1.1\times10^{-3}$ | 0.25 |
| $Pr_4Sn$ | Cyclohexane | No reaction | |
| $Me_4Sn$ | $CCl_4$ | $1.7\times10^{-3}$ | 2.2 |
| $Pr_4Sn$ | $CCl_4$ | $<5\times10^{-4}$ | 10.1 |

## U. Iodinolysis of Organolead(IV) Compounds

Riccoboni and coworkers[393–395] have studied the kinetics of iodinolysis of tetraalkylleads using various solvents in the presence of iodide ion (reaction 149). Through experiments in which the concentration of iodide ion was varied, they were able to establish that reaction 149 in the solvents studied is a bimolecular reaction between

$$R_4Pb + I_2 \longrightarrow RI + R_3PbI \tag{149}$$

tetraalkyllead and iodine (see Section III.A). Rate constants are given in Table 85. Gielen and Nasielski[374] also investigated reaction 149 using an excess of iodide ion and their determined rate constants are also given in Table 85. For the iodinolysis of tetramethyllead in acetonitrile, there is some dispute[374,396] as to whether or not there is a term in $(k_2^b + k_3/K)$ (see Section III.A). Abraham has discussed the experimental results and suggests that there is a small contributing term, attributed[374] to nucleophilic catalysis by $I^-$. Pilloni and Tagliavini[396] extended kinetic studies to acetone and their results are also given in Table 85.

Reaction 149 has also been studied in benzene and carbon tetrachloride[396] in the absence of added iodide ion, and the rate constants for iodinolysis of tetraalkylleads

TABLE 85. Second-order rate constants, $k_2$, and relative second-order rate constants, $k_{rel}$, for the reaction

$$R_4Pb + I_2 \xrightarrow{I^-} RI + R_3PbI$$

| | | $k_2$ ($l\,mol^{-1}\,s^{-1}$) | | | $k_{rel}$ | | | |
|---|---|---|---|---|---|---|---|---|
| Solvent | $T$ (°C) | $Me_4Pb$ | $Et_4Pb$ | $Pr^n{}_4Pb$ | $Me_4Pb$ | $Et_4Pb$ | $Pr^n{}_4Pb$ | Ref. |
| MeCN | 25 | $2.41\times10^5$ | | | 100 | 38.6 | 8.3 | 395 |
| MeOH | 25 | $1.03\times10^4$ | | | 100 | 38.9 | 13.4 | 395 |
| EtOH | 25 | $2.80\times10^3$ | | | 100 | 47.5 | 7.2 | 395 |
| $Pr^nOH$ | 25 | $1.61\times10^3$ | | | 100 | 45.3 | 6.2 | 395 |
| $Me_2CO$ | 35 | $4.4\times10^4$ | | | 100 | 37.6 | | 396 |
| $C_6H_6$[a] | 35 | 1.9 | 22.8 | 7.1 | | | | 396 |
| $CCl_4$[a] | 31 | 0.0355 | 1.08 | 0.43 | | | | 396 |
| MeCN | 20 | $2.7\times10^5$ | | | 100 | 55.5 | | 374 |
| MeOH | 20 | $1.04\times10^4$ | | | 100 | 35.3 | | 374 |
| dmso | 20 | $9.5\times10^3$ | | | 100 | 23.6 | | 374 |
| MeCOOH | 20 | $2\times10^4$ | | | 100 | 56.6 | | 374 |

[a] In absence of added iodide ion.

TABLE 86. Second-order rate constants[19], $k_2$ at 25 °C in carbon tetrachloride for the iodinolysis of tetraalkylleads

| $R_4Pb$ | $k_2$ ($l\,mol^{-1}\,s^{-1}$) |
|---|---|
| $EtPbMe_3$ | 0.10 |
| $Et_2PbMe_2$ | 0.33 |
| $Et_3PbMe$ | 0.76 |
| $Et_4Pb$ | 0.89 |

are given in Table 85. More recently, Fukuzumi and Kochi[19] have reported second-order rate constants for iodinolysis of some mixed methylethyllead compounds in carbon tetrachloride; results are given in Table 86.

As usual for electrophilic substitutions, the phenyl group is cleaved more rapidly than are simple alkyl groups. Nasielski and coworkers[397] report that the second-order rate constant for iodinolysis of phenyltrimethyllead by iodine in methanol at 23 °C is 20 900 $l\,mol^{-1}\,s^{-1}$, so that after taking into account the statistical factor, the Ph—$PbMe_3$ bond is about 8 times as reactive towards iodine in methanol as is the Me—$PbMe_3$ bond. Another value for the Ph—$PbMe_3$/Me—$PbMe_3$ reactivity in methanol is 26, again due to Nasielski's group[374].

## V. Fluorinolysis of Organotin(IV) Compounds

A recent interesting report indicates that fluorinolysis of $Ph_4Sn$ or $PhSnBu_3$ is a preparatively useful route to fluorobenzene. Thus with a deficiency of fluorine in $CFCl_3$ at −78 °C, yields of 15% and 70% fluorobenzene, respectively, were obtained, and with fluorine in $CCl_4$ at 0 °C yields of 56% and 48% were reported[398].

### W. The Fukuzumi–Kochi Analysis of Halogenodemetallations

In the electron-transfer, or charge-transfer, mechanism of electrophilic substitution (see section I.B.5), the organometallic substrate and the electrophile first form a charge-transfer complex or excited ion pair, in which an electron is transferred from the substrate to the electrophile. Because of the Franck–Condon principle, the excited ion pair is not in equilibrium with the solvent, and it may then be supposed that the excited ion pair becomes solvated to yield a 'thermal' ion-pair, that is, an ion pair that is in thermal equilibrium with the solvent. Fukuzumi and Kochi[19,20,22,351] then argue that if such a thermal ion pair is a good model for the transition state in an electrophilic substitution, the calculation of the effect of change of solvent, or of substrate or of electrophile in terms of the thermal ion pair will provide an estimate of such a change on the electrophilic substitution transition state. In Section II.F.14 it was shown that solvent effects on the $R_4Sn/HgCl_2$ reaction could thus be analysed, see equation 150. An analogous equation may be written for the halogenolysis reaction (equation 151); in both of these equations, the $Me_4Sn$ compound is taken as

$$(R_4SnHgCl_2) \xrightarrow{\Delta h\nu_{CT}} (R_4Sn^+HgCl_2^-)^* \xrightarrow{\Delta G_r^s} (R_4Sn^+HgCl_2^-) \qquad (150)$$

$$(R_4SnX_2) \xrightarrow{\Delta h\nu_{CT}} (R_4Sn^+X_2^-)^* \xrightarrow{\Delta G_r^s} (R_4Sn^+X_2^-) \qquad (151)$$

a reference. Fukuzumi and Kochi determined $\Delta h\nu_{CT}$ values for the charge-transfer complexes between tetraalkyltins[19,20,351] and the electrophiles $HgCl_2$, $I_2$, and $Br_2$. Then, in order to obtain relative energies for transformation of $(R_4SnX_2)$ to the thermal ion pair $(R_4Sn^+X_2^-)$, it is necessary to obtain an estimate of the solvation energy, $\Delta G_r^s$. In this way, $\delta\,\Delta G_r^{\neq}$ may be calculated (equation 152) and compared with the observed value (equation 153).

$$\Delta G_r^{\neq}\ (\text{calc}) = \Delta G_r^s + \Delta h\nu_{CT} \qquad (152)$$

$$\Delta G_r^{\neq}\ (\text{obs}) = -RT \ln (k^{R_4Sn}/k^{Me_4Sn}) \qquad (153)$$

There are a number of ways of comparing observed and calculated values; Fukuzumi and Kochi, for example, obtained 'experimental' values of $\Delta G_r^s$ through equation 154 and then compared the $\Delta G_r^s$ values with those calculated using the

$$\Delta G_r^s = \Delta G_r^{\neq}\ (\text{obs}) - \Delta h\nu_{CT} \qquad (154)$$

Born equation[22]. Another approach is to regard $\Delta h\nu_{CT}$ as being made up of an electronic term, given by the ionization potential of the tetraalkyltin, $\Delta I_D$, and a steric term, $\Delta E$, so that $\Delta h\nu_{CT} = \Delta I_D + \Delta E$; again, all quantities refer to values for the tetramethyltin process as a reference standard. Then, knowing $\Delta h\nu_{CT}$ and $\Delta I_D$, the steric term $\Delta E$ can be evaluated[22,351].

We use equations 152 and 153 to compare calculated and observed $\Delta G_r^{\neq}$ values, and in Table 87 are collected results for iodinolysis in several solvents. As pointed out before, there is a pronounced inversion of reactivity with solvent. In the polar solvents the reactivity sequence is $Me_4Sn > Et_4Sn > Pr^n_4Sn \geqslant Bu^n_4Sn \geqslant Bu^i_4Sn > Pr^i_4Sn$, but in the less polar solvents sequences such as the following (in tetrachloromethane) are observed, $Me_4Sn < Et_4Sn > Pr^n_4Sn \approx Bu^n_4Sn > Bu^i_4Sn < Pr^i_4Sn$. Gielen and Nasielski[6] have suggested that these various sequences arise because the transition state in polar solvents and in nonpolar solvents does not have the same structure or charge separation, the mechanism changing from $S_E2$ (open) in polar solvents to $S_E2$ (cyclic) in non-polar solvents. Fukuzumi and Kochi[22], however, attribute the

TABLE 87. Comparison of observed values of $\Delta G_r^{\neq}$ with those calculated by the methods of Fukuzumi and Kochi for the $R_4Sn/I_2$ reaction[a]

| | | MeOH | | | MeCN | | | EtOH | | | MeCOMe | | |
|---|---|---|---|---|---|---|---|---|---|---|---|---|---|
| Compound | $\Delta h\nu_{CT}$ | Obs | Calc | Calc | Obs | Calc | Calc | Obs | Calc | Calc | Obs | Calc | Calc |
| $Me_4Sn$ | 0.0 | 0.0 | 0.0 | 0.0 | 0.0 | 0.0 | 0.0 | 0.0 | 0.0 | 0.0 | 0.0 | 0.0 | 0.0 |
| $Et_4Sn$ | −7.4 | 1.2 | 0.0 | 0.3 | 0.9 | 0.0 | 0.3 | 0.8 | −0.1 | 0.2 | | −0.2 | 0.2 |
| $Pr^n_4Sn$ | −7.4 | 2.5 | 0.7 | 1.6 | 1.9 | 0.7 | 1.6 | 1.9 | 0.6 | 1.5 | 1.6 | 0.5 | 1.4 |
| $Bu^n_4Sn$ | −6.7 | 3.1 | 1.9 | 3.6 | 1.9 | 1.9 | 3.6 | 2.1 | 1.8 | 3.5 | 2.0 | 1.7 | 3.4 |
| $Bu^i_4Sn$ | −8.5 | 2.7 | 0.7 | 1.8 | 2.0 | 0.7 | 1.8 | 3.5 | 0.6 | 1.7 | 2.5 | 0.5 | 1.6 |
| $Pr^i_4Sn$ | −9.0 | 4.4 | 2.3 | 4.4 | 3.6 | 2.3 | 4.4 | 2.7 | 2.2 | 4.2 | | 2.1 | 4.1 |
| $Bu^s_4Sn$ | −10.4 | | 1.1 | 2.2 | 4.5 | 1.1 | 2.2 | 3.2 | 0.9 | 2.0 | | 0.8 | 1.9 |
| $Bu^nSnMe_3$ | −2.3 | 0.1 | 3.1 | | −0.1 | 3.1 | | 0.7 | 3.1 | | −0.2 | 3.0 | |
| $Bu^n_2SnMe_2$ | −5.5 | | 2.7 | | −0.2 | 2.7 | | 0.1 | 2.5 | | | 2.5 | |
| $Pr^i_2SnMe_2$ | −7.4 | | 3.0 | | −0.2 | 3.0 | | 0.3 | 2.8 | | | 2.8 | |
| $Bu^t_2SnMe_2$ | −11.3 | | 2.3 | | 1.2 | 2.3 | | 3.8 | 2.2 | | | 2.0 | |

| | | MeCOOH | | | PhCl | | | $CH_2Cl_2$ | | | $CCl_4$ | | |
|---|---|---|---|---|---|---|---|---|---|---|---|---|---|
| | | Obs | Calc | Calc | Obs | Calc | Calc | Obs | Calc | Calc | Obs | Calc | Calc |
| $Me_4Sn$ | 0.0 | 0.0 | 0.0 | 0.0 | 0.0 | 0.0 | 0.0 | 0.0 | 0.0 | 0.0 | 0.0 | 0.0 | 0.0 |
| $Et_4Sn$ | −7.4 | 0.7 | −1.0 | −0.7 | −1.0 | −1.2 | −0.9 | −1.1 | −0.7 | −0.3 | −2.2 | −3.2 | −3.0 |
| $Pr^n_4Sn$ | −7.4 | 1.9 | −0.4 | 0.4 | 0.2 | −0.6 | 0.2 | −0.6 | 0.0 | 0.8 | −1.9 | −2.7 | −2.3 |
| $Bu^n_4Sn$ | −6.7 | 2.0 | 0.7 | 2.2 | 0.3 | 0.6 | 2.0 | −0.6 | 1.2 | 2.7 | −1.9 | −1.8 | −0.8 |
| $Bu^i_4Sn$ | −8.5 | | −0.5 | | | −0.7 | 0.2 | −0.2 | 0.1 | | −1.4 | −3.2 | −2.6 |
| $Pr^i_4Sn$ | −9.0 | 4.7 | 0.8 | 2.6 | −1.0 | 0.6 | 2.3 | 0.6 | 1.4 | 3.3 | −2.6 | −2.6 | −1.4 |
| $Bu^s_4Sn$ | −10.4 | | | | | −0.7 | 0.3 | 1.3 | 0.1 | 1.1 | −2.5 | −3.8 | −3.2 |
| $Bu^nSnMe_3$ | −2.3 | | | | −0.8 | 2.3 | | −0.3 | 2.7 | | −1.1 | 0.8 | |
| $Bu^n_2SnMe_2$ | −5.5 | | | | −0.9 | 1.4 | | −0.5 | 2.0 | | −1.6 | −0.9 | |
| $Pr^i_2SnMe_2$ | −7.4 | | | | | 1.4 | | −0.8 | 2.0 | | −2.3 | −1.5 | |
| $Bu^t_2SnMe_2$ | −11.3 | | | | −0.1 | 0.2 | | 0.0 | 1.1 | | −1.7 | −3.5 | |

[a] All values in kcal $mol^{-1}$ at 25 °C. Observed quantities found through equation 153 using rate constants in Tables 68, 69, 70 and 73. Calculated quantities found through equation 152; for each solvent the values in the first Calc column were obtained by use of $\Delta G_r^s$ values in acetonitrile[22] and correcting these to other solvents by a Born-type expression[20], and the values in the second Calc column were obtained directly through the modified Born equation of Fukuzumi and Kochi[22].

change in sequence to the variation of $\Delta G_r^s$ with solvent, the transition state throughout being regarded as close to the model ion pair. The reason for the alteration in sequence for iodinolyses, but not for reaction with $HgCl_2$, is due to the much more negative values of $\Delta h\nu_{CT}$ in the former case; compare the results in Tables 39 and 87. With, for example, a very negative $\Delta h\nu_{CT}$ value of −9.0 kcal $mol^{-1}$ for the ($Pr^i_4SnI_2$) charge-transfer complex, relative to ($Me_4SnI_2$), the positive $\Delta G_r^s$ value is large enough to yield a positive $\Delta G_r^{\neq}$ value in polar solvents, but in non-polar solvents $\Delta G_r^s$ is less positive and so ($\Delta h\nu_{CT}+\Delta G_r^s$) becomes negative. However, for the complex ($Pr^i_4SnHgCl_2$) as against ($Me_4SnHgCl_2$), $\Delta h\nu_{CT}$ is −3.0 kcal $mol^{-1}$, so that even in non-polar solvents the term ($\Delta h\nu_{CT}+\Delta G_r^s$) is still positive. Two methods of obtaining $\Delta G_r^s$ values have been advanced[20,22]. The first method was described in Section II.F.14 and applied to the ($R_4SnHgCl_2$) reactions, and exactly the same procedure was used to deduce $\Delta G_r^s$ and then the $\Delta G_r^{\neq}$ (calc) values given in Table 87 (first Calc column under each solvent) through equation 152; in this first method[20] $\Delta G_r^s$ values for $R_4Sn^+$ in acetonitrile were obtained from experiments on outer-sphere oxidation in acetonitrile, and corresponding $\Delta G_r^s$ values in other solvents then calculated from those in acetonitrile by the Born equation. The second method[22] simply used a modified Born equation to calculate $\Delta G_r^s$ using the solvent dielectric constant as the only solvent parameter. In Table 87 (second Calc column under each solvent) are listed the $\Delta G_r^{\neq}$ (calc) values obtained in this way.

As for the ($R_4SnHgCl_2$) processes, the general trends in $\Delta G_r^{\neq}$ are reproduced for the ($R_4SnI_2$) systems using either of the methods of Fukuzumi and Kochi; in particular, the sequence inversion mentioned above can for the first time be accounted for in a unified approach[22]. Since both methods rely on a modified Born equation to obtain relative solvation energies of the $R_4Sn^+$ entities, an exact agreement would not be expected. Indeed, one of the main deficiencies can be seen from results in acetic acid ($\varepsilon = 6.195$) and chlorobenzene ($\varepsilon = 5.62$). Any calculation based on the Born equation will yield similar $\Delta G_r^s$ values in these two solvents, and hence similar $\Delta G_r^{\neq}$ (calc) values—thus the same sequence of reactivity should obtain

TABLE 88. Comparison of observed values of $\Delta G_r^{\neq}$ with those calculated by the methods of Fukuzumi and Kochi for the $R_4Sn/Br_2$ reaction[a]

| | | dmf | | | MeCOOH | | | PhCl | | |
|---|---|---|---|---|---|---|---|---|---|---|
| Compound | $\Delta h\nu_{CT}$ | Obs | Calc | Calc | Obs | Calc | Calc | Obs | Calc | Calc |
| $Me_4Sn$ | 0.0 | 0.0 | 0.0 | 0.0 | 0.0 | 0.0 | 0.0 | 0.0 | 0.0 | 0.0 |
| $Et_4Sn$ | −6.9 | 0.5 | 0.4 | 0.8 | 0.1 | −0.6 | −0.3 | −1.5 | −0.7 | −0.4 |
| $Pr^n{}_4Sn$ | −6.9 | 1.7 | 1.1 | 2.1 | 1.2 | 0.0 | 0.9 | −0.9 | −0.1 | 0.7 |
| $Bu^n{}_4Sn$ | −6.9 | | 1.7 | 3.4 | 1.3 | 0.5 | 2.0 | −0.9 | 0.3 | 1.8 |
| $Bu^i{}_4Sn$ | −7.4 | | 1.9 | 2.9 | | 0.6 | 1.5 | | 0.5 | 1.3 |
| $Pr^i{}_4Sn$ | −11.1 | | 0.3 | 2.3 | 2.2 | −1.3 | 0.4 | −0.7 | −1.5 | 0.2 |
| $Bu^s{}_4Sn$ | −11.8 | | −0.3 | 0.8 | | −1.9 | −0.9 | | −2.1 | −1.1 |
| $Bu^nSnMe_3$ | −1.6 | | | | | | | | | |
| $Bu^n{}_2SnMe_2$ | −5.1 | | | | | | | | | |
| $Pr^i{}_2SnMe_2$ | −8.5 | | | | | | | | | |
| $Bu^t{}_2SnMe_2$ | −9.5 | | | | | | | | | |

| | | $CCl_4$ | | | Hexane | | |
|---|---|---|---|---|---|---|---|
| Compound | $\Delta h\nu_{CT}$ | Obs | Calc | Calc | Obs | Calc | Calc |
| $Me_4Sn$ | 0.0 | 0.0 | 0.0 | 0.0 | 0.0 | 0.0 | 0.0 |
| $Et_4Sn$ | −6.9 | −2.9 | −2.7 | −2.5 | −2.8 | −3.3 | −3.1 |
| $Pr^n{}_4Sn$ | −6.9 | −2.4 | −2.3 | −1.8 | −2.4 | −3.0 | −2.5 |
| $Bu^n{}_4Sn$ | −6.9 | −2.5 | −2.0 | −1.0 | −2.6 | −2.7 | −1.9 |
| $Bu^i{}_4Sn$ | −7.4 | −1.5 | −2.1 | −1.5 | −1.8 | −2.8 | −2.4 |
| $Pr^i{}_4Sn$ | −11.1 | −4.3 | −4.7 | −3.5 | −5.1 | −5.5 | −4.5 |
| $Bu^s{}_4Sn$ | −11.8 | −3.5 | −5.2 | −4.6 | −4.0 | −6.2 | −5.6 |
| $Bu^nSnMe_3$ | −1.6 | −0.4 | 1.5 | | −0.6 | 1.1 | |
| $Bu^n{}_2SnMe_2$ | −5.1 | −1.3 | −0.4 | | −1.2 | −1.1 | |
| $Pr^i{}_2SnMe_2$ | −8.5 | −2.8 | −2.7 | | −3.2 | −3.5 | |
| $Bu^t{}_2SnMe_2$ | −9.5 | −1.0 | −1.7 | | −1.7 | −2.8 | |

[a] All values in kcal mol$^{-1}$ at 25 °C. Observed quantities found through equation 153 using rate constants in Tables 53, 54, 55 and 59. Calculated quantities found through equation 152; for each solvent the values in the first Calc column were obtained by use of $\Delta G_r^s$ values in acetonitrile[22] and correcting these to other solvents by a Born-type expression[20], and the values in the second Calc column were obtained directly through the modified Born equation of Fukuzumi and Kochi)[22].

in these solvents. However, the actual observed $\Delta G_r^{\neq}$ values in acetic acid are $Me_4Sn$ 0, $Et_4Sn$ 0.7, $Pr^n{}_4Sn$ 1.9, $Bu^n{}_4Sn$ 2.0 and $Pr^i{}_4Sn$ 4.7, with the corresponding reactivity sequence $Me_4Sn > Et_4Sn > Pr^n{}_4Sn \geq Bu^n{}_4Sn > Pr^i{}_4Sn$, whereas in chlorobenzene the $\Delta G_r^{\neq}$ values are $Me_4Sn$ 0, $Et_4Sn$ −1.0, $Pr^n{}_4Sn$ 0.2, $Bu^n{}_4Sn$ 0.3 and $Pr^i{}_4Sn$ −1.0, leading to an inverted reactivity sequence $Me_4Sn < Et_4Sn > Pr^n{}_4Sn \approx Bu^n{}_4Sn < Pr^i{}_4Sn$.

An identical procedure to the above may be used for the brominolysis of $R_4Sn$ compounds[22] and details are given in Table 88. Once again, the general trend of $\Delta G_r^{\neq}$ values is reproduced, although the failure of Born-type equations to deal with solvents with similar dielectric constants but dissimilar reactivity sequences (e.g. acetic acid and chlorobenzene) is again evident.

## IV. PROTODEMETALLATION OF ORGANOMETALLIC COMPOUNDS

### A. Protodemetallation of Organolithium Compounds

A stereochemical study[283] showed that the cleavage of the optically active compound (1-methyl-2,2-diphenylcyclopropyl)lithium **25** by methanol in diethyl ether or diethyl ether–dimethoxyethane took place with retention of optical activity and configuration.

### B. Protodemetallation of Organo-magnesium, -zinc and -cadmium Compounds

Emptoz and Huet[399] found that the rates of acidolysis of Group II organometallic compounds with hex-l-yne decreased along the series $Et_2Mg > EtMgX > Et_2Zn > Et_2Cd$. Also, for the cleavage by iso-amyl alcohol, diethylzinc reacted faster than did diethylcadmium.

The acidolysis of dialkylmagnesium compounds with hex-l-yne occurs by two competitive consecutive second-order reactions 155 and 156 in diethyl ether[400,401].

$$R_2Mg + BuC{\equiv}CH \longrightarrow RH + BuC{\equiv}CMgR \tag{155}$$

$$BuC{\equiv}CMgR + BuC{\equiv}CH \longrightarrow RH + (BuC{\equiv}C)_2Mg \tag{156}$$

For diethylmagnesium, a kinetic isotope ratio $k_H/k_D$ of 2.8 for reaction with $BuC{\equiv}CH$ and $BuC{\equiv}CD$ was reported[400]. Rate constants for reactions 155 and 156 are given in Table 89.

Wotiz and coworkers have studied the acidolysis of alkyl Grignard reagents by alk-l-ynes. The reaction between 1 M EtMgBr and 1 M $BuC{\equiv}CH$, to yield ethane, followed overall second-order kinetics and had a rate constant[400] at 35 °C of $1.7 \times 10^{-3}\,l\,mol^{-1}\,s^{-1}$ in diethyl ether. The kinetic isotope ratio $k_H/k_D$ for the reactions between EtMgBr and $BuC{\equiv}CH$ and $BuC{\equiv}CD$ in diethyl ether was 4.4[400]. Relative reactivities of various Grignard reagents RMgBr towards hex-l-yne in diethyl ether have been reported[402] as R = Me 6, $Pr^n$ 59, Et 100, and $Pr^i$ 210. An $S_E2$ (cyclic) type of transition state was proposed, such as **26**, but the actual

$$\left[\begin{matrix} -C{\equiv}C-H \\ -Mg-R \end{matrix}\right]^{\neq}$$

**(26)**

TABLE 89. Second-order rate constants[401], $10^3k_2$, at 33 °C for the reactions (A) and (B) in diethyl ether

$$R_2Mg + BuC{\equiv}CH \longrightarrow RH + BuC{\equiv}CMgR \quad (A)$$

$$BuC{\equiv}CMgR + BuC{\equiv}CH \longrightarrow RH + (BuC{\equiv}C)_2Mg \quad (B)$$

| | $10^3k_2$ ($l\ mol^{-1}\ s^{-1}$) | |
|---|---|---|
| R | (A) | (B) |
| Et[a] | 13 | 2.5 |
| Et | 22 | 3.0 |
| $Pr^n$ | 13 | 2.0 |
| $Pr^i$ | 40 | 8.0 |

[a] Values from ref. 400.

mechanism could not be deduced since the nature of the kinetically active species in the Grignard reagent is not known; at the concentrations used the equilibrium 157 exists[403].

$$RMg(\mu\text{-}X)_2MgR \rightleftharpoons 2RMgX \rightleftharpoons R_2Mg + MgX_2 \rightleftharpoons R_2Mg(\mu\text{-}X)_2Mg \quad (157)$$

Dessy and Salinger[404] investigated reaction 158 in diethyl ether.

$$XC_6H_4MgBr + C_4H_9C{\equiv}CH \longrightarrow C_6H_5X + C_4H_9C{\equiv}CMgBr \quad (158)$$

They concluded that the reaction was second order overall, first order in each reactant; the rate constants are listed in Table 90. A second kinetic study[405], on the same reaction (X = H) carried out in diethyl ether showed that the reaction was only first order, and when tetrahydrofuran was added it then became second order.

TABLE 90. Second-order rate constants[404], $10^4k_2$, at 31.5 °C in diethyl ether for the reaction

$$XC_6H_4MgBr + C_4H_9C{\equiv}CH \longrightarrow C_6H_5X + C_4H_9C{\equiv}CMgBr$$

| X | $10^4k_2$ ($l\ mol^{-1}\ s^{-1}$) |
|---|---|
| $m$-$CF_3$ | 0.22 |
| $m$-Cl | 0.33 |
| $p$-Cl | 0.6 |
| H | 2.8 |
| $p$-Me | 6.22 |

Hashimoto and coworkers[406] concluded that the reaction between ethylmagnesium bromide and hex-l-yne in tetrahydrofuran followed third-order kinetics, and although they proposed a mechanism, they did not take into account equilibrium 157. In diethyl ether, the reaction was first order in each reactant[406].

Deuterium isotope effects were investigated by Pocker and Exner[407]. For the acidolysis of Grignard reagents by water and alcohols in diethyl ether and tetrahydrofuran the values of $k_H/k_D$ were slightly larger than 1, suggesting an initial coordination of the water or alcohol to magnesium, followed by a rapid proton transfer. For reaction with alkynes the isotope ratios were large, indicating a rate-determining proton transfer.

Abraham and Hill[408,409] studied the kinetics of acidolysis of dialkylzincs. The reactions between $Pr^n{}_2Zn$ abd *p*-toluidine or cyclohexylamine at 76 °C in diisopropyl ether were analogous to the acidolysis of dialkylmagnesiums in that they occurred by the competitive consecutive second order reactions 159 and 160 ($R = Pr^n$, $R' = p$-tolyl or cyclohexyl).

$$R_2Zn + H_2NR' \longrightarrow RH + RZnNHR' \tag{159}$$

$$RZnNHR' + H_2NR' \longrightarrow RH + Zn(NHR')_2 \tag{160}$$

Rate constants for reaction 159 were about six times those for step 160. For $Pr^n{}_2Zn$ the rate constants for reaction 159 at 76 °C in diisopropyl ether were $2.1 \times 10^{-2}\,l\,mol^{-1}\,s^{-1}$ for *p*-toluidine and $1.8 \times 10^{-3}\,l\,mol^{-1}\,s^{-1}$ for acidolysis with cyclohexylamine. Relative rate constants for reaction 159 ($R' = p$-tolyl) are given in Table 91, and for the unsymmetrical compounds $EtZnPr^n$ and $Pr^nZnBu^n$ relative rates at 68 °C in diisopropyl ether for reaction with *p*-toluidine were R = Et 100, $Pr^n$ 52, and $Bu^n$ 41[409]. An $S_E2$ (cyclic) mechanism was proposed[409] with a transition state such as **27**.

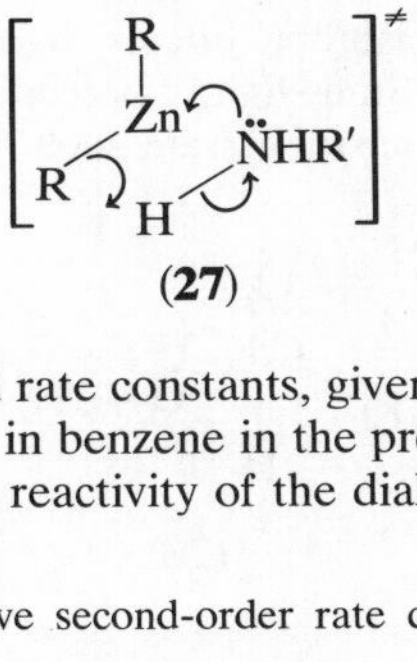

**(27)**

Inoue and Yamada[410] reported rate constants, given in Table 92, for the acidolysis of dialkylzincs by diphenylamine in benzene in the presence of complexing agents. It was suggested that the increased reactivity of the dialkylzincs in the presence of the

TABLE 91. Relative second-order rate constants[a] for the reaction

$$R_2Zn + H_2NR' \longrightarrow RH + RZnNHR' \quad (R' = p\text{-tolyl})$$

| R | | | | | | |
|---|---|---|---|---|---|---|
| Me | Et | $Pr^n$ | $Bu^n$ | $Pr^i$ | Solvent | Temp. (°C) |
| 41 | 100 | 42 | 27 | 61 | $Et_2O$ | 35 |
| | 100 | 33 | 35 | 67 | $Pr^i{}_2O$ | 68 |
| | 100 | | | 71 | Hexane | 69 |
| | 100 | | | 82 | Thf | 66 |

[a] Obtained by competitive experiments, in which two dialkylzincs were allowed to react with a deficiency of *p*-toluidine[409].

TABLE 92. Second-order rate constants[410], $10^6 k_2$, for the acidolysis of $R_2Zn$ by diphenylamine in the presence of complexing agents in benzene at 40 °C

| | $10^6 k_2$ ($l\,mol^{-1}\,s^{-1}$) | |
|---|---|---|
| Complexing agent (0.3 M) | $Et_2Zn$ | $Bu^n{}_2Zn$ |
| None | 4.4 | 4.2 |
| $MeOCH_2CH_2OMe$ | 1.9 | 2.2 |
| $Et_3N$ | 3.1 | |
| tmed[a] | 3.3 | 4.2 |
| py | 27 | |
| dmaq[b] | 89 | |
| bipy | 114 | 97 |

[a] *N,N,N',N'*-Tetramethylethylenediamine.
[b] 8-Dimethylaminoquinoline.

three ligands with conjugated structures was caused by the back-donation from the zinc atom to the ligand reducing the electron density at zinc and thus facilitating the coordination of diphenylamine to zinc in a transition state analogous to **27**.

The acidolysis of diethylcadmium by substituted benzyl alcohols $XC_6H_4CH_2OH$ in diethyl ether followed the overall reaction 161[411,412]; rate constants are given in Table 93.

$$Et_2Cd + 2ArCH_2OH \longrightarrow 2EtH + (ArCH_2O)_2Cd \qquad (161)$$

From substituent and kinetic isotope effects, a mechanism in which nucleophilic coordination of oxygen to cadmium slightly precedes electrophilic attack at the ethyl group was suggested, with a transition state such as **28**.

$$\left[ \begin{array}{c} Et \\ | \\ Cd \cdots \ddot{O}-CH_2Ar \\ Et \quad H \end{array} \right]^{\neq}$$

(**28**)

TABLE 93. Second-order rate constants[411,412], $10^3 k_2$, for the reaction between $Et_2Cd$ and $XC_6H_4CH_2OH$ in diethyl ether at 34.5 °C

| X | $10^3 k_2$ ($l\,mol^{-1}\,s^{-1}$) | X | $10^3 k_2$ ($l\,mol^{-1}\,s^{-1}$) |
|---|---|---|---|
| *p*-OMe | 1.8 | *p*-Cl | 1.1 |
| *p*-Me | 1.5 | *m*-Cl | 1.0 |
| *m*-Me | 1.4 | *m*-$CF_3$ | 0.98 |
| H | 1.3 | *p*-$CF_3$ | 0.85 |
| *m*-OMe | 1.2 | | |

## C. Protodemetallation of Organomercury Compounds

Much of the early work on the protodemetallation of organomercury compounds was concerned with the cleavage of dialkyl- and diaryl-mercury compounds by hydrogen chloride in ethanol. With unsymmetrical compounds, RHgR′, the relative rates of cleavage of the two groups may be obtained by product analyses, see equation 162 (X = Cl).

$$RHgR' + HX \begin{cases} \nearrow RH + R'HgX \\ \searrow R'H + RHgX \end{cases} \tag{162}$$

Kharasch and coworkers[413–418] used qualitative relative rates of cleavage determined in this way to construct an 'electronegativity' series of alkyl and aryl groups, the greater the reactivity of the organic group, the greater was its electronegativity. Abraham[4] has commented on this series and has shown that certainly in the case of alkyl groups there is no substantial theoretical basis for the electronegativity series. Experimentally, the following order of ease of protodemetallation was observed by Kharasch and coworkers:

$$R = Me > Et > Pr^n > Bu^n > Pr^i > Hept^n > Bu^s > Pe^{neo}, Bu^t \tag{163}$$

More recently, there have been a number of kinetic investigations of protodemetallation of alkylmercury compounds, and a stereochemical study by Jensen and coworkers[419], who showed that the protodemetallation of optically active $Bu^s_2Hg$ and of *cis*- and *trans*-di-4-methylcyclohexylmercury involved predominant retention of configuration at the carbon atom undergoing substitution. One of the first kinetic investigations of protodemetallation of alkylmercury compounds was by Winstein and co-workers[14,354], who reported rate constants for protodemetallation by an excess of acetic acid (as solvent); some values are given in Table 94 together with activation parameters. A cyclic transition state was suggested for these uncatalysed protodemetallations. In the presence of perchloric acid, second-order kinetics were observed. A detailed study of the protodemetallation of alkylmercury(II) iodides by aqueous sulphuric and perchloric acids showed that a rate-determining protodemetallation (164) was followed by the fast step (165)[420,421]. Kinetic data and

$$H^+ + RHgI \longrightarrow RH + HgI^+ \tag{164}$$

$$HgI^+ + RHgI \longrightarrow RHg^+ + HgI_2 \tag{165}$$

TABLE 94. First-order rate constants[14,354], $10^6k_1$, at 25 °C and activation parameters for the reaction

$$R_2Hg + AcOH \longrightarrow RH + RHgOAc$$

| *R* | $10^6k_1$ ($s^{-1}$) | $\Delta H^{\neq}$ (kcal $mol^{-1}$) | $\Delta S^{\neq}$ (cal $K^{-1}$ $mol^{-1}$) |
|---|---|---|---|
| Ph | 498 | | |
| $Bu^s$ | 23 | | |
| $Bu^n$ | $2.4^a$ | 20.4 | −16 |
| Camph-4-yl | $0.52^a$ | 17 | −12 |
| Neophyl | $0.036^a$ | 22.6 | −17 |

[a] Extrapolated values.

TABLE 95. Second-order rate constants[420,421], $10^6k_2$, at 110 °C and activation parameters for the reaction with 1 M $H_2SO_4$ in water + 2% methanol

$$H^+ + RHgI \longrightarrow RH + HgI^+$$

| R | $10^6k_2$ (l mol$^{-1}$ s$^{-1}$) | $\Delta H^{\neq}$ (kcal mol$^{-1}$) | $\Delta S^{\neq}$ (cal K$^{-1}$ mol$^{-1}$) |
|---|---|---|---|
| Me | 17.2 | 22.3 | −23.3 |
| Et | 6.8 | 22.8 | −25 |
| $Pr^n$ | 3.8 | 23 | −23 |
| $Hex^n$ | ~2 | | |
| $Pr^i$ | 2.2 | 25 | −21 |
| cyclo-Pr | 13 000 | 20.9 | −11 |
| $Bu^t$ | ~0.14[a] | | |

[a] Solvent water–methanol (96:4).

activation parameters are collected in Table 95. Maguire and Anand[422] have determined second-order rate constants for the protodemetallation of dimethylmercury by dilute aqueous acids at 25 °C and pressures up to 1000 atm, the constants increasing with increase in pressure. At 25 °C and 1 atm, the rate constants were $6 \times 10^{-4}$ l mol$^{-1}$ s$^{-1}$ for protodemetallation by HCl and $2.5 \times 10^{-4}$ l mol$^{-1}$ s$^{-1}$ for protodemetallation by HBr.

Nugent and Kochi[423,424] have investigated the effects of alkyl groups R and R′ on rate constants for the protodemetallation of dialkylmercurys by an excess of acetic acid (reaction 166) and their reported first-order rate constants at 37.5 °C are given

$$RHgR' + HOAc \longrightarrow RH + R'HgOAc \quad (166)$$

in Table 96. With a constant leaving group (HgR′) the alkyl group is cleaved in the

TABLE 96. Pseudo-first-order rate constants[423,424], $10^7k$ and $10^7k'$, at 37.5 °C in excess acetic acid for the reactions

$$RHgR' + HOAc \xrightarrow{k} RH + R'HgOAc$$
$$RHgR' + HOAc \xrightarrow{k'} R'H + RHgOAc$$

| R | R′ | $10^7k$ (s$^{-1}$) | $10^7k'$ (s$^{-1}$) |
|---|---|---|---|
| Me | Me | 3.90[a] | 3.90[a] |
| Me | Et | 23.5 | 14.3 |
| Me | $Pr^i$ | 85.8 | 8.45 |
| Me | $Bu^t$ | 120 | Very slow |
| Et | Et | 81.8[a] | 81.8[a] |
| Et | $Pr^i$ | 255 | 47.0 |
| Et | $Bu^t$ | 378 | ~3 |
| $Pr^i$ | $Pr^i$ | 154[a] | 154[a] |
| $Pr^i$ | $Bu^t$ | 224 | ~10 |
| $Bu^t$ | $Bu^t$ | Very slow | Very slow |

[a] Statistically corrected.

sequence R = Me < Et > Pr$^i$ > Bu$^t$, interpreted by Nugent and Kochi as due to a combination of electronic and steric effects; compare sequence 163 for protodemetallation by hydrogen chloride in ethanol, suggested[4] to arise predominantly from steric effects. All the rate constants obtained by Nugent and Kochi could be accommodated by a single equation (167), in which $k_0$ is the rate constant for

$$\log (k/k_0) = L + C \tag{167}$$

cleavage of MeHgMe and $k$ is that for cleavage of RHgR′. $L$ is the leaving group constant, essentially an electronic effect with a striking relationship to the vertical ionization potential of RHgR′, and $C$ is the cleaved group constant. Kinetic isotope effects showed that the protodemetallation 166 proceeded by a rate-determining proton transfer, and transition state **29** was postulated.

$$\left[\begin{matrix} \text{R} & \\ \vdots & \text{H}\cdots\text{OAc} \\ \text{Hg} & \\ | & \\ \text{R}' & \end{matrix}\right]^{\neq}$$

**(29)**

Both Nesmeyanov *et al.*[425] and Dessy *et al.*[426] have investigated the protodemetallation of a series of organomercury compounds $R_2Hg$, where R = alkyl, aryl, etc. Unfortunately, the experimental conditions are not the same, so that the two sets of results cannot be combined; rate constants are collected in Tables 97 and 98, and it

TABLE 97. Second-order rate constants[425], $10^3k_2$, at 20 °C in 90 vol.-% aqueous dioxane for the reaction

$$R_2Hg + HCl \longrightarrow RH + RHgCl$$

| R | $10^3k_2$ ($l\,mol^{-1}\,s^{-1}$) | R | $10^3k_2$ ($l\,mol^{-1}\,s^{-1}$) |
|---|---|---|---|
| 1,3,5-$Me_3C_6H_2$ | 365 | $C_6H_5$ | 1.77 |
| 2-$MeC_6H_4$ | 47.3[a] | Et | 25.7[b] |
| 3-$MeC_6H_4$ | 12.8[a] | Bu$^n$ | 11.8[b] |
| 4-$MeC_6H_4$ | 10.0 | $PhCH_2$ | 3.54[b] |
| 4-$EtC_6H_4$ | 42.6[a] | α-Naphthyl | 12.8[a] |
| 2-$MeOC_6H_4$ | 3.82 | $CH_2$=CH | 2.77 |
| 3-$MeOC_6H_4$ | 5.52[a] | *cis*-ClCH=CH | 13.8[b] |
| 4-$MeOC_6H_4$ | 125 | *trans*-ClCH=CH | 25.6[b] |
| 4-$EtOC_6H_4$ | 178 | *cis*-MeCH=CH | 166 |
| 2-$MeOCOC_6H_4$ | 2.75[b] | *trans*-MeCH=CH | 213 |
| 2-$MeOOCC_6H_4$ | 13.6[b] | PhCH=CH | 94.8 |
| 4-$MeOOCC_6H_4$ | 24.8[b] | *cis*-PhCH=CPh | 6.94 |
| 3-$FC_6H_4$ | 0.256[a] | *trans*-PhCH=CPh | 136[c] |
| 4-$FC_6H_4$ | 5.82[a] | *cis*-MeC(MeOCO)=CMe | 43.2 |
| 2-$ClC_6H_4$ | 1.84[c] | *trans*-MeC(MeOCO)=CMe | 54.0 |
| 3-$ClC_6H_4$ | 0.209[a] | PhC(MeOCO)=CPh | 28.0[b] |
| 4-$ClC_6H_4$ | 1.23[a] | α-Thienyl | 34.2 |
| 4-$BrC_6H_4$ | 0.846[a] | | |

[a] At 30 °C
[b] At 70 °C.
[c] At 60 °C.

TABLE 98. Second-order rate constants[426], $k_2$, at 40 °C and activation parameters for the reaction[a]

$$R_2Hg + HCl \longrightarrow RHgCl + RH$$

| R | $k_2$ (l mol$^{-1}$ s$^{-1}$) | $E_a$ (kcal mol$^{-1}$) | $\Delta S^{\neq}$ (cal K$^{-1}$ mol$^{-1}$) |
|---|---|---|---|
| cyclo-Pr | $2.5\times10^{-1}$ | 16.5 | −11 |
| Vinyl | $4.2\times10^{-2}$ | 13.6 | −23 |
| $C_6H_5$ | $2.5\times10^{-2}$ | 12.2 | −29 |
| Et | $3.8\times10^{-4}$ | 15.5 | −27 |
| $Pr^i$ | $2.6\times10^{-4}$ | 15.4 | −28 |
| $Pr^n$ | $2.3\times10^{-4}$ | 16.5 | −25 |
| Me | Very slow | | |
| Et[b] | $1.9\times10^{-3}$ | | |

[a] Solvent dimethyl sulphoxide–dioxane (10:1).
[b] Solvent dimethylformamide–dioxane (10:1)[427].

can be seen that not only aryl groups but also vinyl groups are cleaved more easily than are simple alkyl groups. Nesmeyanov *et al.*[425] also determined rate constants for protodemetallation of numerous substituted diphenylmercurys (Table 97), unfortunately at various temperatures. In general, the order of cleavage of $XC_6H_4$ groups (Table 97) is the same as that reported by Dessy and Kim[428] (Table 99) and by Nerdel and Makover[429], who give the series X = 4-Me > 2-Me > 3-Me > H for cleavage of $(XC_6H_4)_2Hg$ by HCl in aqueous dioxane and tetrahydrofuran. A more detailed kinetic study of protodemetallation of an arylmercury compound was earlier reported by Corwin and coworkers[430,431], who showed that in the protodemetallation of diphenylmercury with carboxylic acids, in dioxane, the reaction was approximately third order in the carboxylic acid. When perchloric acid was added, the kinetic form was simplified to second-order kinetics overall, first order in $R_2Hg$ and first order in perchloric acid. Rates of protodemetallation were measured in mixtures of ethanol, dioxane and water, and since these were correlated with the medium acidity function, a rate-determining proton transfer was suggested (see mechanism 168)[431]

$$C_6H_5HgPh + HX \underset{k_{-1}\,(\text{slow})}{\overset{k_1}{\rightleftharpoons}} (C_6H_6Ph)^+ \xrightarrow{k_2\,(\text{fast})} C_6H_6 + PhHgX \qquad (168)$$

TABLE 99. Second-order rate constants, $10^3k_2$, at 32 °C and activation parameters[a] in dimethyl sulphoxide–dioxane (10:1 v/v) for the reaction[428]

$$(XC_6H_4)_2Hg + HCl \longrightarrow C_6H_5X + XC_6H_4HgCl$$

| X | $10^3k_2$ (l mol$^{-1}$ s$^{-1}$) | $E_a$ (kcal mol$^{-1}$) | $\Delta S^{\neq}$ (cal K$^{-1}$ mol$^{-1}$) |
|---|---|---|---|
| 4-MeO | 460[b] | 9.2 | −32 |
| 4-Ph | 26.1 | 11.9 | −30 |
| H | 16.2 | 12.2 | −29 |
| 4-F | 13.6 | 11.4 | −31 |
| 4-Cl | 4.78 | 15.0 | −22 |
| 3-$NO_2$ | 0.14[b] | 20.8 | −10 |

[a] Values to be treated with caution since they were calculated only over 18 °C for each compound.
[b] Extrapolated from data at other temperatures.

TABLE 100. Second-order rate constants, $10^5k_2$, at 70 °C and activation parameters for the protodemetallation of ArHgCl by HCl in 10 vol-% aqueous ethanol[432,433]

| Ar | $10^5k_2$ ($l\,mol^{-1}\,s^{-1}$) | $E_a$ ($kcal\,mol^{-1}$) | $\Delta S^{\neq}$ ($cal\,K^{-1}\,mol^{-1}$) |
|---|---|---|---|
| Ph | 12.0 | 23 | −12 |
| $4\text{-}MeC_6H_4$ | 83.6 | 24 | −5 |
| $3\text{-}MeC_6H_4$ | 28.3 | 25 | −3 |
| $4\text{-}ClC_6H_4$ | 8.04 | 23 | −14 |
| $3\text{-}ClC_6H_4$ | 3.13 | 23 | −16 |
| $4\text{-}MeOC_6H_4$ | 1770 | 19 | −14 |
| $3\text{-}MeOC_6H_4$ | 8.53 | 21 | −18 |
| Fur-2-yl | 49 400 | 18 | −9 |
| Fur-3-yl | 1800 | 16 | −21 |
| Thien-2-yl | 20 600 | 15 | −20 |
| 2-Selenophenyl | 45 600 | 14 | −18 |

Brown *et al.*[432] have studied the kinetics of the hydrochloric acid cleavage of arylmercury(II) chlorides in 10% aqueous ethanol. The process was first order in RHgCl and first order in acid; the second-order rate constants at 70 °C and the associated activation parameters are given in Table 100. The effect of chloride ion is peculiar in that, although the reaction order is zero with respect to chloride ion, the reaction does not occur at all in the absence of $Cl^-$. The mechanism shown in equation 169 was suggested.

$$ArHgCl + H_3O^+ \underset{slow}{\rightleftharpoons} ArHHgCl^+ \overset{Cl^-}{\underset{fast}{\rightleftharpoons}} ArH + HgCl_2 \quad (169)$$

Rate constants were reported for protodemetallation of a number of heterocyclic compounds (Table 100)[433] and as for the arylmercury(II) chlorides, the peculiar effect of chloride ion was again noticed.

The cleavage of phenylmercury(II) bromide by hydrochloric acid[434,435], phosphoric acid[436], and chloroacetic acid[436] in aqueous dioxane was studied in the presence of added sodium iodide to complex the mercury(II) halide produced, since the latter affected the rate constant throughout a given kinetic run. However, this addition of sodium iodide itself accelerated the reaction and also led to other complications. Reutov and coworkers suggested that in 90–95% aqueous dioxane, the reaction took place through an $S_E2$ mechanism, but that in 60–80% aqueous dioxane an $S_E1$ type of process took place[435,436]. The latter seems unlikely, and it is possible that oxygen interferes with these protodemetallations[118,437].

Second-order rate constants for protodemetallation of divinylmercurys (Table 97) show that these compounds are reactive, and Reutov and coworkers[438] further examined the cleavage of *cis*- and *trans*-β-chlorovinylmercury(II) chloride by hydrogen chloride in dioxane and dimethyl sulphoxide. In all cases, the protodemetallation proceeded with complete retention of geometrical configuration, although in dioxane the mechanism was $S_E2$ whereas in dimethyl sulphoxide an $S_E1$ mechanism was observed. Second-order rate constants for the protodemetallation in dioxane at 60 °C were $7.9 \times 10^{-4}\,l\,mol^{-1}\,s^{-1}$ (*trans*) and $8.8 \times 10^{-4}\,l\,mol^{-1}\,s^{-1}$ (*cis*), and in dimethyl sulphoxide at 50 °C the first-order rate constants were $4.13 \times 10^{-3}\,s^{-1}$ (*trans*) and $3.61 \times 10^{-3}\,s^{-1}$ (*cis*).

Earlier work by Winstein and coworkers[439] showed that the allyl—mercury bond in *trans*-crotyl and *trans*-cinnamylmercury(II) halides was rapidly cleaved by a number of acid/solvent systems to yield rearranged products by mechanisms suggested to be of the $S_E2'$ (open) or $S_E2'$ (cyclic) type. No rate constants were obtained[439], but Kreevoy and coworkers[440,441] later investigated the protodemetallation of allylmercury(II) iodides by aqueous acid and iodide ion in water–methanol (96:4) (equation 170). With *trans*-$CH_3CH{=}CHCH_2HgI$, 95% of the rearranged

$$CH_2{=}CHCH_2HgI + H^+ + I^- \longrightarrow CH_2{=}CHCH_3 + HgI_2 \qquad (170)$$

product $CH_2{=}CHCH_2CH_3$ was produced, no doubt by an $S_E'$ process. When the concentrations of allylmercury(II) iodide and added iodide ion were low, reaction 170 was found to follow overall second order kinetics, first order in allylmercury(II) iodide and first order in $H^+$. The rate constant at 25 °C is $1.4 \times 10^{-2}\,l\,mol^{-1}\,s^{-1}$ with $\Delta H^{\neq} = 16.42$ kcal mol$^{-1}$ and $\Delta S^{\neq} = -11.9$ cal K$^{-1}$ mol$^{-1}$. Abraham[4] has estimated that at 110 °C the rate constant will be about $3 \times 10^6$ times greater than that for cleavage of ethylmercury iodide under comparable conditions. Kinetic isotope effects indicated that reaction 170 takes place by a rate-determining proton transfer[440,441], and later work showed that reaction 170 was subject to general acid catalysis[442,443]. Ibrahim and Kreevoy[444] also studied the cleavage of cyclopentadienylmercury(II) bromide by aqueous solutions containing $H^+$ and $Br^-$. General acid catalysis was again observed, and the results were consistent with a mechanism involving fast reversible complexing of bromide ion to form complexes up to and including $[C_5H_5HgBr_4]^{3-}$, followed by protodemetallation of the various complexed species. Nesmeyanov *et al.*[445] have determined rate constants for the acidolysis of allylmercury(II) iodide by HCl in 90% aqueous dioxane; at 30 °C the second order rate constant is $1.94 \times 10^{-2}\,l\,mol^{-1}\,s^{-1}$ with $E_a = 21.1$ kcal mol$^{-1}$ and log A = 13.561.

Protodemetallation of benzylmercury(II) compounds has been the subject of several investigations. The survey of Nesmeyanov *et al.*[425] (see Table 97) indicates that dibenzylmercury is rather less reactive than simple dialkylmercurys towards HCl in aqueous dioxane, and this observation is consistent with the kinetic data of Nerdel and Makover[429] on protodemetallation of a series of $R_2Hg$ compounds by HCl in dioxane at 69 °C: values of the second-order rate constant are $0.45 \times 10^{-3}$ ($R = PhCH_2$), $1.25 \times 10^{-3}$ ($R = PhCH_2CH_2$) and $1.88 \times 10^{-3}\,l\,mol^{-1}\,s^{-1}$ ($R = PhCH_2CH_2CH_2$). Reutov and coworkers have examined a number of protodemetallations of benzylmercury(II) compounds, through both kinetic methods and product analyses. The protodemetallation of benzylmercury(II) chloride by HCl in slightly aqueous dioxane was reported[446,447] to be second order overall: at 70 °C the rate constant was $0.01\,l\,mol^{-1}\,s^{-1}$ with $E_a = 21$ kcal mol$^{-1}$ and $\Delta S^{\neq} = -7$ cal K$^{-1}$ mol$^{-1}$. In the protodemetallation of dibenzylmercury by HCl in dimethyl sulphoxide, dimethylformamide, tetrahydrofuran, *n*-butanol, and aqueous acetonitrile, it was suggested[448] that reaction took place via an $S_E1$ mechanism, since the overall kinetic form was first order (in $R_2Hg$); however, later work[449] indicated a more complex mechanism. Another investigation by Kitching and coworkers[450] showed that when conducted under an atmosphere of nitrogen, the protodemetallation of dibenzylmercury in aqueous acetonitrile was in fact a second-order process, and it seems probable that atmospheric oxidation processes could complicate the kinetics of protodemetallation. In addition, Reutov and coworkers[447,451] have themselves shown that protodemetallation of benzylmercury(II) compounds can lead to anomalous products. Thus, in deuteriodemetallation of benzylmercury(II) chloride with DCl in dioxane, 95% aqueous dioxane, and dimethoxyethane, transfer of the reaction centre was observed, but in dimethylformamide at 70 °C there was only slight transfer of the reaction centre. Reutov and coworkers[451] suggested scheme 171 for the incorporation of deuterium into the aromatic nucleus. In the case of the 2,4,6-trimethylbenzyl compound, reaction seems to proceed in the normal way, by attack at the $\alpha$-carbon atom.

$$C_6H_5CH_2HgCl \xrightarrow{DCl} [\text{cyclic complex: } CH_2\text{–Hg(Cl)}\cdots Cl\text{–D}\cdots \text{ring}] \xrightarrow{-D(H)Cl} 2\text{-D-}C_6H_4CH_2HgCl + C_6H_5CH_2HgCl$$

$$\downarrow -HgCl_2$$

$$[\text{6-D,6-H-methylenecyclohexadiene}] \xrightarrow{DCl} 2\text{-D-}C_6H_4CH_2D + C_6H_5CH_2D \quad (171)$$

A number of other substituted alkyl compounds have been studied kinetically; Coad and Johnson[452] reported that the protodemetallation of 4-pyridiomethylmercury(II) chloride in aqueous perchloric acid proceeded by an anion-catalysed $S_E1$ mechanism, and Coad and Ingold[453] suggested a four-step mechanism for the protodemetallation of $\alpha$-carbethoxybenzylmercury(II) chloride, $PhCH(CO_2Et)HgCl$, by perchloric acid in dioxane–water (70:30), again catalysed by chloride ion.

## D. Protodemetallation of Organoboron Compounds

There have been several studies of the stereochemistry of the protodemetallation reaction. Davies and Roberts[454] showed that the dibutyl ester of optically active 1-phenylethylboronic acid was cleaved by $C_7H_{15}CO_2D$ in boiling diglyme to yield (−)-1-deuterioethylbenzene with 95% retention of configuration, probably through a six-centred $S_E2$ (cyclic) transition state. More recently, Kabalka *et al.*[455] reported that the protodemetallation of *erythro*- and *threo*-$Bu^tCHDCPhDBH_2$ by acetic acid yielded the corresponding *erythro*- and *threo*-$Bu^tCHDCPhDH$ by hydrocarbons, again with retention of configuration at the carbon atom undergoing substitution.

Cleavages in the presence of alkali have also been studied: when the diethanolamine ester of 1-phenylethylboronic acid was treated with $OD^-$ in boiling $D_2O$, cleavage took place with 54% inversion of configuration[454], but the alkaline cleavage of $PhCHMeC(Me)(Ph)BH_2$ took place with retention of configuration at the carbon atom undergoing substitution.

The protodemetallation of triethylboron was examined by Dessy and coworkers[456], who obtained second-order rate constants for cleavage by a range of carboxylic acids in diglyme (Table 101). Since the weaker was the acid, the faster was the cleavage, it seems as though coordination of the acid to the triethylboron must take place (presumably in a prior fast equilibrium step[4]) followed by a slow cleavage step giving rise to the observed $k_H/k_D$ ratio of 3.3[456].

Brown *et al.*[457] have investigated the protodemetallation of boronic acids, $RB(OH)_2$, where R = phenyl, thiophen-2 and thiophen-3, in the presence of perchloric acid. The relative rates of reaction were, in water, R = phenyl 1, thiophen-3

TABLE 101. Second-order rate constants[456], $10^4k_2$, for protodeboronation of $Et_3B$ by various acids in diglyme at 31 °C

| Acid | $10^4k_2$ ($l\,mol^{-1}\,s^{-1}$) | Acid | $10^4k_2$ ($l\,mol^{-1}\,s^{-1}$) |
|---|---|---|---|
| $Bu^tCOOH$ | 81 | $Cl_3CCOOH$ | Very slow |
| $Me(CH_2)_6COOH$ | 59 | $F_3CCOOH$ | Very slow |
| EtCOOH | 48 | $4\text{-}MeOC_6H_4COOH$ | 102 |
| MeCOOH | 47 | $4\text{-}NO_2C_6H_4COOH$ | 8.6 |
| PhCOOH | 42 | $MeCONH_2$ | 10 |
| EtCH(Ph)COOH | 41 | MeCOOD | 1.4 (14)[a] |
| $PhCH_2COOH$ | 36 | Quinolin-8-ol | 8.4 |
| $Ph_2CHCOOH$ | 24 | 2-Pyrrolidone | 1.0 |
| $ClCH_2COOH$ | 5.1 | 2-Pyridone | 460 |
| $Cl_2CHCOOH$ | 0.59 | | |

[a] Values of $k_2$ are given as $4.7\times10^{-3}$ (MeCOOH) and $1.4\times10^{-4}\,l\,mol^{-1}\,s^{-1}$ (MeCOOD), but since the value of $k_H/k_D$ is given as 3.3 for these two acids, the $k_2$ value for MeCOOD should possibly be $1.4\times10^{-3}\,l\,mol^{-1}\,s^{-1}$.

$7.1\times10^3$ and thiophen-2 $8.5\times10^5$, and mechanism 172 was considered the most probable.

$$RB(OH)_2 + H_3O^+ \rightleftharpoons \left[ R\begin{matrix} \cdot\cdot H \\ \cdot\cdot B(OH)_2 \end{matrix} \right]^{\neq} + H_2O \longrightarrow RH + H_3BO_3 + H^+ \qquad (172)$$

Kuivila and Nahabedian[458–461] have studied the protodemetallation of benzeneboronic acids with aqueous sulphuric, phosphoric and perchloric acids. From detailed kinetic studies including solvent isotope effects, they concluded that results were inconsistent with an aromatic $S_E2$ mechanism. Rate constants were obtained for a wide range of substituted benzeneboronic acids using malonic acid–sodium malonate buffer solutions[462,463]; these results have been extensively reviewed[464] and Table 102 gives a selection of rate constants.

TABLE 102. Relative rate constants for protodeboronation of $XC_6H_4B(OH)_2$ with malonate buffers at pH 6.70 and 90 °C[464]

| X | Rel. $k_1$ | X | Rel. $k_1$ |
|---|---|---|---|
| H | 0.145 | 2-F | 11.1 |
| 4-MeO | 0.610 | 3-F | 0.332 |
| 2-MeO | 1.53 | 4-Cl | 0.187 |
| 4-Me | 0.260 | 2-Cl | 8.60 |
| 2-Me | 0.360 | 3-Cl | 0.217 |
| 3-Me | 0.294 | 2,6-Dimethoxy | 18.1 |
| 4-F | 0.250 | | |

## E. Protodemetallation of Organosilicon Compounds

There have been a large number of preparative studies on the protodemetallation of organosilicon compounds by acids and by base-catalysed processes. We first consider these preparative studies and related kinetic investigations, and then deal with the more specialized kinetic and mechanistic studies on the acidic protodemetallations and the base-catalysed processes.

### 1. Preparative and general kinetic studies

Alkyl—silicon bonds are very readily cleaved by hydrogen halides in the presence of aluminium halides (equivalent to the very strong acids $HAlX_4$). For reaction 173 in benzene at 40 °C, the relative rates were[465] $Me \gg Et\,(1.00) > Bu^n\,(0.17) \approx Pr^n\,(0.13) > Pr^i\,(0.04)$, with steric effects outweighing electronic effects. In reaction 174 a methyl

$$\text{R-SiMe}_3 + \text{HBr} \xrightarrow{\text{AlBr}_3} \text{RH} + \text{Me}_3\text{SiBr} \tag{173}$$

$$\text{Me}_3\text{SiCH}_2\text{Cl} + \text{HCl} \xrightarrow{\text{AlCl}_3} \text{CH}_4 + \text{Me}_2(\text{ClCH}_2)\text{SiCl} \tag{174}$$

group is cleaved[466] rather than the chloromethyl group, demonstrating that in the absence of steric requirements, electronic effects govern the course of the protodemetallation. If hexamethyldisilane is used, the reaction temperature can be chosen to obtain the required products[165], with one methyl group being cleaved at 20 °C, two at 55 °C, and three at 90 °C by HCl in the presence of $AlCl_3$.

Cold concentrated sulphuric acid will cleave a methyl group from a wide variety of compounds $Me_3SiR$, according to equation 175. Some of the R groups which have

$$\text{Me}_3\text{SiR} + \text{H}_2\text{SO}_4 \longrightarrow \text{CH}_4 + \text{Me}_2\text{Si(HSO}_4)\text{R} \tag{175}$$

been used are listed in Table 103. The substituted alkyl groups are not cleaved, and from studies[470,483,484] of the acid cleavage of $MeR_2Si(CH_2)_2COOH$ it has been concluded that the ease of cleavage of R—Si bonds falls in the order $R = Am^i \approx Bu^n > Am^n > Me > Pr^n \approx Et$. For R = Et and $Pr^n$, the methyl group is cleaved; for the other R groups the Si—R bond is cleaved.

An excess of fluorosulphonic acid at −47 °C with $Me_4Si$ or $Me_3SiCH_2Cl$ yields methane with reported first-order rate constants of $4.3 \times 10^{-3}$ and $4.6 \times 10^{-5}\ s^{-1}$, respectively[485]. The same reagent at 57.5 °C cleaves $Me_3SiOSO_2F$ and

TABLE 103. Examples of compounds $Me_3SiR$ which have been cleaved by concentrated sulphuric acid to give methane

| R | Ref. | R | Ref. |
|---|---|---|---|
| $CH_2NMe_2$ | 467 | $(CH_2)_4O(CH_2)_4SiMe_3$ | 477 |
| $(CH_2)_2NH_2$ | 468, 469 | $(CH_2)_3Br$ | 475, 478 |
| $(CH_2)_2COOH$ | 468–472 | $(CH_2)_4Br$ | 475 |
| $(CH_2)_2COCH_3$ | 468, 469 | $CHBrCH_2CFClCF_2Br$ | 479 |
| $CH_2Cl$ | 473 | $CH_2NHC_6H_{11}$ | 467 |
| $(CH_2)_2CO(CH_2)_2SiMe_3$ | 469 | $CH_2N(CH_2SiMe_3)_2$ | 467 |
| $(CH_2)_3COMe$ | 474 | $(CH_2)_2CH(NH_2)(CH_2)_2SiMe_3$ | 469 |
| $(CH_2)_3COOH$ | 469, 472, 475 | $(CH_2)_4SiMe_3$ | 480 |
| $CH_2CH(COOH)(CH_2)_2COOH$ | 476 | $(CH_2)_2COC_5H_4FeC_5H_5$ | 481, 482 |

$Me_2Si(CH_2Cl)OSO_2F$ to produce again methane with rate constants of $1.92 \times 10^{-5}$ and $1.90 \times 10^{-6}\ s^{-1}$, respectively[485].

Harbordt and O'Brien[486] later reported the cleavage of compounds $XCH_2SiMe_3$ (X = Br, I) by fluorosulphonic acid, although rate constants could not be obtained. At −78 °C, attack was exclusively at the methyl—silicon bond, producing methane in each case. The subsequent reactions of the products $BrCH_2Me_2SiSO_3F$ and $ICH_2$-$Me_2SiF$ at 30 °C were also studied. In the former case cleavage of both methyl and bromomethyl groups occurred but for the iodo compound there was cleavage only of the $ICH_2$—Si bond. A number of pathways were discussed to account for the complex results.

Hopf and O'Brien[487] investigated the kinetics of the fluorosulphonic acid cleavage of $RMe_2SiF$ (R = Me, Et, $Pr^n$, $Pr^i$, $Bu^t$) in $CH_2Cl_2$, in which competitive cleavage of R—, Me—, and F—Si bonds occurred. The reaction was first order in organosilane and second order in acid, and third-order rate constants were R = Me $2.63 \times 10^{-5}$, Et $4.88 \times 10^{-5}$, $Pr^n$ $3.85 \times 10^{-5}$, $Pr^i$ $1.95 \times 10^{-5}$, and $Bu^t$ $1.82 \times 10^{-5}\ l^2\ mol^{-2}\ s^{-1}$. A mechanism was proposed which involved nucleophilic attack of the acid at silicon, to form the intermediate **30**, followed by electrophilic attack at an $\alpha$-carbon or fluorine

$$-\overset{-}{Si}-\overset{+}{O}(H)-SO_2F$$

**(30)**

to produce the three cleavage products. Because both nucleophilic and electrophilic attack is involved, there is a levelling of steric effects, and the rates for the various alkyl derivatives do not vary much.

Aqueous hydrogen bromide (40%) at 200 °C protodemetallates $PhSiCl_3$ and *p*-$ClC_6H_4SiCl_3$ to give benzene and chlorobenzene, respectively[488]. Anhydrous hydrogen halides also protodemetallate arylsilicon compounds, for example reaction 176.

$$PhSiH_3 + HI \longrightarrow PhH + H_3SiI \qquad (176)$$

The reactivity towards phenylsilane increases along the series HF < HCl < HBr < HI and the reactivity of the organometallic substrate towards a particular hydrogen halide increases in general along the series $PhSiF_3$ < $PhSiCl_3$ < $Ph_2SiCl_2$ < $PhSiHBr_2$ < $Ph_2Si(OH)_2$ < $PhSiH_2Cl$ < $PhSiH_2Br$ < $Ph_2SiHCl$ < $PhSiH_2I$ < $Ph_2SiH_2$ < $PhSiH_3$ < $PhSiMe_3$.

Some trichloromethylsilicon compounds react with water to form chloroform, for example[489,490] reaction 177.

$$Cl_3SiCCl_3 + H_2O \longrightarrow CHCl_3 + Cl_3SiOH \qquad (177)$$

The silicon—chlorine bonds of the other product are usually hydrolysed. Similar protodemetallations are experienced by the compounds $Cl_2Si(CCl_3)_2$[491], $Cl_2MeSiCCl_3$[491,492], $Cl_2PhSiCCl_3$[493], and $(OMe)_3SiCCl_3$[494]. Ethanol at 60 °C will cleave[491] a Si—$CCl_3$ bond in $(OEt)_2Si(CCl_3)_2$ to form chloroform, and ethanolic sodium ethoxide reacts similarly[495] with $(OEt)_3SiCCl_3$. Bromoform is produced when $Me_3SiCBr_3$ is boiled with aqueous acetone[496]. Ethanolic sodium ethoxide will also protodemetallate the compounds $MeCl_2SiCHCl_2$, $Me_2ClSiCHCl_2$, $Me_3SiCHCl_2$, $[Me_2(Cl_2CH)Si]_2O$, and $Me_3SiSiMe_2CHCl_2$, in each case producing $CH_2Cl_2$[490,492,497,498]. Methylene bromide is formed[496] by the action of aqueous acetone containing small amount of alkali on $Me_3SiCHBr_2$.

Methyl chloride is produced by the action of aqueous alcoholic potassium hydroxide on $Ph_3SiCH_2Cl$[499], by the cleavage of $Me_3SiCH_2Cl$ by sodamide in liquid ammonia[467], and by the cleavage[500] of $Me_3SiSiMe_2CH_2Cl$ and $ClCH_2Me_2SiSiMe_2CH_2Cl$ by ethanolic potassium cyanide, and methyl bromide is formed[501] by the protodemetallation of $Me_3SiCH_2Br$ by methanolic potassium cyanide.

As expected, electron-withdrawing groups on the silicon atom make protodemetallations by alcoholic alkalis easier. Thus $p$-$ClC_6H_4Me_2SiCH_2Cl$ reacts faster[502] than $PhMe_2SiCH_2Cl$, to yield $CH_3Cl$, and the compounds $Me_3SiOSiMe_2CH_2Cl$ and $\overline{O(SiMe_2O)_3SiMeCH_2Cl}$ react more readily[498,503] than $Me_3SiCH_2Cl$. Aqueous alkali reacts with $Cl_3SiCF_2CHFCl$ to form[504] mainly the cleavage product $HCF_2CHFCl$. Benzaldehyde (PhCHO) is formed by the protodemetallation of $Ph_3SiCOPh$ by aqueous alcoholic alkali or, with ultraviolet radiation, by neutral or slightly acidic methanol, or by aqueous ethanol. Acyl groups are readily cleaved from other compounds, such as $Me_3SiCOPh$ using similar reagents[505–509].

The esters $R_3SiCH_2COOR'$ and the ketones $R_3SiCH_2COR'$ are cleaved[501,510–512] readily by alkalis and by proton acids to form $CH_3COOR'$ and $CH_3COR'$, respectively, for example reactions 178 and 179[512]. Hot water alone is sufficient to cleave[513]

$$Me_3SiCH_2COOEt + EtOH \longrightarrow CH_3COOEt + Me_3SiOEt \qquad (178)$$

$$Me_3SiCH_2COOEt + HCl \longrightarrow CH_3COOEt + Me_3SiCl \qquad (179)$$

carbon—silicon bonds in $Me_3SiCH_2CN$ and $Me_3SiOSiMe_2CH_2CN$, although acids and alkalis will catalyse the reactions. In each case methyl cyanide is formed. Ethyl cyanide is formed by the alkali cleavage[514] of $Cl_3SiCHMeCN$.

Severe conditions are required to cleave the carbon—silicon bond in $H_3SiCH{=}CH_2$ to form ethylene. The reaction apparently takes place[515] with 30% aqueous sodium hydroxide after 15 h at 170 °C. The presence of halogen groups greatly aids protodemetallation. Thus $Cl_3SiCH{=}CCl_2$ reacts[516,517] with aqueous alkali to form $CH_2{=}CCl_2$ and analogous cleavages occur with $Cl_3SiCCl{=}CH_2$[516] and $Cl_3SiCBr{=}CHBr$[518]. Ethylene is produced by the action of concentrated sulphuric acid on $Me_3SiCH{=}CH_2$[519] or on $Me_2(PhCH_2)SiCH{=}CH_2$[520]. Hydrogen bromide or hydrogen iodide will protodemetallate[281] $(Me_3Si)_2C{=}CH_2$ at low temperatures, as in reaction 180.

$$(Me_3Si)_2C{=}CH_2 + HI \longrightarrow Me_3SiCH{=}CH_2 + Me_3SiI \qquad (180)$$

Sulphuric acid or hot methanolic alkali cleave $Me_3SiCH_2CH{=}CH_2$ to produce propylene[521]. The compounds $Cl_3SiCH_2CH{=}CH_2$ and $Cl_3SiCH_2CH{=}CHMe$ give[522] propylene and but-2-ene with hot aqueous alkali.

### 2. *Kinetic studies on the acidic protodemetallation*

A great deal of work, much of it by Eaborn and coworkers, has been published on the kinetics of protodemetallation of arylsilicon compounds by acids. In the main, aryltrimethylsilanes have been used, and the acids employed have included sulphuric acid[151,523–528], $p$-toluenesulphonic acid[529–531], and perchloric acid[532,533] in aqueous acetic acid. Hydrochloric acid or perchloric acid in aqueous methanol, ethanol, or dioxane[23,152,527,533–540] and anhydrous and aqueous trifluoroacetic acid[25] have also been used as reagents. Solvent isotope studies showed that the cleavage of $p$-$MeOC_6H_4SiMe_3$ was 1.55 times faster[23] in $HCl$–$H_2O$–dioxane than in $DCl$–$D_2O$–

dioxane and 7.3 times faster[25] in $CF_3COOH$–$H_2O$ than in $CF_3COOD$–$D_2O$. These results imply that step 181, the proton transfer to carbon, is rate determining.

$$H_3O^+ + Me_3SiAr \longrightarrow Me_3Si\overset{+}{A}rH + H_2O \quad (181)$$

$$Me_3Si\overset{+}{A}rH \longrightarrow ArH + Me_3Si^+ \quad (182)$$

The Wheland intermediate, **31**, is stabilized compared with **32** because of the electron release of the $Me_3Si$ group and thus arylsilicon compounds are cleaved more

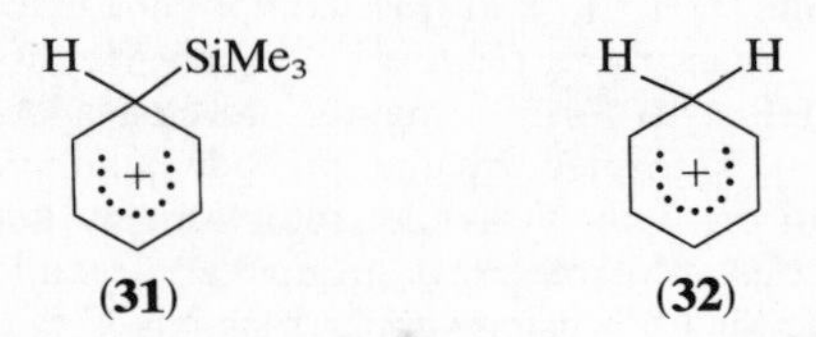


**(31)** **(32)**

readily than simple aryl compounds[23]. In a later study, Eaborn *et al.*[24] measured isotope effects on the cleavage of $XC_6H_4SiMe_3$ ($X = p$-OMe, *p*-Me, H, *p*-Cl, and *m*-OMe) by 1:1 $CF_3COOH$–$CF_3COOD$. The results indicated that there was no significant secondary solvent isotope effect for the proton transfer from the acids. Increasing electron release by the metals Si, Ge, Sn, and Pb results in an increasing ease of cleavage[527] of compounds $PhMEt_3$, M = Si (1) < Ge (36) < Sn ($3.5 \times 10^5$) < Pb ($2 \times 10^8$). Electron-donating substituents in the aryl ring accelerate the cleavage and electron-attracting groups retard reaction. Some kinetic results for compounds $Me_3SiC_6H_4X$ are collected in Tables 104 and 105, and in Table 106 are given the

TABLE 104. Relative rates of cleavage[a] of $Me_3SiC_6H_4X$ to form $C_6H_5X$ using $HClO_4$–MeOH–$H_2O$ at 50 °C

| X | $k_{rel}$ | X | $k_{rel}$ |
|---|---|---|---|
| *o*-OMe | 335 | *p*-Ph | 3.5 |
| *o*-Me | 17 | *p*-F | 0.75 |
| *o*-OPh | 8.7[b] | *p*-Cl | 0.13 |
| *o*-SPh | 1.3[b] | *p*-Br | 0.10 |
| *m*-$Bu^t$ | 3.8 | H | 1.0 |
| *m*-Me | 2.5 | *o*-$CH_2SiMe_3$ | 31 |
| *m*-OMe | 0.5 | *o*-$(CH_2)_2SiMe_3$ | 17 |
| *m*-OPh | 0.36 | *o*-$(CH_2)_3SiMe_3$ | 12 |
| *m*-SMe | 0.19 | *o*-$(CH_2)_4SiMe_3$ | 13 |
| *p*-$NMe_2$ | $\sim 3 \times 10^7$ | *m*-$CH_2SiMe_3$ | 6.6 |
| *p*-OH | 10 700 | *m*-$(CH_2)_2SiMe_3$ | 3.6 |
| *p*-OMe | 1510 | *m*-$(CH_2)_3SiMe_3$ | 3.8 |
| *p*-OPh | 88[b] | *m*-$(CH_2)_4SiMe_3$ | 3.6 |
| *p*-SMe | 78 | *m*-$CH(Pr^n)SiMe_3$ | 8.2 |
| *p*-Me | 21 | *p*-$SiMe_3$ | 1.2 |
| *p*-$Bu^t$ | 15.6 | *p*-$CH_2SiMe_3$ | 315 |
| *p*-SPh | 10.7[b] | *p*-$(CH_2)_2SiMe_3$ | 28 |
| *p*-$PhCH_2$ | 7.8 | *p*-$(CH_2)_3SiMe_3$ | 22 |
| *p*-Et | 19.5 | *p*-$(CH_2)_4SiMe_3$ | 24 |
| *p*-$Pr^i$ | 17.2 | *p*-$CH(Pr^n)SiMe_3$ | 260 |

[a] From refs. 533, 534, 536, 541, and 542.
[b] At 25 °C.

TABLE 105. Relative rates of cleavage[a] of $Me_3SiC_6H_4X$ to form $C_6H_5X$ using $H_2SO_4$–$CH_3COOH$–$H_2O$ at 50 °C

| X | $k_{rel}$ | X | $k_{rel}$ |
|---|---|---|---|
| *p*-OMe | 1000 | *o*-F | 0.07 |
| *p*-$CH_2SiMe_3$ | 200 | *o*-Cl | 0.03 |
| *p*-Me | 18 | *m*-Cl | 0.012 |
| *p*-(4-$MeOC_6H_4$) | 9 | *m*-Br | 0.012 |
| *o*-$SiMe_3$ | 9 | *m*-COOH[b] | $9\times10^{-3}$ |
| *o*-$Bu^t$ | 8 | *o*-$SO_3H$ | $3\times10^{-3}$ |
| *o*-Ph | 6 | *p*-COOH[b] | $2\times10^{-3}$ |
| *p*-Ph | 2.8 | *p*-COOMe | $2\times10^{-3}$ |
| H | 1.0 | *m*-$CF_3$ | $2\times10^{-3}$ |
| *p*-F | 0.95 | *m*-P(O)$(OH)_2$ | $1\times10^{-3}$ |
| *p*-$CH_2P(O)(OEt)_2$ | 0.70 | *p*-$SO_3H$ | $1\times10^{-3}$ |
| *m*-OMe | 0.45 | *p*-$NMe_3^+$ | $4\times10^{-4}$ |
| *m*-Ph | 0.33 | *m*-$NO_2$ | $3\times10^{-4}$ |
| *p*-(4-$NO_2C_6H_4$) | 0.3 | *p*-$NO_2$ | $1\times10^{-4}$ |
| *m*-$CH_2P(O)(OEt)_2$ | 0.21 | *o*-$NO_2$ | $7\times10^{-5}$ |
| *p*-Cl | 0.19 | *p*-$NMe_3^+$ | $4\times10^{-5}$ |
| *p*-Br | 0.10 | *p*-$PMe_3^+$ | $\sim7\times10^{-5}$ |
| *p*-I | 0.10 | *p*-P(O)$Ph_2$ | $\sim4\times10^{-5}$ |

[a] From refs. 151, 523–526, and 541–543.
[b] Values vary with acidity.

relative rates of cleavage of the compounds *p*-$MeOC_6H_4SiR_3$ in which electron-withdrawing groups retard the reaction and electron-donating groups accelerate it, although steric factors are also important (compare, for example, the values for the electronically similar substituents *p*- and *o*-$MeC_6H_4$).

Eaborn and coworkers[544] reported that $PhCH_2SiMe_3$ reacts slowly with $CF_3COOH$ at 70 °C to give a mixture of products, including $Me_3SiCOOCF_3$, $C_6H_5CH_3$, *o*- and *p*-$PhCH_2C_6H_4CH_2SiMe_3$, $C_6H_5CH_2CH_2C_6H_5$, and *o*- and *p*-$PhCH_2C_6H_5CH_3$. However, the cleavage of *m*-$MeOC_6H_4CH_2SiMe_3$ occurred readily, to produce *m*-$MeOC_6H_4CH_3$ only. It was suggested that some of the additional products formed from $PhCH_2SiMe_3$ occur as a result of the reaction between one of the major products, $Me_3SiCOOCF_3$, and $PhCH_2SiMe_3$, and various reaction schemes were postulated.

TABLE 106. Relative rates of cleavage[305,539] of *p*-$MeOC_6H_4SiR_3$ to form *p*-$MeOC_6H_5$ using $HClO_4$–MeOH–$H_2O$ at 50 °C

| $R_3$ | $k_{rel}$ | $R_3$ | $k_{rel}$ |
|---|---|---|---|
| $Me_3$ | 1000 | (*p*-$Me_3SiCH_2C_6H_4)_3$ | 74 |
| $Et_3$ | 490 | (*p*-$MeC_6H_4)_3$ | 36 |
| $Pr^n{}_3$ | 420 | (*p*-$ClC_6H_4)_3$ | 5.4 |
| $Pr^i{}_3$ | 55 | (*o*-$MeC_6H_4)_3$ | ~0.075 |
| $Me_2Ph$ | 330 | $(PhCH_2)_3$ | 88 |
| $MePh_2$ | 74 | $Me_2(ClCH_2)$ | 120 |
| $Ph_3$ | 16 | $Me(ClCH_2)_2$ | 40 |
| (*p*-$MeOC_6H_4)_3$ | ~75 | $Me_2(CH_2NH_2Et^+)$ | ~15 |

### 3. Kinetic studies on the base-catalysed protodemetallation

It has been known for some time that although simple alkyl and aryl groups are not cleaved from silicon in the presence of base, various other organic groups are more or less readily removed to yield the corresponding hydrocarbon. Gilman *et al.*[545] found the following order of ease of removal of the group R from $Ph_3SiR$ by alkali: R = PhC≡C > 1-indenyl, 9-fluorenyl > $Ph_2CH$ > $PhCH_2$ > $m$-$CF_3C_6H_4$, $p$-$ClC_6H_4$ > $\alpha$-$C_{10}H_7$ > Ph, $PhCH_2CH_2$, $n$-$C_6H_{13}$ ≈ 0 and Hauser and Hance[546] found the sequence R = $Ph_3C$ > $Ph_2CH$ > $PhCH_2$.

Eaborn and coworkers have carried out numerous kinetic and mechanistic studies on the more reactive substrates in the above sequence. The base-catalysed protodemetallation of compounds of type $R_3SiC{\equiv}CC_6C_4X$ by aqueous methanolic alkali yields the corresponding $HC{\equiv}CC_6H_4X$ compounds; relative rate constants are given in Table 107[547] and show that electron-withdrawing groups R or X have marked effects on the rates.

Eaborn and coworkers[548] showed that fluoren-9-yl-$SiM_3$ (**33**) is cleaved very readily by aqueous methanolic alkali. They also reported[548] rate constants for the cleavage by aqueous methanolic alkali of some trialkylaryl compounds of silicon, germanium, and tin, which are given in Table 108 and which show that the order of reactivity with respect to the metal atom is Sn > Si > Ge. There is also a possible steric effect exerted by the leaving group, with $Me_3SiCHPh_2$ reacting faster than the triethyl analogue. Relative rates[540,549,550] for the protodemetallations of the compounds $Me_3SiCH_2C_6H_4X$ by aqueous methanolic alkali, producing $CH_3C_6H_4X$, are given in Table 109; the large substituent effects imply that considerable charge must be present on the benzyl group in the transition state. In a later publication[551], relative

TABLE 107. Relative rate constants[547] at 25 °C for the cleavage of $R_3SiC{\equiv}CC_6H_4X$ by aqueous methanolic alkali

| R | X | $k_{rel}$ | R | X | $k_{rel}$ |
|---|---|---|---|---|---|
| Et | H | 1.00 | $m$-$ClC_6H_4$ | H | 3776 |
| Me | H | 280 | Et | $p$-Me | 0.56 |
| $p$-$MeOC_6H_4$ | H | 2.83 | Et | $o$-Me | 0.41 |
| $p$-$MeC_6H_4$ | H | 3.07 | Et | $o$-Cl | 1.81 |
| Ph | H | 11.8 | Et | $p$-Cl | 2.56 |
| $p$-$ClC_6H_4$ | H | 1156 | Et | $m$-Br | 4.5 |

TABLE 108. Second-order rate constants[548], $10^6k_2$, for the cleavage of some organometallic compounds by NaOH in 80% v/v aqueous methanol at 50 °C

| Substrate | $10^6k_2$ ($l\,mol^{-1}\,s^{-1}$) | Substrate | $10^6k_2$ ($l\,mol^{-1}\,s^{-1}$) |
|---|---|---|---|
| $Me_3SnCH_2Ph$ | 10.1 | $Me_3GeCHPh_2$ | 3.7 |
| $Me_3SiCH_2Ph$ | 0.336 | $Et_3SiCHPh_2$ | 1.1 |
| $Me_3SiCPh_3$ | 630 | $Et_3GeCHPh_2$ | 0.05 |
| $Me_3GeCPh_3$ | 29.8 | $Et_3GeCPh_3$ | 0.16 |
| $Me_3SiCHPh_2$ | 480 | | |

TABLE 109. Relative rates of cleavage[540,549,550] of $Me_3SiCH_2C_6H_4X$ by aqueous methanolic alkali at 50 °C

| X | $k_{rel}$ | X | $k_{rel}$ |
|---|---|---|---|
| $p$-$NO_2$ | $1.8\times10^6$ | $m$-$CF_3$ | 100 |
| $o$-$NO_2$ | $5.0\times10^5$ | $o$-Cl | 80 |
| $p$-$PMe_3^+$ | $4.2\times10^5$ | $m$-Cl | 63 |
| $p$-COPh | $1.6\times10^5$ | $p$-Cl | 14 |
| $p$-$SO_2Ph$ | $1.5\times10^5$ | $p$-$SiMe_3$ | 9.6 |
| $p$-$P(O)Ph_2$ | $1.5\times10^4$ | $m$-$CO_2^-$ | 2.1 |
| $m$-$PMe_3^+$ | $4.6\times10^3$ | H | 1.0 |
| $p$-$CONH_2$ | $6.3\times10^2$ | p-Me | 0.2 |
| $o$-I | $1.4\times10^2$ | $p$-OMe | ~0.02 |

TABLE 110. Relative rates of cleavage[551] of $Me_3SiR$ by methanolic sodium methoxide at 25 °C

| R | $k_{rel}$ | $E_a$ (kcal mol$^{-1}$) | Log $A$ |
|---|---|---|---|
| $C_6H_5CH_2$ | 1.0 | 26.1 | 11.2 |
| $m$-$ClC_6H_4CH_2$ | 100 | 22.6 | 10.6 |
| $m$-$NCC_6H_4CH_2$ | 2800 | | |
| $Ph_2CH$ | 4700 | 18.8 | 9.2 |
| $Ph_3C$ | 7600 | 17.2 | 8.3 |
| 3,5-$Cl_2C_6H_3CH_2$ | 8200 | 21.2 | 11.1 |
| $p$-$NCC_6H_4CH_2$ | $6.3\times10^5$ | | |
| 9-Methylfluoren-9-yl | $3.9\times10^5$ | | |
| Fluoren-9-yl | $2.2\times10^7$ | | |

rates were reported for the cleavage of some other trimethylorganosilicon compounds by methanolic sodium methoxide and these, together with some activation parameters, are given in Table 110. Electron withdrawal in the leaving group —$SiR'_3$ aids reaction, as shown by the kinetic data[552] in Table 111. These data also show that steric effects are important.

Values of the product isotope ratios[553,554], $k_H/k_D$, for the cleavage of benzyl—silicon bonds using an equimolar mixture of MeOH and MeOD at 50 °C for the

TABLE 111. Approximate relative rates[552] for cleavage of various compounds by aqueous methanolic alkali

| $R'_3SiCH_2C_6H_4$-$p$-Cl | | $R'_3SiCH_2C_6H_4$-$p$-$COO^-$ | |
|---|---|---|---|
| $R'_3$ | $k_{rel}$ | $R'_3$ | $k_{rel}$ |
| $PhMe_2$ | 100 | $Me_3$ | 100 |
| $p$-$MeC_6H_4Me_2$ | 60 | $Et_3$ | 20 |
| $p$-$MeOC_6H_4Me_2$ | 45 | $Pr^n_3$ | 12 |
| $p$-$ClC_6H_4Me_2$ | 300 | $PhMe_2$ | 480 |
| $Me_3$ | 17 | $Ph_2Me$ | 480 |
| | | $Ph_3$ | 230 |

TABLE 112. Relative rates of cleavage[555] of $XC_6H_4SiMe_3$ by hydroxide ion in 1:6 v/v water–dimethyl sulphoxide at 75 °C

| X | $k_{rel}$ | X | $k_{rel}$ |
|---|---|---|---|
| *m*-Cl | 180 | *m*-OMe | 2.14 |
| *p*-Cl | 30 | H | 1.00 |
| *p*-F | 7.3 | *p*-OMe | 0.31 |
| *p*-SMe | 3.2 | | |

compounds $Me_3SiCH_2C_6H_4X$ were X = H 1.4, *p*-Me 1.5, *m*-Cl 1.6, and *m*-$CF_3$ 1.6, and implied that carbanions were formed from species such as $[Ar—CH_2 \cdots SiMe_3—OMe]^-$. These carbanions then reacted with a solvent molecule to form the products (steps 183 and 184).

$$^{-}OMe + Me_3SiCH_2Ar \longrightarrow Me_3SiOMe + {}^{-}CH_2Ar \tag{183}$$

$$^{-}CH_2Ar + MeOH \longrightarrow CH_3Ar + {}^{-}OMe \tag{184}$$

In another study, Eaborn and coworkers[555] reported on the cleavage of $XC_6H_4MMe_3$ (M = Si, Sn) in 1:6 v/v water–dimethyl sulphoxide containing potassium hydroxide. In all cases, hydrogen transfer from solvent to the aromatic centre was concerted with cleavage of a C—H or C—$MMe_3$ bond by base attack. Electron-releasing substituents assist the electrophilic attack of the solvent. Relative rates of reaction are given in Table 112.

Eaborn and coworkers[556] investigated the base cleavages of $XC_6H_4Me_2SiCH_2C_6H_4Cl$-*p* and $XC_6H_4Me_2SiC_8H_5O$ ($C_8H_5O$ = 2-benzofuryl). Rates were measured in aqueous methanolic KOH and in 1:6 water–dimethyl sulphoxide containing borax, and these are given in Table 113. In each case the rate-determining step was thought to be the breaking of the Si—aryl bond of the pentacoordinate silicon intermediate $[R_3Si(OH)Ar]^-$, which is formed rapidly and reversibly in a pre-equilibrium step.

TABLE 113. Pseudo-first-order rate constants, $10^5k$, at 50.2 °C for cleavage of various compounds in 7.5% aqueous methanol containing 0.82 M potassium hydroxide

| $XC_6H_4Me_2SiCH_2C_6H_4$-*p*-Cl | | $XC_6H_4Me_2SiC_8H_5O^a$ | | |
|---|---|---|---|---|
| X | $10^5k$ ($s^{-1}$) | X | $10^5k$ ($s^{-1}$) | $10^3k$ ($s^{-1}$)$^b$ |
| *m*-Cl | 10.4 | *m*-Cl | 120 | 67 |
| *p*-Cl | 7.4 | *p*-Cl | 67 | 44 |
| H | 2.10 | H | 20.6 | 7.5 |
| *m*-Me | 1.66 | *m*-Me | 15.7 | 6.2 |
| *p*-Me | 1.12 | *p*-Me | 10.1 | 5.1 |
| *p*-OMe | 1.05 | *p*-OMe | 9.3 | 3.6 |

$^a$ $C_8H_5O$ = 2-benzofuryl.
$^b$ Pseudo-first-order rate constants for reaction in 6:1 dimethyl sulphoxide–0.01 M aqueous borax at 30.1 °C.

Roberts and El Kaissi[557] studied the cleavage of trimethyl-3-phenallylsilicon by aqueous alcoholic alkali, which proceeded by the two steps 185 and 186 (R= $CH_2CH$=CHPh).

$$Base^- + Me_3SiR \xrightarrow{slow} Base—SiMe_3 + R^- \quad (185)$$

$$R^- + solvent \xrightarrow{fast} RH + Base^- \quad (186)$$

Since the base is regenerated in step 186, the reaction is actually first order in organosilicon, but second order rate constants may be calculated from $k_2 = k_1^{obs}/[Base^-]$. At 40 °C the values of $k_2$ were $9.7\times10^{-5}$ and $1.43\times10^{-4}$ l mol$^{-1}$ s$^{-1}$ for reaction in 60% EtOH and 60% MeOH, respectively. On the basis of an isotope effect $k(H_2O)/k(D_2O) = 0.5$ for the cleavage of $Me_3SiCH_2CH$=CHPh in aqueous methanol, these workers suggested that bond making and bond breaking in the transition state **34** are of equal importance.

$$\left[\overset{\delta-}{Base}\cdots Si\cdots \overset{\delta-}{C}H_2CH{=}CHPh\right]^{\neq}$$

**(34)**

It was first reported[557] that the hydrocarbon product from reactions 185 and 186 (R = $CH_2CH$=CHPh) was $\beta$-methylstyrene ($CH_3CH$=CHPh), formed without rearrangement, but in a later publication[558] this statement was corrected, and the products obtained in this reinvestigation were 60% $\beta$-methylstyrene and 40% allylbenzene.

In this later paper, Rennie and Roberts[558] reported on the cleavage of benzyl and cinnamyl derivatives of silicon, germanium, and tin in aqueous and alcoholic dimethyl sulphoxide containing a large excess (to ensure psuedo-first order kinetics) of base NaOR′ (R′ = H, Me, Et, Pr, $Bu^t$, $Pe^{neo}$). The observed rate constant depended on the basicity of the medium and the true second order rate constants were obtained by dividing the observed constant by [NaOR′]. In 68.3 mol-% aqueous dmso at 40 °C, relative rates for $Et_3MCH_2CH$=CHPh were M = Sn ($2.8\times10^5$) ≫ Si (50) > Ge (1.0). The reaction rates decreased on passing from aqueous to alcoholic media, and rates for $R_3SiCH_2CH$=CHPh (R = Me, Et) are given in Table 114. The kinetic solvent isotope effect ($k_{H_2O}/k_{D_2O}$) was measured for $R_3SiCH_2CH$=CHPh (R = Me, Et) in aqueous dmso, and the values (less than unity) implied that there was no significant rate-determining proton transfer. Activation parameters were also reported for cleavage of trimethyl- and triethyl-cinnamylsilicon in $H_2O$–dmso and $Bu^tOH$–dmso mixtures.

TABLE 114. Second-order rate constants, $k_2$ (l mol$^{-1}$ s$^{-1}$), for the cleavage of $R_3SiCH_2CH$=CHPh by NaOR′ in R′OH–dmso mixtures ($\chi_{dmso} = 0.5$)[558]

| | R′ | | | | | |
|---|---|---|---|---|---|---|
| R | Me | Et | $Pr^n$ | $Bu^i$ | $Pe^{neo}$ | H |
| Me | 0.341 | 0.442 | 0.447 | 0.424 | 0.492 | 6.93 |
| Et | 0.003 | 0.003 | | | | 0.13 |

The cleavage of 3-phenallyl—silicon bonds by methanolic sodium methoxide was also studied by Eaborn *et al.*[559], who found that *trans*-PhCH=CHCH$_2$SiR$'_3$ (R′ = Me, Et) gave *trans*-PhCH=CHCH$_3$ and PhCH$_2$CH=CH$_2$ in relative amounts which depended on the base concentration. Product– and rate–isotope effects were measured, and the general conclusion was that the cleavage proceeds by an electrophilically assisted mechanism in which the proton transfers from the solvent to either the $C_{(1)}$ or $C_{(3)}$ carbon atoms of the phenallyl group are concerted with the cleavage of the Si—C bond. Rate constants for the cleavage of PhCH=CHCH$_2$SiR$_3$ were R = Me $13.85 \times 10^{-5}$ and Et $6.6 \times 10^{-7}$ l mol$^{-1}$ s$^{-1}$.

## F. Protodemetallation of Organogermanium Compounds

A number of preparative studies have been reported on these protodemetallations, usually under acidic conditions, and there have also been kinetic investigations into acid- and base-catalysed protodemetallations. Cleavage of one, or sometimes two, organo groups from an R$_4$Ge compound by the action of halogen acids often requires the presence of AlCl$_3$ as a catalyst. Reaction 187 has been studied for X = Br,

$$R_4Ge + HX \xrightarrow{AlX_3} RH + R_3GeX \tag{187}$$

R = Me[560], X = Cl, R = Me[561], and X = F, R = Me, Et[562]. Hydrogen bromide in chloroform at room temperature has been shown[563] to cleave a variety of symmetrical and unsymmetrical organogermanium compounds; *m*- and *p*-MeC$_6$H$_4$GePh$_3$ both give Ph$_3$GeBr and PhCH$_3$ while for *p*-MeC$_6$H$_4$Ge (*m*-MeC$_6$H$_4$)$_3$ the *p*-MeC$_6$H$_4$—Ge bond is cleaved. The rate of cleavage increases along the series PhCH$_2$ < Ph < *m*-MeC$_6$H$_4$ < *p*-MeC$_6$H$_4$. Allylgermanium compounds are cleaved by HBr; for example[318] *n*-Bu$_3$GeCH$_2$CH=CH$_2$ gives CH$_3$CH=CH$_2$ and *n*-Bu$_3$GeBr. Allyl and propargyl groups are both rapidly cleaved by carboxylic acids at 20 °C, with vinyl and ethynyl groups being cleaved more slowly[564].

TABLE 115. Relative rates of cleavage[565] of XC$_6$H$_4$GeEt$_3$ by aqueous methanolic perchloric acid at 50 °C

| X | $k_{rel}$ | X | $k_{rel}$ |
|---|---|---|---|
| *p*-NMe$_2$ | ~$3 \times 10^6$ | *m*-Bu$^t$ | 3.3 |
| 2,4,6-Me$_3$[a] | 13 600 | *o*-Ph | 3.22 |
| *p*-OH | 2730 | *p*-Ph | 2.69 |
| *p*-OMe | 540 | *m*-Me | 2.11 |
| *o*-OMe | 207 | 3,4-C$_4$H$_4$[c] | 1.79 |
| *p*-CH$_2$SiMe$_3$ | 162 | H | 1.0 |
| *p*-OPh | 36.7 | *p*-F | 0.93 |
| *p*-Me | 14.0 | *m*-MeO | 0.57 |
| *p*-Et | 13.0 | *p*-Cl | 0.167 |
| *o*-Me | 12.4 | *p*-I | 0.131 |
| *p*-Pr$^i$ | 12.0 | *p*-Br | 0.127 |
| *p*-Bu$^t$ | 11.5 | *m*-Cl | 0.0165 |
| 2,3-C$_4$H$_4$[b] | 6.2 | | |

[a] Denotes mesityltriethylgermane.
[b] Denotes 1-naphthyltriethylgermane.
[c] Denotes 2-naphthyltriethylgermane.

TABLE 116. Relative rates of cleavage[566] of $XC_6H_4GeEt_3$ by $H_2O$–$CH_3COOH$–$H_2SO_4$ at 50 °C

| X | $k_{rel}$ | X | $k_{rel}$ |
|---|---|---|---|
| *p*-OMe | 392 | *m*-Br | 0.019 |
| *o*-OMe | 187.5 | *m*-Cl | 0.019 |
| *p*-Me | 12.4 | *m*-COOH | 0.018 |
| *p*-Ph | 2.43 | *m*-$CF_3$ | $5.4\times10^{-3}$ |
| *m*-Me | 1.78 | *p*-$CF_3$ | $2.5\times10^{-3}$ |
| H | 1.0 | *p*-COOH | $5.2\times10^{-3}$ |
| *m*-OMe | 0.51 | *m*-$NMe_3^+$ | $1.26\times10^{-3}$ |
| *p*-Cl | 0.168 | *p*-$NMe_3^+$ | $1.06\times10^{-3}$ |
| *p*-Br | 0.133 | *m*-$NO_2$ | $8.0\times10^{-4}$ |
| *m*-F | 0.032 | *p*-$NO_2$ | $3.76\times10^{-4}$ |

Eaborn and Pande[565] have reported on the protodemetallation of $XC_6H_4GeEt_3$ by aqueous methanolic perchloric acid, and obtained relative rate constants by the 'overlap method', since a variety of acid concentrations were used. The relative rate constants are given in Table 115 and show that electron-releasing substituents accelerate the cleavage, whereas electron-attracting substituents retard the cleavage, as expected for an aromatic electrophilic substitution. The same workers[566] also obtained relative rate constants for protodemetallation of $XC_6H_4GeEt_3$ using a mixture of acetic and sulphuric acids, and these rate constants are listed in Table 116.

Although simple alkyl- and aryl-germanium compounds are inert towards dilute aqueous alcoholic sodium hydroxide, groups able to support a carbanion are cleaved more readily, and Table 108 includes rate constants reported by Eaborn and coworkers[548] for the base-catalysed protodemetallation of some organogermanium compounds.

## G. Protodemetallation of Organotin Compounds

A number of preparative studies, usually on acidic protodemetallations, have been reported, together with a stereochemical study. Sisido and coworkers[567,568] have shown that protodemetallation of the optically active compound (1-methyl-2,2-diphenylcyclopropyl)trimethyltin (**35**) by HCl or HBr in tetrachloromethane proceeds with retention of optical activity and configuration.

Good yields of trialkyltin salts are reported to be obtained by the action of dry hydrogen chloride in diethyl ether on long-chain ($C_{12}$–$C_{18}$) tetraalkyltins[569] and by the action of carboxylic acids on tetraethyltin or tetra-*n*-propyltin at 275 °C; with tetra-iso-propyltin under the latter conditions, two alkyl groups were removed[570]. Lesbre and Dupont[571] have also studied the cleavage of $R_4Sn$ ($R = Me, Et, Pr^n, Bu^n$) by various carboxylic acids under a variety of conditions, and showed that the ease of cleavage increased in the order $R = Bu^n < Pr^n < Et < Me$. Aryl—tin bonds are broken more easily, and HCl in chloroform will cleave tetraphenyltin[572]; HBr at −78 °C will protodemetallate $Ph_3SnH$, cleaving two Ph—Sn bonds[573]. Above 125 °C, formic and acetic acids will cleave all four bonds in tetraphenyltin[574]. Similarly, all four vinyl groups are cleaved from tetravinyltin under almost the same conditions[575], but with unsymmetrical alkylvinyltins only the vinyl groups are removed by the action of carboxylic acids[332,333].

TABLE 117. Second-order rate constants[577], $10^4k_2$, for the acidolysis of $R_4Sn$ and $R_3SnCl$ by hydrochloric acid in methanol at 50 °C

| R | $10^4k_2$ (l mol$^{-1}$ s$^{-1}$) | |
|---|---|---|
| | $R_4Sn$ | $R_3SnCl$ |
| Me | 2.30 | 0.173 |
| Bu$^n$ | 0.82 | 0.400 |
| Pr$^i$ | 0.68 | 0.192 |

Although tetraalkyltins are not very reactive towards acids, relative rates of the hydrogen chloride cleavage of $R_4Sn$ compounds in dioxane have been reported as R = Me 1 and Et 3; in benzene the relative rates are R = Me 1, Et 7.5, Pr$^n$ 3, and Pr$^i$ 3[576]. Rate constants for the protodemetallation of tetraalkyltins and trialkyltin chlorides by HCl in methanol have been obtained by Bade and Huber[577], and are given in Table 117.

Eaborn and coworkers have studied the acidolysis of aryltin compounds of type $ArSnR_3$ (R = Me or cyclohexyl). Relative rate constants for the protodemetallation of a series of substituted compounds, $XC_6H_4SnR_3$, by aqueous ethanolic perchloric acid were determined by an overlap procedure[578], and are listed in Table 118. Eaborn and coworkers[23,24] also showed from product isotope effects for cleavage by 1:1 mixtures of $CH_3COOH$ and $CH_3COOD$ or by HCl in $H_2O$ and $D_2O$ that the rate-determining stage in these aromatic electrophilic substitutions was proton transfer in a Wheland intermediate. The cleavage of the aryl—tin bond in the compounds *p*-$Me_3MCH_2C_6H_4SnMe_3$ by aqueous ethanolic perchloric acid was found to take place in order of increasing tendency M = Si 1 < Ge 1.35 < Sn 3.21, thus indicating that the order of inductive electron release is $Me_3Si < Me_3Ge < Me_3Sn$[579].

Protodemetallation of a similar series, $ArSnR_3$, has also been investigated by Nasielski and coworkers[580,581], who reported rate constants for cleavage by acetic acid and by HCl in methanol; results are given in Table 119. The general patterns of

TABLE 118. Relative rates of cleavage[578] of $XC_6H_4SnR_3$ by aqueous ethanolic perchloric acid at 50 °C

| X[a] | $k_{rel}$ | X[a] | $k_{rel}$ | X[b] | $k_{rel}$ |
|---|---|---|---|---|---|
| *p*-NMe$_2$ | ~2 × 10$^4$ | *p*-Ph | 1.77 | *p*-Me | 3.4 |
| *p*-MeO | 63 | H | 1.0 | H | 1.0 |
| *p*-Bu$^t$ | 7.0 | *m*-MeO | 0.85 | *p*-Cl | 0.38 |
| *p*-Pr$^i$ | 7.0 | *p*-F | 0.62 | *m*-Cl | 0.18 |
| *p*-Me | 5.6 | *p*-Cl | 0.187 | | |
| *p*-Et | 5.3 | *p*-Br | 0.145 | | |
| *o*-Ph | 1.99 | *m*-Cl | 0.039 | | |
| *m*-Me | 1.85 | *p*-COOH | 0.03 | | |

[a] R = cyclohexyl.
[b] R = Me.

TABLE 119. Second-order rate constants, $k_2$, at 25 °C and activation energies for the protodemetallation of $ArSnR_3$

| Ar | R | $10^2 k_2$ ($l\,mol^{-1}\,s^{-1}$)[a] | $E_a$ ($kcal\,mol^{-1}$)[a] | $10^5 k_2$ ($l\,mol^{-1}\,s^{-1}$)[b] |
|---|---|---|---|---|
| $p$-$BrC_6H_4$ | Me | 1.7 | 16.2 | 1.26 |
| $C_6H_5$ | Me | 3.24 | 16.8 | 4.28 |
| $m$-$MeC_6H_4$ | Me | 5.62 | 16.6 | 6.01 |
| $o$-$MeC_6H_4$ | Me | — | — | 17.6 |
| $p$-$MeC_6H_4$ | Me | 18.6 | 14.7 | 22.1 |
| $p$-$OMeC_6H_4$ | Me | 158 | 13.2 | 304 |
| Naphth-1-yl | Me | 6.31 | 14.4 | |
| Naphth-2-yl | Me | 6.31 | 12.6 | |
| Phenanthr-9-yl | Me | 4.47 | 13.7 | |
| $C_6H_5$ | $Bu^n$ | 1.7 | 14.5 | |
| Naphth-1-yl | $Bu^n$ | 3.8 | 15.1 | |
| Naphth-2-yl | $Bu^n$ | 2.76 | 15.0 | |
| $C_6H_5$ | $Pr^i$ | 0.576 | 14.9 | |
| Naphth-1-yl | $Pr^i$ | 2.24 | 15.3 | |
| Naphth-2-yl | $Pr^i$ | 1.07 | 9.7 | |

[a] Using HCl in methanol[580].
[b] Using acetic acid[581].

substituent effects observed by Nasielski and coworkers[580,581] and Eaborn and Waters[578] are similar, but the former workers showed also that steric effects in the leaving group $SnR_3$ (R = Me, $Bu^n$, $Pr^i$) were small.

Two sets of workers have reported on the protodemetallation of benzyl compounds, $XC_6H_4CH_2SnMe_3$, by aqueous methanolic perchloric acid[582] and by an excess of $CF_3COOH$ in benzene[583]. In both acidic systems, for most substituents, X, there is not a great difference in the rate of cleavage of the benzyl—tin or methyl—tin bonds, but when X = $m$-Me or $m$-OMe, the benzyl—tin cleavage is much faster, indicating a change in mechanism[583] possibly involving ring protonation[582]. For the other substituted compounds, direct electrophilic attack at the benzylic carbon atom was suggested. Rate constants for the $CF_3COOH$–benzene system are given in Table 120 and relative rate constants for the methanolic perchloric acid system in Table 121; in the latter case, these vary with acid concentration and the values given are for the highest acid concentration used. Relative rate constants for cleavage by methanolic perchloric acid were also obtained for the series of compounds $PhCH_2SnMe_3$ 1, $PhCHMeSnMe_3$ 0.96, and $PhCMe_2SnMe_3$ 0.39, consistent again with direct electrophilic attack at the benzylic carbon atom[582].

Kuivila and coworkers[584–586] have studied the hydrogen chloride cleavage of a number of allyl and allenyltin compounds in methanol, aqueous methanol, acetonitrile, dioxane, benzene, and dimethylsulphoxide. In methanol–water (96:4), trimethylallyltin readily reacts with HCl to give $Me_3SnCl$ and propene[584]. The reaction is first order in each reactant; the second-order rate constant is equal to that for cleavage by perchloric acid. The protodemetallation of *cis*- and *trans*-but-2-enyltrimethyltins by HCl in 96% aqueous methanol gave[584] but-1-ene as the major product, together with smaller amounts of *cis*- and *trans*-but-2-enes, the *cis*-isomer

TABLE 120. First-order rate constants[583], $10^5k$, for the cleavage of $XC_6H_4CH_2SnMe_3$ by an excess of $CF_3COOH$ in benzene

| X | $10^5k$ $(s^{-1})^a$ | $10^5k$ $(s^{-1})^b$ |
|---|---|---|
| H | 12.2 | 3.17 |
| *m*-Cl | 3.17 | 0.83 |
| *p*-Cl | 3.67 | 2.17 |
| *m*-$CF_3$ | 1.50 | 2.17 |
| *p*-$CF_3$ | 1.17 | 0.83 |
| *m*-$CH_3$ | 178 | |
| *p*-$CH_3$ | 21.8 | 5.0 |
| *m*-OMe | Very fast | |
| *p*-OMe | 12.8 | 0.83 |

[a] Benzyl—tin cleavage.
[b] Methyl—tin cleavage.

predominating. This led to 188 as being the predominant pathway (mechanism $S_E2'$) with very little reaction by the $S_E2$ mechanism 189.

$$MeCH{=}CHCH_2SnMe_3 + H_2\overset{+}{O}Me \xrightarrow{S_E2'} MeCH_2CH{=}CH_2 + Sn\overset{+}{M}e_3 + HOMe \tag{188}$$

$$MeCH{=}CHCH_2SnMe_3 + H_2\overset{+}{O}Me \xrightarrow{S_E2} MeCH{=}CHCH_3 + Sn\overset{+}{M}e_3 + HOMe \tag{189}$$

Cleavage by a number of other acids was also studied[586] and it was found that the actual distribution of products depended on the acidity of the protonating agent, the solvent polarity, and reagent concentration, with weak acids, polar solvents, and low concentrations favouring the production of but-2-enes, by reaction 189. Transition states such as **36** and **37** were proposed[584], which correspond to mechanisms $S_E2'$ (open) and $S_E2'$ (cyclic).

TABLE 121. Relative rates of cleavage[582] of benzyl—tin bonds in $XC_6H_4CH_2SnMe_3$ by aqueous methanolic perchloric acid at 50 °C

| X | $k_{rel}$ | X | $k_{rel}$ |
|---|---|---|---|
| H | 1.0 | *p*-F | 0.77 |
| *o*-Me | 0.87 | *o*-Cl | 0.41 |
| *m*-Me | 4.4 | *m*-Cl | 0.55 |
| *p*-Me | 1.57 | *p*-Cl | 0.64 |
| *p*-$Bu^t$ | 1.73 | *o*-Br | 0.42 |
| *o*-F | 0.43 | *m*-OMe | 60 |
| *m*-F | 0.60 | 3,5-$Me_2$[a] | 26 |

[a] Denotes 3,5-$Me_2C_6H_3CH_2SnMe_3$.

TABLE 122. Second-order rate constants for the acidolysis of allyltins[584] and allenyltins[585] by HCl in 96% aqueous methanol at 25 °C

| Substrate | $k_2$ (l mol$^{-1}$ s$^{-1}$) | |
|---|---|---|
| | $S_E2'$ | $S_E2$ |
| $CH_2{=}CMeCH_2SnMe_3$ | 24.8 | |
| $CH_2{=}CHCH_2SnMe_3$ | 0.475 | |
| *cis*-$MeCH{=}CHCH_2SnMe_3$ | 0.0508 | |
| *trans*-$MeCH{=}CHCH_2SnMe_3$ | 0.0274 | |
| $CH_2{=}CHCH_2SnPh_3$ | 0.00441 | |
| $MeCH{=}CHCH_2SnPh_3$ | 0.00032 | |
| *trans*-$MeCH{=}C{=}C(Me)SnMe_3$ | 0.223 | 0.296 |
| *trans*-$PhCH{=}C{=}C(Me)SnMe_3$ | 0.067 | 0.0441 |
| *trans*-$MeCH{=}C{=}C(Ph)SnMe_3$ | 0.0188 | 0.1061 |
| *trans*-$MeCH{=}C{=}C(Me)SnEt_3$ | 0.199 | 0.100 |
| *trans*-$MeCH{=}C{=}C(Me)SnPh_3$ | 0.00657 | 0.00082 |

**(36)** **(37)**

Rate constants for HCl cleavage of various allyltins are given in Table 122 together with rate constants for the acidolyses of allenyltin compounds[585]. For the allenyltins the two possible pathways, 190 and 191, yield, respectively, an allene and an alkyne, for example.

$$trans\text{-}MeCH{=}C{=}C(Me)SnMe_3 + H_2\overset{+}{O}Me \xrightarrow{S_E2} cis\text{-}MeCH{=}C{=}CHMe + \overset{+}{Sn}Me_3 + MeOH \quad (190)$$

$$trans\text{-}MeCH{=}C{=}C(Me)SnMe_3 + H_2\overset{+}{O}Me \xrightarrow{S_E2'} MeCH_2C{\equiv}CMe + \overset{+}{Sn}Me_3 + MeOH \quad (191)$$

In Table 122 the overall rate constants for acidolyses of allenyltins have been partitioned into their $S_E2$ and $S_E2'$ components from the proportions of allene and alkyne formed in the reactions.

The cleavages of *cis*- and *trans*-$PhCH{=}CHSnMe_3$ in $CH_3COOD$–MeOD (1:10) were reported[587] to give *cis*- and *trans*-PhCH=CHD, respectively. The rates of cleavage of various $XC_6H_4CH{=}CHSnMe_3$ compounds were measured and showed very little difference between the *cis* and *trans* isomers. Relative rates at 50 °C were X = H 1.0, *p*-OMe 7.0, *p*-Me 2.3, *m*-Cl 0.34, and *m*-Br 0.36. The results were interpreted in terms of a rate-determining proton transfer to the $\beta$-carbon atoms.

TABLE 123. Relative rates of cleavage[587] of $XC_6H_4CH{=}CHSnMe_3$ in aqueous methanolic sodium hydroxide at 50 °C

| X | *trans* : *cis*[a] | $k_{rel}$ |
|---|---|---|
| *m*-Cl | 77:23 | 1.67 |
| *m*-Br | 90:10 | 1.65 |
| H | 100:0 | 1.0 |
| H | 10:90 | 0.77 |
| *p*-MeO | 100:0 | 0.99 |
| *p*-Me | 100:0 | 0.92 |

[a] Composition of starting material.

However, for cleavage of these same compounds by methanolic alkali the product–isotope and rate–isotope effects indicated[587] that there is a rate-determining proton transfer from the solvent to the carbon atom of the C—Sn bond as it breaks as a result of the attack of the $MeO^-$ ion at the tin atom. The small but significant difference in the rates of cleavage of *cis*- and *trans*-$PhCH{=}CHSnMe_3$ were associated with the existence of a substantial amount of carbanionic character at the $\beta$-carbon in the transition state. Kinetic results for the base cleavage are given in Table 123.

The allyl—tin bond is rapidly broken by base-catalysed protodemetallation, and kinetic and mechanistic studies have been reported by Roberts and coworkers[557,558] and Eaborn *et al.*[559]. Roberts and El Kaissi[557] studied the base-catalysed cleavage of triethyl- and tri-*n*-butyl-3-phenallyltin in 60% aqueous MeOH. The second-order rate constants at 40 °C were $4.6 \times 10^{-3}$ and $3.2 \times 10^{-3}\ l\ mol^{-1}\ s^{-1}$, respectively.

In a subsequent paper, Rennie and Roberts[558] reported that the rates of cleavage of $Et_3SnCH_2CH{=}CHPh$ at 40 °C by various bases, NaOR′, in R′OH–dmso mixtures ($\chi_{dmso} = 0.5$) were R′ = Me 0.282, Et 4.22, $Pr^n$ 4.93, $Bu^i$ 7.45, $Pe^{neo}$ 26.6, and H $128\ l\ mol^{-1}\ s^{-1}$. The kinetic solvent isotope effects were measured for $Et_3SnCH_2CH{=}CHPh$ and $Et_3SnCH_2Ph$ in aqueous dmso ($k_{H_2O}/k_{D_2O}$). The values were less than unity, precluding any significant proton transfer to the incipient carbanion in the transition state, but for the benzyl derivative there was thought to be pronounced carbanionic character in the transition state, since the values were only slightly less than unity. Values of $\Delta H^{\neq}$ and $\Delta S^{\neq}$ were also reported for the cleavage of triethylbenzyltin and triethylcinnamyltin in various aqueous dmso mixtures.

Eaborn *et al.*[559] also studied the cleavage of trimethyl- and triethyl-cinnamyltin by methanolic sodium methoxide. The rates at 50 °C were $2.35 \times 10^{-3}$ and $1.44 \times 10^{-3}\ l\ mol^{-1}\ s^{-1}$, respectively, and mixtures of $PhCH{=}CHCH_3$ and $PhCH_2CH{=}CH_2$ were obtained. The results of isotope studies on rates and products led to the conclusion that the cleavage involved an electrophilically assisted mechanism. Proton transfer from the solvent was concerted with bond breaking, and a transition state such as $[MeO{—}SnMe_3 \cdot R \cdot H \cdots OMe]^{\neq -}$ was proposed (R = cinnamyl).

Eaborn and coworkers[548] studied the kinetics of the alkaline cleavage of a number of trimethylbenzyltin compounds, $Me_3SnCH_2C_6H_4X$. The relative second order rate constants for cleavage in 80% v/v aqueous methanol at 50 °C were X = *m*-Cl 32.8, *p*-Br 12.8, *p*-Cl 79, H 1.0, *m*-Me 0.73, *p*-F 0.55, and *p*-Me 0.21, demonstrating that electron-withdrawing substituents aid reaction and suggesting that there is considerable negative charge on the leaving benzyl group in the transition state. It was thought that the actual mechanism of reaction could involve a concerted bond-making and -breaking process, or could be through an intermediate such as that shown in 192, with the rate-determining transition state lying somewhere between the intermediate and the products.

$$MeO^- + Me_3Sn{-}CH_2Ph \longrightarrow \left[ \overset{\delta-}{MeO}\cdots Sn(Me)_3\cdots \overset{\delta-}{CH_2Ph} \right] \longrightarrow MeOSnMe_3 + {}^-CH_2Ph \qquad (192)$$

The carbanion then reacts with the solvent to form the final products (193).

$${}^-CH_2Ph + MeOH \longrightarrow PhCH_3 + {}^-OMe \qquad (193)$$

Product isotope ratios, $k_H/k_D$, for the alkaline cleavage of compounds $Me_3SnCH_2C_6H_4X$ in an equimolar mixture of MeOH and MeOD at 50 °C were[553,554] X=H 2.8, *p*-Me 2.5, *m*-Cl 2.4, and *m*-$CF_3$ 2.0, and these values implied that the carbanion formed in step 192 is not free, but experiences some degree of electrophilic attack by solvent at the benzyl carbon atom, such as **38**, in which a good proportion of the negative charge must be located on the benzyl group.

$$\left[ Ar{-}CH_2 \begin{matrix} \cdots SnMe_3{-}OMe \\ \cdots \underset{(D)}{H}\cdots OMe \end{matrix} \right]^-$$

(**38**)

The base-catalysed protodemetallation of a number of benzyl-type organotin compounds has been further studied by Eaborn and Seconi,[551] who reported rate

TABLE 124. Second-order rate constants, $10^5k_2$, at 25 °C for the cleavage of R—$SnR'_3$ by methanolic soldium methoxide[551]

| R | R′ | [NaOMe] (M) | $10^5k_2$ (l mol$^{-1}$ s$^{-1}$) |
|---|---|---|---|
| $PhCH_2$ | Me | 2 | 0.52[a] |
| *m*-$ClC_6H_4CH_2$ | Me | 1 | 13.7[a] |
| *m*-$CF_3C_6H_4CH_2$ | Me | 2 | 76[a] |
| 3,5-$Cl_2C_6H_3CH_2$ | Me | 0.5 | 575[a] |
| 3,5-$Cl_2C_6H_3CH_2$ | Me | 0.5 | 39 |
| *m*-$NCC_6H_4CH_2$ | Me | 0.5 | 11 |
| *p*-$NCC_6H_4CH_2$ | Me | 0.05 | 7500 |
| $Ph_2CH$ | Me | 0.1 | 450 |
| $Ph_2CH$ | Me | 0.05 | 7250[a] |
| $Ph_3C$ | Me | 0.01 | $159\times10^3$ |
| Fluoren-9-yl | Me | 0.002 | $21\times10^7$ |
| 9-Methylfluoren-9-yl | Me | 0.002 | $89\times10^7$ |
| $Ph_2CH$ | Et | 0.1 | 182 |
| Fluoren-9-yl | Et | 0.002 | $78\times10^6$ |
| $Ph_2CH$ | $Pr^i$ | 0.5 | 7.3 |
| Fluoren-9-yl | $Pr^i$ | 0.001 | $73\times10^4$ |

[a] Rate at 50 °C.

TABLE 125. Activation parameters[551] for the cleavage of $Me_3SnR$ by sodium methoxide in methanol

| R | [NaOMe] (M) | $E_a$ (kcal mol$^{-1}$) | Log *A* |
|---|---|---|---|
| $PhCH_2$ | 2 | 23.8 | 10.7 |
| *m*-$ClC_6H_4CH_2$ | 1 | 21.6 | 10.7 |
| 3,5-$Cl_2C_6H_3CH_2$ | 0.5 | 20.6 | 11.6 |
| $Ph_2CH$ | 0.05 | 21.1 | 13.1 |
| $Ph_3C$ | 0.001 | 17.0 | 12.7 |

TABLE 126. Relative rates of cleavage[555] of $XC_6H_4SnMe_3$ by hydroxide ion in 1:6 v/v $H_2O$–dmso at 75 °C

| X | $k_{rel}$ | X | $k_{rel}$ |
|---|---|---|---|
| *m*-$CF_3$ | 96 | *p*-F | 5.2 |
| *m*-Cl | 58 | *p*-SMe | 3.0 |
| *p*-Cl | 12.6 | H | 1.0 |
| *p*-C≡CH | 11.0 | *p*-OMe | 0.86 |

constants and product isotope ratios for cleavage by methanolic sodium methoxide. The rate constants (Table 124) vary slightly with the sodium methoxide concentration. Some activation parameters are given in Table 125. From the results of this investigation, Eaborn and coworkers felt that two mechanisms were possible. For the majority of the compounds studied reaction took place through a species such as **38**, as they had previously postulated[553,554]. However, for fluoren-9-yl- and 9-methylfluoren-9-yl-$SnR'_3$ ($R' = Me$, Et, $Pr^i$) and for $Ph_3CSnMe_3$, the reaction was thought to involve species such as $[Ar\cdots SnR'_3{-}OMe]^-$ which yields a carbanion which then reacts with solvent to form the final products. This mechanism had previously been postulated for trialkylarylsilicon compounds[553,554]. For *p*-$NCC_6H_4CH_2SnMe_3$ and $Ph_2CHSnMe_3$, the product isotope ratios and kinetic results implied that a mixture of these two mechanisms was in operation[551].

Eaborn and coworkers[555] also reported relative rate constants (given in Table 126) for the cleavage of trimethylaryltin compounds by potassium hydroxide in 1:6 v/v water–dimethyl sulphoxide. Electrophilic attack by hyrogen from the solvent was concerted with bond cleavage by the base.

## H. Protodemetallation of Organolead Compounds

Less work has been carried out on the protodemetallation of organolead compounds than on organic derivatives of silicon, germanium, and tin. However, both acid- and base-catalysed processes have been studied kinetically and mechanistically.

Robinson[588] reported on the protodemetallation of tetraalkylleads by acetic acid, using the latter as solvent, and by perchloric acid in acetic acid. First-order rate constants and activation parameters for the first-order process and second-order rate constants for the perchloric acid cleavage are given in Table 127; in both cases there seems to be a small steric effect of the alkyl group on the protodemetallation.

TABLE 127. First-order rate constants, $10^5 k_1$, at 24.9 °C and activation parameters and second-order rate constants, $10^2 k_2$, at 25 °C for the protodemetallation of $R_4Pb$[588]

| R | $10^5 k_1$ ($s^{-1}$)[a] | $\Delta H^{\neq}$ (kcal $mol^{-1}$)[a] | $\Delta S^{\neq}$ (cal $K^{-1}$ $mol^{-1}$)[a] | $10^2 k_2$ (l $mol^{-1}$ $s^{-1}$)[b] |
|---|---|---|---|---|
| Me | 1.16 | 20.8 | −12 | 22 |
| Et | 0.80 | 20.1 | −15 | 2.4 |
| $Pr^n$ | 0.225 | 20.7 | −15 | 0.89 |
| $Bu^n$ | 0.31 | 21.2 | −13 | 0.68 |
| Isoamyl | 0.30 | 21.2 | −13 | 0.59 |

[a] Using excess acetic acid.
[b] Using perchloric acid in acetic acid.

The acidic protodemetallation of tetraethyllead was also studied by Horn and Huber[589], who used acetic acid in anhydrous toluene as the reagent. Under these conditions, they observed the two competitive consecutive reactions, 194 and 195, whereas Robinson[588] had only observed reaction 194 when the solvent was acetic acid.

$$Et_4Pb + MeCOOH \longrightarrow EtH + Et_3PbOCOMe \tag{194}$$

$$Et_3PbOCOMe + MeCOOH \longrightarrow EtH + Et_2Pb(OCOMe)_2 \tag{195}$$

In anhydrous toluene, reactions 194 and 195 were both second order overall, first order in each reactant, with rate constants at 80 °C $2.18 \times 10^{-5}$ and $1.37 \times 10^{-3}$ l $mol^{-1}$ $s^{-1}$, respectively[589]. As Abraham[4] pointed out, it is remarkable that reaction 195 is faster than reaction 194 in toluene, whereas in acetic acid itself reaction 195 is not observable. In nitrobenzene as solvent, the second-order rate constants for protodemetallation of $Me_4Pb$ and $Me_3PbOCOCD_3$ by $CD_3COOD$ are $4.08 \times 10^{-4}$ and $0.938 \times 10^{-4}$ l $mol^{-1}$ $s^{-1}$, respectively, the first stage now being faster than the second[590].

Rate constants were also obtained by Bade and Huber[577] for the protodemetallation of $R_4Pb$ and $R_3PbCl$ alkyl compounds by HCl in methanol (see Table 128). In all cases the first alkyl group is removed more easily than the second, but whereas

TABLE 128. Second-order rate constants[577], $10^4 k_2$, for the protodemetallation of $R_4Pb$ and $R_3PbCl$ by hydrochloric acid in methanol at 50 °C

| | $10^4 k_2$ (l $mol^{-1}$ $s^{-1}$) | |
|---|---|---|
| R | $R_4Pb$ | $R_3PbCl$ |
| Me | 187.7 | 0.320 |
| Et | 40.8 | 0.370 |
| $Pr^n$ | 23.0 | 1.892 |
| $Bu^n$ | 21.0 | 1.138 |

TABLE 129. Relative rates of cleavage[591] of $XC_6H_4Pb(C_6H_{11})_3$ by aqueous ethanolic perchloric acid at 25 °C

| X | $k_{rel}$ |
|---|---|
| *p*-OMe | 21 |
| *p*-Me | 3.4 |
| H | 1.0 |
| *p*-Cl | 0.32 |
| *m*-Cl | 0.125 |

the protodemetallation of the $R_4Pb$ compounds exhibits a steric sequence with respect to the group R, there is a reverse effect in the cleavage of the $R_3PbCl$ compounds. Bade and Huber[577] also reported second-order rate constants for protodemetallation by HCl in methanol–benzene (20:80) at 50 °C as $Me_4Pb$ 0.0473, $Ph_4Pb$ 0.717, and $(p\text{-}MeOC_6H_4)_4Pb$ 3.28 l mol$^{-1}$ s$^{-1}$, again showing the increased reactivity of aryl—metal bonds compared with alkyl—metal bonds in protodemetallations by acids.

Eaborn and Pande[591] have measured relative rate constants for the acidic protodemetallation of aryl compounds $XC_6H_4Pb(cyclohexyl)_3$ by aqueous ethanolic perchloric acid. The spread of rate constants is smaller than for cleavage of the analogous tin compounds[578] and much smaller than for the cleavage of the corresponding germanium compounds[565], this effect being associated with the greater reactivity of the lead compounds; relative rate constants are given in Table 129.

Comparative studies[427] have shown that the rate of acidic protodemetallation of phenyl compounds increases in the series $Ph_4Sn$ 1 < $Ph_4Pb$ 60 < $Ph_2Hg$ 300, whereas the phenyl group is removed from $PhMEt_3$ by aqueous alcoholic perchloric acid with relative rates[527] $PhSiEt_3$ 1 < $PhGeEt_3$ 36 < $PhSnEt_3$ $(3.5 \times 10^5)$ < $PhPbEt_3$ $(2 \times 10^8)$.

## V. REFERENCES


1. O. A. Reutov and I. P. Beletskaya, *Reaction Mechanisms of Organometallic Compounds,* North Holland, Amsterdam, 1968.
2. F. R. Jensen and B. Rickborn, *Electrophilic Substitution of Organomercurials,* McGraw-Hill, New York, 1968.
3. J. C. Lockhart, *Redistribution Reactions,* Academic Press, New York, 1971.
4. M. H. Abraham, *Electrophilic Substitution at a Saturated Carbon Atom,* in *Comprehensive Chemical Kinetics* (Eds. C. H. Bamford and C. F. H. Tipper), Elsevier, Amsterdam, 1973, Vol. 12.
5. D. S. Matteson, *Organometallic Reaction Mechanisms,* Academic Press, New York, 1974.
6. M. Gielen and J. Nasielski, *Ind. Chim. Belge,* **26,** 1393 (1961); **29,** 767 (1964).
7. C. K. Ingold, *Helv. Chim. acta,* **47,** 1191 (1964); *Rec. Chem. Prog.,* **25,** 145 (1964).
8. O. A. Reutov, *Tetrahedron,* **34,** 2827 (1978); *Pure Appl. Chem.,* **50,** 717 (1978).
9. R. Taylor, in *Aromatic and Heterocyclic Chemistry, Chemical Society Specialist Periodical Reports,* **1,** 176 (1973); **2,** 217 (1974); **3,** 220 (1975); **4,** 227 (1976); see also **6,** 175 (1978) and **7,** 248 (1979).
10. H. B. Charman, E. D. Hughes, and C. K. Ingold, *J. Chem. Soc.,* 2530 (1959).
11. F. R. Jensen and D. D. Davis, *J. Am. Chem. Soc.,* **93,** 4048 (1971).
12. D. A. Slack and M. C. Baird, *J. Am. Chem. Soc.,* **98,** 5539 (1976).
13. J. A. Ross, R. P. Seiders, and D. M. Lemal, *J. Am. Chem. Soc.,* **98,** 4325 (1976).

14. S. Winstein, T. G. Traylor, and C. S. Garner, *J. Am. Chem. Soc.*, **77,** 3741 (1955); S. Winstein and T. G. Traylor, *J. Am. Chem. Soc.*, **77,** 3747 (1955).
15. M. H. Abraham and M. R. Sedaghat-Herati, *J. Chem. Soc., Perkin Trans. 2,* 729 (1978).
16. J. R. Chipperfield, *J. Organomet. Chem.*, **205,** 365 (1981).
17. M. Gielen, J. Nasielski, J. E. Dubois, and P. Fresnet, *Bull. Soc. Chim. Belg.*, **73,** 293 (1964).
18. H. C. Gardner and J. K. Kochi, *J. Am. Chem. Soc.*, **96,** 1982 (1974).
19. S. Fukuzumi and J. K. Kochi, *J. Am. Chem. Soc.*, **102,** 2141 (1980).
20. S. Fukuzumi and J. K. Kochi, *J. Am. Chem. Soc.*, **102,** 7290 (1980).
21. S. Fukuzumi and J. K. Kochi, *J. Phys. Chem.*, **85,** 648 (1981).
22. S. Fukuzumi and J. K. Kochi, *J. Phys. Chem.*, **84,** 2254 (1980).
23. R. W. Bott, C. Eaborn, and P. M. Greasley, *J. Chem. Soc.*, 4804 (1964).
24. C. Eaborn, I. D. Jenkins, and D. R. M. Walton, *J. Chem. Soc., Perkin Trans. 2,* 596 (1974).
25. C. Eaborn, P. M. Jackson, and R. Taylor, *J. Chem. Soc.*, 613 (1966).
26. H. Hashimoto and Y. Morimoto, *J. Organomet. Chem.*, **8,** 271 (1967).
27. L. G. Makarova and A. N. Nesmeyanov, *The Organic Compounds of Mercury,* North-Holland, Amsterdam, 1967.
28. M. Dub, *Organometallic Compounds,* Vols. I (2nd ed. 1966), II (2nd ed. 1967), and III (1962), Springer-Verlag, Berlin and New York.
29. A. E. Goddard and D. Goddard, in *Organic Chemistry* (Ed. J. Newton Friend), Charles Griffin, Vol. XI, Parts I, II, III, and IV (1928).
30. D. Seyferth and L. G. Vaughan, *J. Am. Chem. Soc.*, **86,** 883 (1964).
31. F. R. Jensen and K. L. Nakamaye, *J. Am. Chem. Soc.*, **88,** 3437 (1966).
32. S. W. Breuer, *J. Chem. Soc., Chem. Commun.*, 671 (1972).
33. A. F. Reid and C. J. Wilkins, *J. Chem. Soc.*, 4029 (1955).
34. M. H. Abraham and P. H. Rolfe, *J. Organomet. Chem.*, **8,** 395 (1967).
35. J.-E. Bäckvall and B. Åkermark, *J. Chem. Soc., Chem. Commun.*, 82 (1975).
36. C. L. Wong and J. K. Kochi, *J. Am. Chem. Soc.*, **101,** 5593 (1979).
37. V. D. Nefedov and E. N. Sintova, *Collected Works on Radiochemistry,* Leningrad University Press, Leningrad, 1955, pp. 110–113.
38. V. D. Nefedov, E. N. Sintova, and N. Ya. Frolov, *Zh. Fiz. Khim.*, **30,** 2356 (1956).
39. V. D. Nefedov and E. N. Sintova, *Zh. Neorg. Khim.*, **2,** 1162 (1957).
40. E. N. Sintova, *Zh. Neorg. Khim.*, **2,** 1205 (1957).
41. E. D. Hughes, C. K. Ingold, F. G. Thorpe, and H. C. Volger, *J. Chem. Soc.*, 1133 (1961).
42. E. D. Hughes and H. C. Volger, *J. Chem. Soc.*, 2359 (1961).
43. M. H. Abraham, P. L. Grellier, and M. J. Hogarth, *J. Chem. Soc., Perkin Trans. 2,* 1613 (1974).
44. H. B. Charman, E. D. Hughes, C. K. Ingold, and H. C. Volger, *J. Chem. Soc.*, 1142 (1961).
45. V. Lucchini and P. Wells, *J. Organomet. Chem.*, **92,** 283 (1975).
46. M. H. Abraham, P. L. Grellier, and M. J. Hogarth, *J. Chem. Soc., Perkin Trans. 2,* 221 (1977).
47. M. H. Abraham, D. Dodd, M. D. Johnson, E. S. Lewis, and R. A. More O'Ferrall, *J. Chem. Soc. B,* 762 (1971).
48. T. A. Smolina, Chou Mieh-Ch'u and O. A. Reutov, *Izv. Akad. Nauk SSSR, Ser. Khim.*, 413 (1966); *Engl. Transl.*, 390.
49. T. A. Ansaloni and J. Belmondi, *Ric. Sci.*, **26,** 3315 (1956).
50. I. P. Beletskaya, I. I. Zakharycheva, and O. A. Reutov, *Zh. Org. Khim.*, **5,** 793 (1969); *Engl. Transl.*, 783.
51. I. P. Beletskaya, I. I. Zakharycheva, and O. A. Reutov, *Dokl. Akad. Nauk SSSR,* **195,** 837 (1970); *Engl. Transl.* 861.
52. P. Peringer and P.-P. Winkler, *J. Organomet. Chem.*, **195,** 249 (1980).
53. O. A. Reutov, T. A. Smolina, and V. A. Kalyavin, *Dokl. Akad. Nauk SSSR,* **139,** 389 (1961); *Engl. Transl.*, 697.
54. O. A. Reutov, T. A. Smolina and V. A. Kalyavin, *Zh. Fiz. Khim.*, **36,** 119 (1962); *Engl. Transl.*, 59.

55. O. A. Reutov, T. A. Smolina and V. A. Kalyavin, *Dokl. Akad. Nauk SSSR,* **155,** 596 (1964); *Engl. Transl.*, 273.
56. O. A. Reutov, V. A. Kalyavin, and T. A. Smolina, *Dokl. Akad. Nauk SSSR,* **156,** 95 (1964); *Engl. Transl.*, 460.
57. V. A. Kalyavin, T. A. Smolina, and O. A. Reutov, *Dokl. Akad. Nauk SSSR,* **157,** 919 (1964); *Engl. Transl.*, 762.
58. O. A. Reutov, *Acta Chim. Acad. Sci. Hung.*, **18,** 439 (1959).
59. O. A. Reutov, I. P. Beletskaya, and Yang-Ts'e Wu, *Kinet. Katal., Akad. Nauk SSSR, Sb. Statei,* 43 (1960).
60. O. A. Reutov, V. I. Sokolov, and I. P. Beletskaya, *Dokl. Akad. Nauk SSSR,* **136,** 631 (1961); *Engl. Transl.*, 115.
61. O. A. Reutov, V. I. Sokolov, and I. P. Beletskaya, *Izv. Akad. Nauk SSSR, Otd. Khim. Nauk,* 1217, 1427 (1961); *Engl. Transl.*, 1127, 1328.
62. E. D. Hughes, C. K. Ingold, and R. M. G. Roberts, *J. Chem. Soc.*, 3900 (1964).
63. O. A. Reutov, V. I. Sokolov, and I. P. Beletskaya, *Izv. Akad. Nauk SSSR, Otd. Khim. Nauk,* 1213 (1961); *Engl. Transl.,* 1123.
64. O. A. Reutov, V. I. Sokolov, and I. P. Beleskaya, *Izv. Akad. Nauk SSSR, Otd. Khim. Nauk,* 1561 (1961); *Engl. Transl.*, 1458.
65. O. A. Reutov, V. I. Sokolov, I. P. Beletskaya, and Yu. S. Ryabokobylko, *Izv. Akad. Nauk SSSR, Otd. Khim. Nauk,* 965 (1963); *Engl. Transl.*, 879.
66. O. A. Reutov, B. Praisner, I. P. Beletskaya, and V. I. Sokolov, *Izv. Akad. Nauk SSSR, Otd. Khim. Nauk,* 970 (1963); *Engl. Transl.*, 884.
67. O. A. Reutov, P. Knoll', and U. Ian-Tsei, *Dokl. Akad. Nauk SSSR,* **120,** 1052 (1958); *Engl. Transl.*, 477.
68. A. N. Nesmeyanov, A. E. Borisov, and coworkers, *Izv. Akad. Nauk SSSR, Otd. Khim. Nauk,* 639 (1945); 647 (1946); 445 (1948); 570 (1949); 582 (1949); 402 (1951); 992 (1954); 1008 (1954); 1216 (1959).
69. A. N. Nesmeyanov, A. E. Borisov, and N. V. Novikova, *Dokl. Akad. Nauk SSSR,* **119,** 504 (1958).
70. W. Kitching, M. Bullpitt, P. Sleezer, and S. Winstein, *J. Organomet. Chem.*, **34,** 233 (1972).
71. F. R. Jensen, *J. Am. Chem. Soc.*, **82,** 2469 (1960).
72. A. N. Nesmeyanov, O. A. Reutov, Wu Yang-Ch'ieh, and Lu Ching-Chu, *Izv. Akad. Nauk SSSR, Otd. Khim. Nauk,* 1327 (1958); *Engl. Transl.*, 1280.
73. A. N. Nesmeyanov, O. A. Reutov, and S. S. Poddubnaya, *Izv. Akad. Nauk SSSR, Otd. Khim. Nauk,* 850 (1953); *Engl. Transl.*, 753.
74. O. A. Reutov and E. V. Uglova, *Izv. Akad. Nauk SSSR, Otd. Khim. Nauk,* 1691 (1959); *Engl. Transl.*, 1628.
75. G. F. Wright, *Can. J. Chem.*, **30,** 269 (1952).
76. R. E. Dessy and Y. K. Lee, *J. Am. Chem. Soc.*, **82,** 689 (1960).
77. R. E. Dessy, Y. K. Lee, and J.-Y. Kim, *J. Am. Chem. Soc.*, **83,** 1163 (1961).
78. M. D. Rausch and J. R. van Wazer, *Inorg. Chem.*, **3,** 761 (1964).
79. A. N. Kashin, I. P. Beletskaya, V. A. Milyaev, and O. A. Reutov, *Zh. Org. Khim.*, **10,** 1561 (1974); *Engl. Transl.*, 1577.
80. V. S. Petrosyan, S. M. Sakembayeva, and O. A. Reutov, *Izv. Akad. Nauk SSSR. Ser. Khim.*, 1403 (1973); *Engl. Transl.*, 1366.
81. A. N. Kashin, V. A. Milyaev, I. P. Beletskaya, and O. A. Reutov, *Zh. Org. Khim.*, **15,** 7 (1979), *Engl. Transl.*, 4.
82. A. N. Nesmeyanov and O. A. Reutov, *Dokl. Akad. Nauk SSSR,* **144,** 126 (1962); *Engl. Transl.*, 405.
83. A. N. Nesmeyanov and O. A. Reutov, *Tetrahedron,* **20,** 2803 (1964).
84. R. E. Dessy, W. Kitching, T. Psarras, R. Salinger, A. Chen, and T. Chivers, *J. Am. Chem. Soc.*, **88,** 460 (1966).
85. A. N. Nesmeyanov, O. A. Reutov, and S. S. Poddubnaya, *Dokl. Akad. Nauk SSSR,* **88,** 479 (1953).
86. O. A. Reutov, I. P. Beletskaya, and R. E. Mardaleishvili, *Dokl. Akad. Nauk SSSR,* **116,** 617 (1957); *Engl. Transl.*, 901.

87. O. A. Reutov, I. P. Beletskaya, and R. E. Mardaleishvili, *Zh. Fiz. Khim.*, **33,** 152, 1962 (1959); *Engl. Transl.*, 4, 240.
88. O. A. Reutov, *Rec. Chem. Prog.*, **22,** 1 (1961).
89. O. A. Reutov, *Angew. Chem.*, **72,** 198 (1960).
90. O. A. Reutov, *Dokl. Akad. Nauk SSSR*, **163,** 909 (1965); *Engl. Transl.*, 744.
91. O. A. Reutov, I. P. Beletskaya, and G. A. Artamkina, *Zh. Obshch. Khim.*, **34,** 2817 (1964); *Engl. Transl.*, 2850.
92. F. R. Jensen and B. Rickborn, *J. Am. Chem. Soc.*, **86,** 3784 (1964).
93. F. R. Jensen, B. Rickborn, and J. J. Miller, *J. Am. Chem. Soc.*, **88,** 340 (1966).
94. O. A. Reutov and I. P. Beletskaya, *Dokl. Akad. Nauk SSSR*, **131,** 853 (1960); *Engl. Transl.*, 333.
95. O. A. Reutov, I. P. Beletskaya, and G. A. Artamkina, *Zh. Fiz. Khim.*, **36,** 2582 (1962); *Engl. Transl.*, 1407.
96. O. A. Reutov, I. P. Beletskaya, and G. A. Artamkina, *Izv. Akad. Nauk SSSR, Ser. Khim.*, 1737 (1964); *Engl. Transl.*, 1651.
97. O. A. Reutov, I. P. Beletskaya, and G. A. Artamkina, *Dokl. Akad. Nauk SSSR*, **149,** 90 (1963); *Engl. Transl.*, 181.
98. O. A. Reutov, I. P. Beletskaya and G. A. Artamkina, *Izv. Akad. Nauk SSSR, Otd. Khim. Nauk*, 765 (1963); *Engl. Transl.*, 691.
99. O. A. Reutov, Hu-Hung Weng, T. A. Smolina, and I. P. Beletskaya, *Zh. Fiz. Khim.*, **35,** 2424 (1961); *Engl. Transl.*, 1197.
100. T. A. Smolina, V. A. Kalyavin, and O. A. Reutov, *Izv. Akad. Nauk SSSR, Otd. Khim. Nauk*, 2235 (1963); *Engl. Transl.*, 2070.
101. O. A. Reutov, T. P. Karpov, É. V. Uglova, and V. A. Malyanov, *Tetrahedron Lett.*, **19,** 6 (1960).
102. O. A. Reutov, T. P. Karpov, É. V. Uglova, and V. A. Malyanov, *Dokl. Akad. Nauk SSSR*, **134,** 360 (1960): *Engl. Transl.*, 1017.
103. O. A. Reutov, T. P. Karpov, É. V. Uglova, and V. A. Malyanov, *Izv. Akad. Nauk SSSR, Otd. Khim. Nauk*, 1311 (1960); *Engl. Transl.*, 1223.
104. O. A. Reutov, T. P. Karpov, É. V. Uglova, and V. A. Malyanov, *Izv. Akad. Nauk SSSR, Ser. Khim.*, 1580 (1964); *Engl. Transl.*, 1492.
105. H. B. Charman, E. D. Hughes, C. K. Ingold, and F. G. Thorpe, *J. Chem. Soc.*, 1121 (1961).
106. F. E. Paulik and R. E. Dessy, *Chem. Ind.* (*London*), 1650 (1962).
107. V. S. Petrosyan, S. M. Sakembayeva, V. I. Bakhmutov, and O. A. Reutov, *Dokl. Akad. Nauk SSSR*, **209,** 1117 (1973); *Engl. Transl.*, 309.
108. F. R. Jensen and J. Miller, *J. Am. Chem. Soc.*, **86,** 4735 (1964).
109. I. P. Beletskaya, G. A. Artamkina, and O. A. Reutov, *Dokl. Akad. Nauk SSSR*, **166,** 1347, (1966); *Engl. Transl.*, 242.
110. H. B. Charman, E. D. Hughes, and C. K. Ingold, *J. Chem. Soc.*, 2523 (1959).
111. O. A. Reutov, J. Hun Geng, and T. A. Smolina, *Izv. Akad. Nauk SSSR, Otd. Khim. Nauk*, 559 (1959); *Engl. Transl.*, 534.
112. D. R. Pollard and M. H. Thompson, *J. Organomet. Chem.*, **36,** 13 (1972).
113. R. E. Dessy, F. Kaplan, G. R. Coe, and R. M. Salinger, *J. Am. Chem. Soc.*, **85,** 1191 (1963).
114. T. N. Huckerby, P. H. Lindsay, F. G. Thorpe, and J. C. Podestá, *Tetrahedron*, **31,** 277 (1975).
115. I. P. Beletskaya, K. P. Butin, V. N. Shishkin, V. F. Gomzyakov, I. F. Gun'kin, and O. A. Reutov, *Zh. Org. Khim.*, **10,** 2009 (1974); *Engl. Transl.*, 2027.
116. I. P. Beletskaya, K. P. Butin, V. N. Shiskin, I. F. Gun'kin, and O. A. Reutov, *Zh. Org. Khim.*, **11,** 2483 (1975); *Engl. Transl.*, 2549.
117. C. R. Hart and C. K. Ingold, *J. Chem. Soc.*, 4372 (1964).
118. F. R. Jensen and D. Heyman, *J. Am. Chem. Soc.*, **88,** 3438 (1966).
119. J. R. Kalman, J. T. Pinhey, and S. Sternhell, *Tetrahedron Lett.*, 5369 (1972).
120. R. C. Larock and H. C. Brown, *J. Am. Chem. Soc.*, **92,** 2467 (1970).
121. R. C. Larock and H. C. Brown, *J. Organomet. Chem.*, **26,** 35 (1971).
122. D. S. Matteson and R. A. Bowie, *J. Am. Chem. Soc.*, **87,** 2587 (1965).

123. M. Gielen and R. Fosty, *Bull. Soc. Chim. Belg.*, **83,** 333 (1974).
124. D. E. Bergbreiter and D. P. Rainville, *J. Organomet. Chem.*, **121,** 19 (1976).
125. D. S. Matteson and J. O. Waldbillig, *J. Am. Chem. Soc.*, **85,** 1019 (1963).
126. D. S. Matteson and J. O. Waldbillig, *J. Am. Chem. Soc.*, **86,** 3778 (1964).
127. D. S. Matteson and M. L. Talbot, *J. Am. Chem. Soc.*, **89,** 1119 (1967).
128. D. S. Matteson and M. L. Talbot, *J. Am. Chem. Soc.*, **89,** 1123 (1967).
129. D. S. Matteson, R. A. Bowie, and G. Srivastava, *J. Organomet. Chem.*, **16,** 33 (1969).
130. D. S. Matteson and E. Kramer, *J. Am. Chem. Soc.*, **90,** 7261 (1968).
131. D. S. Matteson and P. G. Allies, *J. Am. Chem. Soc.*, **92,** 1801 (1970).
132. H. G. Kuivila and T. C. Muller, *J. Am. Chem. Soc.*, **84,** 377 (1962).
133. H. Gilman and H. Hartzfield, *J. Am. Chem. Soc.*, **73,** 5878 (1951).
134. F. W. G. Fearon and H. Gilman, *J. Organomet. Chem.*, **10,** 409 (1967).
135. P. J. Morris, F. W. G. Fearon, and H. Gilman, *J. Organomet. Chem.*, **9,** 427 (1967).
136. H. Gilman and D. Aoki, *Chem. Ind.* (*London*), 1619 (1961).
137. D. Seyferth, F. M. Armbrecht, Jr., and E. M. Hanson, *J. Organomet. Chem.*, **10,** P25 (1967).
138. D. Seyferth and T. Wada, *Inorg. Chem.*, **1,** 78 (1962).
139. W. Findeiss, W. Davidsohn, and M. C. Henry, *J. Organomet. Chem.*, **9,** 435 (1967).
140. H. Gilman, *Bull. Soc. Chim. Fr.*, 1356 (1963).
141. A. A. Petrov, V. A. Kormer, and M. D. Stadnichuk, *Zh. Obsch. Khim.*, **31,** 1135 (1961).
142. A. Hofer, H. Kuckertz, and M. Sandler, *Makromol. Chem.*, **90,** 38 (1966).
143. H. Gilman and R. A. Tomasi, *J. Am. Chem. Soc.*, **81,** 137 (1959).
144. G. R. John, L. A. P. Kane-Maguire, and C. Eaborn, *J. Chem. Soc., Chem. Commun.*, 481 (1975).
145. G. Combes, *C. R. Acad. Sci.*, **122,** 622 (1896).
146. Z. Manulkin, *Zh. Obshch. Khim.*, **16,** 275 (1946).
147. A. D. Petrov and L. L. Shchukovskaya, *Zh. Obshch. Khim.*, **25,** 1128 (1955).
148. R. E. DeSimone, *J. Chem. Soc., Chem. Commun.*, 780 (1972).
149. R. A. Benkeser, T. V. Liston, and G. M. Stanton, *Tetrahedron Lett.*, **15,** 1 (1960).
150. R. A. Benkeser, D. I. Hoke, and R. A. Hickner, *J. Am. Chem. Soc.*, **80,** 5294 (1958).
151. F. B. Deans, C. Eaborn, and D. E. Webster, *J. Chem. Soc.*, 3031 (1959).
152. C. Eaborn, Z. Lasocki, and D. E. Webster, *J. Chem. Soc.*, 3034 (1959).
153. D. E. Webster, *Ph.D. Thesis,* University of London, 1958.
154. J. R. Chipperfield, G. D. France, and D. E. Webster, *J. Chem. Soc., Perkin Trans. 2,* 405 (1972).
155. R. M. G. Roberts, *J. Organomet. Chem.*, **12,** 89 (1968).
156. H. Schmidbaur and W. Findeiss, *Angew. Chem., Int. Ed. Engl.*, **3,** 696 (1964).
157. H. Schmidbaur and W. Findeiss, *Chem. Ber.*, **99,** 2187 (1966).
158. H. C. Bell, J. R. Kalman, J. T. Pinhey, and S. Sternhell, *Tetrahedron Lett.*, 3391 (1974).
159. G. Calingaert, H. Soroos, and V. Hnizda, *J. Am. Chem. Soc.*, **62,** 1107 (1940).
160. P. D. George, L. H. Somner, and F. C. Whitmore, *J. Am. Chem. Soc.*, **77,** 1677 (1955).
161. G. A. Russell, *J. Am. Chem. Soc.*, **81,** 4815 (1959).
162. G. A. Russell, *J. Am. Chem. Soc.*, **81,** 4834 (1959).
163. F. H. Pollard, G. Nickless, and P. C. Uden, *J. Chromatogr.*, **19,** 28 (1965).
164. R. O. Sauer and E. M. Hadsell, *J. Am. Chem. Soc.*, **70,** 3590 (1948).
165. H. Sakurai, K. Tominaga, T. Watanabe, and M. Kumada, *Tetrahedron Lett.*, 5493 (1966).
166. H. Sakurai, K. Tominaga, and M. Kumala, *Bull. Chem. Soc. Jpn.*, **39,** 1820 (1966).
167. R. Müller and W. Müller, *Chem. Ber.*, **97,** 1673 (1964).
168. A. Ya Yakubovich and G. V. Motsarev, *Zh. Obshch. Khim.*, **23,** 1414 (1953).
169. R. M. G. Roberts, *J. Organomet. Chem.*, **12,** 97 (1968).
170. F. Rijkens, E. J. Bulten, W. Drenth, and G. J. M. van der Kerk, *Recl. Trav. Chim. Pays-Bas*, **85,** 1223 (1966).
171. E. J. Bulten and J. G. Noltes, *Tetrahedron Lett.* 3471 (1966).
172. J. G. A. Luijten and F. Rijkens, *Recl. Trav. Chim. Pays-Bas,* **83,** 857 (1964).
173. E. J. Bulten and J. G. Noltes, *J. Organomet. Chem.*, **15,** P18 (1968).
174. D. Seyferth and M. A. Weiner, *Chem. Ind.* (*London*), 402 (1959).
175. D. Seyferth and M. A. Weiner, *J. Am. Chem. Soc.*, **83,** 3583 (1961).
176. D. Seyferth and M. A. Weiner, *J. Org. Chem.*, **26,** 4797 (1961).

177. D. Seyferth and M. A. Weiner, *Org. Synth.* **41,** 30 (1961).
178. D. Seyferth and M. A. Weiner, *J. Org. Chem,* **24,** 1395 (1959).
179. D. Seyferth, D. E. Welch, and G. Raab, *J. Am. Chem. Soc.,* **84,** 4266 (1962).
180. D. Seyferth, R. Suzuki, C. J. Murphy, and G. R. Sabet, *J. Organomet. Chem.,* **2,** 431 (1964).
181. D. Seyferth, L. G. Vaughan, and R. Suzuki, *J. Organomet. Chem.,* **1,** 437 (1964).
182. D. Seyferth and H. M. Cohen, *Inorg. Chem.,* **2,** 625 (1963).
183. H. H. Anderson, *Inorg. Chem.,* **1,** 647 (1962).
184. C. Eaborn, K. J. Odell, and A. Pidcock, *J. Chem. Soc., Dalton Trans.,* 357 (1978).
185. C. J. Cardin, D. J. Cardin, M. F. Lappert, and K. W. Muir, *J. Organomet. Chem.,* **60,** C70 (1973).
186. C. J. Cardin, D. J. Cardin, and M. F. Lappert, *J. Chem. Soc., Dalton Trans.,* 767 (1977).
187. R. B. King and F. G. A. Stone, *J. Am. Chem. Soc.,* **82,** 3833 (1960).
188. S. Fukuzumi, C. L. Wong, and J. K. Kochi, *J. Am. Chem. Soc.,* **102,** 2928 (1980).
189. M. H. Abraham and T. R. Spalding, *J. Chem. Soc. A,* 2530 (1968).
190. M. H. Abraham and T. R. Spalding, *J. Chem. Soc. A,* 399 (1969).
191. M. H. Abraham and T. R. Spalding, *J. Chem. Soc. A,* 784 (1969).
192. M. H. Abraham and G. F. Johnston, *J. Chem. Soc. A,* 188 (1970).
193. M. H. Abraham and G. F. Johnston, *J. Chem. Soc. A,* 193 (1970).
194. M. H. Abraham, R. J. Irving, and G. F. Johnston, *J. Chem. Soc. A,* 199 (1970).
195. M. H. Abraham, J. F. C. Oliver, and J. A. Richards, *J. Chem. Soc. A,* 203 (1970).
196. M. H. Abraham, *J. Chem. Soc. A,* 1061 (1971).
197. M. H. Abraham and F. Behbahany, *J. Chem. Soc. A,* 1469 (1971).
198. M. H. Abraham and M. J. Hogarth, *J. Chem. Soc. A,* 1474 (1971).
199. M. H. Abraham and G. F. Johnston, *J. Chem. Soc. A,* 1610 (1971).
200. M. H. Abraham, F. Behbahany, and M. J. Hogarth, *J. Chem. Soc. A,* 2566 (1971).
201. M. H. Abraham and F. J. Dorrell, *J. Chem. Soc., Perkin Trans. 2,* 444 (1973).
202. M. H. Abraham and P. L. Grellier, *J. Chem. Soc., Perkin Trans. 2,* 1132 (1973).
203. M. H. Abraham, D. F. Dadjour, and C. J. Holloway, *J. Organomet. Chem.,* **52,** C27 (1973).
204. M. H. Abraham and D. F. Dadjour, *J. Chem. Soc., Perkin Trans. 2,* 233 (1974).
205. M. H. Abraham, D. F. Dadjour, M. Gielen, and B. de Poorter, *J. Organomet. Chem.,* **84,** 317 (1975).
206. M. H. Abraham and P. L. Grellier, *J. Chem. Soc., Perkin Trans. 2,* 623 (1975).
207. M. H. Abraham, D. F. Dadjour, and M. R. Sedaghat-Herati, *J. Chem. Soc., Perkin Trans. 2,* 1225 (1977).
208. M. H. Abraham, *J. Chem. Soc., Perkin Trans. 2,* 1028 (1977).
209. M. H. Abraham, J. Andonian-Haftvan, and M. R. Sedaghat-Herati, *J. Organomet. Chem.,* **172,** 31 (1978).
210. M. H. Abraham, P. L. Grellier, M. J. Hogarth, T. R. Spalding, M. Fox, and G. R. Wickham, *J. Chem. Soc. A,* 2972 (1971).
211. M. H. Abraham, *Prog. Phys. Org. Chem.,* **2,** 1 (1974).
212. R. M. G. Roberts, *J. Organomet. Chem.,* **32,** 323 (1971).
213. R. M. G. Roberts, personal communication.
214. I. P. Beletskaya, A. N. Kashin, A. Ts. Malkhasyan, and O. A. Reutov, *Zh. Org. Khim.,* **9,** 1089 (1973); *Engl. Transl.,* 1127.
215. A. N. Kashin, I. P. Beletskaya, A. Ts. Malkhasyan, and O. A. Reutov, *Zh. Org. Khim.,* **9,** 1098 (1973); *Engl. Transl.,* 1127.
216. A. N. Kashin, I. P. Beletskaya, and O. A. Reutov, *Zh. Org. Khim.,* **15,** 673 (1979); *Engl. Transl.,* p. 600.
217. A. N. Kashin, I. P. Beletskaya, and O. A. Reutov, *Vestn. Mosk. Univ., Khim.,* **22,** 409 (1981).
218. R. M. G. Roberts, *J. Organomet. Chem.,* **18,** 307 (1969).
219. M. H. Abraham and J. Andonian-Haftvan, *J. Chem. Soc., Perkin Trans. 2,* 1033 (1980).
220. K. L. Jewett, F. E. Brinckman, and J. M. Bellama, in *Organometals and Organometalloids. Occurrence and fate in the Environment,* Advances in Chemistry Series, No. 82, American Chemical Society, Washington, D.C., 1978.

221. A. B. Burg and J. R. Spielman, *J. Am. Chem. Soc.,* **83,** 2667 (1961).
222. W. Gerrard, E. F. Mooney, and R. G. Rees, *J. Chem. Soc.,* 740 (1964).
223. J. E. Burch, W. Gerrard, M. Howarth, and E. F. Mooney, *J. Chem. Soc.,* 4916 (1960).
224. F. E. Brinckman and F. G. A. Stone, *Chem. Ind.* (*London*), 254 (1959).
225. Z. M. Manulkin, *Zh. Obshch. Khim.,* **18,** 299 (1948).
226. J. C. van Egmond, M. J. Janssen, J. G. A. Luijten, G. J. M. van der Kerk, and G. M. van der Want, *J. Appl. Chem.* (*London*), **12,** 17 (1962).
227. A. E. Borisov and N. V. Novikova, *Izv. Akad. Nauk SSSR, Otd. Khim. Nauk,* 1670 (1959).
228. H. Kurasawa, M. Tanaka, and R. Okawara, *J. Organomet. Chem.,* **12,** 241 (1968).
229. M. Tanaka, H. Kurasawa, and R. Okawara, *Inorg. Nucl. Chem. Lett.,* **3,** 565 (1967).
230. M. Tanaka, H. Kurasawa, and R. Okawara, *J. Organomet. Chem.,* **18,** 49 (1969).
231. T. Abe, H. Kurasawa, and R. Okawara, *J. Organomet. Chem.,* **25,** 353 (1970).
232. R. Okawara, *Usp. Khim.,* **41,** 1220 (1972).
233. I. F. Gun'kin, K. P. Butin, I. P. Beletskaya, and O. A. Reutov, *Izv. Akad. Nauk SSSR, Ser. Khim.,* 1016 (1978); *Engl. Transl.,* 879.
234. I. F. Gun'kin, K. P. Butin, I. P. Beletskaya, and O. A. Reutov, *Izv. Akad. Nauk SSSR, Ser. Khim.,* 1297 (1978); *Engl. Transl.,* 1128.
235. M. F. Shostakovskii, B. A. Sokolov, and G. P. Mantsivoda, *Zh. Obshch. Khim.,* **33,** 3779 (1963); *Engl. Transl.,* 3717.
236. V. F. Miranov and A. L. Kravchenko, *Zh. Obshch. Khim.,* **34,** 1356 (1964); *Engl. Transl.,* 1359.
237. H. G. Kuivila, J. E. Dixon, P. L. Maxfield, N. M. Scarpa, T. M. Topka, K.-H. Tsai, and K. R. Wursthorn, *J. Organomet. Chem.,* **86,** 89 (1975).
238. G. Plazzogna, S. Bresadola, and G. Tagliavini, *Inorg. Chim. Acta,* **2,** 333 (1968).
239. G. Plazzogna, V. Peruzzo, S. Bresadola, and G. Tagliavini, *Gazz Chim. Ital.,* **102,** 48 (1972).
240. E. J. Juenge and D. Seyferth, *J. Org. Chem.,* **26,** 563 (1961).
241. H. Gilman and L. A. Woods, *J. Am. Chem. Soc.,* **65,** 435 (1943).
242. C. E. H. Bawn and F. J. Whitby, *Discuss. Faraday Soc.,* No. 2, 228 (1947).
243. C. E. H. Bawn and F. J. Whitby, *J. Chem. Soc.,* 3926 (1960).
244. C. E. H. Bawn and R. Johnson, *J. Chem. Soc.,* 4162 (1960).
245. G. Costa, G. de Alti, and L. Stefani, *Atti Accad. Naz. Lincei, Cl. Sci. Fis. Mat. Nat., Rend.,* **31,** 267 (1961).
246. G. Costa, G. de Alti, L. Stefani, and G. Boscarato, *Ann. Chim.* (*Rome*), **52,** 289 (1962).
247. G. Semerano, L. Riccoboni, and F. Callegari, *Chem. Ber.,* **74,** 1297 (1941).
248. F. Glockling, *J. Chem. Soc.,* 716 (1955).
249. J. E. Spice and W. Twist, *J. Chem. Soc.,* 3319 (1956).
250. F. Glockling and D. Kingston, *J. Chem. Soc.,* 3001 (1959).
251. G. Tagliavini, G. Schiavon, and U. Belluco, *Gazz. Chim. Ital.,* **88,** 746 (1958).
252. U. Belluco, L. Riccoboni, and G. Tagliavini, *Ric. Sci.,* **30,** 1255 (1960).
253. L. Riccoboni, U. Belluco, and G. Tagliavini, *Ric. Sci., Parte 2, Sez. A,* **2,** 323 (1962).
254. G. Tagliavini, U. Belluco, and L. Cattalini, *Ric. Sci., Parte 2, Sez. A,* **2,** 350 (1962).
255. A. N. Nesmeyanov and K. A. Kocheshkov, *Chem. Ber.,* **67,** 317 (1934).
256. M. S. Whelan, *Adv. Chem. Ser.,* **23,** 82 (1959).
257. E. M. Panov and K. A. Kocheshkov, *Dokl. Akad. Nauk SSSR,* **85,** 1037, 1293 (1952).
258. K. A. Kocheshkov and E. M. Panov, *Izv. Akad. Nauk SSSR, Otd. Khim. Nauk,* 711, 718 (1955).
259. G. Tagliavini and U. Belluco, *Ric. Sci., Parte 2, Sez. A,* **2,** 76 (1962).
260. A. E. Goddard and D. Goddard, *J. Chem. Soc.,* **121,** 482 (1922).
261. D. Goddard and A. E. Goddard, *J. Chem. Soc.,* **121,** 256 (1922).
262. V. Mironov and A. L. Kravchenko, *Dokl. Akad. Nauk SSSR,* **158,** 656 (1964).
263. A. E. Goddard, J. N. Ashley, and R. B. Evans, *J. Chem. Soc.,* **121,** 978 (1922).
264. G. Calingaert, M. A. Beatty, and H. Soroos, *J. Am. Chem. Soc.,* **62,** 1099 (1940).
265. H. Gilman and L. D. Apperson, *J. Org. Chem.,* **4,** 162 (1939).
266. M. S. Kharasch, E. V. Jensen, and S. Weinhouse, *J. Org. Chem.,* **14,** 429 (1949).
267. E. Krause and O. Schlöttig, *Chem. Ber.,* **63,** 1381 (1930).
268. E. A. Puchinyan, *Tr. Tashk. Farm. Inst.,* **1,** 310 (1957).

269. H. Weingarten and J. R. van Wazer, *J. Am. Chem. Soc.*, **88,** 2700 (1966).
270. G. Fraenkel and D. T. Dix, *J. Am. Chem. Soc.*, **88,** 979 (1966).
271. G. Fraenkel, D. T. Dix and D. G. Adams, *Tetrahedron Lett.*, 3155 (1964).
272. G. M. Whitesides, M. Witanowski, and J. D. Roberts, *J. Am. Chem. Soc.*, **87,** 2854 (1965).
273. G. M. Whitesides and J. D. Roberts, *J. Am. Chem. Soc.*, **87,** 4878 (1965).
274. M. Witanowski and J. D. Roberts, *J. Am. Chem. Soc.*, **88,** 737 (1966).
275. G. Fraenkel, D. T. Dix, and M. Carlson, *Tetrahedron Lett.*, 579 (1968).
276. M. Gielen and J. Nasielski, *Bull. Soc. Chim. Belg.*, **71,** 32 (1962).
277. N. S. Zefirov, P. P. Kadzyauskas, and Yu. K. Yur'ev, *Zh. Obshch. Khim.*, **36,** 1735 (1966); *Engl. Transl.*, 1731.
278. J.-E. Bäckvall, M. U. Ahmad, S. Uemura, A. Toshimitsu, and T. Kawamura, *Tetrahedron Lett.*, 2283 (1980).
279. C. Eaborn and D. E. Webster, *J. Chem. Soc.*, 4449 (1957).
280. L. M. Stock and A. R. Spector, *J. Org. Chem.*, **28,** 3272 (1963).
281. G. Fritz and J. Grobe, *Z. Anorg. Allg. Chem.*, **309,** 98 (1961).
282. R. D. Chambers, H. C. Clark, and C. J. Willis, *Can. J. Chem.*, **39,** 131 (1961).
283. H. M. Walborsky, F. J. Impastato, and A. E. Young, *J. Am. Chem. Soc.*, **86,** 3283 (1964).
284. D. E. Applequist and A. H. Peterson, *J. Am. Chem. Soc.*, **83,** 862 (1961).
285. D. E. Applequist and G. N. Chmurny, *J. Am. Chem. Soc.*, **89,** 875 (1967).
286. F. R. Jensen and L. H. Gale, *J. Am. Chem. Soc.*, **81,** 1261 (1959).
287. F. R. Jensen, L. D. Whipple, D. K. Wedegaertner, and J. A. Landgrebe, *J. Am. Chem. Soc.*, **81,** 1262 (1959).
288. F. R. Jensen and L. H. Gale, *J. Am. Chem. Soc.*, **82,** 148 (1960).
289. F. R. Jensen, L. D. Whipple, D. K. Wedegaertner, and J. A. Landgrebe, *J. Am. Chem. Soc.*, **82,** 2466 (1960).
290. O. A. Reutov, É. V. Uglova, I. P. Beletskaya, and T. B. Svetlanova, *Izv. Akad. Nauk SSSR, Ser. Khim.*, 1383 (1964); *Engl. Transl.*, 1297.
291. O. A. Reutov, É. V. Uglova, I. P. Beletskaya, and T. B. Svetlanova, *Izv. Akad. Nauk SSSR, Ser. Khim.*, 1151 (1968); *Engl. Transl.*, 1101.
292. Yu. G. Bundel, V. I. Rozenberg, V. K. Piotrovskii, and O. A. Reutov, *Izv. Akad. Nauk SSSR, Ser. Khim.*, 1791 (1971); *Engl. Transl.*, 1679.
293. O. A. Reutov, I. P. Beletskaya, and T. A. Azizyan, *Izv. Akad. Nauk SSSR, Otd. Khim. Nauk,* 424 (1962); *Engl. Transl.*, 393.
294. I. P. Beletskaya, O. A. Reutov, and T. A. Azizyan, *Ivz. Akad. Nauk SSSR, Otd. Khim. Nauk,* 223 (1962); *Engl. Transl.*, 204.
295. I. P. Beletskaya, O. A. Reutov, and T. P. Gur'yanova, *Izv. Akad. Nauk SSSR, Otd. Khim. Nauk,* 2178 (1961); *Engl. Transl.*, 2036.
296. I. P. Beletskaya, T. A. Azizyan, and O. A. Reutov, *Izv. Akad. Nauk SSSR, Ser. Khim.*, 1332 (1963); *Engl. Transl.*, 1208.
297. I. P. Beletskaya, A. V. Ermanson, and O. A. Reutov, *Bull. Acad. Sci. USSR*, 218 (1965).
298. L. M. Sayre and F. R. Jensen, *J. Am. Chem. Soc.*, **101,** 6001 (1979).
299. H. C. Brown and C. F. Lane, *J. Chem. Soc., Chem. Commun.*, 521 (1971).
300. H. G. Kuivila and A. R. Hendrickson, *J. Am. Chem. Soc.*, **74,** 5068 (1952).
301. H. G. Kuivila and L. E. Benjamin, *J. Am. Chem. Soc.*, **77,** 4834 (1955).
302. K. Tamao, J. Yoshida, M. Murata, and M. Kumada, *J. Am. Chem. Soc.*, **102,** 3267 (1980).
303. C. Eaborn and O. W. Steward, *J. Chem. Soc.*, 521 (1965).
304. C. Eaborn and D. E. Webster, *J. Chem. Soc.*, 179 (1960).
305. R. C. Moore, *PhD Thesis,* University of Leicester, 1962.
306. L. H. Sommer, G. M. Goldberg, C. E. Buck, T. S. Bye, F. J. Evans, and F. C. Whitmore, *J. Am. Chem. Soc.*, **76,** 1613 (1954).
307. M. P. Brown and G. W. A. Fowles, *J. Chem. Soc.*, 2811 (1958).
308. C. A. Kraus and E. A. Flood, *J. Am. Chem. Soc.*, **54,** 1635 (1932).
309. C. Eaborn and K. C. Pande, *J. Chem. Soc.*, 3200 (1960).
310. H. H. Anderson, *J. Am. Chem. Soc.*, **74,** 2370 (1952).
311. E. A. Flood, *J. Am. Chem. Soc.*, **54,** 1663 (1932).

312. O. H. Johnson and D. M. Harris, *Inorg. Synth.*, **5,** 74 (1957).
313. O. H. Johnson, W. H. Nebergall, and D. M. Harris, *Inorg. Synth.*, **5,** 76 (1957).
314. P. Mazerolles, *Bull. Soc. Chim. Fr.*, 1907 (1962).
315. P. Mazerolles, M. Lesbre and J. Dubac, *C. R. Acad. Sci.*, **260,** 2255 (1965).
316. P. Mazerolles, J. Dubac and M. Lesbre, *J. Organomet. Chem.*, **5,** 35 (1966).
317. P. Mazerolles and J. Dubac, *C. R. Acad. Sci.*, **257,** 1103 (1963).
318. P. Mazerolles and M. Lesbre, *C. R. Acad. Sci.*, **248,** 2018 (1959).
319. M. Lesbre, P. Mazerolles, and G. Manuel, *C. R. Acad. Sci.*, **255,** 544 (1962).
320. J. Satge, *Ann. Chim. (Paris)*, **6,** 519 (1961).
321. M. Lesbre and J. Satge, *C. R. Acad. Sci.*, **250,** 2220 (1960).
322. I. F. Lutsenko, Yu. I. Baukov, and B. N. Khasapov, *Zh. Ohshch. Khim.*, **33,** 2724 (1963).
323. R. J. Cross and F. Glockling, *J. Chem. Soc.*, 4125 (1964).
324. G. Grüttner and E. Krause, *Chem. Ber.*, **50,** 1802 (1917).
325. J. G. A. Luijten and G. J. M. van der Kerk, *Investigations in the Field of Organotin Chemistry,* Tin Research Institute, Greenford, 1955.
326. S. D. Rosenberg, E. Debreczeni, and E. L. Weinberg, *J. Am. Chem. Soc.*, **81,** 972 (1959).
327. G. J. M. van der Kerk and J. G. A. Luijten, *J. Appl. Chem. (London)*, **6,** 56 (1956).
328. L. N. Snegur and Z. M. Manulkin, *Zh. Obshch. Khim.*, **34,** 4030 (1964).
329. H. Zimmer, I. Hechenbleikner, O. A. Homberg, and M. Danzik, *J. Org. Chem.*, **29,** 2632 (1964).
330. M. M. Koton and T. M. Kiseleva, *Zh. Obshch. Khim.*, **27,** 2553 (1957).
331. G. J. M. van der Kerk and J. G. Noltes, *J. Appl. Chem. (London)*, **9,** 179 (1959).
332. D. Seyferth, *Naturwissenschaften,* **44,** 34 (1957).
333. D. Seyferth, *J. Am. Chem. Soc.*, **79,** 2133 (1957).
334. S. D. Rosenberg and A. J. Gibbons, *J. Am. Chem. Soc.*, **79,** 2138 (1957).
335. D. Seyferth, *J. Org. Chem.*, **22,** 1599 (1957).
336. D. Seyferth, K. Brändle, and G. Raab, *Angew. Chem.*, **72,** 77 (1960).
337. G. Grüttner, E. Krause, and M. Wiernik, *Chem. Ber.*, **50,** 1549 (1917).
338. R. M. G. Roberts, *J. Organomet. Chem.*, **63,** 159 (1973).
339. S. Boué, M. Gielen, and J. Nasielski, *Tetrahedron Lett.*, 1047 (1968).
340. P. Baekelmans, M. Gielen, and J. Nasielski, *Tetrahedron Lett.*, 1149 (1967).
341. A. Rahm and M. Pereyre, *J. Am. Chem. Soc.*, **99,** 1672 (1977).
342. L. F. McGahey and F. R. Jensen, *J. Am. Chem. Soc.*, **101,** 4397 (1979).
343. A. N. Kashin, V. N. Bakunin, V. A. Khutoryanskii, I. P. Beletskaya, and O. A. Reutov, *Zh. Org. Khim.*, **15,** 16 (1979); *Engl. Transl.*, 12.
344. M. Gielen and J. Nasielski, *Bull. Soc. Chim. Belg.*, **71,** 601 (1962).
345. S. Boué, M. Gielen, and J. Nasielski, *J. Organomet. Chem.*, **9,** 443 (1967).
346. M. Gielen and J. Nasielski, *J. Organomet. Chem.*, **1,** 173 (1963).
347. S. Faleschini and G. Tagliavini, *Gazz. Chim. Ital.*, **97,** 1401 (1967).
348. A. N. Kashin, I. P. Beletskaya, A. Ts. Malkhasyan, A. A. Solov'yanov, and O. A. Reutov, *Zh. Org. Khim.*, **10,** 1798 (1974); *Engl. Transl.*, 1812.
349. I. P. Beletskaya, A. N. Kashin, A. Ts. Malkhasyan, A. A. Solov'yanov, E. Yu. Bekhli, and O. A. Reutov, *Zh. Org. Khim.*, **10,** 678 (1974); *Engl. Transl.*, 682.
350. A. N. Kashin, V. A. Khutoryanskii, B. L. Finkel'shtein, I. P. Beletskaya, and O. A. Reutov, *Zh. Org. Khim.*, **12,** 2049 (1976); *Engl. Transl.*, 1997.
351. S. Fukuzumi and J. K. Kochi, *J. Phys. Chem.*, **84,** 2246 (1980).
352. G. Grüttner and E. Krause, *Chem. Ber.*, **49,** 1415 (1916).
353. J. L. Keller, *Thesis,* U.C.L.A., Los Angeles, 1948; quoted in ref. 354.
354. S. Winstein and T. G. Traylor, *J. Am. Chem. Soc.*, **78,** 2597 (1956).
355. A. Lord and H. O. Pritchard, *J. Phys. Chem.*, **70,** 1689 (1966).
356. I. P. Beletskaya, O. A. Reutov, and T. P. Gur'yanova, *Izv. Akad. Nauk SSSR, Otd. Khim. Nauk,* 1589 (1961); *Engl. Transl.*, 1483.
357. I. P. Beletskaya, O. A. Reutov, and T. P. Gur'yanova, *Izv. Akad. Nauk SSSR, Otd. Khim. Nauk,* 1997 (1961); *Engl. Transl.*, 1863.
358. O. A. Reutov, I. P. Beletskaya, and T. P. Fetisova, *Izv. Akad. Nauk SSSR, Ser. Khim.*, 990 (1967); *Engl. Transl.*, 960.

359. O. A. Reutov, I. P. Beletskaya, and T. P. Fetisova, *Dokl. Akad. Nauk SSSR,* **166,** 861 (1966); *Engl. Transl.,* 158.
360. O. A. Reutov, I. P. Beletskaya and T. P. Fetisova, *Dokl. Akad. Nauk SSSR,* **155,** 1095 (1964); *Engl. Transl.,* 347.
361. O. A. Reutov, G. A. Artamkina, and I. P. Beletskaya, *Dokl. Akad. Nauk SSSR,* **153,** 588 (1963); *Engl. Transl.,* 939.
362. P. L. Yadav, V. Ramakrishna, and N. K. Jha, *Indian J. Chem.,* **14A,** 88 (1976).
363. P. L. Yadav, V. Ramakrishna, and N. K. Jha, *Indian J. Chem.,* **16A,** 623 (1978).
364. R. D. Brown, A. S. Buchanan, and A. A. Humffray, *Aust. J. Chem.,* **18,** 1527 (1965).
365. H. G. Kuivila and R. M. Williams, *J. Am. Chem. Soc.,* **76,** 2679 (1954).
366. C. Eaborn, *J. Chem. Soc.,* 2755 (1949).
367. H. C. Clark, J. T. Kwan, and D. Whyman, *Can. J. Chem.,* **41,** 2628 (1963).
368. D. Grafstein, *J. Am. Chem. Soc.,* **77,** 6650 (1955).
369. M. Lesbre and P. Mazerolles, *C. R. Acad. Sci.,* **246,** 1708 (1958).
370. H. H. Anderson, *J. Am. Chem. Soc.,* **73,** 5800 (1951).
371. Z. M. Manulkin, A. B. Kuchkarev, and S. A. Sarankina, *Dokl. Akad. Nauk SSSR,* **149,** 318 (1963).
372. D. F. Seitz, G. L. Tonnesen, S. Hellman, R. N. Hanson, and S. J. Adelstein, *J. Organomet. Chem.,* **186,** C33 (1980).
373. M. H. Abraham, A. T. Broadhurst, I. D. Clark, R. U. Koenigsberger, and D. F. Dadjour, *J. Organomet. Chem.,* **209,** 37 (1981).
374. M. Gielen and J. Nasielski, *J. Organomet. Chem.,* **7,** 273 (1967).
375. J. C. Jungers, L. Sajus, I. de Aguirre, and D. Decroocq, *Rev. Inst. Fr. Pét. Ann. Combust. Liq.,* **20,** 513 (1965); **21,** 285 (1966).
376. G. Martino and J. C. Jungers, *Bull. Soc. Chim. Fr.,* 3392 (1970).
377. N. S. Isaacs and K. Javaid, *Tetrahedron Lett.,* 3073 (1977).
378. J. Mathieu, *Bull. Soc. Chim. Fr.,* 807 (1973).
379. I. P. Beletskaya, A. N. Kashin, A. Ts. Malkhasyan, and O. A. Reutov, *Zh. Org. Khim.,* **9,** 2215 (1973); *Engl. Transl.,* 2231.
380. A. N. Kashin, I. P. Beletskaya, A. Ts. Malkhasyan, and O. A. Reutov, *Zh. Org. Khim.,* **10,** 673 (1974); *Engl. Transl.,* 677.
381. A. N. Kashin, I. P. Beletskaya, A. Ts. Malkhasyan, A. A. Solov'yanov, and O. A. Reutov, *Zh. Org. Khim.,* **10,** 2241 (1974); *Engl. Transl.,* 2257.
382. I. P. Beletskaya, A. N. Kashin, A. Ts. Malkhasyan, A. A. Solov'yanov, and O. A. Reutov, *Zh. Org. Khim.,* **10,** 1566 (1974); *Engl. Transl.,* 1582.
383. R. M. G. Roberts, *J. Organomet. Chem.,* **24,** 675 (1970).
384. M. H. Abraham and J. Andonian-Haftvan, *Bull. Soc. Chim. Belg.,* **89,** 819 (1980).
385. P. Baekelmans, M. Gielen, P. Malfroid, and J. Nasielski, *Bull. Soc. Chim. Belg.,* **77,** 85 (1968).
386. R. W. Bott, C. Eaborn, and J. A. Waters, *J. Chem. Soc.,* 681 (1963).
387. O. Buchman, M. Grosjean, and J. Nasielski, *Helv. Chim. Acta,* **47,** 1679 (1964).
388. O. Buchman, M. Grosjean, and J. Nasielski, *Helv. Chim. Acta,* **47,** 2037 (1964).
389. O. Buchman, M. Grosjean, J. Nasielski, and B. Wilmet-Devos, *Helv. Chim. Acta,* **47,** 1688 (1964).
390. O. Buchman, M. Grosjean, and J. Nasielski, *Bull. Soc. Chim. Belg.,* **72,** 286 (1963).
391. A. Folaranmi, R. A. N. McLean, and N. Wadibia, *J. Organomet. Chem.,* **73,** 59 (1974).
392. G. Redl, B. Altner, D. Anker, and M. Minot, *Inorg. Nucl. Chem. Lett.,* **5,** 861 (1969).
393. L. Riccoboni and L. Oleari, *Ric. Sci., Parte 2., Sez. A,* **3,** 1031 (1963).
394. L. Riccoboni, G. Pilloni, G. Plazzogna, and C. Bernardin, *Ric. Sci., Parte 2, Sez. A,* **3,** 1231 (1963).
395. L. Riccoboni, G. Pilloni, G. Plazzogna, and G. Tagliavini, *J. Electroanal. Chem.,* **11,** 340 (1966).
396. G. Pilloni and G. Tagliavini, *J. Organomet. Chem.,* **11,** 557 (1968).
397. A. Delhaye, J. Nasielski, and M. Planchon, *Bull. Soc. Chim. Belg.,* **69,** 134 (1960).
398. M. J. Adam, B. D. Pate, T. J. Ruth, J. M. Berry, and L. D. Hall, *J. Chem. Soc., Chem. Commun.,* 733 (1981).
399. G. Emptoz and F. Huet, *J. Organomet. Chem.,* **82,** 139 (1974).

400. J. H. Wotiz, C. A. Hollingsworth, and R. E. Dessy, *J. Am. Chem. Soc.*, **79,** 358 (1957).
401. S. K. Podder, E. W. Smalley, and C. A. Hollingsworth, *J. Org. Chem.*, **28,** 1435 (1963).
402. J. H. Wotiz, C. A. Hollingsworth, and R. E. Dessy, *J. Am. Chem. Soc.*, **77,** 103 (1955).
403. E. C. Ashby, *Q. Rev. Chem. Soc.*, **21,** 259 (1967).
404. R. E. Dessy and R. M. Salinger, *J. Org. Chem.*, **26,** 3519 (1961).
405. L. V. Guild, C. A. Hollingsworth, D. H. McDaniel, S. K. Podder, and J. H. Wotiz, *J. Org. Chem.*, **27,** 762 (1962).
406. H. Hashimoto, T. Nakano, and H. Okado, *J. Org. Chem.*, **30,** 1234 (1965).
407. Y. Pocker and J. H. Exner, *J. Am. Chem. Soc.*, **90,** 6764 (1968).
408. M. H. Abraham and J. A. Hill, *Proc. Chem. Soc.*, 175 (1964).
409. M. H. Abraham and J. A. Hill, *J. Organomet. Chem.*, **7,** 23 (1967).
410. S. Inoue and T. Yamada, *J. Organomet. Chem.*, **25,** 1 (1970).
411. A. Jubier, E. Henry-Basch, and P. Fréon, *C. R. Acad. Sci., Ser. C,* **267,** 842 (1968).
412. A. Jubier, G. Emptoz, E. Henry-Basch, and P. Fréon, *Bull. Soc. Chim. Fr.*, 2032 (1969).
413. M. S. Kharasch and L. Chalkley, *J. Am. Chem. Soc.*, **46,** 1211 (1924).
414. M. S. Kharasch and M. N. Grafflin, *J. Am. Chem. Soc.*, **47,** 1948 (1925).
415. M. S. Kharasch and R. Marker, *J. Am. Chem. Soc.*, **48,** 3130 (1926).
416. M. S. Kharasch and A. L. Flenner, *J. Am. Chem. Soc.*, **54,** 674 (1932).
417. M. S. Kharasch, H. Pines, and J. M. Levine, *J. Org. Chem.*, **3,** 347 (1938–9).
418. M. S. Kharasch, R. R. Legault, and W. R. Sprowls, *J. Org. Chem.*, **3,** 409 (1938–9).
419. L. H. Gale, F. R. Jensen and J. A. Landgrebe, *Chem. Ind. (London),* 118 (1960).
420. M. M. Kreevoy, *J. Am. Chem. Soc.*, **79,** 5927 (1957).
421. M. M. Kreevoy and R. L. Hansen, *J. Am. Chem. Soc.*, **83,** 626 (1961).
422. R. J. Maguire and S. Anand, *J. Inorg. Nucl. Chem.*, **38,** 1167 (1976).
423. W. A. Nugent and J. K. Kochi, *J. Am. Chem. Soc.*, **98,** 273 (1976).
424. W. A. Nugent and J. K. Kochi, *J. Am. Chem. Soc.*, **98,** 5979 (1976).
425. A. N. Nesmeyanov, A. E. Borisov, and I. S. Saveleva, *Proc. Acad. Sci. USSR,* **155,** 280 (1964).
426. R. E. Dessy, G. F. Reynolds, and J.-Y. Kim, *J. Am. Chem. Soc.*, **81,** 2683 (1959).
427. R. E. Dessy and J.-Y. Kim, *J. Am. Chem. Soc.*, **83,** 1167 (1961).
428. R. E. Dessy and J.-Y. Kim, *J. Am. Chem. Soc.*, **82,** 686 (1960).
429. F. Nerdel and S. Makover, *Naturwissenschaften,* **45,** 490 (1958).
430. A. H. Corwin and M. A. Naylor, *J. Am. Chem. Soc.*, **69,** 1004 (1947).
431. F. Kaufman and A. H. Corwin, *J. Am. Chem. Soc.*, **77,** 6280 (1955).
432. R. D. Brown, A. S. Buchanan, and A. A. Humffray, *Aust. J. Chem.*, **18,** 1507 (1965).
433. R. D. Brown, A. S. Buchanan, and A. A. Humffray, *Aust. J. Chem.*, **18,** 1513 (1965).
434. I. P. Beletskaya, A. E. Myshkin, and O. A. Reutov, *Izv. Akad. Nauk SSSR, Ser. Khim.*, 240 (1965); *Engl. Transl.*, 226.
435. I. P. Beletskaya, A. E. Myshkin, and O. A. Reutov, *Izv. Akad Nauk SSSR, Ser. Khim.*, 238 (1967); *Engl. Transl.*, 232.
436. I. P. Beletskaya, A. E. Myshkin, and O. A. Reutov, *Izv. Akad. Nauk SSSR, Ser. Khim.*, 245 (1967); *Engl. Transl.*, 239.
437. M. M. Kreevoy and R. L. Hansen, *J. Phys. Chem.*, **65,** 1055 (1961).
438. I. P. Beletskaya, V. I. Karpov, V. A. Moskalenko, and O. A. Reutov, *Dokl. Akad. Nauk SSSR,* **162,** 86 (1965); *Engl. Transl.*, 414.
439. P. D. Sleezer, S. Winstein, and W. G. Young, *J. Am. Chem. Soc.*, **85,** 1890 (1963).
440. M. M. Kreevoy, P. J. Steinwand, and W. V. Kayser, *J. Am. Chem. Soc.*, **86,** 5013 (1964).
441. M. M. Kreevoy, P. J. Steinwand, and W. V. Kayser, *J. Am. Chem. Soc.*, **88,** 124 (1966).
442. M. M. Kreevoy, P. J. Steinwand, and T. S. Straub, *J. Org. Chem.*, **31,** 4291 (1966).
443. M. M. Kreevoy, D. J. W. Goon, and R. A. Kayser, *J. Am. Chem. Soc.*, **88,** 5529 (1966).
444. S. E. Ibrahim and M. M. Kreevoy, *J. Phys. Chem.*, **81,** 2143 (1977).
445. A. N. Nesmeyanov, A. E. Borisov, and I. S. Savel'eva, *Dokl. Akad. Nauk SSSR,* **172,** 1043 (1967); *Engl. Transl.*, 155.
446. O. A. Reutov, Yu. G. Bundel', and N. D. Antonova, *Dokl. Akad. Nauk SSSR,* **166,** 1103 (1966); *Engl. Transl.*, 191.
447. Yu. G. Bundel', V. I. Rosenberg, I. N. Krokhina, and O. A. Reutov, *Zh. Org. Khim.*, **6,** 1519 (1970); *Engl. Transl.*, 1531.

448. O. A. Reutov, I. P. Beletskaya, and L. A. Fedorov, *Dokl. Akad. Nauk SSSR,* **163,** 1381 (1965); *Engl. Transl.,* 794.
449. O. A. Reutov, I. P. Beletskaya, and L. A. Fedorov, *Zh. Org. Khim.,* **3,** 225 (1967); *Engl. Transl.,* 213.
450. B. F. Hegarty, W. Kitching, and P. R. Wells, *J. Am. Chem. Soc.,* **89,** 4816 (1967).
451. Yu. G. Bundel', V. I. Rozenberg, G. V. Gavrilova, E. A. Arbuzova, and O. A. Reutov, *Izv. Akad. Nauk SSSR, Ser. Khim.,* 1155 (1970); *Engl. Transl.,* 1089.
452. J. R. Coad and M. D. Johnson, *J. Chem. Soc. B,* 633 (1967).
453. J. R. Coad and C. K. Ingold, *J. Chem. Soc. B,* 1455 (1968).
454. A. G. Davies and P. B. Roberts, *J. Chem. Soc. C,* 1474 (1968).
455. G. W. Kabalka, R. J. Newton, and J. Jacobus, *J. Org. Chem.,* **44,** 4185 (1979).
456. L. H. Torporcer, R. E. Dessy, and S. I. E. Green, *J. Am. Chem. Soc.,* **87,** 1236 (1965).
457. R. D. Brown, A. S. Buchanan, and A. A. Humffray, *Aust. J. Chem.,* **18,** 1521 (1965).
458. H. G. Kuivila and K. V. Nahabedian, *Chem. Ind. (London),* 1120 (1959).
459. H. G. Kuivila and K. V. Nahabedian, *J. Am. Chem. Soc.,* **83,** 2159 (1961).
460. H. G. Kuivila and K. V. Nahabedian, *J. Am. Chem. Soc.,* **83,** 2164 (1961).
461. H. G. Kuivila and K. V. Nahabedian, *J. Am. Chem. Soc.,* **83,** 2167 (1961).
462. H. G. Kuivila, J. F. Reuwer, and J. A. Mangravite, *Can. J. Chem.,* **41,** 3081 (1963).
463. H. G. Kuivila, J. F. Reuwer, and J. A. Mangravite, *J. Am. Chem. Soc.,* **86,** 2666 (1964).
464. R. Taylor, *Electrophilic Substitution at a Saturated Carbon Atom,* in *Comprehensive Chemical Kinetics* (Eds. C. H. Bamford and C. F. H. Tipper), Elsevier, Amsterdam, 1973, Vol. 13.
465. G. A. Russell and K. L. Nagpal, *Tetrahedron Lett.,* 421 (1961).
466. P. D. George, *US Pat.,* 2 802 852 (1957).
467. J. E. Noll, J. L. Speier, and B. F. Daubert, *J. Am. Chem. Soc.,* **73,** 3867 (1951).
468. L. H. Sommer, N. S. Marans, G. M. Goldberg, J. Rockett, and R. P. Pioch, *J. Am. Chem. Soc.,* **73,** 882 (1951).
469. L. H. Sommer, R. P. Pioch, N. S. Marans, G. M. Goldberg, J. Rockett, and J. Kerlin, *J. Am. Chem. Soc.,* **75,** 2932 (1953).
470. B. N. Dolgov, E. V. Kukharskaya, and D. N. Andreev, *Izv. Akad. Nauk SSSR, Otd. Khim. Nauk,* 968 (1957).
471. L. H. Sommer, W. P. Barie, and J. R. Gould, *J. Am. Chem. Soc.,* **75,** 3765 (1953).
472. L. M. Shorr, H. Freiser, and J. L. Speier, *J. Am. Chem. Soc.,* **77,** 547 (1955).
473. S. Nozakura, *Nippon Kagaku Zasshi,* **75,** 958 (1954).
474. S. I. Sadykh-Zade, I. I. Tsetlin, and A. D. Petrov, *Zh. Obshch. Khim.,* **26,** 1239 (1956).
475. L. H. Sommer, W. D. English, G. R. Ansul, and D. N. Vivona, *J. Am. Chem. Soc.,* **77,** 2485 (1955).
476. L. H. Sommer, G. M. Goldberg, G. H. Barnes, and L. S. Stone, *J. Am. Chem. Soc.,* **76,** 1609 (1954).
477. S. Kohama and S. Fukukawa, *Nippon Kagaku Zasshi,* **80,** 1492 (1959).
478. L. H. Sommer and G. A. Baum, *J. Am. Chem. Soc.,* **76,** 5002 (1954).
479. C. Tomasino, *Diss. Abstr.,* **20,** 103 (1959).
480. L. H. Sommer and G. R. Ansul, *J. Am. Chem. Soc.,* **77,** 2482 (1955).
481. E. V. Wilkus and W. H. Rauscher, *J. Org. Chem.,* **30,** 2889 (1965).
482. E. V. Wilkus and A. Berger, *US Pat.,* 3 326 952 (1967).
483. D. N. Andreev, B. N. Dolgov, and S. V. Butz, *Zh. Obshch. Khim.,* **32,** 1275 (1962).
484. B. N. Dolgov, D. N. Andreev, and V. P. Lyutyi, *Dokl. Akad. Nauk SSSR,* **118,** 501 (1958).
485. D. H. O'Brien and C. M. Harbordt, *J. Organomet. Chem.,* **21,** 321 (1970).
486. C. M. Harbordt and D. H. O'Brien, *J. Organomet. Chem.,* **111,** 153 (1976).
487. D. D. Hopf and D. H. O'Brien, *J. Organomet. Chem.,* **111,** 161 (1976).
488. A. Ya. Yakubovich and G. V. Motsarev, *Zh. Obshch. Khim.,* **29,** 2395 (1959).
489. P. A. DiGiorgio, L. H. Sommer, and F. C. Whitmore, *J. Am. Chem. Soc.,* **70,** 3512 (1948).
490. R. H. Krieble and J. R. Elliott, *J. Am. Chem. Soc.,* **67,** 1810 (1945).
491. G. V. Motsarev, V. R. Rozenberg, and T. Ya. Chashnikova, *Zh. Prikl. Khim.,* **34,** 430 (1961).

492. G. Fritz, J. Grobe, and D. Ksinsik, *Z. Anorg. Allg. Chem.,* **302,** 175 (1959).
493. G. V. Motsarev and V. R. Rozenberg, *Zh. Obshch. Khim.,* **30,** 3011 (1960).
494. W. Zimmerman, *Chem. Ber.,* **87,** 887 (1954).
495. C. Tamborski and H. W. Post, *J. Org. Chem.,* **17,** 1400 (1952).
496. S. Brynolf, *Kgl. Fysiograf. Sallskap. Lund., Forh.,* **29,** 121 (1959).
497. J. L. Speier and B. F. Daubert, *J. Am. Chem. Soc.,* **70,** 1400 (1948).
498. R. H. Krieble and J. R. Elliott, *J. Am. Chem. Soc.,* **68,** 2291 (1946).
499. Chih-Tang Huang and Pao-Jen Wang, *Hua Hsueh Hsueh Pao,* **25,** 341 (1959).
500. M. Kumada, M. Ishikawa and K. Tamao, *J. Organomet. Chem.,* **5,** 226 (1966).
501. C. R. Hauser and C. R. Hance, *J. Am. Chem. Soc.,* **74,** 5091 (1952).
502. C. Eaborn and J. C. Jeffrey, *J. Chem. Soc.,* 137 (1957).
503. C. F. Roedel, *J. Am. Chem. Soc.,* **71,** 269 (1949).
504. R. N. Haszeldine and J. C. Young, *J. Chem. Soc.,* 4503 (1960).
505. A. G. Brook, *J. Am. Chem. Soc.,* **79,** 4373 (1957).
506. A. G. Brook and N. V. Schwartz, *J. Org. Chem.,* **27,** 2311 (1962).
507. H. G. Kuivila and P. L. Maxfield, *Inorg. Nucl. Chem. Lett.,* **1,** 29 (1965).
508. A. G. Brook, W. W. Limburg, D. M. MacRae, and S. A. Fieldhouse, *J. Am. Chem. Soc.,* **89,** 704 (1967).
509. A. G. Brook, R. Kivisikk, and G. E. LeGrow, *Can. J. Chem.,* **43,** 1175 (1965).
510. F. Rijkens, M. J. Janssen, W. Drenth, and G. J. M. van der Kerk, *J. Organomet. Chem.,* **2,** 347 (1964).
511. F. C. Whitmore, L. H. Sommer, J. R. Gold, and R. E. van Strein, *J. Am. Chem. Soc.,* **69,** 1551 (1947).
512. J. R. Gold, L. H. Sommer, and F. C. Whitmore, *J. Am. Chem. Soc.,* **70,** 2874 (1948).
513. M. Prober, *J. Am. Chem. Soc.,* **77,** 3224 (1955).
514. S. Nozakura and S. Konotsune, *Bull. Chem. Soc. Jpn.,* **29,** 326 (1956).
515. S. Tannenbaum, S. Kaye, and G. F. Lewenz, *J. Am. Chem. Soc.,* **75,** 3753 (1953).
516. C. L. Agre and W. Hilling, *J. Am. Chem. Soc.,* **74,** 3895 (1952).
517. G. H. Wagner and A. N. Pines, *J. Am. Chem. Soc.,* **71,** 3567 (1949).
518. C. L. Agre and W. Hilling, *J. Am. Chem. Soc.,* **74,** 3899 (1952).
519. M. Kanazashi, *Bull. Chem. Soc. Jpn.,* **28,** 44 (1955).
520. V. M. Vdovin, N. S. Nametkin, E. Sh. Finkel'shtein, and V. D. Oppengeim, *Izv. Akad. Nauk SSSR, Ser. Khim.,* 458 (1964).
521. L. H. Sommer, L. J. Tyler, and F. C. Whitmore, *J. Am. Chem. Soc.,* **70,** 2872 (1948).
522. D. L. Bailey and A. N. Pines, *Ind. Eng. Chem.,* **46,** 2363 (1954).
523. F. B. Deans and C. Eaborn, *J. Chem. Soc.,* 2299 (1959).
524. F. B. Deans and C. Eaborn, *J. Chem. Soc.,* 2303 (1959).
525. R. W. Bott, B. F. Dowden, and C. Eaborn, *J. Chem. Soc.,* 6306 (1965).
526. R. Baker, R. W. Bott, C. Eaborn, and P. M. Greasley, *J. Chem. Soc.,* 627 (1964).
527. C. Eaborn and K. C. Pande, *J. Chem. Soc.,* 1566 (1960).
528. R. A. Benkeser and F. S. Clark, *J. Am. Chem. Soc.,* **82,** 4881 (1960).
529. R. A. Benkeser and H. R. Krysiak, *J. Am. Chem. Soc.,* **76,** 6353 (1954).
530. R. A. Benkeser, R. A. Hickner, and D. I. Hoke, *J. Am. Chem. Soc.,* **80,** 2279 (1958).
531. R. A. Benkeser, W. Schroeder, and O. H. Thomas, *J. Am. Chem. Soc.,* **80,** 2283 (1958).
532. J. Nasielski and M. Planchon, *Bull. Soc. Chim. Belg.,* **69,** 123 (1960).
533. C. Eaborn and J. A. Sperry, *J. Chem. Soc.,* 4921 (1961).
534. C. Eaborn, *J. Chem. Soc.,* 4858 (1956).
535. C. Eaborn, *J. Chem. Soc.,* 3148 (1953).
536. R. W. Bott, C. Eaborn, and K. Leyshon, *J. Chem. Soc.,* 1971 (1964).
537. C. Eaborn and D. R. M. Walton, *J. Organomet. Chem.,* **3,** 169 (1965).
538. R. Baker, C. Eaborn, and J. A. Sperry, *J. Chem. Soc.,* 2382 (1962).
539. R. W. Bott, C. Eaborn, and P. M. Jackson, *J. Organomet. Chem.,* **7,** 79 (1967).
540. R. W. Bott, C. Eaborn, and B. M. Rushton, *J. Organomet. Chem.,* **3,** 448 (1965).
541. P. M. Greasley, *PhD Thesis,* University of Leicester, 1961.
542. J. R. F. Jaggard, *DPhil Thesis,* University of Sussex, 1965.
543. D. J. Young, *DPhil Thesis,* University of Sussex, 1967.
544. K. A. Andrianov, S. A. Igonina, V. I. Sidorov, C. Eaborn, and P. M. Jackson, *J. Organomet. Chem.,* **110,** 39 (1976).

545. H. Gilman, A. G. Brook, and L. S. Miller, *J. Am. Chem. Soc.,* **75,** 4531 (1953).
546. C. R. Hauser and C. R. Hance, *J. Am. Chem. Soc.,* **73,** 5846 (1951).
547. C. Eaborn and D. R. M. Walton, *J. Organomet. Chem.,* **4,** 217 (1965).
548. R. W. Bott, C. Eaborn, and T. W. Swaddle, *J. Chem. Soc.,* 2342 (1963).
549. C. Eaborn and S. H. Parker, *J. Chem. Soc.,* 126 (1955).
550. R. W. Bott, B. F. Dowden, and C. Eaborn, *J. Chem. Soc.,* 4994 (1965).
551. C. Eaborn and G. Seconi, *J. Chem. Soc., Perkin Trans. 2,* 203 (1979).
552. H. R. Allcock, *PhD Thesis,* University of London, 1956.
553. R. Alexander, C. Eaborn, and T. G. Traylor, *J. Organomet. Chem.,* **21,** 65 (1970).
554. R. Alexander, W. A. Asomaning, C. Eaborn, I. D. Jenkins, and D. R. M. Walton, *J. Chem. Soc., Perkin Trans. 2,* 490 (1974).
555. A. R. Bassindale, C. Eaborn, R. Taylor, A. R. Thompson, D. R. M. Walton, J. Cretney, and G. J. Wright, *J. Chem. Soc. B,* 1155 (1971).
556. B. Bøe, C. Eaborn, and D. R. M. Walton, *J. Organomet. Chem.,* **82,** 327 (1974).
557. R. M. G. Roberts and F. El Kaissi, *J. Organomet. Chem.,* **12,** 79 (1968).
558. W. J. Rennie and R. M. G. Roberts, *J. Organomet. Chem.,* **37,** 77 (1972).
559. C. Eaborn, I. D. Jenkins, and G. Seconi, *J. Organomet. Chem.,* **131,** 387 (1977).
560. L. M. Dennis and W. I. Patnode, *J. Am. Chem. Soc.,* **52,** 2779 (1930).
561. J. E. Griffiths and M. Onyszchuk, *Can. J. Chem.,* **39,** 339 (1961).
562. B. M. Gladshtein, V. V. Rode, and L. Z. Soborovskii, *Zh. Obshch. Khim.,* **29,** 2155 (1959).
563. J. K. Simons, *J. Am. Chem. Soc.,* **57,** 1299 (1935).
564. P. Mazerolles, *Bull. Soc. Chim. Fr.,* 856 (1960).
565. C. Eaborn and K. C. Pande, *J. Chem. Soc.,* 297 (1961).
566. C. Eaborn and K. C. Pande, *J. Chem. Soc.,* 5082 (1961).
567. K. Sisido, S. Kozima, and K. Takizawa, *Tetrahedron Lett.,* 33 (1967).
568. K. Sisido, T. Miyanisi, and T. Isida, *J. Organomet. Chem.,* **23,** 117 (1970).
569. R. N. Meals, *J. Org. Chem.,* **9,** 211 (1944).
570. G. S. Sasin, A. L. Borror, and R. Sasin, *J. Org. Chem.,* **23,** 1366 (1958).
571. M. Lesbre and R. Dupont, *C. R. 78e Congrès des Sociétés Savantes, Paris et Départements, Sect. Sci.,* 429 (1953).
572. G. Bähr, *Z. Anorg. Chem.,* **256,** 107 (1948).
573. G. Fritz and H. Scheer, *Z. Naturforsch., Teil B,* **19,** 537 (1964).
574. M. M. Koton, *Zh. Obshch. Khim.,* **26,** 3581 (1956).
575. A. Henderson and A. K. Holliday, *J. Organomet. Chem.,* **4,** 377 (1965).
576. R. Walraevens, cited by M. Gielen and J. Nasielski, *Recl. Trav. Chim. Pays. Bas.* **82,** 228 (1963).
577. V. Bade and F. Huber, *J. Organomet. Chem.,* **24,** 387 (1970).
578. C. Eaborn, and J. A. Waters, *J. Chem. Soc.,* 542 (1961).
579. R. W. Bott, C. Eaborn, and D. R. M. Walton, *J. Organomet. Chem.,* **2,** 154 (1964).
580. O. Buchman, M. Grosjean and J. Nasielski, *Helv. Chim. Acta,* **47,** 1695 (1964).
581. O. Buchman, M. Grosjean, M. Jeuquet, and J. Nasielski, *J. Organomet. Chem.,* **19,** 353 (1969).
582. R. Alexander, M. T. Attar-Bashi, C. Eaborn, and D. R. M. Walton, *Tetrahedron,* **30,** 899 (1974).
583. C. J. Moore, M. L. Bullpitt, and W. Kitching, *J. Organomet. Chem.,* **64,** 93 (1974).
584. H. G. Kuivila and J. A. Verdone, *Tetrahedron Lett.,* 119 (1964).
585. H. G. Kuivila and J. C. Cochran, *J. Am. Chem. Soc.,* **89,** 7152 (1967).
586. J. A. Verdone, J. A. Mangravite, N. M. Scarpa, and H. G. Kuivila, *J. Am. Chem. Soc.,* **97,** 843 (1975).
587. A. Alvanipour, C. Eaborn, and D. R. M. Walton, *J. Organomet. Chem.,* **201,** 233 (1980).
588. G. C. Robinson, *J. Org. Chem.,* **28,** 843 (1963).
589. H. Horn and F. Huber, *Monatsh. Chem.,* **98,** 771 (1967).
590. V. Bade and F. Huber, *J. Organomet. Chem.,* **24,** 691 (1970).
591. C. Eaborn and K. C. Pande, *J. Chem. Soc.,* 3715 (1961).

The Chemistry of the Metal—Carbon Bond, Volume 2
Edited by F. R. Hartley and S. Patai


CHAPTER **3**

# Homolytic cleavage of metal—carbon bonds: Groups I to V

PHILIP J. BARKER[a]

*Research School of Chemistry, Australian National University, P.O. Box 4, Canberra, A.C.T. 2600, Australia*

JEREMY N. WINTER

*Department of Chemistry, University of York, Heslington, York YO1 5DD, UK*

---



[a] Author to whom correspondence should be addressed.

## I. INTRODUCTION

### A. Historical Survey

Before commencing the chapter, it seems relevant to reflect upon the historical significance of the subject matter not only to the development of organometallic chemistry as a whole, but also to the understanding of free radicals and homolytic reactions. The person most responsible for the development of these areas was Frankland. In his early paper 'On the Isolation of the Organic Radicals'[1], he studied the reaction between zinc metal and ethyl iodide, and isolated a colourless gas, which he believed to be free 'ethyl radicals' and a pyrophoric white crystalline solid, later found to be diethylzinc[2]. The gas was subsequently found to consist mainly of butane and there seems little doubt that the thermally induced homolysis of diethylzinc had been effected.

The future of organometallic chemistry was assured to the extent that only 11 years after his initial discoveries, Frankland was moved to write a 60-page review of the area which described the synthesis of organometallic compounds of every group in the Periodic Table[3] and demonstrated the widespread interest in this new area of chemistry throughout other European Schools.

The general acceptance of the free radical as a chemical entity was, in contrast, very difficult to achieve. Even after Gomberg's work on the triphenylmethyl radical at the turn of the century[4], it was 30 years before Hofeditz and Paneth unequivocally proved the existence of the methyl radical[5]. With relevance to this chapter, it was an organometallic of Group IV used in the demonstration, tetramethyllead.

Advances in instrumentation have now ensured that we can directly observe free radicals produced in organometallic homolyses by electron spin resonance (e.s.r.) and by the use of chemically induced dynamic nuclear polarization (CIDNP), we may infer that a reaction path is homolytic in nature.

A chronological account of the events important to the development of homolytic chemistry with emphasis on organometallic systems may be found in a recent essay by Davies[6], 'The History of Homolysis'.

There has been rapidly expanding interest in the organic chemistry of the main-group elements for 30 years now and, over the past 15 years or so, this general interest has been complemented by an increasing number of studies on the homolytic reactions of these organometallics, and particularly interesting are those reactions where the initial products of metal—carbon bond homolysis, generally free radicals, may be observed directly. In addition to the increase in study of the main group homolysis *per se*, the mechanisms of reactions fundamental to the organic chemist, namely the Grignard reaction, and reactions of organolithium reagents continue to attract much attention. Many of these reactions have been found to involve free-radical intermediates, whether formed by direct metal—carbon bond homolysis or by concerted processes. In this respect, the advances in instrumentation mentioned earlier have been fundamental to the elucidation of both new and known mechanisms.

## B. Scope of Review

From the title of the chapter, the two major types of homolytic reaction implied are outlined in equations 1 and 2.

$$R_nM \longrightarrow R^{\bullet} + R_{n-1}M \tag{1}$$

$$X^{\bullet} + R_nM \longrightarrow R^{\bullet} + XMR_{n-1} \tag{2}$$

Equation 1 indicates a simple unimolecular process which gives a free radical $R^{\bullet}$ and a low-valent metal species, which may or may not be free radical in nature. Such reactions may be achieved by photolysis or thermolysis under relatively mild conditions, or by the use of ionizing radiation (X- or $\gamma$-rays). Equation 2 indicates a bimolecular substitution reaction (designated $S_H2$), in which an incoming radical $X^{\bullet}$ attacks the metal centre with subsequent elimination of $R^{\bullet}$.

Also included in this discussion are those reactions where metal—carbon bond homolysis is initiated by another chemical process or an electron transfer reaction. In the latter case an organometallic substrate gives up an electron to a suitable organic acceptor or inorganic oxidant. The resulting highly unstable organometallic cation rapidly fragments to give a stable cation and an organic free radical. Such reactions are widespread among the groups included in this chapter.

$$R_nM + A \longrightarrow R_nM^{+\bullet} + A^{-\bullet} \tag{3}$$

$$R_nM^{+\bullet} \longrightarrow R^{\bullet} + R_{n-1}M^{+} \tag{4}$$

Of these reaction types, the substitution ($S_H2$) reactions of equation 2 have received the most attention. This interest has arisen from early studies on the autoxidation of organometallics which showed that for many compounds a chain mechanism (reactions 5 and 6) could account adequately for the kinetic and product

$$R^{\bullet} + O_2 \longrightarrow ROO^{\bullet} \tag{5}$$

$$ROO^{\bullet} + R'_nM \longrightarrow ROOMR'_{n-1} + R'^{\bullet} \tag{6}$$

analyses, clearly involving $S_H2$ at the metal atom as a chain propagation step in equation 6[7,8]. Indeed, it is now assured that $S_H2$ at metal centres plays an important part in homolytic reactions of organometals. Thus, results prior to 1970 have been summarized in a valuable monograph by Ingold and Roberts[9] and since then review articles have covered advances fairly comprehensively, among these being general articles[10–13], and those devoted to $S_H2$ reactions of specific metals, for example boron[14], tin[15] and phosphorus[16]. Consequently, the substitution reactions of the more commonly studied elements will not be the focus of attention here, and our discussion will centre on unimolecular reactions, novel and recent advances in substitution reactions, and electron-transfer initiated homolysis. While this last class of reaction has had early results summarized by Kochi[17], it is hoped that the widespread occurrence of these reactions and their importance will be strongly emphasized here.

In discussing the results of the investigation reviewed, much of the data will be from e.s.r. or CIDNP studies, where the radical products (e.s.r.) or the non-radical products from radical reactions (CIDNP) may be detected. However, where product analysis has allowed little ambiguity in mechanistic assignment, these examples are also reviewed. Generally, reactions taking place inside the mass spectrometer, or those requiring very high temperatures or pressures, the mechanistic conclusions of which are derived from kinetic or product studies, are not considered.

## C. Thermochemistry

The facility with which many main-group organometallics lend themselves to homolytic chemistry, whether unimolecular decomposition or bimolecular substitution, is a result of the thermochemical properties of the respective metal—carbon bonds. However, in many cases it is this lability which makes reliable and accurate determination of bond strengths difficult. A review by Steele[18] assessed the reliability of previously published data and outlined the methods by which satisfactory values may be obtained. In addition, the mean bond dissociation energies of several organometallics have been estimated[18] and selected relevant values are listed in Table 1. The method used to obtain these values merely relies on a knowledge of the enthalpy of formation of the gaseous metal ions and the free-radical fragments required.

While giving a good indication of the energy required to transform an organometallic compound into constituent parts, the values give little idea of the

TABLE 1. Mean bond dissociation energies of metal—methyl bonds. Error limits range from $\pm 10$ to $\pm 4\ \text{kJ mol}^{-1}$. From ref. 18

| Bond | $\bar{D}$(M—R) (kJ mol$^{-1}$) | Bond | $\bar{D}$(M—R) (kJ mol$^{-1}$) |
|---|---|---|---|
| Li—Me | 248 | Sn—Me | 226 |
| Zn—Me | 183 | Pb—Me | 168 |
| Cd—Me | 147 | P—Me | 283 |
| Hg—Me | 124 | As—Me | 238 |
| B—Me | 373 | Sb—Me | 224 |
| Al—Me | 286 | Bi—Me | 140 |
| Ga—Me | 252 | | |

importance of the first bond dissociation energy, and it is this value which reflects the ability of an organometallic compound to take part in homolytic reactions. A good example of the difference between first and second bond dissociation energies is provided by those of dimethylmercury[19]. Here, the mean bond dissociation energy is itself a comparatively low 124 kJ $mol^{-1}$ for each metal—carbon bond. However, the value for the first bond is 200–210 kJ $mol^{-1}$ while the second is only 20 kJ $mol^{-1}$. In effect, this corresponds to the pair of decomposition reactions depicted in equations 7 and 8, implying the intermediacy of what must be a very short-lived mercury(I) species.

$$Me_2Hg^{II} \xrightarrow[\text{or } h\nu]{\Delta} MeHg^{I} + Me^{\bullet} \quad (7)$$

$$MeHg^{I} \longrightarrow Hg^{0} + Me^{\bullet} \quad (8)$$

The lifetimes of these low-valent intermediates are discussed in Section V, but in the solid state such species have been confirmed by e.s.r. where decomposition was induced by $\gamma$-radiolysis[20]. Reviews by Skinner[19] and Cox and Pilcher[21] give several first bond dissociation energies for other organometallics (see also Volume 1, Chapter 2).

The relative bond dissociation energies for the alkyl—magnesium bond for several Grignard reagents have been estimated[22,23]. The values are significant in that a good insight may often be gained as to the primary mechanism of a particular Grignard reaction. Thus, as is discussed at greater length later, methylmagnesium bromide, with a relatively high value for $D$(XMg—R) undergoes many reactions via polar mechanisms, whereas under similar conditions *t*-butylmagnesium chloride with a low $D$(XMg—R) often favours the single electron transfer route leading to products resulting from free *t*-butyl radicals reacting outside the solvent cage.

Without doubt, homolytic reactions of Group IV have been studied more than those of other groups during the past decade or so. The need for reliable thermochemical data has become apparent, and the demand has been satisfied by Jackson[24]. In calculating the first bond dissociation energies of several fundamental Group IV compounds from mass spectral appearance potentials and kinetic data, a series of reasonably self-consistent values which seem to reflect the observed chemistries has been obtained. These values are given in Table 2. The first bond dissociation energies of several other organometallics relevant to this discussion are given in Table 3.

TABLE 2. First bond dissociation energies for Group IV organometallics. Calculated from kinetic and appearance Potential Data. From ref. 24

| Bond | $D(Me_3M—Me)$ (kJ $mol^{-1}$) |
|---|---|
| $Me_3C—Me$ | 380 |
| $Me_3Si—Me$ | 380 |
| $Me_3Ge—Me$ | 320 |
| $Me_3Sn—Me$ | 270 |
| $Me_3Pb—Me$ | 180 |

TABLE 3. First bond dissociation energies of some symmetric organometallic compounds. Compiled from refs. 19 and 20

| Compound | $D_1$(kcal mol$^{-1}$) | Compound | $D_1$(kcal mol$^{-1}$) |
|---|---|---|---|
| $Me_2Hg$ | 216 | $Me_3Ga$ | 252 |
| $Et_2Hg$ | 180 | $Me_3In$ | 197 |
| $Pr^i{}_2Hg$ | 113 | $Me_3Tl$ | 113 |
| $Me_2Zn$ | 197 | $Me_3As$ | 265 |
| $Me_2Cd$ | 193 | $Me_3Sb$ | 140 |
| $Me_3Al$ | 273 | $Me_3Bi$ | 185 |

## D. Instrumental Techniques

There are many avenues experimentally available to the organometallic chemist for the study of reaction mechanisms. Here, where discussion is focused on homolytic reactions, those techniques of most assistance include electron spin resonance (e.s.r.), chemically induced dynamic nuclear polarization (CIDNP) and product studies. The first two of these provide much of the experimental data used here, and a brief practical introduction to each is warranted.

### *1. Electron spin resonance*

The principles of e.s.r. have been outlined in several major texts and monographs[25] and are not enlarged upon here. For those to whom the technique is unfamiliar, reference is made to a concise and readable introductory text which has recently become available[26].

There are three major types of radical which may be detected by e.s.r. in solution in organometallic systems. Firstly the free organic radicals and free organometallic (metal-centred) radicals which are the direct products of metal—carbon bond homolysis (except in $S_H2$ reactions where only the alkyl radical is displaced) may be observed. If this is not possible, then 'trapped' radicals may be seen; this term merely refers to the observation of a secondary radical which is a product from reaction of the primary species with another substrate. This other substrate may be an alkene, carbonyl compound, nitroso compound, or nitrone (in 'spin-trapping') and the secondary product is normally (although not always) confirmative of the homolytic pathway. Careful choice of trapping agent enables either fragment to be trapped. Finally, the products of many electron transfer reactions of Grignard reagents, for example with benzophenones or phenanthroline derivatives, are radical ions, but here the spectra are often difficult to analyse and not always indicative that metal—carbon homolysis has taken place.

A large number of examples where radicals are observed directly are initiated photochemically. Decomposition or substitution reactions are initiated by direct irradiation of the samples in the e.s.r. cavity by light from a high-pressure mercury arc. Photolysis is generally carried out at temperatures less than 0 °C to obtain detectable concentrations of the normally very reactive species generated. As an illustration, cyclopentadienyl compounds of tin and mercury have been found to undergo unimolecular photolysis on irradiation at −60 °C in many solvents to give well defined e.s.r. spectra of the cyclopentadienyl radical (reaction 9)[27]. While under

$$R_3Sn(^1\eta{-}C_5H_5) \xrightarrow{h\nu} R_3Sn^{\bullet} + C_5H_5^{\bullet} \qquad (9)$$

these conditions the e.s.r. of the tin species is not observed, its presence is confirmed by repeating the photolysis in an alkene solvent where the tin radical is 'trapped' by the double bond to give a $\beta$-stannylalkyl radical which is readily detected (reaction 10).

$$R_3Sn^{\bullet} + CH_2{=}CH_2 \longrightarrow R_3SnCH_2CH_2^{\bullet} \qquad (10)$$

Direct observation of tin-centred radicals is possible, but often difficult, even at low temperature in solution[28], although in some specific cases very persistent sterically hindered tin radicals may be prepared[29]. The photolysis of cyclopentadienyl compounds of lead also give unimolecular decomposition and this is discussed in Section VII.

Substitution reactions are studied in much the same manner, the major difference being that the photolysis merely serves to generate the attacking radical $X^{\bullet}$ in equation 2. To this effect, Figure 1 shows the e.s.r. spectrum of the radical caused by the $S_H2$ reaction at tin by *t*-butoxy radicals on tri-*n*-propyltin chloride[30].

$$(Bu^tO)_2 \longrightarrow 2Bu^tO^{\bullet} \qquad (11)$$

$$Bu^tO^{\bullet} + Pr^n{}_3SnCl \longrightarrow Bu^tOSn(Pr^n)_2Cl + n\text{-}Pr^{\bullet} \qquad (12)$$

The following discussion will demonstrate the utility of cyclopentadienyl metallics in inducing homolytic chemistry. For instance, the unimolecular photolysis of cyclopentadienyltin compounds has enabled a study of the comparative reactivity of a variety of tin-centred radicals[31]. Photolyses of (alkylcyclopentadienyl)mercury(II) derivatives have allowed detailed study of a series of substituted cyclopentadienyl radicals which are interesting examples of the [*n*]-annulene series which were previously inaccessible[27]. $S_H2$ reactions are also found in some metallocenes and other cyclopentadienyl derivatives.

The method of 'spin-trapping' may also be useful in elucidating homolytic organometallic reaction mechanisms. Here, the experiments are often carried out at

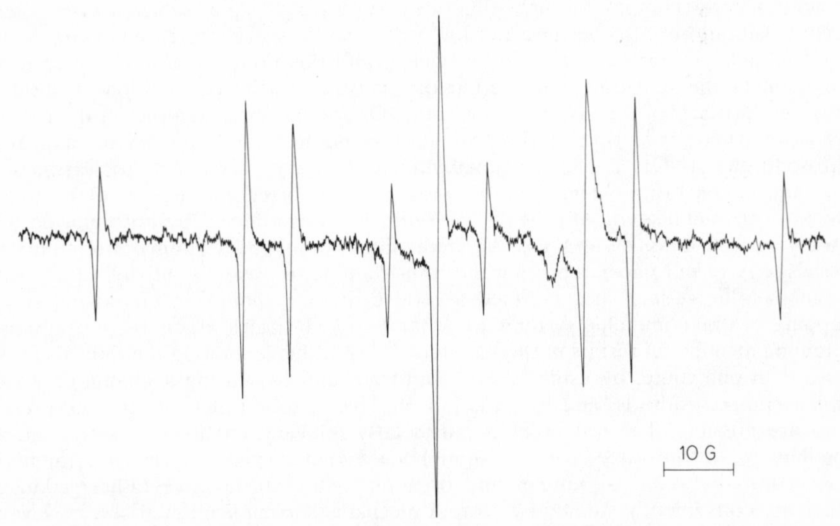


FIGURE 1. E.s.r. spectrum of the *n*-propyl radical, $CH_3CH_2\dot{C}H_2$, formed by the $S_H2$ process in equations 11 and 12. From ref. 30

ambient temperatures which are generally too high to enable direct observation of many types of free-radical. Spin-trapping organic radicals with nitroso compounds or nitrones has achieved much success in purely organic systems. Briefly, a transient radical (alkyl, for example) generated during the course of a reaction in the presence of either of the above compounds reacts to give a more persistent 'spin-adduct' readily detectable by e.s.r.

$$R^{\cdot} + R'N{=}O \longrightarrow RR'NO^{\cdot} \quad (13)$$

$$R^{\cdot} + R'CH{=}N(O)Bu^t \longrightarrow R'CH(R)N(O^{\cdot})Bu^t \quad (14)$$

The subject of spin-trapping has received much attention since its development in the late 1960s and has recently been thoroughly reviewed[32]. In organometallic systems, however, difficulties are encountered when conventional spin-traps are used.

To trap the organometallic fragments of homolytic decompositions a different type of trap is required. Particularly useful for the trapping of tin radicals, for example, is butane-2,3-dione (biacetyl). When trialkyltincyclopentadienes are photolysed in the presence of this substrate, strong e.s.r. spectra of the tin adducts are observed[31]. A conventional trap for alkyl radicals is 2-methyl-2-nitrosopropane (MNP)[33]; however, tin or other metal-centred radicals do not add to this substrate. 2-Methyl-2-nitropropane has been used effectively to trap tin and other low-valent metal species[34]. Confirmation of this was provided by trapping of stannyloxy ($R_3SnO^{\cdot}$) radicals on mnp to give similar adducts to those of stannyl radicals and the nitro compound[35]. Factors governing these contrasting reactivities include the redox potential of the trap, the HOMO and LUMO energies of the trap, the SOMO energy of the radical, etc.

$$R_3Sn^{\cdot} + Bu^tNO_2 \longrightarrow Bu^tN(O^{\cdot})OSnR_3 \quad (15)$$

$$R_3SnO^{\cdot} + Bu^tNO \longrightarrow Bu^tN(O^{\cdot})OSnR_3 \quad (16)$$

In general, experiments involving the use of conventional spin traps in organometallic systems must be carried out with extreme caution. This is not only because nitroso compounds are sometimes sensitive to light, which causes their decomposition, but also because the low-valent metal species often encountered in organometallic systems are strongly reducing, and this property is likely to lead to reduction of the spin-trap. Such reduction may lead to the observation of spectra which are artifacts of the system under study. Of note are two examples in the recent literature where reduction of the two most commonly used spin-traps, mnp and nitrosodurene (ND; 2,3,5,6-tetramethylnitrosobenzene), has led to erroneous mechanistic conclusions. Significantly, both errors occurred where low-valent metal species were implicated, with MNP the reduction was effected by borohydride (or borane radical anion)[36] and the ND reduction by an undetermined titanium(III) species[37]. A recent paper reporting the spin-trapping of cyclopentadienyl radicals in organometallic systems has outlined several cautionary points in employing spin-trapping in organometallic systems in general and illustrates a general strategy for obtaining meaningful results in these systems[38]. Thus, by generating the radicals from more than one route, by using several spin-traps and comparing available data for similar trapped radicals, and by variation of solvent, sound mechanistic interpretations are possible. The last point is particularly relevant and it is stressed that, if possible, use of chlorinated solvents should be avoided in spin-trapping experiments.

Reactions between *o*-quinones and organometallic radicals give radical adducts which are conveniently studied by e.s.r. A mechanistic complication arises, however, as to whether the adducts are formed by addition of the organometallic radical or by a charge transfer from the parent organometallic compound. Thus, while reactions of trialkyltin cyclopentadienes with biacetyl occur only on photolysis[31], (equations 9 and 17), a thermal reaction takes place with *o*-quinones according to equations 18 and 19[34]. However, the mechanistic problem aside, these *o*-quinones are very efficient substrates for metal-centred radical trapping.

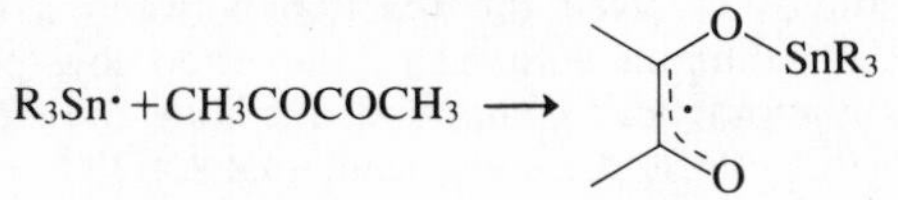

$$R_3Sn^{\cdot} + CH_3COCOCH_3 \longrightarrow \text{(semidione–}SnR_3\text{ adduct)} \qquad (17)$$

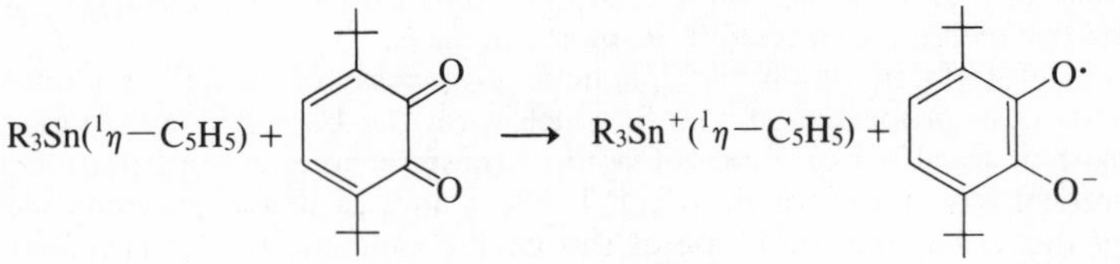

$$R_3Sn(^1\eta{-}C_5H_5) + o\text{-quinone} \longrightarrow R_3Sn^+(^1\eta{-}C_5H_5) + \text{semiquinone radical anion} \qquad (18)$$

$$\longrightarrow \text{(}O\text{-}SnR_3\text{ semiquinone)} + C_5H_5^{\cdot} \qquad (19)$$

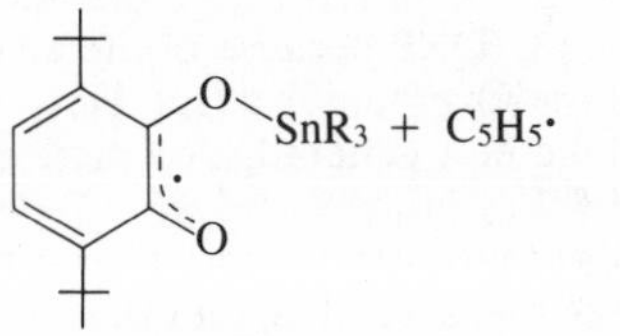

In other charge-transfer reactions conventional traps may be used with safety. Thus, reaction between the inorganic oxidant hexachloroiridate(IV) and bisalkylmercurials gives alkyl radicals which are trapped, and low-valent metal species are generally avoided in these processes (see Section V for a full discussion). E.s.r. measurements in solution, in this way, may yield much more direct mechanistic information.

Homolysis of metal—carbon bonds in rigid matrices is also readily affected and again may yield much useful information, but of a different nature. Homolyses induced by ionizing radiation (usually $^{60}$Co $\gamma$-rays) at 77 K give information on the electronic structure of the organometallic fragments, which may subsequently provide rational explanations for the solution behaviour of these fragments. Such experiments may be carried out on pure samples or in dilute solutions where the mode of ionization (electron gain or loss) may be induced by judicious choice of solvent. Irradiation at 77 K of tetraalkyltin compounds induces homolysis of a metal—carbon bond and not only may the tin(III) fragments be observed, but also the products of reactions between alkyl radiclals and other substrate molecules within the matrix are detected[39]. In the observation of the tin (III) species the e.s.r. data may be used to determine the orbital contribution to the SOMO of the radical. In the matrix isolated trimethylstannyl radical for example, a significant 5*s* orbital contribution is indicated by the large isotropic $^{117/119}$Sn hyperfine coupling of 1550 G, which effectively means a large distortion from planarity is present. This result is in contrast to an earlier result for $H_3Sn^{\cdot}$ where little or no distortion from planarity was assumed from the $^{117/119}$Sn isotropic coupling of 380 G[40].

It should be noted that, with a $\gamma$-source available, radiolysis experiments are very easy to carry out, but at 77 K the resulting e.s.r. spectra are fully anisotropic and therefore, to analyse these spectra fully and draw the correct structural conclusions from them, a detailed knowledge of the theory of the *g*- and *A*-tensors is required.

### 2. CIDNP

Chemically induced dynamic nuclear polarization (CIDNP) has received increasing use for the study of organometallic homolyses since the technique was discovered in 1967. The discovery of the technique was made by two groups simultaneously[41,42], and one of these first studies involved the reaction between *n*-butyllithium and *n*-butyl iodide[42]. Again, in common with e.s.r., the often low bond dissociation energies encountered in organometallic compounds facilitate the study of homolytic reactions by CIDNP. To this effect, it is significant that CIDNP effects have been observed in every group in the Periodic Table covered by this chapter. In addition, CIDNP effects may be observed on simple unmodified n.m.r. spectrometers, a fact which makes the technique accessible to most chemists.

The origins of CIDNP lie in the magnetic properties of nuclei and unpaired electrons, and these properties are those which form the basis for both n.m.r. and e.s.r. The mathematical backgrounds of both of these techniques contribute heavily to the theoretical elucidation of the CIDNP effect and, as in the previous section, discussion of this is beyond the scope of this chapter and the reader is referred to several reviews[43] on the subject. Here only a qualitative description of how the effect arises and becomes evident in the n.m.r. spectra of organometallic systems is given. Initially, the technique was dubbed CIDNP because of the analogy with the dynamic nuclear polarization described as the Overhauser effect. However, the early theoretical models were inadequate, and the now generally accepted 'radical pair model' was proposed by two groups in 1969[44,45].

Qualitatively, when two radicals are formed in close proximity (for example, within a solvent cage) or come together via a random diffusive mechanism, there is an interaction between them. With an unpaired electron on each radical there are three possible spin states describing the system as a whole, $S = 1$, 0 or $-1$. Transitions between these singlet and triplet states occur with nuclear spin dependency, and the products of these encounters have non-Boltzmann spin distributions which are reflected as polarizations in their n.m.r. spectra during the course of the reaction. The polarized products show n.m.r. lines in emission or enhanced absorption which decay as the reaction is completed to give the normal product n.m.r. spectrum. Using the set of rules devised by Kaptein for $|S\rangle$ (singlet), $|T\rangle$ (triplet) and $|F\rangle$ (freely diffusing radicals) precursors, and constants derived from the magnetic properties of the free radicals and products, it is possible to account for and predict the phase of the observed polarizations. Thus, while not observing the intermediates directly, CIDNP provides a powerful tool in determining the observed reaction pathway[46].

A convenient illustration of the utility of the CIDNP effect and Kaptein's rules is provided in a study of the photolysis of acylsilanes in chlorinated solvents. Originally this reaction was thought to proceed via a Norrish type I cleavage of the acyl—Si bond to give triarylsilyl chloride, acetyl chloride, trichloroacetone, and hexachloroethane according to reactions 20 to 24[47].

$$R_3SiC(O)Me \longrightarrow R_3Si^{\bullet} + {}^{\bullet}C(O)Me \quad (20)$$

$$R_3Si^{\bullet} + CCl_4 \longrightarrow R_3SiCl + {}^{\bullet}CCl_3 \quad (21)$$

$${}^{\bullet}C(O)Me + CCl_4 \longrightarrow ClC(O)Me + {}^{\bullet}CCl_3 \quad (22)$$

$${}^{\bullet}CCl_3 + {}^{\bullet}C(O)Me \longrightarrow Cl_3CC(O)Me \quad (23)$$

$$2{}^{\bullet}CCl_3 \longrightarrow Cl_3CCCl_3 \quad (24)$$

A photo-CIDNP study by Porter and Iloff[48], however, showed that this mechanism was inconsistent with the CIDNP results. The CIDNP results showed n.m.r. lines in emission for acetyl chloride, in strongly enhanced absorption for trichloroacetone, and no polarization in the silyl chloride formed. If the above mechanism were valid, polarization in the silyl chloride would be expected (being an escape product from the initial radical pair) and furthermore, from Kaptein's rules, the trichloroacetone formed should exhibit lines in emission. In conclusion, reactions 25–27 were proposed to describe the mechanism more fully, in this case for diphenylacylsilane.

$$CH_3C(O)SiMe_2Ph \xrightarrow[CCl_4]{h\nu} \underset{\mathbf{(2)}}{(CH_3C(O)SiMe_2PhCCl_4)^*} \qquad (25)$$

$$\mathbf{2} \longrightarrow \underset{\mathbf{(3)}}{[CH_3\dot{C}(O)\cdots\dot{C}Cl_3]} + ClSiMe_2Ph \qquad (26)$$

$$\mathbf{3} \longrightarrow \text{products} \qquad (27)$$

This mechanism involves an exciplex, **2**, which undergoes a non-radiative decay to the radical pair **3**. By Kaptein's rules, the cage recombination product, trichloroacetone, should show enhanced absorption (observed) and the escape product, acetyl chloride, should show emission (also observed). This mechanism does have precedent in the literature, where exciplex formation has been postulated[49] in the photolysis of diisopropyl ketone, also in carbon tetrachloride. Photolysis of acylsilanes in other solvents (alcohols, for example) often leads to intermediates of a carbenoid nature, and totally different mechanistic conclusions.

While in fact this work has shown that no simple homolysis of the Si—C bond has taken place, the example serves to demonstrate the utility of simple CIDNP theory and again stresses the fact (see e.s.r. section) that chlorinated solvents sometimes play an important role in determining the mechanistic outcome of homolytic organometallic reactions.

### E. Product Studies

Where instrumental methods are not utilized, product studies may provide valuable information on the course of homolytic reactions. While little comment on the practical aspects of product study is necessary here, some indication of the general approach to experimental strategy available may be made. In the following text there are numerous examples where product studies have been used to gain information on metal—carbon bond homolysis. Approaches most often used include the following:

1. Incorporation of optically active ligands which lose sterochemical discretion on metal—ligand homolysis.

2. Inclusion of a ligand for example 5-hexenyl-, which rearranges to cyclopentylmethyl radicals on homolysis, or neophyl-, which rearranges to 1,1-dimethyl-2-phenylethyl on homolysis. These reactions are particularly useful since kinetic inferences may be made from the product distribution ratios since the rates of both rearrangements are known (see text).

3. Variation of solvent. This is a particularly important point which may dramatically affect the course of homolytic reactions.

If one or all of these points are considered, conclusions on the course of homolytic reactions may be made with confidence.

## II. GROUP IA

Reactions in which homolysis of the metal—carbon bond has been observed or postulated for this group are mainly confined to those involving lithium. Some examples have been included which may be regarded as reactions of carbanions rather than of covalent carbon—lithium bonds. Organometallic compounds of the heavier members of the group are more difficult to study, with the metal—carbon interaction becoming increasingly ionic. Therefore, this section will be devoted to organolithium chemistry with only occasional references to the behaviour of its higher congeners.

### A. Thermal Decompositions

There is little to indicate that free radicals are intermediates in the thermal decomposition of organolithium compounds[50,51], although the formation of organolithium aggregates (commonly dimers, tetramers, and hexamers) often serves to obscure the mechanism of decomposition. The predominant decomposition pathway involves $\beta$-elimination of lithium hydride to give unsaturated products[52]. This type of reaction is common for many classes of alkylmetals.

$$R_2CHCH_2M \longrightarrow R_2C{=}CH_2 + MH \tag{28}$$

Even when $\beta$-hydrogen is not available, lithium—carbon homolysis is unlikely to play more than a minor role compared with other concerted processes; however, evidence suggests that a homolytic pathway should not be discounted in the case of some alkyl-sodium and -potassium compounds[53].

### B. Photolytic Decompositions

The u.v. photolysis of ethyllithium in hydrocarbon solvents yields ethene and ethane but no butane as decomposition products[54]. The absence of butane or monodeuteriated ethane when the photolysis is carried out in perdeuteriated solvents seems to exclude the formation of 'free' ethyl radicals. It is possible though that radicals formed by Li—C homolysis react too rapidly within the ethyllithium aggregate (a hexamer) to allow escape into solution. To this effect, the following mechanism has been proposed.

$$(EtLi)_6^* \longrightarrow {}^{\bullet}C_2H_5(EtLi)_5 + Li \tag{29}$$

$${}^{\bullet}C_2H_5(EtLi)_5 \longrightarrow EtH + {}^{\bullet}C_2H_4Li(EtLi)_4 \tag{30}$$

$${}^{\bullet}C_2H_4Li(EtLi)_4 \longrightarrow C_2H_4 + Li + (EtLi)_4 \tag{31}$$

Non-homolytic pathways are available for the formation of these products, but a CNDO/2 study on idealized tetrameric and hexameric methyllithium indicate that the u.v. absorption of alkyllithium aggregates are associated with charge-transfer transitions from the alkyl groups to give alkyl radicals[55].

$$R^{\delta-}{-}Li^{\delta+} \xrightarrow{h\nu} R^{\bullet} + Li \tag{32}$$

The model also predicts that the transition is higher in energy for the tetramer than the hexamer. The process could also be thought of as a photoinduced electron transfer from a carbanion to an electron acceptor. A further example is found in the photolysis of cyclopentadienyllithium, where a weak e.s.r. spectrum of the cyclopentadienyl radical is observed[27]. Not only does this process occur in dimethoxyethane as solvent, where the compound may be represented as $\eta^5{-}C_5H_5Li$, but also in pentane in the presence of 18-crown-6, where $C_5H_5^-\ Li^+$ may be a more accurate representation. Cyclopentadienyl radicals have also been observed by e.s.r. in the photolysis of cyclopentadienylsodium[56]. A weak e.s.r. spectrum of the pentamethylcyclopentadienyl radical was obtained from an unirradiated suspension of the lithium compound in tetrahydrofuran[57], although this may have been caused by photolysis by laboratory light or by oxidation by traces of oxygen; irradiation, however lead to strong signals from $C_5Me_5^{\bullet}$.

## C. Radical Reactions

The autoxidation of alkyllithium clusters occurs via a free-radical mechanism, similar to that described in the Introduction, involving $S_H2$ by an alkylperoxy radical at a lithium centre. The proposed initiating and propagation steps are outlined in equations 33–35[58].

$$(RLi)_4 + O_2 \longrightarrow R^{\bullet} + (R_3Li_4)^+ + O_2^{\bullet} \quad (33)$$

$$R^{\bullet} + O_2 \longrightarrow ROO^{\bullet} \quad (34)$$

$$ROO^{\bullet}(RLi)_4 \longrightarrow (ROO)R_3Li_4 + R^{\bullet} \quad (35)$$

A study on neophyllithium tetramers confirmed that alkyl radicals are intermediates in the reaction since products derived from 1,1-dimethyl-2-phenylethyl radicals were obtained[58]. Neophyl (i.e. 2,2-dimethyl-2-phenylethyl) radicals rearrange fairly slowly ($k = 10^5\ s^{-1}$ at 100 °C) via a 1,2-phenyl shift to give the more stable tertiary radical from which the products are derived, and rearranged products

$$^{\bullet}CH_2C(Me)_2Ph \xrightarrow{\circ} {}^{\bullet}C(Me)_2CH_2Ph \quad (36)$$

increase with temperature as would be expected. A number of other processes (for example, single electron transfer) could also account for the observations, however, and these deductions are not conclusive. It may be significant that in the autoxidation of neophyllithium dimers, no products from the rearranged radical could be detected[58].

It may well be that radical attack at lithium in organolithium aggregates may be slow compared with radical reactions at other metal centres. The reaction of *t*-butoxyl radicals, generated by photolysis of di-*t*-butyl peroxide appears to abstract hydrogen from methyllithium tetramers rather than give methyl radicals via an $S_H2$ process[59].

The reactions of 'stable' free radicals with organolithium compounds have been studied recently. It has been concluded from product studies that the verdazyl radical[60], **4**, and the di-*t*-butyliminoxyl radical[61], **5**, both react with organolithium compounds via the $S_H2$ processes, outlined in equation 37 (for the former case).

(4) [1,3,5-triphenylverdazyl radical: ring N, N•, N(Ph), $CH_2$, N(Ph), C(Ph)] + 2RLi ⟶ [verdazyl anion: ring N, $N^-$, N(Ph), $CH_2$, N(Ph), C(Ph)] $Li^+$ + RR (37)

R=Ph

It has been noted that the reaction of **5** with alkyllithium compounds does not proceed via the usually electronically symmetric transition state implied by equation 2, but that considerable initial charge transfer from the organolithium to the incoming radical may occur[9,61]. The steric constraints of **5**, the polarizability of the $\sigma$-Li bond and the good donor properties of organolithiums (see the next section) support this rationalization and by all accounts a similar perspective may be adopted for the similar reactions of verdazyl, **4**.

### D. Reactions with Electron Accepttors

Alkyllithium compounds are able to react with electron acceptors (generally with extended $\pi$-systems) to give alkyl radicals and radical anions (reaction 38). Reactions also also take place with $\sigma$-acceptors (alkylbromides and iodides) (reaction 39). The

$$RLi + A \longrightarrow R^{\bullet} + Li^{+}A^{-\bullet} \tag{38}$$

$$RLi + R'X \longrightarrow R^{\bullet} + R'^{\bullet} + LiX \tag{39}$$

reaction with $\pi$-acceptors can be followed by monitoring the appearance of the long-lived radical anions using e.s.r.[62], while the reactions with $\sigma$-acceptors were among the first to be studied by CIDNP[63–65]. Considerable support for the mechanistic conclusions has also been gained from product[66] and stereochemical[67] studies.

An interesting recent example of single electron transfer has been found in the reaction of alkyllithium compounds with titanium tetrachloride. Very shortly after mixing the reagents the e.s.r. spectra of alkyl radicals are detected.[68] This reaction is similar to that with $\sigma$-acceptors, above, where again $R^{\bullet}$ may be detected soon after mixing RLi and alkyl iodides[69,70].

$$(MeLi)_4 + TiCl_4 \xrightarrow[C_6H_6]{Et_2O} {}^{\bullet}CH_3 + LiCl + TiCl_3 \tag{40}$$

As will be seen in the ensuing sections, many other organometallic compounds act as donors in single electron transfer reactions but, surprisingly, a study on alkylsodium and -potassium compounds with alkyl halides failed to provide evidence for alkyl radical formation, under conditions where such processes occur readily with the corresponding lithium derivatives[66].

## III. GROUP IB

Reactions involving the homolysis of Group IB metal—carbon bonds have not previously received the attention of systematic review. Organo-copper, -silver, and -gold compounds appear to undergo homolysis only under exceptional circumstances and have received only isolated attention. However, several such examples are now known and they represent an interesting area of investigation.

### A. Thermal Decompositions

The thermal stability of Group IB organometallics is poor compared with their main group counterparts. Stability is increased if the compounds are prepared in the presence of donor solvents or donor ligands (for example, tetrahydrofuran or triphenylphosphine, respectively) but in many cases it is still not possible to achieve the isolation of these compounds.

Decompositions of these compounds usually proceed to the metal and a mixture of organic products which, superficially, may appear to have resulted from radical termination reactions. Thus, decomposition of ethyl(triphenylphosphine)gold(I) at 124 °C in tetralin gave 7% ethane, 5% ethene, and 73% $n$-butane[71]. These hydrocarbons are produced in the bimolecular termination of ethyl radicals.

$$2CH_3CH_2^{\bullet} \xrightarrow{K_d} CH_3CH_3 + CH_2{=}CH_2 \tag{41}$$

$$2CH_3CH_2^{\bullet} \xrightarrow{K_c} CH_3CH_2CH_2CH_3 \tag{42}$$

The product distribution indicated above does not however, correspond to that normally observed for ethyl radical coupling to disproportionation ($K_c/K_d \approx 5$). Thus, gold—carbon homolysis does not fully account for the observed product ratios. Two other non-homolytic reactions are probably responsible for the decomposition.

$$2[CH_3CH_2Au(PPh_3)] \nearrow CH_3CH_3 + CH_2{=}CH_2 + 2Au + 2PPh_3 \qquad (43)$$

$$2[CH_3CH_2Au(PPh_3)] \searrow n\text{-}C_4H_{10} + 2Au + 2PPh_3 \qquad (44)$$

Both the reductive elimination (43) and coupling (44) involve concerted formation of products, and represent types of decomposition often found in transition metal organometallics. The disproportionation reaction requires the presence of a $\beta$-hydrogen atom and accounts for most of the decomposition products of *sec*- and *t*-alkylgold(I) and alkylcopper(I) compounds which meet this criterion[71,72]. The predominance of reactions 43 and 44 does not exclude the possibility that homolytic reactions make a minor contribution to the decomposition, but some organometallic compounds that do not contain $\beta$-hydrogen also seem reluctant to undergo reductive coupling. Decompositions of these compounds allow a much greater contribution from metal—carbon bond homolysis.

### 1. Decomposition of copper(I) compounds

Tris-*n*-butylphosphine(neophyl)copper(I) is a relatively stable alkyl copper compound which undergoes decomposition in diethyl ether at above 30 °C. If free neophyl radicals are produced as a result of Cu—C homolysis then products containing the 1,1-dimethyl-2-phenylethyl moiety, from the facile rearrangement of neophyl radicals mentioned above, should be observed. This was found to be the case, and these products increased at higher temperatures, which is consistent with the increase in rate of neophyl radical rearrangement[73]. Similar results were obtained from the decomposition of neophylcopper(I) formed from the reaction of neophyl Grignard with copper(I) iodide in tetrahydrofuran[74]. In addition, quantities of products derived from the reaction of radicals with solvent were observed. Neophyl derivatives of other first row transition metals, on the other hand, give no products derived from free radicals but only those from reductive coupling.

Efforts have been made to correlate these results with a general model of orbital symmetry correlation[75]. In a general approach the species assumed to be responsible for the concerted decompositions (e.g. reductive coupling) is the *cis*-dialkylmetal complex and then the metal $d$-orbitals and metal—alkyl bonding orbitals are assigned assuming a $C_{2v}$ symmetry. The metal $d$-orbital energy diagram is outlined in Figure 2.

In general, concerted reductive coupling will occur when electrons from symmetric alkylmetal orbitals correlate with the symmetric dialkyl bonding orbital, and electrons from antisymmetric alkyl metal orbitals correlate with the antisymmetric $d_{x^2}$ orbital in the metal product. If all the orbitals in the reactant complexes correlate with the product orbitals in this way, then reductive coupling is symmetry allowed. Reactant and product orbital classifications and electron allocations are given in Table 4.

Examination of Table 4 reveals that concerted reductive coupling in $d^{10}$ systems is symmetry forbidden in the ground state. Similar tables for $d^8$ to $d^0$ configurations demonstrated that in all these systems the concerted reaction is symmetry allowed.

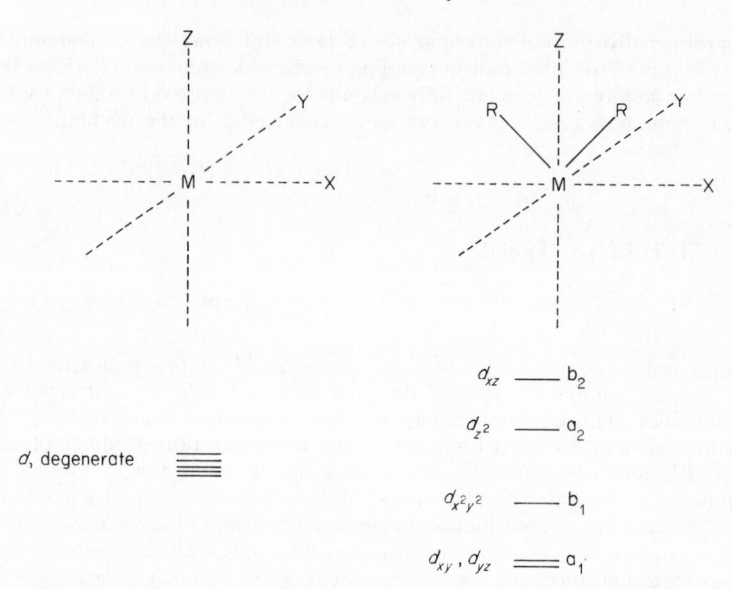


FIGURE 2. Relative $d$-orbital energies in the *cis*-2-coordinate complex and free metal

For our purposes, it is interesting to note that concerted formation of radical pairs is possible from $d^{10}$ excited states and $d^9$ ground states. Although *cis*-dialkylcopper(I) species are unlikely intermediates in concerted radical formation, a similar process is a possibility for *cis*-dialkylcopper(II) decompositions. Addition of neophyl Grignard to copper(II) bromide or copper(I) iodide gives similar product distributions. Therefore, it was considered that a dialkylcopper(II) species may represent a common intermediate[76], formed directly or by disproportionation, respectively.

$$2RMgX + CuBr_2 \longrightarrow R_2Cu^{II} + 2MgXBr \qquad (45)$$

$$2RCu^{I} \longrightarrow R_2Cu^{II} + Cu^{0} \qquad (46)$$

TABLE 4. Symmetry correlations between $d^9$ and $d^{10}$ dialkylmetal valence orbitals and corresponding valence orbitals in eliminated products

| Compound/decomposition configuration | Metal $d$-electron configuration | Number of electrons in symmetry orbitals | | | |
|---|---|---|---|---|---|
| | | $a_1$ | $a_2$ | $b_1$ | $b_2$ |
| $R_2M$ | $d^{10}$ | 6 | 2 | 4 | 2 |
| | $d_{xz}{}^1d^8s^1$ | 7 | 2 | 3 | 2 |
| | $d_{xz}{}^0d^8s^2$ | 8 | 2 | 2 | 2 |
| $R_2+M$ | $d^{10}s^2$ | 8 | 2 | 2 | 2 |
| $2R^{\bullet}+M$ | $d^{10}s^2$ | 7 | 2 | 3 | 2 |
| $R_2M$ | $d_{xz}{}^1d^8$ | 6 | 2 | 3 | 2 |
| | $d_{xz}{}^1d^8s^1$ | 7 | 2 | 2 | 2 |
| $R_2+M$ | $d^{10}s^1$ | 7 | 2 | 2 | 2 |
| $2R^{\bullet}+M$ | $d^{10}s^1$ | 6 | 2 | 3 | 2 |

It was pointed out, however, that the tendency of organocopper compounds to form clusters may relax the symmetry conditions sufficiently to attenuate the energy differences between the ground and excited states. Although this may allow other mechanistic possibilities, straightforward unimolecular homolysis may make a significant contribution to these reactions.

A number of other organocopper(I) compounds which may decompose by Cu—C bond homolysis to a significant extent are also worthy of mention. Trimethylsilylmethylcopper(I) tetramers decompose at 80 °C in toluene to give products including $PhCH_2CH_2SiMe_3$ and $PhCH_2CH_2Ph$ derived from radical attack on the solvent[77].

$$(Me_3SiCH_2)_4Cu^I_4 \longrightarrow (Me_3SiCH_2)_3Cu^I_3Cu^0 + Me_3Si\dot{C}H_2 \tag{47}$$

Methylcopper(I) compounds decompose to give mainly ethane and methane in proportions which vary widely according to the conditions of thermolysis and the presence and nature of the phosphine ligand[78]. The decompositions show first-order kinetics and the most likely mechanism appears to involve methyl radicals[79] which may abstract hydrogen from ligand or substrate to give methane, or terminate bimolecularly to give ethane.

In dimethylformamide or dimethyl sulphoxide at 150 °C perfluoroalkylcopper(I) compounds decompose to give products consistent with formation of perfluoroalkyl radicals[76]. When the same compounds are decomposed in the presence of alkenes not only are radical addition products formed but also polymerizations are readily initiated[80].

Clear evidence for the formation of radicals in the decomposition of arylcopper(I) compounds has not been obtained[81] and the Ullman synthesis almost certainly involves an ionic mechanism[82]. It has been demonstrated that vinylcopper(I) derivatives do not decompose via homolytic mechanisms as the configuration of substituted vinyl radicals does not alter from reactant to the diene products[83]. Vinyl radicals would be expected to lose their configuration by rapid isomerization. Recently, decomposition of more complex vinylcopper clusters in naphthalene at 220 °C has provided evidence, from isomeric product equilibration, for a free radical mechanism[84].

Finally, an interesting comment has been made on the decomposition of ethylcopper(I) in tetrahydrofuran[85]. The reaction showed a marked induction period, at the end of which an e.s.r. spectrum of a postulated copper(0)—copper(I) complex was observed. The induction period can be attributed to autocatalysis by copper (0). This may serve to increase the rate of reductive disproportionation, but it has been suggested that the products could also be the result of an autocatalysed homolytic mechanism (reactions 48–50). Similar selectivity for reaction 49 and the $\beta$-

$$EtCu^I + Cu^0 \longrightarrow EtCu^ICu^0 \longrightarrow Et^{\bullet} + 2Cu^0 \tag{48}$$

$$Et^{\bullet} + CH_3CH_2Cu^I \longrightarrow EtH + {}^{\bullet}CH_2CH_2Cu^I \tag{49}$$

$${}^{\bullet}CH_2CH_2Cu^I \longrightarrow CH_2{=}CH_2 + Cu^0 \tag{50}$$

elimination from $\beta$-metallated alkyl radicals (reaction 50) has precedent in organotin chemistry[28].

### 2. *Decomposition of copper(II) compounds*

Simple alkylcopper(II) compounds are expected to be very unstable species, and as yet have not been isolated. Their formation during the reaction of various organometallic compounds with copper(II) salts has been suggested, however, and available evidence indicates that they may decompose by a homolytic mechanism under certain circumstances. There are two primary mechanisms which may account for the generation of radicals during these reactions, and these may be illustrated with reference to the reaction between tetraethyllead and copper(II) acetate in acetic acid[86], where products obtained from ethyl radical oxidation (ethene, butane, and EtX, where X is from $CuX_2$) were observed.

$$Et_4Pb + Cu^{II}(OAc)_2 \longrightarrow Et_3Pb(OAc) + Cu^{I}(OAc) + Et^{\bullet} \quad (51)$$

$$Et^{\bullet} + Cu^{II}(OAc)_2 \longrightarrow Et_{ox} + Cu^{I}(OAc) + OAc^{-} \quad (52)$$

The ethyl radicals here would be formed by a transalkylation process to give an ethylcopper(II) species which then undergoes homolysis (reactions 53 and 54)

$$Et_4Pb + Cu^{II}(OAc) \longrightarrow Et_3Pb(OAc) + [EtCu^{II}(OAc)] \quad (53)$$

$$[EtCu^{II}(OAc)] \longrightarrow Et^{\bullet} + Cu^{I}(OAc) \quad (54)$$

or a single electron transfer similar to that described in equations 3 and 4 may be operative (reactions 35 and 56).

$$Et_4Pb + Cu^{II}(OAc)_2 \longrightarrow Et_4Pb^{+} + [Cu^{I}(OAc)_2]^{-} \quad (55)$$

$$Et_4Pb^{+} \longrightarrow Et_3Pb^{+\bullet} + Et^{\bullet} \quad (56)$$

In this case the former mechanism (equation 53 and 54) was preferred because studies on methylethylplumbanes demonstrated that reactivities and product distributions mirrored the selectivity of a transalkylation step rather than the mass spectral fragmentation pattern of the plumbane radical cation.

The situation regarding these reactions is by no means clear, however, because the reaction between copper(II) triflate and tetraethyllead in tetrahydrofuran yielded butane as the major product[86] instead of EtX, the major product in the example above. In addition, the reaction between ethylmagnesium bromide with copper(II) chloride gave butane exclusively[87], although it is probable that $EtCu^{II}Cl$ was formed initially. From consideration of these results it has been suggested that reductive coupling of ethylcopper(II) dimers can compete with Cu—C homolysis in suitable situations.

$$2EtCu^{II}X \rightleftharpoons [EtCu^{II}X]_2 \longrightarrow Et{-}Et + 2Cu^{I}X \quad (57)$$

This process may be particularly applicable when high concentrations of organocopper(II) compound are present (i.e. when their formation is rapid).

The reaction of compounds $R_2Hg$ and RHgX compounds with copper(II) bromide gives mainly RBr as product in dimethylformamide[88]. The authors felt that in this particular case, where acylmercury derivatives were involved, electron transfer to give $R_2Hg^{+}$ was less likely (although when R is simple alkyl this occurs readily) than the mechanism involving alkylation of copper(II) with subsequent homolytic decomposition. Further evidence for free radical participation was gained in a study of the reaction between copper(II) salts and organometallic compounds in which the ligands may act as stereochemical probes.

Both *erythro*- and *threo*-$(CH_3)_3CCHDCHDM$ [$M = (^5\eta\text{-}C_5H_5)_2ZrCl$, HgCl] and 2-M-6-acetylbicyclo [2.2.0] heptane [**6**; M = Pd(diphos)Cl, HgCl] react with copper(II) bromide to give products in which all configurational or isomeric integrity is lost[89]. Once again homolysis of an alkylcopper(II) intermediate is thought to be the major pathway for radical production.

### 3. *Decomposition of silver(I) and gold(I) compounds*

In contrast to *n*-alkylcopper(I) compounds, the corresponding silver(I) and gold(I) complexes decompose by reductive coupling to give predominantly *n*-alkyl dimers rather than disproportionation products. There is very little evidence that homolytic mechanisms play anything but a minor role in these decompositions except in the case of some organosilver complexes.

Similar results were found for decomposition of neophyl silver(I)triphenylphosphine to those of the corresponding copper(I) derivatives. The reaction of trialkylsilylmethyl Grignards, $RMe_2SiCH_2MgCl$, with silver(I) chloride or bromide gave, in addition to $(RMe_2SiCH_2)$ and $Me_4Si$, products from radical attack on solvent[90]. It is interesting that it is the 'neopentyl-like' ligands which appear to induce homolytic decomposition in their compounds with Group IB organometallics. This was rationalized in the case of the copper compounds as the lack of $\beta$-hydrogen denying an easier pathway; however, this reductive elimination is not as important for the silver(I) and gold(I) complexes. It may well be that the bulk effect of the 'neopentyl-like' ligands also provides a stereochemical constraint on concerted reactions and perhaps aiding metal—carbon bond homolysis.

There is some evidence to suggest that the 'ate' silver(I) complex $Li[(Bu^n)_2AgPBu^n{}_3]$ may, unlike $[Bu^nAg(PPh_3)]$ decompose via silver—carbon bond homolysis[91].

$$Li[(Bu^n)_2AgPBu^n{}_3] \longrightarrow n\text{-}Bu^{\bullet} + Bu^nLi + Ag^0 + PBu^n{}_3 \qquad (58)$$

The thermolysis in diethyl ether at 0 °C gave *n*-octane (26%), *n*-butane (71%), and but-1-ene (3%) as the principal products. The octane : butene ratio is close to the $K_d/K_c$ ratio for *n*-butyl radical termination and lower yields of *n*-butane were observed in poor hydrogen-donating solvents. Studies on $Li[(Me)(Bu^n)AgPBu^n{}_3]$ demonstrated that products from *n*-butyl groups were formed more rapidly, consistent with an expected preference for the more stable alkyl radical. Crossover experiments with $Li[(n\text{-}C_5H_{11})(Bu^n)AgPBu^n{}_3]$ yielded a statistical distribution of dimeric products (i.e. the *n*-octane : *n*-nonane: *n*-decane ratio was about 1 : 2 : 1). Finally, these decompositions were found to follow first-order kinetics. The homolytic mechanism could not be proved conclusively, but all these observations suggest the formation of free radicals.

## B. Photolytic Decompositions

Photolytic decomposition of Group IB organometallic compounds might be expected to involve metal—carbon bond homolysis as the photosensitivity of these compounds is well known. Unfortunately, however, little effort has been devoted to exploring these reactions.

Photolysis of methyl- and ethyl-(triphenylphosphine)copper(I) in toluene at −20 °C gave mostly methane and ethane, respectively, in contrast to the thermal decompositions[78]. Very little incorporation of deuterium was observed when the photolysis was carried out in deuteriated solvents. If alkyl radicals are produced, they must react rapidly to abstract hydrogen from another substrate molecule.

$$[CH_3Cu^I(PPh_3)] \xrightarrow{h\nu} H_3C^{\bullet} + Cu^0 + PPh_3 \qquad (59)$$

$$H_3C^{\bullet} + [CH_3Cu^I(PPh_3)] \longrightarrow CH_4 + {}^{\bullet}CH_2Cu^I(PPh_3) \qquad (60)$$

The photolysis of $(Me_3SiCH_2Cu)_n$ gave similar results to the thermolysis[77], providing reasonable evidence for Cu–C bond homolysis. Radical intermediates were implicated in the photolysis of *n*-butyl(tri-*n*-butylphosphine)silver(I) in diethyl ether at −60 °C[92]. Butane was the major product but 2-ethoxyhexane was also detected. Methyl(triphenylphosphine)gold(I) complexes have been irradiated under various conditions, but there is little evidence to suggest that the radical reactions which undoubtedly take place (see below) are initiated by photo-induced cleavage of the gold—carbon bond[93,94].

### C. Radical Reactions

The possibility that $\beta$-coppermetallated alkyl radicals may be formed in some alkylcopper(I) decompositions and that these radicals $\beta$-eliminate copper has already been mentioned above. Therefore, the main topic of interest in this brief section concerns the ability of organogold(I) complexes to undergo $S_H2$ reactions (equation 2) with a variety of radicals[95]. Related compounds of platinum(II) were also found to exhibit a similar reactivity towards homolytic substitution[96].

Methyl(triphenylphosphine)gold(I) reacts with benzenethiol at 35 °C in dichloromethane to give $PhSAu^I(PPh_3)$. Methyl radicals were spin-trapped by MNP (see Section I.D.1) to give the adduct $Bu^t(Me)NO^{\bullet}$ which was detected by e.s.r. spectroscopy. The reaction was accelerated when a radical initiator, aibn, was introduced and hindered by the presence of galvinoxyl, a well known inhibitor. It was concluded that the reaction probably proceeds via a radical chain propagated by $S_H2$ attack on the gold(I) alkyl by phenylthiyl radicals[97].

$$PhS^{\bullet} + [MeAu^I(PPh_3)] \longrightarrow [PhSAu^I(PPh_3)] + Me^{\bullet} \qquad (61)$$

Other studies, utilizing the spin-trapping technique and also CIDNP and traditional initiation/scavenging methods, have indicated that $PhCO_2^{\bullet}$[98], $PhSe^{\bullet}$[99], $Me^{\bullet}$[94], $F_3C^{\bullet}$[100], and $(Me)F_2CCF_2^{\bullet}$[93] are all implicated in $S_H2$ reactions at gold(I) centres. No such reactivity is seen at the corresponding gold(III) centres, however (for example, $[Me_3Au^{III}(PPh_3)]$)[97,100].

In addition, a study on the reaction between *n*-butyl(tri-*n*-butylphosphine)silver(I) and the 2,2,6,6-tetramethylpiperidin-1-oxyl (tmpo) radical suggests that this stable radical is able to displace *n*-butyl radicals from silver[92].

## IV. GROUP IIA

While it is true that homolytic chemistry is not common among the members of this group, there are well documented accounts of such reactions in magnesium and beryllium chemistry where free-radical intermediates have been observed directly. Organocalcium compounds have found some use in organic synthesis, and free-radical mechanisms have been inferred in several examples of reactions involving these compounds.

There are numerous accounts of reactions of Grignard reagents with particular substrates where the intermediacy of free radicals has been proposed, and these mechanisms are discussed under a separate subheading.

### A. Thermal Decompositions

There is no direct evidence to indicate that thermal decomposition of organoberyllium compounds proceeds via beryllium—carbon bond homolysis. These thermal decompositions generally proceed via the concerted $\beta$-elimination of metal hydride mentioned earlier. In fact, this route is the most convenient available for the synthesis of compounds RBeH. Thus, heating *t*-butylmethylberyllium at 150 °C for

2 h under nitrogen followed by a further period at 120 °C and $10^{-3}$ mmHg gives methylberyllium hydride[101].

$$\text{Bu}^t\text{BeMe} \longrightarrow \text{HBeMe} + iso\text{-C}_4\text{H}_8 \qquad (62)$$

Although beyond the general scope of this review, it should be mentioned that the mass spectral fragmentation pattern for diphenylberyllium suggests several different beryllium-containing radicals, derived from various bond homolyses[102]. This same work indicates that the mean bond dissociation energy ($\bar{D}$) in $Ph_2Be$ was *ca* 336 kJ $mol^{-1}$, a value higher than those encountered in Section I.C.

Di-*t*-butylmagnesium has been the subject of a detailed investigation in which CIDNP was observed in the thermal decomposition. At 83 °C the products of decomposition are isobutene, isobutane, and 2,2,3,3-tetramethylbutane[103]. Polarizations were observed in the n.m.r. spectra of isobutene and isobutane, suggesting that homolysis of one $Bu^t$—Mg bond gives an initial radical pair from which *t*-butyl radicals escape. From Kaptein's rules only free-diffusing radicals would give rise to the A/E spectra observed. A more detailed account of how such polarizations lead to these conclusions is outlined below.

Surprisingly few other studies have been carried out on the thermal or photolytic stability of alkylberyllium or magnesium compounds. Recent studies on cyclopentadienyl derivatives of beryllium and magnesium have suggested that the compounds are stable towards photolysis[104].

## B. Radical Reactions

The autoxidation of dimethylberyllium was originally thought to proceed via a heterolytic insertion of oxygen into the Be—C bond to give methylberyllium methyl peroxide[105]. However, later it was suggested[9], by analogy with the known mechanisms of autoxidation of a large number of other organometallic compounds, that a free-radical chain reaction was responsible for the observed products, with the product-forming propagation step involving $S_H2$ at the beryllium centre. $S_H2$ reactions have remained unsubstantiated, apart

$$\text{MeOO}^\bullet + \text{BeMe}_2 \longrightarrow \text{MeOOBeMe} + \text{Me}^\bullet \qquad (63)$$

from this postulate until recently when the reactions of *t*-butoxy radicals with $Cp_2Be$, CpBeCl, and CpBeMe were studied by e.s.r.[104]. When solutions of these compounds in toluene or cyclopropane were photolysed in the presence of di-*t*-butyl peroxide in the cavity of an e.s.r. spectrometer, strong e.s.r. spectra of cyclopentadienyl radicals were observed. In the case of CpBeMe both $Cp^\bullet$ and $Me^\bullet$ were directly observable, the relative intensities being temperature dependent. The results with dicyclopentadienylberyllium, in addition to demonstrating again the utility of the cyclopentadienyl ligand to induce homolytic chemistry, provided some insight into a complex structural problem that has been a feature of the inorganic literature for some years. The structure of beryllocene has been a topic for debate since X-ray crystal structure studies[106,107] indicated that a 'slipped sandwich' structure might be preferred over the findings of an early electron diffraction study[108]. However, a later electron diffraction study[107] indicated that a symmetric $C_{5v}$ structure was preferred. The situation was further confused when PRDDO[109] and MNDO[110] calculations indicated that not only were the symmetric $D_{5d}$ or $D_{5h}$ models both considerably lower in energy than the $C_{5v}$ or 'slipped sandwich' models but also that the lowest energy, most stable form of beryllocene would be that containing one $^5\eta$-bonded and one $^1\eta$-bonded cyclopentadienyl ring! Recently, the problem has been reexamined by gas-phase electron diffraction[111], by He(I) photoelectron spectroscopy with *ab initio* M.O. calculations[112], and by Raman spectroscopy in the solid state and as a liquid[113]. In all these papers the results indicated that in the solid state at −160 °C and at 0 °C, as a liquid at 65 °C, and in the gas phase, the molecule exists in a pentahapto/polyhapto (probably $^5\eta:^3\eta$) 'slipped sandwich' structure. The fact that homolytic substitution by *t*-butoxyl radicals takes place at the beryllium centre suggests that any symmetric structure ($C_{5v}$, $D_{5d}$, $D_{5h}$) would be unlikely, purely because no close approach to interact with any metal orbitals is possible by the incoming radical. Schematic diagrams of the highest occupied molecular orbitals of a 'slipped sandwich' structure (with one ring laterally displaced by 1.21 Å) indicate that the required orbitals are available[112] for interaction and that $S_H2$ reactions should be possible, as is experimentally observed. Although few structural studies have been carried out in fluid solution, and in this phase the possibility of a $^5\eta:^1\eta$ structure cannot be altogether dismissed (which would certainly be likely to undergo $S_H2$), the e.s.r. results tend to support the most recent structural studies[108].

The reactions of dialkylberyllium compounds with ketones indicate that charge-transfer processes may occur readily. Thus, mixing di-*t*-butylberyllium with a solution of benzophenone in diethyl ether at −78 °C gives a strongly coloured solution, thought to be due to charge transfer[114].

$$Bu^t_2Be + Ph_2C{=}O \longrightarrow Bu^t_2Be^{+\bullet} + Ph_2CO^- \qquad (64)$$

The colour disappears on warming and isobutene is evolved with the dimeric $(Bu^tBeOCO_2)_2$ being isolated. While no other products (isobutane, 2,2,3,3-ttramethylbutane, etc.) that would have been indicative of free *t*-butyl radicals were observed, the possibility cannot be excluded, given the analogy between this process and similar ones for other metals discussed in this chapter. No attempts to observe e.s.r. spectra or CIDNP effects were made in this work, but such studies may certainly prove useful in characterizing the reactivity of organoberyllium compounds further.

Autoxidations of Grignard reagents were originally thought to proceed via a polar mechanism on the basis of product and kinetic studies[115] which showed that diphenylamine did not inhibit the autoxidation and that butyraldehyde was not co-oxidized when present in the reaction mixture. It was concluded that the autoxidation probably did not involve free radicals, although the possibility of their participation was never completely excluded[116].

Evidence for a free-radical chain mechanism was first provided by the study of 5-hexenylmagnesium bromide[117,118], where an increased percentage of cyclized product (dramatically increased at low oxygen concentrations, where cyclization is able to compete with oxygen quenching) suggested a typical rapid free-radical chain autoxidation, involving initial electron transfer, with $S_H2$ at magnesium being the propagating step.

$$RMgX + O_2 \longrightarrow R^\bullet + XMgO_2^\bullet \qquad (65)$$

$$R^\bullet + O_2 \longrightarrow ROO^\bullet \qquad (66)$$

$$ROO^\bullet + RMgX \longrightarrow ROOMgX + R^\bullet \qquad (67)$$

Confirmation of the ability of dialkylmagnesium compounds and Grignard reagents to undergo $S_H2$ reactions was provided by an e.s.r. study of reactions of these compounds with *t*-butoxy radicals[119]. The course of reaction depends on whether $\alpha$-hydrogen is available in the alkyl group. If $\alpha$-hydrogen atoms are present the two courses shown in reactions 68 and 69 proceed simultaneously.

$$R_2CHMgX \xrightarrow{S_H2} R_2CH^{\bullet} + Bu^tOMgX \quad (68)$$

$$R_2CHMgX \longrightarrow R_2\dot{C}MgX + Bu^tOH \quad (69)$$

Abstraction of $\alpha$-protons by *t*-butoxy radicals, rather than substitution, is widespread among organometallic compounds and in e.s.r. spectra in magnesium systems two signals are superimposed, confirming that both reactions are taking place. While the presence of $R^{\bullet}$ could also be interpreted to imply some unimolecular homolysis of R—Mg, the possibility can be discounted for when the photolysis is carried out in the absence of di-*t*-butyl peroxide, no radicals are observed.

Organocalcium compounds have been studied very little in respect to their ability to undergo calcium—carbon bond homolysis. The compounds are subject to rapid autoxidation, but the mechanistics of the process have not been studied, and while it is likely that free radicals are important the possibility of ligand carbanion autoxidation must also be considered[9]. Free radicals have been considered as intermediates, however, in the reactions between organocalcium halides and 1,3-benzoxathiole and 1,3-benzodioxole derivatives[120]. Here, a product study strongly implicates free radicals in these reactions. The first step appears to be either a simple cleavage of R—Ca to give $R^{\bullet}$ and a calcium species which adds to oxygen, or more likely a charge transfer to give $R\overset{+}{C}aX$ which immediately fragments to $R^{\bullet}$ and $^{+}CaX$, which adds to oxygen. A reaction scheme for where R contains $\beta$-hydrogen atoms is outlined in reaction 70. **8** and **9** then abstract hydrogen from solvent to give the

$$\textbf{(7)} + RCaX \longrightarrow \underset{\textbf{(8)}}{o\text{-}XCaOC_6H_4SCMe_2} + \underset{\textbf{(9)}}{o\text{-}XCaSC_6H_4OCMe_2} \quad (70)$$

corresponding products and products derived from the usual free-radical termination steps. Thus, reaction of **7** with *t*-butylcalcium bromide gave 68% yield of the phenol (from **8**), 7% of the thiophenol (from **9**), and isobutene, isobutane, and 2,2,3,3-tetramethylbutane. These results seem to confirm an earlier report of reactions of organocalcium halides with aliphatic ketones, where free radicals were implicated[121]. In addition, given the facility with which Grignard reagents react with aliphatic thioketones via free-radical pathways (see below), these observations are not surprising. In fact, in an extensive study on reactions of methylcalcium iodide with organic compounds, inspection of some products obtained indicates that other organocalcium reactions may involve free radicals, but no mechanistic studies or conclusions were made[122].

### C. Reactions of Grignard Reagents

While it would be obviously impractical to make any general comments regarding the mechanism of Grignard reactions, it is possible to demonstrate the types of reaction that commonly lead to free radicals via carbon—metal bond homolysis and to indicate the factors which cause the reactions to proceed via polar or electron transfer mechanisms with a view to predicting the cause of reaction.

### *1. Grignard reactions with ketones*

By far the most mechanistic studies on the Grignard reaction have been carried out using ketones as substrates. Several groups of workers have concentrated on these reactions to evaluate many of the factors which contribute to and influence the overall reaction. In a summary of studies to 1974, Ashby *et al.*[123] concluded that an overall reaction scheme such as that outlined in equations 71 and 72 was applicable.

$$\text{R—M} + \text{>C=O} \xrightarrow{\text{Polar}} (\text{>C=O} \cdots \text{R—M}) \longrightarrow \text{>C(R)—OM}$$
$$\text{R—M} + \text{>C=O} \xrightarrow{\text{Set}} (\text{>C=O} \;\; \text{R—M}) \longrightarrow (\text{>}\dot{\text{C}}\text{—O}^- \;\; \dot{\text{R}}\text{—M}^+) \longrightarrow \text{>C(R)—OM} \quad (71)$$

**(10)**

$$\textbf{(10)} \xrightarrow{\text{diffusion}} \text{>}\dot{\text{C}}\text{—OM} + \text{R}\cdot \longrightarrow \text{products} \quad (72)$$

The course which any particular reaction may take was found to depend critically on several factors. The reaction may proceed via the polar mechanism or single electron transfer or even both simultaneously, depending on the redox potentials of both reagents, on solvent polarity, on transition metal impurities in the magnesium used for the reaction, etc. Thus, benzophenone was found to react with methylmagnesium bromide via the polar mechanism, yet *t*-butylmagnesium chloride reacts via single electron transfer. A change from the polar mechanism to single electron transfer is observed for the methylmagnesium bromide when the solvent is changed to hexamethylphosphoramide or when 0.05 mol-% of iron(III) chloride is introduced. In this discussion the reaction is confused further by the fact that if the cage pair **10** does not escape, the product is merely the same as that in the polar route. Therefore, while many studies have shown that electron transfer may be the first step by e.s.r., observation of the ketyl intermediates, products that are diagnostic of Mg—C homolysis, are not necessarily isolated. As a consequence, the examples discussed here only include those where free-radical reactions are caused as a consequence of escape from the solvent cage. In suitable cases this enables both ketyl intermediates (by e.s.r.) and polarized products (CIDNP) which could only (by Kaptein's rules) arise from $|F\rangle$ precursors.

Benzophenone and substituted benzophenone have always been the classical substrates used as mechanistic probes in the study of Grignard reactions. Consequently, while a profusion of data for the system has been available, the most important mechanistic questions remained unanswered until the early and mid-1970s. In fact, a radical mechanism was proposed for this reaction as early as 1929[124], even before the general acceptance of the 'free radical' as a chemical entity which was brought about by the experiments of Hofeditz and Paneth[5]. Pinacol, a product critical to the free radical interpretation, was first observed in reaction products by Arbuzov and Arbuzova[125] and ensuing arguments as to whether this was a minor product formed by impurities[126] did little to clarify the situation. Although

Mosher and Blomberg[127] demonstrated the existence of the benzophenonemagnesium ketyl by e.s.r, the question as to whether this was formed by the main or side reactions remained unclear until the kinetic work of Holm and Crosland[128]. It was found from the product distribution in the reaction between *t*-butylmagnesium chloride and substituted benzophenones in ether that, although the individual rates of formation of pinacols and 1,2-addition, 1,4-addition, and 1,6-addition products were obviously controlled by steric factors, a good Hammett plot was obtained for the sum of the four competing reactions. This observation, and the fact that steric effects have no influence on the *overall* reaction rate, as measured by the disappearance of substrate, strongly indicated that a common rate-limiting step, single electron transfer from Grignard to benzophenone, was involved. This was the first real indication that the radical pathway is not a side reaction but in this case constitutes the major, or indeed only, mechanism. In 1975 a remarkable paper by Savin *et al.*[129] demonstrated fully the complexity of the reaction between benzophenone and *t*-butylmagnesium chloride in dimethoxyethane as solvent. In addition to observing the ketyl radical, formed by the initial electron transfer, by e.s.r. spectroscopy they also observed large polarizations in the n.m.r. spectra of the products isobutane and isobutene during the course of the reaction. No polarizations were observed in the other major products, and using Kaptein's rules it was possible to predict that the observed polarizations could only arise from random encounters of freely diffusing *t*-butyl radicals. The complex scheme in Figure 3 outlines the full mechanistic conclusions of the paper. It is important to note that only the isobutene formed by random encounters has any polarization. In a major series of papers, Savin and coworkers have reported an extensive study of Grignard reactions with many substrates using CIDNP as the main mechanistic probe, and e.s.r. where possible. Thus, by observing polarizations in butane and butene, free butyl radicals are implicated in the reactions between *t*-butylmagnesium chloride and benzophenone[129], dibenzoyl disulphide[130], chloroform[131], benzaldehyde[132], carbon tetrachloride[133], benzoyl chloride[134], and substituted benzoyl chlorides[135] and nitrobenzene[136].

The same workers also examined the reaction between benzylmagnesium chloride and benzoyl peroxide in thf[137]. Here, polarizations in the n.m.r. spectra of the products benzoyl benzoate, benzyl chloride, and dibenzyl were observed. It was deduced that the polarizations of the benzoyl benzoate, which were in emission, were caused by cage recombination of the initial radical pair. However, dibenzyl was formed in 45% total yield and, with enhanced absorption observed in the methylene protons, this must indicate that diffusion from the solvent cage is a major pathway of decomposition of the initial radical pair.

While no e.s.r. spectra of CIDNP effects were observed, kinetic and product studies strongly implicate free-radical mechanisms from the reaction between *t*-butylmagnesium chloride and ethyl cinnamate[138,139]. Here, products from cage reaction and also classical free-radical reactions in solution were detected. Thus according to the reaction scheme, initial single electron transfer occurs to give ester anion and Grignard cation which rearrange to give the initial radical pair, **11** (Scheme 1). As in other similar reactions, the cage products may now collapse to give the expected 1,4-addition product or may diffuse to give the free radicals **12** and **13**. In the absence of radical scavengers, the *t*-butyl radicals can now add directly to the double bond of ethyl cinnamate. This radical may then undergo a further single electron transfer to give, after work-up, the cyclopropanone hemiketal salt **14.**

A very interesting mechanistic development has become evident in a recent study, namely the possibility of a Grignard reaction proceeding apparently via two independent single electron transfer steps[140]. The products generally observed in the reac-

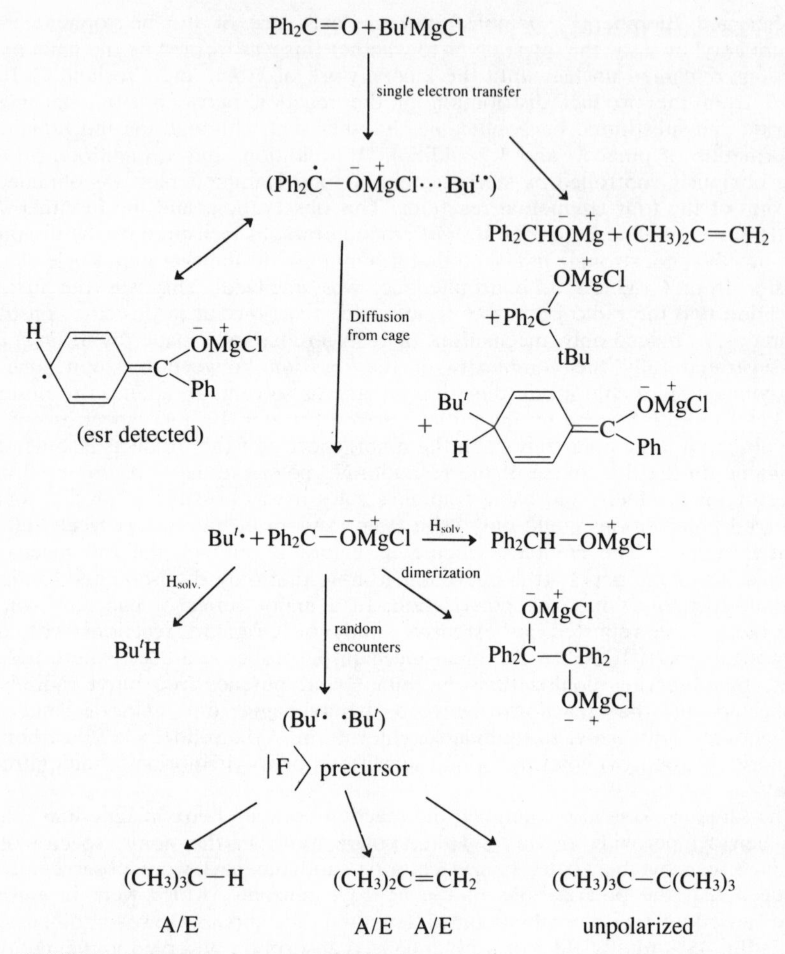


FIGURE 3. Reaction scheme for the Grignard reaction between t-butylmagnesium bromide and benzophenone

tions of carboxylic acids with Grignard reagents are tertiary alcohols. Benzoic acid and substituted benzoic acids are no exception to this rule. The ketyl radicals, products of 'single electron transfer' from alkyl and aryl Grignard reagents to salicyclic and benzoic acids, have been observed by e.s.r., and these intermediates may be stabilized by the presence of nickel(II) ions[141,142]. It has now been demonstrated that reaction between *m*- and *p*-substituted benzoic acids with aryl Grignards in thf. solution under argon, with small amounts of nickel(II) present, has a further single electron transfer step, characterized by e.s.r. observation of the carboxylate anion radicals which are the products of an initial electron transfer. Admittedly these reactions are in the presence of a nickel(II) salt and, as has been mentioned before, traces of transition metals do have dramatic effects on the course of Grignard reaction[123]. However, unlike for example iron(III) impurities which, because of their influence as electron acceptors may catalyse electron transfer, nickel(II) does not

$$\text{Bu}^t\text{MgCl} + \text{PhCH}{=}\text{CHCO}_2\text{Et} \longrightarrow [\text{Bu}^t\overset{+}{\text{Mg}}\text{Cl}\cdot\text{Ph}\dot{\text{C}}\text{HCH}{=}\text{C(O}^-\text{)OEt}]$$

$$\xrightarrow{\circ} [\text{Bu}^{t\cdot}\text{PhCHCH}{=}\text{C(OMgCl)OEt}] \quad \textbf{(11)}$$

$$\textbf{11} \xrightarrow{\text{cage reaction}} \text{Ph}\dot{\text{C}}\text{H(CMe}_3\text{)CH(MgCl)CO}_2\text{Et} \longrightarrow \text{PhCH(CMe}_3\text{)CH}_2\text{Co}_2\text{Et}$$

$$\textbf{11} \longrightarrow \underset{\textbf{(12)}}{\text{Bu}^{t\cdot}} + \underset{\textbf{(13)}}{\text{PhCHCH}{=}\text{C(OMgCl)OEt}}$$

$$\text{Bu}^{t\cdot} + \text{PhCH}{=}\text{CHCo}_2\text{Et} \longrightarrow \text{Ph}\dot{\text{C}}\text{HCH(CMe}_3\text{)Co}_2\text{Et}$$

$$\xrightarrow{\text{set}} \text{PhCH(MgCl)CH(CMe}_3\text{)CO}_2\text{Et}$$

$$\longrightarrow \text{1-Ph-2-CMe}_3\text{-3-OEt-3-OMgCl-cyclopropane} \quad \textbf{(14)}$$

SCHEME 1

readily accept or give up electrons to give Ni(I) or Ni(III) under the conditions of these experiments. It is difficult to visualize the role of this ion being anything other than stabilizing (presumably via chelation) the intermediate carboxylate radical ions. if, subsequently, hmpa is added to the reaction mixture, e.s.r. spectra of the ketyl radicals intermediate in tertiary alcohol formation are observed. after the initial electron transfer the fate of the aryl group from the Grignard appears to be diffusion from the solvent cage, although the required proton for aromatic hydrocarbon formation may well be that from the benzoic acid. While more product and mechanistic studies need to be carried out on this system, the possibilities and implications to mechanism induced by these initial experiments should provide some stimulating and interesting results.

$$\text{ArCOOH} + \text{RMgX} \xrightarrow[\text{thf or Et}_2\text{O}]{\text{(i) Ni)}} \xrightarrow{\text{(ii) set}} \text{Ar}\dot{\text{C}}\text{(O}^-\text{)OMgX} + \text{RH} \qquad (73)$$

$$\text{Ar}\dot{\text{C}}\text{(O}^-\text{)OMgX} \xrightarrow{\text{(i) hmpa, (ii) RMgX}} \xrightarrow{\text{set (?)}} \text{Ar}\dot{\text{C}}\text{(O}^-\text{)R} \longrightarrow \text{products} \qquad (74)$$

This novel system aside, and although there is not space here to include discussion of all mechanistic studies on reactions of carbonyl compounds with Grignards, it is possible to summarize the implications of these studies with the aid of some recent results.

Mechanistic developments involving the reactions of Grignards with carbonyl compounds seem to confirm all previous work in that the nature of the pathway, whether single electron transfer or polar, varies from reaction to reaction with no 'blanket' rule covering any. Thus, 2,2,6,6-tetramethylhepten-4-one was used as a mechanistic probe with alkyl and allyl Grignards[143], the principle being that if the reaction proceeds via single electron transfer, then complete loss of stereochemical integrity in substrate and products would occur. The study showed that while some Grignards (Me, Et, allyl) favoured the polar mechanism, *t*-butyl Grignard caused nearly complete isomerization of both starting materials and products, indicating the single electron transfer pathway. Mechanistically, it appears that from the results available thus far, it is becoming increasingly possible to predict a reaction path for any particular reaction as our knowledge of the factors governing the path by which the reaction may proceed improves. However, it is also evident that a unified mechanistic view of the Grignard reaction may be difficult to achieve given the number of experimental variables in each individual reaction.

### 2. *Reactions with other substrates*

There are many Grignard reactions with substrates other than ketones which proceed via an initial single electron transfer step and lead to participation of free radicals. A few pertinent examples, with a variety of substrates, are discussed.

Reaction of alkylmagnesium bromides with mono-$\gamma$-acetylenic(alkyl) ethers appears to proceed via a free-radical reaction[144].

$$(CH_3)_3CCH_2C{\equiv}CCH_2OR' + RMgBr \longrightarrow (CH_3)_3CCH_2C{\equiv}CCH_2^{\bullet} + R^{\bullet} + R'OMgBr \quad (75)$$

$$(CH_3)_3CH_2\dot{C}{=}C{=}CH_2 \longleftrightarrow (CH_3)_3CH_2C{\equiv}C\dot{C}H_2 \quad (76)$$

All the possible coupled products from the tautomeric intermediate allenyl and $\beta$-acetylenemethyl radicals were isolated. These products were formed in addition to coupled products from the Grignard reagent, $R_2$, and ROMgBr. Significantly, in this study there was very little dependence of product distribution on the nature of R or R′, which may indicate that the driving force for the reaction is formation of the resonance stabilized free-radical intermediates. The total yield of products is always nearly quantitative.

Another recent product study concerns the reaction of neophylmagnesium chloride with carbon tetrachloride or 1,2-dibromoethane, in which the presence of free radicals is implicated, although no e.s.r. or CIDNP studies were carried out[145]. The products of the reaction with carbon tetrachloride are indicative of a purely ionic pathway, but in 1,2-dibromoethane an alternative pathway was proposed. It was suggested that the transition state for single electron transfer with 1,2-dibromoethane is stabilized by homobenzylic conjugation.

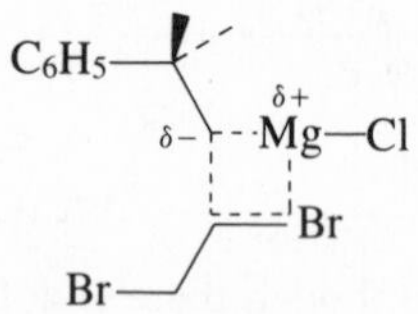

$C_6H_5$ $\delta+$ $\delta-$ Mg—Cl Br Br

**(15)**

Neophyl and 2-bromoethyl radicals may then diffuse from the solvent cage to give the observed variety of products. No products from the rearrangement of neophyl radicals were detected, but this is not surprising as the reactions were carried out at −70 °C, where the rate of rearrangement is very slow compared with that of radical termination steps.

Thioketones react with Grignard reagents to give thioethers according to the general reaction 77. The reaction was subjected to mechanistic investigation and

$$R_2C{=}S \xrightarrow{R'MgX} R_2CHSR' \quad (77)$$

initially, in parallel with studies on carbonyl compounds (see earlier), the reaction between thiobenzophenone and methyl magnesium iodide in ether or thf was examined. Not surprisingly, by analogy with reactions with benzophenone, the e.s.r. spectrum of thiobenzophenone ketyl was observed, confirming single electron transfer to be a primary step[145]. Later it was found that while mechanistic studies on the initial step were difficult, addition of Grignard (wherher polar or single electron transfer) followed by Mg—carbon bond homolysis appear widespread in this reaction.

$$R_1R_2C{=}S + RMgX \longrightarrow R_2R_2C(Mg)SR \qquad (78)$$

$$R_1R_2C(Mg)SR \longrightarrow \underset{(\mathbf{16})}{R_1R_2\dot{C}SR} \longrightarrow \text{products} \qquad (79)$$

It became apparent that the intermediate radicals could be readily observed using e.s.r. Thus, radicals **17**, **18**, **19**, and **20** could be observed from the corresponding thioketones[147].

$C_6H_5\dot{C}(SR)Bu^t$

(**17**)

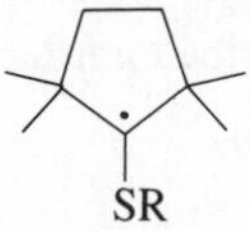

SR

(**18**)

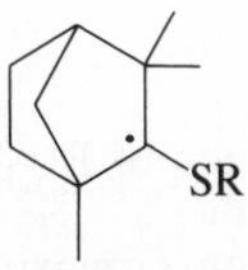

SR

(**19**)

SR

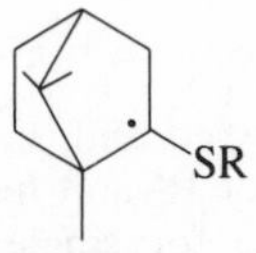

(**20**)

Reaction of vinylmagnesium halides with thioketones again presents evidence for Mg—C bond cleavage from an intermediate Grignard adduct[148]. The general reaction proceeds to give the two possible products according to reaction 80.

$$R_1R_2C{=}S \xrightarrow[\text{several steps}]{XMgCH{=}CH_2} \underset{(\mathbf{21})}{R_1R_2C(SH)CH{=}CH_2} + \underset{(\mathbf{22})}{R_1R_2CHSCH{=}CH_2} \qquad (80)$$

The yield of the products depends on the time of reaction[146]. After 16 h **21** is formed in 100% yield but early quenching gives **22** in 80% yield. A mechanism similar to that in equations 77–79 was again used to present an explanation, with **21** being formed as the kinetic product and **22** is formed from the thermodynamically more stable intermediate radical. Here again e.s.r. was used to observe the intermediate directly, although formation of **21** of course required the intermediacy of a thiyl radical, which is not observable by e.s.r.[149]. A more detailed study on the thiobenzophenone system showed that products arise primarily from the addition of R (from the Grignard) to sulphur. This again implies the intermediacy of the thermodynamically more stable diphenyl(thioalkyl)methyl radical, which in fact is observed by e.s.r.[150].

Significantly, organocadmium compounds undergo similar reactions, although not in such impressive yields.

The reaction between thiobenzophenone and butenylmagnesium bromide demonstrates the utility of the mechanistic interpretation, with the isolation of products which could only arise from a radical cyclization reaction.

$$Ph_2C{=}S + XMgCH_2CH_2CH{=}CH_2 \longrightarrow \underset{\textbf{(23)}}{Ph_2C(MgX)SCH_2CH_2CH{=}CH_2} \quad (81)$$

$$\textbf{23} \longrightarrow \underset{\textbf{(24)}}{Ph_2\dot{C}SCH_2CH_2CH{=}CH_2} \quad (82)$$

$$\textbf{24} \longrightarrow \text{2,2-diphenyl-3-methyltetrahydrothiophene} \quad (83)$$

The major product, formed by abstraction of hydrogen from solvent, was 2,2-diphenyl-3-methyltetrahydrothiophene[150].

From a product study, it has been shown that the stable nitrogen-centred radical 1,3,5-triphenylverdazyl reacts with alkylmagnesium chlorides by an $S_H2$-like process similar to that described earlier for organolithium compounds[60].

$$\text{1,3,5-triphenylverdazyl} + RMgX \longrightarrow \textbf{(25)} + R^{\bullet} \quad (84)$$

**25** is then hydrolysed and, in addition to the dimerization of $R^{\bullet}$, 40–50% yields of coupled product of $R^{\bullet}$ and further verdazyl are isolated.

Direct evidence for single electron transfer between Grignards and nitrogen-containing substrates has recently become available. Thus, 1,10-phenanthroline reacts via single electron transfer with Grignard reagents RMgBr or $R_2Mg$ to give organomagnesium radical complexes[151], $[phen(MgR)]^{\bullet}$, which may be characterized by e.s.r.

$$phen + MgPh_2 \longrightarrow [phen(MgPh)]^{\bullet} + Ph^{\bullet} \quad (85)$$

It may well be that further experiments with ligands such as 1,10-phenanthroline with organometallic compounds will provide more interesting stable paramagnetic complexes, given the ability of this ligand and, for example, 2,2′-bipyridyl to function as negative molecular ions and therefore stabilize metals in formally low oxidation states[152]. Outside the scope of this chapter, it is interesting that electron transfer to phen has recently been observed[153] in reaction with $[(^5\eta\text{-}C_5H_5)Ti^{II}(CO)_2$.

## V. GROUP IIB

It has long been realized that organometallic compounds of this group have been susceptible to homolytic decomposition processes, a major reason being the rather low first bond dissociation energies often encountered in these compounds, which render them labile towards autoxidation, decomposition, and substitution reactions. In fact, both decomposition reactions[154] and substitution reactions[9] of this group have been reviewed at length. Therefore, in this chapter the most significant aspects to be discussed will be some new developments in the decompositions of these compounds when used as precursors to novel, previously unavailable free radicals, and also as organometallic compounds of this group readily undergo electron transfer reactions with a variety of electron acceptors, these reactions will also be included.

## A. Thermal and Photolytic Decompositions

The temperatures at which dialkyl-zinc and -mercury(II) compounds begin to decompose homolytically, 85–140 °C, ensure a suitable rate of unimolecular decomposition for the study of CIDNP effects[103].

Di-*t*-butyl-zinc and -mercury decompose thermally between 82 and 140 °C and 120 and 140 °C, respectively. During these decompositions polarizations are observed in the product n.m.r. spectra. The small negative activation entropies for the reactions (for example, $\Delta S^{\neq} = -3.7$ eu for $Bu^t_2Zn$) and the fact that the $K_D/K_C$ ratio of 3.75 is around that expected for *t*-butyl radicals indicate that the major reaction path is via homolysis of the metal—carbon bond, rather than by other mechanisms (for example, via a four-centre transition state). By Kaptein's rules it can be shown that for the decomposition of the zinc compound that the product polarizations arise from random encounters of $|F\rangle$ precursors, and that simple unimolecular homolysis is preferred.

$$R_2Zn \longrightarrow Zn + 2R^{\bullet} \quad (86)$$

On the other hand, thermal decomposition of the mercury compound is interpreted as a cage reaction of *t*-butyl radicals from singlet precursors, $|S\rangle$, while the photolysis of this compound at ambient temperatures, gives polarizations indicative of cage reactions from $|T\rangle$ precursors, although the possibility of unimolecular decomposition with resulting reactions of $|F\rangle$ precursors cannot be ruled out. These elegant experiments give a good idea of the sensitivity of the CIDNP effect to the rather subtle differences in the possible modes of decomposition.

Similarly, diallylzinc compounds decompose giving A/E product polarizations at temperatures from 85 to 150 °C[155]. In a study of decomposition of diallylzinc, bis(2-methylallyl)zinc, and bis(3-methylallyl)zinc, a common mode of decomposition is indicated, namely unimolecular decomposition, with random encounters of $|F\rangle$ precursors responsible for product polarizations (predicted by Kaptein's rules).

An extensive photo-CIDNP study of organomercury compounds has been undertaken[156]. Not surprisingly, photolysis of asymmetric and symmetric dialkyl- and diaryl-mercury compounds provided no evidence for intermediates $RHg^{\bullet}$, but realistically this could not be expected at the temperatures of the study.

Given the ease with which alkyl-zinc and -mercury compounds undergo homolytic thermal decomposition at temperatures between about 80 and 150 °C, it seems remarkable that homolysis of the metal—carbon bond has not been considered as a possibility in reactions between dialkenylmercury compounds and zinc or magnesium metal in sealed tubes at 120 °C[157]. Thermolysis at 120 °C of dihexenylmercury in the presence of zinc metal gave, after 24 h, 90% of dihexenylzinc and 10% of dicyclopentylmethyl zinc, and after 84 h gave 10% of hexenylzinc and 90% of cyclized product. In the reaction with magnesium metal, only cyclized product was isolated after 24 h. The results were interpreted in terms of a four-centre transition state, **26**, stabilized via a metal—double bond interaction, thus facilitating intramolecular cyclization.

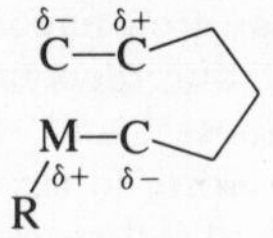

$\delta- \quad \delta+$
C—C
M—C
$/\delta+ \quad \delta-$
R

**(26)**

While similar transition states have been postulated for other systems, it certainly seems possible that free radicals play the major part in this instance. At those temperatures CIDNP measurements should be definitive in proof of this as they have been in the allylmetal, *t*-butylmetal, and other compounds mentioned previously.

Thermal and photolytic decompositions of organomercurials have found considerable application in studying reactions of the free radicals produced, and this is an area which appears to have potential for further expansion. These routes to the free radicals are favoured over others, primarily because they provide a convenient unimolecular pathway which generally gives the free radicals in high yield.

Dibenzylmercury was found to be a convenient source of benzyl radicals when decomposed at 129.8 °C[158]. In a detailed kinetic study, the scheme below was that favoured to describe the decomposition.

$$R_2Hg \xrightarrow{\Delta} RHg^{\bullet} + R^{\bullet} \tag{87}$$

$$RHg^{\bullet} \longrightarrow R^{\bullet} + Hg \tag{88}$$

This mechanism was favoured because the Arrhenius parameters and activation energies were lower than those required if both Hg—C bonds were broken. The reaction is very favourable for the initiation of polymerization, because very little head-to-tail recombination to give substituted aromatics occurs. In a subsequent paper[159], substituent effects on the rate of decomposition were reported. Ten substituted dibenzylmercury compounds were found to decompose with first-order kinetics at 140 °C. Rate constants varied from $5.6 \times 10^5$ mol $l^{-1}$ $s^{-1}$ for the *m*-fluoro compound to $26.3 \times 10^5$ mol $l^{-1}$ $s^{-1}$ for the *p*-methoxy compound. The general method appears suitable for comparing the stabilities of substituted benzyl radicals using Hammett $\rho$ and $\sigma$ values.

Cyclopentadienylmercury compounds undergo unimolecular decomposition on u.v. irradiation to give cyclopentadienyl radicals (from $[(C_5H_5)_2Hg]$) and alkyl-substituted cyclopentadienyl radicals (from $[(RC_5H_4)_2Hg]$)[27]. Previously, the cyclopentadienyl radical had been observed by e.s.r. in single crystals[160] and in neon[161] or adamantane matrices[162] and no convenient route to the radical in solution was available, the reactions of *t*-butoxy or trimethylsiloxy radicals with cyclopentadiene giving products from addition to the diene system, rather than abstraction of hydrogen from the $sp^3$ carbon atom. Thus the solution study, made accessible through the mercury compounds, gave valuable information about these theoretically interesting annulene radicals. The cyclopentadienyl radical, which has two degenerate highest occupied molecular orbitals, was found to be a true planar $\pi$ radical, as deduced by the magnitude of $^{13}C$ hyperfine coupling of 2.69G, with no evidence of a dynamic Jahn–Teller effect, which would increase this value.

The alkyl-substituted cyclopentadienyl radicals, from the corresponding alkylcyclopentadienylmercury compounds, were also studied in detail. It was found that the alkyl substituents[163] and even deuterium[164] could perturb the HOMO orbital degeneracy and that the magnitudes of the observed e.s.r. hyperfine coupling constants precisely reflect the electronic interaction between the substituent and the $\pi$-electron system. The large temperature range over which the photolyses could be carried out (−120 to +10 °C) also enabled a detailed conformational analysis of the temperature-dependent coupling constants and power saturation measurements to be made[27].

The same method of cyclopentadienyl radical generation has been used more recently in a study of the spin-trapping of cyclopentadienyl radicals in organometallic systems[38]. The mercury compounds were particularly useful in this study, since cyclopentadienyl radicals can be produced not only by photolysis, but also by reaction with electron acceptors ($[Ir^{IV}Cl_6]^{2-}$, for example). This study, with a variety of nitroso and nitrone spin-traps, not only clarifies conflicting literature reports[37,165] as to the e.s.r. spectrum of the spin-trapped cyclopentadienyl radical, but also outlines an overall strategy for spin-trapping free radical intermediates in organometallic systems (see Section I.D).

Similar to the cyclopentadienyl radical, the allyl radical may be spin-trapped on 2-methyl-2-nitrosopropane or nitrosodurene when generated thermally or photochemically from bisallylmercury[166]. This is a more convenient route than that previously described involving reaction of hexabutylditin, allyl bromide, and di-*t*-butyl peroxyoxalate in the presence of mnp[167]. It would appear that the use of organomercury compounds as free radical precursors is growing considerably, given the apparent usefulness indicated by these studies.

The discussion as to whether low-valent species are intermediates in photochemical and thermal decompositions of Group IIB dialkylmetals, while having little real effect on the outcome of most reactions, is significant in the general context of very short-lived species, playing important roles in organometallic systems.

$$R_2M^{II} \longrightarrow R^{\bullet} + RM^{I\bullet} \longrightarrow R^{\bullet} + M^0 \quad (89)$$

$$R_2M^{II} \longrightarrow 2R^{\bullet} + M^0 \quad (90)$$

In dialkylmercury compounds, while the second bond dissociation energy is exceptionally low (*ca.* 20 kJ mol$^{-1}$), it is measurable and thus a finite, albeit very short lifetime, in solution should ensure that the intermediate Hg(I) species is important in decomposition. That these species exist has been confirmed by e.s.r. spectroscopy at 77 K, where they may be generated and identified in matrices by irradiation with $\gamma$-rays[20]. An indication of the lifetime of these $RHg^{\bullet}$ species in solution may be estimated from comparison of the lifetimes of the zinc and cadmium analogues. U.v. flash photolysis of dimethylzinc and dimethylcadmium provides direct evidence for the existence of $MeZn^{\bullet}$ and $MeCd^{\bullet}$[168]. Results were obtained from spectra recorded in the ranges from 400 to 445 nm and 264 to 287 nm. From these studies the radical lifetimes were estimated to be of the order of 100 $\mu$s. Indeed, one would expect the mercury

$$Me_2M \longrightarrow MeM^{\bullet} + Me^{\bullet} \quad (91)$$

species to have an even shorter lifetime under similar conditions, owing to the lower second bond dissociation energy. Observation of such briefly existent species in solution is not readily achieved using conventional e.s.r. spectroscopy. There are many other short-lived intermediates, postulated in main group organometallic chemistry, for example the metal cations often encountered in homolyses initiated by charge transfer depicted in equations 3 and 4 (Section I.B). Significantly, Group IV tetraalkylmetal cations are expected to be extremely short-lived and upper limits of 1 ms have been placed on their lifetimes (see Section VII), which is in the same range as those for the Group IIB species mentioned above.

### B. Radical Reactions

All the members of Group IIB are known to undergo $S_H2$ reactions with many substrates[9]. Often, however, it is difficult to distinguish between the products of $S_H2$ reaction and direct decomposition because the conditions necessary to initiate the attacking radical are similar to those under which decomposition is facilitated. Zinc-and cadmium-alkyls both undergo $S_H2$ reactions at the metal centre, with subsequent loss of alkyl radical. Thus, *t*-butoxy radicals will substitute at the metal in diethylzinc or dimethylcadmium to give ethyl and methyl radicals, respectively, the e.s.r. spectra of which may be observed directly when the reaction is initiated in the e.s.r. cavity[169].

Unfortunately, very few investigations have been carried out on $S_H2$ reactions of organic derivatives of Group IIB elements in recent years, despite the increasing importance of these compounds synthetically. Major reviews have appeared in recent years on the use of organocadmium[170] and organomercury reagents[171] in organic synthesis, and thus it is hoped that a concomitant expansion of interest in homolytic reactions will occur.

Dicrotylzinc and dicrotylmagnesium both react with alkyl-substituted cyclohexanones to give dialkylcyclohexanols[172].

$$\text{2-R-cyclohexanone} \xrightarrow[R_2Zn]{R_2Mg} \underset{\textbf{(27)}}{\text{1-[C(CH}_3\text{)CH=CH}_2\text{]-2-R-cyclohexanol}} \text{ or } \underset{\textbf{(28)}}{\text{1-(CH}_2\text{CH=CHCH}_3\text{)-2-R-cyclohexanol}} \quad (92)$$

With the magnesium compound the major product is **27**, but for the zinc compound **28** predominates. While no mechanistic conclusions were made in this study, it is conceivable that the reactions proceed via different mechanisms. With dicrotylzinc, a radical reaction may be postulated with a methylallyl radical intermediate. The observed product is consistent with addition to the carbonyl by the less hindered end of the allyl radical. On the other hand, allyl Grignard reagents are known to favour polar mechanisms in reaction with carbonyl compounds (see Section IV) and thus a concerted polar mechanism with a cyclic transition state may be postulated here.

Cyclopropanes may be prepared via the reaction of alkenes with a 'zinc-carbenoid' reagent prepared from dialkylzincs and *gem*-diiodoalkanes[173].

$$CH_2{=}CHR \xrightarrow[Et_2Zn]{CH_2I_2} \text{R-cyclopropane} \quad (93)$$

The reactive intermediates in these reactions are thought to be $\alpha$-iodoalkylzinc derivatives, which are the carbene transferring agents. It has been found that not only is the reaction accelerated under an oxygen atmosphere but also the yields are improved to near quantitative[174]. The autoxidation of alkylzinc proceeds via an extremely rapid free-radical chain process[175], upon which conventional radical scavengers have little effect (as was the case with the carbene reaction), and therefore the increased efficiency of the carbene reaction was attributed to the increased rate of initial zinc—alkyl cleavage facilitating the rapid formation of the carbene transferring $\alpha$-haloalkylzincs.

### C. Electron Transfer Reactions

The general implications of electron-transfer reactions of organometals with classical inorganic and organic substrates are discussed in Section VII, and are essentially illustrated with reference to Group IV organometals. Reactions of the same electron acceptors with organomercury compounds have proved valuable in mechanistic interpretations of the electron-transfer processes, and these are mentioned below. Zinc- and cadmium-alkyls also react with organic acceptors in synthetically useful reactions.

The compounds $Et_2M$ (M=Zn or Cd) react with orthoquinones in a similar fashion to many main group organometallic derivatives[176].

$$\text{R–(t-Bu)-orthoquinone} + Et_2Zn \longrightarrow \text{R–(t-Bu)-semiquinone}^{\cdot -}\ (\text{A}) + Et_2Zn^{\cdot +} \quad (94)$$

The short-lived metal cation dissociates to give ethyl radicals and the stable (towards further decomposition) cation $RZn^+$, which reacts with the semiquinone anion to give the complex

$$\text{A}\quad Et_2Zn^{\cdot +} \longrightarrow \text{R–(t-Bu)-semiquinone–ZnEt}\ (\text{B}) + Et^{\cdot} \quad (95)$$

This complex is paramagnetic, and its organometallic nature is confirmed by the presence of metal hyperfine coupling ($^{67}Zn$, $I = 5/2$). Very few ethyl radicals escape the solvent cage, as witnessed by the lack of ethane or bibenzyl or other typical termination products in the product mixture. The ethyl radical reacts with the complex within the solvent cage, and the ethylmetal 3,6-di-*t*-butoxy-2-ethoxyphenoxide can be isolated in up to 90% yield.

$$Et^{\cdot} + \text{B} \longrightarrow \text{R–(t-Bu)-C}_6\text{H}_2\text{(OEt)(OZnEt)} \quad (96)$$

The classic inorganic oxidant hexachloroiridate(IV) undergoes electron transfer processes with organomercurials with great facility to liberate alkyl radicals according to the general equation 97[177].

$$R_2Hg \xrightarrow{[Ir^{IV}Cl_6]^{2-}} R_2Hg^+ \longrightarrow RHg^+ + R^{\cdot} \quad (97)$$

In acetic acid with spin-trapping agents present, strong e.s.r. spectra of the appropriate spin-trapped alkyl radicals may be conveniently observed. The mechanism for the reaction is via an inner-sphere electron transfer, which is thought to give a binuclear intermediate. This intermediate may then break down via a pair of mechanistically indistinguishable pathways, the whole scheme being outlined in reactions 98–102.

$$R_2Hg + [Ir^{IV}Cl_6]^{2-} \longrightarrow \underset{\mathbf{(31)}}{[R_2HgClIrCl_5]^{2-}} \quad (98)$$

$$\mathbf{31} \longrightarrow R_2Hg^{+\cdot} + [IrCl_6]^{3-} \quad (99)$$

$$R_2Hg^{+} \longrightarrow R^{\cdot} + RHg^{+} \quad (100)$$

$$\mathbf{31} \longrightarrow R^{\cdot} + \underset{\mathbf{(32)}}{[RHgClIrCl_5]^{2-}} \quad (101)$$

$$\mathbf{32} \longrightarrow \text{other products} \quad (102)$$

The selectivity of R in these reactions when $R_2Hg$ is an unsymmetric dialkyl derivative has helped to formulate the general unified view of charge transfer between inorganic oxidants or electron acceptors and labile alkylmetals, which is discussed more fully in Section VII. Significantly, the reaction has been used as an alternative method in the characterization of the spin-trapped cyclopentadienyl radical as a convenient and efficient route to cyclopentadienyl radicals (reaction 103)[38]. As a general alternative to photolysis for radical generation this route may yet find further general application.

$$(C_5H_5)_2Hg \xrightarrow{[Ir^{IV}Cl_6]^{2-}} C_5H_5^{\cdot} + (C_5H_5)Hg^{+} \text{ etc.} \quad (103)$$

## VI. GROUP III

The metal—carbon bond in Group III organometallic compounds is thermodynamically rather stable towards unimolecular decomposition as the first bond dissociation energies for the earlier members of the group are relatively high, although this value decreases uniformly down the group. There is, however, an extensive homolytic substitution chemistry for members of this group to which the major part of this section is devoted.

### A. Thermal Decompositions

Trialkyl- and triaryl-metal derivatives of Group III elements are thermally stable at ambient temperatures, but the gas-phase decompositions of these compounds, particularly the methyl derivatives, have been studied extensively. For the trimethylmetal derivatives of gallium, indium, and thallium, results obtained by pyrolysis in a toluene carrier flow system are consistent with a mechanism involving stepwise loss of methyl radicals[178].

$$Me_3M \longrightarrow Me_2M^{\cdot} + Me^{\cdot} \quad (104)$$

$$Me_2M \longrightarrow MeM\!: + Me^{\cdot} \quad (105)$$

$$MeM\!: \longrightarrow M + Me^{\cdot} \quad (106)$$

$$n\,MeGa\!: \longrightarrow (MeGa)_n \quad (107)$$

Methylgallium appears to polymerize rather than to decompose further to gallium and methyl radicals. The activation energies for the decompositions were found to be equal to the dissociation energies for the first metal—carbon bond cleavage. The decompositions of trimethylboron and trimethylaluminium appear to be more complex but in both cases the initial step seems likely to involve homolysis of the metal—carbon bond. Thermolysis of trimethylboron in the presence of $D_2$ gave deuteriomethane, probably formed by the $S_H2$ attack of $\alpha$-boronylalkyl radicals at boron[179].

The decomposition of higher boron and aluminium alkyls may also proceed via an elimination mechanism rather than the homolytic process[178,180]. Higher indium- and gallium-alkyls decompose by both homolytic and elimination routes, although long-chain alkylindium compounds prefer the homolytic pathway[181,182].

## B. Photolytic Decompositions

It might be expected that simple Group III organometallic compounds would be susceptible to photo-induced metal—carbon bond homolysis, especially in compounds of the heavier members of the group. However, relatively few studies have been made on this aspect of their chemistry. Care must be taken to ensure that radicals formed under photolytic conditions are not the result of the presence of radical chain carriers or other photolabile impurities, since Group III compounds react rapidly with a variety of radicals via $S_H2$ processes (below). For example, e.s.r. signals of methyl radicals detected during the photolysis of trimethylboron may arise from photolysis of small quantities of peroxide, with subsequent attack of the resulting oxyl radicals on the borane[183]. Nevertheless, u.v. photolysis of trimethylboron was found to give methane as a major product and a mechanism similar to the pyrolytic one was proposed[180].

$$B(CH_3)_3 \xrightarrow{h\nu} {}^{\bullet}CH_3 + {}^{\bullet\bullet}B(CH_3)_2 \qquad (108)$$

$${}^{\bullet}CH_3 + B(CH_3)_3 \longrightarrow {}^{\bullet}CH_2B(CH_3)_2 + CH_4 \qquad (109)$$

Photolysis of Me(Ph)NB(Ph)R (R = Et, $Pr^i$, or benzyl) in carbon tetrachloride at 35 °C gives products consistent with the formation of $R^{\bullet}$ via homolysis of the B—R bond[184]. The low quantum yield for the reaction and the lack of effect of conventional scavengers for trichloromethyl radicals on the quantum yield indicate that a radical chain process is not operative. Both the quantum efficiency for boron—alkyl cleavage and the chemical yield of $RCCl_3$ coupling products decreased in the order R = benzyl > *i*-propyl > ethyl, reflecting the relative stabilities of the respective alkyl radicals.

The photolytically induced homolysis of tri-1-naphthylboron (which has a charge-transfer maximum at 300 nm) was reported to compete with the formation of a monovalent organoboron carbene analogue[185]. Evidence for radical intermediates was obtained in carbon tetrachloride solution although it was not clear whether boron—carbon bond homolysis was the primary photoreaction or a result of $S_H2$ reactions.

An unusual variation on the ubiquitous $S_H2$ reaction was found in the photolysis of 2,2,4,4-tetramethylpentan-3-imine [$(Bu^t)_2C{=}NH$] in the presence of triethylboron[186]. E.s.r. signals of both ethyl radicals and ${}^{\bullet}C(Bu^t)_2CNHBEt_2$ were obtained, apparently paralleling the $S_H2$ reaction of triplet ketones with triethylboron. However, the process was sensitized, rather than desensitized, by triplet quenching agents (e.g. conjugated dienes) and the yield of radicals also increased as the temperature decreased. Direct photolysis of the borane—imine complex formed at low temperatures was implicated.

$$BEt_3 + (Bu^t)_2C{=}NH \rightleftharpoons \underset{\mathbf{(33)}}{(Bu^t)_2C{=}NHBEt_3} \qquad (110)$$

$$\mathbf{33} \xrightarrow{h\nu} Et^{\bullet} + {}^{\bullet}C(Bu^t)_2NHBEt_2 \qquad (111)$$

A recent study has revealed that photolysis of tetralkylcyclobutadienylaluminium chloride complexes in dichloromethane at −80 °C gives good e.s.r. spectra of cyclobutadienyl radical cations[187]. The complexes are formed by the reaction of dialkylacetylenes with aluminium chloride and both X-ray and n.m.r. spectroscopy have indicated that they are σ-complexes. The aluminium—carbon bond is presumably homolysed on photolysis.

$$2RC{\equiv}CR + AlCl_3 \longrightarrow [R_4C_4^{+}]\text{—}\bar{A}lCl_3 \xrightarrow{h\nu} [R_4C_4]^{\cdot +} + \cdot\bar{A}lCl_3 \quad (112)$$

These reactions provide a useful route to substituted cyclobutadienyl radicals which would be otherwise difficult to prepare, and it is interesting to compare this reaction with those of several cyclopentadienyl metallics (see below) as a route to annulene radicals. It must be added though that the cyclopentadienyl boron derivatives **34** and **35** are photostable[27]. N.m.r. evidence indicates that this is due to the bonding mode

$Cp_3B$ **(34)**  Cp—B(OCH$_2$CH$_2$O) **(35)**

of the cyclopentadienyl rings. In these boron derivatives it appears that the cyclopentadiene is bonded to the boron atom at an $sp^2$ carbon 1-$R_2BC_5H_5$; **36**), a bonding mode previously observed in the chloroboranecyclopentadienyl compound $Cl_2BCp$[188] and $Et_2BCp$[189].

Very little appears to be known about the role that homolysis plays in the photolysis of organoaluminium and organogallium compounds, but a study on trialkylindium compounds reported a homolytic mechanism for photoinduced decomposition[190].

The photolability of organothallium compounds has found some synthetic application. Photolysis of arylthallium(III)bis(trifluoroacetates) in benzene gave unsymmetrical biphenyls in good yields, and very small amounts of aryltrifluoroacetates. These results suggest that aryl radicals, generated by photoinduced homolysis of the thallium—carbon bond, add to the solvent or terminate by recombination with trifluoroacetyl radicals, formed by disproportionation of the unstable thallium(II) bis(trifluoroacetate)[191].

$$[ArTl^{III}(CF_3COO)_2] \xrightarrow{h\nu} Ar^{\cdot} + [Tl^{II}(CF_3COO)_2] \quad (113)$$

$$[Tl^{II}(CF_3COO)_2] \longrightarrow [Tl^{I}(CF_3COO)] + {}^{\cdot}OC(O)CF_3 \quad (114)$$

In the presence of cyanide ions in aqueous solution photolysis of the arylthallium(III)bis(trifluoroacetates) gave aromatic nitriles. When the cyanide concentration was reduced, photolysis of the phenyl thallium salt gave benzene, biphenyl, and other products typical of reactions of phenyl radicals[192]. The important aspect of these reactions is the stereospecific introduction of a substituent into the position in the aromatic ring previously occupied by the thallium. In the presence of iodide ion, formation of aryl iodides proceeds almost instantaneously without the need for photolysis[193].

### C. Radical Reactions

Bimolecular homolytic substitution ($S_H2$) reactions at metal centres makes an important contribution to the organometallic chemistry of the lighter members of Group III.

$$X^{\bullet} + MR_3 \longrightarrow XMR_2 + R^{\bullet} \quad (115)$$

Unfortunately, the susceptibility of indium and thallium to these reactions is difficult to assess because many conventional radical precursors react directly with the organometallic compounds of these elements. $S_H2$ reactions have been studied extensively for organoboron compounds. The substantial amount of kinetic data, obtained mostly from e.s.r. spectroscopy, for the reaction of a variety of radicals at boron has been central to a number of books and reviews mentioned previously[9,11,14].

The first $S_H2$ reactions studied extensively were those involving the autoxidation of organometallic compounds (see the Introduction), and both organoboron and organoaluminium compounds are autoxidized by this commonly encountered radical chain process where an alkylperoxy radical displaces an alkyl radical from the metal centre[7]. Other oxygen-centred radicals have been shown to react rapidly with organoboron compounds. Thus, photolysis of di-*t*-butyl peroxide[183] or acetone[194] in the presence of trialkylborons in the cavity of an e.s.r. spectrometer gives intense signals from alkyl radicals.

$$(Bu^tO)_2 \xrightarrow{h\nu} 2Bu^tO^{\bullet} \quad (116)$$

$$Bu^tO^{\bullet} + BR_3 \longrightarrow Bu^tOBR_2 + R^{\bullet} \quad (117)$$

$$Me_2C{=}O \xrightarrow{h\nu} Me_2\dot{C}{-}\dot{O} \quad (118)$$

$$Me_2\dot{C}{-}\dot{O} + BR_3 \longrightarrow {}^{\bullet}CMe_2OBR_2 + R^{\bullet} \quad (119)$$

Reference to Table 5, which gives kinetic parameters for *t*-butoxydealkylation at boron, demonstrates some particularly interesting features of $S_H2$ reactions. The rate of dealkylation decreases in the order $R = Bu^n > Bu^i > Bu^s >$ neopentyl, which suggests that steric crowding at the metal centre hinders the reaction. If the rate of reaction were to depend on the ease of formation of alkyl radicals then the order $R = Bu^s > Bu^n$ might be expected, not only because this is the order of stabilization

TABLE 5. Kinetic parameters for bimolecular homolytic *t*-butoxydealkylation at boron

| Borane | Method[a] | $T$ (°C) | $k$ ($dm^3\ mol^{-1}\ s^{-1}$) | $E_a$ ($kJ\ mol^{-1}$) | A ($dm^3\ mol^{-1}\ s^{-1}$) |
|---|---|---|---|---|---|
| $Bu^n{}_3B$ | 1 | 30 | $3\times10^7$ | 0 | $3.5\times10^7$ |
| $Bu^n{}_3B$ | 2 | 40 | $1\times10^7$ | 0 | $6.1\times10^6$ |
| $Bu^i{}_3B$ | 1 | 30 | $1\times10^6$ | 1.1 | $6.8\times10^6$ |
| $Bu^i{}_3B$ | 2 | 40 | $3\times10^5$ | — | — |
| $Bu^s{}_3B$ | 1 | 30 | $3\times10^5$ | 4.5 | $4.9\times10^8$ |
| $Bu^s{}_3B$ | 2 | 40 | $4\times10^5$ | 5.9 | $5.1\times10^9$ |
| $Me_3CCH_2)_3B$ | 2 | 40 | $4\times10^4$ | 0.7 | $1.1\times10^5$ |
| $PhCH_2)_3B$ | 2 | 25 | $1\times10^7$ | — | — |

[a] Methods 1 and 2 both involve competing *t*-butoxydealkylation with hydrogen abstraction from a suitable donor. If the rate of hydrogen abstraction is known, then relative rates of *t*-butoxydealkylation can be estimated. Method 1 involves an e.s.r. technique; method 2 uses a product analysis

of the alkyl radicals formed, but also because a sterically crowded four coordinate intermediate should collapse more readily. Similar trends have been observed in the autoxidation of organoborons, their reactions with acetone triplets[194], and also in $S_H2$ reactions at tin centres where *n*-butyl radicals are ejected in preference to *t*-butyl radicals in the reaction of *t*-butoxy radicals with $Bu^n(Bu^t)_2Sn^{IV}Cl$[28] (see below).

These observations raise the question as to whether these reactions proceed via a discrete intermediate or not.

$$X^{\bullet} + MR_3 \longrightarrow [X\dot{M}R_3]^{\neq} \longrightarrow XMR_2 + R^{\bullet} \quad (120)$$

$$X^{\bullet} + MR_3 \longrightarrow X\dot{M}R_3 \longrightarrow XMR_2 + R^{\bullet} \quad (121)$$

E.s.r. spectroscopy in solution has been unable to detect intermediates in these reactions, except in the case of phosphorus(III) compounds (see below). However, in a study of optically detected laser flash photolysis of triphenylboron with di-*t*-butyl peroxide the transient boranyl radical **37** was detected as the first intermediate[195].

$$Bu^tO^{\bullet} + Ph_3B \longrightarrow \underset{\mathbf{(37)}}{Ph_3(Bu^tO)B^{\bullet}} \longrightarrow Bu^tOBPh_2 + Ph^{\bullet} \quad (122)$$

The half-life of this species, 10 $\mu$s, is extremely short when compared with that of the corresponding phosphoranyl radical ($t_{1/2} = 900\ \mu s$), but the experiment was the first unambiguous demonstration that this *t*-butoxydephenylation is a stepwise process. Alkylboranyl radicals would be expected to have even shorter lifetimes.

The view that an intermediate is involved in these reactions is consistent with steric control of this process if the rate of the overall reaction is determined by the formation, rather than decomposition, of the boranyl intermediate. It has also been suggested that observations on reactions of aromatic ketone triplets with trialkylboranes might be best explained in terms of an intermediate triplet ketone–borane complex[196]. Significantly in the solid state at 77 K $\gamma$-irradiation of tetramethyltin(IV) leads to formation of among other radicals, the Sn(V) species $Me_5Sn^{\bullet}$, providing a more general interpretation of the stepwise process indicated above[39].

Unlike the simple trialkylborons, boroxines react with *t*-butoxy radicals in a manner which appears less dependent on steric effects, but more dependent on the stability of the displaced alkyl radical. However, the reaction with simple primary alkylboroxines is slower than for the corresponding boranes, presumably because boron–oxygen $\pi$-interactions are disrupted in the four-coordinate transition state or intermediate[197].

Intense e.s.r. spectra of alkyl radicals are also observed in the *t*-butoxydealkylation of trialkylaluminium and trialkylgallium compounds[183].

$S_H2$ reactions at Group III metals have been observed for a variety of different types of attacking radical. The reactions have been observed with radicals centred on oxygen, nitrogen, carbon, sulphur, and the halogens, and several other types of radical could possibly behave similarly. However, very few studies have appeared on the scope of this reaction since the early 1970s although a number of reactions involving $S_H2$ at boron have found synthetic use[198]. An example of this is found in the 1,4-addition of alkylboranes to enones, where hydrolysis of the product **38** gives the extended ketone.

$$R^{\bullet} + CH_2{=}CHCH{=}O \longrightarrow RCH_2CH{=}CHO^{\bullet} \quad (123)$$

$$RCH_2CH{=}CHO^{\bullet} + BR_3 \longrightarrow \underset{\mathbf{(38)}}{RCH_2CH{=}CHOBR_2} + R^{\bullet} \quad (124)$$

A similar radical chain process is found in the reaction of trialkylaluminiums with enones[199].

### D. Chemically Induced Homolytic Decomposition

A problem often encountered in studying $S_H2$ reactions of Group III, particularly the heavier members, is their ready direct reaction with, for example, peroxides to give radicals, which may then take part in $S_H2$ reactions at the metal centre. Thus, trialkylboranes react with hydrogen peroxide to give boronic acids, $RB(OH)_2$, as well as products derived from alkyl radicals, and at temperatures at which homolysis of the peroxide bond would not be expected[10].

$$R_3B + H_2O_2 \longrightarrow R_3\bar{B}\overset{+}{O}(H)OH \quad \textbf{(39)} \tag{125}$$

$$\textbf{39} \longrightarrow R^{\cdot} + R_2BOH + {}^{\cdot}OH \tag{126}$$

$${}^{\cdot}OH + R_3B \longrightarrow R_2BOH + R^{\cdot} \tag{127}$$

The driving force for this reaction, and presumably a contributing factor to the $S_H2$ reactions, is the strength of the Group III metal—oxygen bond. Consequently, the rapid reaction of trialkylboranes and alanes with acylperoxides is envisaged as involving the homolytic breakdown of a bis-organometal peroxide complex[13].

$$2AlR_3 + (PhCOO)_2 \longrightarrow (R_3AlO(Ph)CO)_2 \quad \textbf{(40)} \tag{128}$$

$$\textbf{40} \longrightarrow 2R_3AlO(Ph)CO^{\cdot} \quad \textbf{(41)} \tag{129}$$

$$\textbf{41} \longrightarrow 2R_2AlO(Ph)CO + 2R^{\cdot} \tag{130}$$

This reaction is almost instantaneous at temperatures down to −70 °C with the stoichiometry being 2:1 ($R_3Al$:peroxide) and the final products corresponding to reactions between alkyl radicals and dialkylaluminium benzoate.

Trialkylaluminiums[200] and trialkylthalliums[201] also react rapidly with di-*t*-butyl peroxide to give alkoxides and radical products. A mechanism involving decomposition of a complex could be invoked here also, but the reaction is also reminiscent of the proposed single electron transfer mechanism by which alkyllithium compounds are thought to react with di-*t*-butyl peroxide (Section II).

The reaction of organoaluminiums with *p*-quinones appears to involve a homolytic breakdown of an organoaluminium complex[202]. The reaction of monoalkylaluminium dichlorides with *p*-benzoquinone at −78 °C in diethyl ether gave *p*-alkoxyphenols and benzoquinone oligomers. E.s.r. evidence for an aluminoxysemiquinone was obtained and the mechanism in reactions 131–133 was proposed.

$$p\text{-Benzoquinone}\ \text{(A)} + RAlCl_2 \longrightarrow (A)\cdot Al(R)Cl_2 \quad \textbf{(42)} \tag{131}$$

$$\textbf{42} \longrightarrow p\text{-}Cl_2AlOC_6H_4O^{\cdot} + R^{\cdot} \quad \textbf{(43)} \tag{132}$$

$$\textbf{43} \longrightarrow p\text{-}Cl_2AlOC_6H_4OR \quad \textbf{(44)} \tag{133}$$

The reaction for trialkylaluminium compounds proceeds via a carbanionic mechanism, but here the chlorine substituents are thought to reduce the polarity of the aluminium—carbon bond, thereby favouring the homolytic over the heterolytic mechanism. Similar behaviour was observed in the reactions of $Et_3Al$, $Et_2AlCl$, and $EtAlCl_2$ with *p*-chloranil[203]. Ethyl radical termination products were obtained along with ethyl- and benzyl-toluene when toluene was used as solvent. Interestingly, despite the close analogy between these reactions and those for *o*-quinones with organometals, a charge-transfer mechanism, involving the organoaluminium and uncomplexed quinone, appears improbable here, based on the fact that the reaction between $Et_3(PhCH_2)N^+$ $Et\bar{A}lCl_3$ and *p*-chloranil did not give radical derived products, even though the ethylchloroaluminate anion should be a considerably stronger electron donor than $EtAlCl_2$ itself.

The reactions between triphenyl- and tribenzyl-aluminiums with *p*-quinones also provides evidence for free-radical intermediates[204,205] However, it appears that phenoxyaluminiums are not only much less reactive towards quinones but also less reactive with peroxides[13], seeming to parallel behaviour of the boron analogues of these compounds.

The ready homolytic decomposition of trialkylaluminium compounds using u.v. irradiation or peroxides has been used to effect in catalysis of free radical polymerizations[13,180]. Trialkylaluminiums also effectively catalyse the polymerization of olefins, in this case by the non-homolytic addition of the carbon—aluminium bond across the olefin double bond. The versatility of these organoaluminiums as polymerization catalysts also exemplifies the care that must be taken if organometallic homolysis is to be inferred from the ability of organometallics in general to induce polymerizations.

### E. Electron Transfer Reactions

Both ends of the spectrum of redox behaviour seem to be found in the reaction of Group III organometallics with copper(II) salts, reflecting the higher stability of high oxidation states in light main group metals and lower oxidation states in heavier main group metals. Trialkylboranes react with copper(II) halides at room temperature to give alkyl halides and copper(I) halides[206].

$$R_3B + 2Cu^{II}Br_2 + H_2O \xrightarrow{\text{thf}} RBr + R_2BOH + Cu^{1}{}_2Br_2 + HBr \quad (134)$$

The formation of olefins and alcohols when copper(II) halides were replaced with copper(II) acetate or sulphate indicated that alkyl radicals were formed in this reaction and gave products depending on the nature of the copper(II) salt, similar to the examples illustrated in Section III.A.2. The possible mechanisms have also been described earlier. In this case electron transfer from the borane to copper(II) was preferred because secondary alkyl derivatives were formed from unsymmetrical boranes in considerably greater yield than primary alkyl derivatives, reflecting the relative ease of formation of the corresponding radicals. Without attempting to draw any further conclusions about the details of the oxidation, the reaction may be represented by equations 135 and 136.

$$R^1R^2{}_2BFOH_2 + Cu^{II}X_2 \longrightarrow [R^1R^2{}_2BOH_2]^{+\cdot} + [Cu^{I}X_2]^- \quad (135)$$

$$[R^1R^2{}_2BOH_2]^{+\cdot} \longrightarrow R^1R^2BOH + R^{2\cdot} + H^+ \quad (136)$$

If, alternatively, an alkylcopper(II) intermediate were involved, preference for primary alkyl derivatives would be expected as judged by the trend in migratory aptitudes in the reactions of trialkylboranes with mercuric acetate, i.e. primary > secondary.

In contrast, stereoregular organothallium(III) compounds reacted with copper(I) chloride in acetonitrile to give mixtures of isomeric products in which the thallium had been replaced with chlorine[207]. At the same time, thallium(III) was reduced to thallium(I). At 80 °C the *erythro* : *threo* ratio was 1 : 2 whereas at 60 °C it was 1 : 10, seeming to indicate that an ionic mechanism was responsible for inversion of

$$\underset{erythro}{\text{MeO(D)(H)C(R)(H)–C(Tl(OAc)}_2)} \xrightarrow[\text{KCl}]{\text{CuCl}} \underset{erythro}{\text{MeO(D)(H)(R)(H)C–C(Cl)}} + \underset{threo}{\text{MeO(Cl)(H)(R)C–C(D)(H)}} + [\text{Tl(OAc)}_2]^- \qquad (137)$$

configuration at the lower temperature, with a radical mechanism giving configurational scrambling at higher temperatures. The involvement of radicals was supported by e.s.r. spin-trapping studies, and suggested that radicals were being formed by decomposition of a labile thallium(II) complex formed by reduction of the thallium(III) complex by copper(I).

$$\underset{(\mathbf{45})}{\text{R(MeO)CHCHDTl}^{\text{II}}\text{(OAc)}_2 + \text{Cu}^{\text{I}}\text{X}_2^-} \longrightarrow \underset{(\mathbf{46})}{[\text{R(MeO)CHCHDTl}^{\text{III}}\text{(OAc)}_2^-} + \text{CuX}_2 \qquad (138)$$

$$(\mathbf{47}) \longrightarrow \text{R(MeO)CH}\dot{\text{C}}\text{HD} + [\text{Tl}^{\text{I}}\text{(OAc)}_2]^- \qquad (139)$$

This mechanism is given some support from the behaviour of other organothallium(III) compounds with other reductants. Thus, reaction of optically pure **45** (R = Ph) with ascorbic acid in acetonitrile gave products in which racemization had occurred[208]. When undeuteriated **45** was used, inclusion of perdeuterionitrosodurene gave the e.s.r. spectrum of the radical **48**.

$$\underset{(\mathbf{48})}{2,3,5,6\text{-(CD}_3)_4\text{C}_6\text{HN(O}^\bullet\text{)CH}_2\text{CH(MeO)(Ph)}}$$

The reaction of similar organothallium(III) compounds with *N*-benzyl-1,4-nicotinamide in methanol at room temperature gave good yields of hydrodethallated products[209], again with spin-trapping studies indicating the participation of free radicals. In all these examples, electron transfer to the thallium(III) compound and subsequent homolytic decomposition of the thallium(II) intermediate was thought to provide a reasonable mechanistic interpretation.

Triethyl-Group(III) compounds react with sterically hindered *o*-quinones to give ethoxy phenols in 70–90% yield[210]. While the non-polar mechanism mentioned above is a possibility, the reaction is similar to that of diethylcadmium and diethylzinc with similar substrates, where the charge transfer mechanism was proposed (Section V.C). It is only recently that e.s.r. studies have confirmed the presence of the organoaluminium intermediates in the case of the reaction between $Et_3Al$ and 3,6-di-*t*-butyl-1,2-benzoquinone[211].

## VII. GROUP IV

The homolytic reactions of Group IV have attracted probably the most attention of all other groups under discussion in this chapter. Organosilicon radical chemistry has been extensively studied, although surprisingly the number of reactions in which direct homolysis of silicon—carbon bonds occurs is very few, with many silicon radicals being formed by abstraction of hydrogen from alkylsilanes. Germanium compounds tend to behave similarly.

$$\text{X}^\bullet + \text{R}_3\text{SiH} \longrightarrow \text{XH} + \text{R}_3\text{Si}^\bullet \qquad (140)$$

Organotin and lead compounds are well known to undergo many types of free-radical reactions and while these have been reviewed at length, the past few years have seen some very significant developments in these areas. Therefore, in the subsequent section, these recent developments in (a) the unimolecular cleavage of tin—carbon bonds to give a variety of new tin radicals (b) the reactivity of tin radicals produced in this way, (c) new examples of $S_H2$ reactions at tin, and (d) reactions of tin and lead compounds with electron acceptors will be discussed in depth. As will be seen, the first two of these topics give valuable insight into radical reactivity as a whole and the third is arguably the most important development in the free radical chemistry of Group IV elements. Using alkyltin and lead compounds a unified view of the whole range of previously supposed unrelated aspects of electron-transfer reactions, Marcus theory for outer sphere inorganic processes, inner sphere inorganic processes, and Mulliken charge-transfer organic processes has been made possible. The most recent developments in this field have even led to the selective substitution of some classical inorganic acceptors via outer-sphere electron-transfer mechanisms which involve the intermediacy of free alkyl radicals serived from the tin substrates.

### A. Thermal Decompositions

Organosilicon and organogermanium compounds have rather high bond dissociation energies and thus are not expected to undergo thermal decomposition via metal—carbon bond homolysis under anything less than high-temperature pyrolysis conditions. Even at elevated temperatures homolysis, when it occurs, is not selective and many products are often obtained. In fact, reports of pyrolysis of tetramethylsilane have demonstrated the presence of up to 46 different compounds in the product mixture. As a result, many reports of pyrolyses of alkylsilanes, germanes, and stannanes cannot be subjected to reasonable comparison. The reactions are affected by surface effects, carriers, temperature, and pressure.

A recent account of the pyrolyses of tetramethyl-silicon, -germanium, and -tin in a wall-less reactor[212] outlines many of the difficulties arising in attempting to obtain a unified mechanistic view. Even tetraalkyllead compounds do not decompose readily at 120–140 °C. A CIDNP study has shown that the decomposition of tetraethyllead is solvent dependent, and after appreciable times at 120–140 °C little decomposition occurs in solvents such as decalin or bromobenzene[213]. However, decomposition readily occurs at these temperatures in high-boiling chlorinated solvents such as hexachlorocyclopentadiene or hexachloroacetone. Unfortunately, though, this decomposition does not proceed via simple lead—carbon homolysis, but is thought to involve a concerted bimolecular reaction between the lead compound and the solvent to give a radical pair which may then react in several ways to give the observed n.m.r. product polarizations.

$$Et_4Pb + RCl \longrightarrow Et_3PbCl + R^\bullet + Et^\bullet \tag{141}$$

While the reaction could be construed as involving an initial electron-transfer step (see below) there is little direct evidence to suggest this. The thermal degradation of triphenyllead derivatives has been studied, but the precise nature of the mechanism is not clear[214].

### B. Photolytic Decompositions

Again, the high dissociation energies of the silicon—carbon and germanium—carbon bonds preclude photoinduced homolysis under normal conditions. When short-wave (147 nm) u.v. light is used in the gas-phase photolysis of tetramethylsilane, homolytic decomposition is observed however. Unfortunately, the process is not discrete, like the thermolyses, and at least 16 products are isolated in addition to unidentified polymeric materials[215]. Quenching with nitric oxide provided at least some insight into the nature of the major silicon-containing products, most of these being derived from reactions of $Me_3Si^{\bullet}$ or $Me_3SiCH_2^{\bullet}$. As many as six primary photoreactions were found to be present, although the previously mentioned major product distribution may well arise from reactions 142 and 143.

$$Me_4Si \longrightarrow Me_3Si^{\bullet} + Me^{\bullet} \quad (142)$$

$$Me_3Si^{\bullet} + Me_4Si \longrightarrow Me_3SiH + Me_3SiCH_2^{\bullet} \quad (143)$$

The photolysis of sterically hindered Group IV organometallics **49**, where M may be germanium, tin, or lead, gives homolysis of the metal—carbon bond. Subsequent reaction of the bis(trimethylsilyl)methyl radical at the metal centre of a further unphotolysed molecule gives the extremely persistent radicals **50**[216].

$$\underset{\textbf{(49)}}{[(Me_3Si)_2CH]_2M^{II}} \longrightarrow [(Me_3Si)_2CH]M^{I\bullet} + Me_3Si_2CH^{\bullet} \quad (144)$$

$$(Me_3Si)_2\dot{C}H + \textbf{49} \longrightarrow \underset{\textbf{(50)}}{[(Me_3Si)_2CH]_3M^{III}}. \quad (145)$$

It must be mentioned that the fate of the univalent metal species is uncertain. While it is true that these reactions provide unimolecular routes to some structurally exceptional organometallic radicals, it has very little application in assessing the reactivity of Group IV metal-centred radicals as a whole. However, recently general routes have been discovered to novel organotin radicals which allows the reactivity of many such radicals to be assessed.

Previously, in order to study the reactions of organotin radicals the methods of generation of these radicals outlined below have been utilized.

$$R_3SnSnR_3 \xrightarrow{h\nu} 2R_3Sn^{\bullet} \quad (146)$$

$$Bu^tO^{\bullet} + R_3SnSnR_3 \longrightarrow Bu^tOSnR_3 + R_3Sn^{\bullet} \quad (147)$$

$$Bu^tO^{\bullet} + R_3SnH \longrightarrow Bu^tOH + R_3Sn^{\bullet} \quad (148)$$

$$R_3SnN(Pr^i)_2 \xrightarrow{h\nu} Pr^i_2N^{\bullet} + R_3Sn^{\bullet} \quad (149)$$

Thus, direct photolysis of hexaalkylditins gives trialkylstannyl radicals, although the quantum yield for this reaction is low, mainly owing to the rapid rate of recombination of the tin radicals (when R is not a bulky substituent, the rate constants are near those normally encountered for self-reaction of radicals under diffusion control, i.e. *ca.* $2 \times 10^9$ l mol$^{-1}$ s$^{-1}$). The yield of trialkylstannyl radicals may be considerably improved when *t*-butoxy radicals (from photolysis of di-*t*-butyl peroxide) are present owing to the effective competition of the $S_H2$ process outlined in equation 147[217]. For the study of reactions involving free-radical chain processes, equation 148 is employed, and photolysis of aminotin compounds also gives acceptable yields of organotin radicals[15]. It must be noticed that only $R_3Sn^{\bullet}$ species have been previously available for study because no other radicals, for example $R_2XSn^{\bullet}$, have been readily accessible.

It was first demonstrated in 1978 that cyclopentadienyl derivatives of tin(IV) compounds could be readily photolysed in solution in the cavity of an e.s.r. spectrometer to give strong spectra of the cyclopentadienyl radical[218].

$$R_3Sn(\eta^1{-}C_5H_5) \longrightarrow C_5H_5^{\bullet} + R_3Sn^{\bullet} \quad (150)$$

While the tin radicals could not be detected directly, owing to an interesting physical property of the cyclopentadienyl radical (viz. a reluctance to saturate with increased microwave power), the solutions gave reactions very typical of those containing tin(III) radicals; addition to alkenes, abstraction of halogen, etc., all of which could be identified and characterized by e.s.r. spectroscopy. Significantly, the photolysis could be performed not only on compounds giving $R_3Sn^{\bullet}$ radicals, $(C_5H_5)_4Sn$ and $Bu^n{}_3SnC_5H_5$, but also on $Bu^n{}_2Sn(C_5H_5)_2$ and $(C_5H_5)_3SnCl$, which would obviously lead to the asymmetric tin radicals $Bu^n{}_2(C_5H_5)Sn^{\bullet}$ and $(C_5H_5)_2ClSn^{\bullet}$. These conclusions were confirmed when a subsequent study demonstrated the generality of the reaction for the compounds $R_{3-n}Cl_nSnC_5H_5$, photolysis of which gave a series of radicals $R_{3-n}Cl_nSn^{\bullet}$ which enabled several factors governing reactivity to be studied conveniently for the first time[219].

The presence of these chloroalkyltin radicals could be confirmed by observation of their reaction with butane-2,3-dione (biacetyl) in the e.s.r. spectrometer[31]. Photolysis of dichlorobutyltincyclopentadiene and trichlorotincyclopentadiene in toluene solution in the presence of biacetyl at low temperatures gave similar e.s.r. spectra corresponding to the interaction of one chlorine atom with two non-equivalent methyl groups. At higher temperatures the compounds showed well defined hyperfine coupling of two equivalent chlorine atoms and two equivalent methyl groups, and a complicated pattern thought to arise from three equivalent chlorine atoms and two equivalent methyl groups, respectively.

These results could be interpreted as arising from the radicals **51** and **52**. In these radicals, at low temperature, interaction from only one chlorine in the rigid 5-

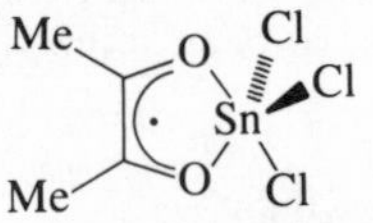

Me Me O O Sn Cl Cl Cl

**(51)**

Me Me O O Sn Bu$^n$ Cl Cl

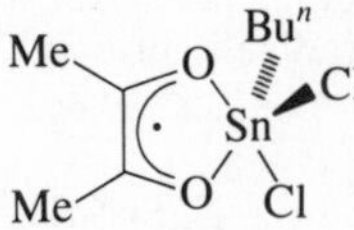

**(52)**

coordinate structure is expected, namely the chlorine in the apical position of the pentagonal bipyramidal structure. At higher temperature rapid pseudorotation renders both the chlorine atoms and methyl groups equivalent on the e.s.r. time scale. The results were confirmed for the dichloro derivative by the generation of the same radical from reaction 151 by proton abstraction of the indicated hydrogen atom by

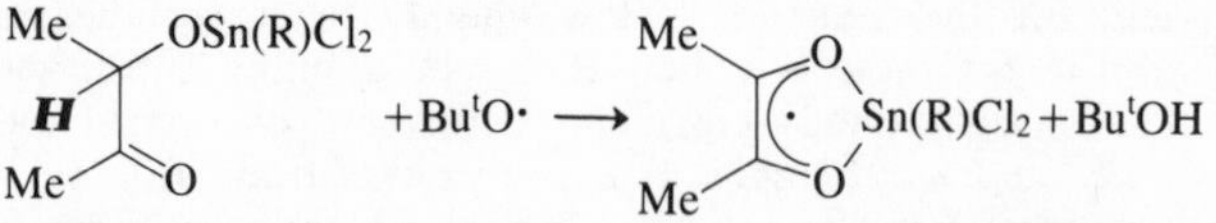

$$\text{Me(}\boldsymbol{H}\text{)C(OSn(R)Cl}_2\text{)C(=O)Me} + Bu^tO^{\bullet} \longrightarrow \text{[Me}_2C_2O_2^{\bullet}]Sn(R)Cl_2 + Bu^tOH \quad (151)$$

*t*-butoxy radicals. Careful analysis of the e.s.r. spectra for the corresponding monochlorotin radical $R_2ClSn^{\bullet}$ and the trialkyltin radical $R_3Sn^{\bullet}$ adducts indicated a *cis*-monodentate structure and a rapidly fluxional monodentate structure, respectively.

Significantly, the reactivities of the tin radicals towards the commonly encountered reactions of Group IV metal radicals were different. Thus, while the tri-*n*-butylstannyl radical readily adds to alkenes and abstracts bromine from alkyl bromides, the monochlorodibutylstannyl radical does not add to alkenes but is able to abstract bromine from alkyl bromides, and the dichlorobutylstannyl radical and trichlorostannyl radical perform neither of these reactions. This order of reactivity could be explained satisfactorily using simple frontier molecular orbital theory, where the consequences of the interaction of the radical SOMO with the substrate HOMO or LUMO are considered. It can be shown particularly well in these cases that alteration of the radical SOMO with respect to the substrate frontier orbitals is able to change the radical reactivity dramatically in an expected and accountable manner[220].

In a more recent study[34] the same method of tin radical generation has been used in the study of the effect of having the cyclopentadienyl group as a ligand on the properties and reactions of tin(III) radicals. Thus, photolysis of a variety of cyclopentadienylalkyl derivatives was carried out to give the corresponding tin radicals.

$$Bu^n_{4-n}(C_5H_5)_nSn \xrightarrow{h\nu} Bu^n_{4-n}(C_5H_5)_nSn^{\bullet} + C_5H_5^{\bullet} \quad (152)$$

Like the correspondingly substituted chlorotin radicals, only the monocyclopentadienylstannyl radical abstracts bromine from alkyl bromides, with the reactivity of the tin radical decreasing with increase in number of cyclopentadienyl ligands. The e.s.r. parameters of the radical adducts with biacetyl suggest a similarity in the effect of the cyclopentadienyl ligand towards reactivity to that of chlorine. The suggestion that electronic effects of the cyclopentadienyl ligand may be responsible for the observed radical reactivities was provided by other physical data, namely Mössbauer and n.m.r. spectroscopy. The Mössbauer spectra of cyclopentadienyltrialkyltin(IV) compounds shows quadrupole splitting (like trialkyltin chlorides), unlike the simple alkyl derivatives.

As can be seen, photoinduced cleavage of the tin—cyclopentadiene bond provides a valuable insight into radical reactivity and the reaction has much more as yet unfulfilled potential for study. It is perhaps significant that trialkyltinmethylcyclopentadienyl derivatives are not subject to efficient homolytic cleavage of the tin—cyclopentadiene bonds under the same conditions as the unsubstituted derivatives[27]. However, this problem was overcome when substituted cyclopentadienyl radicals were required for study by using the ready ability of the bis(cyclopentadienyl)mercury(II) derivatives to undergo unimolecular photolysis[27,163].

In a similar fashion triorgano(cyclopentadienyl)-lead(IV) compounds are readily photolysed to give plumbyl radicals and cyclopentadienyl radicals[221].

$$Ph_3Pb(C_5H_5) \xrightarrow{h\nu} Ph_3Pb^{\bullet} + C_5H_5^{\bullet} \quad (153)$$

To this effect the e.s.r. spectrum of the cyclopentadienyl radical produced in this reaction is shown in Figure 4.

Tetraalkyllead(IV) compounds also undergo direct unimolecular photolysis to alkyl radicals and lead(III) radicals[217]. Thus, photolysis of tetraethyllead gave unimolecular cleavage of the lead—carbon bond and e.s.r. spectra of ethyl radicals were recorded.

$$Et_4Pb \longrightarrow Et_3Pb^{\bullet} + Et^{\bullet} \quad (154)$$

The presence of the lead-centred radicals was confirmed by repeating the photolysis in the presence of allyl bromide, whence e.s.r. spectra of the allyl radical were observed.

$$Et_3Pb^{\bullet} + CH_2{=}CH{-}CH_2Br \longrightarrow \overset{\cdots\cdots\cdot\cdots\cdots}{CH_2CHCH_2} + Et_3PbBr \quad (155)$$

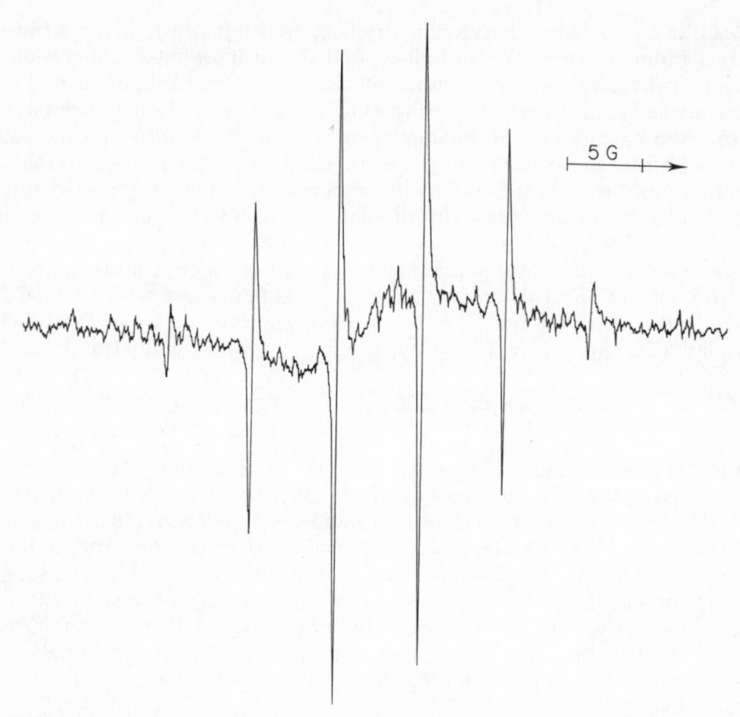


FIGURE 4. E.s.r. spectrum of the cyclopentadienyl radical from the photolysis of $CpPbPh_3$ in toluene at −65 °C. From ref. 221

### C. Radical Reactions

Although the bond energies of alkyl-silanes and -germanes tend to be higher than those of the corresponding tin and lead compounds, in principle they should be more willing to undergo $S_H2$ reactions than saturated carbon centres because of the availability of *d*-orbitals for interaction with the incoming radicals. However, there are still relatively few reports of such processes occurring at these metal centres. In general, abstraction of hydrogen from $\alpha$- or $\beta$-carbon atoms is preferred to substitution at the metal. In illustration, an early report of the reaction of tetramethylgermane with di-*t*-butyl peroxide at 130 °C gave triethyl(*t*-butoxy)germane on product analysis[222]. This, of course, would be the expected product of $S_H2$ at germanium. However, because ethylene was formed in good yields and no ethane was observed, the possibility of ethyl radicals being present as intermediates was rejected. The mechanism was adequately explained by a $\beta$-hydrogen abstraction followed by decomposition of the intermediate $\beta$-germylethyl radical. The subsequent triethyl germyl radical is then able to perform $S_H2$ at oxygen of another peroxide molecule to give the observed product and propagate the radical chain.

$$Bu^tO^{\bullet} + Et_4Ge \longrightarrow Bu^tOH + Et_3GeCH_2CH_2^{\bullet} \tag{156}$$

$$Et_3GeCH_2CH_2^{\bullet} \longrightarrow Et_3Ge^{\bullet} + CH_2{=}CH_2 \tag{157}$$

$$Et_3Ge^{\bullet} + Bu^tOOBu^t \longrightarrow Et_3GeOBu^t + Bu^tO^{\bullet} \tag{158}$$

In the case of the corresponding silane, dimers of the type $(Et_3SiCH_2CH_2)_2$ were formed, providing more evidence for this type of process[223].

Whereas in tin, lead, and mercury analogues the cyclopentadienyl ligand has been shown to induce radical reactions, the corresponding silicon and germanium derivatives show marked reluctance to undergo homolysis of the metal—cyclopentadiene bond whether unimolecularly (as for Sn, Pb, and Hg derivatives) or bimolecularly (in the case of substituted cyclopentadienyl derivatives of tin—see below). In fact, the ability of these compounds to react with *t*-butoxy radicals at the saturated $\alpha$-carbon atom has been used in studies of organometallic substituted cyclopentadienyl radicals[224].

$$5\text{-}R_3MC_5H_5 + Bu^tO^{\bullet} \longrightarrow R_3MC_5H_4^{\bullet} + Bu^tOH \qquad (159)$$

The e.s.r. spectra of these radicals are accountable in the same manner, by using Hückel molecular orbital perturbation theory, as the alkyl cyclopentadienyl radicals mentioned in Section V. The abilities of organogermanium and silicon compounds to undergo bimolecular substitution reactions will not be discussed further here.

### D. Recent Studies of $S_H2$ at Tin

As alluded to previously, homolytic substitution reactions have been studied more extensively for the organic compounds of tin than for any other metal. There are many factors which govern the homolytic substitution reactions at tin centres. Whether the course of a particular reaction favours attack at tin or attack at hydrogen in an alkyl ligand may depend on the relative steric demands of the available processes, the stabilization of side-chain radicals, electronegativity considerations, or other factors. As a consequence, a wide range of reactivity towards a variety of attacking radicals is observed. As an example, *t*-butoxyl radicals generated in the presence of tetralkyltins tend to abstract protons from the alkyl ligand to give $\alpha$- or $\beta$-stannyl substituted alkyl radicals, but under similar conditions alkyltin halides or trialkyltin carboxylates undergo attack at tin to give alkyl radicals or $\beta$-carboxyalkyl radicals, respectively[225]. Although prior to 1976 many examples of $S_H2$ were found, few of the factors governing this difference in reactivity were well understood.

As mentioned above, the normal reaction for *t*-butoxy radicals with, for example, a $\beta$-carbonylalkyltin trichloride proceeds via $S_H2$ at tin to give e.s.r. spectra of the $\beta$-carboxyalkyl radical. However, unlike dialkyltin dichlorides, where usually stronger e.s.r. signals of the alkyl radical are detected than for the monoalkyltin trichlorides, no detectable amounts of $\beta$-carbonylalkyl radicals were obtained from bis-$\beta$-carboxyalkyltin dichlorides[226]. These observations were attributed to the fact that Mössbauer and n.m.r. studies showed that both in solution and in the solid state the complexes are 6-coordinate through chelation by the $\beta$-carbonyl substituent. Consequently, the concentration of 4-coordinate tin in solution is always too low to allow rapid $S_H2$.

Some interesting results were obtained with stannacycloalkanes, which are known to exhibit enhanced reactivity in heterolytic reactions[227,228], but their unusual sensitivity to air indicates a high reactivity to alkoxy and alkylperoxy radicals which facilitates study of homolytic reactions. Thus, in an extensive study of homolytic reactivities of these compounds it was found that *t*-butoxyl, trimethylsiloxyl, benzoyloxyl, and phenylthiyl radicals all react with dialkylstannacyclopentanes to give ring-opened products, the ring-opened radical **54** being observed by e.s.r. spectroscopy[229].

$$\underset{}{Bu^tO^{\bullet}} + \underset{\mathbf{(53)}}{1,1\text{-}R_2\text{-stannacyclopentane}} \longrightarrow \underset{\mathbf{(54)}}{Bu^tO(R)_2Sn(CH_2)_3\dot{C}H_2} \qquad (160)$$

Significantly, the rate constants for the reaction were found to be $1\times10^6$ l $mol^{-1}$ $s^{-1}$ for R=Me and $5.5\times10^5$ l $mol^{-1}$ $s^{-1}$ for R=$Bu^n$ at 213 K, which are nearly two orders of magnitude faster than normal $S_H2$ reactions. A further mechanistic confirmation was made possible by the thermally initiated reaction between phenylthiol and **53** (R=$Bu^n$). At 35 °C, initiated by *t*-butyl hyponitrite, and at 60 °C initiated by AIBN, a smooth chain reaction took place to yield tributyl(phenylthio)tin derivative.

$$RS^{\bullet} + \mathbf{53} \longrightarrow RS(Bu^n)_2Sn(CH_2)_3\dot{C}H_2 \quad (161)$$

$$RS(Bu^n)_2Sn(CH_2)_3\dot{C}H_2 + RSH \longrightarrow (RS)Sn(Bu^n)_3 + RS^{\bullet} \quad (162)$$

The general reaction lends much weight to the 5-coordinate transition state argument proposed for $S_H2$ reactions at tin. By analogy with cyclic phosphorus compounds the stannacyclopentane ring would be expected to be bound to apical and equatorial positions in the trigonal bipyramidal intermediate and thus the preferred cleavage of the apical tin—carbon bond (as in $XPR_3$—see Section VIII) gives the observed product radical, and no $R^{\bullet}$. Stannacyclo-hexanes and -heptanes also show similar behaviour.

$$X^{\bullet} + \text{(1,1-}R_2\text{-stannacyclopentane)} \longrightarrow X\cdots Sn(R_2)\cdots\cdots \longrightarrow X{-}SnR_2(CH_2)_3\dot{C}H_2 \quad (163)$$

Interestingly, the compound with R=$Bu^t$ under similar conditions showed no e.s.r. spectra at all in reaction with *t*-butoxyl radicals. However, when ethyl bromide was incorporated, strong signals of the ethyl radical were observed. This result was explained by the fact that not only would the *t*-butyl groups protect the tin centre but also the $\alpha$-methylene position, therefore directing attack towards the $\beta$-methylene protons. Abstraction from the $\beta$-methylene position would thus yield a $\beta$-metallated alkyl radical, and these are well known to undergo rapid $\beta$-scission so that a tin-centred radical **55** would result. This, of course, would be difficult to detect by e.s.r. under the conditions of the experiment, but would readily abstract bromine from ethyl bromide to give the observed spectrum.

$$Bu^tO^{\bullet} + 1,1{-}Bu^t_2{-}\text{stannacyclopentane} \longrightarrow Bu^t_2Sn\text{(ring, 3-yl radical)} + Bu^tOH \quad (164)$$

**(55)**

$$\mathbf{55} \longrightarrow Bu^t_2\dot{S}nCH_2CH_2CH{=}CH_2 \quad (165)$$

**(56)**

$$\mathbf{56} + EtBr \longrightarrow Bu^t_2Sn(Br)CH_2CH_2CH{=}CH_2 + Et^{\bullet} \quad (166)$$

Another major application of the ability of organotin compounds to undergo $S_H2$ reactions has been in the generation of substituted cyclopentadienyl radicals. From previous sections (V, VII.B, and VII.C) it was noted that bis(cyclopentadienyl)mercury(II) compounds afforded a simple unimolecular route to substituted cyclopentadienyl radicals, and while unsubstituted cyclopentadienyltin(IV) compounds could undergo unimolecular photolysis to give cyclopentadienyl radicals and novel tin radicals, their substituted cyclopentadienyl analogues could not[27]. However, a bimolecular reaction between trialkyltin(alkylcyclopentadienyl) compounds has been used in other studies of cyclopentadienyl radicals[230].

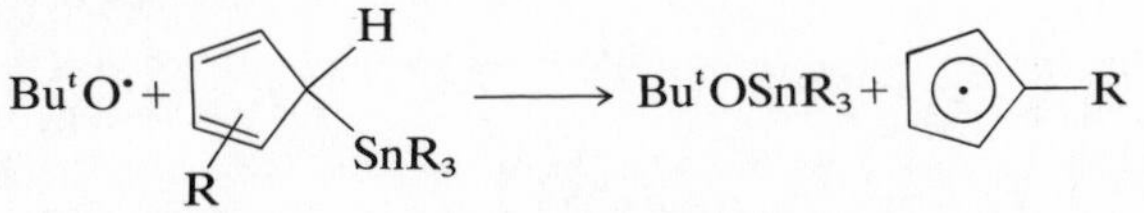

$$Bu^tO^\bullet + (RC_5H_4)SnR_3 \longrightarrow Bu^tOSnR_3 + (C_5H_4^\bullet)\text{—}R \quad (167)$$

One of the major industrial uses of compoununds of the type $R_2SnX_2$ is as a stabilizer of poly(vinyl chloride) (PVC) It is thought that the thermal degradation of PVC, while ultimately complex and possibly involving a number of reaction pathways, may occur via a radical pathway, involving chain-carrying chlorine atoms[231]. In this type of mechanism, the tin compounds may function as chlorine atom traps in simple $S_H2$ processes, given the ability of alkyltin carboxylates to undergo this type of reaction. In a model reaction involving the photolysis of carbon tetrachloride in the presence of di-*n*-butyltin diacetate the following mechanism was proposed from the product study.

$$CCl_4 \xrightarrow{h\nu} Cl^\bullet + {}^\bullet CCl_3 \quad (168)$$

$$Cl^\bullet + Bu_2Sn(OAc)_2 \longrightarrow Bu_2Sn(OAc)Cl + AcO^\bullet \quad (169)$$

$$Cl^\bullet + Bu_2Sn(OAc)Cl \longrightarrow Bu_2SnCl_2 + AcO^\bullet \quad (170)$$

$$AcO^\bullet + Bu_2Sn(OAc)_2 \longrightarrow AcOH + CH_3CH_2\dot{C}HCH_2Sn(OAc)_2Bu \quad (171)$$

$${}^\bullet CCl_3 + Bu_2Sn(OAc)_2 \longrightarrow CHCl_3 + CH_3CH_2\dot{C}HCH_2Sn(OAc)_2Bu \quad (172)$$

$$CH_3CH_2\dot{C}HCH_2Sn(OAc)_2Bu \longrightarrow CH_3CH_2CH{=}CH_2 + Bu\dot{S}n(OAc)_2 \quad (173)$$

$$2Cl_3C^\bullet \longrightarrow Cl_3CCCl_3 \quad (174)$$

## E. Electron Transfer Reactions

There is no doubt that the most rapid expansion in the homolytic reactions of organotin and lead compounds has recently been in studies of reaction between organotin compounds and electron acceptors. These reactions may often be monitored in the e.s.r. cavity to give either structural information on the organometallic fragments or merely to confirm the presence of alkyl radicals which may be used to perform alkylation of particular substrates.

In common with derivatives of Group II and III organotin compounds react with *o*-quinones via electron transfer to give persistent organotin-*o*-semiquinolate radicals and alkyl radicals. In this respect, 3,6-di-*t*-butyl-1,2-benzoquinone has proved very useful as a trap for organotin radicals[34,232]. It must be noted that unlike reactions discussed earlier between organotin radicals and biacetyl, these reactions occur thermally on mixing the reactants and photolysis is not required. Thus, when cyclopentadienyltin trichloride is mixed with the quinone in toluene solution at −45 °C the e.s.r. spectrum in Figure 5 is observed, derived from radical **56** in equation 176. The cyclopentadienyl radical is not observed under these conditions.

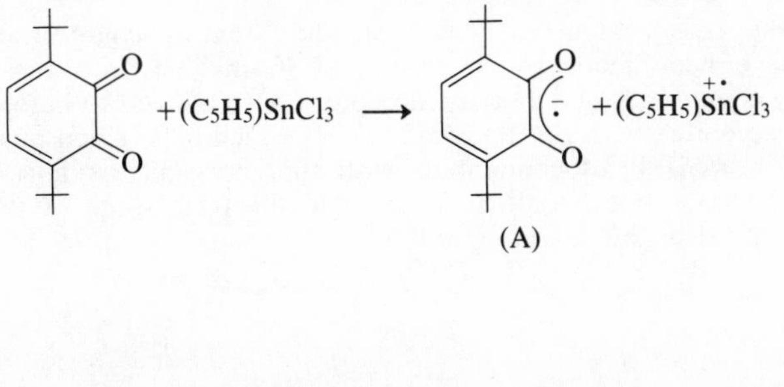

$$\text{3,6-di-}t\text{-butyl-1,2-benzoquinone} + (C_5H_5)SnCl_3 \longrightarrow \text{semiquinone radical anion} + (C_5H_5)\overset{+\bullet}{Sn}Cl_3 \quad (175)$$

(A)

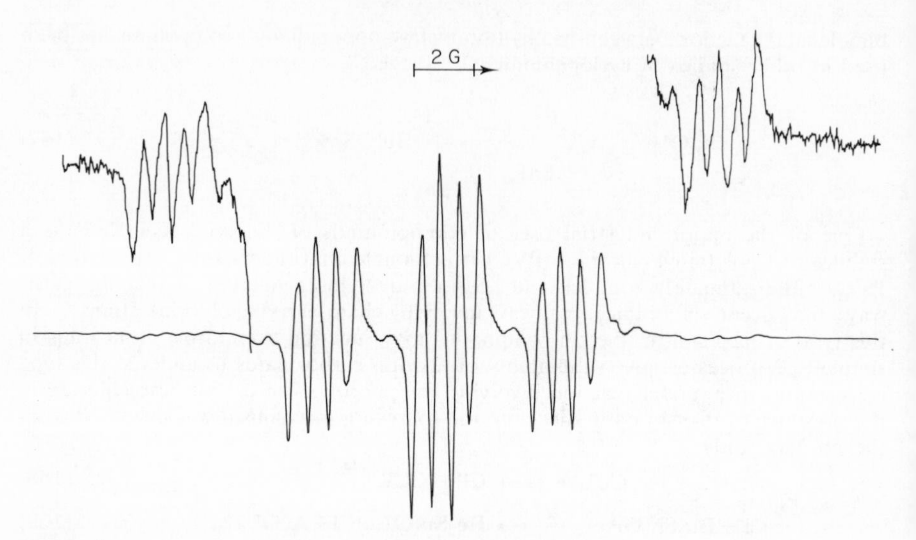


FIGURE 5. E.s.r. spectrum of the $Cl_3Sn$ derivative of 3,6-di-*t*-butyl-*o*-semiquinone from mixing $CpSnCl_3$ and the quinone in toluene at −45 °C. Insets are $^{119/117}Sn$ satellites. From ref. 221

$$A + \overset{+}{Sn}Cl_3 + (C_5H_5)^{\cdot} \longrightarrow \textbf{(56)} \qquad (176)$$

**(56)**

The spectrum, showing hyperfine coupling apparently from two equivalent protons and one chlorine atom, suggests that a bidentate cyclic structure similar to **52** for biacetyl is present, again with only one apical chlorine atom coupling. The tin satellite lines are well resolved and may be used to estimate the degree of interaction between the tin nucleus and the unpaired electron. As in the biacetyl case, the monochlorodibutyltin radical acts as a monodentate ligand, fixed at low temperatures and fluxional at higher temperatures. For comparison, the rate constants for the exchange process have been estimated to be $2.8 \times 10^6\ s^{-1}$ at 333 K for $Bu^n{}_2\dot{S}nCl$[34] and $2.5 \times 10^6\ s^{-1}$ at the same temperature for $Me_2SnCl$[232].

Perhaps the most extensive studies in the area of charge-transfer induced homolysis have been made by the group of Kochi. The reactions between organomercury compounds and electron acceptors were alluded to before, but it is the reactions of acceptors with tin- and lead- alkyls which have given most results. The reactions of tetraalkyllead compounds with the inorganic oxidant hexachloroiridate(IV) and with tetracyanoethylene serve to illustrate some of the mechanistic subtleties involved in this work.

Tetraalkyllead(IV) compounds react with hexachloroiridate(IV) via an initial electron transfer followed by cleavage of alkyl according to equations 177 and 178[233].

$$PbR_4 + [Ir^{IV}Cl_6]^{2-} \longrightarrow {}^{+\bullet}PbR_4 + [Ir^{III}Cl_6]^{3-} \quad (177)$$

$$R_4Pb^{+\bullet} \longrightarrow R_3Pb^{+} + R^{\bullet} \quad (178)$$

The alkyl radical may be detected by e.s.r. using the spin-trapping technique with 2-methyl-2-nitrosopropane. In the overall reaction $R^{\bullet}$ either abstracts hydrogen from solvent or (more generally) reacts with another hexachloroiridate(IV) molecule to give alkyl halide and an iridium(III) species. The overall stoichiometry of the reaction in acetic acid is shown in equations 177 and 178 and it is interesting that it is the same as that in equations 179 and 180 for the reaction between tetraalkyllead with chlorocuprate(II)[234]. Both reactions require 2 equivalents of metal complex and give alkyl chloride as a product.

$$Me_4Pb + [Ir^{IV}Cl_6]^{2-} \xrightarrow{HOAc} Me_3PbOAc + MeCl + [Ir^{III}Cl_6]^{3-} + [Ir^{III}Cl_5]^{2-} \quad (179)$$

$$Me_4Pb + 2[Cu^{II}Cl_3]^{-} \longrightarrow Me_3PbCl + MeCl + 2[Cu^{I}Cl_2]^{-} \quad (180)$$

However, equations 179 and 180 are thought to proceed via an *electrophilic* pathway[234]. Obviously, alkyl transfers involving electrophilic pathways and those which involve electron transfer have subtle differences, and the pathway of each particular reaction must be subjected to careful experimental analysis before mechanistic conclusions can be obtained.

In contrast, electron transfer between tetraalkyllead and TCNE leads to insertion of tcne into a lead—alkyl bond[233]. However, while proceeding via a similar lead cation as in reaction 181, no free alkyl radical is observed, which indicates that the reaction between $R_4Pb^{+}$ and $TCNE^{-}$ must be a concerted process.

$$R_4Pb + TCNE \longrightarrow [R_4Pb^{+} \cdot TCNE^{-}] \longrightarrow R_3PbC\text{—}C\text{—}R \quad (181)$$

One of the most important applications of the elegant mechanistic interpretations of these types of reaction has been in presenting a more unified view of electron-transfer reactions in general than has previously been accepted.

The reactions of tetraalkyltin compounds with tris(phenanthroline)iron(III) proceeds via equations 182 and 183, again involving the breakdown of an intermediate $R_4Sn^{+\bullet}$ radical cation[235].

$$[Fe^{III}(phen)_3]^{3+} + R_4Sn \longrightarrow R_4Sn^{+\bullet} + [Fe^{II}(phen)]^{2+} \quad (182)$$

$$R_4Sn^{+\bullet} \longrightarrow R_3Sn^{+} + R^{\bullet} \quad (183)$$

The reactions of the same compounds with hexachloroiridate(IV) proceed via the apparently similar mechanism (via $R_4Sn^{+\bullet}$) outlined for the analogous lead complexes in equations 177 and 178. Similarly, the reaction with TCNE leads to insertion of TCNE into an alkyltin bond[236]. All these processes occur via a rate-limiting single electron transfer from alkylmetal to oxidant or acceptor. Significantly, however, the reactions with $[Fe^{III}(phen)_3]^{3+}$ and $[Ir^{IV}Cl_6]^{2-}$ differ in more than one way. Thus, while the reactions of $R_4Sn$ with the Fe(III) complexes exhibit behaviour fully predictable by Marcus theory, and follow the Marcus linear free energy relationship with the predicted Brönsted slope ($\alpha = 0.5$) for a classical outer-sphere mechanism[237,238], the reactions of $R_4Sn$ with hexachloroiridate do not; the major differences are that the observed rate of electron transfer is $10^7$–$10^9$ faster than those predicted by the Marcus theory, and there is a uniform scatter of points in a similar plot of the free energy relationship. It is implicit from the results that both $[Ir^{IV}Cl_6]^{2-}$ and TCNE react via inner-sphere mechanisms[239], and that steric effects greatly affect the rates of electron transfer (whereas in the outer-sphere process they do not). Then, by defining a steric term for each alkylmetal with reference to $Me_4Sn$ as standard, it was possible to calculate a steric correction factor for the observed rate constants for electron transfer, with the result that the plot of log $K$ *vs.* ionization potential for a number of alkyl metals now gave a straight line of Brönsted slope $\alpha = 1$ ($\alpha = 0.5$ for outer-sphere processes) corresponding to the Marcus prediction for the limit of an inner-sphere mechanism. From a combination of Marcus theory for outer sphere processes and Mulliken charge-transfer theory, a method for obtaining accurate predictions of rate constants of electron transfer for inner-sphere processes has been developed[239], where neither theory was previously adequate individually to describe them. For a fully detailed account of the discussion the reader is referred to several recent papers[235,236,239].

The principles outlined above have been used to study the homolytic aromatic substitution of phenanthroline as coordinated ligand in the iron(III) derivatives[240]. Thus, the stoichiometry of the overall reaction proceeds via equation 184.

$$Me_4Sn + 2[Fe(phen)_3]^{3+} \longrightarrow [Me(phen)Fe^{II}(phen)]^{2+} + [Fe^{II}(phen)_3]^{2+} + Me_3Sn^+ \quad (184)$$

The methyl-substituted complex was identified by proton n.m.r. of the iron(II) complex and by isolation of the free ligand after hydrolysis. Significantly, an identical product was obtained when diacetyl peroxide was thermally decomposed in the presence of $[Fe^{III}(phen)_3]^{3+}$, confirming the nature of the homolytic reaction. The substitution was shown to be positionally consistent with the 4 and 7 ligand positions being progressively substituted.

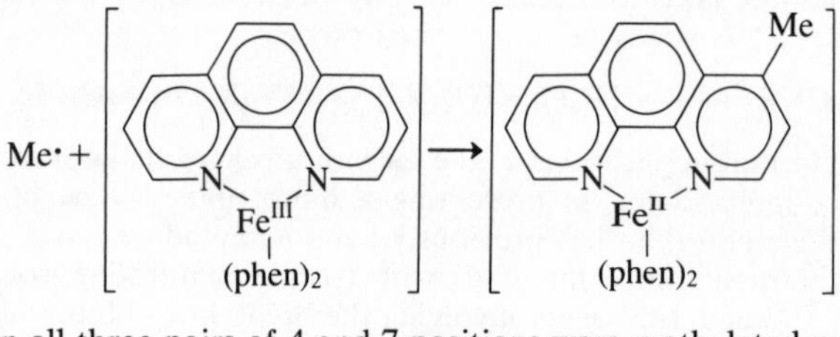

$$Me^{\bullet} + \left[ \text{(phen)} \; Fe^{III}(phen)_2 \right] \longrightarrow \left[ \text{(4-Me-phen)} \; Fe^{II}(phen)_2 \right] \quad (185)$$

Only when all three pairs of 4 and 7 positions were methylated was position 6 found to substitute.

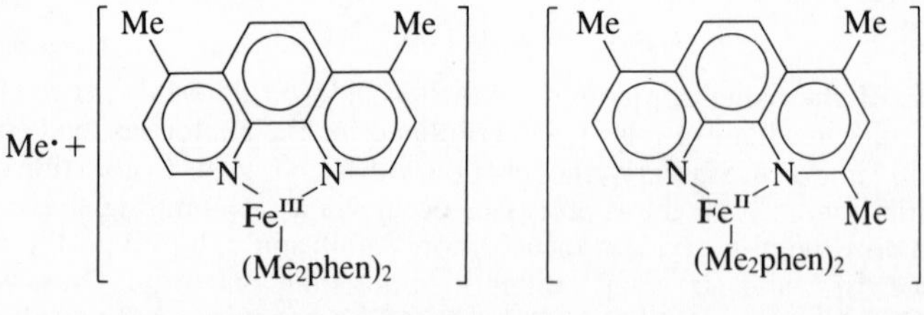

$$Me^{\bullet} + \left[ \text{(4,7-Me}_2\text{phen)} \; Fe^{III}(Me_2phen)_2 \right] \quad \left[ \text{(4,6,7-Me}_3\text{phen)} \; Fe^{II}(Me_2phen)_2 \right] \quad (186)$$

Efficient alkylation was effected by ethyl, *n*-propyl, and isobutyl radicals but with *t*-butyl radicals, products obtained from the *t*-butyl cation were obtained in 95% yield[241]. This indicates that oxidation of the radical is favoured to ligand substitution as depicted in equation 187.

$$Bu^{t\bullet} + [Fe^{III}(phen)_3]^{3+} \longrightarrow [Fe(phen)_3]^{2+} + Bu^{t+} \quad (187)$$

It appears that isopropyl, benzyl, and cyclohexyl radicals all favour this alternative. Developing the theories of steric effects in electron transfer, it can be found that the radical oxidation process (rates essentially independent of steric effects) occur via outer-sphere mechanisms, where as the substitution reactions are favoured by an inner-sphere process. Studies were also made on ruthenium and osmium analogues and similar reactivities were found although the rates of ligand substitution increased in the order $Os^{III} < Fe^{III} < Ru^{III}$. It was confirmed that the alkyl radicals were all derived from the homolysis of alkyltin derivatives via an initial electron transfer step to give the $R_4Sn^{+\bullet}$ cation.

Another demonstration of electron transfer induced homolysis of a tin—carbon bond has been found in the reaction of optically active tetraorganotin with dicarbonylcyclopentadienylferrate, $[Cp(CO)_2Fe]^-$[242]. When an optically pure tin compound was used, racemized products were obtained and this was explained by a rapid inversion of an intermediate tin radical.

$$R^1R^2R^3SnR^4 \xrightarrow{[Cp(CO)_2Fe]^-} [(R^1R^2R^3Sn)Fe(CO)_2Cp] \tag{188}$$

$$R^1 = Me,\ R^2 = neophyl,\ R^3 = phenyl$$

Comparing the results with others obtained in similar systems,[243] the following mechanism was proposed.

$$R^1R^2R^3SnR^4 \longrightarrow \underset{\mathbf{(57)}}{[R^1R^2R^3Sn^{\bullet}R^{4-}Fe(CO)_2Cp]} \tag{189}$$

$$\mathbf{57} \xrightarrow{\text{diffusion}} R^1R^2R^3Sn^{\bullet} + R^- + {}^{\bullet}Fe(CO)_2Cp \tag{190}$$

$$\xrightarrow{\text{dimerization}} [(R^1R^2R^3Sn)Fe(CO)_2Cp] \tag{191}$$

A similar explanation was afforded to account for the cleavage of the cobalt—tin bond without any stereoselectivity by $[Cp(CO)_2Fe]^-$[244].

## F. Miscellaneous.

### 1. *The lifetime of $R_4M^{+\bullet}$*

In the electron-transfer processes discussed above, a common intermediate for all the organometals is the cation radical of the organometal formed on electron loss. Two recent developments have assisted in characterizing such intermediates. In the first paper, the electrochemistry of the mercury- and Group IV-alkyls used in the chemical studies was undertaken and perhaps the most significant result was the double-step chronamperometry of $R_4Sn$ which indicated a maximum upper limit for the lifetime of the $R_4Sn^{+\bullet}$ cation of 1 ms[245]. By comparison it might be expected that on thermochemical grounds $R_2Hg^{+\bullet}$ and $R_4Pb^{+\bullet}$ might have similar lifetimes whereas $R_4Si^{+\bullet}$ and $R_4Ge^{+\bullet}$ may be longer lived.

Significantly, recently $R_4Si^{+\bullet}$[246], $R_4Ge^{+\bullet}$[246], and $R_4Sn^{+\bullet}$[247] have been characterized by e.s.r. when the parent compounds were irradiated with $^{60}Co$ $\gamma$-rays in halocarbon solution at 77 K. This solvent is known to cause electron loss by substrate and exceptionally well resolved spectra of the cations $Me_4Si^+$ and $Me_4Ge^+$ were recorded.

### 2. Toxicity

All these findings may have some relevance to the toxicity particularly of mercury), tin-, and lead-alkyls. Both tin and lead compounds owe their toxicity in mammals and man to formation of $R_3M^+$ cations and mercury to the RM species[248]. The variety of illnesses and diseases caused by these species is astounding. Considering the discussion in the previous sections it would be reasonable to conclude that the mechanism causing breakdown of the tetraorgano-tin or -lead compounds and the dialkylmercurials would be via electron-transfer processes with suitable oxidants or electron acceptors, of which there are many in living systems, to give the appropriate organometal cation and alkyl radical.

## VIII. GROUP V

Homolysis of the Group V metal—carbon bond has been studied in some detail, when it occurs in thermolytic, photolytic, or radical reactions, but there appear to be few examples outside these areas of investigation. The decomposition of phosphoranyl radicals, however, represents a field of study without parallel in organometallic chemistry and must be a major inclusion in any general discussion of homolytic reactions of Group V.

As a prelude, the bond dissociation energies of several Group V metal—carbon bonds are listed in Tables 1 and 3 (Section I.C), and inspection of these yields the usual trend of decreasing bond strength with increasing atomic weight of the metal.

### A. Thermal Decompositions

The decompositions of trimethyl-arsine, -stibine, and -bismuthine have been studied in the gas phase and the results obtained appear to be consistent with initial loss of methyl radicals[178].

$$Me_3M \longrightarrow Me_2M^{\bullet} + R^{\bullet} \qquad (192)$$

Indeed, the decomposition of the bismuth derivative probably involves a stepwise loss of all three methyl radicals. The thermal stability of these compounds decreased, as expected, in the order $Me_3As > Me_3Sb > Me_3Bi$. A similar order of stability was found for decomposition of the triphenyl analogues but although biphenyl is a product of these decompositions in an inert atmosphere, and benzene was produced under hydrogen, it is not clear whether direct metal—carbon bond homolysis is involved[249].

Thermolysis of tris(trifluoromethyl)-arsine and -stibine was found to yield products consistent with the formation of trifluoromethyl radicals[250,251]. The thermal decompositions of higher alkyl Group V organometallics have not been studied in detail, although there is a recent report that triethylstibine decomposes via a free-radical mechanism[252].

Organometallic compounds of Group V metals in the +5 oxidation state have also received some attention. Once again the bismuth analogues are far less stable than their higher analogues. Decomposition of pentaalkyl compounds of phosphorous and arsenic proceed via ylide formation[253].

$$Me_5As \longrightarrow Me_3As{=}CH_2 + CH_4 \qquad (193)$$

However, the ylides are less stable for the heavier members of the group and reductive elimination and coupling become more important[254].

$$Me_5M \longrightarrow Me_3M + CH_3CH_3 \qquad (194)$$

Reductive elimination is probably responsible for the decomposition of pentaaryl derivatives[255], but the presence of benzene among the products and the observation that heating pentaphenylantimony in chloroform decolorizes added diphenylpicrylhydrazyl may be indicative of some contribution of metal—carbon bond homolysis towards the overall mechanism[256].

More recently, studies on the thermal decomposition of unsymmetrical organoantimony(V) compounds have been made. The decomposition of hydroxytetraarylantimony(V) in *p*-xylene at 50–70 °C in the dark was thought to proceed via homolysis of an antimony—aryl bond[257]. A radical chain mechanism finally accounted for the antimony oxide and benzene among the products.

$$Ph_4SbO^{\bullet} \longrightarrow Ph_3Sb{=}O + Ph^{\bullet} \tag{195}$$

$$Ph_4SbOH + Ph^{\bullet} \longrightarrow Ph_4SbO^{\bullet} + PhH \tag{196}$$

If substituted phenyl groups were present in these compounds an approximate order for the ease of formation of the corresponding substituted phenyl radicals was established: *m*-nitrophenyl > *p*-nitrophenyl > *m*-chlorophenyl > *p*-chlorophenyl ≈ *p*-methoxyphenyl ≈ *m*-methoxyphenyl > *p*-tolyl > phenyl > *m*-tolyl[258].

Heating $(p\text{-}MeOC_6H_4S)SbPh_4$ in chloroform–carbon tetrachloride or benzene in the presence of 2-methyl-2-nitrosopropane in the cavity of an e.s.r. spectrometer gave spectra of the spin-trapped phenyl radical. Triphenylstibine, phenyl (*p*-methoxyphenyl) sulphide, benzene, and biphenyl were found among the products of the decomposition in chloroform–carbon tetrachloride and a concerted homolysis was preferred to a radical chain decomposition mechanism[259].

$$(p\text{-}MeOC_6H_4S)SbPh_4 \longrightarrow Ph_3Sb + Ph^{\bullet} + pMeOC_6H_4S^{\bullet} \tag{197}$$

## B. Photolytic Decompositions

Examples of photoinduced homolytic decomposition have been observed for organic compounds of all the elements of Group V, but are more common for the heavier members. A comparative study of the photodissociation reactions of the triphenyl derivatives has been made at 77 and 300 K using both e.s.r. and spectrophotometric methods[260]. With the aid of the spin-trap phenyl *t*-butyl nitrone, quantum yields for the generation of phenyl radicals from the arsenic (0.05), antimony (0.1), and bismuth (0.15) derivatives could be estimated. However, the spin-trapped phenyl radical could not be detected during photolyses of the phosphorus or nitrogen derivatives.

$$Ph_3M \xrightarrow{h\nu} Ph_2M^{\bullet} + Ph^{\bullet} (M = As, Sb, Bi) \tag{198}$$

$$Ph^{\bullet} + PhCH{=}\overset{+}{N}(\bar{O})(Bu^t) \longrightarrow Ph_2CHN(\dot{O})(Bu^t) \tag{199}$$

There is some evidence to suggest that phenyl radicals are formed on photolysis of phenylphosphines in other e.s.r. experiments[261]. In addition, photolysis of single crystals of triphenylphosphine gave e.s.r. signals of both the diphenylphosphinyl radical, $Ph_2P^{\bullet}$, and the phenyl radical[262]. The corresponding diphenylarsinyl radical has also been observed by e.s.r. on photolysis of triphenylarsine in ethanol glasses at 97 K, together with the triphenylarsine cation, $Ph_3As^+$. Unfortunately, under these conditions the corresponding phosphorus radicals could not be detected, owing to interaction with the solvent[263].

The photolysis of triphenylbismuth in substituted benzenes gave mixed biaryls, bismuth metal, and polymeric materials. It was thought that all three phenyl groups could be lost by a stepwise photoinduced homolysis and that the polymer was probably due to reactions of the carbene type species **:BiPh**[264].

Kinetic absorption and emission spectroscopy was used, in a flash photolysis study on trimethylbismuth [$\varepsilon_{max}(211.5\ nm) = 1.65 \times 10^4\ l\ mol^{-1}\ cm^{-1}$] and trimethylantimony [$\varepsilon_{max}(225\ nm) = 1 \times 10^4\ l\ mol^{-1}\ cm^{-1}$], to investigate the nature of excited metal species thought to be formed by stepwise loss of methyl radicals in a multiphoton process[265]. One species detected in the $BiMe_3$ system which decayed rapidly during the flash but persisted for *ca.* $3 \times 10^{-4}$ s in the dark was tentatively assigned to **:BiMe**.

Photolysis of pentaphenylantimony in benzene gave biphenyl as the major product. To demonstrate that this was not formed by a concerted elimination (the preferred mechanism for thermal decomposition) the compound was $^{14}C$-labelled at the $C_{(1)}$ position. The biphenyls isolated contained unlabelled phenyl groups derived from solvent and labelled phenyl groups from $Ph_5Sb$, indicating that phenyl radicals had indeed been formed in the reaction[255].

Finally, mention might be made of the photolytic decomposition of aroyldiphenylphosphines ($ArCOPPh_2$). Products obtained upon u.v. irradiation of *o*-$MeOC_6H_4COPPh_2$ in benzene included diphenyl and *o*-$MeOC_6H_4COOPh_2$ in benzene included diphenyl and *o*-$MeOC_6H_4CHO$, suggesting that not only had phosphorus—phenyl bond homolysis occurred, whether by direct photodissociation or homolytic substitution, but also that a Norrish type I fragmentation of the photoexcited ketone had occurred[266].

$$Ph_2P(Ar)C{=}O \xrightarrow{h\nu} Ph_2P(Ar)\dot{C}{-}\dot{O} \qquad (200)$$

$$Ph_2P(Ar)\dot{C}{-}\dot{O} \longrightarrow Ar\dot{C}{=}O + Ph_2P^{\bullet} \qquad (201)$$

### C. Radical Reactions

Radical substitution reactions at Group III metal centres were considered to be $S_H2$ processes, although a radical intermediate has been characterized in one instance, and more such examples may come to light in the future. The $S_H2$ description seems justified if, as is likely, any intermediates are too short-lived to affect the course and rate of the overall reaction.

Radical substitution at trivalent Group V metal centres is also commonly encountered but, in the case of the phosphorus(III) and arsenic(III) compounds, the radical intermediates often have lifetimes which enable them to be readily characterized by both optical and e.s.r. spectroscopy. Under these circumstances, it is the chemistry of the intermediates that becomes most important in determining the nature of the substitution reactions.

$$R_3M + X^{\bullet} \longrightarrow R_3(X)M^{\bullet} \longrightarrow R_2MX + R^{\bullet} \qquad (202)$$

The phosphorus-containing intermediates, phosphoranyl radicals, have recently been the subject of an excellent review[16]. Here, the first consideration is of factors affecting the $\alpha$-scission of carbon–phosphorus bonds in phosphoranyl radicals; however, it will be seen that this might be considered as a direct homolytic cleavage or, alternatively, as a redistribution of the electrons from a three electron bond.

#### 1. *α-Scission of phosphoranyl radicals*

E.s.r. spectroscopy has proved to be particularly useful in the investigation of the factors which influence the $\alpha$-scission of phosphorus—carbon bonds in phosphoranyl radicals. Much of the work has involved photolysis of di-*t*-butyl peroxide in the presence of trialkylphosphorus(III) compounds in solution at low temperatures. Thus, under these conditions trimethylphenylphosphine reacts with *t*-butoxyl radicals to give e.s.r. spectra of both the intermediate phosphoranyl radical, and the displaced methyl radical[267].

$$Bu^tO^{\bullet} + Me_3P \longrightarrow Me_3(Bu^tO)P^{\bullet} \longrightarrow Bu^tO(Me)_2P + Me^{\bullet} \qquad (203)$$

If larger alkyl groups are present in the phosphine, the intermediate phosphoranyl radical is not observed (except for $Et_3(Bu^tO)P^{\bullet}$) and even at the lowest accessible temperatures only the displaced alkyl radicals are detected. It is likely, however, that phosphoranyl radicals are real intermediates in these processes, but they are too short-lived to be detected by e.s.r.

*t*-Butoxydealkylation of mixed trialkylphosphines is relatively unselective in the nature (primary, secondary, or tertiary) of the alkyl radical displaced, with the exception of $Bu^tPMe_2$, which gave only *t*-butyl radicals[268]. This suggests that, as with the *t*-butoxydealkylation of the trialkylboranes, the course of the reaction is controlled by the formation of the radical intermediate rather than its decay.

In contrast with the above, both the intermediate phosphoranyl radical and the displaced alkyl radical may be readily detected in *t*-butoxydealkylation reactions of dialkyl(*t*-butoxy)phosphines[268,269].

$$Bu^tO^{\bullet} + R_2(Bu^tO)P \longrightarrow \underset{\mathbf{(58)}}{R_2(Bu^tO)_2P^{\bullet}} \qquad (204)$$

$$\mathbf{58} \longrightarrow R^{\bullet} + R(Bu^tO)_2P \qquad (205)$$

The decay of **58** could easily be measured and the observed values for $k_{205}$ increased in the order $R = Me < Bu^t < Pr^n$, $Et < Pr^i <$ allyl; with the exception of *t*-butyl, this order clearly reflects the ease of formation of the corresponding alkyl radicals. The anomalous position of $R = Bu^t$ can be explained by considering the non-equivalence of ligand sites in the phosphoranyl radical.

Phosphoranyl radicals have traditionally been considered as having trigonal-bipyramidal structures, with two ligand sites (apical and equatorial), the unpaired electron occupying an equatorial position, **59**.

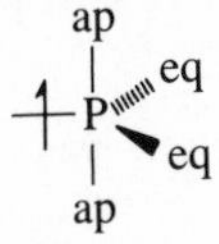

ap
eq
P
eq
ap

**(59)**

Electronegative ligands (e.g. $Cl^-$, $Bu^tO^-$) prefer to occupy apical sites, while alkyl groups prefer equatorial sites. Exchange between sites is very rapid, however, and small equilibrium concentrations of **59** with alkyl ligands occupying apical sites can be envisaged. This provides the basis for an explanation of the order of *t*-butoxydalkylation described above if phosphorus—carbon bond homolysis takes place preferentially from an apical site, i.e. if $\alpha$-scission is itself apical site selective.

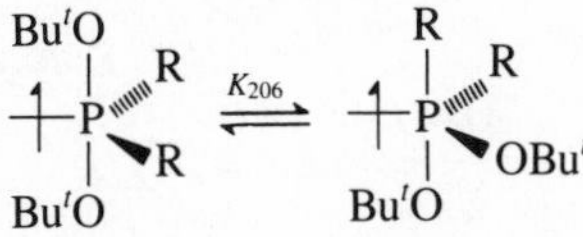

$$\underset{\mathbf{(60)}}{(Bu^tO)_{ap}(Bu^tO)_{ap}PR_2^{\bullet}} \overset{K_{206}}{\rightleftharpoons} \underset{\mathbf{(61)}}{(R)_{ap}(Bu^tO)_{ap}P(R)(OBu^t)^{\bullet}} \longrightarrow R^{\bullet} + (Bu^tO)_2PR \qquad (206)$$

The steric interaction between alkyl groups is greater in **61** than in **60**, so $k_{206}$ will decrease as the bulk of the alkyl group increases. In the case of the stereochemically less active alkyl ligands, the overall rate of $\alpha$-scission ($k_\alpha$) may mainly depend on $k_{206}$, which would be expected to increase with decreasing strength of the phosphorus—carbon bond. Thus $k_\alpha$ increases in the observed order Me < Pr$^i$ < allyl. However, the steric interaction of the two $t$-butyl groups in **61** may reduce $K_\alpha$ more than enough to offset the further increase in $k_{206}$. This unfavourable steric interaction is absent in the radical (Bu$^t$)Me(Bu$^t$O)$_2$P$^\bullet$, which was found to decay to give $t$-butyl radicals more rapidly than **60** (R = Pr$^i$) under similar conditions.

There is further evidence that $\alpha$-scission is apical site selective. The phosphoranyl radicals R$_2$(Bu$^t$O)$_n$(EtO)$_{2-n}$P$^\bullet$ ($n$ = 0, 1, 2) exhibit decreased stability towards loss of R$^\bullet$ as the number of ethoxy ligands is increased[269]. Ethoxy ligands are apparently less apicophilic than $t$-butoxy ligands, thereby allowing a higher proportion of phosphoranyl radical isomers with apical alkyl groups.

The rate of decay of the intermediate phosphoranyl radicals in the $t$-butoxydealkylation of trialkylphosphines would be expected to be particularly fast (as seems likely) since these radicals cannot avoid possessing an apical alkyl ligand. It is also found that the nature of R′ does not affect the rate of decay significantly in R$_3$(R′O)P$^\bullet$[268].

Ligand exchange in Me$_3$(Bu$^t$O)P$^\bullet$ is apparently still more rapid than its decay by $\alpha$-scission. To explain this result it has been suggested that $\alpha$-scission, like ligand exchange, is mediated by a $\sigma^*$ intermediate in which the unpaired electron resides in an antibonding phosphorus ligand $\sigma$-orbital[270]. The $t$-butoxydealkylation reaction could then be rationalized in the following way.

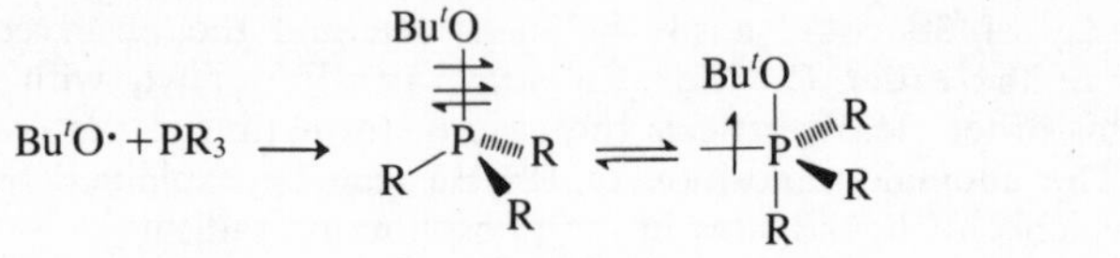

$$\mathrm{Bu^tO^\bullet + PR_3 \longrightarrow [Bu^tO\text{–}PR_3]^{\sigma^*} \rightleftharpoons [Bu^tO(R)_2P\text{–}R]} \qquad (207)$$

$$\longrightarrow \mathrm{[Bu^tO(R)_2P\text{–}R]^{\sigma^*} \longrightarrow R^\bullet + Bu^tO\,PR_2} \qquad (208)$$

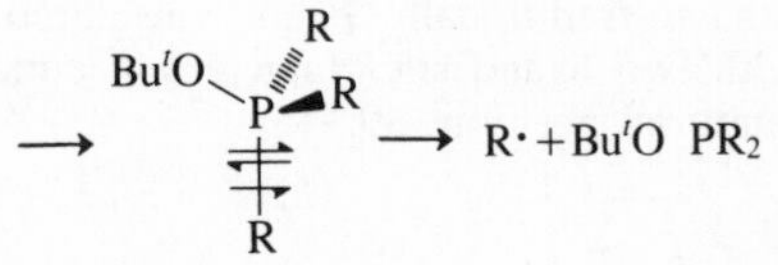

The bond that is broken in the $\sigma$ structure was thus originally an apical bond in the trigonal bipyramidal structure.

$\alpha$-Scission is not the only process via which phosphoranyl radicals may decompose. $\beta$-Scission to give the thermodynamically stable P=O bond is energetically very favourable.

$$\mathrm{R'O^\bullet + R_3P \longrightarrow R_3(R'O)P^\bullet \longrightarrow R_3P{=}O + R^\bullet} \qquad (209)$$

Thus, the reaction of $t$-butoxyl radicals with tri-$n$-butylphosphine at 403 K gives 80% $\alpha$-scission products and 20% $\beta$-scission products[271].

$$\mathrm{Bu^tO^\bullet + Bu^n{}_3P \longrightarrow Bu^n{}_3(Bu^tO)P^\bullet \xrightarrow{\alpha} Bu^{n\bullet} + Bu^n{}_2POBu^t} \qquad (210)$$

$$\mathrm{Bu^n{}_3(Bu^tO)P^\bullet \xrightarrow{\beta} Bu^{t\bullet} + Bu^n{}_3P{=}O} \qquad (211)$$

Even though oxidation is 100 kJ mol$^{-1}$ more exothermic than displacement, the $\alpha$-scission is much faster at this temperature and the exclusive mode of decay of the intermediate phosphoranyl radical at lower temperatures. Hence, the activation energy for $\alpha$-scission is less than that for $\beta$-scission. If the phosphorus—carbon bond strength is increased as would be expected in $Ph_3(RO)P^{\bullet}$ radicals then $\beta$-scission predominates. Phosphoranyl radicals formed by addition of thiyl radicals to phosphines usually decay by $\beta$-scission unless a very weak phosphorus—carbon bond is also present (for example, in benzylphosphoranyl radicals)[272].

The chemistry of arsanyl radicals, $X_4As^{\bullet}$, resembles that of the phosphoranyl radicals but comparatively few studies have been performed on these species. A number of arsanyl radicals have been detected and characterized by e.s.r. spectroscopy. Intermediate arsanyl radicals could not be detected in the reaction of trimethylarsine with *t*-butoxyl radicals under similar conditions which did allow observation of the phosphorus analogue[273]. However, strong signals of methyl radicals were observed, as were alkyl radicals when higher trialkylarsines were employed.

E.s.r. signals of $Ph_3(Bu^tO)As^{\bullet}$ and $Ph_2(Bu^tO)(RO)As^{\bullet}$ could be detected in the reactions between *t*-butoxyl radicals and the corresponding arsines[274]. Both radicals decay by $\alpha$-scission to give phenyl radicals which were spin-trapped on phenyl *t*-butyl nitrone. This behaviour contrasts with that of the phosphorus analogues but is consistent with weakening of the metal—carbon bond for heavier members of the group.

### 2. Radical reactions of antimony and bismuth

E.s.r. signals of ethyl radicals were detected when *t*-butoxy radicals were reacted with triethylstibine and bismuthine[9,10], but intermediate stibyl or bismuthyl radicals could not be detected. Radicals of the type $X_4Sb^{\bullet}$ and $X_4Bi^{\bullet}$ have not been observed in fluid solution but in all probability are very short-lived. Radical reactions at antimony and bismuth probably have more of the character of radical reactions at Group(III) metal centres.

### 3. Miscellaneous reactions

E.s.r. spectra attributed to radicals **62** and **63** were obtained when triphenylbismuth was heated with the corresponding *o*-quinones at 160 °C[275]. Apparently the charge-transfer process ubiquitous in reactions of organometals of Groups IIb, III and IV with *o*-quinones is again operative, and many other Group V derivatives exhibit similar behaviour[276].

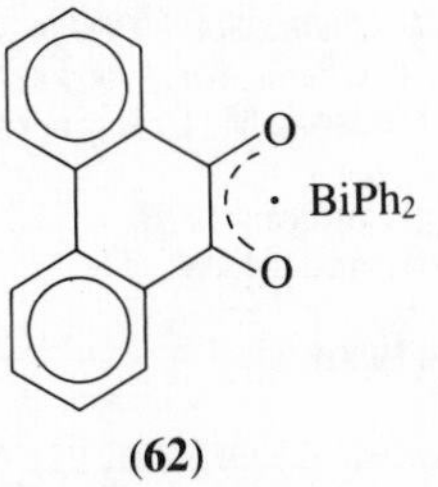

**(62)** 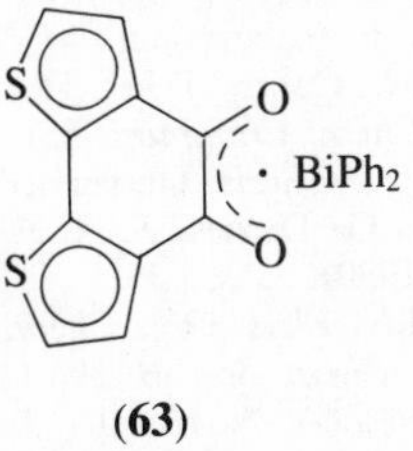 **(63)**

## IX. ACKNOWLEDGEMENTS

The authors thank Professors Alex McAuley, and Paul Tordo and Dr Paul R. West for advice and guidance. They also acknowledge in general the support and friendship of their colleagues at University College, London, and York in the UK, at the University of Victoria (B.C.), at the Université de Provènce, and at the Australian National University, and in particular special thanks are due to Dr D. H. Roberts, Dr J. A.-A. Hawari, and Mr D. M. Whitfield. In addition, both authors are indebted to and grateful for the constant encouragement, support, and confidence of Professor Alwyn G. Davies.

## X. REFERENCES

1. E. Frankland, *J. Chem. Soc.*, **2,** 263 (1849).
2. E. Frankland, *J. Chem. Soc.*, **3,** 44 (1850).
3. E. Frankland, *J. Chem. Soc.*, **13,** (1860).
4. M. Gomberg, *J. Am. Chem. Soc.*, **22,** 757 (1900).
5. W. Hofeditz and K. Paneth, *Chem. Ber.*, **62,** 1335 (1929).
6. A. G. Davies, *J. Chem. Phys. Soc. Univ. Coll., London,* **4,** 63 (1976).
7. A. G. Davies, *Organic Peroxides*, Butterworths, London, 1960.
8. A. G. Davies, in *Organic Peroxides* (Ed. D. Swern), Vol. II, Wiley, New York, 1970.
9. K. U. Ingold and B. P. Roberts, *Free Radical Substitution Reactions,* Wiley, New York, 1971.
10. A. G. Davies and B. P. Roberts, in *Free Radicals*, (Ed. J. K. Kochi), Vol. I, Wiley, New York, 1973.
11. A. G. Davies and B. P. Roberts, *Acc. Chem. Res.*, **5,** 387 (1972).
12. A. G. Davies, *J. Organomet. Chem.,* **200,** 87 (1980).
13. E. B. Milovskaya, *Russ. Chem. Rev.*, **42,** 384 (1973).
14. A. G. Davies, *Pure Appl. Chem.,* **39,** 497 (1974).
15. A. G. Davies, in *Organotin Compounds: New Chemistry and Applications* (Ed. J. J. Zuckermann), *Adv. Chem. Ser.,* No. **157,** American Chemical Society, Washington, DC, 1976.
16. B. P. Roberts, *Adv. Free Radical Chem.,* **6,** 225 (1980).
17. J. K. Kochi, *Organometallic Mechanisms in Catalysis,* Wiley, New York, 1978.
18. W. V. Steele, *Chem. Soc. Ann. Rep.,* **71A,** 103 (1974).
19. H. A. Skinner, *Adv. Organomet. Chem.,* **2,** 49 (1964).
20. B. W. Fullam and M. C. R. Symons, *J. Chem. Soc., Dalton Trans.,* 1086 (1974).
21. J. D. Cox and G. Pilcher, *Thermochemistry of Organic and Organometallic Compounds,* Academic Press, New York, 1970.
22. T. Holm, *J. Organomet. Chem.,* **56,** 87 (1973).
23. T. Holm, *J. Organomet. Chem.,* **77,** 27 (1974).
24. R. A. Jackson, *J. Organomet. Chem.,* **166,** 17 (1979).
25. See, for example, J. E. Wertz and J. R. Bolton, *Electron Spin Resonance: Elementary Theory and Practical Applications,* McGraw-Hill, New York, 1972.
26. M. C. R. Symons, *Chemical and Biological Aspects of E.S.R. Spectroscopy,* Halsted Press, New York, 1978.
27. P. J. Barker, A. G. Davies, and M.-W. Tse, *J. Chem. Soc., Perkin Trans 2,* 941 (1980).
28. A. G. Davies, B. P. Roberts, and M.-W. Tse, *J. Chem. Soc., Perkin Trans 2,* 145 (1978).
29. J. D. Cotton, C. S. Cundy, D. H. Harris, A. Hudson, M. F. Lappert, and P. W. Lednor, *J. Chem. Soc., Chem. Commun.,* 651 (1974).
30. B. Muggleton, *PhD Thesis,* University College London, 1976.
31. P. J. Barker, A. G. Davies, J. A.-A. Hawari, and M.-W. Tse, *J. Chem. Soc., Perkin Trans. 2,* 1488 (1980).
32. M. J. Perkins, *Adv. Phys. Org. Chem.,* **17,** 1 (1980).
33. M. J. Perkins, *J. Chem. Soc. B,* 395 (1970).
34. A. G. Davies and J. A.-A. Hawari, *J. Organomet. Chem.,* **201,** 221 (1980).

35. D. Rehorek, personal communication.
36. C. Lai and L. H. Piette, *Tetrahedron Lett.*, 775 (1979).
37. Z.-T. Tsai and C. H. Brubaker, *J. Organomet. Chem.*, **166,** 199 (1978).
38. P. J. Barker, S. R. Stobart, and P. R. West, *Can. J. Chem.*, unpublished results.
39. S. A. Fieldhouse, A. R. Lyons, H. C. Starkie, and M. C. R. Symons, *J. Chem. Soc., Dalton Trans.*, 1966 (1974).
40. G. S. Jackel and W. Gordy, *Phys. Rev.*, **176,** 443 (1968).
41. J. Bargon and H. Fischer, *Z. Naturforsch., Teil A,* **22,** 1556 (1967).
42. H. R. Ward and R. G. Lawler, *J. Am. Chem. Soc.,* **89,** 5518 (1967).
43. A. R. Lepley and G. L. Closs (Eds.), *Chemically Induced Dynamic Nuclear Polarization,* Wiley, New York, 1973.
44. R. Kaptein and J. L. Oosterhoff, *Chem. Phys. Lett.,* **4,** 195 and 214 (1969).
45. G. L. Closs, *J. Am. Chem. Soc.,* **91,** 4552 (1969).
46. R. Benn, *Rev. Chem. Intermediates,* **3,** 45 (1979).
47. A. G. Brook, P. J. Dillon, and R. Pearce, *Can. J. Chem.,* **49,** 133 (1971).
48. N. A. Porter and P. M. Iloff, *J. Am. Chem. Soc.,* **96,** 6200 (1974).
49. J. A. Den Hollander, R. Kaptein and P. A. T. M. Brand, *Chem. Phys. Lett.,* **10,** 430 (1971).
50. D. Bryce-Smith, *J. Chem. Soc.,* 1712 (1955).
51. R. H. Finnegan and H. W. Kutta, *J. Org. Chem.,* **30,** 4139 (1965).
52. W. H. Glaze, T. H. Brewer, R. Hatch, and J. Nathan, in *Decomposition of Organometallics to Refractory Ceramics, Metals and Metal Alloys,* (Ed. K. S. Mazdiyashi), University of Dayton Press, Dayton, Ohio, 1968, pp. 187–194.
53. A. A. Morton and E. J. Lampher, *J. Org. Chem.,* **20,** 839 (1955).
54. W. H. Glaze and T. H. Brewer, *J. Am. Chem. Soc.,* **91,** 4490 (1969).
55. J. B. Smart, R. Hogan, P. A. Scherr, L. Ferrier, and J. P. Oliver, *J. Am. Chem. Soc.,* **94,** 8371 (1972).
56. A. G. Davies, J. R. M. Giles, and J. Luzstyk, *J. Chem. Soc., Perkin Trans. 2,* 747 (1981).
57. A. G. Davies and J. Luzstyk, *J. Chem. Soc., Perkin Trans. 2,* 692 (1981).
58. E. J. Panek and G. M. Whitesides. *J. Am. Chem. Soc.,* **94,** 8768 (1972).
59. K. S. Chen, F. S. Bertini, and J. K. Kochi, *J. Am. Chem. Soc.,* **95,** 1340 (1973).
60. Y. Miura, Y. Monmoto, and M. Kinoshita, *Bull. Chem. Soc. Jpn.,* **49,** 1715 (1976).
61. C. Schenk and Th. J. de Boer, *Tetrahedron,* **35,** 2119 (1976).
62. G. A. Russell, E. G. Janzen, and E. T. Strom, *J. Am. Chem. Soc.,* **86,** 1807 (1964).
63. H. R. Ward, R. G. Lawler, and R. A. Copper, in ref. 43.
64. W. A. Nugent, F. S. Bertini, and J. K. Kochi, *J. Am. Chem. Soc.,* **96,** 4945 (1979).
65. X. Creary, *J. Am. Chem. Soc.,* **99,** 7632 (1977).
66. D. Bryce-Smith, *J. Chem. Soc.,* 1603 (1956).
67. H. D. Zook and R. N. Goldey, *J. Am. Chem. Soc.,* **75,** 3975 (1953).
68. G. A. Russell and D. W. Lamson, *J. Organomet. Chem.,* **156,** 17 (1978).
69. G. A. Russell and D. W. Lamson, *J. Am. Chem. Soc.,* **91,** 3967 (1969).
70. H. Fischer, *J. Phys. Chem.,* **73,** 3834 (1969).
71. A. Tamaki and J. K. Kochi, *J. Organomet. Chem.,* **61,** 441 (1973).
72. G. M. Whitesides, E. J. Panek, C. P. Casey, and J. San Filippo, *J. Am. Chem. Soc.,* **92,** 1426 (1970).
73. G. M. Whitesides, E. J. Panek, and E. R. Stedronsky, *J. Am. Chem. Soc.,* **94,** 232 (1972).
74. B. Akermark and A. Ljungquist, *J. Organomet. Chem.,* **182,** 47 (1979).
75. B. Akermark and A. Ljungquist, *J. Organomet. Chem.,* **182,** 59 (1979).
76. U. C. R. McLoughlin and J. Thrower, *Tetrahedron,* **25,** 5921 (1969).
77. M. F. Lappert and R. Pearce, *J. Chem. Soc., Chem Commun.,* 24 (1973).
78. A. Miyashita, T. Yamamoto, and A. Yamamoto, *Bull. Chem. Soc. Jpn.,* **50,** 1109 (1977).
79. A. E. Jukes, *Adv. Organomet. Chem.,* **12,** 215 (1979).
80. P. L. Coe and N. E. Milner, *J. Organomet. Chem.,* **39,** 395 (1972).
81. A. Cairncross and W. A. Sheppard, *J. Am. Chem. Soc.,* **93,** 247 (1971).
82. P. E. Fanta, *Synthesis,* **1,** 9 (1974).
83. G. M. Whitesides and C. P. Casey, *J. Am. Chem. Soc.,* **88,** 4541 (1966).
84. R. W. M. Tan Hoedt, G. van Koten, and J. G. Noltes, *J. Organomet. Chem.,* **201,** 327 (1980).

85. K. Wada, M. Tamura, and J. K. Kochi, *J. Am. Chem. Soc.*, **92,** 6650 (1970).
86. N. A. Clinton and J. K. Kochi, *J. Organomet. Chem.*, **42,** 241 (1972).
87. M. Tamura and J. K. Kochi, *J. Organomet. Chem.*, **42,** 205 (1972).
88. I. P. Beletskaya, Y. A. Artemkina, and O. A. Reutov, *J. Organomet. Chem.*, **99,** 343 (1975).
89. A. A. Budnik and J. K. Kochi, *J. Organomet. Chem.*, **116,** C3 (1976).
90. K. Yamamoto, K. Nakanishi, and M. Kumada, *J. Organomet. Chem.*, **7,** 197 (1967).
91. D. E. Berbreiter and T. J. Lynch, *J. Org. Chem.*, **46,** 727 (1981).
92. G. M. Whitesides, D. E. Bergbreiter, and P. E. Kendall, *J. Am. Chem. Soc.*, **96,** 2806 (1974).
93. C. M. Mitchell and F. G. A. Stone, *J. Chem. Soc., Dalton Trans.*, 103 (1972).
94. P. W. N. M. Van Leeuwen, R. Kaptein, R. Huis, and C. F. Roobeek, *J. Organomet. Chem.*, **104,** C44 (1976).
95. R. J. Puddephatt, *The Chemistry of Gold,* Elsevier, Amsterdam, 1978, Chapter 9.
96. M. F. Lappert and P. W. Lednor, *Adv. Organomet. Chem.*, **14,** 345 (1976).
97. A. Johnson and R. J. Puddephatt, *J. Chem. Soc., Dalton Trans.*, 115 (1975).
98. R. Kaptein, P. W. N. M. Van Leeuwen, and R. Huis, *J. Chem. Soc., Chem. Commun.*, 568 (1975).
99. R. J. Puddephatt and P. J. Thompson, *J. Organomet. Chem.*, **117,** 395 (1976).
100. A. Johnson and R. J. Puddephatt, *J. Chem. Soc., Dalton Trans.*, 1360 (1976).
101. G. E. Coates, D. L. Smith, and R. C. Srivastava, *J. Chem. Soc., Dalton Trans.*, 618 (1973).
102. F. Glockling, R. J. Morrison, and J. W. Wilson, *J. Chem. Soc., Dalton, Trans.*, 94 (1973).
103. R. Benn, *Chem. Phys.*, **15,** 369 (1976).
104. P. J. Barker, A. G. Davies, and J. Lusztyk, unpublished observations.
105. T. G. Brilkina and V. A. Shushunov, *Reactions of Organometallic Compounds with Oxygen and Peroxides,* Illiffe, London, 1969.
106. E. O. Fischer and H. P. Hoffmann, *Chem. Ber.*, **92,** 482 (1959).
107. A. Almenningen, O. Bastiansen, and A. Haaland, *J. Chem. Phys.*, **40,** 3434 (1964).
108. T. J. Kealy and P. L. Pauson, *Nature (London),* **168,** 1039 (1951).
109. D. S. Marguick, *J. Am. Chem. Soc.*, **99,** 1436 (1977).
110. M. J. S. Dewar and H. S. Rzepa, *J. Am. Chem. Soc.*, **100,** 777 (1978).
111. A. Almenningen, A. Haaland, and J. Lusztyk, *J. Organomet. Chem.*, **170,** 271 (1979).
112. R. Gleiter, M. C. Böhm, A. Haaland, R. Johansen, and J. Lusztyk, *J. Organomet. Chem.*, **170,** 285 (1974).
113. J. Lusztyk and K. B. Starowieski, *J. Organomet. Chem.*, **170,** 293 (1979).
114. R. A. Andersen and G. E. Coates, *J. Chem. Soc., Dalton Trans.*, 1171 (1974).
115. C. Walling and S. A. Buckler, *J. Am. Chem. Soc.*, **77,** 6032 (1955).
116. C. Walling, *Free Radicals in Solution,* Wiley, New York, 1957.
117. R. C. Lamb, P. W. Ayers, M. K. Torey, and J. F. Garst, *J. Am. Chem. Soc.*, **88,** 4261 (1966).
118. C. Walling and A. Cioffiari, *J. Am. Chem. Soc.*, **92,** 6609 (1970).
119. K. S. Cheng, J.-P. Battioni, and J. K. Kochi, *J. Am. Chem. Soc.*, **95,** 4439 (1973).
120. S. Cabbidu, S. Melis, E. Marongui, P. P. Piras, and G. Fodda, *J. Organomet. Chem.*, **111,** 249 (1976).
121. M. Chastrette and R. Gauthier, *J. Organomet. Chem.*, **66,** 219 (1974).
122. N. Kawabata, H. Nakamura, and S. Yamashita, *J. Org. Chem.*, **38,** 3403 (1973).
123. E. C. Ashby, J. Laemmle, and H. M. Neumann, *Acc. Chem. Res.*, **7,** 272 (1974).
124. F. F. Blicke and L. D. Powers, *J. Am. Chem. Soc.*, **51,** 3378 (1929).
125. A. E. Arbuzov and I. A. Arbuzova, *J. Gen. Chem. USSR,* **2,** 388 (1932).
126. M. S. Kharasch and S. Weinhouse, *J. Org. Chem.*, **1,** 209 (1936).
127. H. S. Mosher and C. Blomberg, *J. Organomet. Chem.*, **13,** 519 (1968).
128. T. Holm and I. Crosland, *Acta Chem. Scand.*, **25,** 59 (1971).
129. V. I. Savin, I. D. Temyachev, and F. D. Yambushev, *J. Org. Chem. USSR,* **11,** 1227 (1975).
130. V. I. Savin and Yu. P. Kitaev, *J. Org. Chem. USSR,* **12,** 262 (1976).
131. V. I. Savin and Yu. P. Kitaev, *J. Org. Chem. USSR,* **12,** 273 (1976).

132. V. I. Savin and Yu. P. Kitaev, *J. Org. Chem. USSR,* **11,** 2622 (1975).
133. V. I. Savin, A. G. Abulkhunov, and Yu. P. Kitaev, *J. Org. Chem. USSR,* **12,** 479 (1976).
134. V. I. Savin, *J. Org. Chem. USSR.,* **12,** 1818 (1976).
135. V. I. Savin and Yu. P. Kitaev, *J. Org. Chem. USSR,* **13,** 1137 (1977).
136. V. I. Savin, *J. Org. Chem. USSR,* **14,** 1936 (1978).
137. V. I. Savin, I. D. Temyachev, and F. D. Yambushev, *J. Org. Chem. USSR,* **12,** 268 (1976).
138. I. Crosland, *Acta Chem. Scand.,* **29B,** 468 (1975).
139. T. Holm, I. Crosland, and J. O. Madsen, *Acta, Chem. Scand.,* **23B,** 754 (1978).
140. A. Stasko, L. Malik, E. Mat'usova, and A. T. Kac, *Org. Magn. Reson.,* **17,** 74 (1981).
141. A. Stasko, A. T. Kac, R. Prikyl, and L. Malik, *J. Organomet. Chem.,* **92,** 253 (1975).
142. A. Stasko, L. Malik, A. Tkac, V. Adamcik, and M. Hronec, *Org. Magn. Reson.,* **9,** 269 (1977).
143. E. C. Ashby and T. L. Wiesmann, *J. Am. Chem. Soc.,* **100,** 3101 (1978).
144. G. M. Mkryan, S. M. Gasparyan, M. A. Papazian, and E. S. Voskanian, *J. Org. Chem. USSR,* **8,** 1397 (1972).
145. A. Ljungquist, *J. Organomet. Chem.,* **159,** 1 (1978).
146. M. Dagonneau, J.-F. Hemidy, D. Cornet, and J. Vialle, *Tetrahedron Lett.,* 3003 (1972).
147. M. Dagonneau and J. Vialle, *Tetrahedron Lett.,* 3017 (1973).
148. M. Dagonneau, *J. Organomet. Chem.,* **80,** 1 (1974).
149. M. C. R. Symons, *J. Chem. Soc.,* 1618 (1974).
150. M. Dagonneau and J. Vialle, *Tetrahedron,* **30,** 3119 (1974).
151. W. Kaim, *J. Organomet. Chem.,* **222,** C17 (1981).
152. W. R. McWhinnie and J. D. Miller, *Adv. Inorg. Chem. Radiochem.,* **12,** 135 (1969).
153. D. R. Corbin, W. S. Willis, E. N. Duesler, and G. D. Stucky, *J. Am. Chem. Soc.,* **102,** 5969 (1980).
154. K. C. Bass, *Organomet. Chem. Rev.,* **1,** 391 (1966).
155. R. Benn, E. G. Hoffmann, H. Lehmkuhl, and H. Nehl, *J. Organomet. Chem.,* **146,** 103 (1978).
156. F. J. J. De Kanter, *Org. Magn. Reson.,* **8,** 129 (1976).
157. J. St. Denis, J. P. Oliver, T. W. Dolzine, and J. B. Smart, *J. Organomet. Chem.,* **71,** 315 (1974).
158. R. A. Jackson and D. W. O'Neill, *J. Chem. Soc., Perkin Trans. 2,* 509 (1978).
159. S. Dincturk, R. A. Jackson, and M. Townson, *J. Chem. Soc., Chem. Commun.,* 172 (1979).
160. G. R. Leibling and H. M. McConnell, *J. Chem. Phys.,* **42,** 3931 (1965).
161. E. Hedaya, *Acc. Chem. Res.,* **12,** 367 (1979).
162. R. E. Linder and A. C. Ling, *Can. J. Chem.,* **50,** 3982 (1972).
163. P. J. Barker, A. G. Davies, and J. D. Fisher, *J. Chem. Soc., Chem. Commun.,* 584 (1974).
164. P. J. Barker and A. G. Davies, *J. Chem. Soc., Chem. Commun.,* 815 (1979).
165. E. Klahne, C. Giannotti, H. Marquet-Ellis, G. Folcher, and R. D. Fischer, *J. Organomet Chem.,* **201,** 399 (1980).
166. P. J. Barker and A. L. J. Beckwith, unpublished observations.
167. E. M. Flesia and J.-M. Surzur, *Tetrahedron Lett.,* 123 (1974).
168. P. J. Young, R. K. Gosavi, J. Connor, O. P. Strausz, and H. E. Gunning, *J. Chem. Phys.,* **58,** 5280 (1973).
169. A. G. Davies and B. P. Roberts, *J. Organomet Chem.,* **19,** C17 (1969).
170. P. R. Jones and P. J. Desio, *Chem. Rev.,* **78,** 491 (1978).
171. A. J. Bloodworth, in *The Chemistry of Mercury* (Ed. C. A. MacAuliffe), Macmillan, London, 1977.
172. D. Abenhaim, *J. Organomet. Chem.,* **92,** 275 (1975).
173. J. Furukawa, N. Kawabata, and J. Nishimura, *Tetrahedron,* **24,** 53 (1968).
174. S. Miyano and H. Hashimoto, *Bull. Chem. Soc. Jpn.,* **46,** 892 (1973).
175. A. G. Davies and B. P. Roberts, *J. Chem. Soc. B,* 1074 (1968).
176. E. N. Gladyshev, P. Ya. Bayushkin, G. Abakumov, and E. S. Klimov, *Izv. Akad. Nauk SSSR, Ser. Khim.,* 176 (1978).

177. J. Y. Chen, H. C. Gardner, and J. K. Kochi, *J. Am. Chem. Soc.*, **98,** 6150 (1976).
178. S. J. W. Price, in *Comprehensive Chemical Kinetics* (Ed. C. H. Bamford and C. F. H. Tipper), Vol. 4, Elsevier, Amsterdam, 1972.
179. M. P. Brown, A. K. Holliday, G. M. Way, R. B. Whittle, and G. M. Woodward, *J. Chem. Soc., Dalton Trans.*, 1862 (1977).
180. T. Mole and E. A. Jeffrey, *Organoaluminium Compounds,* Elsevier, Amsterdam, 1972.
181. Yu. A. Alexandrov, O. N. Druzhkov, Yu. Baryshnikov, T. K. Polznikova, G. I. Makin, and B. I. Kozyrkin, *J. Gen. Chem. USSR,* **50,** 2642 (1980).
182. Yu. A. Alexandrov, G. I. Makin, O. N. Druzhkov, Yu. Baryshnikov, T. K. Polznikova, and B. I. Kozyrkin, *J. Gen. Chem. USSR,* **51,** 70 (1981).
183. P. J. Krusic and J. K. Kochi, *J. Am. Chem. Soc.,* **91,** 3942 (1969).
184. K. G. Hancock and D. A. Dickinson, *J. Am. Chem. Soc.,* **95,** 280 (1973).
185. B. G. Ramsey and D. M. Anjo, *J. Am. Chem. Soc.,* **99,** 3182 (1977).
186. J. C. Scaiano and K. U. Ingold, *J. Chem. Soc., Chem. Commun.,* 878 (1975).
187. Q. B. Broxtermann, H. Hoogeveen, and D. M. Kok, *Tetrahedron Lett.,* 173 (1981).
188. B. Lockman and T. Onak, *J. Org. Chem.,* **38,** 2552 (1973).
189. H. Grandke and P. I. Paetzold, *Chem. Ber.,* **104,** 1134 (1971).
190. G. A. Razuvaev, G. G. Pelukhov, U. I. Scherbakov, O. N. Druzhkov, and S. F. Zhil'tsov, *J. Gen. Chem. USSR,* **37,** 1516 (1967).
191. E. C. Taylor, F. Kienzle, and A. McKillop, *J. Am. Chem. Soc.,* **92,** 6088 (1970).
192. E. C. Taylor, H. W. Altland, R. Danforth, G. McGillivray, and A. McKillop, *J. Am. Chem. Soc.,* **92,** 3520 (1970).
193. A. McKillop, J. D. Hunt, M. J. Zelesko, J. S. Fowler, E. C. Taylor, G. McGillivray, and F. Kienzle, *J. Am. Chem. Soc.,* **93,** 4841 (1971).
194. A. G. Davies, B. P. Roberts, and J. C. Scaiano, *J. Chem. Soc. B,* 2171 (1971).
195. D. Griller, K. U. Ingold, L. K. Pattison, J. C. Scaiano, and R. D. Small, *J. Am. Chem. Soc.,* **101,** 3780 (1979).
196. A. G. Davies and J. C. Scaiano, *J. Chem. Soc., Perkin Trans. 2,* 2234 (1972).
197. A. G. Davies, D. Griller, and B. P. Roberts, *J. Chem. Soc. B,* 1823 (1971).
198. H. C. Brown and M. M. Maitland, *Angew. Chem., Int. Ed. Engl.,* **11,** 692 (1972).
199. G. W. Kaballa and R. F. Daly, *J. Am. Chem. Soc.,* **95,** 4428 (1973).
200. A. G. Davies and B. P. Roberts, *J. Chem. Soc., Chem. Commun.,* 384 (1973).
201. S. F. Zhil'stov, U. I. Scherbakov, and O. N. Druzhkov, *J. Gen. Chem. USSR,* **39,** 1327 (1969).
202. Z. Floryanczyk, W. Kuran, S. Pasenkiewicz, and G. Kuas, *J. Organomet. Chem.,* **112,** 21 (1976).
203. Z. Floryanczyk, W. Kuran, S. Pasenkiewicz, and A. Krasnicka, *J. Organomet. Chem.,* **145,** 21 (1978).
204. A. Alberola, A. M. Gonzalez-Nogal, and F. J. Pulido, *Ann. Quim.,* **74,** 1147 (1978).
205. A. Alberola, A. M. Gonzalez-Nogal, and F. J. Pulido, *Ann. Quim., Ser. C.* **76,** 119 (1980).
206. C. F. Lane, *J. Organomet. Chem.,* **31,** 421 (1971).
207. J. E. Baekvall, M. U. Ahmed, S. Uemura, A. Toshimitsu, and T. Kawamura, *Tetrahedron Lett.,* **21,** 2283 (1980).
208. H. Kurosawa and M. Yashuda, *J. Chem. Soc., Chem. Commun.,* 716 (1978).
209. H. Kurosawa, H. Okada, and M. Yashuda, *Tetrahedron Lett.,* **21,** 959 (1980).
210. G. A. Ruzuvaev, G. A. Abakumov, E. S. Klimar, E. N. Gladyshev, and P. Yu, Bayushkin, *Izv. Akad. Nauk SSSR, Ser. Khim.,* 1128 (1976).
211. A. G. Davies and J. Lusztyk, unpublished observations.
212. J. E. Taylor and T. S. Milazzo, *J. Phys. Chem.,* **82,** 247 (1978).
213. P. W. N. M. Van Leeuwen, R. Kaptein, R. Huis, and W. I. Kalisvaart, *J. Organomet. Chem.,* **93,** C5 (1975).
214. I. P. Malysheva, V. A. Varyukkin, and O. S. D'yachkouskaya, *Khimiya Elementoorgan. Soedin. (Gorkii),* 51 (1978); *Ref. Zh., Khim.,* **12,** ll (1979).
215. L. Gammie, C. Sandorfy, and O. P. Strausz, *J. Phys. Chem.,* **83,** 2075 (1979).
216. A. Hudson, M. F. Lappert, and P. W. Lednor, *J. Chem. Soc., Dalton Trans.,* 2369 (1976).

217. J. Cooper, A. Hudson, and R. A. Jackson, *J. Chem. Soc., Perkin Trans. 2,* 1056 (1973).
218. A. G. Davies and M.-W. Tse, *J. Chem. Soc., Chem. Commun.*, 353 (1978).
219. P. J. Barker, A. G. Davies, and J. A.-A. Hawari, *J. Organomet. Chem.*, **187,** C7 (1980).
220. A. G. Davies, *Proceedings A.M.P.E.R.E. Colloq. CNRS, Organic Free Radicals, Aix-en-Provence,* 1977.
221. J.A.-A. Hawari, *PhD Thesis,* University College, London, 1981.
222. N. S. Uyazankin, E. N. Gladyshev, and G. A. Ruzavaev, *Dokl. Akad. Nauk SSSR,* **153,** 104 (1963).
223. G. A. Razuvaev, N. S. Uyazankin, and O. S. D'yachkovskaya, *J. Gen. Chem. USSR,* **32,** 2161 (1962).
224. M. Kira, M. Watanabe, and H. Sakurai, *J. Am. Chem. Soc.*, **99,** 7760 (1977).
225. A. G. Davies and J. C. Scaiano, *J. Chem. Soc., Perkin Trans 2,* 1777 (1973).
226. R. M. Haigh, A. G. Davies, and M.-W. Tse, *J. Organomet, Chem.*, **174,** 163 (1979).
227. B. C. Pant, *J. Organomet. Chem.*, **66,** 321 (1974).
228. D. Hanssgen and E. Odenhauser, *J. Organomet. Chem.*, **124,** 143 (1977).
229. A. G. Davies, B. P. Roberts, and M.-W. Tse, *J. Chem. Soc., Perkin Trans. 2,* 1499 (1977).
230. M. Kira, M. Watanabe, and H. Sukurai, *J. Am. Chem. Soc.*, **102,** 5202 (1980).
231. G. Ayrey, F. P. Man, and R. C. Poller, *J. Organomet. Chem.*, **173,** 171 (1979).
232. S. G. Kuker, A. I. Prokof'ev, N. N. Bubnov, S. P. Solodnikov, E. D. Korniets, D. N. Kravtsov, and M. I. Kabachnik, *Dokl. Akad. Nauk SSSR,* **229,** 877 (1976).
233. H. C. Gardner and J. K. Kochi, *J. Am. Chem. Soc.*, **97,** 1855 (1975).
234. N. A. Clinton and J. K. Kochi, *J. Organomet. Chem.*, **56,** 243 (1973).
235. C. L. Wong and J. K. Kochi, *J. Am. Chem. Soc.*, **101,** 5593 (1979).
236. S. Fukuzumi, K. Mochida, and J. K. Kochi, *J. Am. Chem. Soc.*, **101,** 5961 (1979).
237. R. A. Marcus, *J. Chem. Phys.*, **23,** 966 (1956).
238. R. A. Marcus, *J. Chem. Phys.*, **26,** 867 (1957).
239. S. Fakuzami, C. L. Wong, and J. K. Kochi, *J. Am. Chem. Soc.*, **102,** 2428 (1980).
240. K. L. Rollick and J. K. Kochi, *J. Org. Chem.*, **47,** 435 (1982).
241. K. L. Rollick and J. K. Kochi, *J. Am. Chem. Soc.*, **104,** 1319 (1982).
242. M. Gielen and I. Van Eynde, *J. Organomet. Chem.*, **218,** 315 (1981).
243. P. J. Krusic, P. J. Fagan, and J. S. San Filippo, *J. Am. Chem. Soc.*, **99,** 250 (1977).
244. M. Gielen and I. Van Eynde, *Transition Met. Chem.*, **6,** 344 (1981).
245. R. J. Klinger and J. K. Kochi, *J. Am. Chem. Soc.*, **102,** 4740 (1980).
246. B. W. Walther and F. Williams, *J. Chem. Soc., Chem. Commun.*, 270 (1982).
247. M. C. R. Symons, personal communication.
248. J. S. Thayer, *J. Organomet. Chem.*, **76,** 265 (1976).
249. W. Ipatiew and G. Rasuwajew, *Chem. Ber.*, **63,** 1110 (1930).
250. P. B. Ayscough and H. J. Emeleus, *J. Chem. Soc.*, 3381 (1954).
251. J. W. Dale, H. J. Emeleus, R. N. Hazeldine, and J. H. Moss, *J. Chem. Soc.*, 3708 (1957).
252. N. N. Travkin, B. K. Skachkov, I. G. Tonoyan, and B. I. Kosyrkin, *J. Gen. Chem. USSR,* **48,** 276 (1978).
253. G. O. Doak and L. D. Freedman, *Organometallic Compounds of Arsenic, Antimony and Bismuth,* Wiley, New York, 1970.
254. K.-H. Mitschke and H. Schmidbauer, *Chem. Ber.*, **106,** 3645 (1973).
255. K. Shen, W. E. McEwan, and A. P. Wolf, *J. Am. Chem. Soc.*, **91,** 1283 (1969).
256. G. A. Razuvaev, N. A. Osanove, N. P. Shulov, and B. M. Tsigin, *J. Gen. Chem. USSR,* **37,** 216 (1967).
257. W. E. McEwan and F. L. Chupka, *Phosphorus Sulphur,* **1,** 227 (1972).
258. F. L. Chupka, J. W. Knapczyk, and W. E. McEwan, *J. Org. Chem.*, **42,** 1399 (1977).
259. J. L. Wardell and D. W. Grant, *J. Organomet Chem.*, **149,** C13 (1978).
260. S. G. Smirnov, A. N. Rodinov, K. L. Royozhin, O. P. Syalkina, E. M. Parnov, D. N. Shigorin, and K. A. Kocheskev, *Izv. Akad. Nauk SSSR, Ser. Khim.*, 335 (1976).
261. J. R. M. Giles and B. P. Roberts, *J. Chem. Soc., Perkin Trans. 2,* 1211 (1981).
262. W. T. Cook, J. S. Vincent, I. Bernal, and F. Ramirez, *J. Chem. Phys.*, **61,** 3479 (1974).
263. J. D. Preer, F.-D. Tsay, and H. B. Gray, *J. Am. Chem. Soc.*, **94,** 1875 (1972).
264. D. H. Hey, G. A. Shingleton, and G. H. Williams, *J. Chem. Soc.*, 5612 (1963).

265. J. S. Connor, P. J. Young, and O. P. Strausz, *J. Am. Chem. Soc.*, **93,** 822 (1971).
266. M. Dankowski, K. Praefic, J.-Si. Lee, and S. C. Nyburg, *Phosphorus Sulphur,* **8,** 359 (1980).
267. J. K. Kochi and P. J. Krusic, *J. Am. Chem. Soc.*, **91,** 3944 (1969).
268. J. W. Cooper and B. P. Roberts, *J. Chem. Soc., Perkin Trans. 2,* 808 (1976).
269. A. G. Davies, R. W. Dennis, and B. P. Roberts, *J. Chem. Soc., Perkin Trans. 2,* 1101 (1974).
270. R. S. Hay and B. P. Roberts, *J. Chem. Soc., Perkin Trans. 2,* 770 (1978).
271. S. A. Buckler, *J. Am. Chem. Soc.*, **84,** 3093 (1962).
272. W. G. Bentrude, E. R. Hansen, W. A. Khan, T. B. Min, and P. E. Rogers, *J. Am. Chem. Soc.*, **95,** 2286 (1973).
273. A. G. Davies, D. Griller, and B. P. Roberts, *J. Organomet. Chem.*, **38,** C8 (1972).
274. E. Furimsky, J. A. Howard, and J. R. Morton, *J. Am. Chem. Soc.*, **95,** 6574 (1973).
275. A. Alberti and A. Hudson, *J. Organomet. Chem.*, **182,** C49 (1979).
276. A. Alberti, A. Hudson, A. Maccioni, G. Fodda, and G. F. Pedulli, *J. Chem. Soc., Perkin Trans. 2,* 1274 (1981).

The Chemistry of the Metal—Carbon Bond, Volume 2
Edited by F. R. Hartley and S. Patai


CHAPTER **4**

# Insertions into main group metal–carbon bonds

J. L. WARDELL and E. S. PATERSON
*Department of Chemistry, University of Aberdeen, Meston Walk, Old Aberdeen AB9 2UE, Scotland, UK*

---

## I. INTRODUCTION

This chapter considers insertion reactions into carbon—metal bonds of main group organometallics and is divided into sections: reactions with carbon—oxygen, carbon—sulphur, carbon—carbon, and carbon—nitrogen multiple bonds. A final section is concerned with miscellaneous insertions, which do not fit into the earlier categories. This includes, for example, insertions by elements from Group VI and sulphur oxides.

Insertion reactions include some of the most frequently used laboratory reactions (e.g. additions to aldehydes and ketones as well as to other carbonyl-containing compounds) and some of the more important industrial reactions, e.g. polymerization of alkenes and dienes, while others only have a curiosity value.

Although many of the reactions involve a polar mechanism, i.e. attack by carbanionic grouping, R, from RM, with possibly additional involvement of the metal to give a cyclic transition state, others proceed via radical and electron-transfer mechanisms. Evidence for the electron-transfer mechanism has been increasingly found in the past few years. All the reactions included here have the one common feature of an insertion as a critical step, whether it be 1,2-, 1,4-, 1,6-, or any other 1,$n$- formal addition. Subsequent reactions or work-up procedures, of course, may mask this initial reaction.

Only where necessary is mention made of transition metal catalysed reactions as a detailed survey could take this chapter deep into transition metal chemistry. However, it appeared to the author useful to make some passing reference in the chapter, for example, to conjugate (1,4-) additions to $\alpha,\beta$-unsaturated systems of cuprates, generated from organo-lithium or -magnesium reagents, and also to Ziegler–Natta catalysts.

Individual reactions for particular metals are covered more fully in appropriate books and articles; these should be consulted in particular for further examples. One review[1] published in 1967 was concerned solely with insertion reactions; it's coverage, however, included both transition metal and main group compounds in insertion reactions involving any metal—element bond.

## II. INSERTIONS INTO CARBON—OXYGEN MULTIPLE BONDS

Main group organometallics react with carbonyl-containing compounds, including aldehydes, ketones, carboxylic acids, esters, and isocyanates, as well as carbon oxides.

### A. Reactions with Aldehydes and Ketones

#### 1. *Scope*

A general review on carbonyl additions was published in 1966[2]. These reactions are arguably the most frequently used organometallic reactions, being among the most valuable preparations of alcohols available to organic chemists.

$$\mathrm{RM} + \mathrm{R'(H)R''C{=}O} \longrightarrow \mathrm{R'{-}C(R)((H)R''){-}O^-\ M^+} \xrightarrow{\mathrm{H_3O^+}} \mathrm{R'{-}C(R)((H)R''){-}OH} \qquad (1)$$

Examples of additions to aldehydes and ketones are known for most main group elements. References for alkali metals, beryllium, magnesium, zinc, cadmium, and Group III elements, for example, are to be found in general discussions or reviews on the organometallic chemistry of the particular element[3,4], while typical references for other elements are calcium, strontium, and barium[5], boron[6,7], mercury[8], silicon[9–13], and tin[11–23].

Only a relatively few reactions are known for the less electropositive elements: mercury, silicon, and tin. For these metals, it is only the most reactive organic derivatives that take part in additions to carbonyl groups, e.g. compounds having $XCH_2$-metal bonds ($X = CO_2R$, COR, $CONR_2$, or CN; metal = mercury[8], silicon[13], or tin[15]), allyl—metal bonds (metal = silicon[9–12] or tin[11,14,18–23]), polyhalophenyl—silicon[16] and ethynyl—tin[17] bonds (e.g. equations 2–4). Even then, a Lewis acid catalyst (e.g. $BF_3$, $GaCl_3$, $ZnCl_2$, or $TiCl_4$) may be necessary for reactions with aldehydes and ketones not having strong electron-withdrawing groups. Base catalysts have also been used for $R_3SiCH_2X$–$R_2CO$ reactions[13]. Although only a limited number of carbon—tin and carbon—silicon bonds react, these reactions (especially of the allylic derivatives) have proved to be of great synthetic value. One advantage of using these organometallics is that a large number of functional groups can be tolerated. The tin and silicon alkoxides can be readily isolated; hydrolyses to alcohols are conveniently achieved using acids. As found with other organometallics, aldehydes are more reactive than ketones.

$$Hg(CH_2COMe)_2 \xrightarrow[\text{(ii) } H_3O^+]{\text{(i) RCHO}} RCH(OH)CH_2COMe \qquad (2)^{8}$$

$$Me_3SiCH_2CH{=}CHR' + RCHO \xrightarrow[-25^\circ C]{AlCl_3, CH_2Cl_2} RCH(OSiMe_3)CHR'CHCH{=}CH_2 \qquad (3)^{9}$$

$$Me_3SnCH_2CH{=}CH_2 + m\text{-}O_2NC_6H_4CHO \xrightarrow[-78\,^\circ C,\ 1/2h]{BF_3 \cdot OEt_2, CH_2Cl_2} m\text{-}O_2NC_6H_4CH(OSnMe_3)CH_2CH{=}CH_2 \qquad (4)^{22}$$

The most studied reactions are those involving organo-magnesium and -lithium reagents, and to a lesser extent organoaluminium compounds. The complete range of organometal derivatives of these elements react with carbonyl compounds. As these insertion reactions are used in the main for preparations of alcohols, the initial alkoxide insertion products are not usually isolated but are hydrolysed to the alcohols. Moreover, the organo-magnesium or -lithium reagents, prepared for example from organic halides or hydrocarbons, are not isolated either and so the various stages (preparation of organometallic through to alcohol) can be performed successively in the same reaction pot. Procedures (including the Barbieri synthesis) have also been published for the simultaneous addition of the organic halide and carbonyl compound to the metal[24]. As given in equation 1, the reaction is simply shown as a 1,2-addition reaction. In reality, the situation can be appreciably more complex[25,26]. For appropriate combinations of RM and carbonyl compounds, formal 1,4- and/or 1,6-additions or other 1,*n*-, e.g. 1,3-additions[27], can also arise. This becomes important for sterically hindered compounds (see Scheme 1[25]). The greater the hindrance about the carbonyl group, the more chance of 1,4- and/or 1,6-additions. Apparently organo-lithiums provide more 1,2-additions to hindered carbonyls than do the corresponding Grignard reagents.

$$RCOPh \xrightarrow[\text{(ii) } H_3O^+]{\text{(i) MeMgBr}} RC(OH)(Me)Ph + \underset{\text{1,4-addition}}{RCOC_6H_4Me\text{-}o} + \underset{\text{1,6-addition}}{RCOC_6H_4Me\text{-}p}$$

$R = 2,3,5,6\text{-}Me_4C_6H$

$$\left[\xrightarrow{-H_2O} RC({=}CH_2)Ph\right]$$

SCHEME 1

Replacement of groups *ortho* or *para* to the carbonyl groups can also occur[28–30], e.g. equations 5 and 6.

$$o\text{-}XC_6H_4COR \xrightarrow[\text{(ii) } H_2O]{\text{(i) } PhCH_2MgX} o\text{-}(PhCH_2)C_6H_4COR \qquad (5)^{28}$$

$X = Br$ or $OMe$
$R = 2,4,6\text{-}Me_3C_6H_2$

TABLE 1. Products of reaction of alkyl-Grignards[2] with $Pr^i_2CO$

| Alkyl-Grignard | Enolization (%) | Reduction (%) | Addition (%) |
|---|---|---|---|
| MeMgX | 0 | 0 | 95 |
| EtMgX | 2 | 21 | 77 |
| PrMgX | 2 | 60 | 36 |
| $Pr^iMgX$ | 29 | 65 | 0 |
| $Me_3CCH_2MgX$ | 90 | 0 | 4 |

$$p\text{-}XC_6H_4COR \xrightarrow[\text{(ii) } H_2O]{\text{(i) } R^1MgX} p\text{-}R^1C_6H_4COR \qquad (6)^{30}$$

X = CN or OMe
$R = 2,3,5,6\text{-}Me_4C_6H$
$R^1 = PhCH_2$ or $Bu^t$

Other by-products can also arise from competing reactions, such as:
(i) *enolization*[31,32], when $\alpha$-hydrogens are present in the carbonyl compound, e.g. equation 7. Enolizations have been found to increase in the order metal = potassium > sodium > lithium > magnesium.

$$RCOCH_2R^1 + R^2MgX \longrightarrow \left[ \text{cyclic transition state: R, C=O, Mg–X, H–C(R}^1\text{), H}\cdots\text{R}^2 \right] \longrightarrow RC(=CHR^1)OMgX + R^2H \qquad (7)$$

(ii) *reductions to hydrols*, when $\beta$-hydrogens are present in the organometallics, e.g. equation 8. Variations of ratios of enolization, reduction, and 1,2-addition products of alkyl-Grignard reactions with $Pr^i_2CO$ are shown in Table 1. Organolithiums give much less enolization and reduction products than do Grignard reagents. Organoaluminium reagents, especially those with $\beta$-hydrogens, give appreciable amounts of reduction products.

$$RCOR^1 + R^2CH_2CH_2MgX \longrightarrow \left[ \text{cyclic transition state: R, R}^1\text{C–O}\cdots\text{Mg(X)–CH}_2\text{, H}\cdots\text{C(H)(R}^2\text{)} \right] \longrightarrow RR^1CHOMgX + R^2CH{=}CH_2 \qquad (8)$$

and (iii) formation of pinacols, $RR^1C(OH)C(OH)RR^1$.

### 2. Mechanisms

The most extensive mechanistic study has been made on Grignard reactions. These will be considered first.

a. *Mechanism of Grignard addition to ketones*[33]. For a complete understanding of the reaction, regard has to be paid to the composition of the Grignard reagent. This is because different species have different reactivities. Evidence for various species and equilibria in solution is well documented. For example, in $Et_2O$ solution, RMgCl

$$\text{Higher aggregates} \rightleftharpoons 2/3\ \text{Trimers} \rightleftharpoons \text{Dimers} \rightleftharpoons 2\text{RMgX} \rightleftharpoons R_2Mg + MgX_2 \quad (9)$$

exists as dimers even up to 3.5 M, while for RMgBr and RMgI increasing aggregation results with increasing concentration. In thf solution, monomers of RMgX (X=R, Cl, or Br) are found, while RMgF and RMgOR are dimeric.

Two distinct mechanistic types have been recognized; these are a polar mechanism (a four-centre pericyclic concerted mechanism)[34] and a single electron transfer (s.e.t.) mechanism (see Scheme 2). The latter mechanism accounts for the 1,4- and 1,6-adducts, the pinacol products, and other free radical derived products.

As pointed out by Ashby[33], several factors are important in deciding the mechanistic type. These are (i) the nature of the organic group in the Grignard reagent, (ii) the nature of the ketone, (iii) the purity of the magnesium, and (iv) the solvent. A collection of factors which favour a polar pathway are Grignard reagents that are difficult to oxidize (e.g. primary alkyl Grignards), ketones that are difficult to reduce (e.g. dialkyl ketones), solvents of low polarity (e.g. $Et_2O$), and Grignards made from pure magnesium. Conversely, factors which favour a s.e.t. mechanism are easily oxidized Grignards (e.g. tertiary-alkyl Grignards), easily reduced ketones (e.g. diaryl

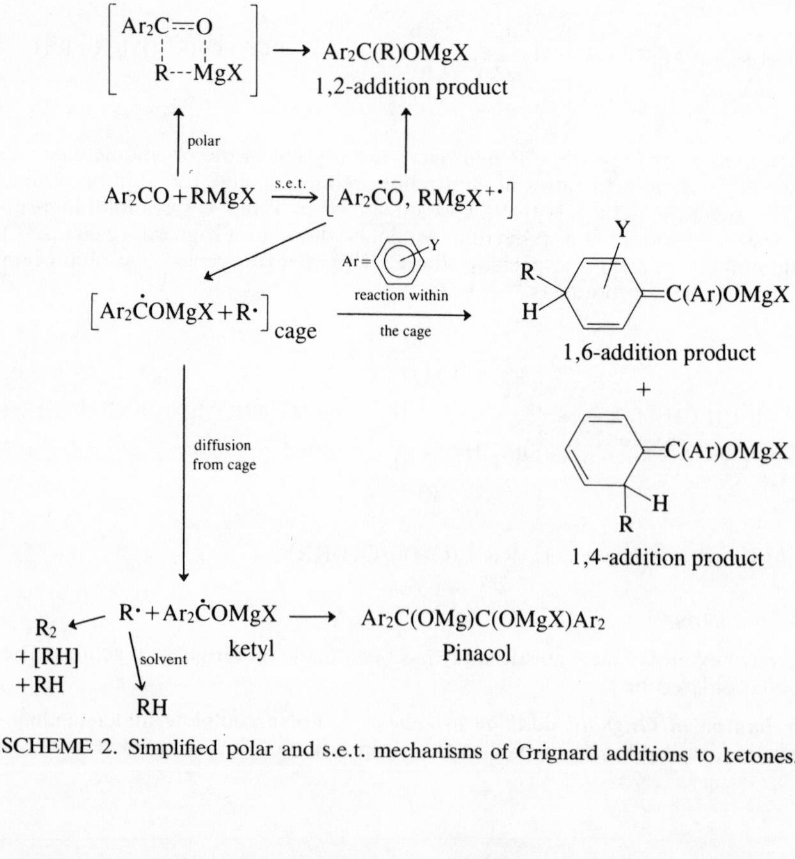

$[Ar_2C{=}O \cdots R{-}{-}{-}MgX] \rightarrow Ar_2C(R)OMgX$ (1,2-addition product)

polar

$Ar_2CO + RMgX \xrightarrow{\text{s.e.t.}} [Ar_2\dot{C}\bar{O}, RMgX^{+\cdot}]$

Ar = $C_6H_4Y$

$[Ar_2\dot{C}OMgX + R\cdot]_{\text{cage}}$ —reaction within the cage→ 1,6-addition product + 1,4-addition product

(R, H)(Y)C=C(Ar)OMgX

1,6-addition product

+

=C(Ar)OMgX, H, R

1,4-addition product

diffusion from cage

$R\cdot + Ar_2\dot{C}OMgX \longrightarrow Ar_2C(OMg)C(OMgX)Ar_2$

ketyl Pinacol

$R_2$ + [RH] + RH ← R·

solvent → RH

SCHEME 2. Simplified polar and s.e.t. mechanisms of Grignard additions to ketones.

ketones), more polar solvents e.g. thf or hmpa), and Grignards made from magnesium containing a first-row transition metal impurity. The methyl-Grignard–$Ph_2CO$ and $Bu^tMgBr$–$Me_2CO$ reactions are both affected by transition metal impurities, while the $Bu^tMgBr$–$Ph_2CO$ (always s.e.t.) and MeMgBr–$Me_2CO$ (always polar addition) reactions are unaffected[35].

In Scheme 3 is presented a fairly comprehensive mechanism of the reaction of a ketone with the Grignard reagent from methyl bromide under conditions where monomeric species prevail and where only 1,2-addition occurs. Such a system for example pertains[36] using excess $PhCOC_6H_4Me$-*o*, methyl-Grignard (less than 0.1 M) made from single-crystal magnesium (>99.9995% pure) and $Et_2O$ as the reaction medium (equation 10). Under these conditions, the order in Grignard reagent was

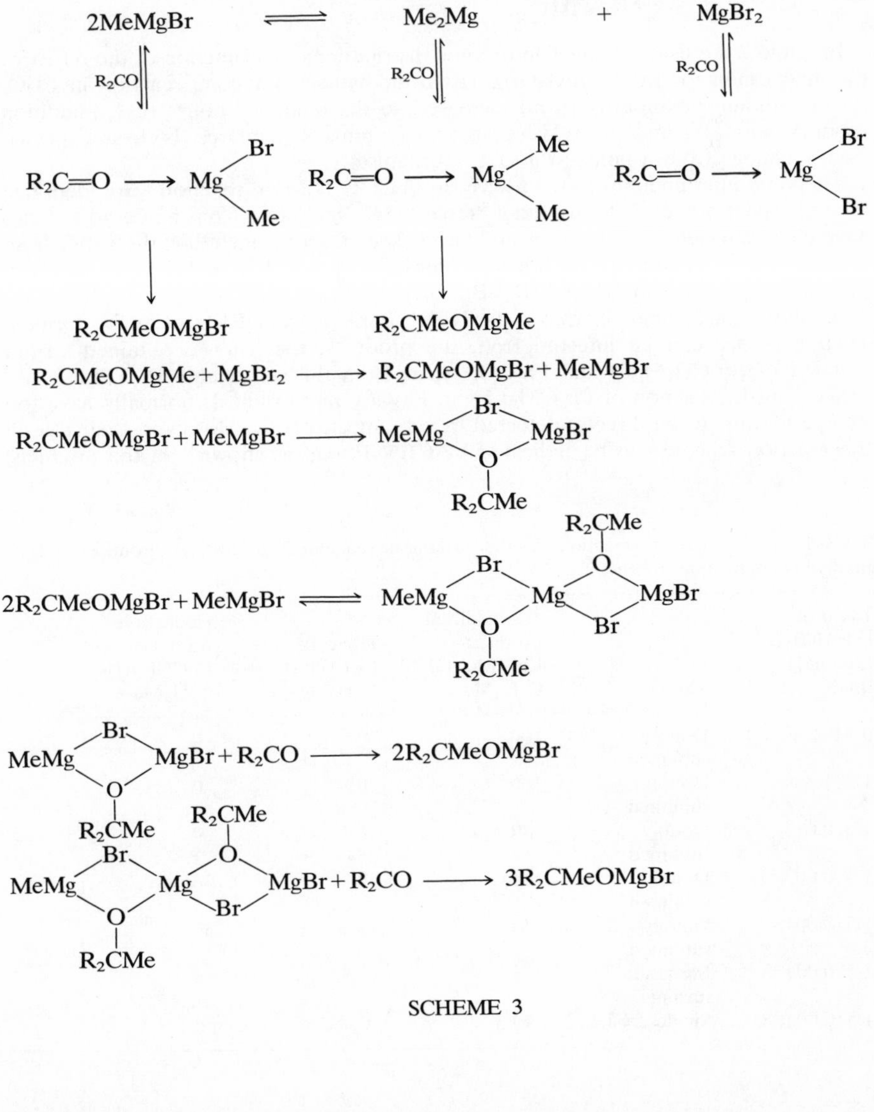

$$2MeMgBr \rightleftharpoons Me_2Mg + MgBr_2$$

$$2MeMgBr \xrightleftharpoons{R_2CO} R_2C{=}O \rightarrow Mg(Br)(Me) \longrightarrow R_2CMeOMgBr$$

$$Me_2Mg \xrightleftharpoons{R_2CO} R_2C{=}O \rightarrow Mg(Me)(Me) \longrightarrow R_2CMeOMgMe$$

$$MgBr_2 \xrightleftharpoons{R_2CO} R_2C{=}O \rightarrow Mg(Br)(Br)$$

$$R_2CMeOMgMe + MgBr_2 \longrightarrow R_2CMeOMgBr + MeMgBr$$

$$R_2CMeOMgBr + MeMgBr \longrightarrow MeMg(\mu\text{-}Br)(\mu\text{-}OCMeR_2)MgBr$$

$$2R_2CMeOMgBr + MeMgBr \rightleftharpoons MeMg(\mu\text{-}Br)(\mu\text{-}OCMeR_2)Mg(\mu\text{-}OCMeR_2)(\mu\text{-}Br)MgBr$$

$$MeMg(\mu\text{-}Br)(\mu\text{-}OCMeR_2)MgBr + R_2CO \longrightarrow 2R_2CMeOMgBr$$

$$MeMg(\mu\text{-}Br)(\mu\text{-}OCMeR_2)Mg(\mu\text{-}OCMeR_2)(\mu\text{-}Br)MgBr + R_2CO \longrightarrow 3R_2CMeOMgBr$$

SCHEME 3

$$PhCOC_6H_4Me\text{-}o \xrightarrow[\text{(ii) } H_2O]{\text{(i) MeBr,Mg,Et}_2\text{O}} PhC(OH)MeC_6H_4Me\text{-}o \qquad (10)$$

found to be unity; both MeMgBr and $Me_2Mg$ were involved in the addition. Although the reactivity of $Me_2Mg$ is *ca.* ten times that of MeMgBr, there is *ca.* ten times less of it; hence there is about an equal amount of reaction via these two species.

Complexation of ketones by magnesium species[36,37] can be detected by the u.v. or i.r. spectral changes, e.g. see equation 11.

$$\underset{\nu_{CO}\ 1695\ cm^{-1}}{RMgBr + R^1COMe} \underset{K=14\ l\ mol^{-1}}{\overset{Et_2O, 25\,°C}{\rightleftharpoons}} \underset{\underset{\nu_{CO} 1658\ cm^{-1}}{Br}}{RMg} \longleftarrow O{=}C(Me)R^1 \qquad (11)^{37}$$

R = cyclopentyl
$R^1 = p\text{-}MeSC_6H_4$

In Table 2 are listed changes in product distributions as a function of the purity of the magnesium for the methyl-Grignard–*o*-methylbenzophenone reaction in $Et_2O$. As the amount of impurity (iron) increases, so the relative amount of 1,2-addition product decreases and the relative amount of pinacol increases, both as expected from a change from a polar to a s.e.t. mechanism.

The ketyl intermediates in the MeMgBr–$PhCOC_6H_4Me$-*o* reaction were identified by e.s.r. spectroscopy[38]. Many other stable ketyls generated from hindered ketones have been detected[26,30,39] by e.s.r. and visible spectroscopy, including the ketyls from reactions of 2,4,6-$Me_3C_6H_2COPh$ with PhMgBr and of 2,3,5,6-$Me_4C_6HCOC_6H_4Y$ (Y = OMe or CN) with RMgCl (R = $Bu^t$ or $PhCH_2$) in thf.

As shown in Scheme 2, free radicals, $R^\bullet$, derived from RMgX can be formed. Their presence can be inferred from the products, e.g. $Bu^tMe$ obtained[40] from reaction of $Bu^tCH_2MgCl$ and $Ph_2CO$. Apart from the very stable $R^\bullet = Ph_3C^\bullet$, detected[41] in the reaction of $Ph_3CMgCl$ and $Ph_2CO$, these radicals normally have too short a lifetime to be directly detected by e.s.r spectroscopy. However, their role in the reaction scheme can be indicated by CIDNP, e.g. as shown[42] in the products,

TABLE 2. Products of reaction of the Grignard reagent from methyl bromide and *o*-methylbenzophenone in ether[33]

| Ratio of [MeMgBr]: [ketone] ($mol\ l^{-1}$) | Mg | 1,2-Addition product: PhC(Me)(OH)-$C_6H_4Me$-*o* | Pinacol: (—C(Ph)(OH)-$C_6H_4Me$-*o*$)_2$ | Reduction product: PhCH(OH)-$C_6H_4Me$-*o* |
|---|---|---|---|---|
| 0.01:1 | Doubly sublimed | 100 | 0 | 0 |
| 1.5:1.5 | Doubly sublimed | 100 | 0 | 0 |
| 1.5:0.015 | Doubly sublimed | 89.5 | 1.5 | 9 |
| 1.5:0.00375 | Doubly sublimed | 62 | 2 | 36 |
| 1.5:0.00188 | Doubly sublimed | 40 | 4 | 56 |
| 1.5:0.00188 | Mg grade turnings | 55 | 18.5 | 8 |
| 1.5:0.00188 | Single crystal | 99.5 | 0 | 0.5 |

$$H_2C{=}CHCH_2CH_2CMe_2CH_2MgCl + Ph_2CO \xrightarrow[\text{s.e.t.}]{\text{Et}_2\text{O, room temp}} \left[\begin{array}{c} Ph_2\dot{C}OMgX \\ H_2C{=}CH(CH_2)_2CMe_2CH_2^{\bullet} \end{array}\right]$$

etc.

R = 3,3-dimethylcyclopentyl

$$Ph_2C(OMgX)CH_2R \longleftarrow \left[\begin{array}{c} Ph_2\dot{C}{-}OMgX, \\ \dot{C}H_2R \end{array}\right]$$

SCHEME 4

$Me_2C{=}CH_2$ and $Me_3CH$, from reactions of equimolar $Bu^tMgBr$ and PhCOPh in dme, as well as among the products of reactions of $Bu^tMgCl$ and PhCHO. The cyclised products formed[43] in reactions of the primary Grignard, $H_2C{=}CH(CH_2)_2CMe_2CH_2MgCl$, and of the tertiary Grignard, $H_2C{=}CH(CH_2)_3CMe_2MgCl$, with $Ph_2CO$ also point to radical intermediates and hence to s.e.t. mechanisms even for primary Grignard reagents, e.g. Scheme 4.

The reaction of isomeric 2-norbornylmagnesium bromides and $H_2CO$ proceeds with retention of configuration, even in the presence of 5% $FECl_3$. It would thus appear that the s.e.t. mechanism does not apply in reactions of simple aliphatic aldehydes[44]. Ashby *et al.*[45] have shown that addition of salts can lead to a significant rate enhancement in $R_2Mg$–$R_2^1CO$ reactions; $LiClO_4$ is particularly effective. In addition a MeLi–$LiCuMe_2$ combination is about $10^3$ times more effective than MeLi alone[45].

b. *Mechanism of organolithium additions to ketones.* One area of study of organolithium additions to ketones has been concerned with establishing the aggregation state of the active organometallic species. Methyllithium is tetrameric in diethyl ether solution. Smith *et al.*[46], however, reported that reaction between $(MeLi)_4$ and 2,4-dimethyl-4′-(methylthio)benzophenone in $Et_2O$ was first order in ketone and one quarter order in methyl-lithium. This indicates that the reaction proceeds with monomeric methyllithium.

$$(MeLi)_4 \rightleftharpoons 4MeLi \tag{12}$$

$$MeLi + ArCOAr' \longrightarrow \text{products} \tag{13}$$

Another study in benzene solution involved butyllithium (hexameric) and *p*-$MeSC_6H_4COMe$. The reaction rate increased rapidly as the concentration of butyllithium was increased from 0.014 to 0.1 M, but beyond 0.1 M the rate remained fairly constant. While this could be interpreted as indicating active BuLi monomers, an alternative explanation involving a complexed hexamer, e.g. $(BuLi)_6{\cdot}O{=}CArR$, could not be dismissed[47].

The arrangements of groups in the transition state has also been investigated. The major rotamer obtained from reaction of *o*-tolyllithium and $Bu^t_2CO$ was found[48] to be antiplanar (ap); it was assumed that this is consistent with a four-centred transition state having the *o*-tolyl ring in the same plane as the carbonyl group and bisecting the $Bu^t$—C—$Bu^t$ angle. The alternative route to $Bu^t_2C(C_6H_4Me\text{-}o)OH$

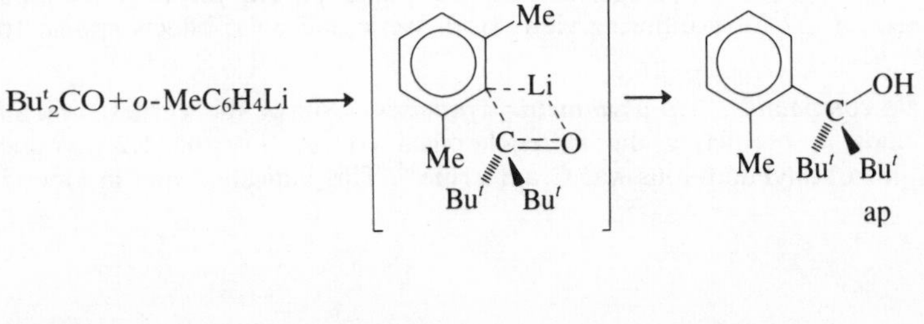

$$Bu^t_2CO + o\text{-}MeC_6H_4Li \longrightarrow [\text{transition state: } o\text{-tolyl (Me), Li}\cdots C{=}O, Bu^t, Bu^t] \longrightarrow o\text{-}MeC_6H_4C(OH)Bu^t_2 \ \text{(ap)} \tag{14}$$

using $Bu^tCOC_6H_4Me$-*o* and $Bu^tLi$ provides the sp rotamer. In the latter case, the dihedral angle between the carbonyl group and *o*-tolyl ring is 40° with the carbonyl group being close to the *o*-methyl group.

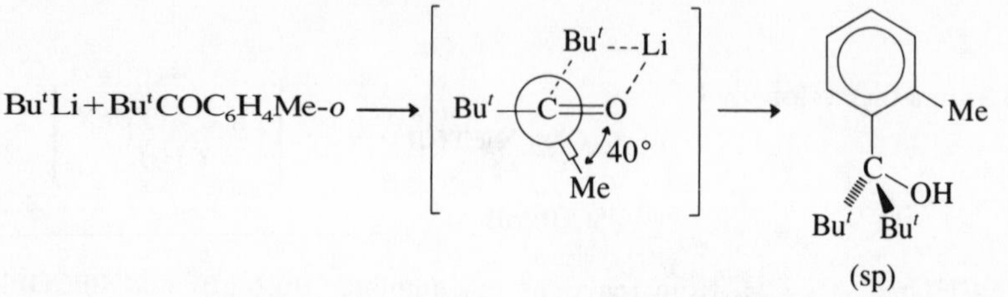

$$Bu^tLi + Bu^tCOC_6H_4Me\text{-}o \longrightarrow \left[ Bu^t\text{—C=O} \cdots Li\text{—}Bu^t,\ 40° \right] \longrightarrow (sp) \quad (15)$$

Bu^t---Li; Bu^t; C=O; 40°; Me; Me; C; OH; Bu^t Bu^t; (sp)

c. *Mechanism of organoaluminium additions to ketones*[49]. Triorganoaluminiums are monomeric in polar solvents (e.g. $Et_2O$) and exist as monomers and dimers in benzene solution.

The benzophenone–$Me_3Al$ reaction in $Et_2O$ solution was established as being first order in each component. An intermediate complex, $Ph_2CO:AlMe_3$ could be detected by u.v. However, in benzene solution, the reaction rate was considerably increased as the concentration of $Me_3Al$ increased from a 1:1 to a 2:1 $Me_3Al:Ph_2CO$ ratio. Both $Me_3Al$ molecules were involved in the transition (Scheme 5). Additions of salts, such as LiX ($X = ClO_4$, Br, I, or $OBu^t$) and MX (N = Na, K, or $Bu_4N$; X = halide), lead to rate reductions as a consequence of complexation with the ketone[45,50].

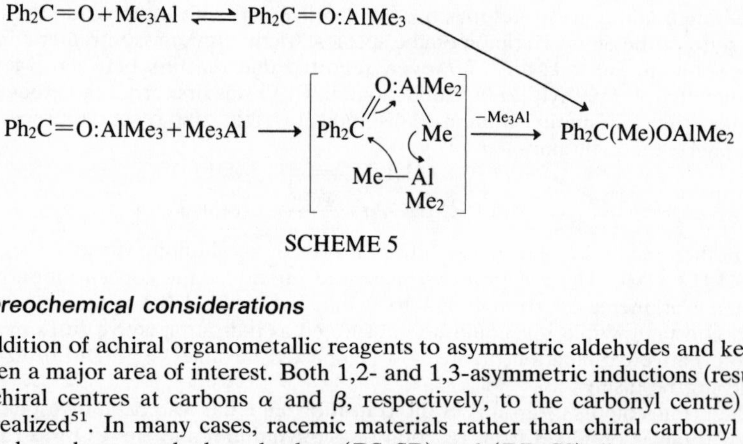

$$Ph_2C{=}O + Me_3Al \rightleftharpoons Ph_2C{=}O{:}AlMe_3 \longrightarrow Ph_2C(Me)OAlMe_2$$

$$Ph_2C{=}O{:}AlMe_3 + Me_3Al \longrightarrow \left[ Ph_2C(O{:}AlMe_2)\cdots Me\cdots Me\text{—}AlMe_2 \right] \xrightarrow{-Me_3Al} Ph_2C(Me)OAlMe_2$$

SCHEME 5

### 3. Stereochemical considerations

The addition of achiral organometallic reagents to asymmetric aldehydes and ketones has been a major area of interest. Both 1,2- and 1,3-asymmetric inductions (resulting from chiral centres at carbons $\alpha$ and $\beta$, respectively, to the carbonyl centre) have been realized[51]. In many cases, racemic materials rather than chiral carbonyl compounds have been used; these lead to (*RS*, *SR*) and (*RR*, *SS*) diastereomeric mixtures with basically the same information being gained. Considerable efforts have been spent on establishing models and experimental conditions for asymmetric induction. Almost all theories concerning nucleophile attack on carbonyl compounds consider attack to occur orthogonally to the plane of the carbonyl compound; however, see ref. 52 for a differing view. Both steric and polar effects appear to be important.

a. *Acyclic compounds: 1,2-asymmetric syntheses.* One of the earliest and more important aids in predicting the stereochemical course (i.e. the 1,2-asymmetric induction) in carbonyl additions was Cram's rule[53]. This indicated 'that in kinetically

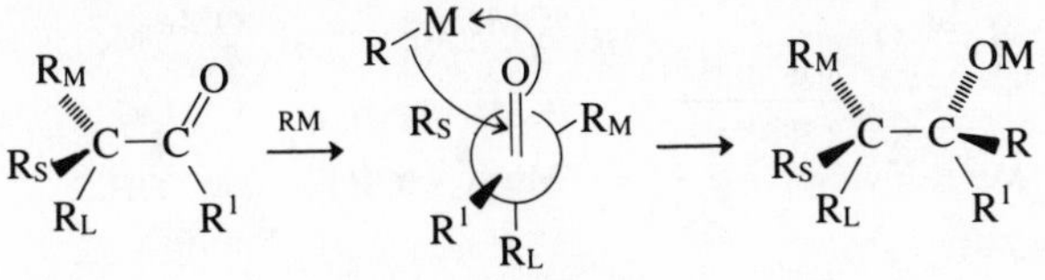

Favoured pathway

S, M, and L refer to small, medium, and large groups

SCHEME 6

controlled reactions the diastereomer will dominate which would be formed by the approach of the entering nucleophile from the less hindered side of the C═O bond when the rotational configuration of the C—C bond is such that the double bond is flanked by the two least hindered bulky groups, e.g. $R_S$ and $R_M$ in Scheme 6, attached to the asymmetric centres'. These reactions will have open transition states [open model]. An example[54] is given in equation 16.

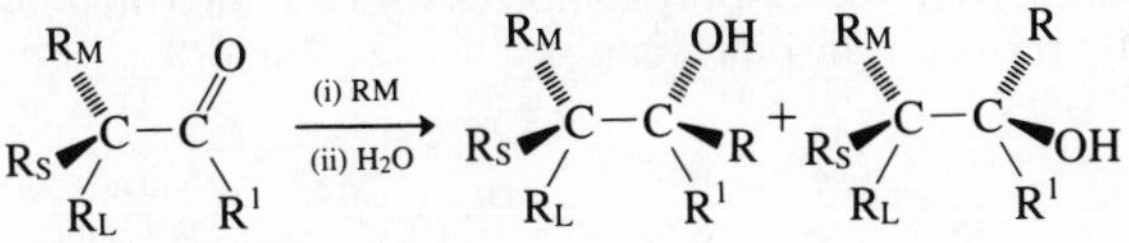

(i) RM; (ii) $H_2O$ → **(1)** + **(2)** (16)[54]

$R_S$=Me, $R_M$=Et, $R_L$=Ph

$R^1$=Me, RM=EtLi [**1**] : [**2**]=2.3 : 1

$R^1$=Et, RM=MeLi [**1**] : [**2**]=10.0 : 1

When a good donor group such as $NR_2$, or OR is present the situation may be altered as a consequence of the additional complexation between RM and the donor group, e.g. Scheme 7 and equations 17 and 18. In these cases a cyclic transition state occurs [cyclic model].

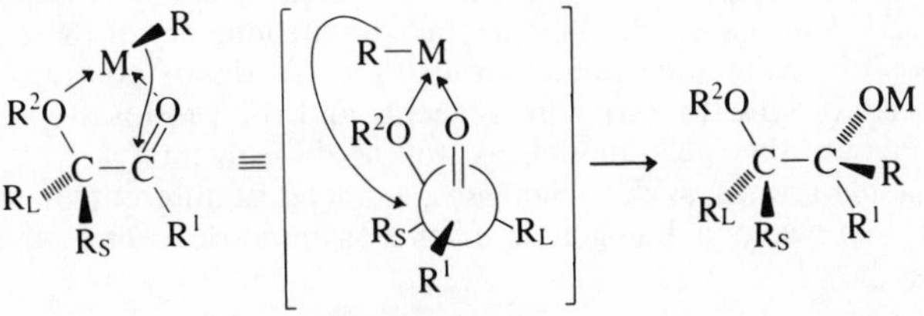

Favoured pathway

SCHEME 7

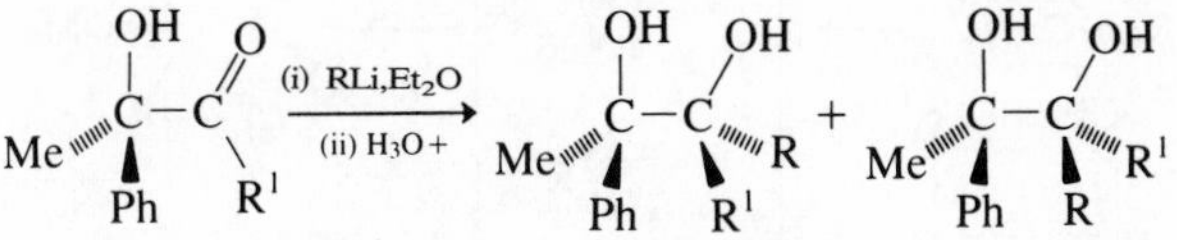

(i) RLi,$Et_2O$; (ii) $H_3O^+$ → **(3)** + **(4)** (17)[55]

$R^1$=Me, R=Ph [**3**] : [**4**]=7 : 1

$R^1$=Ph, R=Me [**3**] : [**4**]=11 : 1

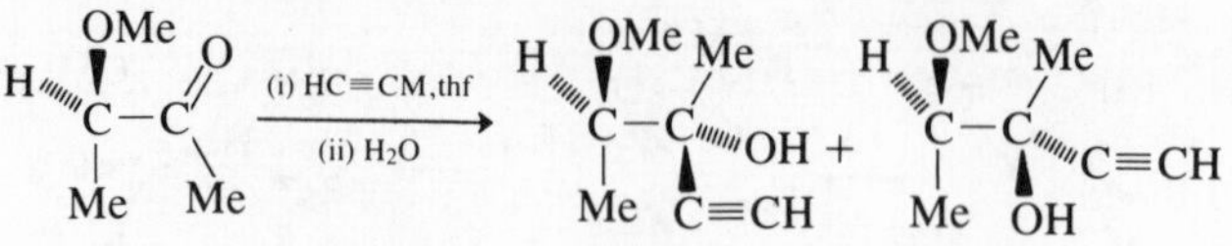

$$\text{MeCH(OMe)COMe} \xrightarrow[\text{(ii) } H_2O]{\text{(i) } HC{\equiv}CM\text{, thf}} \text{MeCH(OMe)C(Me)(OH)C}{\equiv}\text{CH } \mathbf{(5)} + \mathbf{(6)} \qquad (18)^{56}$$

$M = MgX$ [**5**] : [**6**] = 83 : 17
$M = Li$ [**5**] : [**6**] = 90 : 10
$M = Na$ [**5**] : [**6**] = 75 : 25

Examples with neighbouring amino groups have recently been reported by Lattes and coworkers[57], equation 19. The (*RR* + *SS*) diastereoisomer is favoured in benzene

$$\text{RNMeCHMeCOMe} \xrightarrow[\text{(ii) } H_2O]{\text{(i) } R^1MgX} \text{RNMeCHMeC(OH)}R^1\text{Me} \quad (RR+SS) \text{ and } (RS+SR) \qquad (19)$$

solution, in which a cyclic transition state occurs. In contrast in the more powerful coordinating solvents, e.g. thf and $NEt_3$, the (*RS* + *SR*) diastereomer dominates. In the latter case the solvent successfully competes with the built-in donor group for the magnesium and an open transition state results, see **7** and **8**.

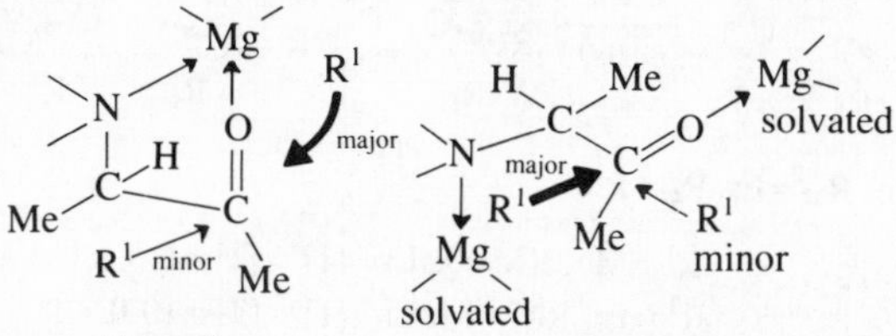

Mg, N, H, C, Me, O, C, $R^1$, major, $R^1$ minor, Me

(**7**) cyclic

H, Me, Mg, solvated, N, C, O, C, major, $R^1$, Mg, solvated, Me, $R^1$, minor

(**8**) open

As expected from the cyclic model, reaction of cyclo-$C_6H_{11}$NMeCHMeCOMe and PhMgBr in benzene produces more (*RR* + *SS*)-cyclo-$C_6H_{11}$NMeCHMeC(OH)PhMe, whereas from the cyclo-$C_6H_{11}$NMeCHMeCOPh–MeMgBr reaction the major diastereomer is the (*RS* + *SR*) form. Many other examples of similar compounds are to be found in ref. 51. Factors which influence the stereochemistry of these reactions of carbonyl compounds, containing donor groups, are the donor ability of the group, the halide associated with the Grignard reagent, and the presence of salts such as $MgBr_2$ (which favours the cyclic model) as well as the solvent.

Highly polarizable groups, such as halogens, can lead to different situations. It has been proposed that when a halogen is at the asymmetric centre, the important

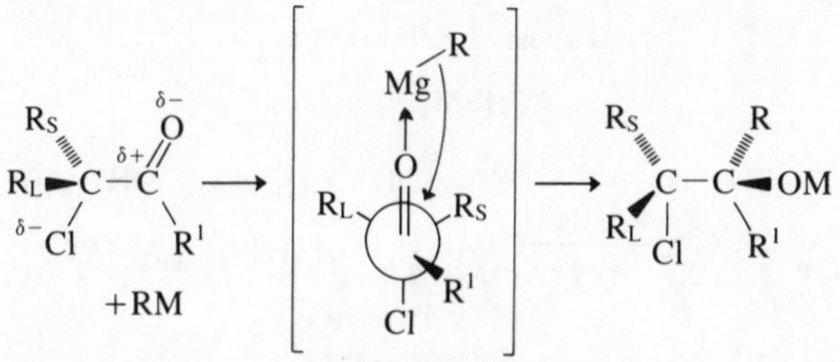

$R_S$, $R_L$, C, C, $\delta-$ O, $\delta+$, $\delta-$ Cl, $R^1$, + RM → [ R, Mg, O, $R_L$, $R_S$, $R^1$, Cl ] → $R_S$, R, C–C–OM, $R_L$, Cl, $R^1$

Favoured pathway

SCHEME 8

conformation in the addition reaction is the *trans* dipolar one [dipolar model][58], Scheme 8 and equation 20.

$$\mathrm{R_L(H)(Cl)C{-}CHO} \xrightarrow[\text{(ii) } H_2O]{\text{(i) RM}} \mathrm{R_L(H)(Cl)C{-}C(H)(R)OH}\ (\mathbf{9}) + \mathrm{R_L(H)(Cl)C{-}C(H)(OH)R}\ (\mathbf{10}) \qquad (20)^{58}$$

$R_L$=Bu; RM=EtMgBr [**9**] : [**10**]=70 : 30

$R_L$=Et; RM=BuMgBr mainly [**9**]

A useful discussion of the application of the various models has been made by Morrison and Mosher[51]. Clearly the extent of the 1,2-asymmetric induction depends on the differences in the steric bulk of the groups at the chiral centre. Designation of the groups as $R_S$, $R_M$, and $R_L$ is not completely straightforward. For the purpose of using such models, phenyl is considered larger than alkyl groups, including $PhCH_2$ and $Pr^i$ and probably $Bu^t$ also. Temperature, reagents, and solvents are factors which also influence the stereoselectivity.

Modifications of Cram's rule and other developments have been subsequently made by Karabatsos[59], Felkin and coworkers[60], and Perez-Ossorio and co-workers[61], among others. There are an infinite number of conformations of an enantiomer of $R_SR_MR_LCCOR'$. Cram considered one having staggered carbonyl and eclipsed R groups. The approach of Karabatsos, a semi-empirical treatment, made the assumptions (i) that little bond breaking and making occurred in the transition state, i.e. the transition state is reactant-like with eclipsed carbonyl and staggered R, and (ii) the diastereomeric transition states that control product stereospecificity have the smallest group, $R_S$, closest to the incoming reagent group. The Felkin approach made the assumptions (i) the transition states are reactant like, (ii) torsional strain (Pitzer strain) involving partial bonds in transition states represents a substantial fraction of the strain between fully formed bonds, even when the degree of bonding is low, (iii) the important transition state interactions involve RM (the entering reagent group) and R′ (the achiral group attached to the carbonyl carbon) rather than the interaction of the group on the chiral centre with the carbonyl oxygen (as envisaged by both Cram and Karabatsos), and (iv) polar effects stabilize those transition states in which the separation between R and the electronegative groups ($R_L$, $R_M$ and $R_S$) is greatest and destabilize others. These assumptions lead to a conformation in which all groups are staggered. Perez-Ossorio and coworkers also considered reactant-like transition states. However they considered all possible (Karabatsos-like) conformers and calculated for each conformer the molar ratio of carbinol products. The overall stereoselectivity was based on the summation from each conformer. All transition states were considered *ab initio* but the evaluation of their differential energy contents involved *a posteriori* selection. Figure 1 illustrates typical transition state arrangements for each of the three approaches.

These approaches, particularly that of Perez-Ossorio and coworkers, have been successfully used with reactions[62] such as those of RCHMeCHO(R= $Bu^t$, Ph or $PhCH_2$) and PhCHMeCOMe, e.g. equation 21.

$$\mathrm{PhCHMeCHO + PhMgBr} \xrightarrow[\text{(ii) } H_2O]{\text{(i) } Et_2O} \mathrm{PhCHMeCPhHOH} \qquad (21)$$

78% (*RS*+*SR*)+22% (*RR*+*SS*)

The Perez-Ossorio approach predicted 62% (*RS*+*SR*); the three most favoured transition state conformations are **11–13**.

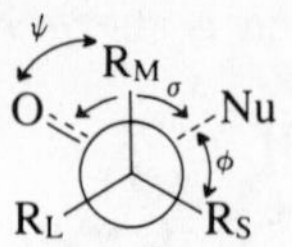

FIGURE 1. Angles in transition states.

| | $\psi$ | $\sigma$ | $\phi$ |
|---|---|---|---|
| Karabatsos[59] | 0 | 90 | 30 |
| Felkin[60] | 30 | 90 | 60 |
| Perez-Ossorio[61] | 15 | 90 | 45 |

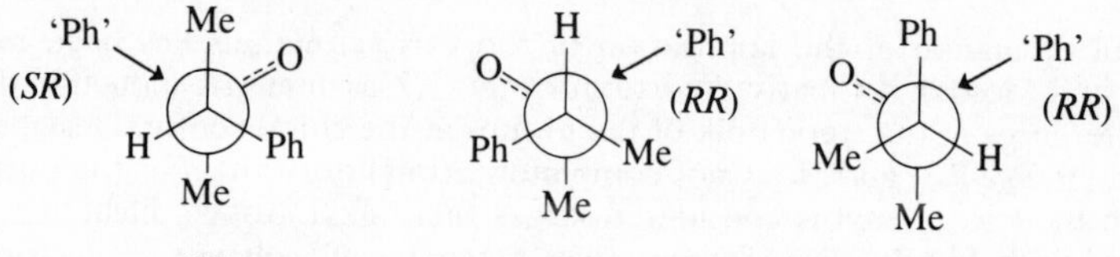

(11) (*SR*) (12) (*RR*) (13) (*RR*)

The Karabatsos and Felkin approaches considered just **11** and **12**, respectively. However success was not met with the alkyl aryl ketones, $PhCHMeCOC_6H_4X$-*p*. For reactions of these compounds, four-centre transition states were considered with considerable metal–oxygen coordination. This resulted in deviations from reactant-like transition states to give tetrahedral rather than trigonal geometry. Hence the model does not work that well[34].

b. *Acyclic compounds:* 1,3-*asymmetric inductions.* There are various known 1,3-asymmetric inductions, i.e. with the asymmetric centre $\beta$ to the carbonyl group, e.g. equations 22 and 23. The examples[62,63] also include carbonyls having polar groups.

$$Bu^tCHPhCH_2CHO \xrightarrow[\text{(ii) } H_2O]{\text{(i) PhMgBr, 30 °C, } Et_2O \text{ or thf}} Bu^tCHPhCH_2CHPhOH \quad (22)^{62}$$

68% (*RS* + *SR*) + 32% (*RR* + *SS*)

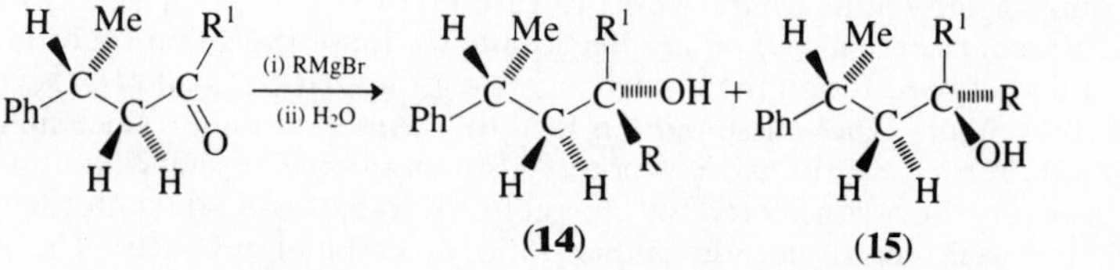

(23)[63] (i) RMgBr; (ii) $H_2O$ → (14) + (15)

$R^1 = Me$, R = Ph, T = −110 °C   [(14)]:[(15)] = 83:17
$R^1 = Ph$, R = Me, T = −78 °C   [(14)]:[(15)] = 59:41

c. *Alicyclic compounds*[64–68]. In Tables 3 and 4 are listed the stereochemical results for addition of organometallics to cyclic ketones. The problem of the stereochemistry of additions to cycloalkanones has attracted much attention. One of the most frequently studied cycloalkanones is 4-*tert*-butylcyclohexane (**16**). Two factors or two different

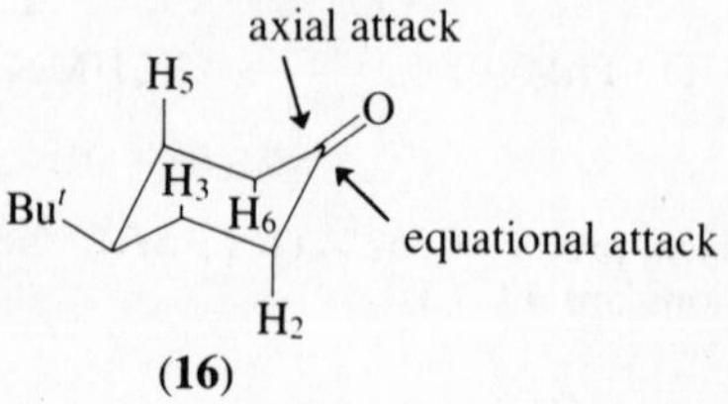

axial attack

equational attack

(16)

TABLE 3. Stereochemistry of reactions of organometallics, RM, with 4-*tert*-butylcyclohexanone

$Bu^t$–C$_6$H$_9$=O $\xrightarrow[\text{(ii) } H_2O]{\text{(i) RM}}$ $Bu^t$–(OH, R) + $Bu^t$–(R, OH)

Axial alcohol; Equatorial alcohol

| RM | Solvent conditions | Axial alcohol | Equatorial alcohol | Ref. |
|---|---|---|---|---|
| MeLi | $Et_2O$ | 65 | 35 | 64 |
| MeLi + $Me_4NI$ | $Et_2O$, 2 h, −78 °C | 70 | 30 | 46 |
| MeLi + LiBr | $Et_2O$, 2 h, −78 °C | 76 | 24 | 46 |
| MeLi + LiI | $Et_2O$, 2 h, −78 °C | 81 | 19 | 46 |
| MeLi + $LiClO_4$ | $Et_2O$, 2 h, −78 °C | 92 | 8 | 46 |
| MeLi + $LiCuMe_2$ | $Et_2O$, 2 h, −78 °C | 93 | 7 | 46 |
| MeLi + $LiCuMe_2$ | thf | 65 | 35 | 46 |
| $n$-$C_6H_{13}Li$ | PhH–$C_6H_{14}$ | 75 | 25 | 64 |
| $Bu^tLi$ | $Et_2O$, 2 h, −78 °C | 100 | 0 | 46 |
| (1,3-dithiane ring: S, S; Ph, Li) | thf, RT | 100 | 0 | 65 |
| | $C_6H_{14}$ | 82 | 18 | 65 |
| (1,3-dithiane ring: S, S; H, Li) | $C_6H_{14}$ | 49 | 51 | 65 |
| | thf | 40 | 60 | 65 |
| | tmed | 58 | 42 | 65 |
| PhLi | $Et_2O$ | 58 | 42 | 64 |
| $Me_2Mg$ | $Et_2O$ | 62 | 38 | 64 |
| | $NEt_3$ | 76 | 24 | 64 |
| $Me_2Mg$ + $LiClO_4$ | $Et_2O$, 2 h, −78 °C | 65 | 35 | 46 |
| $Pr_2Mg$ | $Et_2O$ | 75 | 25 | 64 |
| MeMgI | $Et_2O$ | 53 | 47 | 64 |
| MeMgBr | $Et_2O$ | 60 | 40 | 64 |
| | PhH | 62 | 38 | 64 |
| MeMgCl | $Et_2O$ | 59 | 41 | 64 |
| $MeMgOPr^i$ | $Et_2O$ | 79 | 21 | 64 |
| EtMgBr | $Et_2O$ | 71 | 29 | 64 |
| $Pr^iMgBr$ | $Et_2O$ | 82 | 18 | 64 |
| $Bu^tMgBr$ | $Et_2O$ | 100 | 0 | 13 |
| PhMgBr | $Et_2O$ | 49 | 51 | 64 |
| $Me_3Al$ | | | | |
| [$Me_3Al$ : $R_2CO$ = 1 : 1] | $Et_2O$ | 85 | 15 | 64 |
| [$Me_3Al$ : $R_2CO$ = 3 : 1] | $Et_2O$ | 87 | 13 | 64 |
| [$Me_3Al$ : $R_2CO$ = 1 : 1] | PhH | 76 | 24 | 64 |
| [$Me_3Al$ : $R_2CO$ = 2 : 1] | PhH | 17 | 83 | 64 |
| [$Me_3Al$ : $R_2CO$ = 3 : 1] | PhH | 12 | 88 | 64 |
| $Ph_3Al$ | | | | |
| [$Ph_3Al$ : $R_2CO$ = 1 : 1 (or 3 : 1)] | $Et_2O$ | 44 | 56 | 64 |
| [$Ph_3Al$ : $R_2CO$ = 1 : 1] | PhH | 51 | 49 | 64 |
| [$Ph_3Al$ : $R_2CO$ = 2 : 1] | PhH | 27 | 73 | 64 |
| [$Ph_3Al$ : $R_2CO$ = 4 : 1] | PhH | 8 | 92 | 64 |
| MeMgI + $CdCl_2$ | $Et_2O$ | 38 | 62 | 64 |
| $Me_2Cd$ + $MgBr_2$ | $Et_2O$ | 49 | 51 | 64 |

TABLE 3. Continued

$Bu^t$-cyclohexanone $\xrightarrow[\text{(ii) } H_2O]{\text{(i) RM}}$ $Bu^t$-cyclohexanol (OH axial, R equatorial) + $Bu^t$-cyclohexanol (R axial, OH equatorial)

Axial alcohol — Equatorial alcohol

| RM | Solvent conditions | Axial alcohol | Equatorial alcohol | Ref. |
|---|---|---|---|---|
| $LiMgMe_3$ | $Et_2O$ | 69 | 31 | 64 |
| $MeLiBMe_4$ | PhH | 70 | 30 | 64 |
| $LiAlMe_4$ | $Et_2O$ | 42 | 58 | 64 |
| $CH_2=CHCH_2Li$ | thf, −20 °C | 35 | 65 | 66 |
| $CH_2=CHCH_2Na$ | thf, −20 °C | 35 | 65 | 66 |
| $CH_2=CHCH_2K$ | thf, −20 °C | 37 | 63 | 66 |
| $(CH_2=CHCH_2)_2Mg$ | $Et_2O$ | 44 | 56 | 64 |
| $CH_2=CHCH_2MgBr$ | thf, 50 °C | 45 | 55 | 66 |
| | $Et_2O$, −40 °C | 46.7 | 53.3 | 6 |
| | $Et_2O$, 0 °C | 48.5 | 51.5 | 6 |
| | $Et_2O$, 35 °C | 50.3 | 49.7 | 6 |
| | $Et_2O$, refluxing | 54.3 | 45.7 | 6 |
| $(CH_2=CHCH_2)_2Zn$ | $Et_2O$ | 84 | 16 | 64 |
| $CH_2=CHCH_2ZnBr$ | thf, 50 °C | 85 | 15 | 66 |
| $CH_2=CHCH_2BR_2$ | $Et_2O$, refluxing | 54.8 | 45.2 | 6 |
| $CH_2=CHCH_2AlBr_2$ | thf, 50 °C | 68 | 32 | 66 |
| $Bu_3SnCH_2CH=CH_2$ | | 92 | 8 | 22 |
| $EtO_2CCH_2Li$ | $n$-$C_5H_{12}$, 60 °C | 63 | 37 | 67 |
| | $n$-$C_5H_{12}$–hmpt (40:60), −60 °C | 46 | 54 | 67 |
| $EtO_2CCH_2AlBr_2$ | $Et_2O$, 35 °C | 41 | 59 | 67 |
| $EtO_2CCH_2ZnBr$ | $(MeO)_2CH_2$, 0 °C | 57 | 43 | 67 |
| | $(MeO)_2CH_2$–dmso (3:1), 0 °C | 33 | 67 | 67 |

TABLE 4. Products of reactions of methyllithium with cycloalkanones in ether[68]

$$\text{(cyclo)C=O} \xrightarrow[\text{(ii) } H_2O]{\text{(i) MeLi,}Et_2O} \text{(cyclo)C(Me)(OH)}$$

| Ketone | % *trans* or *endo* alcohol |
|---|---|
| 4-*tert*-Butylcyclohexanone | 35 |
| 3-Methylcyclohexanone | 66 |
| 3-*tert*-Butylcyclohexanone | 78.5 |
| 2-Methylcyclohexanone | 16 |
| 3,5,5-Trimethylcyclohexanone | 99.7 |
| 2-Methylcyclopentanone | 33 |
| 2,5-Dimethylcyclopentanone | 11.6 |
| 2,5,5-Trimethylcyclopentanone | 3 |
| 2,4,4-Trimethylpentanone | 67 |
| Bicyclo[2.2.1]heptan-2-one | 99.3 |
| 1-Methylbicyclo[2.2.1]heptan-2-one | 98 |
| 1,7,7-Trimethylbicyclo[2.2.1]heptan-2-one | 2 |

interactions must be involved. One factor directs the nucleophile into the axial position. It appears that there is general agreement that if the reaction has an early, i.e. reactant-like, transition state, the steric strain between the nucleophile and the $C_{(3)}$ and $C_{(5)}$ axial hydrogens destabilizes the axial transition state and so directs the nucleophile into the equatorial position. However, another (non-steric) factor counters this to give axial attack. Suggestions as to the nature of this influence have included thermodynamic stability and frontier orbitals. Two approaches that have gained much acceptance[64,69] have been (i) that the equatorial transition state is actually more destabilized than the axial one by torsional strain: in the equatorial attack, the incipient bond is eclipsing the axial $C_{(2)}$— and $C_{(6)}$—hydrogen bonds (a destablizing interaction)[60] and (ii) that the axial transition state is stabilized by interaction with the antibonding orbitals ($\sigma^*$) orbitals of the axial $C_{(2)}$— and $C_{(6)}$—hydrogen bonds[70].

However, a more recent approach by Cieplak[71] has a different rationalization, namely that the axial attack is favoured by electron donation from the cyclohexanone $\sigma_{CC}$ and $\sigma_{CH}$ bonds into the low-lying vacant orbitals $\sigma^*_{\#}$ associated with the $\sigma$-bond being formed in the reaction. Consequently, nucleophile structure, metal cation complexing the carbonyl oxygen, solvent, and counter ions or other solutes may influence the stereoselectivity of the reaction by changes in $\Sigma(\sigma^*_{\#})$. Electron-withdrawing substituents in the nucleophile should lower $\Sigma(\sigma^*_{\#})$ and so favour axial attack. Donation by solvent or solute, such as LiX, will increase the $\sigma^*_{\#}$ energy level and should thus lead to an increase in the relative yield of the product of equatorial attack. As seen from entries in the Tables 3 and 4, changes in the stereoselectivity are brought about changes in nucleophile, counter ion, etc.

Gaudemar[66] has accounted for the stereochemical results of reaction of allylic organometallics with 4-*tert*-butylcyclohexanone in terms of the hardness and softness of the nucleophiles. It has been argued that hard nucleophiles, if steric effects are not too significant, will attack in an axial direction, i.e. under change control. This attack is governed by the disymmetry of the LUMO. Soft nucleophiles will be under orbital control and for these, equatorial attack is preferred. Hence allylic alkali metal compounds, which are hard nucleophiles, give mainly the equatorial alcohol via axial attack, whereas $CH_2{=}CHCH_2ZnBr$, a soft nucleophile, provides mainly the axial alcohol (see Table 3).

Other cycloalkanones can be similarly discussed. 2-Methylcyclohexanone is attacked from the equatorial side to a much larger extent than is 4-*tert*-butyl cyclohexanone; 2-, 3-, and 4-methylcyclohexanones with organo-alkali metal compounds indicate clearly the loss of the influence of a substituent as it becomes more remote from the reaction site. With 3,3,5-trimethylcyclohexanone, 100% equatorial attack occurs; camphor also provides a single product, the *exo* alcohol.

d. *Asymmetric synthesis using chiral ligands.* A variety of chiral ligands has been used in asymmetric additions of organometallics, in particular RLi and RMgX, to carbonyl compounds. Recent examples[72–75] of ligands are (2*S*,2′*S*)-2-hydroxymethyl-1-[(1-alkylpyrrolin-2-yl)methyl]pyrrolidines[72], e.g. **17**, (−)-1,2,3,4-$(MeO)_4$butane, (−)-1,4-$(MeO)_2$-*N*,*N*,*N*′,*N*′-$Me_4$-butane-2,3-diamine[73], and compounds such as **18**, **19**, **20**, and **21**.

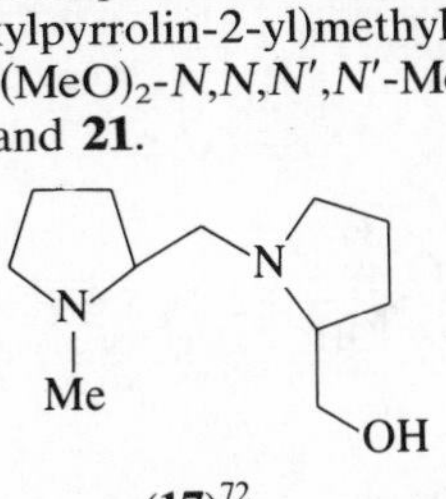

(**17**)[72]

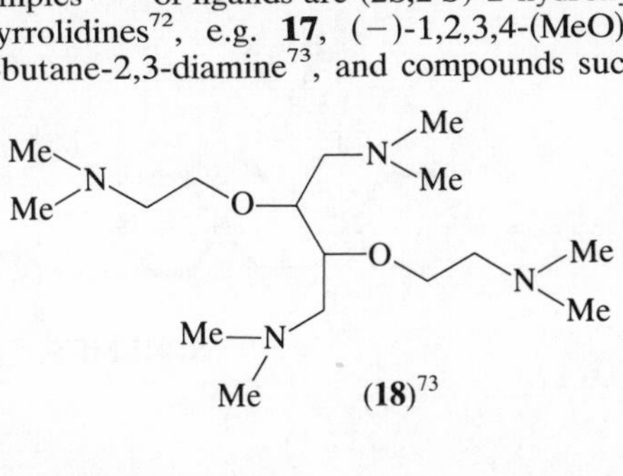

(**18**)[73]

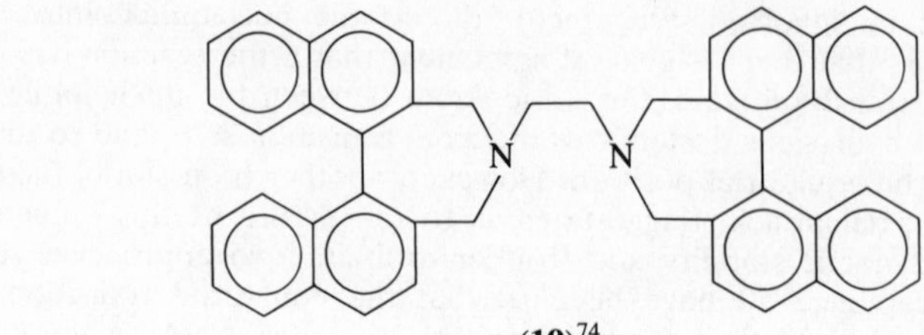

(19)[74]

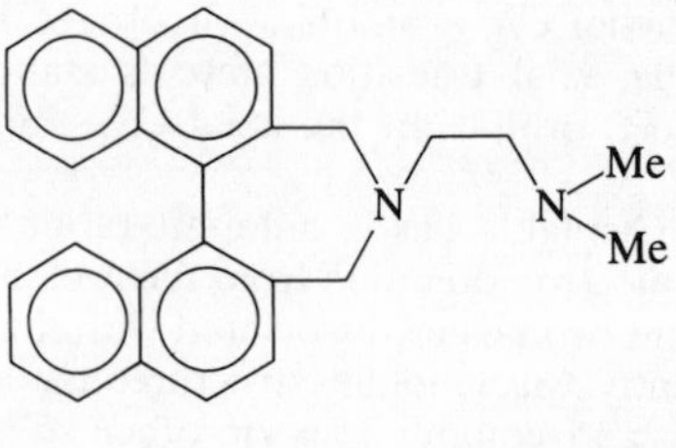

(20)[74]

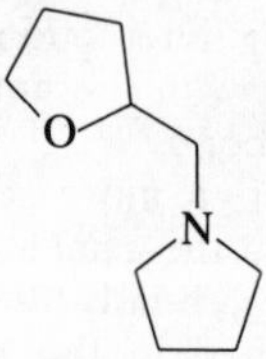

(21)[75]

Particularly good optical yields have been achieved using **19** (up to 92% e.e.) and **17** (up to 95% e.e.), see Schemes 9 and 10. There are considerable solvent effects.

Using **19** and **20**, it was found that the optical yields are greater (i) with reactants with higher steric requirements, (ii) the more highly shaped and sterically confining the catalyst, and (iii) using PhCHO rather than BuCHO. In many cases, it was found that the lower the temperature, the higher were the optical yields.

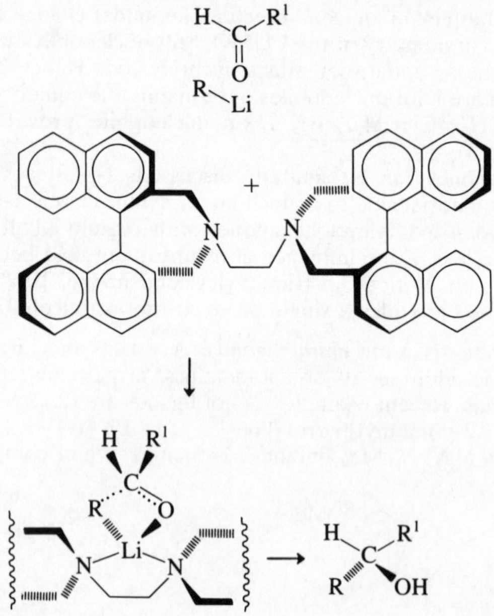

SCHEME 9

$$BuLi + PhCHO \xrightarrow[\text{(ii) } H_2O]{\text{(i) } (R,R)\text{—}\mathbf{19},\ Et_2O,\ -120\,^\circ C} (R)\text{-PhCBuHOH} \quad 92\%\ \text{e.e.}$$

$$BuLi + PhCHO \xrightarrow[\text{(ii) } H_2O]{\text{(i) } (2S,2'S)\text{—}\mathbf{17},\ (MeO)_2CH_2\text{—}Me_2O,\ -123\,^\circ C} (S)\text{-PhCBuHOH} \quad 95\%\ \text{e.e.}$$

$$Bu_2Mg + PhCHO \xrightarrow[\text{(ii) } H_2O]{\text{(i) Li salt of } (2S,2S')\text{—}\mathbf{17},\ PhMe,\ -110\,^\circ C,\ 1\,h} (R)\text{-PhCBuHOH} \quad 88\%\ \text{e.e.}$$

$$PhLi + BuCHO \xrightarrow[\text{(ii) } H_2O]{\text{(i) } (R,R)\text{—}\mathbf{19},\ Et_2O,\ -120\,^\circ C} (S)\text{-PhCBuHOH} \quad 47\%\ \text{e.e.}$$

SCHEME 10

e. *Stereochemistry of allylic—metal compound additions to aldehydes.* For certain allyl derivatives, (e.g. of tin[18,19] (equation 24), boron (equation 25), and aluminium[76] (equation 26), additions to aldehydes are both highly stereoselective (or even stereospecific) and regiospecific reactions. (*E*) and (*Z*)-allyl-metal compounds produce *threo* and *erythro* products, respectively; complete allylic rearrangement occurs; (*Z*)- and (*E*)-mixtures can also react stereospecifically.

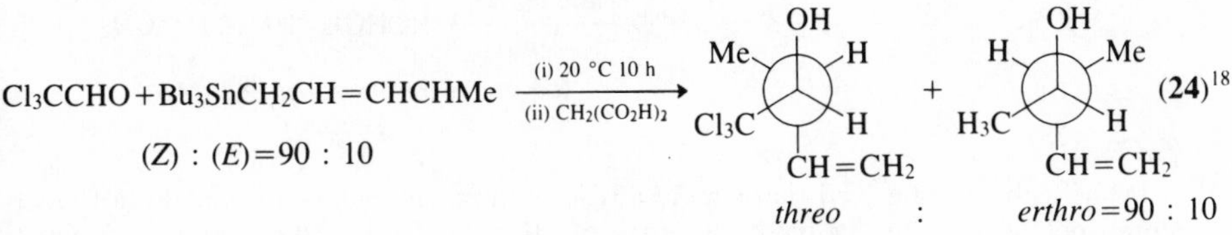

$$Cl_3CCHO + Bu_3SnCH_2CH{=}CHCHMe \xrightarrow[\text{(ii) } CH_2(CO_2H)_2]{\text{(i) } 20\,^\circ C\ 10\,h} \text{threo} + \text{erthro} \qquad (24)^{18}$$

(*Z*) : (*E*)=90 : 10

threo (Newman projection: OH, Me, H, $Cl_3C$, H, $CH{=}CH_2$) : erthro (Newman projection: OH, H, Me, $H_3C$, H, $CH{=}CH_2$) = 90 : 10

$$\text{Me–CH=CH–CH}_2\text{–B(pinacolato)} + RCHO \longrightarrow \text{R–CH(O–B(pinacolato))–CH(Me... )–CH=CH}_2 \qquad (25)^{7}$$

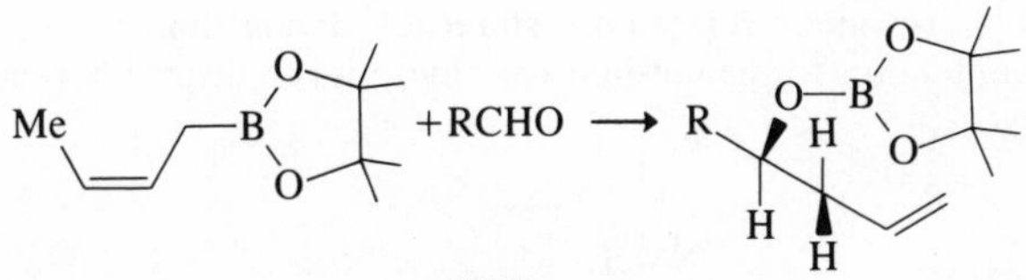

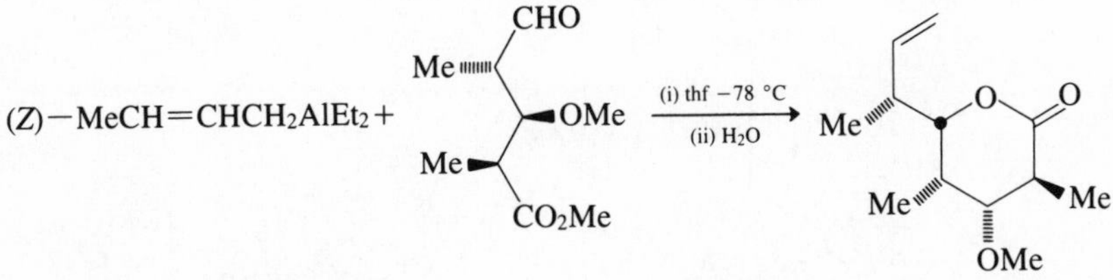

$$(Z)\text{–}MeCH{=}CHCH_2AlEt_2 + \text{CHO–CH(Me)–CH(OMe)–CH(Me)–CO}_2\text{Me} \xrightarrow[\text{(ii) } H_2O]{\text{(i) thf } -78\,^\circ C} \text{lactone} \qquad (26)^{76}$$

For simple allyl derivatives (e.g. crotyl) of lithium[77], magnesium[78], zinc[79], or cadmium[79] and for some allyl-boron derivatives[6], of unknown geometry, the stereoselectivity is poor; however, the stereoselectivity increases slightly as the steric bulk of the substituents on the allyl group increases.

It is generally accepted that these highly stereoselective reactions proceed via a cyclic transition state (**22**) in which the metal cation can interact with the oxygen of the carbonyl group.

It has been shown more recently that in the presence of a Lewis acid, e.g. $BF_3$, high yields of *erythro* products result from either (*Z*)- or (*E*)-crotyltin derivatives[20]. Clearly, the presence of $BF_3$ alters the mechanism

$$\underset{(Z):(E)=40:60}{Bu_3SnCH_2CH{=}CHMe} \xrightarrow[\text{(ii) } H_3O^+]{\text{(i) PhCHO,BF}_3} \underset{\textit{erythro}:\textit{threo}=96:4}{PhCHOHCHMeCH{=}CH_2} \qquad (27)$$

and non-cyclic transition states must occur with no tin–carbonyl oxygen interaction, e.g. **23** and **24**. Of interest the greater affinity of silicon for oxygen apparently results in

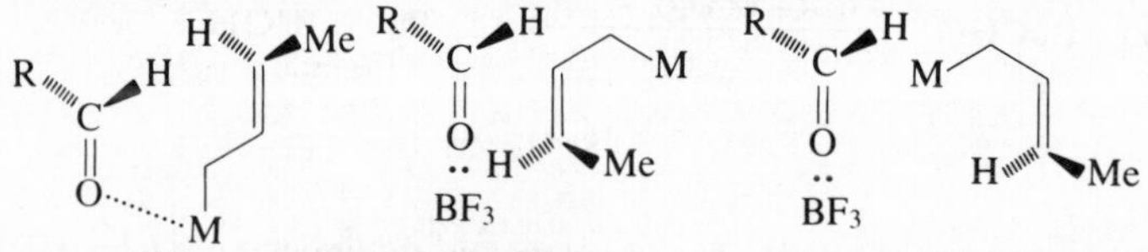

(22) (23) (24)

cyclic transition states in related organo-silicon reactions even in the presence of a Lewis acid catalyst such as $TiCl_4$. *Threo*-selective reactions have also been found using allylic boronates[80], e.g. equation 28.

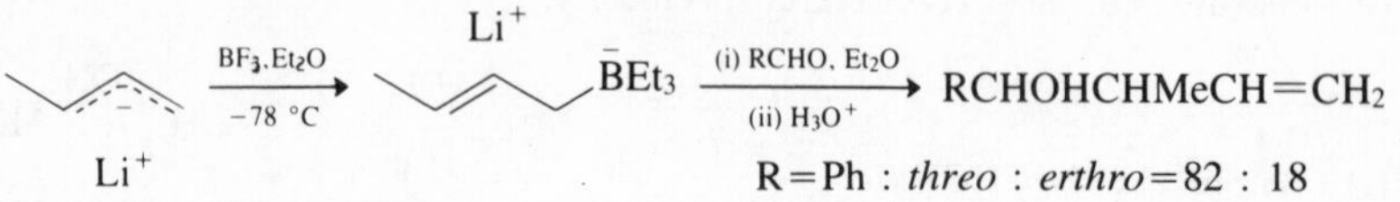

$$\text{(25)} \xrightarrow[-78\ ^\circ C]{BF_3.Et_2O} Li^+\ \text{CH}_3\text{CH=CHCH}_2\bar{B}Et_3 \xrightarrow[\text{(ii) } H_3O^+]{\text{(i) RCHO, Et}_2O} RCHOHCHMeCH{=}CH_2 \qquad (28)$$

R=Ph : *threo* : *erthro*=82 : 18

It has to be emphasized that simple allylic-lithiums and -magnesiums do not react stereospecifically, e.g. **25** in the absence of $BF_3$ gives a 1:1 *threo*:*erythro* product mixture. More rigid allyl-lithiums, such as **26**, however, can react[81] stereoselectively, equation 29; the (*Z*)-isomers react more stereoselectively than do the (*E*)-isomers. Changing the counter ion to magnesium (or zinc) makes little difference.

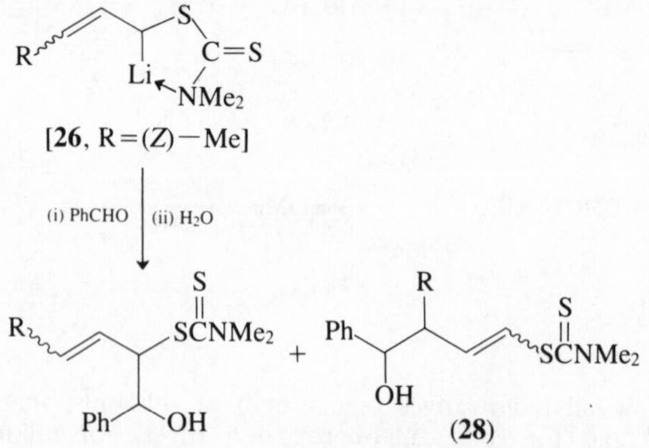

[**26**, R=(*Z*)−Me]

(i) PhCHO (ii) $H_2O$ (29)

(**27**) + (**28**)

Overall yield 85%

[**27**] : [**28**]:=1 : 9

### 4. Reversible additions

A number of additions to carbonyl compounds have been shown to be reversible[82–85], especially those involving stable carbanions. This is particularly so for metal allyls (e.g. metal = magnesium, zinc, lithium or tin). In equation 30 as the reaction time and temperature increase, so the amount of **29**, the thermodynamic product, increases at the expense of **30**, the kinetic product. A cross-over experiment

$$\mathrm{Me_2C{=}CHCH_2M + Pr^i{}_2C{=}O \rightleftharpoons \underset{(\mathbf{30})}{CH_2{=}CHCMe_2CPr^i{}_2OM} + \underset{(\mathbf{29})}{Me_2C{=}CHCH_2CPr^i{}_2OM}} \tag{30}$$

M = MgX, ZnBr, or Li

of two alkoxides has also been performed, equation 31. The exchanges clearly arise from the reformation of ketones and allyl-Grignards[84]. Other reversible reactions have been shown with Grignards from propargylic halides[86] and 2-Li-2-phenyl-1,3-dithiane[65].

$$\mathrm{Bu^tC(OMgBr)(Pr^i)(MeCHCH{=}CH_2) + Bu^tC(OMgBr)(Bu^t)(CH_2CH{=}CH_2) \xrightarrow{\Delta,\ thf} \underset{(Z)\ and\ (E)}{Bu^tC(OMgBr)(Pr^i)(CH_2CH{=}CHMe)} + Bu^tC(OMgBr)(Pr^i)(CH_2{=}CH{-}CH_2) + \underset{(Z)\ or\ (E)}{Bu^tC(OMgBr)(Bu^t)(CH_2CH{=}CHMe)}} \tag{31}$$

### 5. Reductions[87]

As indicated earlier, reductions of carbonyls by alkyl-metals can occur if $\beta$-hydrogens are present. Although this topic is outside the scope of this chapter, some space is given to it here since it is a frequently met side-reaction in Grignard reactions with carbonyl species. The usually accepted view is that reduction occurs via six-membered cyclic transition states and that appreciable C—H bond stretching occurs in the rate-determining step. Hydrogen isotope effects, $k_H/k_D$, have been measured, e.g. $k_H/k_D = 2$ for reduction of *p*-$MeSC_6H_4COMe$ by cyclopentyl- and 2,2,5,5,5-$D_4$-cyclopentyl-Grignards in $Et_2O$[37] and $k_H/k_D = 1.46$ for reduction of $Ph_2CO$ by *iso*-butyl Grignards in $Et_2O$[88].

The Grignard reagent from 2-*exo*-chloro-3-*exo*-deuterionorbornane is a mixture of *exo*- and *endo*-isomers. Using this isomeric mixture in reaction with $Ph_2CO$ in $Et_2O$ solution, it was concluded that reduction proceeded via *cis*-*exo*-eliminative transfer of D and MgCl, since only the *exo*-Grignard reacted and that deuterium transfer had occurred[89]. Fauvergue[90] had earlier reported that the reduction of $Pr^iCOPh$ with the Grignard reagent from $\alpha$-exo-deuterioisobornyl chloride proceeded with preferential but not total transfer of deuterium; *cis*-*exo*-elimination of D and MgCl was the major component with a minor contribution from *cis*-*endo*-transfer.

Ashby most recently produced the first direct evidence for the involvement of an s.e.t. mechanism in reduction of diaryl ketones by primary, secondary, or tertiary Grignard reagents[91]. Asymmetric reduction of ArCOR has been achieved using Grignards with chiral centres[92] at the $\beta$-carbon, e.g. using $Ar'CH(Et)CH_2MgCl$, $PhCHDCD_2MgBr$ or $MeCH_2CMeHCH_2MgCl$. Evidence has been provided to show that small quantities of magnesium hydride are also obtained in the preparation of Grignards and that such species may lead to reduction of carbonyls, see Table 2.

### 6. *Oxophilic additions*

Addition of the organic part of an organometallic compound to the oxygen of certain carbonyl groups has also been reported. Such reactions (oxophilic reactions)

$$RM + {>}C{=}O \longrightarrow {>}C(OR)(M) \xrightarrow{H_2O} {>}C(OR)(H) \qquad (32)$$

with phenanthraquinone[93,94], tetraphenylcyclopentadienone[95], and quinol acetates[96] provide ethers. While phenanthraquinone reacts with vinyl- and phenyl-magnesium halides [to give 9-vinyloxy- and 9-phenyloxy-10-hydroxy-9,10-dihydrophenanthrenes as well as the normal (carbophilic) addition products, e.g. equation 33], reactions with vinylic lithium compounds, as well as methyl- and ethyl-magnesium bromide, occur solely to give the carbophilic product, e.g. equation 34.

9,10-Phenanthraquinone

$\xrightarrow[\text{(ii) } H_2O]{\text{(i) } CH_2{=}CHMgBr(4eq),thf}$ 9,10-dihydroxy-9,10-divinyl-9,10-dihydrophenanthrene
22%

+9-vinyloxy-10-hydroxy-9,10-dihydrophenanthrene (33)
43%

9,10-Phenanthraquinone

$\xrightarrow[\text{(ii) } H_2O]{\text{(i) } CH_3CH{=}CHLi,thf}$ 9,10-dihydroxy-9,10-di(prop-1-enyl)-9,10-dihydrophenanthrene
63% (34)

There is predominant retention of the geometry of the organic group on oxophilic addition to phenanthraquinone. This rules out a predominant reaction via free radicals but does not distinguish between mechanisms involving a caged radical and direct nucleophilic addition. The latter process was preferred[94] for addition of vinylic magnesium halides to phenanthraquinone.

However, one-electron transfer is considered to occur in oxophilic addition to quinol acetates, e.g. equation 35[96a] and Scheme 11.

Oxophilic additions occur in reaction[95] of $Bu^tMgCl$ (but not, however, of vinyllithium)[94] to tetraphenylcyclopentadiene.

Heterophilic reactions have been reported much more frequently with C=S bonded compounds (see Section III.B), and to a much reduced extent with C=N.

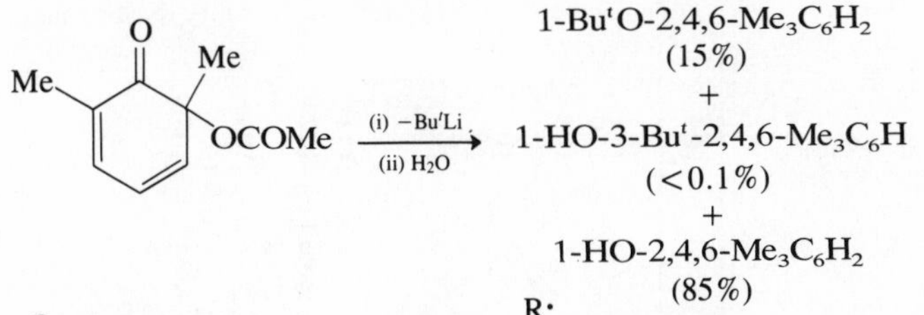

$$\text{2,6-Me}_2\text{-6-OCOMe-cyclohexa-2,4-dienone} \xrightarrow[\text{(ii) H}_2\text{O}]{\text{(i) }-\text{Bu}^t\text{Li}} \begin{array}{c} \text{1-Bu}^t\text{O-2,4,6-Me}_3\text{C}_6\text{H}_2 \\ (15\%) \\ + \\ \text{1-HO-3-Bu}^t\text{-2,4,6-Me}_3\text{C}_6\text{H} \\ (<0.1\%) \\ + \\ \text{1-HO-2,4,6-Me}_3\text{C}_6\text{H}_2 \\ (85\%) \end{array} \quad (35)$$

$$\text{2,4,6-Me}_3\text{-6-OCOMe-cyclohexa-2,4-dienone} \xrightarrow{\text{RM}} \text{2,4,6-Me}_3\text{C}_6\text{H}_2\text{O}\cdot + \text{R}\cdot + \text{MO}_2\text{CMe} \longrightarrow \text{1-RO-2,4,6-Me}_3\text{C}_6\text{H}_2$$

$$\text{2,4,6-Me}_3\text{C}_6\text{H}_2\text{O}\cdot \longrightarrow \text{1-HO-2,4,6-Me}_3\text{C}_6\text{H}_2 \quad \text{and} \quad \text{1-HO-3-R-2,4,6-Me}_3\text{C}_6\text{H}$$

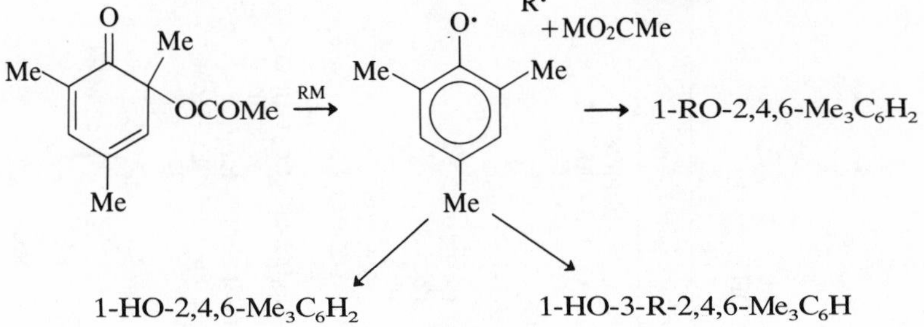

SCHEME 11

## B. Reactions with Carboxylic Acids, Esters, and Related Carbonyl Compounds

Reactions of organic derivatives of the more electropositive main group elements, in particular lithium, magnesium, zinc, cadmium, and aluminium, with carboxylic acids and their derivatives, e.g. esters, anhydrides, acyl chlorides, and amides, as well as ketenes and isocyanates, have been variously reported[3,4]. Enolizations and reductions, as well as other side reactions may complicate matters but these reactions constitute useful syntheses of carbonyl compounds[97,98] (ketones or aldehydes) and carbinols (secondary or tertiary); some examples are shown in Table 5. The general scheme for these reactions of acyl derivatives is given in Scheme 12. The initial step (step i) is a 1,2-addition to give RR′CX(OM), hydrolysis of which provides a ketone (or aldehyde), RR′CO; if a subsequent and further addition of R′M proceeds then $RR'_2COM$ is obtained. This will give the tertiary (or secondary) alcohol, $RR'_2COH$, on hydrolysis. Whether a carbinol or carbonyl compound is formed depends on several factors; these include (i) the leaving group ability of X, (ii) steric hindrance, (iii) the temperature, (iv) the molar ratio of the reactants, (v) solvent, and (vi) the hydrolytic conditions.

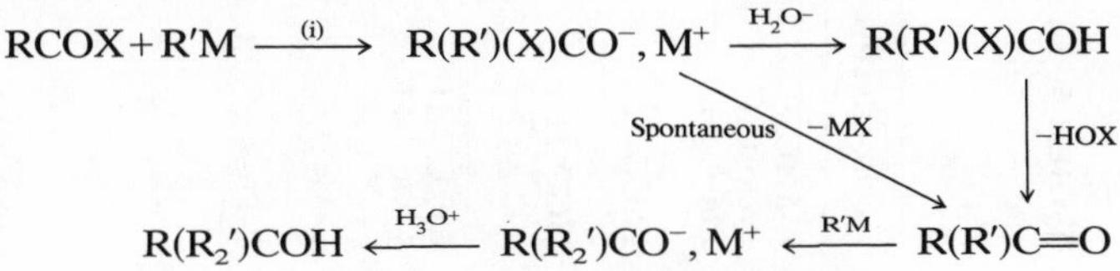

$$\text{RCOX} + \text{R}'\text{M} \xrightarrow{\text{(i)}} \text{R(R}')\text{(X)CO}^-, \text{M}^+ \xrightarrow{\text{H}_2\text{O}-} \text{R(R}')\text{(X)COH}$$

$$\text{R(R}')\text{(X)CO}^-, \text{M}^+ \xrightarrow[\text{Spontaneous}]{-\text{MX}} \text{R(R}')\text{C=O}; \quad \text{R(R}')\text{(X)COH} \xrightarrow{-\text{HOX}} \text{R(R}')\text{C=O}$$

$$\text{R(R}_2')\text{COH} \xleftarrow{\text{H}_3\text{O}^+} \text{R(R}_2')\text{CO}^-, \text{M}^+ \xleftarrow{\text{R}'\text{M}} \text{R(R}')\text{C=O}$$

X = OLi, Cl, OR, OCOR, or $NR_2$

SCHEME 12

TABLE 5. Reactions of acyl derivatives with organometallics, R'M

$$RCOX + \xrightarrow[(ii)\ H_2O]{(i)\ R'M} RCOR' \text{ or } RR'_2COH$$

| Acyl derivative | RM | Product (yield, %) | Ref. |
|---|---|---|---|
| $p$-$MeC_6H_4CO_2Li$ | PhLi | $p$-$MeC_6H_4COPh$ (69) | 102 |
| cyclo-$C_6H_{11}CO_2H$ | MeLi | cyclo-$C_6H_{11}COMe$ (91) | 120 |
| $PhCO_2H$ | $PhLi/Et_2O$ 2 equiv. | $Ph_3COH$ (25) + $Ph_2CO$ (54) | 101 |
| $PhCO_2Li$ | $PhLi/Et_2O$ 1 equiv. | $Ph_2CO$ (85) | 101 |
| $HCO_2H$ | PhCH=CHMgBr (**E**):(**Z**) = 89:11 in thf | PhCH=CHCHO (**E**):(**Z**) = 87:13 (67) | 103 |
| $CF_3CO_2H$ | PhLi | $CF_3COPh$ (75) | 121 |
| 1-Me-1-$HO_2C$-2,2-$Ph_2$-cyclopropane (+) | PhLi | 1-Me-1-PhCO-2,2-$Ph_2$-cyclopropane (−) (64) | 122 |
| 2-$MeO_2C$-furan | 1-Li-2-$LiO_2C$-benzene | (2-furyl)(2-$HO_2CC_6H_4$)CO (40) | 109 |
| 2-Ph-4-$EtO_2C$-furan | $EtMgI.2Et_2O + NEt_3$ | 2-Ph-4-$EtO_2C$-furan + 2-Ph-4-$HOCEt_2$-furan (7) | 107a |
| $PhCO_2Et$ | 2-Me-2-Li-1,3-dithiane [structure: S, S ring with Me, Li] | (2-Me-1,3-dithian-2-yl)$_2$CPhOH [structure: (S, S ring with Me)$_2$ CPhOH] (95) | 123 |
| $HCO_2Et$ | PrMgBr | $Pr_2CHOH$ (63) | 124 |

| | | | |
|---|---|---|---|
| $HCO_2Et$ | $o$-$MeC_6H_4MgBr$ | $o$-$MeC_6H_4CHO$ (45) | 125 |
| $PhCO_2Et$ | $p$-$MeC_6H_4SOCH_2Li$ +(−) sparteine | $p$-$MeC_6H_4^*SOCH_2COPh$ (80% chemical yield; 15% optical yield) | 126 |
| $PhCO_2Et$ | PhCaI | $Ph_3COH$ | 127 |
| $PhCO_2Me$ | $Et_3Al$ | $PhEtCHOH + PhCEt_2OH$ 2,7 : 1 | 128 |
| $MeCO_2Me$ | $(H_2C{=}CHCH_2)_2Cd$ | $(CH_2{=}CHCH_2)_2CMeOH$ (65) | 129 |
| $MeO_2C(CH_2)_3CON(OMe)Me$ | PhMgBr | $Ph_2C(OH)(CH_2)_3COPh$ | 99 |
| $PhCH_2CH_2COS$(2-pyridyl) | cyclo-$C_6H_{11}MgBr$ | $PhCH_2CH_2CO$-cyclo-$C_6H_{11}$ (95) | 130 |
| PrCOCl | $Bu^tMgX$ | $Bu^tCOPr$ (82) | 107 |
| $(MeCO)_2O$ | $PhCH_2MgX$ | $PhCH_2COMe$ (52) | 131 |
| $(MeCO)_2O$ | $CH_2{=}CH{-}CH_2ZnBr$ | $(CH_2{=}CHCH_2)_2CMeOH$ (77) | 132 |
| piperidine-1-carbaldehyde | 1-M-cyclopropane | cyclopropane-1-carbaldehyde M = MgBr : yield 80% M = Li : yield 75% | 100 |
| 2-PhCON(Me)-pyridine | EtMgBr | PhCOEt (83) | 133 |
| γ-Butyrolactone | PhMgX | $HO(CH_2)_3CPh_2OH$ | 134 |
| γ-Butyrolactone | $BrMg(CH_2)_4MgBr$ | 1-(1-hydroxypropyl)cyclopentanol | 135 |

The poorer the leaving group ability of X, in the acyl derivative RCOX, the greater is the chance of limiting the reaction to a single addition and hence to carbonyl formation. Compared with X = Cl, OR″ or OCOR″, both $X = NR''_2$[4,99,100] and OM (M = MgX or Li)[98,101–103] are poor leaving groups; indeed, $X = NR''_2$ and X = OLi are considered only to leave on hydrolysis, i.e. $RR'CNR''_2(OM)$ and $RR'CN(OM)_2$ (M = MgX or Li) are stable under the reaction conditions and do not break down spontaneously to RR′C=O as do other RR′CXOM compounds (e.g. X = Cl, OR″, or OCOR″). Such stability of $RR'CNR''_2(OM)$ and $RR'C(OM)_2$ in the reaction mixture clearly prohibit any additional reactions. They do give RR′CO on hydrolysis. Compounds, $PhCR(OLi)_2$ (R = Ph, p-$MeC_6H_4$, p-$MeOC_6H_4$) have been reported to be even stable in refluxing diethyl ether for 96 h[102]. Thus there have been a number of published routes to ketones (or aldehydes) from amides and salts of carboxylic acids[98], see equations 36–38.

$$p\text{-}MeC_6H_4CO_2Li + PhLi \xrightarrow[96\,h]{Et_2O,\Delta} p\text{-}MeC_6H_4CPh(OLi)_2 \xrightarrow{H_2O} p\text{-}MeC_6H_4COPh \quad 69\% \qquad (36)^{102}$$

$$2\text{-}PhCH{=}CHMgBr + HCO_2H \xrightarrow[(ii)\ H_2O]{(i)\ thf,\ 0\ ^\circ C} PhCH{=}CHCHO \qquad (37)^{103}$$

$(E):(Z) = 89:11$ → $(E):(Z) = 87:13$

67% overall yield

$$RCON(Me)OMe + BuM \xrightarrow[(ii)\ H_2O]{(i)\ thf} RCOBu \qquad (38)$$

R = cyclohexyl; BuM (1.5 equiv)

M = MgCl 97%
M = Li 94%

There have been several reports[101,104] that all the RLi should be destroyed or used up prior to hydrolysis in $RCO_2Li$–RLi reaction mixtures if high yields of RR′CO are required. It appears that RLi can survive the addition of water and is thereby able to react with the liberated RCOR′ to give the further addition product $R_2R'COLi$. Good reagents for removing excess RLi prior to hydrolysis of the $RR'C(OLi)_2$ are aniline and HCHO[104]. It appears better to use performed $RCO_2Li$ rather than form it *in situ* from $RCO_2H$ and RLi, since some $RCO_2H$ may still survive to protonate the $RR'C(OLi)_2$, and thereby form RR′CO[101].

For other leaving groups, X, the milder the reaction conditions and the greater the control of the quantity of the organometallic, the better are the prospects of obtaining good yields of RR′CO. Other factors leading to more RR′CO are greater steric hindrance[105,106] in the reagents, see equation 39 and 40, and use of $NEt_3$ and thf (equations 41 and 42)[107] at least as part of the solvent system. Reaction 40 should be compared with the less hindered reaction 43. Esters in particular have found considerable use in the syntheses of carbinols[108].

2,6-$(MeO)_2C_6H_3Li$ + $ClCO$–$C_6H_3(OMe)_2$ $\xrightarrow[(ii)\ H_2O]{(i)\ Et_2O,\ -70\ ^\circ C}$ 2,6-$(MeO)_2C_6H_3$–C(=O)–$C_6H_3(OMe)_2$ **(39)**[105]

$$Bu^tCO_2Et + Bu^tLi \longrightarrow Bu^tCOBu^t \qquad \mathbf{(40)}^{106}$$

$$\mathrm{PhMgX + PrCOCl \xrightarrow{thf,\ -78\ ^\circ C} PhCOPr} \quad (\mathbf{41})^{107b}$$

71%

$$\mathrm{RCO_2Me + MeMgBr \xrightarrow[5-10\ ^\circ C]{Et_3N} RCOMe} \quad (\mathbf{42})^{107a}$$

R=bicyclo[2.22]oct-l-yl 80%

$$\mathrm{MeCO_2Et + 2\text{-}Li\text{-}furan \longrightarrow (furyl)_2CMeOLi} \quad (\mathbf{43})^{108}$$

94%

Scheme 13 indicates both mono- and di-additions of PhLi to phthalic anhydride. The mono-addition is favoured using 1 equivalent of PhLi whereas 2 equivalents of PhLi and a reverse addition favours the di-addition product[109].

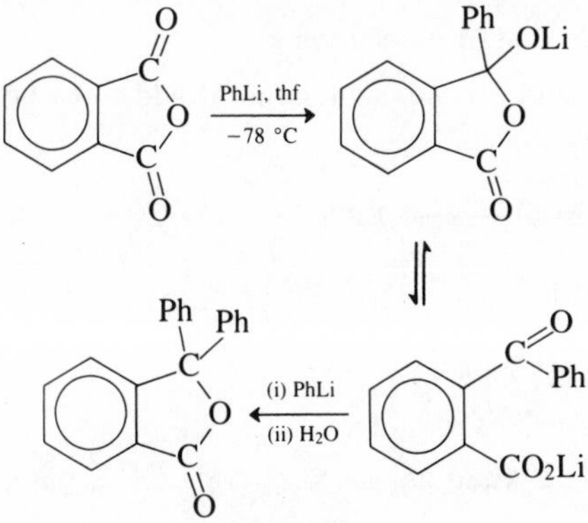

PhLi, thf, −78 °C; (i) PhLi (ii) $H_2O$

SCHEME 13

Reactions 39–41 illustrate the formation of ketones from (i) hindered acyl chloride and organolithiums and (ii) using thf at low temperature. Particularly useful general procedures to ketones from acyl halides have been developed for organo-cadmium[110], -zinc[111] and -copper compounds[112].

Isocyanates also react with organometallics, e.g. organo-lithium[3,4], -magnesium[3,113] (equation 44), -aluminium[3,114] (equation 45), -boron[1], -antimony[1], -indium, and gallium[115] derivatives.

$$\mathrm{R'MgX + RNC{=}O \longrightarrow RN{=}C(R)OMgX \xrightarrow{H_2O} RNHCOR} \quad (44)^{113}$$

$$\mathrm{Et_3Al} \begin{cases} \xrightarrow{Bu^tNCO} \mathrm{Et_2AlOCEt{=}NBu^t} \\ \xrightarrow{Bu^sNCO} \mathrm{Et_2AlN(Bu^s)CEt(O)} \end{cases} \quad (45)^{114}$$

Ketenes, too, react with electropositive metal derivatives (e.g. organo-lithiums[3,4] (equation 46)[116], -magnesiums[3,4] and -zincs[3]), as well as with derivatives of mercury[3,117] (equation 47) and silicon[114] (equation 48).

$$\mathrm{(R_3M)_2C{=}C{=}O + BuLi \longrightarrow (R_3M)_2C{=}C(Bu)OLi \xrightarrow{EtOH} (R_3M)_2CHCOBu} \quad (46)^{116}$$

M = Si or Ge

$$RHgX + CH_2{=}C{=}O \longrightarrow CH_2{=}C(R)OHgX \xrightarrow{HX} CH_3COR \quad (47)^{117}$$

R = furyl or phenyl

$$Me_3SiC{\equiv}CNEt_2 + Ph_2C{=}C{=}O \longrightarrow \underset{68\%}{Me_3SiOC({=}CPh_2)C{\equiv}CNEt_2} \quad (48)^{118}$$

Evidence for a s.e.t. mechanism in a hindered system has been obtained[119]. Russian workers have reported CIDNP effects in the reaction of $Bu^tMgCl$ and $o$-$BrC_6H_4COCl$ in dme solution.

## C. Additions to Carbon Oxides

### 1. *Carbon suboxide ($C_3O_2$) insertions*

Organolithiums[136] and organomagnesiums[137] add to carbon suboxide as shown in equations 49 and 50.

$$2RLi + O{=}C{=}C{=}C{=}O \longrightarrow LiO(R)C{=}C{=}C{=}C(R)OLi \xrightarrow{H_2O} RCOCH_2COR \quad (49)^{136}$$

R = Ph 60%
R = Bu 67%

$$3RMgX + 3O{=}C{=}C{=}C{=}O \longrightarrow 3(R)LiOC{=}C{=}C{=}O \xrightarrow{H_2O} 2,4,6\text{-}(RCO)_3\text{-phloroglucinol} \quad (50)^{137}$$

R = Me or cyclohexyl

### 2. *Carbon monoxide insertions: carbonylations*

Reactions of CO have been reported with organo-lithium[138–141] -magnesium[142,143], -zinc[144], and -boron[145] as well as with organo-mercury[146] reagents, if palladium compounds are also present[147–50]. The initial products of reaction of CO with organolithiums, RLi (e.g. R = Ph or $Bu^t$) appear to be the insertion products[138] RCOLi. Subsequent reactions depend on the conditions and other reagents present, e.g. diphenylalkylcarbinols (and acyloins) can be easily formed by adding an alkyl bromide (R = Bu, $Pr^i$ or $Bu^t$) to a solution of PhLi in thf at −78 °C and exposing the

$$PhLi + RBr + CO \xrightarrow[\text{(ii) } H_2O]{\text{(i) thf, } -78\,°C} Ph_2RCOH + PhCOC(OH)HPh \quad (51)$$

| | | |
|---|---|---|
| e.g. R = Bu: | 80% | 15%. |
| R = $Bu^t$: | 20% | 38%. |

solution to a carbon monoxide atmosphere, equation 51. Other products may occur, e.g. when R = $Bu^t$ some $Me_2CHCPh_2COBu^t$ and $Ph_2CHOBu^t$ are formed. Such complex product mixtures are unattractive features of these organolithium reactions. The mechanism for formation of $Ph_2RCOH$ in equation 51 is shown in Scheme 14.

$$PhLi + CO \rightleftharpoons PhCOLi \xrightarrow{RBr} PhCOR \xrightarrow{PhLi} Ph_2C(R)OLi$$

SCHEME 14

Triphenylsilyl chloride has been used to trap $Bu^tCOLi$ (as $Bu^tCOSiMe_3$)[139].

$$Bu^tLi \xrightarrow{CO} Bu^tCOLi \xrightarrow{Me_3SiCl} \underset{16\%}{Bu^tCOSiMe_3} \quad (52)$$

In the absence of a trapping agent, a variety of products are obtained, see equation 53; the particular yields depend[140] on the reaction conditions. Whitesides and

$$PhLi \xrightarrow[\text{(ii) } H_2O]{\text{(i) CO (1 atm)}/Et_2O} PhCOPh + PhCOCH_2Ph_2 + PhCOCH_2OH +$$
$$+ PhCOCHOHCPh_2OH + \text{traces of } PhCOOCPh + PhCOCOHPh_2 + PhCHOHPh \quad (53)$$

coworkers[140] found spectral evidence for the intermediacy of the dilithium benzophenone dianion, $[Ph_2CO^{2-}, 2Li^+]$. It appears that only derivatives of strongly basic anions react with CO, e.g. $Bu^tLi$, BuLi, MeLi, and cyclopropyllithium (as well as $R_2NLi$[141] and dimsylsodium). In contrast, $PhC{\equiv}CLi$, $BuC{\equiv}CLi$, and CpLi do not react with CO.

The carbonylation of Grignard reagents is catalysed by hmpt[142], equation 54; the presence of a base (e.g. $KOBu^t$) appears necessary for carbonylation[144] of dialkylzincs, equation 55.

$$EtMgBr \xrightarrow[\text{(ii) } H_2O]{\text{(i) CO (50 psi), hmpt (1 equiv)}} \underset{36\%}{EtCOCHEt_2} \quad (54)$$

$$Bu_2Zn + CO\ (1\ atm) \xrightarrow[\text{(ii) } H_2O]{\text{(i) } KOBu^t,\ 15\ °C} \underset{42\%}{BuCHOHCOBu} \quad (55)$$

Primary alkyl-Grignards normally produce $RCOCHR_2$ derivatives; other organomagnesium halides give mainly the acyloins, RCHOHCOR. The reactivity of RMgX decreases in the sequence R = primary group > *sec*- or *tert*-group > aryl or aralkyl group. It has not been found normally possible to trap RCOMgX (R = alkyl) since presumably it's further reactions are too fast; however, PhCOMgBr was trapped at 0–5 °C as PhCHO (20%), on hydrolysis. In addition, the reaction of BuMgBr with CO in neat hmpt and in the presence of $NaBH_4$ gave the alcohol $BuCH_2OH$. Scheme 15 was devised to account for the great variety of possible products in the carbonylation of Grignard reagents. Nickel complexes[143] also have been used to catalyse the formation of $R^1COR^2$ from the reaction of $R^1MgX$, CO and $R^2X$.

a. *Carbonylation of organoboranes.* With characteristic skill and enthusiasm, Brown[145] has developed the carbonylation of organoboranes into a series of valuable organic syntheses. Trialkylboranes react with CO at high pressure (or lower pressures

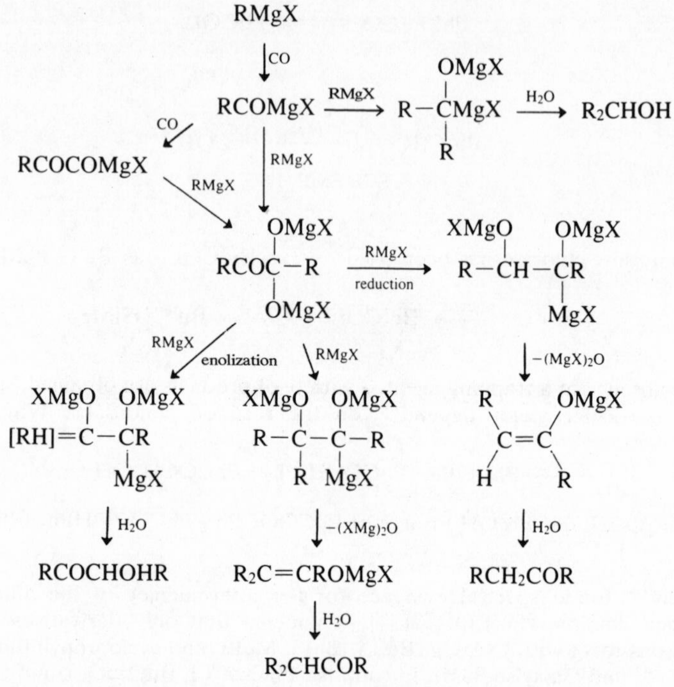

$$
\begin{array}{l}
\text{RMgX} \xrightarrow{\text{CO}} \text{RCOMgX} \xrightarrow{\text{RMgX}} \text{R-C(OMgX)(MgX)-R} \xrightarrow{H_2O} R_2CHOH \\
\text{RCOMgX} \xrightarrow{\text{CO}} \text{RCOCOMgX} \xrightarrow{\text{RMgX}} \text{RCOC(OMgX)}_2\text{-R} \\
\text{RCOMgX} \xrightarrow{\text{RMgX}} \text{RCOC(OMgX)}_2\text{-R} \\
\text{RCOC(OMgX)}_2\text{-R} \xrightarrow[\text{reduction}]{\text{RMgX}} \text{R-CH(OMgX)-CR(OMgX)(MgX)} \xrightarrow{-(MgX)_2O} \text{RHC=C(OMgX)R} \xrightarrow{H_2O} RCH_2COR \\
\text{RCOC(OMgX)}_2\text{-R} \xrightarrow[\text{enolization}]{\text{RMgX}} \text{[RH]=C(OMgX)-CR(OMgX)(MgX)} \xrightarrow{H_2O} \text{RCOCHOHR} \\
\text{RCOC(OMgX)}_2\text{-R} \xrightarrow{\text{RMgX}} \text{R-C(OMgX)(R)-C(OMgX)(MgX)-R} \xrightarrow{-(XMg)_2O} R_2C{=}CROMgX \xrightarrow{H_2O} R_2CHCOR
\end{array}
$$

SCHEME 15

on heating, e.g. to 100–125 °C in diglyme) to give the sequence of reactions shown in Scheme 16. All transfers of R groups from boron to carbon occur in an intramolecular manner. A modification involves carbonylation in the presence of $(CH_2OH)_2$, equation 56.

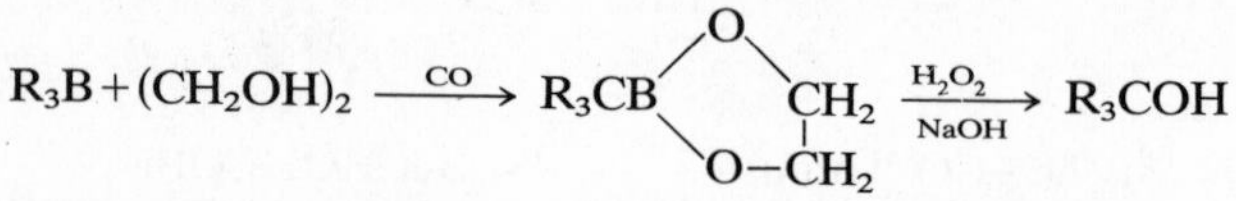

$$R_3B + (CH_2OH)_2 \xrightarrow{CO} R_3CB\langle O\text{-}CH_2\text{-}CH_2\text{-}O\rangle \xrightarrow[\text{NaOH}]{H_2O_2} R_3COH \tag{56}$$

e.g. $R = Me_2CHCH_2$, 90% yield

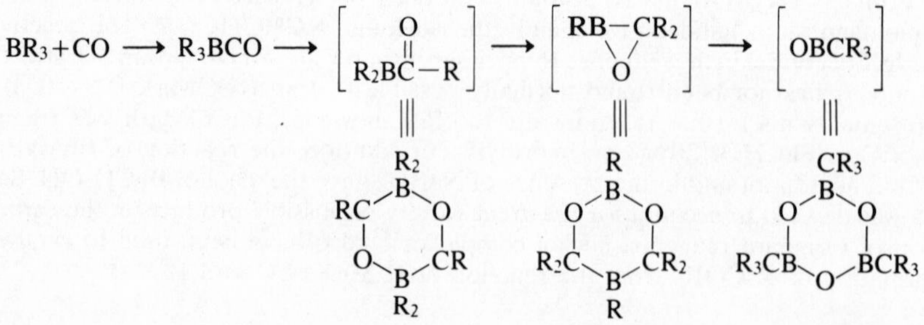

$$BR_3 + CO \longrightarrow R_3BCO \longrightarrow \left[ R_2B\overset{O}{\overset{\|}{C}}-R \right] \longrightarrow \left[ RB\underset{O}{\diagdown\!\diagup}CR_2 \right] \longrightarrow [OBCR_3]$$

(31): cyclic dimer of $[R_2BC(O)R]$ — ring $R_2B$–O–CR=…, i.e. $RC$/$O$/$R_2B$ ring with $CR$ and $BR_2$

(32) 2,5-diboradioxane: ring of $RB$, O, $CR_2$, $RB$, O, $CR_2$

(33) boroxine: ring of $B(CR_3)$, O, $BCR_3$, O, $R_3CB$, O

SCHEME 16

Procedures have been devised to trap the intermediate species shown in Scheme 16, e.g. addition of hydroxide to the carbonylation medium enables the 2,5-diboradioxane to be trapped, equation 57, whereas on addition of $LiBH_4$ the product from transfer of one R group is obtained, equation 58.

$$BR_3 + CO \xrightarrow{H_2O} (\mathbf{32}) \xrightarrow{NaOH} RB(OH)_2 + R_2CHOH \tag{57}$$

$$BR_3 + CO \longrightarrow (\mathbf{31}) \xrightarrow[\text{(ii) } H_2O]{\text{(i) } LiBH_4 [\text{or } LiAlH(OMe)_3]} RCHO \tag{58}$$

The groups transferred need not be the same, e.g. equation 59.

$$RR'_2B \xrightarrow[\text{(ii) } H_2O_2, NaOH, H_2O]{\text{(i) } CO, H_2O} RCOR' \tag{59}$$

e.g. R′ = cyclohexyl; R = MeEtCH, yield of RCOR′ = 66%

In addition, cyclic ketones can be obtained, equation 60.

$$\begin{array}{l} Me{-}CH{-}CH_2 \\ \quad | \\ \quad CH_2{-}CH_2 \end{array}\!\!\!> B{+\!\!+} \xrightarrow[\text{(ii) } Na_2O_2,\ NaOH,\ H_2O]{\text{(i) } CO,\ H_2O} \underset{40\%}{\text{3-methylcyclopentanone}} \tag{60}$$

b. *Palladium-catalysed alkoxycarbonylations of organomercury compounds.* Uncatalysed carbonylations of organomercury compounds require high pressures and temperatures and then only poor yields of products are obtained[146]. Ethylmercury chloride is carbonylated at atmospheric pressure in the presence of palladium chloride to give a low yield of propionic acid on hydrolysis. Arylmercurials are also carbonylated at atmospheric pressure. The products of these reactions depend on the solvent and catalyst, for example, using $PdCl_2$, acid chlorides are formed in MeCN solution whereas esters are obtained in alcoholic solvents[147]. The carbon ylations proceed stereospecifically as shown with cyclohexylmercury chloride[148] and vinylmercurials[149], equation 61.

$$(E)\text{-}RCH{=}CHHgCl \xrightarrow[-HgCl_2]{[PdCl_4]^{2-}, CO, ROH} (E)\text{-}RCH{=}CHCO_2R + Pd^0 + 2Cl^- + HCl \tag{61}$$

Divinyl and diaryl ketones have been formed from reaction of the appropriate organomercury salts and carbon monoxide at room temperature using $[Rh(CO_2)_2Cl]_2$ as catalyst[150], equation 62.

$$(E)\text{-}RCH{=}CHHgCl + CO \xrightarrow[\text{LiCl, room temp.}]{[Rh(CO)_2Cl]_2} [(E)\text{-}RCH{=}CH]_2CO \tag{62}$$

In all these reactions intermediate organopalladium compounds are formed.

$$RHgX + PdX_2 \xrightarrow{-HgX_2} [RPdX] \tag{63}$$

$$[RPdX] + CO \longrightarrow RCOPdX \xrightarrow{ROH} RCO_2R + HX + Pd \tag{64}$$

c. *Reactions with complexed carbonyls*[151]. Metal carbonyls have been used as reagents with main group organometallics, especially lithium compounds. The initial reaction products have been reacted with various electrophiles; for example, they have been alkylated to yield carbene complexes as with $[Cr(CO)_6]$, equation 65, or hydrolysed to give aldehydes, as with $[Fe(CO)_5]$, equation 66; in addition, acyloins are formed using $[Ni(CO)_4]$, equation 67.

$$[(CO)_6Cr] + MeLi \xrightarrow{0\,°C} \left[(CO)_5Cr{:}C\begin{matrix}Me\\OLi\end{matrix}\right] \xrightarrow[\text{(ii) } H_2O]{\text{(i) } Me_3O^+,\ BF_4^-} \left[(CO)_5Cr{-}C\begin{matrix}OMe\\Me\end{matrix}\right] \quad (65)$$

$$ArLi \xrightarrow[\text{(ii) } H_2O]{\text{(i) } [Fe(CO)_5]} ArCHO \quad (66)$$

$$ArLi \xrightarrow[\text{(ii) aq. MeOH, HCl}]{\text{(i) } [Ni(CO)_4]} ArCOCHOHAr \quad (67)$$

### *3. Carbon dioxide insertions*

Insertion of carbon dioxide into carbon—metal bonds has been achieved for many metals[152], including all the alkali metals, alkaline earth elements, zinc, aluminium, indium, and thallium. No example has yet been found for compounds of the Group IVB metals and mercury.

Most study has been with organo-lithium and -magnesium compounds. Ready reactions occur with all organic derivatives of lithium[4] and magnesium, including hindered ones—an exception perhaps being pentabromophenyllithium[4]. The initial products are the carboxylates, which are readily hydrolysed to the carboxylic acids. This is one of the most useful methods for preparing carboxylic acids. If these

$$RM + CO_2 \longrightarrow RCO_2M \xrightarrow{H_2O} RCO_2H \quad (68)$$

reactions are used as preparations of $RCO_2H$, care must be taken to ensure that an excess of carbon dioxide is present (i.e. pour RM solution on to crushed solid $CO_2$) to avoid further reaction of the metal carboxylate, $RCO_2M$, with the organometal, RM. Low temperatures are also desirable. In addition hydrolysis should be delayed until all the RM is used up (see also Section II.B). Without these precautions, products $R_2CO$ and $R_3COH$ would also be formed in appreciable amounts. An illustration of this has been provided by Soloski and Tamborski[153] (Table 6). Another example is the formation of the alcoholate which resulted from bubbling $CO_2$ into excess trialkylaluminium at room temperature[154], equation 69. In contrast, if $Et_3Al$ is added dropwise to a saturated solution of carbon dioxide, the carboxylate is

$$Et_3Al + CO_2 \longrightarrow Et_2AlO_2CEt \xrightarrow{2Et_3Al} Et_2AlOCEt_3 + Et_2AlOAlEt_2 \quad (69)$$

formed[154]. As mentioned earlier, attack by $CO_2$ on sterically hindered organolithiums still proceeds at a reasonable rate; however, subsequent reaction of the lithium carboxylate with the hindered organolithium is seriously retarded and so less secondary products are obtained.

A further problem regarding the formation of $RCO_2M$ with the more electropositive metals (e.g. sodium, potassium, and magnesium)[152] is that $\alpha$-metallation of the initial carboxylate may occur. This leads to the formation of substituted malonic acids (e.g. Scheme 17).

TABLE 6. Products of carboxylation of $p\text{-}CF_3C_6H_4Li$ (ArLi) under different reaction conditions[153]

| Conditions | Yield (%) | | |
|---|---|---|---|
| | $ArCO_2H$ | $Ar_2CO$ | $Ar_3COH$ |
| ArLi solution poured on to crushed $CO_2$, then hydrolysed | 64 | 30 | 6 |
| $CO_2$ bubbled into ArLi solution at 0 °C, then hydrolysed | 0 | 89 | 11 |
| $CO_2$ bubbled into ArLi solution at −78 °C, then hydrolysed | 43 | 50 | 7 |

$$RCH_2Na \xrightarrow{CO_2} RCH_2CO_2Na \xrightarrow{RNa} RCHNaCO_2Na$$

$$RCHNaCO_2Na \xrightarrow{CO_2} RCH(CO_2Na)_2$$

$$RCH(CO_2Na)_2 \xrightarrow{H_2O} RCH(CO_2H)_2$$

SCHEME 17

As well as for synthesis of carboxylic acids, carboxylations have been used to identify the site and the extent of metallation. With care, almost quantitative conversion into carboxylic acids can be achieved. This use can be a general one, since carboxylations have been shown to proceed with retention, e.g. for *trans*-4-*t*-butylcyclohexyllithium[155], vinyllithiums[156], substituted cyclopropyllithiums[157], PhCHLiSOMe[158] and alkenylaluminium derivatives[159,160], e.g. equation 70.

$$trans\text{-}BuCH{=}CHAlBu^i_2 \xrightarrow{MeLi} \underset{(\mathbf{34})}{trans\text{-}BuCH{=}CHAlBu^i_2(Me)Li} \xrightarrow[(ii)\ H_2O]{(i)\ CO_2} trans\text{-}BuCH{=}CHCO_2H \quad (70)^{159a}$$

Reaction of allyl-Grignards occurs with allylic rearrangement, e.g. equation 71[160]; propargylic–allenylic rearrangements occur in reaction of propargylmagnesium halides[161] and diallenylzinc compounds[162] with $CO_2$.

$$CH_2{=}CHCHMeCH_2CH{=}CMeCH_2MgX \xrightarrow[(ii)\ H_2O]{(i)\ CO_2} CH_2{=}CHCHMeCH_2CH(CO_2H)CMe{=}CH_2 \quad (71)$$

The reactivities towards carbon dioxide are varied[152]. Dimethylberyllium may inflame in a $CO_2$ atmosphere; in ether solution $(MeCO_2)_2Be$ can be formed. Simple dialkylzincs react only very slowly with $CO_2$, unless temperatures of 150–160 °C are used or a coordinating base, such as *N*-methylimidazole, tmed, or $\alpha,\alpha'$-bipyridyl is also present[163]. Allenylzinc compounds are much more reactive[162]. Sequences of reactivity have been established as $R_2Zn > R_2Cd > R_2Hg$[152] and $R_3Al > R_3Ga > R_3Tl$[164].

As shown in equation 70, the vinyl—aluminium bond is more reactive than an alkyl—aluminium bond, which in turn is more reactive than an aryl—aluminium bond. All complexes, such as **34**, are generally more reactive than the corresponding triorganoalanes and indeed in many cases (but not all)[159b] their formation may be required for reaction. Only one organo—aluminium bond in triorganoalanes reacts under normal conditions; however, at high temperatures (180–200 °C) all four metal—carbon bonds in $MAlR_4$ can react.

The mechanism(s) of the $CO_2$ insertion are probably similar to those for reaction with carbonyl compounds, i.e. by polar and s.e.t. mechanisms. An electron-transfer mechanism has been discovered for the hindered 1-triptycyllithium–carbon dioxide reaction (Scheme 18). It is of interest that the reaction of 1-triptycyllithium with ketones do not proceed via electron transfer, since no 1-(1-triptycyl)ethanol is formed[165].

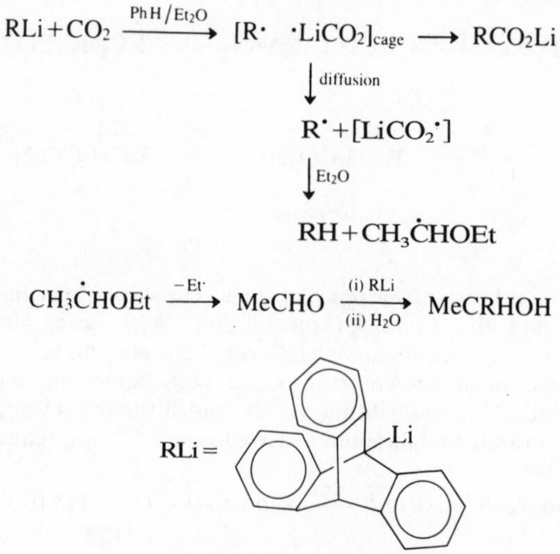

$$RLi + CO_2 \xrightarrow{PhH/Et_2O} [R^{\cdot}\ \ {}^{\cdot}LiCO_2]_{cage} \longrightarrow RCO_2Li$$

$$[R^{\cdot}\ \ {}^{\cdot}LiCO_2]_{cage} \xrightarrow{\text{diffusion}} R^{\cdot} + [LiCO_2^{\cdot}] \xrightarrow{Et_2O} RH + CH_3\dot{C}HOEt$$

$$CH_3\dot{C}HOEt \xrightarrow{-Et^{\cdot}} MeCHO \xrightarrow[\text{(ii) } H_2O]{\text{(i) RLi}} MeCRHOH$$

RLi = (1-triptycyllithium)

SCHEME 18

## D. Additions to β-unsaturated Carbonyl Compounds

Additions to $\alpha,\beta$-enones and -ynones are considered separately since 1,2-and 1,4-(conjugated or Michael) additions may occur.

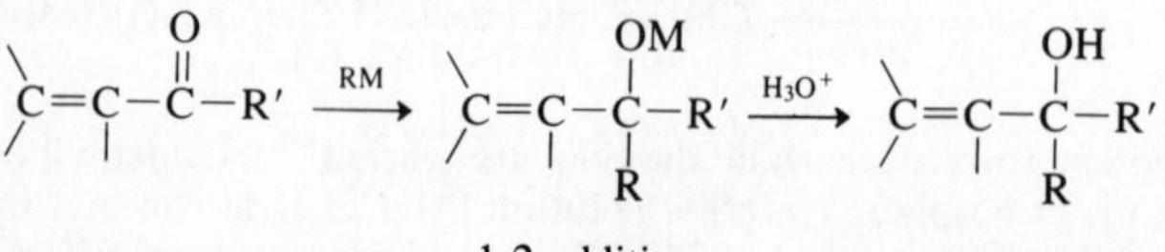

$$\rangle C{=}C{-}\overset{O}{\overset{\|}{C}}{-}R' \xrightarrow{RM} \rangle C{=}C{-}\underset{R}{\overset{OM}{C}}{-}R' \xrightarrow{H_3O^+} \rangle C{=}C{-}\underset{R}{\overset{OH}{C}}{-}R' \qquad (72)$$

1,2-addition

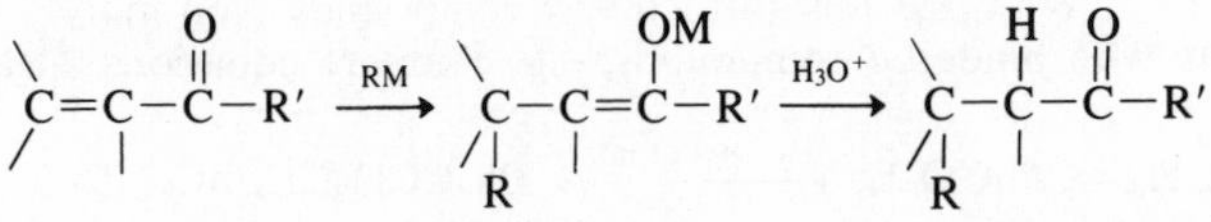

$$\begin{array}{c} \backslash \quad \ \ O \\ C{=}C{-}\overset{\|}{C}{-}R' \\ / \quad | \end{array} \xrightarrow{RM} \begin{array}{c} \backslash \quad \ \ OM \\ C{-}C{=}\overset{|}{C}{-}R' \\ /| \quad | \\ R \end{array} \xrightarrow{H_3O^+} \begin{array}{c} \backslash \quad H \ \ O \\ C{-}\overset{|}{C}{-}\overset{\|}{C}{-}R' \\ /| \quad | \\ R \end{array} \qquad (73)$$

1,4-addition

R′=alkoxy, alkyl, aryl, $NR_2$, etc.

Generally, organolithiums and other organo-alkali metals tend to give more 1,2-addition than organomagnesiums (although this is not always upheld). A good example is shown[166] in Scheme 19, while Table 7 contains data for a variety of phenyl metal derivatives with benzalacetophenone[167].

$$PhC(O)C{\equiv}CC(O)Ph \xrightarrow{PhLi} Ph_2C(OLi)C{\equiv}CC(OLi)Ph_2 \xrightarrow{H_3O^+} Ph_2C(OH)C{\equiv}CC(OH)Ph_2$$

$$PhC(O)C{\equiv}CC(O)Ph \xrightarrow{PhMgX} PhC(OMgX){=}C(Ph)C(Ph){=}C(OMgX)Ph \xrightarrow{H_3O^+} PhCOCH(Ph)CH(Ph)COPh$$

SCHEME 19

Additions of $\alpha,\beta$-unsaturated aldehydes tend to be 1,2-additions (even for RMgX and $R_3Al$ as well as RLi), whereas with Grignard reagents and $\alpha,\beta$-unsaturated alkyl ketones both 1,2- and 1,4-additions are obtained. $Me_3SiCH_2MgCl$ and $\alpha,\beta$-unsaturated aldehydes provide 1,2-adducts, even in the presence of stoichiometric CuBr; $Me_3SiCH_2Li$ and $\alpha,\beta$-unsaturated ketones also give 1,2-adducts, in contrast to $Me_3SiCH_2MgCl$, which provides mainly 1,4-additions[168a]. It is of interest that (*E*)-$RCOCH{=}CHSiMe_3$ (R = Me, Et, $Pr^i$, or Ph) gives 1,2-products with

TABLE 7. Additions of phenyl-metals to benzalacetophenone

$$PhCH{=}CHCOPh \xrightarrow[\text{(ii) } H_3O^+]{\text{(i) PhM}} \underset{\text{1,2-addition}}{PhCH{=}CHC(OH)Ph_2} + \underset{\text{1,4-addition}}{Ph_2CHCH_2COPh}$$

| | Yield (%) | |
|---|---|---|
| Phenylmetal | 1,2-Addition | 1,4-Addition |
| PhLi | 69 | 13 |
| PhNa | 39 | 3.5 |
| PhK | 52 | |
| PhMgBr | | 94 |
| PhCaI | 52 | |
| $Ph_2Be$ | | 90 |
| $Ph_2Zn$ | | 91 |
| $Ph_2Cd$ | | 100 |
| $Ph_3Al$ | | 94 |

$Me_3SiCH_2MgCl$[168b]. With $\alpha,\beta$-unsaturated aryl compounds even more 1,4-additions occur, especially with hindered compounds, e.g. compare equations 74 and 75.

$$CH_2{=}CPhCO_2Et \xrightarrow[\text{(ii) } H_3O^+]{\text{(i) } Ph_3CLi,\text{ thf}} Ph_3CCH_2CHPhCO_2Et \qquad (74)^{169a}$$

$$CH_2{=}CMeCO_2Et \xrightarrow[\text{(ii) } H_3O^+]{\text{(i) } PhLi,\ Et_2O} CH_2{=}CMeCPh_2OH \qquad (75)^{169b}$$

Polymerization may become a considerable nuisance in reactions between RMgX and $CH_2{=}CH_2CO_2Et$; however, good yields of the conjugate adduct may be obtained if low temperatures are employed[170a]. Frequently the conjugate additions are carried out with excess of the organometallic; with deficit organometallic reagent complex reaction mixtures may ensue[170b].

As can be seen from the above discussion, conjugate additions by organolithiums are not frequently found. However, Michael or conjugate addition can become the dominant reaction under certain circumstances, e.g. (i) when considerable steric hindrance is met, as exemplified by equation 74, as well as in reactions with *cis*-chalcones and cyclopropenones[4]; (ii) using stabilized carbanions[171–188]; (iii) reactions with unsaturated anides[189–191], e.g. reaction 76;

$$PhCH{=}CHCON(C_6H_{11}\text{-cyclo})_2 \xrightarrow[\text{(ii) } H_2O]{\text{(i) } PhLi} Ph_2CHCH_2CON(C_6H_{11}\text{-cyclo})_2 \qquad (76)^{189}$$

(iv) reactions in the presence of hmpt[172,188,192–194], e.g. Scheme 20;

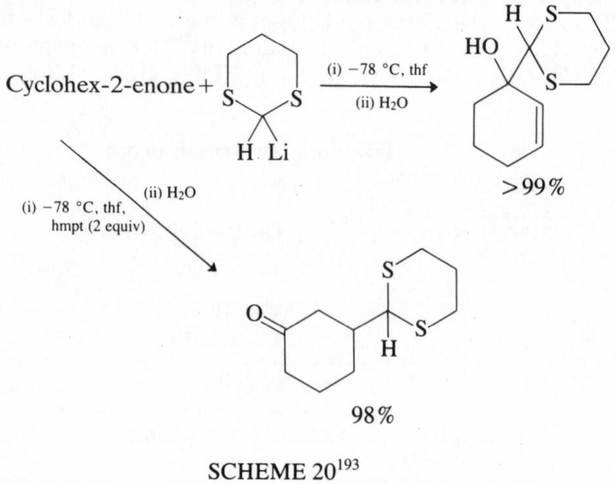


SCHEME 20[193]

(v) under thermodynamic control[178,195], e.g. Scheme 21, and (vi) via cuprates[196] (conjugate addition of organocuprates, obtained by treatment of $Cu^I$ salts with RMgX or RLi, to $\alpha$-enones is a well established reaction).

Conjugate additions of the following stabilized organolithiums have been reported: $Ph_3CLi$[169,173], $Me_3SiLi$[174], $RS(RSO)CR'Li$[176], $(PhS)_3CLi$[177], $(MeS)_2C(MMe_3)Li$ (M = Si or Sn)[171], $(RS)_2CHLi$[181], $(RSe)_2CHLi$[181], 2-Li-2-R-1,3-dithianes[172,178–180] ester

Cyclohex-2-enone + Li Ph S S (i) thf, −78 °C (ii) $H_2O$ → HO Ph S S 65% + O S Ph S (A) (35%)

(i) thf, −78 °C→25 °C (ii) $H_2O$ → A 93%

SCHEME 21[178]

enolates, $X\bar{C}HCO_2R, Li^+$ (X = PhO[183], ArS[183,185,192] or PhSe[184]), $ArS\bar{C}HCN, Li^+$[186], 2-$LiCH_2$-isoquinoline[187] and $RXCH{=\!=\!=}\overset{(-)}{C}H{=\!=\!=}CH_2, Li^+$ (X = S or Se)[188], Scheme 22. These compounds undergo direct conjugate addition or can be made to do so by an increase in reaction temperature and/or solvent polarity. The outstanding reagent for inducing 1,4-addition is hmpt, a reagent which solvates the $Li^+$ particular well[197]. A number of stabilized carbanions, which normally react by 1,2-addition with $\alpha$-enones in thf alone, e.g. 2-lithio-1,3-dithiane[193], yield the 1,4-adducts in presence of hmpt, either under conditions of thermodynamic[183,184] or kinetic control[181,191,192]. Higher temperatures also can favour 1,4-addition[184].

Conjugate additions of Grignard reagents to $\alpha,\beta$-unsaturated amides have been known for sometime; references for 1,4-addition of organolithiums to these compounds are more limited[189–191]. Recently, 1,4-addition to $\alpha,\beta$-unsaturated thioamides have been reported; these are discussed in Section III.A.2.

Cyclopent-2-enone + MeS $Li^+$ (i) thf, −78 °C (ii) $H_3O^+$ → HO SMe (43%) (diastereomeric ratio 3 : 2) + HO SMe (44%) (diastereomeric ratio 5 : 3)

(i) thf, hmpt, −78 °C (ii) $H_3O^+$ → O SMe (89%) (diastereomeric ratio 3 : 2) + O SMe *ca.* 5%

SCHEME 22[188]

Conjugate additions have been reported with $\alpha,\beta$-unsaturated lactones[198], e.g. equation 77, and with complexed $\alpha,\beta$-unsaturated esters[199], e.g.

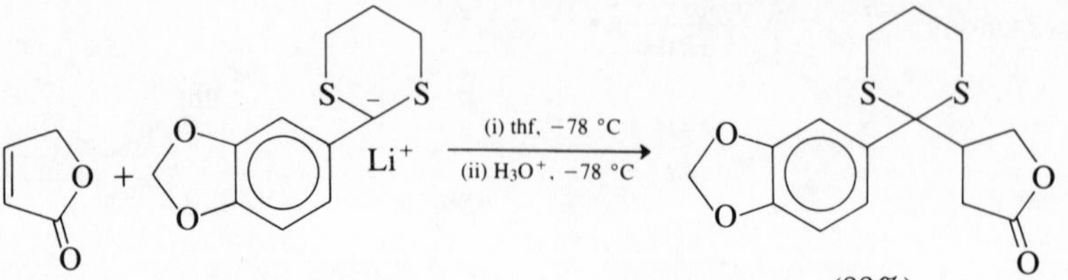

$$\text{2(5H)-furanone} + \text{2-lithio-2-(3,4-methylenedioxyphenyl)-1,3-dithiane } (Li^+) \xrightarrow[\text{(ii) } H_3O^+,\ -78\ ^\circ C]{\text{(i) thf},\ -78\ ^\circ C} \text{4-[2-(3,4-methylenedioxyphenyl)-1,3-dithian-2-yl]butanolide}\ (88\%) \qquad (77)^{198}$$

$[(CH_2{=}CHCO_2Me)Fe(CO)_4]$, Scheme 23. Of interest, it appears that a CO insertion can also be worked into these syntheses (e.g. step b in Scheme 23).

$$[CH_2{=}CHCO_2Me \cdot Fe(CO)_4] + NaCH(CO_2Et)_2$$

$$\xrightarrow[\text{(ii) } CF_3CO_2H]{\text{(i) thf, } 0\rightarrow 25\ ^\circ C} (EtO_2C)_2CHCH_2CH_2CO_2Me\ \ 92\%$$

$$\xrightarrow[\text{(i) thf, } 0\rightarrow 25\ ^\circ C;\ \text{(ii) MeI}]{\text{(b)}} (EtO_2C)_2CHCH_2CH(COMe)CO_2Me\ \ 85\%$$

SCHEME 23

As mentioned earlier, organocuprates are very good and general reagents for conjugate addition. However, there is one exception: alkynylcuprates are poor reagents, probably owing to the strength of the RC≡C—Cu bond. Hence there is a need for a reagent to transfer alkynyl groups. Alkynyl-alanes[200] and -boranes[201] do conjugatively add to enones, but only to those capable of achieving an *S-cis*-conformation (a *transoid* geometry leads to 1,2-addition). However, a way around this has recently been established; this employs [Ni(acac)$_2$] and $Bu^i_2AlH$[202] as additional reagents, equation 78.

$$Me_2AlC{\equiv}CBu^t + \text{cyclopent-2-enone} \xrightarrow[\text{(ii) } 0\ ^\circ C, Et_2O]{\text{(i) [Ni(acac)}_2\text{]–(Bu}^i\text{)}_2\text{AlH}} \text{1-(Me}_2\text{AlO)-3-(Bu}^t\text{C}{\equiv}\text{C)-cyclopentene}\ \ 60\% \qquad (78)$$

Stoichiometric copper (II) acetate has been used[203] to catalyse the conjugate additions of alkylalanes to $\alpha,\beta$-unsaturated ketones or aldehydes, e.g. equation 79. Free-radical initiators have been found to catalyse conjugate additions of trialkylalanes[204], equation 80; normally for $R_3Al$, considerable amounts of polymer and 1,2-adducts are obtained in reactions with $\alpha,\beta$-unsaturated carbonyls.

$$\left(2RCH{=}CH_2 \xrightarrow{LiAlH_4}\right) LiAlH_2(CH_2CH_2R)_2 \xrightarrow[\text{(ii) } H_3O^+]{\text{(i) } CH_2{=}CHCOR',\ Cu(OAc)_2\text{(stoich.)}} R(CH_2)_4COR' \qquad (79)$$

R = H or Me

$$Pr_3Al + \text{cyclohex-2-enone} \xrightarrow[\text{(ii) } H_3O^+]{\text{(i) initiator, } O_2} \text{3-propylcyclohexanone} \qquad (80)$$

As shown in equation 78, for organoalanes the alkynyl group is transferred in preference to alkyl groups. Alkenyl groups also preferentially react (and with retention of configuration)[205,206], e.g. equations 81 and 82.

$$(E)\text{-BuCH{=}CHAlBu}^i{}_2 + \text{PhCH{=}CHCOMe} \xrightarrow[\text{(ii) } H_2O]{\text{(i) hydrocarbon}} (E)\text{-BuCH{=}CHCH(Ph)CH}_2\text{COMe} \quad (81)^{205}$$

69%

$$(E)\text{-C}_6\text{H}_{13}\text{CH{=}CHAlBu}^i{}_2\text{Me,Li}^+ + 2\text{-[(EtO}_2\text{C(CH}_2)_6]\text{-cyclopent-2-enone}$$

$$\xrightarrow[\text{(ii) } H_2O]{\text{(i) hydrocarbon}} \text{2-(CH}_2)_6\text{CO}_2\text{Et-3-[(}E\text{)-CH{=}CHC}_6\text{H}_{13}]\text{-cyclopentanone} \quad (82)^{206}$$

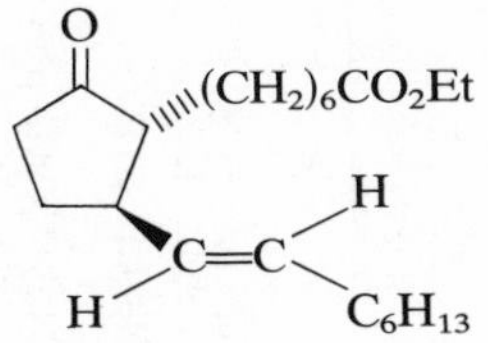

(76%)

In this section we have concentrated on organo-lithium and -magnesium compounds, with some space also given over to aluminium compounds. This is appropriate since these are the most used derivatives. However, some systematic study with other metals have been undertaken[207]. For example, diorganocadmiums, prepared from RMgX and $CdX_2$, tend to give more 1,2-addition to conjugated aldehydes and more 1,4-addition to unsaturated ketones. The primary 1,4-enolates (but not the 1,2-enolates), however, can also conjugatively add to the unsaturated ketone to give further products, Scheme 24.

$$(E)\text{—MeCH{=}CHCOMe} \xrightarrow{\text{Me}_2\text{Cd, MgBr}_2} \text{Me}_2\text{CHCH{=}CMeOCdMe}^+ + \text{MeCH{=}CHCMe}_2\text{OCdMe} \; (<1\%)$$

$$\text{Me}_2\text{CHCH{=}CMeOCdMe}^+ \xrightarrow{(E)\text{—MeCH{=}CHCOMe}} \text{Me}_2\text{CHCH(COMe)—MeCHCH{=}CMeOCdMe}$$

$$\xrightarrow{\text{H}_3\text{O}^+} \text{Me}_2\text{CHCH(COMe)—MeCHCH}_2\text{COMe} \; (83\%)$$

SCHEME 24

Conjugate additions by organo-magnesiums[170b] and -lithiums[177c] (including ate complexes, $Me_3ZnLi$) have been carried out in the presence of chiral ligands, such as (−)-sparteine[170b] and $(Me_2NCH_2CHOMe)_2$[177c]; however, only low optical yields were obtained, e.g. cyclohex-2-enone and $(PhS)_3CLi$ in the presence of $(Me_2N$-$CH_2CHOMe)_2$ gave chiral 3-$(PhS)_3C$-cyclohexanone with an optical yield of only 7.6%[177c]. Table 8 contains further examples of conjugate additions.

Some theoretical consideration of additions to unsaturated carbonyl compounds has been given by Anh and coworkers[70,208] and by Lefour and Loupy[197]. It has been argued that reactions at carbon 2 (hard site) are under charge control while reactions

TABLE 8. Conjugate or Michael addition to $\alpha,\beta$-unsaturated carbonyl compounds

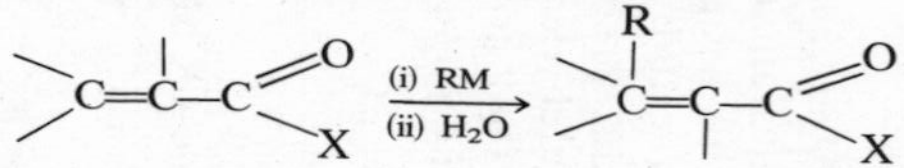

$$>C{=}C{-}C({=}O)X \xrightarrow[\text{(ii) } H_2O]{\text{(i) RM}} >C(R){-}C{=}C({=}O)X$$

| 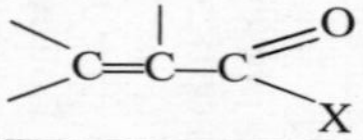 $>C{=}C{-}C({=}O)X$ | RM, conditions | Product (yield, %) | Ref. |
|---|---|---|---|
| PhCH=CHCHO | (i) $p$-$MeOC_6H_4CHLiC{\equiv}N$ thf, −78 °C, hmpt<br>(ii) $H_3O^+$, 78 °C | PhCH—$CH_2CHO$<br>\|<br>$NCCHC_6H_4OMe$-$p$<br>(40)<br>+1,2-adduct<br>(40) | 192 |
| 2-methylcyclopent-2-enone | (i) 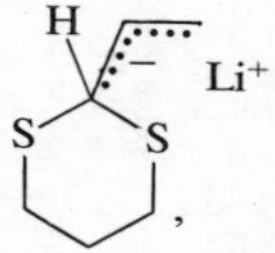 H, S, S, $Li^+$, thf, −78 → +25 °C.<br>(ii) $H_2O$ | 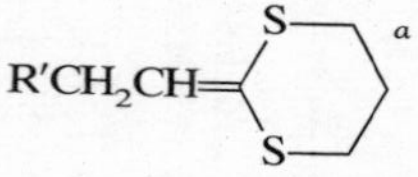 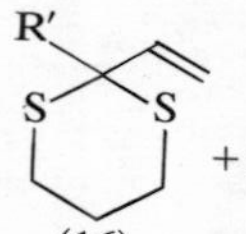 $R'CH_2CH{=}$ S, S (60) [a]<br>R′, S, S (16) +1,2-adduct (24) [a] | 172 |
| | (i) 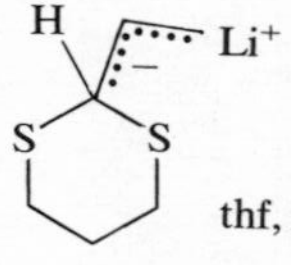 H, S, S, $Li^+$, thf, hmpt −78 → +25 °C<br>(ii) $H_2O$ | 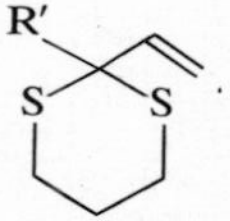 R′, S, S (100) [a] | 172 |

| | | | |
|---|---|---|---|
| cyclopent-2-enone | (i) $(MeS)_2C(SnMe_3)Li$<br>(ii) $H_3O^+$ | $R''C(SMe)_2SnMe_3$ [b]<br>(70) | 171 |
| cyclohex-2-enone | (i) $Me_3SiCH_2Li$<br>(ii) $H_3O^+$ | $R'''CH_2SiMe_3$ [c]<br>(70) no catalyst<br>(97) with $Cu_2Br_2$ | 168a |
| cyclohex-2-enone | (i) $(PhS)_3CLi$, $(Me_2NCH_2CHOMe)_2$<br>(ii) $H_2O$ | $R'''C(SPh)_3$ [c]<br>(7.6 optical yield) | 177c |
| $CH_2{=}CHCOMe$ | (i) $[(C_6H_{13})_2AlH_2]Li$<br>(ii) $H_2O$ | $C_6H_{13}CH_2CH_2COMe$<br>(54) | 203 |
| $CH_2{=}CHCO_2Et$ | (i) $CH_2{=}CHMgBr$, −35 °C<br>CuCl<br>(ii) $H_3O^+$ | $CH_2{=}CHCH_2CH_2CO_2Et$<br>(55) | 170a |
| (**E**)-$Cl(CH_2)_3CH{=}CHCO(CPPh_3)CO_2Et$ | (i) MeLi<br>(ii) $H_2O$ | O, $\overset{+}{P}Ph_3$, C—$\bar{C}$′, $CO_2Et$, Me<br>(82) | 209a |
| 1-[(**E**)-$MeCH{=}CHC(O)$]piperidine | (i) $Bu^tLi$<br>(ii) PrBr<br>(iii) $H_3O^+$ | 1-[$MeCH(Bu^t)CH(Pr)C(O)$]piperidine<br>(85) | 191 |
| (**E**)-$MeCH{=}CHC(O)NMe_2$ | (i) PhMgBr<br>(ii) $H_3O^+$ | $MeCH(Ph)CH_2C(O)NMe_2$<br>(50) | 191 |

TABLE 8. Continued

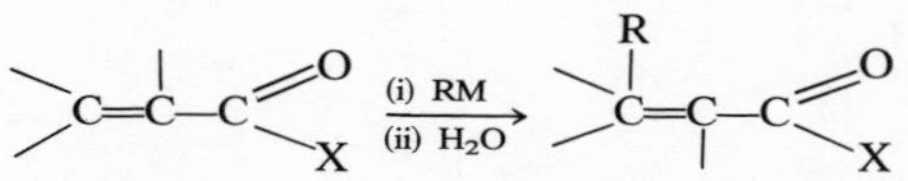

$$>C{=}C(-)-C({=}O)X \xrightarrow[\text{(ii) } H_2O]{\text{(i) RM}} >C(R)-C(-)-C({=}O)X$$

| 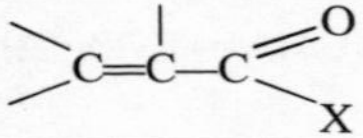 $>C{=}C-C({=}O)X$ | RM, conditions | Product (yield, %) | Ref. |
|---|---|---|---|
| $MeCH{=}CHC(O)NMe_2$ | (i) 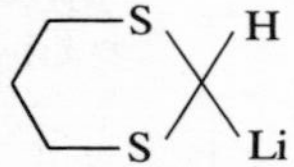 2-lithio-1,3-dithiane; (ii) 3,4-dimethoxybenzaldehyde; (iii) $H_3O^+$ | 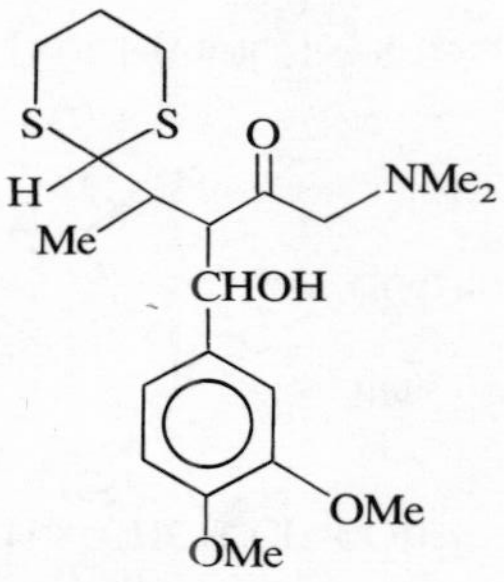 2-(1,3-dithian-2-yl)(H)CH(Me)–CH(CHOH–C$_6$H$_3$(OMe)$_2$-3,4)–C(O)CH$_2$NMe$_2$ (80) | 191 |
| $CH_2{=}CHC(O)NHPh$ | (i) BuLi (ii) $H_2O$ | $BuCH_2CH_2CONHPh$ (90) | 190 |
| $(E)$-$PhCH{=}CHCOMe$ | (i) $PhHgCl,Cl,PdCl_2,$ $CH_2Cl_2,HCl,Bu_4NCl$ | $Ph_2CHCH_2COMe$ (85) | 209b |
| $CH_2{=}CHCOMe$ | (i) $(Z)$-$C_6H_{13}CH{=}CHB$(cyclic borane) (ii) $H_2O$ 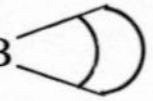 | $(E)$-$C_6H_{13}CH{=}CHCH_2CH_2COMe$ (87) | 209c |

[a] R′ = 2-Me-3-oxocyclopentyl; [b] R″ = 3-oxocyclopentyl; [c] R″ = 3-oxocyclohexyl.

at carbon 4 (soft site) are under frontier control. From this it could be reasoned that organic derivatives of hard metals, e.g. alkali metals, attack at $C_{(2)}$, and those of soft metals, e.g. cadmium and copper, attack at $C_{(4)}$. Derivatives of intermediate metals, e.g. magnesium and aluminium, produce both 1,2- and 1,4-adducts. The complexation of $M^+$ with the carbonyl or its association with $R^-$ can also be important roles played by the cation of the organometallic, RM.

Okubo[30] has detected ketyls at room temperature in the reactions of sterically hindered chalcones, e.g. 2,4,6-$Me_3C_6H_2CH{=}CHCOC_6H_2Me_3$-2,4,6, with *o*-$MeC_6H_4MgBr$. Okubo suggested that the strength of the C—Mg linkage, the conformation of the $\alpha,\beta$-enones, and the spin distribution in the ketyl radical all are important in deciding the mode of addition to $\alpha,\beta$-enones. The generally considered view is that the transition states for conjugate addition have *cis*-enone structures.

## III. ADDITIONS TO CARBON—SULPHUR DOUBLE BONDS[94,210]

Organometallic compounds can add to carbon—sulphur double bonds to give initial products of nucleophilic attack at sulphur (thiophilic addition) and/or at carbon (carbophilic addition). The relative importance of these additions depend upon the

$$>C{=}S + RM \longrightarrow \begin{cases} M{-}\overset{|}{\underset{|}{C}}{-}SR & \text{thiophilic addition} \\ M^+,\bar{S}{-}C{-}SR & \text{carbophilic addition} \end{cases} \qquad (83)$$

type of carbon—sulphur double bonded compound, the organometallic compound, the solvent, and other factors. The hard–soft acid–base (HSAB) approach has been applied to these reactions. It has been argued that hard reagents would attack at carbon and soft reagents at sulphur[211]; thus the extent of thiophilic addition depends on the softness of the nucleophile. However, the solvent effects and lack of metal cation effects in certain thiophilic additions are not at all expected for a complete HSAB interpretation. Molecule orbital calculations[212] have indicated a slight negative character to carbon in thiocarbonyls (compare **35** with **36**). However, other factors must be present to account for the range of behaviour met with in these reactions.

$$>\overset{-0.13}{C}{=}S^{+0.13} \quad (\mathbf{35}) \qquad\qquad >\overset{+0.24}{C}{=}O^{-0.24} \quad (\mathbf{36})$$

### A. Carbophilic Additions

Examples are known of initial carbophilic addition to a dithioester[213], a thioester[214], thioamides[215], and generally to carbon disulphide[210,216–221], equation 84, carbon oxysulphide (COS)[210], e.g. equation 85, and isothiocyanates[1,114,222], equation 86.

$$RM + CS_2 \longrightarrow RC(S)SM \xrightarrow[EY]{E^+ \text{or}} RC(S)SE \qquad (84)$$

e.g. $E^+ = H^+$, $EY = MeI$

$$p\text{-MeC}_6\text{H}_4\text{MgBr} + \text{COS} \xrightarrow[\text{(ii) H}_2\text{O}]{\text{(i) thf}} p\text{-MeC}_6\text{H}_4\text{C(O)SH} \quad (50\%) \tag{85}$$

$$\text{RN{=}C{=}S} \xrightarrow{\text{R'M}} \text{RN{=}C(SM)R'} \xrightarrow{\text{H}_3\text{O}^+} \text{RNHC(S)R'} \tag{86}$$

In addition, certain organometal–thioketone combinations produce carbophilic products, e.g. benzyl- and benzylhydryl-lithiums and -sodiums and thiobenzophenone[94].

### 1. *Carbophilic additions to carbon disulphide*

These reactions are known for at least RLi, RNa, RMgX, and $R_3Al$. As shown in equation 84 and Table 9, they may be used as synthetic routes to dithiocarboxylic acids and their derivatives. Recently it was reported that copper(I) bromide was a most useful catalyst[218]. Treatment of the initial $CS_2$ adduct with further RLi, before reaction with an alkylating agent, has become a useful method of producing ketene thioketals[216,221], Scheme 25.

$$\text{RCH}_2\text{Li} \xrightarrow{\text{CS}_2} \text{RCH}_2\text{C(S)SLi} \xrightarrow{\text{R}^1\text{Li}} \text{RCH{=}C(SLi)}_2 \xrightarrow{\text{R}^2\text{X}} \text{RCH{=}C(SR}^2)_2$$

SCHEME 25

TABLE 9. Reactions of carbon disulphide with organometallic compounds

$$\text{RM} + \text{CS}_2 \longrightarrow \text{RCS}_2\text{M} \xrightarrow{\text{E}^+} \text{RCS}_2\text{E}$$

| RM | $E^+$ | Product (yield,%) | Ref. |
|---|---|---|---|
| BuMgBr | $H_2O$ | $BuCS_2H$ (40) | 210 |
| MeMgCl | EtBr | $MeCS_2Et$ (70) | 219 |
| EtMgCl | $CH_2{=}CHCH_2Br$ | $EtCS_2CH_2CH{=}CH_2$ (80) | 219 |
| 2-BrMg-thiophene | MeI | $2\text{-MeS}_2\text{C}$-thiophene (68) | 219 |
| $Bu^tMgCl$, CuCl | MeI | $Bu^tCS_2Me$ (95) | 218 |
| $Me_2C{=}CHMgCl$ | MeI | $Me_2C{=}CHCS_2Me$ (93) | 218 |
| *cyclo*-$C_6H_{11}MgCl$,CuCl | MeI | cyclo-$C_6H_{11}CS_2Me$ (100) | 218 |
| 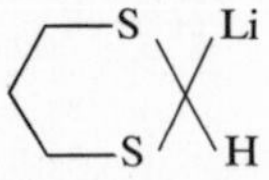 2-lithio-1,3-dithiane (ring with S, S; Li, H on C-2) | (i) $CS_2$ (ii) BuLi (iii) MeI | 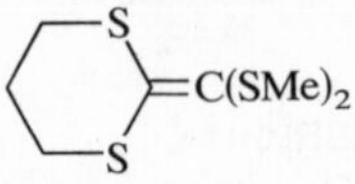 2-[$=C(SMe)_2$]-1,3-dithiane | 217 |
| 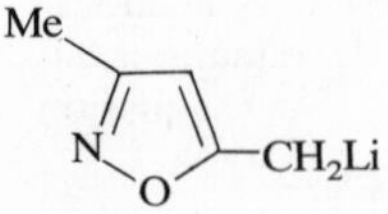 3-Me-isoxazol-5-yl-$CH_2Li$ | (i) $\frac{1}{2}CS_2$ (ii) MeI | 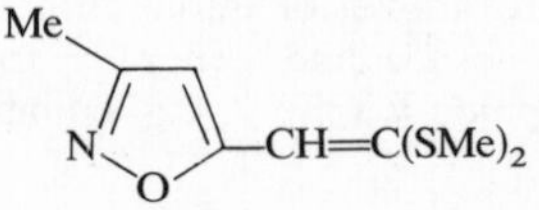 3-Me-isoxazol-5-yl-$CH{=}C(SMe)_2$ (75) | 221 |
| $Et_3Al$ | | $Et_2AlSAlEt_2$ and $Et_2CSAlEt_2$ | 220 |

$$\mathrm{RCHLiCO_2^-, Li^+ \xrightarrow[-75\,^\circ C]{CS_2} Li^+, {}^-SC(S)CH(R)CO_2^-, Li^+ \xrightarrow[\text{(ii) 2MeLi}]{\text{(i) 50}\,^\circ C} RCH{=}C(SMe)_2}$$

$$\mathrm{Li^+, {}^-SC(S)CH(R)CO_2^-, Li^+ \xrightarrow{MeI\ at\ -75\,^\circ C} RCH(CS_2Me)CO_2Li}$$

$$\mathrm{RCH_2CS_2Me \xleftarrow{\Delta} RCH(CS_2Me)CO_2H \xleftarrow{H_3O^+} RCH(CS_2Me)CO_2Li}$$

SCHEME 26

Scheme 26 illustrates[217] some products of reaction of $RCHLiCO_2Li$ with $CS_2$; good yields of each product can be obtained.

### 2. *Carbophilic conjugate addition to α,β-unsaturated thioamides*

Conjugate additions to α,β-unsaturated tertiary thioamides occur with both organo-lithium and -magnesium reagents[223], e.g. equation 87.

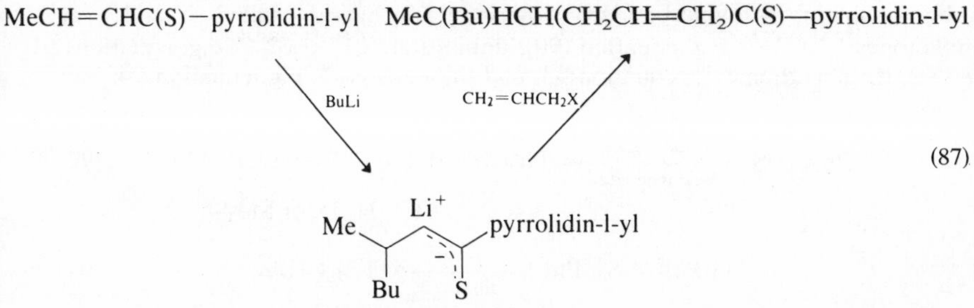

$$\mathrm{MeCH{=}CHC(S){-}pyrrolidin\text{-}l\text{-}yl \xrightarrow{BuLi} [MeCH(Bu){-}CH{=}C(S^-)\text{-}pyrrolidin\text{-}l\text{-}yl,\ Li^+] \xrightarrow{CH_2{=}CHCH_2X} MeC(Bu)HCH(CH_2CH{=}CH_2)C(S){-}pyrrolidin\text{-}l\text{-}yl} \quad (87)$$

Organolithiums, but not organomagnesiums, also conjugatively add to α,β-unsaturated secondary thioamides. However, by using *N*-alkyl-*N*-trimethylsilylamido derivatives, it is possible to obtain[224] 1,4-addition even with Grignard reagents, and since the *N*-trimethylsilyl group can be easily removed on hydrolysis, the desired product can be obtained, e.g. equation 88.

$$\mathrm{Me{\sim}CH{=}CHC(S)N(Me)SiMe_3 \xrightarrow[\text{(ii) 2NHCl}]{\text{(i) RM}} MeCH(R)CH_2C(S)NHMe} \quad (88)$$

| | |
|---|---|
| RM = BuLi | 80% yield |
| $CH_2{=}CHMgBr$ | 79% |
| $Pr^iMgBr$ | 87% |
| $Bu^tO_2CCH_2Li$ | 88% |

## B. Thiophilic Additions

The carbophilic conjugate additions to α,β-unsaturated thioamides discussed in the last section are analogous to those to α,β-unsaturated amides, discussed in Section II.D, but contrast[225] with the thiophilic 1,2- and 1,4-additions of alkylmagnesium bromide to α,β-unsaturated thioketones, e.g. equation 89. Apparently aryl-, vinyl-, and alkynyl-magnesium bromides do not react with the unsaturated thioketones[225].

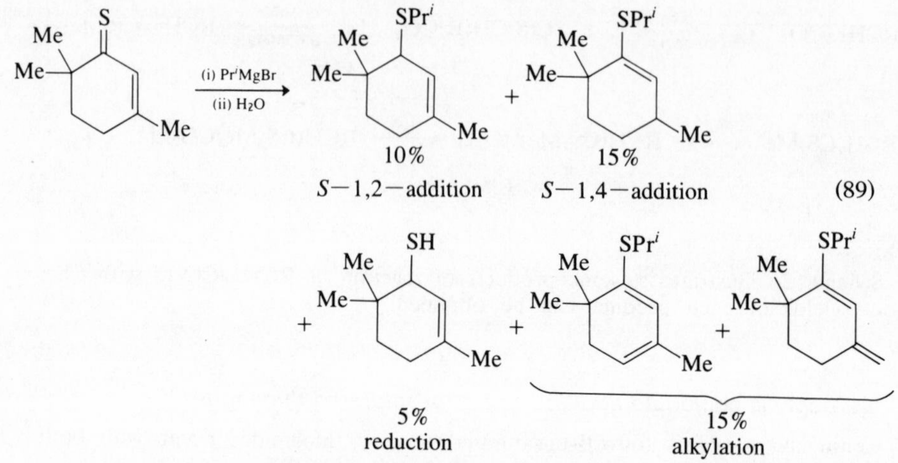

$$\text{(structures: 3,3,5-trimethylcyclohex-2-ene-1-thione} \xrightarrow[\text{(ii) } H_2O]{\text{(i) } Pr^iMgBr} \text{S-1,2-addition product (10\%)} + \text{S-1,4-addition product (15\%)} + \text{reduction product (SH, 5\%)} + \text{alkylation products (15\%))} \quad (89)$$

Me, Me, S, Me → SPr$^i$ 10% S−1,2−addition + SPr$^i$ 15% S−1,4−addition (89)

+ SH 5% reduction + SPr$^i$ + SPr$^i$ 15% alkylation

Products of thiophilic addition have been obtained from a variety of other carbon—sulphur double-bonded species, including alkyl- and aryl-thioketones[77,94,226–228], e.g. equation (90), dithioesters[77,94,211,228,229], e.g. equations 91 and 92, trithiocarbonate[94], equation 93, and thioketenes[230], e.g. equation 94.

$$Ph_2C{=}S \xrightarrow[\text{Et}_2\text{O, room temp.}]{\text{PhLi(2,2 equiv)}} Ph_2C(SPh)Li \xrightarrow{E^+} Ph_2C(E)SPh \qquad (90)^{228}$$

$$E = H, D, \text{ or } Me_3Si$$

$$PhS(Ph)C{=}S + PhLi \xrightarrow[\text{(ii) } H_2O]{\text{(i) } -78\,^\circ C} (PhS)_2CHPh \quad 49\% \qquad (91)^{228}$$

$$EtC(S)SMe + Pr^iMgX \xrightarrow[\text{(ii) } H_2O]{\text{(i) } -17\,^\circ C\text{, thf}} EtCH(SPr^i)Me \quad 84\% \qquad (92)^{229g}$$

$$(PhS)_2C{=}S + PhLi \xrightarrow[\text{(ii) } H_2O]{\text{(i) } -78\,^\circ C} (PhS)_3CH \quad 66\% \qquad (93)^{228}$$

$$R(R_1)C{=}C{=}S \xrightarrow[\text{(ii) } H_2O]{\text{(i) } R^2Li,\ -78\,^\circ C} R(R^1)C{=}CHSR^2 \qquad (94)^{230}$$

$$R = Pr^i, Bu^t;\ R^1 = Bu^t;\ R^2 = Ph$$

Most studies have involved organo-lithium and -magnesium compounds. Products from thioketones can be numerous, especially from enolizable compounds, and include the following:

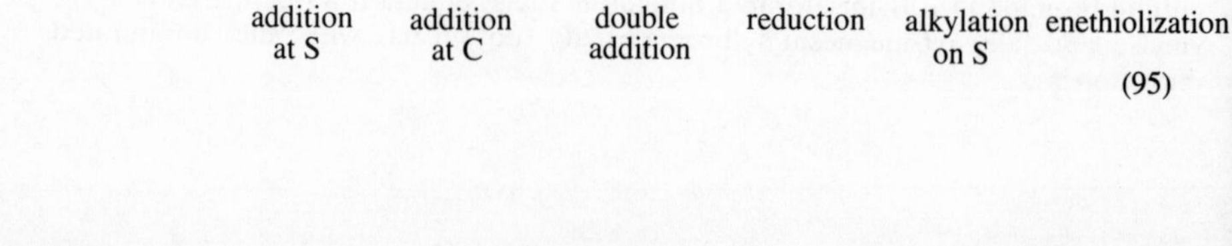

$$-\overset{S}{\overset{\|}{C}}-CH\!\!\diagdown \xrightarrow[\text{(ii) } H_2O]{\text{(i) RM}} -\underset{SR}{CH}CH + -\overset{R}{\underset{SH}{C}}-CH + -\overset{R}{\underset{SR}{C}}-CH + -\underset{SH}{CH}-CH + -\underset{SR}{C}{=}C + -\overset{SH}{C}{=}C \qquad (95)$$

addition at S; addition at C; double addition; reduction; alkylation on S; enethiolization

Not, all, of course, will be realised for every RM–thioketone combination[210], as shown in the thiocamphor (equation 96) and the Bu$^t$CSMe (equations 97 and 98) examples. The difference in products of reactions of Bu$^t$CSMe with MeMgI in $Et_2O$

$$\text{thiocamphor} \xrightarrow[\text{(ii) } H_2O]{\text{(i) EtMgBr}} \text{(SH, H)} + \text{(SEt, H)} + \text{(SEt, alkene)} \quad (96)$$

64% (*endo* : exo = 45 : 55); 10% (*endo* : *exo* = 49 : 51); 11%

$$Bu^tC(S)Me \xrightarrow[\text{(ii) } H_2O]{\text{(i) MeMgI, } Et_2O} \underset{10\%}{Bu^tC(SH){=}CH_2} + \underset{90\%}{Bu^tC(SH)Me_2} \quad (97)$$

$$Bu^tC(S)Me \xrightarrow[\text{(ii) } H_2O]{\text{(i) MeMgBr, thf}} \underset{50\%}{Bu^tC(SMe){=}CH_2} + \underset{33\%}{Bu^tCH(SMe)Me} + \underset{17\%}{Bu^tC(SMe)Me_2} \quad (98)$$

and with MeMgBr in thf also clearly indicates the effect of solvent changes as well as changes in the organometallic. The temperature is also an important factor, for example compare the products of reactions 93 and 99. The effect of metal cation changes is not always pronounced but some differences have been found for RMgX and RLi reactions with thiobenzophenone[210], equations 100–102.

$$(PhS)_2C{=}S \xrightarrow[\text{(ii) } H_2O]{\text{(i) PhLi, 25°C}} \underset{27\%}{(PhS)_2CHPh} + \underset{10\%}{Ph(PhS)C{=}C(SPh)_2} \quad (99)^{228}$$

$$MeMgI \xrightarrow[\text{(ii) } H_2O]{\text{(i) } Ph_2C{=}S\text{, } Et_2O\text{, 16h}} \underset{95\%}{Ph_2C{=}CPh_2} \quad (100)$$

$$MeLi \xrightarrow[\text{(ii) } H_2O]{\text{(i) } Ph_2C{=}S\text{, } Et_2O\text{, 16h}} \underset{30\%}{Ph_2CHSMe} + \underset{70\%}{Ph_2C{=}CH_2} \quad (101)$$

$$EtMgBr \xrightarrow[\text{(ii) } H_2O]{\text{(i) } Ph_2C{=}S\text{, } Et_2O\text{, 16h}} \underset{67\%}{Ph_2CHSEt} + \underset{\text{unreacted } 29\%}{Ph_2C{=}S} \quad (102)$$

As can be gathered from the above, a variety of processes are possible, including many secondary reactions. Processes after the initial formation of the addition product (the $\alpha$-thiaorganometallic) have been variously considered to lead to free radicals[211,229a–e], episulphides[231], alkenes[119g,231], enethiol esters[225,229b,c,e], and double addition products[211,216,229d,e].

Similar complexities can occur in $RCS_2R'$ and $(PhS)_2C{=}S$ reactions; in the reactions of the latter, carbene intermediates, $(PhS)_2C$:, can be obtained by elimination from the initial product $(PhS)_3CLi$[94,210]. Secondary reactions may mask the initial thiophilic reaction. 2,3-Sigmatropic rearrangements of the initial thiophilic addition products have been suggested to lead to the isolated carbon—carbon bonded

$$Bu^tC(S)Me + RCH{=}CHCH_2MgBr \longrightarrow [Bu^tC(Me)(MgBr)SCH_2CH{=}CHR] \xrightarrow{\sigma[2,3]} Bu^tC(SMgBr)CHRCH{=}CH_2 \quad (103)$$

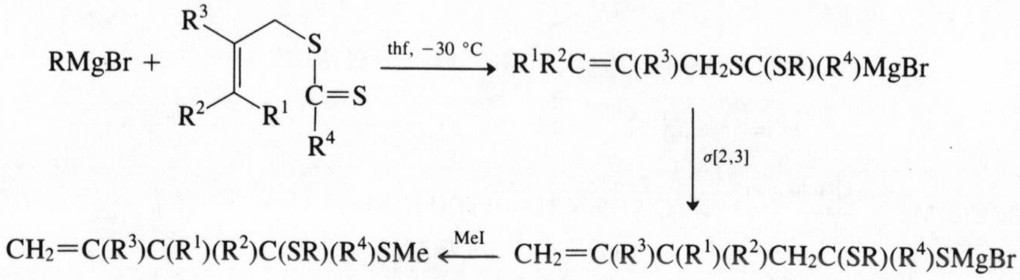

$$RMgBr + \text{(2-}R^3\text{-3-}R^1\text{-3-}R^2\text{-allyl)}S{-}C(=S)R^4 \xrightarrow{\text{thf, } -30\,°C} R^1R^2C{=}C(R^3)CH_2SC(SR)(R^4)MgBr$$

$$\xrightarrow{\sigma[2,3]} CH_2{=}C(R^3)C(R^1)(R^2)CH_2C(SR)(R^4)SMgBr \xrightarrow{MeI} CH_2{=}C(R^3)C(R^1)(R^2)C(SR)(R^4)SMe$$

SCHEME 27

products, in the allylic organometallic reactions with thioadamantane[226b] and with $Bu^tCSMe$[232], equation 103; see also Scheme 27[233] for a variation of this.

In contrast to these sulphur to carbon migrations of allyl groups, the addition of vinyl Grignard, $CH_2{=}CHMgX$, to PhCSR (R = Ph or $Bu^t$) was considered to involve a carbon to sulphur migration of a vinyl group[234], equation 104. Free radicals

$$PhRC{=}S + CH_2{=}CHMgX \longrightarrow [PhC(R)(CH{=}CH_2)SMgX] \longrightarrow PhC(Bu^t)(CH{=}CH_2)MgX \quad (104)$$

$PhBu^t\dot{C}SCH{=}CH_2$, were detected by e.s.r. and suggested the presence of these as intermediates in the migration. However 1-propenylmetallics add to $Ph_2C{=}S$ primarily with retention of configuration[94], equation 105. The small amount of

$$\underset{(E):(Z) = 94.7:5.3}{MeCH{=}CHLi} \xrightarrow[\text{(ii) } H_2O]{\text{(i) } Ph_2C{=}S} \underset{(E):(Z) = 86.9:13.1}{Ph_2CHSCH{=}CHMe} \quad (105)^{94}$$

apparent isomerization in the product was considered to arise from different reactivities of the isomeric Grignards. This predominant retention of configuration must rule out predominant involvement of free radicals, but neither caged radicals nor a direct nucleophilic reaction. The latter was preferred since the yields of thiophilic addition of RLi to $Ph_2C{=}S$ are inversely proportional to the ability of RLi to transfer an electron; the yields for RM are in the sequence R = vinyl ≈ phenyl > butyl > benzyl ≈ benzhydyl[227] (no thiophilic additions occur with the last two derivatives). For Grignard reactions[94], the extent of thiophilic addition does not follow a regular order; in addition a pronounced solvent effect can be observed. Alkyl-Grignards add directly to sulphur of $Ph_2C{=}S$, as do phenyl metals (metal = lithium or magnesium) [from the reaction[94,228] of $(C_6D_5)_2C{=}S$ and $C_6H_5M$; $(C_6D_5)_2CHSC_6H_5$ was obtained which suggested it was formed directly by attack of $C_6H_5M$ on S, rather than from a rearrangement of $(C_6D_5)_2(C_6H_5)CSM$, which would have provided scrambled phenyl groups].

$$Ph_2C{=}S + PhCH_2MgCl \longrightarrow Ph_2CHSH + Ph_2C(SH)CH_2Ph + Ph_2C(CH_2SPh)SCH_2Ph \quad (106)$$

The extent of free radical involvement in these thiophilic reactions remains essentially unanswered. There are clear signs that free radicals play some role, however. E.s.r. signals have been detected in a number of reactions, e.g. see refs. 229a and b. Formation of $Ph_2CHSH$ in reactions of $Ph_2C{=}S$ with $MeCH{=}CHMgBr$ also suggests some radical behaviour[94]. In addition, the double addition products are difficult to rationalize by an ionic mechanism. The thiophilic additions have been used in the synthesis of carbonyl compounds[235], equation 107.

$$RC(S)SEt \xrightarrow[\text{(ii) E}^+]{\text{(i) EtMgI}} RC(E)(SEt)_2 \xrightarrow{H_3O^+} RCOE \qquad (107)$$

There are a few examples[229f,236–238] known where an organometallic compound prefers thiophilic attack at a carbon—sulphur double bond rather than carbophilic addition at a carbon—oxygen double bond in compounds containing both groups, e.g. $MeC(S)CMe_2CO_2Et$[229f] and 2,2,4,4-tetramethyl-1,3-cyclobutanonethione[237], equations 108 and 109, respectively.

$$MeC(S)CMe_2CO_2Et + Bu^tM \xrightarrow{\text{thf(Et}_2\text{O)}} \underset{10\%\ (30\%)}{MeCH(SH)CMe_2CO_2Et}$$

$$+ \underset{30\%\ (30\%)}{MeCH(SBu^t)CMe_2CO_2Et} + \underset{30\%\ (10\%)}{CH_2{=}C(SBu^t)CMe_2CO_2Et} \qquad (108)$$

$$\text{2,2,4,4-tetramethyl-3-thioxocyclobutanone (S=C, C=O; Me, Me, Me, Me)} \xrightarrow[\text{(ii) H}_2\text{O}]{\text{(i) RMgX}} \text{3-(H)(SR)-2,2,4,4-tetramethylcyclobutanone} \qquad (109)$$

The ester functions are not attacked even with excess of RMgX. But-2-en-1-ylmagnesium bromide attacks initially at sulphur in $Bu^tCSCO_2Et$ to give at −20 °C $Bu^tC(MgBr)(SCHMeCH{=}CH_2)CO_2Et$; at higher temperatures this undergoes a σ[2, 3] rearrangement[236] to $Bu^tC(CH_2CH{=}CHMe)(SMgBr)CO_2Et$, equation 110.

$$Bu^tC(S)CO_2Et \xrightarrow{CH_2{=}CHCHMeMgBr,\ -20\,^{\circ}C} Bu^tC(MgX)(SCHMeCH{=}CH_2)CO_2Et \xrightarrow{\sigma[2,3]} Bu^tC(SMgX)(CO_2Et)CH_2{=}CHCHMe \qquad (110)$$

Stereospecific formations of cyclopropane rings have been achieved with 1,3-onethione compounds[238], equation 111; a *cis*-concerted addition results.

$$\text{XMg—R} + \text{O=C–C–C(=S)X} \xrightarrow{\text{thf}} \text{cyclopropane bearing OMgX, SR, X} \qquad (111)$$

## IV. ADDITIONS TO ALKENES AND ALKYNES

### A. Introduction

Additions of organic derivatives of main group elements, especially those of the more electropositive elements, to carbon—carbon multiple bonds are well established. Most studies have been with lithium, magnesium, and aluminium, with other alkali metals, boron, and zinc attracting considerable but reduced attention. Various references to these elements will be made in the remainder of this section. Fewer examples have been noted for other electropositive elements, such as calcium, strontium, barium[239], beryllium[240], indium, and gallium[241].

Examples of insertions into carbon—metal bonds of the less electropositive metals are rare. However insertion of carbon—carbon multiple bonds into silicon—carbon bonds of strained cyclic compounds have been reported[242,243], e.g. equations 112 and 113. Reaction 113 contrasts with the reaction with less reactive alkenes, which provide dimethylsilene ($SiMe_2$) trapped products.

$$\text{(silacyclobutane, } SiMe_2) + EtO_2CC{\equiv}CCO_2Et \xrightarrow{Pd^{II}\text{complex}} \text{1,1-dimethyl-2,3-bis(ethoxycarbonyl)silacyclohex-2-ene} \qquad (112)^{242}$$

$$\text{hexamethylsilirane } (Me_2C\text{—}CMe_2\text{—}SiMe_2) \xrightarrow{\Delta} \left[\begin{array}{c} Me_2C-\dot{S}iMe_2 \\ | \\ Me_2\dot{C} \end{array}\right] \xrightarrow{PhRC{=}CH_2} \text{1,1,2,2,3,3-hexamethyl-5-Ph-5-R-silacyclopentane} \qquad (113)^{243}$$

To these reactions can be added tetracyanoethylene insertions. These are considered in Section IV. Tetracyanoethylene is a powerful electron acceptor and these insertion reactions proceed via charge-transfer complexes.

### B. General

Organolithiums are generally less reactive towards carbon—carbon multiple bonds than organo-aluminium and -boron compounds, but more so than organo-magnesium and -zinc species. As in other reactions, the presence of coordinating solvents or reagents leads to more reactive organolithium species. This contrasts with the decreased reactivity of organo-aluminium and -magnesium species under similar circumstances.

Lehmkuhl *et al.*[244a] have reported differences in the regioselectivity of reactions of RLi, RMgCl, and $R_3Al$ ($R = Pr^i$ or $Bu^t$) with alk-1-enes; for example, RLi adds solely to $C_{(1)}$ whereas $(Bu^t)_3Al$ and $Bu^tMgCl$ only add to $C_{(2)}$. For the isopropyl derivatives of aluminium and magnesium addition to both $C_{(1)}$ and $C_{(2)}$ (major) result. Differences are also noted in the regioselectivity of additions of organo-lithium, -magnesium, and -zinc compounds to the enyne BuCH=CHC≡CH; diallyl zinc adds to the triple bond to give, after hydrolysis, $BuCH{=}CHC(CH_2CH{=}CH_2){=}CH_2$ as the major product with also a small amount of $BuCH{=}CHC(CH_2CH{=}CH_2)_2CH_3$ being formed. The major reaction of allylmagnesium bromide is addition to the double bond (1,2-addition) to give

$CH_2{=}CHCH_2CHBuCH_2C{\equiv}CH$ as well as some 1,4-adduct, $CH_2{=}CHCH_2CHBuCH{=}C{=}CH_2$. Organolithiums, RLi (R = allyl or Bu), solely give 1,2-addition[244b]. Organo-calcium, -strontium and -barium compounds also mainly if not solely add 1,4- to conjugated enynes to give allenic products[239b]. Some differences in stereoselectivity have been also indicated, e.g. in reactions of allyl-Grignard and butyllithium to bicycloalkenols[245].

Particularly reactive organo—metal bonds in general are allyl—metal bonds. Benzyl—metal bonds are also reactive, while it is usually found that *tert*-alkyl—metal bonds are more reactive than secondary and primary alkyl—metal bonds. Greater reactivity has been shown by some alkenes and alkynes which contain coordinating groups, such as hydroxy, alkoxy, and amino groups. Conjugated alkenes, e.g. buta-1,3-diene and styrene, are also considerably more reactive than simple isolated alkenes. Strained cycloalkenes are another class of compound which react readily with organometallic derivatives of main group elements.

Vinyl–metal(oid) compounds, e.g. $CH_2{=}CHX$, $X = SiR_3$, $GeR_3$, and $AsR_2$, (as well as SeR, and SR), also show greater reactivity, towards at least organolithiums, than do simple alkenes. In addition, intramolecular additions involving alkenyl and alkynyl metal compounds proceed more readily than intermolecular reactions of RM to isolated carbon—carbon multiple bonds. These various and general effects will be illustrated in the more specific discussions of individual elements. One of the more important features of these insertions by carbon—carbon multiple bonded compounds is polymerization. Indeed, the greatest industrial use of organolithiums is as polymerization initiators. Conditions for the polymerization of monomers and dienes are well established, especially using organolithiums, including its tmed complexes[246a], and organoaluminium–transition metal combinations (Ziegler–Natta catalysts)[246b].

### C. Insertions into Carbon—Lithium Bonds

Organolithium compounds do add to isolated carbon—carbon multiple bonds but forcing conditions may be necessary. Ethylene, for example, reacts with ethyl-or *n*-butyl-lithium (RLi) under pressure to give simple addition compounds, $RCH_2CH_2Li$ and low molecular weight polymers[247]. Secondary and tertiary alkyllithiums add much more readily, especially in the presence of a donor such as an ether or an amine[248]. With a particularly powerful donor, polymerization may result; ethylene

$$R'CH{=}CH_2 \xrightarrow{RLi} RR'CHCH_2Li \longrightarrow RR'CHCH_2[CH(R)CH_2]_nLi \quad (114)$$

can be polymerized by BuLi in the presence of dabco (1,4-diazabicyclo[2.2.2]octane), tmed, or $MeOCH_2CH_2OMe$[4,246a]. No further consideration is given here to polymerization.

Additions to strained alkenes, e.g. bicyclo[3.3.1]non-1-ene[249], equation 115, bicyclo[2.2.1]hept-2-ene[245], equation 116, and methylcyclopropenes[250], e.g. equation 117, occur more readily.

$$\text{bicyclo[3.3.1]non-1-ene} \xrightarrow[\text{(ii) } H_2O]{\text{(i) MeLi, } Et_2O\text{, 4 d, room temp.}} \underset{54\%}{\text{1-methylbicyclo[3.3.1]nonane}} \quad (115)$$

$$\underset{(\mathbf{37})}{\text{bicyclo[2.2.1]hept-2-ene}} \xrightarrow[\text{(ii) } H_2O]{\text{(i) BuLi, tmed, hexane, 24 h, room temp.}} \text{2-butylbicyclo[2.2.1]heptane} \quad (116)$$

$$\text{3-methylcyclopropene} \xrightarrow[\text{(ii) } H_2O]{\text{(i) } Et_2O,\ PhLi} \text{trans-1-Me-2-Ph-cyclopropane} \quad (117)$$

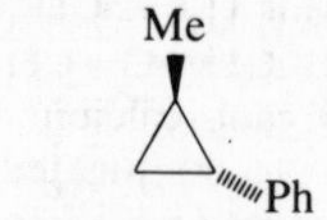

overall yield 30.5%
*trans*, 94%

Intramolecular additions[251–255] have been reported with alkenyl- and alkynyllithiums; typical examples are shown in equations 118–121.

$$Ph_2C{=}CHCH_2CH_2Li \underset{Et_2O}{\overset{THF}{\rightleftharpoons}} Ph_2CLi\text{-cyclopropane} \quad (118)^{253}$$

(**38**)

$$CH_2{=}CHCH_2CH_2CH(Li)R \xrightarrow{Et_2O,\ RT,\ 1h} \text{1-R-2-LiCH}_2\text{-cyclopentane} \quad (119)^{251,252}$$

$$RC{\equiv}C(CH_2)_4Li \longrightarrow \text{cyclopentylidene}{=}CRLi \quad (120)^{255}$$

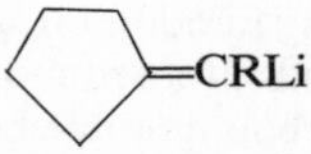

$$CH_2{=}C(Me)(CH_2)_5Li \xrightarrow{\text{tmed}} \text{1-Me-1-LiCH}_2\text{-cyclohexane} \quad (121)^{252}$$

These intramolecular additions occur under much milder conditions than are required for the intermolecular analogues. A noteworthy feature is that whenever there is a choice of forming rings of different sizes by additions to multiple bonds, the smaller ring is most often obtained.

As shown in reaction 118, equilibria or reversible reactions may arise. Of interest, the cyclopropylmethyl analogue of **38** is stable to ring opening in all the solvents studied, whereas the magnesium analogue always exists as the acyclic species. Sodium shows similar intermediate behaviour to that of lithium[253]. In some other cases, reversible behaviour can be detected by isotropic labelling, e.g. equation 122;

$$CH_2{=}CMeCH_2CD_2Li \rightleftharpoons \left[ \text{cyclo-}(CD_2CH_2)C(Me)CH_2Li \right] \rightleftharpoons CH_2{=}CMeCD_2CH_2Li \quad (122)^{256}$$

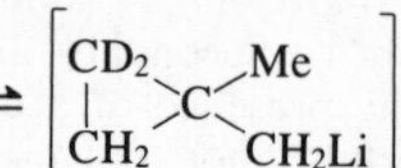

only the acyclic species can be detected but the cyclic species must be an intermediate[256]. More elaborate equilibria may be set up, e.g. Scheme 28[254].

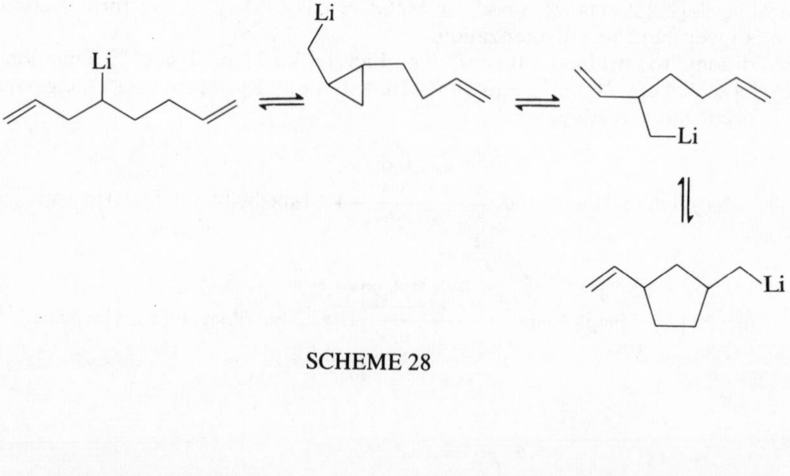

SCHEME 28

A strategically placed internally donor group (OH, OR, $NR_2$, or SR) in the unsaturated compound has facilitated additions of organolithiums[245,257–259]. Allylic alcohols, propargylic alcohols, and *tert*-amino-, alkoxy-, and alkylthio-substituted allylic, homoallylic compounds and propargylic species are more reactive towards organolithium than their hydrocarbon counterparts, e.g. equations 123–125. The presence of tmed may, however, still be necessary for reaction to occur[250].

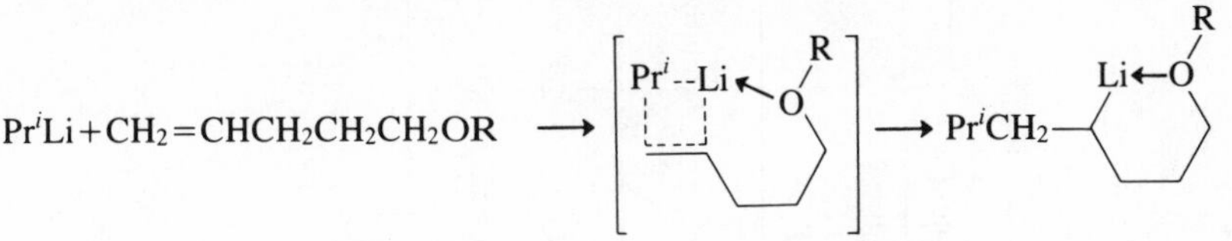

$$Pr^iLi + CH_2{=}CHCH_2CH_2CH_2OR \longrightarrow [\text{cyclic transition state: } Pr^i\text{--}Li \leftarrow O(R)] \longrightarrow Pr^iCH_2\text{-(cyclic Li}\leftarrow O(R)\text{)} \qquad (123)$$

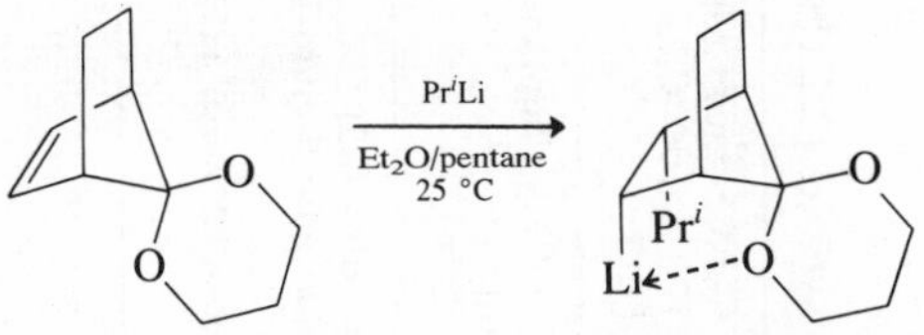

$$\xrightarrow[\text{Et}_2\text{O/pentane, 25 °C}]{Pr^iLi} \qquad (124)^{257a}$$

*cis*-addition

$$CH_2{=}CHCH_2OH \xrightarrow{\text{2RLi, tmed}} LiCH_2CHRCH_2OLi \qquad (125)^{257b}$$

Richey and coworkers[245] carried out an elaborate study with hydroxy-bicyclo[2.2.1]hept-2-enes. The results are given in Table 10. It was argued that a metallated hydroxyl group which does not assist an addition of RM actually results in a rate retardation[245] (compare **39** with **43**). This holds for additions by BuLi as well as by allyl-Grignards. It was also concluded that addition of BuLi to **42** is not hydroxyl-assisted, in contrast to the addition by allyl-Grignard. Indeed, in the assisted additions by allyl-Grignard, **42** is more reactive than **40**, the reverse of the finding for BuLi. Attachment of the allyl group from allyl MgX to **42** is *exo* and to **40** is *endo*, i.e. in assisted additions of allyl Grignards, there is a preferential attachment of the allyl group to the face of the double bond *nearer* to the hydroxyl group. A further significant factor is that *exo* attack is usually faster than *endo* attack on bicycloalkenes. This accounts for the relative reactivity **42**>**40** towards the allyl-Grignard.

For the butyllithium reactions, it appears that an assisted addition (to these homoallylic alcohols) leads preferentially to attachment of the organic group to the face of the double bond *further* from the hydroxyl group (i.e. opposite to the Grignard situation). So assisted reaction to **40** provides the *exo* isomer in a fast reaction. The assisted reaction to **42**, if this occurred, would give the *endo* isomer. The slowness of this reaction allows a faster but unassisted reaction to occur instead to give the *exo* product. Richey and coworkers[245] further argued that assisted addition to allylic and homo-allylic alcohols may proceed with different selectivities.

Vinyl-silanes[260], -arsines[261], -germanes, -phosphines[262], -halides[263], sulphides[262], and selenides[261] are also reactive towards organolithiums, e.g. equations 126 and 127.

$$CH_2{=}CHSiMe_3 \xrightarrow{Bu^tLi} Bu^tCH_2CH(Li)SiMe_3 \qquad (126)^{260}$$

$$CH_2{=}CHAsPh_2 \xrightarrow{C_6H_{13}Li,\ \text{thf, room temp.}} \underset{95\%}{C_6H_{13}CH_2CH(Li)AsPh_2} \qquad (127)^{261}$$

TABLE 10. Products of additions of BuLi and $CH_2$=$CHCH_2MgCl$ to hydroxybicyclo[2.2.1]heptenes

| | | (i) BuLi, TMED, 24 h room temp. (ii) $H_2O$ | | (i) $CH_2CH$=$CH_2MgCl$, thf 120 h, 100 °C (ii) $H_2O$ | |
|---|---|---|---|---|---|
| No. | Bicyclo[2.2.1]hept-2-ene | Bicyclo[2.2.1]heptane product (yield, %) | Relative rate | Bicyclo[2.2.1]heptane product (yield, %) | Relative rate |
| **39** | Parent compound | 2-*exo*-Bu (30) | 1 | 2-*exo*-allyl (25) | 1 |
| **40** | 5-*endo*-HO | 2-*exo*-Bu-5-*endo*-HO (51) | 10[a] | 2-*endo*-allyl-5-*endo*-HO | 20[a] |
| **41** | 5-*exo*-HO | 2-exo-Bu-5-*exo*-HO (47) | [b] | — | [c] |
| **42** | 7-*syn*-HO | 2-*exo*-Bu-7-*syn*-HO (26) | <0.1[b] | 2-*exo*-allyl-7-*syn*-HO (>52) | 20[a] |
| **43** | 7-*anti*-HO | 2-exo-Bu-7-*anti*-HO (2) | <0.1[b] | — | [c] |

[a] Assistance; [b] no assistance; [c] no reaction.

Addition to conjugated carbon—carbon multiple bonds occur readily and polymerization may result, as shown by styrene, buta-1,3-diene, and isoprene. For a discussion on aspects of these polymerisations, see refs. 4 and 246. The reactivity of RLi in THF towards styrene[265] has been shown to be R = alkyl > benzyl > allyl > phenyl > vinyl > $Ph_3C$. Under controlled conditions, e.g. using low temperatures or with hindered systems, mono-addition products can be obtained, with $PhCR{=}CH_2$, buta-1,3-diene, and related compounds[265–269], e.g. equations 128–131.

$$Bu^tLi + CH_2{=}CHCH{=}CH_2 \xrightarrow{\text{hydrocarbon}} Bu^tCH_2CH{=}CHCH_2Li \qquad (E):(Z) = 3:1 \qquad (128)^{266}$$

$$Bu^tLi + p\text{-}X\text{-}C_6H_4C({=}CH_2)Me \xrightarrow[-78\,°C,\ NR_3\ \text{or}\ R_2O]{\text{cyclopentane}} X\text{-}C_6H_4\text{-}C^-(Me)(CH_2Bu^t)\ Li^+ \qquad (129)^{267}$$

$$m\text{-}PhC({=}CH_2)C_6H_4CPh{=}CH_2 \xrightarrow[1\ h,\ 20\ °C]{2Bu^sLi} [m\text{-}PhC(CH_2Bu^s)C_6H_4CPhCH_2Bu^{2-}]2Li^+ \qquad (130)^{268}$$

$$\text{hexamethyl-4-methylenecyclohexa-2,5-diene} \xrightarrow[\text{hydrocarbon},\ L = R_2O\ \text{or}\ NR_3]{RLi,\ 20\ °C} \text{pentamethyl(CH}_2\text{R)cyclohexadienyl anion}\ Li^+(L)_n \qquad (131)^{269}$$

R = $Bu^s$, $Bu^t$, or $Bu^n$

The rate of addition in reaction 131 is helped by ligands which reduce the size of the organolithium aggregate[269]. Organolithiums also add to diphenylacetylenes, especially if tmed is present, but metallation also results[270], equation 132.

$$PhC{\equiv}CPh \xrightarrow[\text{tmed}]{Bu^nLi} o\text{-}LiC_6H_4C(Bu){=}C(Li)Ph\text{-}(\mathbf{E}) \qquad (132)$$

An intramolecular addition to a diarylacetylene is shown[271] in equation 133.

$$2\text{-Li-}2'\text{-}(PhC{\equiv}C)\text{biphenyl} \rightarrow 9\text{-}(Ph(Li)C{=})\text{fluorene} \qquad (133)$$

The magnesium analogue reacts similarly.

Alkyllithiums add to aromatic rings, e.g. of naphthalene[272,273], equation 134. As

$$\text{naphthalene} \xrightarrow[165\ °C]{Bu^tLi} \text{1-H-1-Bu}^t\text{-4-H-4-Li-1,4-dihydronaphthalene} \longrightarrow \text{1-Bu}^t\text{-naphthalene} \qquad (134)^{272}$$

LiH is rapidly lost from the adduct, the overall reaction is alkylation. Addition to benzene also can occur at 165 °C. Milder conditions have been employed for polynuclear aromatic hydrocarbons (including azulene[274], equation 135).

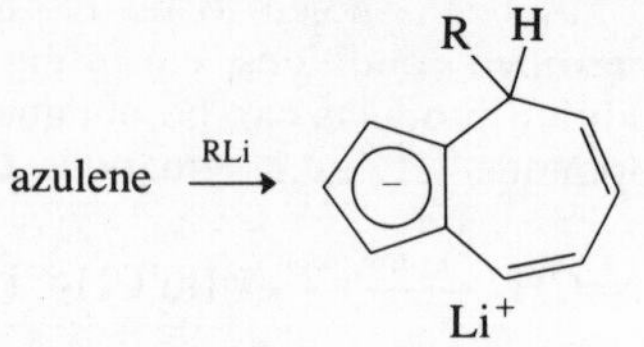

$$\text{azulene} \xrightarrow{\text{RLi}} \text{(R, H-substituted dihydroazulenide anion)}\ Li^+ \qquad (135)^{274}$$

## D. Insertions into Carbon—Magnesium Bonds

As already indicated, organomagnesium compounds are less reactive towards alkenes and alkynes than are organolithium reagents. Intramolecular additions of alkenyl and alkynyl magnesiums have, however, been extremely well studied and Hill has most thoroughly reviewed this area[275,276]. It is with these intramolecular reactions that we shall begin this section of insertions into carbon—magnesium bonds.

### 1. *Intramolecular additions and the reverse reactions*[275,276]

Since the initial publications by Robert *et al.*[277] that cyclopropylmethyl-Grignard underwent ring cleavage to 3-butenyl-Grignards, many related organomagnesium

$$\text{cyclopropyl-}CH_2MgX \longrightarrow CH_2{=}CHCH_2CH_2MgX \qquad (136)$$

rearrangements have been reported. These have involved both cleavage of three-, four-, and five-membered rings and the reverse reaction, cyclization, which form three- to six-membered rings, equation 137.

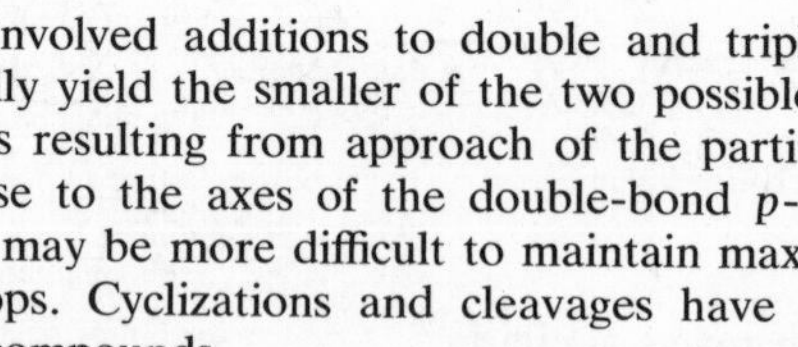

$$R^3\text{(ring)}C(R^1){=}...R^1R^2,\ C(R^4)(R^5)\text{–}MgX \rightleftarrows \text{cyclic } C(R^3)(CR^1R^2MgX)\text{–}C(R^4)(R^5) \qquad (137)$$

**(44)**

Cyclizations have involved additions to double and triple bonds as well as to allenes. These normally yield the smaller of the two possible rings. This preference has been explained as resulting from approach of the partially carbanionic carbon along a direction close to the axes of the double-bond $p$-orbital. If the terminal carbon is attached, it may be more difficult to maintain maximum $\pi$ overlap as the new $\sigma$-orbital develops. Cyclizations and cleavages have also been observed in bicyclic and tricyclic compounds.

There are several general features of these rearrangements:

(i) Rearrangements occur more readily in less polar solutions, e.g. they are faster in $Et_2O$ than in thf; the presence of hmpt increases the rate relative to that in thf.

(ii) Rearrangements follow first-order kinetics and the rate is largely independent of the total organomagnesium concentration in dilute solution (0.5–1.0 M). At higher concentrations, the rate is often found to increase linearly with the concentration.

(iii) The rate is relatively insensitive to magnesium purity.

(iv) Reactivity of RMgX is in the order $X = R > Cl > Br > I$.

(v) Relative reactivities of (cycloalkyl)methyl magnesium halides in ring cleavages (ring containing $n$ carbons) are in the sequence of ring size $n = 3 > 4 \gg 5, 6$, while for ring closure to various ring sizes, the reactivities follow the order $n = 5 > 4 > 6$.

(vi) Alkyl groups substitution at the double bond (i.e. $R^1$, $R^2$, or $R^3$ in equation 137) appears to decrease the rate of ring closure. The order of reactivity for **44** is $R^3 = R^4 = \text{alkyl} < R^3 = H, R^4 = \text{alkyl} > R^3 = R^4 = H$ (this may be a consequence of the opposing effects of steric repulsion on one hand and Grignard stabilization on the other, which is in the sequence primary alkyl > secondary alkyl > tertiary alkyl).

In ring cleavages, alkyl substitution at $R^1$ has little effect on the rate, whereas an alkyl group at $R^4$ decreases the rate. Halogen substitution at $R^1$ or $R^2$ or a phenyl group at $R^3$ decreases the rate of cyclization. Phenyl substitution at $R^1$ or $R^2$ produces a sizeable decrease in the cyclization rate for three-membered ring formation but slight increases for the five-membered ring. Electron-withdrawing substituents in an aryl ring at $R^1$ increase the cyclization rate.

a. *Mechanism.* If a single general mechanism were to apply to all the various reactions of this type, then a concerted four-centre mechanism, possibly one involving a $\pi$-complex intermediate, would be the one favoured by Hill. A coplanar transition state was considered to be unfavoured (equation 138); instead, the preferred attack by the C—Mg bond on the carbon—carbon double bond was considered to lead to the perpendicular arrangement **45**.

(138)

(**45**)

b. *Examples.* With some 3-butenylmagnesium compounds (and for which the cyclopropylmethyl analogues are too unstable to be detected), the occurrence of an isotopic scrambling or rearrangement of substituents indicates exchange reactions, e.g. equation 139[278]; the *cis–trans* equilibration at the double bond and isotropic scrambling occurred at the same rate.

$$R^1R^2C{=}CHCH_2CD_2MgCl \rightleftharpoons \left[ \text{cyclopropane}(D_2)\text{-}CR^1R^2MgX \right] \rightleftharpoons R^1R^2C{=}CHCD_2CH_2MgCl \quad (139)$$

An interesting 3-butylmagnesium halide rearrangement[279] involving a cyclopropylmethylmagnesium species is shown in equation 140.

(140)

(**46**) (**47**)

Maercker and Geuss[279b] calculated that **47** (R = H) was 15.5 kJ $mol^{-1}$ more stable than 46 (R = H). The position of equilibria depends on the solvent, e.g. $K$ = [**45** (R = H)]/[**44** (R = H)] = 5.7 in $Et_2O$/hmpt compared with 9.3 in $Et_2O$. Complete cleavages of cyclopropylmethyl and cyclobutylmethyl compounds to 3-butenyl and 4-pentenyl derivatives, respectively, do not always result; for example, the relative stability of three- or four-membered rings can be increased appreciably by alkyl substitution; compare reaction 136 with equation 141.

$$\underset{(\mathbf{48})}{Me_2C(MgCl)CMe_2CH{=}CH_2} \underset{}{\overset{\text{thf, room temp.}}{\rightleftharpoons}} \underset{(\mathbf{49})\ >99.9\%}{\text{1-ClMgCH}_2\text{-2,2,3,3-Me}_4\text{-cyclopropane}} \qquad (141)$$

One driving force in equilibrium 141 is the formation of a primary Grignard reagent **48** from a tertiary one **49**. Another factor[280] is the stabilizing influence of *gem*-dimethyl groups in **48**—the Thorpe–Ingold effect. The values of the equilibrium constants for equilibria 142–144[281–282] also show the stabilizing influence of alkyl substitution (although an important factor in equilibrium 143 may be the conversion of a secondary to a primary alkyl Grignard).

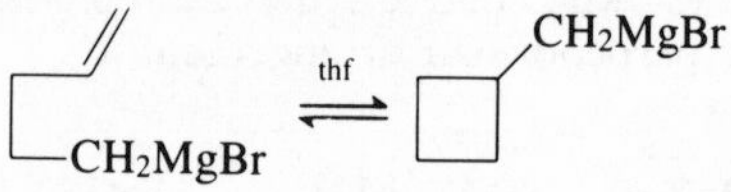

$CH_2{=}CHCH_2CH_2CH_2MgBr \overset{\text{thf}}{\rightleftharpoons}$ cyclobutyl-$CH_2MgBr$ (142)[281]

$K \leqslant 9 \times 10^{-5}$ at 100 °C

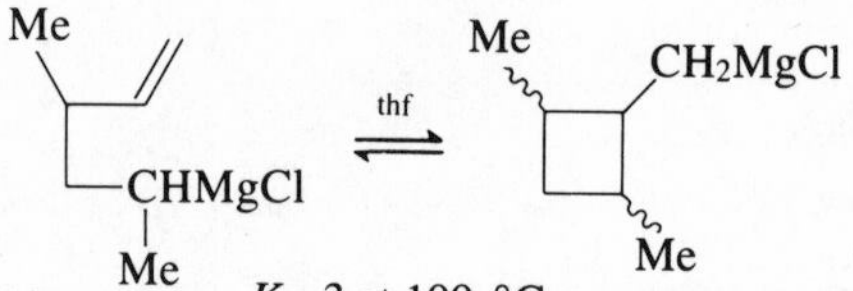

$CH_2{=}CHCH(Me)CH_2CH(Me)MgCl \overset{\text{thf}}{\rightleftharpoons}$ 2,4-dimethylcyclobutyl-$CH_2MgCl$ (143)[282]

$K = 3$ at 100 °C

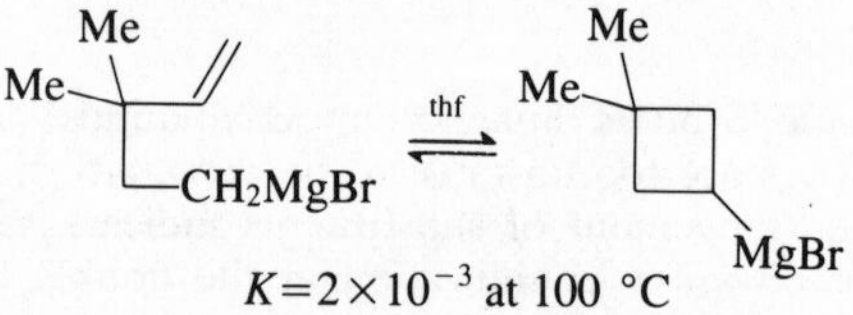

$CH_2{=}CHCMe_2CH_2CH_2MgBr \overset{\text{thf}}{\rightleftharpoons}$ 3,3-dimethylcyclobutyl-MgBr (144)[283]

$K = 2 \times 10^{-3}$ at 100 °C

To these equilibria can be added the quantitative conversion of $CH_2{=}CHCMe_2CH_2CMe_2MgCl$ to 2,2,4,4-tetramethylcyclobutylmethylmagnesium chloride on[284] heating at 70 °C. A more elaborate rearrangement[285] of a cyclobutylmethyl Grignard derivative, the 2-bicyclo[3.2.0]heptyl-Grignard reagent, is shown in Scheme 29. A primary and a secondary Grignard reagent are formed, the latter slowly rearranging on further heating.

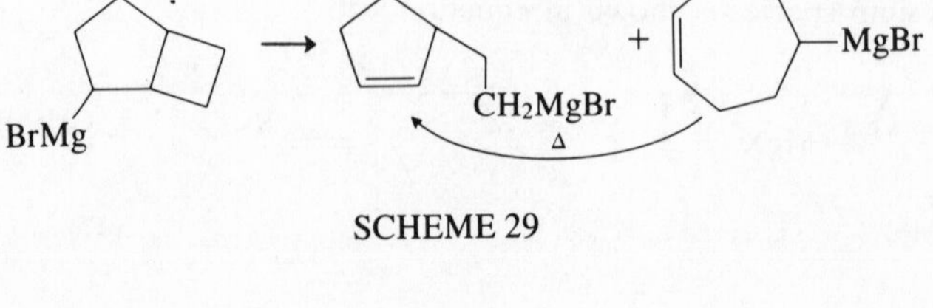

2-bicyclo[3.2.0]heptyl-MgBr (BrMg) → cyclopent-2-enyl-$CH_2CH_2MgBr$ (CH₂MgBr) + cyclohept-3-enyl-MgBr; the latter → former (Δ)

SCHEME 29

Examples of rearrangements giving five-membered rings are known[286,287], e.g. equation 145[286]. Such cyclizations occur much more readily than do the corresponding cyclizations to cyclohexylmethyl compounds; for example, the relative reactivities

$$XMgCH(R)(CH_2)_3CH{=}CH_2 \xrightarrow{\text{thf}} \text{1-XMgCH}_2\text{-2-R-cyclopentane} \quad \text{mainly } trans \qquad (145)^{286}$$

of 5-hexenyl- and 6-heptenyl-Grignard reagents in cyclization reactions have been[288] calculated to be about $3 \times 10^3$. In both these reactions *trans*-isomers dominate. Whilst cyclization of 5-hexenyl-Grignard reagents normally occurs readily, ring strain in bicycloalkenyl analogues may result in easy ring cleavage. Felkin *et al.*[289] have reported the stereospecific cyclization of the non-conjugated diene **50**, equation 146.

MgBr → MgBr (146)

(**50**) *cio*

## 2. *Intermolecular additions*

Except under forcing conditions, RMgX are generally unreactive towards intermolecular addition to unactivated and non-terminal alkynes. An increased reactivity is found for allylmagnesium halides; benzyl-, secondary alkyl-, and tertiary alkyl-Grignards are also more reactive than primary alkyl-Grignard reagents[290]. Reactions occur more readily in cyclohexane and in $Et_2O$ than in thf solution[291].

For reaction between diethylmagnesium and ethylene in diethyl ether (giving $Bu_2Mg$) a pressure of 50 atm and a temperature of 100 °C were required[292]. Milder conditions have been reported, however, for reactions of benzyl-[293], *tert*-butyl-[290], isopropyl[294] and allyl-Grignards[295] (equation 147).

$$CH_2{=}CHCH_2MgCl + CH_2{=}CH_2 \xrightarrow[\text{cyclohexane}]{\text{30 atm, 25–75 °C}} CH_2{=}CHCH_2CH_2CH_2MgCl \quad 41\% \qquad (147)$$

Considerable attention has been paid to the additions to alk-1-enes, conjugated alkenes, and strained cyclic alkenes, especially by Lehmkuhl and coworkers[290,291,296–303].

The addition of allylmagnesium halides to alkenes was found to be first order in Grignard reagent and in alkene. The following reactivity sequences were established[300]:

(i) The relative reactivity towards $CH_2{=}C(Me)CH_2MgX$ was norbornene: styrene: oct-1-ene = 53.1 : 4.1 : 0.3 (see equations (148–150) for the initial reactions). This sequence illustrates well the increased reactivity of conjugated and strained alkenes compared with isolated unstrained species.

(ii) the relative reactivity of $CH_2{=}C(Me)CH_2MgX$ decreases as X = $CH_2{=}C(Me)CH_2 > Cl > Br$.

(iii) the relative reactivity towards norbornene was $MeCH{=}CHCH(Me)MgX > CH_2{=}C(Me)CH_2MgX > CH_2{=}CHCH_2MgX > MeCH{=}CHCHCH_2MgX > Me_2C{=}CHCH_2MgX$.

(iv) the rate of reaction decreased in the solvent order $Et_2O > dioxane \approx thf$.

$$XMgCH_2C(Me){=}CH_2 + CH_2{=}CHC_6H_{13} \longrightarrow CH_2{=}C(Me)CH_2CH(CH_2MgX)C_6H_{13} \quad (148)$$

$$XMgCH_2C(Me){=}CH_2 + CH_2{=}CHPh \longrightarrow CH_2{=}C(Me)CH_2CH(CH_2MgX)Ph + CH_2{=}C(Me)CH_2CH(MgX)Ph \quad (149)$$

$$XMgCH_2C(Me){=}CH_2 + \text{bicyclo[2.2.1]hept-2-ene} \longrightarrow \text{2-}[CH_2{=}CMeCH_2]\text{-3-XMg-bicyclo[2.2.1]heptane} \quad (150)$$

The regioselectivity of the initial additions to alk-1-enes, $RCH{=}CH_2$, by Grignard reagents, $R^1R^2R^3CMgX$, has also been investigated[299]. A correlation was found between the ratio of $Mg \rightarrow C_{(2)}$ addition to $Mg \rightarrow C_{(1)}$ addition and the inductive effects of the $R^1$, $R^2$, and $R^3$ groups on the Grignard reagents. An accumulation of methyl groups ($+I$ effect) leads to up to 99% addition of metal to $C_{(2)}$. Successive replacement of the methyl groups by H, vinyl, or Ph groups ($-I$ effects) causes an increase in the $Mg \rightarrow C_{(1)}$ addition, e.g. 99% addition of magnesium to $C_{(1)}$ in oct-1-ene or non-1-ene by allylmagnesium halide (equation 148) or by $PhCH_2MgX$.

Addition of $R^1R^2R^3CMgX$ to styrene occurs mainly to give the $Mg{-}C_{(2)}$ bonded isomer. Additions of alk-2-enyl-Grignard reagents to styrene are also reversible[299,301], with $PhCH(allyl)CH_2MgX$ slowly rearranging to the thermodynamically more stable benzylic compound, $PhCH(MgX)CH_2$-allyl.

These initial adducts of allylmagnesium halides to the alkenes are ideally set up for the rearrangement reactions mentioned in Section IV.D.1. While it is possible to isolate the initial monoadducts, it has been found that with longer reaction times and higher reaction temperatures, further addition and or rearrangements occur. Examples of such processes are given in Schemes 30–33. Further reaction also results with the initial $Bu^tMgCl$–butadiene adduct[304], Scheme 34.

It was shown that additions by allyl-Grignard reagents to 3,3-dimethylcyclopropene(I) occur in a stereoselective *cis*-manner at 0–20 °C (equation 151). Additions by 1-, 3-, or 4-alkenyl-Grignard are less selective; for other additions to cyclopropenes, see ref. 305.

$$XMgCH_2CR^1{=}CR^2R^3 \xrightarrow{(I)} \text{cyclopropane bearing } CH_2{=}CR^1CR^2R^3,\ XMg,\ Me,\ Me \quad (151)^{303}$$

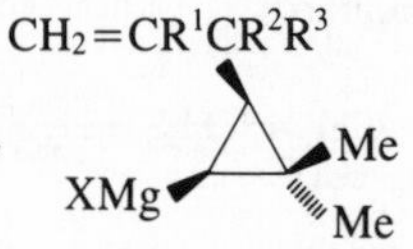

a. *Assisted additions.* Alkenes and alkynes containing donor groups such as hydroxyl, alkoxyl, or amino groups may react faster with Grignard reagents than the corresponding species without the donor group[245b,306–308]. In some cases the increase in the reactivity is appreciable, as shown with various allylic, homoallylic (e.g. equation 152), propargylic, and homopropargylic alcohols as well as allenols. The

$$CH_2{=}CHCH_2MgCl + CH_2{=}CHC_6H_{13} \longrightarrow C_6H_{13}CH(CH_2MgCl)CH_2CH{=}CH_2$$

$$\downarrow CH_2{=}CHCH_2MgCl$$

$$\text{1-}C_6H_{13}\text{-3,5-}(ClMgCH_2)_2\text{-cyclohexane} \longleftarrow C_6H_{13}CH(CH_2MgCl)CH_2CH(CH_2MgCl)CH_2CH{=}CH_2$$

SCHEME 30. Reaction of $CH_2{=}CHCH_2MgCl$ with oct-1-ene[291].

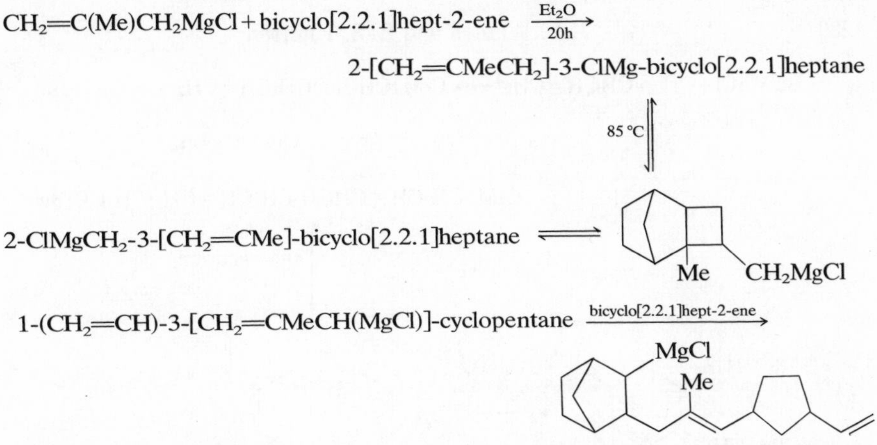

$CH_2{=}C(Me)CH_2MgCl + \text{bicyclo[2.2.1]hept-2-ene} \xrightarrow[20h]{Et_2O}$

2-[$CH_2{=}CMeCH_2$]-3-ClMg-bicyclo[2.2.1]heptane

⇅ 85 °C

2-ClMg$CH_2$-3-[$CH_2{=}CMe$]-bicyclo[2.2.1]heptane ⇌ (tricyclic cyclobutane with Me and $CH_2MgCl$)

1-($CH_2{=}CH$)-3-[$CH_2{=}CMeCH(MgCl)$]-cyclopentane $\xrightarrow{\text{bicyclo[2.2.1]hept-2-ene}}$ (product with MgCl, Me)

SCHEME 31. Reaction of 2-methyallymagnesium chloride with norbornene (bicyclo[2.2.1]hept-2-ene[291].

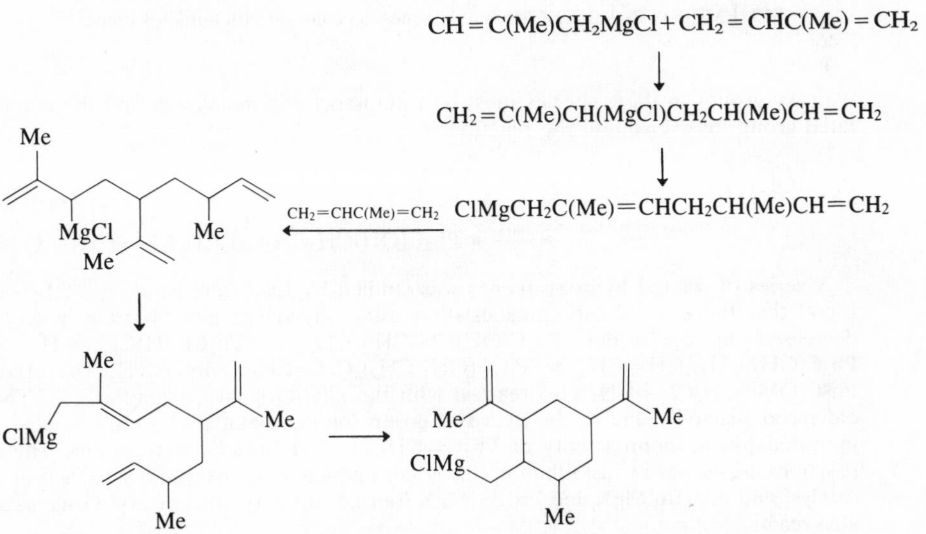

$CH{=}C(Me)CH_2MgCl + CH_2{=}CHC(Me){=}CH_2$

↓

$CH_2{=}C(Me)CH(MgCl)CH_2CH(Me)CH{=}CH_2$

↓

$ClMgCH_2C(Me){=}CHCH_2CH(Me)CH{=}CH_2 \xrightarrow{CH_2=CHC(Me)=CH_2}$ (structure with Me, MgCl, Me, Me)

↓

(ClMg, Me, Me, Me structure) ⟶ (cyclohexane with Me, Me, ClMg, Me)

SCHEME 32. Reaction of 2-methylallyl-Grignard with buta-1,3-diene[304].

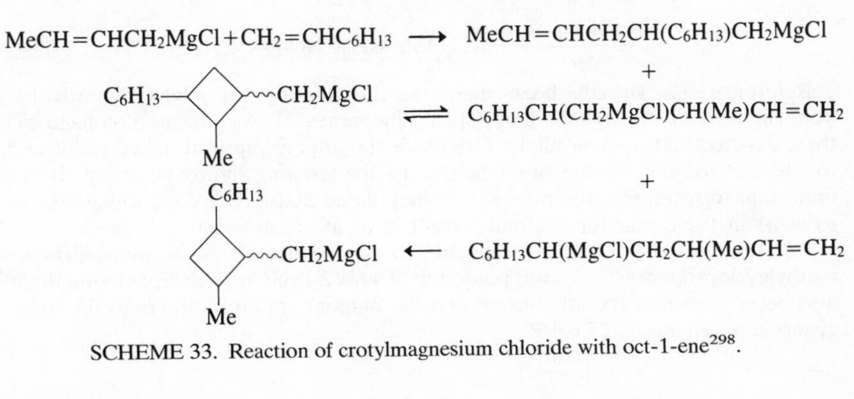

$MeCH{=}CHCH_2MgCl + CH_2{=}CHC_6H_{13} \longrightarrow MeCH{=}CHCH_2CH(C_6H_{13})CH_2MgCl$

\+

$C_6H_{13}$–(cyclobutane, Me)~$CH_2MgCl$ ⇌ $C_6H_{13}CH(CH_2MgCl)CH(Me)CH{=}CH_2$

\+

$C_6H_{13}$ (cyclobutane, Me)~$CH_2MgCl$ ⟵ $C_6H_{13}CH(MgCl)CH_2CH(Me)CH{=}CH_2$

SCHEME 33. Reaction of crotylmagnesium chloride with oct-1-ene[298].

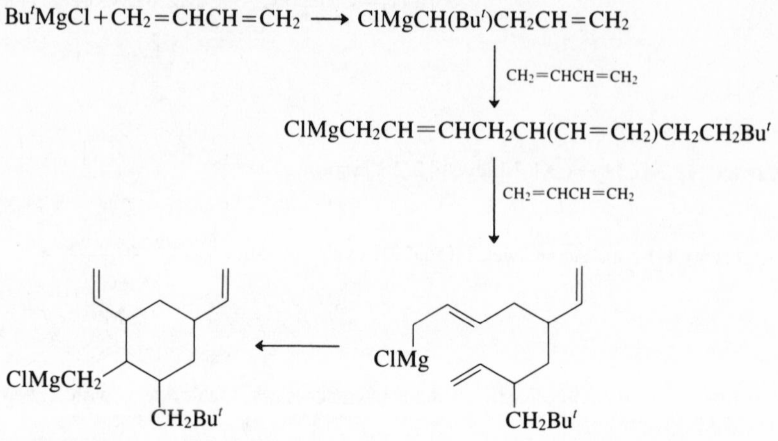

$$Bu^tMgCl + CH_2{=}CHCH{=}CH_2 \longrightarrow ClMgCH(Bu^t)CH_2CH{=}CH_2$$

$$\downarrow CH_2{=}CHCH{=}CH_2$$

$$ClMgCH_2CH{=}CHCH_2CH(CH{=}CH_2)CH_2CH_2Bu^t$$

$$\downarrow CH_2{=}CHCH{=}CH_2$$

ClMgCH2 CH2Bu$^t$ ⟵ ClMg CH2Bu$^t$

SCHEME 34. Reaction of *tert*-butylmagnesium chloride with buta-1,3-diene[304].

hydroxy groups in these species must be instantaneously metallated and the metallated group must facilitate the reaction.

$$Ph_2C(OH)CH_2CH{=}CH_2 \xrightarrow[\text{(ii) } H_3O^+]{\text{(i) } CH_2{=}CHCH_2MgBr, Et_2O, reflux} Ph_2C(OH)CH_2CH_2CH_2CH_2CH{=}CH_2 \quad (152)$$

A series of related hydroxyalkenes was studied by Eisch and Merkley[307]. It was found that the ease of carbomagnesiation using allylmagnesium bromide in $Et_2O$ decreased in the order $Ph_2C(OH)CH_2CH{=}CH_2 > Ph_2C(OH)CH{=}CH_2 \gg Ph_2C(OH)(CH_2)_2CH{=}CH_2 \ggg Ph_2C(OH)(CH_2)_3CH{=}CH_2$ (unreactive). The ether $Ph_2C(OMe)CH_2CH{=}CH_2$ also reacted with the allyl-Grignard, equation 153. The enhanced reactivity due to the hydroxyl group (or the metallated group) is clearly indicated by the nonreactivity of $Ph_2C{=}CHCH{=}CH_2$ in similar reactions. These reactions are slower in $Et_2O$ than in thf. Other findings were that $[Ni(acac)_2]$ acts as a catalyst and that $Bu^tMgX$ and $PhCH_2MgX$ (but not primary alkyl or aryl Grignards) also react.

$$Ph_2C(OMe)CH_2CH{=}CH_2 \xrightarrow[PhMe]{CH_2{=}CHCH_2MgBr} Ph_2C(OMe)CH_2CH(MgX)CH_2CH_2CH{=}CH_2 \quad (153)$$

Reference has already been made to the addition of allyl-Grignards to the homoallylic alcohols, hydroxybicyclo[2.2.1]heptenes[245b]. As indicated on page 271, in these assisted additions of allylic Grignards the allyl group is attached preferentially to the face of the double bond nearer to the existing hydroxyl group. It is also important to recall that the presence of metallated hydroxyl groups which are unable to assist in these reactions actually result in a rate retardation.

Additions of allyl Grignard reagents to other homoallylic alcohols, 3-hydroxymethylcyclopropenes[306], 3-cyclopentenol[308], and 2-cyclohexenyldiphenylmethanol[308] also occur such that the allyl group and the magnesium are both *cis* to the hydroxyl group, e.g. equations 154–155.

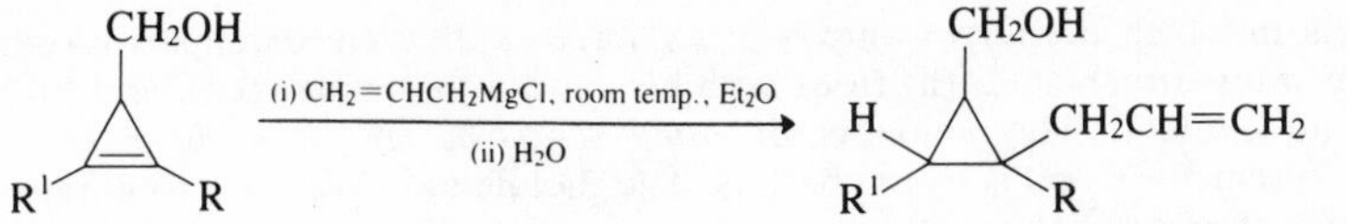

$$\textbf{(51)} \xrightarrow[\text{(ii) } H_2O]{\text{(i) } CH_2{=}CHCH_2MgCl\text{, room temp., } Et_2O} \text{cyclopropane with } CH_2OH,\ H,\ CH_2CH{=}CH_2,\ R^1,\ R \qquad (154)^{306}$$

$R^1 = R{=}Et$; $R^1 = H$, $R = Pr$

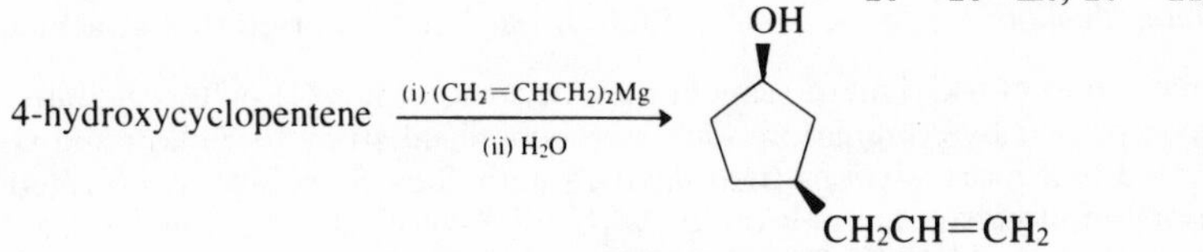

$$\text{4-hydroxycyclopentene} \xrightarrow[\text{(ii) } H_2O]{\text{(i) } (CH_2{=}CHCH_2)_2Mg} \text{cyclopentanol with } OH \text{ and } CH_2CH{=}CH_2 \qquad (155)^{308}$$

Predominant, if not exclusive, allylic rearrangement occurs. Surprisingly, considering the reactivity of simple alkylcyclopropenes towards Grignard reagents, no addition to **51** occurs with PhMgBr, Bu$^t$MgBr or MeMgI (all in $Et_2O$) or $CH_2{=}CHMgBr$ in thf.

All these stereochemical findings for homoallylic alcohols, $>C_\delta{=}C_\delta{-}C_\beta{-}C_\alpha{-}OH$ are consistent with the mechanism proposed by Eisch for addition to cyclohex-2-enyldiphenylmethanol, which involved the $OMgCH_2CH{=}CH_2$ grouping (Scheme 35). This is an electrophilic reaction of allylmagnesium at a carbon-carbon $\pi$ bond

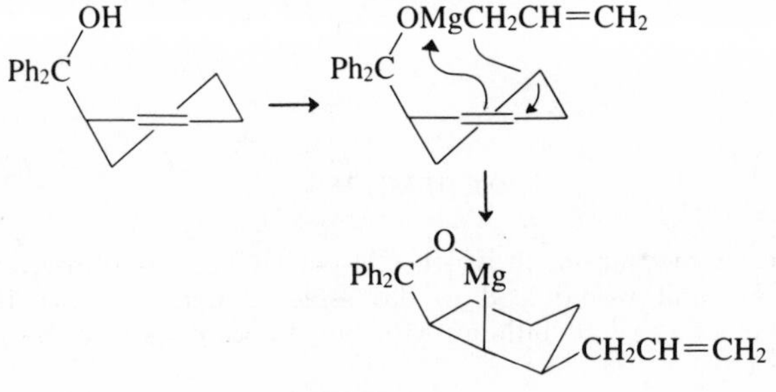

OH, $Ph_2C$ → $OMgCH_2CH{=}CH_2$, $Ph_2C$ → $Ph_2C$, O, Mg, $CH_2CH{=}CH_2$

SCHEME 35

and is similar to carboalumination of alkenes[309] and alkynes[310]. It was strongly argued that it was the allylmagnesium group bonded to oxygen in these homoallylic alcohols which was the one which added to the C=C intramolecularly to give *cis*-addition. Attack by a second group (an intermolecular allyl-Grignard) would lead to *anti*-addition. Results of additions to other types of alkenols and alkynols indicate different situations and have been summarized by Richey and Wilkins[245b].

i. *Homopropargylic alcohols.* $-C_\delta{\equiv}C_\gamma{-}C_\beta{-}C_\alpha{-}OH$. Addition to these by allyl-Grignards results in attachment of the allyl group to either $C_\gamma$ or $C_\delta$. Products in which the allyl group is at $C_\gamma$ have resulted exclusively from *cis*-attachment of the allyl and magnesium entities, e.g. equation 156.

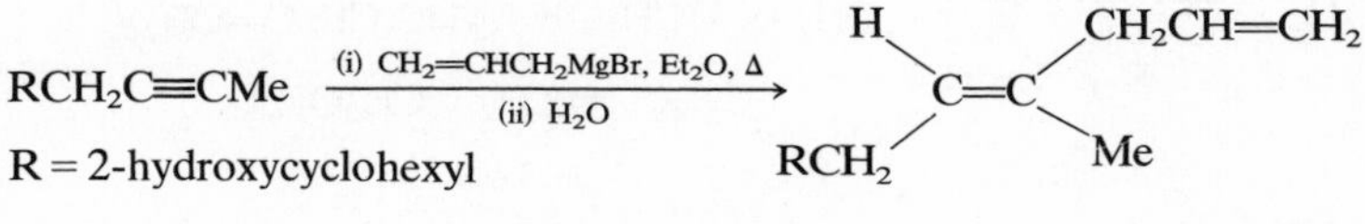

$$RCH_2C{\equiv}CMe \xrightarrow[\text{(ii) } H_2O]{\text{(i) } CH_2{=}CHCH_2MgBr,\ Et_2O,\ \Delta} (H)(RCH_2)C{=}C(CH_2CH{=}CH_2)(Me) \qquad (156)^{308}$$

$R$ = 2-hydroxycyclohexyl

Products in which the allyl group is at $C_\delta$ have in different examples resulted (i) only from *trans*-attachment, (ii) from both *cis*- and *trans*-attachment, and (iii) from only *cis*-attachment. The amounts of *trans*-$\gamma$, *cis*-$\delta$, and *trans*-$\delta$ products are sensitive to structural and solvent changes. The metallated hydroxyl group must play a role in all of these reactions, since the hydroxyalkynes are more reactive than the parent alkyne.

ii. *Allylic alcohols.* $>C_\gamma{=}C_\beta{-}C_\alpha{-}OH$. It has been reported that attachment of the organic group of a Grignard reagent may occur to either $C_\beta$ or $C_\gamma$, the latter also being accompanied by elimination. One stereochemical study by Felkin and coworkers revealed that more *erythro*- than *threo*-3-methylhex-5-en-2-ol was formed[311] in the reactions of allylmagnesium bromide with but-3-enol. The mechanism in Scheme 36 was proposed in which the $CH_2{=}CHCH_2{-}$ and the OMgX groups are kept as far

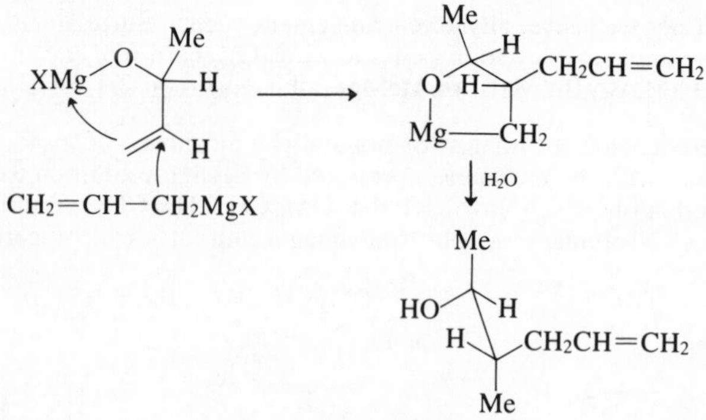

SCHEME 36

apart as possible. A mechanism in which $CH_2{=}CHCH_2{-}$ is transferred from the $OMgCH_2CH{=}CH_2$ unit would lead to the *threo*-adduct, equation 157. It is of interest that the reaction of allyllithium with but-3-enol does give the *threo*-isomer predominantly[257c].

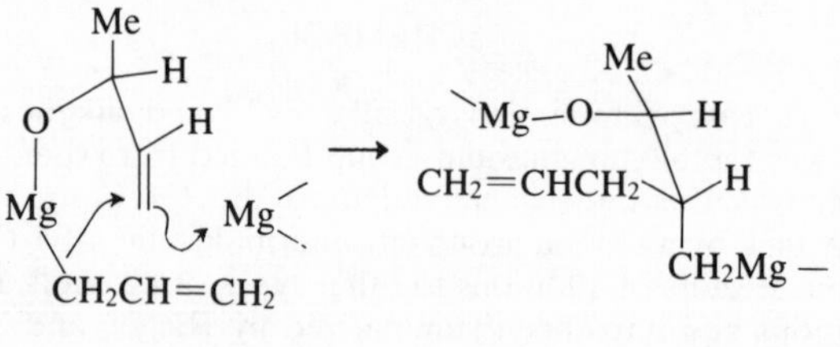

(157)

iii. *Propargylic alcohols.* $-C_\gamma{\equiv}C_\beta{-}C_\alpha{-}OH$. Additions to these compounds also fit in with Felkin's mechanism. Products in which the organic group is attached to $C_\beta$ result from *trans*-addition of the organo group and magnesium[312].

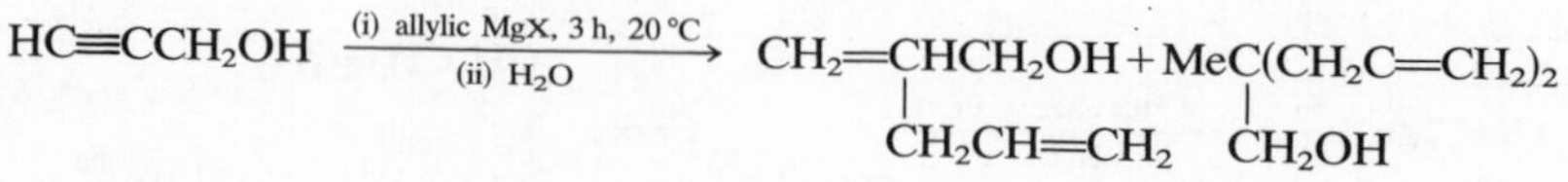

(158)

74%

iv. *Other hydroxy compounds.* Addition of allyl Grignards also occurs to *o*-allylphenol, equation 159. This reaction is faster than either those of *o*-$MeOC_6H_4CH_2CH{=}CH_2$ or $PhCH_2CH{=}CH_2$. Hence the metallated hydroxy group

$$CH_2{=}CHCH_2MgCl + RCH_2CH{=}CH_2 \xrightarrow[Et_2O]{100\,°C,\ 3\,h} CH_2{=}CHCH_2CH(MgCl)CH_2R + ClMgCH_2CH(CH_2CH{=}CH_2)CH_2R \quad (159)$$

R = *o*-hydroxyphenyl 58%

is giving assisatnce although it is remote from the double bond (i.e. the double bond is $\gamma$ to the hydroxyl group)[313]. No reaction occurs with BuLi–tmed in hexane even at 100 °C.

v. *Additions to unsaturated amines.* Allylic Grignard reagents also add to unsaturated amines such as *N*-allylaniline, equation 160, *N,N*-dimethylcinnamylamine, equation 161, and 3-dimethylamino-1-phenylprop-1-yne (but not, however, to others such as allylamine, diallylamine, or *N,N*-dimethylaniline[314]).

$$CH_2{=}CHCH_2NHPh \xrightarrow{CH_2{=}CHCH_2MgCl,\ reflux} ClMgCH_2CH(CH_2CH{=}CH_2)CH_2NHPh \quad (160)$$

$Et_2O$, 22 h: 32%

thf, 66 h: 44%

$$PhCH{=}CHCH_2NMe_2 \xrightarrow[40\,h]{CH_2{=}CHCH_2MgCl,\ PhMe,\ reflux} PhCH(MgCl)CH(CH_2CH{=}CH_2)CH_2NMe_2 \quad (161)$$

44%

The enhanced reactivity of the aminoalkenes suggests that the tertiary amino and metallated primary and secondary amino functions can assist the addition of Grignards; a sequence of effectiveness was established as $XMgO > XMg(R)N > R_2N$. The greater effectiveness of the metallated hydroxyl groups than of the amino groups is consistent with the idea that the effectiveness in assisting reactions increases with the decreasing basicity of the metallated assisting group. Mornet and Gouin[315] have studied additions of Grignard reagents to substituted alkynes, $R_2NCH_2C{\equiv}CCH_2X$ ($X = OH, OR', SR''$, or $NR'_2$), e.g. equation 162. Additions[316] to alkenylsulphones also occur, e.g. equation 163.

$$R_2NCH_2C{\equiv}CCH_2OR' \xrightarrow[(ii)\ H_2O]{(i)\ R^2MgX} (\mathbf{E})\text{-}R'OCH_2C(R^2){=}CHCH_2NR_2 + CH_2{=}C{=}CR^2CH_2NR_2 \quad (162)$$

$$(E)\text{-}BrCH_2CH{=}CHSO_2Ph \xrightarrow{RMgX} BrCH_2CHRCH(SO_2Ph)MgX \quad (163)$$

R = allyl, proparglylic, aryl, and benzylic but not alkyl

b. *Transition metal-catalysed additions.* There have been a number of reports that transition metal complexes such as $[(Ph_3P)_2NiCl_2]$ catalyse additions of Grignard reagents to alkynes and conjugated dienes. The $[(PPh_3)_2NiCl_2]$-catalysed alkyne addition reactions are regiospecific and stereospecific reactions in ether solution[317]. A reduced stereoselectivity occurs in benzene solution. Non-reducible alkyl Grignards

$$\text{PhMgBr} + \text{PhC}\equiv\text{CR} \xrightarrow[\text{(ii) } H_2O]{\text{(i) } [(Ph_3P)_2NiCl_2],\ Et_2O} \underset{R}{\overset{Ph}{}}\!\!>\!C{=}C\!<\!\!\underset{Ph}{\overset{H}{}} \tag{164}$$

$R = Et$ (100%)

may also be used. *Syn*-addition to an *o*-aminophenylalkyne was also achieved using $[(Ph_3P)_2NiCl_2]$ catalysis[318].

In contrast to the *syn*-addition in the presence of $[(Ph_3P)_2NiCl_2]$, equimolar $[(Ph_3P)_3RhBr]$ apparently leads to *trans*-addition[319] of MeMgBr to PhC≡CPh.

$$\underset{}{\text{PhC}\equiv\text{CPh}} + \underset{\text{excess}}{\text{MeMgBr}} \xrightarrow[\text{(ii) } H_2O]{\text{(i) } [(Ph_3P)_3RhBr]} \underset{Me}{\overset{Ph}{}}\!\!>\!C{=}C\!<\!\!\underset{Ph}{\overset{H}{}} \tag{165}$$

With $Ni^{II}$ catalysis, additions of aryl-Grignards to enynes, RC≡CCH=$CH_2$, occur at the triple bond, to give after hydrolysis ArCR=CHCH=$CH_2$[320]. In some cases, oligomerization[321], e.g. equation 166, or polymerization may occur; indeed, RMgX has been used as components of Ziegler–Natta type catalysts.

$$2CH_2{=}CHCH{=}CH_2 + Pr^iMgBr \xrightarrow[Et_2O,\ \text{room temp.}]{[(Ph_3P)_2NiCl_2]} CH_2{=}CH(CH_2)_3CH{=}CHCH_2MgBr + MeCH{=}CH_2 \tag{166}$$

## E. Insertions Into Carbon—Zinc Bonds

Organozinc compounds follow the same trends as observed with organo-lithiums and -magnesiums. However, even the most reactive organozinc derivatives, e.g. allylzinc[322] and *tert*-butylzinc[323] compounds, add to ethylene only under vigorous conditions, e.g. for $(Bu^t_2)Zn$ a pressure of 60 atm and temperatures of 50–70 °C are required to give reaction (94% yield). Dicrotylzinc undergoes allylic rearrangement in reactions with ethylene. Additions of $(Bu^t_2)Zn$[323] and bisallylzincs[322,324] to alk-1-enes and to buta-1,3-diene (equation 167) and styrene have been reported. For the alk-1-enes, RCH=$CH_2$, the major products have metal bonded to $C_{(2)}$.

$$CH_2{=}CHCH{=}CH_2 + Bu^tZn{-} \xrightarrow{253\text{–}283\ K} (Bu^tCH_2CH{=}CHCH_2)Zn{-} \tag{167}$$

1,4-addition; *cis* : *trans* = 1 : 2

+

$$-ZnCH_2CH(CH{=}CH_2)CH(CH{=}CH_2)CH_2Bu^t$$

The regioselectivity of the addition of $(CH_2{=}CMeCH_2)_2Zn$ to *p*-$RC_6H_4CH{=}CH_2$ at 20 °C in benzene varies[325] with the group R; electron-withdrawing groups favour **52**, e.g. [**53**]:[**52**] = 6.0 for R = MeO and 0.6 for $CF_3$.

$$p\text{-}RC_6H_4CH{=}CH_2 + CH_2{=}C(Me)CH_2Zn{-}$$

$$\longrightarrow \underset{\mathbf{(52)}}{p\text{-}RC_6H_4CH(Zn{-})CH_2CH_2C(Me){=}CH_2} + \underset{\mathbf{(53)}}{p\text{-}RC_6H_4CH[CH_2C(Me){=}CH_2]CH_2Zn{-}} \tag{168}$$

Di(*tert*-butyl)zinc adds[326] to alk-1-ynes, in refluxing thf but not $Et_2O$, to give *cis*, *trans* mixtures of $Bu^tCH{=}CRZn$—. Additions to terminal alkynes carrying donors groups, e.g. $HC{\equiv}CCH_2Y$ (Y = OR or $NH_2$), have been featured in other studies; a variety of organozincs have been used, including allyl-[327] and $MeC(CO_2Et)_2$-zinc[328] compounds. The allylic reactions can be reversible, as shown in Scheme 37. With increasing reaction time and temperature **54** increases at the expense of **55**.

$$HC{\equiv}CCH_2Y + RCH{=}CHCH_2ZnBr \xrightleftharpoons{thf} BrZnCH{=}C(CH_2Y)(RCHCH{=}CH_2) \quad (\mathbf{55})$$

$$\downarrow thf$$

$$BrZnCH{=}C(CH_2Y)(CH_2CH{=}CHR) \quad (\mathbf{54})$$

Y = OR or $NR_2$

SCHEME 37

Allylzinc compounds undergo spontaneous oligomerisations[329], equation 169.

$$XZnCH_2CH(CH_2CH{=}CHR)CHRZnX \longleftarrow XZnCH_2CH{=}CHR + RCH{=}CHCH_2ZnX \tag{169}$$

X = Br, solvent thf or dme, Δ
X = $CH_2{=}CHCH_2$; solvent $Et_2O$, 20–50 °C

*Trans*-Additions of $(Bu^t)_2Zn$ to $HC{\equiv}CCH{=}CHCH_2R$ (R = $Bu^t$, $NEt_2$, etc.) in $Et_2O$ at 35 °C occur at the triple bond. In refluxing thf, the reactions are still regioselective but are not stereoselective, a consequence of a series of rapid equilibrations[330].

## F. Insertions into Carbon—Boron Bonds

Reactions of trialkylboranes with alkenes require high temperatures; at these temperatures the most frequently met reactions are alkene exchanges. These exchanges result from elimination–addition reactions. Few simple insertions have been noted. However, $CH_2{=}CH_2$ and dec-1-ene do insert into a carbon—boron bond of $Et_3B$ to give $EtCH_2CH_2BEt_2$ and $Et(C_8H_{17})CHCH_2BEt_2$, respectively[331].

$$B(C_nH_{2n+1})_3 + 3C_mH_{2m} \longrightarrow B(C_mH_{2m+1})_3 + 3C_nH_{2n} \qquad n \geqslant 3;\ m \geqslant 2 \tag{170}$$

It is only with activated alkenes, e.g. alkoxyalkenes, and strained alkenes, such as cyclopropenes and activated organoboranes, that reactions proceed at all readily. The most frequently studied species are allylic boranes[332]. The reactions with alkoxyalkenes require temperatures above 100 °C; at such temperatures, the initial adducts are themselves unstable, equation 171. More ready reaction occurs with 1-methylcyclopropene, equation 172. All three bonds of triallylic boranes can react. The addition occurs in a *cis*-manner and with allylic rearrangement[333a].

Trialkylboranes, such as tripropylborane, also react with 1-methylcyclopropene, but the course of the reaction is clearly different[333b], equation 173. While it is not a

(171)

(172)

$B(C_3H_4Me)_2$ 85–90%

$Pr_3B$ + 1-methylcyclopropene —(−50 to −20 °C)→ [ $Pr_2B$ … Me … Pr ] —(1-methylcyclopropene)→ … $BPr_2$ (173)

carbon—carbon multiple insertion, the initial reaction is still an insertion and thus is appropriate to be included in this chapter.

Very few examples of thermal insertion into alkyl- or aryl-boranes by alkenes have been reported. One is the reaction of $PhBCl_2$ with norbornadiene[334]; both $Ph_2BCl$ and $Ph_3B$ are unreactive, though. However, sensitized photoinsertions have been reported[335], equation 174.

1-ethylcyclohexene —($R_3B$, $h\nu$ (254 nm), *p*-xylene, 48 h)→ … $BR_2$ —($H_2O_2$, $OH^-$)→ … OH (174)

R = Et 71%
R = Bu 68%

Triallylboranes react with alkoxyalkynes even below room temperature; all three carbon—boron bonds can react[336]. As shown in reaction 175, allylic rearrangement occurs in these *cis*-additions.

(175)

Terminal alkynes, RC≡CH, also react in a similar way[337]. The initial products are also prone to further reaction, see Scheme 38.

Few insertions of alkynes into alkyl- or aryl-boranes have been reported[338,339]. Trialkylboranes and borocycloalkanes have been shown[338] to react with $R_3SnC{\equiv}CSnR_3$ (R = Me or Bu) to give *cis*-adducts, e.g. equation 176.

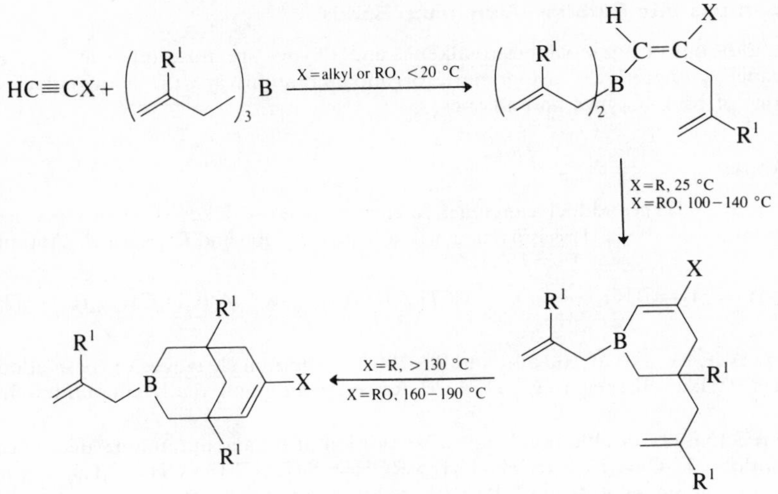

SCHEME 38

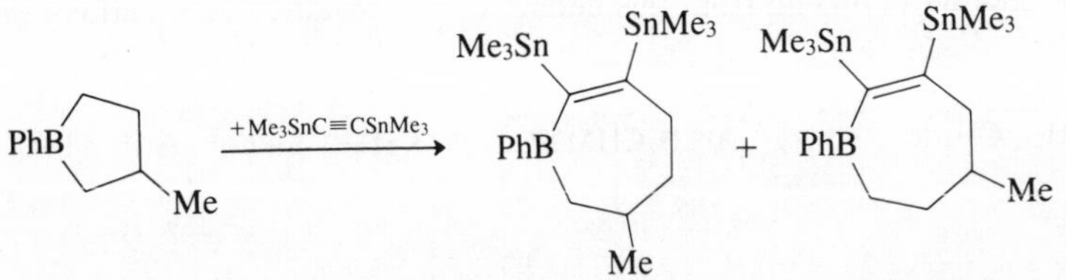

$$\text{PhB(3-methylborolane)} \xrightarrow{+Me_3SnC{\equiv}CSnMe_3} \text{PhB(...)} + \text{PhB(...)} \qquad (176)$$

In the reaction of 1-phenylborocyclohexane with $Me_3SnC{\equiv}CSnMe_3$, insertion into the phenyl—boron bond results instead of into the ring.

Allenes[337] also react with triallylboranes; a temperature of >150 °C is required to initiate the reaction. This results in the reaction proceeding beyond the initial addition stage (Scheme 39).

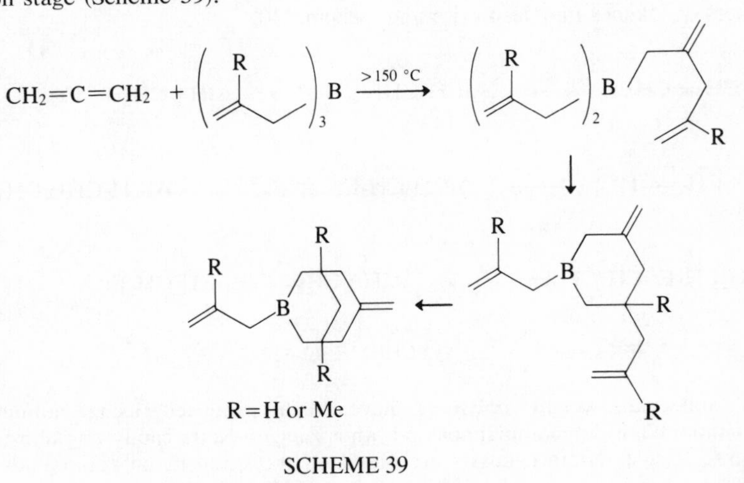

R=H or Me

SCHEME 39

## G. Insertions into Carbon—Aluminium Bonds

The additions of organoalanes to alkenes and alkynes are most important reactions in organic synthesis[340,341]. Monomeric organoaluminiums are the active species in reactions of both alkynes and alkenes.

### 1. Alkenes

The initial alkene adduct can react further with more alkenes to produce long-chain aluminium alkyls. This is particularly so with ethylene and $Et_3Al$ or $Pr_3Al$ (but[342]

$$>Al{-}R + CH_2{=}C(R')_2 \longrightarrow >AlCH_2CR_2'R \longrightarrow >Al(CH_2CR_2')_nR \qquad (177)$$

not, however, $Me_3Al$) at 100 atm and 90–200 °C. Thermal cleavages of these adducts provides higher alkenes (e.g. $n = 4$–20 for ethylene) and dialkylaluminium hydrides.

The reactivity of double bonds towards addition of organoaluminiums decreases in the order $H_2C{=}CH_2 > RCH{=}CH_2 > RCH{=}CHR > R_2C{=}CH_2$. The carboalumination of monosubstituted alkenes requires higher temperatures than are needed for $CH_2{=}CH_2$ and gives predominantly the mono addition product. The major reaction leads to the aluminium at the terminal position of the originally double bond; thermal loss of an alkene from this adduct occurs readily, see equations 178 and 179.

$$>AlPr + CH_2{=}CHMe \xrightarrow{140\ ^\circ C} \left[>AlCH_2CHMePr\right] \longrightarrow CH_2{=}CMePr + >AlH \qquad (178)^{343}$$

$$\text{1,2-dihydronaphthalene} \xrightarrow{Ph_3Al} \left[\text{(1,2,3,4-tetrahydronaphthalene with H, AlPh}_2\text{ at C-2 and H, Ph at C-3)}\right] \xrightarrow{-Ph_2AlH} \text{3-phenyl-1,2-dihydronaphthalene} \qquad (179)^{309b}$$

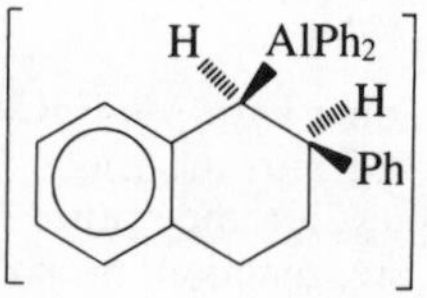

Oligomers of alkenes may also arise, e.g. Scheme 40.

$$Et_3Al + CH_2{=}CHEt \xrightarrow{150\ ^\circ C} >AlCH_2CHEt_2 \longrightarrow >AlH + CH_2{=}CEt_2$$

$$>AlH + CH_2{=}CHEt \longrightarrow >AlCH_2CH_2Et \xrightarrow{CH_2=CHEt} >AlCH_2CHEtCH_2CH_2Et$$

$$>AlCH_2CH(Et)CH_2CH_2Et \xrightarrow{\Delta} >AlH + CH_2{=}C(Et)CH_2CH_2Et$$

SCHEME 40

High molecular weight polymers have been produced using aluminium-alkyl/titanium chloride combinations and other Ziegler–Natta catalyst systems[344,345]. In general, Ziegler–Natta catalysts are based on transition metal compounds normally formed by reaction of a transition metal halide, alkoxide, alkyl, or aryl derivative with a main group alkyl or haloalkyl compound. The variety of combinations patented is vast. Ziegler catalyses involves rapid polymerization of ethylene and $\alpha$-alkenes at low pressures and temperatures up to *ca.* 120 °C. With monoenes, stereoregular structures (isotactic and syndiotactic rather than atactic) can be realized. 1,4-*Cis*-, 1,4-*trans*-, and 1,2-polybutadienes and 3,4-polyisoprene can also be produced.

Side reactions, other than polymerization and $R_2AlH$ eliminations from the initial adducts, have been observed. These include (i) *cis–trans* isomerization of alkenes, e.g. the conversion of *cis*- to *trans*-PhCH=CHPh at 175 °C in refluxing mesitylene in the presence of $Ph_3Al$, (ii) metallation of vinylic alkenes, e.g. *cis*-PhCH=CHPh by $Ph_3Al$ at 200 °C, and (iii) decarboalumination with C—C bond scission as exemplified in the interaction of $Ph_3CCH{=}CH_2$ with $Ph_3Al$, equation 180[309b].

$$Ph_3CCH{=}CH_2 + Ph_3Al \longrightarrow [Ph_3CCHPhCH_2AlPh_2] \longrightarrow PhCH{=}CH_2 + Ph_3CAlPh_2 \quad (180)$$

As with other organometallic compounds, additions to strained and conjugated alkenes occur under more moderate conditions. The reactivity of a series of alkenes (both strained and conjugated) towards $Ph_3Al$ was found to be in the order norbornadiene > *cis*-PhCH=CHMe > *trans*-PhCH=CHMe ≈ 1,2-dihydronaphthalene ≈ 1,1-dimethylindene > *cis*-PhCH=CHPh > 3,3,3-triphenylpropene > *trans*-1,2-diarylethylenes ≈ phenylcyclopropane[309b]. As shown in equation 181,

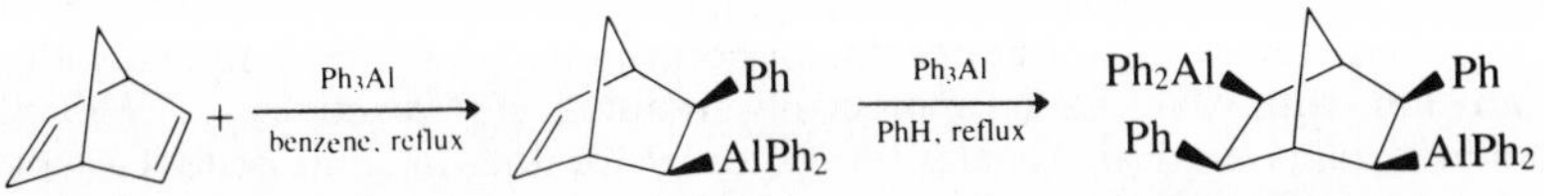

(181)[309b]

*syn,exo*-additions to norbornadiene occur in refluxing PhH at both double bonds. *cis*-Additions are generally considered to occur with alkenes and alkynes.

The relative reactivity of 6-substituted benzonorbornadiene and regioselectivity of their reactions with $Ph_3Al$ were also studied[309a], equation 182.

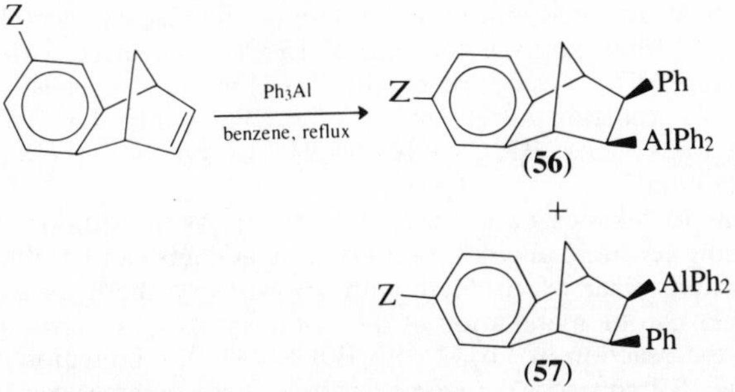

(182)

The relative reactivities were in the sequence Z = Me > H > F > Cl. In addition, the logarithm of the relative rates correlated with $\frac{1}{2}(\sigma_p + \sigma_m)$ with $\rho = -0.66$. The ratios of [**56**]:[**57**] were found to be 53:47 and 30:70 for Z = Me and F, respectively. Both the stereochemistry (*cis-syn*) and the substituent effects point to electrophilic attack by monomeric $Ph_3Al$ on the alkene in the rate determining step.

Intramolecular additions also occur readily[346–348]. Interesting examples are found in hydroalumination reactions of diene, e.g. equations 183 and 184.

$$\text{Me}_2\text{-cyclohexene (H, H; vinyl, vinyl)} + \text{HAl} \longrightarrow [\,\cdots\text{Al}\,] \longrightarrow \text{bicyclic product–CH}_2\text{Al} \qquad (183)^{346}$$

$$CH_2{=}CMeCH_2CH_2CH{=}CH_2 + Et_2AlH$$
$$\downarrow$$
$$[Et_2AlCH_2CHMeCH_2CH{=}CH_2 + CH_2{=}CMeCH_2CH_2CH_2CH_2AlEt_2]$$
$$\downarrow \qquad (184)^{348}$$

Me-cyclopentyl–$CH_2AlEt_2$ + 1-Me-1-($CH_2AlEt_2$)cyclopentane

*cis* : *trans* 4·1 : 95·9

It should be noted that such cyclizations do not occur in the presence of ethers, which on coordination with the alane appreciably reduces it's reactivity. 1-(Hex-5-enyl)aluminium compounds are ideally set up for cyclization, in contrast to 1-(hept-6-enyl)- and 1-(oct-7-enyl)-aluminium species[251]. Evidence has been gained from spectral data for $\pi$-interaction of the double bond with the aluminium centre[251].

## 2. Alkynes

Carboalumination of carbon—carbon triple bonds also occurs readily[349]. Indeed, additions to alkynes proceed more readily than to alkenes, e.g. acetylene reacts with $Et_3Al$ at 40–60 °C whereas a temperature of 150 °C is required with ethylene[349]. In addition, PhCH═CHPh does not react with $Ph_3Al$ in refluxing mesitylene even after 40 h, while 100% reaction occurs with PhC≡CPh within 4 h[309b,310]. An internal competition in 3-en-1-ynes, RR′C═CR′C≡CH, also points to the increased reactivity of the triple bond[350].

The additions to alkynes occur in a *cis* manner. With addition to unsymmetric alkynes, including terminal alkynes, two isomeric adducts can be obtained. The lack of regiospecificity is clearly a problem with regard to synthetic uses of the reaction. Other drawbacks can be metallation at the acidic hydrogen of terminal alkynes, e.g. as happens in the reaction of $Ph_3Al$ with BuC≡CH[351], elimination of $R_2AlH$ from adducts having $\beta$-hydrogens, oligomerizations and polymerizations.

Di-insertion of alkynes may result, e.g. equation 185 and Scheme 41.

$$Et_3Al + EtC{\equiv}CEt \longrightarrow [Et_2C{=}CEt(AlEt_2)] \xrightarrow{EtC\equiv CEt} Et_2C{=}C(Et){-}C(Et){=}C(Et)AlEt_2 \qquad (185)$$

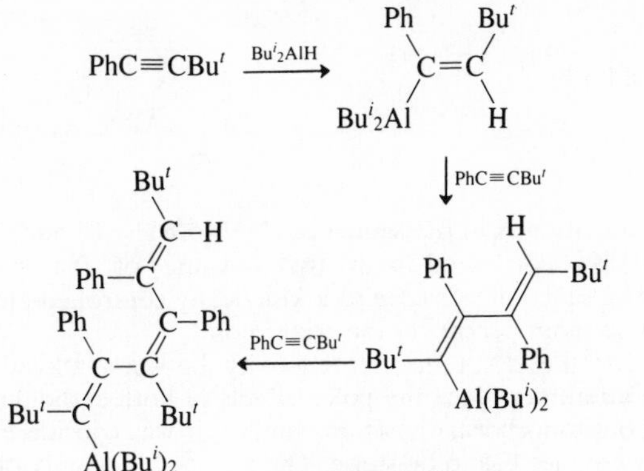

$$PhC{\equiv}CBu^t \xrightarrow{Bu^i_2AlH} (Bu^i_2Al)(Ph)C{=}C(Bu^t)(H) \xrightarrow{PhC{\equiv}CBu^t} \text{dienylalane} \xrightarrow{PhC{\equiv}CBu^t} \text{trienylalane}$$

SCHEME 41[352]

An interesting continuation of this is the formation of arenes from alkynes in the presence[353] of catalytic $Bu^i_2AlH$, equation 186.

$$3PhC{\equiv}CMe \xrightarrow[\text{cat. } (Bu^i)_2AlH]{140-150\ ^\circ C} \text{1,2,4-trimethyl-3,5,6-triphenylbenzene} \quad (186)$$

Alkylalanes can react with alkynes with retention of configuration, e.g. equation 187[354].

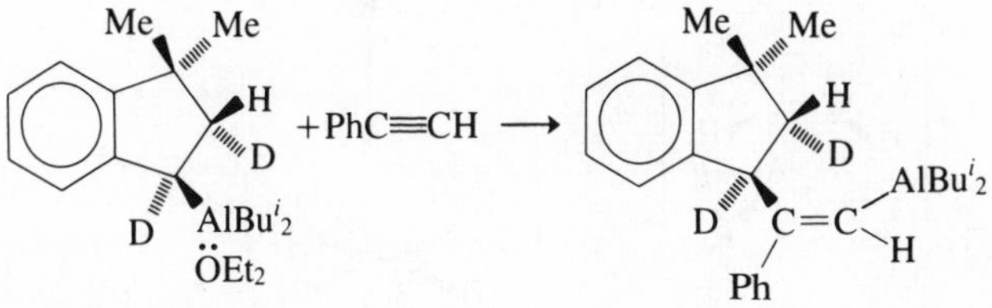

$$\text{indanyl-}AlBu^i_2\cdot OEt_2 + PhC{\equiv}CH \longrightarrow \text{indanyl-}C(Ph){=}C(H)AlBu^i_2 \quad (187)$$

The regioselectivity of the alkyne additions are governed by both steric and polar effects[353,355]. Steric control is clearly seen in the reactions[355] of $PhC{\equiv}CMe$ and $PhC{\equiv}CBu^t$ with $Ph_3Al$, equations 188 and 189.

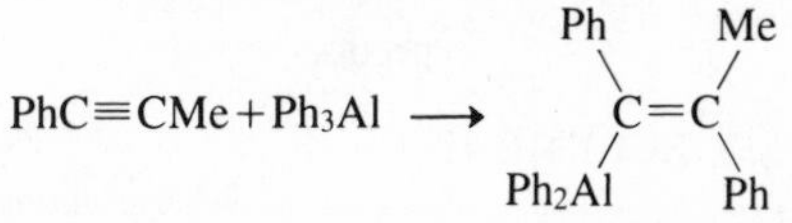

$$PhC{\equiv}CMe + Ph_3Al \longrightarrow (Ph_2Al)(Ph)C{=}C(Me)(Ph) \quad 98\% \quad (188)$$

$$PhC{\equiv}CBu^t + Ph_3Al \xrightarrow{90-100\ ^\circ C} (Ph_2Al)(Bu^t)C{=}C(Ph)(Ph) \quad 95\% \quad (189)$$

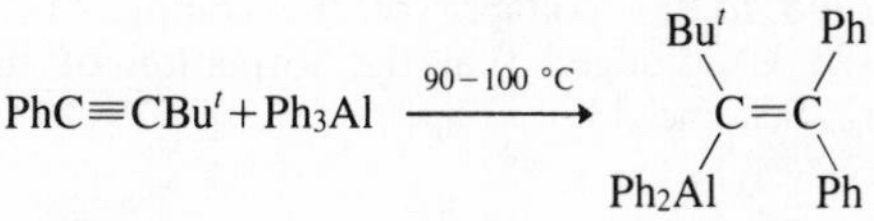

Polar effects are shown in the reaction of $p\text{-}ZC_6H_4C{\equiv}CPh$ with $Ph_3Al$; both *syn* isomers were formed[310], equation 190.

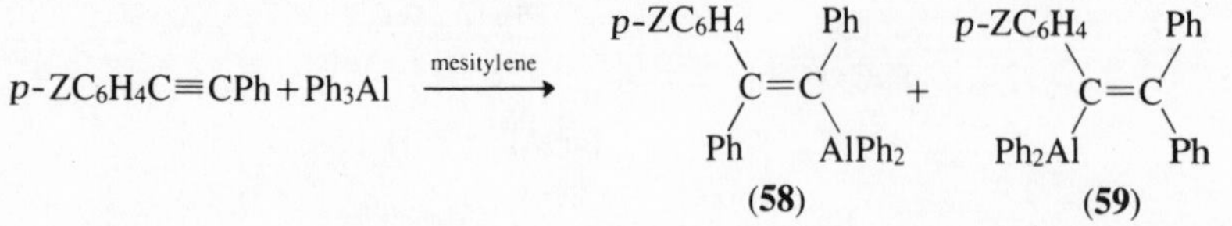

$$p\text{-}ZC_6H_4C{\equiv}CPh + Ph_3Al \xrightarrow{\text{mesitylene}} \underset{(\mathbf{58})}{(p\text{-}ZC_6H_4)(Ph)C{=}C(Ph)(AlPh_2)} + \underset{(\mathbf{59})}{(p\text{-}ZC_6H_4)(Ph_2Al)C{=}C(Ph)(Ph)} \qquad (190)$$

The relative reactivity was in the sequence Z = Me > H > Cl and the logarithm of the ratio of [**58**]:[**59**] correlated with the $\sigma$-value of the *para*-substituent (Z) ($\rho = -0.713$). The reaction was seen as a kinetically controlled electrophilic attack of the trivalent aluminium centre on the triple bond.

The reason for the difference in reactivity between alkynes and alkenes has attracted some attention. Since the polar effects in both carboalumination of diarylacetylenes and benzonorbornadienes are similar, it was considered that the origin of the different reactivities has to be steric. The reduced reactivity of alkenes is thought to arise from steric factors that destabilize the $\pi$-complex between $R_3Al$ and the alkene and that cause a trapezoidal-like transition state to be rate determining, Scheme 42.

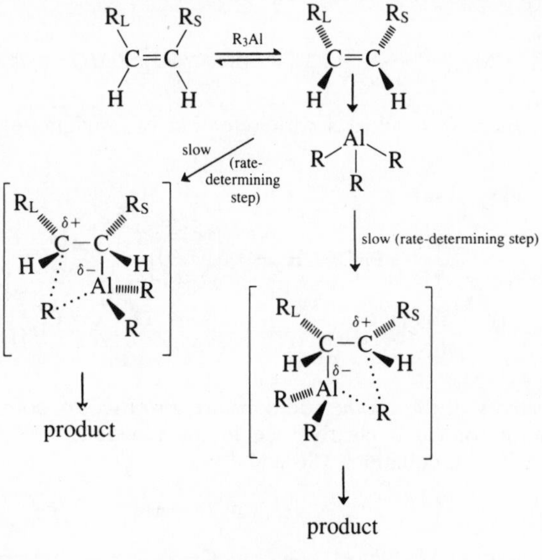

SCHEME 42

The higher reactivity of alkynes is ascribed to less steric hindrance both to $\pi$-complexation with $Ph_3Al$ and to the collapse of the complex via a trapezoidal configuration. For alkynes it is adjudged that the formation of the intimate $\pi$-complex is rate determining, Scheme 43.

### 3. *Catalysed reactions*

Use has been made of combinations of organoalane–transition metal compounds. Controlled additions to alkynols and alkynes have been reported among others by Thompson's group[356] and Negishi's group[357,358]. Some, but not all, of these systems

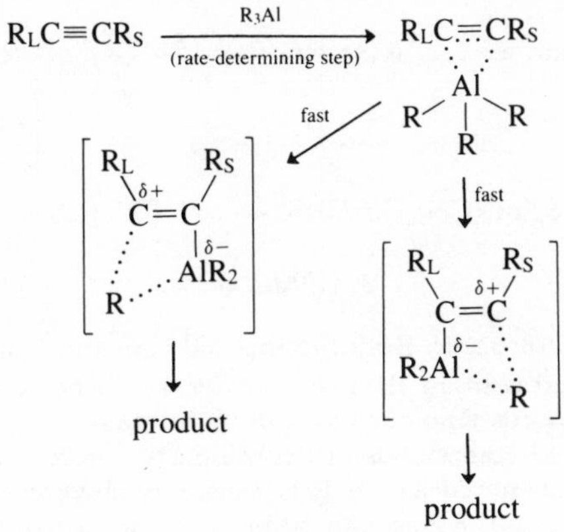

$$R_LC{\equiv}CR_S \xrightarrow[\text{(rate-determining step)}]{R_3Al} R_LC{\equiv}CR_S \cdots AlR_3$$

fast → $[R_L(\delta+)C{=}C(R_S)\text{—}AlR_2(\delta-) \cdots R]$ → product

fast → $[R_L\text{—}C{=}C(\delta+)(R_S)\text{—}R_2Al(\delta-) \cdots R]$ → product

SCHEME 43

lead to regio- or stereo-selective reactions. Thus, use of $TiCl_4$–$Me_3Al$ with hex-3-yn-1-ol provides (*Z*)-4-methylhex-3-en-1-ol, after hydrolysis[356], in marked contrast to the non-selective reaction with but-3-en-1-ol. Negishi and coworkers[357,358] have used $[Cp_2ZrCl_2]$–organoalane combinations with alkynes, propargyl, and homoproparyl derivatives and obtained alkynylmetal derivatives selectivity in good yields, equations 191 and 192.

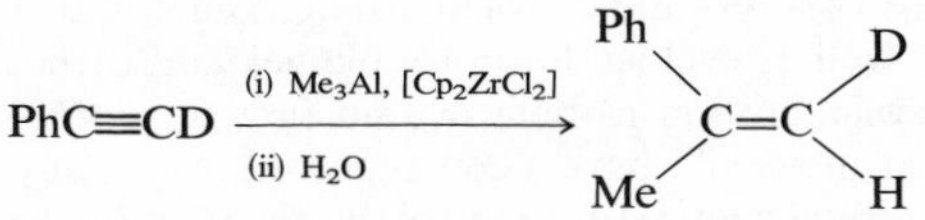

$$PhC{\equiv}CD \xrightarrow[\text{(ii) } H_2O]{\text{(i) } Me_3Al,\ [Cp_2ZrCl_2]} (Ph)(Me)C{=}C(D)(H) \tag{191}$$

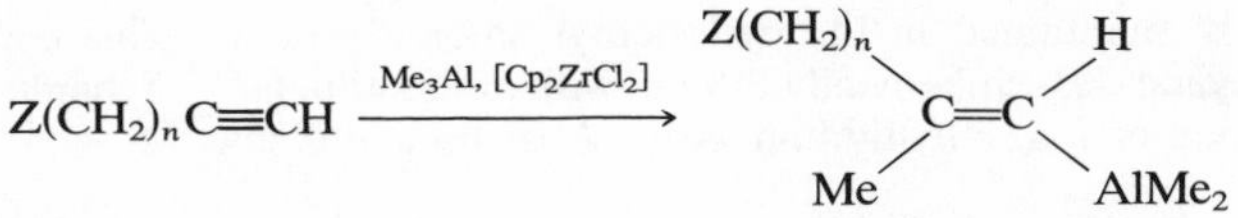

$$Z(CH_2)_nC{\equiv}CH \xrightarrow{Me_3Al,\ [Cp_2ZrCl_2]} (Z(CH_2)_n)(Me)C{=}C(H)(AlMe_2) \tag{192}$$

[358]

Regioselectivity >90%

## H. Tetracyanoethylene Insertions[359–362]

Insertions of tetracyanoethylene into carbon—metal bonds occurs for both electropositive metals such as magnesium[362] and for metals such as tin, lead, and mercury[359–361]. Tetracyanoethylene is a powerful electron acceptor and forms complexes with both $\sigma$- and $\pi$-donors. The mechanism of the tcne insertions into the less

$$RM + (NC)_2C{=}C(CN)_2 \longrightarrow RC(CN)_2C(CN)_2M \tag{193}$$

electropositive element—carbon bonds has been shown to involve 1:1 charge-transfer complexes with the metal—carbon bond acting as the $\sigma$-donor. For tetraalkyltin compounds, $R_4Sn$, the disappearance of the charge-transfer complex follows overall second-order kinetics and leads to the insertion product, $R_3SnC(CN)_2C(CN)_2R$. The same product is obtained in a unit quantum yield by the direct irradiation at the charge transfer maximum at low temperatures, when the

$$R_4Sn + tcne \overset{K_{CT}}{\rightleftharpoons} [R_4Sn.tcne] \xrightarrow[\text{or } \Delta]{h\nu_{CT}} [R_4Sn^{+\bullet}, tcne^{-\bullet}]$$

$$\downarrow \text{fast}$$

$$R_3SnC(CN)_2C(CN)_2R \longleftarrow [R^{\bullet}, R_3Sn^{+}, tcne^{-\bullet}]$$

SCHEME 44

thermal reaction does not occur. Both the thermal activation and the photochemical activation are associated with an electron transfer, see Scheme 44. The reactivity of tetraalkylstannanes towards tcne decreases in the sequence $Bu_4Sn > Pr_4Sn > Et_4Sn > Me_4Sn$. Tetraorganolead compounds are considerably more reactive than both the tin analogues and dialkylmercurials. If R is a primary alkyl group, further reactions, particularly of the lead and magnesium adducts, occur, equation 194.

$$R'_3PbC(CN)_2C(CN)_2CH_2R \xrightarrow{-HCN} [R'_3PbC(CN)_2C(CN){=}CHR]$$

$$\downarrow \text{tcne, } -HCN$$

$$R'_3Pb^{+} + (NC)_2C{=}C(CN)\bar{C}(R)C(CN){=}C(CN)_2 \quad (194)$$

## I. Cycloaddition Reactions

Various [3+2]cycloadditions of allyl- and 2-azaallyl-magnesium and -alkali metal compounds are known[363]. As well as cycloaddition to alkenes and dienes[364], to give five-membered carbocycles, cycloaddition to heterocumulenes (e.g. RNCS, RNCO, $CS_2$ and carbodiimides)[365] and nitriles[366] have been reported, especially for 2-azaallyllthiums. Conservation of orbital symmetry predicts the reaction of the allyl anion to carbon—carbon multiple bonds to be a $[\pi^4s + \pi^2s]$ process in which the alkene configuration is maintained in the cyclopentyl anionic product. This occurs in the reaction of 2-cyano-1,3-diphenylallyllithium with *trans*-stilbene[367], Scheme 45, as well as in the reaction of 2-azaallyllithium with *cis*- or *trans*-stilbene[368].

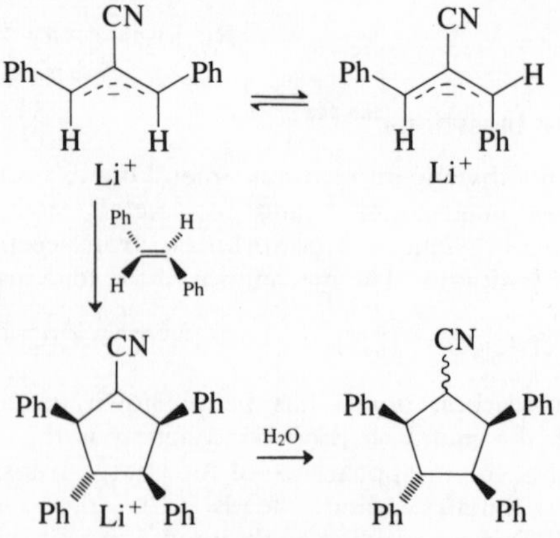

SCHEME 45

Only two of the ten possible isomers was detected in the 2-cyano-1,3-diphenylallyl-lithium–*trans*-stilbene reaction, which points to (but does not prove) a concerted reaction.

Two features which promote the anionic [3+2]cycloaddition reactions are (i) electron-withdrawing groups on the C=C bond and (ii) stabilization of charge in the cyclopentyl anion product by a suitable substituent, e.g. nitrogen as in the 2-azaallyl-lithium or a carbonyl, cyano, or aryl group. Whereas 2-cyano-1,3-diphenylallyl-lithium[369] and 1,3-diphenyl-2-azaallyllithium[370] cycloadd to acenaphthylene, $CH_2$=$CPhCH_2MgOPh \cdot 2hmpt$ gives only a simple addition product,[363b], Scheme 46.

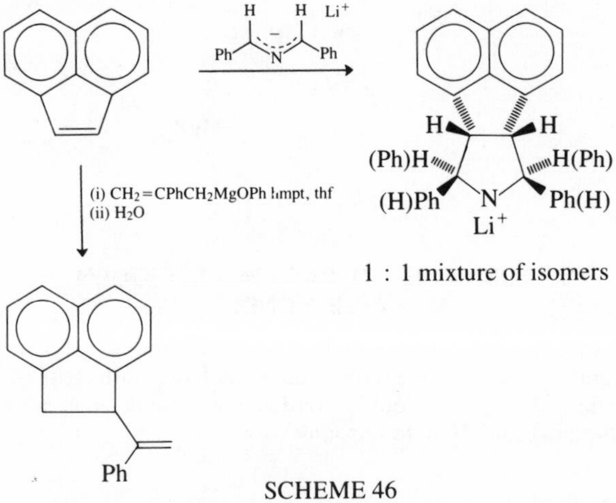

(i) $CH_2$=$CPhCH_2MgOPh \cdot hmpt$, thf
(ii) $H_2O$

1 : 1 mixture of isomers

SCHEME 46

No cycloaddition products are obtained using 2-phenylallylmagnesium phenoxide in the absence of hmpt or [2.1.1]cryptate[371]. A cycloaddition reaction of PhC≡C$\bar{C}$HPh, $Li^+$ has also been reported[363].

Cycloadditions to benzyne have been reported. Thus benzyne and covalent $\eta'$-$C_5H_5MgBr$ in thf produces the benzonorbornadienyl-Grignard[372], equation 195. In

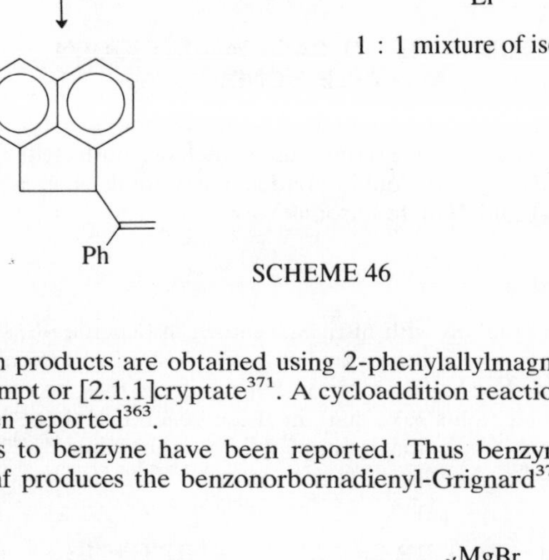

(195)

**(60)**

contrast, the ionic CpM (M = Li, N, or K) in thf just undergoes nucleophilic addition to give 1-[$o$-M-$C_6H_4$]-cyclopenta-2,4-diene, **61**. It is of interest that $\eta'$-$C_5H_5MgBr$ in the presence of hmpt gave both **60** and **61**[373].

Allyl-Grignard reagents react with benzyne by three pathways[374], namely (i) nucleophilic addition, (ii) [2+2]cycloaddition, and (iii) [3+2]cycloaddition, see Scheme 47.

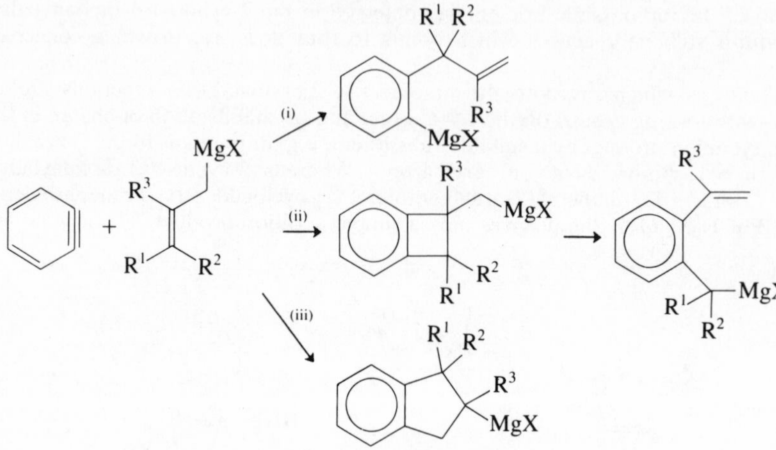

SCHEME 47

## V. ADDITIONS TO CARBON—NITROGEN MULTIPLE BONDS[3,4]

A number of carbon—nitrogen multiple bonded compounds react with organometallic compounds. The reactive compounds include both triple-bonded compounds, i.e. nitriles, RCN, and double-bonded compounds, e.g. azomethines (aldimines and ketimines) and *N*-heteroaromatics.

### A. Additions to Nitriles

The simple addition reaction with nitriles is shown in Scheme 48, step (i). Nitriles in general are much less reactive than acid chlorides, esters and even amides and so longer reaction times and/or higher reaction temperatures may be required. Organic derivatives of various elements take part in these reactions; these include organic derivatives of lithium[4a,376–378], magnesium[4b,379–383], aluminium[4c,384–387], zinc[388–394], alkaline earth metals[5a,395], beryllium[396], and boron[397,398].

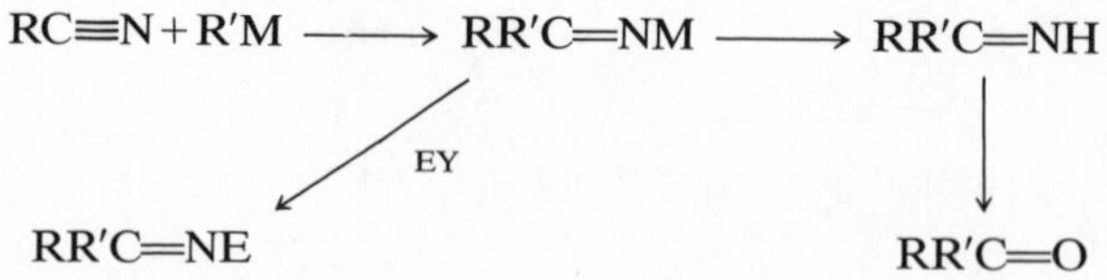

$$RC{\equiv}N + R'M \longrightarrow RR'C{=}NM \longrightarrow RR'C{=}NH \longrightarrow RR'C{=}O$$

$$RR'C{=}NM \xrightarrow{EY} RR'C{=}NE$$

$EY = R_3MCl$, M = Si, Ge, or Sn[375]; $MeOSO_3Me$ (ref. 4)

SCHEME 48

The insertion products, the metallated imines, may be studied, either in solution or in the solid state, derivatized on reaction with electrophiles, hydrolysed carefully to ketimines, RR′C=NH (e.g. using MeOH[380]), or hydrolysed more forcefully to ketones, RR′C=O. Examples of these reactions are listed in Table 11.

The metallated imines have been studied[399,400] for lithium[401], beryllium, magnesium[402], zinc[388], boron, and aluminium[384,387] derivatives. These compounds appear to be associated species, with the nitrogens acting as bridging groups; it has been established, for example, that beryllium, magnesium, zinc, boron, and aluminium compounds are dimers, e.g. **62**.

$$\begin{array}{c} Ph_2C{=}N(\mu\text{-}ZnR)_2N{=}CPh_2 \end{array}$$

**(62)**

It also appears that coordination compounds are formed prior to the insertion process. These adducts have been detected, for example, for organoaluminium, organoberyllium[403], and organozinc[388] compounds at ambient or even lower temperatures; strong heating may be necessary to cause the insertion, e.g. equations 196 and 197. In equation 197, the mono-insertion product undergoes complete dispro-

$$\tfrac{1}{2}(Me_3Al)_2 + PhCN \xrightarrow{\text{room temp.}} Me_3Al{\cdot}NCPh \xrightarrow[150\text{–}220\,^{\circ}C]{\Delta} \underset{60\%}{(PhCMe{=}NAlMe_2)_2} \qquad (196)$$

$$\underset{\nu_{CN}=2229\ cm^{-1}}{Ph_2Zn + PhCN} \xrightarrow{\text{room temp.}} \underset{\nu_{CN}=2259\ cm^{-1}}{PhCN{\cdot}ZnPh_2} \xrightarrow{100\,^{\circ}C}$$

$$[(Ph_2C{=}N)ZnPh]_2 \longrightarrow \underset{\nu_{CN}=1607\ cm^{-1}}{(Ph_2C{=}N)_2Zn} + Ph_2Zn \qquad (197)$$

portionation at the reaction temperature. Not all adducts give rise to insertion products; an example[403b] is $Bu^t{}_2Be{\cdot}NCBu^t$. Four-centre transition states are envisaged for these insertions, e.g. equation 198.

$$R{-}C{\equiv}N{\rightarrow}AlEt_2(Et) \longrightarrow R(Et)C{=}N{-}AlEt_2 \qquad (198)$$

Higher yields of the metallated imines[385,386] have been obtained using 2:1 $R_3Al{:}R'CN$ molar ratios rather than 1:1 ratios. As with other organoaluminium reactions, $[Ni(acac)_2]$ acts as a catalyst in the methylation of nitriles, enabling ambient temperatures to be employed instead of the 140 °C required in the absence of the catalyst. Methyl group transfer also occurs to give $(PhCMe{=}NAlMeCl)_2$ on heating the complex $Me_2AlCl{:}NCPh$[384].

The mechanism of the reaction between benzonitrile and the Grignard reagent from methyl bromide was studied by Ashby *et al.*[404]. Under conditions where both MeMgBr and $Me_2Mg$ are monomeric (i.e. <0.1 M in diethyl ether), the reaction was found to be an overall second order reaction, first order in each component. The

TABLE 11. Reactions of organometallics with nitriles

| Organometallic compound | Nitrile | Conditions | Product (yield, %) | Ref. |
|---|---|---|---|---|
| $Me_3Al$ | MeCN | 120–150 °C | $(Me_2C{=}NAlMe_2)_2$ (15–20) | 384 |
| $(CH_2{=}CHCH_2)_3B$ | $CH_2{=}CMeCN$ | $CH_2Cl_2$, Δ | $CH_2{=}CMe{-}C(CH_2CH{=}CH_2){=}NB(CH_2CH{=}CH_2)_2$ (23) | 398 |
| pyrid-2-yl-MgBr | EtCN | (i) $Et_2O$, PhH (ii) MeOH, HCl | pyrid-2-yl-C(Et)=NH·HCl (41) | 380 |
| thien-2-yl-MgBr | 2-cyanotoluene | (i) $Et_2O$, PhH (ii) MeOH, HCl | thien-2-yl-C(2-tolyl)=NH·HCl (61) | 380 |
| $RCOCMe_2ZnBr$ | R′CN | (i) EtOAc, $HgCl_2$ (trace) (ii) $H_2O$ | $RCOCMe_2CR'{=}NH$ | 390 |
| BuLi | PhCN | (i) $Et_2O$, 20 °C, 3 h (ii) $H_2O$ | PhCOBu (75) | 406b |
| fur-2-yl-Li | PhCN | (i) $Et_2O$, −20 °C (ii) aq. HCl | fur-2-yl-COPh (87) | 377 |
| 2-Me-2-Li-1,3-dithiane [structure: S, S ring; Me; Li] | PhCN | (i) thf, −78 °C, 40 min (ii) $H_2O$ | 2-Me-2-(C(=O)Ph)-1,3-dithiane [structure: S, S ring; Me; CPh=O] (82) | 123 |
| $(EtCHMeCH_2)_2Be$ | PhCN | (i) PhCN, 0 °C (ii) $H_2O$ | $PhCOCH_2CHMeEt$ (93), PhCHO (7) } (32) | 396 |
| $MeCH{=}CHCH_2MgBr$ | MeCN | (i) $Et_2O$, −10 °C (ii) aq. HCl | $MeCH{=}CHCH_2COMe$ (38); $CH_2{=}CHCHMeCOMe$ (7) | 324b |

| | | | | |
|---|---|---|---|---|
| PrMgBr | MeCN | (i) $Et_2O$, 20 °C (ii) $H_2O$ | MeCOPr (2) + condensation product | 382 |
| | | (i) $Et_2O$, 20 °C, $LiClO_4$ (ii) $H_2O$ | MeCOPr (20) + condensation product | 382 |
| $PhCH_2MgCl$ | MeCN (0.5 equiv.) | (i) PhH (ii) $H_2O$ | $MeCOCH_2Ph$ (48) | 383 |
| cyclohexyl-MgBr | cyanocyclohexane | (i) PhH (ii) $H_2O$ | dicyclohexylketone (68) | 383 |
| PhMgBr | $CF_3CN$ | (i) $Et_2O$ (ii) $H_2O$ | $PhCOCF_3$ (82) | 379 |
| $(PhC{\equiv}C)_2M$ M = Ca, Sr, or Ba | PhCN | (i) (ii) $H_2O$ | $PhC{\equiv}CCOPh$ | 395 |
| MeCaI | PhCN | (i) thf, −30 °C (ii) $H_2O$ | PhCOMe (20) | 5[a] |
| $CH_2{=}CHCH_2ZnBr$ | MeCN | (i) (ii) $H_2O$ | $CH_2{=}CHCH_2COMe$ (52) | 393 |
| $EtMeCHO_2CCHMeZnBr$ | $C_5H_{11}CN$ | (i) PhH (ii) $H_3O^+$ | $C_5H_{11}COCHMeCO_2CHMeEt$ | 394 |
| $Me_2C{=}CHCH_2Br$ + Zn/Ag | PhCN | (i) thf (ii) $H_2O$ | $CH_2{=}CHCMe_2COPh$ (67) | 389 |
| $EtCH(CO_2H)ZnBr$ | $PhCH_2CN$ | (i) thf (ii) $H_2O$ | $EtCH(CO_2H)COCH_2Ph$ (77) | 392 |
| $[Me_3Si(CH_2)_3]_3Al$ | EtCN (0.5 equiv.) | (i) PhH, 5 h, 80 °C (ii) $H_2O$ | $Me_3Si(CH_2)_3COEt$ (68) | 385 |
| $Me_3Al$ | PhCN | (i) PhH, $[Ni(acac)_2]$ 20 °C, 70 h (ii) dil. HCl | PhCOMe (80) | 386 |
| (a) BuMgBr (b) BuLi | 2-cyanotetra-hydrofuran | (i) reaction (ii) $H_2O$ | 2-($H_2NCBu_2$)-tetrahydrofuran (72) | 408 |
| $EtOCH_2CN$ | $CH_2{=}CHCH_2MgBr$ | (i) $Et_2O$ (ii) $H_2O$ | $EtOCH_2C(CH_2CH{=}CH_2)_2NH_2$ (79) | 409 |

purity of magnesium is important, with single-crystal magnesium providing 100% addition. In contrast, triply sublimed magnesium (99.95%) pure) in excess provides up to 13% of by-product. The reaction involves both $Me_2Mg$ and MeMgBr, see Scheme 49, with the relative reactivity ($k_2 : k_1$) being 17.7:1.3 at 25 °C. Insertion into the second methyl group in dimethylmagnesium occurs much less readily than insertion into the first, i.e. $k_3 < k_2$. Earlier work[405] in thf had shown that only *ca.* 50% of the ethyl groups in diethylmagnesium reacted with benzonitrile. This may be due to $k_3$ in thf being exceptionally small.

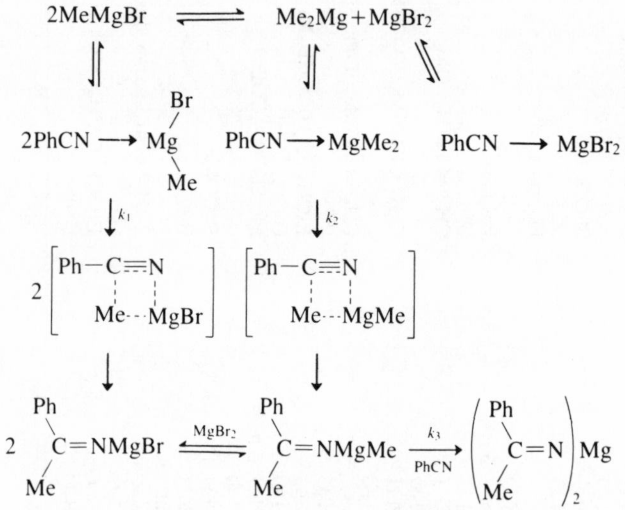

$$2MeMgBr \rightleftharpoons Me_2Mg + MgBr_2$$

$$2PhCN \rightarrow Mg(Br)(Me) \qquad PhCN \rightarrow MgMe_2 \qquad PhCN \rightarrow MgBr_2$$

$$\downarrow k_1 \qquad\qquad \downarrow k_2$$

$$2\left[\begin{array}{c}Ph-C\equiv N\\ \vdots \quad \vdots\\ Me\cdots MgBr\end{array}\right] \qquad \left[\begin{array}{c}Ph-C\equiv N\\ \vdots \quad \vdots\\ Me\cdots MgMe\end{array}\right]$$

$$2\ Ph(Me)C{=}NMgBr \underset{}{\overset{MgBr_2}{\rightleftharpoons}} Ph(Me)C{=}NMgMe \underset{PhCN}{\overset{k_3}{\longrightarrow}} \left(Ph(Me)C{=}N\right)_2Mg$$

SCHEME 49

These cyanide reactions are counted among the more general routes to ketones; especially good yields of ketones are obtained from aryl cyanides. Particularly good yields of ketones have been produced in toluene containing 1 mol of ether rather than in ether alone[383]. However, the reactions do suffer, in varying degrees, from side reactions, such as the following:

(i) $\alpha$-metallations when there are $\alpha$-hydrogens present, e.g. as in $CH_3CN$, equation 199;

$$CH_3CN + R'M \rightleftharpoons R'H + MCH_2CN \text{ (or } CH_2{=}C{=}NM) \qquad (199)$$

(ii) deprotonation of products, especially in the presence of excess of organometallic;
(iii) reductions of RCN to RCH=NM, especially noted with organoaluminium reagents; and
(iv) further condensations or insertions of RCN into the N—M bonds of adducts.

Much less $\alpha$-metallation of cyanides has been found for organolithium than for Grignard reactions. The presence of lithium perchlorate has been shown to reduce the amount of $\alpha$-metallation with Grignard reagents[382]. Even with nitriles having reactive $\alpha$-hydrogens, such as $CH_3CN$ or $PhCH_2CN$, yields of ketones were increased from 2 to 20%.

Use of organometallics having $\beta$-hydrogens can lead to reductions. As well as organoaluminium reagents, organoberylliums are noted in giving reductions. Six-centred transition states have been proposed, e.g. equation 200.

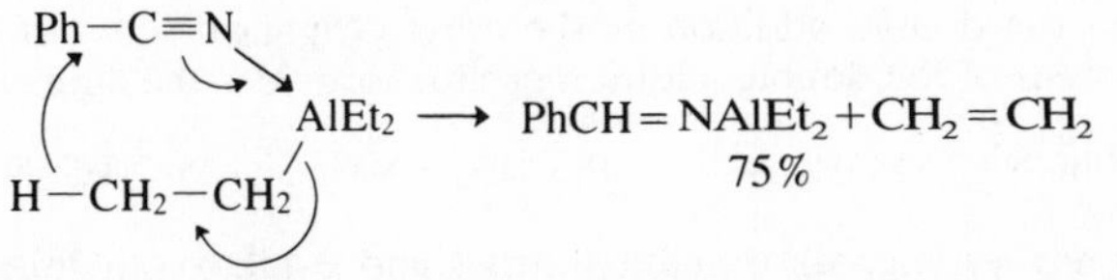

$$Ph-C\equiv N \cdots AlEt_2 (H-CH_2-CH_2) \longrightarrow PhCH=NAlEt_2+CH_2=CH_2 \quad 75\% \qquad (200)$$

For organoberylliums, higher temperatures have been shown to lead to more reduction[397].

Further insertions are particularly important when excess of nitrile is present[388,406]; products include enamidines, triazine derivatives, and other condensation products. These reactions involve successive insertions of the nitrile into initially metal—carbon bonds and subsequently into nitrogen—metal bonds, e.g. equation 201.

$$PhCN+Me_2Zn \longrightarrow MePhC{=}NZnMe \longrightarrow MePhC{=}NCPh{=}NZnMe \longrightarrow \text{chain polymers and/or cyclic oligomers} \qquad (201)$$

Reactions of triallylborane with nitriles can lead to transfer of one or two allyl groups from boron to carbon[397], e.g. Scheme 50. Hydrolysis of the product from transfer of the two allyl groups provides a primary amine.

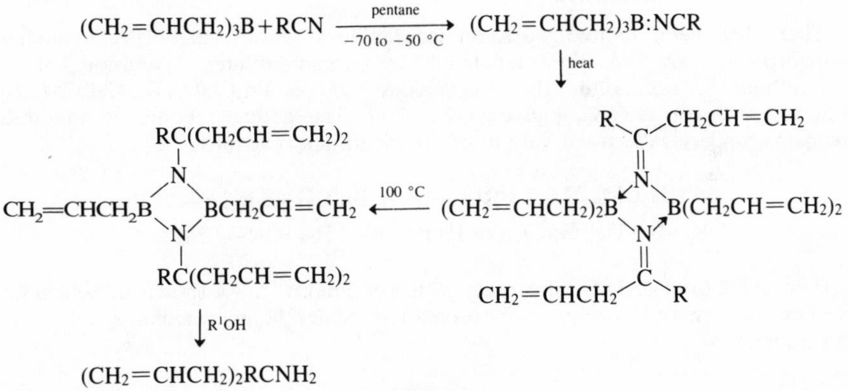

$$(CH_2{=}CHCH_2)_3B+RCN \xrightarrow[-70\text{ to }-50\ ^\circ C]{\text{pentane}} (CH_2{=}CHCH_2)_3B{:}NCR$$

$$\xrightarrow{\text{heat}} (CH_2{=}CHCH_2)_2B\langle N(=C(R)CH_2CH{=}CH_2)\rangle_2 B(CH_2CH{=}CH_2)_2$$

$$\xrightarrow{100\ ^\circ C} CH_2{=}CHCH_2B\langle N(RC(CH_2CH{=}CH_2)_2)\rangle_2 BCH_2CH{=}CH_2 \xrightarrow{R^1OH} (CH_2{=}CHCH_2)_2RCNH_2$$

SCHEME 50

The formation of primary amines would clearly be a valuable reaction. Unfortunately, double additions to nitriles by organo-magnesiums or -lithiums are not frequently met. Clearly the metallated imines, RR′C=NM (where M is magnesium or lithium), must be deactivated towards RM compared with ketimines, which as shown in Section V.C are reactive towards a variety of organometallics. It was pointed out in Section IV.D.2 that a metallated hydroxyl group can reduce the reactivity of an alkenol. However, use of refluxing toluene as the reaction medium

$$RC{\equiv}N+R'M \longrightarrow RRC'{=}NM \xrightarrow{R'M} RR'_2CNM_2 \qquad (202)$$

has allowed the double addition of the ethyl Grignard[407] to benzonitrile, equation 203. Hydrolysis of the double addition product provides the amine, $PhCEt_2NH_2$. Use

$$PhCN + EtMgBr \xrightarrow{PhMe,\Delta} PhC(Et){=}NMgBr + PhC(Et)_2N(MgBr)_2 \qquad (203)$$

of reactive nitriles, e.g. alkoxymethylnitriles and $\beta$-alkoxyethylnitriles[409,410], and the reactive allyl organometallics also leads to double addition, e.g. equation 204.

$$EtOCH_2CN + CH_2{=}CHCH_2MgBr \xrightarrow{Et_2O,\ reflux} EtOCH_2C(CH_2CH{=}CH_2)_2N(MgBr)_2$$

$$\downarrow H_2O \qquad (204)^{409}$$

$$EtOCH_2C(CH_2C{=}CH_2)_2NH_2$$
$$(77\%)$$

An additional means of achieving double addition was devised by Gauthier *et al.*[408]. This involves the successive use of two organometallics; the first may be any organomagnesium and the second an aliphatic, an unhindered, or an allylic lithium reagent, equation 205.

$$MeOCH_2CN \xrightarrow{RMgX} MeOCH_2CR{=}NMgX \xrightarrow{R'LI} MeOCH_2CRR'NLi(MgX) \qquad (205)$$

## B. Additions to isonitriles

There has been limited work on isocyanide reactions. Isonitriles, containing $\alpha$-hydrogens, are readily metallated by organolithiums. Addition of organolithiums[411] occurs to other isocyanides such as $Ph_3CNC$, $Bu^tCH_2CMe_2NC$, PhCMeEtNC, or $Bu^tNC$, e.g. equation 206. The adducts, being organolithium reagents, undergo the usual variety of organolithium reactions.

$$Bu^tCH_2CMe_2N{=}C + RLi \longrightarrow Bu^tCH_2CMe_2N{=}C(Li)R$$
$$R = Bu,\ Bu^s,\ Bu^t,\ Et,\ \text{or}\ Ph\ \text{but not}\ CH_2{=}CH\ \text{or}\ PhC{\equiv}C \qquad (206)$$

Triorganoboranes, $R_3B$, also react with isocyanides[412]. As shown in Scheme 51, several transfers of R groups from boron are possible; the intermediate products may be trapped.

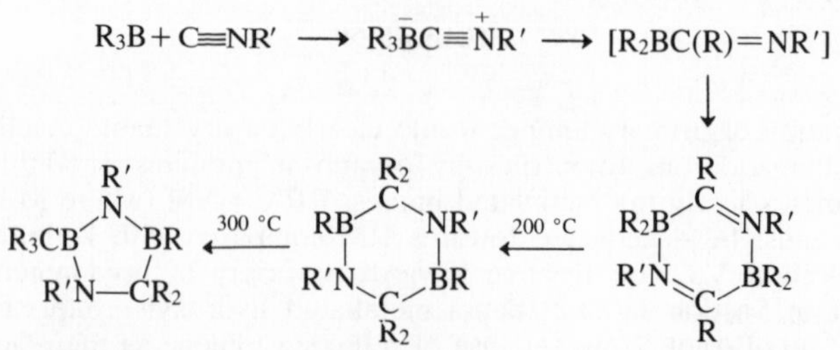

$$R_3B + C{\equiv}NR' \longrightarrow R_3\overset{-}{B}C{\equiv}\overset{+}{N}R' \longrightarrow [R_2BC(R){=}NR']$$

$$\downarrow$$

$$\text{[5-membered ring: } R_3CB,\ N(R'),\ BR,\ CR_2,\ R'N] \xleftarrow{300\,^{\circ}C} \text{[6-membered ring: } RB,\ C(R_2),\ NR',\ BR,\ C(R_2),\ R'N] \xleftarrow{200\,^{\circ}C} \text{[6-membered ring: } R_2B,\ C(R),\ NR',\ BR_2,\ C(R),\ R'N]$$

SCHEME 51

## C. Additions to Azomethines

Organometallics may add to azomethines (aldimines, RCH=NR′ or ketimines, $RR'C{=}NR^2$), equation 207. Organic derivatives of the following main group elements

$$RR'C{=}NR^2 + R^3M \longrightarrow RR'R^3CNR^2M \xrightarrow{H_2O} RR'R^3CNR^2H \quad (207)$$

react with azomethines: lithium[4], beryllium[403a], magnesium[4,413], alkaline earth metals[414], zinc[413,415–417], cadmium[413b] (especially in the presence of metal halides), boron[399] (e.g. for triallylboron), and aluminium[410,418].

Allyl derivatives are, as usual, especially reactive derivatives[324b]; they react with allylic rearrangement, equation 208. A further feature of these allyl metal reactions

$$PhCH{=}NR + MeCH_2{=}CHCH_2M \longrightarrow PhCH(NHR)CHMeCH{=}CH_2 + PhCH(NHR)CH_2CH{=}CHMe \quad (208)$$

$$M{=}MgBr \text{ or } Li$$

is the reaction's reversibility with the kinetic and thermodynamic products being rearranged and unrearranged derivatives, respectively. Hence longer reaction times and higher temperatures favour the unrearranged allylic derivatives. Prior coordination complexes have been detected in $Me_2Be$ reactions[403a], e.g. equation 209. A four-centred transition state is envisaged for the insertion reaction.

$$Me_2Be + PhCH{=}NPh \xrightarrow{-78\,°C} Me_2Be{:}N(Ph){=}CHPh \xrightarrow{\text{room temp.}} MeBeNPh{=}CHMePh \quad (209)$$

Methyllithium is more reactive than MeMgX towards aldimines, e.g. MeLi but not MeMgX reacts[419] with the hindered aldimine, $PhCH{=}NBu^t$. Alkylmagnesium halides are, however, sufficiently reactive towards less hindered Schiff bases, e.g. PhCH=NR (R = Me or Et), equations 210 and 211.

$$PhCH{=}NEt + PrMgBr \xrightarrow[\text{(ii) dil. HCl}]{\text{(i) } Et_2O,\ \Delta} PhCHPrNHEt \quad (210)^{420}$$

$$PhCH{=}NMe + PhCH_2MgCl \xrightarrow[\text{(ii) } H_2O,\ HCl]{\text{(i) } Et_2O,\ 35\,°C,\ 2\,h} PhCH(CH_2Ph)NHMe{\cdot}HCl \quad 91\text{–}96\% \quad (211)^{421}$$

Salt effects on Grignard reactions have been studied by Thomas[422]. Magnesium bromide activates addition of ethyl-Grignards to PhCH=NPh whereas lithium salts have no (or just a slight deactivating) effect. Thomas[413] also determined $\rho$ values for reaction of $XC_6H_4CH{=}NPh$ (+0.28 and +0.84 in $Et_2O$ and thf, respectively) and of $PhCH{=}NC_6H_4X$ (+0.39 and +0.90 in $Et_2O$ and thf, respectively).

Reactions other than insertions may also occur in the azomethine–organometallic systems, as follows:

(i) Metallations, for azomethines having $\alpha$-protons[4], e.g. equation 212.

$$Me_2CHCH{=}NBu^t \xrightarrow[\text{THF},\ \Delta]{EtMgBr} Me_2C{\overset{\text{---}}{=}}\bar{C}H{\overset{\text{---}}{=}}NBu^tMgX^+ \quad (212)^{423}$$

(ii) Reductions for organometallics having β-hydrogens[4,403a,414], e.g. $Bu^t_2Be$ reduces PhCH=NMe to dimeric $(PhCH_2NMeBeBu^t)_2$ and PhCH=CHCH=NPh to $(Bu^tBeNPhCH_2CH{=}CHPh)_2$. However, less reductions are found for dialkylberyllium reactions with aldimines than with ketones.
(iii) Reactions leading to coupled products, e.g. equation 213.

$$PhCH{=}NEt \xrightarrow{Bu^tMgCl} EtNHCH(Ph)CH(Ph)NHEt \qquad (213)^{424}$$

Additions of organometallics (e.g. organo-lithium, -magnesium, and -zinc derivatives) to azirines (three-membered cyclic compounds containing a nitrogen—carbon double bond) have also been reported[425].

In several reactions a preference for *threo*-products has been shown, e.g. in reactions of organolithiums with PhCH=NCHR′Ph[426a], of R′CH=C=CHMgBr with PhCH=NR, equation 214[426b], of RCH(ZnBr)CONR$'_2$ with R″CH=NR‴[417], and of allylmagnesium or -lithium (but not -aluminium) with PhCH=NR[324b], especially with bulky R groups.

$$\underset{\underset{[MeCHBrC{\equiv}CH+Mg]}{\uparrow}}{MeCH{=}C{=}CHMgBr} \xrightarrow[\text{(ii) } H_2O]{\text{(i) PhCH=NR}} \underset{threo}{Ph-\overset{H}{\underset{RNH}{C}}-\overset{C\equiv CH}{\underset{H}{C}}-Me} \quad \underset{erythro}{Ph-\overset{H}{\underset{RNH}{C}}-\overset{H}{\underset{C\equiv CH}{C}}-Me} \qquad (214)$$

| | threo | | erythro |
|---|---|---|---|
| R=Me | 80 | : | 20 |
| R=Et | 90 | : | 10 |

However, some stereochemical differences have been found in reactions of *N*-benzyl-D-glyceraldimine (and its 2,3-*O*-isopropylidine derivatives) with phenyl—metal derivatives; with phenyl-lithium or -calcium derivatives, the *threo*-product, **63**, dominates while for phenyl-beryllium, -magnesium, or -zinc species the major isomer is the *erythro*-product, **64**[427].

(63): Fischer projection — Ph at top; PhCH₂NH on left of C-2; OH on right of C-3; $CH_2$OH (—OH) at bottom.

(64): Fischer projection — Ph at top; NHCH₂Ph on right of C-2; OH on right of C-3; $CH_2$OH (—OH) at bottom.

Various organometallics add 1,2- to $Ph_2C{=}NPh$ in ether; these include RLi, RK, RNa, RCaI, RBaX, RSrX, and allyl-MgBr. In contrast, MeMgI, PrMgBr, and PhMgBr do not react even in refluxing ether. However, in toluene–ether at 90–105 °C, PhMgBr does add but in a 1,4-sense, equation 215[428].

$$Ph_2C{=}NPh + PhMgBr \xrightarrow[\text{(ii) } H_2O]{\text{(i) PhMe, } Et_2O} PhCH(NHPh)C_6H_4Ph\text{-}o \qquad (215)$$

1,4-Additions of allyl—metal (metal = zinc, magnesium, or lithium) can occur with PhCH=CHCH=$NPr^i$. Whether 1,2- or 1,4-addition by BuLi occurs to PhCH=CHCH=NR depends on the bulk of R; for R = Me only 1,2-addition results, whereas for R = $Bu^t$ or $Pr^i$ some 1,4-addition is also obtained[429].

Initial reactions of diazabutadienes, RN=CHCH=NR, with organometallics can also lead to 1,2-addition, e.g. with $Me_3Al$[418], equation 216, or to 1,4-addition, e.g. with $Et_2Zn$[430], equation 217.

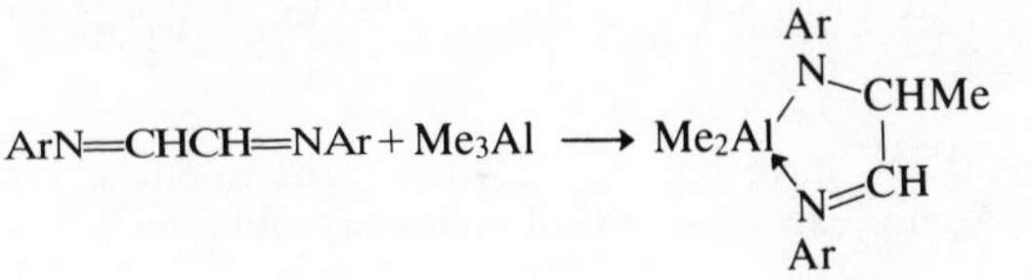

$$ArN{=}CHCH{=}NAr + Me_3Al \longrightarrow Me_2Al(N(Ar)CHMe{-}CH{=}NAr) \qquad (216)$$

$$Bu^tN{=}CHCH{=}NBu^t + Et_2Zn \longrightarrow EtZn(N(Bu^t)CH{=}CHN(Bu^t)Et) \qquad (217)$$

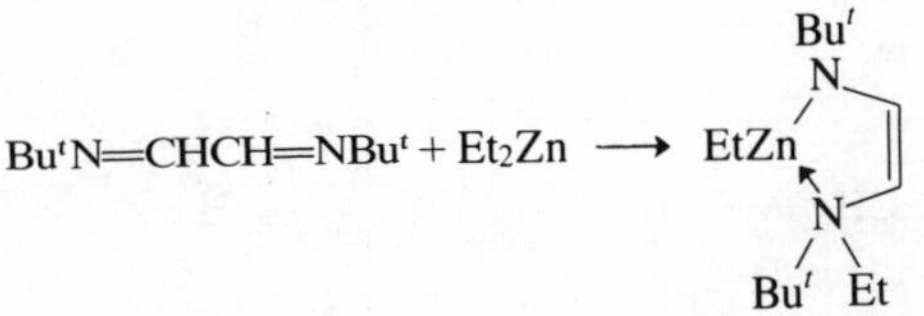

A number of 1,4-additions of organometallics to chiral $\alpha,\beta$-unsaturated imines have been reported[431], e.g. Scheme 52. Complete 1,4-addition occurs with Grignard reagents for $R' = Bu^t$; some 1,2- also occurs for $R' = Pr^i$. In contrast, BuLi adds completely 1,2- to **65** to give chiral unsaturated amines, e.g. equation 218.

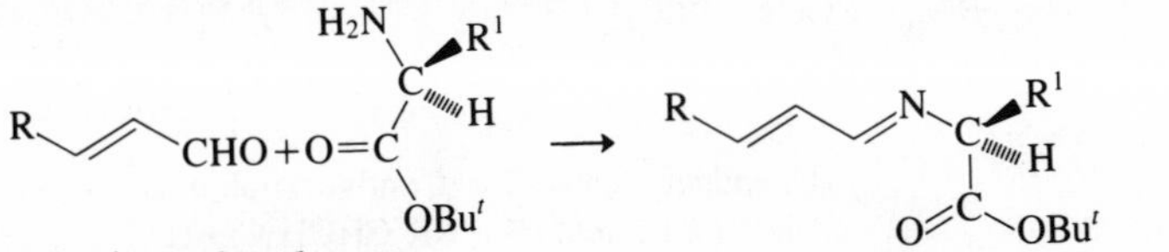

R–CH=CH–CHO + O=C(OBu$^t$)–C(H)(R$^1$)–$H_2N$ ⟶ R–CH=CH–CH=N–C(H)(R$^1$)–C(=O)OBu$^t$

(L)-*tert*-butyl ester of *tert*-leucine

(**65**)

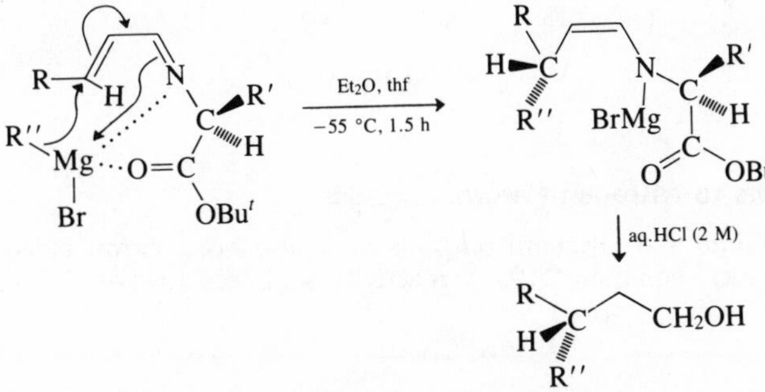

R″–Mg(Br)···O=C(OBu$^t$)–C(H)(R′)–N=CH–CH=CH(H)–R ⟶ (Et$_2$O, thf, −55 °C, 1.5 h) R(H)(R″)C–CH=CH–N(BrMg)–C(H)(R′)–C(=O)OBu$^t$

aq.HCl (2 M)

R–C(H)(R″)–$CH_2OH$

| R | R′ | R″ | Chemical Yield (%) | Optical Yield (%) | Configuration |
|---|---|---|---|---|---|
| Me | Bu$^t$ | Ph | 52 | 91 | (*R*) |
| Ph | Bu$^t$ | Et | 56 | 95 | (*S*) |
| Me | Bu$^t$ | Bu | 40 | 98 | (*S*) |
| Me | Pr$^i$ | Ph | 41 | 63 | (*R*) |

SCHEME 52

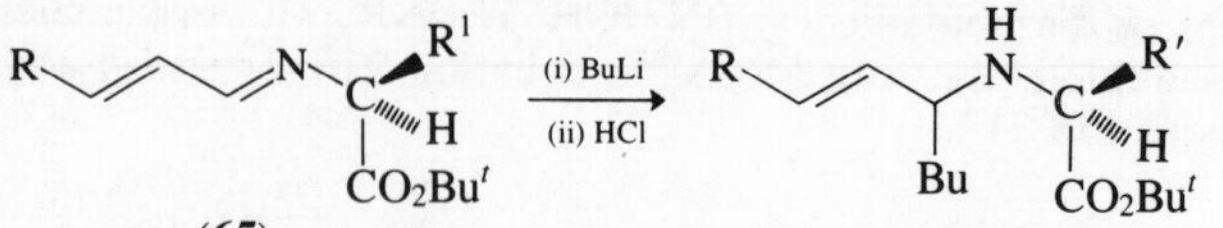

$$R\text{-}CH{=}CH\text{-}CH{=}N\text{-}C(R^1)(H)(CO_2Bu^t)\ \textbf{(65)} \xrightarrow[\text{(ii) HCl}]{\text{(i) BuLi}} R\text{-}CH{=}CH\text{-}CH(Bu)\text{-}NH\text{-}C(R')(H)(CO_2Bu^t) \qquad (218)$$

Meyers and Whitten[432,433] have reported 1,4-additions to chiral 1-alkenyl-oxazolines, Scheme 53[432]. Again, excellent optical yields are obtained.

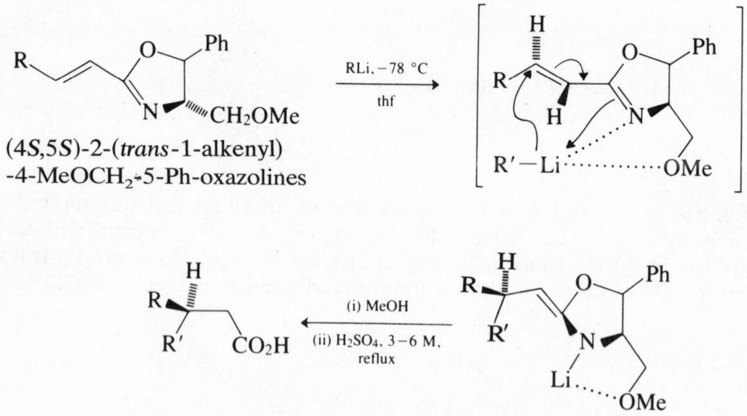

(4*S*,5*S*)-2-(*trans*-1-alkenyl)-4-$MeOCH_2$-5-Ph-oxazolines $\xrightarrow[\text{thf}]{\text{RLi, }-78\ ^\circ\text{C}}$ [R′—Li chelated intermediate with OMe] $\longrightarrow$ adduct $\xrightarrow[\text{(ii) } H_2SO_4,\ 3-6\ M,\ \text{reflux}]{\text{(i) MeOH}}$ $RR'CHCH_2CO_2H$

| R | R′ | Chemical Yield (%) | Optical Yield (%) | Configuration of $RR'CHCH_2CO_2H$ |
|---|---|---|---|---|
| Me | Et | 30 | 92 | (*R*) |
| Me | Ph | 34 | 98 | (*S*) |
| Et | Ph | 31 | 92 | (*S*) |
| Ph | Et | 66 | 97 | (*R*) |

SCHEME 53

## D. Additions to Nitrogen Heteroaromatics

Organolithium and Grignard reagents have also been shown to add to nitrogen heteroaromatics, equation 219. A reactivity sequence towards RMgX was estab-

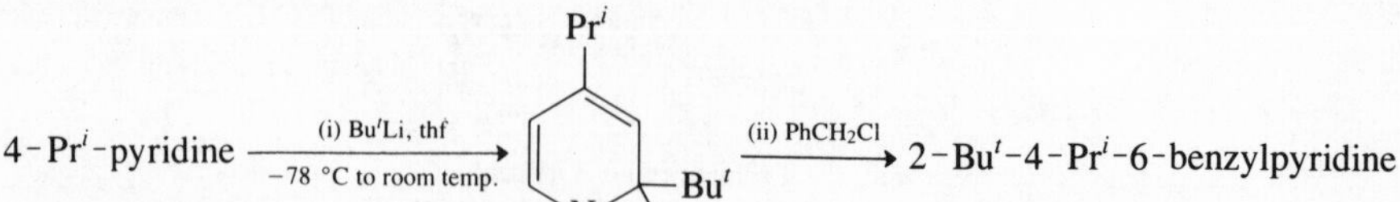

$$4\text{-}Pr^i\text{-pyridine} \xrightarrow[-78\ ^\circ\text{C to room temp.}]{\text{(i) }Bu^tLi\text{, thf}} \text{4-}Pr^i\text{-2-}Bu^t\text{-2H-1-lithio-1,2-dihydropyridine} \xrightarrow{\text{(ii) }PhCH_2Cl} 2\text{-}Bu^t\text{-}4\text{-}Pr^i\text{-}6\text{-benzylpyridine} \qquad (219)$$

lished as acridine ≈ phenanthridine > isoquinoline ≈ quinoline > pyridine[435], while 2,2′-bipyridine was found to be more reactive than pyridine towards RLi[436]. The adducts may be (i) aromatized either by loss of LiH or by oxidation of the dihydroaromatic, formed by hydrolyses, or (ii) derivatized using allylating agents

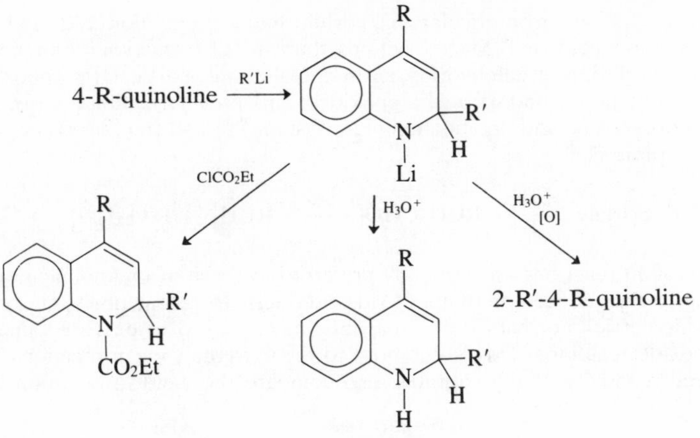

4-R-quinoline $\xrightarrow{R'Li}$ [1-Li-1,2-dihydro-2-R'-4-R-quinoline]

ClCO₂Et → [1-CO₂Et-1,2-dihydro-2-R'-4-R-quinoline]

$H_3O^+$ → [1,2-dihydro-2-R'-4-R-quinoline, N–H]

$H_3O^+$, [O] → 2-R'-4-R-quinoline

SCHEME 54

(Scheme 54); see also step (ii) in equation 219. These reactions provide good syntheses of alkyl-heteroaromatics.

Organolithiums add solely 1,2- i.e. to the carbon—nitrogen double bond aza-aromatics such as pyridine or quinoline[437,438]. However, for Grignard reagents, the situation appears not to be so clear cut, with for example both 1,2- and 1,4-addition to pyridines being reported.

Bryce-Smith *et al.*[438] have established that the presence of the free metal, as would occur in *in situ* reactions involving pyridine, magnesium, and the alkyl halide, leads to the formation of 4-substituted products, possibly via free-radical routes. With no free magnesium present, reactions of aza-aromatics and Grignard reagents produce only 2-alkylated species.

## VI. MISCELLANEOUS INSERTIONS

### A. Reactions with Epoxides and Related Compounds[4,439]

#### 1. *Epoxides*

Ring openings of epoxides can occur on reaction with organometallic compounds, e.g. organo-lithium[123,411,440,441], -beryllium[442], -magnesium[443], and -aluminium[444–447] reagents. The products, after hydrolysis, are $\beta$-substituted ethanols. In additions to

$$\text{oxirane} + RM \longrightarrow RCH_2CH_2OM \xrightarrow{H_2O} RCH_2CH_2OH \qquad (220)$$

unsymmetric epoxides, the organic group of an organolithium becomes preferentially attached to the least electron-rich carbon atom of the ring. For Grignard reactions, the situation is not so clear cut; $S_N2$ type reactions occur with alkyl substituted epoxides while $S_N1$ openings occur with aryl-substituted epoxides, e.g. equation 221.

$$\text{2-phenyloxirane} \xrightarrow[H_2O]{\text{(i) } Me_2Mg} PhCHMeCH_2OH \qquad (221)$$

Several by-products may be formed. These include products arising from rearrangements of the epoxides to aldehydes. These isomerizations are catalysed by electrophilic metal halides, in particular magnesium halides, equation 222, and are very much more prevalent in R'MgX reactions than in R'Li reactions. Other side reactions are (i) formation of halohydrins; (ii) $\alpha$-metallations, particularly important with basic organometallics and/or acidic epoxides; (iii) polymerizations, especially significant with organozinc and organoaluminium systems[448]; and (iv) reductions, e.g. using triethylaluminium.

$$\text{2-R-oxirane} \xrightarrow{MgX_2} RCH_2CHO[\xrightarrow{R'M} RCH_2CHR'OM] \qquad (222)$$

The organolithium reactions are generally preferred to those of organomagnesium reagents owing to the reduced amounts of side products. In recent times, cuprates, either as stoichiometric[449] or catalytic[450] reagents, have proved to be very valuable reagents in epoxide reactions. This is due both to the extremely clean reactions and to the enhanced reactivity of the cuprates, e.g. compare the yield in equation 223

$$\text{1,2-epoxycyclohexane} \xrightarrow[\text{thf, 0 °C, 2 h; (ii) } H_2O]{\text{(i) PhMgCl (1.5 equiv.), CuI (0.15 equiv.)}} \text{trans-2-phenylcyclohexanol (82\%)} \qquad (223)$$

with the 3% yield in the absence of the copper salt. However, no catalyst is required for benzyl- or allyl-Grignard reactions. *Trans*-addition usually occurs, as shown in equation 223. Allylic rearrangement is found in reactions of allylic metal derivatives[324b,451], see entries in Table 12.

The findings with organoaluminium–epoxide reactions match those from other reactions of organoaluminium, i.e. (i) alkenyl—aluminium bonds are more reactive than alkyl—aluminium bonds, (ii) 'ate' complexes, e.g. **66**, are more reactive than the corresponding triorganoaluminium, e.g. (*E*)-EtCH=CHAl$(Bu^i)_2$, and (iii) retention of configuration occurs[444,445], as shown in equation 224. The anti-Markownikov

$$(E)\text{-EtCH=CHAl}(Bu^i)_2BuLi\ \textbf{(66)} \xrightarrow[\text{(ii) } H_2O]{\text{(i) 2-R-oxirane}} (E)\text{-EtCH=CHCH}_2\text{CHROH} \qquad (224)^{445}$$

adduct is formed in equation 224. However, in the reaction of triethylaluminium with propene epoxide, the Markownikov product, $MeCHEtCH_2OH$, is obtained[447]. Reactions with substituted epoxides are more sluggish. As shown in equation 225, reactions of borates are significantly different[452].

$$(\mathbf{E})\text{-}R^1CH{=}CHBR_3^-, Li^+ \xrightarrow[\text{Et}_2\text{O, 10 h, room temp.}]{\text{2-}R^2\text{-oxirane}} \text{R}_2\text{B-oxaborinane (R, } R^1, R^2 \text{ substituted)} \qquad (225)$$

With $\alpha,\beta$-unsaturated epoxides, 1,2- or 1,4-additions may occur, see equations 226 and 227. Particularly good yields of conjugate additions are obtained with cuprates. With the cyclohexadiene epoxide, 1,2-addition and arrangements arise. As

TABLE 12. Reactions of organometallic compounds with epoxides

| Organometallic | Epoxide | Product, after hydrolysis (yield, %) | Ref. |
|---|---|---|---|
| $Bu^tCH_2CMe_2N{=}CLiBu$ | oxirane | $BuCOCH_2CH(OH)Me$ (90) | 411 |
| S, S, Me, Li | 2-Ph-oxirane | S, S, Me, OH, Ph (70) | 123 |
| $EtCH{=}\bar{C}H{=}NMe,Li^+$ | 2-Meoxirane | 2-MeNH-3-Et-5-Me-tetrahydrofuran (73) | 440 |
| PhLi | OLi, O | 2-phenylcyclohexa-2--enone (63) | 441 |
| BuMgBr | oxirane | $BuCH_2CH_2OH$ (60) | 443 |
| $(MeCH{=}CHCH_2)_2Zn$ | 2-Ph-oxirane | $PhCH(CH_2OH)CH(Me)CH{=}CH_2$ (75) | 324b |
| $CH_2{=}\overset{(-)}{C}H{=}CH_2$, $Na^+$ | 1,2-epoxycyclohexane | OH, $CH_2CH{=}CH_2$ (85) | 324b |
| $Bu^tCH_2{-}CH{=}\overset{(-)}{C}H{=}CH_2$, $Li^+$ | 2,3-$Me_2$-oxirane | $Bu^tCH_2CH{=}CHCH_2{-}CH(Me)CH(Me)OH$ (*trans* : *cis* = 80 : 11) + $H_2C{=}CHCH(CH_2Bu^t)CH(Me)CH(Me)OH$ | 451 |
| (**Z**)-$Et_2AlCH{=}CHEt$ | oxirane | (**Z**)-$HOCH_2CH_2CH{=}CHEt$ (95) | 446 |
| $Me_2AlC{\equiv}CR$ | $(CH_2)_2OH$, HO, O | HO, $CH_2CH_2OH$, $C{\equiv}CR$, HO (60) | 447 |
| $CH_2{=}CHCH_2Li$ | 3,4-epoxycyclooctene | $CH_2$, OH (>99) | 456 |
| $Me_2C{=}C{=}CHLi$ | 2-allyloxirane | $Me_2C{=}C{=}CHCH(OH){-}CH{=}CH_2$ (95) | 453 |
| $Bu^tLi$, tmed | 2-allyl-2-methyloxirane | $Bu^tCH_2CH{=}C(Me)CH_2OH$ $(Z):(E) = 96:4$ | |

$$\text{2-allyloxirane} \xrightarrow[\text{(ii) } H_2O]{\text{(i) BuM, } Et_2O\ -10\ ^\circ C} BuCH_2CH{=}CHCH_2OH + CH_2{=}CHCH_2CH(Bu)OH + CH_2{=}CHCH(Bu)CH_2OH \quad (226)^{454}$$

M=MgBr: 56% (*trans* : *cis*=78 : 22); 19%; 3%

M=CuBuLi: 93% (*trans* : *cis*=86 : 14)

(227)[455]

M=MgMe 95%
M=MgBr 70% 19%
M=Li 37% 63%

shown in both examples, reactions involving magnesium halides produce products derived from the aldehyde isomers of the epoxides.

## 2. Related compounds

The sulphur analogues, episulphides, in the main do not give inserted products with organometallics, RM, but provide instead alkenes and thiolates[457]. The nitrogen analogues, aziridines, are generally unreactive towards ring openings. However, an *N*-tosylatoaziridine was shown to react[458] with MeMgBr, equation 228.

$$\text{2-phenyl-}N\text{-Tos-aziridine} \xrightarrow[\text{(ii) } H_2O]{\text{(i) MeMgBr (4 equiv.)}} PhCHMeCH_2NH\text{-}Tos \quad 46\% \quad (228)$$

Other heterocycles which react with organometallics include oxetanes[450,459], equation 229, thietones[460], thiacyclohexanes[461], e.g. equation 230, 1,3-benzoxathioles, and 1,3-benzodioxoles[462], e.g. equation 231.

$$\text{oxetane} \xrightarrow[M=MgX \text{ or } Li]{RM} R(CH_2)_3OM \quad (229)$$

$$\text{2,2-diphenylthiane} \xrightarrow[-10\ ^\circ C,\ thf]{Bu^tLi,\ tmed} Bu^t(CH_2)_4CPh_2Li \quad (230)$$

$$\text{1,3-benzoxa(Y)ole (CH}_2\text{)} \xrightarrow[\text{(ii) } H_2O]{\text{(i) } MeCH{=}CHCH_2MgBr} CH_2{=}CHCH(Me)CH_2YC_6H_4OH\text{-}o \quad (231)$$

Y=O or S

### B. Reactions of Organometallic Compounds with Oxygen[463]

Reactions of RM with oxygen under controlled conditions may lead to oxygen-containing products, e.g. peroxides, ROOM, and alkoxides (or aryloxides) ROM (see Table 13) and/or products derived from the free radical $R^{\bullet}$. A wide range of reactivity towards oxygen is exhibited, e.g. from those compounds which ignite in air at ambient temperature (e.g. lower alkyls of Groups I, II, and III) to those which show a considerable resistance to any oxidation under normal conditions (e.g. derivatives of silicon, germanium, and tin). Derivatives of lead and mercury are also classed as being difficult to oxidize; however, secondary and tertiary alkyl derivatives of these elements are more prone to oxidation, albeit slowly, than often thought or described.

Among the factors which influence the reactivity towards oxygen include the relative strengths of the metal—carbon and metal—oxygen bonds, the Lewis acidity (or complexing ability) of the organometallic and the electronegativity of the metal. The oxygen-containing products are obtained by oxygen insertion into the metal—carbon bond. These dominate for simple alkyl and vinyl derivatives. The peroxide products stem from the initial oxygen insertion, equation 232, whereas the alkoxide (or aryl oxide) result from a subsequent redox reaction of the metal alkyl peroxide, equation 233. For polyorganometallic compounds, $R_mM$ ($m = 2$–4), the reactions

$$RM + O_2 \longrightarrow ROOM \qquad (232)$$

$$ROOM + RM \longrightarrow 2ROM \qquad (233)$$

with oxygen can in principle lead to all possible peroxy-alkoxy derivatives, e.g. from $R_2M$, compounds RMOOR, $M(OR)_2$, $M(OOR)_2$ could be formed. In addition, the redox reactions of the peroxide can be either intramolecular or intermolecular processes, equations 234 and 235, respectively.

$$RMOOR \longrightarrow M(OR)_2 \qquad (234)$$

$$RMOOR + R_2M \longrightarrow 2RMOR \qquad (235)$$

Successive oxygen insertions into carbon—metal bonds of $R_mM$ occur at decreasing rates, as shown with $R_2Cd$[464] and $R_3Al$[465]. Different reactivities are shown by secondary and primary alkyl derivatives, e.g. tricyclohexylborane, a typical secondary alkylborane, reacts with 2 mol of oxygen at −78 °C to give $(\text{cyclohexyl-OO})_2B$-cyclohexane while a primary alkylborane will only take up one molecule of oxygen at −78 °C, the second insertion requires a higher temperature (0 °C). The presence of $Et_2O$, however, renders these boranes more reactive[466], equation 236. Donors do not

$$\text{cyclo-}C_6H_{11}BCl_2 \xrightarrow[\text{(ii) } H_2O]{\text{(i) } O_2,\ Et_2O,\ -18\,^\circ C} \underset{93\%}{\text{cyclo-}C_6H_{11}OOH} \qquad (236)$$

always lead to increased reactivity, e.g. complexation of trialkylboranes by nitrogen donors results in greater stability towards oxygen[463]. The relative amounts of peroxides and alkoxides among the reaction products depend on factors such as the particular metal and organic group, the molar ratios of RM and $O_2$, the temperature, and the mode of addition (i.e. oxygen added to a solution of the organometallic or *vice versa*).

Peroxides, for example, have been obtained in good yield at room temperature from organo-zincs[467,468], -cadmiums[468], and -boranes[466], but only at low temperatures, e.g. *ca.* −78 °C, for organo-magnesiums[469,470] and -lithiums[468]. At ambient

TABLE 13. Controlled reaction of organometallics, RM, with oxygen

| RM | Oxidation conditions | Product (yield, %) | Ref. |
|---|---|---|---|
| BuLi | (i) $Et_2O$, room temp. $O_2$<br>(ii) $H_2O$ | BuOH (75) | 473 |
| BuLi | (i) $Et_2O$, $O_2$, −75 °C, 45 min<br>(ii) $H_2O$ | BuOOH (31) | 468 |
| $Me_2Be$ | $O_2$, $Et_2O$, −78 to +25 °C | $Be(OMe)_2$ + peroxide (<7) | 475 |
| $PhCH_2MgCl$ | (i) Inverse addition of RM to $O_2$-saturated $Et_2O$ solution at −78 °C<br>(ii) $H_2O$ | $PhCH_2OOH$ (30) | 469 |
| $Bu^tMgCl$ | (i) Inverse addition of RM to $O_2$-saturated $Et_2O$ solution at −78 °C<br>(ii) $H_2O$ | $Bu^tOOH$ (86) | 469 |
| Me Me Me H MgCl | (i) Inverse addition, $Et_2O$, −75 °C<br>(ii) $H_2O$ | Me Me Me OOH (90 overall)<br>(*l*)-bornyl-OOH: (*d*)-isobornyl-OOH = 56 : 44 | 469 |
| $Bu^iMgBr$ | (i) $O_2$, room temp.<br>(ii) $H_2O$ | $Bu^iOH$ (74) | 471 |
| $n$-$C_8H_{17}MgBr$ | (i) $O_2$ bubbled through $Et_2O$ solution at 0 °C<br>(ii) $H_2O$ | $n$-$C_8H_{17}OH$ (85) | 472 |
| $Et_2Zn$ | $O_2$ bubbled through $Et_2O$ solution, 5 h | $Zn(OOEt)_2$ | 467 |
| $Et_2Zn$ | (i) Aerial oxidation of anisole solution, room temp. 6 weeks<br>(ii) dil. $H_2SO_4$ | EtOH (87)<br>EtOOH (10) | 467 |
| $Bu_2Zn$ | (i) Added to $O_2$-saturated $Et_2O$ solution at 0 °C<br>(ii) $H_2O$ | BuOOH (88) | 468 |
| $BrZnCH_2CH{=}CH_2$ | (i) $O_2$ at −75 °C<br>(ii) dil. HCl | $CH_2{=}CHCH_2OOH$ (50) | 476 |
| $Bu_2Cd$ | (i) $O_2$ at 10 °C<br>(ii) $H_3O^+$ | BuOOH (95) | 468 |
| $PhCH_2CdCl$ | $O_2$ at −5 °C | $PhCH_2OOCdCl$ (86) | 468 |
| $Bu_3B$ | (i) $O_2$, thf, 5 min, 0 °C<br>(ii) NaOH | BuOH (94) | 477 |
| (cyclo-$C_6H_{11})_3B$ | (i) $O_2$, thf, 5 min, 0 °C<br>(ii) NaOH | cyclo-$C_6H_{11}OH$ (80) | 477 |
| (cyclo-$C_6H_{11})_3B$ | (i) $O_2$, thf, −78 °C<br>(ii) aq. $H_2O_2$ | cyclo-$C_6H_{11}OOH$ (82) | 478 |
| (norbornyl)$_3$B | (i) $O_2$, thf, 0 °C | norborneol (91)<br>*exo* : *endo* = 86 : 14 | 477 |
| $(PrCHMeCH_2)_3Al$ | (i) air, <−70 °C<br>(ii) dil. HCl | $PhCMeHCH_2OH$ (60) | 474 |
| $Et_3Tl$ | (i) $O_2$, $n$-octane, −70 °C | $Et_2TlOEt$ | 479 |

temperature, alkoxides are the major products from alkyl-magnesium[471,472] and -lithium[473] reactions with oxygen. Only poor yields of peroxides are obtained from organo-aluminium[468,474] and -beryllium[475] reactions, even at low temperatures. It is of interest that $Et_2TlOOEt$, unlike both $Et_2BOOEt$ and $Et_2AlOOEt$, shows little tendency to rearrange to the $EtM(OEt)_2$ compound at room temperature[479]. Inverse addition, i.e. addition of a solution of the organometallic to an oxygen-saturated solution, has been recommended especially for the formation of peroxides from organomagnesium reagents[469]. Examples are quoted in Table 13. The alkynyl—metal bond (e.g. metal = magnesium or sodium) is particularly resistant to oxidation, e.g. less than 1 mol-% of oxygen is taken up by $C_6H_{13}C{\equiv}CMgBr$ after 1 h at ambient temperature[469].

Mechanistic studies of oxidations have been widely made. No single mechanism has been universely confirmed but that shown (simplistically) in Scheme 55 appears to

*Initiation*: $RM + O_2 \xrightarrow{k_i} R^{\bullet} + {}^{\bullet}OOM$

*Chain propagation*: $R^{\bullet} + O_2 \longrightarrow ROO^{\bullet}$

$ROO^{\bullet} + RM \longrightarrow ROOM + R^{\bullet}$

*Termination*: $2ROO^{\bullet} \xrightarrow{2k_t}$ inactive products

*Other reactions*: $R^{\bullet} + SH \longrightarrow RH$

$R^{\bullet} + R^{\bullet} \longrightarrow RR$

$R^{\bullet} + R^{\bullet} \longrightarrow RH + [RH]$

$ROOM + RM \longrightarrow 2ROM$

SH = solvent or other proton source

SCHEME 55

satisfy many of the conclusions from lithium[480], magnesium, zinc[481], cadmium[464,481], boron[482] and aluminium[481] oxidations.

As radical scavengers, such as galvinoxyl, react with organolithiums, organomagnesiums and dialkylzincs, such reagents have little use as probes for free-radical chain mechanisms for these derivatives. On the other hand, galvinoxyl does not react with $R_2Cd$[483], RZnOOR, or $R_3B$[484] and has been used to confirm the free-radical chain character of oxidations of these compounds. The similarity of the product distributions in oxidations of norbornyl derivatives of magnesium and boron indicates similar, i.e. radical, mechanisms.

Russian workers have concluded that at least for $R_2Cd$ (at low temperature)[485] and also $R_2Hg$[486], there is a prior coordination of oxygen to the organometallic. The initiation step is often written as shown in Scheme 55; a detailed study of the oxidation of $Me_2Cd$, however, reveals a more elaborate step with a second-order dependence on $Me_2Cd$[464].

The reaction of simple alkyl radicals with oxygen are normally diffusion controlled[487]. However, with stabilized free radicals, $R^{\bullet}$, e.g. aryl radicals, or when the amount of oxygen is limited, reactions other than with oxygen may ensue. These reactions will give rise to non-oxygen-containing products, such as RR, RH and [RH]; in some cases such products dominate.

Products of oxidation of aryl metals, e.g. metal = Li, Mg, Zn, Ca, Tl, and Al, include phenols, arenes, biaryls, and others obtained by $Ar^{\bullet}$ attack on the solvent; little if any peroxide, ArOOM or ArOOH, is obtained[488]. A typical result is that with $Ph_2Zn$; $Ph_2Zn$ in benzene absorbed 0.5 mol of $O_2$ to give $Ph_2$ (17–21%), PhOH (17%), and other products. When the oxidation was carried out in $C_6D_6$ solution, the biphenyl was found to contain deuterium. Chemiluminescence has been observed in the oxidation of aryl—metallics[463]. Other stabilized free radicals are obtained from oxidation of benzylic lithium compounds[489,490], e.g. equations 237 and 238, pentadienyllithium[491], and from the aluminate compound[492] **67**, equation 239, all these reactions provide considerable amounts of hydrocarbon products.

$$\text{pyrid-2-yl-CH}_2\text{Li} \xrightarrow[\text{(ii) dil. HCl}]{\text{(i) O}_2\text{, Et}_2\text{O}} (\text{pyrid-2-yl-CH}_2\text{—})_2 + \text{pyrid-2-ylmethanol} \qquad (237)^{489}$$

$$\text{Ph–C}^{-}\text{(Me)–CH}_2\text{Bu}^t\ \text{Li}^+ \xrightarrow[-30\,^\circ\text{C}]{\text{O}_2\text{, isooctane}} [\text{Ph–C}^{\bullet}\text{(Me)–CH}_2\text{Bu}^t] \longrightarrow \underset{58\%}{\text{PhC(CH}_2\text{Bu}^t)\text{=CH}_2} + \underset{40\%}{\text{PhCH(Me)CH}_2\text{Bu}^t} \qquad (238)^{490}$$

$$\text{Na}^+\ \textbf{(67)} \xrightarrow{\text{O}_2} \text{naphthalene} \qquad (239)^{492}$$

**(67)**

A number of non-oxygen-containing products have been obtained in the oxidation of tetrametric neophyll-lithium in hydrocarbon solutions (equation 240); included

$$\begin{aligned}(\text{PhCMe}_2\text{CH}_2\text{Li})_4 \xrightarrow[\text{(ii) H}_2\text{O}]{\text{(i) O}_2\text{, pentane, 25 °C}} & \text{PhCH}_2\text{CMe}_2\text{OH} + \text{PhCMe}_2\text{CH}_2\text{OH} + \text{PhCMe}_3 \\ & + \text{PhCH}_2\text{CHMe}_2 + \text{PhCH}_2\text{C(Me)=CH}_2 \\ & + \text{PhCH=CMe}_2 \\ & + (\text{PhCMe}_2\text{CH}_2)_2 + (\text{PhCH}_2\text{CMe}_2)_2 \\ & + \text{PhCMe}_2\text{CH}_2\text{CMe}_2\text{CH}_2\text{Ph} \qquad (240)\end{aligned}$$

among the products were those derived from the rearrangement of the neophyll radical, i.e. $PhCMe_2CH_2^{\bullet} \longrightarrow PhCH_2CMe_2^{\bullet}$. This rearrangement is sufficiently slow so not to compete with reactions within the solvent cage. The rearranged products become more important when the amount of oxygen is limited[493]. Of some interest is the fact that oxidations either in ethereal solutions or in mixed hydrocarbon–ether solutions, in which neophyll-lithium dimers are present, the rearranged radical products are largely suppressed. This suggests different mechanisms for oxidation of the dimer and tetramer and that Scheme 55 is too simple. Radical rearrangements also result in the oxidation of hex-5-enylmagnesium bromides[494], see Scheme 56.

In keeping with the radical nature of the reactions, the oxidations are not stereospecific, as illustrated in Schemes 57 and 58. *Endo*- and *exo*-norbornylmagnesium halides, as well as tris(norbornyl)boranes, give the same isomeric mixture of *exo*- and *endo*-norbornyl peroxides[482] (i.e. 76:24). This clearly indicates a common intermediate—the norbornyl radical. A final illustration here is the autoxi-

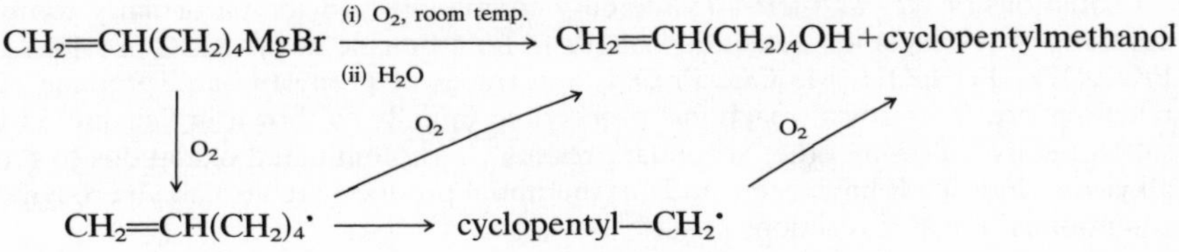

$$CH_2{=}CH(CH_2)_4MgBr \xrightarrow[\text{(ii) } H_2O]{\text{(i) } O_2\text{, room temp.}} CH_2{=}CH(CH_2)_4OH + \text{cyclopentylmethanol}$$

$$CH_2{=}CH(CH_2)_4MgBr \xrightarrow{O_2} CH_2{=}CH(CH_2)_4^{\bullet} \longrightarrow \text{cyclopentyl—}CH_2^{\bullet}$$

($CH_2{=}CH(CH_2)_4^{\bullet} \xrightarrow{O_2} CH_2{=}CH(CH_2)_4OH$; cyclopentyl—$CH_2^{\bullet} \xrightarrow{O_2}$ cyclopentylmethanol)

SCHEME 56

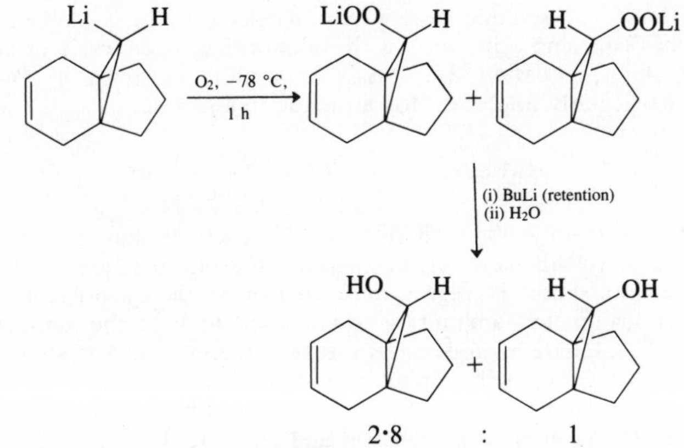

(Li, H substituted tricyclic alkene) $\xrightarrow[\text{1 h}]{O_2,\ -78\ ^\circ C,}$ (LiOO, H isomer) + (H, OOLi isomer) $\xrightarrow[\text{(ii) } H_2O]{\text{(i) BuLi (retention)}}$ (HO, H isomer) + (H, OH isomer)

2·8 : 1

SCHEME 57[495]

Et(H)C=C(Ph)Li $\xrightarrow{O_2,\ -78\ ^\circ C}$ Et(H)C=C$^{\bullet}$(Ph) $\xrightarrow{\text{slow}}$ Et(H)C=C(Ph)$O_2Li$ $\xrightarrow{\text{reduction with retention}}$

⇅ fast

Et(H)C=C(Li)Ph $\xrightarrow{O_2,\ -78\ ^\circ C}$ Et(H)C=C$^{\bullet}$Ph $\xrightarrow{\text{slow}}$ Et(H)C=C($O_2Li$)Ph $\xrightarrow{\text{reduction with retention}}$

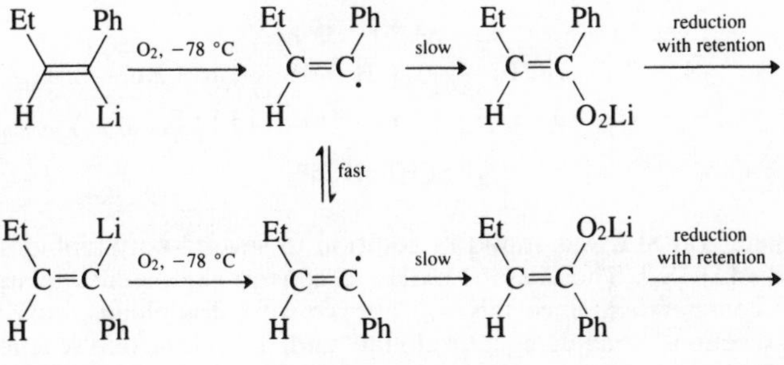

SCHEME 58[496]

dation of chiral $MeCHPhB(OH)_2$ which provided the racemic product[484] $MeCHPhOOB(OH)_2$.

The redox reactions, e.g. equations 234 or 235, proceed with retention, as shown for lithium as shown in Schemes 57 and 58 and for boron derivatives, e.g. equation 241. Such findings rule out a radical nature to the redox reactions, unless very tight radical cages are present.

$$endo\text{-norbornyl—B—O—OR} \longrightarrow endo\text{-norbornyl—O—B—OR} \qquad (241)$$

R = norbornyl

Oxidations of *sec*- and *tert*-alkylmercury compounds provide particularly complex mixtures of products, e.g. oxidation of $(Pr^i)_2Hg$ in nonane at 60–85 °C provides Hg, $Pr^iHgOPr^i$, $Pr^iHgOH$, $Me_2CO$, $Pr^iOH$, and traces of propylene and propane. The reactions are free-radical reactions, proceeding initially as shown in Scheme 55 but subsequently following other secondary routes[497]. Photoinitiated oxidations of tetra-alkyl-tins and -leads have been studied; the initial products are also alkylperoxymetal compounds. Further reactions ensue[498].

## C. Reactions with Sulphur, Selenium, and Tellurium

The reaction[499] of organic derivatives of electropositive elements, especially organo-magnesiums and -lithiums, with sulphur has been used as a method of synthesizing thiols, equation 242. Some examples are given in Table 14. The synthesis is particularly important for aromatic thiols.

$$RM + \tfrac{1}{8}S_8 \longrightarrow RSM \xrightarrow[\text{LiAlH}_4]{\text{H}_3\text{O}^+ \text{ or}} RSH \qquad (242)$$

Alkanethiols are readily obtained from other reactions and, except for tertiary-alkanethiols and cycloalkanethiols, the sulphur insertion reaction is inferior to these methods. For thiol synthesis, regard must be paid to the quantity of sulphur used; less than a stoichiometric amount is usually used to limit the amounts of dialkyl mono- and poly-sulphide byproducts, $RS_nR$ ($n = 1, 2, 3, \ldots$). A study has been made of the reaction of $Bu^sLi$ with sulphur at different molar ratios of reagents[500]. The following initial reactions (Scheme 59) were proposed by the authors [neglecting the aggregation of $Bu^sLi$ in benzene solution and any role that $(Bu^sLi)_n$ may play].

$$Bu^sLi + S_8 \longrightarrow Bu^s\text{—}S_8\text{—}Li$$

$$Bu^s\text{—}S_8\text{—}Li + Bu^sLi \longrightarrow Bu^s\text{—}S_x\text{—}Bu^s + Li_2S_y$$

$$x + y = 8$$

$$Bu^s\text{—}S_x\text{—}Bu^s + Bu^sLi \longrightarrow Bu^s\text{—}S_z\text{—}Bu^s + Bu^s\text{—}S_{(x-z)}\text{—}Li$$

$$Bu^sLi + Li_2S_y \longrightarrow Bu^sSLi + Li_2S_{(y-1)}$$

SCHEME 59

The thiolate, $Bu^sSLi$, was found in addition to several polysulphides at all molar ratios of $[Bu^sLi]:[S_8]$. The use of $LiAlH_4$ to convert the metal thiolate to the thiol has been recommended, since this will also convert disulphides, etc., to thiols. An interesting reaction[501], involving a disulphide and, as an insertion reaction, within the remit of this chapter, is shown in equation 243.

$$\text{1,2-dithiacyclohexane} + RLi \longrightarrow RS(CH_2)_4SLi \qquad (243)$$

$$R = Bu \text{ or } Ph$$

Selenium[502–509] and tellurium[504,508–510] also readily react with organo-magnesium and -alkali metal compounds to give the insertion products, metal salts of selenols (RSeM), and tellurols (RTeM). Care must be taken to exclude air during the insertion reactions, since these salts are very readily oxidized. The metal salts can be hydrolysed to selenols and tellurols or derivatized, see Table 13.

Sulphur and trialkylaluminiums react, at temperatures of 40–60 °C and $[R_3Al]:[S_8]$ more ratios of 8 : 1, to give distillable products, $R_2AlSR$, in good yields[510], e.g. 83 and 86% for R = Et and $Bu^i$, respectively. Refluxing toluene has

TABLE 14. Insertions of sulphur, selenium, and tellurium into carbon—metal bonds

$$RM + X_8 \longrightarrow RXM \xrightarrow{E^+ \text{ or } EX} RXE$$

$$X = S, Se, \text{ or } Te$$

| Organometallic RM | X | $E^+$ or EX | Product (yield, %) | Ref. |
|---|---|---|---|---|
| $Bu^tMgCl$ | S | $H_3O^+$ | $Bu^tSH$ (70–75) | 499 |
| $Ph_3CMgBr$ | S | $H_2O$ | $Ph_3CSH$ (70) | 499 |
| PhLi | S | $H_3O^+$ | PhSH (62) | 499 |
| $p$-$FC_6H_4MgBr$ | S | $H_3O^+$ | $p$-$FC_6H_4SH$ (50) | 499 |
| 2-M-thiophene M = Li or MgBr | S | $H_3O^+$ | 2-mercaptothiophene (65–70) | 499 |
| $PhCH_2MgCl$ | Se | HCl | $PhCH_2SeH$ (45) | 502 |
| PhMgBr | Se | HCl | PhSeH (57–71) | 503 |
| PhLi | Se or Te | $[(\pi\text{-}C_5H_5)_2NbCl_2]$ | $[(\pi\text{-}C_5H_5)_2Nb(XPh)_2]$ X = Se or Te (75–80) | 504 |
| PhLi | Te | $CH_2I_2$ | $(PhTe)_2CH_2$ (7) | 510 |
| $p$-$C_5H_{11}C_6H_4MgBr$ | Te | air | $(p\text{-}C_5H_{11}C_6H_4Te)_2$ (92) | 506 |
| $p$-$PhC_6H_4MgBr$ | Se | (i) $H_3O^+$ (ii) $O_2$ | $(p\text{-}PhC_6H_4Se)_2$ (31) | 506 |
| $p$-$C_5H_{11}C_6H_4MgBr$ | Se | $p$-$(n\text{-}C_8H_{17}O)C_6H_4COCl$ | $p$-$C_5H_{11}C_6H_4Se$—$CO$—$C_6H_4(OC_8H_{17}\text{-}n)$-$p$ (structure: $p$-$C_5H_{11}C_6H_4Se$ over $p$-$(n\text{-}C_8H_{17}O)C_6H_4CO$) | 506 |
| $o$-$(EtO)_2CHC_6H_4Li$ | Te | $BrCH_2CH(OEt)_2$ | $o$-$(EtO)_2CHC_6H_4TeCH_2CH(OEt)_2$ (50) | 511 |
| fur-2-yllithium | Se | MeI | fur-2-ylmethylselenide (49) | 507 |
| 1,8-dilithio-naphthalene | S, Se or Te | | 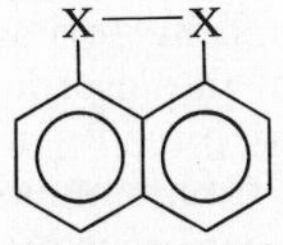 [naphthalene with X—X bridge at 1,8-positions] X = S (35–40) X = Se (18–22) X = Te (8–22) | 508 |
| PrSC≡CNa | S, Se, or Te | MeI | PrSC≡CXMe X = S, Se or Te | 509 |

also been used as a medium for S and Se insertions into carbon—aluminium bonds. Organometallic compounds of silicon, germanium, tin, and lead also react with sulphur, selenium, and even tellurium[513]. However, these reactions have no value in organic synthesis. Tetraphenyltin reacts with sulphur on heating at 200 °C, the isolated products being $Ph_2S$ and $(Ph_2SnS)_3$. Nucleophilic attack on $S_8$ was proposed. Schumann and Schmidt[513] considered the initial step to be similar to those in the reactions of organolithiums, i.e. an insertion step, equation 244, which gives $Ph_3SnS_8Ph$ and a subsequent reaction to give $Ph_3SnSPh$, equation 245.

$$Ph_4Sn + S_8 \longrightarrow Ph_3SnS_8Ph \quad (244)$$

$$Ph_3SnS_8Ph + Ph_4Sn \longrightarrow Ph_3SnSPh + [Ph_3SnS_7Ph], \text{ etc.} \quad (245)$$

At the temperature employed, $Ph_3SnSPh$ reacts further with sulphur to give $Ph_2Sn(SPh)_2$, which is thermally labile and decomposes to $Ph_2S$ and $(Ph_2SnS)_3$. Higher temperatures and excess sulhur lead to cleavages of further phenyl—tin bonds.

Tetraalkyltins are more reactive towrds sulphur, e.g. $Bu_4Sn$ reacts at 150 °C to give $(Bu_2SnS)_3$ and $Bu_2S$. Tetraphenyl-silicon and -germanium require higher temperatures to react initially with sulphur; at these temperatures, the isolated products are $MS_x$ ($x = 1$ or 2) and $Ph_2S$. Tetraorganolead compounds react with sulphur at 150 °C, via free-radical reactions.

An interesting ring expansion reaction occurs with the germacyclic compound dibutylgermacyclobutane[514], equation 246. The reaction of selenium with $Ph_4Sn$ occurs

$$Bu_2Ge\langle\diamond\rangle + \tfrac{1}{8}S_8 \xrightarrow{220-250\ °C} Bu_2Ge\text{(five-membered ring containing S)} \quad (246)$$

at 200 °C to give $Ph_3SnSePh$, which can be isolated at this reaction temperature. Tellurium requires a higher temperature to react (>240 °C); after several days only $Ph_2Te$ and tin tellurides are isolated.

## D. Sulphur Dioxide Insertions

There have been numerous reports of sulphur dioxide insertions into carbon—metal bonds of main group compounds, equation 247. A representative listing is

$$RM + SO_2 \longrightarrow RSO_2M \xrightarrow{H_3O^+} RSO_2H \quad (247)$$

given in Table 15. The topic was reviewed generally by Kitching and Fong[515] in 1970, while a more recent review was concerned solely with organotin compounds[516]. Hydrolysis of the insertion product, the metal sulphinate, $RSO_2M$, leads to a sulphinic acid and provides a useful method of preparation of these compounds[517], particularly from organo-magnesium[518] and -lithium[519] compounds. A drawback with the Grignard reaction is reported to be the formation of symmetrical sulphoxides as by-products.

Not all main group elements have as yet been studied but there is little doubt that, given the correct conditions, all would react with sulphur dioxide. Even organosilanes[520] and -germanes[521], which are among the least reactive main group organometallics, have been shown to react, e.g. equation 248. The rates of insertion and the

$$R_2M\langle\diamond\rangle + SO_2 \longrightarrow R_2M\text{(six-membered ring containing O–S(=O))} \quad (248)$$

TABLE 15. Insertion of sulphur dioxide into carbon—metal bonds

$$SO_2 + RM \longrightarrow RSO_2M \xrightarrow{H_2O} RSO_2H$$

| Organometal compound, RM | Conditions | Product (yield, %) | Ref. |
|---|---|---|---|
| $n$-$C_{12}H_{25}MgBr$ | (i) $Et_2O$ solution, $SO_2$ bubbled in at −35 to −40 °C (ii) aq. $NH_4Cl$ solution | $(n$-$C_{12}H_{25}SO_2)_2Mg{\cdot}2H_2O$ (80) | 518 |
| p-$MeC_6H_4MgBr$ | (i) $Et_2O$, 0 °C (ii) HCl | p-$MeC_6H_4SO_2H$ (18) + (p-$MeC_6H_4)_2SO$ | 518 |
| p-($n$-$C_{12}H_{25})C_6H_4Li$ | (i) $Et_2O$ solution, 0 °C (ii) $H_2O$ | p-($n$-$C_{12}H_{25})C_6H_4SO_2H$ (63) | 519 |
| $Me_2Zn$ | $SO_2$ equimolar | MeZnOSOMe | 528 |
| | excess $SO_2$ | $Zn(OSOMe)_2$ | |
| $Et_2Hg$ | refluxing liquid $SO_2$, 16 h | EtHgOSOEt (98 crude) | 536 |
| $Ph_2Hg$ | liquid $SO_2$, 23 °C | PhHgOSOPh (49) | 532 |
| $(C_{12}H_{25})_3Al$ | toluene solution, −45 °C to room temp. | $Al(O_2SC_{12}H_{25})_3$ (88) | 523 |
| $Et_3Ga$ | liquid $SO_2$, −50 °C | $(Et_2GaOSOEt)_2$ | 525 |
| $Me_3In$ | liquid $SO_2$, 1 min, −50 °C | $(Me_2InOSOMe)_2$ (90) | 526 |
| | liquid $SO_2$, −45 °C to room temp., 1 h | $[(MeSO_2)_3In]_n$ (*ca.* 100) | 526 |
| $Ph_3In$ | liquid $SO_2$, $\frac{1}{2}$ h | $Ph_2InOSOPh$ (98) | 526 |
| $Me_3Tl$ | liquid $SO_2$, excess | $(Me_2TlOSOMe)_2$ (92) | 527 |
| $Et_2Ge$ (four-membered ring) | −10 °C | $Et_2Ge$ (ring, O—SO) | 521 |
| $Ph_3SnMe$ | $SO_2$, 60 °C | $PhMeSn(O_2SPh)_2$ | 529 |
| $Me_4Sn$ | $SO_2$-bipy, −30 °C | $Me_2Sn[OS(O)Me]_2$ | |
| $Et_4Pb$ | benzene solution, 0 °C | $Et_3PbOS(O)Et$ | 531 |
| | room temp. | $Et_2Pb[OS(O)Et]_2$ | |
| $Ph_3PbCH_2C_6H_4F$-$p$ | $SO_2$ | $(PhSO_2)_3PbCH_2C_6H_4F$-$p$ | 531 |

necessary conditions for insertion vary significantly with the metal and the organic group.

Insertion into all the carbon—metal bonds in reactive compounds can occur. Thus all three carbon—aluminium bonds in trialkylaluminiums react[154,522–524], equation 249; even in alkylaluminium halides, $R_2AlX$ and $RAlX_2$, are all carbon—aluminium

$$R_3Al + SO_2 \longrightarrow (RSO_2)_3Al \qquad (249)$$

$$R = \text{alkyl}$$

bonds cleaved. In contrast, only one carbon—metal bond of triethylgallium[525], triphenylindium[526], and trimethylthallium[527] reacts. Controlled conditions can result in the insertion of sulphur dioxide into only one of the bonds in trimethylindium[522,526], e.g. a reaction time of 1 min at −50 °C in liquid sulphur dioxide provides dimeric $[Me_2InOS(O)Me]_2$, while allowing a solution of trimethylindium in liquid sulphur dioxide to warm up from −45 °C to room temperature over a period of 1 h gives the polymeric trisulphinate $[In\{OS(O)Me\}_3]_n$. The methyl—zinc bonds in dimethylzinc can be cleaved sequentially.[528] It is possible to cleave more than one bond in organotin[516,529,530] and -lead compounds[531].

Reactivity sequences reflect the electrophilic nature of the sulphur dioxide insertions, e.g. the reactivity of carbon—Group IV metal bonds decreases in the sequence C—Pb > C—Sn ≫ C—Ge, C—Si. Most of the mechanistic and rate studies have been performed with organotin compounds. However, relative reactivities have been gathered from study of other elements, such as mercury[515,532] and lead[515,531,533]. For organotin compounds, the ease of insertions decreases in the sequences allyl, allenyl, propargyl > phenyl > benzyl > methyl[515] and aryl > vinyl > benzyl > alkyl ≫ perfluoroalkyl[516].

The sulphur dioxide reaction in methanol is an overall second-order reaction, being first order in each component. The rate constants for reaction[534] of *p*- and m-$X\text{-}C_6H_4SnMe_3$ with sulphur dioxide in methanol solution (equation 250) at 30 °C

$$XC_6H_4SnMe_3 + SO_2 \longrightarrow XC_6H_4S(O)OSnMe_3 \qquad (250)$$

correlate with $\sigma^*$ with a $\rho$ value of −1.87; in comparison, $\rho$ values for cleavages by HCl, $CH_3CO_2H$, $Br_2$, and $I_2$ in methanol[535] are −2.17, −2.24, −2.58 and −2.54, respectively. Kitching and Fong concluded from the low $\rho$ value and the slight positive salt effect that a $S_Ei$ mechanism operates, with a four-centred transition state having some polar character, e.g. **68** in which the aryl—carbon bond is still essentially $sp^2$ hybridized. In contrast, Kunze[516] has argued for a $S_E$ mechanism with an open transition state, e.g. **69** for the $SO_2$ insertions.

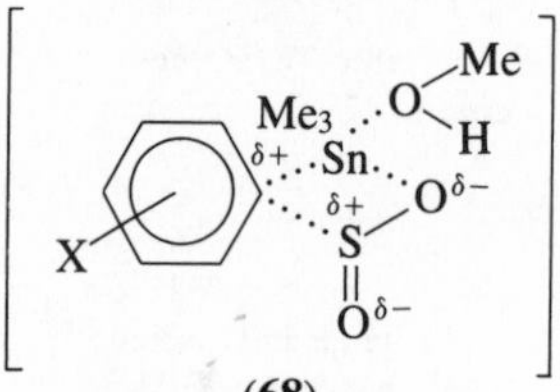

**(68)** 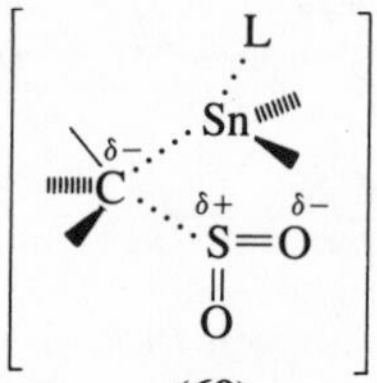 **(69)**

Towards organotin compounds, $SO_2$ can be placed in the following reactivity sequence: $Br_2 \gg I_2 > HCl > SO_2 > MeCO_2H$. The reactivity of $SO_2$ is enhanced in the presence of a donor, such as bipyridyl[530]. In addition, the reactivity is greater in methanol than in benzene solutions. Insertion of sulphur dioxide into substituted allyl—tin proparglyl—tin and allenyl—tin bonds lead to completely rearranged products[536], equations 251–253.

$$Me_3SnCH_2CH{=}CHMe \xrightarrow{SO_2} Me_3SnOS(O)CHMeCH{=}CH_2 \qquad (251)$$

$$Me_3SnCH_2C{\equiv}CH \xrightarrow{SO_2} Me_3SnOS(O)CH{=}C{=}CH_2 \qquad (252)$$

$$Me_3SnCH{=}C{=}CH_2 \xrightarrow{SO_2} Me_3SnOS(O)CH_2C{\equiv}CH \qquad (253)$$

The kinetic product of sulphur dioxide insertion into $Me_3SnCH_2CH{=}CHPh$, namely

$Me_3SnOS(O)CHPhCH{=}CH_2$, rapidly rearranges to the thermodynamic product $Me_3SnOS(O)CH_2CH{=}CHPh$.

Vinyl—tins react with sulphur dioxide with retention, e.g. equation 254. The

$$(E)\text{-}Me_3SnCH{=}CHPh \xrightarrow[-25\,^\circ C,\ 3\,h]{\text{liquid } SO_2} (E)\text{-}Me_3SnOS(O)CH{=}CHPh \qquad (254)$$

retention of configuration for the styryltin compounds and the rearrangement for the allylic, propargylic, and allenic compounds are compatible but not exclusively so, with an $S_Ei$ reaction.

Sulphinates can in principle be M—O or M—S bonded. Only for mercury have any M—S bonded sulphinates been established among main group sulphinates. Phenylmercury benzenesulphinate[532], PhHgOS(O)Ph, is monomeric in the solid state and in solution and exists in both Hg—O and Hg—S bonded forms. The compound obtained from chloroform–pentane is the *O*-sulphinate; recrystallization of this thermodynamically more stable Hg—O bonded form from acetone, methyl ethyl

| PhHgOS(O)Ph | $PhHgS(O)_2Ph$ |
|---|---|
| **70** | **71** |
| $\nu_{(SO)}$ 1048 and 838 $cm^{-1}$ | $\nu_{(SO)}$ 1175 and 1049 $cm^{-1}$ |

ketone, or methanol leads to the Hg—S form. Ethylmercury ethanesulphinate is Hg—S bonded in the solid state but is Hg—O bonded in solution. On the other hand, benzylmercury phenylmethanesulphinate, $PhCH_2HgOS(O)CH_2Ph$, is Hg—O bonded in both the solid state and in solution.

Zinc sulphinates appear from their i.r. spectra to be zinc—oxygen bonded[528], they are also highly aggregated as is $Me_2TlOSOMe$[527]. The diethylgallium ethanesulphinate, $Et_2GaOSOEt$[525], and dimethylindium methanesulphinate[527] are metal—oxygen bonded dimers with bridging O—SR—O groups. Group IVB sulphinates are M—O bonded compounds. Tin sulphinates, $R_3SnOS(O)R$, in benzene and $Me_3PbO$-SOMe exist as small aggregates[515]. However, $Me_3SnOSO(O)Ph$ is reported[532] to be slightly dissociated in methanol.

## E. Sulphur Trioxide Insrtions

Sulphur trioxide is a more powerful electrophilic reagent than sulphur dioxide. Possibly as a consequence of this reactivity, it's insertion reactions have not been so often reported. Most studies have been concerned with Group III[537–539] and IV[540–546] organometallic compounds. Organomercury compounds[547] have also been studied. Indeed, as far back as 1870 Otto[548] mentioned that $SO_3$ reacted violently with diphenylmercury to cleave both carbon—mercury bonds to give mercury bisbenzenesulphonate. Metal sulphonates are the products, usually in high yield, see Table 16. Organometal sulphates, $(R_3MO)_2SO_2$, and sulphonic acid anhydrides, $RSO_2OSO_2R$, could be detected as by-products in the reactions of tetraorganometallic derivatives of Group IVB. Disulphonation of tetraalkyl-tin[543] and -lead[546] compounds arises using excess of sulphur trioxide.

Hydrolyses of the sulphonate insertion products provides sulphonic acids, e.g. equation 255[540]. The overall conversion of arylmetal to arenesulphonic acid occurs

$$p\text{-}Me_3SiC_6H_4SiMe_3 \xrightarrow[CCl_4]{SO_3} p\text{-}Me_3SiC_6H_4SO_2OSiMe_3 \xrightarrow{H_2O} p\text{-}Me_3SiC_6H_4SO_3H \qquad (255)$$

TABLE 16. Sulphur trioxide insertions into carbon—metal bonds

| Organometallic compound, RM | Conditions | Product (yield, %) | Ref. |
|---|---|---|---|
| $Me_2Hg$ | $CH_2Cl_2$ solution, −70 °C, equimolar $SO_3$ | $MeHgOSO_2Me$ (quant.) | |
| $Me_3M$ M = Al, Ga, In or Tl | equimolar $SO_3$, −30 to −50 °C | $Me_2MOSO_2Me$ M = Al (65) Ga (70) In (77) Tl (85) | 537 |
| $(n\text{-}C_8H_{17})_3Al$ | $ClCH_2CH_2Cl$ soln., −30 °C | $n\text{-}C_8H_{17}SO_2OH$ (good) | 538 |
| $(Me_3SiCH_2)_3Al{\cdot}OEt_3$ | 3 equiv. $SO_3$, $CH_2Cl_2$ solution | $[Me_3SiCH_2SO_3]_3Al$ | 539 |
|  $Me_2Si$ (cyclic, four-membered ring) $SiMe_2$ | −78 °C in $CH_2Cl_2$ solution, $\frac{1}{2}$ h | 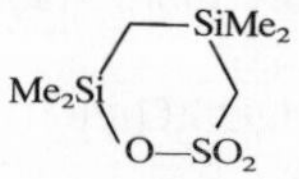 $Me_2Si$ (ring: $SiMe_2$, O—$SO_2$) | 542 |
|  $R_2M$ (cyclic, five-membered ring) M = Si; R = Me, Et or Bu M = Ge; R = Et or Bu | −70 °C in $CH_2Cl_2$ solution | 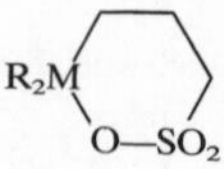 $R_2M$ (ring: O—$SO_2$) | 544 |
| $Me_4Si$ | 0 °C, $\frac{1}{2}$ h | $Me_3SiOSO_2Me$ (84) | 543 |
| $PhSiMe_3$ | 0 °C, 1 h | $Me_3SiOSO_2Ph$ (82) | 543 |
| $Me_3SiCH_2SiMe_3$ | 0–5 °C, 1 h | $Me_3SiCH_2SiMe_2OSO_2Me$ (80) | 543 |
| *o*-, *m*-, or *p*- $Me_3SiCH_2C_6H_4SiMe_3$ | | *o*-, *m*-, or *p*- $Me_3SiCH_2C_6H_4SO_3SiMe_3$ | 541 |
| $Me_4Ge$ | −35 °C, 45 min | $Me_3GeOSO_2Me$ (92) | 543 |
| $p\text{-}(Et_3Ge)_2C_6H_4$ | $SO_3$ in $CCl_4$ | $p\text{-}Et_3GeC_6H_4SO_3GeEt_3$ | 545 |
| $Me_4Sn$ | 1:1 $SO_3$ | $Me_3SnOSO_2Me$ (72) | 543 |
| | 1:2 $SO_3$ | $Me_2Sn(OSO_2Me)_2$ (67) | 543 |

regioselectively and hence is a useful route to sulphonic acids (and to their derivatives) under mild conditions (the direct sulphonation of arenes, ArH, by sulphur trioxide is far from a selective process). Substituent effects for $SO_3$ insertion into aryltrimethylsilanes were briefly reported on. By means of competition experiments, it was shown that substituent effects were small but were in the order expected for an electrophilic aromatic substitution reaction, equation[549] 256. Eaborn and

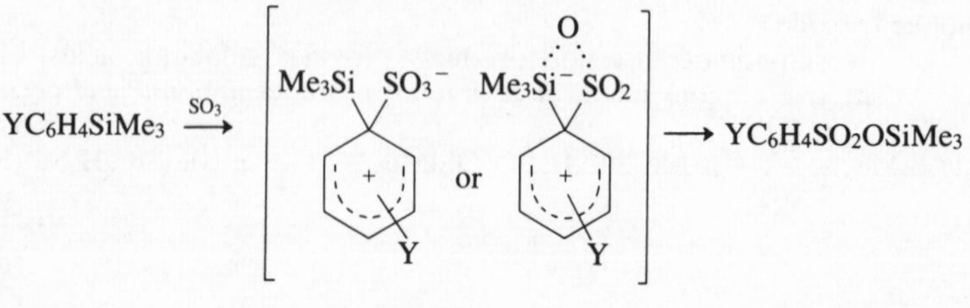

$$YC_6H_4SiMe_3 \xrightarrow{SO_3} \left[ \text{Me}_3\text{Si}\ \ \text{SO}_3^- \text{ (arenium, Y) or Me}_3\text{Si}^-\ \text{SO}_2\text{ (O-bridged arenium, Y)} \right] \longrightarrow YC_6H_4SO_2OSiMe_3 \qquad (256)$$

coworkers[540,541,545] also showed that aryl—silicon bonds are cleaved by sulphur trioxide more readily than methyl—silicon and benzyl—silicon bonds.

## F. Other Insertions

Main group organometallic compounds also undergo insertion reactions with various other compounds, as follows.

(i) other S=O bonded compounds, e.g. sulphoxides[550], equation 257.

$$Ph_2S{=}O + PhMgBr \longrightarrow Ph_3SOMgBr \qquad (257)$$

(ii) N=N bonded compounds, e.g. azo compounds[551], equation 258.

$$PhN{=}NPh + Me_2Be \longrightarrow (PhNMeNPh)_2Be \qquad (258)$$

(iii) N=O bonded compounds, e.g. nitric oxide[552], equation 259;

$$RM + NO \longrightarrow MONR \xrightarrow{NO} MONRN{=}O \qquad (259)$$

nitrous oxide[553], equation 260;

$$R_2CHLi + N_2O \longrightarrow R_2CHN{=}NOLi \qquad (260)$$

nitroso compounds, equations 261 and 262;

$$RO(R')NN{=}O + R^2M \longrightarrow [RO(R')NNR^2OM] \xrightarrow{-ROM} R'N{=}NOR^2 \qquad (261)^{554}$$

$$2,4,6\text{-}Bu^t{}_3C_6H_2N{=}O \xrightarrow{RMgX}$$

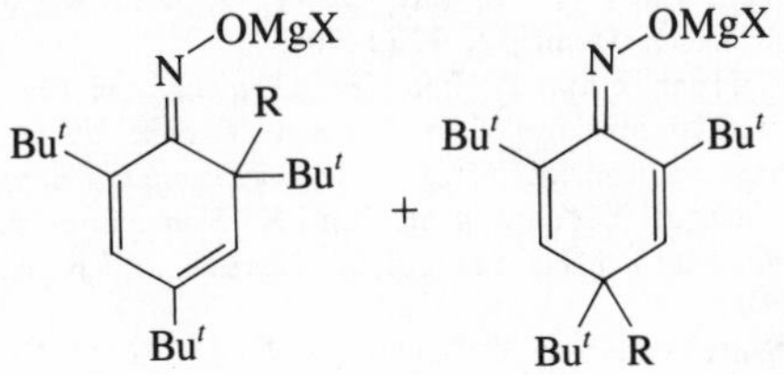

(262)[555]

and nitroarenes[556], equation 263

$$C_6H_5NO_2 \xrightarrow{RMgX}$$

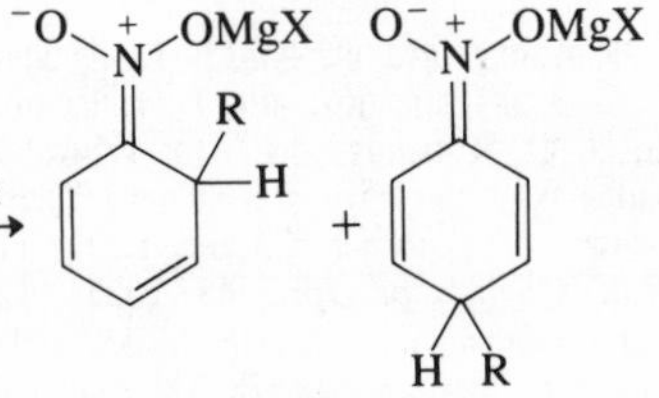

(263)

## VII. REFERENCES


1. M. F. Lappert and B. Prokai, *Adv. Organomet. Chem.*, **5**, 225 (1967).
2. T. Eicher, in *The Chemistry of the Carbonyl Group* (Ed. S. Patai), Wiley–Interscience, London, 1966, Chapter 13.
3. See, for example, the pertinent sections or volumes in the following: *Houben-Weyl Methoden der Organischen Chemie,* Georg Thieme Verlag, Stuttgart, Band 13; D. N. Jones (Ed.), *Comprehensive Organic Chemistry,* Pergamon Press, Oxford, 1979, Vol. 3;

A. N. Nesmeyanov and K. A. Kochestkov, (Eds.), *Methods of Elemento-Organic Chemistry,* North-Holland Press, Amsterdam.

4. (a) B. J. Wakefield, *The Chemistry of Organolithium Compounds,* Pergamon Press, Oxford, 1974; (b) M. S. Kharasch and O. Reinmuth, *Grignard Reactions of Non-Metallic Substances,* Prentice Hall, Englewood Cliffs, N.J., 1954; (c) T. Mole and E. A. Jeffrey, *Organoaluminium Compounds,* Elsevier, Amsterdam, 1972.
5. (a) N. Kawabata, H. Nakamura, and S. Yamashita, *J. Org. Chem.,* **38,** 3403 (1973); (b) B. G. Gowenlock, W. E. Lindsell, and B. Singh, *J. Organomet. Chem.,* **101,** C37 (1975); (c) M. A. Zemlyanichenko, N. I. Sheverdina, V. A. Chernoplekova, and K. A. Kocheshkov, *J. Gen. Chem. USSR (Engl. Transl.)* **42,** 832 (1972) (d) G. Santini, M. Le Blanc, and J. G. Reiss, *J. Organomet. Chem.,* **140,** 1 (1977); (e) L. F. Kozhemyakina, N. I. Sheverdina, I. E. Paleeva, V. A. Chernoplekova, and K. A. Kocheshkov, *Bull. Acad. Sci. USSR, Div. Chem. Sci.,* **25,** 413 (1976).
6. G. W. Kramer and H. C. Brown, *J. Org. Chem.,* **42,** 2292 (1977).
7. R. W. Hoffmann and H. J. Zeiss, *Angew. Chem., Int. Ed. Engl.,* **18,** 306 (1979); *J. Org. Chem.,* **46,** 1309 (1981); R. W. Hoffmann and W. Ladner, *Tetrahedron Lett.,* 4653 (1979); R. W. Hoffmann and T. Herold, *Chem. Ber.,* **114,** 375 (1981).
8. G. A. Artamkina, I. P. Beletskaya, and O. A. Reutov, *J. Organomet. Chem.,* **42,** 17C (1972).
9. G. Deleris, J. Dunoques and R. Calas, *Tetrahedron Lett.,* 2449 (1976).
10. R. Calas, J. Dunoques, G. Deleris, and F. Pisciotti, *J. Organomet. Chem.,* **69,** C15 (1974); G. Deleris, J. Dunoques, and R. Calas, *J. Organomet. Chem.,* **93,** 43 (1975).
11. E. W. Abel and R. J. Rowley, *J. Organomet. Chem.,* **84,** 199 (1975).
12. A. Hosomi and H. Sakurai, *Tetrahedron Lett.,* 1295 (1976).
13. L. Birkover, A. Ritter, and H. Wieden, *Chem. Ber.,* **95,** 971 (1962).
14. K. König and W. P. Neumann, *Tetrahedron Lett.,* 495 (1967).
15. S. V. Ponomarev and I. F. Lutsenko, *J. Gen. Chem. USSR (Engl. Transl.),* **34,** 3492 (1964); J. G. Noltes, H. M. J. C. Creemers, and G. J. M. van der Kerk, *J. Organomet. Chem.,* **11,** P 21 (1968); J. G. Noltes, F. Verbeck, and H. M. J. C. Creemers, *Organomet. Synth.,* **1,** 57 (1970).
16. P. F. Hudrlik and D. Peterson, *Tetrahedron Lett.,* 1785 (1972).
17. R. G. Mirshov and V. M. Vlasov, *J. Gen. Chem. USSR (Engl. Transl.)* **36,** 581 (1966).
18. C. Servens and M. Pereyre, *J. Organomet. Chem.,* **35,** C20 (1972).
19. H. Yatagai, Y. Yamamoto, and K. Maruyama, *J. Am. Chem. Soc.,* **102,** 4548 (1980).
20. Y. Hamamoto, H. Yatagai, Y. Naruta, and K. Maruyama, *J. Am. Chem. Soc.,* **102,** 7107 (1980).
21. K. Maruyama and Y. Naruta, *J. Org. Chem.,* **43,** 3796 (1978).
22. Y. Naruta, S. Ushida, and K. Maruyama, *Chem. Lett.,* 919 (1979).
23. V. Peruzzo, G. Tagliavini, and A. Gambaro, *Inorg. Chim. Acta,* **34,** L263 (1979); A. Gambaro, D. Marton, V. Peruzzo, and G. Tagliavini, *J. Organomet. Chem.,* **204,** 191 (1981).
24. For Mg, C. Blomberg and F. A. Hartzog, *Synthesis,* 18 (1977); for Mg or Ca, A. V. Bogatskii, N. G. Luk'yanenko, and L. I. Lyamtseva, *Dokl. Akad. Nauk SSSR,* **25,** 851 (1981); for Zn, J. C. Ruppert and J. D. White, *J. Org. Chem.,* **41,** 550 (1976); for Li, G. G. Cameron and A. J. S. Milton, *J. Chem. Soc., Perkin Trans. 2,* 378 (1976).
25. R. C. Fuson, *Adv. Organomet. Chem.,* **1,** 221 (1964).
26. M. Okubo, *Bull. Chem. Soc. Jpn.,* **48,** 1057, 1327, 2057 (1975).
27. L. Jalander, *Acta Chem. Scand., Ser. B,* **35,** 419 (1981); T. Holm, I. Crossland, and J. O. Madsen, *Acta Chem. Scand. Ser. B,* **32,** 754 (1978); I. G. C. Coutts and M. Hamblin, *J. Chem. Soc., Chem. Commun.,* 58 (1976).
28. R. C. Fuson and S. B. Speck, *J. Am. Chem. Soc.,* **64,** 2446 (1942).
29. R. C. Fuson and W. S. Friedländer, *J. Am. Chem. Soc.,* **75,** 5410 (1953).
30. M. Okubo, *Bull. Chem. Soc. Jpn.,* **50,** 2379 (1977).
31. B. F. Landrum and C. T. Lester, *J. Am. Chem. Soc.,* **76,** 5797 (1954).
32. W.I. O'Sullivan, F. W. Swamer, W. J. Humphlett, and C. R. Hauser, *J. Org. Chem.,* **26,** 2306 (1961).
33. E. C. Ashby, *Pure Appl. Chem.,* **52,** 545 (1980).

34. M. Lasperas, A. Perez-Rubalcaba, and M. L. Quiroga-Feijoo, *Tetrahedron,* **36,** 3403 (1980).
35. E. C. Ashby and T. L. Wiesemann, *J. Am. Chem. Soc.,* **100,** 189 (1978).
36. E. C. Ashby, J. Laemmle, and H. M. Neumann, *J. Am. Chem. Soc.,* **94,** 542 (1972).
37. S. E. Rudolph, L. F. Charbonneau, and S. G. Smith, *J. Am. Chem. Soc.,* **95,** 7083 (1973).
38. I. G. Lopp, J. D. Buhler, and E. C. Ashby, *J. Am. Chem. Soc.,* **97,** 4966 (1975).
39. T. Holm and I. Crossland, *Acta Chem. Scand.,* **25,** 59 (1971).
40. C. Blomberg, R. M. Salinger, and H. S. Mosher, *J. Org. Chem.,* **34,** 2385 (1969).
41. M. P. Ponomarchuk, L. F. Kasukhin, and V. D. Pokhodenko, *J. Gen. Chem. USSR (Engl. Transl.),* **41,** 40 (1971).
42. V. I. Savin, I. D. Temyachev, and F. D. Yambushev, *J. Org. Chem. USSR* (*Engl. Transl.*), **11,** 1227 (1975); V. I. Savin and Yu. P. Kitaev, *J. Org. Chem. USSR* (*Engl. Transl.*), **11,** 2622 (1975).
43. E. C. Ashby, J. R. Bowers, and R. Depriest, *Tetrahedron Lett.,* 3541, (1980); E. C. Ashby and J. R. Bowers, *J. Am. Chem. Soc.,* **103,** 2242 (1981).
44. D. E. Bergbreiter and O. M. Reichert, *J. Organomet. Chem.,* **125,** 119 (1977).
45. E. C. Ashby and S. A. Noding, *J. Org. Chem.,* **44,** 4371 (1979); E. C. Ashby, J. J. Lin, and J. J. Watkins, *Tetrahedron Lett.,* 1709 (1977).
46. S. G. Smith, L. F. Charbonneau, D. P. Novak, and T. L. Brown, *J. Am. Chem. Soc.,* **94,** 7059 (1972).
47. L. F. Charbonneau and S. G. Smith, *J. Org. Chem.,* **41,** 808 (1976).
48. J. S. Lomas, P. K. Luong, and J. E. Dubois, *J. Org. Chem.,* **42,** 3394 (1977).
49. E. C. Ashby and J. T. Laemmle, *J. Org. Chem.,* **33,** 3398 (1968); E. C. Ashby, J. T. Laemmle, and H. M. Neumann, *J. Am. Chem. Soc.,* **90,** 5179 (1968).
50. M. Chastrette and R. Amouroux, *J. Organomet. Chem.,* **70,** 323 (1974).
51. J. D. Morrison and H. S. Mosher, *Asymmetric Organic Reactions,* Prentice Hall, Englewood Cliffs, N.J., 1971, Chapter 3.
52. H. B. Bürgi, J. D. Dunitz, J. M. Lehn, and G. Wipff, *Tetrahedron,* **30,** 1563 (1974).
53. D. J. Cram and F. A. A. Elhafez, *J. Am. Chem. Soc.,* **74,** 5828 (1952).
54. D. J. Cram and J. D. Knight, *J. Am. Chem. Soc.,* **74,** 5835 (1952).
55. D. J. Cram and D. R. Wilson, *J. Am. Chem. Soc.,* **85,** 1245 (1963).
56. D. Guillerm-Dron, M. L. Capmau, and W. Chodkiewicz, *Bull. Soc. Chim. Fr.,* 1417 (1973).
57. A. Gaset, M. T. Maurette and A. Lattes, *J. Appl. Chem. Biotechnol.,* **25,** 1 (1975); A. Gaset, P. Audoye, and A. Lattes, *J. Appl. Chem. Biotechnol.,* **25,** 13 (1975); P. Audoye, A. Gaset, and A. Lattes, *J. Appl. Chem. Biotechnol.,* **25,** 19 (1975); *J. Organomet. Chem.,* **88,** 303 (1975).
58. J. W. Cornforth, R. H. Cornforth, and K. K. Mathew, *J. Chem. Soc.,* 112 (1959).
59. G. J. Karabatsos, *J. Am. Chem. Soc.,* **89,** 1367 (1967).
60. M. Cherest, H. Felkin and N. Prudent, *Tetrahedron Lett.,* 2199 (1968); M. Cherest and H. Felkin, *Tetrahedron Lett.,* 2205 (1968); M. Cherest, *Tetrahedron,* **36,** 1593 (1980).
61. F. Fernandez-Gonzalez and R. Perez-Ossorio, *An. Quim.,* **68,** 1411 (1972).
62. O. Arjona, R. Perez-Ossorio, A. Perez-Rubalcaba, and M. L. Quiroga, *J. Chem. Soc., Perkin Trans. 2,* 597 (1981), and references cited therein.
63. T. J. Leitereg and D. J. Cram, *J. Am. Chem. Soc.,* **90,** 4011 (1968).
64. E. C. Ashby and J. T. Laemmle, *Chem. Rev.,* **75,** 521 (1975), and references cited therein.
65. E. Juaristi and E. L. Eliel, *Tetrahedron Lett.,* 543 (1977).
66. M. Gaudemar, *Tetrahedron,* **32,** 1689 (1976).
67. M. Bellassoued, F. Dardoize, F. Gaudemar-Bardone, M. Gaudemar, and N. Gasdoue, *Tetrahedron,* **32,** 2713 (1976).
68. M.-H. Rei, *J. Org. Chem.,* **44,** 2760 (1979).
69. W. T. Wipke and P. Gund, *J. Am. Chem. Soc.,* **98,** 8107 (1976).
70. N. T. Anh, *Top. Curr. Chem.,* **88,** 145 (1980).
71. A. S. Cieplak, *J. Am. Chem. Soc.,* **103,** 4540 (1981).
72. T. Mukaiyama, K. Soai, T. Sato, H. Shimizu, and K. Suzuki, *J. Am. Chem. Soc.,* **101,** 1455 (1979).
73. D. Seebach, H. O. Kalinowski, B. Bastini, G. Crass, H. Daum, H. Dörr, N. P. Dupreez,

V. Ehrig, W. Langer, C. Nussler, H. A. Oei, and M. Schmidt, *Helv. Chim. Acta,* **60,** 301 (1977); D. Seebach, G. Crass, E. M. Wilka, D. Hilvert, and E. Brunner, *Helv. Chim. Acta,* **62,** 2695 (1979).
74. J. P. Mazaleyrat and D. J. Cram, *J. Am. Chem. Soc.,* **103,** 4585 (1981).
75. J. K. Whitesell and B. R. Jaw, *J. Org. Chem.,* **46,** 2798 (1981).
76. D. B. Collum, J. H. McDonald, and W. C. Still, *J. Am. Chem. Soc.,* **102,** 2117 (1980).
77. V. Rautenstrauch, *Helv. Chim. Acta,* **57,** 496 (1974).
78. H. Felkin, Y. Gault, and G. Roussi, *Tetrahedron,* **26,** 3761 (1970).
79. D. Abenhaim and E. Henri-Basch, *C.R. Acad. Sci., Ser. C,* **267,** 87 (1968).
80. Y. Yamamoto, H. Yatagai, and K. Maruyama, *J. Chem. Soc., Chem. Commun.,* 1072 (1980); *J. Am. Chem. Soc.,* **103,** 1969 (1981).
81. T. Hayashi, N. Fujitaka, T. Oishi, and T. Takeshima, *Tetrahedron Lett.,* 303 (1980).
82. F. Barbot and Ph. Miginiac, *Bull. Soc. Chim. Fr.,* 113 (1977).
83. R. Couffignal and M. Gaudemar, *J. Organomet. Chem.,* **60,** 209 (1973).
84. R. A. Benkeser and M. P. Siklosi, *J. Org. Chem.,* **41,** 3212 (1976).
85. R. A. Benkeser, M. P. Siklosi, and E. C. Mozdzen, *J. Am. Chem. Soc.,* **100,** 2134 (1978).
86. J. L. Moreau, *Bull. Soc. Chim. Fr.,* 1248 (1975).
87. E. C. Ashby and A. B. Goel, *J. Am. Chem. Soc.,* **103,** 4983 (1981).
88. T. Holm, *Acta Chem. Scand.,* **27,** 1552 (1973).
89. J. D. Morrison and G. Lambert, *J. Org. Chem.,* **37,** 1034 (1972).
90. J. F. Fauvergue, *C.R. Acad. Sci., Ser. C,* 1053 (1971).
91. J. Capillon and J. P. Guette, *Tetrahedron,* **35,** 1817 (1979).
92. J. D. Morrison, J. E. Tomaszewski, H. S. Mosher, J. Dale, D. Miller, and R. L. Elsenbaumer, *J. Am. Chem. Soc.,* **99,** 3167 (1977); H. S. Mosher, J. E. Stevenott, and D. O. Kimble, *J. Am. Chem. Soc.,* **78,** 4374 (1956).
93. (a) B. Eistert and L. Klein, *Chem. Ber.,* **101,** 900 (1968); (b) D. Wege, *Aust. J. Chem.,* **24,** 1531 (1971); (c) C. Blomberg, H. H. Grootveld, T. H. Gerner, and F. Bickelhaupt, *J. Organomet. Chem.,* **24,** 549 (1970).
94. P. Beak, J. Yamamoto, and C. J. Upton, *J. Org. Chem.,* **40,** 3052 (1975).
95. K. Dimroth and J. von Laufenberg, *Chem. Ber.,* **105,** 1044 (1972).
96. (a) F. Wessely and J. Kotlan, *Monatsh. Chem.,* **84,** 124 (1953); (b) B. Miller, *J. Chem. Soc., Chem. Commun.,* 750 (1974); (c) B. Miller, *J. Org. Chem.,* **42,** 1402, 1408, (1977); (d) B. Miller, E. R. Matjeka, and J. G. Haggerty, *Tetrahedron Lett.,* 323 (1977); (e) B. Miller, *J. Am. Chem. Soc.,* **95,** 8458 (1973).
97. M. Cais and A. Mandelbaum, in *Chemistry of the Carbonyl Group* (Ed. S. Patai), Wiley, London, 1966, p. 303.
98. M. J. Jorgensen, *Org. React.,* **18,** 1 (1970).
99. S. Nahm and S. M. Weinreb, *Tetrahedron Lett.,* **22,** 3815 (1981).
100. G. A. Olan and M. Arvanaghi, *Angew. Chem., Int. Ed. Engl.,* **20,** 878 (1981).
101. R. Levine and M. J. Karten, *J. Org. Chem.,* **41,** 1176 (1976); R. Levine, M. J. Karten, and W. M. Kadunce, *J. Org. Chem.,* **40,** 1770 (1975).
102. P. Hodge, G. M. Perry, and P. Yates, *J. Chem. Soc., Perkin Trans. 1,* 680 (1977).
103. F. Sato, K. Oguro, H. Watanabe, and M. Sato, *Tetrahedron Lett.,* 2869 (1980).
104. D. E. Nicodem and M. L. P. F. C. Marchiori, *J. Org. Chem.,* **46,** 3928 (1981).
105. H. D. Locksley and I. G. Murray, *J. Chem. Soc., C,* 1392 (1970).
106. J. E. Dubois, B. Leheup, F. Hennequin, and P. Bauer, *Bull. Soc. Chim. Fr.,* 1150 (1967).
107. (a) I. Kikkawa and T. Yorifuji, *Synthesis,* 877 (1980); (b) F. Sato, M. Inoue, K. Oguro, and M. Sato, *Tetrahedron Lett.,* 4303 (1979).
108. V. Ramanathan and R. Levine, *J. Org. Chem.,* **27,** 1216 (1962).
109. W. E. Parham and R. M. Piccirilli, *J. Org. Chem.,* **41,** 1268 (1976).
110. J. Cason, *Chem. Rev.,* **40,** 15 (1947); D. A. Shirley, *Org. React.,* **8,** 28 (1954).
111. R. G. Jones, *J. Am. Chem. Soc.,* **69,** 2350 (1947).
112. D. E. Bergbreiter and J. M. Killough, *J. Org. Chem.,* **41,** 2750 (1976).
113. K. A. Parker and E. G. Gibbons, *Tetrahedron Lett.,* 981 (1975).
114. H. Reinheckel, D. Jahncke, and G. Kretzschmar, *Chem. Ber.,* **99,** 11 (1966).
115. H. V. Schwering and J. Weidlein, *Chimia,* **27,** 535 (1973).
116. S. A. Lebedev, S. V. Ponomarev, and I. F. Lutsenko, *J. Gen. Chem. USSR (Engl. Transl.).* **42,** 643 (1972).

117. J. H. Robson and G. F. Wright, *Can. J. Chem.*, **38,** 1 (1960).
118. G. Himbert, *Angew. Chem., Int. Ed. Engl.*, **15,** 51 (1976).
119. V. I. Savin and Yu. P. Kitaev, *J. Org. Chem. USSR* (*Engl. Transl.*), **13,** 1137 (1977).
120. T. M. Bare and H. O. House, *Org. Synth.*, **49,** 81 (1969).
121. T. F. McGrath and R. Levine, *J. Am. Chem. Soc.*, **77,** 3656 (1955).
122. F. J. Impastato and H. M. Walborsky, *J. Am. Chem. Soc.*, **84,** 4838 (1962).
123. E. J. Corey and D. Seebach, *Angew. Chem., Int. Ed. Engl.*, **4,** 1075, 1077 (1965).
124. F. C. Whitmore and C. E. Lewis, *J. Am. Chem. Soc.*, **64,** 2964 (1942).
125. L. Gattermann, *Justus Leibigs Ann. Chem.*, **393,** 215 (1912).
126. N. Kunieda and M. Kinoshita, *Phosphorus Sulphur,* **10,** 383 (1981).
127. D. Bryce-Smith and A. C. Skinner, *J. Chem. Soc.*, 577 (1963).
128. Y. Baba, *Bull. Chem. Soc. Jpn.*, **41,** 1022 (1968).
129. D. Abenhaim, E. Henry-Basch, and P. Freon, *Bull. Chim. Soc. Fr.*, 4038 (1969).
130. T. Mukaiyama, M. Araki, and H. Takei, *J. Am. Chem. Soc.*, **95,** 4763 (1973).
131. M. S. Newman and S. Blum, *J. Org. Chem.*, **29,** 1416 (1964).
132. M. Gaudemar, *Bull. Soc. Chim. Fr.*, 1475 (1963).
133. A. J. Meyers and D. L. Comins, *Tetrahedron Lett.*, 5179 (1978).
134. J. F. Vozza, *J. Org. Chem.*, **24,** 720 (1959).
135. P. Canonne, G. Foscolos, and G. Lemay, *J. Chem. Soc., Chem. Commun.*, 691 (1979).
136. L. B. Dashkevich and L. N. Kuzmenkov, *J. Gen. Chem. USSR* (*Engl. Transl.*), **29,** 2330 (1959).
137. J. H. Billman and C. M. Smith, *J. Am. Chem. Soc.*, **61,** 457 (1939); **74,** 3174 (1952).
138. N. S. Nudelman and A. A. Vitale, *J. Org. Chem.*, **46,** 4625 (1981).
139. P. Jutzi and F. W. Schröder, *J. Organomet. Chem.*, **24,** 1 (1970).
140. L. S. Trzupek, T. L. Newirth, E. G. Kelly, N. E. Sbarbati, and G. M. Whitesides, *J. Am. Chem. Soc.*, **95,** 8118 (1973).
141. P. Jutzi and F. W. Schröder, *Angew. Chem., Int. Ed. Engl.*, **10,** 339 (1971).
142. W. J. J. M. Sprangers and R. Louw, *J. Chem. Soc., Perkins Trans. 2,* 1895 (1976), and references cited therein.
143. T. Tamamoto, T. Kohara, and A. Yamamoto, *Chem. Lett.*, 1217 (1976).
144. M. W. Rathke and H. Yu, *J. Org. Chem.*, **37,** 1732 (1972).
145. H. C. Brown, *Acc. Chem. Res.*, **2,** 65 (1969); H. C. Brown, in *Boranes in Organic Synthesis,* Cornell University Press, Ithaca, N.Y., 1972, p. 343; A. Pelter and K. Smith, in *Comprehensive Organic Chemistry,* ed. D. N. Jones, Pergamon Press, Oxford, Vol. 3, p. 824 (1979).
146. B. K. Nefedov, N. S. Sergeeva and Ya. T. Eidus, *Bull. Acad. Sci. USSR, Div. Chem. Sci.*, **21,** 1694, 2429 (1972).
147. P. M. Henry, *Tetrahedron Lett.*, 2285 (1968).
148. J. K. Stille and P. K. Wong, *J. Org. Chem.*, **40,** 335 (1975).
149. R. C. Larock, *J. Org. Chem.*, **40,** 3237 (1975); see also S. M. Brailovski, G. A. Kabalina, V.S. Shestakova, and O. N. Temkin, *J. Org. Chem. USSR (Engl. Transl.),* **13,** 1066 (1977).
150. R. C. Larock and S. S. Hershberger, *J. Org. Chem.*, **45,** 3840 (1980).
151. M. Ryang, *Organomet. Chem. Rev.*, **5A,** 67 (1970).
152. M. E. Volpin and I. S. Kolomnikov, *Organomet. React.*, **5,** 313 (1975).
153. E. J. Soloski and C. Tamborski, *J. Organomet. Chem.*, **157,** 373 (1978).
154. K. Ziegler, F. Krupp, K. Weyer, and W. Larbig, *Justus Liebigs Ann. Chem.*, **629,** 251 (1960); K. Ziegler, *Angew. Chem.*, **68,** 721 (1956).
155. W. H. Glaze and C. M. Selman, *J. Organomet. Chem.*, **11,** P3 (1968).
156. D. Seyferth and L. G. Vaughan, *J. Am. Chem. Soc.*, **86,** 883 (1964); D. Y. Curtin and W. J. Koehl, Jr., *J. Am. Chem. Soc.*, **84,** 1967 (1962).
157. H. M. Walborsky and F. J. Impastato, *J. Am. Chem. Soc.*, **81,** 5835 (1959).
158. K. Nishihata and M. Nishio, *Tetrahedron Lett.*, 1695 (1976).
159. (a) G. Zweifel and R. B. Steele, *J. Am. Chem. Soc.*, **89,** 2754 (1967); (b) J. J. Eisch and M. W. Foxton, *J. Organomet. Chem.*, **11,** P7 (1968).
160. F. Barbot and P. Miginiac, *J. Organomet. Chem.*, **145,** 269 (1978).
161. J. H. Wotiz, *J. Am. Chem. Soc.*, **73,** 693 (1951), and earlier papers.
162. J. Pansard and M. Gaudemar, *Bull. Soc. Chim. Fr.*, 3332 (1968).

163. S. Inoue and Y. Yokoo, *J. Organomet. Chem.*, **39,** 11 (1972).
164. J. Weidlein, *J. Organomet. Chem.*, **49,** 257 (1973).
165. Y. Kawada and H. Iwamura, *J. Org. Chem.*, **46,** 3357 (1981).
166. R. E. Lutz, C. L. Dickerson, W. J. Welstead, and R. G. Bass, *J. Org. Chem.*, **28,** 711 (1963).
167. H. Gilman and R. H. Kirby, *J. Am. Chem. Soc.*, **63,** 2046 (1941).
168. (a) R. T. Taylor and J. G. Galloway, *J. Organomet. Chem.*, **220,** 295 (1981); (b) J. P. Pilot, J. Dunoques, and R. Calas, *J. Chem. Res. (S),* 268 (1977).
169. (a) E. M. Kaiser, M. Chung-Ling, C. F. Hauser, and C. R. Hauser, *J. Org. Chem.*, **35,** 410 (1970); (b) H. Schreiber, *Makromol. Chem.*, **36,** 86 (1960).
170. (a) S. H. Liu, *J. Org. Chem.*, **42,** 3209 (1977); (b) R. A. Kretchmer, *J. Org. Chem.*, **37,** 2744 (1972).
171. R. Bürstinghaus and D. Seebach, *Chem. Ber.*, **110,** 841 (1977); D. Seebach and R. Bürstinghaus, *Angew. Chem., Int. Ed. Engl.*, **14,** 57 (1975).
172. F. E. Ziegler and C. C. Tam, *Tetrahedron Lett.*, **49,** 4717 (1979).
173. R. A. Lee and W. Reusch, *Tetrahedron Lett.*, 969 (1973).
174. W. C. Still and A. Mitra, *Tetrahedron Lett.*, 2659 (1978).
175. B.-T. Gröbel and D. Seebach, *Synthesis,* 357 (1977).
176. (a) J. L. Herrmann, J. E. Richman, and R. H. Schlessinger, *Tetrahedron Lett.*, 3271, 3275 (1973); (b) K. Ogura, M. Yamashita, and G.-I. Tsuchihashi, *Tetrahedron Lett.*, 1303 (1978).
177. (a) A. R. B. Manas and R. A. J. Smith, *J. Chem. Soc., Chem. Commun.*, 216 (1975); (b) T. Cohen and S. M. Nolan, *Tetrahedron Lett.*, 3533 (1978); (c) W. Langer and D. Seebach, *Helv. Chim. Acta,* **62,** 1710 (1979).
178. P. C. Ostrowski and V. V. Kane, *Tetrahedron Lett.*, 3549 (1977).
179. Y. Köksal, P. Raddatz, and E. Winterfeldt, *Angew. Chem, Int. Ed. Engl.*, **19,** 472 (1980).
180. J. A. Thomas and C. H. Heachcock, *Tetrahedron Lett.*, 3235 (1980).
181. J. Lucchetti, W. Dumont, and A. Krief, *Tetrahedron Lett.*, 2695 (1979).
182. J. Mulzer, G. Hartz, U. Kühl, and G. Brüntrup, *Tetrahedron Lett.*, 2949 (1978).
183. (a) S. Yamagiwa, N. Hoshi, H. Sato, H. Kosugi, and H. Uda, *J. Chem. Soc., Perkin Trans. 1,* 214 (1978); (b) A. G. Schultz and Y. K. Yee, *J. Org. Chem.*, **41,** 4044 (1976).
184. J. Luchetti and A. Krief, *Tetrahedron Lett.*, 2697 (1978).
185. (a) R. Sauvetre and J. Seyden-Penne, *Tetrahedron Lett.*, 3949 (1976); (b) R. Sauvetre, M. C. Roux-Schmidt, and J. Seyden-Penne, *Tetrahedron,* **34,** 2135 (1978); (c) B. Deschamps, M.-C. Roux-Schmidt, and L. Wartski, *Tetrahedron Lett.*, **34,** 1377 (1979).
186. N.-Y. Wang, S.-S. Su, and L.-Y. Tsai, *Tetrahedron Lett.*, 1121 (1979).
187. E. M. Kaiser, P. L. Knutson, and J. R. McClure, *Tetrahedron Lett.*, 1747 (1978).
188. M. R. Binns and R. K. Haynes, *J. Org. Chem.*, **46,** 3790 (1981).
189. G. Gilbert and B. F. Aycock, *J. Org. Chem.*, **22,** 1013 (1957).
190. J. E. Baldwin and W. A. Dupont, *Tetrahedron Lett.*, **21,** 1881 (1980).
191. G. B. Mpango, K. K. Mahalanabis, Z. Mahdvi-Damghani, and V. Snieckus, *Tetrahedron Lett.*, **21,** 4823 (1980).
192. L. Wartski, M. El-Bouz, J. Seyden-Penne, W. Dumont, and A. Krief, *Tetrahedron Lett.*, 1543 (1979).
193. C. A. Brown and A. Yamaichi, *J. Chem. Soc., Chem. Commun.*, 100 (1979).
194. M. El-Bouz and L. Wartski, *Tetrahedron Lett.*, **21,** 2897 (1980).
195. L. Wartski, M. El-Bouz, and J. Seyden-Penne, *J. Organomet. Chem.*, **177,** 17 (1979).
196. G. H. Posner, *Org. React.*, **19,** 1 (1972).
197. J.-M. Lefour and A. Loupy, *Tetrahedron,* **34,** 2597 (1978).
198. F. E. Ziegler and J. A. Schwartz, *J. Org. Chem.*, **43,** 985 (1978).
199. B. W. Roberts, M. Ross, and J. Wong, *J. Chem. Soc., Chem. Commun.*, 428 (1980).
200. P. W. Collins, E. Z. Dajani, M. S. Bruhn, C. H. Brown, J. R. Palmer, and R. Pappo, *Tetrahedron Lett.*, 4217 (1975).
201. M. S. Bruhn, C. H. Brown, P. W. Collins, J. R. Palmer, E. Z. Dajani, and R. Pappo, *Tetrahedron Lett.*, 235 (1976).
202. R. T. Hansen, D. B. Carr, and J. Schwartz, *J. Am. Chem. Soc.*, **100,** 2244 (1978).
203. F. Sato, T. Oikawa, and M. Sato, *Chem. Lett.*, 167 (1979).
204. G. W. Kabalka and R. F. Daley, *J. Am. Chem. Soc.*, **95,** 4428 (1973).

205. J. Hooz and R. B. Layton, *Can. J. Chem.*, **51,** 2098 (1973).
206. K. F. Bernady and M. J. Weiss, *Tetrahedron Lett.*, 4083 (1972).
207. M. Gocmen and M. G. Soussan, *J. Organomet. Chem.*, **61,** 19 (1973); M. Gocmen, M. G. Soussan, and P. Freon, *Bull. Soc. Chim. Fr.*, 562, 1310 (1973).
208. M. Cossentini, B. Deschamps, N. T. Anh, and J. Seyden-Penne, *Tetrahedron,* **33,** 409 (1977).
209. (a) M. P. Cooke, Jr., *Tetrahedron Lett.*, 2199 (1979); (b) S. Cacchi, D. Misiti, and G. Palmieri, *Tetrahedron,* **37,** 2941 (1981); (c) P. Jacob and H. C. Brown, *J. Am. Chem. Soc.*, **98,** 7832 (1976).
210. P. Paquer, *Bull. Soc. Chim. Fr.*, 1439 (1975).
211. M. Dagonneau, *C.R. Acad. Sci., Ser. C,* **276,** 1683 (1973).
212. A. Julg, M. Bonnet, and Y. Ozias, *Theor. Chim. Acta,* **17,** 49 (1970); see also N. C. Baird and J. R. Swenson, *J. Phys. Chem.*, **77,** 277 (1973).
213. C. E. Loader and H. J. Anderson, *Can. J. Chem.*, **49,** 45 (1971).
214. H. Gilman, J. Robinson, and N. J. Beaber, *J. Am. Chem. Soc.*, **48,** 2715 (1926).
215. H. Wuyts and A. Lacourt, *Bull. Soc. Chim. Belg.*, **45,** 445 (1946).
216. D. M. Baird and R. D. Bereman, *J. Org. Chem.*, **46,** 458 (1981).
217. D. A. Konen, P. E. Pfeffer, and L. S. Silbert, *Tetrahedron,* **32,** 2507 (1976).
218. H. Westmijze, H. Kleijn, J. Meijer, and P. Vermeer, *Synthesis,* 432 (1979);
219. J. Meijer, P. Vermeer, and L. Brandsma, *Recl. Trav. Chim. Pays-Bas,* **92,** 601 (1973).
220. H. Reinheckel and D. Jahnke, *Chem. Ber.*, **99,** 23 (1966).
221. R. G. Micetich, *Can. J. Chem.*, **48,** 2006 (1970).
222. e.g. for Li, H. Gilman and F. Breuer, *J. Am. Chem. Soc.*, **55,** 1262 (1933); for B. M. F. Lappert and B. Prokai, *J. Chem. Soc.*, 4223 (1963).
223. Y. Tamaru, T. Harada, H. Iwamoto, and Z.-I. Yoshida, *J. Am. Chem. Soc.*, **100,** 5221 (1978).
224. Y. Tamuru, M. Kagotani, and Z.-I. Yoshida, *Tetrahedron Lett.*, **22,** 3409 (1981).
225. P. Metzner and J. Vialle, *Bull. Soc. Chim. Fr.*, 1703 (1973).
226. (a) A. Schönberg, E. Singer, E. Frese, and K. Praefcke, *Chem. Ber.*, **98,** 3311 (1965); (b) A. Ohno, K. Nakamura, M. Uohama, S. Oka, T. Yamabe, and S. Nagata, *Bull. Chem. Soc. Jpn.*, **48,** 378 (1975).
227. M. Dagonneau and J. Vialle, *Bull. Soc. Chim. Fr.*, 2067 (1972).
228. P. Beak and J. W. Worley, *J. Am. Chem. Soc.*, **92,** 4142 (1970); **94,** 597 (1972), and references cited therein.
229. (a) M. Dagonneau, J.-F. Hemidy, D. Cornet, and J. Vialle, *Tetrahedron Lett.*, 3003 (1972); (b) M. Dagonneau and J. Vialle, *Tetrahedron Lett.*, 3017 (1973); (c) M. Dagonneau, P. Metzner, and J. Vialle, *Tetrahedron Lett.*, 3675 (1973); (d) M. Dagonneau, D. Paquer, and J. Vialle, *Bull. Soc. Chim. Fr.*, 1699 (1973); (e) D. Paquer and R. Pou, *Bull. Soc. Chim. Fr.*, 3887 (1972); (f) J. C. Wesdorp, J. Meijer, P. Vermeer, H. J. T. Bos, L. Brandsma, and J. F. Arenes, *Recl. Trav. Chim. Pays-Bas.*, **93,** 184 (1974); (g) L. Leger and M. Saquet, *Bull. Soc. Chim. Fr.*, 657 (1975).
230. E. Schaumann and W. Walter, *Chem. Ber.*, **107,** 3562 (1974).
231. A. Schönberg, A. Rosenbach, and O. Schültz, *Justus Liebigs Ann. Chem.*, **454,** 37 (1927).
232. M. Dagonneau and J. Vialle, *Tetrahedron,* **30,** 415 (1974).
233. L. Leger, M. Saquet, A. Thiullier, and S. Julia, *J. Organomet. Chem.*, **96,** 313 (1975).
234. M. Dagonneau, *J. Organomet. Chem.*, **80,** 1 (1974).
235. A. I. Meyers, T. A. Tait, and D. L. Comins, *Tetrahedron Lett.*, 4657 (1978).
236. P. Metzner, J. Vialle, and A. Vibet, *Tetrahedron,* **34,** 2289 (1978); *Tetrahedron Lett.*, 4295 (1976).
237. D. Paquer and S. Smadja, *Recl. Trav. Chim. Pays-Bas,* **95,** 175 (1976).
238. J.-L. Burgot, J. Masson, P. Metzner, and J. Vialle, *Tetrahedron Lett.*, 4297 (1976); J.-L. Burgot, J. Masson, and J. Vialle, *Tetrahedron Lett.*, 4775 (1976); J. Masson, P. Metzner, and J. Vialle, *Tetrahedron,* **33,** 3089 (1978).
239. (a) H. Gilman, R. N. Meals, G. O'Donnell, and L. Woods, *J. Am. Chem. Soc.*, **65,** 268 (1943); K. A. Allen, B. G. Gowenlock, and W. E. Lindsell, *J. Polym. Sci., Polym. Chem.*, **12,** 1131 (1974); (b) L. N. Cherkasov and R. S. Razina, *J. Org. Chem. USSR* (*Engl. Transl.*), **9,** 14 (1973); L. N. Cherkasov, *J. Gen. Chem. USSR* (*Engl. Transl.*), **41,** 1565 (1971); **42,** 1520 (1972); *J. Org. Chem. USSR* (*Engl. Transl.*), **10,** 1555 (1974); L. N.

Cherkasov, G. I. Pis'mennaya, Kh. V. Bal'yan, and A. A. Petrov, *J. Org. Chem. USSR* (*Engl. Transl.*), **9,** 870 (1973); L. N. Cherkasov, G. I. Pis'mennaya, Kh. V. Bal'yan, and A. A. Petrov, *J. Org. Chem. USSR (Engl. Transl.),* **7,** 2530 (1971); L. N. Cherkasov, S. I. Radchenko, and Kh. V. Bal'yan, *J. Org. Chem., USSR* (*Engl. Transl.*), **10,** 2635 (1974); L. N. Cherkasov, *J. Org. Chem. USSR* (*Engl. Transl.*), **9,** 12 (1973).
240. K. Ziegler, *Chem. Abstr.,* **52,** 1203 (1958); Br. Pat. 763,824 (1956).
241. J. J. Eisch, *Organomet. Chem. Rev. B,* **4,** 331 (1968); T. Spencer and F. G. Thorpe, *J. Organomet. Chem.,* **99,** C8 (1975).
242. H. Sakarai and T. Imai, *Chem. Lett.,* 891 (1975).
243. D. Seyferth and D. C. Annarelli, *J. Organomet. Chem.,* **117,** C51 (1976).
244. (a) H. Lehmkuhl, O. Olbrysch, D. Reinehr, G. Schomberg, and D. Henneberg, *Justus Liebigs Ann. Chem.,* 145 (1975); (b) B. Mauzé, *J. Organomet. Chem.,* **131,** 321 (1977).
245. (a) H. G. Richey, Jr., C. W. Wilkins, Jr., and R. M. Bension, *J. Org. Chem.,* **45,** 5042 (1980), and references therein; (b) H. G. Richey, Jr., and C. W. Wilkins, Jr., *J. Org. Chem.,* **45,** 5027 (1980), and references cited therein.
246. (a) A. F. Halasa, D. N. Schulz, D. P. Tate, and V. D. Mochel, *Adv. Organomet. Chem.,* **18,** 55 (1980); (b) H. Sinn and W. Kaminsky, *Adv. Organomet. Chem.,* **18,** 99 (1980).
247. K. Ziegler and H.-G. Gellert, *Justus Liebigs Ann. Chem.,* **567,** 195 (1950).
248. P. D. Bartlett, S. J. Tauber, and W. P. Weber, *J. Am. Chem. Soc.,* **91,** 6362 (1969).
249. J. A. Marshall and H. Faubl, *J. Am. Chem. Soc.,* **92,** 948 (1970).
250. S. Wawzonek, B. J. Studnicka, and A. R. Zigman, *J. Org. Chem.,* **34,** 1316 (1969); see also J. G. Welch, and R. M. Magid, *J. Am. Chem. Soc.,* **89,** 5300 (1967).
251. T. W. Dolzine and J. P. Oliver, *J. Organomet. Chem.,* **78,** 165 (1974).
252. V. N. Drozd, Yu. A. Ustynyuk, M. A. Tseleva, and L. B. Dmitriev, *J. Gen. Chem. USSR* (*Engl. Transl.*), **38,** 2047 (1968); *J. Gen. Chem. USSR* (*Engl. Transl.*), **39,** 1951 (1969).
253. A. Maercker and J. D. Roberts, *J. Am. Chem. Soc.,* **88,** 1742 (1966).
254. S. E. Wilson, *Tetrahedron Lett.,* 4651 (1975).
255. D. J. N. Keyton, *Diss. Abstr. (B),* **29,** 3262 (1969).
256. A. Maercker and K. Weber, *Justus Liebigs Ann. Chem.,* **756,** 43 (1972).
257. Alkenols: (a) M. Kool and G. W. Klumpp, *Tetrahedron Lett.,* 1873 (1978); (b) J. K. Crandall and A. C. Clark, *Tetrahedron Lett.,* 325 (1969); *J. Org. Chem.,* **37,** 4236 (1972); (c) H. Felkin, G. Swierczewski, and A. Tambuté, *Tetrahedron Lett.,* 707 (1969).
258. Alkenyl ethers: A. H. Veefkind, F. Bickelhaupt, and G. W. Klumpp, *Recl. Trav. Chim. Pays-Bas.,* **88,** 1058 (1969).
259. Amines: (a) H. G. Richey, Jr., W. F. Erickson, and A. S. Heyn, *Tetrahedron Lett.,* 2187 (1971); (b) A. H. Veefkind, J. van der Schaaf, F. Bickelhaupt, and G. W. Klumpp, *Chem. Commun.,* 722 (1971) (also alkenyl sulphides).
260. T. H. Chan and E. Chang, *J. Org. Chem.,* **39,** 3264 (1974); see also C. R. Buell, R. Corriu, G. Guerin, and L. Spialter, *J. Am. Chem. Soc.,* **92,** 7424 (1970).
261. T. Kauffmann, H. Ahlers, H.-J. Tilhard, and A. Woltermann, *Angew. Chem., Int. Ed. Engl.,* **16,** 710 (1977).
262. D. J. Peterson, *J. Org. Chem.,* **31,** 950 (1966).
263. J. D. Park, T. S. Croft, and R. W. Anderson, *J. Organomet. Chem.,* **64,** 19 (1974).
264. W. E. Parham and R. F. Motter, *J. Am. Chem. Soc.,* **81,** 2146 (1959); see also *J. Org. Chem.,* **27,** 2628 (1962).
265. R. Waack and M. A. Doran, *J. Org. Chem.,* **32,** 3395 (1967).
266. W. H. Glaze, J. E. Hanicak, M. L. Moore, and J. Chaudhuri, *J. Organomet. Chem.,* **44,** 39 (1972).
267. G. Fraenkel and M. J. Geckle, *J. Am. Chem. Soc.,* **102,** 2869 (1980); G. Fraenkel, M. J. Geckle, A. Kaylo, and D. W. Esters, *J. Organomet. Chem.,* **197,** 249 (1980).
268. G. Schulz and H. Höcker, *Angew. Chem. Int. Ed. Engl.,* **19,** 219 (1980).
269. M. Hallden-Abberton, C. Engelman, and G. Fraenkel, *J. Org. Chem.,* **46,** 538 (1981); G. Fraenkel and M. Hallden-Abberton, *J. Am. Chem. Soc.,* **103,** 5657 (1981).
270. J. E. Mulvaney and D. J. Newton, *J. Org. Chem.,* **34,** 1936 (1969).
271. S. A. Kandil and R. E. Dessy, *J. Am. Chem. Soc.,* **88,** 3027 (1966).
272. J. A. Dixon, D. H. Fishman, and R. S. Dubinyak, *Tetrahedron Lett.,* 613 (1964).
273. J. A. Dixon and D. H. Fishman, *J. Am. Chem. Soc.,* **85,** 1356 (1963); D. J. Schaeffer and

H. E. Ziegler, *J. Org. Chem.*, **34,** 3958 (1969); H. E. Ziegler, *J. Org. Chem.*, **31,** 2977 (1966).
274. K. Hafner and H. Weldes, *Justus Liebigs Ann. Chem.*, **606,** 90 (1957).
275. E. A. Hill, *J. Organomet. Chem.*, **91,** 123 (1975).
276. E. A. Hill, *Adv. Organomet. Chem.*, **16,** 131 (1977).
277. J. D. Roberts, E. R. Trumbull, Jr., W. Bennett and R. Armstrong, *J. Am. Chem. Soc.*, **72,** 3116 (1950); J. D. Roberts and R. H. Mazur, *J. Am. Chem. Soc.*, **73,** 2509 (1951).
278. H. G. Richey, Jr. and H. S. Veale, *Tetrahedron Lett.*, 615 (1975).
279. (a) E. A. Hill and G. E.-M. Shih, *J. Am. Chem. Soc.*, **95,** 7764 (1973); (b) A. Maercker and R. Guess, *Chem. Ber.*, **106,** 773 (1973).
280. A. Maercker, P. Güthlein, and H. Wittmayr, *Angew. Chem., Int. Ed. Engl.*, **12,** 774 (1973).
281. E. A. Hill, H. G. Richey, Jr., and T. C. Rees, *J. Org. Chem.*, **28,** 2161 (1963); E. A. Hill and H.-R. Ni, *J. Org. Chem.*, **36,** 4133 (1974).
282. E. A. Hill and M. M. Myers, *J. Organomet. Chem.*, **173,** 1 (1979).
283. E. A. Hill, D. C. Link, and P. Donndelinger, *J. Org. Chem.*, **46,** 1177 (1981).
284. H. Lehmkuhl and W. Bergstein, *Justus Liebigs Ann. Chem.*, 1876 (1978).
285. E. A. Hill, R. J. Theissen, C. E. Cannon, R. Miller, R. B. Guthrie, and A. T. Chen, *J. Org. Chem.*, **41,** 1191 (1976).
286. H. G. Richey, Jr. and H. S. Veale, *J. Am. Chem. Soc.*, **96,** 2641 (1974).
287. H. G. Richey, Jr. and A. M. Rothman, *Tetrahedron Lett.*, 1457 (1968).
288. W. C. Kossa, Jr., T. C. Rees, and H. G. Richey, Jr., *Tetrahedron Lett.*, 3455 (1971).
289. H. Felkin, J. D. Umpleby, E. Hagaman, and E. Wenkert, *Tetrahedron Lett.*, 2285 (1972).
290. H. Lehmkuhl and D. Reinehr, *J. Organomet. Chem.*, **34,** 1 (1972).
291. H. Lehmkuhl, D. Reinehr, G. Schomberg, D. Henneberg, H. Damen and G. Schroth, *Justus Liebigs Ann. Chem.*, 103 (1975).
292. H. E. Podall and W. E. Foster, *J. Org. Chem.*, **23,** 1848 (1958).
293. H. Lehmkuhl, D. Reinehr, J. Brandt, and G. Schroth, *J. Organomet. Chem.*, **57,** 39 (1973).
294. L. H. Shepherd, Jr., *U.S. Pat.* 3 597 488 (1971); *Chem. Abstr.*, **75,** 88751 (1971).
295. H. Lehmkuhl and D. Reinehr, *J. Organomet. Chem.*, **25,** C47 (1970).
296. H. Lehmkuhl and E. Jannssen, *Justus Liebigs Ann. Chem.*, 1854 (1978).
297. H. Lehmkuhl, D. Reinehr, D. Henneberg, and G. Schroth, *J. Organomet. Chem.*, **57,** 49 (1973).
298. H. Lehmkuhl, D. Reinehr, D. Henneberg, G. Schomberg, and G. Schroth, *Justus Liebigs Ann. Chem.*, 119 (1975).
299. H. Lehmkuhl, W. Bergstein, D. Henneberg, E. Janssen, O. Olbrysch, D. Reinehr, and G. Schomberg, *Justus Liebigs Ann. Chem.*, 1176 (1975).
300. H. Lehmkuhl and E. Janssen, *Justus Liebigs Ann. Chem.*, 1854 (1978).
301. H. Lehmkuhl and W. Bergstein, *Justus Liebigs Ann. Chem.*, 1436 (1978).
302. H. Lehmkuhl and D. Reinehr, *J. Organomet. Chem.*, **57,** 29 (1973).
303. H. Lehmkuhl and K. Mehler, *Justus Liebigs Ann. Chem.*, 1841 (1978).
304. H. Lehmkuhl, D. Reinehr, K. Mehler, G. Schomburg, H. Kötter, D. Henneberg, and G. Schroth, *Justus Liebigs Ann. Chem.*, 1449 (1978).
305. M. Yu. Lukina, T. Yu. Rudavshevskaya, and O. A. Nesmeyanova, *Dokl. Akad. Nauk. SSSR*, **190,** 1109 (1970).
306. H. G. Richey, Jr. and R. M. Bension, *J. Org. Chem.*, **45,** 5036 (1980), and references cited therein.
307. J. J. Eisch and J. H. Merkley, *J. Am. Chem. Soc.*, **101,** 1148 (1979), and references cited therein.
308. J. J. Eisch, J. H. Merkley, and J. E. Galle, *J. Org. Chem.*, **44,** 587 (1979).
309. (a) J. J. Eisch and N. E. Burlinson, *J. Am. Chem. Soc.*, **98,** 753 (1976); (b) J. J. Eisch, N. E. Burlinson, and M. Boleslawski, *J. Organomet. Chem.*, **111,** 137 (1976).
310. J. J. Eisch and C. K. Hordis, *J. Am. Chem. Soc.*, **93,** 2974, 4496 (1971).
311. M. Cherest, H. Felkin, C. Frajerman, C. Lion, G. Roussi and G. Swierczewski, *Tetrahedron Lett.*, 875 (1966).
312. Y. Frangin and M. Gaudemar, *Bull. Soc. Chim. Fr.*, 1173 (1976).

313. H. G. Richey, Jr., and M. S. Domalski, *J. Org. Chem.*, **46**, 3780 (1981).
314. H. G. Richey, Jr., L. M. Moses, M. S. Domalski, W. F. Erickson, and A. S. Heyn, *J. Org. Chem.*, **46**, 3773 (1981), and references cited therein.
315. G. Bouet, R. Mornet and L. Gouin, *J. Organomet. Chem.*, **135**, 151 (1977); R. Mornet and L. Gouin, **86**, 57, 297 (1975).
316. J. J. Eisch and J. E. Galle, *J. Org. Chem.*, **44**, 3277 (1979).
317. J. G. Duboudin and B. Jousseaume, *J. Organomet. Chem.*, **162**, 209 (1978); **44**, Cl (1972); *C.R. Acad. Sci., Ser. C*, **276**, 1421 (1973).
318. R. W. T. Ten Hoedt, G. van Koten, and J. G. Noltes, *J. Organomet. Chem.*, **170**, 131 (1979).
319. M. Michman and M. Balog, *J. Organomet. Chem.*, **31**, 395 (1971).
320. L. M. Zubritskii, L. N. Cherkasov, T. N. Fomina, and Kh. V. Bal'yan, *J. Org. Chem. USSR* (*Engl. Transl.*), **11**, 204 (1975); see also L. M. Zubritskii, T. N. Fomina, and Kh. V. Bal'yan, *J. Org. Chem. USSR* (*Engl. Transl.*), **17**, 63 (1981).
321. H. Felkin, L. D. Kwart, G. Swierczewski, and J. D. Umpleby, *J. Chem. Soc., Chem. Commun.*, 242 (1975).
322. H. Lehmkuhl and H. Nehl, *J. Organomet. Chem.*, **60**, 1, (1973).
323. H. Lehmkuhl and O. Olbrysch, *Justus Liebigs Ann. Chem.*, 1162 (1975).
324. (a) G. Courtois, B. Mauzé, and L. Miginiac, *J. Organomet. Chem.*, **72**, 309 (1974); (b) G. Courtois and L. Miginiac, **69**, 1 (1974).
325. H. Lehmkuhl and R. McLane, *Justus Liebigs Ann. Chem.*, 736 (1980).
326. G. Courtois, B. Mauzé, and L. Miginiac, *C.R. Acad. Sci., Ser. C.*, **285**, 207 (1977).
327. F. Bernadou and L. Miginiac, *Tetrahedron Lett.*, 3083 (1976).
328. M. T. Bertrand, G. Courtois and L. Miginiac, *C.R. Acad. Sci. Ser. C*, **280**, 999 (1975); D. Mesnard and L. Miginiac, *J. Organomet. Chem.*, **117**, 99 (1976).
329. (a) G. Courtois and L. Miginiac, *J. Organomet. Chem.*, **52**, 241 (1973); (b) H. Lehmkuhl, I. Döring, and H. Nehl, *J. Organomet. Chem.*, **221**, 7 (1981); H. Lehmkuhl, I. Döring, R. McLane, and H. Nehl, *J. Organomet. Chem.*, **221**, 1 (1981).
330. J. Auger, G. Courtois and L. Miginiac, *J. Organomet. Chem.*, **133**, 285 (1977); G. Courtois and L. Miginiac, *J. Organomet. Chem.*, **195**, 13 (1980).
331. R. Koster, *Justus Liebigs Ann. Chem.*, **618**, 31 (1958).
332. B. M. Mikhailov, *Organomet. Chem. Rev.*, **8A**, 1 (1972).
333. (a) Yu. N. Bubnov, O. A. Nesmeyanova, T. Yu. Rudashevskaya, B. M. Mikhailov, and B. A. Kazanskii, *Tetrahedron Lett.*, 2153 (1971); *J. Gen. Chem. USSR* (*Engl. Transl.*), **43**, 125, 132 (1973); (b) B. M. Mikhailov, Yu, N. Bubnov, O. A. Nesmeyanova, V. K. Kiselev, T. Yu. Rudashevskaya, and B. A. Kazanskii, *Tetrahedron Lett.*, 4627 (1972).
334. F. Joy, M. F. Lappert, and B. Prokai, *J. Organomet. Chem.*, **5**, 506 (1966).
335. N. Miyamoto, S. Isiyama, K. Utimoto, and H. Nozaki, *Tetrahedron*, **29**, 2365 (1973); *Tetrahedron Lett.*, 4597 (1971).
336. B. M. Mikhailov, Yu. N. Bubnov, and M. Sh. Grigoryan, *J. Gen. Chem. USSR* (*Engl. Transl.*), **44**, 2425, 2669 (1974).
337. B. M. Mikhailov and V. N. Smirnov, *Bull. Acad. Sci. USSR, Div. Chem. Sci.*, **23**, 1079 (1974); B. M. Mikhailov, Yu. N. Bubnov, M. Sh. Grigoryan, and V. S. Bobdanov, *J. Gen. Chem. USSR* (*Engl. Transl.*), **44**, 2669 (1974).
338. B. Wrackmeyer and H. Nöth, *J. Organomet. Chem.*, **108**, C21 (1976); L. Killian and B. Wrackmeyer, *J. Organomet. Chem.*, **132**, 213 (1977).
339. M. F. Lappert and B. Prokai, *J. Organomet. Chem.*, **1**, 384 (1964), J. J. Eisch and L. J. Gonsoir, quoted in *Adv. Organomet. Chem.*, **16**, 67 (1977).
340. G. Bruno, *The Use of Aluminium Alkyls in Organic Synthesis*, Ethyl Corp., Baton Rouge, La., 1970, p. 75.
341. K. Ziegler, *Adv. Organomet. Chem.*, **6**, 1 (1968); K. Ziegler, H.-G. Gellert, K. Zosel, E. Holzkamp, J. Schneider, M. Söll, and W.-R. Kroll, *Justus Liebigs Ann. Chem.*, **629**, 121 (1960); K. Ziegler, H.-G. Gellert, E. Holzkamp, G. Wilke, E. W. Duck, and W.-R. Kroll, *Justus Liebigs Ann. Chem.*, **629**, 172 (1960).
342. K. Ziegler, H.-G. Gellert, H. Martin, K. Nagel, and J. Schneider, *Justus Liebigs Ann. Chem.*, **589**, 91 (1954).
343. K. Ziegler, F. Krupp, and K. Zosel, *Justus Liebigs Ann. Chem.*, **629**, 241 (1960).
344. H. Sinn and W. Kamińsky, *Adv. Organomet. Chem.*, **18**, 99 (1980).

345. J. P. Kennedy and A. W. Langer, Jr., *Fortschr. Hochpolym. Forsch.*, **3**, 508 (1964); K. J. Ivin, J. J. Rooney, C. D. Stewart, M. L. H. Green, and R. Mahtab, *J. Chem. Soc., Chem. Commun.*, 604 (1978).
346. R. Schimpf and P. Heimbach, *Chem. Ber.*, **103**, 2122 (1970).
347. G. Hata and A. Miyake, *J. Org. Chem.*, **28**, 3237 (1963).
348. A. Stefani, *Helv. Chim. Acta*, **57**, 1346 (1974).
349. G. Wilke and H. Müller, *Justus Liebig Ann. Chem.*, **629**, 222 (1960).
350. A. M. Caporusso, G. Giacomelli, and L. Lardicci, *J. Chem. Soc., Perkin Trans. 1*, 1900 (1981).
351. J. J. Eisch and W. C. Kaska, *J. Organomet. Chem.*, **2**, 184 (1964); T. Mole and J. R. Surtees, *Chem. Ind.* (*Lond*), 1727 (1963).
352. J. J. Eisch and R. Amtmann, *J. Org. Chem.*, **37**, 3410 (1972).
353. J. J. Eisch and W. C. Kaska, *J. Am. Chem. Soc.*, **88**, 2213, 2976 (1966).
354. J. J. Eisch and K. C. Fichter, *J. Am. Chem. Soc.*, **96**, 6815 (1974).
355. J. J. Eisch, R. Amtmann, and M. W. Foxton, *J. Organomet. Chem.*, **16**, P55 (1969).
356. M. D. Schiavelli, J. J. Plunkett and D. W. Thompson, *J. Org. Chem.*, **46**, 807 (1981), and references cited therein.
357. T. Yoshida and E.-I. Negishi, *J. Am. Chem. Soc.*, **103**, 4985 (1981), and references cited therein.
358. C. L. Rand, D. E. van Horn, M. W. Moore, and E.-I. Negishi, *J. Org. Chem.*, **46**, 4093 (1981).
359. O. A. Reutov, V. I. Rozenberg, G. V. Gavrilova, and V. A. Nikanorov, *J. Organomet. Chem.*, **177**, 101 (1979).
360. S. Fukuzumi, K. Mochida, and J. K. Kochi, *J. Am. Chem. Soc.*, **101**, 5961 (1979).
361. H. C. Gardiner and J. K. Kochi, *J. Am. Chem. Soc.*, **98**, 2460 (1976).
362. H. C. Gardiner and J. K. Kochi, *J. Am. Chem. Soc.*, **98**, 558, (1976).
363. e.g. see (a) T. Kauffmann, *Angew. Chem., Int. Ed. Engl.*, **13**, 627 (1974); (b) G. F. Luteri and W. T. Ford, *J. Org. Chem.*, **42**, 820 (1977), and references cited therein.
364. T. Kauffmann and R. Eidenschink, *Chem. Ber.*, **110**, 645 (1977).
365. T. Kauffmann and R. Eidenschink, *Angew. Chem., Int. Ed. Engl.*, **12**, 568 (1973); *Chem. Ber.*, **110**, 651 (1977).
366. T. Kauffmann, A. Busch, K. Habersaat and E. Köppelmann, *Angew. Chem., Int. Ed. Engl.*, **12**, 569 (1973).
367. W. T. Ford and G. F. Luteri, *J. Am. Chem. Soc.*, **99**, 5330 (1977).
368. T. Kauffmann and E. Köppelmann, *Angew. Chem., Int. Ed. Engl.*, **11**, 290 (1972); T. Kauffmann, K. Habersaat, and E. Köppelmann, *Angew. Chem., Int. Ed. Engl.*, **11**, 291 (1972).
369. G. Boche and D. Martens, *Angew. Chem., Int. Ed. Engl.*, **11**, 724 (1972); G. Boche, D. Martens, and H.-U. Wagner, *J. Am. Chem. Soc.*, **98**, 2668 (1976).
370. T. Kauffmann, K. Haberssat, and E. Köppelmann, *Chem. Ber.*, **110**, 638 (1977).
371. G. F. Luteri and W. T. Ford, *J. Organomet. Chem.*, **105**, 139 (1976).
372. W. T. Ford, *J. Org. Chem.*, **36**, 3979 (1971).
373. G. R. Buske and W. T. Ford, *J. Org. Chem.*, **41**, 1995 (1976).
374. J. G. Duboudin, B. Jousseaume, and M. Pinet-Vallier, *J. Organomet. Chem.*, **172**, 1 (1979); *J. Chem. Soc., Chem. Commun.*, 454 (1977).
375. L.-H. Chan and E. G. Rochow, *J. Organomet. Chem.*, **9**, 231 (1967).
376. D. J. Berry and B. J. Wakefield, *J. Chem. Soc., C*, 642 (1971).
377. A. D. Petrov, E. B. Sokolova, and C. L. Kao, *Zh. Obshch. Khim.*, **30**, 1107 (1960).
378. J. A. Gautier, M. Miocque, and L. Mascrier-Demagny, *Bull. Soc. Chim. Fr.*, 1551 (1967).
379. R. L. Jones, D. E. Pearson, and M. Gordon, *J. Org. Chem.*, **37**, 3369 (1972).
380. P. L. Pickard and T. L. Tolbert, *J. Org. Chem.*, **26**, 4886 (1961).
381. H. Edelstein and E. I. Becker, *J. Org. Chem.*, **31**, 3375 (1966).
382. M. Chastrette, R. Amouroux, and M. Subit, *J. Organomet. Chem.*, **99**, C41 (1975).
383. P. Canonne, G. B. Foscolos, and G. Lemay, *Tetrahedron Lett.*, **21**, 155 (1980).
384. J. R. Jennings, J. E. Lloyd, and K. Wade, *J. Chem. Soc.*, 2662, 5083 (1965).
385. (a) G. Sonnek, K.-G. Baumgarten, H. Reinheckel, S. Pasynkiewicz, and K. B. Starowieyski, *J. Organomet. Chem.*, **150**, 21 (1978); (b) S. Pasynkiewicz and S. Maciaszek, *J. Organomet. Chem.*, **15**, 301 (1968).

386. L. Bagnell, E. A. Jeffrey, A. Meisters, and T. Mole, *Aust. J. Chem.*, **27,** 2577 (1974).
387. L. Lardicci and G. P. Giacomelli, *J. Organomet. Chem.*, **33,** 293 (1971).
388. I. Pattison and K. Wade, *J. Chem. Soc., A,* 57 (1968).
389. G. Rousseau and J. M. Conia, *Tetrahedron Lett.*, **22,** 649 (1981).
390. F. G. Saitkulova, G. G. Abashev, and I. I. Lapkin, *J. Org. Chem. USSR* (*Engl. Transl.*), **12,** 978 (1976).
391. H. B. Kagan and Y.-H. Suen, *Bull. Soc. Chim. Fr.*, 1819 (1966).
392. M. Bellassoued and M. Gaudemar, *J. Organomet. Chem.*, **81,** 139 (1974).
393. M. Gaudemar, *Bull. Soc. Chim. Fr.*, 974 (1962).
394. K. L. Rinehart, *Org. Synth.*, Coll. Vol. IV, 120 (1963).
395. H. Gilman and L. A. Woods, *J. Am. Chem. Soc.*, **67,** 520 (1945).
396. G. P. Giacomelli and L. Lardicci, *Chem. Ind.* (*London*), 689 (1972).
397. B. M. Mikhailov, *Organomet. Chem. Rev.*, **8A,** 23 (1972).
398. A. Meller and W. Gerger, *Monatsh. Chem.*, **105,** 684 (1974).
399. R. Snaith, K. Wade, and B. K. Wyatt, *J. Chem. Soc., A,* 380 (1970); C. Summerford and K. Wade, *J. Chem. Soc., A,* 2010 (1970); C. Summerford, K. Wade, and B. K. Wyatt, *J. Chem. Soc., A,* 2016 (1970); B. Samuel, R. Snaith, C. Summerford, and K. Wade, *J. Chem. Soc., A,* 2019 (1970).
400. R. Snaith, C. Summerford, K. Wade, and B. K. Wyatt, *J. Chem. Soc., A,* 2635 (1970).
401. I. Pattison, K. Wade, and B. K. Wyatt, *J. Chem. Soc., A,* 837 (1968).
402. E. C. Ashby, L. C. Chao, and H. M. Neumann, *J. Am. Chem. Soc.*, **95,** 5186 (1973).
403. (a) R. A. Anderson and G. E. Coates, *J. Chem. Soc., Dalton Trans.*, 1171 (1974); (b) B. Hall, J. B. Farmer, H. M. M. Shearer, J. D. Sowerby, and K. Wade, *J. Chem. Soc., Dalton Trans.*, 102 (1979).
404. E. C. Ashby, L.-C. Chao, and H. M. Neumann, *J. Am. Chem. Soc.*, **95,** 4896 (1973).
405. S. J. Storfer and E. I. Becker, *J. Org. Chem.*, **27,** 1868 (1962).
406. (a) A. A. Scala, N. M. Bikales, and E. I. Becker, *J. Org. Chem.*, **30,** 303 (1965); (b) L. S. Cook and B. J. Wakefield, *J. Chem. Soc., Perkin Trans. 1,* 2392 (1980); *Tetrahedron Lett.*, 147 (1976).
407. G. Alvernhe and A. Laurent, *Tetrahedron Lett.*, 1057 (1973).
408. R. Gauthier, G. P. Axiotis, and M. Chastrette, *J. Organomet. Chem.*, **140,** 245 (1977); *Tetrahedron Lett.*, 23 (1977).
409. B. B. Allen and H. R. Henze, *J. Am. Chem. Soc.*, **61,** 1790 (1939); H. R. Henze and T. R. Thompson, *J. Am. Chem. Soc.*, **65,** 1422 (1943); H. R. Henze, B. B. Allen, and W. B. Leslie, *J. Am. Chem. Soc.*, **65,** 87 (1943); H. R. Henze and L. R. Swett, *J. Am. Chem. Soc.*, **73,** 4918 (1951).
410. C. E. Rehberg and H. R. Henze, *J. Am. Chem. Soc.*, **63,** 2785 (1941).
411. G. E. Niznik, W. H. Morrison, III, and H. M. Walborsky, *J. Org. Chem.*, **39,** 600 (1974), and references cited therein; N. Hirowatari and H. M. Walborsky, *J. Org. Chem.*, **39,** 604 (1974); G. E. Niznik and H. M. Walborsky, *J. Org. Chem.*, **39,** 608 (1974); M. P. Periasamy and H. M. Walborsky, *J. Org. Chem.*, **39,** 611 (1974).
412. J. Casonova, in *Isonitrile Chemistry* (Ed. I. Ugi), Academic Press, New York, 1971, p. 169.
413. (a) J. Thomas, *Bull. Soc. Chim. Fr.*, 1296 (1973); (b) 1300 (1973).
414. H. Gilman, A. H. Haubein, G. O'Donnell, and L. A. Woods, *J. Am. Chem. Soc.*, **67,** 922 (1945).
415. N. Goasdoue and M. Gaudemar, *J. Organomet. Chem.*, **125,** 9 (1977).
416. H. Gilman and M. Speeter, *J. Am. Chem. Soc.*, **65,** 2255 (1943).
417. F. Dardoize and M. Gaudemar, *Bull. Soc. Chim. Fr.*, 1561 (1976).
418. J. M. Klerks, D. J. Stufkens, G. van Koten, and K. Vrieze, *J. Organomet. Chem.*, **181,** 271 (1979).
419. B. L. Emling, R. J. Horvath, A. J. Saraceno, E. F. Ellermeyer, L. Haile, and L. D. Hudac, *J. Org. Chem.*, **24,** 657 (1959).
420. K. N. Campbell, C. H. Helbing, M. P. Florkowski, and B. K. Campbell, *J. Am. Chem. Soc.*, **70,** 3868 (1948).
421. R. B. Moffat, *Org. Synth.*, Coll. Vol. V., 605 (1963).
422. J. Thomas, *Bull. Soc. Chim. Fr.*, 209 (1975).
423. G. Stork and S. R. Dowd, *J. Am. Chem. Soc.*, **85,** 2178 (1963).

424. H. Thies and H. Schoenenberger, *Chem. Ber.*, **89,** 1918 (1956).
425. S. Arseniyadis, J. Goré, A. Laurent, and M.-L. Roumestant, *J. Chem. Res. (S),* 416 (1978); B. Kryczka, A. Laurent, and B. Marquet, *Tetrahedron,* **34,** 3291 (1978).
426. (a) L. I. Sivova, N. A. Sivov, R. A. Gracheva, and V. M. Potapov, *J. Org. Chem. USSR (Engl. Transl.),* **14,** 731 (1978); (b) J.-L. Moreau, and M. Gaudemar, *Bull. Soc. Chim. Fr.,* 1211 (1975).
427. Y. Ohgo, Y. Konda, and J. Yoshimura, *Bull. Chem. Soc. Jpn.,* **46,** 1892 (1973).
428. H. Gilman and J. J. Eisch, *J. Am. Chem. Soc.,* **79,** 2150 (1957).
429. B. Mauzé, and L. Miginiac, *Bull. Soc. Chim. Fr.,* 1078, 1082 (1973).
430. J. T. B. H. Jastrzebski, J. M. Klerks, G. van Koten, and K. Vrieze, *J. Organomet. Chem.,* **210,** C49 (1981).
431. S.-I. Hashimoto, S. I. Yamada, and K. Koga, *Chem. Pharm. Bull.,* **27,** 771 (1979); *J. Am. Chem. Soc.,* **98,** 7450 (1976).
432. A. I. Meyers and C. E. Whitten, *J. Am. Chem. Soc.,* **97,** 6266 (1975).
433. A. I. Meyers and C. E. Whitten, *Tetrahedron Lett.,* 1947 (1976); *Heterocycles,* **4,** 1687 (1976).
434. C. S. Giam, T. E. Goodwin, K. F. Rion, and S. D. Abbott, *J. Chem. Soc., Perkin Trans. 1,* 3082 (1979).
435. H. Gilman, J. J. Eisch, and T. Soddy, *J. Am. Chem. Soc.,* **79,** 1245 (1957).
436. T. Kauffmann, J. König, and A. Woltermann, *Chem. Ber.,* **109,** 3864 (1976).
437. (a) C. E. Crawforth, O. Meth-Cohn, and C. A. Russell, *J. Chem. Soc., Perkin Trans. 1,* 2807 (1972); (b) D. I. C. Scopes and J. A. Joule, *J. Chem. Soc., Perkin Trans. 1,* 2810 (1972); (c) J. J. Eisch and D. R. Comfort, *J. Organomet. Chem.,* **38,** 209 (1972).
438. D. Bryce-Smith, P. J. Morris, and B. J. Wakefield, *J. Chem. Soc., Perkin Trans. 1,* 1977 (1976), and references cited therein.
439. R. E. Parker and N. S. Issacs, *Chem. Rev.,* **59,** 737 (1959); J. K. Crandall, J. P. Arrington, and J. Hen, *J. Am. Chem. Soc.,* **89,** 6208 (1967).
440. J. F. Le Borgne, *J. Organomet. Chem.,* **122,** 139 (1976).
441. P. A. Wender, J. M. Erhardt, and L. J. Letendre, *J. Am. Chem. Soc.,* **103,** 2114 (1981).
442. G. E. Coates and A. H. Fishwick, *J. Chem. Soc., A,* 477 (1968).
443. E. E. Preger, *Org. Synth.,* Coll. Vol. 1, 2nd ed., 306 (1964).
444. S. Warwel, G. Schmitt, and B. Ahlfaenger, *Synthesis,* 632 (1975).
445. E.-I. Negishi, S. Baba, and A. O. King, *J. Chem. Soc., Chem. Commun.,* 17 (1976).
446. D. B. Malpass, S. C. Watson, and G. S. Yeargin, *J. Org. Chem.,* **42,** 2712 (1977).
447. E. I. Negishi, *Organometallics in Organic Synthesis,* Vol. 1, Wiley, New York, 1980.
448. Y. Yasuda and H. Hani, *J. Polym. Sci., Polym. Chem.,* **11,** 3103 (1973); C. Coulon, N. Spassky, and P. Sigwalt, *Polymer,* **17,** 821 (1976).
449. J. P. Marino and D. M. Floyd, *Tetrahedron Lett.,* 675 (1979); R.-D. Acker, *Tetrahedron Lett.,* 3407 (1977).
450. C. Huynh, F. Derguini-Boumechal, and G. Linstrumelle, *Tetrahedron Lett.,* 1503 (1979).
451. W. H. Glaze, D. P. Duncan, and D. J. Berry, *J. Org. Chem.,* **42,** 694 (1977).
452. K. Utimoto, K. Uchida, and H. Nozaki, *Tetrahedron Lett.,* 4527 (1973).
453. G. Balme, A. Doutheau, J. Gore, and M. Malacria, *Synthesis,* 508 (1979).
454. R. J. Anderson, *J. Am. Chem. Soc.,* **92,** 4978 (1970); R. W. Herr and C. R. Johnson, *J. Am. Chem. Soc.,* **92,** 4979 (1970).
455. D. M. Wieland and C. R. Johnson, *J. Am. Chem. Soc.,* **93,** 3047 (1971); J. Staroscik and B. Rickborn, *J. Am. Chem. Soc.,* **93,** 3046 (1971).
456. M. Apparu and M. Barrelle, *Bull. Soc. Chim. Fr.,* 947 (1977).
457. F. G. Bordwell, H. M. Andersen, and B. M. Pitt, *J. Am. Chem. Soc.,* **76,** 1082 (1954).
458. A. P. Kozikowski, H. Ishida, and K. Isobe, *J. Org. Chem.,* **44,** 2788 (1979).
459. S. Searles, *J. Am. Chem. Soc.,* **73,** 124 (1951).
460. M. Morton and R. E. Kammereck, *J. Am. Chem. Soc.,* **92,** 3217 (1970).
461. J. F. Biellmann, J. B. Ducep, J. L. Schmitt, and J. J. Vicens, *Tetrahedron,* **32,** 1061 (1976).
462. P. Beltrame, G. Gelli, A. Lai, and M. Monduzzi, *J. Org. Chem.,* **41,** 580 (1976); S. Cabiddu, S. Melis, E. Marongiu, P. P. Piras, and G. Podda, *J. Organomet. Chem.,* **111,** 249 (1976); S. Cabiddu, E. Marongiu, S. Melis, and F. Sotgiu, *J. Organomet. Chem.,* **116,** 275 (1976).

463. T. G. Brilkina and V. A. Shushunov, in *Reactions of Organometallic Compounds with Oxygen and Peroxides,* Iliffe Books, London, 1969; A. G. Davies, in *Organic Peroxides,* Butterworths, London, 1961; G. Sosnovsky and J. H. Brown, *Chem. Rev.,* **66,** 529 (1966).
464. Yu. A. Aleksandrov, S. A. Lebedev, and N. V. Kuznetsova, *Zh. Obshch. Khim.,* **48,** 1659 (1978); Yu. A. Aleksandrov, S. A. Lebedev, N. V. Kuznetsova, and G. A. Razuvaev, *Dokl. Akad. Nauk. SSSR,* **245,** 842 (1979).
465. K. Ziegler, F. Krupp, and K. Zosel, *Angew. Chem.,* **629,** 241 (1960); see also M. A. Margulis, L. K. Putilina, A. T. Menyailo, and I. M. Krikun, *Russ. J. Phys. Chem.,* **46,** 1005 (1972).
466. M. M. Midland and H. C. Brown, *J. Am. Chem. Soc.,* **95,** 4069 (1973).
467. M. H. Abraham, *J. Chem. Soc.,* 4130 (1960).
468. H. Hock and F. Ernst, *Chem. Ber.,* **92,** 2716 (1959).
469. C. Walling and S. A. Buckler, *J. Am. Chem. Soc.,* **75,** 4372 (1953); **77,** 6032 (1955).
470. H. Hock and F. Ernst, *Chem. Ber.,* **92,** 2732 (1959).
471. F. C. Whitmore and A. R. Lux, *J. Am. Chem. Soc.,* **54,** 3449 (1932).
472. M. T. Goebel and C. S. Marvel, *J. Am. Chem. Soc.,* **55,** 1693 (1933).
473. E. Müller and T. Töpel, *Chem. Ber.,* **72,** 273 (1939).
474. A. M. Sladkov, V. A. Merkevich, I. A. Yavich, V. N. Chernov, and L. K. Luneva, *Dokl. Akad. Nauk SSSR,* **119,** 1159 (1958); A. M. Sladkov and L. K. Luneva, *J. Gen. Chem. USSR,* **28,** 2894 (1958).
475. R. Masthoff, *Z. Anorg. Allg. Chem.,* **336,** 252 (1965).
476. H. E. Seyfarth, J. Henkel, and A. Rieche, *Angew. Chem., Int. Ed. Engl.,* **4,** 1074 (1965).
477. H. C. Brown, M. M. Midland, and G. W. Kabalka, *J. Am. Chem. Soc.,* **93,** 1024 (1971).
478. H. C. Brown, *Organic Syntheses via Boranes,* Wiley, New York, 1975.
479. Yu. A. Aleksandrov, G. N. Figurova, V. A. Dodonov, and G. A. Razuvaev, *Dokl. Chem.* (Engl. Translation), **195,** 895 (1970).
480. E. J. Panek and G. M. Whitesides, *J. Am. Chem. Soc.,* **94,** 8768 (1972).
481. A. G. Davies and B. P. Roberts, *J. Chem. Soc., B,* 1074 (1968).
482. A. G. Davies and B. P. Roberts, *J. Chem. Soc., B,* 311, 317 (1969).
483. A. G. Davies and J. E. Packer, *J. Chem. Soc.,* 3164 (1959).
484. A. G. Davies and B. P. Roberts, *J. Chem. Soc., B,* 17 (1967).
485. Yu. A. Aleksandrov, G. N. Figurova, and G. M. Razuvaev, *J. Organomet. Chem.,* **57,** 71 (1973).
486. I. P. Beletskaya, G. A. Artamkina, A. Yu. Karmitov, and A. L. Buchachenko, *Izv. Akad. Nauk SSSR, Ser. Khim.,* 2409 (1980).
487. G. A. Russell, in *Peroxide Reaction Mechanisms* (Ed. J. O. Edwards), Interscience, New York, 1962, p. 107.
488. J. F. Garst, C. D. Smith, and A. C. Farrar, *J. Am. Chem. Soc.,* **94,** 7707 (1972); H. A. Pacevitz and H. Gilman, *J. Am. Chem. Soc.,* **61,** 1513 (1939); R. L. Bardsley and D. M. Hercules, *J. Am. Chem. Soc.,* **90,** 4545 (1968); G. A. Razuvaev, G. G. Petukhov, R. F. Galiulina, and N. N. Shabanova, *J. Gen. Chem. USSR (Engl. Transl.),* **34,** 3863 (1964).
489. A. M. Jones and C. A. Russell, *J. Chem. Soc., C,* 2246 (1969).
490. G. Fraenkel and M. J. Geckle, *J. Chem. Soc., Chem. Commun.,* 55 (1980).
491. G. di Maio, A. Colantoni, and E. Zeuli, *Gazz. Chim. Ital.,* **103,** 477 (1973).
492. H. Lehmkuhl, *Justus Liebigs Ann. Chem.,* **719,** 20 (1968).
493. E. J. Panek and G. M. Whitesides, *J. Am. Chem. Soc.,* **94,** 8768 (1972).
494. G. M. Whitesides, E. J. Panek, and E. R. Stedronsky, *J. Am. Chem. Soc.,* **94,** 232 (1972).
495. P. Warner and S. L. Wu, *J. Org. Chem.,* **41,** 1459 (1976).
496. E. J. Panek, L. R. Kaiser, and G. M. Whitesides, *J. Am. Chem. Soc.,* **99,** 3708 (1977).
497. G. A. Razuvaev, S. F. Zhilt'sov, O. N. Druzhkov, and G. G. Petukhov, *Dokl. Akad. Nauk SSSR,* **152,** 633 (1963); Yu. A. Aleksandrov, O. N. Druzhkov, S. F. Zhilt'sov and G. A. Razuvaev, *Dokl. Akad. Nauk SSSR,* **157,** 1395 (1964); *J. Gen. Chem. USSR* (*Engl. Transl.*), **35,** 1444 (1965); F. R. Jensen and D. Heyman, *J. Am. Chem. Soc.,* **88,** 3438 (1966).
498. Yu. A. Aleksandrov, B. A. Radbil', and V. A. Shushunov, *J. Gen. Chem. USSR (Engl. Transl.),* **37,** 190 (1967); Yu. A. Aleksandrov and B. A. Radbil', *J. Gen. Chem. USSR (Engl. Transl.),* **37,** 2230 (1967).
499. J. L. Wardell, in *The Chemistry of the Thiol Group* (Ed. S. Patai), Wiley, London, 1974, p. 211, and references cited therein.

500. J. F. Boscato, J.-M. Catala, E. Franta, and J. Brossas, *Tetrahedron Lett.*, 1519 (1980); *Makromol. Chem.*, **180,** 1571 (1979).
501. M. Hallensleben, *Makromol. Chem.*, **175,** 3315 (1974).
502. E. P. Painter, *J. Am. Chem. Soc.*, **69,** 229 (1947).
503. D. G. Foster, *Org. Synth.*, Coll. Vol. 3, 771 (1955).
504. M. Sato and T. Yoshida, *J. Organomet. Chem.*, **87,** 217 (1975).
505. E. S. Gould and J. D. McCullough, *J. Am. Chem. Soc.*, **73,** 1109 (1951).
506. B. Kohne, W. Lohner, K. Praefcke, H. J. Jakobsen, and B. Villadsen, *J. Organomet. Chem.*, **166,** 373 (1979).
507. E. Niwa, H. Aoki, H. Tanaka, K. Munakata, and M. Namiki, *Chem. Ber.*, **99,** 3215 (1966).
508. J. Meinwald, D. Dauplaise, F. Wudl, and J. J. Hauser, *J. Am. Chem. Soc.*, **99,** 255 (1977).
509. S. I. Radchenko and A. A. Petrov, *J. Org. Chem. USSR* (*Engl. Transl.*), **13,** 36 (1977).
510. D. Seebach and A. K. Beck, *Chem. Ber.*, **108,** 314 (1975).
511. J.-L. Piette, R. Lysy, and M. Renson, *Bull. Soc. Chim. Fr.*, 3559 (1972).
512. L. I. Zakharkin and V. V. Gavrilenko, *Bull. Acad. Sci. USSR* (*Engl. Transl.*), 1260 (1960); A. P. Kozikowski and A. Ames, *J. Org. Chem.*, **43,** 2735 (1978).
513. H. Schumann and M. Schmidt, *Angew. Chem., Int. Ed. Engl.*, **4,** 1007 (1965).
514. P. Mazerolles, J. Dubac, and M. Lesbre, *J. Organomet. Chem.*, **5,** 35 (1966).
515. W. Kitching and C. W. Fong, *Organomet. Chem. Rev.*, **5A,** 281 (1970).
516. U. Kunze, *Rev. Silicon, Germanium, Tin and Lead Compounds,* **2,** 251 (1977).
517. W. E. Truce and A. M. Murphy, *Chem. Rev.*, **48,** 69 (1951).
518. C. S. Marvel and R. S. Johnson, *J. Org. Chem.*, **13,** 822 (1948).
519. W. E. Truce and J. F. Lyons, *J. Am. Chem. Soc.*, **73,** 126 (1951).
520. J. Dubac, P. Mazerolles, M. Joly, W. Kitching, C. W. Fong, and W. H. Atwell, *J. Organomet. Chem.*, **34,** 17 (1972).
521. J. Dubac and P. Mazerolles, *C.R. Acad. Sci., Ser. C,* **267,** 411 (1968).
522. E. B. Baker and H. H. Sisler, *J. Am. Chem. Soc.*, **75,** 5193 (1953).
523. A. V. Kuchin, D. J. Akhmetov, V. P. Yuveev, and G. A. Tolstikov, *Zh. Obshch. Khim.*, **49,** 401 (1979); **48,** 420 (1978).
524. J. Weidlein, *J. Organomet. Chem.*, **24,** 63 (1970).
525. J. Weidlein, *Z. Anorg. Allg. Chem.*, **366,** 22 (1969).
526. A. T. T. Hsieh, *J. Organomet. Chem.*, **27,** 293 (1971).
527. A. G. Lee, *J. Chem. Soc., A,* 467 (1970).
528. N. A. D. Carey and H. C. Clark, *Can. J. Chem.*, **46,** 649 (1968).
529. J. D. Koola and U. Kunze, *J. Organomet. Chem.*, **77,** 325 (1974).
530. E. Lindner and D. W. R. Frembs, *J. Organomet. Chem.*, **49,** 425 (1983).
531. F. Huber and F. J. Padberg, *Z. Anorg. Allg. Chem.*, **351,** 1 (1967).
532. G. B. Deacon and P. W. Felder, *Aust. J. Chem.*, **22,** 549 (1969).
533. C. W. Fong and W. Kitching, *J. Organomet. Chem.*, **21,** 365 (1970).
534. C. W. Fong and W. Kitching, *J. Organomet. Chem.*, **59,** 213 (1973).
535. J. Nasielsky, O. Buchman, M. Grosjean, and M. Jauquet, *J. Organomet. Chem.*, **19,** 353 (1969).
536. C. W. Fong and W. Kitching, *J. Organomet. Chem.*, **22,** 107 (1970).
537. H. Olapinski, J. Weidlein, and H.-D. Hausen, *J. Organomet. Chem.*, **64,** 193 (1974).
538. H. Olapinski and J. Weidlein, *J. Organomet. Chem.*, **35,** C53 (1972).
539. G. Sonnek, G. Müller, and K.-G. Baumgarten, *J. Organomet. Chem.*, **194,** 9 (1980).
540. C. Eaborn and T. Hashimoto, *Chem. Ind.* (*London*), 1081 (1961).
541. R. W. Bott, C. Eaborn, and T. Hashimoto, *J. Organomet. Chem.*, **3,** 442 (1965).
542. H. Schmidbaur, L. Sechser, and M. Schmidt, *Chem. Ber.*, **102,** 376 (1969).
543. H. Schmidbaur, L. Sechser, and M. Schmidt, *J. Organomet. Chem.*, **15,** 77 (1968).
544. J. Dubac and P. Mazerolles, *J. Organomet. Chem.*, **20,** P5 (1969).
545. R. W. Bott, C. Eaborn, and T. Hashimoto, *J. Chem. Soc.*, 3906 (1963).
546. R. Gelius and R. Müller, *Z. Anorg. Allg. Chem.*, **351,** 42 (1967).
547. K. A. Salib and J. B. Senior, *J. Chem. Soc., Chem. Commun.*, 1259 (1970).
548. R. Otto, *J. Prakt. Chem.*, **(2) 1,** 183 (1870).

549. C. Eaborn and R. W. Bott, in *Organometallic Compounds of the Group IV Elements* (Ed. A. G. MacDiamid), Vol. 1, Part 1, Marcel Dekker, New York, 1968, p. 425.
550. B. S. Wildi, S. W. Taylor and H. A. Potuatz, *J. Am. Chem. Soc.*, **73,** 1965 (1951); N. A. Nesmeyanov, V. A. Kalyarin, and O. A. Reutov, *Dokl. Chem.* (*Engl. Transl.*), **230,** 651 (1976).
551. H. Gilman and P. Schulze, *Recl. Trav. Chem. Pays-Bas,* **48,** 1129 (1929); G. Wittig and A. Schuhmacher, *Chem. Ber.,* **88,** 234 (1955); E. M. Kaiser and G. J. Bartling, *Tetrahedron Lett.,* 4357 (1969).
552. M. H. Abraham, J. H. N. Garland, J. A. Hill, and L. F. Larkworthy, *Chem. Ind.* (*London*), 1615 (1962).
553. R. Meier, *Chem. Ber.,* **86,** 1483 (1953); R. Meier and W. Frank, *Chem. Ber.,* **89,** 2747 (1956); R. Meir and K. Rappold, *Angew. Chem.,* **65,** 560 (1953).
554. A. C. M. Meesters, H. Rüeger, K. Rajeswari, and M. H. Benn, *Can. J. Chem.,* **59,** 264 (1981).
555. Y. Inagaki, R. Okazaki, and N. Inamoto, *Bull. Chem. Soc. Jpn.,* **48,** 621 (1975).
556. G. Bartoli, M. Bosco, A. Melandri, and A. C. Boicelli, *J. Org. Chem.,* **44,** 2087 (1979), and references cited therein.

The Chemistry of the Metal—Carbon Bond, Volume 2
Edited by F. R. Hartley and S. Patai


CHAPTER **5**

# Insertions into transition metal—carbon bonds

JOHN J. ALEXANDER
*Department of Chemistry, University of Cincinnati, Cincinnati, Ohio 45221, USA*

---

## I. INTRODUCTION

Insertion reactions, in which small molecules X—Y react with a transition metal—alkyl or —aryl complex to form products of 1,1- or 1,2-addition, are ubiquitous in organometallic chemistry.

MX(Y)C (1,1-addition product) (**1**)

MXYC (1,2-addition product) (**2**)

Equation 1 provides an example of 1,1-addition, the well known carbonylation reaction. Alkyne insertion (equation 2) gives a 1,2-addition product.

$$[MeMn(CO)_5] + CO \longrightarrow [MeCOMn(CO)_5] \quad (1)$$

$$trans\text{-}[PtCl(Me)(PMe_2Ph)_2] + MeCO_2C{\equiv}CCO_2Me \longrightarrow trans\text{-}[PtCl(PMe_2Ph)_2\{C(CO_2Me){=}C(Me)(CO_2Me)\}] \quad (2)$$

1,4-Additions, although less frequently observed, are also known.

$$[CpCr(NO)_2Me] + (CN)_2C{=}C(CN)_2 \longrightarrow [CpCr(NO)_2\{N{=}C{=}C(CN)C(Me)(CN)_2\}] \quad (3)$$

Insertion reactions often are reversible and the reverse is termed an elimination or extrusion reaction.

The name insertion reaction has no necessary mechanistic significance and refers only to the structure of the product. The molecules X—Y are unsaturated species capable of increasing the number of groups coordinated to X and/or Y. Examples are

$CO$, $SO_2$, $SO_3$, olefins, isocyanates, and $CO_2$. Oxidative addition reactions such as equation 4 might be thought of as an insertion of a metal-containing fragment into a

$$[Ir(CO)Cl(PPh_3)_2] + MeI \longrightarrow [MeIr(CO)(PPh_3)_2ClI] \quad (4)$$

C—I bond, for example. However, conventionally, the term insertion is employed for reactions in which the metal does not increase in oxidation state or in coordination number as in equations 1–3. The occurrence of insertion reactions is not limited to metal—carbon bonds[1] and has been observed *inter alia* for M—P, M—N, M—Cl and M—O bonds[2], where M may be either a transition or main-group metal. However, this chapter is concerned only with insertions into transition metal—carbon bonds.

Repeated insertions offer a mechanism for (*inter alia*) polymerization of small molecules; transition metal alkyl and aryl complexes have been actively studied as oligomerization and polymerization catalysts[3]. Insertion of small molecules followed perhaps by rearrangement or reaction of the insertion product also may afford a variety of other organic products on cleavage of the metal-containing fragment[4]. These may be catalytic or stoichiometric reactions. This chapter mentions few studies in which intermediate products of insertion in the reaction scheme have not been isolated. Only information obtained from isolable insertion products has been reported and analysed in order to limit the scope to manageable size. Emphasis is on results since the latest review articles in the literature, but background information has been provided.

Two important classes of insertion reactions are those in which the inserting molecule X—Y behaves either as a nucleophile or as an electrophile toward the metal—alkyl complex (in order to avoid lengthy qualification, alkyl will be understood to mean both alkyl and aryl unless otherwise specified). The paradigmatic cases are CO (nucleophilic behaviour) and $SO_2$ (electrophilic behaviour), and these will be treated first.

## II. CO INSERTION AND EXTRUSION. CARBONYLATION AND DECARBONYLATION

### A. General

The carbon monoxide insertion or carbonylation reaction converts metal alkyls to metal acyls (equation 1), and was discovered about 25 years ago. Several authoritative reviews on CO insertion are available[5]. This is probably the most thoroughly studied insertion reaction, and a great deal of what is known has resulted from studies on $[RMn(CO)_5]$ and $[CpFe(CO)_2R]$ systems, which may be regarded as paradigms for discussion of carbonylation and decarbonylation. Results on these complexes will be presented first. Aspects of carbonylation of square planar complexes will then be discussed. This will be followed by recent results whose significance is more apparent in the context thus set. A theoretical treatment of CO insertion with many excellent references has been given by Berke and Hoffmann[6].

### B. Studies on Mn and Fe Complexes

#### 1. *Intramolecular nature of insertion*

Labelling studies (equation 5) show that the insertion is intramolecular: the inserted CO is one already coordinated to the metal[7,8].

$$[MeMn(CO)_5] + {}^*CO \longrightarrow [(MeCO)Mn({}^*CO)(CO)_4] \qquad (5)$$

This suggests that carbonylation can also be effected by other nucleophiles[8] L Manganese complexes are rather insensitive to the nature of L, undergoing insertions on reaction with $I^-$, amines, phosphines, stibines, arsines, etc. Iron systems are more selective. Another feature of equation 5 deserves comment: the acyl product has the entering group (whether *CO or other[7]) *cis* to the acyl group. The initially produced *cis* isomer may subsequently rearrange to a *cis–trans* mixture[8,9].

### *2. Alkyl migration vs. 'CO insertion'*

Decarbonylation of *cis*-[(MeCO)Mn($^{13}$CO)(CO)$_4$] (the product in equation 5) proceeds with loss of a terminal CO to give isomers **3**, **4**, and **5** in a 2:1:1 ratio[7].

**(3)** **(4)** **(5)**

This result is consistent with production of a stereochemically rigid square pyramid and migration of the methyl group to occupy the position vacated by the departing CO. By the principle of microscopic reversibility, the carbonylation reaction must also involve alkyl migration. The alternative possibility is the 'CO insertion' mechanism, in which a coordinated CO moves to insert itself into the Mn—C bond is excluded by the decarbonylation experiment. If the CO insertion mechanism were operative, decarbonylation would be expected to lead only to **3**.

These results are consistent with the photochemically induced decarbonylation of [CpFe(CO)$_2$($^{13}$COMe)][10] in hexane which affords exclusively [CpFe(CO)($^{13}$CO)Me] with complete label retention, indicating loss of a terminal rather than the 'inserted' CO. If the decarbonylation is carried out in the presence of PPh$_3$, [CpFe(CO)($^{13}$CO)Me] and [CpFe(CO)(PPh$_3$)($^{13}$COMe)] result. This is consistent with a competition for the coordination position vacated by CO loss between methyl migration and PPh$_3$ attack. However, see Section II.B.3 for an instance of 'CO insertion'.

### *3. Stereochemical studies*

Retention of stereochemistry at the $\alpha$-carbon is observed in carbonylation reactions. Equilibration of [(+)$_D$-(PhCH$_2$*CH(Me)CO)Mn(CO)$_5$] with the alkyl in a CO atmosphere over several hours led to no change in optical rotation[11]. More recently, Whitesides and coworkers[12] have shown that carbonylation of [(*threo*-(Me$_3$CCHDCHD)Fe(CO)$_2$Cp] with L = PPh$_3$ or Bu$^t$NC leads to [(*threo*-(Me$_3$CCHDCHDCO)Fe(CO)LCp] with configuration retention.

As mentioned previously, insertion in octahedral Mn complexes leads to a kinetic product having the entering nucleophile and the acyl group *cis*. A number of investigations of the stereochemistry about chiral iron in pseudo-tetrahedral complexes have shown sometimes retention and sometimes inversion. Attig and Wojcicki[13] showed that the photochemical decarbonylation of diastereomerically related enantiomers of $[(\eta^5\text{-MeC}_5H_4)^*Fe(CO)(PPh_3)(COMe)]$ proceeded with >84% stereospecificity. Davison and Martinez[14] demonstrated that the decarbonylation of equation 6 led to inversion of stereochemistry at Fe in a topological sense. Consistent with this result is cleavage of a terminal CO followed by alkyl migration.

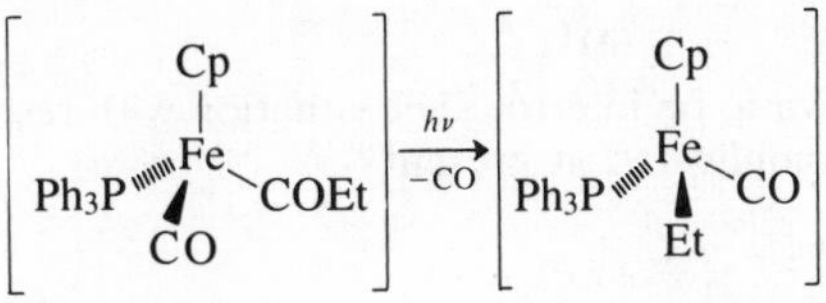

$$[\mathrm{Cp(Ph_3P)(CO)Fe{-}COEt}]\ (+)_{546}{-}R \xrightarrow[-\mathrm{CO}]{h\nu} [\mathrm{Cp(Ph_3P)(CO)Fe{-}Et}]\ (-)_{546}{-}S \qquad (6)$$

Thermal carbonylation of optically pure $Cp^*Fe(CO)[P(OCH_2)_3CH_3]Et$ and $Cp^*Fe(CO)(PPh_3)Et$ by CO and $C_6H_{11}NC$ was found to occur with widely varying stereospecificity and with stereochemistry corresponding to *formal* CO migration (producing **7a,b,c,d**) or to *formal* alkyl migration (producing **7e,f,g,h**) depending on

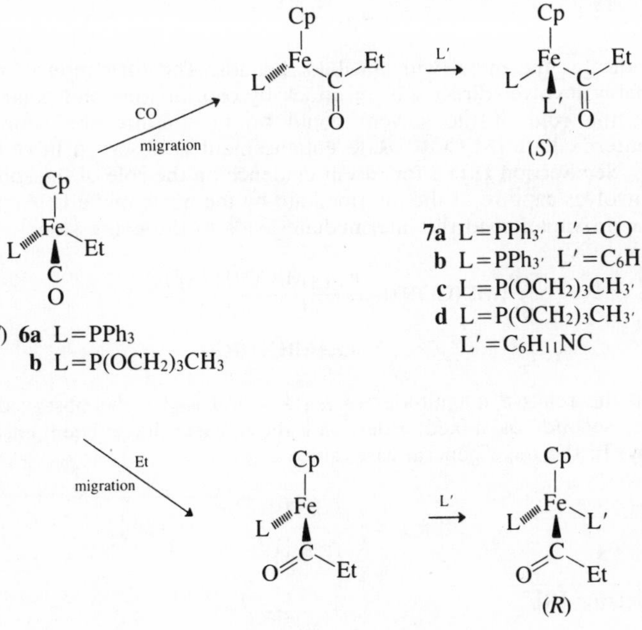

$$(S)\ \mathbf{6}:\ \mathrm{CpFe(L)(CO)Et} \xrightarrow{\text{CO migration}} \mathrm{CpFe(L)(COEt)} \xrightarrow{L'} \mathrm{CpFe(L)(L')(COEt)}\ (S) \qquad (7)$$

$$(S)\ \mathbf{6}:\ \mathrm{CpFe(L)(CO)Et} \xrightarrow{\text{Et migration}} \mathrm{CpFe(L)(COEt)} \xrightarrow{L'} \mathrm{CpFe(L)(L')(COEt)}\ (R)$$

(*S*) **6a** L=$PPh_3$
**b** L=$P(OCH_2)_3CH_3$

**7a** L=$PPh_3$, L′=CO
**b** L=$PPh_3$, L′=$C_6H_{11}NC$
**c** L=$P(OCH_2)_3CH_3$, L′=CO
**d** L=$P(OCH_2)_3CH_3$, L′=$C_6H_{11}NC$

**7e** L=$PPh_3$, L′=CO
**f** L=$PPh_3$, L′=$C_6H_{11}NC$
**g** L=$P(OCH_2)_3CH_3$, L′=CO
**h** L=$P(OCH_2)CH_3$, L′=$C_6H_{11}NC$

solvent[15a]. Carbonylation of optically active $CpFe(CO)[PPh_2N(CH_3)$-(S)-*$CH(CH_3)Ph]CH_3$, on the other hand, gives almost completely epimerized acetyl product **8**[15b]. An earlier assignment[15c] of product stereochemistry which indicated

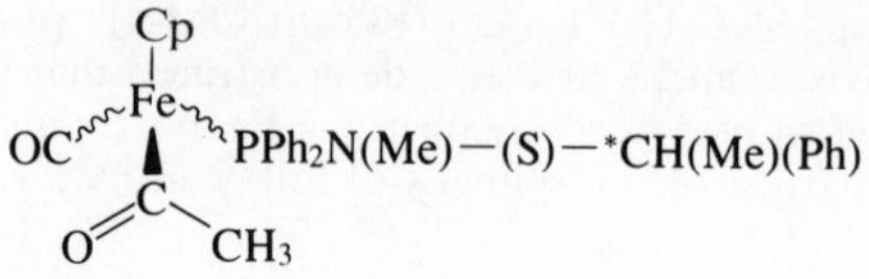

Cp
Fe
OC PPh₂N(Me)—(S)—*CH(Me)(Ph)
C
O CH₃

(**8**)

retention at Fe was shown to be in error. The situation with respect to stereospecificity at Fe seems quite complicated at present[16,17].

### 4. Kinetic studies

Kinetic studies which have been made on reactions of both iron and manganese alkyls with nucleophiles L are consistent with the mechanism shown in Scheme 1,

$$[M](CO)R + S \underset{k_{-1}}{\overset{k_1}{\rightleftharpoons}} [M](COR)(S) \quad S = \text{solvent}$$

$$[M](COR)(S) + L \xrightarrow{k_2} [M]L(COR) + S$$

SCHEME 1

where [M] indicates the metal with ancillary ligands. The formation of the intermediate probably involves direct solvent attack by coordinating and polar solvents. Alternatively, the role of the solvent could be to stabilize the coordinatively unsaturated intermediate [M]COR. Rate enhancement is observed in coordinating polar solvents. See Section II.D.2 for recent evidence on the role of the solvent. The second step involves capture of the intermediate by the nucleophile L. Applying the steady-state approximation to the intermediate leads to the expression

$$\frac{d}{dt}[ML(COR)] = \frac{k_1 k_2 [M(CO)R][S][L]}{k_{-1} + k_2[L]} = k_{obs}[M(CO)R] \quad (8)$$

Depending on the relative magnitudes of $k_1$, $k_{-1}$ and $k_2[L]$, the observed kinetics could be first, second, or mixed order. All these cases have been encountered experimentally. In the most general case, since

$$k_{obs} = \frac{k_1 k_2 [L][S]}{k_{-1} + k_2[L]} \quad (9)$$

we have, neglecting [S],

$$\frac{1}{k_{obs}} = \frac{k_{-1}}{k_1 k_2} \cdot \frac{1}{[L]} + \frac{1}{k_1} \quad (10)$$

and a so-called double inverse plot of $1/k_{obs}$ *vs.* $1/[L]$ allows extraction of $k_1$ and the ratio $k_{-1}/k_2$. However, the solvent dependence cannot be extracted from the form of

TABLE 1. Rate data for some carbonylation reactions

$[RCH_2Mn(CO)_5] + L \rightarrow [RCH_2C(O)Mn(CO)_5]$:

| R | $\sigma^*(RCH_2)$ | L | $10^5 k_1$ ($s^{-1}$) | $10^5 k_{obs}$ ($l\ mol^{-1}\ s^{-1}$) | Conditions |
|---|---|---|---|---|---|
| Et | −0.115 | CO | | (14 000)[a] | 30 °C; β,β′-diethoxy diethyl ether |
| Me | −0.10 | | | (12 000)[a] | |
| cyclo-$C_6H_{11}$ | −0.06 | | | ~2500 ± 300[a] | |
| H | 0.00 | | | 1200 ± 190[a] | |
| MeO | +0.52 | | | 25 ± 4.0[a] | |
| Ph | +0.22 | | | 12 ± 2.5[a] | |
| HOOC | +1.05 | | | ≤5 ± 3.0 | |
| p-$C_6H_4Me$ | | n-$C_4H_9NC$ | 7.24 ± 0.11[b] | [c] | 26 °C; thf |
| Ph | +0.22 | | 5.32 ± 0.38[b] | | |
| p-$C_6H_4Cl$ | | | 3.43 ± 0.30[b] | | |
| p-$C_6H_4NO_2$ | | | ~0 | | |
| Me | −0.10 | $C_6H_{11}NH_2$ | 96.3[d] | | 25.5 °C; thf |

$[CpFe(CO)_2R] + L \rightarrow [CpFe(CO)LCOR]$:

| R | $\sigma^*(R)$ | L | $10^6 k_1$ ($s^{-1}$) | $10^6 k_{-1}$ ($s^{-1}$) | $10^6 k_2$ ($l\ mol^{-1}\ s^{-1}$) | Conditions |
|---|---|---|---|---|---|---|
| Me | 0.0 | $PPh_3$[e] | 1.5 | 20 | — | 29 °C; dmso |
| Et | −0.10 | | 21.8 | 40.3 | 35 | |
| n-Pr | −0.12 | | 19.4 | 46.3 | — | |
| n-Bu | −0.13 | | 17.6 | 43.9 | 94 | |
| n-$C_6H_{13}$ | | | 18.8 | 45.7 | — | |
| i-Bu | | | 91 | 91 | 136 | |
| i-Pr | −0.19 | | 103 | 414 | 751 | |
| s-Bu | | | 780 | 220 | 420 | |
| $Me_3CCH_2$ | | | 910 | 200 | 330 | |
| $(Me_3Si)_2CH$ | | | Very large | Very large | Very large | |
| $PhCH_2$ | | +0.22 | ~0 | | | |
| $C_6H_{11}CH_2$ | | $PPh_3$[f] | 230 | 320 | 50 | 37 °C; dmso |
| Et | | $PPh_3$[g] | 13.2 | | | 47.5 °C; MeCN |
| | | $PMe_2Ph$[g] | 22.2 | | | |
| | | $PMePh_2$[g] | 17.8 | | | |
| | | $P(OMe)_3$[g] | 18.4 | | | |
| | | $P(OPh)_3$[g] | 13.5 | | | |

| R | L | $10^5 k_{obs}$ ($s^{-1}$) | Conditions |
|---|---|---|---|
| $C_6H_{11}CH_2$ | $PPh_3$[f] | 5.9 | 37 °C; MeCN |
| | | 5.4 | 37 °C; dmso |
| | | 3.6 | 37 °C; $CH_2Cl_2$ |
| | | 1.7 | 37 °C; $C_6H_6$ |
| | | 1.3 | 37 °C; heptane |

[a] J. N. Cawse, R. A. Fiato, and R. L. Pruett, *J. Organomet. Chem.*, **172,** 405 (1979).
[b] D. W. Kuty, *PhD Thesis*, University of Cincinnati, 1976.
[c] $10^3 k_{-1}/k_2 = 4.85 \pm 0.65$.
[d] R. J. Mawby, F. Basolo, and R. G. Pearson, *J. Am. Chem. Soc.*, **86,** 3994 (1964).
[e] J. D. Cotton, G. T. Crisp, and L. Latif, *Inorg. Chim. Acta*, **47,** 171 (1981).
[f] K. Nicholas, S. Raghu, and M. Rosenblum, *J. Organomet. Chem.*, **78,** 133 (1974).
[g] M. Green and D. J. Westlake, *J. Chem. Soc. A*, 367 (1971).

the rate law. In a very few cases for manganese (but more frequently for other metals) a second-order term $k_3$[M(CO)R][L] representing direct attack by L may be added to the rate law. When [L] is large, $k_{obs} = k_1$, and this is the situation generally observed for both manganese and iron. When L = CO, manganese complexes may display mixed order kinetics at low pressures. At high pressures of CO the rate becomes independent of the gas pressure.

Table 1 reports a selection of rate constants for carbonylation of [RMn$(CO)_5$] and [CpFe$(CO)_2$R] complexes by various nucleophiles L. The data are far from comprehensive, but are representative. The most striking feature is the rate acceleration by electron-donating alkyl groups. Electron-withdrawing groups such as $CF_3$ and p-$NO_2C_6H_4CH_2$ are inert to insertion. This effect presumably arises because electron-withdrawing R groups lead to stronger M—C bonds and retard the rate of R migration, which involves M—C bond breaking. This same effect should also strengthen M—COR bonds, but the magnitude is smaller. Thus, the increase in activation energy is only partially offset. A linear free energy relationship exists[18] between rate and $\sigma^*$ for the Mn reactions with CO having $\rho^* = -8.7$. CNDO calculations have recently shown that the preferred insertion pathway for [RMn$(CO)_5$] is alkyl migration. Further, the migration can be regarded as an intramolecular nucleophilic attack on coordinated CO and should decrease in rate with decreasing electron density on alkyl C in the order Et > Me > $CH_2F$ > $CF_3$[19], as observed.

In the study of iron complexes in dmso bulky R groups migrated more readily to form the coordinatively unsaturated intermediate than did less bulky groups of similar electron-withdrawing ability. This is consistent with the importance of M—C bond breaking in influencing insertion rates. As might be expected, the influence of the entering ligand L on the rate is very small. When corrected for the expected difference between R = $PhCH_2$ and R = Me, reaction rates are about the same with isocyanide and amine ligands for manganese. For the carbonylation of Fe complexes in dmso, the values of $k_2$ span a range of 10 while the values of $k_1$ span a range of *ca.* $10^3$ as R varies. The last set of data in Table 1 illustrate the same effect for variation of phosphine ligands.

It must be noted that the interpretation of kinetic data on carbonylation of [CpFe$(CO)_2$R] in dmso by Cotton and coworkers presented in Table 1 would not command universal agreement. It was observed some time ago by Rosenblum and coworkers (whose data are also presented in Table 1) that the intermediates [CpFe(CO)(COR)(dmso)] could be detected by i.r. and n.m.r. in this solvent. However, Rosenblum and colleagues observed no specific increase in rate for carbonylation in dmso and suggested that the detectable species was not the intermediate in Scheme 1, but rather the product of a dead-end equilibrium. According to this view, the reactive intermediate in Scheme 1 would probably be unsolvated since the overall rate does not vary much with large changes in solvent donicity. Cotton and colleagues assumed that [CpFe(CO)(COR)(dmso)] *is* the intermediate on the reaction path and that values of $k_1$ and $k_{-1}$ could be extracted from rates of formation and equilibrium constants for the solvated species in dmso. $k_2$ values were obtained by adding $PPh_3$ to equilibrated solutions and monitoring the appearance of product. Regardless of the correctness of these assumptions, the relative values of '$k_1$' and '$k_{-1}$' should give an indication of the migratory abilities of various R.

Thermal decarbonylation has been investigated kinetically only for manganese compounds since [CpFe$(CO)_2$(COR)] complexes do not lose CO under thermal conditions and [CpFe(CO)L(COR)] (L = phosphine) complexes do so only at inconveniently high temperatures. The mechanistic scheme for decarbonylation may be represented as in Scheme 2.

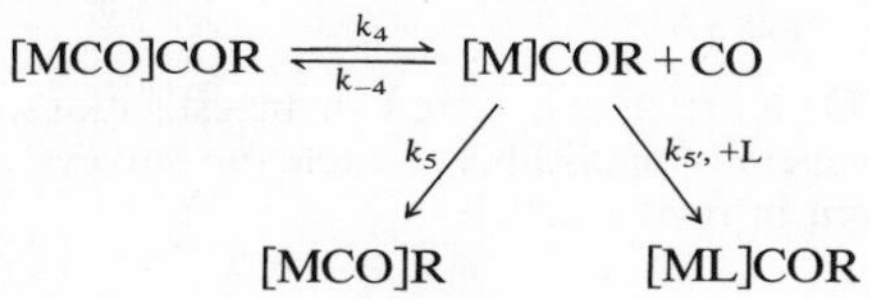

$$[MCO]COR \underset{k_{-4}}{\overset{k_4}{\rightleftharpoons}} [M]COR + CO$$

$$[M]COR \xrightarrow{k_5} [MCO]R \qquad [M]COR \xrightarrow{k_{5'}, +L} [ML]COR$$

SCHEME 2

The intermediate is the same one as for carbonylation. The $k_{5'}$ path relates ligand substitution by L to the decarbonylation. Applying the steady state treatment to the $k_5$ path gives

$$\frac{d}{dt}[M(CO)R] = \frac{k_4 k_5 [M(CO)(COR)]}{k_{-4}[CO] + k_5} \tag{11}$$

Because $k_{-4}[CO]$ is usually small, the decarbonylation rate is $k_4[M(CO)(COR)]$. When $k_{5'}[L]$ is large, ligand substitution has the same rate as the decarbonylation. In practice, it is usually smaller.

Table 2 presents data for the thermal decarbonylation of acyl manganese pentacarbonyls. Surprisingly, loss of CO is retarded by electron-withdrawing R. This may be because the loss of two electrons donated by CO becomes less favourable as the metal becomes more depleted of electrons. The same trend was noted[20] in a study of the decarbonylation rates of $[(RCO)Co(CO)_4]$. This explanation could also account for the observation[21] that [CpFe(CO)L(COR)] complexes may be thermally decarbonylated when L = phosphine or phosphite but not when L is the better $\pi$-acid CO.

The data in Table 1 allow a ranking of values of $k_5$ for [CpFe(CO)(COR)] which could be expected to parallel $k_{-1}$ values if the decarbonylation were thermally accessible. The migration rate constants from the acyl group to iron with solvent displacement are in the order $i$-Pr > $Me_3CCH_2$ ≈ $(Me_3Si)_2CH$ ≫ $i$-Bu > $n$-Pr ≈ $n$-$C_6H_{13}$ ≈ $n$-Bu ≈ Et > Me. The non-reactivity of Fe benzyl species restricts the range of $\sigma^*$ values severely. Consequently, it is difficult to say anything very definitive about the promotion or inhibition of alkyl migration to iron as a function of electron-withdrawing ability. The ease of migration by bulky groups again is seen.

TABLE 2. Decarbonylation rates for $[RCOMn(CO)_5]$ at 80 °C in benzene

| R | $\sigma^*$ | $10^4 k_4$ ($s^{-1}$) |
|---|---|---|
| cyclo-$C_6H_{11}CH_2$[a] | | ⩾250 ± 20 |
| Me[a] | | 250 ± 15 |
| Me[b] | | 290 |
| $PhCH_2$[a] | | 159 ± 12.0 |
| $MeOCH_2$[a] | 0.64 | 58 ± 4.5 |
| EtOOC[a] | | 28 ± 3.1 |
| MeOC[b] | | 21 |
| $PhCH_2OC$[b] | | 20 |
| $CF_3$[a] | | 2.8 ± 0.9 |

[a] J. N. Cawse, R. A. Fiato, and R. L. Pruett, *J. Organomet. Chem.*, **172,** 405 (1979).

[b] C. P. Casey, C. A. Bunnell, and J. Calabrese, *J. Am. Chem. Soc.*, **98,** 1166 (1976).

### 5. Conclusion

The aspects of CO insertion uncovered in investigations of these paradigmatic systems are rather generally applicable. Systematic surveys of reactions by central metal have been given in review articles[5].

## C. Square-Planar Complexes

An extensive study of over sixty square-planar phosphine complexes of nickel, palladium, and platinum by Garrou and Heck[22] has shown that the mechanism in Scheme 3 can account for all kinetic results obtained.

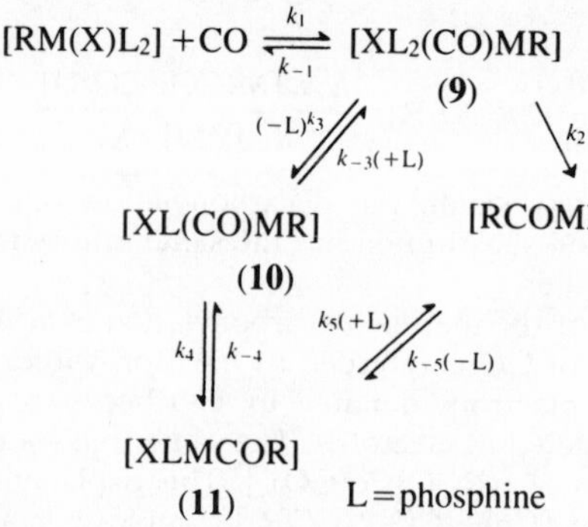

$$[RM(X)L_2] + CO \underset{k_{-1}}{\overset{k_1}{\rightleftharpoons}} [XL_2(CO)MR] \; (\mathbf{9})$$

$$[XL_2(CO)MR] \; (\mathbf{9}) \xrightarrow{k_2} [RCOML_2X]$$

$$[XL_2(CO)MR] \; (\mathbf{9}) \underset{k_{-3}(+L)}{\overset{(-L)k_3}{\rightleftharpoons}} [XL(CO)MR] \; (\mathbf{10})$$

$$[XL(CO)MR] \; (\mathbf{10}) \underset{k_{-4}}{\overset{k_4}{\rightleftharpoons}} [XLMCOR] \; (\mathbf{11})$$

$$[XLMCOR] \; (\mathbf{11}) \underset{k_{-5}(-L)}{\overset{k_5(+L)}{\rightleftharpoons}} [RCOML_2X]$$

L = phosphine

SCHEME 3

As with octahedral complexes, prior coordination of the inserting CO occurs, followed by migration of the alkyl group. Two pathways involve migration in the five-coordinate intermediate or in a three-coordinate (possibly solvated) intermediate produced by phosphine loss. The predominant path is via the three-coordinate intermediate **11**—the dissociative path. Different complexes had differing values of the rate constants and, consequently, the kinetics may be dominated by any step in a particular case. This made it impossible to extract all rate constants or to compare activation parameters meaningfully over the entire series. However, in groups of complexes where comparisons were possible, electron-withdrawing substituents on R = aryl were found to retard the migration rate while electron-donating substituents enhanced it. These effects were smaller than in octahedral manganese complexes. The rate constants for carbonylation of *trans*-$[(p\text{-}XC_6H_4)Pt(PPh_3)_2I]$ were in the ratio 0.005 : 1.0 : 4.0 for X = $NO_2$, H, and $OCH_3$. It has also been demonstrated[23] for a series of platinum complexes that less electron-withdrawing R groups increase the importance of the dissociative path relative to migration in the five-coordinate intermediate **9**. Presumably, this occurs by stabilization of **11**.

Some further aspects of the Garrou and Heck mechanism (Scheme 3) have recently been uncovered by Anderson and Cross and coworkers. There are three possible isomers of **10** (R = Ph, L = $PPh_3$, X = Cl): **12**, **13**, and **14**. Only **12** reacted with CO, giving the dimer **15** (X = Cl) (both isomers)[24,25].

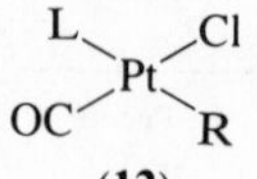

| L, Cl / Pt / OC, R | 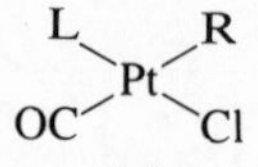 L, R / Pt / OC, Cl |
|---|---|
| (**12**) | (**13**) |

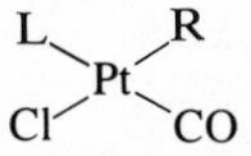

**(14)** 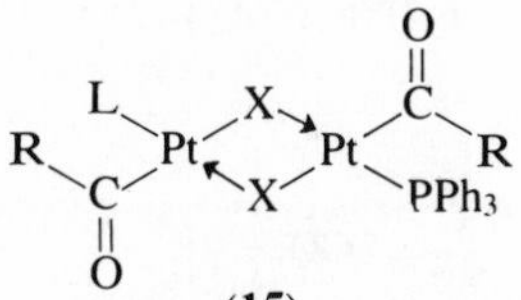 **(15)**

Presumably, the reaction occurs via a three-coordinate acyl **11** which subsequently dimerizes. The lack of reactivity of **13** can be attributed to the fact that CO and Ph are *trans*. Even in the presence of excess $PPh_3$, **13** and **14** do not carbonylate, but undergo CO displacement. The ease of R migration in **12** would then be related to the value of $K$ for the **12** $\rightleftharpoons$ **15** equilibrium. When L = $PMe_2Ph$ the $K$ values decrease in the order R = Et ($K = 14$) > Ph > Me > $CH_2Ph$ ($k = 0$). For substituted phenyl groups $C_6H_4R^1$ the values of $K$ (38 °C) decrease in the order $R^1$ = $p$-$NMe_2$ ($\infty$) > $p$-OMe (2170) > $p$-Me (360) > $p$-Cl (20) > $m$-Cl (3.3) > $p$-CN, $o$-Me, $o$-OMe (0). These orders are similar to those for the *rates* of carbonylation of $[RMn(CO)_5]$ and suggests that the manganese reactions reflect relative rates of migration. Values of $\Delta H^0$ and $\Delta S^0$ were measured for the substituted phenyl complexes. $\Delta H^0$ were $<0$ and $\Delta S^0$ very negative as expected for a dimerization. Although activation parameters reflect a contribution from the formation of $\mu$-Cl bonds as well as energy differences between Pt—COR and Pt—R bonds, the former two may be relatively constant in the series. If so, then the less negative $\Delta H^0$ values for electron-withdrawing R would reflect an increase in Pt—R bond strength inhibiting migration.

R migration may be promoted by the high *trans*-influence of the phosphine. For various phosphines *trans* to R = Ph, $K$ decreases in the order[26] $PEt_3$ > $PMe_2Ph$ > $PMePh_2$ > $PPh_3$ > $PCy_3$ $\gg$ $P(o\text{-}MeC_6H_4)_3$, $AsMePh_2$, $AsPh_3$ (0). This order is consistent with the dominance of electronic effects in weakening the M—C bond up to a critical cone angle when steric effects destabilize **11**. Values of $K$ are independent of solvent, suggesting that **11** is not solvated.

A difficulty in studying the geometry of **10** under carbonylation conditions is that the L released in its formation leads to its conversion to product preventing accumulation. By adding elemental sulphur to the solution, released phosphine could be converted to the phosphine sulphide. $^{31}P$ NMR studies[27] on solutions of *trans*-$[PhPt(X)L_2]$ containing $S_8$ where L was a variety of phosphines revealed that only isomers **13** and **14** were detectably present on addition of CO. Presumably, these must isomerize to **12** in order for insertion to occur.

In polar solvents and with nucleophilic phosphines ionic species $[PhPt(CO)L_2]^+X^-$ were detected by $^{31}P$ NMR. Conversely, in non-polar solvents and with less electron-donating phosphines, five-coordinate $[PhPt(CO)L_2X]$ complexes were detectable[27]. Further evidence for the importance of carbonylation via migration in the five-coordinate intermediate was obtained. In the presence of CO and elemental sulphur, $[PhPt(PMePh_2)_2Cl]$ afforded no $Ph_2MePS$, indicating no phosphine dissociation to produce **10**. Nevertheless, *trans*-$[(PhCO)Pt(PMePh_2)_2Cl]$ forms slowly. In light of these results, Scheme 3 can be expanded to Scheme 4 for Pt-Ph complexes. The isomers of the five-coordinate intermediate have been labelled **9a–g**. Isomer **9b** has arbitrarily been assumed to be the reactive species in the $k_2$ path and isomerization has been assumed rapid compared to ligand loss or alkyl migration.

Scheme 4 elucidates some other aspects of the chemistry of *trans*-$[RPtL_2X]$ complexes. The reaction $[RPtL_2X] \rightarrow$ **9** $\rightarrow$ **13** or **14** is tantamount to substitution of L by CO. In some cases the reaction stops at this stage. For example, *trans*-$[Ph_2Pt(PPh_3)_2]$ gives only *cis*-$[Ph_2Pt(CO)(PPh_3)]$ (structure **13** X = Ph, L = $PPh_3$) on

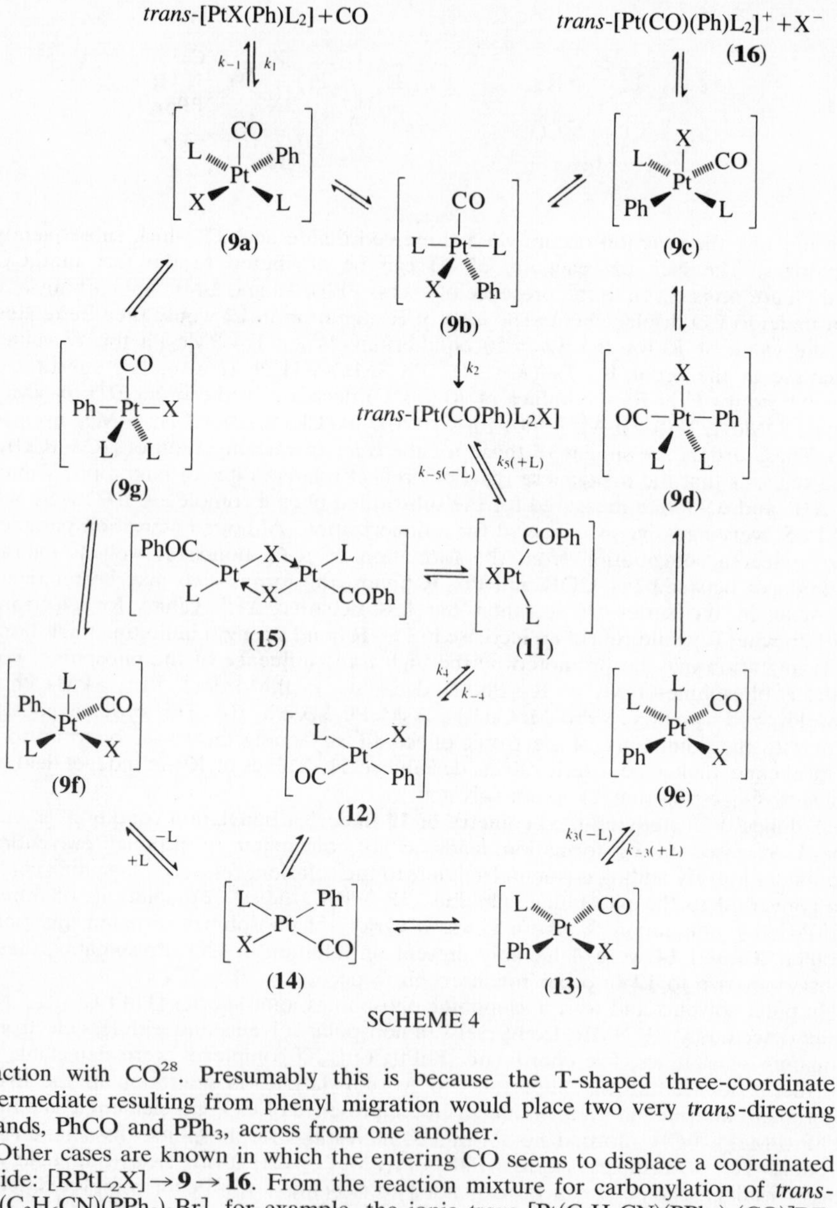

trans-[PtX(Ph)$L_2$]+CO

$k_{-1}$ $k_1$

(9a)

(9b)

(9c)

$k_2$

trans-[Pt(COPh)$L_2$X]

trans-[Pt(CO)(Ph)$L_2$]$^+$+$X^-$ (16)

(9d)

(9g)

$k_{-5}(-L)$ $k_5(+L)$

(15)

(11)

(9f)

$k_4$ $k_{-4}$

(12)

(9e)

−L +L

$k_3(-L)$ $k_{-3}(+L)$

(14)

(13)

SCHEME 4

reaction with CO[28]. Presumably this is because the T-shaped three-coordinate intermediate resulting from phenyl migration would place two very *trans*-directing ligands, PhCO and $PPh_3$, across from one another.

Other cases are known in which the entering CO seems to displace a coordinated halide: [$RPtL_2X$]→**9**→**16**. From the reaction mixture for carbonylation of *trans*-[Pt($C_3H_6CN$)($PPh_3$)$_2$Br], for example, the ionic *trans*-[Pt($C_3H_6CN$)($PPh_3$)$_2$(CO)]$BF_4$ could be isolated[29].

Recent studies[30,31] of the rate enhancement by X = $SnCl_3$ on the carbonylation of *trans*-[$PhPtL_2X$] and the decarbonylation of [(RCO)$PtL_2X$] have shown the importance of the ionic species **16**. In the absence of excess CO (*not* the usual carbonylation conditions) it has been shown that the important equilibria are those in Scheme 5.

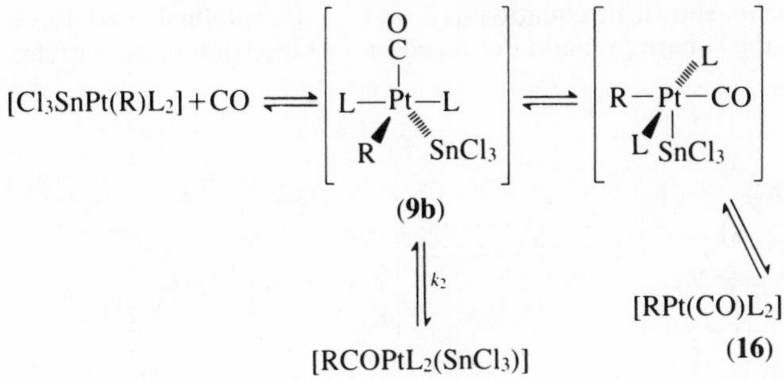

$$[Cl_3SnPt(R)L_2] + CO \rightleftharpoons \left[ L{-}Pt(CO)(R)(SnCl_3){-}L \right] \rightleftharpoons \left[ R{-}Pt(L)(L)(SnCl_3){-}CO \right]$$

(**9b**)

$\downarrow k_2$ from (**9b**); ⇌ from the second intermediate

$[RCOPtL_2(SnCl_3)]$ $\quad$ $[RPt(CO)L_2]^+SnCl_3^-$ (**16**)

SCHEME 5

The function of the $SnCl_3^-$ is[31] to serve as a good leaving group to produce **16**, which does not undergo further reaction with CO (except for exchange). Slow re-attack by $SnCl_3^-$ affords a five-coordinate species which is stabilized by this ligand and gives the product via the $k_2$ path. The dissociative path proceeding via L loss is suppressed. $I^-$ functions similarly but is a worse leaving group. Other $X^-$ act to replace CO.

Acyl nickel complexes are relatively rare since, once formed, they undergo reductive elimination, disproportionation, and ligand displacement. Some stable complexes of the type *trans*-$[(RCO)NiL_2Cl]$ ($R = CH_2CMe_2Ph$, $CH_2SiMe_3$; $L = PMe_3$, $PMe_2Ph$) have recently been prepared by bubbling CO through solutions of the alkyls[32]. For $L = PMe_3$, the carbonylation is not thermally reversible. $[Ni(COCH_2SiMe_3)(PMe_3)_2L]$ was found *not* to have $\eta^2$-acyl coordination which would lead to an 18-electron configuration. Another series of composition *cis*-$[(RCO)Ni(bipy)X]$ ($R = Me$, Et) was synthesized. Depending on the nature of R and X, reductive elimination to give RCOX occurs, presumably aided by the *cis* position of acyl and X[33].

Studies on the kinetics of thermal decarbonylation of square-planar acyl complexes are not numerous because of interference by simultaneously occurring decomposition, reductive elimination, substitution, and disproportionation. In contrast to decarbonylation of manganese octahedral acyls which involves relatively large positive $\Delta S^{\neq}$ connected with prior CO dissociation, thermal decarbonylation of $[(PhCO)Ni(acac)(PPh_3)]$ was recently found[34] to have $\Delta S^{\neq} = -19.2$ cal mol$^{-1}$K$^{-1}$. This is probably connected with the migration of Ph to nickel followed by loss of CO. Negative entropies of activation of similar magnitude were observed in the acyl→alkyl rearrangement of $[(XCH_2CO)Ir(PPh_3)_2Cl_2]$[35] where the CO remains coordinated to the alkyl product.

We now present some significant recent developments which are better understood in the context of the foregoing results.

## D. Recent Results on Carbonylation

### 1. The 'CO insertion' mechanism

Evidence has begun to accumulate that the 'CO insertion' mechanism involving motion of a coordinated CO to insert into an M—C bond is sometimes operative. Pankowski and Bigorgne[36] have demonstrated that CO insertion occurs in *cis*-$[MX(Me)(CO)_2(PMe_3)_2]$($M = Fe$, Ru; $X = I$, Me, CN, $CN \leftarrow BPh_3$). The results of labelling experiments shown in equations 12 and 13 are intelligible on the basis of this mechanism if the square pyramid produced on CO insertion does not rearrange.

$$[(L)_2M(CO)_2(I)(Me)] \underset{\text{argon; } -30\,°C}{\overset{^{13}CO;\ -30\,°C,\ \text{hexane}}{\rightleftharpoons}} [(L)_2M(CO)(^{13}CO)(I)(C(O)Me)] \quad (12)$$

$$[(L)_2M(^{13}CO)_2(I)(Me)] \underset{\text{argon; } -30\,°C}{\overset{CO;\ -30\,°C,\ \text{hexane}}{\rightleftharpoons}} [(L)_2M(^{13}CO)(CO)(I)(^{13}C(O)Me)] \quad (13)$$

If reaction 12 is allowed to occur at higher temperatures, isomerization occurs, giving a product with acetyl and iodo ligands *trans*, which could lead to the erroneous assignment of a methyl migration mechanism. The possibility that a similar isomerization leads to erroneous assignment of the insertion mechanism for [$MeMn(CO)_5$] has been ruled out by following the reaction by $^{13}C$ NMR at $-115\,°C$[8] as well as because *cis* → *trans* isomerization of the acetyl is slow even at room temperature[8].

The related complexes **17** give **18** on carbonylation consistent with CO insertion[37]. **18** rearrange on standing to the isomer with *cis*-carbonyls when X = Cl, Br, I.

$$(Me_3P)_2Fe(CO)_2(Me)(X) \quad (\mathbf{17}) \qquad (Me_3P)_2Fe(CO)_2(COMe)(X) \quad (\mathbf{18})$$

X = Cl, Br, I, NCS, NCO, CN, OMe, Me

(**17**) (**18**)

The equilibrium described by reaction 14 involving a trigonal bipyramidal intermediate has been found to take place in solution. The isomeric composition of the acetyl product mixture depends on the nature of L′[38].

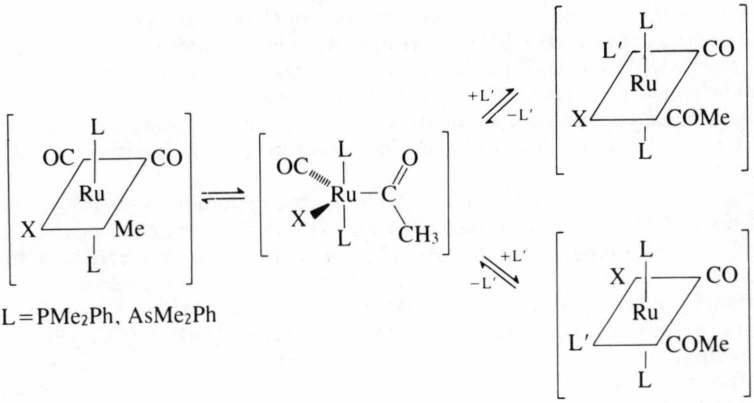

$$[\text{trans-}RuMe(CO)_2XL_2] \rightleftharpoons [Ru(COCH_3)(CO)XL_2] \xrightleftharpoons[-L']{+L'} [RuL'(COMe)(CO)XL_2] \quad (14)$$

L=$PMe_2Ph$, $AsMe_2Ph$

L′=CO, phosphines, arsines

These results are consistent with either methyl migration or CO insertion. The crystal structure of [Ru(CO-*p*-tolyl)(CO)$(PPh_3)_2$I] showed $\eta^2$-coordination of the acyl group[39].

## 2. Acceleration of alkyl migration

A dramatic acceleration of methyl migration was observed[40] on oxidizing [CpFe$(CO)_2$Me] with Ce(IV) in acetonitrile at −78 °C. The green cation radical $[CpFe(CO)(MeCN)COMe]^{+\bullet}$ was isolated as the triflate salt.

Lewis acids have been shown by Shriver and coworkers[41] to induce alkyl migration in the absence of CO. $AlBr_3$ promotes migration in [$L_n$MR] to form the products [$L_n$M{C(OAlBrBr$_2$)R}], where [$L_n$MR] can be [MeMn$(CO)_5$], [$PhCH_2Mn(CO)_5$], [CpFe$(CO)_2$Me], and [CpMo$(CO)_3$Me]. The reactions also occur with $AlCl_3$ and $BF_3$. The structure of a manganese compound is shown as **19**.

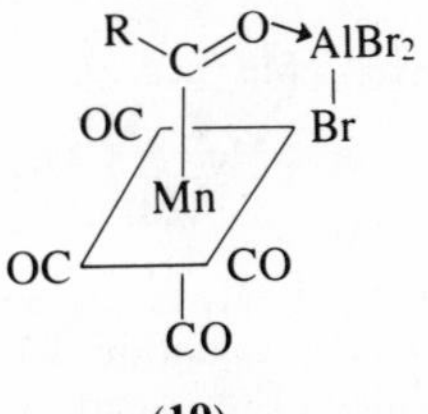

$$[Mn(CO)_4\{C(R)=O\rightarrow AlBr_2(\mu\text{-}Br)\}]$$

**(19)**

The structure shows that the Lewis acid stabilizes the acyl group and occupies the vacated coordination site. The coordinated halide can be replaced by CO. It is noteworthy that normally unstable molybdenum acetyls are stabilized by Lewis acid coordination. I.r. evidence indicates that Me migration with CO loss occurs with [MeMn$(CO)_5$] and [CpFe$(CO)_2$Me] adsorbed on alumina[41].

A kinetic study of the reactions of $AlCl_3$ and $AlBr_3$ with [MeMn$(CO)_5$] shows[42] that carbonylation is accelerated by Lewis acids by a factor of $10^8$, indicating that their function is to accelerate alkyl migration rather than to capture the unsaturated acyl intermediate.

$BF_3$ was recently found[15b] to accelerate carbonylation of optically active complexes such as **6a** and $Cp^*Fe(CO)L(CH_3)$($L = PPh_2N(Me)$-(*S*)-*CH(Me)Ph, $PPh_2N(CH_2C_6H_5)$-(*S*)*CH(Me)Ph) as well as to increase stereoselectivity. The observed stereochemistry corresponds to *formal* alkyl migration. However, a wide variety of intermediates is possible in the presence of $BF_3$ and the stereochemistry of the product may, therefore, not really be indicative of the mechanism. See also Section II.B.2.

Protonic acids were also shown[43] to accelerate methyl migration in $[MeMn(CO)_5]$ in the presence of CO, but not by as much as Lewis acids. The rate enhancement was proportional to acid strength, but metal—alkyl bond cleavage occurred with the strongest acids. Previously Collman and coworkers[44] found rate enhancement of alkyl migration in $[RFe(CO)_4]^-$ by small cations which formed Lewis acid ion pairs.

In contrast to the corresponding ruthenium complexes, $[OsR(X)(CO)_2(PPh_3)_2]$ (R = *p*-tolyl, Me) are resistant to migratory insertion. In accord with results quoted in Section II.B.4 and II.C, the osmium complex having less electron-withdrawing R = Et inserts CO on reaction with CO, CNR, and $S_2CNEt_2^-$($X = ClO_4^-$)[45]. In a similar vein, a kinetic study of the $PPh_3$-promoted carbonylation of $[CpMo(CO)_3CH_2C_6H_4X]$ showed[46] the reaction to proceed by a mechanism like that in Scheme 1. The rate constants increased with the electron-donating ability of X. $\rho$ for these reactions was found to be $-0.97$.

Wax and Bergman[47] have now shown that the solvent participation indicated in Scheme 1 actually occurs in the carbonylation of $[CpMo(CO)_3Me]$ with $PMePh_2$. By using substituted thf solvents, these workers were able to show that for a series of solvents of about the same dielectric constant, $k_1$ decreased with steric bulk of the solvent while $k_3$ (representing L attack) remained the same, as would be expected. This constitutes evidence that the role of the solvent in this carbonylation (and presumably in others) is one of direct attack on the metal.

### 3. Novel modes of CO insertion

Treatment of $[(\eta^5\text{-}Me_5C_5)TaMe_4]$ with CO at $-78\,°C$ in ether was found[48] to afford **20**, the product of addition of two methyl groups to CO.

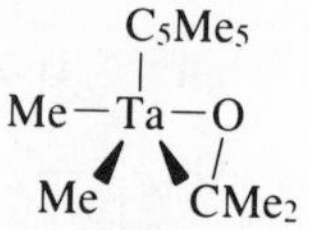

C5Me5
|
Me—Ta—O
Me CMe2

**(20)**

The $\eta^2$-acetone ligand is reduced by $H_2$ giving an $—OCHMe_2$ ligand. An interesting case of double insertion of CO is equation 15 where the function of BuLi is to produce a carbanion nucleophile at the $\alpha$-carbon in the thioether ligand, which then attacks coordinated CO[49].

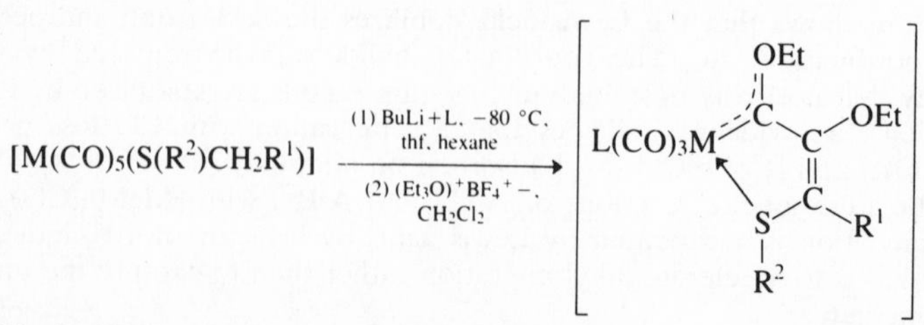

$$[M(CO)_5(S(R^2)CH_2R^1)] \xrightarrow[\text{(2) }(Et_3O)^+BF_4^-,\ CH_2Cl_2]{\text{(1) BuLi + L, }-80\,°C,\ \text{thf, hexane}} \left[ L(CO)_3M \text{ chelate: } M{=}C(OEt)\text{–}C(OEt){=}C(R^1)\text{–}S(R^2)\to M \right] \qquad (15)$$

M = Cr, W; L = isocyanide, phosphine, phosphite

Other examples of intramolecular nucleophilic attack on CO are known. The reaction in Scheme 6 is acid catalysed and features attack of a deprotonated OH on CO to give a lactone[50].

$$[M]^- + ClCH_2C{\equiv}C(CH_2)_nCRR'OH \longrightarrow [M]CH_2C{\equiv}C(CH_2)_nC(R)(R')OH$$

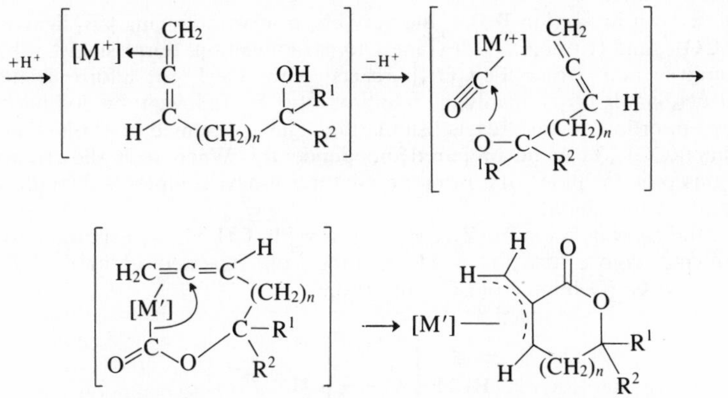

$\xrightarrow{+H^+}$ [ $[M^+] \leftarrow CH_2{=}C{=}C(H)(CH_2)_nC(OH)R^1R^2$ ] $\xrightarrow{-H^+}$ [ $[M'^+]$, $O{=}C$, $CH_2{=}C{=}C{-}H$, $(CH_2)_n$, $^-O{-}C(R^1)(R^2)$ ] $\longrightarrow$

[ $H_2C{=}C{=}C(H)(CH_2)_nC(R^1)(R^2){-}O{-}C({=}O){-}[M']$ ] $\longrightarrow$ $[M']$—lactone (H, H, O, $(CH_2)_n$, $R^1$, $R^2$)

$[M] = CpMo(CO)_3, Mn(CO)_5$
$[M'] = [M]{-}CO$

$n=0$; $R^1=R^2=H$
$n=0$; $R^1=H$, $R^2=Ph$
$n=0$; $R^1=R^2=Me$
$n=1$; $R^1=R^2=H$

SCHEME 6

Another example is equation 16[51].

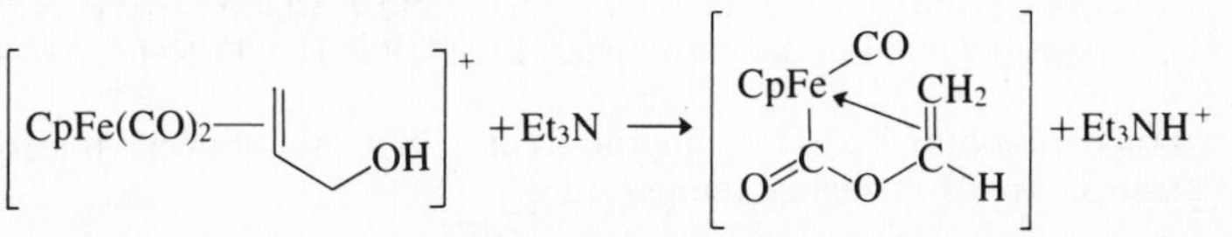

$$\left[CpFe(CO)_2{-}(CH_2{=}CHCH_2OH)\right]^+ + Et_3N \longrightarrow \left[CpFe(CO)(CH_2{=}C(H){-}O{-}C({=}O))\right] + Et_3NH^+ \qquad (16)$$

An oft-repeated theme in the carbonylation chemistry of the transition metals is the attempt to synthesize a metal formyl complex by CO insertion into an M—H bond. So far, this has not been achieved although formyl complexes have been synthesized by other means[52]. Organoactinide hydrides[53] have now been discovered to yield formyl complexes.

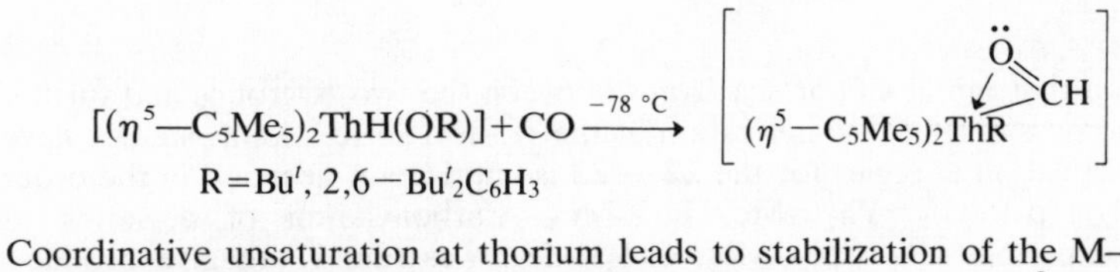

$$[(\eta^5{-}C_5Me_5)_2ThH(OR)] + CO \xrightarrow{-78\ °C} \left[(\eta^5{-}C_5Me_5)_2ThR(\ddot{O}{=}CH)\right] \qquad (17)$$

$R = Bu^t, 2,6{-}Bu^t_2C_6H_3$

Coordinative unsaturation at thorium leads to stabilization of the M—O bond.

Coordinative unsaturation at titanium accounts for the $\eta^2$-acyl structure of $[Cp_2Ti(CO\text{-}p\text{-}MeC_6H_4]$, a 17-electron species made by carbonylation of the alkyl[54]. The acyl reacts with HCl giving $[Cp_2TiCl(CO\text{-}p\text{-}MeC_6H_4)]$ and with PhSSPh to give $[Cp_2Ti(SPh)(CO\text{-}p\text{-}MeC_6H_4)]$. Other early transition metal acyls having $\eta^2$-acyl coordination are $[\{(C_3H_2Ph_3)CO\}V(CO)_3(Ph_2AsCH_2PPh_2)]$[55] and $[(\mu\text{-}Cl)Mo(CO)_2(COCH_2SiMe_3)]_2$[56]. Caulton and coworkers have noted the significance of such coordination in unsaturated transition metal complexes[57]. They attribute the order of decreasing equilibrium constants for CO insertion into $[Cp_2ZrMeX]$ (X = Me > Cl > OEt) to competition between $\pi$-donor orbitals on acetyl and X for the empty metal orbitals.

As pointed out in Section II.B.4, the very electron-withdrawing $CF_3$ ligand makes $[CF_3Mn(CO)_5]$ and $[CpFe(CO)_2CF_3]$ inert to carbonylation. Surprisingly, it has now been found[58] that $[(diars)Fe(CO)_3]$ reacts with $CF_3I$ to afford isomers of $[(CF_3CO)Fe(diars)(CO)_2I]$ involving addition of $CF_3$ followed by iodide-induced migratory insertion. MeI reacts similarly. The presumed intermediate *fac*-$[MeFe(diars)(CO)_3]^+$ can be prepared independently. When it is allowed to react with various pseudohalides, the isomeric mixture of acyl complexes depends on the solvent and on the anion.

Finally, the carbonylation of $[Zr(CH_2R)_4]$ (R = Ph, CH=CH$_2$) is found[59] to afford only a polymer whose structure is **21** resulting from repeated addition of $ZrCH_2R$ across the acyl C=O bond from CO insertion.

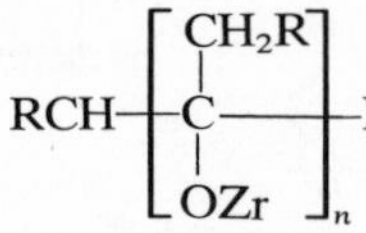

(**21**)

### *4. Other recent results*

In keeping with the theme of insertion of an already coordinated CO, the intermediates $[(PhCH_2)_4Ti(CO)_2]$ and $[(PhCH_2)_3\{PhCH_2(O)C\}Ti(CO)]$ have been detected by i.r. during the low-temperature carbonylation of $[(PhCH_2)_4Ti]$ to $[\{PhCH_2(O)C\}_2Ti(CH_2Ph)_2]$[60]. Likewise, the intermediate $[(PhCH_2)_4Ti(CO)\text{-}\{NH(C_6H_{11})_2\}]$ was detected in the carbonylation of $[(PhCH_2)_4Ti\{NH(C_6H_{11})_2\}]$ to the monoacyl.

The carbonylation of $[Cp_2ZrR_2]$ compounds at −78 °C was shown[61] to produce **22**, which rearranges to **23** at higher temperature.

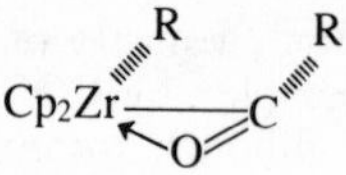

(22) 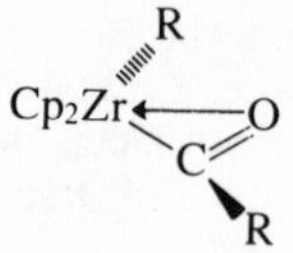 (23)

This must mean that initial CO attack is *not* between the two R groups and further emphasizes the possibility of isomerization at the $\eta^2$-acyl stage. N.m.r. studies have shown[62] that activation energies for the **22** → **23** isomerization decrease in the order $R_2 = (p\text{-anisyl})_2 \approx (p\text{-tolyl})_2 \approx Ph_2 > Me$, $Ph \gg Me_2$. Carbonylation of a series of $[Cp_2ZrRX]$ compounds ($R = CH_2CMe_3$, $CH_2SiMe_3$) was found[63] to give products $[Cp_2Zr(COR)X]$ with $\eta^2$-acyls. When $R = CH(SiMe_3)_2$ and X = Me the CO inserted into the Zr—C bond. This effect is presumably related to the stereochemical results just cited. Attack of CO adjacent to X = Me would push the bulky R group too near the Cp rings for stability.

**24** was found to insert CO into both Hf—C bonds at 20 °C and 1 atm CO[64]. The

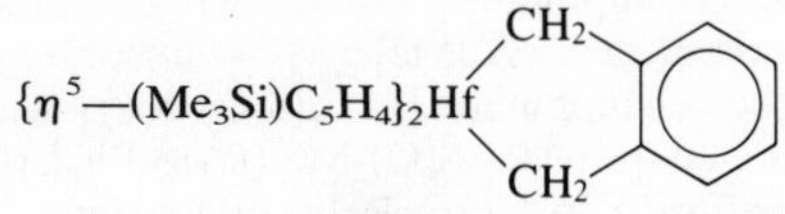

$\{\eta^5-(Me_3Si)C_5H_4\}_2Hf$ $CH_2$ $CH_2$

**(24)**

first example of CO insertion into a Re—C bond has been discovered[65,66]. Under more vigorous conditions, CO can replace coordinated $AlBr_3$. Finally, the unsatu-

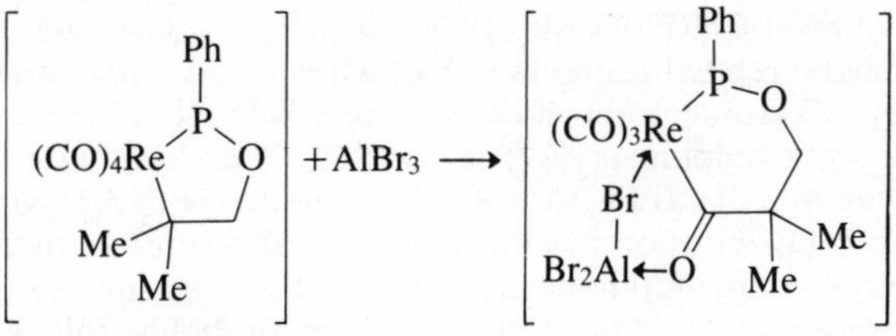

$$[(CO)_4Re\text{—}P(Ph)\text{—}O\text{—}CH_2\text{—}CMe_2] + AlBr_3 \longrightarrow [(CO)_3Re(Br\text{—}AlBr_2)(P(Ph)\text{—}O\ldots CMe_2\text{—}C{=}O\rightarrow AlBr_2)] \tag{18}$$

rated ligand in equation 19 leads to enhanced reactivity toward CO. Perhaps this is due to ring strain in the reactant[67].

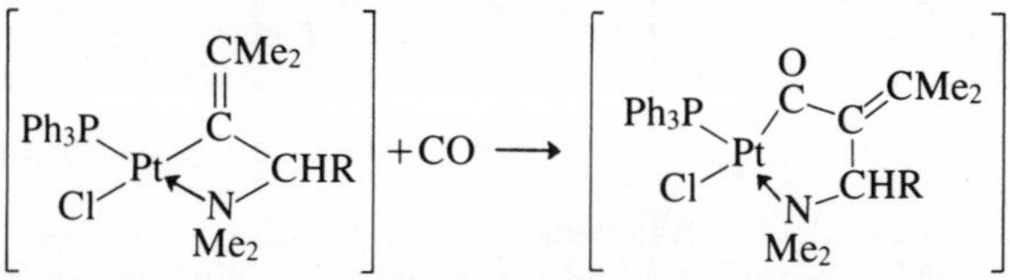

$$[(Ph_3P)(Cl)Pt\text{—}C(=CMe_2)\text{—}CHR\text{—}NMe_2\rightarrow Pt] + CO \longrightarrow [(Ph_3P)(Cl)Pt\text{—}C(O)\text{—}C(=CMe_2)\text{—}CHR\text{—}NMe_2\rightarrow Pt] \tag{19}$$

## III. INSERTION OF SULPHUR DIOXIDE AND RELATED ELECTROPHILES

### A. Insertion into Metal—Alkyl Bonds

#### *1. Scope and mechanism*

Sulphur dioxide insertion reactions may be represented as

$$[L_nMR] + SO_2 \longrightarrow [L_nM(SO_2R)] \tag{20}$$

An excellent review of $SO_2$ insertions was written in 1974 by Wojcicki[68], whose group has done much work on these reactions. An immediate contrast with CO insertion is apparent in that the reactant molecule is the one which is inserted. Various modes of attachment for the inserted $SO_2$ are possible, the ones commonly observed being **25** and **26** with **25** the more usual. This is the mode of coordination

| $MS(O)_2R$ | $MOS(O)R$ | M(O)(O)SR |
|---|---|---|
| *S*-sulphinate | *O*-sulphinate | *O,O′*-sulphinate |
| **(25)** | **(26)** | **(27)** |

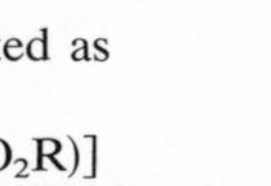

which would be anticipated for class *b* metals. Some i.r. studies are useful in distinguishing structures. *O*-Sulphinates are known for titanium[69], zirconium[70,71], and, recently, nickel[72] complexes. [AuMeL] (L = phosphine) has been found[73] to react with $SO_2$ giving the *S*-sulphinate [LAu{S(O)$_2$Me}] while [AuMe$_3$(PMe$_2$Ph)] inserts 1 mol of $SO_2$ giving *cis*-[AuMe$_2${S(O)$_2$Me}(PMe$_2$Ph)]. [Cp$_3$UR] (R = Me, Bu$^n$) also react with $SO_2$(l) to afford [Cp$_3$U(SO$_2$R)] products whose structures are, as yet, unestablished[74].

In contrast to CO, $SO_2$ is not readily eliminated from the sulphinate products. Two exceptions to this generalization are the thermal and photochemical extrusion of $SO_2$ from [CpFe(CO)$_2${S(O)$_2$C$_6$F$_5$}] only in toluene[75] and thermal elimination from solid [CpFe(CO){P(OPh)$_3$}{S(O)$_2$CH(Ph)(SiMe$_3$)}] *in vacuo*[75]. In some cases, the inability to eliminate $SO_2$ may be related to the fact that, when a coordination site is vacated by ligand expulsion, alkyl migration does not occur (cf., decarbonylation, Section II.A.4.) Instead an *O,O'*-sulphinate (**27**) is formed. Complexes of this type were found[76] to result when *trans*-[Pt(PPh$_3$)$_2$Cl(SO$_2$R)] is treated with $Ag^+$ to remove $Cl^-$.

The mechanism of $SO_2$ insertion has received a good deal of study and seems to depend on the electron configuration of the complex. For coordinatively unsaturated systems prior coordination of $SO_2$ to the metal is probably followed by 'alkyl migration' or '$SO_2$ insertion'. A recent observation[77] of retention of stereochemistry in a gold complex is consistent with this view. The alternative would be attack

$$\left[\begin{array}{c} CD_3 \\ | \\ Me_3-Au-PMe_3 \\ | \\ Me \end{array}\right] + SO_2 \longrightarrow \left[\begin{array}{c} S(O)_2CD_3 \\ | \\ Me-Au-PMe_3 \\ | \\ Me \end{array}\right] + \left[\begin{array}{c} CD_3 \\ | \\ Me-Au-PMe_3 \\ | \\ S(O)_2Me \end{array}\right] \quad (21)$$

on Me. Such an attack followed by formation of an ion pair, $[Me_2Au(PMe_3)]^+O_2SMe^-$, would also be consistent with configuration retention if the ion pair were to recombine faster than rearrangement occurs. However, the regiospecificity of $SO_2$ insertion in platinum and gold complexes as well as in titanium complexes[78] might be noted. In mixed Me,Ph complexes, $SO_2$ invariably inserts into the M—CH$_3$ bond. If ligand attack rather than coordination to the metal were the first step, the stabilization provided by a Wheland–type intermediate (**28**) might be expected to lead to preferential insertion at the M—C$_6$H$_5$ bond.

$O_2S^-$, M— (Wheland-type cyclohexadienyl cation)

(**28**)

For 18-electron complexes, prior coordination of $SO_2$ is not likely. The most studied system with regard to mechanism is [CpFe(CO)$_2$R]. Similar mechanistic features apply to [CpMo(CO)$_3$R], [CpW(CO)$_3$R], [RMn(CO)$_5$], and [RRe(CO)$_5$] with appropriate modification.

A study by Wojcicki and coworkers[79] produced n.m.r. and i.r. evidence for the intermediacy of [M]OS(O)R ([M] = CpFe(CO)$_2$, CpMo(CO)$_3$, Mn(CO)$_5$, and Re(CO)$_5$; R = Me, CH$_2$Ph) which rearranges to the [M]S(O)$_2$R product. Trapping experiments with $I^-$ indicated that a small fraction of the intermediate must be present as the ion pair $[M]^{+-}O_2SR$. Kinetic studies by Jacobson and Wojcicki[80] on the insertion by [CpFe(CO)$_2$R] in $SO_2$(l) have shown that reactivity decreases in the order Et > Me > CH$_2$OMe > CH$_2$CN. As the bulkiness of R increases, the rate decreases in the order Me > CH$_2$CHMe$_2$ > CH$_2$CMe$_3$ > CMe$_3$. These results reflect electrophilic attack by $SO_2$. Moreover, the rates in organic solvents were shown[81] to be first order in $SO_2$. The observation[12] that *threo*-[CpFe(CO)$_2$CHDCHDCMe$_3$]

reacts with $SO_2$(l) giving *erythro*-[$CpFe(CO)_2S(O)_2CHDCHDCMe_3$] shows that inversion occurs at $\alpha$-C. This behaviour indicates backside attack on R by electrophilic $SO_2$. The currently accepted mechanism is shown in Scheme 7. An orbital analysis of this mechanism was given by Fukui and coworkers[82].

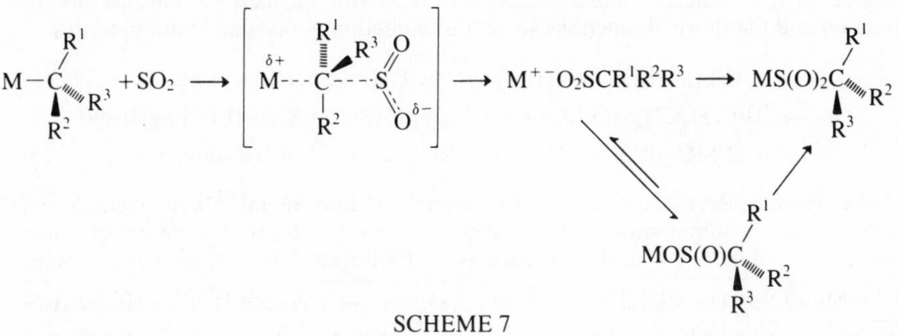

$$M-CR^1R^2R^3 + SO_2 \longrightarrow \left[ \overset{\delta+}{M}\cdots CR^1R^2R^3\cdots S(=O)\cdots O^{\delta-} \right] \longrightarrow M^{+\,-}O_2SCR^1R^2R^3 \longrightarrow MS(O)_2CR^1R^2R^3$$

$$M^{+\,-}O_2SCR^1R^2R^3 \rightleftharpoons MOS(O)CR^1R^2R^3 \longrightarrow MS(O)_2CR^1R^2R^3$$

SCHEME 7

Consistent with this mechanism are the $\Delta H^{\neq}$ values of 2.9–7.8 kcal mol$^{-1}$ and $\Delta S^{\neq}$ of −62 to −43 cal K$^{-1}$ mol$^{-1}$ in $SO_2$(l)[80]. Similarly, [$CpFe(CO)_2CHMe_2$] was found to have $\Delta H^{\neq} = 8.7 \pm 0.6$ kcal mol$^{-1}$ and $\Delta S^{\neq} = -45 \pm 2$ cal K$^{-1}$ mol$^{-1}$ for insertion in $CHCl_3$[81]. The recent observations of Stanley and Baird[75] that $SO_2$ insertion by [$CpFe(CO)\{P(OPh)_3\}CH(Ph)SiMe_3$] in $CDCl_3$ has $\Delta H^{\neq} \approx 20$ kcal mol$^{-1}$ and $\Delta S^{\neq} \approx 0$ is not inconsistent with the mechanism of Scheme 7 since the bulky R group may prevent attack by $SO_2$ until Fe—C bond breaking has proceeded relatively far in the transition state, leading to larger $\Delta H^{\neq}$. Further, the less negative $\Delta S^{\neq}$ may reflect release of steric strain and increased molecular motion on going from the alkyl to the sulphinate. It is conceivable that the mechanism of $SO_2$ insertion may differ in $SO_2$(l) from that in organic solvents since some changes in reactivity order as a function of R have been found[80,81]. However, it has been shown[75] that insertion of $SO_2$ into [$CpFe(CO)_2CHDCHDPh$] proceeds with configuration inversion at $\alpha$-C in $CHCl_3$ as well as in $SO_2$(l), pointing to a similar mechanism.

The observation that electron-withdrawing R groups retard insertion rates[80,81,83] explains the lack of reactivity of perfluoroalkyl complexes toward $SO_2$ insertion. The first example of such an insertion has now been reported[84] (equation 22).

$$cis\text{-}[FeR_FI(CO)_4] + SO_2 + 4dmf \xrightarrow{20\,^{\circ}C} [Fe(SO_2R_F)(dmf)_4I] + 4CO \qquad (22)$$

$$R_F = CF_3,\ C_2F_5,\ C_6F_{13}$$

Interestingly, this reaction does not occur in $SO_2$(l) or hexane. The structure of the product was not reported.

Tungsten compounds are extremely unreactive toward $SO_2$(l) at reflux[83,85]. However, the compounds [$CpW(CO)_3\{S(O)_2R\}$] (R = Me, Et, $CH_2Ph$) have been prepared by bomb reactions at 50–55 °C[85]. In another approach, Lewis acids $BF_3$ or $SbF_5$ were added to $SO_2$(l) to increase its electrophilicity by formation of species such as $OSO \rightarrow BF_3$. The reaction at −40 °C with [$CpW(CO)_3R$] (R = Me, $CH_2Ph$) affords an *O*-sulphinate (**29**).

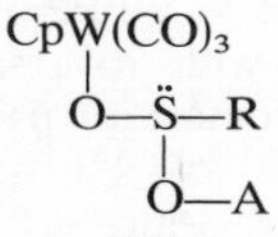

$$CpW(CO)_3-O-\ddot{S}(R)-O-A \qquad A = BF_3,\ SbF_5$$

(**29**)

Removal of the Lewis acid with ammonia or aniline results in rearrangement to the *S*-sulphinate[86].

The first example of a radical chain mechanism has been reported for alkylcoboloximes. The fact that cross-products are obtained in reaction 23 indicates the presence of free radicals[87]. Also, reactions of $SO_2$ with alkylcoboloximes are photocatalysed and the thermal reactions subject to induction periods and variable rates.

$$\begin{aligned} &[BrC_6H_4CH_2Co(dmgH)_2py] + [PhCH_2Rh(dmgH)_2py] + SO_2 \\ &\longrightarrow [\{BrC_6H_4CH_2S(O)_2\}Co(dmgH)_2py] + [\{PhCH_2S(O)_2\}Rh(dmgH)_2py] \\ &+ [\{PhCH_2S(O)_2]Co(dmgH)_2py] + [\{BrC_6H_4CH_2S(O)_2\}Rh(dmgH)_2py] \end{aligned} \quad (23)$$

Other recent developments in $SO_2$ insertions into metal alkyls include the synthesis of *S*-sulphinates with chelate structures from the corresponding alkyls by Lindner and coworkers. Produced in this way were $[(CO)_5MnS(O)_2(CH_2)_3PPh_2]$[66,68], $[CpFe(CO)\{S(O)_2(CH_2)_3PPh_2\}]$[88,89], $[(CO)_4ReS(O)_2(CH_2)_3OPPh_2]$[90], $[(CO)_4MnS(O)_2(CH_2)_nPPh_2]$ ($n = 2,4$), $[(CO)_4ReS(O)_2(CH_2)_nPPh_2]$ ($n = 2–4$)[91], and $[CpNiS(O)_2(CH_2)_3PPh_2]$[89]. The metallacyclopentane complex $[(dppe)Pd(CH_2)_4]$ inserts two molecules of $SO_2$, affording an *S*-sulphinate product[92]. $[(PhCH_2)_2Mn]$ was prepared and found to insert $SO_2$ into both Mn—C bonds[93].

## 2. Stereochemistry

As mentioned in Section III.A.1, the configuration inversion at $\alpha$-C during $SO_2$ insertion seems well established. Inversion has also been shown for some other systems[94].

$$threo\text{-}[CpFe(CO)_2CHDCHDPh] + SO_2(l) \longrightarrow erythro\text{-}[CpFe(CO)_2\{S(O)_2CHDCHDPh\}] \quad (24)$$

$$cis\text{-}[\{threo\text{-}PhCHDCHD\}Mn(CO)_4(PEt_3)] + SO_2(l) \longrightarrow cis\text{-}[\{erythro\text{-}PhCHDCHDS(O)_2\}Mn(CO)_4(PEt_3)] \quad (25)$$

$$trans\text{-}[\{threo\text{-}PhCHDCHD\}W(CO)_2(PEt_3)Cp] + SO_2(l) \longrightarrow trans\text{-}[\{erythro\text{-}PhCHDCHDS(O)_2\}W(CO)_2(PEt_3)Cp] \quad (26)$$

In the light of the free-radical mechanism demonstrated[83] for benzylcoboloximes, the stereospecific inversions seen in equations 27 and 28 are interesting[95].

$$[1(e)\text{-}Me\text{-}4(e)\text{-}\{Co(dmgH)_2(py)\}cyclohexane] + SO_2(l) \longrightarrow [1(a)\text{-}Me\text{-}4(e)\text{-}\{S(O)_2Co(dmgH)(py)\}cyclohexane] \quad (27)$$

$$[1(a)\text{-}Me\text{-}4(e)\text{-}\{Co(dmgH)_2(py)\}cyclohexane] + SO_2(l) \longrightarrow [1(e)\text{-}Me\text{-}4(e)\text{-}\{S(O)_2Co(dmgH)(py)\}cyclohexane] \quad (28)$$

A good deal of work has been done to establish the stereochemistry at the metal. Results so far show stereospecificity with configuration retention for iron and titanium. In an early study, diastereomers of $[CpFe(CO)(PPh_3)CH_2CH(Me)Ph]$ were prepared and found to undergo $SO_2$ insertion with 90% stereospecificity[96]. This result is interesting because it requires even the contact ion pair of Scheme 7 to have great stereochemical stability. Diastereomers of $[\{\eta^5\text{-}(Me)(Ph)C_5H_3\}Fe(CO)(PPh_3)Me]$ were shown by n.m.r. to insert $SO_2$ with >95% stereoselectivity in

$CH_2Cl_2$ and 79% in $SO_2$(l)[97]. The epimerization seen in $SO_2$(l) may be due to dissociation of the tight ion pair. Diastereomers of **30** were prepared and their degree of resolution established by use of n.m.r. shift reagents.

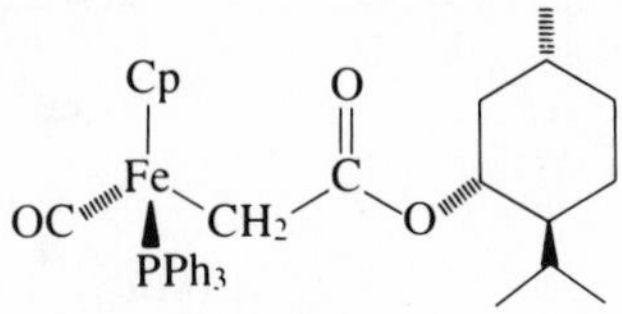

(**30**)

$SO_2$ insertion was found to occur in organic solvent with >90% stereospecificity and comparison of the circular dichroism spectrum with that of a compound of known absolute configuration indicated retention at iron[98].

To avoid problems associated with possible asymmetric induction in diastereomers, enantiomers of [Cp*Fe(CO)($PPh_3$)R] were prepared and found to undergo $SO_2$ insertion in organic solvents stereospecifically at iron as determined by n.m.r.[99]. However, insertion into Fe—$CH_2$-cyclo-$C_3H_5$ leads to [CpFe(CO)($PPh_3$)-{S(O)$_2$$CH_2CH_2CH$=$CH_2$}] with only 40% stereospecificity at iron. An X-ray crystal structure of $(-)_{578}$-[CpFe(CO)($PPh_3$){S(O)$_2$$CH_2CHMe_2$}] prepared from $(+)_{578}$-(*S*)-[CpFe(CO)($PPh_3$)$CH_2CHMe_2$] showed that the sulphinate also has the *S* configuration, thus establishing retention of stereochemistry at iron[100].

French workers have established that reaction of $SO_2$ with [Cp{$\eta^5$-Ph(Me)CHC$_5$H$_4$}Ti(Me)($C_6F_5$)] proceeds regioselectively with insertion into the Ti—$CH_3$ bond to give an *O*-sulphinate and stereoselectively at titanium involving retention. The stereochemistry has been established by chemical correlation of configuration[101] and n.m.r.[78]. The *O*-sulphinate product has chiral sulphur and diastereomers can be detected by n.m.r[102]. *O*-Sulphinates of the type [CpCp$^1$Ti{OS(O)R}$_2$] (Cp$^1$ = substituted cyclopentadienyl) have two chiral sulphur atoms. Racemic- and *meso*-sulphinates can be detected and inversion at sulphur can be followed by n.m.r.[103].

### B. Reactions with Metal Alkenyl and Propargyl Complexes

When the R group contains a centre of unsaturation, several types of reaction may occur with $SO_2$. For an allyl complex (**31**), products of simple insertion may have unrearranged (**32**) or rearranged (**33**) allyl groups. [M] represents a metal with its ancillary ligands.

$$[M]CH_2CR^1{=}CR^2R^3 \quad (\mathbf{31}) \qquad [M]S(O)_2CH_2CR^1{=}CR^2R^3 \quad (\mathbf{32})$$

$$[M]S(O)_2C(R^2)(R^3)C(R^1){=}CH_2 \quad (\mathbf{33})$$

These results may be accounted for by a mechanism first proposed by Giering and Rosenblum[104] shown in Scheme 8, where EN represents a molecule having an electrophilic part (E) and a nucleophilic part (N). Attack on the electron-rich double bond generates the zwitterion **34**. This ion can suffer any of several fates, depending on the identity of EN (Scheme 8). Attack on $CH_2$ gives a cyclic product. Rearrangement to **36**, which is in equilibrium with **37**, gives rearranged and unrearranged insertion products, respectively.

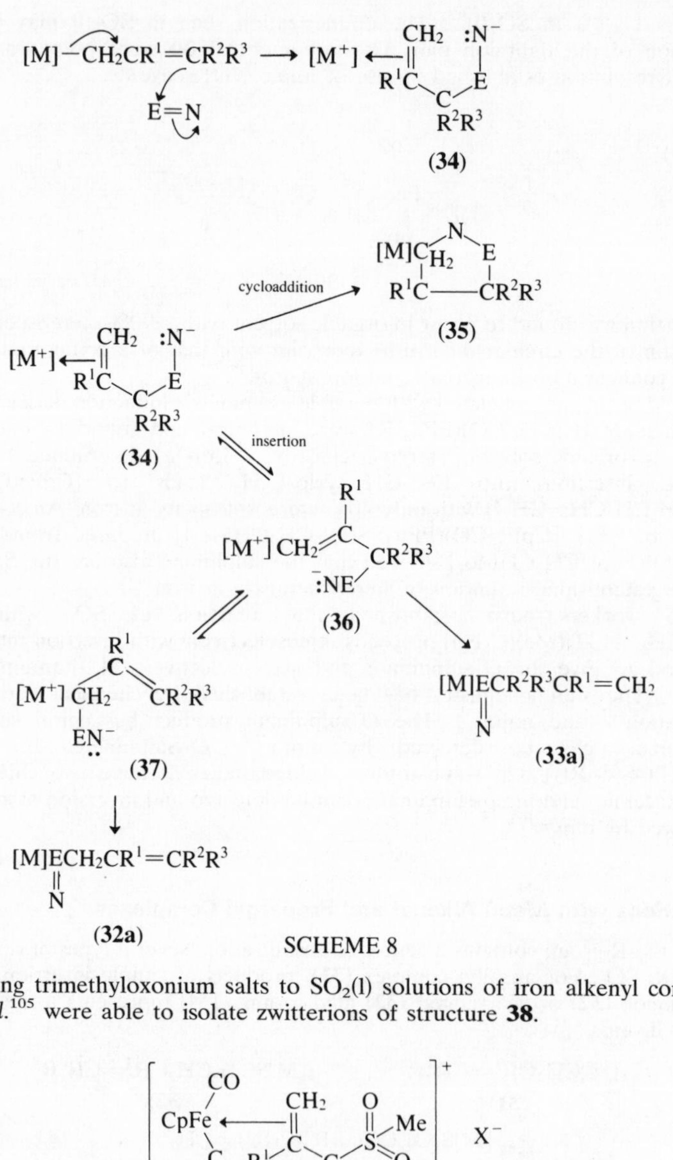

$[M]-CH_2CR^1=CR^2R^3 + E=N \longrightarrow [M^+] \leftarrow$ (34)

(34) $\xrightarrow{\text{cycloaddition}}$ $[M]CH_2$–N–E, $R^1C$—$CR^2R^3$ (35)

(34) $\rightleftharpoons$ (insertion) $[M^+]\ CH_2=C(R^1)CR^2R^3\,:\bar{N}E$ (36) $\longrightarrow$ $[M]ECR^2R^3CR^1=CH_2$ (with $E=N$) (33a)

(36) $\rightleftharpoons$ $[M^+]\ CH_2C(R^1)=CR^2R^3$, $EN^-$ (37) $\longrightarrow$ $[M]ECH_2CR^1=CR^2R^3$ (with $E=N$) (32a)

SCHEME 8

By adding trimethyloxonium salts to $SO_2$(l) solutions of iron alkenyl complexes, Chen *et al.*[105] were able to isolate zwitterions of structure **38**.

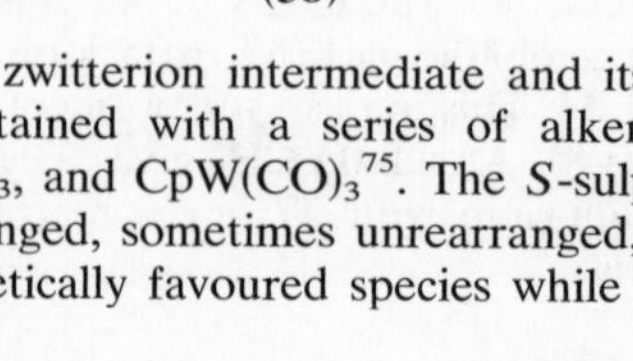

$$\left[ CpFe(CO)_2 \leftarrow CH_2=C(R^1)C(R^2)(R^3)S(O)_2Me \right]^+ X^-$$

**(38)**

The existence of the zwitterion intermediate and its equilibration to **36** and **37** account for results obtained with a series of alkenyl complexes where [M] = $CpFe(CO)_2$, $CpMo(CO)_3$, and $CpW(CO)_3$[75]. The *S*-sulphinate products were sometimes completely rearranged, sometimes unrearranged, and sometimes consisted of mixtures; **36** is the kinetically favoured species while **37** is less sterically crowded and, hence, thermodynamically favoured. Less sterically demanding substituents favour **36** and lead to larger yields of rearranged products. Parallel results were obtained on a series of ring-substituted alkenyl complexes where $[M] = (\eta^5\text{-}MeC_5H_4)Fe(CO)_2$, $(\eta^5\text{-}1,3\text{-}Ph_2C_5H_3)CpFe(CO)_2$, $Mo(CO)_2\{P(OPh)_3\}$[106].

The complex $[(\eta^3\text{-}C_3H_5)Pd(PPh_3)(\eta^1\text{-}C_3H_5)]$ inserts $SO_2$ at −30 °C giving[107] $[(\eta^3\text{-}C_3H_5)Pd(PPh_3)\{S(O)_2CH_2CH{=}CH_2\}]$. With propargyl ($[M]CH_2C{\equiv}CR$) complexes $SO_2$ reacts via a mechanism similar to that of Scheme 8 to afford cycloaddition products $\overline{[M]{-}C{=}C(R)S(O)OCH_2}$ containing a sultine ring instead of an insertion product[108].

## C. Insertion of Species Isoelectronic with $SO_2$

*N*-Sulphinylsulphonamides, $R^2S(O)_2N{=}S{=}O$, and disulphonylsulphur diimides, $R^2S(O)_2N{=}S{=}NS(O)_2R^2$, were also found to insert into $Fe{-}R^1$ bonds to afford products of structures **39** and **40**, respectively. Compound **39** can be transformed thermally to **41**.

$CpFe(CO)_2N[S(O)R^2]S(O)_2R^1$ (**39**)

$CpFe(CO)_2N[S(O)_2R^2]S(R^1)NS(O)_2R^2$ (**40**)

$CpFe(CO)_2S(O)(R^1){=}NS(O)_2R^2$ (**41**)

These insertions are similar to those of $SO_2$, since they both occur with inversion of configuration at α-C. Moreover, **39**, the initial product with *N*-sulphinylsulphonamides, also has the hard base bonded to iron, but in this case is isolable[109,110] in contrast to analogous *O*-sulfonates.

$SeO_2$ has been found to insert into $[CpFe(CO)_2Me]$ and $[CpMo(CO)_3Me]$, affording *Se*-selinate products[111]. $[(\eta^7\text{-}C_7H_7)Mo(CO)_2Me]$ reacts with $SO_2$, $SeO_2$, and $TeO_2$ to give products of structure $[(\eta^7\text{-}C_7H_7)Mo(CO)_2\{E(O)_2Me\}]$ (E = S, Se, Te)[112,113].

## D. Insertion of Other Electrophiles

### 1. Tetracyanoethylene

Tetracyanoethylene is discussed in this section while tetrafluoroethylene is arbitrarily placed in Section IV.A.2 with olefins. Tcne displays a variety of behaviour with transition metal carbonyl alkyl complexes giving products of structures **42**, **43**, and **44**[114].

$[M]C(CN)_2C(CN)_2R$
Metal cyanoalkyl
(**42**)

$[M]N{=}C{=}C(CN)C(CN)_2R$
metal keteniminate
(**43**)

$(NC)_2C{-}C(CN)_2$ bridged by $[M]C(O)R$
tcne acyl
(**44**)

For $[M] = CpMo(CO)_2L$, products of structure **42** result when $L = PPh_3$ and $P(OPh)_3$, but no reaction occurs for L = CO, reflecting the electrophilic nature of the cleavage by tcne. The products **42** can be converted thermally to **43**.

For $[CpFe(CO)_2R]$, the order of reactivity toward tcne is $R = CH_2Ph > CHMePh > Me, Et, Pr^n \gg Ph$ which is different from the reactivity order toward $SO_2$. This reflects a free-radical path for insertion and, indeed, free radicals have been detected by e.s.r.

for mixtures of tcne and benzyl complexes[115]. For [M] = $CpFe(CO)_2$, isomers **42** and **43** are both produced and do not interconvert, reflecting simultaneous parallel paths for their production. Similar results were seen[116] for $[CpCr(NO)_2R]$. For [M] = CpFe(CO)L (L = phosphine, phosphite), reactions with tcne occur rapidly in $CH_2Cl_2$. For R = $CH_2Ph$, a product of structure **43** results whereas tcne acyls (**44**) are produced for R = Me, Et, and $Pr^n$. These acyl complexes rearrange thermally to **43** or lose tcne to give starting material[109].

$[(\eta^4\text{-}2,3\text{-}Me_2C_4H_4)Ru(CO)_2L]$, which has a $\sigma$-component of the Ru—C bond inserts tcne giving **45** when L = $PPh_3$ or $P(OMe)_3$. No reaction occurs when L = CO[117].

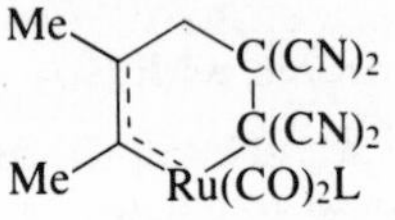

Me — C(CN)$_2$ — C(CN)$_2$ — Ru(CO)$_2$L — Me

(**45**)

With alkenyl and propargyl complexes tcne and related electrophilic olefins give cycloaddition products via Scheme 8[118].

### 2. $SnX_2$ and $GeX_2$

In 1970, Nesmeyanov and coworkers[119] found that $GeCl_2$ reacts with $[CpFe(CO)_2R]$ (R = Me, Et, $Pr^n$, $Pr^i$, $CH_2Ph$, Ph) in dioxane or thf to give $[CpFe(CO)(GeCl_2R)]$. No reaction was observed with electron-withdrawing R = $C_6F_5$, $COCF_3$, COMe. Also, $SnCl_2$ was found to react with $[CpFe(CO)_2Me]$ to give $[CpFe(CO)_2(SnCl_2Me)]$. With $SnBr_2$, the products were $[CpFe(CO)_2Br]$ and $[CpFe(CO)_2SnBr_3]$. $[CpFe(CO)_2Et]$ gave $[CpFe(CO)_2(SnCl_2Et)]$, $[CpFe(CO)_2SnCl_3]$, and $[\{CpFe(CO)_2\}_2SnCl_2]$ with $SnCl_2$[120]. $[Sn\{CH(SiMe_3)_2\}_2]$ inserts into the M—C bonds of $[CpMo(CO)_3Me]$ and $[CpFe(CO)_2Me]$[121].

Recent kinetic studies following reactions by n.m.r. indicated a free-radical chain mechanism for insertion of $SnCl_2$ and $GeCl_2$ into Fe—C bonds in $[CpFe(CO)_2R]$. The reactions were photocatalysed and sometimes showed induction periods; 1% 1,1-diphenyl-2-picrylhydrazyl inhibited the reactions completely. Qualitative observations indicate a rate decrease in the order R = $Pr^i > Pr^n >$ Et > Me > $CH_2Ph \gg CF_3$, Ph for $GeCl_2$ insertion consistent with electrophilic behaviour of the inserting species[122].

An investigation of reactions of metal alkenyl complexes $[CpFe(CO)_2CH_2C(R^1){=}CR^2R^3]$ with $SnCl_2$ showed insertion to give $[CpFe(CO)_2\{SnCl_2CH_2C(R^1){=}CR^2R^3\}]$ with no allylic rearrangement. It was proposed that the mechanism of Scheme 8 could account for these results[123]. However, recent evidence indicates that a free-radical mechanism may also be operative in these reactions and that the kinetic product is the rearranged isomer which is converted back to unrearranged isomer at a rate proportional to the excess of $MX_2$ in solution[124].

### 3. Hexafluoroacetone

Hexafluoroacetone, $CF_3C(O)CF_3$, generally reacts with alkenyl and propargyl complexes via the mechanism of Scheme 8 to give cycloaddition products. However, with $[CpFeCH_2C(Me){=}CH_2]$, the zwitterionic intermediate **34** apparently dissociates and attack occurs at iron by oxygen giving an insertion product **46**, which reacts with another mole of hfac to give **47**[125].

$$CpFe(CO)_2OC(CF_3)_2CH_2C(Me){=}CH_2$$
**(46)**

$$CpFe(CO)_2OC(CF_3)_2CH_2C({=}CH_2)CH_2C(CF_3)_2OH$$
**(47)**

Attack on double bonds coordinated to the metal is presumably responsible for insertion of hfac into **48** giving **49**[126], as well as relating insertions involving $[(cod)_2Pt]$[127] and $[(\eta^5\text{-indenyl})Rh(isoprene)]$[128].

$[Pt(cod)(R)_2]$
$R = CH_2{=}CHCOCH_3$

**(48)**

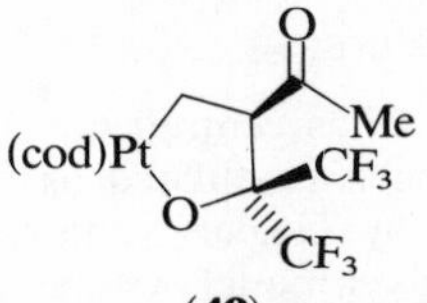

(cod)Pt, O, O, Me, $CF_3$, $CF_3$

**(49)**

## IV. INSERTION OF ALKENES AND ALKYNES

### A. General

A vast number of reactions are known that involve insertions of alkenes and alkynes into transition metal—carbon bonds. Hence, this section cannot attempt comprehensive coverage. Instead, major types of reactions and recent results will be emphasized. Most of the work done is synthetic and structural. Except for labelling studies, information on mechanisms is largely postulated from the structures of products. However, features similar to those of carbonyl insertion often seem to be operative: vacation of a coordination position thermally or photochemically (not necessary if the metal has less than 18 electrons), coordination of the unsaturated molecule, and insertion which may involve alkyl migration or attack on alkyl by the inserting ligand. With electron-withdrawing molecules such as perfluorinated olefins and alkynes, a dipolar pathway also is possible, as represented in Scheme 9. The

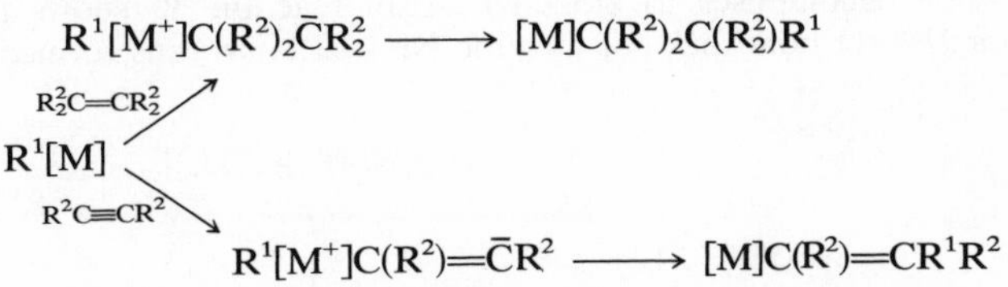

$$R^1[M^+]C(R^2)_2\bar{C}R^2_2 \longrightarrow [M]C(R^2)_2C(R^2_2)R^1$$

$$R^1[M] \xrightarrow{R^2_2C{=}CR^2_2} \quad ; \quad R^1[M] \xrightarrow{R^2C{\equiv}CR^2}$$

$$R^1[M^+]C(R^2){=}\bar{C}R^2 \longrightarrow [M]C(R^2){=}CR^1R^2$$

SCHEME 9

formation of a Lewis acid–base adduct places negative charge on the carbon originally at the terminus of the multiple bond followed by migration of $R^1$. This pathway is possible for 18-electron metals even without prior loss of a ligand with its complement of electrons.

Attack could be on the metal itself or on an unsaturated ligand. Many examples of multiple insertions are known. Mono- or multiple insertion may be followed by rearrangement via hydride shift or other paths.

Perfluorinated alkenes and alkynes constitute a large fraction of the inserting molecules known. Some questions exist as to whether their insertions are really very typical of alkenes and alkynes since relatively so few insertions are known for the hydrocarbon analogues. Nevertheless, because the structures of insertion products are similar, we shall discuss both types of unsaturated reactants together. The order of discussion is from simple to more complex inserting molecules. We treat complexes in order of electron number: 16-electron complexes followed by 18 within each inserting ligand class.

## B. Alkenes

### 1. General features

Much interest has centred on alkene insertion into M—C bonds since repeated insertion has been postulated as the step responsible for chain growth in alkene polymerization by Ziegler–Natta catalysts. Until recently, however, no well characterized metal alkyl model systems were known to undergo insertion of $C_2H_4$. Now, reaction 29 has been found to occur by Evitt and Bergman[129].

$$[CpCo(PPh_3)Me_2] + 2CH_2{=}CH_2 \longrightarrow [CpCo(C_2H_4)(PPh_3)] + CH_4 + CH_3CH{=}CH_2 \quad (29)$$

Labelling studies showed that the products must arise from ethylene insertion rather than from a metallacycloubtane as in **50**[130].

CpCo

PPh$_3$

**(50)**

Insertion of simple olefins into M—H bonds is far more common than into M—C bonds. Two examples must suffice to illustrate this. $[RCo(PMe_3)_2(C_2H_4)]$ complexes were prepared for R = H and Ph[131]. In solution the hydride complex catalyses isomerization of pentene via successive insertions and $\beta$-eliminations. The phenyl complex merely decomposes to biphenyl. Also, reaction 30 shows that ethylene inserts into a Nb—H bond but not into the Nb—Et bond thus formed[132].

$$[Cp_2Nb(H)(C_2H_4)] \xrightarrow{C_2H_4} [Cp_2Nb(Et)(C_2H_4)] \not\longrightarrow [Cp_2Nb(C_4H_9)(C_2H_4)] \quad (30)$$

Several theoretical treatments of ethylene insertion are available, including a CNDO study of insertion into Ti—Me bonds[133]. A review of olefin insertion into M—C bonds of zinc and magnesium alkyls has been published[134].

A review of olefin insertion in catalytic processes[135] has documented the regiospecificity of insertions of unsymmetrical alkenes.

### *2. Insertion of isolated double bonds*

The calculations of Thorn and Hoffmann on Pt—H insertion[133] have revealed features which seem relevant to Pt—C insertions of both alkenes and alkynes. In particular, no low activation energy pathway was found for insertion into five-coordinate platinum(II) complexes. Four-coordinate complexes having H and $C_2H_4$ *cis* displayed a moderately facile insertion path. That *cis* orientation is not a sufficient condition for insertion is shown by the fact that *cis*-$[(diars)PtMe(C_2H_4)]^+$ does not insert[136]. Platinum(II) complexes have a greater tendency to insert activated alkenes and, especially, alkynes.

$C_2F_4$ is known to insert into M—C bonds in *trans*-$[PtMeL_2X]$ complexes[137], $[MeAu(PPhMe_2)]$[138], and, with photochemical activation, $[MeAu(PMe_3)]$[139]. Another 16-electron complex **51** was found to insert in solution giving an allyl complex, **52**[140].

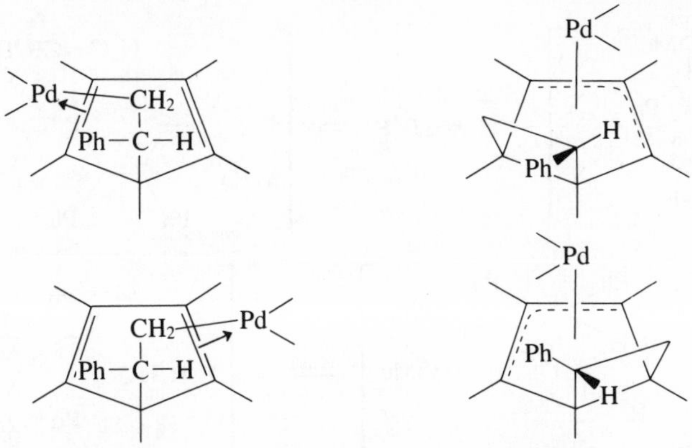

**(51)** **(52)**

Because $\eta^3$-allyl complexes generally participate in equilibria analogous to that in equation 31, insertions can occur into metal—allyl bonds. Norbornene was found[141]

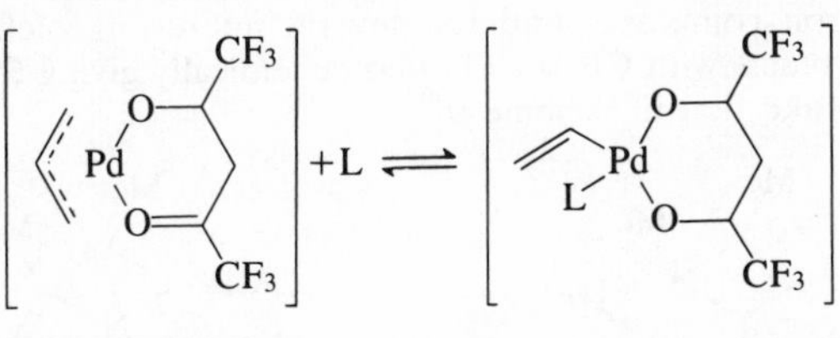

$$(\eta^3\text{-allyl})Pd(hfacac) + L \rightleftharpoons (\eta^1\text{-allyl})PdL(hfacac) \quad (31)$$

to insert into $[(\eta^3\text{-allyl})M(hfacac)]$ (M = Pd, Pt) giving *cis-exo*-products **53**. Similarly, NaOAc reacts with $[Ni\{\eta^3\text{-}CH_2C(Me)CH_2\}(\eta^2\text{-norbornene})Cl]$ giving **54**[142].

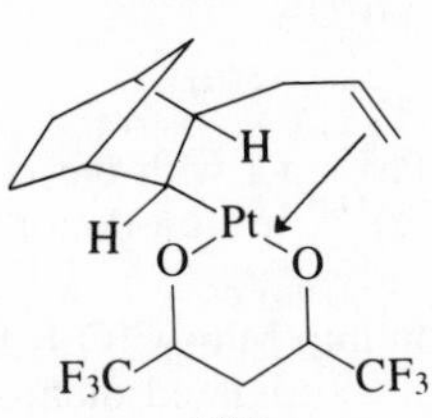

**(53)** 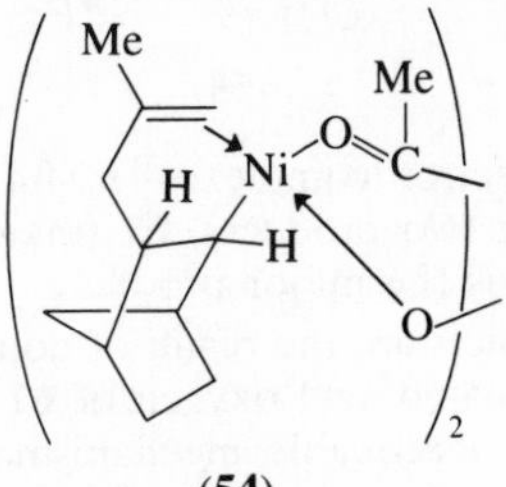 **(54)**

The *endo*-cyclobutenyl complexes **55** exist in equilibrium in solution with the ring-opened form[143] **56**, which insert olefins to give **57** via a route shown in Scheme 10 involving a hydride shift[144].

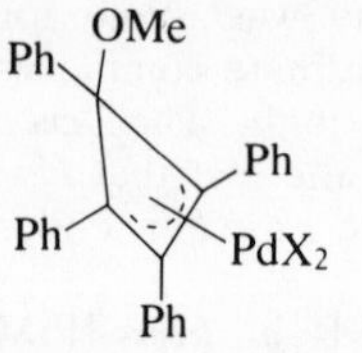

$X_2 = Cl_2$, acac, hfacac

**(55)**

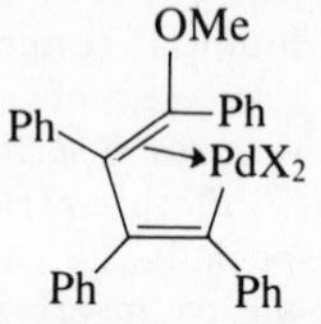

X = $Cl_2$, acac, hfacac

**(56)**

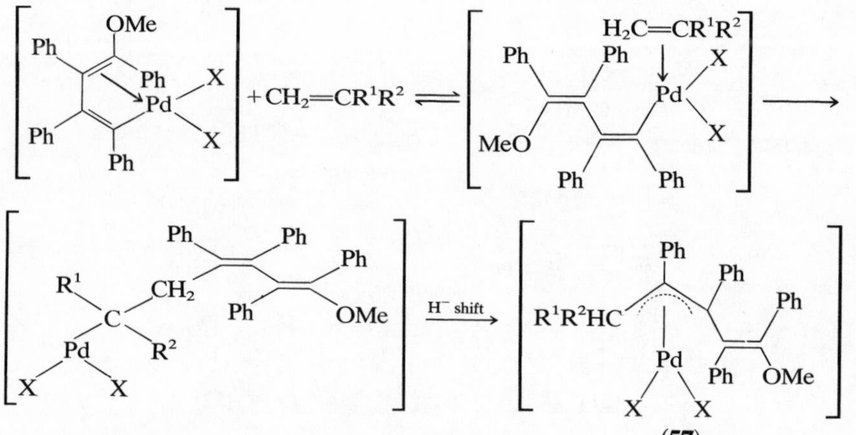

H⁻ shift

**(57)**

Eighteen-electron complexes undergo several interesting olefin insertions. [($\eta^4$-$Me_4C_4$)$Fe(CO)_3$] reacts with $CF_2$=CFH photochemically giving **58** and **59**, probably via a mechanism like that in Scheme 9[145].

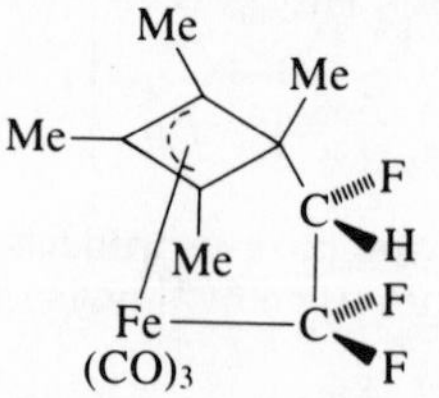

**(58)**

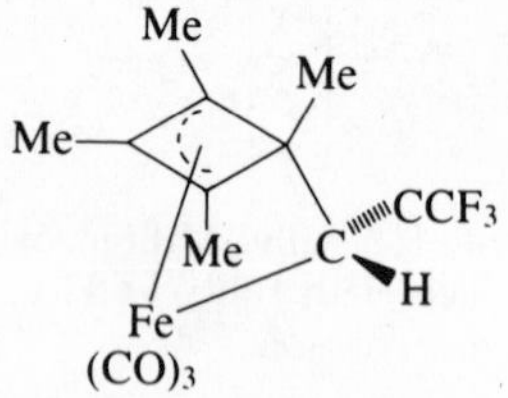

**(59)**

The 18-electron complexes [$RMn(CO)_5$] (R = Me, Ph) react with dicyclopentadiene (**60**) giving two products, **61** (major) and **62** (minor)[146], in non-donor solvents. In MeCN **62** is the major product.

Both products are the result of double bond insertion into Mn—C(O)R bonds. In **62** the coordinated acyl oxygen of **61** is displaced by a coordinated olefin. Scheme 11 represents a probable mechanism where only the double bond function of dicyclopentadiene is represented.

(60) 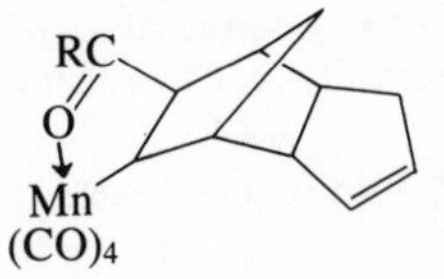 (61) 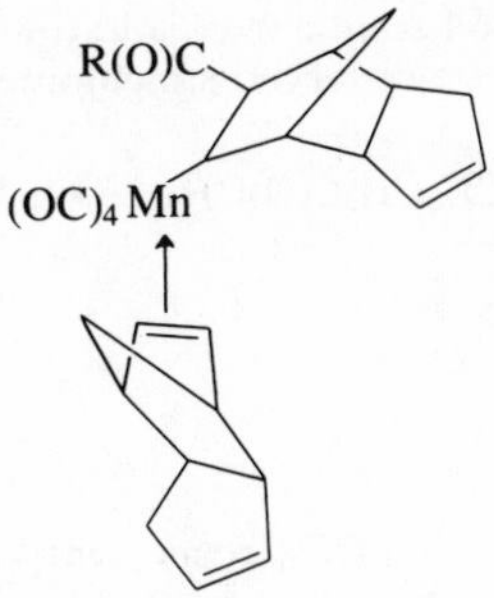 (62)

$\equiv R'Mn(CO)_4 \leftarrow R''$

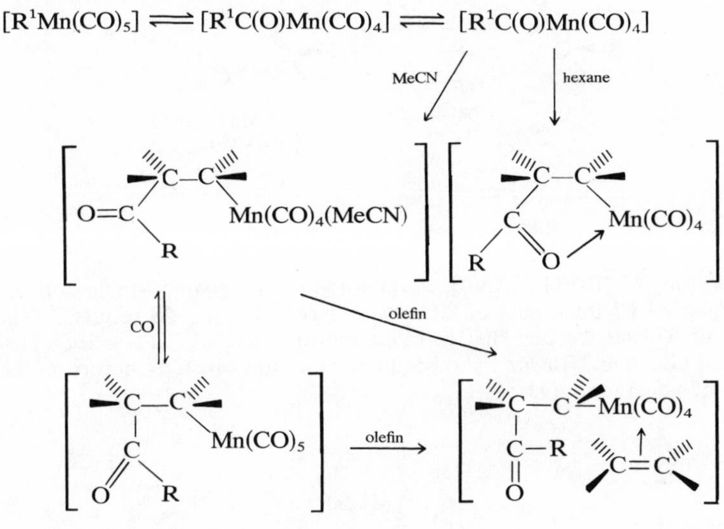

$$[R^1Mn(CO)_5] \rightleftharpoons [R^1C(O)Mn(CO)_4] \rightleftharpoons [R^1C(O)Mn(CO)_4]$$

MeCN hexane CO olefin olefin

SCHEME 11

In MeCN a small amount of acyl **63** resulting from carbonylation of **62** is isolated. No reactions occurred with 1,5-cod, norbornene, cyclohexene or *cis*-but-2-ene.

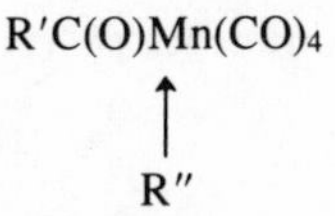

$R'C(O)Mn(CO)_4 \leftarrow R''$

R′, R″ as in **62**

**(63)**

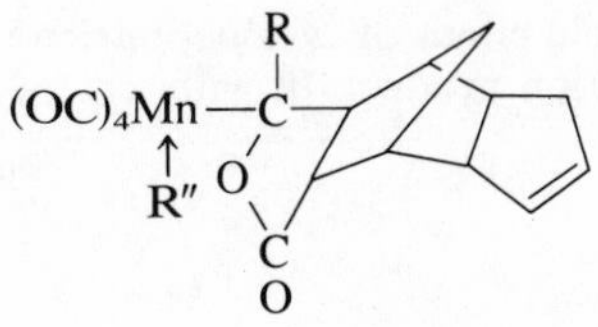

R″ as in **62**

**(64)**

$[RC(O)Mn(CO)_5]$ (R = Me, Ph) afford **61** and **62** in hexane. However, in MeCN an inseparable mixture of **63** and lactone **64** results[147] when R = Me and only **64** when

R = Ph. Compound **64** results from addition of C—Mn across >C=O in **63** (R = Ph). This addition has been observed stoichiometrically as shown in equation 32.

$$Mn(CO)_5^- + ClC(O)CH_2CH_2C(O)CH_3 \longrightarrow [(CO)_5MnC(O)CH_2CH_2C(O)CH_3] \downarrow [(CO)_5MnR] + [(CO)_5MnC(O)R] \quad (32)$$

R =

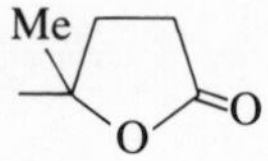

$C_2F_4$ and $CF_3CF{=}CFCF_3$ react with $[\{\eta^3\text{-}CH_2C(R)CH_2\}Co(CO)_2L]$ (L = CO, phosphines) giving monoinsertion products **65** with equatorial double bonds for R = H. When R = Me, $CF_3CF{=}CF_2$ regioselectively gives **66**[148] [L = $P(OMe)_3$]. With

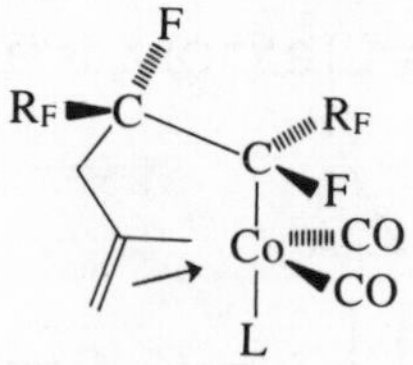

$R_F$ = F, $CF_3$

**(65)**

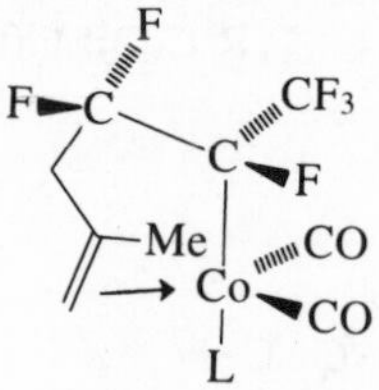

**(66)**

$[\{\eta^3\text{-}CH_2C(Me)CH_2\}Ir(PPh_3)_2(CO)]$, tetrafluoroethylene gives a stable adduct (**67**). On treatment of **67** for 4 days at 80 °C with excess of $C_2F_4$, **68** results[149]. The low reactivity of **67** and the fact that coordination of a second $C_2F_4$ to iridium seems unlikely led Green and Taylor [149] to postulate that attack by $C_2F_4$ occurs on the allyl ligand of **67** generating **69**.

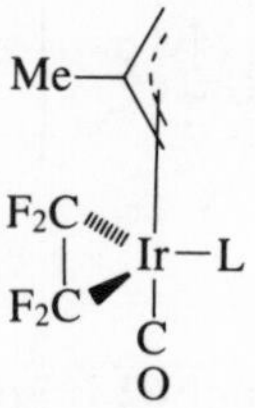

**(67)** 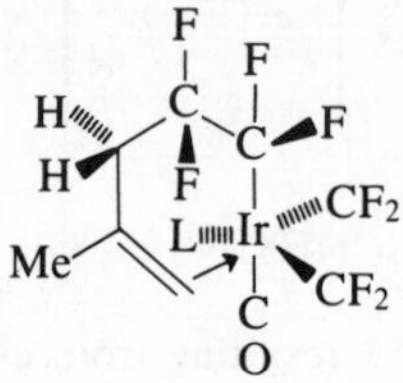 **(68)** 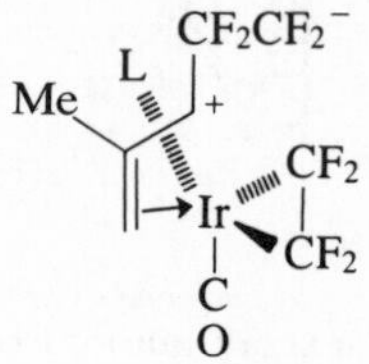 **(69)**

One of the double bonds of cycloheptatriene was found to react with $[PhMn(CO)_5]$ affording an insertion product **70** with loss of CO[150].

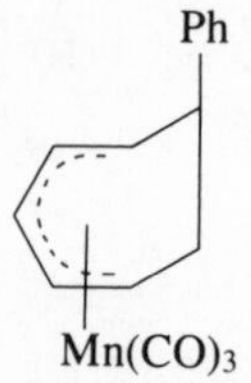

**(70)**

### 3. Dienes

Allenes (1,2-dienes) insert into Pt—C bonds in cationic complexes giving 2-allyl complexes, i.e. R adds to $C_{(2)}$. This reaction seems to require a cationic complex for allene activation. The insertion into *trans*-$[PtMe(\eta^2\text{-}CH_2{=}C{=}CH_2)(PMe_2Ph)_2]^+$ to afford[151] $[\{\eta^3\text{—}CH_2C(CH_3)CH_2\}Pt(PMe_2Ph)_2]^+$ was first order in the platinum complex and was retarded by excess of allene, indicating that a five-coordinated species is not involved in the insertion as expected from the results of Lauher and Hoffmann[133]. *trans*-$[Pt(Ph)(PEt_3)_2Cl]$ must be converted into a cationic complex (which also has a vacant site for coordination of allene) by addition of $AgBF_4$ in the presence of allene before insertion occurs giving $[\{\eta^3\text{-}CH_2C(Ph)CH_2\}Pt(PEt_3)_2]BF_4$[152].

Hughes and Powell have investigated the reactions of allene with more labile palladium complexes. The insertions depicted in Scheme 12 show that palladium—norbornenyl complexes behave[153] like allyl complexes (see Scheme 13)[154].

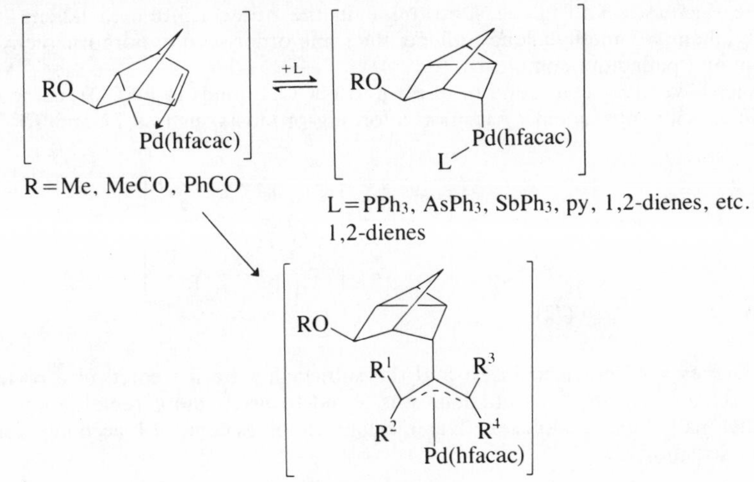

R=Me, MeCO, PhCO

L=$PPh_3$, $AsPh_3$, $SbPh_3$, py, 1,2-dienes, etc.

SCHEME 12

$X_2$=$Cl_2$, acac, hfacac

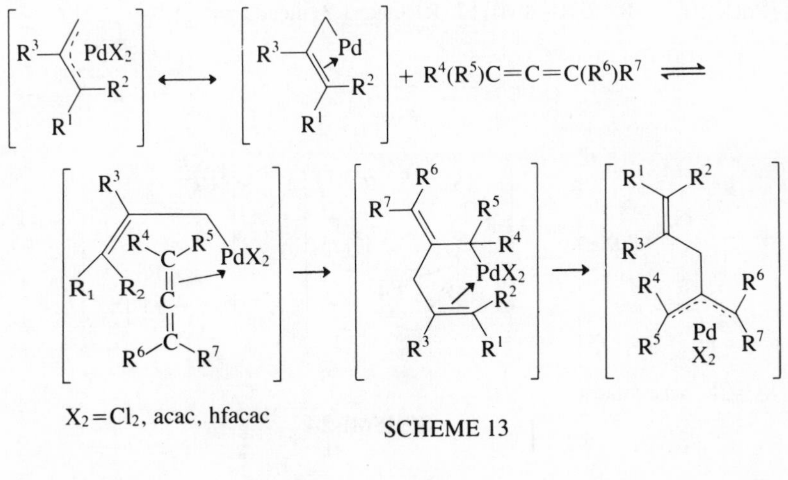

SCHEME 13

Allenes also react[144] with **55** via the allylic form **56**, giving products of structure **71** where the allene numbering corresponds to that in Scheme 13.

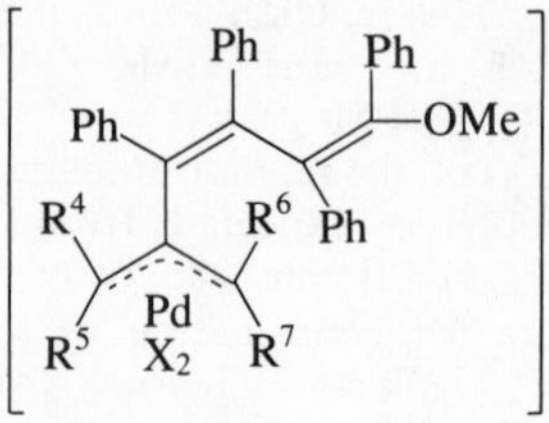

**(71)**

Kinetic studies have shown that the reaction is first order both in palladium complex and allene. Electron-withdrawing groups $R^1$, $R^2$, $R^3$, or X speed up the insertion, and the rate decreases with allene substitution in the order 1,3-dimethylallene >1,1-dimethylallene > 1-methylallene > allene, the same order seen in norbornadienyl and cyclobutenyl palladium complexes.

Allenes have also been found to insert into FeCOR bonds in $Fe(CO)_4$ compounds sometimes with subsequent cyclization affording products such as **72** and **73**[155].

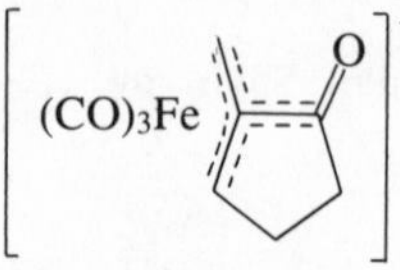

**(72)** 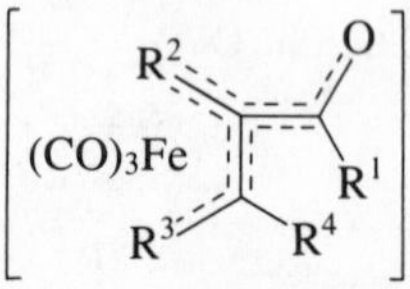 **(73)**

1,3-Dienes are known to insert and the subject has been treated in a review of metal—diene chemistry[156]. Butadiene was found to insert more rapidly than allene into allyl palladium complexes. The mechanism of Scheme 14 accounts for the product structure[157].

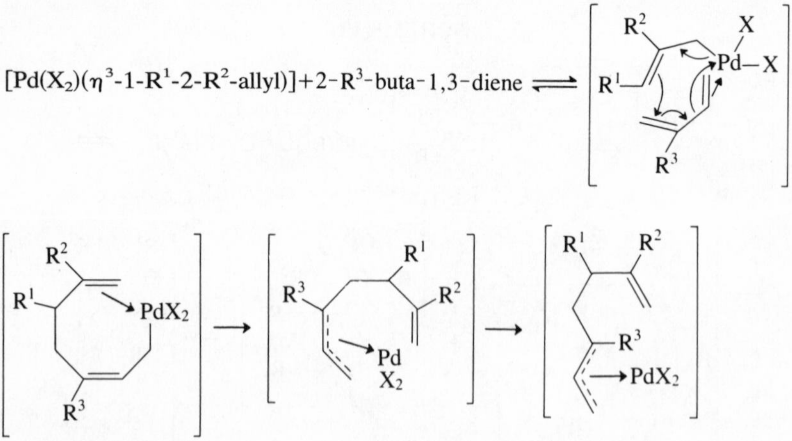

$[Pd(X_2)(\eta^3\text{-}1\text{-}R^1\text{-}2\text{-}R^2\text{-allyl})] + 2\text{-}R^3\text{-buta-}1,3\text{-diene} \rightleftharpoons$

$X_2 = Cl_2$, acac, hfacac

SCHEME 14

Similar results were also found for the structure of the allylic product when dienes $CH_2{=}CHC(R^2){=}CH_2$ were allowed to insert into the dimers $[\{\eta^3\text{-}CH_2C(R^1)CH_2\}PdCl]_2$. Treatment of the dimeric product with $PPh_3$ and $AgClO_4$ gave cations of structure **74**[158] where the isolated double bond is in the coordination

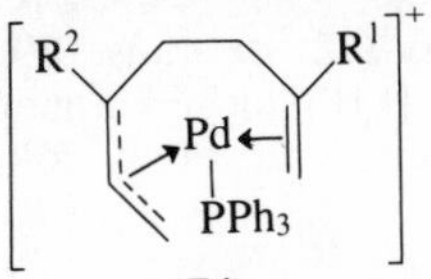

$R^2$ $R^1$ Pd $PPh_3$

**(74)**

sphere of palladium 1,3-Dienes are also known to insert into the cyclobutenyl complexes **55**[144].

Palladium allyls containing unsymmetrical diketonate and Schiff base ligands insert 1,3-dienes. When **75** is allowed to react with butadiene, the product has the uncoordinated but-1-enyl-2-methyl substituent *trans* to sulphur (**76**)[159]. $[\{\eta^3\text{-}CH_2C(Cl)CH_2\}Pd(\text{salicyclaldiminato})]$ inserts isoprene. However, the 2-H compound

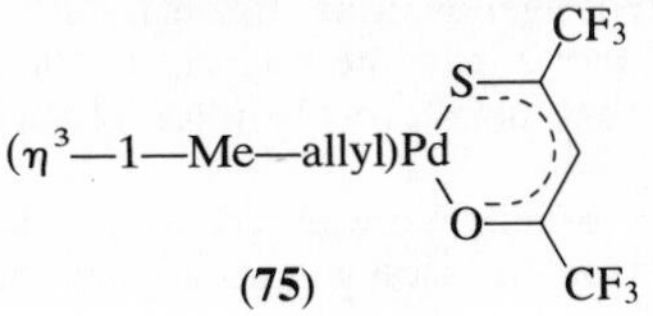

$(\eta^3\text{—1—Me—allyl})Pd$ (S, O, $CF_3$, $CF_3$) **(75)**

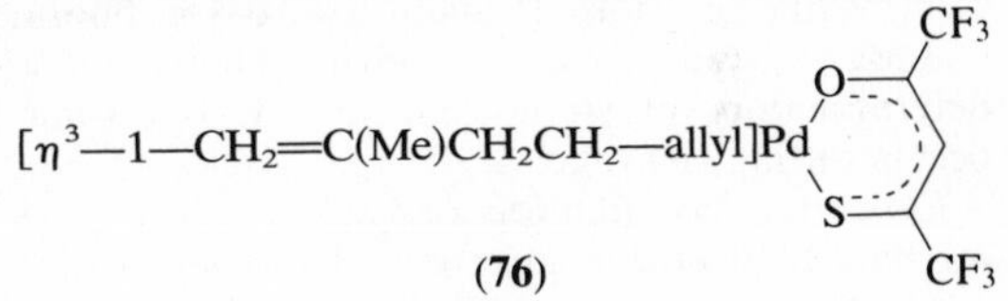

$[\eta^3\text{—1—}CH_2{=}C(Me)CH_2CH_2\text{—allyl}]Pd$ (O, S, $CF_3$, $CF_3$) **(76)**

does not insert. The lower reactivity of the Schiff base complex compared with that of the hfacac complex is in keeping with the results noted above that insertion is facilitated by electron-withdrawing groups on palladium[160].

Another example of insertion of an olefin into Mn—COR bonds is seen in the reaction of butadiene with $[RMn(CO)_5]$ (R = Me, Ph). Whether *cis*-or *trans*-butadiene is employed the product is stereospecifically **77** with a *syn* acyl group.

$$[Mn(CO)_4(\eta^3\text{-1-Me-3-COR-allyl})]$$

**(77)**

As Scheme 15 shows, this can be accounted for by carbonylation induced by olefin coordination followed by *cis*-1,2-addition of Mn—COR across a double bond (giving the wrong allyl isomer) and 1,4-hydride shift via the metal. This mechanism is in accord with a labeling experiment using $CD_2{=}CHCH{=}CD_2$ for the benzoyl complex[161].

$[RMn(CO)_5]$ + buta-1,3-diene ⟶ $CH_2{=}CHCH{=}CH_2$ ↓ $(CO)_4MnCOPh$ $\xrightarrow{cis\text{-1,2-addition}}$

$[Mn(CO)_4(\eta^3\text{-1-}CH_2COPh\text{-allyl})]$ ⟶ $[HMn(CO)_4(\eta^2\text{-1-COPh-buta-1,3-diene})]$

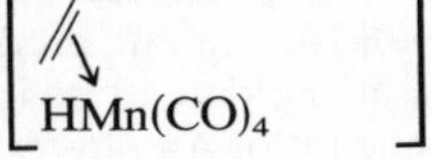

⟶ [COPh; $HMn(CO)_4$] ⟶ $[Mn(CO)_4(\eta^3\text{-1-COPh-3-}CH_3\text{-allyl})]$ **(77)**

SCHEME 15

## C. Alkynes

### 1. General features

The products of insertion of actylenes $R^2C{\equiv}CR^2$ into [M]—$R^1$ are alkenyl (vinyl) complexes which may have *cis* (**78**) or *trans* (**79**) stereochemistry of [M] and $R^1$ about the double bond where [M] indicates a metal with its ancillary ligands. The designation of these structures as *E* or *Z* follows the usual rules of organic nomenclature.

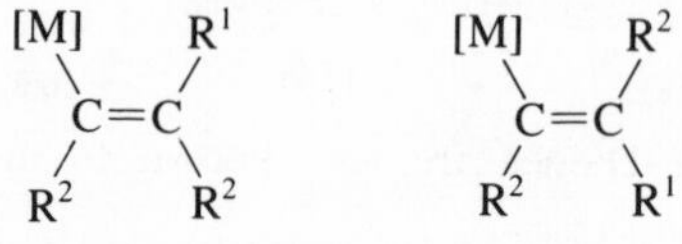

```
[M]      R¹          [M]      R²
   \    /               \    /
    C=C                  C=C
   /    \               /    \
R²       R²          R²       R¹
   (78)                 (79)
```

The stereochemistry of the adduct is often determined by n.m.r. chemical shift and/or coupling constant data. Determination of stereochemistry via coupling constant values is more straightforward for fluorinated ligands than for, say, $R^2 = CO_2Me$. Chemical cleavage giving known olefins and X-ray methods also allow determinations of stereochemistry. It is assumed that no stereochemical change occurs during cleavage.

Concerted *cis* addition of [M]—R to alkynes is usually considered to be the mechanistic course of insertions. However, with electron-withdrawing acetylenes the mechanism of Scheme 9 is an attractive possibility, especially since a carbanionic intermediate could attack other alkynes and account for cases of multiple insertion. Present evidence does not rule out concerted *trans*-addition, $\alpha$-hydride migration to give alkylidene intermediates, or attack on coordinated alkyne. Products of exclusive *cis*-addition and of exclusive *trans*-addition have been observed as well as stereorandom products. In some cases, a metal complex gives *cis*-insertion for one alkyne and *trans* for another.

Most insertions occur with activated alkynes such as $CF_3C{\equiv}CCF_3$, PhC≡CPh, and $MeO_2CC{\equiv}CCO_2Me$ containing electron-withdrawing substituents. Reviews of the preparation of fluorinated compounds have appeared[162] and also a review of alkyne complexes which includes insertions, especially into M—H bonds[163].

All alkynes discussed in the following sections are monoalkynes.

### 2. Fourteen- and sixteen-electron complexes

$[MeAu(PMe_3)]$[139] and $[MeAu(PMe_2Ph)]$[138] react with $CF_3C{\equiv}CCF_3$ giving an adduct containing two molecules of [MeAuL] per acetylene. With excess of alkyne these give *cis*-$[\{Me(CF_3)C{=}C(CF_3)\}AuL]$ (L = phosphine). Photochemical reaction of $[Cp_2TiMe_2]$ with PhC≡CPh or $C_6F_5C{\equiv}CC_6F_5$ give *cis* addition products $[Cp_2TiMe\{C(R){=}C(R)Me\}]$ (R = Ph, $C_6F_5$)[164]. An addition of the unactivated alkyne MeC≡CMe to *trans*-$[NiPh(L_2)Br]$ gives an insertion product[165] having *cis* stereochemistry at the double bond.

Huggins and Bergman[166] have studied the insertion of a wide variety of unactivated alkynes with [NiMe(acac)Cl]. The vinyl products were formed regiospecifically with the sterically larger substituent near the nickel atom and methyl migration to the less hindered carbon atom. Depending on the alkyne, kinetic products displayed all *cis*, all *trans*, or a mixture of stereochemistries at the double bond. Further, the kinetic isomer ratio differed from the thermodynamic ratio. An interesting result was that both the reaction of [NiMe(acac)Cl] with PhC≡CPh and the reaction of

[NiPh(acac)Cl] with PhC≡CMe give the same product, namely [Ni{*Z*-C(Ph)=C(Ph)Me}(acac)Cl]. This result (and those of other experiments) can be rationalized as shown in Scheme 16 on the assumption that exclusive *cis*-addition

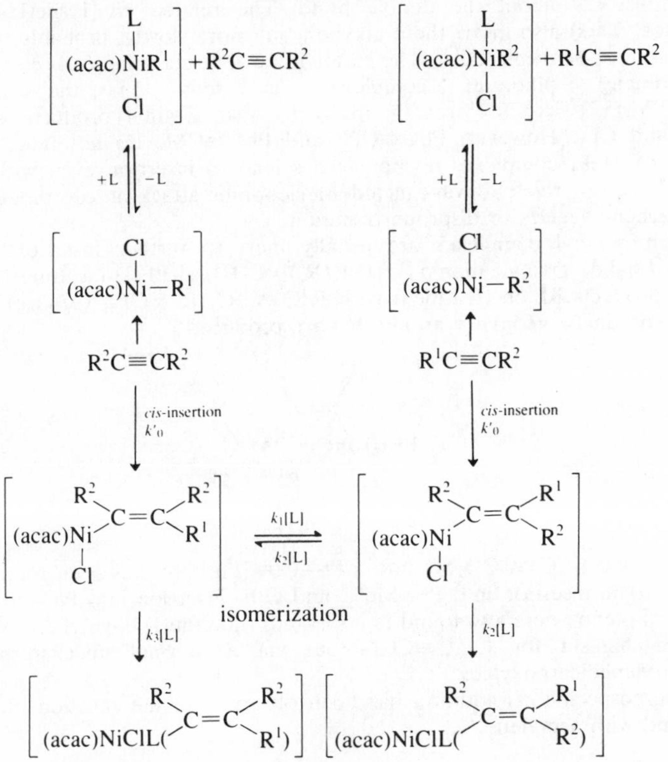

$$\left[(\text{acac})\text{Ni}(\text{L})\text{R}^1\text{Cl}\right] + \text{R}^2\text{C}{\equiv}\text{CR}^2 \qquad \left[(\text{acac})\text{Ni}(\text{L})\text{R}^2\text{Cl}\right] + \text{R}^1\text{C}{\equiv}\text{CR}^2$$

$$\Big\Updownarrow {+\text{L}}\ {-\text{L}} \qquad\qquad \Big\Updownarrow {+\text{L}}\ {-\text{L}}$$

$$\left[(\text{acac})\text{Ni}(\text{Cl}){-}\text{R}^1\right] \leftarrow \text{R}^2\text{C}{\equiv}\text{CR}^2 \qquad \left[(\text{acac})\text{Ni}(\text{Cl}){-}\text{R}^2\right] \leftarrow \text{R}^1\text{C}{\equiv}\text{CR}^2$$

$$\Big\downarrow \textit{cis}\text{-insertion}\ k'_0 \qquad\qquad \Big\downarrow \textit{cis}\text{-insertion}\ k'_0$$

$$\left[(\text{acac})\text{Ni}(\text{Cl}){-}\text{C}(\text{R}^2){=}\text{C}(\text{R}^2)\text{R}^1\right] \underset{k_2[\text{L}]}{\overset{k_1[\text{L}]}{\rightleftharpoons}} \left[(\text{acac})\text{Ni}(\text{Cl}){-}\text{C}(\text{R}^2){=}\text{C}(\text{R}^1)\text{R}^2\right]$$

isomerization

$$\Big\downarrow k_3[\text{L}] \qquad\qquad \Big\downarrow k_2[\text{L}]$$

$$\left[(\text{acac})\text{NiClL}(\text{C}(\text{R}^2){=}\text{C}(\text{R}^2)\text{R}^1)\right] \qquad \left[(\text{acac})\text{NiClL}(\text{C}(\text{R}^2){=}\text{C}(\text{R}^1)\text{R}^2)\right]$$

SCHEME 16

occurs to give an unsaturated species which can undergo isomerization to the *trans*-species at a rate competitive with phosphine readdition to the metal, forming product. A possible mode of isomerization would involve attack of free phosphine at carbon giving **80**. Such a unified mechanism is very attractive and may be very generally applicable.

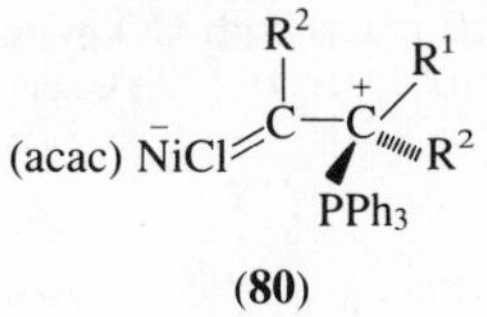

$$(\text{acac})\,\overset{-}{\text{Ni}}\text{Cl}{=}\text{C}(\text{R}^2){-}\overset{+}{\text{C}}(\text{R}^1)(\text{R}^2)(\text{PPh}_3)$$

(**80**)

The product of reaction of $MeO_2CC{\equiv}CCO_2Me$ with *trans*-$[PtMeX(PMe_2Ph)_2]$, which formerly was thought to result from insertion into the Pt—Me bond, has now been correctly identified as a $\beta$-chlorovinyl complex resulting from a free-radical process[167]. An insertion product *trans*-[Pt{*Z*—(C(R)=C(R)Me)}$(PMe_2Ph)_2X$] ($R = CO_2Me$; X = Cl, I) does result from treatment of $[PtMe(PMe_2Ph)_2(acetone)]^+$

with the acetylene followed by LiX. This is keeping with activation of olefins by coordination to cationic platinum species noted previously. Also, the cationic complexes *cis*-[PtMe(diphos) (acetone)]$^+$ and *cis*-[PtMe(diars) (acetone)]$^+$ react with the activated alkynes $CF_3C{\equiv}CCF_3$ and $MeO_2CC{\equiv}CCO_2Me$ giving products with *cis* stereochemistry[136] about the double bond. The related *cis*-[PtMe($L_2$)($NO_3$)] $L_2$ = diphos, diars) also insert these alkynes, but more slowly, probably because the nitrate ion must be displaced[136]. The reactivity of four-coordinate (as compared with five-coordinate) platinum complexes is seen in the fact that [($R^1_2Bpz_2$)PtMe($R^2C{\equiv}CR^2$)] readily insert to give *E*-vinyl products when $R^2$ = $MeO_2C$ and $CF_3$. However, PhC≡CPh and PhC≡CMe do not react[136,168]. The reactions of MePt complexes do not always lead to insertion even with activated acetylenes. Alternative pathways include nucleophilic attack on coordinated alkynes to give carbene species or disproportionation.

Platinum σ-vinyl complexes are usually inert to further insertion. However, removal of halide giving cationic [Pt{*E*-C($R^1$)=C($R^2$)H}($PEt_3$)$_2$(acetone)]$^+$ results in insertion products **81** on treatment with $R^3C{\equiv}CR^3$ ($R^3$ = $CF_3$, $CO_2Me$). Both *cis*- and *trans*-products (geometry around Pt) are produced[169].

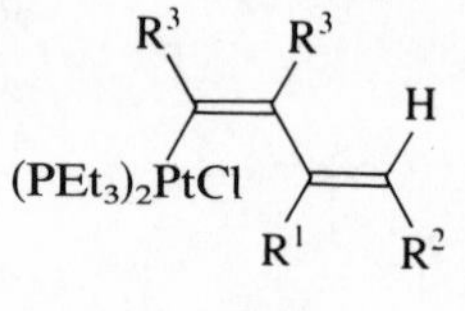

**(81)**

Insertion of $MeO_2CC{\equiv}CCO_2Me$ into a Pt—C≡CPh bond has been reported[170], as well as benzyne insertion into Pt—Me[171] and $C_4F_6$ insertion into Pt—allyl[172].

Activated olefins were also found to insert into rhodium(I) σ-vinyl complexes via a dipolar mechanism for $CF_3C{\equiv}CCF_3$ but via a 'normal' mechanism for dimethylacetylenedicarboxylate[173].

99% stereospecific *cis*-addition has been observed in the reaction of the Cu—$C_7H_{15}$ bond with acetylene[174].

### 3. *Eighteen-electron complexes*

Some recently discovered novel reactions of acetylenes with 18-electron complexes will be treated first, followed by a survey of results organized by Periodic Group as in other parts of Section IV.

a. *Insertion into metal—acyl bonds.* Irradiation of [CpW(CO)$_3$Me] with HC≡CH was found[175] to afford **82** (M = W; $R^1$ = Me, $R^2$ = H) and the 16-electron complex [CpW(CO)Me(HC≡CH)] which reacts with CO giving the vinyl ketone complex **82** resulting from carbonylation and insertion[176]. A second resonance form **83** can be

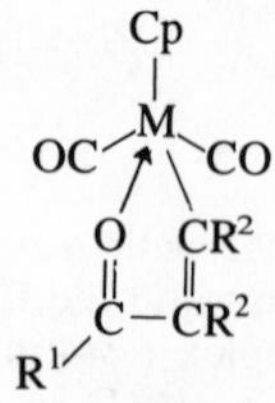

**(82)** 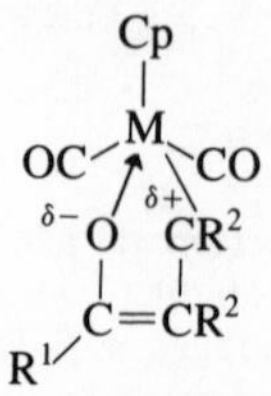 **(83)**

written for **82**, suggesting the nucleophilic property of the ring carbon. Basic $PMe_3$ reacts[177] with **83** giving a P-ylid for M = Mo and W and $R^2$ = H. The reaction does not occur if $R^2$ is electron-donating.

[$CpM(CO)_3R^1$] (M = Mo, W, $R^1$ = $COCF_3$, Me; M = Mo, $R^1$ = $CH_2Ph$) reacted with MeC≡CMe giving complexes **82** ($R^2$ = Me), the molybdenum compounds thermally and tungsten compounds photochemically[178]. The products **82** react with L = Bu$^t$NC, $C_6H_{11}NC$, $PPh_3$, CO to give stereoisomers of the lactone **84**, which is a product of CO insertion into the M—alkyl bond and nucleophilic attack by the δ-C oxygen on the α-C. The unsaturation permits formation of an $\eta^3$-allylic ligand in contrast to **64** and equation 32. When M = Mo and $R^1$ = $COCF_3$, **85** arises from multiple alkyne insertion followed by carbonylation and rearrangement. The structure was established by X-ray crystallography.

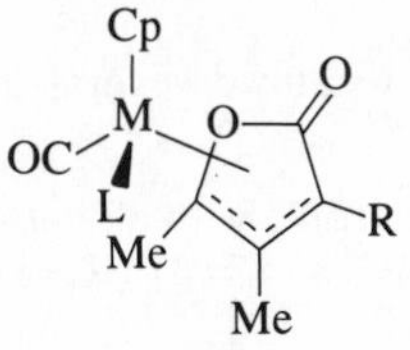

**(84)** 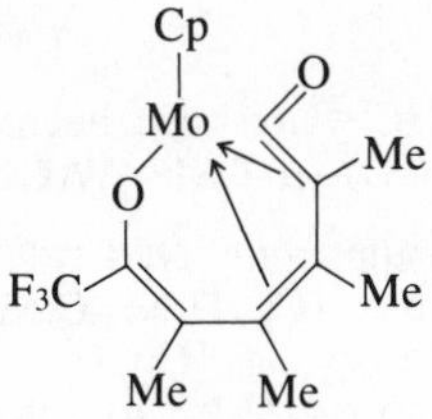 **(85)**

With ω-alkynes, **86** can be prepared thermally (Mo) or photochemically (W) from [$CpM(CO)_3\{(CH_2)_nC{\equiv}CCH_3\}$] via the carbonylated product[179]. Indenyl complexes are even more reactive than Cp[180,181] and indenyl analogues of **82** and **84** are readily prepared at room temperature for M = Mo.

b. *Insertion of electron-rich alkynes into metal—carbene bonds.* Dötz and coworkers have prepared a number of interesting compounds by insertion of $R^3C{\equiv}CNEt_2$ into the metal—C bond of Fischer-type carbenes [$(CO)_5M{=}C(R^1)(R^2)$] of structure **87** (M = Cr, $R^3$ = H, Me, $R^1$ = Me, Ph, $R^2$ = OMe[182]; M = Cr, Mo, W, $R^3$ = H, Me, $R^1$ = Me, Ph, $R^2$ = OMe[183,184]; M = W, $R^3$ = H, Me, $R^1$ = $R^2$ = Ph[183,184]; $R^3$ = $NEt_2$, $R^1$ = Ph, $R^2$ = OMe[185]).

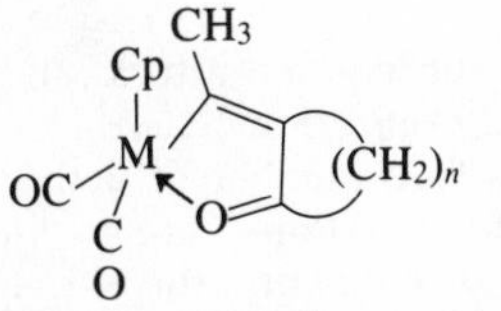

$n$ = 3,4,5

**(86)** **(87)** 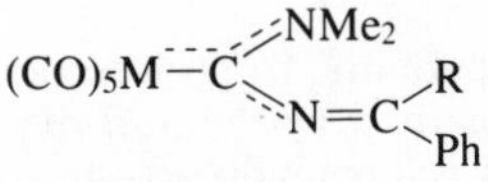 **(88)**

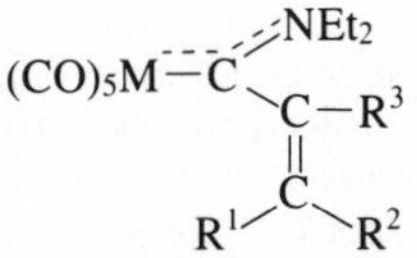

The reactions are stereoselective with the metal—carbene and OR substituents predominantly *trans* to each other. [$(\eta^5\text{-}MeC_5H_4)(CO)_2Mn{=}C(R^1R^1)$] was found to insert MeC≡CNEt giving products of 10:1 *trans*:*cis* insertion[186]. The products are in keeping with the usual chemistry of Fischer-type carbenes which undergo attack at the carbene C by nucleophiles.

Isoelectronic with $R^1C{\equiv}CNR^2_2$ is the cyanamide $N{\equiv}CNMe_2$. This has also been found to react with Fischer-type carbenes giving products of structure **88** (M = Cr, R = OMe; M = W, R = OMe, Ph, $p\text{-}XC_6H_4$, where X = OMe, Me, H, Br, $CF_3$)[187,188,189]. Kinetic studies[187] showed that the reaction was first order both in carbene and cyanamide. Electron-withdrawing substituents on the carbene increase the reaction rate consistent with nucleophilic attack on the carbene.

c. *Insertion into metal—μ-alkylidene bonds*. Reaction of RC≡CR with $[Cp_2M_2(CO)_2(\mu\text{-}CO)(\mu\text{-}CH_2)]$ gives insertion products **89** (M = Fe, Ru, R = H; M = Fe, R = $CO_2Me$)[190].

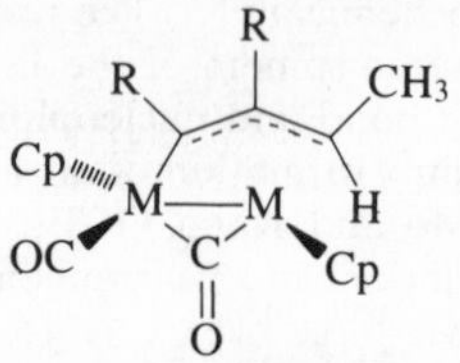

**(89)**

MeC≡CMe and HC≡CH have been found to afford several interesting insertion products with $[W(CO)_5\text{-}\mu\text{-}CMe_2\text{-}W(CO)_5]$[191].

d. *Survey by periodic group*. MeC≡N, isoelectronic with alkynes, was found to give the insertion products $[Cp_2Ti\{N{=}C(Me)CHRCH{=}CH_2\}]$ (R = H, Me) on reaction with the 17-electron species $[Cp_2Ti(\eta^3\text{-}RCHCHCH_2)]$[192].

$[MeMn(CO)_5]$ inserts $C_4F_6$ to afford an *E*-alkenyl product[193]. $[R^1Mn(CO)_5]$ $R^1$ = Me, MeCO, Ph) afford the products **90** with $R^2C{\equiv}CR^3$ = HC≡CPh, $MeO_2CC{\equiv}CCO_2Me$, $HO_2CC{\equiv}CCO_2H$, $HC{\equiv}CO_2Me$, and HC≡CHO. As with Ni—Me complexes, the carbon with largest substituent is coordinated to manganese. Excess of CO inhibits product formation. With HC≡CPh, a 2:1 product **91** results,

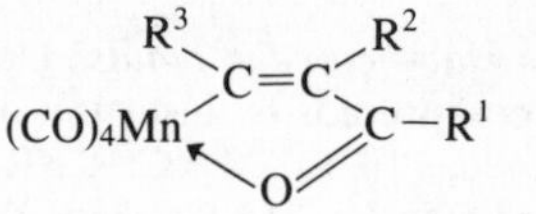

**(90)** 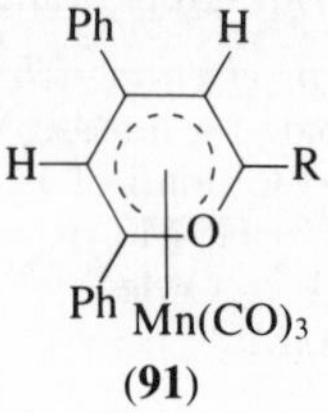 **(91)**

apparently from insertion of a second acetylene and subsequent attack at the α-carbon by the acyl oxygen followed by loss of CO and cyclization[194].

Cyclopentadienyl—iron complexes have been treated with a number of activated acetylenes. Insertion of $MeO_2CC{\equiv}CCO_2Me$ into *erythro*- or *threo*-$[CpFe(CO)_2CHDCHDCMe_3]$ occurs with 80% retention of configuration at α-C[12]. Photochemical reactions of $[CpFe(CO)_2R]$ (R = Me, $CH_2Ph$) with $CF_3C{\equiv}CH$ gave **92** (M = Fe, L = CO), **93**, and **94**. The last two products arise from carbonylation

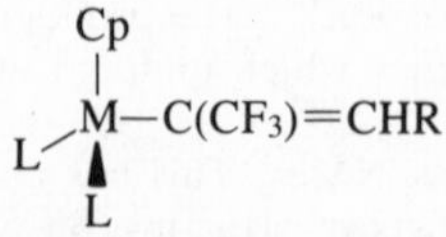

**(92)** 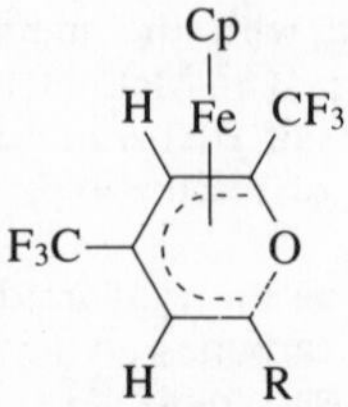 **(93)** 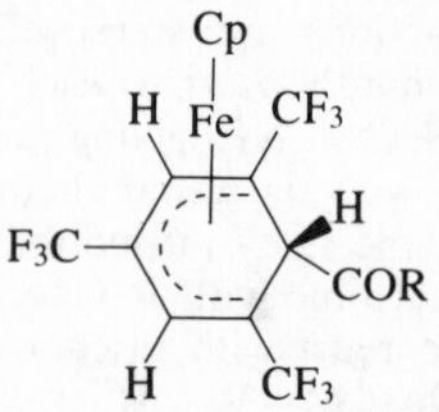 **(94)**

followed by a second and third insertion, respectively, and cyclization. When R = allyl only **95** (R = H) was obtained. *Cis*-insertions occur in **92** and **95** again with the larger group closer to the metal[195].

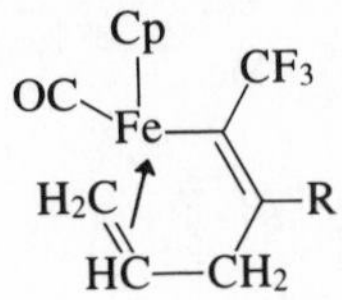

(**95**) 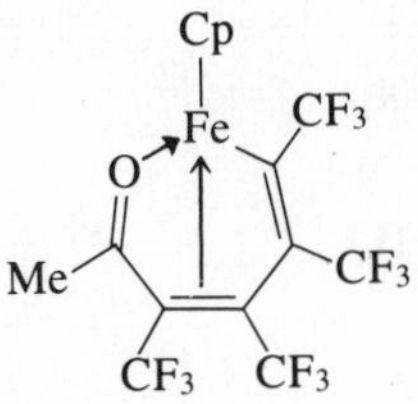 (**96**) 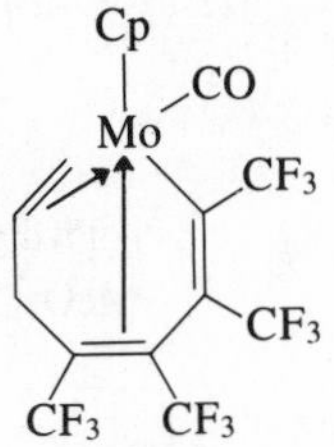 (**97**)

Reaction of $C_4F_6$ (again photochemical) with $[CpFe(CO)_2R]$ (R = allyl) gives **95** ($R = CF_3$)[193]. For R = Me, **96** results from bis insertion and coordination of double bond and acyl oxygen in place of CO. Again, all additions are *cis*[193]. In contrast to the iron—allyl complex, $[CpMo(CO)_3(\eta^1\text{-allyl})]$ affords **97**, the product of two *cis*-insertions, with $C_4F_6$[193]. $[CpFe(CO)_2(\eta^1\text{-}C_5H_5)]$ forms a product of addition to the $\eta^1$-cyclopentadienyl ring with $C_4F_6$[193].

The chemistry of cyclopentadienyl—ruthenium complexes has received much attention. $[CpRu(PPh_3)_2Me]$ reacts thermally with $C_4F_6$ and $MeO_2CC{\equiv}CCO_2Me$ giving **98** ($R = CF_3$, $MeO_2C$, respectively). With $C_4F_6$ in diglyme **99** is isolated, which contains a CO ligand derived from solvent and is the apparent product of *trans*-addition[196].

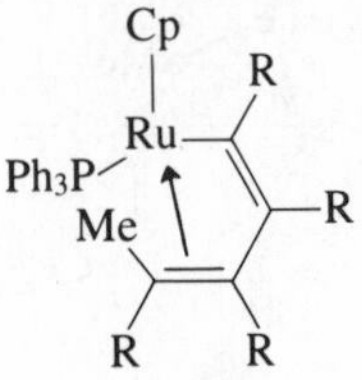

(**98**) 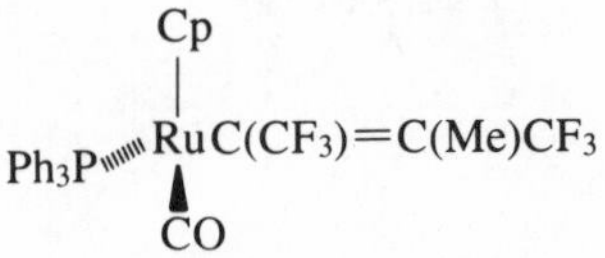 (**99**)

In refluxing decalin $[CpRu(PPh_3)_2Me]$ gives the orthometallated $[CpRu\overline{(PPh_3)(C_6H_4}PPh_2)]$, which inserts 2 mol of $C_4F_6$ into the Ru—C bond giving a compound analogous to **98**[197].

The alkenyl complex **100** ($L = CO$, $R = CF_3$) photochemically inserts $C_4F_6$ producing **101**[193]. In contrast, **100** ($L = PPh_3$, $R = CO_2Me$) inserts $C_4F_6$ into the C—H bond of the ligand via a dipolar intermediate resulting from attack on the vinyl ligand[198].

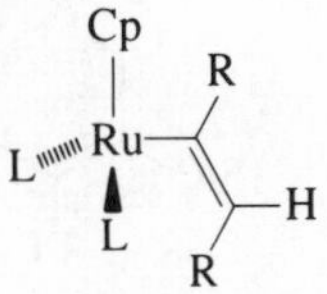

(**100**) 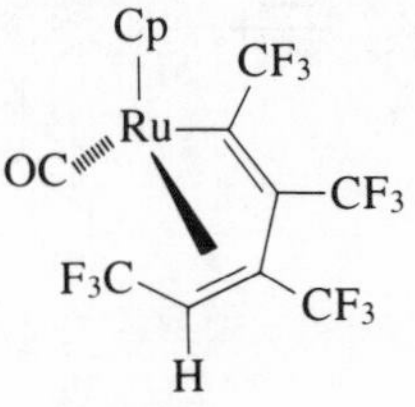 (**101**)

Unsymmetrical acetylenes led to a head-to-tail insertion with ruthenium—vinyl complexes made by acetylene insertion into $[CpRu(PPh_3)_2H]$. Reaction of **102** (R = H) with MeC(O)C≡CH yields **103**[199].

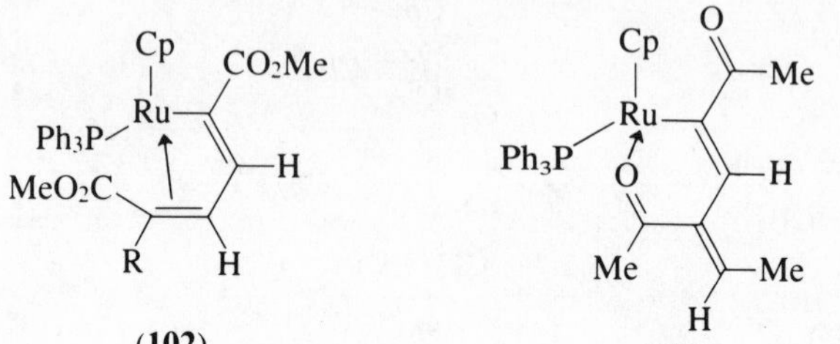

(102)

(103)

$[CpRu(PPh_3)_2Me]$ inserts 2 mol of $MeO_2CC{\equiv}CH$ to produce **102** (R = Me)[199]. Allenic products were also isolated from the reactions of $[CpRu(PPh_3)_2R]$ (R = H, Me) with unsymmetrical alkynes[199].

Iron—vinyl complexes containing coordinated SMe groups also underwent photochemical insertion. For example, **104** gave **105**[200] with $CF_3C{\equiv}CH$ as well as the product of insertion into the S—C bond.

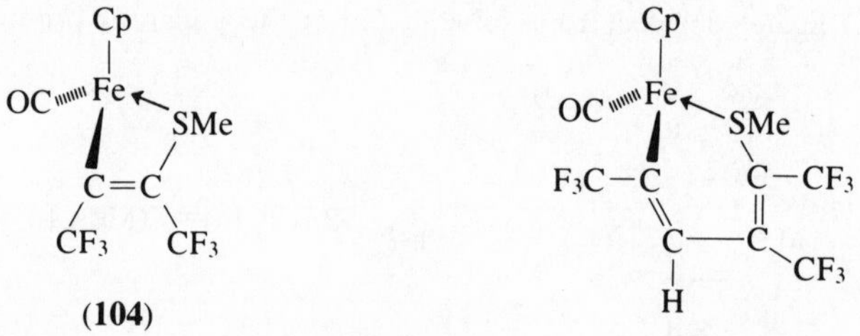

(104)

(105)

The product of $C_4F_6$ reaction with $[(\eta^3\text{-allyl})Co(CO)_3]$ complexes[148] has now been formulated as **106**, resulting from attack on the coordinated double bond as shown in Scheme 17[195].

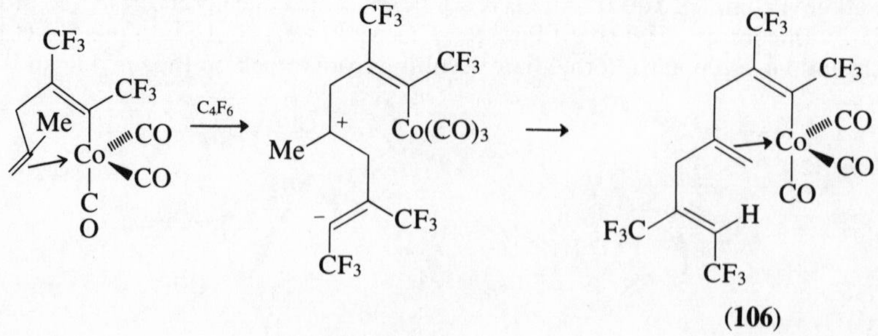

(106)

SCHEME 17

[{$\eta^3$-$CH_2C(R)CH_2$}$Ir(PPh_3)_2(CO)$] complexes (R = H, Me) afforded isomers **107** and **108** (L = CO, $PPh_3$; R = H, Me) with $C_4F_6$ at −30 °C in toluene. The *trans*-insertion product was also formed[201]. The reaction is solvent dependent. In 9:1

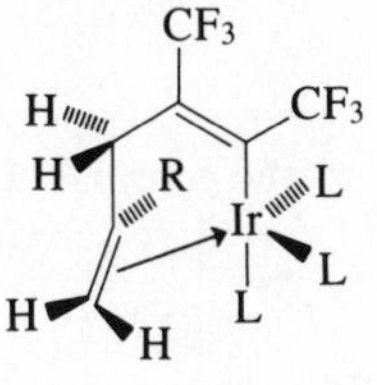

(107) 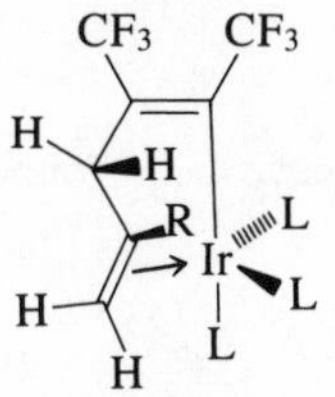 (108) 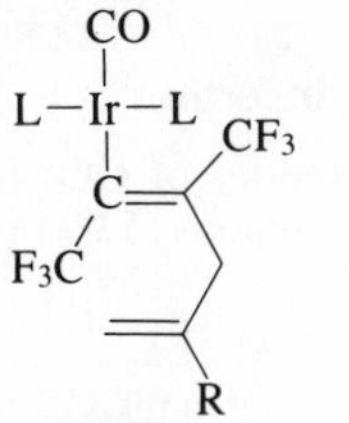 (109)

toluene–methanol only an adduct [{$\eta^3$-$CH_2C(R)CH_2$}$Ir(PPh_3)(CO)(C_4F_6)$] can be isolated. This is apparently a precursor to the final products since it reacts with $PPh_3$ in refluxing benzene giving **107**, **108**, and **109**, probably via a dipolar intermediate from attack on the allyl. [($\eta^3$-$CH_2CMeCH_2$)Ir(CO)(dppe)] gave the analogue of **107** and **110**[201]. The diazobenzene complex **111** (and substituted analogues) gives **112** with $C_4F_6$. This can be thermally decarbonylated to *ortho*-metallated **113**[202].

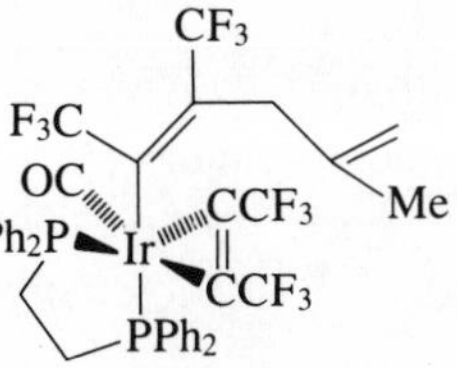

(110)

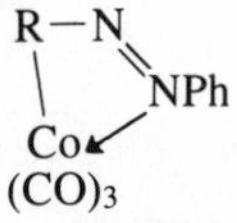

(111) 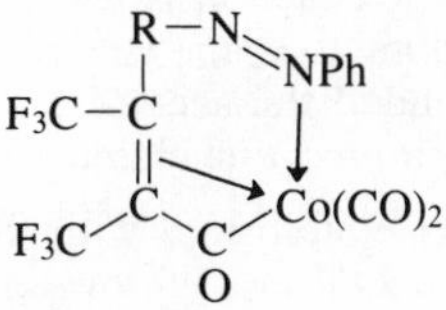 (112) 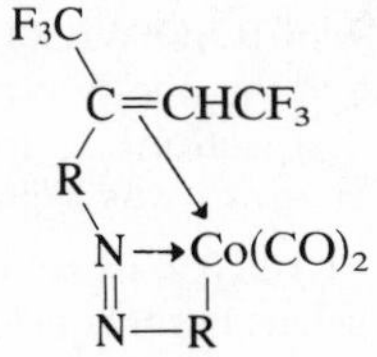 (113)

R = *o*—$C_6H_4$

## V. INSERTION OF CARBON DIOXIDE, CARBON DISULPHIDE, AND ISOELECTRONIC SPECIES

### A. General

$CO_2$, $CS_2$, COS, RNCO, and RNCS are known to insert into metal—carbon bonds. However, they often display a preference for insertion into metal—nitrogen or metal—halogen bonds. A recent example is the reaction of $CO_2$ with the complexes [$R_2M_2(NMe)_4$] (R = Me, Et, $Bu^n$, $Bu^i$, $CH_2CMe_3$, $CH_2SiMe_3$, $Bu^t$; M = Mo, W) containing metal—metal triple bonds[203]. Only insertion into M—N bonds was observed. Interest in $CO_2$ insertions stems in part from the possibility of employing $CO_2$ as a feedstock in catalytic processes including reduction to alcohols and aldehydes.

Excellent recent reviews of $CO_2$[204] and $CS_2$[205] complexes are available which treat insertion reactions. Our coverage will be confined to work not covered in these.

## B. $CO_2$ Insertion[68]

Two modes of $CO_2$ insertion have been found leading to the 'normal' **114**, **115** and 'anomalous' **116** products.

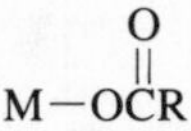

M—OC(=O)R **(114)** M(κ²-O₂C—R) **(115)** M—C(=O)OR **(116)**

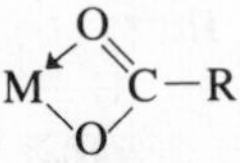

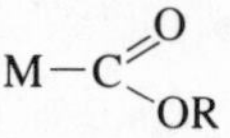

$[Cp_2TiR]$ (R = Me, Bu$^t$) insert $CO_2$ giving products of structure **115**[206]. 'Normal' complexes are also found when R = $\eta^3$-$CHMeCHCH_2$[192,207] and CMe=CHMe[208]. Treatment of $[Cp_2Ti(p\text{-}MeC_6H_4)_2]$ and $[Cp_2Ti(m\text{-}MeC_6H_4)_2]$ with $CO_2$ followed by $BF_3$ in MeOH gave only *m*- and *p*-methyltoluates, showing that the intermediates formed in decomposition had Ti—C bonds only *ortho* to the ring carbon bonded to $CO_2$[209].

Vanadium complexes give reaction 33[210]. With excess of $CO_2$, $[V(OCOC_6F_5)_3]$ can also be obtained[210]. The paramagnetic complex $[Cr(o\text{-}CH_2C_6H_4NMe_2)_3]$ inserts only one molecule of $CO_2$, even at 75 °C, giving a chelated structure (**115**)[211].

$$2[R_3V]\cdot n\text{thf} + CO_2 \xrightarrow[\text{thf}]{} [V(OCOR)_2] + [R_4V]\cdot n\text{thf} \tag{33}$$

$$R = C_6F_5,\ n = 2;\ R = CH_2SiMe_3,\ n = 0$$

Darensbourg and Rokicki[212] have discovered that $CO_2$ reacts with $[MeW(CO)_5]^-$ giving $[\{MeC(O)O\}W(CO)_5]^-$. The reaction is accelerated by $Li^+$ probably by complexation at the non-bonded acetate oxygen. The insertion also occurs with COS and $[HCr(CO)_5]^-$, affording an S-bonded thioacetate.

$[(PhCH_2)_2Mn]$ was reported to give a chelated complex on reaction with $CO_2$[93].

$[Fe(PMe_3)_4]$ exists in solution in equilibrium with the metallated $[HFe(CH_2PMe_2)(PMe_3)_3]$. This latter species gives **117** and, (at low temperatures) with excess of $CO_2$, **118**[213].

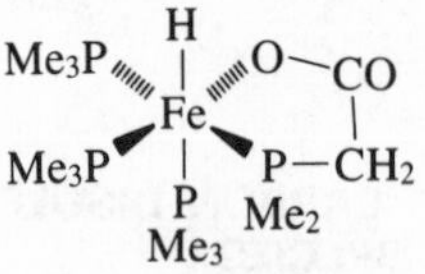

**(117)** 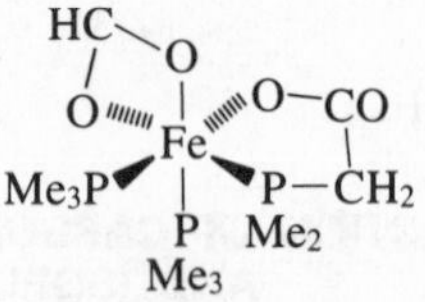 **(118)**

A similar reaction occurs on heating the $CO_2$ adduct of $[Ir(Me_2PCH_2CH_2PMe)_2]Cl$ producing $[(Me_2PCH_2CH_2Me_2)(H)(Ir\{OC(O)(CH_2P(Me)CH_2CH_2PMe_2\}]Cl$, which may have a polymeric structure[214]. $[Et_2Ni(bipy)]$ inserts $CO_2$ at 40–50 °C in benzene, the compound then decomposing to diethyl ketone[215]. $[(\eta^3\text{-Allyl})Pd(PR_3)(\eta^1\text{-allyl})]$ was also found to yield the product of insertion into the Pd—$\eta^1$-allyl bond[107].

Insertion products also have been reported with $CO_2$ and $[PhC{\equiv}CCu(PBu_3)_3]$[216]

as well as $[RCuL_2]$ ($R = Me$, Et, $Pr^n$, $Bu^i$; $L = PPh_3$, $PMePh_2$, $PBu^n{}_3$, $PEt_3$, $P(C_6H_{11})_3$)[217]. The triphenylphosphine complexes reversibly coordinate additional $CO_2$ giving $[(RCOO)Cu(CO_2)_2(PPh_3)_2]$. PhAg coordinates but does not insert carbon dioxide[218].

Many insertions of $CO_2$ are thermally reversible. However, carboxylate complexes prepared by other routes often undergo thermal decarboxylation, especially those having electron-withdrawing R groups in basic solvents. $[(RCO_2)_2Ni(phen)]\cdot H_2O$ ($R = p$-$MeOC_6H_4$, $p$-$EtOC_6H_4$) and $[(RCO)_2Ni(bipy)]$ ($R = C_6F_5$, $p$-$MeOC_6H_4$, $p$-$EtOC_6H_4$) give $[R_2NiL]$ (L = phen, bipy) on heating[219]. In pyridine, $[(RCOO)Rh(PPh_3)_2(CO)]$ lose $CO_2$ with rates decreasing in the order $R = C_6F_5 > p$-$MeOC_6F_4 > p$-$HC_6F_4 > m$-$HC_6F_4 > 4,5$-$H_2C_6F_3 > 3,5$-$H_2C_6F_3$[220]. $[Cu(O_2CCH_2CN)_2]$ undergoes decarboxylation and reduction producing $Cu(CH_2CN)$ at 50 °C in dmf[221]. $[\{CF_3CH(CF_3)CO_2\}Ag]$[222] and $[C_6F_5CO_2AuPPh_3]$[223] also give the corresponding alkyls on heating. The carboalkoxy species $[Pt(CO_2R)ClL_2]$ ($L = PPh_3$, $PPh_2Me$; $R = CH_2CHCH_2$, $CH_2C(Me){=}CH_2$, $CH_2CH{=}CHMe$, $CH(Me)CH{=}CH_2$) decarboxylate in refluxing benzene or on removal of $Cl^-$ by $Ag^+$ affording $[(\eta^3$-allyl$)PtL_2]Cl$[224].

## C. $CS_2$ Insertion[68]

$[Cp_2Ti\{C(Me){=}CHMe\}]$ inserts $CS_2$ as well as $CO_2$[208]. Although $CS_2$ inserts into Ta—N bonds in $[Me_3Ta(NMe_2)_2]$, it produces $[(MeCS_2)_3TaCl_2]$ when allowed to react with $[Me_3TaCl_2]$. Presumably the insertion stops when CN (coordination number) 8 is reached for tantalum[225].

$o$-$MeC_6H_4Cu$ and $CS_2$ give $[o$-$MeC_6H_4CS_2Cu]_{\sim 5}$; however, other aryl copper compounds afford only unstable products. $[(ArCu)_nPPh_3]$ inserts $CS_2$ providing air stable complexes $[ArCSSCu(PPh_3)_2]$. Tetrahedral geometry around copper was established for $(o$-$MeC_6H_4C(S)S)Cu(PPh_3)_2$ by X-ray crystallography[226]. Similar dithiocarboxylates containing alkyl groups exist: $[(RCSS)Cu(PPh_3)_2]$ ($R = Me$, Et, $Pr^n$) and $[(RCSS)_2Cu_2(Ph_2PCH_2CH_2PPh_2)_3]$. In contrast to the analogous $CO_2$ complexes, these species did not coordinate additional molecules of $CS_2$ on further treatment[217].

## D. Insertion of Isoelectronic Species[68]

$[Cp_2TiR]$ ($R = Me$, $Bu^n$) affords **119** with PhNCO[206]. This is the usual mode of insertion: the organic group is attached to isocyanate carbon while oxygen forms a dative bond to the metal. When $R = \eta^3$-$MeCHCHCH_2$, **119** is the product with PhNCO while $PhN{=}CHPh$ and $Me_2C{=}O$ give **120** and **121**, respectively[192]. PhNCO

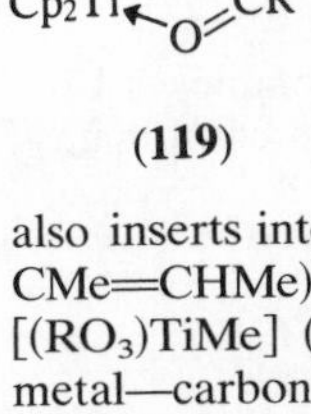

**(119)** 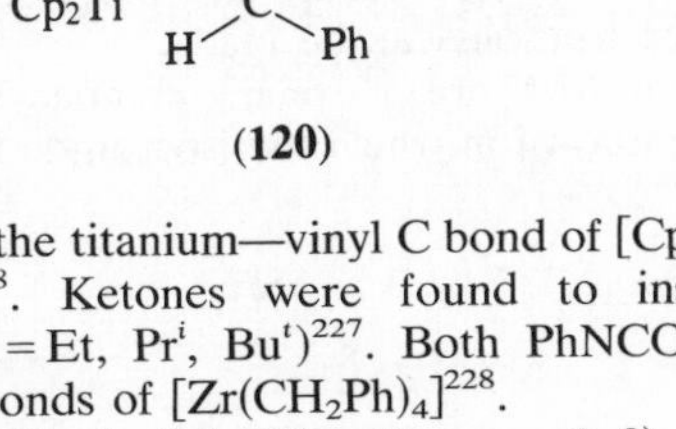 **(120)** 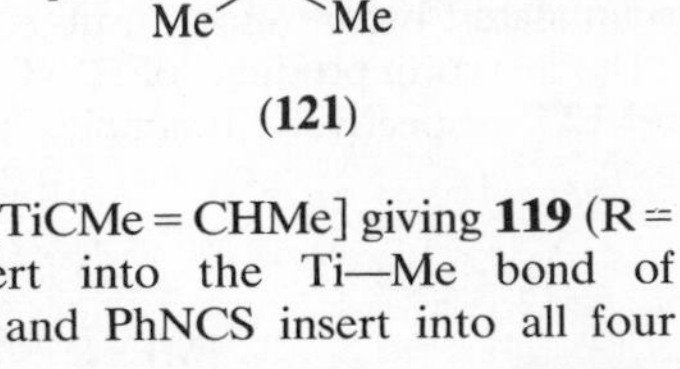 **(121)**

also inserts into the titanium—vinyl C bond of $[Cp_2TiCMe = CHMe]$ giving **119** ($R = CMe{=}CHMe$)[208]. Ketones were found to insert into the Ti—Me bond of $[(RO_3)TiMe]$ ($R = Et$, $Pr^i$, $Bu^t$)[227]. Both PhNCO and PhNCS insert into all four metal—carbon bonds of $[Zr(CH_2Ph)_4]$[228].

With $[Me_xMCl_{5-x}]$ (M = Nb, Ta; $x = 1$–$3$), MeNCO, PhNCO[229], MeNCS, PhNCS[230], and $RN{=}C{=}NR$ ($R = Pr^i$, $Bu^t$, $C_6H_{11}$, $p$-$MeC_6H_4$)[231] insert into Me—M bonds. All bonds do not necessarily undergo insertion and the stoichiometry of stable products seems to be governed by the achievement of six- or seven-coordination around the metal. The reactivity order is $[MeMCl_4] > [Me_2MCl_3] > [M_3MCl_2]$ and MeNCO > PhNCO suggesting initial formation of donor–acceptor complexes. No insertions into M—Cl bonds were observed in these reactions. The modes of coordination for the products of isothiocyanate and carbodiimide reactions are shown in **122** and **123**.

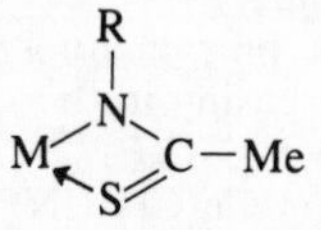

(**122**)

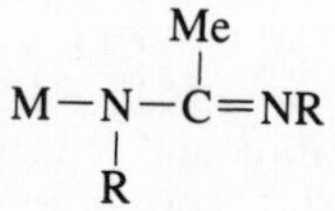

(**123**)

When PhNCS reacts with $[CpNiL\{CH(CN)_2\}]$ ($L = PBu_3$, $PPh_3$) the products have structure **124** instead of **122**[232].

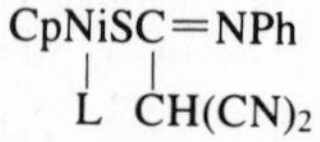

(**124**)

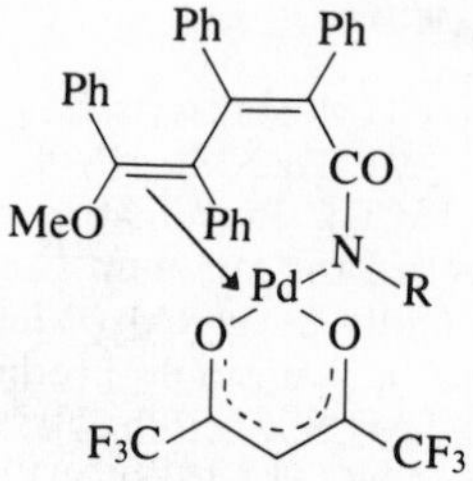

(**125**)

The palladium complex (**55**) ($X_2$ = hfacac) inserts aryl isocyanates affording **125** with a double bond coordinated to palladium. Electron-withdrawing substituents on the aryl group accelerate the reactions[233], in apparent contrast to the niobium and tantalum complexes discussed previously.

## VI. INSERTION OF ISOCYANIDES

### A. General

Isocyanides (RNC) are isoelectronic with CO and, hence, might be expected to exhibit similar features with respect to insertion reactions. Many similarities are known. However, some surprising differences occur, and more data need to be accumulated before many firm generalizations can be made.

The insertion products of $R^2NC$ into $MR^1$ are $\eta^1$- or $\eta^2$- iminoacyl complexes **126** and **127**, respectively. Examples are known in which the isocyanide first coordinates

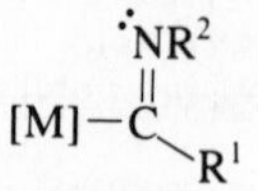

(**126**)

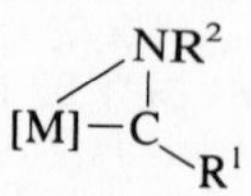

(**127**)

to the mètal followed by alkyl migration or motion of the isocyanide. In other cases, direct insertion apparently occurs. In still other instances, isocyanides behave as nucleophiles on reaction with metal carbonyl alkyls yielding metal acyl isocyanide complexes **128**, especially when $R^2$ is bulky. Sometimes iminoacyls **129** are co-products. No cases are known in which alkyl migration occurs from the acyl group of **128** to isocyanide to give **129**.

$$\underset{(\mathbf{128})}{R^1C(O)[M]CNR^2} \qquad \underset{(\mathbf{129})}{OC[M]C(R^1){=}NR^2}$$

Again in contrast to CO, multiple insertions are fairly common with RNC. As with CO, insertions generally do not occur when the metal is bonded to alkyls with electron-withdrawing groups such as perfluoroalkyls. However, $CNC_6H_{11}$ has been found[234] to insert regiospecifically into the Ti—$C_6F_5$ bond of $[Cp_2TiMe(C_6F_5)]$ while CO inserts regiospecifically into the Ti—Me bond.

Two excellent reviews of isocyanide chemistry which treat insertion reactions are available[235]. Except insofar as results mentioned in these sources are germane to recent developments, only subsequent work is included here.

## B. Square-Planar Complexes

Some reactions of isocyanides with palladium complexes can be rationalized by a pathway analogous to that depicted for carbonyl insertion in Scheme 4. Crociani and coworkers[236] reported that reaction 34 occurs. This suggests the operation of a

$$cis\text{-}[Pd(PPh_3)(CNPh)Cl_2] + YPh \longrightarrow [Pd(PPh_3)(CNPh)(Ph)Cl] \quad (34)$$

$$\downarrow$$

$$\left[ (Ph_3P)(Ph(C{=}NPh))Pd(\mu\text{-}Cl)_2Pd((C{=}NPh)Ph)(PPh_3) \right]$$

Y = PhHg, $Ph_3Pb$, etc.

mechanism involving a three-coordinate intermediate which can dimerize. The bridge bonds in the dimer can be cleaved by $PPh_3$ giving *trans*-$\{Pd(PPh_3)_2[C({=}NPh)Ph]Cl\}$. Similar results were found with *trans*-$[RPd(Bu^tNC)_2I]$ (R = Me, MeCO) which underwent insertion at 11 °C and presumably formed a dimer which reacted with L = $Bu^tNC$, $PPh_3$ to give a product having L *cis* to an iminoacyl[237].

Otsuka and Ataka[238] studied the slower rates of dimerization of the benzyl complexes *trans*-$[Pd(CH_2Ph)(Bu^tNC)_2X]$ (X = Cl, Br, I) to produce $\{Pd[C({=}NBu^t)CH_2Ph](Bu^tNC)X\}_2$ and found that the rates decreased in the order Cl > Br > I, which is opposite to the *trans*-effect order. Moreover, the rates of reaction of *trans*-$[Pd(Me)(Bu^tNC)_2I]$ with L = $CNBu^t$, $PPh_3$, or $P(OPh_3)_3$ to give **130** were first order in palladium complex and independent of the nature of L. The rates were very slow in hexane and fast in polar organic solvents. This suggests the presence of a three-coordinate iminoacyl intermediate **131** which (in contrast to **11**) is solvent-stabilized and which can dimerize or react with L. This intermediate can be detected at low temperatures in toluene by n.m.r. in the absence of L and gives the dimer on raising the temperature.

$$\underset{\mathbf{(130)}}{Bu^tNCPd(I)(L)C(Me){=}NBu^t} \qquad \underset{\mathbf{(131)}}{Bu^tPd(I)C(Me){=}NBu^t}$$

For the L-induced insertion $\Delta H^* = +9.4$ kcal mol$^{-1}$ and $\Delta S^* = 22$ cal mol$^{-1}$K$^{-1}$, values similar to those for solvent-assisted carbonylations such as that of $[MeMn(CO)_5]$.

The authors suggest[238] that these results can be rationalized on the assumption that the role of the *trans*-halide ligand is to polarize the Pd—C bond rather than to exert a *trans*-effect. The more carbanionic the migrating alkyl and the more electrophilic the isocyanide terminus, the faster is the production of the iminoacyl three-coordinate species. The requirement seems to be only for alkyl and isocyanide in *cis*-position.

A similar mechanism surely prevails for reaction 35 which further shows that the inserted isocyanide is the one pre-coordinated to the metal[239].

$[(R)(NMe_2)Pd(\mu\text{-}Cl)]_2 + 2R^1NC \longrightarrow 2\,[(R)(NMe_2)Pd(CNR^1)Cl]$ (35)

$\xrightarrow{\Delta} [(R{-}C{=}NR^1)(NMe_2)Pd(\mu\text{-}Cl)]_2 \xrightarrow{2R^2NC} 2\,[(R{-}C{=}NR^1)(NMe_2)Pd(CNR^2)Cl]$; also $[(R)(NMe_2)Pd(CNR^1)Cl] \xrightarrow{R^2NC} [(R{-}C{=}NR^1)(NMe_2)Pd(CNR^2)Cl]$

$R = o{-}C_6H_4$

*trans*-$[PdMeL_2I]$ [L = $PPh_3$, $PPhMe_2$, $PBu_3$, $PPh_2(C_6H_{11})$, $PPh_2Me$, $PPhMe_2$] were found to react with $C_6H_{11}NC$ to produce iminoacyl insertion products having *trans* phosphine ligands. *trans*-$[PdMe(PPhMe_2)_2I]$ behaved similarly with $Bu^tNC$ and $PhCH_2NC$, as did *trans*-$[PdMe(PMe_3)_3I]$ with $Bu^tNC$ and *trans*-$[Pd(o\text{-}MeC_6H_4)(PPhMe_2)_2I]$ with $C_6H_{11}NC$[240]. These reactions are analogous to the carbonylation depicted in Scheme 4 in the sense that CNR is the entering group and the products have *trans* stereochemistry. The reactions could proceed via the mechanism outlined in Scheme 4 involving five-coordinate species containing CNR rather than CO. Alternatively, an ionic species *trans*-$[PdR^1(CNR^2)L_2]^+I^-$ may be formed initially which undergoes insertion on reattack by $I^-$. Such a species has been shown to be involved in isocyanide insertions in platinum complexes (see below), but has not as yet been implicated as leading to carbonylated products. The importance of steric effects in isocyanide insertion is emphasized by the facts that reactions of 2,6-$Me_2C_6H_3NC$ with *trans*-$[PdMe(PPhMe_2)_2I]$ and of $Bu^tNC$ with *trans*-$[Pd(o\text{-}MeC_6H_4)(PPhMe_2)_2I]$ do not occur[240].

Bis-insertion products **132** were obtained on reaction of $C_6H_{11}NC$ with *trans*-$[PdMeL_2I]$ (L = $PPh_3$, $PPh_2Me$, $PPhMe_2$, $PMe_3$, $PBu^n_3$ in a 2:1 molar ratio as well as by treatment of the mono-insertion products (L = $PPhMe_2$, $PMe_3$) with excess of

$C_6H_{11}NC$, showing that, in some cases at least, insertion proceeds stepwise. This is not invariably the case, however. Even though *trans*-[PdMe(PPh$_3$)$_2$I] reacts at 5 °C with a 2:1 molar ratio of $C_6H_{11}NC$ producing **132** (L = PPh$_3$), the mono-insertion product *trans*-[Pd{C(=NC$_6$H$_{11}$)Me}(PPh$_3$)$_2$I] does *not* react with a second mole of cyclohexyl isocyanide at room temperature. It does react at 70 °C, however, to give **132** (L = PPh$_3$). These results suggest the operation of an additional pathway besides stepwise insertion.

Reaction of *trans*-[PdMe(PPh$_2$Me)$_2$I] with a 3:1 molar ratio of $C_6H_{11}NC$ affords the product of tris-insertion **133** (L = PPh$_2$Me)[240].

L(I)Pd(L)–C(=NC$_6$H$_{11}$)C(=NC$_6$H$_{11}$)Me

**(132)**

L(I)Pd–C(=NC$_6$H$_{11}$)–C(=NC$_6$H$_{11}$)–C(Me)=N(C$_6$H$_{11}$)→Pd

**(133)**

Isocyanides were also found to insert into the Pt—C bonds of *trans*-[PtR(PPh$_3$)$_2$X] (R = Me, Ph; X = Cl, Br, I)[241,242], *trans*-[PtMe(PEt$_3$)$_2$I] and *trans*-[Pt(CH$_2$Ph)-(PPhMe$_2$)$_2$I][242–244]. These insertions are slower than those of palladium complexes and involve ionic intermediates $[R^1Pt(CNR^2)L_2]^+X^-$ which may be isolated and converted thermally to iminoacyl insertion products, which presumably result from reattack by $X^-$. The structure of one of these insertion products, *trans*-[Pt{C(=NC$_6$H$_4$Cl)Me}(PEt$_3$)$_2$I], was investigated by X-ray crystallography and found to contain an $\eta^1$-iminoacyl group[245].

Complexes of the type $[R^1_2PtL_2]$ were found[246] to afford insertion products on reaction with isocyanides when L was a basic phosphine such as PEt$_3$ or PPhMe$_2$. With less basic ligands, phosphine replacement giving $[R^1_2Pt(CNR^2)L]$ occurs.

Bu$^t$NC and *p*-MeC$_6$H$_4$NC were found to react with the carboxyvinyl complexes **134** (M = Pd, Pt) to give isolable ionic intermediates **135**, which afford insertion products **136** with retention of stereochemistry both at the double bond and at the

Ph$_3$P(Br)M(PPh$_3$)–CH=CHCO$_2$Me

**(134)**

[Ph$_3$P(RNC)M(PPh$_3$)–CH=CHCO$_2$Me]$^+$ Br$^-$

**(135)**

Ph$_3$P(Br)M(PPh$_3$)–C(=NR)–CH=CHCO$_2$Me

**(136)**

metal[238]. Presumably, the cation of **135** must rearrange to a species having the isocyanide and carboxyvinyl ligands *cis* in order for insertion to occur. In keeping with this view is the fact that [(dppe)Pd(CH=CHCO$_2$Me)Br] in which the dppe ligand occupies *cis* positions readily inserts three molecules of Bu$^t$NC giving [(dppe)Pd{(C=NBu$^t$)$_3$CH=CHCO$_2$Me}Br] again with retention of stereochemistry at both palladium and the double bond.

In contrast, **137** ($M = Pd$, Pt) dimerizes to give iminoacyl products **138** containing only *trans*-double bonds when $M = Pd$ and a mixture of *cis*- and *trans*-double bonds in a slower reaction when $M = Pt$[238].

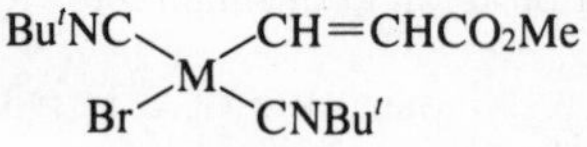

$$\mathrm{Bu^tNC}\diagdown\mathrm{M}(\mathrm{Br})(\mathrm{CNBu^t})\mathrm{CH{=}CHCO_2Me}$$

**(137)**

$$[\mathrm{MeO_2CHC{=}CHC({=}NBu^t)M(Br)(CNBu^t)}]_2$$

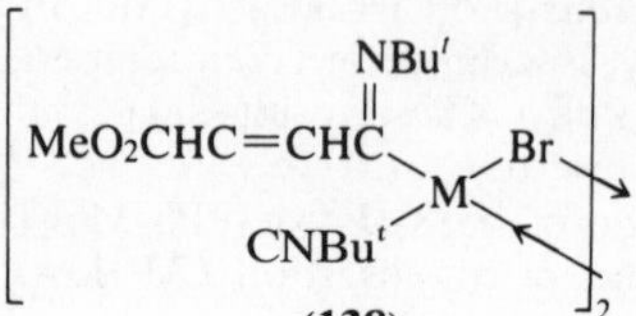

**(138)**

This can be rationalized on the basis of arguments presented previously regarding dimerization of palladium alkyl isocyanide complexes. The reactions involve polarization of the M—C bond by the *trans*-bromide and migration of a carbanionic ligand to an electrophilic isocyanide. When the migrating ligand contains a double bond, a contribution from structure **139** ($L = Bu^tNC$) could lead to double-bond isomeriza-

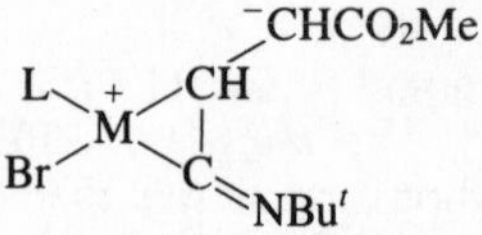

$$\mathrm{L(Br)M^{+}{-}CH(^{-}CHCO_2Me){-}C{=}NBu^t}$$

**(139)**

tion. That no isomerization occurs when $L = PPh_3$ indicates the importance of ancillary ligands in determining the carbanionic character of $\alpha$-C and the electrophilicity of RNC.

## C. Copper Complexes

A number of copper complexes **140** ($n = 1$, $R^1 = H$, $R^2 = C_6H_{11}$; $n = 1$, $R^1 = Me$, $R^2 = C_6H_{11}$, $Bu^t$; $n = 0$, $R^1 = H$, $R^2 = C_6H_{11}$; $L = PPh_3$, $CNC_6H_{11}$) have been prepared by reaction of isocyanide with the copper—alkyl complex followed by addition of L[247].

$$\text{1-R}^1\text{-3-C(=NR}^2\text{)CuL-4-(CH}_2)_n\text{NMe}_2\text{-benzene}$$

**(140)**

## D. Seventeen- and Eighteen-Electron Complexes

$[Cp_2TiR]$ ($R = Me$, Et, $Bu^n$, $Bu^s$, $Bu^t$, Ph, $o$-$MeC_6H_4$) were found[206,248,249] to insert $2,6\text{-}Me_2C_6H_3NC$ to afford paramagnetic iminoacyl products. In some cases intermediate adducts can be isolated in which $\nu_{CN}$ is lowered by *ca.* 40 $cm^{-1}$ compared with the free ligand. This is evidence for back-donation. Migration of alkyl groups to these electron-rich ligands is in contrast to the behaviour of diamagnetic square-planar palladium and platinum complexes. $Y_2 = I_2$, PhSSPh oxidize these iminoacyl complexes to titanium(IV) species $[Cp_2Ti(Y)\{C(R) = N(2,6\text{-}Me_2C_6H_3)\}]$. $[(\eta^3\text{-}C_3H_5)TiCp_2]$ also reacts[192,250] with $2,6\text{-}Me_2C_6H_3NC$ affording **141**, which has an

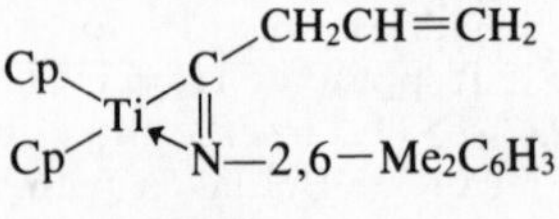

$$\mathrm{Cp_2Ti}\leftarrow\mathrm{N(2,6\text{-}Me_2C_6H_3){=}C(CH_2CH{=}CH_2)}$$

**(141)**

$\eta^2$-iminoacyl ligand, as do insertion products of other titanium(III) alkyls mentioned above. In contrast, the aryl complexes were found to have $\eta^1$-iminoacyl ligands by examination of their i.r. spectra[248].

Even with excess of isocyanide, $[Cp_2TiMe_2]$ gives only $[Cp_2TiMe(C{=}NC_6H_{11})Me]$[251] or $[Cp_2Ti\{C({=}NMe)Me\}_2]$[252]. The zironium(IV) complexes $[Cp_2ZrR^1R^2]$ [$R^1 = CH(SiMe_3)C$, $R^2 = Cl$, Me, $CH_2(SiMe_3)$; $R^1 = CH_2(SiMe_3)$; $R^1 = R^2 = CH_2(CMe_3)$] insert 1 mol of $CNC_6H_4Me$ (or CO) into the Zr—C bond of the bulkier $R^1$ ligand, yielding $\eta^1$-iminoacyl (or acyl) complexes[253]. Steric bulk may aid in breaking the Zr—C bond. MeNC inserts into all four of the Zr—C bonds of $[Zr(CH{=}CMe_2)_4]$[252]. However, $[Zr\{CH_2(CMe_3)\}_4]$ is reported to yield a product of the unusual composition, $[Zr\{CH_2(CMe_3)\}_3(CNBu^t)\{C({=}NBu^t)CH_2CMe_3\}]$, with $Bu^tNC$, probably for steric reasons[252]. The complexes $[R_2Hf\{N(SiMe_3)_2\}_2]$ (R = Me, Et) insert one molecule of $Bu^tNC$ into *both* Hf—C bonds. $CO_2$, on the other hand, inserts into both Hf—N bonds[254]. All three of the Ta—C bonds in $[Me_3TaCl_2]$ are reported[252] to insert MeNC.

Adams and Chodosh have investigated the reactions of $[CpMo(CO)_{3-n}(CNR)_n]^-$ (R = Me, $n = 1,2$; R = Ph, $n = 1$) with MeI. The course of the reactions is outlined in Scheme 18[255].

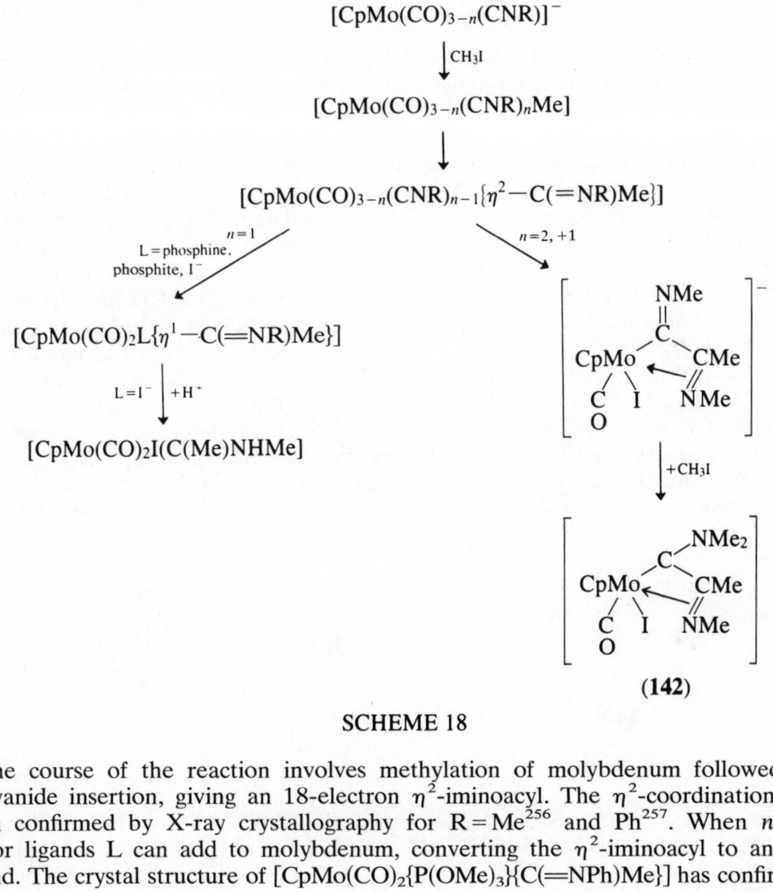

$[CpMo(CO)_{3-n}(CNR)]^-$

↓ $CH_3I$

$[CpMo(CO)_{3-n}(CNR)_nMe]$

↓

$[CpMo(CO)_{3-n}(CNR)_{n-1}\{\eta^2{-}C({=}NR)Me\}]$

$n = 1$, L = phosphine, phosphite, $I^-$ ↙ ↘ $n = 2$, +I

$[CpMo(CO)_2L\{\eta^1{-}C({=}NR)Me\}]$

↓ $L = I^-$, $+H^+$

$[CpMo(CO)_2I(C(Me)NHMe]$

$[CpMo(CO)I\{C({=}NMe)C(Me){=}NMe\}]^-$ (NMe, C, CpMo, CMe, NMe, C, O, I)

↓ $+CH_3I$

$[CpMo(CO)I\{C(NMe_2)C(Me){=}NMe\}]$ (NMe₂, C, CpMo, CMe, NMe, C, O, I)

**(142)**

SCHEME 18

The course of the reaction involves methylation of molybdenum followed by isocyanide insertion, giving an 18-electron $\eta^2$-iminoacyl. The $\eta^2$-coordination has been confirmed by X-ray crystallography for R = Me[256] and Ph[257]. When $n = 1$, donor ligands L can add to molybdenum, converting the $\eta^2$-iminoacyl to an $\eta^1$-ligand. The crystal structure of $[CpMo(CO)_2\{P(OMe)_3\}\{C({=}NPh)Me\}]$ has confirmed the mode of coordination[257]. If $L = I^-$, the anionic complex can be protonated at nitrogen giving a carbene ligand. When $n = 2$, an $I^-$-induced insertion of a second isocyanide occurs, giving an anionic complex which is not isolated but is methylated *in situ* giving **142**. As is the case with CO, the corresponding tungsten complexes are resistant to insertion. In the special case where M = W, $R^1 = R^2 = Me$ and $L = Bu^tNC$, the product of reaction 15 is **143**, resulting from isocyanide insertion and proton shift[40].

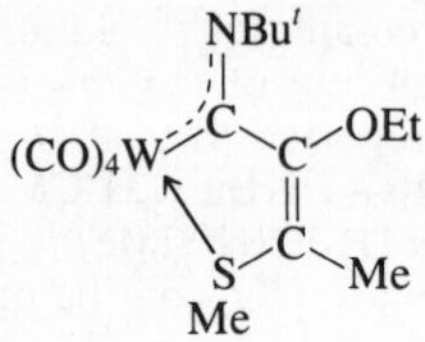

**(143)** 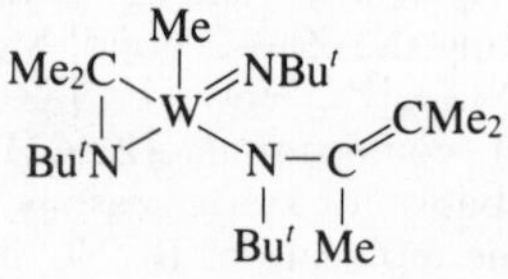 **(144)**

A very unusual product, **144**, results from treatment of $[WMe_6]$ with $Bu^tNC$ at −7 °C. Its formation must involve isocyanide insertion and subsequent methyl transfers[252]. $[ReMe_6]$ affords a more straightforward product, $[Re(CNBu^t)_2\{C(=NBu^t)Me\}_3]$, which contains trigonal bipyramidal rhenium[252].

Yamamoto, Yamazaki and coworkers have carried out extensive studies of reactions of isocyanides with transition metal complexes, including insertions. This group found a number of interesting results with cyclopentadienyl—iron complexes which may be compared with those in Section II.B on carbonylation. Benzyl and phenyl complexes seem to insert RNC more readily than alkyl complexes. Pre-coordinated isocyanide molecules can undergo insertion, but 'direct insertion' of entering RNC seems unlikely on the basis of data so far accumulated. Thus, room-temperature reaction of $[CpFe(CO)_2Me]$ with $C_6H_{11}NC$ affords the acyl $[CpFe(CO)(CNC_6H_{11})(COMe)]$[258]. On the other hand, CO reacts under pressure with $[CpFe(CO)(CNR^2)R^1]$ ($R^1 = CH_2Ph$, $p$-$ClC_6H_4CH_2$, $p$-$ClC_6H_4$; $R^2 = C_6H_{11}$, $CH_2Ph$) giving preferential insertion of isocyanide which produces the iminoacyls $[CpFe(CO)_2\{C(=NR^2)R^1\}]$[259]. When $R^2 = Bu^t$, no reaction occurred, apparently for steric reasons. The insertion rates were noted qualitatively to decrease in the order $R^2 = PhCH_2 > C_6H_{11} \gg Bu^t$ and $R^1 = p\text{-}ClC_6H_4CH_2 > PhCH_2 > p\text{-}ClC_6H_4$. The complex $[CpFe(CO)_2\{C(=NC_6H_{11})CH_2Ph\}]$ could be decarbonylated photochemically.

Photolysis of $[CpFe(CO)(CNBu^t)Me]$ in the presence of $C_6H_{11}NC$ gave a bis(imino) complex **145** ($R^1 = Bu^t$, $R^2 = C_6H_{11}$), suggesting insertion of $Bu^tNC$, thereby creating a vacant site for coordination of $C_6H_{11}NC$ which subsequently inserts[260].

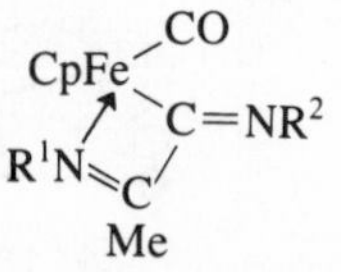

**(145)** 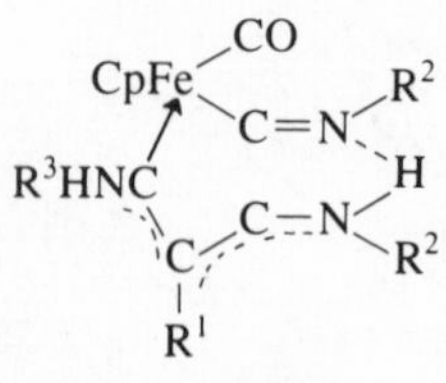 **(146)**

Reflux in thf or benzene of $C_6H_{11}NC$ with $[CpFe(CO)(CNC_6H_{11})CH_2R]$ (R = H, Ph, $p$-$ClC_6H_4$) gives products **146** ($R^2 = R^3 = C_6H_{11}$). These cyclic carbene complexes are the result of insertion of three molecules of isocyanide and shift of the benzyl protons onto nitrogen[261]. These were previously incorrectly formulated[258] as tris(imino) complexes **147** ($R^2 = C_6H_{11}$). Insight into the structure of the carbene

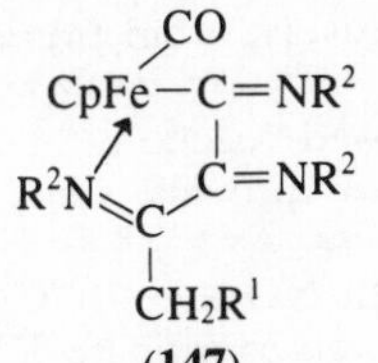

**(147)**

complexes was provided by X-ray determination of the crystal structure of $[CpFe(CO)\{(C{=}NC_6H_{11})(CNHC_6H_{11})CH\{CNHBu^t\}\}]$[262] (**146**) ($R^1 = H$, $R^2 = C_6H_{11}$, $R^3 = Bu^t$), which was prepared from the reaction of **145** ($R^1 = R^2 = C_6H_{11}$) with $Bu^tNC$.

Other compounds **146** containing two different alkyl groups bonded to nitrogen were synthesized from the reactions of $[CpFe(CO)(CNR^2)CH_2R^1]$ ($R^1 = H$, Ph, $R^2 = Bu^tNC$; $R^1 = Ph$, $R^2 = CH_2Ph$) with $C_6H_{11}NC$ ($= R^3NC$). The structures of compounds **146** show that the last isocyanide molecule inserts into a C—H bond of **145**[261]. However, treatment of $[CpFe(CO)(CNBu^t)Me]$ with $CNBu^t$ gives only a mono-insertion product for steric reasons.

Aryl iron complexes $[CpFe(CO)(CNR^2)R^1]$ ($R^1 = Ph$, $p$-$ClC_6H_4$; $R^2 = C_6H_{11}$, $Bu^t$) have no benzylic H and react with $R^3NC$ ($R^3 = C_6H_{11}$) to give tris(imino) complexes **148**[261].

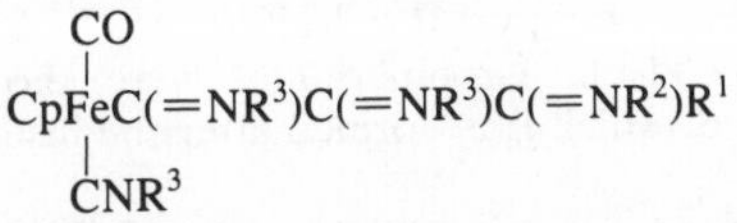

**(148)** 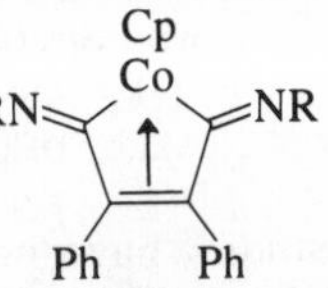 **(149)**

These complexes catalyse the polymerization of isocyanides.

$[CpCo(PPh_3)(\eta^2\text{-}PhC{\equiv}CPh)]$ has been found[263] to insert RNC (R = Ph, $p$-$MeC_6H_4$, 2,5-$MeC_6H_3$, $Bu^t$) into both Co—C bonds, giving the metallocycles **149**. The double bond may be displaced from the cobalt coordination sphere by additional RNC when R is an aryl group. $[CpNi(PPh_3)R]$ (R = Ph, 2,4,6-$Me_3C_6H_2$, $(\eta^5\text{-}C_5H_4)Mn(CO)_3$) were reported[264] to insert $p$-$MeC_6H_4NC$ producing $[CpNi(CN\text{-}p\text{-}MeC_6H_4)\{C({=}N\text{-}p\text{-}MeC_6H_4)R\}]$.

## VII. INSERTION OF OTHER UNSATURATED MOLECULES

### A. Nitric Oxide[68]

Nitric oxide is a paramagnetic molecule which seems to react with organometallic substrates to form diamagnetic species. Thus, 2 mol of NO react with diamagnetic metal alkyls forming $N$-alkyl-$N$-nitrosohydroxylamine ligands **150**.

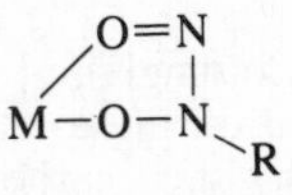

**(150)** 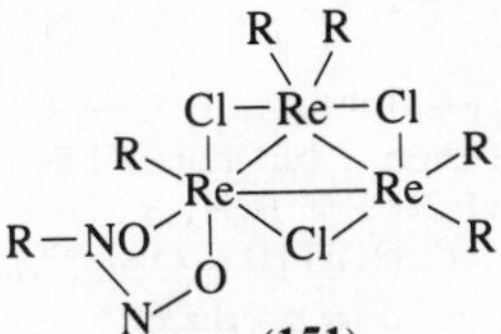 **(151)**

When several M—C bonds are present, reaction continues until the metal is coordinatively saturated. When the organometallic complex itself is also paramagnetic, only one NO inserts. Alternatively, atom transfer reactions may occur without insertion.

McCleverty has recently reviewed reactions of NO with transition metal complexes, including insertions[265]. We provide some more recent examples of the behaviour noted above.

$[Re_3Cl_3(CH_2SiMe_3)_6]$ reacts with NO at −78 °C producing **151** ($R = CH_2SiMe_3$) in which one rhenium atom is seven-coordinate[266]. The N—N bond distance indicates delocalization. The compounds [CpCo(NO)R] (R = Me, Et) were prepared and found to react with $PPh_3$ at 0 °C to give $[CpCo(PPh_3)\{N(O)R\}]$ at 0 °C. β-Elimination does not compete with insertion when R = Et[267]. $[ReOMe_4]$ produces $[Me_2ReO_2]$ + MeN = NMe rather than an insertion product on reaction with NO[268].

## B. Dioxygen

Results on the insertion of $O_2$ into M—C bonds giving peroxo complexes have been summarized by Wojcicki[68]. Both thermal and photochemical activation of these insertions have been observed. The photoinduced insertions into the Co—C bonds of $[RCo(dmgH)_2L]$ (R = Me, L = $H_2O$; R = Et, L = 3-cyanopyridine; R = $Pr^n$, L = 2-methylpyridine; R = $Bu^n$, L = 4-cyanopyridine; R = $(CH_2)_4Me$, L = 3-bromopyridine; R = $Bu^i$, L = 4-bromopyridine; R = $C_6H_{11}$, L = py; R = 2-HO($C_6H_{10}$), L = 4-Me-pyridine; R = $Pr^i$, L = morpholine; R = $CH_2Ph$, L = piperidine) were shown[269] to involve initial cleavage of the Co—L bond, which is re-formed after the insertion has occurred.

Kinetic studies have been conducted on insertion of $O_2$ into $[RCo(dmgH)_2L]$ (L = py, $H_2O$)[270]. Both thermal and photochemical reactions showed first-order dependence on the cobalt complex and on $O_2$. Consistent with the results reported above, excess of L retards the reaction in ethanol and water. For the thermal insertion, reaction rates increase in the order $R = Me_2CH \ll PhCH_2 < R^1R^2C{=}CHCH_2 < PhCHR$. The results were consistent with a free-radical (but not a radical chain) mechanism. For the photochemical reaction, the rates increase in the order R = Me < alkyl < allenyl < propargyl < benzyl < allyl. However, the variation of rate with R is not large, suggesting that the role of $h\nu$ is *not* to induce Co—R homolysis. Again, a radical chain mechanism is ruled out by the rate law.

An unambiguous demonstration that an optically active R racemizes during $O_2$ insertion has been given for (−)-*R*-[(2-octyl)Co$(dmgH)_2$py]. The octan-2-ol cleaved from the product is racemic[271].

Photochemical insertion of $O_2$ into Co—$CH_2$ bonds of $[PhCH_2Co(CN)_5]^{3-}$ and [RCo(tpp)] (R = Et, Pr, $Bu^n$, $Bu^i$, $C_6H_{11}$, $CH_2CH_2Ph$; tpp = tetra-phenylporphinato) 7 have been reported[2,273].

## C. Sulphur

Insertions of tetrasulphur into Co—C bonds of [RCo(dmgH)py] were reviewed previously[68]. Since then a full paper has been published on these reactions[274].

Reaction of MeNCS, EtNCS or PhNCS with the carbene complexes $[W(CO)_5\{C(Ph)(p\text{-}RC_6H_4)\}]$ (R = OMe, Me, H, Br, $CF_3$) gives insertion of S into the tungsten—carbene bond to produce the thioketone complexes $[W(CO)_5\{S{=}C(Ph)(p\text{-}RC_6H_4)\}]$[275,276].

## D. Carbenes

The tantalum—alkylidene complex [$Cp_2Ta({=}CHMe)Me$] decomposes to give [$Cp_2Ta(\eta^2$-$CH_2{=}CHMe)H$], which can be rationalized as involving insertion of the alkylidene into the Ta—Me bond to give [$Cp_2Ta$-$CHMe_2$] followed by $\beta$-elimination[277].

The complexes [$Cp_2Nb(R)({=}CH{-}O{-}Zr(H)(\eta^5$-$Me_5C_5)_2$] (R = H, Me, $CH_2Ph$, $CH_2C_6H_4OMe$-*p*, Ph) can be prepared. In the presence of a trapping ligand L = $PhC{\equiv}CPh$, these complexes undergo carbene insertion into the Nb—R bond, giving an unisolated [$Cp_2NbCH(R)OZr(H)(\eta^5$-$Me_5C_5)_2$] which undergoes $\beta$-elimination to produce [$Cp_2Nb(H)L$] and [$(\eta^5$-$Me_5C_5)_2Zr\{(H)O{-}CH{=}CHR\}$] (R = H, Ph, *p*-$OMeC_6H_4$)[278].

The cationic [$Cp_2WMe_2]^+$ has been found to give [$Cp_2WMe({=}CH_2)]^+$ on treatment with $Ph_3C$. This complex undergoes methylene insertion giving unisolated [$Cp_2WCH_2CH_3]^+$, which can be trapped as [$Cp_2WLEt]^+$ by L = $PPh_2Me$ or, alternatively, undergoes $\beta$-elimination, producing [$Cp_2W(H)(\eta^2$-$C_2H_4)]^+$[279].

## E. Nitrogen

$IN_3$ reacts with [$Re_2(CO)_{10}$] to insert N into an Re—C bond, affording [$Re(CO)_4(\mu$-$NCO)]_2$[280].

## VIII. REFERENCES


1. For a recent review of insertions into transition metal—carbon bonds, see A. Wojcicki, *Ann. N.Y. Acad. Sci.*, **239,** 100 (1974).
2. For a review, see M. F. Lappert and B. Prokai, *Adv. Organomet. Chem.*, **5,** 225 (1967).
3. See, for example, R. F. Heck, *Organotransition Metal Chemistry, A Mechanistic Approach,* Academic Press, New York, 1974; M. M. Taqui Khan and A. E. Martell, *Homogeneous Catalysis by Transition Metal Complexes,* Vol. II, Academic Press, New York, 1974, P. W. Jolly and G. Wilke, *The Organic Chemistry of Nickel,* Vol. II, Academic Press, New York, 1975, P. C. Wailes, R. S. P. Coutts, and H. Weigold, *Organometallic Chemistry of Titanium, Zirconium and Hafnium,* Academic Press, New York, 1974, Chapter V.
4. See, for example, P. M. Maitlis, *The Organic Chemistry of Palladium,* Vol. II, Academic Press, New York, 1971, I. Wender and P. Pino, *Organic Synthesis via Metal Carbonyls,* Wiley–Interscience, New York, Vol. 1, 1968, and Vol. 2, 1977; *Adv. Organomet. Chem.*, **17,** 1979; J. Tsuji, *Organic Synthesis by Means of Transition Metal Complexes,* Springer-Verlag, New York, 1975; P. N. Rhylander, *Organic Syntheses with Noble Metal Catalysts,* Academic Press, New York, 1973; H. Alper (Ed.), *Transition Metal Organometallics in Organic Synthesis,* Academic Press, New York, Vol. I, 1976, Vol. II, 1978.
5. A. A. Wojcicki, *Adv. Organomet. Chem.,* **11,** 87 (1973); F. Calderazzo, *Angew. Chem., Int. Ed. Engl.,* **16,** 299 (1977); E. J. Kuhlmann and J. J. Alexander, *Coord. Chem. Rev.,* **33,** 195 (1980).
6. H. Berke and R. Hoffmann, *J. Am. Chem. Soc.,* **100,** 7224 (1978).
7. T. H. Coffield, J. Kozikowski, and R. D. Closson, *Chem. Soc. Spec. Publ.*, No. 13, 126 (1959).
8. K. Noack and F. Calderazzo, *J. Organomet. Chem.*, **10,** 101 (1967).
9. T. C. Flood, J. E. Jensen, and J. A. Statler, *J. Am. Chem. Soc.*, **103,** 4410 (1981).
10. J. J. Alexander, *J. Am. Chem. Soc.*, **97,** 1729 (1975).
11. F. Calderazzo and K. Noack, *Coord. Chem. Rev.,* **1,** 118 (1966).
12. P. L. Bock, D. J. Boschetto, J. R. Rasmussen, J. P. Demers, and G. M. Whitesides, *J. Am. Chem. Soc.*, **96,** 2814 (1974).
13. T. G. Attig and A. Wojcicki, *J. Organomet. Chem.*, **82,** 397 (1974).
14. A. Davison and N. Martinez, *J. Organomet. Chem.,* **74,** C17 (1974).

15. (a) T. C. Flood, K. D. Campbell, H. H. Downs, and S. Nakanishi, *Organometallics*, **2,** 1590 (1983); (b) H. H. Brunner, B. Hammer, I. Bernal and M. Draux, *Organometallics*, **2,** 1595 (1983); (c) H. Brunner and H. Vogt, *Angew. Chem., Int. Ed., Engl.* **20,** 405 (1981).
16. H. Brunner and H. Vogt, *Chem. Ber.*, **114,** 2186 (1981).
17. G. Fachinetti, S. DelNero, and C. Floriani, *J. Chem. Soc., Dalton Trans.*, 203 (1976).
18. J. N. Cawse, R. A. Fiato, and R. L. Pruett, *J. Organomet. Chem.*, **172,** 405 (1979).
19. D. Saddei, H. J. Freund, and G. Hohlneicher, *J. Organomet. Chem.*, **186,** 63 (1980).
20. R. F. Heck, *J. Am. Chem. Soc.,* **85,** 651 (1963).
21. S. R. Su and A. Wojcicki, *J. Organomet. Chem.*, **27,** 231 (1971).
22. P. E. Garrou and R. F. Heck, *J. Am. Chem. Soc.*, **98,** 4115 (1976).
23. N. Sugita, J. Minkiewicz, and R. F. Heck, *Inorg. Chem.*, **17,** 2809 (1978).
24. G. K. Anderson and R. J. Cross, *Chem. Commun.*, 819 (1978).
25. G. K. Anderson and R. J. Cross, *J. Chem. Soc., Dalton Trans.*, 1246 (1979); R. J. Cross and J. Gemmill, *J. Chem. Soc., Dalton Trans.*, 2317 (1981).
26. G. K. Anderson and R. J. Cross, *J. Chem. Soc., Dalton Trans.*, 712 (1980).
27. G. K. Anderson and R. J. Cross, *J. Chem. Soc., Dalton Trans.*, 1434 (1980).
28. G. K. Anderson, H. C. Clark, and J. A. Davies, *Inorg. Chem.,* **20,** 3607 (1981).
29. R. Ros, J. Renaud, and R. Roulet, *Helv. Chim. Acta,* **58,** 133 (1975).
30. M. Kubota, D. A. Phillips, and J. E. Jacobsen, *J. Coord. Chem.*, **10,** 125 (1980).
31. G.K. Anderson, H. C. Clark, and J. A. Davies, *Organometallics,* **1,** 64 (1982).
32. E. Carmona, F. Gonzalez, M. L. Poveda, J. L. Atwood, and R. D. Rogers, *J. Chem. Soc., Dalton Trans.*, 2108 (1980).
33. T. Yamamoto, T. Kohara, and A. Yamamoto, *Bull. Chem. Soc. Jpn.,* **54,** 2161 (1981).
34. K. Marayuma, T. Ito, and A. Yamamoto, *J. Organomet. Chem.,* **157,** 463 (1978).
35. M. Kubota, D. M. Blake, and S. A. Smith, *Inorg. Chem.,* **10,** 1430 (1971).
36. M. Pankowski and M. Bigorgne, *Abstracts of Papers, VIIIth International Conference on Organometallic Chemistry, Kyoto, Japan,* 1977, 5B02.
37. G. Cardaci, G. Bellachioma, and G. Reichenbach, *Congr. Naz. Chim. Inorg., (Atti), 13th,* 1980; *Chem. Abstr.,* **95,** 97951q.
38. C. F. J. Barnard, J. A. Daniels, and R. J. Mawby, *J. Chem. Soc., Dalton Trans.,* 1331 (1979).
39. W. R. Roper, G. E. Taylor, J. M. Waters, and L. J. Wright, *J. Organomet. Chem.,* **182,** C46 (1979).
40. R. H. Magnuson, S. Zulu, W.-M. Tsai, and W. P. Giering, *J. Am. Chem. Soc.,* **102,** 6887 (1980).
41. S. B. Butts, S. H. Strauss, E. M. Holt, R. E. Stinson, N. W. Alcock, and D. F. Shriver, *J. Am. Chem. Soc.,* **102,** 5093 (1980); F. Correa, R. Nakamura, R. E. Stinson, R. L. Burwell, Jr., and D. F. Shriver, *J. Am. Chem. Soc.,* **102,** 5112 (1980).
42. T. G. Richmond, F. Basolo, and D. F. Shriver, *Inorg. Chem.,* **21,** 1272 (1982).
43. S. B. Butts, T. G. Richmond, and D. F. Shriver, *Inorg. Chem.,* **20,** 278 (1981).
44. J. P. Collman, R. G. Finke, J. N. Cawse, and J. I. Brauman, *J. Am. Chem. Soc.,* **100,** 4766 (1978).
45. K. R. Grundy and W. R. Roper, *J. Organomet. Chem.,* **216,** 255 (1981).
46. J. D. Cotton, G. T. Crisp, and V. A. Daly, *Inorg. Chim. Acta,* **47,** 165 (1981).
47. M. J. Wax and R. G. Bergman, *J. Am. Chem. Soc.,* **103,** 7028 (1981).
48. C. D. Wood and R. R. Schrock, *J. Am. Chem. Soc.,* **101,** 5421 (1979).
49. H. G. Raubenheimer, S. Lotz, H. E. Swanpoel, H. W. Viljoen, and J. C. Rautenbach, *J. Chem. Soc., Dalton Trans.,* 1701 (1979).
50. J. Benaim and F. Giulien, *J. Organomet. Chem.,* **165,** C28, (1979).
51. A. Cutler, D. Entholt, P. Lennon, K. Nicholas, D. F. Marten, M. Madavarao, S. Ragu, A. Rosan, and M. Rosenblum, *J. Am. Chem. Soc.,* **97,** 3149 (1975).
52. For a review of transition metal formyl complexes, see J. A. Gladysz, *Adv. Organomet. Chem.,* **20,** 1, (1982).
53. P. J. Fagan, K. G. Moloy, and T. J. Marks, *J. Am. Chem. Soc.,* **103,** 6959 (1981).
54. E. J. M. de Boer, L. C. Ten Cate, A. G. J. Starling, and J. H. Teuben, *J. Organomet. Chem.,* **181,** 61 (1979).
55. U. Franke and E. Weiss, *J. Organomet. Chem.,* **165,** 329 (1979).

56. E. C. Guzman, G. Wilkinson, R. D. Rogers, W. E. Hunter, M. J. Zaworotko, and J. L. Atwood, *J. Chem. Soc., Dalton Trans.*, 229 (1980).
57. J. A. Marsella, K. G. Moloy, and K. G. Caulton, *J. Organomet. Chem.*, **201,** 389 (1980).
58. C. R. Jablonski, *Inorg. Chem.*, **20,** 3940 (1981).
59. C. J. Attridge, B. Dobbs, and S. J. Maddock, *J. Organomet. Chem.*, **57,** C55 (1973).
60. A. Röder, K. H. Thiele, G. Palyi, and L. Marko, *J. Organomet. Chem.*, **199,** C31 (1980).
61. G. Erker and F. Rosenfeldt, *Angew. Chem., Int. Ed. Engl.*, **17,** 605 (1978).
62. G. Erker and F. Rosenfeldt, *J. Organomet. Chem.*, **188,** Cl (1980).
63. J. Jeffrey, M. F. Lappert, N. T. Luong-Thi, M. Webb, J. L. Atwood, and W. E. Hunter, *J. Chem. Soc., Dalton Trans.*, 1593 (1981).
64. M. F. Lappert, T. R. Martin, J. L. Atwood, and W. E. Hunter, *Chem. Commun.*, 476 (1980).
65. E. Lindner and G. von Au, *Angew. Chem., Int. Ed. Engl.*, **19,** 824 (1980).
66. E. Lindner, G. von Au, H.-J. Eberle, and S. Hoehne, *Chem. Ber.*, **115,** 513 (1982).
67. A. De Renzi, A. Panunzi, M. Scalone, and A. Vitigliano, *J. Organomet. Chem.*, **192,** 129 (1980).
68. A. Wojcicki, *Adv. Organomet. Chem.*, **12,** 31 (1974).
69. P. C. Wailes, H. Weigold, and A. P. Bell, *J. Organomet. Chem.*, **33,** 181 (1971).
70. P. C. Wailes, G. W. A. Fowles, and D. A. Rice, *J. Organomet. Chem.*, **74,** 417 (1974).
71. P. C. Wailes, H. Weigold, and A. P. Bell, *J. Organomet. Chem.*, **34,** 155 (1972).
72. C. Mealli and P. Stoppioni, *J. Organomet. Chem.*, **175,** C19 (1979); C. Mealli, M. Perruzini, and P. Stoppioni, *J. Organomet. Chem.*, **192,** 437 (1980).
73. A. Johnson and R. J. Puddephatt, *J. Chem. Soc., Dalton Trans.*, 1384 (1977).
74. L. Arnaudet, G. Folcher, and H. Marquet-Ellis, *J. Organomet. Chem.*, **214,** 215 (1981).
75. R. L. Downs and A. Wojcicki, *Inorg. Chim. Acta*, **27,** 91 (1978); K. Stanley and M. C . Baird, *J. Am. Chem. Soc.*, **99,** 1808 (1977).
76. M. Kubota, R. K. Rothrock, M. R. Kernan, and R. B. Haven, *Inorg. Chem.*, **21,** 2491 (1982).
77. R. J. Puddephatt and M. A. Stalteri, *J. Organomet. Chem.*, **193,** C27 (1980).
78. A. Dormond, C. Moise, A. Dalchour, J. C. Leblanc, and J. Tirouflet, *J. Organomet. Chem.*, **177,** 191 (1979).
79. S. E. Jacobson, P. Reich-Rohrwig, and A. Wojcicki, *Inorg. Chem.*, **12,** 717 (1973).
80. S. E. Jacobson and A. Wojcicki, *J. Am. Chem. Soc.*, **95,** 6962 (1973).
81. S. E. Jacobson and A. Wojcicki, *Inorg. Chim. Acta*, **10,** 229 (1974).
82. S. Inagaki, H. Fujimoto, and K. Fukui, *J. Am. Chem. Soc.*, **98,** 4693 (1976).
83. S. E. Jacobson and A. Wojcicki, *J. Organomet. Chem.*, **72,** 113 (1974).
84. K. von Werner and H. Blank, *J. Organomet. Chem.*, **195,** C25 (1980).
85. J. O. Kroll and A. Wojcicki, *J. Organomet. Chem.*, **66,** 95 (1974).
86. R. G. Severson and A. Wojcicki, *J. Am. Chem. Soc.*, **101,** 877 (1979).
87. A. E. Crease and M. D. Johnson, *J. Am. Chem. Soc.*, **100,** 8013 (1978).
88. E. Lindner, G. Funk, and S. Hoehne, *Angew. Chem., Int. Ed. Engl.*, **18,** 535 (1979).
89. E. Lindner, G. Funk, and F. Bouachir, *Chem. Ber.*, **114,** 2653 (1981).
90. E. Lindner and G. von Au, *J. Organomet. Chem.*, **202,** 163 (1980).
91. E. Lindner and G. Funk, *J. Organomet. Chem.*, **216,** 393 (1981).
92. P. Diversi, G. Ingrosso, and A. Lucherini, *Chem. Commun.*, 735 (1978).
93. K. Jacob and K.-H. Thiele, *Z. anorg. allg. Chem.*, **455,** 3 (1979).
94. D. Dong, D. A. Slack, and M. C. Baird, *J. Organomet. Chem.*, **153,** 219 (1978).
95. J. D. Cotton and G. T. Crisp, *J. Organomet. Chem.*, **186,** 137 (1980).
96. P. Reich-Rohrwig and A. Wojcicki, *Inorg. Chem.*, **13,** 2457 (1974).
97. T. G. Attig and A. Wojcicki, *J. Am. Chem. Soc.*, **96,** 262 (1974); T. G. Attig and A. Wojcicki, *J. Am. Chem. Soc.*, **101,** 619 (1979).
98. T. C. Flood and D. L. Miles, *J. Am. Chem. Soc.*, **95,** 6460 (1973).
99. T. C. Flood, F. J. Di Santi, and D. L. Miles, *Inorg. Chem.*, **15,** 1910 (1976).
100. S. L. Miles, D. L. Miles, R. Bau, and T. C. Flood, *J. Am. Chem. Soc.*, **100,** 7278 (1978).
101. A. Dormond, C. Moise, A. Dahchour, and J. Tirouflet, *J. Organomet. Chem.*, **168,** C53 (1979).
102. A. Dormond, C. Moise, A. Dahchour, and J. Tirouflet, *J. Organomet. Chem.*, **177,** 181 (1979).

103. A. Dormond, A. Dahchour, and J. Tirouflet, *J. Organomet. Chem.*, **216,** 49 (1981).
104. W. P. Giering and M. Rosenblum, *J. Am. Chem. Soc.,* **93,** 5299 (1971).
105. L. S. Chen, S. R. Su, and A. Wojcicki, *J. Am. Chem. Soc.,* **96,** 5655 (1974).
106. D. A. Ross and A. Wojcicki, *Inorg. Chim. Acta,* **28,** 59 (1978).
107. T. Hung, P. W. Jolly, and G. Wilke, *J. Organomet. Chem.*, **190,** C5 (1980).
108. J. E. Thomasson, P. W. Robinson, D. A. Ross, and A. Wojcicki, *Inorg. Chem.*, **10,** 2130 (1971); M. Churchill, T. Wormald, D. A. Ross, J. E. Thomasson, and A. Wojcicki, *J. Am. Chem. Soc.,* **92,** 1795 (1970).
109. R. G. Severson and A. Wojcicki, *J. Organomet. Chem.*, **149,** C66 (1978).
110. R. G. Severson, T. W. Leung, and A. Wojcicki, *Inorg. Chem.*, **19,** 915 (1980).
111. I.-P. Lorenz, *Angew. Chem., Int. Ed. Engl.*, **17,** 53 (1978).
112. W. Dell and M. Ziegler, *Angew. Chem., Int. Ed. Engl.*, **20,** 471 (1981).
113. W. H. Dell and M. L. Ziegler, *Z. Naturforsch.*, **37B,** 1 (1981).
114. S. R. Su and A. Wojcicki, *Inorg. Chem.*, **14,** 89 (1975).
115. P. J. Krusic, H. Stoklosa, L. E. Manzer, and P. Meakin, *J. Am. Chem. Soc.,* **97,** 667 (1975).
116. J. L. Hanna and A. Wojcicki, *Inorg. Chim. Acta,* **9,** 55 (1974).
117. K. L. Amos and N. G. Conelly, *J. Organomet. Chem.*, **194,** C57 (1980).
118. See, for example, M. Rosenblum, *Acc. Chem. Res.*, **7,** 122 (1974); A. Cutler, D. Entholt, W. P. Giering, P. Lennon, S. Raghu, A. Rosen, M. Rosenblum, J. Tancrede, and D. Wells, *J. Am. Chem. Soc.,* **98,** 3495 (1976); S. R. Su and A. Wojcicki, *Inorg. Chim. Acta,* **8,** 55 (1974); J. P. Williams and A. Wojcicki, *Inorg. Chem.*, **16,** 2506 (1977); J. P. Williams and A. Wojcicki, *Inorg. Chem.*, **16,** 3116 (1977); A. Davison and J. P. Solar, *J. Organomet. Chem.*, **166,** C13 (1979).
119. A. N. Nesmeyanov, N. E. Kolobava, K. N. Asimov, and F. S. Denisov, *Proc. Acad. Sci. USSR,* **192,** 395 (1970).
120. B. J. Cole, J. D. Cotton, and D. McWilliams, *J. Organomet. Chem.*, **64,** 223 (1974).
121. J. D. Cotton, P. J. Davidson, D. E. Goldberg, M. F. Lappert, and K. M. Thomas, *Chem. Commun.*, 893 (1974); J. D. Cotton, P. J. Davidson, and M. F. Lappert, *J. Chem. Soc., Dalton Trans.*, 2275 (1976).
122. J. D. Cotton and G. A. Morris, *J. Organomet. Chem.*, **145,** 245 (1978).
123. C. V. Magotti and W. P. Giering, *J. Organomet. Chem.*, **73,** 85 (1974).
124. J. D. Cotton, *J. Organomet. Chem.*, **159,** 465 (1978).
125. D. W. Lichtenberg and A. Wojcicki, *Inorg. Chem.*, **14,** 1295 (1975).
126. M. Green, J. A. K. Howard, P. Mitrprachachon, M. Pfeffer, J. L. Spencer, F. G. A. Stone, and P. Woodward, *J. Chem. Soc., Dalton Trans.*, 306 (1979).
127. M. Green, J. A. K. Howard, A. Laguna, L. E. Smart, J. L. Spencer, and F. G. A. Stone, *J. Chem. Soc., Dalton Trans.*, 278 (1977).
128. P. Caddy, M. Green, J. A. K. Howard, J. Squire, and N. J. White, *J. Chem. Soc., Dalton Trans.*, 400 (1981).
129. E. R. Evitt and R. G. Bergman, *J. Am. Chem. Soc.,* **101,** 3973 (1979).
130. For an example of dimerization via a metallocyclic path, see J. D. Fellmann, G. A. Rupprecht, and R. R. Schrock, *J. Am. Chem. Soc.,* **101,** 5099 (1979).
131. H.-F. Klein, R. Hammer, J. Gross, and U. Schubert, *Angew. Chem., Int. Ed. Engl.*, **19,** 809 (1980).
132. F. N. Tebbe and G. W. Parshall, *J. Am. Chem. Soc.,* **93,** 3793 (1971).
133. J. W. Lauher and R. Hoffmann, *J. Am. Chem. Soc.,* **98,** 1729 (1976), and references cited therein; D. L. Thorn and R. Hoffmann, *J. Am. Chem. Soc.,* **100,** 2079 (1978); A. Dedieu, *Inorg. Chem.*, **20,** 2083 (1981); P. Cassoux, F. Crasnier, and J.-F. Labarre, *J. Organomet. Chem.*, **165,** 303 (1979).
134. H. Lehmkuhl, *Bull. Soc. Chim. Fr.*, 87 (1981).
135. G. Henrici-Olivé and S. Olivé, *Top. Curr. Chem.*, **67,** 107 (1976).
136. H. C. Clark, C. R. Jablonski, and K. von Werner, *J. Organomet. Chem.*, **82,** C51 (1974).
137. H. C. Clark and R. J. Puddephatt, *Inorg. Chem.*, **9,** 2671 (1970).
138. A. Johnson, R. J. Puddephatt, and R. J. Quirk, *Chem. Commun.*, 938 (1972).
139. A. Johnson and R. J. Puddephatt, *J. Chem. Soc., Dalton Trans.*, 1384 (1977).
140. D. J. Mabbot, P. M. Bailey, and P. M. Maitlis, *Chem. Commun.*, 521 (1975).
141. R. P. Hughes and J. Powell, *J. Organomet. Chem.*, **60,** 387 (1973).

142. M. G. Gallazi, T. L. Hanlon, G. Vitulli, and L. Pori, *J. Organomet. Chem.*, **33,** C45 (1971).
143. T. R. Jack, C. J. May, and J. Powell, *J. Am. Chem. Soc.*, **100,** 5057 (1978).
144. C. J. May and J. Powell, *J. Organomet. Chem.*, **184,** 385 (1980).
145. A. Bond, M. Green, and S. Taylor, *Chem. Commun.*, 112 (1973).
146. B. L. Booth, M. Gardner, and R. N. Haszeldine, *J. Chem. Soc., Dalton Trans.*, 1856 (1975).
147. B. L. Booth, M. Gardner, and R. N. Haszeldine, *J. Chem. Soc., Dalton Trans.*, 1863 (1975).
148. A. Greco, M. Green, and F. G. A. Stone, *J. Chem. Soc. A*, 3476 (1971); M. Bottrill, R. Goddard, M. Green, and P. Woodward, *J. Chem. Soc., Dalton Trans.*, 1671 (1979).
149. M. Green and S. H. Taylor, *J. Chem. Soc., Dalton Trans.*, 1128 (1975).
150. J. C. Burt, S. A. R. Knox, R. J. McKinney, and F. G. A. Stone, *J. Chem. Soc., Dalton Trans.*, 1 (1977).
151. M. H. Chisholm and W. S. Johns, *Inorg. Chem.*, **14,** 1189 (1975).
152. R. Stevens and G. D. Shier, *J. Organomet. Chem.*, **21,** 495 (1970).
153. R. P. Hughes and J. Powell, *J. Organomet. Chem.*, **34,** C51 (1972).
154. R. P. Hughes and J. Powell, *J. Organomet. Chem.*, **60,** 409 (1973).
155. A. Guinot, P. Cadiot, and J. L. Roustan, *J. Organomet. Chem.*, **166,** 379 (1979); J. L. Roustan, A. Guinot, P. Cadiot, and A. Forgues, *J. Organomet. Chem.*, **194,** 179 (1980); J. L. Roustan, A. Guinot, and P. Cadiot, *J. Organomet. Chem.*, **194,** 191, 357, 367 (1980).
156. M. I. Lobach and V. A. Kormer, *Russ. Chem. Rev.*, **48,** 759 (1979).
157. R. P. Hughes and J. Powell, *J. Am. Chem. Soc.*, **94,** 7723 (1972).
158. Y. Takahashi, S. Sakai, and Y. Ishii, *Inorg. Chem.*, **11,** 1516 (1972).
159. J. A. Sadownick and S. J. Lippard, *Inorg. Chem.*, **12,** 2659 (1973).
160. I. D. Rae, B. E. Reichert, and B. O. West, *J. Organomet. Chem.*, **81,** 227 (1974).
161. M. Green and R. I. Hancock, *J. Chem. Soc. A*, 109 (1968).
162. M. I. Bruce and F. G. A. Stone, *Prep. Inorg. React.*, **4,** 177 (1968); M. I. Bruce, and W. R. Cullen, *Fluorine Chem. Rev.*, **4,** 79 (1969).
163. S. Otsuka and A. Nakamura, *Adv. Organomet. Chem.*, **14,** 245 (1976).
164. W. H. Boon and M. D. Rausch, *Chem. Commun.*, 397 (1977).
165. S. J. Tremont and R. G. Bergman, *J. Organomet. Chem.*, **140,** C12 (1977).
166. J. M. Huggins and R. Bergman, *J. Am. Chem. Soc.*, **101,** 4410 (1979); **103,** 3002 (1981).
167. T. G. Appleton, M. H. Chisholm, H. C. Clark, and K. Yasufuku, *J. Am. Chem. Soc.*, **96,** 6600 (1974).
168. H. C. Clark and K. von Werner, *J. Organomet. Chem.*, **101,** 347 (1975).
169. H. C. Clark, C. R. C. Milne, and C. S. Wong, *J. Organomet. Chem.*, **136,** 265 (1977).
170. Y. Tohda, K. Sonogashira, and N. Hagihara, *Chem. Commun.*, 54 (1975).
171. T. G. Appleton, M. A. Bennett, A. Singh, and T. Yoshida, *J. Organomet. Chem.*, **154,** 369 (1978).
172. T. G. Appleton, H. C. Clark, R. C. Poller, and R. J. Puddephatt, *J. Organomet. Chem.*, **39,** C13 (1972).
173. H. Eshtiagh-Hosseini, J. F. Nixon, and J. S. Poland, *J. Organomet. Chem.*, **164,** 107 (1979).
174. A. Alexakis, G. Cahiez, and J. F. Normant, *J. Organomet. Chem.*, **177,** 293 (1979).
175. H. G. Alt, *J. Organomet. Chem.*, **127,** 349 (1977).
176. H. G. Alt and J. A. Schwärtzle, *J. Organomet. Chem.*, **155,** C65 (1978).
177. H. G. Alt, J. A. Schwärtzle, and F. R. Kreisel, *J. Organomet. Chem.*, **152,** C57 (1978).
178. J. L. Davidson, M. Green, J. Z. Nyathi, C. Scott, F. G. A. Stone, A. J. Welch, and P. Woodward, *Chem. Commun.*, 714 (1976); M. Green, J. Z. Nyathi, C. Scott, F. G. A. Stone, A. J. Welch, and P. Woodward, *J. Chem. Soc., Dalton Trans.*, 1067 (1978).
179. P. L. Watson and R. G. Bergman, *J. Am. Chem. Soc.*, **101,** 2055 (1979).
180. H. G. Alt, *Z. Naturforsch.*, **32B,** 1139 (1977).
181. M. Bottrill and M. Green, *J. Chem. Soc., Dalton Trans.*, 820 (1979).
182. K. H. Dötz and C. G. Kreiter, *J. Organomet. Chem.*, **99,** 309 (1975).
183. K. H. Dötz, *Chem. Ber.*, **110,** 78 (1977).
184. K. H. Dötz and I. Pruskil, *Chem. Ber.*, **111,** 2059 (1978).

185. K. H. Dötz and C. G. Kreiter, *Chem. Ber.*, **109,** 2026 (1976).
186. K. H. Dötz and I. Pruskil, *J. Organomet. Chem.*, **132,** 115 (1977).
187. H. Fischer, *J. Organomet. Chem.*, **197,** 303 (1980).
188. H. Fischer and U. Schubert, *Angew. Chem. Int. Ed. Engl.*, **20,** 461. (1981).
189. H. Fischer, U. Schubert, and R. Märkl, *Chem. Ber.*, **114,** 3412 (1981).
190. A. F. Dyke, S. A. R. Knox, P. J. Naish, and G. E. Taylor, *Chem. Commun.*, 803 (1979).
191. J. Levisalles, F. Rose-Munch, and H. Rudler, *Chem. Commun.* 152 (1981).
192. E. Klei, J. H. Teuben, H. J. DeLiefde Meijer, E. Kwak, and A. P. Bruins, *J. Organomet. Chem.*, **224,** 327 (1982).
193. J. L. Davidson, M. Green, F. G. A. Stone, and A. J. Welch, *J. Chem. Soc., Dalton Trans.*, 2044 (1976).
194. B. L. Booth and R. G. Hargreaves, *J. Chem. Soc. A*, 308 (1970).
195. M. Bottrill, M. Green, E. O'Brien, L. E. Smart, and P. Woodward, *J. Chem. Soc., Dalton Trans.*, 292 (1980).
196. M. I. Bruce, R. C. F. Gardner, and F. G. A. Stone, *J. Chem. Soc., Dalton Trans.*, 906 (1976).
197. M. I. Bruce, R. C. F. Gardner, and F. G. A. Stone, *J. Chem. Soc., Dalton Trans.*, 81 (1976).
198. T. Blackburn, M. I. Bruce, and F. G. A. Stone, *J. Chem. Soc., Dalton Trans.*, 106 (1974).
199. M. I. Bruce, R. C. F. Gardner, J. A. K. Howard, F. G. A. Stone, M. Welling, and P. Woodward, *J. Chem. Soc., Dalton Trans.*, 621 (1977).
200. F. Y. Petillon, F. LeFloch-Perennou, J. E. Guerchais, and D. W. A. Sharp, *J. Organomet. Chem.*, **173,** 89 (1979).
201. M. Green and S. H. Taylor, *J. Chem. Soc., Dalton Trans.*, 1142 (1975).
202. M. I. Bruce, B. L. Goodall, and F. G. A. Stone, *J. Chem. Soc., Dalton Trans.*, 1651 (1975).
203. M. H. Chisholm and D. A. Hartko, *J. Am. Chem. Soc.*, **101,** 6784 (1979).
204. I. S. Kolomnikov and M. Kh. Grigoryan, *Russ. Chem. Rev.*, **47,** 334 (1978); M. E. Vol'pin and I. S. Kolomnikov, *Organomet. React.*, **5,** 313 (1975).
205. P. V. Yaneff, *Coord. Chem. Rev.*, **23,** 183 (1977); I. S. Butler and A. E. Fenster, *J. Organomet. Chem.*, **66,** 161 (1974).
206. E. Klei, J. H. Telgen, and J. H. Teuben, *J. Organomet. Chem.*, **209,** 297 (1981).
207. F. Sato, S. Iijima, and M. Sato, *Chem. Commun.*, 181 (1981).
208. E. Klei and J. H. Teuben, *J. Organomet. Chem.*, **222,** 79 (1981).
209. M. Kh. Grigoryan, I. S. Kolomnikov, E. G. Berkovich, T. V. Lysak, V. B. Shur, and M. E. Vol'pin, *Izv. Akad. Nauk SSSR, Ser. Khim.*, 1177 (1978).
210. G. A. Razuvaev, V. N. Latyaeva, L. I. Vyshinskaya, and V. V. Drobtinko, *J. Organomet. Chem.*, **208,** 169 (1981).
211. L. E. Manzer, *J. Organomet. Chem.*, **135,** C6 (1977); *J. Am. Chem. Soc.*, **100,** 8068 (1978).
212. D. J. Darensbourg and A. Rokicki, *J. Am. Chem. Soc.*, **104,** 349 (1982).
213. H. H. Karsch, *Chem. Ber.*, **110,** 2213 (1977).
214. T. Herskovitz, *J. Am. Chem. Soc.*, **99,** 2391 (1977).
215. T. Yamamoto and A. Yamamoto, *Chem. Lett.*, 615 (1978).
216. T. Tsuda, Y. Cujo, and T. Saegusa, *Chem. Commun.*, 963 (1975).
217. A. Miyashita and A. Yamamoto, *J. Organomet. Chem.*, **113,** 187 (1976).
218. T. Ikariya and A. Yamamoto, *J. Organomet. Chem.*, **72,** 145 (1974).
219. P. G. Cookson and G. B. Deacon, *Aust. J. Chem.*, **25,** 2095 (1972).
220. G. B. Deacon, S. J. Faulks, and J. M. Miller, *Transition Met. Chem.*, **5,** 305 (1980).
221. T. Tsuda, T. Nakatsuka, T. Hiruyama, and T. Saegusa, *Chem. Commun.*, 557 (1974).
222. V. R. Polishchuk, L. A. Federov, P. O. Okulevich, I. S. German, and I. L. Knunyants, *Tetrahedron Lett.*, 3933 (1970).
223. C. M. Mitchell and F. G. A. Stone, *Chem. Commun.*, 1263 (1970).
224. H. Kurosawa, *Inorg. Chem.*, **14,** 2148 (1975).
225. C. Santini-Scampucci and G. Wilkinson, *J. Chem. Soc., Dalton Trans.*, 807 (1976).
226. A. Camus, N. Marsich, and G. Nardin, *J. Organomet. Chem.*, **188,** 389 (1980).

227. C. Blandy and D. Gervais, *Inorg. Chim. Acta,* **52,** 79 (1981).
228. J. F. Clarke, G. W. A. Fowles, and D. A. Rice, *J. Organomet. Chem.,* **74,** 417 (1974).
229. J. D. Wilkins, *J. Organomet. Chem.,* **67,** 269 (1974).
230. J. D. Wilkins, *J. Organomet. Chem.,* **65,** 383 (1974).
231. J. D. Wilkins, *J. Organomet. Chem.,* **80,** 349 (1974).
232. F. Sato, J. Noguchi, and M. Sato, *J. Organomet. Chem.,* **118,** 117 (1976).
233. C. J. May and J. Powell, *J. Organomet. Chem.,* **209,** 131 (1981).
234. A. Dormond and A. Dahchour, *J. Organomet. Chem.,* **193,** 321 (1980).
235. Y. Yamamoto and H. Yamazaki, *Coord. Chem. Rev.,* **8,** 225 (1972); P. M. Treichel, *Adv. Organomet. Chem.,* **11,** 73 (1973).
236. B. Crociani, M. Nicolini, and T. Boschi, *J. Organomet. Chem.,* **33,** C81 (1971); B. Crociani, M. Nicolini, and R. L. Richards, *J. Organomet. Chem.,* **104,** 259 (1976).
237. S. Otsuka, A. Nakamura, and T. Yoshida, *J. Am. Chem. Soc.,* **91,** 7196 (1969).
238. S. Otsuka and K. Ataka, *J. Chem. Soc., Dalton Trans.,* 327 (1976).
239. Y. Yamamoto and H. Yamazaki, *Inorg. Chim. Acta,* **41,** 229 (1980).
240. Y. Yamamoto and H. Yamazaki, *Inorg. Chem.,* **13,** 438 (1974).
241. P. M. Treichel and R. W. Hess, *J. Am. Chem. Soc.,* **92,** 4731 (1970).
242. Y. Yamamoto and H. Yamazaki, *Bull. Chem. Soc. Jpn.,* **43,** 3634 (1970).
243. Y. Yamamoto and H. Yamazaki, *Bull. Chem. Soc. Jpn.,* **43,** 2653 (1970).
244. Y. Yamamoto and H. Yamazaki, *Bull. Chem. Soc. Jpn.,* **44,** 1873 (1971).
245. K. P. Wagner, P. M. Treichel, and J. C. Calabrese, *J. Organomet. Chem.,* **71,** 299 (1974).
246. P. M. Treichel and K. P. Wagner, *J. Organomet. Chem.,* **61,** 415 (1973).
247. G. van Kotten and J. G. Noltes, *Chem. Commun.,* 59 (1972).
248. E. J. M. DeBoer and J. H. Teuben, *J. Organomet. Chem.,* **166,** 193 (1979).
249. E. Klei and J. H. Teuben, *J. Organomet. Chem.,* **188,** 97 (1980).
250. B. Klei, J. H. Teuben, and H. J. de L. Meijer, *Chem. Commun.* 343 (1981).
251. R. J. H. Clark, J. A. Stockwell, and J. D. Wilkins, *J. Chem. Soc., Dalton Trans.,* 120 (1976).
252. K. W. Chiu, R. A. Jones, G. Wilkinson, A. M. R. Gales, and M. B. Hursthouse, *J. Chem. Soc., Dalton Trans.,* 2088 (1981).
253. M. F. Lappert, N. T. L. Thi, and C. R. C. Milne, *J. Organomet. Chem.,* **174,** C35 (1979).
254. R. A. Andersen, *Inorg. Chem.,* **18,** 2928 (1979).
255. R. D. Adams and D. F. Chodosh, *J. Am. Chem. Soc.,* **98,** 5391 (1976); R. D. Adams and D. F. Chodosh, *J. Am. Chem. Soc.,* **99,** 6544 (1977).
256. R. D. Adams and D. F. Chodosh, *J. Organomet. Chem.,* **122,** C11 (1976).
257. R. D. Adams and D. F. Chodosh, *Inorg. Chem.,* **17,** 41 (1978).
258. Y. Yamamoto and H. Yamazaki, *Inorg. Chem.,* **11,** 211 (1972).
259. Y. Yamamoto and H. Yamazaki, *Inorg. Chem.,* **13,** 2145 (1974).
260. Y. Yamamoto and H. Yamazaki, *J. Organomet. Chem.,* **90,** 329 (1975).
261. Y. Yamamoto and H. Yamazaki, *Inorg. Chem.,* **16,** 3182 (1977).
262. K. Aoki and Y. Yamamoto, *Inorg. Chem.,* **15,** 48 (1976).
263. H. Yamazaki, K. Aoki, Y. Yamamoto, and Y. Wakatsuki, *J. Am. Chem. Soc.,* **97,** 3546 (1975).
264. A. N. Nesmeyanov, L. I. Leont'eva, and K. I. Khomik, *Bull. Acad. Sci. USSR,* **25,** 1571 (1976).
265. J. A. McCleverty, *J. Mol. Catal.,* **13,** 309 (1981).
266. P. Edwards, K. Mertis, G. Wilkinson, M. B. Hursthouse, and K. A. M. Malik, *J. Chem. Soc., Dalton Trans.,* 334 (1980).
267. W. P. Weiner, M. A. White, and R. G. Bergman, *J. Am. Chem. Soc.,* **103,** 3612 (1981).
268. A. R. Middleton and G. Wilkinson, *J. Chem. Soc., Dalton Trans.,* 1888 (1980).
269. C. Giannotti, C. Fontaine, and B. Septe, *J. Organomet. Chem.,* **71,** 107 (1974).
270. C. Bied-Charreton and A. Gaudemer, *J. Organomet. Chem.,* **124,** 299 (1977).
271. J. Deniau and A. Gaudemer, *J. Organomet. Chem.,* **191,** C1 (1980).
272. A. Vogler and R. Hirschmann, *Z. Naturforsch.,* **31B,** 1082 (1976).
273. M. Peree-Fauvet, A. Gaudemer, P. Couchy, and J. Devynk, *J. Organomet. Chem.,* **120,** 439 (1976).

274. C. Giannotti and G. Merle, *J. Organomet. Chem.*, **113,** 45 (1976).
275. H. Fischer, *J. Organomet. Chem.*, **222,** 241 (1981).
276. H. Fischer and R. Märkl, *Chem. Ber.*, **115,** 1349 (1982).
277. P. R. Sharp and R. R. Schrock, *J. Organomet. Chem.*, **171,** 43 (1979).
278. R. S. Threlkel and J. E. Bercaw, *J. Am. Chem. Soc.*, **103,** 2650 (1981).
279. J. C. Hayes, G. D. N. Pearson, and N. J. Cooper, *J. Am. Chem. Soc.*, **103,** 4648 (1981).
280. K. Dehnicke and R. Dübgen, *Z. anorg. allg. Chem.*, **444,** 61 (1978).

The Chemistry of the Metal—Carbon Bond, Volume 2
Edited by F. R. Hartley and S. Patai


CHAPTER **6**

# Nucleophilic attack on transition metal organometallic compounds

LOUIS S. HEGEDUS
*Department of Chemistry, Colorado State University, Fort Collins, Colorado 80523, USA*

The reactions of nucleophiles with transition metal organometallic complexes constitute some of the most broadly useful procedures for both the preparation of these complexes and, more importantly, for their application in synthetically useful transformations of metal-coordinated organic functional groups. These reactions can be divided into two general categories, those which involve nucleophilic attack at the metal centre itself, and those which involve nucleophilic attack at a metal-coordinated organic ligand. The former process is of major importance for the preparation of a variety of transition metal organometallic complexes and for a number of transition metal-catalysed coupling reactions involving 'transmetallation' reactions. The latter is of major consequence in the important and rapidly evolving field of the use of transition metal complexes in organic synthesis. It is also the topic which constitutes the bulk of this review. No attempt at a comprehensive treatment of either area is made in this chapter. Rather, important topics will be introduced and the general features of specific processes will be presented. Existing major reviews will be briefly summarized, and this chapter will focus on current progress in the subject areas. The literature is covered up to December 1981.

# I. NUCLEOPHILIC ATTACK AT THE METAL IN TRANSITION METAL ORGANOMETALLIC COMPLEXES

## A. Formation of Metal—Carbon σ-Bonds

### 1. *Metal carbonyls, isonitriles, and cyanides*

Carbon monoxide is the most ubiquitous carbon ligand in organotransition metal chemistry[1,2]. Examples of compounds with at least one CO ligand are known for all the transition metals. A number of homoleptic metal carbonyls, e.g. $[Ni(CO)_4]$, $[Fe(CO)_5]$, can be prepared by the direct reaction of carbon monoxide with the finely divided metal (equations 1–3), or by reaction with the corresponding metal salt under

$$\mathrm{Ni + CO \xrightarrow[1\ atm]{30\,^{\circ}C} [Ni(CO)_4]} \tag{1}$$

$$\mathrm{Co + CO \xrightarrow[30\text{–}40\ atm]{150\,^{\circ}C} [Co_2(CO)_8]} \tag{2}$$

$$\mathrm{Fe + CO \xrightarrow[1\text{–}200\ atm]{25\text{–}200\,^{\circ}C} [Fe(CO)_5]} \tag{3}$$

reducing conditions. Carbon monoxide, like most common ligands in organotransition metal chemistry, also undergoes facile ligand exchange reactions and can displace non-carbon ligands from metals (equation 4). All of these reactions can be

$$\mathrm{[RhCl(PPh_3)_3] + CO \longrightarrow [RhCl(CO)(PPh_3)_2] + PPh_3} \tag{4}$$

viewed as proceeding by nucleophilic attack on the metal by the lone pair of electrons of the carbon monoxide, much like other ligand exchange processes involving simple donor ligands such as phosphines or amines (see below). However, once coordinated, carbon monoxide is a strong $\pi$-acceptor ligand, and can remove excess electron density from a metal atom and thereby stabilize low oxidation states[3]. Thus, exposure of virtually any low-valent metal complex to carbon monoxide will result in nucleophilic attack on the metal by the CO, resulting in CO incorporation in some manner.

The isonitrile ligand, RNC: is formally analogous to carbon monoxide, but is a stronger $\sigma$-donor and weaker $\pi$-acceptor than is carbon monoxide and thus can stabilize metal complexes in higher oxidation states[4]. Most transition metal isonitrile complexes are prepared by direct reaction (nucleophilic attack) of the isonitrile with a metal complex to effect either a ligand exchange (equations 5[5] and 6[6]) or the displacement of a molecule of coordinated solvent from a cationic species (equation 7[7]), or both (equation 8[6]). Both coordinated isonitriles and carbon monoxide are

$$[PtCl_2(Ph_3P)_2] + MeNC \xrightarrow[25\,°C]{PhH} [PtCl_2(Ph_3P)(MeNC)] + Ph_3P \tag{5}$$

$$\left[R^1R^2Pd\begin{matrix}Cl\\NHEt_2\end{matrix}\right] + MeNC \xrightarrow{Et_2O} \left[R^1R^2Pd\begin{matrix}Cl\\CNMe\end{matrix}\right] + Et_2NH \tag{6}$$

$R^1R^2 = $ (chelating $CH_2CH_2C(O)$ chain bonded through $NEt_2$ and C)

$$[Pt(CH_3)(MeOH)(Ph_3P)_2]^+PF_6^- + RNC \xrightarrow{MeOH} [Pt(CH_3)(RNC)(Ph_3P)_2]^+PF_6^- \tag{7}$$

$$\left[R^1R^2Pd\begin{matrix}thf\\NHEt_2\end{matrix}\right]^+ BF_4^- + MeNC \xrightarrow{thf} \left[R^1R^2Pd\begin{matrix}CNMe\\CNMe\end{matrix}\right]^+ BF_4^- + Et_2NH \tag{8}$$

$R^1R^2$ as in equation 6

themselves subject to attack by nucleophiles. This process forms the basis of a number of synthetically useful procedures and will be discussed below.

The cyanide ligand, $CN^-$, is also isoelectronic with carbon monoxide, but is a much stronger nucleophile and has considerably less $\pi$-acceptor ability than does carbon monoxide. Most metals in the *d* group form cyanide complexes[8], and cyanide anion is often capable of displacing *all* of the ligands from a given transition metal complex forming a homoleptic complex (e.g. $[Fe(CN)_6]^{4-}$). Perhaps because of this overwhelming ability to complex, the cyanide ligand has found little use in organotransition metal reactions.

### 2. $\eta^5$-Cyclopentadienyl- and $\eta^3$-allyl-metal complexes

The cyclopentadienyl group, $C_5H_5$, is central to much of organometallic chemistry, both because of its important role in the early development of the field (the discovery and characterization of ferocene, $[(\eta^5\text{-}C_5H_5)_2Fe]$[9]), and because virtually all transition metals form stable $\eta^5$-$C_5H_5$ complexes, many of which have important catalytic activity or illustrate an important feature of structure or bonding[10–15]. The most general method for the preparation of $\eta^5$-cyclopentadienyl complexes is the reaction of the cyclopentadienyl anion with a transition metal complex having displaceable ligands (equations 9[9], 10[13], and 11[14]). This reaction again can be viewed as a

$$\text{cyclo-}C_5H_6 + Na \longrightarrow Na^+Cp^- + FeCl_2 \longrightarrow [FeCp_2] \tag{9}$$

$$Na^+Cp^- \xrightarrow{W(CO)_6} [WCp(CO)_3] + 3CO \tag{10}$$

$$Na^+Cp^- + ZrCl_4 \longrightarrow [ZrCl_2Cp_2] \tag{11}$$

nucleophilic attack of the carbanion on the metal. Although in most cases the cyclopentadienyl ligand is pentahapto and occupies three coordination sites, some examples of $\sigma$-bonded monohapto cyclopentadienyl complexes are known[16]. An example of this type of complex is $[(\eta^5\text{-}C_5H_5)_3Zr(\eta^1\text{-}C_5H_5)]$[17], in which both types of cyclopentadienyl groups are present. This ability to bond either $\eta^5$ or $\eta^1$ and to fill either three or one coordination sites is important to some reactions catalysed by cyclopentadienyl metal complexes in that the $\eta^5$-bound ligand can 'release' coordination sites when required for catalysis by becoming $\eta^1$-bound.

The $\eta^3$-allyl ligand is another ubiquitous carbon ligand in organometallic chemistry, and a variety of $\eta^3$-allylmetal complexes are of central importance in a number of synthetically useful processes[18,19] The most general approach to homoleptic $\eta^3$-allylmetal complexes is the reaction of allylic Grignard reagents with the corresponding metal halide (equation 12). Homoleptic $\eta^3$-allyl complexes are known for Ni, Pd,

$$n(\text{allylMgX}) + MX_n \longrightarrow [(\eta^3\text{-allyl})_nM] + n(MgX_2) \tag{12}$$

Pt, Co, Rh, Ir, Fe, Cr, Mo, W, V, Nb, Ta, Ti, Zr, Hf, and Th[20]. Mixed systems containing both $\eta^3$-allyl groups and other ligands are similarly prepared (equation 13)[21,22]. These reactions involve direct nucleophilic attack on the metal by the allyl

$$[\text{RhCl(cod)}]_2 + 2CH_2{=}CHCH_2MgCl \longrightarrow 2[\text{Rh}(\eta^3\text{-allyl})(\text{cod})] \tag{13}$$

anion. (There are many other ways of making $\eta^3$-allylmetal complexes that do not involve nucleophilic attack on a transition metal, but they are outside the scope of this review.)

### 3. $\sigma$-Alkyl, aryl, and vinylmetal complexes

a. *Preparation.* $\sigma$-Alkyl, aryl, and vinyl transition metal complexes are among the most important types of complexes of use in synthetic organic chemistry. They can be prepared in a number of ways[23], but the most common involves nucleophilic attack of organo-lithium or -magnesium halide species on halogenometal complexes (equation 14). Many $\sigma$-alkylmetal complexes are relatively unstable, and initially it was

$$\underset{}{[L_nMX]} + \underset{(\text{MgX})}{\text{RLi}} \longrightarrow [L_nMR] + \underset{(MgX_2)}{\text{LiX}} \tag{14}$$

thought that metal—carbon $\sigma$-bonds were inherently weak. However, it was subsequently shown that this instability was a kinetic problem and was due to the existence of facile $\alpha$- or $\beta$-hydride elimination processes (equations 15 and 16) (see also

$$\text{M}{-}\text{C}({-}\text{H}) \rightleftharpoons \underset{\text{H}}{\underset{|}{\text{M}}}{=}\text{C} \tag{15}$$

$$\text{M}{-}\text{C}{-}\text{C}{-}\text{H} \rightleftharpoons \underset{\text{H}}{\underset{|}{\text{M}}}{-}\|(\text{C}{=}\text{C}) \tag{16}$$

Chapter 8). Thus, alkyl groups lacking $\beta$-hydrogens, or unable to generate stable olefins by $\beta$-elimination, are often fairly stable. Examples of this are aryl compounds (equations 17[24] and 18[25]) and acetylides (equations 19[25] and 20[26]) which lack

$$cis\text{-}[PtCl_2(PEt_3)_2] + PhLi \longrightarrow cis\text{-}[PtPh_2(PEt_3)_2] \quad (17)$$

$$trans\text{-}[IrCl(CO)(PPh_3)_2] + PhLi \text{ or } PhMgX \longrightarrow trans\text{-}[IrPh(CO)(PPh_3)_2] \quad (18)$$

$$\xrightarrow{PhC\equiv CLi} trans\text{-}[Ir(C\equiv CPh)(CO)(PPh_3)_2] \quad (19)$$

$$[FeCl(CO)_2(Cp)] + PhC\equiv CMgBr \longrightarrow [Fe(C\equiv CPh)(CO)_2(Cp)] \quad (20)$$

$$RLi, R = \text{bicyclo[2.2.1]hept-1-yl} + CrCl_3 \longrightarrow \xrightarrow{-e^-} [CrR_4] \quad (21)$$

$\beta$-hydrogens, and bridgehead norbornyl complexes (equation 21)[27] which cannot form a stable olefin. Other alkyl groups lacking $\beta$-hydrogens, such as methyl, neopentyl, trimethylsilylmethyl, and benzyl also form relatively stable $\sigma$-alkylmetal complexes.

Metallacyclic species whose ring strain prevents the achievement of a M—C—C—H dihedral angle of 0° required for $\beta$-elimination are also relatively stable compared to corresponding acyclic dialkylmetal complexes (equation 22)[28]. Although most $\sigma$-alkylorganometallic complexes have other ligands, homoleptic metal alkyl complexes are known for many members of the transition series[28–30].

$$cis\text{-}[PtCl_2(PPh_3)_2] + Li(CH_2)_4Li \longrightarrow cis\text{-}\left[(Ph_3P)_2Pt\langle\text{cyclo-}C_4H_8\rangle\right] \quad (22)$$

b. *Alkylation reactions of alkylrhodium(I) complexes.* Although stable $\sigma$-alkylmetal complexes have played an important role in the elucidation of structural and bonding characteristics of this large class of compounds, many of the less stable members of the family have found synthetic utility because of their reactivity. For example, Wilkinson's complex **1** reacts with methyllithium or methylmagnesium bromide to form the $\sigma$-alkylrhodium complex **2** (equation 23). Although unstable to decomposition via an oxidative addition/*ortho*-metallation/reductive elimination sequence, it can, with care, be isolated[31–33]. It is, however, reactive towards a number of organic compounds. Aryl and alkenyl halides undergo a facile oxidative addition/reduction elimination process to result in overall methylation of the organic (equation 24)[34].

$$\underset{(\mathbf{1})}{[RhCl(PPh_3)_3]} + CH_3Li \longrightarrow \underset{(\mathbf{2})}{[RhMe(PPh_3)_3]} \xrightarrow{\text{ox. addn}} \left[o\text{-}(Ph_2P)C_6H_4\text{-}Rh(H)(Me)(PPh_3)_2\right] \xrightarrow[\text{elim.}]{\text{red.}} CH_4 + \left[o\text{-}C_6H_4(PPh_2)\text{-}Rh(PPh_3)_2\right] \quad (23)$$

$$[RhMe(Ph_3P)_3] + RX \xrightarrow[60-90\ °C]{dmf} [Rh(X)(R)(Me)(PPh_3)_3] \longrightarrow RCH_3 + [RhX(PPh_3)_3] \quad (24)$$

R = Ph, *p*-tolyl, *p*-CNPh, *p*-$NO_2$Ph, *p*-MeCOPh, *p*-PhPh, *o*-$CO_2$MePh; X = Br, I

Diphenylacetylene undergoes an apparent insertion process to produce *trans*-$\alpha$-methylstilbene after hydrolysis (equation 25a). This stoichiometric reaction required severe conditions, in contrast to the related $[RhCl(PPh_3)_3]$-catalysed reaction of methylmagnesium halide with diphenylacetylene (equation 25b), which gives the

$$[RhCl(PPh_3)_3] + CH_3MgX \xrightarrow{(a)} [RhMe(PPh_3)_3] \xrightarrow[(2)\ H^+]{(1)\ PhC{\equiv}CPh,\ 130\,°C,\ 0.5\,h} trans\text{-}PhCH{=}C(Me)Ph \qquad (25)$$

$$(\mathbf{1}) \xrightarrow{(b)\ PhC{\equiv}CPh,\ 10\,°C,\ H^+} trans\text{-}PhCH{=}C(Me)Ph \qquad (\mathbf{2})$$

same product. Since the methylmagnesium halide does not itself react with diphenylacetylene, and since complex **2** is made from methylmagnesium bromide and $[RhCl(PPh_3)_3]$, it was assumed that both the stoichiometric and catalytic reactions proceded via the same reactive complex, **2**. However, the drastic disparity between minimum conditions for the two processes belies this assumption, and the two reactions must proceed by different mechanisms. This is one of many cases for which the *stoichiometric* reaction of a complex assumed to be the reactive intermediate in a catalytic process in fact shows different behaviour than the catalytic system itself.

Less stable alkyl- and aryl-rhodium(I) complexes result from the reaction of Grignard reagents or organolithium complexes with $[RhCl(CO)(PPh_3)_2]$, **3**. Although both the phenyl- (4) and methyl-rhodium(I) (**5**) complexes were unstable in the absence of solvent, evidence for their formation was obtained from solution i.r. spectroscopy[36]. Reaction of **3** with methyl- or phenyl-lithium produced complexes that had CO absorptions at 1962 and 1969 $cm^{-1}$, respectively, indicative of an alkylrhodium(I) species {The CO band of $[Rh(Cl)(CO)(PPh_3)_2]$ itself appears at 1980 $cm^{-1}$ under these conditions}. Reaction of the methyl complex **5** with CO led to the development of i.r. absorptions at 1983 and 1955 $cm^{-1}$ $[Rh^{(I)}CO]$, as well as a band at 1679 $cm^{-1}$ characteristic of an acetylrhodium(I) species (equation 26).

$$\underset{\nu_{CO}\ 1980\ cm^{-1}}{\underset{(\mathbf{3})}{trans\text{-}[RhCl(CO)(PPh_3)_2]}} + CH_3Li \xrightarrow[-78\,°C]{thf} \underset{\nu_{CO}\ 1962\ cm^{-1}}{\underset{(\mathbf{5})}{trans\text{-}[Rh(CO)(Me)(PPh_3)_2]}} \underset{}{\overset{CO}{\rightleftharpoons}} \underset{\nu_{CO}\ 1983,\ 1955,\ 1679\ cm^{-1}}{[Rh(COCH_3)(CO)_2(PPh_3)_2]} \qquad (26)$$

Although these complexes could not be isolated, they were synthetically useful nonetheless. The parent chloro compound **3** did not react with acid chlorides. Replacement of the chloro group by an (electron-donating) aryl or alkyl group (**4** or **5**) increased the reactivity of the complex, which then participated in an oxidative addition/reductive elimination process with acid halides to produce ketones in excellent yield (equation 27). $\sigma$-Alkyl complexes of secondary or tertiary alkyl groups were even less stable, and underwent a facile $\beta$-hydride elimination even at −78 °C. Readdition of the resulting rhodium(I) hydride to the olefin in the opposite sense led to the rearranged primary alkylrhodium(I) complex, which reacted with acid chlorides in the usual fashion. The existence of a rhodium(I) hydride species was demonstrated by trapping it with excess of hexene (equation 28).

$$\begin{array}{l} trans\text{-}[RhCl(CO)(PPh_3)_2] + RM \longrightarrow trans\text{-}[Rh(CO)(R)(PPh_3)_2] \\ \quad (\mathbf{3}) \qquad\qquad\qquad\qquad \downarrow R'COCl \\ \qquad\qquad\qquad trans\text{-}[RhCl(COR')(CO)(R)(PPh_3)_2] \\ \qquad\qquad\qquad\qquad \downarrow \\ \qquad\qquad\qquad\qquad \mathbf{3} + RCOR' \end{array} \tag{27}$$

R = Me, Ph, $n$-Bu, allyl, Et

R′ = $n-C_{11}H_{23}$, Ph, *trans*-PhCH=CH, Me, $ClCH_2$, $Bu^s$, ($S$)—(+)—$C_3H_7CH(CH_3)$, $n$-$C_6H_{16}CH(CH_3)CH_2$,

[structure: tetrahydroisoquinolinone with Ph, N–Me and C=O substituents]

$$\begin{array}{l} \text{2-butyllithium} + trans\text{-}[RhCl(CO)(PPh_3)_2] \\ \longrightarrow [MeCH_2CH(Me)Rh(CO)(PPh_3)_2] \xrightarrow{PhCOCl} CH_3CH(COPh)CH_2Me \\ [MeCH_2CH(Me)Rh(CO)(PPh_3)_2] \rightleftharpoons CH_2{=}CHCH_2Me + [RhH(CO)(PPh_3)_2] \\ [RhH(CO)(PPh_3)_2] \xrightarrow{CH_2=CH(CH_2)_3CH_3} [Rh\{(CH_2)_5Me\}(CO)(PPh_3)_2] \\ PhCO(CH_2)_5Me \xleftarrow{PhCOCl} [Rh\{(CH_2)_5Me\}(CO)(PPh_3)_2] \\ [RhH(CO)(PPh_3)_2] \rightleftharpoons [Rh(Bu)(CO)(PPh_3)_2] \longrightarrow PhCO(CH_2)_3Me \end{array} \tag{28}$$

c. *Alkylcopper complexes and copper-catalysed Grignard reactions.* Of all the $\sigma$-alkylmetal complexes available by the nucleophilic attack of carbanions on transition metal halogen complexes, organocopper(I) complexes, in all of their manifestations, are by far the most widely used in organic synthesis, as evidenced by the 80–100 papers that appear annually in this area. Most of these alkylcopper(I) complexes are not stable and rarely, if ever, are they isolated before use. Rather, they are generated *in situ,* and used immediately. The types of organocopper reagents extant are summarized in Table 1, along with their methods of preparation[37]. Each type of organocopper species has its own chemical reactivity, and the behaviour of one type towards a specific substrate may be different from that of another.

Organocopper species are extremely useful for two types of reactions: the alkylation of organic halides and tosylates (substitution reactions)[38] (equation 29) and the 1,4-alkylation of conjugated enones (equation 30)[39].

$$2RLi + CuX \xrightarrow{<0\,°C} R_2CuLi \xrightarrow{R'X} RR' \tag{29}$$

R = 1°, 2°, 3° alkyl, vinyl, aryl, heteroaryl

R′ = 1° > 2° ⋙ 3° alkyl, vinyl aryl, allyl, acyl

X = I > Br > Cl, OTs

TABLE 1. Summary of organocopper reagents

| Type | Preparation |
|---|---|
| RCu | $CuX + RMgX$ or $RLi \longrightarrow RCu + LiX$ or $MgX_2$ |
| $RCu{\cdot}L$ [$L = R_3P, R_2S, (RO)_3P$] | $LCuX + RMgX$ or $RLi \longrightarrow RCuL$ |
| $R_2CuLi$ or $R_2CuMgX$ | $CuX + 2RMgX$ or $2RLi \longrightarrow R_2CuMgX(Li)$ |
| $RR'CuLi$ or $RR'CuMgX$ | $CuX + RLi \longrightarrow RCu + R'Li$ or $R'MgX \longrightarrow RR'CuMgX(Li)$ |
| $R(Y)CuLi$ or $R(Y)CuMgX$ $Y = CN, OR', SR'$ | $CuY + RMgX$ or $RLi \longrightarrow R(Y)CuMgX(Li)$ |
| $R_nCuLi_{n-1}$ | $nRLi + CuX \longrightarrow R_nCuLi_{n-1} + LiX$ |

$$2RLi + CuX \xrightarrow{<0\,°C} R_2CuLi \xrightarrow{CH_2{=}CHCOY} RCH_2CH_2COY \quad (30)$$

$R = 1°, 2°, 3°$ alkyl, vinyl, aryl, allyl, heteroaryl

$Y = R', OR'$

The dialkylcuprates, $R_2CuLi$, have been the most extensively studied in this regard, but recently the other reagents have come under closer scrutiny. The area has been the subject of numerous recent reviews[40–47] and the interested reader is referred to these for more detailed information.

Much of the current work involving stoichiometric reactions of organocopper complexes had its genesis in the early observation that copper salts catalysed the 1,4-alkylation of conjugated enones by Grignard reagents[48]. It is now clear that many of these catalytic reactions proceed *through* organocopper complexes formed by nucleophilic attack of the carbanion on the copper metal centre and many copper-catalysed Grignard reactions are reproduced using stoichiometric organocopper reagents. At the same time, copper-catalysed Grignard reactions often behave differently from the same reaction using stoichiometric copper reagents[41]. For example, propargyl alcohols reacted with Grignard reagents in the presence of a copper(I) iodide catalyst to give an *anti* addition product (equation 31)[49], whereas the corresponding stoichiometric reagent gave the *syn* addition product (equation 32)[50]. The reasons for these differences are not clear. The course of the reaction

$$RC{\equiv}CCH_2OH \; (H) \xrightarrow[3\%\ CuI]{R'MgX} \text{[R(H), R' on C=C; Mg–O–CH}_2\text{ ring]} \xrightarrow{CO_2} \text{[R(H), R' on C=C; lactone O=C–O–CH}_2\text{]} \quad (31)$$

*anti*

$$HC{\equiv}CCH_2OH + RCu, MgX_2 \longrightarrow \xrightarrow{H_3O^+} RCH{=}CHCH_2OH \quad (32)$$

*syn*

depends heavily on substrate, Grignard reagent, counter ion, solvent, and temperature. The field is in a state of flux with a great deal of empirical work going on, but with little clear understanding of the important factors involved. A few of the more interesting and unusual copper-catalysed Grignard reactions are presented below.

The bromohydrin of isoprene reacted with Grignard reagents in the presence of copper(I) iodide to produce 2-alkylpen-4-en-2-ols in a process that involves an unexpected rearrangement (equation 33)[51]. In contrast, in the absence of a copper

$$CH_2{=}CHC(Me)(OH)CH_2Br + RMgX \xrightarrow[\text{cat.}]{\text{CuI}} CH_2{=}CHCH_2C(Me)(OH)R \quad (33)$$

(via R—Cu ... OH···Mg ... Br ?)

catalyst, 2-methyl-4-alkylbut-2-en-1-ols were produced, probably by an $S_N2'$ displacement of an intermediate epoxide (equation 34)[52]. Optically active propargyl ethers,

$$CH_2{=}CHC(Me)(OH)CH_2Br \xrightarrow{\text{RMgX}} [R^- \ \text{epoxide}] \longrightarrow RCH_2CH{=}C(Me)CH_2OH \quad (34)$$

esters, sulphones, and amines reacted with *n*-hexylmagnesium bromide in the presence of copper(I) bromide to give optically active allenes via an $S_N2'$ type displacement of the propargyl leaving group (equation 35)[53].

$$CH_3{-}C(OR)(H){-}C{\equiv}CH + n{-}C_6H_{13}MgBr \xrightarrow{\text{CuBr}} (Me)(H)C{=}C{=}C(H)(C_6H_{13}) \quad (35)$$

R = Me, Ac, $MeSO_2$, $NEt_2$
ee = 16, 79, 66, 16%

This propargyl-to-allene displacement is common both for organocuprate and copper-catalysed Grignard reactions. Propargyl[54] and allylic $\beta$-lactones[55] reacted similarly, undergoing an $S_N2'$ ring opening when exposed to Grignard reagents in the presence of copper catalysts (equation 36 and 37). In addition, the literature is rife with copper-catalysed displacement and conjugate addition reactions, as well as applications of these reactions to the synthesis of complex organic compounds. These areas are outside the scope of this chapter[47].

$$\text{2-ethynyl-4-oxyoxetane} + RMgX \xrightarrow{\text{CuI}} RHC{=}C{=}C(H)(CO_2H) \quad (36)$$

72–97%

R = Me, $Bu^n$, $sBu^s$, $Bu^t$, Ph, vinyl

$$\text{2-vinyl-4-oxyoxetane} + RMgX \xrightarrow{\text{CuI}} RCH_2CH{=}CHCH_2CO_2H \quad (37)$$

R = Et, Me, $Bu^n$, $Bu^s$, $Bu^i$, $Bu^t$, Ph, vinyl

d. *Metal-catalysed Grignard reactions.* Although copper-catalysed Grignard reactions have been known and studied for a long time, it is only recently that other transition metals have been examined for their effects on the course of the Grignard reaction[56]. Nickel(II) complexes have been investigated most extensively[57], and are best known for catalysing the cross-coupling of alkyl, aryl, and alkenyl Grignard reagents with aryl and alkenyl halides[58]. The reaction is thought to proceed by the sequence shown in Scheme 1, in which a bis-phosphine nickel complex undergoes

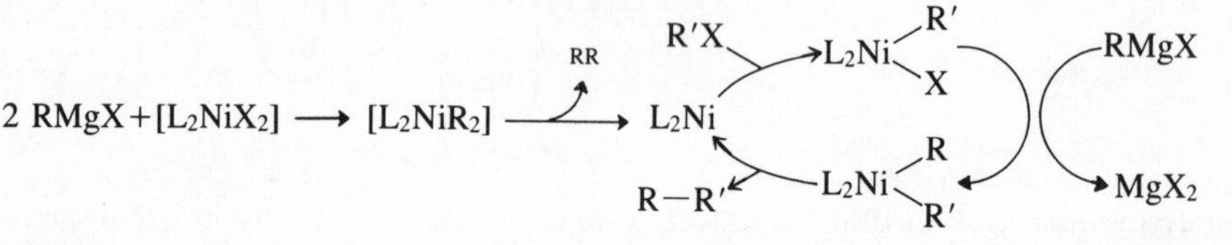

$$2\ RMgX + [L_2NiX_2] \longrightarrow [L_2NiR_2] \xrightarrow{-RR} L_2Ni \xrightarrow{R'X} L_2Ni(R')X \xrightarrow{RMgX,\ -MgX_2} L_2Ni(R)R' \longrightarrow R{-}R' + L_2Ni$$

SCHEME 1. Nickel-catalysed cross-coupling of Grignard reagents with halides.

direct alkylation (nucleophilic attack) by the Grignard reagent to give a dialkylnickel(II) complex, which decomposes to a nickel(0) species by reductive elimination of alkane. This nickel(0) species then undergoes an oxidative addition of the aryl or alkenyl halide to produce the catalytically active monoalkyl species $[L_2Ni(R')X]$. This again undergoes a nucleophilic displacement of halide by the Grignard reagent producing an unsymmetrical dialkylnickel(II) species. Reductive elimination forms the cross-coupled product and regenerates the nickel(0) complex to carry the catalytic cycle. This last step may be assisted by the incoming organic halide[59]. The key step, nucleophilic attack of the Grignard reagent on the $[L_2Ni(R')X]$ species is well documented in stoichiometric reactions of stable $\sigma$-arylnickel and palladium complexes with carbanions, and has been used in the synthesis of natural products (e.g. equation 38)[60].

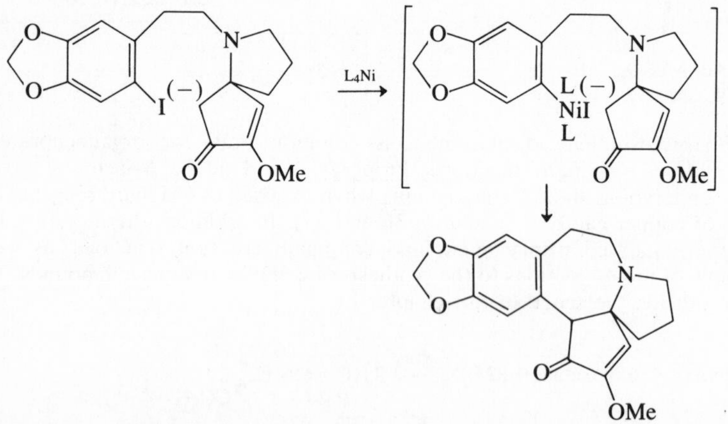

$$\xrightarrow{L_4Ni} \qquad (38)$$

The choice of the specific nickel—phosphine complex depends on the Grignard reagent and the organic halide substrate used. The complex $[NiCl_2\{Ph_2P(CH_2)_3PPh_2\}]$ was most effective for the coupling reactions of primary and secondary alkyl and aryl Grignard reagents, whereas the more basic complex $[NiCl_2\{Me_2P(CH_2)_2PMe_2\}]$ was the most suitable catalyst for use with allylic and vinylic Grignard reagents, and

$[NiCl_2(PPh_3)_2]$ was best for reactions of sterically hindered Grignard reagents or substrates. This is an extremely general process, with over 70 cases cited in the original paper[58] and hundreds more in the intervening period. One of the more interesting uses of this process is in the alkylation of heteroaromatic halides such as chloroquinolines (equation 39)[61] or chloroisoquinolines[62], chloropyridines (equation 40)[63], and chloropyridazines (equation 41)[64]. The use of bis-Grignard reagents with aromatic dihalides resulted in a cyclocoupling (equation 42)[65]. Bromoenol ethers reacted similarly, resulting in a two-carbon homologation of the Grignard reagent (equation 43)[66].

$$\text{2-chloroquinoline} + \text{RMgX} \xrightarrow{[L_2NiCl_2]} \text{2-R-quinoline} \quad (39)$$

$$\text{2-chloropyridine} + \text{RO(CH}_2)_n\text{MgX} \xrightarrow{[L_2NiCl_2]} \text{2-(CH}_2)_n\text{OR-pyridine} \quad (40)$$

$$\text{dichloropyridazine} + \text{RMgX} \xrightarrow{[L_2NiCl_2]} \text{R-X-pyridazine} \quad (41)$$

X = Me, Ph, MeO, piperid-1-yl

R = Me, Et, Ph, $\alpha$-naphthyl, $\alpha$-thienyl

$$\text{Cl–C}_4\text{H}_3\text{Y–Cl} + \text{XMg(CH}_2)_n\text{MgX} \xrightarrow[\text{dppe}]{\text{NiCl}_2} \text{C}_4\text{H}_3\text{Y(CH}_2)_n \quad (42)$$

Y = N; $n = 6-10$
Y = CH; $n = 8, 9, 10, 12$

$$\text{RMgBr} + \text{BrCH=CHOEt} \xrightarrow{[(dppe)NiCl_2]} \text{RCH=CHOEt} \quad (43)$$

Palladium(0) species and, less frequently, palladium(II) complexes have also been used to catalyse the reactions of alkenyl halides[67,68] and aryl halides[69,70] with aryl, benzyl, alkenyl, alkynyl, and methyl Grignard reagents. Palladium complexes are less useful than nickel salts for the coupling of alkyl Grignard reagents because alkylpalladium(II) complexes tend to undergo $\beta$-hydride elimination processes faster than do nickel(II) complexes. Nickel- and palladium-catalysed systems differ in other regards also. For example, with allylic halides as substrates, nickel catalysts gave exclusive alkylation at the more substituted position while palladium catalysts gave substitution at the less substituted position, regardless of the position of the halide in the starting substrate (equation 44)[71]. Palladium catalysts have found a number of useful synthetic applications recently (equations 45–47)[72–74].

$$\left.\begin{array}{l}\text{RCH=CHCH}_2\text{X}\\ \text{CH}_2\text{=CHCH(R)X}\end{array}\right\} + \text{PhMgBr} \begin{array}{l}\xrightarrow{[L_2NiX_2]} \text{CH}_2\text{=CHCH(R)Ph}\\ \xrightarrow{[L_2PdX_2]} \text{RCH=CHCH}_2\text{Ph}\end{array} \quad (44)$$

$$\left.\begin{array}{l}\text{R}^1\text{R}^2\text{C(X)C}\equiv\text{CH}\\ \text{R}^1\text{R}^2\text{C=C=CHX}\end{array}\right\} \xrightarrow[\text{Pd(0) cat}]{\text{RMgX}} \text{R}^1\text{R}^2\text{C=C=CHR} \quad (45)$$

$$R^1R^2C{=}CR^3X + \begin{matrix} RLi \\ \text{or} \\ RMgX \end{matrix} \xrightarrow{3\%[L_4Pd]\text{cat.}} R^1R^2C{=}CR^3R \qquad (46)$$

R = Me, Bu$^n$, *p*-tolyl, PhS, EtS, 3-furyl, 2-dimethylanilino 80–90%

99–100% isomeric purity

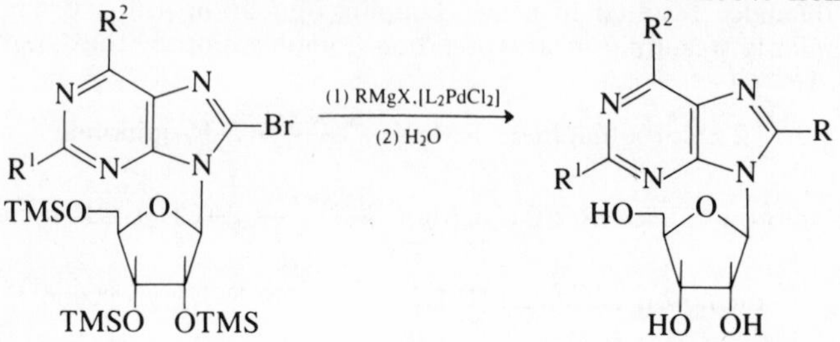

$$\text{(8-bromopurine nucleoside: } R^2, R^1, \text{TMSO, TMSO, OTMS)} \xrightarrow[(2)\ H_2O]{(1)\ RMgX.[L_2PdCl_2]} \text{(8-R purine nucleoside: } R^2, R^1, \text{HO, HO, OH)} \qquad (47)$$

15–35%

R = Me, allyl, Ph, $p$-MeOC$_6$H$_4$, $p$-FC$_6$H$_4$

A particularly useful modification of the nickel-catalysed coupling of Grignard reagents with halides involves the use of optically active phosphine ligands to induce chirality in the coupling process. The most extensively studied system was the coupling of vinyl bromide with the 1-phenethyl Grignard reagent (equation 48)[75], for

$$PhCH(Me)MgCl + CH_2{=}CHBr \xrightarrow[L^*]{NiCl_2} PhCH(Me)CH{=}CH_2 \qquad (48)$$

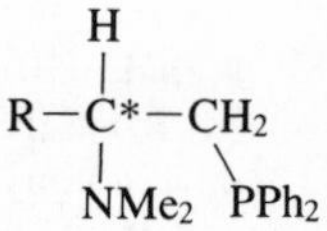

83–98% yield

up to 63% optical purity

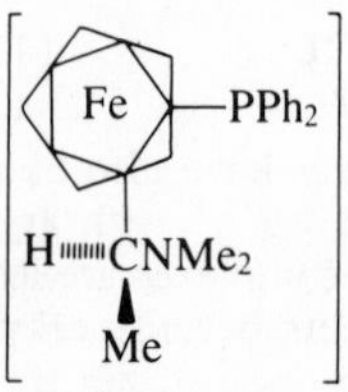

(**6**) (*S*)(*R*): [Fe, —PPh$_2$, H''''CNMe$_2$, Me]

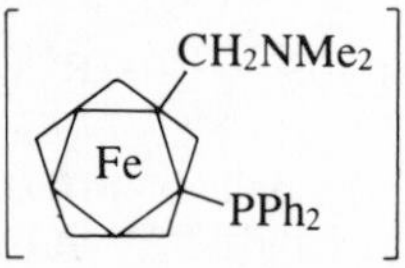

(**7**): [Fe, CH$_2$NMe$_2$, PPh$_2$]

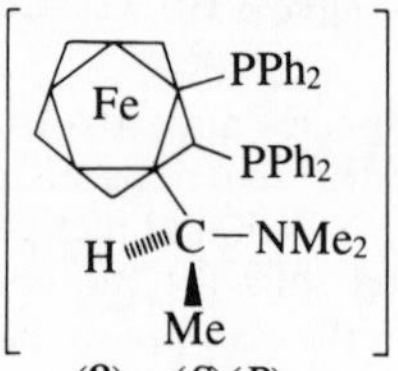

(**8**) (*S*)(*R*): [Fe, PPh$_2$, PPh$_2$, H''''C—NMe$_2$, Me]

(**9**): $R{-}\overset{H}{C^*}{-}CH_2PPh_2$ with NMe$_2$ on C*

which the unusual chiral ferrocenylphosphine ligands[76] **6–8** were most effective and gave products with optical purities of up to 63% (R or S depending on the ligand). Even higher optical purities (up to 94% ee) were obtained using chiral $\beta$-dimethylaminoalkyl phosphines such as **9**[77]. These reactions are actually kinetic resolutions of racemic Grignard reagents. Since the Grignard reagent undergoes inversion rapidly relative to coupling, high yields of chiral products can be obtained. Using this procedure, optically active $\alpha$-curcumene was synthesized (equation 49)[78].

Phosphine—nickel(II) complexes also catalyse the reaction between allylic alcohols and Grignard reagents lacking $\beta$-hydrogens. From the product distribution obtained with unsymmetrical allylic alcohols it was concluded that the reaction proceeded via a $\pi$-allylnickel intermediate as shown in Scheme 2[56a]. This proposed mechanism involves the rather unusual 'oxidative addition' of Grignard reagent to the nickel(0) species to give the key reactive intermediate [L$_2$Ni(R) MgX]. Alternatively, oxidative addition of the magnesium alkoxide to the nickel(0) complex, followed by nucleophilic attack of Grignard on the resulting $\pi$-allylnickel complex would result in the same overall process (equation 50). Whatever the case, the reaction is useful in synthesis (equation 51)[56a].

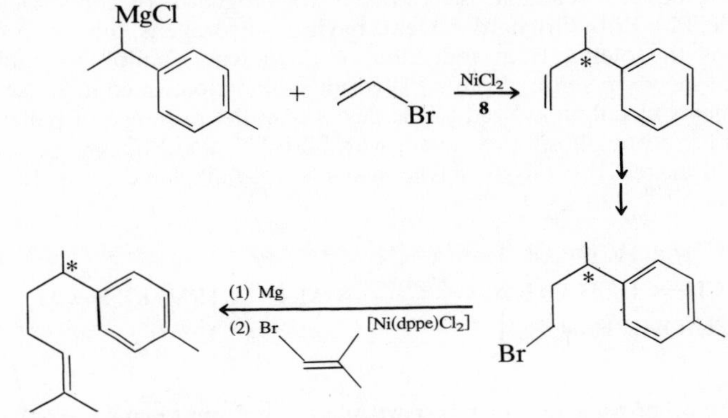

MgCl + Br —$NiCl_2$, **8**→ * ↓ ↓ * Br —(1) Mg (2) Br [Ni(dppe)$Cl_2$]→ *

55% optical purity

(49)

$[L_2NiX_2] + 2RMgX \longrightarrow [L_2NiR_2] \longrightarrow L_2Ni$ (−RR)

$L_2Ni$ + RMgX → $L_2Ni(R)MgX$

$L_2Ni(R)MgX$ → $R'CH{=}CHCH(OMgX)R''$ / $L_2Ni(R)MgX$ → $MgO + MgX_2$

$R'CH(R)CH{=}CHR'' + R'CH{=}CHCH(R)R''$

RMgX

$R'CH(R)CH{=}CHR'' + R'CH{=}CHCH(R)R''$

R′—(allyl)—R″, Ni, $L_2$, R

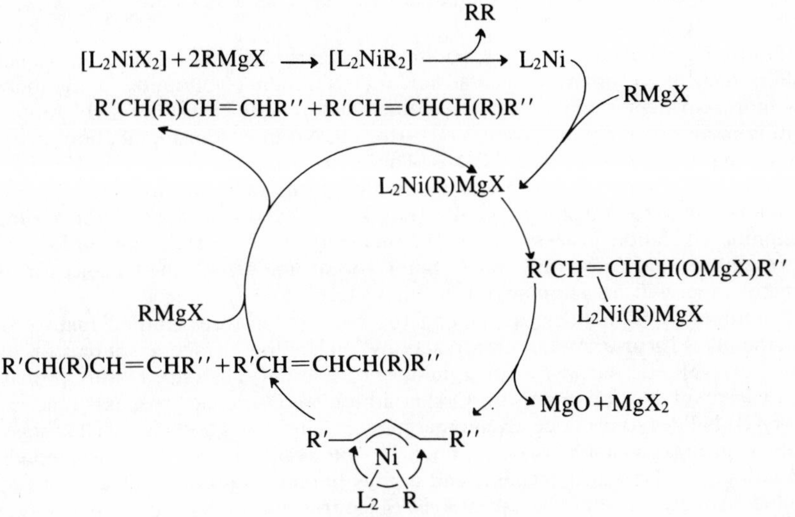

SCHEME 2

$$[NiL_2] + CH_2{=}CHCH_2OMgX \longrightarrow [Ni(L)(OMgX)(\eta^3\text{—allyl})] \xrightarrow{RMgX} [Ni(L)(R)(\eta^3\text{—allyl})] \quad (50)$$

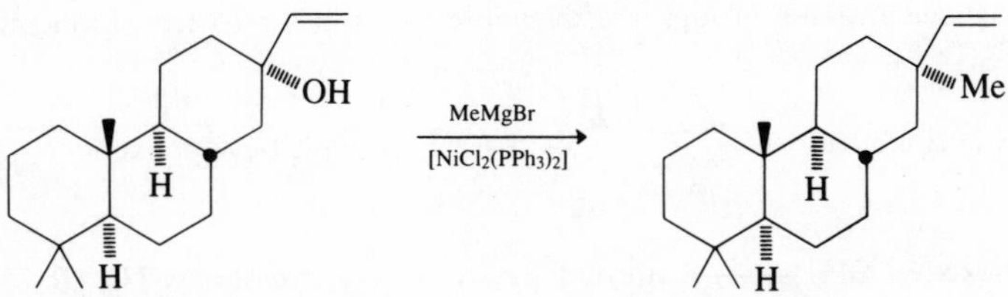

OH, H, H —MeMgBr, [$NiCl_2(PPh_3)_2$]→ Me, H, H (51)

65%

The nickel-catalysed reaction is restricted to Grignard reagents lacking $\beta$-hydrogens (Me, Ph). With Grignard reagents having $\beta$-hydrogens, the sole product of the reaction is the olefin, from reduction of the allylic alcohol, probably by a nickel–hydride species resulting from a $\beta$-hydride elimination. In contrast, secondary Grignard reagents cleanly alkylated allylic alcohols in the presence of palladium(II) complexes of ferrocenylphosphines, again with both $S_N2$ and $S_N2'$ regiochemistry[79]. By using chiral nickel(II) catalysts, allylic alcohols were alkylated in modest optical yield (equation 52)[80,81].

$$\left.\begin{array}{r} trans\text{-}MeCH_2CH{=}CHCH_2OH \\ cis\text{-}MeCH_2CH{=}CHCH_2OH \\ MeCH_2CH(OH)CH{=}CH_2 \end{array}\right\} \xrightarrow[\text{[NiCl}_2\{(-)\text{diop}\}]]{\text{MeMgBr}} \begin{array}{c} MeCH_2CH(Me)CH{=}CH_2 + \text{isomers} \\ 15\%\ ee \end{array} \tag{52}$$

A number of other transition metals, particularly iron[82] and manganese[83], catalyse various Grignard reactions. However, they have been little studied and the role of nucleophilic attack on transition metals in these systems is unclear.

e. *Transmetallation reactions.* In addition to clearly anionic main group organometallics such as Grignard reagents and organolithium compounds, a number of much more covalent main group organometallics readily transfer alkyl groups to transition metals (usually of Group VIII—Rh, Pd, Ni) in a 'transmetallation' process of increasing utility in organic synthesis. The main group metals include Hg, Sn, B, Zn, Al, and $d^0$ complexes of Zr and Ti. Although the mechanism of alkyl transfer from main group to transition metal is not known, and is probably not a simple nucleophilic alkylation process, the gross chemistry of these transmetallation reactions resembles that presented above. Some recent and significant advances in this very active area will be summarized in this section.

Organomercury(II) halides are among the most extensively studied main group organometallics for transmetallation reactions[84]. They are attractive sources of alkyl groups (for synthetic purposes) for a number of reasons. They are readily prepared from a variety of starting materials. Organolithium and Grignard reagents react with mercury(II) halides to produce organomercury(II) halides in good yield. The range of functional groups available is determined by the availability of the corresponding organolithium or Grignard reagent, and is thus limited to groups stable to strongly basic and strongly nucleophilic carbanions. Since transition metals directly catalyse a number of synthetically useful reactions of organolithium and Grignard reagents, the preparation of organomercury(II) halides from these precursors is of use only for processes in which the organolithium or Grignard reagents themselves reduce the transition metal rather than alkylate it. A much more generally useful source of organomercurials is organoboranes made by hydroboration of alkenes or alkynes. Organoboranes undergo a rapid reaction with mercury(II) salts to produce organomercurials in good yield. Since both the hydroboration and the mercuration reactions tolerate a wide range of functional groups, and are often regio- and stereo-specific, highly functionalized organic groups are available for transfer to transition metals (equations 53[85], 54[86], and 55[87]).

$$\text{5-vinylcyclohexene} \xrightarrow[\substack{(2)\ Hg(OAc)_2 \\ (3)\ NaCl}]{(1)\ HB(cyclo\text{-}C_6H_{11})_2} \text{5-}(CH_2CH_2HgCl)\text{cyclohexene} \tag{53}$$

$$3\ \text{cyclohexene} + BH_3 \longrightarrow B(\text{cyclohexyl})_3 \xrightarrow{Hg(OR)_2} \text{cyclohexyl-HgOR} \tag{54}$$

$$RC\equiv C(H \text{ or } R') + R''BH \longrightarrow RHC{=}C(H \text{ or } R')BR'' \xrightarrow{HgX_2} RHC{=}C(H \text{ or } R')HgX \quad (55)$$

$R'' =$ (1,2-dioxyphenylene, catechol-derived ring)

*β*-Substituted organomercurials are *directly* available from alkenes or alkynes by 'solvomercuration reactions'. Some useful examples are given in equations 55a–58.

$$RCH{=}CH_2 + HgX_2 + YH \longrightarrow RCH(Y)CH_2HgX \quad (55a)$$

$Y = Cl$[88], OAc[89], OR′[90], $NR_2'$[91]

$$MeC\equiv CMe \xrightarrow{Hg(OAc)_2} MeC(OAc){=}C(Me)HgOAc \quad (56)$$[92]

*cis* and *trans*

$$R_2C(OH)C\equiv CH \xrightarrow[NaCl]{HgCl_2} R_2C(OH)C(Cl){=}CHHgCl \quad (57)$$[93]

$$RC\equiv CCOX \xrightarrow{HgCl_2} RC(Cl){=}C(HgCl)COX \quad (58)$$[94]

Aromatic mercurial halides are available by direct electrophilic aromatic substitution (equation 59)[95], a process particularly useful for the preparation of heteroaromatic derivatives (equation 60)[96].

$$PhX + HgY_2 \longrightarrow XC_6H_4HgY + HY \quad (59)$$

(1-R-uracil, 5-H) $\xrightarrow{HgCl_2}$ (1-R-uracil, 5-HgCl) (60)

All of these organomercurial complexes transfer their organic groups to a variety of transition metals, forming alkyl or aryl transition metal complexes of synthetic utility. One of the earliest uses of transmetallation from mercury to palladium was the palladium(II)-assisted coupling of arylmercury(II) halides to biaryls (equation 61)[97]. The reaction is thought to proceed by a double arylation of palladium followed

$$ArHgX + PdX_2 \longrightarrow ArPdX + HgX_2 \xrightarrow{ArHgX} ArPdAr + HgX_2 \longrightarrow ArAr + Pd(0) \quad (61)$$

by reductive elimination of the biaryl. Mercurated pyrimidines couple in a similar manner, but two positional isomers are obtained, the ratio depending on counter ion used (equation 62)[98]. Vinylmercury(II) halides couple to form 1,3-dienes resulting

$$\text{(1,3-dimethyluracil-5-yl)HgX} \xrightarrow{PdX_2} \text{5,5'-bi(1,3-dimethyluracil)} + \text{isomeric bi(1,3-dimethyluracil)} \qquad (62)$$

X=OAc 89% 11%
X=Cl 28% 72%

from head-to-head or head-to-tail coupling, depending on reaction conditions (equation 63)[99]. Vinyl carboxylates form in the palladium-catalysed coupling of vinyl-

$$RC{\equiv}C(H \text{ or } R') + R''_2BH + Hg(OAc)_2 \xrightarrow{NaCl} RHC{=}C(H \text{ or } R')HgCl$$

$$RHC{=}C(H \text{ or } R')HgCl \xrightarrow{PdCl_2,\ PhH} RHC{=}C(H \text{ or } R')CH{=}CH(H \text{ or } R')$$

$$RHC{=}C(H \text{ or } R')HgCl \xrightarrow[hmpa]{Li_2[PdCl_4]} RHC{=}C(H \text{ or } R')C(H \text{ or } R'){=}CHR \qquad (63)$$

mercurials with mercury carboxylates (equation 64)[100]. Organomercurials can also transmetallate to methyl rhodium(III) complexes, resulting in methyl transfer (equation 65)[101].

$$RR''C{=}C(R')HgX + Hg(O_2CR''')_2 \xrightarrow{Pd(OAc)_2} RR''C{=}C(R')O_2CR''' \qquad (64)$$

$R = Bu^t, Ph, OAc, Bu^n$
$R' = H, Me, Ph, p\text{-tolyl}, Et$
$R'' = H, Et, OAc$
$R''' = Me, Ph, n\text{-Pr}$

$$[MeRhX_2L_n] + R'HgX \longrightarrow [MeRh(R')XL_n] \longrightarrow MeR' + [L_nRhX] \qquad (65)$$

$R' = PhCH{=}CH, Ar, Bu^tC{\equiv}C$

Aryl- and vinyl-palladium complexes formed from organomercury(II) halides undergo a number of synthetically useful insertion reactions, leading to functionalization of the organic group. Vinylmercury(II) halides, available from alkynes both regio- and stereo-specifically, are readily carbonylated by reaction with carbon monoxide in the presence of a palladium(II) catalyst. The key steps in the process are the transmetallation from mercury to palladium(II) and insertion of carbon monoxide into the $\sigma$-alkylpalladium complex (equation 66)[102]. Propargyl alcohols can be converted to butenolides using this chemistry (equation 67)[103].

$$\mathrm{RCH{=}CHHgCl + Pd(II) \longrightarrow [RCH{=}CHPdCl] \xrightarrow{CO} [RCH{=}CHC(O)PdCl] \xrightarrow{MeOH} Pd(0) + RCH{=}CHC(O)OMe}$$

(Pd(0) → Pd(II): oxidize) (66)

$$\mathrm{R_2C(OH)C{\equiv}CH \xrightarrow[NaCl\ H_2O]{HgCl_2} R_2C(OH)C(Cl){=}CHHgCl \xrightarrow[Li_2[PdCl_4]]{CO}} \text{4-chloro-5,5-dialkylcyclopent-2-enone-type product (Cl, R, R, =O)} \quad (67)$$

Olefins also readily insert into palladium—carbon $\sigma$-bonds, permitting alkylation of olefins by organomercurials. This chemistry is normally restricted to aryl- or vinyl-mercurials since alkyl groups having $\beta$-hydrogens undergo a competitive $\beta$-hydride elimination on transmetallation to palladium. Arylmercury(II) halides readily arylate olefins, particularly electrophilic ones, in the presence of palladium(II) complexes by a transmetallation/insertion/elimination sequence (equation 68)[104]. With allylic halides or acetates, the halide or acetate is preferentially eliminated, leading to arylation with allylic transposition (equation 69)[105]. This reaction has

$$\mathrm{ArHgX + PdCl_2 \longrightarrow ArPdX \xrightarrow{CH_2{=}CHY} ArCH_2CH(Y)PdCl \xrightarrow{-PdH} ArCH{=}CHY + Pd(0) + HCl} \quad (68)$$

$$\mathrm{ArHgX + PdCl_2(catalyst) \longrightarrow ArPdCl + R'CH{=}CHCH_2X \longrightarrow R'CH(Ar)CH(PdX)CH_2X \xrightarrow{-PdX_2} R'CH(Ar)CH{=}CH_2} \quad (69)$$

found extensive application in nucleoside and related heterocyclic chemistry, since direct mercuration of these substrates is possible. Thus, a number of 5-substituted nucleosides, normally difficult to synthesize, have been prepared in good yield by transmetallation to palladium(II) followed by insertion. Saturated side-chains are obtained under stoichiometric (in palladium) conditions with a reductive isolation (equation 70a)[106], while unsaturated side-chains are introduced under catalytic conditions using copper(I) as a re-oxidant (equations 70b[107] and 70c[108]). Allylic halides and acetates also insert (equation 70d[109]), as do enol ethers and acetates (equation 70e[110]). Mercurated purines undergo similar insertion reactions[110d]. *N*-Methylisoquinolines were alkylated in the 4-position using this chemistry (equation 71)[111].

Organotin compounds also readily transmetallate to palladium(II) complexes, providing another source of active alkyl groups for organopalladium chemistry. A very efficient conversion of acid chlorides to ketones, involving transmetallation from tin to palladium as a key step, has been developed (equation 72)[112]. This process goes in very high yield, and tolerates virtually all functional groups (including aldehydes), making it a very general process. Any substrate which can undergo

$$\text{RHgCl} \xrightarrow[\text{(2) NaBH}_4/\text{H}_2]{\text{(1) CH}_2\text{=CH}_2/\text{Li}_2[\text{PdCl}_4]} \text{REt} \quad (70a)$$

$$\text{RHgCl} \xrightarrow[\text{Li}_2[\text{PdCl}_4]]{\text{CH}_2\text{=CO}_2\text{Me}} \text{RCH=CHCO}_2\text{Me} \quad (70b)$$

$$\text{RHgCl} \xrightarrow[\text{Li}_2[\text{PdCl}_4]\text{CuCl}]{\text{YC}_6\text{H}_4\text{CH=CH}_2} \text{RCH=CHC}_6\text{H}_4\text{Y} \quad (70c)$$

(70d)

(70e)

R = (R″)HN ... O, N, R′

R′ = Me, deoxyribose

(71)

(1) $Hg(OAc)_2$
(2) Pd(II), $CH_2{=}CHX$

CH=CHX

$X = CO_2Me$, Ar    41–78%

oxidative addition to palladium(0) complexes can in principle be alkylated by organotin compounds via the same type of transmetallation process. Aromatic[113,114] and benzylic[113] halides were cleanly alkylated by methyl-, phenyl-, benzyl-, vinyl-, *n*-butyl-, and phenylacetylide-tin reagents in the presence of palladium catalysts; allyl bromides reacted with allyltin reagents to give cross-coupled products without allylic transposition in the allyl halide partner but with predominate allylic rearrangement from the tin partner (equation 73)[115]. Although all of these processes were thought to proceed by the oxidative addition/transmetallation/reductive elimination sequence presented in equation 72, recent mechanistic studies[116] have shown the process to be considerably more complex, with several competing transmetallations, and a reductive elimination step preferentially from a palladium(IV) intermediate!

$$\text{L}_2\text{Pd(0)} + \text{RCOCl} \xrightarrow[\text{addn.}]{\text{ox.}} \text{RC(O)Pd(L}_2\text{)Cl} \xrightarrow{\text{R}'_4\text{Sn}} \text{RC(O)Pd(L}_2\text{)R}' \xrightarrow{\text{red. elim.}} \text{L}_2\text{Pd(0)} + \text{RCOR}' \quad (72)$$

$$RCH_2C(Me){=}CHCH_2Br + MeCH{=}CHCH_2SnR'_3 \xrightarrow{\text{Pd cat.}} RCH_2C(Me){=}CHCH_2CH(Me)CH{=}CH_2 \quad (73)$$

Transmetallation from tin to palladium or nickel has also been used to prepare ketones from aryl and benzyl halides and $\alpha$-bromoesters (equation 74)[117,118]. This

$$RX + M(0) \longrightarrow RMX \xrightarrow{CO} RC(O)MX \xrightarrow{Me_4Sn} RC(O)MMe \longrightarrow M(0) + RCOMe \quad (74)$$

R = Ph, *p*-tolyl, *p*-anisyl, *p*-$CO_2EtPh$, *p*-CNPh, 2-thiophene, $PhCHCH_3$, PhCHEt, $CH_3CHCO_2Et$; M(0) = $[L_2PdCl_2]$ + CO, $[Ni(CO)_4]$, $[LNi(CO)_3]$, $[L_2Ni(CO)_2]$, $[L_2NiCl_2]$ + CO

process involves an oxidative addition/CO insertion/transmetallation/reductive elimination sequence (in all cases such as this, for which zerovalent metal catalysts are required, but divalent catalyst precursors are used, reduction of the metal by carbon monoxide or the main group organometallic is a presumed first step). Olefin insertion into a palladium—alkyl species formed by transmetallation from tin has been used to form carbocyclic ring systems (equations 75 and 76)[119]. In these cases, the olefin is a

Sme, $SnMe_3$ — Pd(II) cat. → [SMe, Pd, Cl]; $HgX_2$ → $Hg(SMe)_2$; ↓ [MeS, PdCl]; −PdSMe; ↓ 90%

SMe, $SnMe_3$ — $PdCl_2$, $HgCl_2$ → $CH_2$, 59% (76)

vinyl sulphide and a palladium–sulphur complex is eliminated in the final step. To prevent the loss of catalyst by tight complexation to sulphur, mercury salts were added as a sulphur scavenger.

One synthetic limitation of the tin-to-palladium transmetallation process is the difficulty in preparing functionalized tin reagents. The development of procedures permitting direct transmetallation from boron to palladium has greatly expanded the utility of these transmetallation reactions because of the wide availability of organoboranes from hydroboration reactions. Thus, vinylboranes, from hydroboration of alkynes, couple with vinyl halides in the presence of palladium(0) catalyst (equation 77)[120,121]. Aryl halides react in a similar fashion to form styrenes[122], and benzyl

$$Br(R^3)C{=}CR^4R^5 + [L_4Pd] \longrightarrow [PdBr(L_2)\{C(R^3){=}CR^4R^5\}] \quad (77)$$

$$[PdBr(L_2)\{C(R^3){=}CR^4R^5\}] \xrightarrow{R^1R^2C{=}CHBX_2} [Pd(L_2)\{CH{=}CR^1R^2\}\{C(R^3){=}CR^4R^5\}]$$

$$[Pd(L_2)\{CH{=}CR^1R^2\}\{C(R^3){=}CR^4R^5\}] \longrightarrow R^1R^2C{=}CHC(R^3){=}CR^4R^5 \; (\to [L_4Pd])$$

>98% isomeric purity
42–86% yield

$R^1$ = Bu$^n$, H, Ph, cyclohex-1-enyl
$R^2$ = H, Bu$^n$
$R^3$ = H, Ph
$R^4$ = H, Ph
$R^5$ = Ph, H, *n*-hexyl

halides form allylbenzenes[123]. Allylic halides couple to form non-conjugated dienes (equation 78)[124]. Biaryls are available by the palladium(0)-catalysed coupling of aryl

$$RC{\equiv}CH + R'_2BH \longrightarrow RHC{=}CHBR'_2 \xrightarrow[Pd(OAc)_2]{CH_2{=}CHCH_2Cl} RHC{=}CHCH_2CH{=}CH_2 \quad (78)$$

halides with phenylboronic acid (equation 79)[125]. In all of these reactions, transmetallation from boron to an alkylpalladium halide complex is a key step. Vinylboranes can be carbonylated by transmetallation to palladium chloride itself, followed by a carbon monoxide insertion reaction (equation 80)[126].

$$PhB(OH)_2 + ZC_6H_4Br \xrightarrow[aq.\ Na_2CO_3]{[PdL_4]} PhC_6H_4Z \quad (79)$$

Z = Me, Cl, α-naphthyl; 40–95%

$$RC{\equiv}CR' + R''H \longrightarrow RHC{=}C(R')R''$$

$$RHC{=}C(R')R'' \xrightarrow[\text{CO, MeOH}]{\text{PdCl}_2\text{ cat., NaOAc, benzoquinone}} RHC{=}C(R')CO_2Me \quad (80)$$

60–95%

R = Bu$^n$, *n*-hexyl, Et, Ph, HC≡C$(CH_2)_4$
R′ = H, Et, tms

R″ = catecholboryl ($C_6H_4O_2B$–)

Aluminium alkyls have also been used as alkyl sources in transmetallation reactions. Enol phosphates are alkylated by trialkylaluminium compounds in the presence of a palladium(0) catalyst, probably by a mechanism similar to that observed in the reactions of vinyl halides with palladium(0) catalysts and other main group organometallics discussed above (equation 81)[126]. When this procedure is applied to mixed O—S acetals, a formal conversion of aldehydes to ketones can be achieved (equation 82). This chemistry also provides a procedure for ketone transposition (equation 83)[127].

$$\mathrm{RC({=}CH_2)OP(O)(OPh)_2 + [PdL_4] \longrightarrow [Pd(L_2)\{OP(O)(OPh)_2\}\{RC({=}CH_2)\}] \xrightarrow{R'_3Al} [Pd(L_2)(R')\{RC({=}CH_2)\}] \longrightarrow [PdL_4] + \underset{90\%}{RC({=}CH_2)R'}} \quad (81)$$

R = *n*-decyl, Ph, 4-$Bu^t$-cyclohexen-1-yl
R′ = Me, Et, PhC≡C, *n*-$C_5$C≡C, *n*-$C_5$CH=CH

$$\mathrm{R'CH{=}C(OP(O)(OPh)_2)(SPh) \xrightarrow[\text{[PdL}_4\text{]cat.}]{R_3Al} \underset{60\text{–}80\%}{R'CH{=}C(R)(SPH)} \longrightarrow \longrightarrow R'CH_2COR} \quad (82)$$

R′ = $Pr^n$, Ph; R = Me, Et, PhC≡C, *n*-$C_5$—C≡C

$$\text{4-Bu}^t\text{-cyclohexan-1-one} \xrightarrow[\text{(2) NaH, ClP(O)(OPh)}_2]{\text{(1) LDA, PhSSPh}} \text{1-[OP(O)(OPh)}_2\text{]-2-SPh-4-Bu}^t\text{-cyclohex-1-ene} \xrightarrow[\text{[PdL}_4\text{]cat.}]{\text{Me}_3\text{Al}} \text{1-Me-2-SPh-4-Bu}^t\text{-cyclohex-1-ene} \xrightarrow{\text{TiCl}_4 \,|\, \text{H}_2\text{O}} \text{2-Me-5-Bu}^t\text{-cyclohexan-1-one} \quad (83)$$

Vinylalanes resulting from hydroalumination or zirconium-catalysed carboalumination of *terminal* acetylenes[128] couple to aryl halides[129], vinyl halides (equation 84)[130], and allyl halides (equation 85) in excellent yield in the presence of palladium(0) catalysts. The stereochemistry of the vinyl groups and both the regio- and stereo-chemistry of the allyl groups are maintained. However, the reaction failed utterly when internal alkynes were used as alane precursors. This difficulty can be

$$\mathrm{RC{\equiv}CH + (Bu^i)_2AlH \longrightarrow \underset{(\mathit{trans})}{RHC{=}CHAl(Bu^i)_2} \xrightarrow[\text{[PdL}_4\text{]}]{R'CH{=}CHI} RHC{=}CHCH{=}CHR'} \quad (84)$$

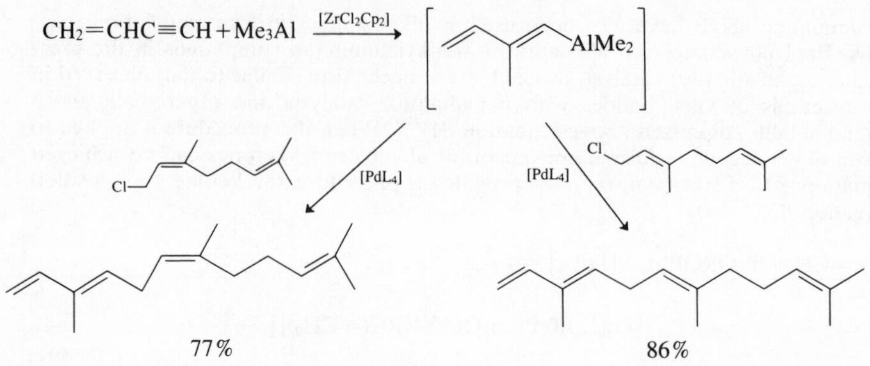


overcome by another transmetallation reaction, this time from aluminium to zinc. For reasons as yet unclear, organozinc halides are by far the most efficient main group organometallic compounds in transmetallation reactions to palladium(II)—alkyl complexes. Since the requisite zinc reagents are themselves readily available from most other organometallic reagents (e.g. Mg, Li, Al, Zr) this is an extremely useful discovery. Thus aryl, alkynyl, and alkenyl halides couple readily to disubstituted vinylalanes in the presence of zinc chloride and palladium(0) catalyst (equation 86)[129]. Allyl benzenes are available either from benzyl halides and vinyl alanes or from vinyl halides and benzylzinc halides (equation 87)[132], styrenes from aryl halides

$$\mathrm{RC{\equiv}CR' + Bu^i{}_2AlH \longrightarrow RHC{=}C(R')AlBu^i{}_2 \xrightarrow{ZnCl_2} RHC{=}C(R')ZnCl}$$

$$\mathrm{R''X + [PdL_4] \longrightarrow [Pd(L_2)(R'')] \longrightarrow \downarrow} \qquad (86)$$

$$\mathrm{RHC{=}CR'R'' + [PdL_4] \longleftarrow [Pd(L_2)(R'')\{C(R'){=}CHR\}]}$$

R″ = aryl, vinyl, alkynyl

$$\mathrm{ArCH_2ZnX + R^1C(X){=}CR^2R^3 \xrightarrow{[PdL_4]} \underset{90\%}{R^1C(CH_2Ar){=}CR^2R^3}} \qquad (87)$$

$$\mathrm{ArCH_2X + R^1C(R_2Al){=}CR^2R^3 \xrightarrow[{[PdL_4]}]{ZnCl_2} \uparrow}$$

and vinylcuprates (equation 88)[133], 1,3-dienes from vinyl halides and vinyl cuprates (equation 89)[134], alkynylpyrimidines from iodopyrimidines and zinc acetylides (equation 90)[135], $\beta,\gamma$-unsaturated esters from vinyl halides and Reformatsky reagents (equation 91)[136], and substituted allenes from propargyl or allenyl halides and alkylzinc chlorides (equation 92)[137].

$$\underset{\text{(cis)}}{\mathrm{(RCH{=}CH)_2CuLi}} \xrightarrow[\text{(2) 5\%[PdL}_4\text{]; (3) ArI}]{\text{(1) ZnBr}_2\text{/thf}} \underset{65\text{–}80\%}{\mathrm{RCH{=}CHC_6H_4Y}} \qquad (88)$$

Y = H, *p*-Br, *p*-OMe, *p*-$NO_2$, *p*-$CO_2Me$, *o*-$CO_2Me$

$$\mathrm{(R^1R^2C{=}CH)_2CuLi + XCH{=}CR^3R^4 \xrightarrow[5\%[PdL_4]]{ZnX_2} R^1R^2C{=}CHCH{=}CR^3R^4}$$

82–94%
>97% isomeric purity (89)

$$RC{\equiv}CZnCl + \text{(5-iodouracil, N-P)} \xrightarrow[\text{or Pd(0) cat.}]{\text{Ni(0)}} \text{(5-(C}{\equiv}\text{CR)uracil, N-P)} \quad 35{-}90\% \qquad (90)$$

$$RCH{=}CHBr + BrZnCH_2CO_2Et \xrightarrow[{[PdL_4]}]{[NiL_4]\text{ or}} RCH{=}CHCH_2CO_2Et \quad 70{-}96\% \qquad (91)$$

R=H, Me, Ph

$$RR^1C{=}C{=}C{\sim\sim}X + RZnCl \text{ or } RR^1C(X)C{\equiv}CH \xrightarrow[\text{cat.}]{[PdL_4]} RR^1C{=}C{=}CHR \quad 80{-}90\% \qquad (92)$$

A particularly appealing feature of these reactions is their tolerance to a variety of functional groups, and the specificity of the coupling reactions. These features have permitted them to be used to synthesize a number of polyunsaturated natural products (equations 93[138,139], 94[140], and 95[141]).

The final transmetallation to be discussed is that from zirconium to palladium. It is not at all clear that this is nucleophilic process, but it is included here for completeness. It is important because of the generality of the hydrozirconation process, which generates vinylzirconates from alkynes[142]. Vinyl zirconates couple with vinyl halides

(93): R–CH₂CH₂C≡CH → Me₃Al, [Cp₂ZrCl₂]; I₂ → vinyl iodide; [L₄Pd], tms—C≡C–CH₂CH₂ZnCl → enyne-tms product, 90%

(94): polyenyl bromide → (1) Mg, thf (2) ZnBr₂ (3) Br-butenolide → butenolide product, 62%

Me$_3$Al [$Cp_2ZrCl_2$] AeMe$_2$ I$_2$ I [$L_4Pd$], $ZnCl_2$, 50–60% Otms ClZn [$L_4Pd$] Otms 80–85% OH (95)

in the presence of palladium(0) catalysts[129], in the same fashion as do vinylalanes. This reaction is fairly tolerant of functional groups (equation 96)[143]. $\pi$-Allyl-

$$[ZrCl(Cp)_2(trans\text{-}CH{=}CHCH_2OR)] + trans\text{-}BrCH{=}C(Me)CO_2Me \xrightarrow{[PdL_4]} trans\text{-}ROCH_2CH{=}CHCH{=}C(Me)CO_2Me \quad (96)$$

R = tetrahydropyran-2-yl

palladium complexes react with vinyl zirconates to result in coupling (equation 97)[144]. The regiochemistry of this reaction is controlled by the ligands on palladium, with phosphines leading to alkylation at the most substituted allyl position and maleic anhydride at the least. This reaction has been used to introduce steroidal side-chains[144b].

$$R^1CH{=}C(R^2)CH_2X \xrightarrow[Pd(0)]{thf} [PdX(\eta^3\text{-}1\text{-}R^1\text{-}2\text{-}R^2\text{-allyl})]_2 \xrightarrow{[ZrCl(Cp)_2(CH{=}CHR^3)]} [PdX(CH{=}CHR^3)(\eta^3\text{-}1\text{-}R^1\text{-}2\text{-}R^2\text{-allyl})] \longrightarrow R^1CH{=}C(R^2)CH_2CH{=}CHR^3 + CH_2{=}C(R^2)CH(R^1)CH{=}CHR^3 \quad (97)$$

## B. Formation of Transition Metal Hydrides by Nucleophilic Attack by Hydride

One of the first syntheses of transition metal hydride complexes was the reaction of the corresponding metal halide with sodium borohydride[145]. Innumerable transition metal hydrides have since been made by this procedure and the reaction ostensibly is a displacement of halide by hydride (e.g. equation 98)[146]. However, it is

$$[NbCl_2(\eta^5\text{-}C_5H_5)_2] + NaBH_4 + PMe_2Ph \longrightarrow [Nb(H)(PMe_2Ph)(\eta^5\text{-}C_5H_5)_2] \quad (98)$$

$$[NbCl_2(\eta^5\text{-}C_5H_5)_2] \longrightarrow [Nb(BH_4)(\eta^5\text{-}C_5H_5)_2] \xrightarrow{PMe_2Ph} [Nb(H)(PMe_2Ph)(\eta^5\text{-}C_5H_5)_2]$$

now clear that most, if not all, of these reactions proceed through the formation of an intermediate complex containing the borohydride ligand itself, which subsequently reacts with a Lewis base to liberate borane ($BH_3$) and leave the transition metal hydride (equation 98). Other active main group hydrides also convert transition metal halo complexes to hydride complexes. Again, no clear case for direct hydride transfer has been made, and many or all of these complexes may be formed by more complicated mechanisms. Selected examples are presented in equations 99–102.

$$[NiBr(PPh_3)(\eta^3\text{-}C_3H_5)] \xrightarrow[-135°]{NaHBMe_3} [NiH(PPh_3)(\eta^3\text{-}C_3H_5)] \qquad (99)^{146b}$$

$$[Mn(CO)_5(PPh_3)]PF_6 \xrightarrow[MeOH]{NaBH_3CN} [MnH(CO)_4(PPh_3)] \qquad (100)^{146c}$$

$$[FeCl_2(Et_2PCH_2CH_2PEt_2)_2] \xrightarrow[thf]{LiAlH_4} [Fe(H)Cl(Et_2PCH_2CH_2PEt_2)_2] \qquad (101)^{147}$$

$$[ZrCl_2(\eta^5\text{-}C_5H_5)_2] \xrightarrow[PhH]{Na[AlH_2(OCH_2CH_2OCH_3)_2]} [Zr(H)Cl(\eta^5\text{-}C_5H_5)] \qquad (102)^{148}$$

Recently, a number of methods which involve reaction of boron or aluminium hydrides with transition metal salts for the reduction of a range of organic functional groups have been developed[149]. Most of these assume that transition metal hydrides are the reactive species, but this has rarely been demonstrated. For example, the reagent resulting from the reaction of $LiAlH_4$ with four equivalents of copper(I) iodide is very efficient for the 1,4-reduction of conjugated enones[150]. However, a detailed study of this system showed that the reactive species was actually *not* a copper hydride complex, but rather $AlH_2I$. Discrete copper hydride 'ate' complexes of composition $Li_nCuH_{n+1}$ for which $n = 1–5$ can be made by the reaction of $LiAlH_4$ with the corresponding lithium alkylcuprate complexes, $Li_nCu(CH_3)_{n+1}$. The chemistry of these reagents depends very much on their constitution, and their utility as reducing agents has been examined[151].

One system which is likely to involve a transition metal hydride, in this case formed by a transmetallation reaction from tin, is the recently reported reduction of acid chlorides to aldehydes by trialkyltin hydrides, catalysed by palladium(0) complexes (equation 103)[153]. This is a generally useful reduction which tolerates aryl bromide and nitro groups.

$$RCOCl + [PdL_4] \longrightarrow [PdCl(L_2)(COR)] \xrightarrow{R_3SnH} [PdH(L_2)(COR)] \longrightarrow [PdL_4] + RCHO \qquad (103)$$

## C. Ligand Exchange Reactions

One of the most fundamental processes which involves nucleophilic attack on transition metal complexes is the process of ligand exchange, whereby an external ligand, usually having a lone pair of electrons, displaces a ligand coordinated to the metal. These reactions are of central importance to transition metal chemistry. Most of the complexes discussed in this chapter are synthesized by ligand exchange reactions, and a frequent ploy in the use of transition metals in organic synthesis involves the modification of the reactivity of a coordinated organic functional group by changing one or more of the ancillary ligands on the complex involved. Finally, a major (undesired) competing reaction of metal-bound organic functional groups is the displacement of that functional group by the nucleophile, in which attack occurs at the metal rather than at the organic functional group. Having stated the overwhelming importance of ligand exchange processes, the topic will not be further discussed here, and the interested reader is referred to standard texts on the subject[153].

## II. NUCLEOPHILIC ATTACK ON TRANSITION METAL COORDINATED ORGANIC LIGANDS

### A. Cleavage of Metal—Carbon Bonds

The cleavage of metal—carbon sigma bonds figures extensively in a multitude of organometallic reactions, particularly those used in organic synthesis. Many of these cleavages involve nucleophilic attack, although as a class they have been little studied and hence little understood mechanistically. A common but initially unexpected type of nucleophilic cleavage involves a nucleophilic displacement of the metal from an alkylmetal complex by another metal. In this process one metal acts as the nucleophile and the other as the leaving group (the significance of this will become clear shortly). One example of this type of process is the equilibration of a cobalt(III) glyoximatomethyl complex with a similar cobalt(I) glyoximato complex (equation 104)[154]. This can be viewed as a nucleophilic attack of the cobalt(I) species on the

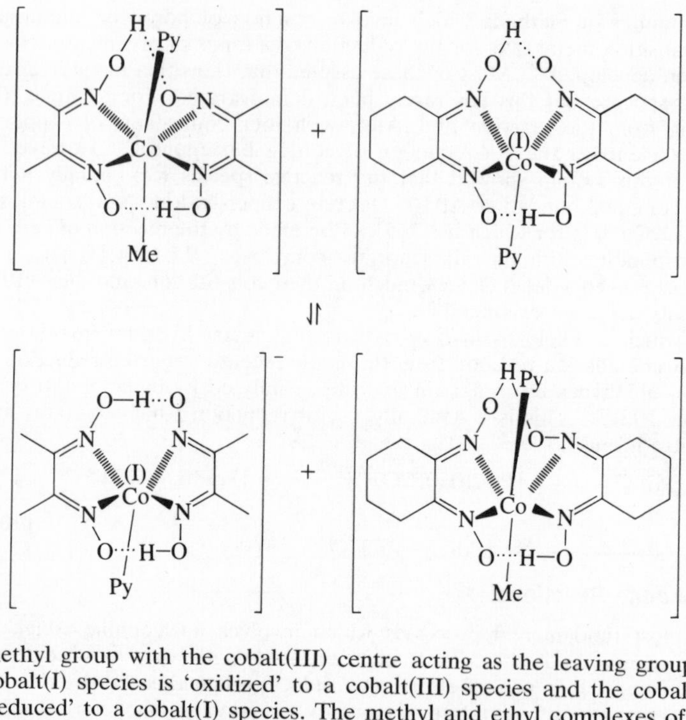

(104)

methyl group with the cobalt(III) centre acting as the leaving group. *Formally* the cobalt(I) species is 'oxidized' to a cobalt(III) species and the cobalt(III) species is 'reduced' to a cobalt(I) species. The methyl and ethyl complexes of the two cobalt species shown in fact equilibrated fairly readily. However, the process was sensitive to steric hindrance on the alkyl group, and neither secondary alkyl complexes nor isobutyl complexes of either cobalt species underwent this exchange. A similar exchange was observed between neutral rhodium(III) alkyl species and the related rhodium(I) complex (equation 105)[155]. Again the self-displacement reaction was strongly dependent on steric effects and the methyl complexes equilibrated more rapidly than the ethyl complexes.

This self-exchange gains significance when reactions of chiral substrates are considered, since it provides a mechanism for racemization of chiral alkylmetal

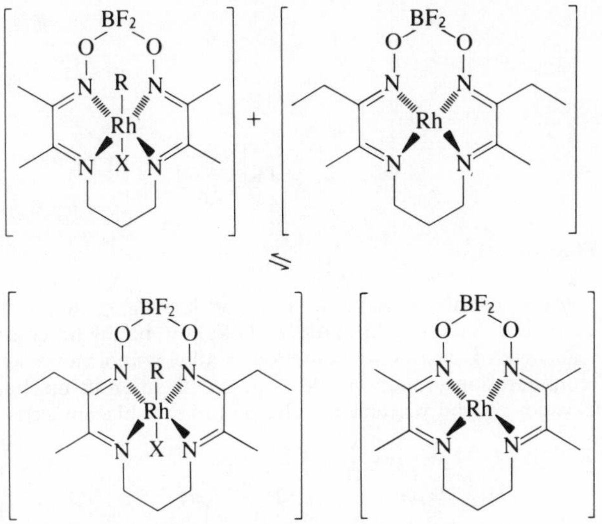

(105)

complexes. Indeed, in studies of the stereochemical course of oxidative addition reactions of palladium(0) complexes with chiral benzyl halides, some racemization of the chiral alkylpalladium(II) complex was observed[156,157]. This was rationalized by invoking the 'self-displacement' of palladium(II) by palladium(0) in a nucleophilic cleavage of a metal—carbon $\sigma$ bond (equation 106).

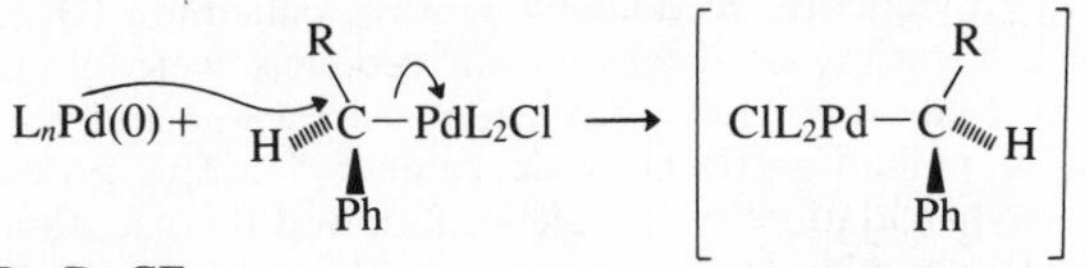

$$L_nPd(0) + H\text{—}C(R)(Ph)\text{—}PdL_2Cl \longrightarrow [ClL_2Pd\text{—}C(R)(Ph)\text{—}H] \qquad (106)$$

$R = D, CF_3$

Self-exchange reactions are probably common because the overall process is 'thermally neutral' in that the starting materials and products have the same net energy. However, displacement of metals from $\sigma$-alkylmetal complexes by other types of nucleophiles is much less common, since the metals involved are generally poor leaving groups. However, it has recently been recognized[158] that oxidation of the metal weakens the metal—carbon bond and makes the metal a better leaving group, and the resulting oxidized alkylmetal complex becomes reactive towards nucleophilic substitution. Thus, a number of oxidatively induced nucleophilic metal—carbon bond cleavages are now recognized, and some synthetically useful processes have been developed using this phenomenon.

One of the earliest systems studied was the oxidative cleavage of alkyl [$Co^{III}$(dimethylglyoximato)$_2$] complexes such as shown in equation 104, in which the alkyl group was benzyl, (+)-*sec*-octyl, or (+)-*sec*-butyl. Oxidation of the benzyl complex with $IrCl_6^{2-}$ or ICl in the presence of excess of chloride produced benzyl chloride in excellent yield[159]. Similar results were obtained using $Cl_2$ or $I_2$ as the oxidant[160]. Oxidation of the optically active complex containing the (+)-*sec*-octyl group with $[IrCl_6]^{2-}/Br^-$ or bromine in acetic acid produced *sec*-octyl bromide with *clean* inversion, indicating an $S_N2$-like displacement[161]. Inversion was also observed in the $Br_2$-induced cleavage of cobalt(III) complexes of cholestane (equation 107)[162].

$$\left[ (Py)(L)Co^{III}\text{-steroid} \right] \xrightarrow[CH_2Cl_2,\ -78\,^{\circ}C]{Br_2} \text{Br-steroid} \tag{107}$$

L = salen, $(dmgh)_2$

Again, oxidation of the metal to make it a better leaving group, followed by displacement of the oxidized metal by bromide, is thought to be the process involved. Finally, oxidative cleavage of cyclopentadienyliron $\sigma$-alkyl complexes also occurred with inversion of configuration (equation 108)[163]. Several different mechanisms for these oxidative cleavages can be written, but the $S_N2$-like displacement is the most likely.

$$Fe(CO)_2Cp \xrightarrow[I_2]{Br_2 \text{ or}} X \tag{108}$$

(X = Br or I)

Oxidative cleavage of metal—carbon $\sigma$-bonds has been used extensively in the synthesis of important organic compounds. In variations on the commercially important Wacker oxidation of ethylene to acetaldehyde using palladium(II) catalysts under oxidizing conditions, a variety of disubstitution products were obtained[164]. Ethylene was converted to 2-chloroethanol by reaction with water and copper(II) chloride in the presence of palladium(II) chloride catalysts[165]. The process was thought to involve initial 'oxypalladation' of the olefin, followed by oxidation of the resulting $\sigma$-alkylpalladium(II) complex by the copper(II) salt, followed by displacement of the oxidized palladium by chloride (equation 109). This reaction was carried

$$CH_2{=}CH_2 + PdCl_2 \rightleftharpoons \tfrac{1}{2}[PdCl_2(\eta^2\text{-}CH_2{=}CH_2)]_2 \xrightarrow{H_2O} \left[\text{>}PdCl(CH_2CH_2OH)\right] \xrightarrow{Cl^-,\ CuCl_2} ClCH_2CH_2OH \tag{109}$$

out with *trans*-dideuterioethylene to establish the stereochemistry of the hydroxypalladation step in the Wacker process. Since $CuCl_2$-cleavage of the palladium—carbon bond was known to proceed with inversion, the hyroxypalladation step in the process that led to chlorohydrin occurred with *trans* stereochemistry (equation 110)[166]. By inference, the hydroxypalladation step in the production of acetaldehyde was claimed to proceed in a *trans* fashion, although there was not universal agreement on this point.

Olefins were converted to a number of glycol derivatives, including glycol mono- and di-acetates and chloroacetates using variations of this oxypalladation (acetoxypalladation)–oxidative cleavage process. Conjugated dienes led to mixtures of 1,4-

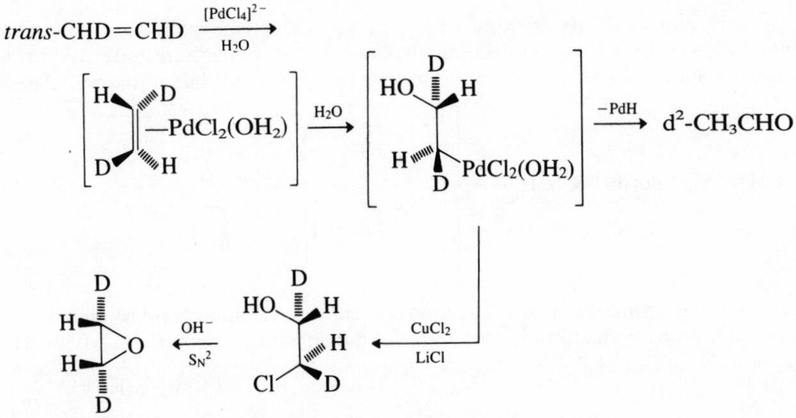

$$trans\text{-}CHD{=}CHD \xrightarrow[H_2O]{[PdCl_4]^{2-}} \left[ \text{(H)(D)C=C(D)(H)—}PdCl_2(OH_2) \right] \xrightarrow{H_2O} \left[ \text{HO(D)(H)C—C(H)(D)}PdCl_2(OH_2) \right] \xrightarrow{-PdH} d^2\text{-}CH_3CHO$$

$$\left[ \text{HO(D)(H)C—C(H)(D)}PdCl_2(OH_2) \right] \xrightarrow[LiCl]{CuCl_2} \text{HO(D)(H)C—C(H)(D)Cl} \xrightarrow[S_N2]{OH^-} \text{epoxide (H, D; H, D)}$$

(110)

and 1,2-disubstitution products[164]. An example of an unexpected rearrangement resulting from oxidative cleavage of a palladium—carbon $\sigma$-bond is seen in equation 111[167]. A similar rearrangement of a stable $\sigma$-alkylpalladium(II) complex under

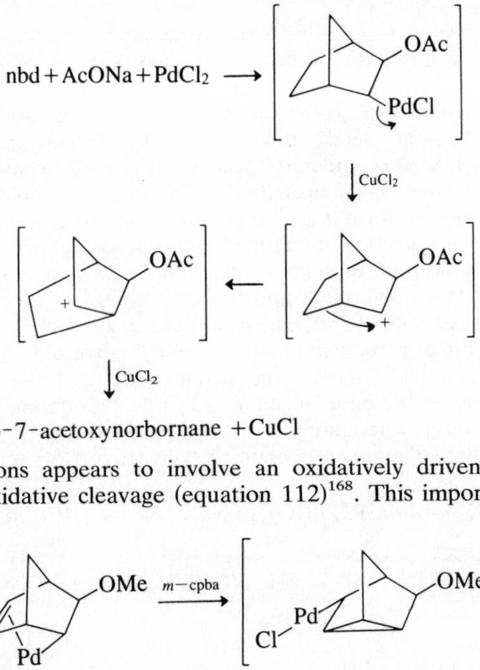

$$nbd + AcONa + PdCl_2 \longrightarrow [\text{OAc, PdCl}] \xrightarrow{CuCl_2} [\text{OAc}, +] \longrightarrow [\text{OAc}, +] \xrightarrow{CuCl_2}$$

(111)

2-chloro-7-acetoxynorbornane + CuCl

oxidizing conditions appears to involve an oxidatively driven olefin insertion in addition to an oxidative cleavage (equation 112)[168]. This important process will be discussed below.

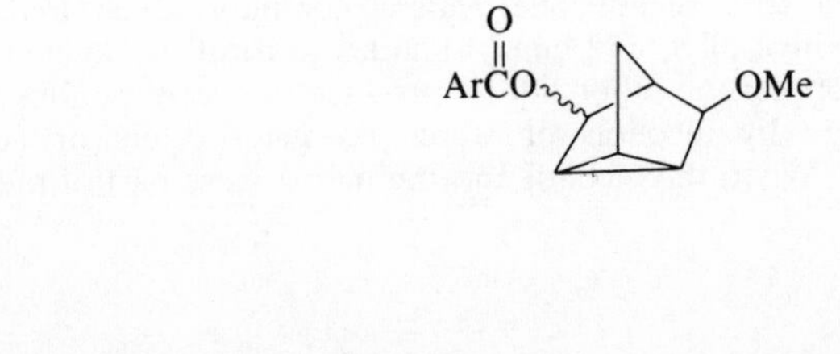

$$\text{(OMe, Pd)} \xrightarrow{m\text{-cpba}} [\text{OMe, Pd, Cl}] \xrightarrow{ArCO_2^-} \text{ArC(=O)O—(OMe)}$$

(112)

Olefins were converted to aziridines by a process involving palladium-promoted amination followed by an intramolecular displacement of palladium under oxidizing conditions (equation 113)[169]. Oxidative cleavage of the initial aminopalladation

$$\mathrm{RCH{=}CH_2 + PdCl_2 + R'NH_2} \longrightarrow \left[\text{R–CH(NHR')–CH}_2\text{–Pd–Cl}\right] \xrightarrow{\mathrm{Br_2}} \text{aziridine (N–R', C–R)} \qquad (113)$$

$$\mathrm{R = Ph,\ } n\mathrm{C_6}$$

complex resulting from the use of secondary amines as nucleophiles resulted in overall oxamination or diamination of the olefinic substrate (equation 114)[170]. The

$$\mathrm{RCH{=}CH_2 + PdCl_2 + R'_2NH} \longrightarrow \left[\text{R–CH(NR}_2\text{)–CH}_2\text{–Pd–Cl}\right] \begin{cases} \xrightarrow[\mathrm{AcO^-}]{\text{ox.}} \mathrm{RCH(NR_2)CH_2OAc} \\ \xrightarrow[\mathrm{R_2'NH}]{\text{ox.}} \mathrm{RCH(NR_2)CH_2NR_2} \end{cases} \qquad (114)$$

process was stereospecific, proceeding with overall *cis* stereochemistry resulting from a *trans* amination of the olefin, followed by displacement of the palladium with inversion. Terminal olefins underwent this reaction in fair to good yield whereas internal olefins reacted with lower yields. A variety of oxidizing agents were effective, including bromine, lead tetraacetate, and *N*-bromosuccinimide.

Metal—acyl $\sigma$-bonds also undergo facile cleavage by nucleophiles to produce carboxylic acid derivatives. Since virtually all $\sigma$-alkylmetal complexes can be made to insert carbon monoxide to form $\sigma$-acyl complexes, this type of cleavage is central to literally thousands of transition metal-mediated carbonylation reactions. An early example is the reaction of organic halides normally reactive in $S_N2$ processes (primary and secondary alkyl, allyl, and benzyl halides, $\alpha$-haloesters) with sodium tetracarbonylcobaltate, $Na[Co(CO)_4]$, to produce a $\sigma$-alkylcobalt complex, which, in the presence of 1 atm of carbon monoxide, rapidly converted to the corresponding $\sigma$-acylcobalt complex. Cleavage by an alcohol produced the ester and, in the presence of a tertiary amine base, regenerated the cobalt anion, making the system catalytic in cobalt complex (equation 115)[171]. Epoxides reacted with hydridocobalt tetracarbonyl in a similar manner, producing $\beta$-hydroxyesters (equation 116).

$$\mathrm{Na[Co(CO)_4] + RX \longrightarrow [RCo(CO)_4] + NaX \xrightarrow{CO} [RC(O)Co(CO)_4] \xrightarrow{R'OH} [HCo(CO)_4] + RC(O)OR' \xrightarrow{R_3N} R_3\overset{+}{N}HX^-} \ (+\ \mathrm{Na[Co(CO)_4]}) \qquad (115)$$

$$\mathrm{Na[Co(CO)_4] + oxirane + CO + MeOH \xrightarrow[40\ psi]{45\,^\circ C} HOCH_2CH_2CO_2Me} \qquad (116)$$

Nickel carbonyl is a reactive zerovalent complex which undergoes oxidative addition to aryl, vinyl, allyl, and benzyl halides to form $\sigma$-alkylnickel(II) complexes which readily insert carbon monoxide to produce the corresponding $\sigma$-acylnickel(II) complex. Cleavage by alcohols or water produces esters or acids respectively (equation 117)[172]. When the alcohol and the halide were part of the same molecule,

$$\mathrm{RX + [Ni(CO)_4] \longrightarrow [RNi(CO)_2X] \xrightarrow{CO} [RC(O)Ni(CO)_2X] \xrightarrow[\text{or } H_2O]{R'OH} RC(O)OR' \text{ or } RC(O)OH} \quad (117)$$

R = aryl, vinyl, allyl, benzyl

lactones were formed. Using $\omega$-hydroxyvinyl halides as substrates resulted in the formation of $\alpha$-methylene lactones (equations 118–120)[173].

$$\mathrm{HO(CH_2)_nC({=}CH_2)Br \xrightarrow[Ph_3P,\ base]{[Ni(CO)_4]} (CH_2)_nC({=}CH_2)C{=}O} \text{ (ring closed through O)} \quad (118)$$

$$\mathrm{2{-}[C({=}CH_2)Br]cyclohexan\text{-}l\text{-}ol}\ (\textit{cis} \text{ or } \textit{trans}) \xrightarrow[Ph_3P,\ base]{[Ni(CO)_4]} \quad (119)$$

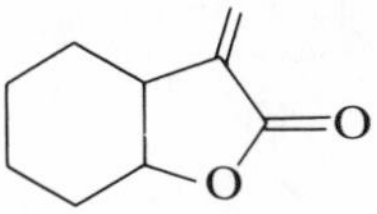

$$\mathrm{2{-}[CH_2C({=}CH_2)Br]cyclohexan\text{-}l\text{-}ol}\ (\textit{cis} \text{ or } \textit{trans}) \xrightarrow[Ph_3P,\ base]{[Ni(CO)_4]} \quad (120)$$

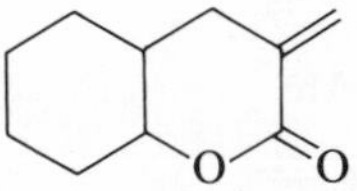

By far the most extensively used metal for carbonylation reactions is palladium, because palladium forms $\sigma$-alkylcomplexes from a wide variety of organic substrates, and because both carbon monoxide insertion and $\sigma$-acyl—palladium bond cleavage by nucleophiles are very facile processes. Palladium(0) complexes undergo oxidative addition reactions with aryl and vinyl halides and the resulting $\sigma$-alkylpalladium(II) complexes undergo facile carbon monoxide insertion and $\sigma$-acyl—metal bond cleavage reactions (equation 121)[174]. [Note: when palladium(II) salts are used as catalysts reduction to palladium(0) by carbon monoxide is assumed to generate the catalytically active species.] With acetylides as nucleophile, acetylenic ketones were formed in reasonable yields (equation 122)[175], whereas acid cyanides were formed when cyanide was the nucleophile (equation 123)[176]. This type of $\sigma$-acyl cleavage reaction has been used to synthesize complex intermediates in the course of the total synthesis of natural products, as exemplified in equations 124[177] and 125[178].

$$\mathrm{ArX + Pd(0) \longrightarrow ArPdX \xrightarrow{CO} ArC(O)PdX \xrightarrow{R'OH} ArC(O)OR' + HX + Pd(0)} \quad (121)$$

$$\left.\begin{array}{c}\mathrm{ArX}\\ \text{het-ArX}\\ \mathrm{CH_2{=}CHX}\end{array}\right\} + \mathrm{CO + HC{\equiv}CR'} \xrightarrow[\text{Pd(II) catalyst}]{\mathrm{R_3N,\ 120\ ^\circ C,\ 80\ atm}} \left\{\begin{array}{c}\mathrm{ArC(O)C{\equiv}CR'}\\ \text{het-}\mathrm{ArC(O)C{\equiv}CR'}\\ \mathrm{CH_2{=}CHC(O)C{\equiv}CR'}\end{array}\right. \quad (122)$$

47–95%

$$\mathrm{ArX + CO + KCN \xrightarrow[100\ ^\circ C]{[PdI(Ph)(L_2)]} ArCOCN} \quad (123)$$

45–92%

Ar = Ph, *p*-OMePh, *p*-tolyl, fur-2-yl

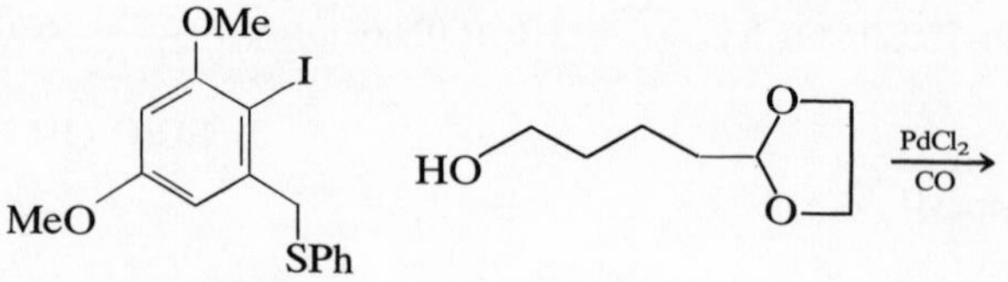

$$2\text{-}[C(O)OCH(Me)(CH_2)_3R']R \qquad (124)$$

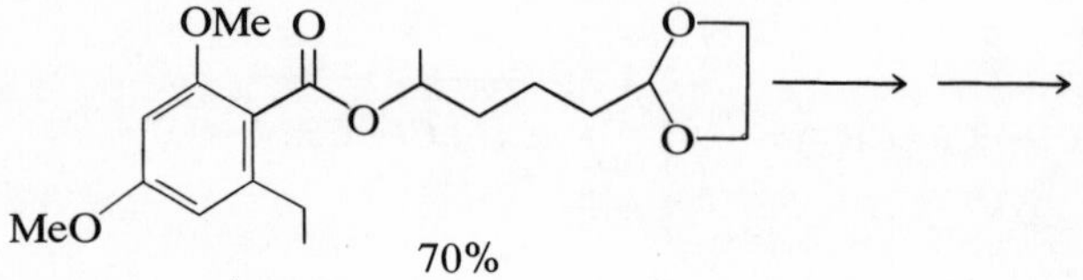

→ → (±)-zearalenone

70%

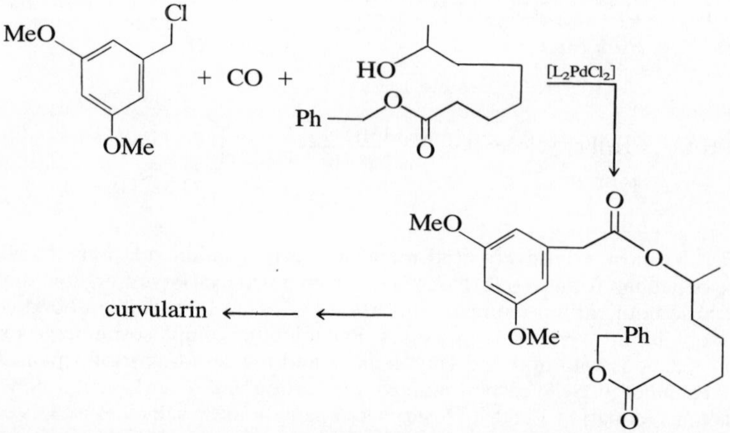

$$+ CO + \xrightarrow{[L_2PdCl_2]} \qquad (125)$$

curvularin ← ←

Intramolecular versions of this process have been especially useful for the synthesis of a variety of lactones and lactams, including some rather complex molecules. A variety of four- and five-membered lactones were available from haloalcohols by this oxidative addition/CO insertion/nucleophilic cleavage process, using palladium salts as catalysts (equations 125a–128)[179a]. Lactones were prepared in a simplar process from haloamines (equation 129).[179b]. Even $\beta$-lactams could be formed by this process (equations 130[179c] and 131[179d]).

$$1\text{-X-2-}(CH_2OH)C_6H_4 \xrightarrow[CO]{Pd(0)} \text{phthalide} \quad 88\% \qquad (125a)$$

$$HOCH_2CH{=}CHCH_2Cl \xrightarrow[CO]{Pd(0)} 3\text{-}({=}CHMe)\text{oxetan-2-one} \quad 52\% \qquad (126)$$

$$\xrightarrow[CO]{Pd(0)} \qquad (127)$$

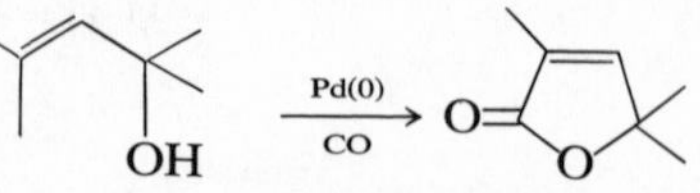

3,5,5-trimethyloxol-3-en-2-one

$$1\text{-OH-2-I-4,6-}(OMe)_2C_6H_4 \xrightarrow[CO]{Pd(0)} 4,6\text{-dimethoxyphthalate} \qquad (128)$$

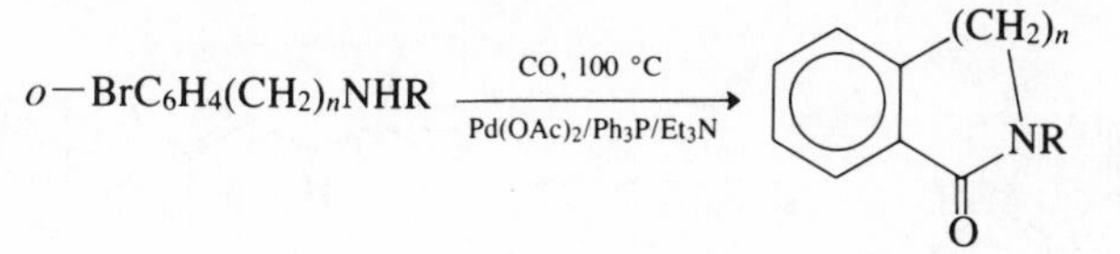

$$o\text{-BrC}_6\text{H}_4(\text{CH}_2)_n\text{NHR} \xrightarrow[\text{Pd(OAc)}_2/\text{Ph}_3\text{P/Et}_3\text{N}]{\text{CO, 100 °C}} \text{(CH}_2)_n \ \text{NR} \ \text{O} \quad (129)$$

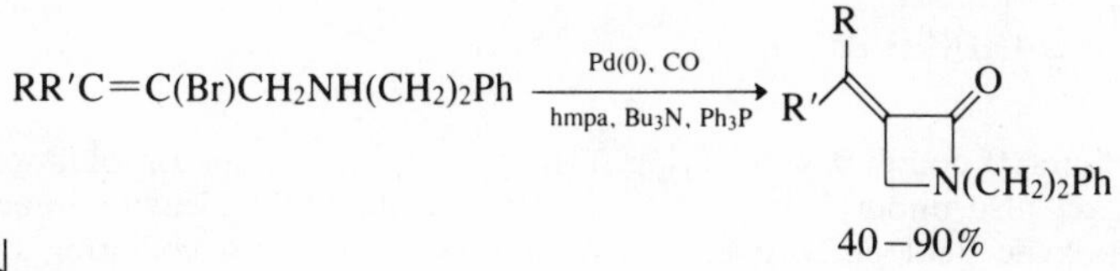

$$\text{RR}'\text{C}{=}\text{C(Br)CH}_2\text{NH(CH}_2)_2\text{Ph} \xrightarrow[\text{hmpa, Bu}_3\text{N, Ph}_3\text{P}]{\text{Pd(0), CO}} \text{R} \ \text{R}' \ \text{O} \ \text{N(CH}_2)_2\text{Ph} \quad (130)$$

40–90%

$$\text{Br} \ \text{NH} \ \text{OCH}_2\text{Ph} \ \text{CO}_2\text{CH}_2\text{Ph} \xrightarrow[\text{Ph}_3\text{P, Bu}_3\text{N}]{\text{CO, Pd(OAc)}_2} \text{O} \ \text{NCH(CO}_2\text{CH}_2\text{Ph)C}_6\text{H}_4(\text{OCH}_2\text{Ph)}{-}p \quad (131)$$

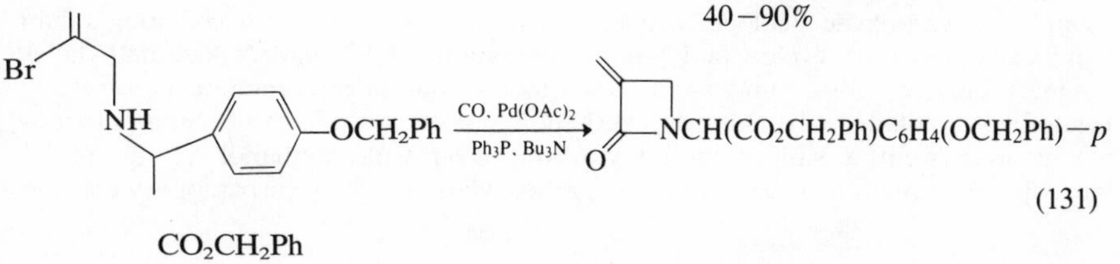

Phthalimides (equation 132), isoquinolones (equation 133), and quinolones (equation 134) were prepared by this procedure[179e]. This chemistry was also used to prepare a number of more complex polycyclic heterocyclic systems, as exemplified by equations 135[180], 136[181], and 137[182].

$$o\text{-BrC}_6\text{H}_4\text{C(O)NHR} \xrightarrow[\text{Bu}_3\text{N, CO}]{\text{Pd(OAc)}_2\text{, Ph}_3\text{P}} N\text{-R-phthalimide} \quad (132)$$

$$o\text{-BrC}_6\text{H}_4(\text{CH}_2)_2\text{NHAc} \longrightarrow N\text{-acetylisoquinol-l-one} \quad (133)$$

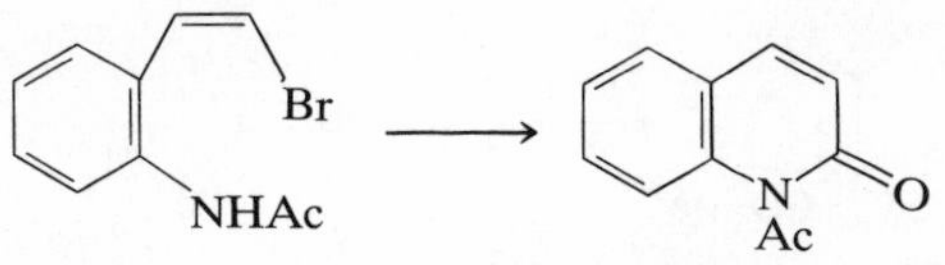

$$\text{Br} \ \text{NHAc} \longrightarrow \text{N} \ \text{Ac} \ \text{O} \quad (134)$$

$$\text{R}^2 \ \text{R}^1 \ \text{NH} \ \text{Br} \ \text{R}^4 \ \text{R}^3 \xrightarrow[\text{Bu}_3\text{N, CO}]{\text{Pd(OAc)}_2\text{, Ph}_3\text{P}} \text{R}^2 \ \text{R}^1 \ \text{N} \ \text{O} \ \text{R}^4 \ \text{R}^3 \quad (135)$$

40–50%

$$\text{MeO} \ \text{PhCH}_2\text{O} \ (\text{CH}_2)_2\text{NHCH}_2\text{C}_6\text{H}_4\text{OMe}{-}p \ \text{Br} \xrightarrow[\text{Ph}_3\text{P, Bu}_3\text{N}]{\text{CO, Pd(OAc)}_2} \text{MeO} \ \text{PhCH}_2\text{O} \ \text{NCH}_2\text{C}_6\text{H}_4\text{OMe}{-}p \ \text{O} \quad (136)$$

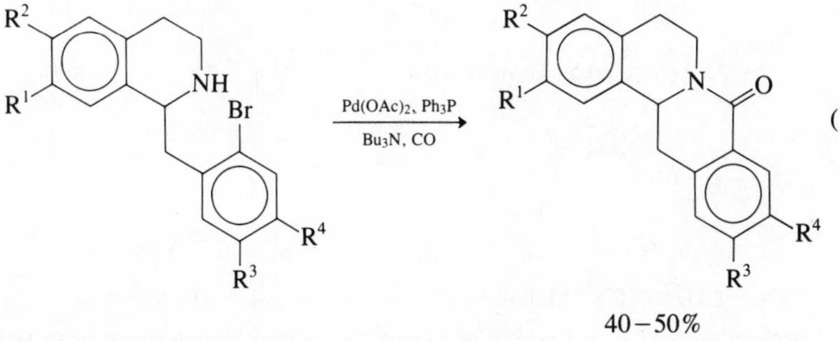

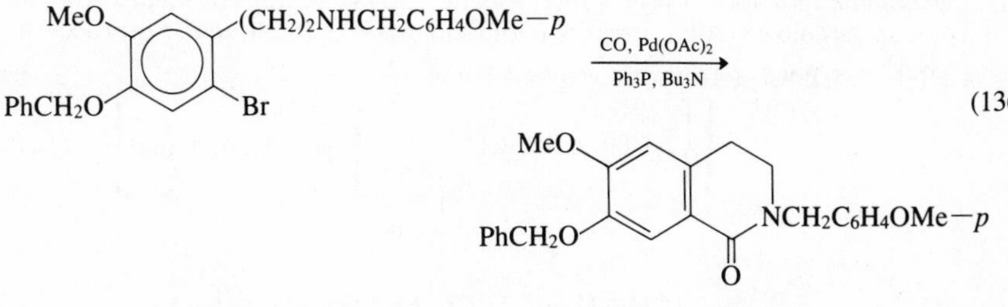

$$\text{(indole-fused tetrahydropyridine with NH and } CH_2C_6H_4Br\text{-}o\text{)} \xrightarrow[Ph_3P,\ Bu_3N,\ 100\ °C]{CO,\ Pd(OAc)_2} \text{lactam product, 55\%} \quad (137)$$

σ-Alkylpalladium(II) complexes resulting from nucleophilic attack on olefinpalladium(II) complexes also undergo facile CO insertion/nucleophilic cleavage reaction to produce carboxylic acid derivatives. Early studies focused on chelating olefin complexes, and both cycloocta-1,5-diene (equation 138)[183] and *N,N*-dimethylallylamine (equation 139)[184] underwent clean methoxylation/carbomethoxylation when treated with palladium(II) chloride, methanol, and carbon monoxide. Simple mono-olefins underwent a similar methoxycarbonylation with methanol as the initial nucleophile (equation 140)[185], and a 'carboacylation' when carbanions were the

$$[PdCl_2(cod)_2] \xrightarrow{MeOH} [\text{OMe-cyclooctenyl-}PdCl_2] \xrightarrow{CO} [\text{OMe-cyclooctenyl-C(O)-Pd(Cl)}]_2 \xrightarrow{MeOH} \text{5-methoxy-6-(methoxycarbonyl)cyclooctene (OMe, } CO_2Me\text{), 77\%} \quad (138)$$

$$CH_2{=}CHCH_2NMe_2 + PdCl_2 \longrightarrow [\text{chelate } PdCl_2\text{–N}] \xrightarrow{MeOH,\ CO} Me_2NCH_2C(OMe)CH_2CO_2Me \quad (139)$$

$$RCH{=}CH_2 + PdCl_2 + MeOH + CO \longrightarrow RCH(OMe)CH_2CO_2Me \quad (140)$$

initial nucleophile (equation 141)[186]. Indoleacetic acid esters were produced from an intramolecular 'aminoacylation' process (equation 142)[187]. In all of these processes,

$$CH_2{=}CHNHAc + PdCl_2 + \text{1-lithoxycyclohex-1-ene} \longrightarrow \left[\left\{\text{2-CH(NHAc)}CH_2\text{—Pd—}\right\}\text{cyclohexan-1-one}\right] \xrightarrow{CO,\ MeOH} \text{2-[CH(NHAc)}CH_2CO_2Me\text{]cyclohexan-1-one} \quad (141)$$

$$o\text{-MeNHC}_6\text{H}_4\text{CH}_2\text{C(Me)}{\equiv}\text{CH}_2 \xrightarrow{\text{Pd(II)}} \text{(2-methyl-2Pd-CH}_2\text{)indoline} \xrightarrow{\text{CO} \mid \text{MeOH}} \text{2-Me-2-CH}_2\text{CO}_2\text{Me-indoline} \quad (142)$$

the σ-acyl complex was rapidly cleaved by methanol to produce the corresponding methyl esters.

σ-Alkylpalladium(II) complexes from orthopalladation reactions also undergo facile CO insertion/nucleophilic cleavage reactions. Since orthopalladation is a very general process, it provides an efficient procedure for introduction of carboxylic acid functionality into aromatic rings. Examples are seen in equations 143[188], 144[188], and 145[189]. In what *may* be a related process, the σ-alkylpalladium complex shown in equation 146 underwent cleavage by organolithium reagents[190]. The mechanism of this process has not been reported.

$$\text{C}_6\text{H}_5\text{C(Me)}{=}\text{NNMe}_2 \xrightarrow{\text{Pd(OAc)}_2} [o\text{-\{C(Me)}{=}\text{NNMe}_2\}\text{C}_6\text{H}_4\text{Pd(OAc)}]_2 \xrightarrow[100\,^\circ\text{C}]{\text{CO}} \text{(isoindolinone, N-NMe}_2\text{)} \quad (143)$$

$$\text{C}_6\text{H}_5\text{CH}_2\text{NHMe} \xrightarrow{\text{Pd(OAc)}_2} [o(\text{CH}_2\text{NHMe})\text{C}_6\text{H}_4\text{Pd(OAc)}]_2 \xrightarrow{\text{CO}} o\text{-(CH}_2\overset{+}{\text{N}}\text{H}_2\text{Me)C}_6\text{H}_4\text{CO}_2^- \quad (144)$$

63%

$$\text{1-R}^1\text{-2-R}^2\text{-C}_6\text{H}_3\text{-4-NHC(O)R}^3 \xrightarrow{\text{Pd(OAc)}_2} [\text{orthopalladated complex: R}^1, \text{R}^2, \text{N–C(O)R}^3, \text{Pd, AcO}] \quad (145)$$

60–90%

$\xrightarrow{\text{CO} \mid \text{EtOH}}$ 1-R$^1$-2-R$^2$-4-NHC(O)R$^3$-5-CO$_2$Et—C$_6$H$_2$

40–90%

R$^1$ = H, Me, OMe, Cl, CO$_2$Et, COMe
R$^2$ = H, Me, OMe, Cl, CO$_2$Me
R$^3$ = Me, Et, Pr$^i$

$$[\text{Y-C}_6\text{H}_3(\text{CH}{=}\text{NMe})\text{Pd(Cl)}]_2 \xrightarrow[\text{(2) RLi} \; \text{(3) H}_3\text{O}^+]{\text{(1) 4PPh}_3} \text{2-Y-6-R-benzaldehyde} \quad (146)$$

60–99%

R = Me, Pr$^n$, Bu$^n$, Ph

Not all σ-acyl complexes are easily cleaved by nucleophiles such as methanol, and many of the more stable ones required an oxidatively induced cleavage, the mechanism of which has been little studied. An example is seen in equation 147, in which aminopalladation/CO insertion of simple olefins led to chelating β-aminoacyl palladium(II) complexes which were very resistant to cleavage by nucleophiles[191]. This stability was thought to be due to the presence of a five-membered chelate ring, particularly stable for organopalladium complexes. Treatment with excess of dppe (to displace the coordinated amino group and destroy the stabilization-by-chelation effect) or with bromine led to cleavage by methanol. The very useful acyliron complexes, produced from the reaction of sodium tetracarbonylferrate with organic halides, also require an oxidative cleavage, usually by halogen (equation 148)[192]. The mechanism of oxidative cleavage of $\sigma$-acylmetal complexes by halogens is not well understood. It may proceed by oxidation of the metal, making it a better leaving group, followed by displacement of the metal by the nuelcophile (halide or methanol). Alternatively, the halogen may 'oxidatively add' to the metal, and subsequently an acid halide may form by a reductive elimination (equation 149). An

$$CH_2{=}CHR + PdCl_2 \xrightarrow{R'_2NH} [\text{R-CH(NR}'_2\text{)CH}_2\text{-Pd(Cl)(NHR}_2)] \xrightarrow{CO} [\text{chelated acyl-Pd(Cl)(NHR}_2)] \text{ (stable)} \xrightarrow[\text{or } Br_2/CH_2Cl_2/MeOH]{Ph_2PCH_2CH_2PPh_2/MeOH} MeOC(O)CH_2CH(R)NR'_2 \qquad (147)$$

$$Na_2[Fe(CO)_4] + RX \longrightarrow [RFe(CO)_4]Na \xrightarrow{L} [RC(O)Fe(CO)_3L]Na \xrightarrow{Br_2} RC(O)Br \xrightarrow{MeOH} RCO_2Me \qquad (148)$$

$$RC(O)M^n \xrightarrow{X_2} RC(O)M^{n+e} \xrightarrow{X^-} RCOX + M^n$$
$$RC(O)M^n \xrightarrow{X_2} RC(O)M^{n+2}X_2 \longrightarrow RCOX + M^nX \qquad (149)$$

interesting example for which oxidation drives both a carbon monoxide insertion and a nucleophilic cleavage of the resulting $\sigma$-acyl complex is seen in equation 150. This

$$[FeCp(CO)_2(\eta^2{-}RCH{=}CH_2)]^+ + PhCH_2NH_2 \longrightarrow [FeCp(CO)_2\{CH_2CH(R)NHCH_2Ph\}] \xrightarrow{Ag_2O} [\text{Cp(OC)Fe}^+\text{-C(O)CH}_2\text{CH(R)NHCH}_2\text{Ph}] \longrightarrow \text{1-PhCH}_2\text{-4-R-azetidin-2-one} \qquad (150)$$

process provided the elegant synthesis of $\beta$-lactams seen here[193]. A similar oxidatively induced cleavage is probably responsible for the formation of $\beta$-lactones in the study of the mechanism of the Wacker process (equation 151)[194].

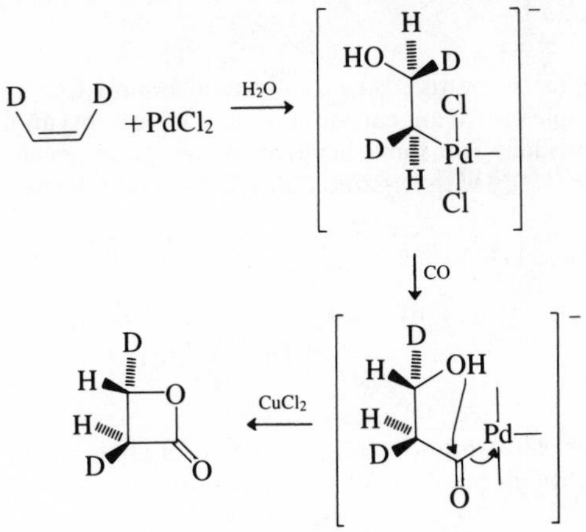

(151)

## B. Nucleophilic Attack on Transition Metal—Carbon Monoxide and Isonitrile Complexes

Coordination of carbon monoxide to most transition metals activates it towards attack by a wide range of nucleophiles, including hydride, carbanions, alkoxides, hydroxide, and amines[195]. The most direct approach to the long-sought (for study as Fischer–Tropsch model compounds) but only recently synthesized metal formyl complexes is direct attack of hydride on a metal-bound carbon monoxide. The most efficient hydride source found was a series of alkoxyborohydrides, which reacted with a variety of transition metal carbonyl compounds to produce metal formyls in reasonable yield (equation 152)[196–198]. In this manner, $[(CO)_4FeCHO]^-$, $[(CO)_5CrCHO]^-$, $[(CO)_5WCHO]^-$, $[(Ph_3P)(CO)_3FeCHO]^-$, $[(Ph_3P)(CO)_4CrCHO]^-$, and $[(Ph_3P)(CO)_4WCHO]^-$ were prepared and characterized by n.m.r. spectroscopy ($\delta$ CHO $\approx$ 15 ppm). Metal carbonyl acyls reacted in a similar fashion to give acyl–formyl complexes (equation 153)[199]. These metal formyl complexes have several

$$\underset{(K)}{NaHB(OR)_3} + [ML_n(CO)] \longrightarrow [ML_n(CHO)] \quad (152)$$

$$[RC(O)M(CO)_n] + LiEt_3BH \longrightarrow [RC(O)M(CO)_{n-1}CHO] \quad (153)$$

M = Mn, Fe, Mo

R = Ph, $MeOCH_2$, MeOCO, Me

interesting properties. Many of them are strong hydride sources themselves, and reduce ketones and alkyl halides. In addition, some metal formyl complexes can transfer their formyl hydride group to another metal carbonyl, generating a new formyl complex in the process (equation 154)[200].

$$[Fe\{(PhO)_3P\}(CO)_3(CHO)]^- + [Re_2(CO)_{10}] \rightleftharpoons$$

$$[Fe\{(PhO)_3P\}(CO)_4\} + [(Re_2(CO)_9(CHO)]^- \underset{}{\overset{Fe(CO)_5}{\rightleftharpoons}} [Fe(CO)_4(CHO)]^- + Re_2(CO)_{10} \quad (154)$$

Alkyl- and aryl-lithium reagents also attack metal-bound CO groups, producing acylmetal anionic complexes. Iron carbonyl reacted with organolithium reagents (equation 155)[201] to produce the same acyliron carbonylate species available from the reaction of $Na_2[Fe(CO)_4]$ with organic halides discussed above.

$$RLi + [Fe(CO)_5] \longrightarrow [RC(O)Fe(CO)_4]^- Li^+ \quad (155)$$

Although these reagents have found extensive use in organic synthesis[192,202,203] they are rarely prepared via the organolithium route, the iron anion route being considerably more convenient[192].

Nickel carbonyl also reacts with organolithium reagents to form acylnickel carbonylate species, which have not been well characterized. However, the general formula $[Ni(RCO)(CO)_3]^-$ has been assigned to them, and much of their chemistry is consistent with this assumption. For example, treatment of the acylnickel carbonylate resulting from aryllithium reagents and nickel carbonyl with ethanolic hydrogen chloride produced $\alpha$-diketones whereas those resulting from alkyllithium reagents gave symmetrical ketones (equation 156)[204]. Although the nickel complexes produced from the reaction of nickel carbonyl with aryllithium reagents were relatively unreactive, those from alkyllithium reagents reacted with allylic halides at −50 °C to produce $\beta,\gamma$-unsaturated ketones (equation 157)[205]. Acylation occurred predominantly to exclusively at the halogen-bearing carbon, with little or no allylic transposition being noted. Aryl halides, benzoyl chloride, and alkyl halides including primary iodides were unreactive toward these reagents. In contrast, vinyl halides were fairly reactive, undergoing facile acylation (equation 158). With relatively

$$RLi + [Ni(CO)_4] \xrightarrow{-70\,^{\circ}C} [RC(O)Ni(CO)_3]^- Li^+ \begin{cases} \xrightarrow[R = Ar]{HCl} ArC(O)C(O)Ar \\ \xrightarrow[R = alkyl]{HCl} RC(O)R \end{cases} \quad (156)$$

$$RLi + [Ni(CO)_4] \xrightarrow[-50\,^{\circ}C]{Et_2O} [RC(O)Ni(CO)_3]^- Li^+ \xrightarrow{R'CH{=}CHCH_2X} R'CH{=}CHCH(CH_3)C(O)R \quad (157)$$

unsubstituted vinyl halides, the product conjugated enone reacted further with the acylnickel carbonylate to produce 1,4-diketones (equation 159). This was a general

$$[MeC(O)Ni(CO)_3]^-Li^+ + Ph_2C{=}CHBr \xrightarrow[-78\,^\circ C]{Et_2O} Ph_2C{=}CHCOMe \tag{158}$$

$$[MeC(O)Ni(CO)_3]^-Li^+ + trans\text{-}PhCH{=}CHBr \longrightarrow trans\text{-}PhCH{=}CHCOMe \downarrow MeOCCH(Ph)CH_2COMe \tag{159}$$

reaction for these species, and a variety of conjugated enones were acylated by this reagent (equation 160)[260]. Monosubstituted alkynes reacted in a similar fashion (equation 161)[207]. The mechanism of these acyl transfer reactions has not been studied.

$$RLi + [Ni(CO)_4] + CH_2{=}CHCOMe \longrightarrow ROCCH_2CH_2COMe \tag{160}$$

$$[ArC(O)Ni(CO)_3]^-Li^+ + RC{\equiv}CH \longrightarrow [RCH{=}CHC(O)Ar] \longrightarrow ArC(O)CH(R)CH_2C(O)Ar \tag{161}$$

Group VI metal carbonyls also react with organolithium reagents to form acylmetal carbonylates in which the oxygen atom of the acyl group is nucleophilic and can be alkylated by reactive agents such as trialkyloxonium salts or $CH_3OSO_2F$. The resulting complexes are heterocarbene complexes (equation 162)[208] and have interesting chemistry in their own right. Other reactive nucleophiles form similar carbenes by nucleophilic attack on metal carbonyl species (equations 163[209], 164[210], and 165[211]). Other metal carbonyl complexes, including $[Mn_2(CO)_{10}]$, $[Re_2(CO)_{10}]$,

$$[M(CO)_6] + RLi \longrightarrow [(CO)_5MC(O)R] \xrightarrow{R'_3O^+} [(CO)_5M{=}C(OR')R] \tag{162}$$

M=Cr, Mo, W

$$[Cr(CO)_6] + LiNEt_2 \longrightarrow [(CO)_5CrC(O)NEt_2] \xrightarrow{Et_3O^+} [(CO)_5Cr{=}C(OEt)NEt_2] \tag{163}$$

$$[Cr(CO)_6] + \text{(1,3-dithian-2-yl)Li} \longrightarrow \xrightarrow{Et_3O^+} \left[(CO)_4Cr{=}C(OEt)C(OH){=}C(\text{-S-}(CH_2)_3\text{-S-})\right] \tag{164}$$

$$[W(CO)_6] + LiPMe_2 \longrightarrow \xrightarrow{Et_3O^+} [(CO)_4W\{{=}C(OEt)PMe_2\}_2] \tag{165}$$

$[Fe(CO)_5]$, and $[Ni(CO)_4]$, react with carbanions to form carbene complexes, but they are considerably less stable than those of Group VI metals, and have been much less studied[208d].

Carbene complexes themselves are reactive towards nucleophiles at the carbene carbon. This reaction provides a rapid entry into heterocarbenes having an element other than oxygen as its heteroatom. Alkoxycarbene complexes behave somewhat like very reactive esters. For example, chromium—carbene complexes underwent facile aminolysis when treated with amines (equation 166)[208d]. Ammonia and primary and secondary amines reacted, but the reaction was sensitive to the steric bulk of the secondary amines studied. This chemistry provides a general route to aminocarbene complexes.

$$(CO)_5Cr{=}C(OMe)R \;\; R'\ddot{N}H_2 \longrightarrow [(CO)_5Cr{=}C(R)NHR']$$
$$\updownarrow$$
$$(CO)_5\bar{C}r{-}\overset{+}{C}(OMe)R \qquad (166)$$

Other nucleophiles behave in a similar fashion. Thiols cleanly replaced alkoxides in a two-step process (equation 167)[212]. More important, at low temperatures organolithium reagents displaced the alkoxide group in a tungsten alkoxycarbene complex to produce a diarylcarbene complex, which was stable only at low temperature (equation 168)[213]. This reaction provided an important synthetic route to dialkylcarbene complexes for use in the study of the olefin metathesis reaction[214].

$$[(CO)_5M{=}C(OMe)Ph] \xrightarrow{p\text{-}RC_6H_4S^-} [(CO)_5MC(OMe)(Ph)SC_6H_4R\text{-}p]^- \xrightarrow{H^+} [(CO)_5M{=}C(SC_6H_4R\text{-}p)Ph] \qquad (167)$$

M = Cr, Mo, W
R = Br, Me, OMe

60–90%

$$[(CO)_5M{=}C(OMe)Ph] \xrightarrow[-78\,^\circ C]{RLi} [(CO)_5MC(OMe)(R)Ph]^- \xrightarrow[-78\,^\circ C]{HCl \text{ or } SiO_2} [(CO)_5M{=}C(R)Ph] \qquad (168)$$

M = Cr, W
R = Ph, $p$-$MeC_6H_4$, $p$-$CF_3C_6H_4$

The analogy of carbene complexes to esters is further exemplified by the conjugate addition of nucleophiles to $\alpha,\beta$-unsaturated carbene complexes observed with both amines (equation 169)[215] and carbanions (equation 170)[216].

$$[(CO)_5W{=}C(OEt)C{\equiv}CPh] + Me_2NH \longrightarrow [(CO)_5W{=}C(NMe_2)CH{=}C(Ph)NMe_2] \qquad (169)$$

$$[(CO)_5Cr{=}C(OMe)CH{=}CHPh] + Ph_2CuLi \longrightarrow \xrightarrow{HCl} [(CO)_5Cr{=}C(OMe)CH_2CHPh_2] \qquad (170)$$

30%

Carbyne complexes also react with nucleophiles, undergoing addition to the carbyne carbon to produce carbene complexes (equation 171)[208d]. Since carbyne complexes are normally synthesized from carbene complexes, this process seems redundant. However, it can lead to carbene complexes inaccessible by standard approaches.

$$[(CO)_5M{\equiv}CR] + Nuc^- \longrightarrow [(CO)_5M{=}C(R)Nuc] \qquad (171)$$

$M = Mn, Re, Cr$

$R = Ph, NMe_2$

$Nuc^- = CN, NCS, NMe_2, NCO, RO$

Metal carbonyl complexes undergo nucleophilic attack by water or hydroxide to produce unstable carboxylate complexes which readily lose $CO_2$, producing a metal hydride complex (equation 172)[217]. This may further decompose, resulting in overall

$$MCO + HO^- \longrightarrow [MCO_2H] \longrightarrow MH + CO_2 \qquad (172)$$

reduction of the metal. The susceptibility of metal carbonyl ligands to nucleophile attack, in general, is inversely dependent upon the extent to which they serve as a $\pi$-acceptor, and can be predicted from their infrared stretching frequencies[218]. Thus, water is sufficiently nucleophilic to attack cationic (electron-poor) metal carbonyl complexes, whereas hydroxide is required for most neutral metal carbonyls. These differences are reflected in the conditions required for the reactions in equations 173–175[219]. This process is important in the currently popular 'water-gas shift

$$[Mn(CO)_6]^+BF_4^- \xrightarrow[CH_3CN]{H_2O} [HMn(CO)_5] + CO_2 \qquad (173)$$

$$[Fe(CO)_5] \xrightarrow[MeOH]{NaOH,\ H_2O} Na[HFe(CO)_4] + CO_2 \qquad (174)$$

$$[Cr(CO)_6] \xrightarrow[thf,\ 50\,^\circ C]{KOH/H_2O/MeOH} K[HCr_2(CO)_{10}] \qquad (175)$$

reaction'[220] for the production of hydrogen from carbon monoxide and water (equation 176) and for the overall reduction of palladium(II) complexes to palladium(0) species for use in catalysis (equation 177).

$$\begin{array}{ccccc} CO + H_2O & \overset{M\ cat.}{\rightleftharpoons} & CO_2 + H_2 & & \\ \downarrow & & \nwarrow & \nwarrow H^+ & \\ [MCO] & \xrightarrow{H_2O} & [MCO_2H] & \longrightarrow & [MH] \end{array} \qquad (176)$$

$$PdCl_2 + CO \longrightarrow [PdCl_2(CO)_2]_n \xrightarrow{H_2O} CO_2 + \text{`}[HPdCl]\text{'} \xrightarrow{-HCl} Pd(O) \qquad (177)$$

Amine *N*-oxides similarly react with metal carbonyl compounds to remove one or more CO ligands by oxidizing them to $CO_2$ (equation 178)[221]. This process has proved particularly useful as a mild method for the removal of the iron tricarbonyl fragment from 1,3-diene iron tricarbonyl complexes (see Section F).

$$MCO + R_3NO^- \longrightarrow [MC(O)ONR_3] \longrightarrow M + CO_2 + NR_3 \qquad (178)$$

Alkoxides and amines also attack metal-bound carbon monoxide groups to give alkoxycarbonyl and carbamoyl complexes, respectively[222]. Again, the reactivity of a metal carbonyl towards nucleophilic attack by alkoxides or amines correlates with the CO stretching frequency, with electron-deficient, cationic metal carbonyls being more reactive than electron-rich, neutral, or anionic carbonyl compounds. A variety of phosphino-palladium(II) and -platinum(II) halide complexes reacted with CO, methanol, and triethylamine to give stable alkoxycarbonyl complexes (equation 179)[223]. Vaska's compound, $[IrCl(CO)L_2]$, reacted in a similar fashion, but both

$$[M(X_2)L_2] + CO \rightarrow [M(X)(CO)L_2]^+X^- \xrightarrow[Et_3N]{MeOH} [M(X)(CO_2Me)(L_2)] + Et_3NHCl \quad (179)$$

M = Pd, Pt

L = $Ph_3P$, $PMePh_2$, $PMe_2Ph$

X = Cl, Br

$[RhCl(CO)(PPh_3)_2]$ and $[CoCl(CO)_2(PPh_3)_2]$ were inert under these conditions. These reactions were assumed to proceed by nucleophilic attack of the alcohol on a cationic carbonyl complex formed *in situ* by replacement of a chloride ligand by carbon monoxide. (Diene)platinum(II) alkoxycarbonyl complexes were formed by the reaction of alcohols with unstable neutral carbonyl complex intermediates (equation 180)[224].

$$[(RO\text{-diene})Pt(Cl)(CO)] + R'OH \longrightarrow [(\text{diene})Pt(CO)(COR')] + ROH \quad (180)$$

Alkoxycarbonyl complexes have been implicated in a number of transition metal-catalysed carboxylation [mostly palladium(II)] procedures[225]. Simple monoolefins were converted to $\alpha,\beta$-unsaturated esters[226] or 1,2-diesters[227] when treated with carbon monoxide, an alcohol, and palladium(II) complexes as catalysts. The key step was thought to be insertion of the olefin into a $\sigma$-acylpalladium(II) complex (equation 181). In the absence of olefinic substrate, oxalate esters were formed by

$$RCH{=}CH_2 + CO + R'OH \xrightarrow[CuCl_2]{PdCl_2} RCH{=}CHCO_2R' + RCH(CO_2R')CH_2CO_2R' \quad (181)$$

$$\text{via } PdCl_2: [ClPdCO_2R'] \xrightarrow[RCH=CH_2]{insert} [ClPdCH(R)CH_2CO_2R'] \xrightarrow{CO} [ClPdCOCH(R)CH_2CO_2R']$$

$$[ClPdCH(R)CH_2CO_2R'] \xrightarrow{-PdH} RCH{=}CHCO_2R'; \quad [ClPdCOCH(R)CH_2CO_2R'] \xrightarrow{R'OH} RCH(CO_2R')CH_2CO_2R'$$

coupling of two alkoxycarbonyl groups (equation 182)[228]. Alkynes underwent a similar carboalkoxylation (equation 183). When acetylenic alcohols were subjected to similar reaction conditions, $\alpha$-methylene lactones were formed (equation 184). The mechanism of this process has been carefully studied and the alkoxycarbonylpalladium complex was shown to be the key intermediate in this reaction[229]. Another

$$[Pd(OAc)_2(PPh_3)_2] \xrightarrow{CO,\ MeOH} [Pd(CO_2Me)(OAc)(PPh_3)_2] \xrightarrow{CO,\ MeOH} [Pd(CO_2Me)_2(PPh_3)_2] \longrightarrow [Pd(PPh_3)_n] + (CO_2Me)_2 \quad (182)$$

$$RC{\equiv}CH + CO + MeOH \xrightarrow{Pd(II)\ cat.} RCH{=}CHCO_2Me + CH_2{=}C(R)CO_2Me \quad (183)$$

$$[PdX_2L_2]+CO \longrightarrow [Pd(CO)LX_2] \xrightarrow{HOCH_2CH_2C\equiv CH} \left[ LX_2Pd-C(=O)-OCH_2CH_2C\equiv CH \right]^- \quad (184)$$

$$\longrightarrow \left[ LX_2Pd(X)C=C(\text{lactone}) \right]^- \xrightarrow{H^+} \text{(α-methylene lactone)} + LX_2Pd$$

$$R'O^-Na^+ + [Ni(CO)_4] \longrightarrow [R'OC(O)Ni(CO)_3]^- \xrightarrow{trans-RCH=CHBr} trans-RCH=CHCO_2R' \quad (185)$$

reaction likely to involve an alkoxycarbonyl complex as the reactive intermediate is the carboxylation of vinyl halides by nickel carbonyl and alkoxides (equation 185)[230]. However, this reaction has not been subjected to a careful mechanistic study. With sensitive substrates this may be the method of choice for carboxylation of vinyl halides (equation 186)[231].

(186)

Me, I → $CO_2Me$

$Mg/CO_2/MeOH$ 10%

$MeO^-$, $[Ni(CO)_4]$, MeOH 85%

Carbamoyl complexes are available from the reaction of amines with metal carbonyl complexes[222]. Ammonia and primary and secondary amines reacted smoothly, but aromatic amines did not produce carbamoyl complexes. Unusual platinum(II) and palladium(II) carbamoyl complexes in which $\alpha$-amino acids were the amine component have been made and characterized (equation 186a)[232]. Car-

$$[M(L_2)(X_2)]+CO \xrightarrow{H_2NCHR'CO_2R} [M(L_3)(CONHCHR'CO_2R)] \quad (186a)$$

M = Pd, Pt; R = Et, Me;

$L = Ph_3P$; R′ = H, Me

bamoyl complexes of nickel were thought to be involved in the conversion of vinyl halides to amides, as shown in equation 187[230,233]. They are also likely intermediates in most transition metal-catalysed carbonylations of amines to formamides (e.g. equation 188)[234], isocyanates, and ureas.

$$[Ni(CO)_4]+LiNMe_2 \xrightarrow[thf]{PhC\equiv CH} Me_2NC(O)CH(Ph)CH_2C(O)NMe_2 \quad (187)$$

$$[Ni(CO)_4]+LiNMe_2 \xrightarrow{PhCH=CHBr} PhCH=CHC(O)NMe_2$$

$$\begin{matrix} RNH_2 \\ \text{or} \\ R_2NH \end{matrix} + CO \xrightarrow[180\,^\circ C,\ 750\ psi]{[Ni(CO)_4]} \begin{matrix} RNHCHO \\ \text{or} \\ R_2NCHO \\ \sim 50\% \end{matrix} \tag{188}$$

Isonitriles coordinated to transition metals also react readily with nucleophiles, forming metal carbenes which are usually fairly stable (equation 189)[208a,d]. These

$$[L_nMC{\equiv}NR] + R'XH \longrightarrow [L_nM{=}C(XR')NHR] \tag{189}$$

$$RXH = MeOH, RNH_2, ArNH_2, RSH$$

types of carbenes are most common for palladium and platinum, but those of other metals (Fe, Mo) are also known. Cationic carbene complexes of platinum(II) and palladium(II) were prepared by the reaction of the corresponding isonitrile complexes with amines (equation 190)[235]. Nickel, palladium, and platinum carbene

$$[PtCl(PEt_3)_2CNR]^+ + R'R''NH \longrightarrow [PtCl(PEt_3)_2\{{=}C(NHR)NR'R''\}] \tag{190}$$

complexes containing chelating fluorocarbon ligands were prepared in a similar manner (equation 191)[236]. Alcohols also reacted with palladium isocyanide complexes to produce alkoxycarbene complexes (equation 192)[237]. The kinetics of

$$\left[\left(\begin{matrix} CF_2 \\ \\ CF_2 \end{matrix}\right) M(CNR)_2\right] + R'_2NH \longrightarrow \left[\left(\begin{matrix} CF_2 \\ \\ CF_2 \end{matrix}\right) M(CNR)\{{=}C(NHR)NR'_2\}\right] \tag{191}$$

$M = Pd, Pt;\ R = Bu^t, Pr^i$

$$[PdCl_3(CNPh)]^- + MeOH \longrightarrow [PdCl_3\{{=}C(OMe)NHPh\}] \tag{192}$$

carbene formation from *cis*-$[PdCl_2(CNR)(PPh_3)]$ and substituted anilines were consistent with direct nucleophilic attack of the amine on the coordinated isonitrile to form a stable intermediate which underwent a subsequent proton transfer to form the observed carbene complex (equation 193)[238].

$$cis\text{-}[PdCl_2(L)(CNC_6H_4X\text{-}p)] + p\text{-}RHNC_6H_4Y \rightleftharpoons [\bar{P}dCl_2(L)\{C({=}NC_6H_4X\text{-}p)\overset{+}{N}H(R)C_6H_4Y\text{-}p\}] \tag{193}$$

$$\swarrow$$

$$[PdCl_2(L)\{{=}C(NHC_6H_4X\text{-}p)N(R)C_6H_4Y\text{-}p\}]$$

## C. Nucleophilic Attack on Metal-Complexed Olefins

### 1. *With chelating olefin complexes*

Complexation of an olefin to a transition metal, particularly a Group VIII metal in a higher oxidation state (+2), often activates the olefin to undergo nucleophilic attack. Chelating diolefin complexes of palladium(II) and platinum(II) have long been known to undergo nucleophilic attack at one of the coordinated olefins to produce $\sigma$-alkylmetal complexes[239]. Both the platinum(II) and palladium(II) complexes of cycloocta-1,5-diene and dicyclopentadiene reacted with methanol in high yield to give the $\sigma$-alkyl-$\pi$-olefin metal(II) complexes (equation 194). The same

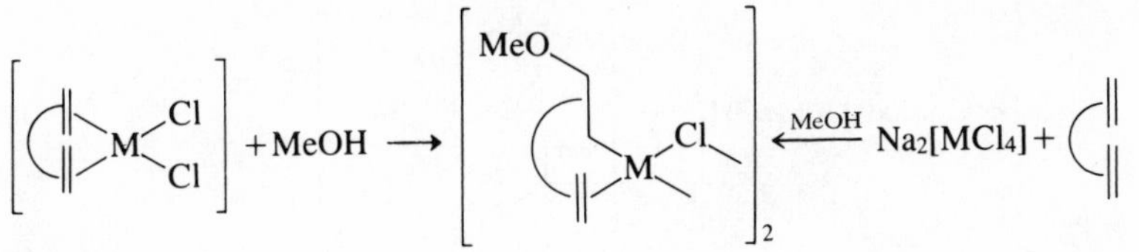


(194)

M = Pd, Pt

 = cod, dcpd, nbd

complexes were formed by direct reaction of the diolefin and chloro-palladate or -platinate in methanol[240]. These complexes were something of a curiosity for a number of years. However, later studies showed that they underwent several interesting further reactions. The complex resulting from alkoxypalladation of dicyclopentadiene was acylated by treatment with acetyl chloride, or by exposure to carbon monoxide followed by electrophilic cleavage (equation 195)[241]. The alkoxypalladation product of cyclooctadiene reacted with isonitriles by an insertion process to produce a ketenimine (equation 196)[242]. The $\beta$-lactone linkage was appended to cyclooctadiene by a hydroxypalladation/CO insertion/acyl cleavage reaction sequence (equation 197)[243].

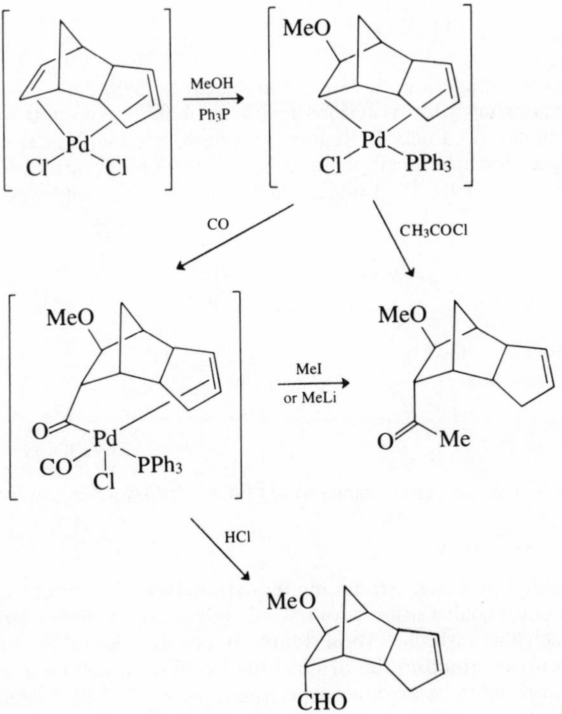


(195)

$$[PdCl_2(cod)] + MeOH \longrightarrow [\text{(MeO-cyclooctenyl)Pd}(\mu\text{-Cl})]_2 \tag{196}$$

RNC, dbu

5-OMe-6-(=C=NR)cyclooct-l-ene

$$[PdCl_2(cod)] \xrightarrow{H_2O} [\text{(HO-cyclooctenyl)Pd}(\mu\text{-Cl})]_2 \xrightarrow{CO} [\text{(HO-cyclooctenyl-C(O))Pd}(\mu\text{-Cl})]_2 \longrightarrow \text{bicyclic }\beta\text{-lactone } (74\%) \tag{197}$$

Acetate also attacks chelating diolefin complexes of palladium. A number of complex rearrangement/insertion reactions result, probably promoted by oxidation, and unusual products are obtained. Palladium acetate reacted with hexa-1,5-diene to produce 1-methylene-3-acetoxycyclopentane by a nucleophilic attack–olefin insertion sequence (equation 198)[244]. Longer chain dienes did not cyclize. Bicyclic

$$[\text{(hexa-1,5-diene)Pd(OAc)}_2] + AcO^- \longrightarrow [\text{AcO-hexenyl-Pd(OAc)}] \tag{198}$$

1-methylene-3-acetoxycyclopentane (64%) ⟵ [1-$CH_2$PdOAc-3-acetoxycyclopentane]

materials were obtained by either irradiating (equation 199)[245] or oxidizing (equation 200)[246] the initial acetoxypalladation product of cycloocta-1,5-diene. Bridged polycyclic olefins in which the carbon–carbon double bonds are rigidly held close to the $\sigma$-alkylpalladium centre formed in the original nucleophilic attack by acetate lead to even more complex products, as evidenced by equations 201[247], 202[248], and 203[249].

$[Pd(OAc)_2(cod)] \xrightarrow{AcO^-}$ [OAc ... Pd OAc]$_2$ (199)

insertion

OAc — $h\nu$ — OAc — $Pb(OAc)_4$ — OAc — Pd (200)

OAc

Pd(OAc)$_2$ — AcO — Pd →

(201)

AcO — −PdH — AcO — H — Pd

CO$_2$R — CO$_2$R — PdCl$_2$/CuCl$_2$ AcONa — OAc — CO$_2$R — O — O (202)

74%

PdCl$_2$/CuCl/HOAc NaOAc — OAc — Cl (203)

70%

A variety of other nucleophiles also react with chelating diolefin complexes of palladium and platinum. Amines reacted with the platinum and palladium complexes of cyclooctadiene and dicyclopentadiene to give the expected amination products, whereas the norbornadiene complexes decomposed to metallic palladium on treatment with benzylamine[250]. The dicyclopentadiene complexes of platinum also reacted with alkoxides, aniline, thiocyanate, and thiophenol in a similar fashion (equation 204). The sulphur nucleophiles also displaced the halide ligands on

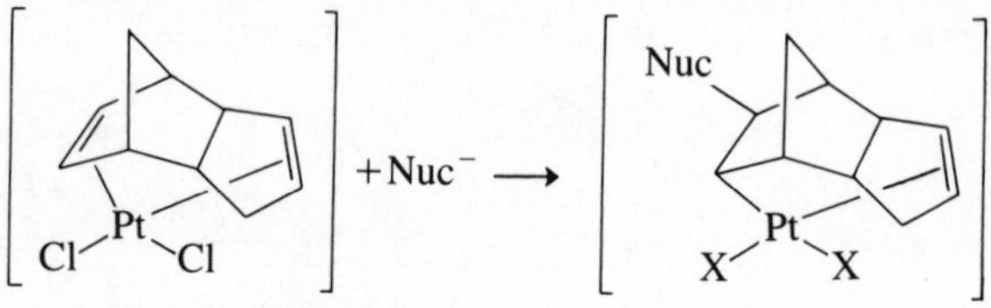

$$[\text{Pt}(\text{dicyclopentadiene})Cl_2] + \text{Nuc}^- \longrightarrow [\text{Pt}(\text{Nuc-dicyclopentadienyl})X_2] \qquad (204)$$

$Nuc^- = RO^-, PhNH^-, SCN^-, PhS^-$
$X = Cl, Py, PPh_3, PhNH, SPh$
$Y = Cl, Py, SCN, SPh$

platinum. Surprisingly, the remaining olefin in the initially formed $\sigma$-alkyl complex was also subject to nucleophilic attack, provided that the chloro ligands were replaced by the bipyridyl ligand, generating a cationic complex (equation 205). A number of generalizations emerged from these studies. With alkoxides as nucleophiles, the order of reactivity was $MeO > Pr^nO > Bu^tO$, indicating a strong steric effect in the nucleophilic attack process. Sulphur nucleophiles attack both the olefin and the metal[251].

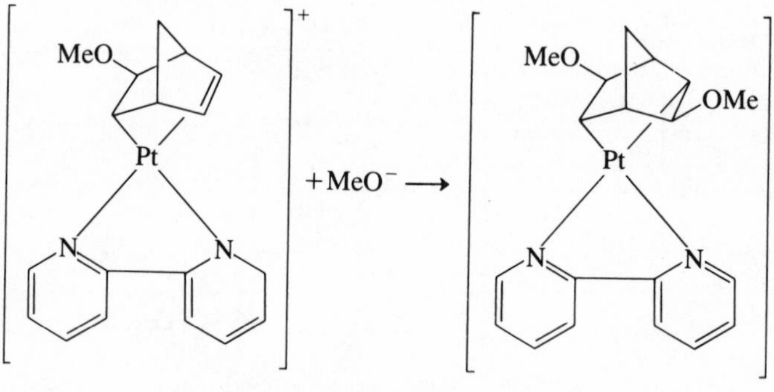

$$[\text{Pt}(\text{MeO-C}_{10}H_{12})(\text{bipy})]^+ + MeO^- \longrightarrow [\text{Pt}(\text{MeO,OMe-C}_{10}H_{12})(\text{bipy})]$$

The cyclooctadiene complex of palladium reacted with carbamates to produce aminated cyclooctene (equation 206)[252], while reaction with secondary amines followed by carbonylation produced a mixture of bicyclic amino esters and ketones from insertion of the remaining olefin into the palladium acyl $\sigma$-bond (equation 207)[253].

$$[PdCl_2(cod)] + H_2NCO_2Et \longrightarrow \underset{58\%}{\text{5-(NHCO}_2\text{Et)cyclooct-1-ene}} \qquad (206)$$

Stabilized carbanions also reacted with chelating diolefin complexes of palladium and platinum. One of the earliest systems studied was the reaction of cyclooctadiene

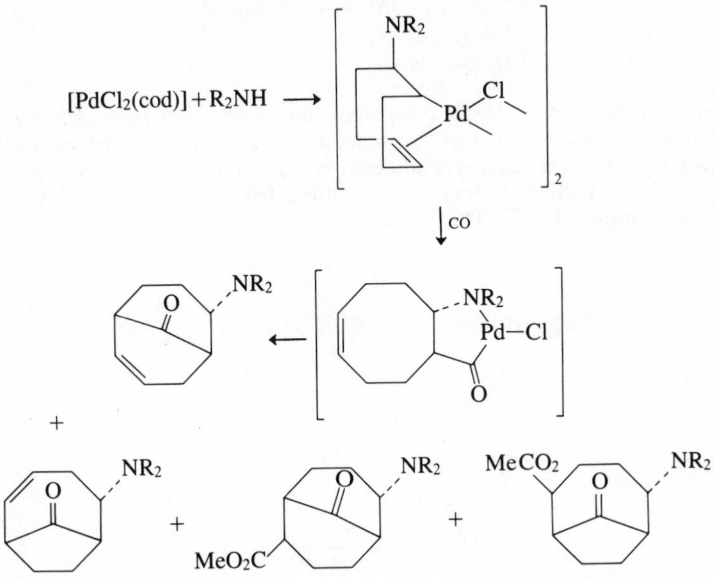

$$[PdCl_2(cod)] + R_2NH \longrightarrow \quad (207)$$

complexes with diethyl malonate and acetylacetone (equation 208)[253–256]. The remaining double bond of the cyclooctadiene resisted further attack, and the $\sigma$-alkyl complexes were fairly stable. However, they could be forced to undergo further reactions (equation 208). A careful study of the reaction of the platinum and palladium complexes of dicyclopentadiene with diethyl malonate, acetylacetone, and ethyl acetoacetate showed that this reaction occurred in a *trans,exo* fashion at the 5,6-double bond of the diolefin, without skeletal rearrangement of the dicyclopentadiene skeleton. This alkylation reaction was sensitive to the nature of both the chelating diolefin and the carbanion. The corresponding norbornadiene complexes of both palladium and platinum produced only metallic palladium or platinum on reaction with these carbanions[250]. Less stabilized carbanions reacted with the cyclooctadiene complex of palladium and platinum to give stable dialkylmetal species from nucleophilic attack on the metal rather than the olefin (equation 209)[257,258].

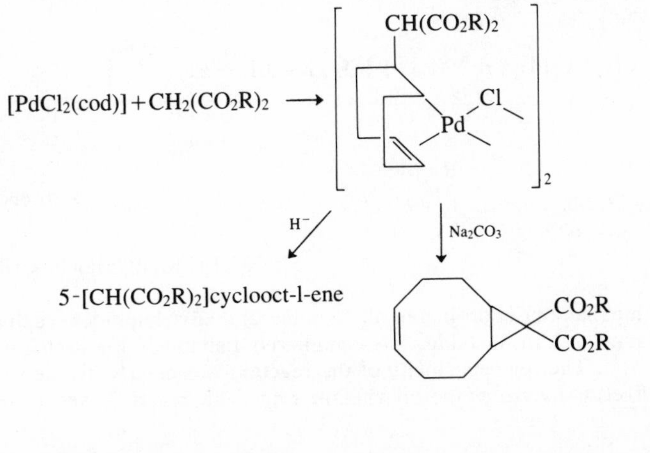

$$[PdCl_2(cod)] + CH_2(CO_2R)_2 \longrightarrow \quad (208)$$

5-[CH(CO$_2$R)$_2$]cyclooct-1-ene

$$[PdCl_2(cod)] + RM \longrightarrow [PdR_2(cod)] \quad (209)$$

$RM = MeMgX, PhMgX, PhSO_2CH_2Li$

Although chelating diolefin complexes have been the most extensively studied, other types of chelating olefins also react with nucleophiles. Allylamines were found to react with lithium chloropalladate in methanol to give a stable five-membered cyclic $\sigma$-alkylpalladium(II) complex resulting from alkoxypalladation of the allylamine (equation 210)[259]. The ability to form a chelate, *five*-membered ring was

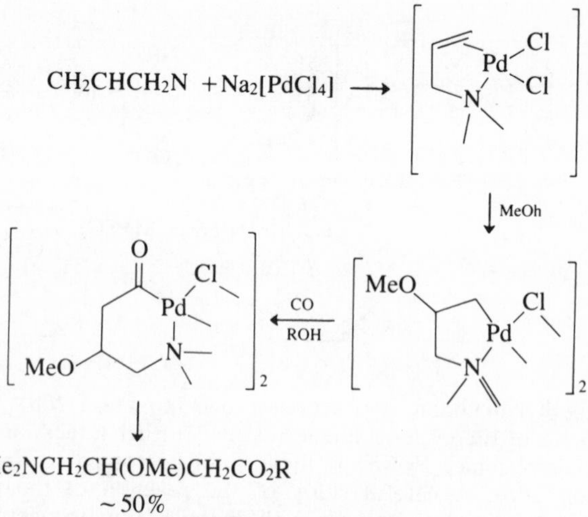

$CH_2CHCH_2N + Na_2[PdCl_4] \longrightarrow$ [chelate complex] $\xrightarrow{MeOh}$ $[\ldots]_2$ $\xrightarrow{CO,\ ROH}$ $[\ldots]_2$ $\longrightarrow$ $Me_2NCH_2CH(OMe)CH_2CO_2R$ ~50% (210)

crucial to both the success of the initial nucleophilic attack and to the stability of the resulting $\sigma$-alkylpalladium complex. If a five-membered chelate ring could not be formed, nucleophilic attack did not occur. If a larger ring was formed by an insertion reaction, the resulting complex decomposed. For example, exposure of the methoxypalladation product of *N,N*-dimethylallylamines to carbon monoxide resulted in the formation of a $\gamma$-amino ester[260]. The expected six-membered $\gamma$-aminoacylpalladium complex intermediate could not be detected, attesting to its instability. Conjugated enones also inserted into the $\sigma$-alkylpalladium complex, leading to further functionalization of the starting allylamine (equation 211)[261].

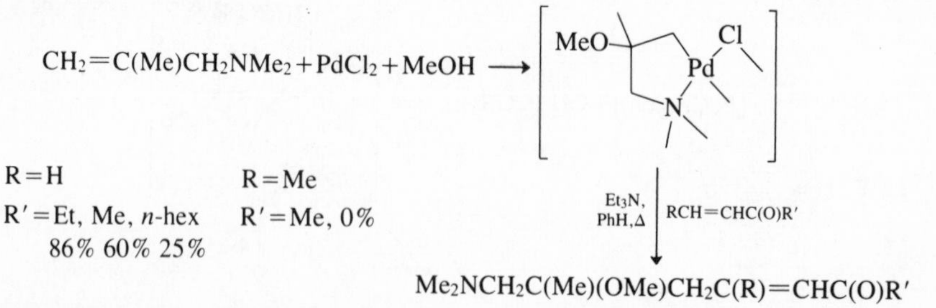

$CH_2{=}C(Me)CH_2NMe_2 + PdCl_2 + MeOH \longrightarrow$ [complex] $\xrightarrow{RCH=CHC(O)R',\ Et_3N,\ PhH,\Delta}$ $Me_2NCH_2C(Me)(OMe)CH_2C(R){=}CHC(O)R'$

R = H
R' = Et, Me, *n*-hex
86% 60% 25%

R = Me
R' = Me, 0%

Both palladium-coordinated allyl amines and allyl sulphides reacted with stabilized carbanions to form stable, five-membered palladiacycles in high yield (equation 212)[262,263]. The regiospecificity of the reaction was clearly dictated by the formation of a five(*vs.* four)-membered chelate ring. The reaction was restricted to olefins

$$XCH_2CH{=}CH_2 + Li_2[PdCl_4] \longrightarrow \left[\text{(chelate)}PdCl_2\right] \xrightarrow[81-94\%]{R^-} \left[\text{R-alkyl-Pd(Cl) chelate with X}\right]_2 \quad (212)$$

X=$NMe_2$, $SCHMe_2$

$R^-$=$CH(CO_2Et)_2$, $CH(COPh)_2$, $CH_3COCHCO_2Et$, $CH_3COCHCOPh$, $C_2H_5C(CO_2Et)_2$, 2−($CO_2Et$)cyclopentan-l-one

which could chelate. Thus, allyl alcohols, allyl phenyl ether, ethyl acrylate, and oct-1-ene (none of which coordinate as strongly to palladium as do nitrogen- or sulphur-containing ligands) failed to undergo the alkylation process. Only stabilized carbanions reacted cleanly. Ketone enolates, organocopper, and Grignard reagents produced metallic palladium and intractable mixtures of organic compounds. Homoallylic amines and sulphides also underwent this 'carbopalladation' reaction with stabilized carbanions (equation 213)[264]. In these cases alkylation occurred at the less

$$X(CH_2)_2CH{=}CH_2 + Li_2[PdCl_4] \longrightarrow \left[\text{(chelate)}PdCl_2\right] \xrightarrow{R^-} \left[\text{R-alkyl-Pd(Cl) chelate with X}\right]_2 \quad (213)$$

X=$NMe_2$, $SCHMe_2$

$R^-$=$CH(CO_2Et)_2$, 2−($CO_2Me$)cyclopentan-l-one

($MeO^-$ reacted similarly)

substituted olefinic carbon, in contrast to the situation with allylic systems, for which attack at the *most* substituted olefin carbon occurred. However, both systems reacted to produce the stable five-membered chelate $\sigma$-alkylpalladium complex. *O*-Allyl-*N,N*-dimethylhydroxylamines also reacted with $PdCl_2$ and stabilized carbanions to form five-membered chelate $\sigma$-alkylpalladium complexes (equation 214)[265]. Further

$$Me_2C{=}CH(CH_2)_2C(Me)(ONMe_2)CH{=}CH_2 + [PdCl_2(MeCN)_2] + NaC(R^1)(COR^2)COR^3 \quad (214)$$

$$\longrightarrow \left[Me_2C{=}CH(CH_2)_2\text{-C(Me)(O-N-Pd ring)-}CH_2C(R^1)(COR^2)COR^3\right]$$

$$\xrightarrow{TMSCl} Me_2C{=}CH(CH_2)_2C(Me){=}CHCH_2C(R^1)(COR^2)COR^3 \quad 56-73\%$$

$R^1$=H, Me, $Bu^n$

$R^2$=MeO, Me, Et

$R^3$=MeO, EtO

reaction with trimethylsilyl chloride removed both the palladium and the ether group to produce the alkylated olefin.

The synthetic utility of these processes lies in the further transformations of the $\sigma$-alkylpalladium complexes produced by the nucleophilic attack. Hydrogen or hydride reducing agents such as $NaBH_4$ and $NaBH_3CN$ reduced these complexes to give the saturated compounds corresponding to the formal addition of the stabilized keto ester across the double bond[263,264]. Conjugated enones inserted into the palladium—carbon $\sigma$-bond (equation 215)[263]. This chemistry has been used as an elegant synthesis of a key prostaglandin precursor (Corey's lactone)[266]. Unfortunately, full experimental details of this series of complex transformations have not yet been published (1982).

$$\left[\text{R-CH(CH}_2\text{X)CH}_2\text{-Pd(Cl)}\right]_2 \xrightarrow[NaBH_4]{H_2} XCH_2CH(R)Me$$

$$\xrightarrow{CH_2{=}CHC(O)Me,\ Et_3N} XCH_2CH(R)CH_2CH{=}CHC(O)Me \qquad (215)$$

In all of the above cases, formation of a five-membered chelate ring was essential to the reaction. When the chelating ligand was changed from nitrogen or sulphur to phosphorus, the situation changed. The reactions of the ligands $CH_2{=}CH(CH_2)_nPPh_3$ ($n = 1, 2, 3$) with both palladium(II) and platinum(II) salts were investigated. The but-3-enylphosphine ($n = 2$) chelated to both platinum(II) and palladium(II), the pent-4-enylphosphine ($n = 3$) chelated only to platinum(II) and the allylphosphine ($n = 1$) chelated to neither. Methoxide and acetate reacted with the ligands which did chelate to product six- or seven-membered chelate $\sigma$-alkylplatinum complexes[267].

## 2. With monolefin complexes

In contrast to chelating olefin complexes of palladium(II), which react with a wide range of nucleophiles, simple monoolefin complexes were much more restricted in their reactions until very recently. The main difficulty was a highly competitive displacement of the olefin by the nucleophile. Oxygen nucleophiles were among the first reagents to be used successfully to attack simple monoolefinpalladium(II) complexes, since oxygen ligands do not coodinate particularly well to palladium(II), and displacement of the olefin does not compete. Reaction of olefinpalladium(II) complexes with water is the basis of the industrially important Wacker process for the 'oxidation' of ethylene to acetaldehyde (equation 216). Since palladium(II) was required for the nucleophilic attack step and palladium(0) was produced in the last step, reoxidation was required to make the system catalytic in palladium. Copper(II) salts proved to be the most effective, and are widely used for this process[268]. The key step in the Wacker process is clearly nucleophilic attack of water on the metal-bound olefin. There is still some controversy regarding the question of nucleophilic attack by an external nucleophile *vs.* prior coordination of the nucleophile[269], although model studies suggested that *trans* attacked by non-coordinated water was occurring[270,271]. Other oxygen nucleophiles, such as acetate and alcohols, also reacted well

$$PdCl_2 + CH_2{=}CH_2 \rightleftharpoons [PdCl_2(\eta^2{-}CH_2{=}CH_2)]_2 \rightleftharpoons [(H_2O)(CH_2{=}CH_2)PdCl_2] \xrightarrow{H_2O:} [HOCH_2CH_2PdCl_2] \rightarrow HCl + Pd(0) + MeCHO \quad (216)$$

$CuCl_2, O_2$ (HCl + Pd(0) → $PdCl_2$)

in Wacker-type reactions with olefins, leading to vinyl acetates, acetals, and glycol ethers. Many of these have been commercialized[268]. All of them are thought to involve nucleophilic attack on a palladium-bound olefin, elimination of Pd(0) from the $\sigma$-alkylpalladium(II) complex, and reoxidation of Pd(0) to Pd(II) by air and copper salts, or some other suitable oxidant.

Longer chain olefins are invariably oxidized to ketones, through nucleophilic attack on the most substituted olefinic carbon. This also is suggestive of external *trans* attack by water rather than a *cis* insertion, since external nucleophilic attack occurs predominantly at the most substitued carbon, whereas insertion processes tend to place the nucleophile at the less substituted carbon[272]. For longer chain olefins, terminal double bonds reacted faster than internal olefins, and polar solvents such as dmf, sulpholane, or *N*-methylpyrrolidone were preferred to water[268b]. This oxidation has found extensive use in organic synthesis when coupled to the palladium-catalysed telomerization reactions of butadiene, discussed below. Recently it has been shown that palladium nitro complexes were very efficient for the oxidation of terminal olefins to methyl ketones. The key step was proposed to be the intramolecular nucleophilic attack of a nitro-group oxygen on the palladium-bound olefin (equation 217)[273].

$$RCH{=}CH_2 + [PdCl(NO_2)(MeCN)_2] \xrightarrow{PhMe} RC(O)Me \quad (217)$$

catalyst

via [L(Cl)Pd(olefin)(N(+)(=O)O⁻)] → [L(Cl)Pd–CH₂–CH(R)(H)–O–N(+)(=O)] → RC(O)Me

Reactions of monoolefinpalladium(II) complexes with other classes of nucleophiles proved to be a much more sensitive and difficult reaction. With amines as nucleophiles, displacement of the olefin by the amine nucleophile was the major difficulty. By carrying out the amination reaction at temperatures below −50 °C, this pathway was suppressed, and amination of olefins by secondary amines was achieved in excellent yield[274]. The reaction proceeded best with nonhindered secondary amines and terminal olefins. Primary amines and internal olefins reacted to give lower (*ca.* 50%) yields of amination. Trisubstituted olefins reacted in very low yield, as did ammonia. The stereochemistry of amination was cleanly *trans*, indicating the nucleophile attacked from the face opposite the metal, without prior coordination[275]. Regioselective attack at the most substituted carbon, also consistent with external nucleophilic attack, was observed. The reaction was not catalytic in palladium, since a reduction step was needed to prevent the formation of unstable enamines by $\beta$-elimination.

This reaction has several features of general importance to the reactions of nucleophiles with olefinpalladium(II) complexes. Three equivalents of amine were required for reasonable yields and no amination had occurred after addition of the first equivalent. This suggested that nucleophilic attack was occurring on a palladium–olefin–amine complex, rather than on the olefin–palladium dimer itself. A mechanism consistent with the experimental observations is shown in equation 218.

$$\tfrac{1}{2}[PdCl_2(\eta^2\text{-}RCH{=}CH_2)]_2 + R'_2NH \longrightarrow [PdCl_2(NHR'_2)(\eta^2\text{-}RCH{=}CH_2)]$$

$$[PdCl_2(NHR'_2)(\eta^2\text{-}RCH{=}CH_2)] \longrightarrow [PdCl(NHR'_2)_2(\eta^2\text{-}RCH{=}CH_2)]^+ [PdCl_3(\eta^2\text{-}RCH{=}CH_2)]^-$$

$$[PdCl_2(NHR'_2)(\eta^2\text{-}RCH{=}CH_2)] \text{ or } [PdCl(NHR'_2)_2(\eta^2\text{-}RCH{=}CH_2)]^+ \xrightarrow{R'_2NH} [RCH(NR'_2)CH_2Pd(Cl)(NHR'_2)] + R_2NH_2Cl \qquad (218)$$

$$[RCH(NR'_2)CH_2Pd(Cl)(NHR'_2)] \xrightarrow{H_2 \text{ or } H^-} MeCH(R)NR'_2$$

$$[RCH(NR'_2)CH_2Pd(Cl)(NHR'_2)] \xrightarrow{CO} [RCH(N R'_2)CH_2C(=O)Pd(Cl)(NHR'_2)]$$

Either the neutral olefinpalladium amine complex or the cationic portion of the complex ion pair reacted with the second equivalent of amine to produce the relatively unstable $\beta$-aminoalkylpalladium complex, which has been isolated and characterized by n.m.r. spectroscopy at temperatures below $-20\,°C$[276]. Reduction of this complex with hydrogen or hydride reducing agents gave the observed amination product. The unstable complex was also treated with carbon monoxide to produce the very stable $\beta$-aminoacylpalladium(II) complex, which was fully characterized by an X-ray crystal structure[277]. This palladium-assisted amination of olefins was also the key step in the aminoacylation, oxamination, and aziridine-forming reactions of olefins discussed above. [Olefinplatinum(II) complexes also underwent amination with amines. In this case, all of the intermediates were stable, and were isolated and fully characterized[278].] Recent uses of the palladium assisted amination of olefins are described in equations 219[279] and 220[280].

$$PhNHR + CH_2{=}CHZ \xrightarrow{Pd^{II}} PhN(R)CH{=}CHZ \qquad (219)$$

$$Z = CO_2Me,\ COMe,\ CN \qquad 50\text{–}70\%$$

$$RNa + CH_2{=}CH_2 + [PdCl_2(CH_3CN)_2] \longrightarrow [R(CH_2)_2Pd] \qquad (220)$$

$$[R(CH_2)_2Pd] \xrightarrow{CO,\ MeOH} R(CH_2)_2CO_2Me; \quad \xrightarrow{H_2} REt; \quad \longrightarrow RCH{=}CH_2$$

$$R = \text{indol-1-yl}$$

The use of carbanions as nucleophiles for addition to olefinpalladium(II) complexes posed different problems. In most cases, the major reaction of carbanions with olefinpalladium(II) complexes appeared to be reduction of the palladium(II) to metallic palladium and concurrent oxidative coupling of the carbanion. For example, methyllithium reacted with styrene in the presence of palladium(II) chloride to give a mere 3% yield of $\beta$-methylstyrene. With palladium acetate, the yield rose to 75% and with palladium(II) acetylacetone it rose to 90%[280] [this increase in yield parallels an increase in the resistance to reduction of the palladium(II) salt]. By the use of specifically deuteriated styrenes, this alkylation was shown to proceed by initial nucleophilic attack at the metal, followed by a *cis*-insertion and a *cis*-elimination (equation 221). The regiochemistry (alkylation of the *less* substituted carbon) was

$$\left[-\overset{|}{\underset{|}{Pd}}(\eta^2-PhCH{=}CH_2)\right] + CH_3Li \xrightarrow[25\ °C]{thf} [PdL_2(Me)(\eta^2\text{-}PhCH{=}CH_2)]$$

*cis*-insertion (221)

$$\left[\begin{array}{c} PdL_2 \quad CH_3 \\ H\cdots C{-}C\cdots H \\ Ph \qquad\quad H \end{array}\right] \longrightarrow \left[\begin{array}{c} PdL_2 \quad H \\ H\cdots C{-}C\cdots CH_3 \\ Ph \qquad\quad H \end{array}\right]$$

$\beta$-elimination

$$HX + Pd(0) + trans\text{-}PhCH{=}CHMe$$

also consistent with this mechanism. In contrast, the stabilized carbanion of dimethyl malonate reacted in very low yield (18%) and the regiochemistry was opposite to that observed for methyllithium, suggesting a change in mechanism.

A more general and efficient alkylation of olefins by carbanions was based on observations reported for the related amination of olefins discussed above. In that reaction, the key reactive intermediate appeared to be an olefinpalladium(II) amine species of some type. Whereas reaction of olefinpalladium(II) chloride complexes with stabilized carbanions resulted in no alkylation of the olefin, addition of two equivalents of triethylamine to the olefinpalladium(II) complex prior to addition of the carbanion led to high yields of alkylation with a range of olefins and stabilized carbanions (equation 222)[282]. Triethylamine was most effective for promoting this

$$RCH{=}CH_2 + [PdCl_2(CH_3CN)_2] + 2Et_3N + R'\bar{C}(X)Y \xrightarrow[-60\ °C]{thf} \left[-\overset{|}{\underset{|}{Pd}}CH_2CH(R)C(R')(X)Y\right]$$

$\beta$-elimination; $H_2$; CO, MeOH (222)

$$CH_2{=}C(R)C(R')(X)Y \qquad MeCH(R)C(R')(X)Y \qquad MeOC(O)CH_2CH(R)C(R')(X)Y$$

$R = H$, Me, Et, $Bu^n$, NHAc  $X = CO_2Et$, $CO_2Me$, $CO_2Bu^t$, COMe
$R' = H$, Me, *n*-hex  $Y = CO_2Et$, $CO_2Me$, COMe, Ph

reaction (90% yields). Terminal monoolefins were alkylated in almost quantitative yield, with alkylation at the most substituted carbon predominating. The electron-rich olefin *N*-vinylacetamide reacted in high yield, but the electron-deficient olefin methyl acrylate was not alkylated under thse conditions. Internal olefins were alkylated in only 30–40% yields, and cyclohexane and isobutene did not react under these conditions. With $\alpha$-acetamidomalonic esters as the carbanion, $\alpha$-amino acid derivatives were prepared (equation 223)[283].

$$CH_2{=}CH_2 + [PdCl_2(MeCN)_2] + Et_3N + AcNHC(CO_2Et)_2 \longrightarrow \xrightarrow{H_2} MeCH_2C(NHAc)(CO_2Et)_2 \quad (223)$$

This alkylation process was restricted to the use of stabilized carbanions ($pK_a$ 10–18) as nucleophiles. Less stabilized carbanions reduced the complexes without alkylating them. However, addition of hmpa to the olefinpalladium(II) complex prior to addition of the triethylamine and the carbanion permitted the use of much less stabilized carbanions[282]. Under these conditions ketone and ester enolates, oxazoline anions (carboxylic acid carbanion equivalents), protected cyanohydrin anions (acyl anion equivalents), and even benzylmagnesium chloride alkylated olefins in fair to excellent yields. With non-stabilized carbanions, propene reacted almost exclusively at the 2-position, hexene at the 1-position, and styrene at both positions, the product distribution depending on the specific carbanion used. The $\sigma$-alkylpalladium intermediates in these alkylation reactions were carbonylated to result in the overall carboacylation of olefins (equation 222)[284]. Stable $\sigma$-alkylpalladium(II) complexes resulted from the alkylation of cationic cyclopentadienylpalladium(II) olefin complexes (equation 224)[285].

$$[Pd(PR_3)(Cp)(\eta^2\text{-}CH_2{=}CH_2] + X\bar{C}HY \longrightarrow [Pd(PR_3)(Cp)\{CH_2CH_2CH(X)Y\}] \quad (224)$$

Intramolecular versions of the above palladium-assisted nucleophilic attack reactions on olefins have found extensive use in the synthesis of heterocyclic compounds. The reactions all involve the general features of olefin activation by complexation to palladium(II), but in some instances the regiochemistry is more sensitive to the structural features of the substrate and to the reaction conditions than are the corresponding intermolecular processes. In most cases the intramolecular processes were more facile than the same intermolecular reaction. Benzofurans were produced by the intramolecular oxypalladation of 2-allylphenols (equation 225)[286]. With substitution on the olefin, both benzofurans and chromenes were obtained (equation 226)[287]. The product distribution was affected by the addition of sodium acetate and

$$o\text{-}HOC_6H_4CH_2CH{=}CHR \xrightarrow[Cu(OAc)_2,\ O_2]{Pd(OAc)_2} 2\text{-}(CH_2R)\text{benzofuran} \quad (225)$$

20–60%

$$o\text{-}HOC_6H_4CH_2CH{=}CMe_2 \xrightarrow[NaOAc]{PdCl_2} 2\text{-}(CHMe_2)\text{benzofuran} \quad (226)$$

+

2-[C(Me)=CH$_2$]-2,2-dihydrobenzofuran

+

2,2-dimethylchromene

by changes in substrate concentration[288]. Using a chiral $\pi$-allylpalladium complex as a catalyst led to the production of 2-vinyldihydrobenzofurans in up to 20% optical yield (equation 227)[289]. Flavones were formed by the palladium(II)-catalysed cyclization of 2-hydroxychalcones (equation 228)[290], and 2-vinyltetrahydrofurans were formed from γ,δ-unsaturated alcohols (equation 229)[291]. This chemistry was elegantly applied to the synthesis of brevicomin (equation 230)[292].

$$o\text{-}HOC_6H_4CH_2CH{=}CHMe + [(-)\text{ catalyst}]_2 \xrightarrow[Cu(OAc)_2]{O_2} S(+) \quad (227)$$

$$o\text{-}HOC_6H_3(R)COCH{=}CHAr \xrightarrow[CH_3CN]{Li_2[PdCl_4]} \quad (228)$$

65–85%

$$HOC(R^1)(R^2)(CH_2)_2CH{=}CHMe \xrightarrow[O_2/Cu(OAc)_2]{Pd(OAc)_2} \quad (229)$$

$R^1$=Ph, Me
$R^2$=H, Me, Et, Ph

20–50%

$$CH_2{=}CHCH(R^1)CH_2CH_2CH(OH)CH(OH)R^2 \xrightarrow{PdCl_2/CuCl_2} \quad (230)$$

Carboxylate salts were also capable of attacking π-olefinpalladium complexes. Thus, o-allylbenzoic acids were converted to isocoumarins by reaction with sodium carbonate and palladium(II) chloride (equation 231)[293]. Five-membered lactones resulted from the reaction of 3- and 4-alkenoic acids with $Li_2[PdCl_4]$ (equation 232)[294], while conjugated dienoic acids were converted to pyrones in good yield

$$o\text{-}(CO_2H)C_6H_3(X)CH_2CH{=}CHR \xrightarrow[thf,\ Na_2CO_3]{[PdCl_2(CH_3CN)_2]} \quad (231)$$

60–96%
R=Me, Et, $Pr^i$
X=H, 6—Cl, 7—OMe

$$RCH{=}CHCH_2CO_2H \text{ or } RCH{=}CHCH_2CH_2COOH \xrightarrow{Li_2[PdCl_4]} R(CH_2)_n\text{-butenolide} \quad (232)$$

32–38%
$n$=0, 1

(equation 233)[295]. α,β-Unsaturated oximes were converted to isoxazoles in fair to good yield (equation 234)[296].

$$R^2CH{=}CHC(R^1){=}CHCO_2H \xrightarrow[H_2O]{Li_2[PdCl_4]} 4\text{-}R^1\text{-}6\text{-}R^2\text{-pyran-2-one} \quad (233)$$

$R^1$=H, Me, Ph 65–75%

$R^2$=H, Me, Ph

$$HON{=}C(R^1)C(R^2){=}CHR^3 + [L_2PdCl_2] + PhO^- \xrightarrow[\Delta]{PhH} 3\text{-}R^1\text{-}4\text{-}R^2\text{-}5\text{-}R^3\text{-isoxazole} \quad (234)$$

$R^1, R^3 = Ph, R^2 = H$ 95%

$R^1 = Bu^t, R^2 = H, R^3 = Ph$ 50%

$R^1 = Ph, R^2 = Me, R^3 = Pr^i$ 15%

Nitrogen heterocycles could also be formed in this fashion. Indoles were formed in the palladium(II)-catalysed cyclization of *o*-allylanilines (equation 235)[297]. Substituted allyl groups cyclized by attack at the most substituted carbon (equation 236). These

$$o\text{-}H_2NC_6H_3(X)CH_2CH{=}CH_2 \xrightarrow[\text{benzoquinone, thf, }\Delta]{[PdCl_2(CH_3CN)_2]\text{ cat.}} \quad (235)$$

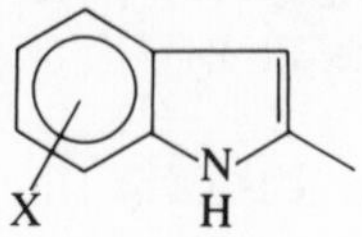

60–90%

$$o\text{-}H_2NC_6H_4CH_2CH{=}CMe_2 \longrightarrow \quad (236)$$

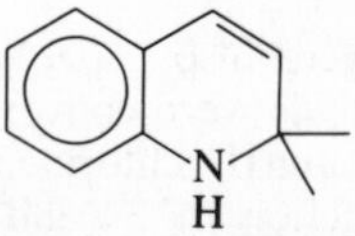

reactions proceeded through unstable σ-alkylpalladium complexes, which could be trapped by CO (equation 237) or conjugated enones[298]. Tricyclic material was formed in an intramolecular version of the olefin insertion reaction (equation 238).

$$o\text{-}H_2NC_6H_4CH_2C(Me){=}CH_2 \xrightarrow{Pd^{II}} \quad (237)$$

(R)

H (R) Pd

$CH_2{=}CHC(O)R'$ CO, MeOH

N H $CH_2CH{=}CHC(O)R$

R=Me, OMe

N R $CH_2CO_2Me$

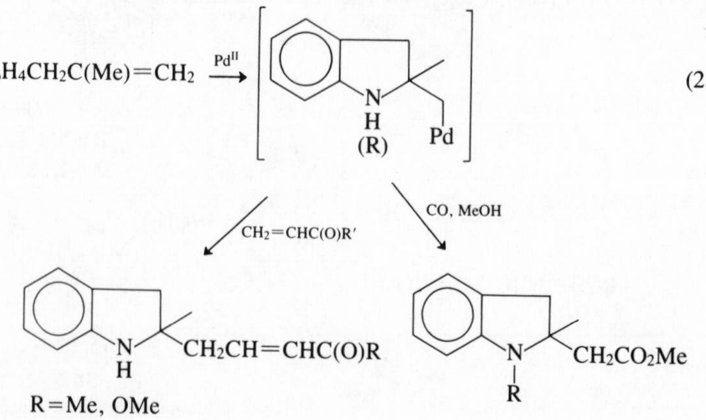

$$o\text{-}[NHC(O)CH{=}CH_2]C_6H_4CH_2C(Me){=}CH_2 \xrightarrow{[PdCl_2(CH_3CN)_2]} \quad (238)$$

83%

The change in regiochemistry was thought to be due to the intramolecular nature of the reaction. Seven-membered rings were formed from an aniline having a dienic side-chain (equation 239)[280]. Azepines were formed from an aminohexatrienyl system (equation 240)[299].

$$cis\text{-}o\text{-}H_2NC_6H_4CH{=}CHCH_2CH{=}CH_2 \xrightarrow{[PdCl_2(MeCN)_2]} \quad (239)$$

CO, MeOH; $H_2$

$CH_2CO_2Me$

34%     36%

$Pd^{II}$, $Na_2CO_3$ → $H_2$ →

NHAc, $CO_2Et$ → NAc, $CO_2Et$

54%

(240)

As the last example shows, amides were sufficiently nucleophilic to attack olefin–palladium(II) complexes. 2-Allylbenzamides cyclized to isoquinolones when treated with palladium(II) chloride and sodium hydride (equation 241)[293]. 2-Amidostilbenes cyclized in a similar manner (equation 242)[300]. Amides of alka-2,4-dienoic acids

$$o\text{-}(CH_2CH{=}CH_2)C_6H_4C(O)NHR + NaH + [PdCl_2(MeCN)_2] \longrightarrow \quad (241)$$

R=H, 68%
R=Me, 91%

$$o\text{-}(CONH_2)C_6H_4CH{=}CHR \xrightarrow{Li_2[PdCl_4]} 1\text{-}OH\text{-}3\text{-}R\text{-isoquinoline} \quad (242)$$

R = *p*-tolyl, PhCO, CN    37–52%

cyclized to pyrid-2-ones (equation 243)[301]. Hydrazides of $\alpha,\beta$-unsaturated acids cyclized to pyrazol-3-ones (equation 244)[302], and unsaturated urea derivatives cyclized to uracil (equation 245)[303]. *N,N*-Diallyl amides of conjugated acids reacted to give a number of cyclic amides (equation 246)[304].

$$RCH{=}CHCH{=}CHCONH_2 + Et_3N + Li_2[PdCl_4] \xrightarrow[125\ ^\circ C]{CH_3CN} 6\text{-}R\text{-pyridin-2-one} \quad (243)$$

R = H, Me, Ph    60–65%

$$RCH{=}C(R')C(O)NHNH_2 + Et_3N + Li_2[PdCl_4] \xrightarrow[Et_3N]{CH_3CN} \text{[pyrazol-3-one: } R^1, R, NH, N\text{–H]} \quad (244)$$

35–42%

$$CH_2{=}CHC(O)NHCONH_2 \xrightarrow{PdCl_2} \text{[uracil]} + Pd(0) + H$$

$$(CH_2{=}CHCH_2)_2NC(O)CH{=}CH_2 + PdCl_2 \text{ catalyst} \xrightarrow[\Delta,\ 5\ h]{MeCH(OH)CH_2Me} \text{[N-}CH_2CH{=}CH_2\text{ dimethylpyrrolinone]}$$

(30%)

+

[MeCH₂CH= pyrrolinone, N–H] 10% + [methyl methylene pyrrolidinone, N–$CH_2CH{=}CH_2$] 30%

Aliphatic nitrogen compounds proved to be considerably more difficult to cyclize because of the high basicity and coordinating ability of aliphatic amines relative to aromatic amines or amides. $\omega$-Olefinic amines cyclized *very* slowly (8–67 days) when treated with $[PtCl_4]^{2-}$ in aqueous acid solution (equation 247)[305]. An improvement

$$CH_2{=}CH(CH_2)_2CH(Me)NH_2 \xrightarrow[11\ days]{H^+,\ [PtCl_4]^{2-}} \text{[cis-2,5-dimethylpyrrolidine]} + \text{[trans-2,5-dimethylpyrrolidine]} \quad 90\% \quad (247)$$

40 : 60

on this process was realized by reacting trifluoromethanesulphonate salts of these amines with palladium(II) chloride, followed by neutralization of the salts with triethylamine (equation 248)[306]. Alternatively, the tosamides of these amines readily

$$CH_2{=}CH(CH_2)_nNH_3^+CF_3SO_3^- + [PdCl_2(PhCN)_2] \xrightarrow{-60\ ^\circ C} \xrightarrow{2Et_3N} \text{2-methyl cyclic amine } (CH_2)_n\text{, NH}$$

$n=2$ 0%
$n=3$ 65%
$n=4$ 76%

cyclized when heated to reflux with a catalytic (1%) amount of palladium(II) chloride, using benzoquinone as the reoxidant (equations 249–251)[307]. Reduction of the resulting tosylated enamine ($H_2$, Pd/C) followed by photolytic detosylation gave the desired cyclic amines in good yields.

$$\text{1-(NHtos)-2-}(CH_2CH{=}CH_2)\text{cyclopentane} \xrightarrow[\text{benzoquinone, thf}, \Delta, 2\text{ h}]{1\%[PdCl_2(CH_3CN)_2]} \text{N-tos bicyclic enamine} \quad (249)$$

90%

(1) $H_2$, Pd/C (2) $h\nu$, ROH

N–H bicyclic amine

$$CH_2{=}CHCH_2CH(Ph)CH_2NHtos \longrightarrow \text{1-tosyl-2-methyl-4-phenyl-4,5-dihydropyrrole}$$

85% (250)

$$\text{1-}(CH_2NHtos)\text{-1-}(CH_2CH{=}CH_2)\text{cyclohexane} \longrightarrow \text{Ntos bicyclic product} \quad (251)$$

Nucleophilic attack on iron–olefin complexes also has been extensively developed. The most generally useful olefin complexes are the cationic olefin species containing the $[(\text{cyclopentadienyl})Fe(CO)_2]^+$ species. They are easily made, react with a wide variety of nucleophiles, and the resulting σ-alkyliron complexes undergo further useful reactions. Alkoxides, mercaptans, phosphites, and phosphines attacked $[CpFe(CO)_2(\text{olefin})]^+$ complexes to give relatively stable adducts (equation 252)[309].

$$[Fp(\eta^2\text{-}CH_2{=}CHR)]^+ + Nuc \longrightarrow [Fp\{CH_2CH(R)Nuc\}] \quad (252)$$

R = H, Me, Ph, CHO, $CH_2OMe$; Nuc = $MeO^-$, $Bu^tS^-$, $Ph_3P$, $(EtO)_3P$, $R_2NH$

Secondary amines reacted in a similar fashion, but primary amines underwent a double addition (equation 253). The regiochemistry of the reaction reflected nucleophilic attack at the most substituted carbon of the olefin. The resulting σ-alkyliron complexes required chemical decomposition to free the alkyl ligand and it was not always possible to isolate the organic product. A notable exception to this was the observation that oxidation of the aminated olefin complexes led to a carbonyl insertion reaction. This formed the basis of an elegant approach to $\beta$-lactams, utilizing amination of an iron-bound olefin, followed by oxidatively induced CO insertion and subsequent ring closure to the $\beta$-lactam (equation 254)[310]. With $\omega$-olefinic amines as substrates, bicyclic $\beta$-lactams were obtained (equation 255). Bicyclic $\beta$-lactams were also available from unsaturated ketones, as shown in equation 256[311].

$$[Fp(\eta^2\text{—}CH_2\text{=}CH_2]^+ + MeNH_2 \longrightarrow [Fp\{(CH_2)_2MeNH(CH_2)_2\}Fp]^+ \quad (253)$$

$$Fp(\eta^2\text{—}CH_2\text{=}CHR] + PhCH_2NH_2 \longrightarrow [Fp\{CH_2CH(R)NHCH_2Ph]$$

(254)

Ag2O; Cp, Fe+, CO, O, PhCH2NH, R; R, N, PhCH2, O; 47%

$$[Fp\{\eta^2\text{—}CH_2\text{=}CH(CH_2)_4NH_3^+\}] \xrightarrow{Bu_3N} [Fp(CH_2\text{-pipedrid-2-yl})] \quad (255)$$

Ag2O; N, O; 30% overall

$$[Fp\{\eta^2\text{—}CH_2\text{=}CH(CH_2)_2C(O)Me\}]^+ \xrightarrow{NH_3}$$

FpCH2, N; 68%

(256)

NaBH4; H, FpCH2, HN; 52%; Δ; Cp, N, Fe, CO, O; 30%; Ag2O; N, O; 72%

A wide variety of carbanions also attacked iron-olefin complexes[312]. Enolates of nitromethane, acetoacetates, malonates, and cyanoacetates, as well as the enamines of isobutyraldehyde, cyclopentanone, and cyclohexanone, reacted with $[Fp(olefin)]^+$ complexes of ethylene, propene, styrene, cyclopentene, cyclohexene, allene, and butadiene (equation 257). The regiospecificity depended on both the olefin and the

$$[Fp(\eta^2\text{—}CH_2{=}CH_2)]^+ + R^- \longrightarrow Fp(CH_2CH_2R)] \tag{257}$$

anion. With propene the regioselectivity was low, but with styrene, alkylation of the benzylic position predominated. The butadiene complex underwent clean 1,2-addition followed by a spontaneous CO insertion (equation 258). Allene alkylated

$$[Fp(\eta^2\text{—}CH_2{=}CHCH{=}CH_2)]^+ + MeO_2C\bar{C}HCO_2Me \longrightarrow$$

$$[Fp\{CH_2CH(CH{=}CH_2)CH(CO_2Me)_2] \longrightarrow \left[\text{Cp(OC)Fe(C=O)} \ldots CH(CO_2Me)_2\right] \tag{258}$$

cleanly at the terminal carbon. Non-stabilized carbanions such as Grignard reagents and organolithium reagents reduced the complexes rather than alkylating them. In contrast, organocuprates reacted in considerably better yield. The phosphonium ylid of ethyl bromoacetate alkylated the ethylene complex nicely, and the resulting adduct reacted as a typical Wittig reagent (equation 259). The related iron complexes

$$[Fp(\eta^2\text{—}CH_2{=}CH_2)]^+ + Ph_3P{=}CHCO_2Et \longrightarrow [Fp\{(CH_2)_2CH(CO_2Et)\overset{+}{P}Ph_3]$$

$$\xrightarrow{(1)\ OH^-\ (2)\ PhCHO} [Fp\{(CH_2)_2C(CO_2Et){=}CHPh] \tag{259}$$

of enol ethers have also been prepared, and they react with carbanions in a number of useful ways. These complexes can be viewed as vinyl cation equivalents, and offer a procedure for the introduction of vinyl groups $\alpha$ to carbonyl groups (equation 260)[313,314]. The complex of the enol ether of pyruvic ester was used in a synthesis of

$$1\text{-}R^1\text{-}2\text{-}OLi\text{-}3\text{-}R^2\text{-cyclohex-}1\text{-ene} + [Fp(\eta^2\text{—}R^3CH{=}CHOEt)]^+ \xrightarrow[\text{thf}]{-78\ ^\circ C} \text{2-}R^2\text{-6-}R^1\text{-6-}[CH(OEt)CH(Fp)R^3]\text{cyclohexanone}$$

$$\xrightarrow{H^+} \text{2-}R^2\text{-6-}R^1\text{-6-}[CH{=}CHR^3\cdot Fp^+]\text{cyclohexanone} \xrightarrow{I^-} \text{2-}R^2\text{-6-}R^1\text{-6-}(CH{=}CHR^3)\text{cyclohexanone}; \xrightarrow{R^1=H} \text{2-}R^2\text{-6-}(=CHR^{3\prime})\text{cyclohexanone} \tag{260}$$

$\alpha$-methylene lactones (equation 261)[315]. Electron-deficient olefins also complexed to the $Fp^+$ group, and were activated to undergo nucleophilic attack. This was used to promote the first step of a classical Robinson annellation (equation 262)[316].

1-lithoxycyclohex-1-ene + $[Fp\{\eta^2-CH_2=C(OEt)CO_2Et\}]^+ \longrightarrow$ 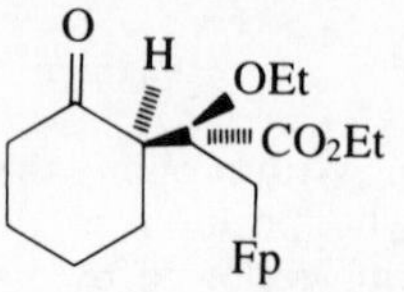 (261)

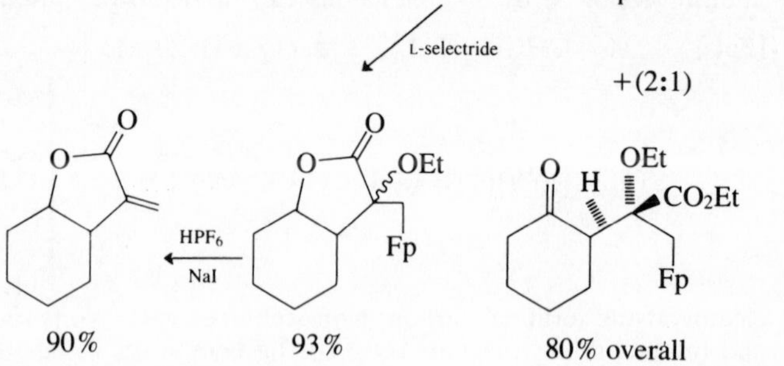

L-selectride

HPF$_6$ / NaI

+(2:1)

90% 93% 80% overall

$[Fp\{\eta^2-CH_2=CHC(O)Me\}]^+$ + 1-lithoxycyclohex-1-ene (262)

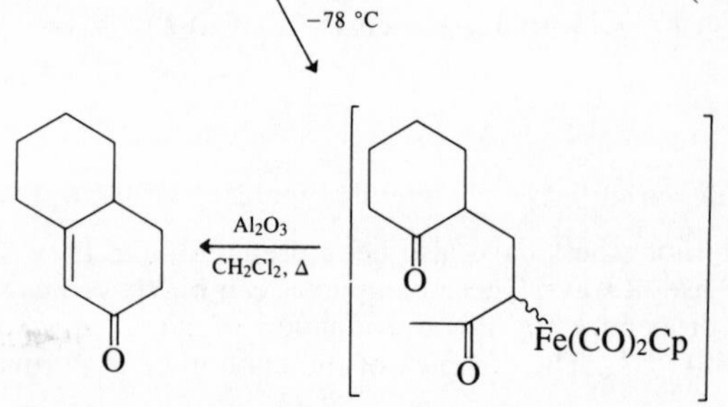

−78 °C

$Al_2O_3$ / $CH_2Cl_2$, $\Delta$

$Fe(CO)_2Cp$

Olefin complexes of iron tetracarbonyl reacted with stabilized carbanions to produce alkylation product after oxidative removal of the iron (equation 263)[317]. The instability of simple olefiniron tetracarbonyl complexes precluded their effective use in this process, but complexes of conjugated enones underwent alkylation in good yield. Difunctionalization was effected by treatment of the $\sigma$-alkyliron complex with organic halides (equation 264)[318]. Dialkylation also resulted when $\alpha$-halo-$\alpha,\beta$-unsaturated esters were used as substrates (equation 265)[319].

$$[Fe(CO)_4(\eta^2-CH_2=CHR] + R''CH(CO_2R')_2 \longrightarrow [Fe(CO)_4\{CH(R)CH_2C(R'')(CO_2R')_2\}]^- \quad (263)$$

(1) $CF_3COOH$ (2) $H^+$, $H_2O_2$ | (3) $Ce^{4+}$

$$\downarrow$$

$$R(CH_2)_2C(R'')(CO_2R')_2$$

R = H, $CO_2Me$
R′ = Me, Et
R″ = H, Me

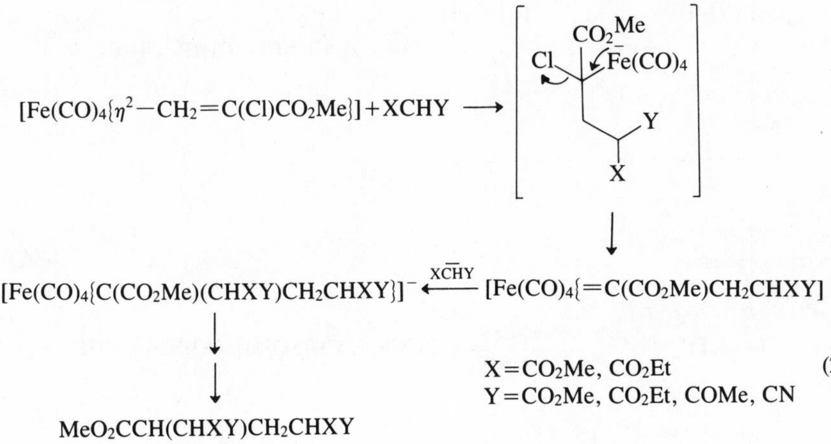

$$[Fe(CO)_4\{\eta^2-CH_2=C(Cl)CO_2Me\}] + XCHY \longrightarrow \left[ \text{Cl, } CO_2^-Me, Fe(CO)_4, Y, X \right]$$

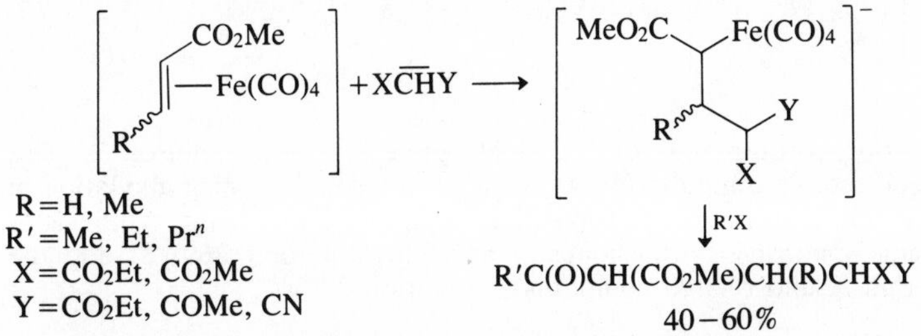

$$\downarrow$$

$$[Fe(CO)_4\{C(CO_2Me)(CHXY)CH_2CHXY\}]^- \xleftarrow{X\bar{C}HY} [Fe(CO)_4\{=C(CO_2Me)CH_2CHXY]$$

$$\downarrow$$

$$\downarrow$$

$$MeO_2CCH(CHXY)CH_2CHXY$$

$X = CO_2Me$, $CO_2Et$ (264)
$Y = CO_2Me$, $CO_2Et$, COMe, CN

$$\left[ R' \text{—}CO_2Me \text{—}Fe(CO)_4 \right] + X\bar{C}HY \longrightarrow \left[ MeO_2C, Fe(CO)_4, R, Y, X \right]^- \xrightarrow{R'X} R'C(O)CH(CO_2Me)CH(R)CHXY \quad (265)$$

R = H, Me
R′ = Me, Et, $Pr^n$
$X = CO_2Et$, $CO_2Me$
$Y = CO_2Et$, COMe, CN

$R'C(O)CH(CO_2Me)CH(R)CHXY$
40–60%

## D. Nucleophilic Attack on π-Allylmetal Complexes

A variety of π-allylmetal complexes undergo attack by nucleophiles at the π-allyl ligand. By far the most extensively studied and most highly developed chemistry is that of π-allylpalladium halide complexes. This chemistry has been the subject of hundreds of papers, and a number of recent detailed reviews[320–327]. For this reason only the basic principles of the process, selected examples of its use, and current studies will be presented here.

π-Allylpalladium halide complexes are yellow, air stable, crystalline solids which are generally prepared directly from olefins by reaction with palladium(II) salts under a variety of conditions[328]. They are rather inert to a variety of chemical reagents, and for this reason had been little studied until it was observed that, in the presence of strongly coordinating ligands such as phosphines or dmso, nucleophiles readily attacked the π-allyl ligand (equation 265a). Early studies used stabilized carbanions

$$[PdCl(\eta^3\text{-allyl}]_2 + \text{excess } L + Nuc^- \longrightarrow Nuc\text{-}CH_2CH{=}CH_2 + [L_nPd] \quad (265a)$$

such as malonate[329] or ketone enamines[330] as nucleophiles. In more recent studies, a range of stabilized carbanions ($pK_a \approx 12$–$18$) have been used in this allylic alkylation process[331]. With unsymmetrical π-allylpalladium complexes the regioselectivity of attack was strongly dependent on the specific structure of the complex, the nature of the carbanion, and the specific reaction conditions employed. Usually, however, alkylation at the less substituted allyl terminus predominated (equation 266). With cyclic π-allyl complexes, attack at the exocyclic position predominated (equation 267), and the carbanion was shown to attack from the face opposite the metal,

$$\underbrace{Pr^nC(CH_2CH(CO_2R)_2)=CHEt \;+\; Pr^nC(CH_2CH(CO_2R)_2)=CHEt}_{8} \;+\; Pr^nC(=CH_2)CH(Et)CH(CO_2R)_2 \quad 8:1$$

$$\uparrow \text{NaCH(CO}_2\text{R)}_2 \;|\; \text{thf, PPh}_3 \tag{266}$$

$$[PdCl(\eta^3\text{-1-Et-2-Pr}^n\text{-allyl})]_2 \xrightarrow[\text{Ph}_3\text{P, thf}]{\text{MeSO}_2\bar{\text{C}}\text{HCO}_2\text{Me}} Pr^nC[CH_2CH(SO_2Me)CO_2Me]=CHEt$$

$$[(\text{cycloalkenyl-}\pi\text{-allyl})PdCl]_2 + L + XCHY \longrightarrow (\text{cycloalkene})CH_2CH(X)Y \tag{267}$$

indicating that prior coordination of the nucleophile was not required[331]. This chemistry has been used in a number of interesting syntheses, including alkylation of cholestanone and testosterone at the 6-position (equation 268)[332], the synthesis of steroids possessing abnormal stereochemistry at C-20 (equation 269)[333], and the synthesis of vitamin A and related compounds (equation 270)[334].

$$[\text{steroid 4,5,6-}\pi\text{-allyl-PdCl complex}]_2 \xrightarrow[\text{(2) LiI, dmf}]{\text{(1) }\bar{\text{C}}\text{H(CO}_2\text{Me)}_2\text{, dmso}} \text{6-CH}_2\text{CO}_2\text{Me-}\Delta^4\text{-3-one} \tag{268}$$

R = $C_8H_{17}$, OH

~70% 6:1 $\alpha:\beta$

$$[\text{3-MeO-estratrienyl 17,20,21-}\pi\text{-allyl-PdCl complex}] \xrightarrow[\text{Ph}_2\text{P(CH}_2)_2\text{PPh}_2]{\text{X}\bar{\text{C}}\text{HCO}_2\text{Me}} \text{MeO-steroid-}\Delta^{16}\text{-20-CH(X)CO}_2\text{Me}$$

X = $CO_2Me$ 81%
X = $PhSO_2$ 82%

$[PdCl(\eta^3\text{-}1\text{-}CH_2OAc\text{-}2\text{-}Me\text{-allyl})]_2$ + (R-allyl-SO₂Ph anion) $\xrightarrow{Ph_3P,\ dmf}$ R-…-SO₂Ph-…-OAc 52% (270)

R = (2,6,6-trimethylcyclohexenyl)–CH=CH–

Until recently, this reaction was restricted to the use of stabilized carbanions (p$K_a$ 12–17). Non-stabilized carbanions tended to destroy the $\pi$-allylpalladium complexes, but produced none of the desired alkylation product. However, as is often the case in organometallic chemistry, both the success and course of these reactions were strongly dependent on the reaction conditions. Recently, $\pi$-allylpalladium complexes have been alkylated by ketone enolates, in a reaction shown to occur by attack of the carbanion from the face opposite the palladium (equation 271)[335]. In contrast, alkylation (by the methyl Grignard reagent) of a $\pi$-allylpalladium complex which could not undergo elimination to a diene occurred from the *same* face occupied by the palladium, indicating a change in mechanism (equation 272)[336]. Finally, reaction of $\pi$-allylpalladium complexes with $\alpha$-branched ester enolates led to nucleophilic attack at the *central* carbon of the $\pi$-allyl group, producing cyclopropanes (equation 273)[337]. Clearly this reaction proceeded by a unique mechanism.

$[\text{(OMe, H-cyclohexenyl)PdCl}]_2 + 4PPh_3 + MeC(OK){=}CH_2 \xrightarrow{thf}$ (OMe-substituted cyclohexene product) 60% (271)

$[\text{(bicyclic allyl)PdCl}]_2 + CH_3MgI \longrightarrow$ (bicyclic product with H, $CH_3$) 90% (272)

$[PdCl(\eta^3\text{-}2\text{-}D\text{-allyl})]_2 +$ (1-CO₂Me-cyclohexyl anion) $Li^+ + 3Et_3N \xrightarrow{thf/hmpa}$ (1-MeO₂C-1-(1-D-cyclopropyl)cyclohexane) 70% (273)

Other π-allylmetal complexes also have been reported to react with carbanions at the π-allyl ligand. π-Allyliron dicarbonylnitrosyl complexes reacted with stabilized carbanions in much the same fashion as did π-allylpalladium halide complexes, leading to mixtures of allylic alkylation products (equation 274)[338]. Cationic π-

$$[Fe(CO)_2(NO)(\eta^3\text{-1-R-allyl})] + X\overline{C}HY \xrightarrow{CO} \longrightarrow RCH{=}CHCH_2CHXY + CH_2{=}CH(R)CHXY \quad (274)$$

R = H, Me, Ph    77–95%

X = COMe, $CO_2Et$, $CO_2Me$, CN

Y = $CO_2Et$, $SO_2Ph$

allyliron tetracarbonyl complexes reacted with dialkylcadmium reagents and with acetoacetate anion to produce allylic alkylation products[339], whereas cationic cyclopentadienylmolybdenum nitrosyl carbonyl complexes reacted with nucleophiles to give neutral π-olefin complexes (equation 275)[340]. Finally, cationic cyclopentadienyl-

$$[MoCp(CO)(NO)(\eta^3\text{-allyl})]^+ + Nuc^- \longrightarrow [MoCp(CO)(NO)(\eta^2\text{-}CH_2{=}CHCH_2Nuc] \quad (275)$$

$Nuc^- = H^-, C_5H_5^-, MeO^-, AcO^-$

π-allyl-molybdenum and -tungsten complexes reacted with methyllithium at the central carbon to produce a stable metallacyclobutane complex (equation 276)[341]. This pattern of reactivity resembles that of equation 273, although it is not known if there is, in fact, any correspondence in mechanism[342].

$$[MCp_2(\eta^3\text{-allyl}]^+ + Nuc^- \longrightarrow \left[ Cp_2M\diamond\text{—}Nuc \right] \quad (276)$$

M = Mo, W
$Nuc^- = H^-, CH_3^-$

Nucleophiles other than carbanions also react with π-allylmetal complexes. Amines have been most extensively studied, again with π-allylpalladium complexes. Simple π-crotylpalladium complexes reacted with dimethylamine in the presence of phosphines to produce the corresponding allylamine in good yield[343]. The stereochemistry of this process was shown to be cleanly *trans*, as shown in equation 277[344].

[π-cyclohexenyl(NMe₂)Pd(Me₂NH)Cl complex] $\xrightarrow[AgBF_4]{Me_2NH,\ PPh_3}$ 3,6-bis(dimethylamino)cyclohexene (Me₂N, NMe₂) (277)

The mechanism of both alkylation and amination of π-allylpalladium halide complexes had been assumed to involve formation of a cationic π-allyl species by displacement of chloride by phosphine, followed by nucleophilic attack at the π-allyl ligand. Evidence for this mechanism in the alkylation process came from the observation that the product distribution from the alkylation of a π-allylpalladium chloride complex and the corresponding pre-formed *bis*-phosphine cationic complex were identical (equation 278)[331]. However, in a careful study of the amination of π-allylpalladium complexes in the presence of excess of phosphine, no cationic complex was detected by conductimetric measurements in thf. N.m.r. spectra of $\pi$-allylpalladium–phosphine mixtures also showed no cationic species, but rather an equilibrating $\sigma$-$\pi$-allyl system when greater than two equivalents per palladium were added. In contrast, addition of $AgBF_4$ to $\pi$-allylpalladium chloride complexes followed by addition of two equivalents of phosphine clearly generated the desired cationic $\pi$-allylpalladium complexes. The regiochemistry of amination of $\pi$-allylpalladium chloride complexes in the presence of added phosphine was different from that observed in the amination of the preformed cationic species, and different mechanisms for the two processes were proposed[345].

½ [(π-allyl)Pd(μ-Cl)]$_2$ (cyclohexenyl-methyl allyl) + $2Ph_3P$

[(π-allyl)Pd($Ph_2$P–P$Ph_2$)]$^+$ $BF_4^-$

$\xrightarrow{MeSO_2\overline{C}HCO_2Me}$ 1-[$CH_2CH(SO_2Me)CO_2Me$]cyclohex-1-ene 25% + 1-methylene-2-[$CH(CO_2Me)(SO_2Me)$]cyclohexane 75% (278)

½[PdCl($\eta^3$-1,1-dimethylallyl)]$_2$

$\xrightarrow{AgBF_4,\ 1L}$ [PdL($\eta^3$-1,1-dimethylallyl)]$^+$ + $BF_4^-$ $\xrightarrow{Me_2NH}$ $Me_2NCH_2CH{=}CMe_2$ >90%

$\xrightarrow{4L}$ [PdCl($L_2$)($CH_2CH{=}CMe_2$)] $\xrightarrow{Me_2NH,\ S_N2'}$ $CH_2{=}CHC(Me)_2NMe_2$ >90% (279)

The reactions discussed above are useful for the direct allylic alkylation and amination of olefins, but suffer from requiring stoichiometric amounts of expensive palladium salts. A related and potentially more useful process is the palladium(0)-catalysed nucleophilic substitution of allylic compounds (equation 280)[346,347]. A wide

$$CH_2{=}CHCH_2X + [L_4Pd(0)]\text{catalyst} + \text{Nuc} \longrightarrow CH_2{=}CHCH_2\text{—Nuc}$$

X = OAc, OOCR, OPh, OH, $NR_2$, $SO_2Ph$, oxirane (280)

range of allylic leaving groups including acetates and other esters, ethers, alcohols, amines, sulphinates, and epoxides were found to react in this system and the most common nucleophiles were stabilized carbanions and amines. With chiral allylic substrates the net process was found to proceed with *retention* of configuration, presumably from two inversions (see below). The reaction was thought to proceed by oxidative addition of the allyl substrate to the palladium(0) complex to form the same cationic $\pi$-allyl complex proposed in the stoichiometric reactions discussed above. Nucleophilic attack on this species produces the observed allylic substitution product and regenerates the palladium(0) catalyst (equation 281)[348]. This mechanism

$$\begin{array}{c} CH_2{=}CHCH_2X + [L_4Pd(0)] \longrightarrow [PdX(L_2)(CH_2CH{=}CH_2) \\ \searrow \\ [L_4Pd(0)] + \text{Nuc-}CH_2CH{=}CH_2 \xleftarrow{\text{Nuc}^-} [Pd(L_2)(\eta^3\text{-allyl}]^+X^- \end{array} \tag{281}$$

fits well the observed stereochemistry, net retention, resulting from an inversion in the oxidative addition step, and another one in the nucleophilic attack step. However, when allyl acetate reacted with $[Pd(Ph_3P)_4]$, *no* $\pi$-allylpalladium acetate was formed. Indeed, no apparent reaction occurred, although later studies showed in fact a 1,3-shift of acetate had occurred. Thus, if the first step of the proposed mechanism is correct, it must be readily reversible and $\pi$-allylpalladium species must be present in only very small concentrations {the more basic palladium complex $[(Cy_3P)_4Pd]$ did form $\pi$-allylpalladium complexes when reacted with allylic acetates}[349]. It was also observed that the pre-formed chiral complex, [$\pi$-cyclohexenylpalladium{(+)-diop}]$^+BF_4^-$, reacted with diethyl malonate to give lower optical yields than the corresponding catalytic reaction of cyclohexenyl acetate with [Pd{(+)-diop}$_2$] and diethyl malonate[350]. It was claimed that this catalytic reaction proceeded via $S_N2$ or $S_N2'$ displacement of $\sigma$-allylpalladium intermediates, rather than by direct nucleophilic attack on a cationic $\pi$-allylpalladium complex. In turn, this interpretation has recently been questioned[351], and the controversy continues.

Whatever the mechanism, this reaction has been used extensively in organic synthesis[320–327]. Steroids having the natural configuration at C-20 were synthesized using a palladium(0)-catalysed side-chain elaboration involving two inversions (equation 282)[333]. Similar chemistry was used to elaborate the side-chain of the steroid

$$\text{H}\blacktriangleleft\overset{\text{Me}}{\underset{\text{R}}{\text{C}}}\text{—OAc} \xrightarrow[\text{[PdL}_4\text{] cat.}]{^-\text{CH(CO}_2\text{Me)}_2} \text{H}\blacktriangleleft\overset{\text{Me}}{\underset{\text{R}}{\text{C}}}\text{—CH(CO}_2\text{Me)}_2 \tag{282}$$

R = [steroid skeleton: 3-MeO aromatic A-ring, 13-Me, side chain at C-17]

ecdysone[352]. Since the carbon—carbon bond-forming step in the catalytic process ostensibly is the same as in the stoichiometric reactions discussed above, the catalytic process shares many of the features of the stoichiometric process. Thus, the regioselectivity of the catalytic process depended on the nature of the carbanion and on the structure of the substrate. With cyclic allylic acetates, attack at the *exo* terminus of the allyl system predominated[348]. The reaction was specific for allylic acetates, and primary bromides in the same molecule did not react[348]. Allylic acetates of enol ethers underwent clean alkylation by stabilized carbanions without loss of or reaction at the sensitive enol ether group[353]. Allyl acetates reacted in preference to allylic alcohols, a feature used to an advantage in the synthesis of chrysanthemic acid (equation 283)[354a]. The complex $[Fe(CO)_3NO]^-Na^+$ also catalysed the allylic alkylation of simple allyl acetates and formates by malonic esters[354b].

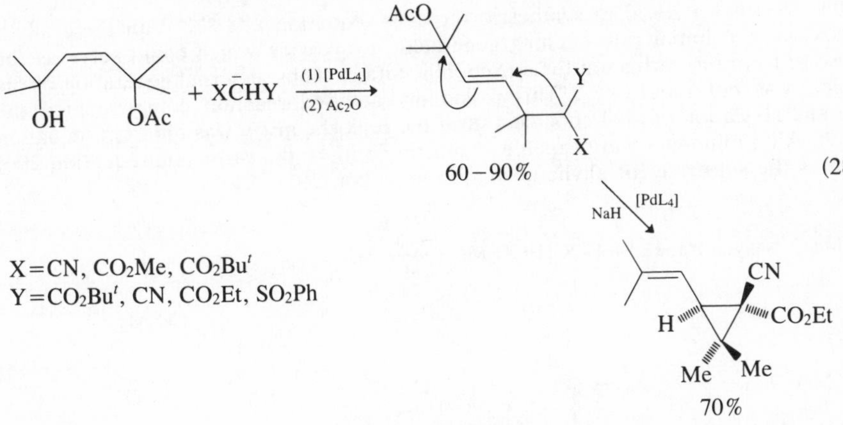

$$\text{(1) [PdL}_4\text{]; (2) Ac}_2\text{O} \quad 60-90\% \qquad \text{NaH, [PdL}_4\text{]} \quad 70\% \tag{283}$$

X = CN, $CO_2Me$, $CO_2Bu^t$
Y = $CO_2Bu^t$, CN, $CO_2Et$, $SO_2Ph$

More highly functionalized allylic substrates also underwent clean allylic alkylation with stabilized carbanions. Vinylsilanes were produced from the palladium(0)-catalysed allylic alkylation of tms-containing allylic acetates (equation 284)[355], regardless of the initial position of the trimethylsilyl group. Enamines were also acceptable nucleophiles (equation 285). Cyanohydrin acetates of allylic aldehydes

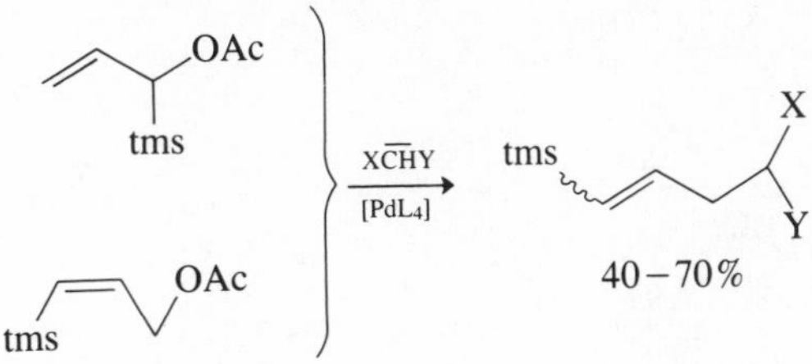

$$\xrightarrow[\text{[PdL}_4\text{]}]{X\overline{C}HY} \quad 40-70\% \tag{284}$$

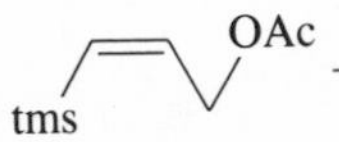

+ 1-(pyrrolidin-1-yl)-6-R-cyclohex-1-ene

$$\xrightarrow{\text{[PdL}_4\text{]}} \tag{285}$$

2-($CH_2CH{=}CH$-tms)-6-R-cyclohexan-2-one
65–71%

underwent clean $S_N2'$ allylic alkylation (equation 286)[356]. Allylic epoxides underwent a similar ($S_N2'$) alkylation when catalysed by palladium(0) complexes (equation 287).

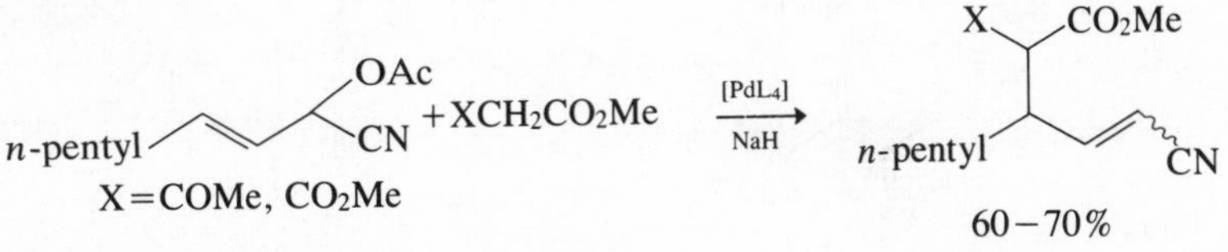

$$n\text{-pentyl-CH=CH-CH(OAc)CN} + XCH_2CO_2Me \xrightarrow[\text{NaH}]{\text{[PdL}_4\text{]}} \quad 60-70\% \tag{286}$$

X = COMe, $CO_2Me$

$$\text{1-R-2-vinyloxirane} + \text{Nuc} \xrightarrow{[PdL_4]} \text{RCH(OH)CH=CHCH}_2\text{—Nuc} \qquad (287)$$

Nuc = stabilized $C^-$ or pyrrolidine 64–92%

This process was used to synthesize terpenes (equation 288)[357]. With these allylic epoxides, the initial ring opening generated an alkoxide which could serve as the base to form the carbanion for nucleophilic attack. Thus, external generation of the anion was not required[358]. This palladium-catalysed reaction complemented the normal alkylation of allyl epoxides, since the regiochemistry was different (equation 289). Allyl sulphones were capable of reacting both as the carbanion (equation 290) and as the substrate for allylic alkylation (equation 291)[359].

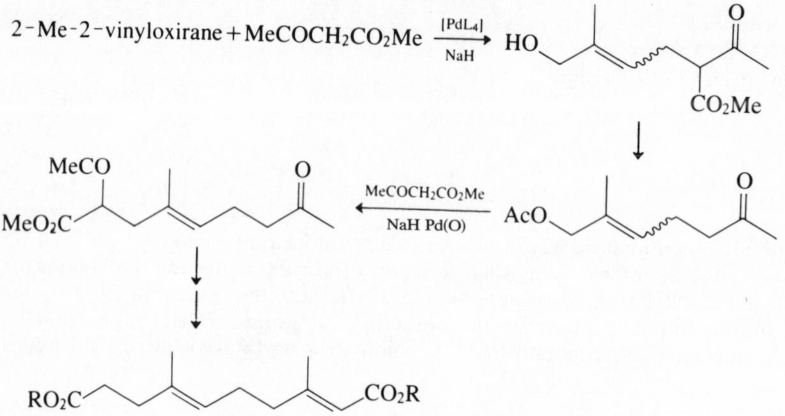

2-Me-2-vinyloxirane + $MeCOCH_2CO_2Me$ $\xrightarrow[NaH]{[PdL_4]}$ HO … $CO_2Me$

(288)

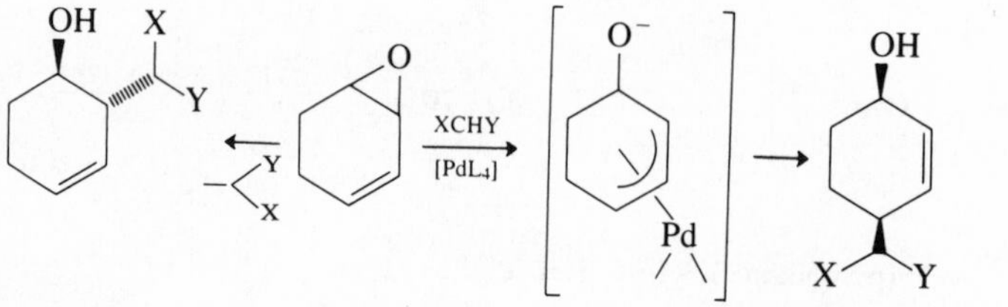

MeCO, $MeO_2C$ … O $\xleftarrow[NaH\ Pd(O)]{MeCOCH_2CO_2Me}$ AcO … O

$RO_2C$ … $CO_2R$

OH X Y ← X, Y; O; $\xrightarrow[{[PdL_4]}]{XCHY}$ [O⁻, Pd] → OH; X Y (289)

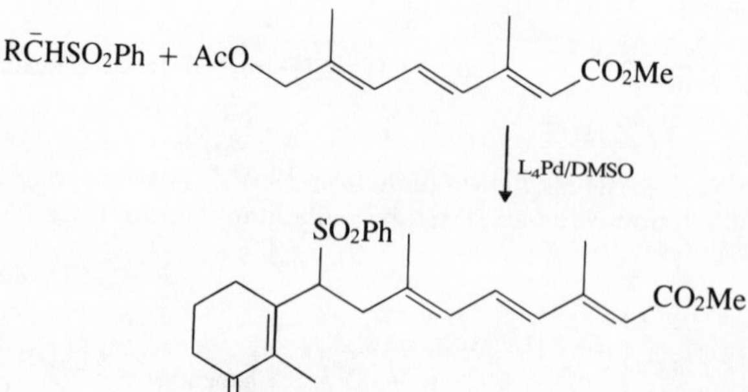

$R\bar{C}HSO_2Ph$ + AcO … $CO_2Me$

$L_4Pd/DMSO$ (290)

$SO_2Ph$ … $CO_2Me$; O

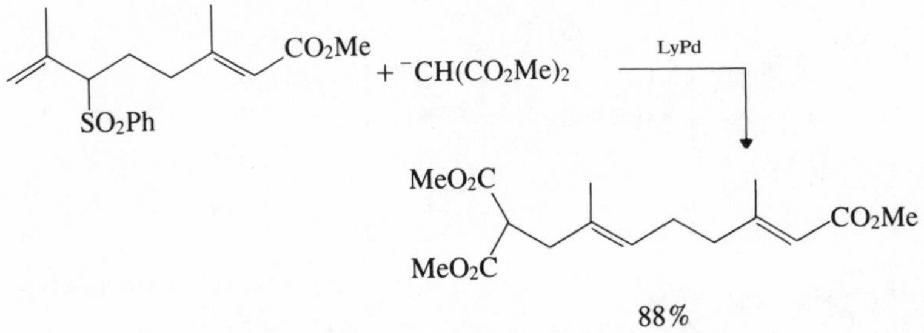

$$+ \; ^{-}CH(CO_2Me)_2 \xrightarrow{L_yPd} \quad (291)$$

88%

Studies of this palladium-catalysed allylic alkylation process have mostly been restricted to stabilized carbanions (equation 292)[360]. However, by changing the catalyst from $[PdL_4]$ to a (dppe) palladium catalyst generated *in situ* from $[Pd(dba)_2]$, simple ketone enolates reacted well (equation 293)[361].

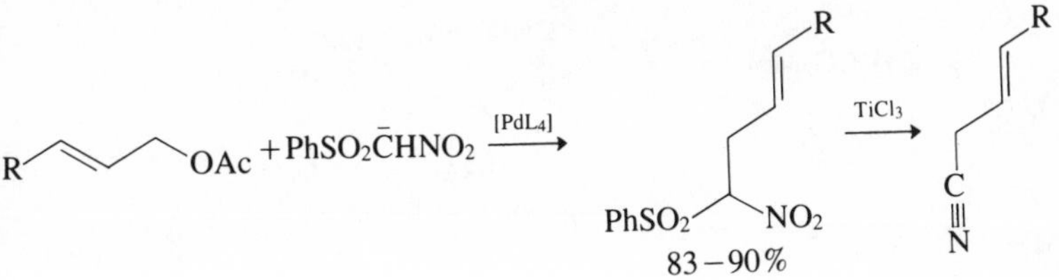

$$R\text{—}CH{=}CH\text{—}CH_2OAc + PhSO_2\bar{C}HNO_2 \xrightarrow{[PdL_4]} \xrightarrow{TiCl_3} \quad (292)$$

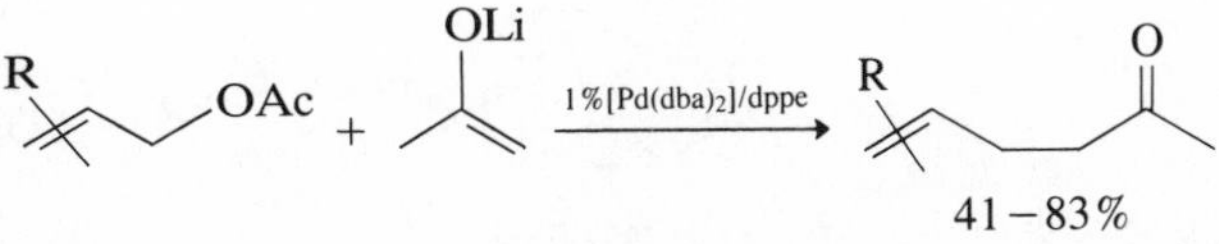

83–90%

$$\xrightarrow{1\%[Pd(dba)_2]/dppe} \quad (293)$$

41–83%

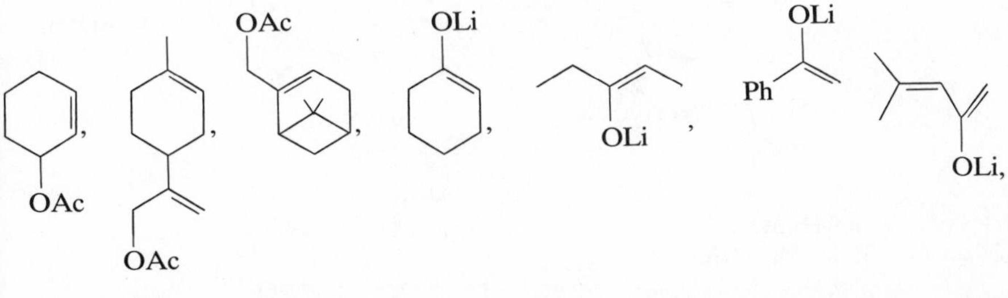

Control of stereochemistry in acylic systems is among the most difficult of problems faced in organic synthesis. The stereospecific nature of both the formation of $\pi$-allylpalladium complexes and their reactions with carbanions has been used to relay the stereochemistry of one chiral centre to a remote position in conformationally mobile systems[362]. Thus, organopalladium chemistry was used to transfer the chirality of a vinyl lactone moiety to the remote vinylic carbon, producing an intermediate having two chiral centres in a 1,5 relationship (equation 294). For the process to succeed, formation of the $\pi$-allylpalladium complex must proceed *only* from the conformation shown, the $\pi$-allyl complex must maintain the stereochemistry, and the nucleophile must attack regiospecifically at the terminal carbon of the $\pi$-allyl system. All of these criteria were met, and the reaction went in 90% yield with greater than 95% stereoselectivity. This chemistry was used to synthesize the

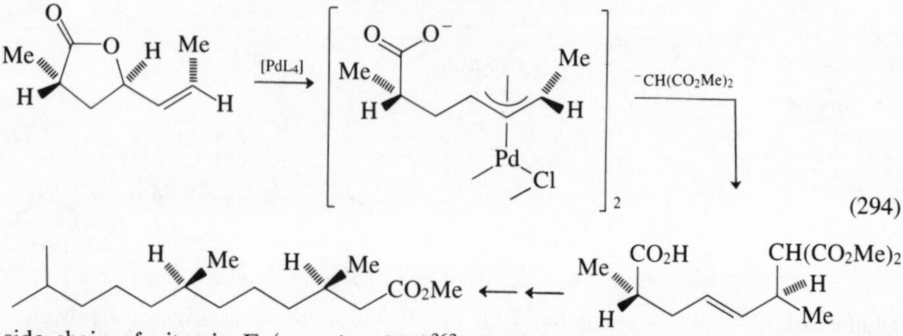

O O H Me Me H H $\xrightarrow{[PdL_4]}$ [ O O⁻ Me H Me H Pd Cl ]$_2$ $\xrightarrow{^{-}CH(CO_2Me)_2}$

(294)

H Me H Me $CO_2Me$ ← ← Me $CO_2H$ H $CH(CO_2Me)_2$ H Me

side-chain of vitamin E (equation 295)[363]. Similar chemistry was used to transfer chirality in the palladium(0)-catalysed rearrangement of vinylogous lactones to cyclopentanones or cycloheptanones (equation 296)[304].

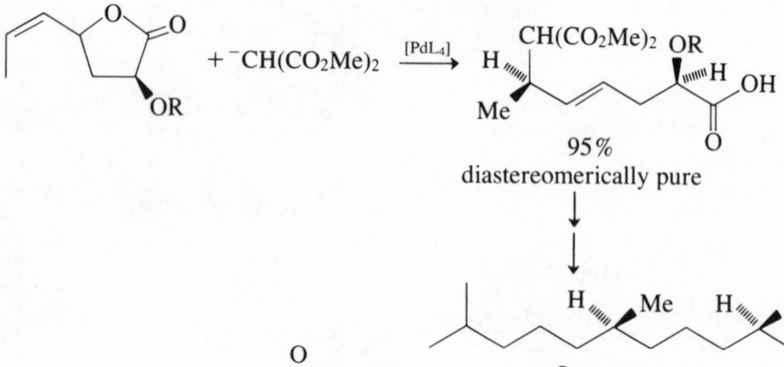

O O OR $+ ^{-}CH(CO_2Me)_2 \xrightarrow{[PdL_4]}$ $CH(CO_2Me)_2$ OR H H OH Me O

95%
diastereomerically pure

↓
↓

(295)

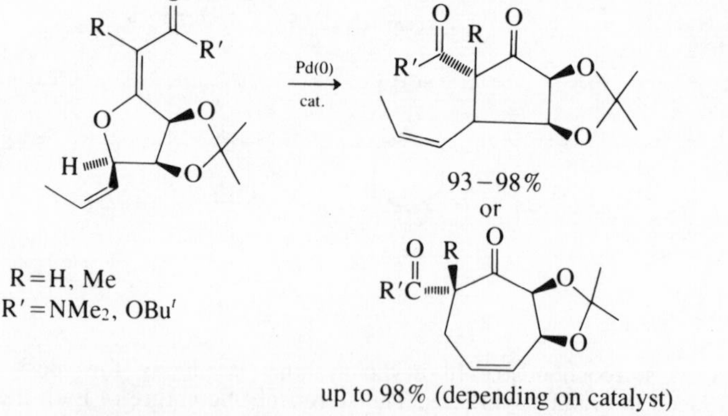

H Me H Me $CO_2Me$

R O R' O O H O $\xrightarrow{Pd(0), cat.}$ O R O R' O O

(296)

93–98%
or

$R = H, Me$
$R' = NMe_2, OBu^t$

O R O R'C O O

up to 98% (depending on catalyst)

Intramolecular palladium(0)-catalysed allylic alkylation reactions have been developed for the synthesis of cyclic compounds[322–327]. Medium-sized rings were produced in good yields from the allylic acetates shown in equations 297–299, and

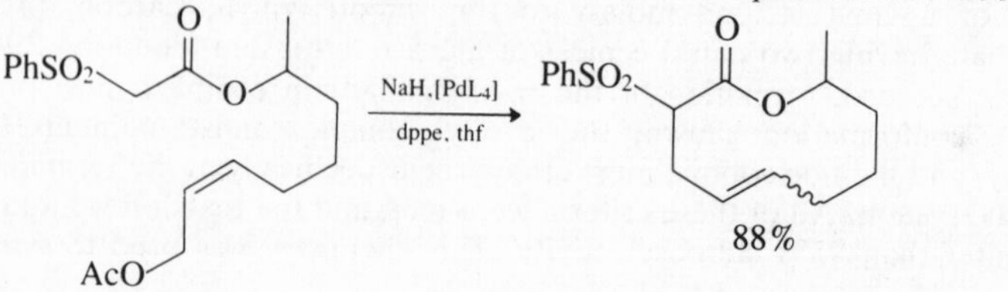

O $PhSO_2$ O AcO $\xrightarrow{NaH, [PdL_4], dppe, thf}$ O $PhSO_2$ O

(297)

88%

$$\text{MeO}_2\text{C–CH(SO}_2\text{Ph)–CH}_2\text{C(O)O(CH}_2)_2\text{CH=CHCH}_2\text{OAc} \xrightarrow[\text{thf}]{\text{NaH},[\text{PdL}_4]} \text{PhSO}_2\text{, CO}_2\text{Me-substituted macrolactone} \quad (298)$$

$$\text{PhSO}_2\text{CH}_2\text{C(O)O(CH}_2)_2\text{CH=CH}_2\text{CH}_2\text{OAc} \xrightarrow[\text{dppe, thf}]{\text{NaH},[\text{PdL}_4]} \text{(8-membered lactone, SO}_2\text{Ph)} + \text{(vinyl 6-membered lactone, SO}_2\text{Ph)} \quad (299)$$

formation of the *larger* of the two possible ring sizes was always favoured[365]. This corresponded *in all cases* to alkylation at the least substituted (most favoured) terminus of the $\pi$-allyl ligand, indicating, as could be expected, that the metal was exercising primary control over the reaction.

By proper choice of starting materials virtually any ring size was accessible by this chemistry. Thus, 6,4-fused ring systems, bicyclooctane [2.2.2] ring systems and 6,5-fused ring systems were synthesized by this approach. Macrolide lactones including exaltolide, recifeiolide, and other twelve-, fourteen-, and sixteen-membered ring compounds, and the eleven-membered carbocyclic sesquiterpene humulene[366], were synthesized by this procedure[369].

In contrast, with allyl phenyl ethers as substrates, palladium-catalysed allylic alkylation resulted in the formation of the smaller of the two possible ring sizes, resulting from nucleophilic attack at the *most* substituted allyl terminus (equations 300 and 301)[367]. This cyclization was used to synthesize highly substituted five- and six-membered rings as intermediates for the synthesis of steroids[323,324,368]. The five-membered $\beta$-carboxy-$\alpha$-methylene cyclopentanone sarkomycin was synthesized using this procedure.

$$\text{(CO}_2\text{Me, Pe}^n\text{ keto allyl phenyl ether)} \xrightarrow[\text{EtCN}]{\text{Pd(OAc)}_2/\text{PPh}_3} \text{2-CO}_2\text{Me-2-Pe}^n\text{-3-vinylcyclopentanone } (87\%) \quad (300)$$

$\text{Pe}^n = n\text{-pentyl}$

$$\text{MeO}_2\text{CCH}_2\text{C(O)(CH}_2)_3\text{CH=CHCH}_2\text{OPh} \xrightarrow{\text{Pd(OAc)}_2,\ \text{Ph}_3\text{P}} \text{2-CO}_2\text{Me-3-vinylcyclohexan-1-one} \quad 62\% \quad (301)$$

Reaction of the bifunctional allylic acetate shown in equation 302 was claimed to produce a trimethylenemethane complex, which reacted with conjugated enones to result in a cycloaddition. This was thought to proceed by a combined nucleophilic–electrophilic attack sequence[325,370]. With unsymmetrical allyl acetate trimethylsilanes, the most stable (least substituted) $\pi$-allylpalladium complex formed (equation 303)[371].

$$\text{tms-CH}_2\text{C(=CH}_2\text{)CH}_2\text{OAc} \xrightarrow{[L_4Pd]} [\text{Pd}^+(L)_2(\eta^3\text{-2-}\bar{C}H_2\text{-allyl})] \xrightarrow{\text{dimethyl norbornenedicarboxylate}} \text{cycloadduct} \quad (302)$$

via

$$\text{tms-CH}_2\text{C(=CH}_2\text{)CH(OAc)Me} + [\text{PdL}_4] \longrightarrow [\overset{+}{\text{Pd}}(\text{L}_2)(\eta^3\text{-2-}\bar{\text{C}}\text{HMe-allyl})]$$

not

$$[\overset{+}{\text{Pd}}(\text{L}_2)(\eta^3\text{-1-Me-2-}\bar{\text{C}}\text{H}_2\text{-allyl})] \quad (303)$$

chromen-2-one

Nucleophiles other than carbanions have been used in palladium-catalysed allylic substitution reactions. Acetate itself was a sufficiently strong nucleophile to attack the $\pi$-allylpalladium complex of a sulphone (equation 304)[372]. Palladium-catalysed rearrangements of allyl acetates probably occur by an intramolecular nucleophilic attack (equation 305)[373]. With chiral allyl acetates this rearrangement occurred stereospecifically (equation 306)[374]. Allyl acetates were converted to tosylates by a palladium-catalysed reaction with the sodium salt of $p$-toluenesulphonic acid (equation 307)[375].

$$[\text{PdCl}(\eta^3\text{-1-R-3-SO}_2\text{tol-allyl})]_2 \xrightarrow{\text{AcO}^-} \text{RCH(OAc)CH=CHSO}_2\text{tol} \quad (304)$$

60–80%

$\text{R} = \text{Ph, H, Me, } n\text{—C}_{19}$

$$\text{R}^3(\text{R}^4)\text{C=C(R}^2)\text{C(R}^1)(\text{CN})\text{OAc} \xrightarrow[\text{thf, 25 °C}]{[\text{PdL}_4]} \text{AcOCCR}^3(\text{R}^4)\text{C(R}^2)\text{=C(R}^1)\text{CN} \quad (305)$$

50–80%

$$\text{C}_5\text{H}_{11}\text{CH=CHCH(OAc)R} \xrightarrow{\text{Pd}^{\text{II}}\text{ cat.}} \text{C}_5\text{H}_{11}\text{CH(OAc)CH=CHR} \quad (306)$$

stereospecific
76% yield

R = (bromo, OBz-substituted spiro bicyclic group)

$$Me_2C{=}CH(CH_2)_2C(Me)(OAc)CH{=}CH_2 + tosNa \cdot H_2O \xrightarrow[\text{thf/MeOH}]{[PdL_4]\ \text{cat.}} Me_2C{=}CH(CH_2)_2C(Me){=}CHCH_2tos \quad 84\% \tag{307}$$

Palladium-catalysed aminations of allyl acetates has been used to synthesize the basic ring systems of the actinabolamine (equation 308), ibogamine (equation 309), and mesembrine (equation 310) alkaloids[376]. This chemistry was used in an elegant synthesis of ibogamine (equation 311)[377] and the closely related alkaloid catharanthine[378]. In these latter systems, the bicyclic amine was formed by an allylic amination, while the cyclization at the 2-position involved nucleophilic attack (or insertion) on a palladium–olefin complex[379]. Spirocyclic amines were synthesized in a

3-OAc-5-(NHCH₂Ph)-cyclohex-1-ene $\xrightarrow[\text{cat.}]{[PdL_4]}$ N-Ph (67%) (308)

3-OAc-4-(CH₂NHCH₂R)-cyclohex-1-ene $\xrightarrow{[PdL_4]\ \text{cat.}}$ R N (60%) (309)

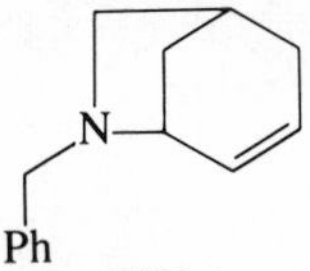

3-OAc-4-[(CH₂)₂NHCH₂Ph]-cyclohex-1-ene $\xrightarrow{[PdL_4]\ \text{cat.}}$ H, H, N, Ph (>50%) (310)

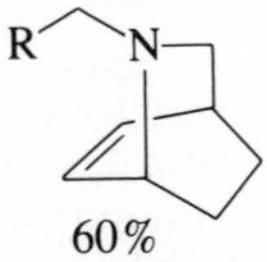

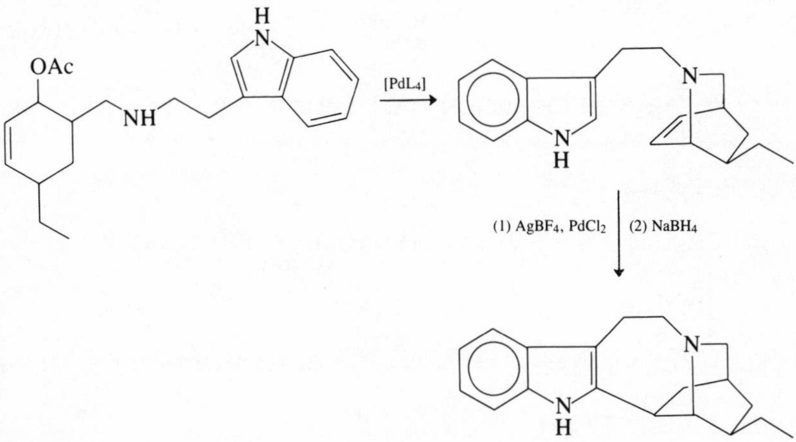

OAc, NH, H N $\xrightarrow{[PdL_4]}$ N, N H $\xrightarrow{(1)\ AgBF_4,\ PdCl_2\ \ (2)\ NaBH_4}$ N, N H (311)

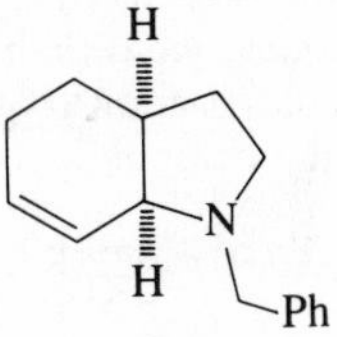

similar manner (equation 312) although few data concerning the purity of the products were given[380]. A variety of pyridine-containing diarylallylamines were synthesized using this same type of reaction (equation 313)[381].

$$\text{2-[(CH}_2)_4\text{NHCH}_2\text{Ph]-6-OAc-cyclohex-1-ene} \xrightarrow[\text{Et}_3\text{N, MeCN}]{[\text{PdL}_4]\text{ cat.}} \quad >95\% \qquad (312)$$

$$\xrightarrow[\text{Me}_2\text{NH, thf, 55 °C}]{[\text{Pd(acac)}_2]\text{, dppe}} \quad 30-80\% \qquad (313)$$

3-Py, 4-Py, 2-Py
X=H, 4-F, 4-Cl, 4-Br, 4-OMe, 2-Br

Conjugated dienes react with palladium salts to produce 1-substituted $\pi$-allylpalladium complexes, which can react further with nucleophiles to lead to some interesting and useful compounds. Acetoxypalladation of butadiene, followed by amination of the resulting $\pi$-allylpalladium complex with secondary amines, produced 1-acetoxy-4-dialkylaminobut-2-enes, whereas the use of primary amines gave pyrroles in modest yield (equation 314)[382]. Dienes were dialkylated or methoxyalkylated in a similar fashion (equation 315)[383]. The stereochemistry of the introduction of the second nucleophile could be controlled by appropriate choice of reaction

$$\text{R'CH=C(CR}^2\text{)C(R}^3\text{)=CHR}^4 + \text{PdCl}_2 + \text{HOAc} + \text{CuCl} \longrightarrow [\text{PdCl}\{\eta^3\text{-1-R}^1\text{-2-R}^2\text{-3-(R}^3\text{)CH(R}^4\text{)OAc-allyl}]_2 \quad 70\text{–}100\%$$

$$\xrightarrow[\text{Me}_2\text{NH}]{\text{R}^1=\text{R}^4=\text{H, AgBF}_4} \text{Me}_2\text{NCH}_2\text{CH=CHCH}_2\text{OAc}$$

$$\xrightarrow[\text{Cu(BF}_4)_2]{\text{RNH}_2\text{, AgBF}_4\text{, Ph}_3\text{P}} \text{1-R-2-R}^1\text{-3-R}^2\text{-4-R}^3\text{-5-R}^4\text{-Pyrrole} \quad 12\text{–}45\% \qquad (314)$$

$$\text{buta-1,3-diene} + \text{PdCl}_2 \longrightarrow [\text{PdCl}(\eta^3\text{-1-CH}_2\text{Cl-allyl})]_2 \xrightarrow[\text{tmeda}]{\text{R}\bar{\text{C}}\text{(X)Y}} \text{RC(X)(Y)CH}_2\text{CH=CHCH}_2\text{C(X)(Y)R} \quad 35\text{–}90\%$$

$$[\text{PdCl}(\eta^3\text{-1-CH}_2\text{Cl-allyl})]_2 \xrightarrow{\text{MeOH}} [\text{PdCl}(\eta^3\text{-1-CH}_2\text{OMe)allyl})]_2 \xrightarrow{\overline{\text{C}}\text{H(CO}_2\text{Et)}_2} \text{MeOCH}_2\text{CH=CHCH}_2\text{CH(CO}_2\text{Et)}_2 \quad 60\%$$

X, Y = $CO_2Et$
R = H, Me, Ph, $PhCH_2$, OHCNH (315)

conditions[384]. $\pi$-Allylpalladium complexes underwent amination by secondary amines in good yield (equation 316)[385]. Dienes were 1,4-acylated-alkylated via nucleophilic attack on $\pi$-allylcobalt complexes (equation 317)[386].

$$\mathrm{MeCH{=}CHBr + Pd(0) \longrightarrow MeCH{=}CHPdBr \xrightarrow{CH_2{=}CHCH(OMe)_2} MeCH{=}CHCH_2CH(PdBr)CH(OMe)_2}$$

$$\downarrow$$

$$\mathrm{\underset{73\%}{MeCH(piperid\text{-}1\text{-}yl)CH{=}CHCH_2CH(OMe)_2} \xleftarrow{piperidine} \left[ \pi\text{-allyl(Me)}\text{-}CH_2CH(OMe)_2 \; PdBr \right]} \tag{316}$$

$$\mathrm{Na[Co(CO)_4] + RX \longrightarrow [RCo(CO)_4] \xrightarrow{CO} [RC(O)Co(CO)_4]}$$

$$\downarrow \text{buta-1,3-diene} \tag{317}$$

$$\mathrm{XCH(Y)CH_2CH{=}CHCH_2COR \xleftarrow{X\overline{C}HY} [Co(CO)_3(\eta^3\text{-}1\text{-}CH_2COMe\text{-}allyl)]}$$

## E. Palladium-Catalysed Telomerization Reactions of Conjugated Dienes

Conjugated dienes react with nucleophiles in the presence of palladium acetate/triphenylphosphine catalysts to produce dimers with incorporation of the nucleophile (equation 318). Nucleophiles such as water, alcohols, carboxylate salts, ammonia,

$$\text{buta-1,3-diene} + \mathrm{YH} \xrightarrow{\mathrm{Pd(0)\ cat.}} \underset{\text{(major)}}{\mathrm{CH_2{=}CH(CH_2)_3CH{=}CHCH_2Y}} + \underset{\text{(minor)}}{\mathrm{CH_2{=}CH(CH_2)_3CH(Y)CH{=}CH_2}} \tag{318}$$

amines, enamines, nitroalkanes, and stabilized carbanions participate in this reaction. The mechanism has not been studied in detail but is thought to involve a palladium(0)-catalysed dimerization of butadiene, followed by nucleophilic attack on the resulting $\pi$-allyl complex (equation 319).

$$\mathrm{Pd^{2+} \longrightarrow Pd^0 \xrightarrow{\text{buta-1,3-diene}} [Pd\ bis\text{-}\pi\text{-allyl}] \longrightarrow [Pd\ complex;\ Y^-,\ H^+] \longrightarrow products} \tag{319}$$

Both the telomerization reaction itself and the chemistry of the resulting products have been extensively reviewed[322–324,387–390], and only minimal coverage will be presented here. With acetate as the nucleophile, the resulting telomer was converted to a 1,5-diketone (equation 320), which was used as a bis-annelation reagent to synthesize steroids[323]. The other (major) isomer of this telomerization was used to synthesize macrocyclic lactones, including diplodialide C, lasiodiplodin, zearalone[322,391], and dehydroxy-*trans*-resorcyclide[391]. With carbanions as nucleophiles, a variety of telomers were prepared (equation 321)[327], and were used further

$$\text{buta-1,3-diene} + \text{AcOH} \xrightarrow{\text{Pd cat.}} CH_2{=}CH(CH_2)_3CH(OAc)CH{=}CH_2 \longrightarrow CH_2{=}CH(CH_2)_3C(O)CH{=}CH_2$$

$$CH_2{=}CH(CH_2)_3C(O)CH{=}CH_2 \xrightarrow{Y^-} CH_2{=}CH(CH_2)_3C(O)(CH_2)_2Y \xrightarrow[O_2]{PdCl_2/CuCl} MeC(O)(CH_2)_3C(O)(CH_2)_2Y \quad (320)$$

Y = enolate of C, D-steroid ring system

$$\text{buta-1,3-diene} + ACH(X)Y \xrightarrow[Ph_3P]{Pd(OAc)_2} CH_2{=}CH(CH_2)_3CH{=}CHCH_2C(X)(Y)A \quad (321)$$

X = Y = $CO_2Et$, A = H
X = $CO_2Et$, Y = COMe, A = H
X = $SO_2Ph$, Y = $CO_2Me$, A = H
X = Y = $CO_2Me$, A = NHAc
X = COMe, Y = NHAc, A = R
X = COR, Y = OH, A = R
X = $NO_2$, Y = $CH_3$, A = H

in natural product syntheses[389]. With amines as nucleophiles, tertiary amines usually resulted. Thus ammonia incorporated three octadienyl chains, as mixtures of regioisomers, primary amines incorporated two octadienyl chains, and secondary amines incorporated one octadienyl chain[389].

This telomerization reaction, and the products of it, continue to be exploited in synthetic chemistry. The telomer from butadiene and phenol provided the substrates for the cyclization reactions of allylphenyl ethers discussed above (equation 300 and 301), as shown in equation 322[367]. It also served as the basis for a number of synthetic approaches to steroids[323,324,390].

$$\text{buta-1,3-diene} + PhOH \xrightarrow[Ph_3P]{Pd(OAc)_2} CH_2{=}CH(CH_2)_3CH{=}CHCH_2OPh \xrightarrow{PdCl_2/CuCl,\ O_2} MeC(O)(CH_2)_3CH{=}CHCH_2OPh$$

$$MeC(O)(CH_2)_3CH{=}CHCH_2OPh \longrightarrow \longrightarrow MeO_2CCH_2C(O)(CH_2)_3CH{=}CHCH_2OPh \quad (322)$$

Glycols reacted with butadienes and palladium catalysts to give octa-2,7-dienyl monoethers of the glycol (equation 323)[392]. Monodienols were synthesized by the cotelomerization of butadiene, isoprene, and water (equation 324)[393]. In both cases, mixtures of isomers were obtained.

OH OH + buta-1,3-diene $\xrightarrow[Ph_3P]{[Pd(acac)_2]}$ OH $O{-}CH_2CH{=}CH(CH_2)_3CH{=}CH_2$

\+

OH $O{-}CH(CH{=}CH_2)(CH_2)_3CH{=}CH_2$ (323)

\+ + $H_2O$ $\xrightarrow[Ph_3P,\ CO_2]{[Pd(acac)_2]}$ OH

\+

OH

OH + (324)

\+

OH

Ammonia and butadiene telomerized to give mixtures of tertiary octadienylamines (equation 325)[394]. Small amounts of water were required by the reaction. Isoprene and ammonia telomerized to give an incredible mixture of amines (equation 326)[395]. Butadiene and secondary amines or carboxylic acids telomerized with a polymer-supported palladium(0) catalyst to give almost exclusive terminal amination (equation 327)[396]. Secondary amines and butadiene telomerized in the presence of nickel catalysts containing chiral phosphites of sugars to give amine telomers with some optical activity. In normal cases, however, six to ten different products were obtained[397].

$$\text{buta-1,3-diene} + NH_3 \xrightarrow[Ph_3P]{Pd(OAc)_2} [CH_2{=}CH(CH_2)_3CH{=}CHCH_2]_3N$$
$$+ [CH_2{=}CH(CH_2)_3CH{=}CHCH_2]_2NCH(CH{=}CH_2)(CH_2)_3CH{=}CH_2 \quad (325)$$

$$NH_3 + CH_2{=}CHC(Me){=}CH_2 \xrightarrow[P(OR)_3]{[Pd(acac)_2]} RC(Me)(NH_2)CH{=}CH_2$$
$$+ RCH(NH_2)C(Me){=}CH_2RCH{=}C(Me)CH_2NH_2$$
$$+ RCH{=}C(Me)CH_2NHCH(R)C(Me){=}CH_2 + [RCH{=}C(Me)CH_2]_2NH$$
$$+ [RCH{=}C(Me)CH_2]_2NH + [RCH{=}C(Me)CH_2]_3N \quad (326)$$
$$R = CH_2{=}C(Me)CH_2CH_2$$

40–60% overall

$$\text{buta-1,3-diene} + R_2NH + \underset{\substack{|\\ Pd(0)\\ cat.}}{\overset{\substack{(70\%\ \text{P-loading})\\ |}}{Ⓟ}} \longrightarrow \begin{array}{c} CH_2{=}CH(CH_2)_3CH{=}CHCH_2NR_2 \\ 92\text{–}100\% \\ CH_2{=}CH(CH_2)_3CH(NR_2)CH{=}CH_2 \\ <7\% \end{array} \quad (327)$$

## F. Nucleophilic Attack on Cationic $\pi$-Dienyl Complexes

Iron carbonyl reacts with conjugated dienes to produce remarkably stable diene iron tricarbonyl complexes. Once complexed in this manner, the reactivity of the diene is drastically altered. For example, the diene in ($\eta^4$-butadiene) tricarbonyliron is inert to catalytic hydrogenation, and unreactive in a Diels–Alder reaction towards maleic anhydride. The complex undergoes Friedel–Crafts acylation reactions without suffering decomplexation from the metal. Complexes of this type are rather inert to nucleophilic attack, although reactive carbanions have recently been reported to attack $\pi$-cyclohexadiene iron complexes, producing alkylated cyclohexenes on protonation[418]. However, diene complexes having allylic hydrogens undergo a hydride abstraction process to produce cationic dienyliron complexes which are fairly reactive towards nucleophiles[398,399].

Cyclohexadienyl complexes have been most extensively studied. Since substituted cyclohexadienes are readily available by Birch reduction of substituted aromatics, a variety of substrates are readily available. Both cyclohexa-1,3- and -1,4-diene reacted with iron pentacarbonyl to produce tricarbonyl(cyclohexa-1,3-diene)iron in fair yield. Treatment with triphenylmethyl tetrafluoroborate produced the cationic dienyl complex in excellent yield, by abstraction of an allylic hydride (equation 328). This is a general approach to these compounds and cyclohexadienyl complexes containing methoxy, methyl and carbomethoxy groups have been made in a similar manner.

$$\text{cyclohexa-1,3-diene or cyclohexa-1,4-diene} + [Fe(CO)_5] \longrightarrow [(\text{cyclohexadiene})\text{—}Fe(CO)_3] \xrightarrow{Ph_3C^+} [(\text{cyclohexadienyl})\text{—}Fe(CO)_3]^+ \quad (328)$$

These cationic complexes were generally reactive towards nuclephiles, regenerating the cyclohexadieneiron complex in the process. The nucleophiles ranged from organo-lithium[400], -copper[401], -cadmium, and -zinc[399] reagents (equation 329) through ketone enolates (equation 330)[399] and nitroalkyl anions (equation 331)[402], to nucleophilic aromatic compounds such as indoles (equation 332)[399] and di- and trimethoxybenzenes (equation 333)[403]. Phthalimide also attacked these complexes (equation 334)[404]. In all cases, nucleophilic attack occurred at the terminus of the

$$[FeR(CO)_3]^+ + R'M \longrightarrow [(5\text{-}R'\text{-cyclohexa-1,3-diene})\text{—}Fe(CO)_3] \quad (329)$$

$[FeR(CO)_3]^+$ + cyclohexanone $\xrightarrow[\Delta]{EtOH}$ [ O —$Fe(CO)_3$ ]

70%

$\downarrow$ $FeCL_3$ | HCl (330)

5-(cyclohexan-1-one-2-yl)cyclohexa-1,3-diene

75%

R as in equation 328

$[FeR(CO)_3]^+ + CH_3NO_2 \xrightarrow{NaH}$ [ $O_2N$— —$Fe(CO)_3$ ]

87%

$\downarrow$ Zn | HCl/HOAc (331)

[ $H_2N$— —$Fe(CO)_3$ ]

81%

R as in equation 328

$[FeR(CO)_3]^+$ + 2-methylindole $\longrightarrow$ [ —$Fe(CO)_3$ N H ] (332)

R as in equation 328

$[FeR(CO)_3]^+$ + 1,2,4-trimethoxybenzene $\longrightarrow$ [ MeO OMe —$Fe(CO)_3$ MeO ] (333)

R as in equation 328

$[FeR(CO)_3]^+$ + $(phthalimide)^-$ $\longrightarrow$ [ O N —$Fe(CO)_3$ O ] (334)

R as in equation 328

dienyl system, and from the face *opposite* the metal. The resulting diene could, in principle, be released from the tricarbonyliron fragment by treatment with an amine oxide.

The tricarbonyl(2-methoxycyclohexadienyl)iron complex has proved especially useful for organic synthesis, as it undergoes nucleophilic attack at the terminus *remote* from the methoxy group, and is thus a synthetic equivalent to a cyclohexenone γ-cation. The 1-methoxy-4-methylcyclohexadienyl complex reacted exclusively at the methyl-bearing position with a large number of stabilized carbanions (equation 335)[405]. Complexes of bicyclic ring systems behaved in a similar manner, permitting the introduction of angular alkyl groups in the absence of steric hindrance (equation 336). When substitution in the non-complexed ring blocked the angular position, attack at the other terminus of the dienyl system was observed (equation

OMe, $-Fe(CO)_3$, Me, $]^+$ + $R^-$ ⟶ [OMe, $Fe(CO)_3$, R, Me] $\xrightarrow[(2)\ H_3O^+]{(1)\ Me_3NO}$ O, R, Me (335)

$R^- = CN^-$, $^-CH(CO_2R)_2$, $^-CH(COR)(CO_2R')$, etc.

[MeO, $(CH_2)_n$, $Fe(CO)_3$$]^+$ $+R^-$ ⟶ [R, $(CH_2)_n$, MeO, $Fe(CO)_3$] (336)

[O, O, MeO, $Fe(CO)_3$$]^+$ $+R^-$ ⟶ [R, MeO, $Fe(CO)_3$] (337)

[OMe, $-Fe(CO)_3$, $CO_2Me$, O$]^+$ $\xrightarrow[CH_2Cl_2,\ -78\ °C]{Et_3N}$ [OMe, $Fe(CO)_3$, $CO_2Me$, O] $\xrightarrow[(2)\ H_3O^+]{(1)\ Me_3NO}$ O, $CO_2Me$, O

(338)

X=Y=CN
X=CN, Y=$CO_2Me$

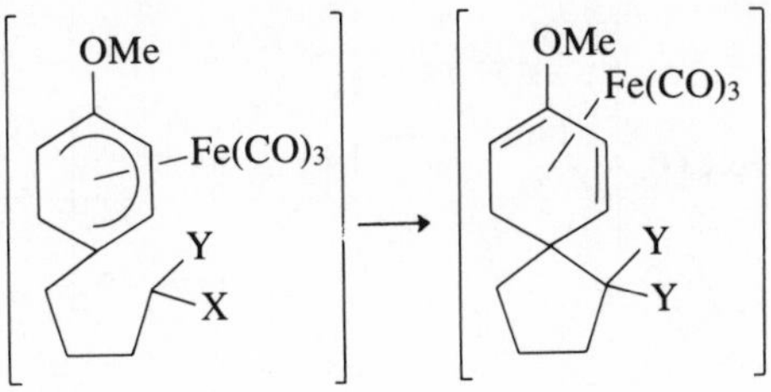

(339)

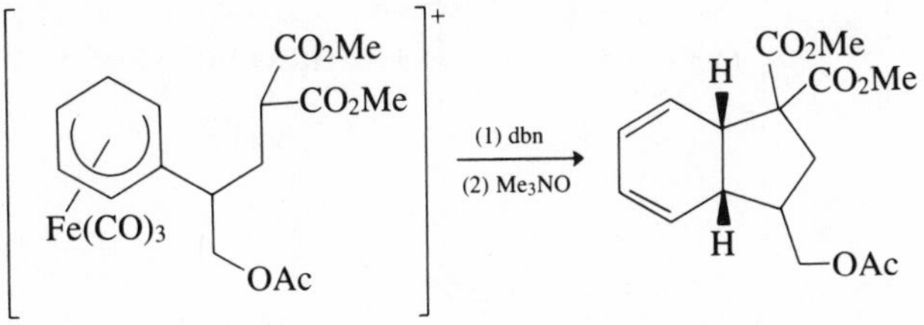

(1) dbn
(2) $Me_3NO$

(340)[405]

337). Intramolecular alkylation reactions of these complexes were used to make spirocyclic compounds (equations 338–340)[398].

Other related complexes have also been studied. The isomeric methoxy complex in equation 341 underwent alkylation by a variety of carbanions including trimethylsilylenol esters to give alkylated cyclohexenones[406]. The 2-*O*-trimethylsilyl cyclohexadiene complex formed a cationic dienyl complex which underwent facile alkylation (equation 342), whereas the 1-*O*-trimethyl-silyl complex simply gave a stable,

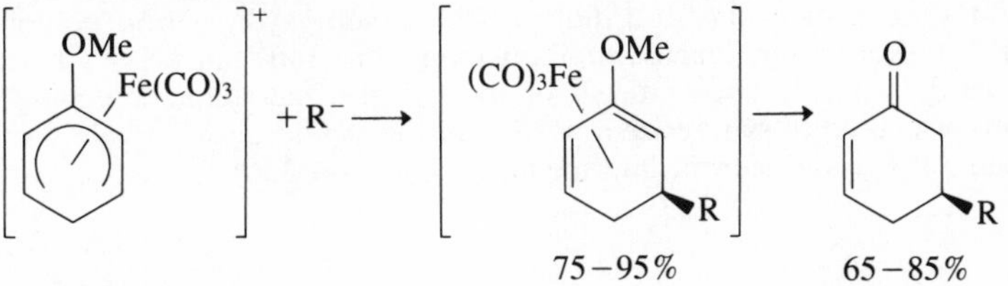

75–95%   65–85%

(341)

$R^-$ = $^-Bu^t$, $Me_2CH_2COCH$, $Me_2C$=COtms, 2-(−)-cyclopentan-1-one,
1,2-di(Otms)-cyclopent-1-ene

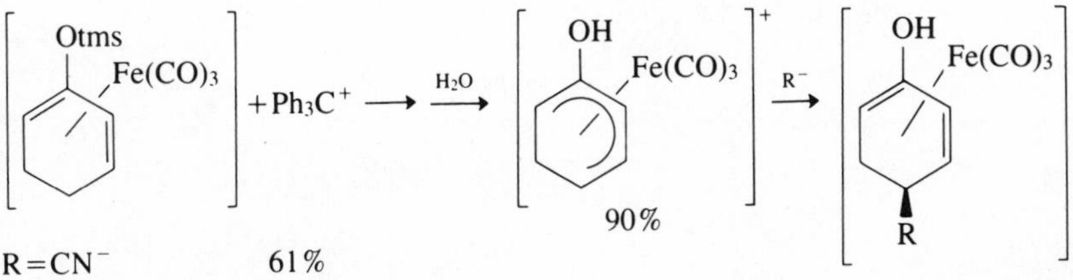

(342)

R=$CN^-$ 61%
R=$^-CH(CO_2Me)_2$ 40%

neutral dienone complex upon reaction with triphenylmethyl cation (equation 343)[407]. The absolute configurations of the iron tricarbonyl complexes of 1-methoxy- and 1-methoxy-4-methylcyclohexadiene were assigned by a process of conversion to the cationic dienyl complexes, stereospecific alkylation, and degradation of the product to compounds of known absolute configuration (equation 344)[408,409].

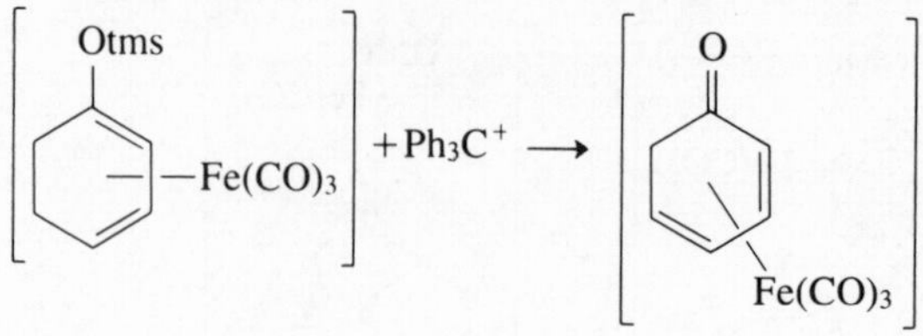

(343)

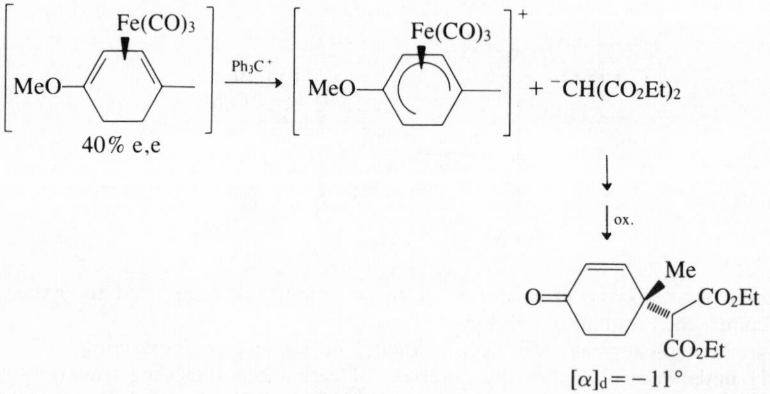

(344)

Cationic tricarbonyl cyclohexadienyliron complexes have been extensively used in organic synthesis[410]. An early example was the synthesis of a benzofuran derivative via the reaction of a cyclohexane-1,3-dione with a methoxydienyliron complex (equation 345)[411]. β-Keto esters reacted similarly (equations 346[412] and 347[413]). This chemistry was used in an elegant synthesis of the tricothecenes (equation 348)[414]. Similar chemistry was used to synthesize a portion of the aspidosperma alkaloid ring system (equation 349)[415] and steroid intermediates (equation 350)[416,417].

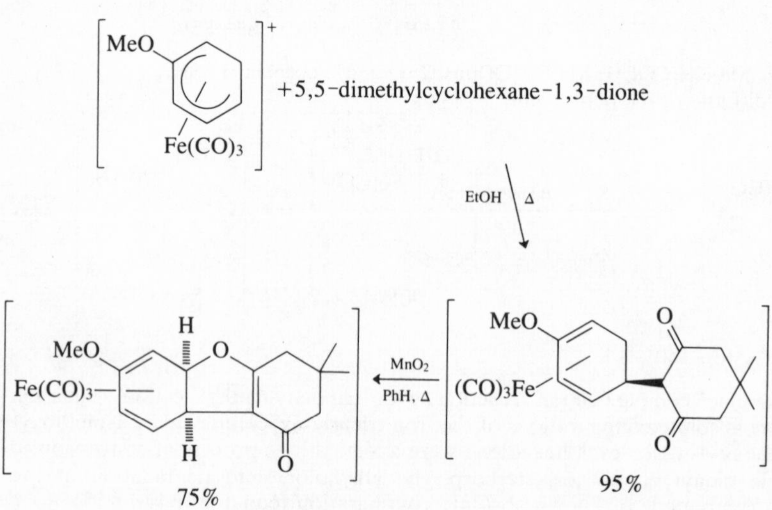

(345)

90% (1:1 mixture of epimers)

$Me_3NO$

84%

(346)

$NaBH_4$

ttfa

(347)

A number of other cationic complexes containing unsaturated hydrocarbon ligands are known, and virtually all of them react with nucleophiles. Since many of these have two or more sites available for nucleophilic attack, the regiochemistry of attack is variable. Recently, three rules have been proposed (and substantiated by numerous examples) that permit the prediction of the most favourable position of nucleophilic attack on 18-electron organo-transition metal cations containing unsaturated hydroarbon ligands, including olefin, $\pi$-allyl, diene, dienyl, and arene ligands[342]. Ligands were classified as to whether an even or odd number of ligand carbons were attached to the metal, and as to whether the carbon ligand was cyclically conjugated (closed) or not (open) (Fig. 1). The rules were: (1) nucleophilic attack occurs preferentially at *even* coordinated polyenes which have no unpaired electrons in the HOMOs; (2) nucleophilic addition to open coordinated polyenes is preferred to addition to closed polyenes; (3) for *even open* polyenes, nucleophilic attack at the

[MeO Fe(CO)3]+ + O CO2Me → [Fe(CO)3 MeO O Me CO2Me] + [MeO O Fe(CO)3 Me CO2Me] (348)

[MeO O Fe(CO)3 Me CO2Me] —(1) NaBH4 (2) p-tosOH (3) SOCl2→ [Fe(CO)3 MeO Me CO2Me]

↓

O O Me OMe → → H O H O Me OMe

12,13-epoxy-14-methoxytricothecene

[MeO Fe(CO)3]+ + $^{-}CH(CO_2Me)_2$ → —$-CO_2$→ [Fe(CO)3 MeO CO2Me]

(1) dibah (2) tosCl (3) NaCN (4) $Me_3NO$ (5) $LiAlH_4$ (349)

MeO NH2 —$H^+$→ O H H N

[MeO Fe(CO)3]+ + OMe MeO2C $O^-$ (350)

→ [MeO Fe(CO)3 O CO2Me OMe] → → Me OMe H O O

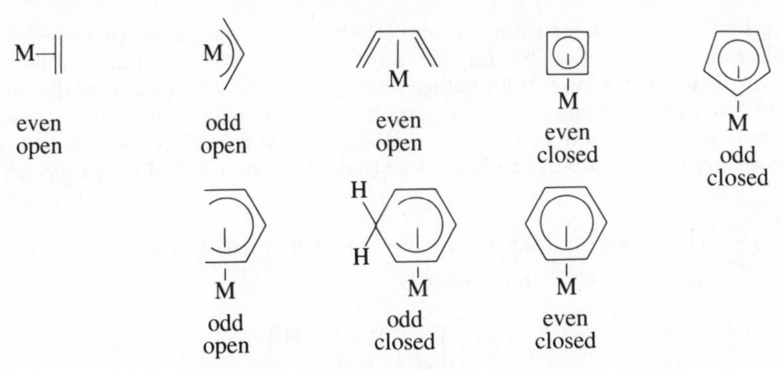


FIGURE 1. Classification of unsaturated hydrocarbon ligands

terminal carbon atom is always preferred; for *odd open* polyenyls, attack occurs at the terminal carbon only if $ML_n^+$ is a strong electron-withdrawing group. These rules were illustrated with over 100 examples.

## G. Nucleophilic Attack on $\pi$-Arene Complexes[419]

The complexation of an arene to a transition metal has a profound influence on the chemistry of that arene. The chemical consequences of complexation are summarized in Fig. 2[420], and most arise from the strong electron-withdrawing abilities of the metal

Enhanced nucleophilic substitution
Enhanced solvolysis
Enhanced acidity → H
X X
C—C
H H
M
Enhanced acidity
Steric hindrance

FIGURE 2. Effects of complexation of arenes to transition metals

fragment, usually an $M(CO)_3$ group. Of particular interest for this review is the enhanced reactivity of the aryl ring and $\alpha$ and $\beta$ side-chain positions toward nucleophilic attack, and the steric hindrance of the face of the arene coordinated to the metal.

Illustrative of the possibilities for nucleophilic attack on arenetricarbonylmetal complexes is the behaviour of the cationic manganese complex shown in equation 351. Depending on the nature of the nucleophile, attack can occur on the metal

$[Mn(CO)_2(CONHR)(\eta^6\text{-}C_6H_6)]$ $\xleftarrow{RNH_2}$ $[Mn(CO)_3(\eta^6\text{-}C_6H_6)]^+$ $\xrightarrow{Ph_3P}$ $[Mn(CO)_2(PPh_3)(\eta^6\text{-}C_6H_6)]^+$

$[Mn(CO)_3(\eta^6\text{-}C_6H_6)]^+ \xrightarrow{R^-}$ [(R)(H)-cyclohexadienyl $Mn(CO)_3$]

$[Mn(CO)_3(\eta^6\text{-}C_6H_6)]^+ \xrightarrow{I^-,\ CH_3CN} C_6H_6$ (351)

($PPh_3$, $I^-$, $CH_3CN$), on a carbonyl group ($RNH_2$), or on the arene ring itself, giving a cyclohexadienyl complex. By far the most common site for nucleophilic attack, however, is the arene ring. With halogenated aromatics, substitution of the halide by the nucleophile normally occurs (equation 352), whereas simple arene complexes react with nucleophiles to produce $\eta^5$-cyclohexadienyl complexes, which give nucleophilic aromatic substitution products after (oxidative) removal of a hydride (equation 353)[420,421].

$$[Fe(Cp)(\eta^6\text{-}C_6H_5X)]^+ + R^- \longrightarrow [Fe(Cp)(\eta^6\text{-}C_6H_5R)]^+ + X^- \quad (352)$$
$$R = RO^-, RS^-, (\text{phthalimide})^-$$

$$[Fe(Cp)(\eta^6\text{-}C_6H_6)]^+ + R^- \longrightarrow \left[\eta^5\text{-}C_6H_6R\,FeCp\right] \xrightarrow[-H^-]{\text{nbs}} [Fe(CP)(\eta^6\text{-}C_6H_5R)] \quad (353)$$
$$R^- = Me, Et, PhCH_2, C_5H_5$$

By far the most extensively studied complexes are the $\eta^6$-arenetricarbonylchromium complexes. The tricarbonylchromium complex of chlorobenzene reacted with sodium methoxide in methanol to produce the corresponding anisole complex in excellent yield, an overall nucleophilic aromatic substitution of alkoxide for halide. The rate of this reaction approximated that of uncomplexed *p*-nitrochlorobenzene[422]. The complex of fluorobenzene reacted with sodium phenoxide and aniline to produce diphenyl ether and diphenylamine, respectively[423]. The kinetics of the reaction suggested initial rapid formation of a $\eta^5$-cyclohexadienyl complex followed by a rate-limiting loss of fluoride from the *endo* face of the ring (equation 354)[424].

$$[(\eta^6\text{-}C_6H_5F)Cr(CO)_3] + R_2NH \xrightarrow{\text{fast}} [(\eta^5\text{-}C_6H_5(F)(NHR_2^+))\,{}^-Cr(CO)_3] \xrightarrow{\text{slow}} [Cr(CO)_3(\eta\text{-}C_6H_5NR_2] + HF \quad (354)$$

A wide range of carbanions performed a similar substitution reaction, and the alkylated aromatic was freed from complexation by oxidation with iodine (equation 355). However, two types of carbanions failed to react in this process. Simple methyl,

$$[(\eta^6\text{-}C_6H_5X)Cr(CO)_3] + R^- \longrightarrow [Cr(CO)_3(\eta^6\text{-}C_6H_5R] \xrightarrow{I_2} C_6H_5R \quad (355)$$

$$R = CH(CO_2Me)_2,\ C(Me)_2CN,\ CN,\ C(Me)_2CO_2Et,\ C(Ph)(OR)(CN),$$

allyl, phenyl, and *t*-butyl Grignard reagents and lithium dimethylcuprate failed to react at low temperatures and led to intractable materials at higher temperatures. In contrast, unbranched primary ester enolate and cyano-stabilized carbanions reacted even at −78 °C, but alkylated the aryl ring *ortho* and *meta* to the halogen producing

$\eta^5$-cyclohexadienyl complexes unable to lose halogen and, in contrast to similar complexes from branched carbanions, unable to rearrange to the $\eta^5$-cyclohexadiene complex which could lose chloride (equation 356)[425].

$$[Cr(CO)_3(\eta^6\text{-}C_6H_5Cl)] + R^- \longrightarrow \left\{ [\text{R,H-adduct, Cl ortho}]^- + [\text{R,H-adduct, Cl meta}]^- + [\text{R,Cl-adduct}]^- \right\}$$

$$\xrightarrow{\text{oxidize}} 1\text{-R-2-Cl-}C_6H_4 + 1\text{-R-3-Cl-}C_6H_4 \qquad \xrightarrow{-Cl^-} [Cr(CO)_3(\eta^6\text{-}C_6H_5R)] \quad (356)$$

R = $CH_2CO_2Me$, $C(Me)_2CO_2Me$, $CH_2CO_2Bu^t$, $CH_2CN$, 1,3-dithian-2-yl

Although this feature limited the utility of chromium complexes of halobenzenes, it pointed out that direct alkylation of unfunctionalized aromatic rings was possible. This process has been developed into a very useful reaction for organic synthesis[426]. Reaction of the benzenetricarbonylchromium complex with a wide range of carbanions (Table 2) produced an $\eta^5$-cyclohexadiene complex (characterized by X-ray diffraction analysis). Oxidation gave the free alkylated aromatic compound, whereas protonation ($CF_3COOH$) followed by oxidation gave the corresponding cyclohexadiene (equation 357)[426a]. Attempts to perform a hydride abstraction with a variety of electrophiles so as to make this reaction catalytic in chromium regenerated the original arene complex[427].

In general, substituted arenetricarbonylchromium complexes displayed both the reactivity and the regioselectivity expected for a nucleophilic aromatic substitution reaction[428]. Thus, the order of reactivity towards carbanions was $PhCl > PhCH_3 > PhOCH_3$. Methoxy groups were powerful *meta* directors, and very little *ortho* (<5%) and no *para* substitution was observed in the alkylation of chromium complexed anisole. Methyl groups were also *meta* directing, but less strongly so (43–92% *meta*

TABLE 2. Reactivity of carbanions in equation 357[426a]

| Unreactive | Successful | Metallation |
|---|---|---|
| $LiCH(CO_2R)_2$ | $LiCH_2CO_2R$ | $Bu^nLi$ |
| $LiCH_2COR$ | $LiCH_2CN$ | MeLi |
| MeMgBr | $KCH_2CO_2Bu^t$ | $Bu^sLi$ |
| $Bu^tMgBr$ | LiCH(CN)(OR) | |
| $Me_2CuLi$ | $LiCH_2SPh$ | |
| $LiCH{=}CH_2$ | $LiCH_2[\overline{S(CH_2)_3S}]$ | |
| | LiPh | |
| | $LiC{\equiv}CR$ | |
| | $LiCH_2CH{=}CH$ | |
| | $LiBu^t$ | |

$$[Cr(CO)_3(\eta^6\text{-}C_6H_6)] + R^- \longrightarrow \left[ \text{(R)(H)}C_6H_6\ldots Cr(CO)_3,\ 38.6° \right]^- \begin{cases} \xrightarrow{I_2} C_6H_5R \\ \xrightarrow{(1)\,H^+,\ (2)\,I_2} \text{1-R-cyclohexa-1,3-diene} \\ \xrightarrow{E^+} RE + [Cr(CO)_3(\eta^6\text{-}C_6H_6)] \end{cases} \tag{357}$$

($E^+ = MeI, Ph_3C^+, Et_3P, Ph_2CO$)

substitution was observed, depending on the carbanion). Surprisingly, even chlorine favored *meta* and *ortho* alkylation over *para* alkylation. In contrast, trimethylsilyl and trifluoromethyl groups were *para* directors.

With anisoletricarbonylchromium complexes, alkylation followed by protonation and oxidation provided a route to substituted cyclohexenones[429]. By careful control of the reaction conditions either isomer was available (equation 358).

$$[Cr(CO)_3(\eta^6\text{-}C_6H_5OMe] + LiC(CN)Me_2 \longrightarrow \left[ \text{(OMe)}C_6H_5\text{(H)(CMe}_2\text{CN)}\ Cr(CO)_3 \right] \tag{358}$$

$\downarrow$ $H^+$ fast

$[Cr(CO)_3\{\eta^4\text{-1-OMe-5-C(CN)Me}_2\text{-cyclohexa-1,3-diene}\}]$

slow $\rightleftharpoons$ $\qquad\qquad\searrow$

$[Cr(CO)_3\{\eta^4\text{-2-OMe-4-C(CN)Me}_2\text{-cyclohexa-1,3-diene}\}]$

5-[C(CN)Me$_2$]cyclohex-1-en-3-one

$\downarrow$

1-[C(CN)Me$_2$]-cyclohex-1-en-3-one

The alkylation of arenes was not restricted to simple monocyclic aromatic compounds. The naphthalenechromium complex reacted with similar carbanions, undergoing clean alkylation at the $\alpha$-position (equation 359)[430]. Replacement of the carbon monoxode ligands with phosphines caused a decrease in yield. The spiro-fused indane system in equation 360 underwent alkylation predominantly at the 7-position (79%) with only minor amounts of substitution at C-4 (3%) or C-6

$$[Cr(CO)_3(\eta^6\text{-naphthalene})] + R^- \longrightarrow \xrightarrow{I_2} \text{1-R-naphthalene} \quad 54\text{–}94\% \tag{359}$$

$R = CMe_2CN, CH_2CN, CH[\overline{S(CH_2)_3S}], OMe$

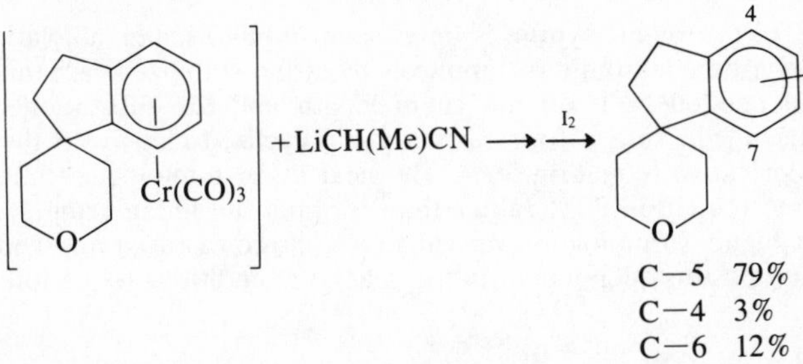

$Cr(CO)_3$ + LiCH(Me)CN $\longrightarrow \xrightarrow{I_2}$ (360)

4, 5, 6, 7, O, CN

C—5 79%
C—4 3%
C—6 12%

(12%)[431]. Indole complexed to chromium exclusively at the benzene ring, and the resulting arenechromium tricarbonyl complex underwent alkylation with nitrile-stabilized carbanions and with dithianes (equation 361). Initially, exclusive attack at the 7-position was reported for both anions[432]. However, subsequent studies showed that with nitrile-stabilized carbanions, the major product resulted from attack at the 4-position[418], in the absence of undue steric crowding. 1,2-Dihydropyridines were stabilized by complexation to chromium[433]. Cyano-stabilized carbanions alkylated the resulting complexes (equations 362 and 363)[434].

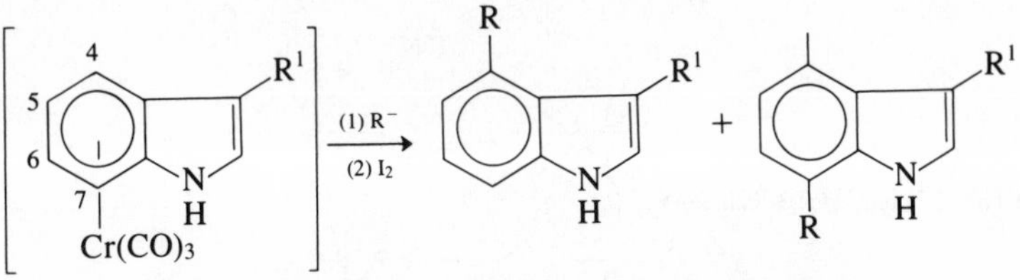

4, 5, 6, 7, $R^1$, N, H, $Cr(CO)_3$ $\xrightarrow[(2)\ I_2]{(1)\ R^-}$ R, $R^1$, N, H + $R^1$, N, H, R (361)

$R^1$=H, tms$CH_2$; $R^-$=$CH_2CN$, $C(Me_2)CN$, $CMe_2CO_2Bu^t$, dithane

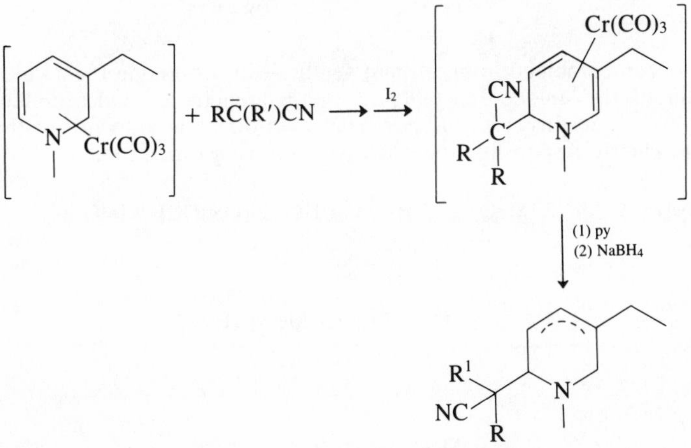

N, $Cr(CO)_3$ + R$\bar{C}$(R′)CN $\longrightarrow \xrightarrow{I_2}$ $Cr(CO)_3$, CN, C, N, R′, R $\xrightarrow[(2)\ NaBH_4]{(1)\ py}$ $R^1$, NC, R, N (362)

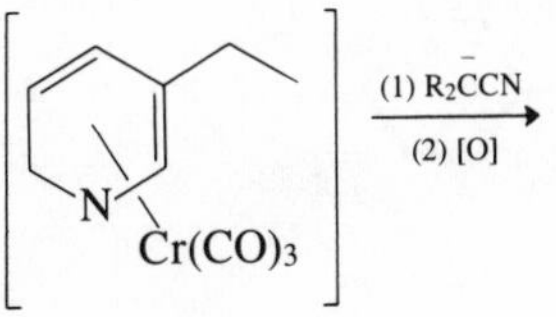

N, $Cr(CO)_3$ $\xrightarrow[(2)\ [O]]{(1)\ R_2\bar{C}CN}$ 3-Et-5-C(R)(R′)CN-pyridine (363)

Particularly useful in organic synthesis have been *intramolecular* alkylation reactions of $\pi$-arenetricarbonylchromium complexes by cyano-stabilized carbanions. The product formed depended both on the chain length and the substituents on the aromatic ring. The nitrile with a four carbon chain cyclized cleanly to the corresponding tetralin derivative (equation 364). The next lower homologue dimerized to the metacyclophane (equation 365) rather than forming the indane ring system. In contrast, the next higher homologue cyclized to a mixture of fused and spiro fused products whose composition depended on the reaction conditions (equation 366)[435].

$$[Cr(CO)_3(\eta^6\text{-}C_6H_5(CH_2)_4CN] \xrightarrow[(2)\ I_2]{(1)\ \text{Litmp},\ -78\ ^\circ C} \text{1-cyanotetrahydronaphthalene} \quad (364)$$

(Litmp = lithium tetramethylpipyridide) 84%

$$[Cr(CO)_3(\eta^6\text{-}C_6H_5(CH_2)_3CN] \xrightarrow[(2)\ I_2]{(1)\ \text{lda},\ 0\ ^\circ C} \quad (365)$$

CN

CN

$$[Cr(CO)_3(\eta^6\text{-}C_6H_5(CH_2)_5CN] \longrightarrow \quad + \quad (366)$$

NC

CN

3–72% 28–97%

This chemistry has been combined in an elegant synthesis of acorenone B in which two key steps utilized the unique reactivity of the $\pi$-arenetricarbonylchromium system (equation 367)[436]. In this case, the powerful *meta*-directing influence of the methoxy group was clearly responsible for exclusive spiro ring formation.

$$[Cr(CO)_3(\eta^6\text{-1-Me-2-MeO-}C_6H_4)] + Me_2CHC(Li)(CN)OCH(Me)OEt$$

(1) $I_2$ (2) $OH^-$ (3) $H^+$ (367)

7 steps

$$\text{1-Me-2-MeO-4-C(O)CHMe}_2\text{-}C_6H_3$$

92% (one isomer only)

CN MeO Me $Cr(CO)_3$

(1) lda, −78 °C (2) $CF_3COOH$ (3) $NH_4OH$

MeO CN Me

→ → Me O Me

The *meta*-directing ability of the methoxy group was similarly used in the synthesis of the precursor to the antibiotic aklavinone (equation 368)[418].

$$\left[\text{(OMe)(OH)-arene-}Cr(CO)_3\text{ nitrile}\right] \xrightarrow[\text{(2) } LiNR_2 \text{ (3) } I_2]{\text{(1) MeI}} \text{OMe, OMe-tetralin-CN} \quad (368)$$

Complexation of an arene to chromium also effected the chemistry of substituents on the aromatic ring. Again, as a consequence of the electron-withdrawing abilities of the tricarbonylchromium group, complexed styrenes underwent alkylation at the $\beta$-position of the olefin to produce a chromium-stabilized benzyl anion which could be trapped with a variety of electrophiles (equation 369)[437]. Peculiarly, the tricarbonylchromium group also stabilized benzyl cations, permitting the facile amidation of chromium-complexed benzyl alcohols (equation 370)[438].

$$[Cr(CO)_3\{\eta^6\text{-}C_6H_5C(R){=}CH_2\}] + R'(-) \longrightarrow [Cr(CO)_3\{\eta^6\text{-}C_6H_5C(R)CH_2R'\}] \quad (369)$$

$$\downarrow \text{(1) } E^+ \text{ (2) oxidize}$$

$$C_6H_5C(R)(E)CH_2R'$$

R = H, Me, SEt

R' = $CMe_2CN$, $CMe_2CO_2Bu^t$, C(CN)(OR)Me, $Bu^n$, Ph, Me, dithiane

$E^+ = H^+$, MeI, MeCOCl, PhSSPh

$$[Cr(CO)_3\{\eta^6\text{-}p\text{-R-}C_6H_4C(R')(R'')OH\}] + R'''CN \xrightarrow[-15\,^\circ C]{H_2SO_4} [Cr(CO)_3\{\eta^6\text{-}p\text{-R-}C_6H_4C(R')(R'')NHC(O)R'''] \quad (370)$$

80–90%

R = Me, MeO, H; R' = H, Me, Ph; R'' = H, Me; R''' = Me, Ph

The face of the aryl ring occupied by the $Cr(CO)_3$ group is sterically hindered, and virtually all nucleophilic reactions of arenetricarbonylchromium complexes occur by attack from the *exo* face of the complex. This permits an unprecedented degree of steric control in reactions of planar aryl systems, and has found use in a number of ways[439]. In addition, arene complexes containing different *ortho* or *meta* substituents are chiral and can be resolved, permitting the synthesis of chiral compounds. For example, reaction of complexed benzaldehydes with Grignard reagents, or of complexed aryl ketones with potassium borohydride, led to the same mixture of diastereoisomeric alcohols, which were separated by chromatography (equation 371). Decomplexation gave optically active alcohols with 100% optical purity[440,441]. Racemic indan-1-ol reacted with hexacarbonylchromium to produce the complex with chromium exclusively on the same face of the indanol as the hydroxyl group. Resolution and oxidation of these complexes was followed by alkylation of one enantiomer with a Grignard reagent. This reaction occurred exclusively *exo* and, after decomplexation, the indanol was obtained with 100% optical purity (equation 372)[442]. Similar chemistry was carried out with tetralones[443]. The reduction of

R[1] H C OH R[2] $Cr(CO)_3$

R[1] CHO $Cr(CO)_3$ $R^2MgX$ $KBH_4$ R[1] O $CR^2$ $Cr(CO)_3$ (371)

R[1] H C $R^2$ OH $Cr(CO)_3$

OH + $[Cr(CO)_6]$ ⟶ resolve ⟶ $(CO)_3Cr$ OH

oxidize (372)

R OH (one enantiomer) ⟵ (1) RMgX (2) $h\nu$, air $(CO)_3Cr$ O

complexed substituted indanones with borohydrides always occurred in an *exo* fashion, producing the *endo* complexed indanols. In this manner, a variety of substituted indanols were prepared stereospecifically[444].

## H. Other Reactions of Nucleophiles with Transition Metal Complexes

Alkynes form very stable complexes with octacarbonyldicobalt. Propargyl alcohols reacted with this reagent to produce the corresponding cobalt complex. Treatment with fluoroboric acid produced a very stable complexed propargyl cation. This species was alkylated with $\beta$-dicarbonyl compounds[445], enol ethers[446], and enol acetates[447] in fair yield (equation 373). Organomagnesium, aluminium, cadmium,

$$R^1C{\equiv}CC(R^2)(R^3)OH \xrightarrow{[Co_2(CO)_8]} \left[ \begin{array}{c} R^1{-}C{-}C{-}C(R^2)(R^3){-}OH \\ (CO)_3Co{-}Co(CO)_3 \end{array} \right] \xrightarrow{HBF_4} \left[ \begin{array}{c} R^1{-}C{-}C{-}C^+(R^2)(R^3) \\ (CO)_3Co{-}Co(CO)_3 \end{array} \right]$$

$$\xrightarrow{(1)\ \text{pentane-2,4-dione};\ (2)\ Fe^{3+}} R^1C{\equiv}CC(R^2)(R^3)CH[C(O)Me]_2$$

$$\xrightarrow{CH_2{=}C(Me)OZ,\ Z{=}R,\ Ac} R^1C{\equiv}CC(R^2)(R^3)CH_2C(O)Me$$

$$\xrightarrow{R^4C(O)CH_2R^5} R^1C{\equiv}CC(R^2)(R^3)CH(R^5)C(O)R^4$$

silver, tin, iron, and silicon reagents alkylated this species in low yield[448]. This same propargyl alcohol complex reacted with acetonitrile in the presence of sulphuric acid to produce the *N*-propargylamide. (equation 374). Since the cobalt could be removed from these complexes by oxidation, this procedure represents a method for the alkylation of propargyl systems without the formation of allenic products which often plagues classical displacement reactions of propargyl derivatives.

$$[Co_2(CO)_6(\eta^2\text{-}CH{\equiv}CCH_2OH)] \xrightarrow[MeCN]{H_2SO_4} [Co_2(CO)_6\{\eta^2\text{-}CH{\equiv}CCH_2NHC(O)Me\}] \quad 35\% \qquad (374)$$

$\alpha,\alpha'$-Dihaloketones reacted with $[Fe_2(CO)_9]$ to give species which, although not isolated and characterized, behaved as if they were iron-stabilized oxallyl cations[449]. For example, one reacted with methanol to give a methyl ether, apparently by nucleophilic attack on the complexed cationic species (equation 375). These iron-stabilized oxallyl cations underwent a number of synthetically useful 'cycloaddition' reactions. With 1,3-dienes, a formal $\pi^2+\pi^4$ (a 3+4 cyclocoupling) cycloaddition occurred to produce seven-membered ring systems (equation 376)[450]. This reaction was used to synthesize nezukone, $\beta$-thujaplicin, $\alpha$-thujaplicin[451], and *C*-nucleosides[452] from cycloaddition to furans, and tropane alkaloids including tropine, scopine, and hyoscymine from cycloaddition to pyrroles[453]. Since this process was shown to be a concerted process, it is not formally a nucleophilic reaction at all[454]. However, this same iron-stabilized oxallyl cation also reacted with aromatic olefins[455]

$$\text{2,6-dibromo-2,6-diisopropylcyclohexanone} \xrightarrow{[Fe_2(CO)_9]} \left[ \text{OFeBr-stabilized oxallyl cation} \right] \xrightarrow{MeOH} \text{2-methoxy-2,6-diisopropylcyclohexanone (OMe)} \quad 61\% \qquad (375)$$

$$R^2R^1C(Br){-}C(O){-}C(Br)R^1R^2 + CH_2{=}C(R^3)C(R^4){=}CH_2 \xrightarrow[PhH,\ 60\ ^\circ C]{[Fe_2(CO)_9]} \text{2,2,7,7-}(R^1,R^2)\text{-4-}R^3\text{-5-}R^4\text{-cyclohept-4-enone} \qquad (376)$$

and enamines[456] in a formal (3+2) (forbidden) cycloaddition, which in fact proceeded stepwise and did involve nucleophilic attack as a first step (equations 377 and 378). The result of this attack was an iron enolate, a nucleophile in its own right, and a cation. Ring closure occurred by nucleophilic attack of the enolate on this cation. The reactions were stereospecific because, although stepwise, free rotation in the intermediate was prevented by strong attraction between the cationic and anionic portions of the intermediate[457].

$R^1$, $R^2$, Br; $+[Fe_2(CO)_9] \longrightarrow$ [OFe, $R^2$, $R^1$]$^+$ $\xrightarrow{CH_2{=}CHAr}$ [OFe, $R^1$, $R^2$, Ar, +] $\longrightarrow$ O, $R^1$, $R^2$, Ar (377)

R, Br, O + morpholine enamine ($R^1$, $R^2$, $R^3$) $\xrightarrow{[Fe_2(CO)_9]}$ O, R, N, $R^3$, $R^2$, $R^1$ $\longrightarrow$ O, R, $R^1$, $R^2$, $R^3$ (378)

A conceptually related cycloaddition process occurs in the reactions of cyclopentadienyldicarbonylmonohaptoallyliron complexes with electron-deficient olefins such as tcne, ddq (dichlorodicyanobenzoquinone), $\beta,\beta$-dicyano-$o$-chlorostyrene, dimethyl methylene malonate, and chlorosulphonyl isocyanates[458]. The reaction was proposed to proceed by initial nucleophilic attack of the $\gamma$-position of the $\pi$-allyl group on the electron-deficient olefin. This produced a stabilized carbanion and a cationic iron–olefin complex known to be reactive towards nucleophiles. Attack of the carbanion centre on this metal-bound olefin completed the cycloaddition and gave the observed product (equation 379)[459,460]. Related tungsten and molybdenum complexes behaved in a similar manner[460]. This reaction was used in a very clever synthesis of the hydroazulene ring system in which two different iron complexes alternately behaved as nucleophile and electrophile in the process (equation 380)[461].

[$(CO)_2CpFe$–$CH_2CH{=}CH_2$] + $X_2C{=}CX_2$ $\longrightarrow$ [$(CO)_2CpFe^+$ (X, X, X, X)] $\longrightarrow$ [$(CO)_2CpFe$–(cyclopentyl, X, X, X, X)]

(379)

[$(CO)_2CpFe$–$CH_2CH{=}CH_2$] + [$C_7H_7Fe(CO)_3$]$^+$ $\longrightarrow$ [$Fe(CO)_3$ H … $^+FeCp(CO)_2$] $\longrightarrow$ [H, H, $Fe(CO)_3$, $FeCp(CO)_2$]$^+$ $\xrightarrow{(1)\ NaBH_4;\ (2)\ Ce^{IV}}$ $CO_2Me$

(380)

Recently this cycloaddition reaction has been expanded to include a wider variety of electron-deficient olefins (equation 381)[462] and to the use of ($\eta^1$-3-methoxyallyl)iron (equation 382)[463] and ($\eta^1$-2-methoxyallyl)iron complexes (equation 383)[464]. Dimethylacetylene dicarboxylate reacted in a similar fashion.

$$[FeCp(CO)_2(CH_2CH{=}CH_2)] + R^1(R^2)C{=}CHR^3 \longrightarrow [(CO)_2CpFe\text{–}(R^1, R^2, H, R^3)]$$

$$\xrightarrow{Ce^{IV},\ ROH} RO_2C\text{–}(R^1, R^2, H, R^3)$$

(381)

| $R^1$ | $R^2$ | $R^3$ | % |
|---|---|---|---|
| $CO_2Et$ | CN | $CO_2Et$ | 81 |
| $CO_2Et$ | $CO_2Et$ | H | 70 |
| $CO_2Me$ | $CO_2Me$ | $CO_2Me$ | 50 |
| CN | CN | $CO_2Et$ | 67 |
| CN | $CO_2Me$ | CN | 94 |
| $CO_2Me$ | $CO_2Me$ | H | 64 |

$$[FeCp(CO)_2(CH_2CH{=}CHOMe)] + R^1(R^2)C{=}CHR^3 \longrightarrow [(CO)_2CpFe\text{-cyclopentyl}(OMe, R^1, R^2, R^3, H)] + [(CO)_2CpFe\text{-cyclopentyl}(OMe, R^1, R^2, R^3, H)] \quad (382)$$

$R^1, R^2, R^3 = CO_2Me, CN$

85–89%

$$[FeCp(CO)_2\{CH_2C(OMe){=}CH_2\}] + R^1(R^2)C{=}C(R^3)R^4 \longrightarrow [(CO)_2CpFe^{+}{=}C(OMe)CH_2C(R^3)(R^4)C^{-}(R^1)(R^2)] \quad (383)$$

$$\longrightarrow [(CO)_2CpFe\text{-}C(MeO)\text{-cyclopentyl}(R^1, R^2, R^3, R^4)]$$

$$\longrightarrow [(CO)_2CpFeCH_2C(MeO){=}C(C(R^3)(R^4)C(R^1)(R^2))_nH]$$

## III. REFERENCES


1. I. Wender and P. Pino (Eds.), *Organic Synthesis via Metal Carbonyls,* Vol. I, Interscience, New York, 1968.
2. I. Wender and P. Pino (Eds.), *Organic Synthesis via Metal Carbonyls,* Vol. II, Interscience, New York, 1977.
3. J. P. Collman and L. S. Hegedus, *Principles and Applications of Organotransition Metal Chemistry,* University Science Books, Mill Valley, CA, 1980, pp. 81–89.
4. F. Bonate and L. Malatesta, *Isocyanide Complexes of Metals,* Wiley, New York, 1969.
5. P. M. Treichel, W. J. Knebel, and R. W. Hess, *J. Am. Chem. Soc.,* **93,** 5424 (1971).
6. L. S. Hegedus, O. P. Anderson, K. Zetterberg, G. Allen, K. Surala-Hansen, D. J. Olsen, and A. B. Packard, *Inorg. Chem.,* **16,** 1887 (1977).
7. H. C. Clark and L. E. Manzer, *Inorg. Chem.,* **11,** 503 (1972).

8. A. G. Sharpe, *The Chemistry of Cyano Complexes of Transition Metals,* Academic Press, New York, 1976.
9. M. Rosenblum, *The Iron-Group Metallacenes: Ferrocene, Ruthenocene, and Osmocene,* Wiley, New York, 1965.
10. P. C. Bharara, *J. Organomet. Chem. Libr.,* **5,** 259 (1977).
11. E. Maslowsky, Jr., *J. Chem. Educ.,* **55,** 276 (1978).
12. J. T. Sheats, *Organomet. Chem. Rev.,* **7,** 1 (1979).
13. F. A. Cotton and G. Wilkinson, *Advanced Inorganic Chemistry,* Wiley, New York, 1980, pp. 1163–1166.
14. R. B. King, *Transition Metal Organometallic Chemistry,* Wiley, New York, 1969, pp. 14–20, 51–53.
15. J. P. Collman and L. S. Hegedus, *Principles and Applications of Organotransition Metal Chemistry,* University Science Books, Mill Valley, CA, 1980, pp. 119–133.
16. A. J. Campbell, *Inorg. Chem.,* **15,** 1326 (1976).
17. R. D. Rogers, *J. Am. Chem. Soc.,* **100,** 5238 (1978).
18. R. Baker, *Chem. Rev.,* **73,** 487 (1973).
19. L. S. Hegedus, *J. Organomet. Chem. Libr.,* **1,** 329 (1976).
20. G. Wilke, B. Bogdanovic, P. Hardt, P. Heimbach, W. Keim, M. Kröner, W. Oberkirch, K. Tanaka, E. Steinrucke, D. Walter, and H. Zimmerman, *Angew. Chem., Int. Ed. Engl.,* **5,** 151 (1966).
21. A. Kasahara and K. Tanaka, *Bull. Chem. Soc. Jpn.,* **39,** 634 (1966).
22. For a general review on the synthesis, structure and occurrence of $\eta^3$-allylmetal complexes, see M. L. H. Green and P. L. I. Hagy, *Adv. Organomet. Chem.,* **2,** 325 (1964).
23. J. P. Collman and L. S. Hegedus, *Principles and Applications of Organotransition Metal Chemistry,* University Science Books, Mill Valley, CA, 1980, pp. 536–602.
24. R. J. Cross, *Organomet. Chem. Rev.,* **2,** 97 (1962).
25. M. D. Rausch and G. A. Moser, *Inorg. Chem.,* **13,** 11 (1974).
26. M. L. H. Green, *Organometallic Compounds,* Vol. II, Methuen, 1968, p. 200.
27. B. K. Bower and H. G. Tennet, *J. Am. Chem. Soc.,* **94,** 2512 (1972).
28. R. R. Schrock and G. W. Parshall, *Chem. Rev.,* **76,** 243 (1976).
29. P. J. Davidson, M. F. Lappert, and R. Pearce, *Acc. Chem. Res.,* **7,** 209 (1974).
30. P. J. Davidson, M. F. Lappert, and R. Pearce, *Chem. Rev.,* **76,** 214 (1976).
31. W. Keim, *J. Organomet. Chem.,* **14,** 179 (1968).
32. W. Keim, *J. Organomet. Chem.,* **19,** 161 (1969).
33. K. C. Dewhirst, W. Keim, and C. A. Reilly, *Inorg. Chem.,* **7,** 546 (1968).
34. M. F. Semmelhack and L. Ryono, *Tetrahedron Lett.,* 2967 (1973).
35. M. Michman and M. Balog, *J. Organomet. Chem.,* **31,** 395 (1971).
36. L. S. Hegedus, P. M. Kendall, S. M. Lo, and J. R. Sheats, *J. Am. Chem. Soc.,* **97,** 5448 (1975).
37. G. H. Posner, *An Introduction to Synthesis Using Organocopper Reagents,* Wiley, New York, 1980.
38. G. H. Posner, *Org. React.,* **22,** 253 (1975).
39. G. H. Posner, *Org. React.,* **19,** 1 (1972).
40. J. F. Normant, *Synthesis,* 63 (1972).
41. J. F. Normant, *J. Organomet. Chem. Lib.,* **1,** 219 (1976).
42. J. F. Normant, *Pure Appl. Chem.,* **50,** 709 (1978).
43. A. E. Jukes, *Adv. Organomet. Chem.,* **12,** 215 (1974).
44. T. Kauffman, *Angew. Chem., Int. Ed. Engl.,* **13,** 291 (1974).
45. R. G. R. Bacon and H. A. O. Hill, *Q. Rev. Chem. Soc.,* **19,** 95 (1965).
46. P. E. Fanta, *Synthesis,* 9 (1974).
47. The synthetic applications of organocopper reagents are covered in an annual survey by L. S. Hegedus entitled "Transition Metals in Organic Synthesis" which appears annually in regular issues of *J. Organomet. Chem.*
48. M. S. Kharasch and P. O. Tawney, *J. Am. Chem. Soc.,* **63,** 2308 (1941).
49. B. Jousseaume and J. Duboudin, *J. Organomet. Chem.,* **91,** C1 (1975).
50. F. W. von Rein and M. G. Richey, *Tetrahedron Lett.,* 3780 (1971).
51. Y. Butsugan, I. Kadosuba, and S. Araki, *Chem. Lett.,* 527 (1979).

52. S. Araki and Y. Butsugan, *Chem. Lett.*, 185 (1980).
53. L.-I. Olsson and A. Claesson, *Acta Chem. Scand. B,* **33,** 679 (1979).
54. T. Sato, M. Kawashima, and T. Fujisawa, *Tetrahedron Lett.*, **22,** 2375 (1981).
55. T. Sato, M. Takeguchi, T. Itoh, M. Kawashima, and T. Fujisawa, *Tetrahedron Lett.*, **22,** 1817 (1981).
56. For reviews see (a) H. Felkin and G. Swierczewski, *Tetrahedron,* **31,** 2735 (1975); (b) K. Tamao and M. Kumada, in *Organometallic Reactions and Synthesis,* (Ed. E. I. Becker and M. Isutsui), Vol. 7, Plenum Press, New York, 1982.
57. M. Kumada, *Pure Appl. Chem.*, **52,** 669 (1980).
58. K. Tamao, K. Sumitani, Y. Kiso, M. Zembayashi, A. Kujioka, S.-I. Kodama, I. Nakajimi, A. Minato, and M. Kumada, *Bull. Chem. Soc. Jpn.*, **49,** 1958 (1976).
59. D. G. Morrell and J. K. Kochi, *J. Am. Chem. Soc.*, **97,** 7262 (1975).
60. M. F. Semmelhack, B. P. Chong, R. D. Stauffer, T. D. Rogerson, A. Chong, and L. D. Jones, *J. Am. Chem. Soc.*, **97,** 2507 (1975).
61. E. D. Thorsett and F. R. Stermitz, *J. Heterocycl. Chem.*, **10,** 243 (1973).
62. L. N. Pridgen, *J. Heterocycl. Chem.*, **17,** 1289 (1980).
63. O. Piccolo and T. Martinengo, *Synth. Commun.*, **11,** 497 (1981).
64. (a) A. Ohsawa, Y. Abe, and H. Igeta, *Chem. Pharm. Bull.*, **26,** 2550 (1978); (b) H. Yamanaka, K. Edo, F. Snoji, K. Konno, T. Sakamoto, and M. Mizugaki, *Chem. Pharm. Bull.*, **26,** 2160 (1978).
65. K. Tamao, S.-I. Kodama, T. Nakatsutka, Y. Kiso, and M. Kumada, *J. Am. Chem. Soc.*, **97,** 4405 (1975).
66. K. Tamao, M. Zembayashi, and M. Kumada, *Chem. Lett.*, 1237 (1976).
67. M. Yamamura, I. Mortani, and S.-I. Murahashi, *J. Organomet. Chem.*, **91,** C39 (1975).
68. H. P. Dung and G. Linstrumelle, *Tetrahedron Lett.*, 191 (1978).
69. A. Sekiya and N. Ishikawa, *J. Organomet. Chem.*, **118,** 349 (1976).
70. A. Sekiya and N. Ishikawa, *J. Organomet. Chem.*, **125,** 281 (1977).
71. T. Hayashi, M. Konishi, K.-I. Yakota, and M. Kumada, *J. Chem. Soc., Chem. Commun.*, 313 (1981).
72. T. Jeffrey-Luong and G. Linstrumelle, *Tetrahedron Lett.*, **21,** 5019 (1980).
73. S.-I. Murahashi, M. Yamamura, K.-I. Yanagesawa, N. Mita, and K. Kondo, *J. Org. Chem.*, **44,** 2408 (1979).
74. N. Cong-Danh, J. P. Beaucourt, and L. Pichat, *Tetrahedron Lett.*, 3159 (1979).
75. (a) T. Hayashi, M. Tajika, K. Tamao, and M. Kumada, *J. Am. Chem. Soc.*, **98,** 3718 (1976); (b) K. Tamao, H. Matsumoto, H. Yamamoto, and M. Kumada, *Tetrahedron Lett.*, 7155 (1979).
76. T. Hayashi, T. Mise, M. Fukushima, M. Kagotani, N. Nagashima, Y. Hamada, A. Matsumoto, S. Kawakami, M. Konishi, K. Yamamoto, and M. Kumada, *Bull. Chem. Soc. Jpn.*, **53,** 1138 (1980).
77. T. Hayashi, N. Nagashima, and M. Kumada, *Tetrahedron Lett.*, **21,** 79 (1980).
78. K. Tamao, T. Hayashi, H. Matsumoto, H. Yamamoto, and M. Kumada, *Tetrahedron Lett.*, 2155 (1979).
79. T. Hayashi, M. Konishi, and M. Kumada, *J. Organomet. Chem.*, **186,** C1 (1980).
80. G. Consiglio, F. Morandini, and O. Piccolo, *Helv. Chim. Acta.*, **63,** 987 (1979).
81. M. Cherest, H. Felkin, J. D. Umpleby, and S. G. Davies, *J. Chem. Soc., Chem. Commun.*, 681 (1981).
82. S. M. Newmann and J. Kochi, *J. Org. Chem.*, **40,** 599 (1975).
83. G. Cahiez, D. Bernard, and J. F. Normant, *Tetrahedron Lett.*, 3155 (1976).
84. For reviews see (a) R. C. Larock, *J. Organomet. Chem. Libr.*, **1,** 257 (1976); (b) R. C. Larock, *Angew. Chem., Int. Ed. Engl.*, **17,** 27 (1978).
85. R. C. Larock and H. C. Brown, *J. Am. Chem. Soc.*, **92,** 2467 (1970).
86. R. C. Larock, *J. Organomet. Chem.*, **67,** 353 (1974); **72,** 35 (1974).
87. R. C. Larock, S. K. Gupta, and H. C. Brown, *J. Am. Chem. Soc.*, **94,** 4371 (1972).
88. H. Staub, K. P. Zeller, and H. Leditsche, in *Houben-Weyl, Methoden der Organische Chemie,* 4th ed., Vol. 13/2b, Thieme, Stuttgart, 1974, pp. 205*ff.*
89. H. Staub, K. P. Zeller, and H. Leditsche, in *Houben-Weyl, Methoden der Organische Chemie,* 4th ed., Vol. 13/2b, Thieme, Stuttgart, 1974, pp. 138*ff.*
90. H. C. Brown and M.-H. Rei, *J. Am. Chem. Soc.*, **91,** 5646 (1969).

91. H. K. Hall, Jr., J. P. Schaefer, and R. J. Spanggard, *J. Org. Chem.*, **37**, 3069 (1972), and references cited therein.
92. A. E. Borisov, V. P. Vil'chevskaga, and A. N. Nesmeyanov, *Izv. Akad. Nauk SSSR, Otd. Khim. Nauk,* 1008 (1954).
93. R. C. Larock and B. Riefling, *Tetrahedron Lett.*, 4661 (1976).
94. (a) A. N. Nesmeyanov, N. K. Kochetkov, and V. M. Dashunin, *Izv. Akad. Nauk SSSR, Otd. Khim. Nauk,* 77 (1950); (b) A. N. Nesmeyanov and N. K. Kochetkov, *Izv. Akad. Nauk SSSR, Otd. Khim. Nauk,* 305 (1949).
95. H. Staub, K. P. Zeller, and H. Leditsche, in *Houben-Weyl, Methoden der Organische Chemie,* 4th ed., Vol. 13/2b, Thieme, Stuttgart, 1974, pp. 28*ff.*
96. D. E. Bergstrom and J. L. Ruth, *J. Am. Chem. Soc.*, **98,** 1587 (1976).
97. R. F. Heck, *Proc. Welch Foundation,* **17,** 53 (1973).
98. I. Arai, R. Hanna, and G. D. Davies, Jr., *J. Am. Chem. Soc.*, **103,** 7684 (1981).
99. R. C. Larock and B. Riefling, *J. Org. Chem.*, **43,** 1468 (1978).
100. R. C. Larock, K. Oertle, and K. M. Beatty, *J. Am. Chem. Soc.*, **102,** 1966 (1980).
101. R. C. Larock and S. S. Hershberger, *Tetrahedron Lett.*, **22,** 2443 (1981).
102. R. C. Larock, *J. Org. Chem.*, **40,** 3237 (1975).
103. R. C. Larock, B. Riefling, and C. A. Fellows, *J. Org. Chem.*, **43,** 131 (1978).
104. (a) R. F. Heck, *J. Am. Chem. Soc.*, **90,** 5518 (1968); **91,** 6707 (1969); **93,** 6896 (1971); (b) R. F. Heck, *J. Organomet. Chem.*, **37,** 389 (1972).
105. R. C. Larock, J. C. Bernhardt, and R. J. Driggs, *J. Organomet. Chem.*, **156,** 45 (1978).
106. D. E. Bergstrom and J. L. Ruth, *J. Am. Chem. Soc.*, **98,** 1587 (1976).
107. J. L. Ruth and D. E. Bergstrom, *J. Org. Chem.*, **43,** 2870 (1978).
108. C. F. Bigge, P. Kalarites, J. R. Peck, and M. P. Mertes, *J. Am. Chem. Soc.*, **102,** 2033 (1980).
109. D. E. Bergstrom, J. L. Ruth, and P. E. Warwick, *J. Org. Chem.*, **46,** 1432 (1981).
110. (a) I. Arai and G. D. Davies, Jr., *J. Heterocycl. Chem.*, **15,** 351 (1978); (b) *J. Org. Chem.*, **43,** 4110 (1978); (c) *J. Am. Chem. Soc.*, **100,** 287 (1978); (d) D. E. Bergstrom, A. J. Brattesani, M. K. Ogawa, and M. J. Schweickert, *J. Org. Chem.*, **46,** 1423 (1981).
111. S. F. Dyke and M. J. McCartney, *Tetrahedron,* **37,** 431 (1981).
112. D. Milstein and J. K. Stille, *J. Org. Chem.*, **44,** 1613 (1979).
113. D. Milstein and J. K. Stille, *J. Am. Chem. Soc.*, **101,** 4992 (1979).
114. A. N. Kashin, I. G. Bumagina, N. A. Bumagina, and D. P. Beletskya, *J. Org. Chem. USSR,* **17,** 18 (1981).
115. (a) J. Godschalx and J. K. Stille, *Tetrahedron Lett.*, **21,** 2599 (1980); (b) B. M. Trost and E. Keinan, *Tetrahedron Lett.*, **21,** 2591, 2595 (1980).
116. D. Milstein and J. K. Stille, *J. Am. Chem. Soc.*, **101,** 4981 (1979).
117. T. Kobayashi and M. Tanaka, *J. Organomet. Chem.*, **205,** C27 (1981).
118. M. Tanaka, *Synthesis,* 47 (1981).
119. B. M. Trost and Y. Tanigawa, *J. Am. Chem. Soc.*, **100,** 4743 (1979).
120. N. Miyaura, K. Yamada, and A. Suzuki, *Tetrahedron Lett.*, 3437 (1979).
121. N. Miyaura, H. Suginome, and A. Suzuki, *Tetrahedron Lett.*, **22,** 127 (1981).
122. N. Miyaura and A. Suzuki, *J. Chem. Soc., Chem. Commun.*, 866 (1979).
123. N. M. Miyaura, T. Yano, and A. Suzuki, *Tetrahedron Lett.*, **21,** 2865 (1980).
124. H. Yatagai, *Bull. Chem. Soc. Jpn.*, **53,** 1670 (1980).
125. N. Miyaura, T. Yanagi, and A. Suzuki, *Synth. Commun.*, **11,** 513 (1981).
126. K. Takai, K. Oshima, and H. Nozaki, *Tetrahedron Lett.*, **21,** 2531 (1980).
127. M. Sato, K. Takai, K. Oshima, and H. Nozaki, *Tetrahedron Lett.*, **22,** 1609 (1981).
128. For a review of zirconium-catalysed hydroalumination, and transmetallation reactions to nickel and palladium complexes see E.-I. Negishi, *Pure Appl. Chem.*, **53,** 2333 (1981); E.-I. Negishi, *Aspects Mech. Organomet. Chem.*, 285 (1978); E.-I. Negishi, *Acc. Chem. Res.*, **15,** 340 (1982).
129. E.-I. Negishi, N. Okukado, A. O. King, D. E. Van Horn, and B. I. Spiegel, *J. Am. Chem. Soc.*, **100,** 2254 (1978).
130. S. Baba and E.-I. Negishi, *J. Am. Chem. Soc.*, **98,** 6729 (1976).
131. H. Matsushita and E.-I. Negishi, *J. Am. Chem. Soc.*, **103,** 2282 (1981).
132. E.-I. Negishi, H. Matsushita, and N. Okukado, *Tetrahedron Lett.*, **22,** 2715 (1981).
133. N. Jabri, A. Alexakis, and J. F. Normant, *Tetrahedron Lett.*, **22,** 3581 (1981).

134. N. Jabri, A. Alexakis, and J. F. Normant, *Tetrahedron Lett.*, **22,** 959 (1981).
135. P. Vincent, J.-P. Beaucourt, and L. Pichat, *Tetrahedron Lett.*, **22,** 945 (1981).
136. J. F. Fauvurque and A. Jutland, *J. Organomet. Chem.*, **209,** 109 (1981).
137. K. Ruitenberg, H. Kleijn, C. J. Elsevier, J. Meyer, and P. Vermeer, *Tetrahedron Lett.*, **22,** 1451 (1981).
138. E.-I. Negishi, L. F. Valenti, and M. Kobayashi, *J. Am. Chem. Soc.*, **102,** 3298 (1980).
139. C. L. Rand, D. E. Van Horn, M. W. Moore, and E.-I. Negishi, *J. Org. Chem.*, **46,** 4093 (1981).
140. M. Kabayashi, and E.-I. Negishi, *J. Org. Chem.*, **45,** 5223 (1980).
141. A. O. King, *PhD Dissertation,* Syracuse University, 1979.
142. (a) J. Schwartz, *J. Organomet. Chem. Libr.*, **1,** 461 (1976); (b) N. Okukado, D. E. Van Horn, W. Klima, and E.-I. Negishi, *Tetrahedron Lett.*, 1027 (1978).
143. (a) E.-I. Negishi and D. E. Van Horn, *J. Am. Chem. Soc.*, **99,** 3168 (1977); (b) N. Okukado, D. E. Van Horn, W. Klima, and E.-I. Negishi, *Tetrahedron Lett.*, 1027 (1978).
144. (a) Y. Hayasi, M. Riedike, J. S. Temple, and J. Schwartz, *Tetrahedron Lett.*, **22,** 2629 (1981); (b) M. Riediker and J. Schwartz, *Tetrahedron Lett.*, **22,** 4655 (1981).
145. For reviews on transition metal hydrides see (a) E. L. Muetterties (Ed.), *Transition Metal Hydrides,* Marcel Dekker, New York, 1971; (b) H. D. Kaesz and R. B. Saillant, *Chem. Rev.*, **72,** 231 (1972); (c) D. Giusto, *Inorg. Chim. Acta Rev.*, **6,** 91 (1972); (d) D. M. Roundhill, *Adv. Organomet. Chem.*, **13,** 273 (1975); (e) H. D. Kaesz, *Inorg. Synth.*, **17,** 52 (1977).
146. (a) C. R. Lucas, *Inorg. Synth.*, **16,** 109 (1976); (b) H. Bonnemann, *Angew. Chem., Int. Ed., Engl.*, **9,** 736 (1970); (c) T. Bodnar, E. Coman, S. LaCroce, C. Lambert, K. Menard, and A. Cutler, *J. Am. Chem. Soc.*, **103,** 2741 (1981).
147. M. J. Mays and B. E. Prater, *Inorg. Synth.*, **15,** 23 (1974).
148. D. B. Carr and J. Schwartz, *J. Am. Chem. Soc.*, **101,** 3521 (1979).
149. J. P. Collman and L. S. Hegedus, *Principles, and Applications of Organotransition Metal Chemistry,* University Science Books, Mill Valley, CA, 1980, pp. 405–411.
150. E. C. Ashby, J. J. Lin, and R. Kovar, *J. Org. Chem.*, **41,** 1939 (1976).
151. E. C. Ashby, J. J. Lin, and A. B. Goel, *J. Org. Chem.*, **43,** 183 (1978).
152. P. Four and F. Guibe, *J. Org. Chem.*, **46,** 4439 (1981).
153. (a) F. A. Cotton and G. Wilkinson, *Advanced Inorganic Chemistry,* 4th ed., Wiley, New York, 1980, pp. 1185–1217; (b) C. H. Langford and H. B. Gray, *Ligand Substitution Reactions,* W. A. Benjamin, Menlo Park, Ca., 1965; (c) F. Basolo and R. G. Pearson, *Mechanisms of Inorganic Reactions,* Wiley, New York, 1967.
154. D. Dodd and M. D. Johnson, *Chem. Commun.*, 1371 (1971).
155. J. P. Collman and M. R. Maclaury, *J. Am. Chem. Soc.*, **96,** 3019 (1974).
156. K. S. Y. Lau, J. K. Stille, and R. W. Fries, *J. Am. Chem. Soc.*, **96,** 4983 (1974).
157. J. K. Stille and K. S. Y. Lau, *J. Am. Chem. Soc.*, **98,** 5832 (1976).
158. G. W. Daub, *Prog. Inorg. Chem.*, **22,** 409 (1977).
159. S. N. Anderson, D. H. Ballard, J. Z. Chrzastowski, D. Dodd, and M. D. Johnson. *J. Chem. Soc., Chem. Commun.*, 685 (1972).
160. S. N. Anderson, D. H. Ballard, and M. D. Johnson, *J. Chem. Soc., Perkin Trans. 2,* 311 (1972).
161. D. Dodd and M. D. Johnson, *J. Chem. Soc., Chem. Commun.*, 571 (1971).
162. J. P. Collman, unpublished results.
163. G. M. Whitesides and D. J. Boschetto, *J. Am. Chem. Soc.*, **93,** 1529 (1971).
164. J. Tsuji, *Organic Synthesis with Palladium Compounds,* Springer-Verlag, Berlin, 1980, pp. 5–22.
165. (a) P. M. Henry, *J. Org. Chem.*, **32,** 2575 (1967); (b) H. Stangl and R. Jira, *Tetrahedron Lett.*, 3589 (1970).
166. J. E. Backvall, B. Åkermark, and S. O. Ljunggren, *J. Chem. Soc., Chem. Commun.*, 264 (1977); *J. Am. Chem. Soc.*, **101,** 2411 (1979).
167. W. C. Baird, Jr., *J. Org. Chem.*, **31,** 2411 (1966).
168. I. J. Harvie and F. J. McQuillin, *J. Chem. Soc., Chem. Commun.*, 369 (1976).
169. J. E. Backvall, *J. Chem. Soc., Chem. Commun.*, 413 (1977).
170. J. E. Backvall and E. E. Bjorkman, *J. Org. Chem.*, **45,** 2893 (1980).

171. R. F. Heck, in *Organic Synthesis via Metal Carbonyls,* (Ed. I. Wender and P. Pino), Vol. I, Wiley, New York, 1968, pp. 373–403.
172. T. A. Weil, L. Cassar, and M. Foa, *Organic Synthesis via Metal Carbonyls,* (Ed. I. Wender and P. Pino), Vol. II, Wiley, New York, 1977, pp. 517–543.
173. M. F. Semmelhack and S. J. Brickner, *J. Org. Chem.,* **46,** 1723 (1981).
174. (a) R. F. Heck, *Pure Appl. Chem.,* **50,** 691 (1978); (b) J. K. Stille and P. K. Wong, *J. Org. Chem.,* **40,** 532 (1975); A. Schoenberg and R. F. Heck, *J. Org. Chem.,* **39,** 3327 (1974).
175. T. Kobayashi and M. Tanaka, *J. Chem. Soc., Chem. Commun.,* 333 (1981).
176. M. Tanaka, *Bull. Chem. Soc. Jpn.,* **54,** 637 (1981).
177. T. Takahashi, H. Ikeda, and J. Tsuji, *Tetrahedron Lett.,* **22,** 1363 (1981).
178. T. Takahashi, H. Ikeda, and J. Tsuji, *Tetrahedron Lett.,* **21,** 3885 (1980).
179. (a) A. Cowell and J. K. Stille, *J. Am. Chem. Soc.,* **102,** 4193 (1980); *Tetrahedron Lett.* 133 (1979); (b) M. Mori, K. Chiba, and Y. Ban, *J. Org. Chem.,* **43,** 1684 (1978); (c) M. Mori, K. Chiba, M. Okita, and Y. Ban, *J. Chem. Soc. Chem. Commun.,* 698 (1979); (d) K. Chiba, M. Mori, and Y. Ban, *J. Chem. Soc.,* 770 (1980); (e) M. Mori, K. Chiba, N. Ohta, and Y. Ban, *Heterocycles,* **13,** 329 (1979).
180. G. D. Pandey and K. P. Tiwari, *Synth. Commun.,* **9,** 895 (1979); *Tetrahedron,* **37,** 1213 (1981).
181. M. Mori, K. Chiba, and Y. Ban, *Heterocycles,* **6,** 1841 (1977).
182. G. Pandey and K. P. Tiwari, *Synth. Commun.,* **10,** 523 (1980).
183. J. K. Stille and L. F. Hines, *J. Am. Chem. Soc.,* **94,** 485 (1972).
184. D. Medema, R. Van Helden, and C. F. Kohle, *Inorg. Chim. Acta,* **3,** 255 (1969).
185. J. K. Stille, D. E. James, and L. F. Hines, *J. Am. Chem. Soc.,* **95,** 5062 (1973).
186. L. S. Hegedus and W. H. Darlington, *J. Am. Chem. Soc.,* **102,** 4980 (1980).
187. L. S. Hegedus, G. Allen, and D. J. Olsen, *J. Am. Chem. Soc.,* **102,** 3583 (1980).
188. J. M. Thompson and R. F. Heck, *J. Org. Chem.,* **40,** 2667 (1975).
189. H. Horino and N. Inoue, *J. Org. Chem.,* **46,** 4416 (1981).
190. S.-I. Murahashi, Y. Tamba, M. Yamamura, and N. Yoshimura, *J. Org. Chem.,* **43,** 4099 (1978).
191. L. S. Hegedus, O. P. Anderson, K. Zetterberg, G. Allen, K. Sürala-Hansen, D. J. Olsen, and A. P. Packard, *Inorg. Chem.,* **16,** 1887 (1977).
192. J. P. Collman, *Acc. Chem. Res.,* **8,** 342 (1975).
193. (a) P. K. Wong, M. Madhavaro, D. F. Marten, and M. Rosenblum, *J. Am. Chem. Soc.,* **99,** 2823 (1977); (b) R. S. Berryhill and M. Rosenblum, *J. Org. Chem.,* **45,** 1984 (1980).
194. J. K. Stille, and R. Divakaruni, *J. Am. Chem. Soc.,* **100,** 1303 (1978); *J. Organomet. Chem.,* **72,** 127 (1974).
195. J. P. Collman and L. S. Hegedus, *Principles and Applications of Organotransition Metal Chemistry,* University Science Books, Mill Valley, CA, 1980, pp. 299–303.
196. (a) C. P. Casey and S. M. Newmann, *J. Am. Chem. Soc.,* **98,** 5395 (1978); (b) C. P. Casey, M. A. Andrews, and J. E. Rinz, *J. Am. Chem. Soc.,* **101,** 741 (1979).
197. S. R. Winter, G. W. Cornett, and E. A. Thompson, *J. Organomet. Chem.,* **133,** 339 (1977).
198. (a) J. A. Gladysz and W. Tam, *J. Am. Chem. Soc.,* **100,** 2545 (1978); (b) W. Tam, W. K. Wong, and J. A. Gladysz, *J. Am. Chem. Soc.,* **101,** 1589 (1979).
199. J. A. Gladysz and J. C. Selover, *Tetrahedron Lett.,* 319 (1978).
200. C. P. Casey and S. M. Newmann, *J. Am. Chem. Soc.,* **100,** 2544 (1978).
201. E. O. Fischer and A. Massbol, *Angew. Chem., Int. Ed. Engl.,* **3,** 580 (1964).
202. M. Ryang and S. Tsutsumi, *Synthesis,* 55 (1971).
203. M. Ryang, *Organomet. Chem. Rev. A,* **5,** 67 (1970).
204. S. K. Myeong, Y. Sawa, M. Ryang, and S. Tsutsumi, *Bull. Chem. Soc. Jpn.,* **38,** 330 (1965).
205. L. S. Hegedus, *PhD Thesis,* Harvard University, 1970, pp. 85–87.
206. E. J. Corey and L. S. Hegedus, *J. Am. Chem. Soc.,* **91,** 4926 (1969).
207. Y. Sawa, I. Hashimoto, M. Ryang, and S. Tsutsumi, *J. Org. Chem.,* **33,** 2159 (1968).
208. For reviews concerning transition metal carbene complexes see (a) F. A. Cotton and C. M. Lukehart, *Prog. Inorg. Chem.,* **16,** 487 (1972); (b) E. O. Fischer, *Adv. Organomet.*

*Chem.*, **14,** 1 (1976); (c) C. P. Casey, in *Transition Metals in Organic Synthesis,* (Ed. H. Alper), Vol. 1, Academic Press, New York, 1976, pp. 189–233; (d) F. J. Brown, *Prog. Inorg. Chem.*, **27,** 1 (1980).
209. E. O. Fischer and H. J. Kollmeier, *Angew. Chem., Int. Ed. Engl.*, **9,** 309 (1970).
210. H. G. Raubenheimer and E. O. Fischer, *J. Chem. Soc. Chem. Commun.*, 732 (1976).
211. E. O. Fischer, F. Kreissl, C. G. Kreiter, and E. W. Meineke, *Chem. Ber.*, **105,** 2558 (1972).
212. E. O. Fischer, M. Leupald, C. G. Kreiter, and J. Müller, *Chem. Ber.*, **105,** 150 (1972).
213. (a) C. P. Casey and T. J. Burkhardt, *J. Am. Chem. Soc.*, **95,** 5883 (1973); (b) E. O. Fischer, W. Held, and F. Kreissl, *Chem. Ber.*, **110,** 3842 (1977); (c) E. O. Fischer, W. Held, F. R. Kriessl, A. Frank, and G. Hultner, *Chem. Ber.*, **110,** 656 (1977).
214. For reviews on the olefin metathesis reaction see R. H. Grubbs, *Prog. Inorg. Chem.*, **24,** 1 (1978); T. J. Katz, *Adv. Organomet. Chem.*, **16,** 283 (1977).
215. E. O. Fischer and F. R. Kreissl, *J. Organomet. Chem.*, **35,** C47 (1972).
216. W. R. Brunswald and C. P. Casey, *J. Organomet. Chem.*, **77,** 345 (1974).
217. (a) N. Grice, S. C. Kao, and R. Petit, *J. Am. Chem. Soc.*, **101,** 1627 (1979); (b) M. Catellani and J. Halpern, *Inorg. Chem.*, **19,** 566 (1980).
218. D. J. Darensbourg, B. J. Baldwin, and J. A. Froelich, *J. Am. Chem. Soc.*, **102,** 4688 (1980).
219. J. R. Norton, in *Organic Reactions and Methods,* (Ed. J. R. Zuckerman), Springer-Verlag, Berlin, in press.
220. J. P. Collman and L. S. Hegedus, *Principles and Applications of Organotransition Metal Chemistry,* University Science Books, Mill Valley, CA, 1980, pp. 418–467.
221. (a) Y. Shvo and E. Hazum, *Chem. Commun.*, 336 (1974); (b) D. J. Blumer, K. W. Barnett, and T. L. Brown, *J. Organomet. Chem.*, **173,** 71 (1979).
222. R. L. Angelici, *Acc. Chem. Res.*, **5,** 335 (1972).
223. H. C. Clark and K. von Werner, *Synth. React. Inorg. Mt.-Org. Chem.*, **4,** 355 (1974).
224. A. Vitagliano, *J. Organomet. Chem.*, **81,** 261 (1974).
225. J. Tsuji, *Organic Synthesis with Palladium Compounds,* Springer-Verlag, Berlin, 1980, pp. 22, 77, 81, 159.
226. T. Yukawa and S. Tsutsumi, *J. Org. Chem.*, **34,** 738 (1969).
227. (a) D. E. James, L. F. Hines, and J. K. Stille, *J. Am. Chem. Soc.*, **98,** 1806 (1976); (b) J. K. Stille, D. E. James, and L. F. Hines, *J. Am. Chem. Soc.*, **95,** 5062 (1973); (c) D. E. James and J. K. Stille, *J. Am. Chem. Soc.*, **98,** 1810 (1976).
228. F. RuVette and U. Romano, *J. Organomet. Chem.*, **154,** 323 (1978).
229. (a) T. F. Muuray aad J. R. Norton, *J. Am. Chem. Soc.*, **101,** 4107 (1979); (b) T. F. Murray, E. G. Samsel, V. Varma, and J. R. Norton, *J. Am. Chem. Soc.*, **103,** 7520 (1981).
230. E. J. Corey and L. S. Hegedus, *J. Am. Chem. Soc.*, **91,** 1233 (1969).
231. E. J. Corey, H. A. Kirst, and J. A. Katzenellenbogen, *J. Am. Chem. Soc.*, **92,** 6314 (1970).
232. W. Beck and B. Purucker, *J. Organomet. Chem.*, **112,** 361 (1976).
233. S. Fukuoka, M. Ryang, and S. Tsutsumi, *J. Org. Chem.*, **33,** 2973 (1968).
234. W. E. Martin and M. F. Farona, *J. Organomet. Chem.*, **206,** 393 (1981).
235. L. Busetto, A. Palazzi, B. Crociani, U. Belluco, E. M. Badley, B. J. L. Kilby, and R. L. Richards, *J. Chem. Soc., Dalton Trans.*, 1800 (1972).
236. C. H. Davies, C. H. Game, M. Green, and F. G. A. Stone, *J. Chem. Soc., Dalton Trans.*, 357 (1974).
237. T. Boschi, B. Crociani, M. Nicolini, and U. Belluco, *Inorg. Chim. Acta,* **12,** 39 (1975).
238. B. Crociani, P. Uguagliati, and U. Belluco, *J. Organomet. Chem.*, **117,** 189 (1976).
239. P. M. Maitlis, *The Organic Chemistry of Palladium,* Vol. 1, Academic Press, New York, 1971, pp. 163–168.
240. (a) J. Chatt, L. M. Valdarino, and L. M. Venanzi, *J. Chem. Soc.*, 3413 (1957); (b) J. K. Stille, R. A. Morgan, D. D. Whitehurst, and J. R. Doyle, *J. Am. Chem. Soc.*, **87,** 3282 (1965).
241. (a) G. Carturan, M. Graziani, R. Ros, and U. Belluco, *J. Chem. Soc., Dalton Trans.*, 262 (1972); (b) J. K. Stille and L. F. Hines, *J. Am. Chem. Soc.*, **92,** 1798 (1970); **94,** 485 (1972).

242. Y. Ito, T. Hirao, N. Ohta, and T. Saegusa, *Tetrahedron Lett.*, 1009 (1977).
243. (a) J. K. Stille and D. E. James, *J. Am. Chem. Soc.*, **97,** 674 (1975); (b) J. K. Stille and D. E. James, *J. Organomet. Chem.*, **108,** 401 (1976).
244. N. Adachi, K. Kikukawa, M. Takagi, and T. Matsuda, *Bull. Chem. Soc. Jpn.*, **48,** 521 (1975).
245. C. B. Anderson and B. J. Burresson, *Chem. Ind.* (*London*), 620 (1967); M. Akbarzadch and C. B. Anderson, *J. Organomet. Chem.*, **197,** C5 (1980).
246. P. M. Henry, M. Davis, G. Ferguson, S. Phillips, and R. Restive, *Chem. Commun.*, 112 (1974).
247. M. Sakai, *Tetrahedron Lett.*, 347 (1973).
248. T. Saski, K. Kanematsu, and A. Kondon, *J. Chem. Soc., Perkin Trans. 1,* 2516 (1976).
249. A. Heumann, M. Reglier, and B. Waegell, *Angew. Chem., Int. Ed. Engl.*, **18,** 866, 867, (1979).
250. J. K. Stille and D. B. Fox, *J. Am. Chem. Soc.*, **92,** 1274 (1970).
251. R. N. Haszeldine, R. V. Parish, and D. W. Robbins, *J. Chem. Soc., Dalton Trans.*, 2355 (1976).
252. S. Ozaki and A. Tamaki, *Bull. Chem. Soc. Jpn.*, **51,** 3391 (1978).
253. B. F. G. Johnson, J. Lewis, and M. S. Subramanian, *Chem. Commun.*, 117 (1966).
254. B. F. G. Johnson, J. Lewis, and M. S. Subramanian, *J. Chem. Soc., A,* 1993 (1968).
255. J. Tsuji and H. Takahashi, *J. Am. Chem. Soc.*, **87,** 3275 (1965).
256. J. Tsuji and H. Takahashi, *J. Am. Chem. Soc.*, **90,** 2387 (1968).
257. C. R. Kistner, J. H. Hutchinson, J. R. Doyle, and J. C. Storbe, *Inorg. Chem.*, **2,** 1255 (1963).
258. (a) M. Julia and L. Saussine, *Tetrahedron Lett.*, 3443 (1974); (b) L. Benchekroun, P. Herpin, M. Julia, and L. Saussine, *J. Organomet. Chem.*, **128,** 275 (1977).
259. A. C. Cope, J. M. Kliegman, and E. C. Friedrich, *J. Am. Chem. Soc.*, **89,** 287 (1967).
260. D. Medema, R. van Helden, and C. F. Kohle, *Inorg. Chim. Acta,* **3,** 256 (1969).
261. R. A. Holton and R. A. Kjonaas, *J. Organomet. Chem.*, **133,** C5 (1977).
262. Y. Takahashi, A. Tokuda, S. Sakai, and Y. Ishii, *J. Organomet. Chem.*, **35,** 415 (1972).
263. R. A. Holton and R. A. Kjonaas, *J. Am. Chem. Soc.*, **99,** 4177 (1977).
264. R. A. Holton and R. A. Kjonaas, *J. Organomet. Chem.*, **142,** C15 (1977).
265. K. Hirai, N. Ishii, H. Suzuki, Y. Moro-Oka, and T. Ikawa, *Chem. Lett.*, 1113 (1979).
266. R. A. Holton, *J. Am. Chem. Soc.*, **99,** 8083 (1977).
267. R. N. Haszeldine, R. J. Lunt, and R. V. Parish, *J. Chem. Soc., A,* 3705 (1971).
268. For reviews concerning the Wacker Process and updated oxypalladations see (a) P. M. Maitlis, *The Organic Chemistry of Palladium,* Academic Press, New York, 1971, Vol. I, pp. 133–138; Vol. II, pp. 77–108; (b) J. Tsuji, *Organic Synthesis with Palladium Compounds,* Springer-Verlag, Berlin, 1980, pp. 6–22, and references cited therein; (c) R. Jira and W. Freisleben, *Organomet. React.*, **3,** 5 (1972).
269. P. M. Henry, *Adv. Organomet. Chem.*, **13,** 363 (1975).
270. J. E. Backvall, B. Åkermark, and S. O. Ljunggren, *J. Am. Chem. Soc.*, **101,** 2411 (1979).
271. J. K. Stille and R. Divakaruni, *J. Am. Chem. Soc.*, **100,** 1303 (1978); *J. Organomet. Chem.*, **169,** 239 (1979).
272. J. P. Collman and L. S. Hegedus, *Principles and Applications of Organotransition Metal Chemistry,* University Science Books, Mill Valley, CA, 1980, pp. 603–625.
273. M. A. Andres and K. P. Kelly, *J. Am. Chem. Soc.*, **103,** 2894 (1981).
274. B. Åkermark, J. E. Backvall, K. Sürala–Hansen, K. Sjoberg, L. S. Hegedus and K. Zetterberg, *J. Organomet. Chem.*, **72,** 127 (1974).
275. B. Åkermark, J. E. Backvall, K. Sürala-Hansen, K. Sjoberg, and K. Zetterberg, *Tetrahedron Lett.*, 1363 (1974).
276. L. S. Hegedus, B. Åkermark, and K. Zetterberg, submitted, *J. Am. Chem. Soc.*
277. L. S. Hegedus, O. P. Anderson, K. Zetterberg, G. Allen, K. Sürala-Hansen, D. J. Olsen, and A. B. Packard, *Inorg. Chem.*, **16,** 1887 (1977).
278. (a) A. Panunzi, A. DeRenzi, R. Palumbo, and G. Paiaro, *J. Am. Chem. Soc.*, **91,** 3879 (1969); (b) A. Panunzi, A. DeRenzi, and G. Paiaro, *J. Am. Chem. Soc.*, **92,** 3488 (1970).
279. J. J. Bozell and L. S. Hegedus, *J. Org. Chem.*, **46,** 2561 (1981).
280. L. S. Hegedus, P. M. Winton, and S. Varaprath, *J. Org. Chem.*, **46,** 2215 (1981).
281. S.-I. Murahashi, M. Yamamura, and N. Mitra, *J. Org. Chem.*, **42,** 2870 (1977).

282. L. S. Hegedus, R. E. Williams, M. A. McGuire, and T. Hayashi, *J. Am. Chem. Soc.*, **102,** 4973 (1980).
283. J. P. Haudegoud, Y. Chairvin, and D. Commereuc, *J. Org. Chem.*, **44,** 3063 (1979).
284. L. S. Hegedus and W. H. Darlington, *J. Am. Chem. Soc.*, **102,** 4980 (1980).
285. H. Kurosawa, T. Majima, and N. Asada, *J. Am. Chem. Soc.*, **102,** 6996 (1980).
286. T. Hosokawa, H. Ohkata, and I. Moritani, *Bull. Chem. Soc. Jpn.*, **48,** 1533 (1975).
287. T. Hosokawa, S. Yamashita, S.-I. Murahashi, and A. Sonoda, *Bull. Chem. Soc. Jpn.*, **49,** 3662 (1976).
288. T. Hosokawa, S. Miyagi, S.-I. Murahashi, and A. Sonoda, *J. Org. Chem.*, **43,** 2752 (1978).
289. T. Hosokawa, T. Uno, S. Inui, and S.-I. Murahashi, *J. Am. Chem. Soc.*, **103,** 2318 (1981).
290. A. Kasahara, I. Izumi, and M. Ooshima, *Bull. Chem. Soc. Jpn.*, **47,** 2526 (1974).
291. T. Hosokawa, M. Hirata, S.-I. Murahashi, and A. Sonoda, *Tetrahedron Lett.*, 1821 (1976).
292. N. T. Bryon, R. Grigg, and B. Kongathip, *J. Chem. Soc., Chem. Commun.*, 216 (1976).
293. D. E. Korte, L. S. Hegedus, and R. K. Wirth, *J. Org. Chem.*, **42,** 1329 (1977).
294. A. Kasahara, T. Izumi, K. Sato, K. Maemura, and T. Hayasaka, *Bull. Chem. Soc. Jpn.*, **50,** 1899 (1977).
295. T. Izumi and A. Kasahara, *Bull. Chem. Soc. Jpn.*, **48,** 1673 (1975).
296. K. Maeda, T. Hosokawa, S.-I. Murahashi, and I. Moritani, *Tetrahedron Lett.*, 5075 (1973).
297. L. S. Hegedus, G. F. Allen, J. J. Bozell, and E. L. Waterman, *J. Am. Chem. Soc.*, **100,** 5800 (1978).
298. L. S. Hegedus, G. F. Allen, and D. J. Olsen, *J. Am. Chem. Soc.*, **102,** 3583 (1980).
299. S. Hatano, M. Saruwatari, K. Isomura, and H. Taniguchi, *Heterocycles,* **15,** 747 (1981).
300. A. Kasahara, T. Izumi, and O. Saito, *Chem. Ind.* (*London*), 666 (1980).
301. A. Kasahara and T. Saito, *Chem. Ind.* (*London*), 745 (1975).
302. A. Kasahara, *Chem. Ind.* (*London*), 1032 (1976).
303. A. Kasahara and N. Fukuda, *Chem. Ind.* (*London*), 485 (1976).
304. E. Schmitz, R. Urban, U. Hueck, G. Zimmermann, and E. Grundemann, *J. Prakt. Chem.*, **318,** 185 (1976).
305. J. Ambwehl, P. S. Pregosin, L. M. Venanzi, G. Consiglio, F. Bachechi, and L. Zambonelli, *J. Organomet. Chem.*, **181,** 255 (1979).
306. B. Pugin and L. M. Venanzi, *J. Organomet. Chem.*, **214,** 125 (1981).
307. L. S. Hegedus and J. M. McKearin, *J. Am. Chem. Soc.*, **104,** 2444 (1982).
308. For a recent review on the reactions of $CpFe(CO)_2(olefin)^+$ species with nucleophiles see M. Rosenblum, T. C. Chang, B. M. Foxman, S. B. Samuels, and C. Stockman, *Org. Synth. Today, Tomorrow, Proc. IUPAC Symp. Org. Synth., 3rd, 1980,* Pergamon Press, Oxford, 1980, pp. 47–54.
309. P. Lennon, M. Madhavaro, A. Rosan, and M. Rosenblum, *J. Organomet. Chem.*, **108,** 93 (1976).
310. M. Madhavarao, D. F. Marten, and M. Rosenblum, *J. Am. Chem. Soc.*, **99,** 2823 (1977).
311. R. S. Berryhill and M. Rosenblum, *J. Org. Chem.*, **45,** 1984 (1980).
312. P. M. Lennon, A. M. Rosan, and M. Rosenblum, *J. Am. Chem. Soc.*, **99,** 8426 (1977).
313. T. C. T. Chang, M. Rosenblum, and S. B. Samuels, *J. Am. Chem. Soc.*, **102,** 5930 (1980).
314. T. C. T. Chang and M. Rosenblum, *J. Org. Chem.*, **46,** 4103 (1981).
315. T. C. T. Chang and M. Rosenblum, *J. Org. Chem.*, **46,** 4676 (1981).
316. A. Rosan and M. Rosenblum, *J. Org. Chem.*, **40,** 3622 (1975).
317. B. W. Roberts and J. Wong, *J. Chem. Soc., Chem. Commun.*, 20 (1977).
318. B. W. Roberts, M. Ross, and J. Wong, *J. Chem. Soc., Chem. Commun.*, 428 (1980).
319. M. A. Baar and B. W. Roberts, *J. Chem. Soc., Chem. Commun.*, 1129 (1979).
320. For a recent general treatment of the reactions of π-allylpalladium complexes with nucleophiles, see J. P. Collman and L. S. Hegedus, *Principles and Applications of Organotransition Metal Complexes,* University Science Books, Mill Valley, CA, 1980, pp. 679–692.
321. J. Tsuji, *Organic Synthesis with Palladium Compounds,* Springer-Verlag, Berlin, 1980, pp. 45–51.

322. J. Tsuji, *Top. Curr. Chem.*, **91,** 30 (1980).
323. J. Tsuji, *Pure Appl. Chem.*, **53,** 2371 (1981).
324. J. Tsuji, *Pure Appl. Chem.*, **54,** 197 (1982).
325. B. M. Trost, *Acc. Chem. Res.*, **13,** 385 (1980); *Pure Appl. Chem.*, **53,** 2357 (1981).
326. B. M. Trost, *Tetrahedron,* **33,** 2615 (1977).
327. L. S. Hegedus, in *Comprehensive Carbanion Chemistry,* (Ed. T. Durst and E. Buncell), Elsevier, Lausanne, (1984).
328. (a) B. M. Trost, P. E. Strege, L. Weber, T. J. Fullerton, and T. J. Dietsch, *J. Am. Chem. Soc.,* **100,** 3407 (1978); (b) B. M. Trost and P. Metzner, *J. Am. Chem. Soc.,* **102,** 3572 (1980).
329. (a) J. Tsuji, H. Takahashi, and M. Morikawa, *Tetrahedron Lett.*, 4387 (1965); (b) J. Tsuji, *Bull. Chem. Soc. Jpn.*, **46,** 1896 (1973).
330. H. Onoue, I. Moritani, and S.-I. Murahashi, *Tetrahedron Lett.*, 121 (1973).
331. B. M. Trost, L. Weber, P. E. Strege, T. J. Fullerton, and T. J. Dietsche, *J. Am. Chem. Soc.,* **100,** 3416 (1978).
332. D. J. Collins, W. R. Jackson, and R. N. Timms, *Tetrahedron Lett.*, 495 *(1976).*
333. B. M. Trost and T. R. Verhoeven, *J. Am. Chem. Soc.,* **100,** 3435 (1978).
334. P. S. Marchand, H. S. Wong, and J. F. Blount, *J. Org. Chem.*, **43,** 4769 (1978).
335. B. Åkermark and A. Jutland, *J. Organomet. Chem.*, **217,** C41 (1981).
336. Y. Castanet and F. Petit, *Tetrahedron Lett.*, 3221 (1979).
337. L. S. Hegedus, W. H. Darlington, and C. E. Russell, *J. Org. Chem.*, **45,** 5193 (1980).
338. J. L. A. Roustan and F. Houlihan, *Can. J. Chem.*, **57,** 2790 (1979).
339. (a) T. H. Whitesides, R. W. Arhart, and R. W. Slaven, *J. Am. Chem. Soc.*, **95,** 5792 (1973); (b) A. J. Birch and I. P. Jenkins, in *Transition Metal Organometallics in Organic Synthesis,* (Ed. H. Alper), Vol. 1, Academic Press, New York, 1973, p. 13.
340. N. A. Bailey, W. G. Kita, J. McCleverty, A. J. Murray, B. E. Mann, and N. W. Walker, *Chem. Commun.*, 592 (1974).
341. (a) M. Ephritikhine, M. L. H. Green, and R. E. MacKenzie, *J. Chem. Soc., Chem. Commun.*, 619 (1976); (b) M. Ephritikhine, B. R. Francis, M. L. H. Green, R. E. MacKenzie, and M. S. Smith, *J. Chem. Soc., Dalton Trans.*, 1131 (1977).
342. For a review concerning nucleophilic addition to organotransition metal cations see S. G. Davies, M. L. H. Green, and D. M. P. Mingos, *Tetrahedron,* **34,** 3047 (1978).
343. B. Åkermark and K. Zetterberg, *Tetrahedron Lett.*, 3733 (1975).
344. B. Åkermark, J. E. Backvall, A. Lowenborg, and K. Zetterberg, *J. Organomet. Chem.*, **166,** C33 (1979).
345. B. Åkermark, G. Åkermark, L. S. Hegedus, and K. Zetterberg, *J. Am. Chem. Soc.*, **103,** 3037 (1981).
346. K. Takahashi, A. Miyaki, and G. Hata, *Bull. Chem. Soc. Jpn.*, **45,** 230 (1972).
347. K. E. Atkins, W. E. Walker, and R. M. Manyik, *Tetrahedron Lett.*, 3821 (1970).
348. B. M. Trost and T. R. Verhoeven, *J. Am. Chem. Soc.*, **102,** 4730 (1980), and references cited therein.
349. T. Yamamoto, O. Saito, and A. Yamamoto, *J. Am. Chem. Soc.*, **103,** 5600 (1981).
350. J. C. Fiaud and J. L. Malleron, *Tetrahedron Lett.*, **22,** 1399 (1981).
351. B. M. Trost and N. R. Schmuff, *Tetrahedron Lett.*, **22,** 2999 (1981).
352. B. M. Trost and Y. Matsumura, *J. Org. Chem.*, **42,** 2036 (1977).
353. B. M. Trost and F. W. Gowland, *J. Org. Chem.*, **44,** 3448 (1979).
354. (a) J.-P. Genet and F. Piau, *J. Org. Chem.*, **46,** 2414 (1981); (b) J. L. Roustan, J. Y. Mérour, and F. Houlihan, *Tetrahedron Lett.*, 3721 (1979).
355. T. Hirao, J. Enda, Y. Ohshiro, and T. Agawa, *Tetrahedron Lett.*, **22,** 3079 (1981).
356. J. Tsuji, H. Ueno, Y. Kobayashi, and H. Okumoto, *Tetrahedron Lett.*, **22,** 2573 (1981).
357. J. Tsuji, H. Kataoka, and Y. Kobayashi, *Tetrahedron Lett.*, **22,** 2575 (1981).
358. B. M. Trost and G. A. Molander, *J. Am. Chem. Soc.*, **103,** 5969 (1981).
359. B. M. Trost, N. R. Schmuff, and M. J. Miller, *J. Am. Chem. Soc.*, **102,** 5979 (1980).
360. P. A. Wade, H. R. Hinney, N. V. Amin, P. D. Vail, S. D. Morrow, S. A. Hardinger, and M. S. Saft, *J. Org. Chem.*, **46,** 765 (1981).
361. J. C. Fiaud and J. L. Malleron, *J. Chem. Soc., Chem. Commun.*, 1159 (1981).
362. B. M. Trost and T. P. Klun, *J. Am. Chem. Soc.*, **101,** 6756 (1979).
363. B. M. Trost and T. P. Klun, *J. Am. Chem. Soc.*, **103,** 1864 (1981).

364. B. M. Trost and T. A. Runge, *J. Am. Chem. Soc.*, **103,** 2485 (1981).
365. B. M. Trost and T. R. Verhoeven, *J. Am. Chem. Soc.*, **102,** 4743 (1980).
366. Y. Kitagawa, A. Itoh, S. Hashimoto, H. Yamamoto, and H. Nozaki, *J. Am. Chem. Soc.*, **99,** 3864 (1977).
367. J. Tsuji, Y. Kobayashi, H. Kataoka, and T. Takahashi, *Tetrahedron Lett.*, **21,** 1475 (1980).
368. J. Tsuji, Y. Kobayashi, H. Katoaka, and T. Takahashi, *Tetrahedron Lett.*, **21,** 3393 (1980).
369. Y. Kobayashi and J. Tsuji, *Tetrahedron Lett.*, **22,** 4295 (1981).
370. B. M. Trost and D. M. T. Chan, *J. Am. Chem. Soc.*, **102,** 6361 (1980); **101,** 6432 (1979); **101,** 6429 (1979).
371. B. M. Trost and D. M. T. Chan, *J. Am. Chem. Soc.*, **103,** 5972 (1981).
372. K. Ogura, N. Shibuya, and H. Iida, *Tetrahedron Lett.*, **22,** 1519 (1981).
373. T. Mandai, S. Hashio, J. Goto, and M. Kawada, *Tetrahedron Lett.*, **22,** 2187 (1981).
374. P. A. Grieco, P. A. Tuthill, and H. L. Shain, *J. Org. Chem.*, **46,** 5005 (1981).
375. K. Inomata, T. Yamamoto, and H. Kotake, *Chem. Lett.*, 1357 (1981).
376. B. M. Trost and J. P. Genet, *J. Am. Chem. Soc.*, **98,** 8516 (1976).
377. B. M. Trost, S. A. Godleski, and J. P. Genet, *J. Am. Chem. Soc.*, **100,** 3930 (1978).
378. B. M. Trost, S. A. Godleski, and J. L. Belletire, *J. Org. Chem.*, **44,** 2052 (1979).
379. B. M. Trost and J. M. D. Fortunak, *Organometallics,* **1,** 7 (1980).
380. S. A. Godleski, J. D. Meinhart, D. J. Miller, and S. Van Wallendael, *Tetrahedron Lett.*, **22,** 2247 (1981).
381. J.-E. Backvall, R. E. Nordberg, J. E. Nystrom, T. Hogberg, and B. Ulff, *J. Org. Chem.*, **46,** 3479 (1981).
382. J.-E. Backvall and J. E. Nystrom, *J. Chem. Soc., Chem. Commun.*, 59 (1981).
383. B. Åkermark, A. Ljungquist, and M. Panunzio, *Tetrahedron Lett.*, **22,** 1055 (1981).
384. J.-E. Backvall and R. E. Nordberg, *J. Am. Chem. Soc.*, **103,** 4959 (1981).
385. (a) B. A. Patel, J.-i.-i. Kim, D. D. Bender, L. C. Kao, and R. F. Heck, *J. Org. Chem.*, **46,** 1061 (1981); (b) R. F. Heck, *Pure Appl. Chem.*, **53,** 2323 (1981).
386. L. S. Hegedus and Y. Inoue, *J. Am. Chem. Soc.*, **104,** 4917 (1982).
387. J. Tsuji, *Acc. Chem. Res.*, **6,** 8 (1973).
388. J. Tsuji, *Adv. Organomet. Chem.*, **17,** 141 (1979).
389. J. Tsuji, *Organic Synthesis with Palladium Compounds,* Springer-Verlag, Berlin, 1980, pp. 90–124.
390. J. Tsuji, *Ann. N.Y. Acad. Sci.*, **333,** 250 (1980).
391. T. Takahashi, I. Minami, and J. Tsuji, *Tetrahedron Lett.*, **22,** 2651 (1981).
392. U. M. Dzhemilev, R. V. Kunakova, N. Z. Baibulatova, and G. A. Tolstikov, *Izv. Akad. Nauk SSSR, Ser. Khim.*, 1837 (1981).
393. J.-P. Bianchini, B. Waegell, S. M. Gaydow, H. Tzehak, and W. Keim, *J. Mol. Catal.*, **10,** 247 (1981).
394. J. Tsuji and M. Takahashi, *J. Mol. Catal.*, **10,** 107 (1981).
395. W. Keim and M. Roper, *J. Org. Chem.*, **46,** 3702 (1981).
396. K. Kaneda, H. Kurosaki, M. Terasawa, I. Imanaka, and S. Teranishi, *J. Org. Chem.*, **46,** 2356 (1981).
397. U. Dzhemilev, R. N. Fakhreldinov, A. G. Telin, G. A. Tolstikov, A. A. Panasenko, and E. V. Vasil-eva, *Bull. Akad. Nauk SSSR,* **29,** 1943 (1980).
398. For current reviews see A. J. Pearson, *Acc. Chem. Res.*, **13,** 463 (1980); *Transition Met. Chem.*, **6,** 67 (1981).
399. (a) For a review covering the literature through 1975 see A. J. Birch and I. D. Jenkins, in *Transition Metal Organometallics in Organic Synthesis,* (Ed. H. Alper), Vol. 1, Academic Press, New York, 1976, pp. 1–76. (b) For a review on the alkylation of these complexes see L. S. Hegedus, in *Comprehensive Carbanion Chemistry,* (Ed. T. Durst and E. Buncel), Elsevier, Lausanne, (1984).
400. B. M. R. Bandara, A. J. Birch, and T. C. Khor, *Tetrahedron Lett.*, **21,** 3625 (1980).
401. A. J. Pearson, *Aust. J. Chem.*, **29,** 1101 (1976); *Aust. J. Chem.*, **30,** 345 (1977).
402. B. F. G. Johnson, J. Lewis, D. G. Parker, and G. R. Stephenson, *J. Organomet. Chem.*, **204,** 221 (1981).
403. G. R. John, L. A. Kane-Maguire, *Inorg. Chim. Acta,* **48,** 179 (1981).

404. A. J. Birch, A. J. Liepa, and G. R. Stephenson, *Tetrahedron Lett.*, 3565 (1979).
405. A. J. Pearson and M. Chandler, *Tetrahedron Lett.*, **21**, 3933 (1980).
406. L. F. Kelly, P. Dahler, A. S. Narula, and A. J. Birch, *Tetrahedron Lett.*, **22**, 1433 (1981).
407. F. Effenberger and M. Keil, *Tetrahedron Lett.*, **22**, 2151 (1981).
408. A. J. Birch, W. D. Ravertz, and G. R. Stephenson, *J. Org. Chem.*, **46**, 5166 (1981).
409. A. J. Birch and G. R. Stephenson, *Tetrahedron Lett.*, **22**, 779 (1981).
410. A. J. Birch, B. M. Ratnayake, *et al. Tetrahedron,* **37**, 289 (1981).
411. A. J. Birch and D. H. Williamson, *J. Chem. Soc., Perkin Trans. 1*, 1892 (1973).
412. A. J. Pearson and P. R. Rathby, *J. Chem. Soc., Perkin Trans. 1*, 395 (1980).
413. A. J. Pearson and C. W. Ong, *J. Chem. Soc., Perkin Trans. 1*, 1614 (1981).
414. A. J. Pearson and C. W. Ong, *J. Am. Chem. Soc.*, **103**, 6686 (1981).
415. A. J. Pearson, *Tetrahedron Lett.*, **22**, 4033 (1981).
416. E. Mincione, A. J. Pearson, P. Bovicelli, M. Chandler, and G. C. Heywood, *Tetrahedron Lett.*, **22**, 2929 (1981).
417. A. J. Pearson and G. Heywood, *Tetrahedron Lett.*, **22**, 1645 (1981).
418. M. F. Semmelhack, *Pure Appl. Chem.*, **53**, 2379 (1981).
419. For a summary of $\eta^6$-arenetricarbonyl chromium chemistry see J. P. Collman and L. S. Hegedus, *Principles and Applications of Organotransition Metal Chemistry,* University Science Books, Mill Valley, CA, 1980, pp. 653–670.
420. This Figure was adapted from an excellent review on the chemistry of arenemetal tricarbonyl complexes by M. F. Semmelhack, *J. Organomet. Chem. Libr.*, **1**, 361 (1976).
421. J. C. Boutonnet and E. Rose, *J. Organomet. Chem.*, **221**, 157 (1981).
422. S. J. Rosca and S. Rosca, *Rev. Chim.*, **25**, 461 (1974).
423. C. A. L. Mahaffy and P. L. Pauson, *J. Chem. Res.*, 128 (1979).
424. J. F. Bunnett and H. Hermann, *J. Org. Chem.*, **36**, 4081 (1971).
425. M. F. Semmelhack and H. T. Hall, Jr., *J. Am. Chem. Soc.*, **96**, 7091; 7092 (1974).
426. For reviews on the reactions of carbon nucleophiles with $\eta^6$-arenechromium tricarbonyl complexes see (a) M. F. Semmelhack, G. R. Clark, J. L. Garcia, J. J. Harrison, Y. Thebtaranonth, W. A. Wulff, and A. Yamashita, *Tetrahedron,* **37**, 3957 (1981); (b) Ref. 418; (c) M. F. Semmelhack, *Org. Synth., Today, Tomorrow, Proc. IUPAC Symp. Org. Synth., 3rd, 1980,* Pergamon Press, Oxford, 1980, pp. 63–69. (d) M. F. Semmelhack, *Ann. N.Y. Acad. Sci.*, **295**, 36 (1977).
427. (a) M. F. Semmelhack, H. T. Hall, Jr., M. Yoshifuji, and G. R. Clark, *J. Am. Chem. Soc.*, **97**, 1247 (1975); (b) M. F. Semmelhack, H. T. Hall, Jr., R. Farina, M. Yoshifuji, G. R. Clark, T. Bargar, K. Hirotsu, and J. Clardy, *J. Am. Chem. Soc.*, **101**, 3535 (1979).
428. (a) M. F. Semmelhack, G. R. Clark, R. Farina, and M. Saeman, *J. Am. Chem. Soc.*, **101**, 217 (1979); (b) M. F. Semmelhack and G. R. Clark, *J. Am. Chem. Soc.*, **99**, 1675 (1977).
429. M. F. Semmelhack, J. J. Harrison, and Y. Thebtaranonth, *J. Org. Chem.*, **44**, 3275 (1979).
430. V. Desobry and E. P. Kundig, *Helv. Chim. Acta*, **64**, 1288 (1981).
431. J. C. Boutonnet, L. Mordente, E. Rose, O. LeMartret, and G. Precigoux, *J. Organomet. Chem.*, **221**, 147 (1981).
432. A. P. Kozikowski and K. Isobe, *J. Chem. Soc., Chem. Commun.*, 1076 (1978).
433. J. P. Kutney, R. A. Badger, W. R. Cullen, R. Greenhouse, M. Noda, V. E. Ridara-Sanz, Y. H. So, A. Zanarotti, and B. R. Worth, *Can. J. Chem.*, **57**, 300 (1979).
434. J. P. Kutney, M. Noda, and B. R. Worth, *Heterocycles,* **12**, 1269 (1979).
435. M. F. Semmelhack, V. Thebtaranonth, and L. Keller, *J. Am. Chem. Soc.*, **102**, 3275 (1979).
436. M. F. Semmelhack and A. Yamashita, *J. Am. Chem. Soc.*, **102**, 5926 (1980).
437. M. F. Semmelhack, W. Seufert, and L. Keller, *J. Am. Chem. Soc.*, **102**, 6586 (1980).
438. S. Top and G. Jaouen, *J. Org. Chem.*, **46**, 78 (1981).
439. For a review see G. Jaouen, in *Transition Metal Organometallics in Organic Synthesis,* (Ed. H. Apler), Vol. II, Academic Press, New York, 1978, pp. 65–120.
440. J. Besancon, J. Tirouflet, A. Card, and Y. Dusausoy, *J. Organomet. Chem.*, **59**, 267 (1973).
441. A. Meyer, *Ann. Chim.* (*Paris* [*1Y*]), **8**, 397 (1973).
442. A. Meyer and G. Jaouen, *J. Chem. Soc., Chem. Commun.*, 787 (1974).
443. G. Jaouen and A. Meyer, *J. Am. Chem. Soc.*, **98**, 4667 (1975).

444. B. Caro and G. Jaouen, *J. Organomet. Chem.*, **220,** 309 (1981).
445. H. D. Hodes and K. M. Nicholas, *Tetrahedron Lett.*, 4349 (1978).
446. K. M. Nicholas, M. Mulvaney, and M. Bayer, *J. Am. Chem. Soc.*, **102,** 2508 (1980).
447. S. Padmanabhan and K. M. Nicholas, *Synth. Commun.*, **10,** 503 (1980).
448. S. Padmanabhan and K. M. Nicholas, *J. Organomet. Chem.*, **212,** 115 (1981).
449. R. Noyori, *Acc. Chem. Res.*, **12,** 61 (1979).
450. H. Takaya, S. Makino, Y. Hawakawa, and R. Noyori, *J. Am. Chem. Soc.*, **100,** 1768 (1978).
451. H. Takaya, Y. Hayakawa, S. Makino, and R. Noyori, *J. Am. Chem. Soc.*, **100,** 1778 (1978).
452. R. Noyori, T. Sato, and Y. Hayakawa, *J. Am. Chem. Soc.*, **100,** 2561 (1978).
453. Y. Hayawaka, Y. Baba, S. Makino, and R. Noyori, *J. Am. Chem. Soc.*, **100,** 1786 (1978).
454. R. Noyori, F. Shimizu, K. Fukuta, H. Takaya, and Y. Hayakawa, *J. Am. Chem. Soc.*, **99,** 5196 (1977).
455. Y. Hayakawa, K. Yokoyama, and R. Noyori, *J. Am. Chem. Soc.*, **100,** 1791 (1978).
456. Y. Hayakawa, K. Yokoyama, and R. Noyori, *J. Am. Chem. Soc.*, **100,** 1799 (1978).
457. Y. Hayakawa, K. Yokoyama, and R. Noyori, *Tetrahedron Lett.*, 4347 (1976).
458. M. Rosenblum, *Acc. Chem. Res.*, **7,** 122 (1974).
459. S. Raghu and M. Rosenblum, *J. Am. Chem. Soc.*, **95,** 3060 (1973).
460. A. Cutler, D. Entholt, W. P. Giering, P. Lennon, S. Raghu, A. Rosan, M. Rosenblum, J. Tancredi, and D. Wells, *J. Am. Chem. Soc.*, **98,** 3495 (1976).
461. N. Genco, D. Marten, S. Raghu, and M. Rosenblum, *J. Am. Chem. Soc.*, **98,** 848 (1976).
462. T. S. Abram, R. Baker, C. M. Exon, and V. B. Rao, *J. Chem. Soc., Perkin Trans. 1,* 285 (1982).
463. R. Baker, C. M. Exon, V. B. Rao, and R. W. Turner, *J. Chem. Soc., Perkin Trans. 1,* 295 (1982).
464. T. S. Abram, R. Baker, C. M. Exon, V. B. Rao, and R. W. Turner, *J. Chem. Soc., Perkin Trans. 1,* 301 (1982).

The Chemistry of the Metal—Carbon Bond, Volume 2
Edited by F. R. Hartley and S. Patai


CHAPTER 7

# Electrophilic attack on transition metal $\eta^1$-organometallic compounds

M. D. JOHNSON
*Department of Chemistry, University College London, 20 Gordon Street, London WC1H 0AJ, UK*

## I. INTRODUCTION

In order to define the mechanisms of a chemical reaction in solution accurately and completely it is necessary to know, in detail, the nature of the reactants in that solution, the composition of the transition state, the character of all intermediates on the reaction path, and the nature of the products. To obtain all such information, even for a single set of reagents in a single solvent, requires careful study of many aspects of that reaction, including pre-equilibria, stereochemistry, kinetics, solvent reagent, and substrate isotope effects and, because more than one mechanism may operate, the influence of temperature on the above quantities.

It is not surprising, therefore, that only a very small proportion of the known chemical reactions have been studied in this detail. Examples in organic chemistry include nucleophilic aliphatic substitution and electrophilic aromatic substitution[1] and examples in inorganic chemistry include ligand substitution in octahedral complexes of cobalt and square-planar complexes of platinum[2]. Yet even in these examples there remain many uncertainties of detail, especially where changes of solvent are contemplated.

Therefore, we tend with good reason to use the principles enunciated for one set of reactions studied in detail to *rationalize* the course of related but less well studied reactions. The simpler the system and the closer it lies to the well studied system, the more likely is the rationalization to be correct, but there is always the possibility that a change of conditions, especially solvent or temperature, will bring about a change of mechanism. Our concept of 'the mechanism' may thus change with experience.

In the field of organotransition metal complexes, the study of reaction mechanisms is still in its infancy. A few specific processes of industrial importance, such as the oxidative addition reaction concerned in the manufacture of acetic acid[3], have been the subject of some elegant studies, but the rapid development of the field has meant that, for most reactions, a study of reagents and products alone has sufficed. There are thus rationalizations based on direct evidence as well as rationalizations based on indirect evidence from other complexes of the same metal or even from complexes of other metals.

One of the major problems with organotransition metal complexes is their complexity. There may be present in a single complex a variety of functional groups on the organic ligand and/or the other ligands attached to the metal. Atoms of widely different electronegativity and groups of widely differing reactivity may thus be combined within the same molecule. Moreover, ligands are not necessarily rigidly fixed and one has to consider the possible reactivities of the parent complex, of the dissociated complex and of the free ligand(s). Organotransition metal complexes are thus akin to the more complex of organic molecules and a true understanding of their reactivity requires an understanding of the reactivities of the several component parts. Any extrapolation from a complex with ligand L to one with ligand L′, or from a complex in one solvent to the same complex in another solvent, or particularly from one metal to another, even in the same triad, adds a considerable measure of uncertainty to any conclusion about the reaction mechanism.

## A. Classification of Reagents

With the limited amount of evidence at our disposal, any classification of the reactions of organotransition metal complexes with electrophiles will be imperfect. Although certain types of reaction can be defined, there will be many reactions of uncertain mechanism the classification of which may change with new evidence. We have chosen as the basis of our classification the *site of reaction* and the *extent to which that reaction influences the carbon—metal bond,* i.e. the immediate consequences of the attack at the particular site. As will be evident in the following discussion, it is not always easy to define a particular site because of subsequent rapid processes. We must therefore qualify our classification by referring to the first site of attack that can be either detected directly or inferred from subsequent processes.

### 1. Electrophilic reagents

We have considered that the nature of the electrophile should be used only as a secondary means of classification, because of the several different roles that any electrophile can play. Ingold[1] defined an electrophile as 'a reagent which acts by aquiring electrons, or a share in electrons, which previously belonged exclusively to a foreign molecule'. Thus the term electrophile embraces oxidizing agents, which act by removal of one or more electrons from the substrate, Lewis acids which act by coordination to the substrate, and substituting reagents which, bonding to the substrate, cause the synchronous expulsion of some other electrophilic species. Certain electrophiles, such as Hg(II) and halogens, may behave in all three of the above ways; others, such as the proton (a very common reagent since many reactions of electrophiles are carried out in acidic solution) and cerium(IV), tend to act in only one of the above ways.

Electrophiles may also be classified according to type, i.e. (i) *metallic electrophiles,* such as Hg(II), Tl(III), Ag(I), Hg(I), Au(III), Pt(IV), Ce(IV), Ir(IV), and Cu(II), the majority of which have an accessible oxidation state one or two units lower in value; it is important to note that the character of such electrophiles is markedly influenced by both solvation and ligation; (ii) *non-metallic electrophiles,* such as halogens, oxyacids, $H_3O^+$, $NO^+$, $NO_2{}^+$, $CO_2$, $SO_2$, $(SCN)_2$ and BrSCN; and (iii) *organic electrophiles,* such as $R_3O^+$, $Ph_3C^+$, RX (where X = I, Br, $OSO_2F$, etc.), tetracyanoethylene, hexafluoroacetone, and related unsaturated molecules.

Whilst a considerable amount of attention has been given to understanding the nature of nucleophilicity and basicity[4], and appropriate nucleophilicity scales have been drawn up for reaction a wide range of nucleophiles at a variety of electrophilic centres, the same consideration has not been given to scales of electrophilicity except in relation to aromatic substitution reactions. However, from our knowledge of the variation of nucleophilicity and basicity we should not expect any direct correlation between acidity, electrophilicity, and oxidizing power.

### 2. Prediction of sites of attack

Because of the molecular complexity, it is often difficult to predict the site of attack of an electrophile on an organotransition metal complex. The choice of site will necessarily be dictated *solely* by the relative energies of the various possible transition states. These transition states may resemble the products of the reaction, they may resemble the reagents themselves or they may have characteristics mid-way between reagents and products. One approach to the prediction of reaction paths has been to assume that the transition state is reagent-like so that consideration of the LUMO of the electrophile and HOMO of the substrate may indicate where attack will take place; the closer the energies of the two orbitals of appropriate symmetry, the greater the stabilization produced on interaction and the more likely the reaction is to be orbital-controlled[5]. This 'Frontier Orbital Approach' has its devotees, but the problem is that resemblance of transition state to reagents may be slight and appreciable orbital energy perturbations may occur on approach of the reagent to the substrate. In short, prediction by this method at this stage is fraught with danger except with hindsight of the nature of the products! A clear example is shown in equation 6, where attack is at neither of the three highest occupied molecular orbitals of the substrate, but at the olefinic carbon.

### 3. Types of reaction

With the above qualifications in mind we can divide the various mechanisms into the several main classes. (1) Those which take place at sites remote from the metal on the organic ligand or other ligands, without any immediate substantial change in the character of the carbon—metal bond. Modification of the carbon—metal bond *may* occur in reactions of this class, *provided* it does so in *subsequent* steps. (2) Reactions at the organic ligand which cause substantial modifications, but not cleavage, of the carbon metal bond. For example, reactions in which there is a change of hapticity of the organic ligand. (3) Electron transfer reactions (i) to the electrophile *from* the substrate and (ii) from other reagents *to* the substrate, thereby rendering the reduced substrate liable to attack by electrophiles.* (4) Direct reaction of the electrophile with the metal centre. (5) Attack at the organic ligand with synchronous or near synchronous expulsion of the metal.

## II. LIGAND MODIFICATION WITHOUT SUBSTANTIAL CHANGE OF THE CARBON—METAL BOND

There are a large number of reactions of organotransition metal complexes which take place under forcing conditions but which have little influence on the carbon—metal bond. These are, in general, where there are nucleophilic sites remote from the metal on the organic ligand or on other ligands. Besides illustrating the fact that the carbon—metal bond is by no means always the most reactive centre, these reactions may serve either as a useful means of minor structural modification of the complex or as unfortunate competing side reactions which prevent or complicate the study of reactions at the carbon—metal bond. It is for these reasons that it is useful to consider such reactions first, since many of them, especially several reversible processes where the electrophile is the proton, will be relevant to subsequent sections of this chapter.

### A. Reversible Prototropic Equilibria

A few examples illustrate the prototropic equilibria of organic and other ligands in aqueous and non-aqueous solvents. The reversible protonation of the heteroatom of pyridylmethylmetal (**1–3**) complexes in aqueous solution (equation 1) does not

$$\left[\text{N}\bigcirc\!\!-\text{CH}_2\text{M}\right] \xrightleftharpoons{H_3O^+} \left[\text{HN}\bigcirc\!\!-\text{CH}_2\text{M}\right]^+ \qquad (1)$$

(**1**) $M = Fe(CO)_2(Cp)$ (**1a**) $pK_a = 8.0$
(**2**) $M = Co(CN)_5^{3-}$ (**2a**) $pK_a = 9.7$
(**3**) $M = Mn(CO)_5$ (**3a**) $pK_a = 8.0$

* It might be considered that reaction of the reduced organometallic substrate would fall within one of the other categories.

apparently change the nature of the carbon—metal bond, but has a profound influence on the rates of subsequent reactions of the complex with other electrophilic reagents[6]. Moreover, the high value of the p$K_a$ of the pyridinium ion formed (e.g. **1a–3a**) gives a good indication of the marked inductive and hyperconjugative interaction between the metal—carbon bond and the aromatic ring[7].

Reversible protonation of alkylbis(dimethylglyoximato)pyridinecobalt(III) complexes can take two forms. For example, on addition of trifluoroacetic acid to a solution of the ethyl complex (**4**) in methylene chloride, one of the dioximato ligands becomes protonated to give **4a**[8]. Such protonation, whether in aqueous or non-aqueous solvents, not only causes a marked decrease in the reactivity of the complex to subsequent attack by metallic electrophiles such as $Hg^{2-}$ (see Section VI), but also causes a marked weakening of the carbon—metal bond, such that homolysis also takes place more readily, for example with the $\alpha$-phenylethylcobaloxime complex[9]. On addition of more trifluoroacetic acid to **4a**, the pyridine ligand becomes reversibly detached and protonated. It is presumed that the vacant coordination site becomes filled by the trifluoroacetate ion as in **4b**. The influence of the latter change is such as markedly to promote intramolecular rearrangements of certain organic ligands[10]. A similar reversible interaction of the dioximato ligands of methylcobaloxime with iron(III) ion has also been studied[11].

The reversible proton- and mercury(II)-induced removal of the 3,5-dimethylbenzimidazole ligand of coenzyme $B_{12}$ and of methylcobalamin (shown in abbreviated form in equation 3) also has a profound influence on subsequent reactions of those substrates (see Section VI). Detailed studies have been made of the equilibrium constants for such reactions because of their importance in biological systems[12].

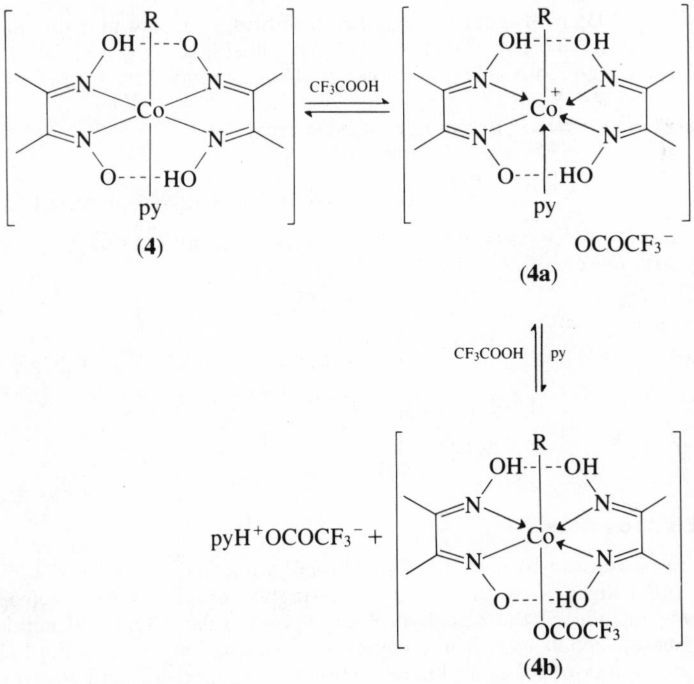

R, OH---O, N, Co, N, O---HO, py

(**4**)

$CF_3COOH$

R, OH---OH, N, $Co^+$, N, O---HO, py, $OCOCF_3^-$

(**4a**)

$CF_3COOH$ py (2)

$pyH^+OCOCF_3^-$ +

R, OH---OH, N, Co, N, O---HO, $OCOCF_3$

(**4b**)

$$\left[\begin{array}{c} R \\ | \\ Co \\ \uparrow \\ N \end{array}\right] \underset{}{\overset{X^{n+}}{\rightleftharpoons}} \left[\begin{array}{c} R \\ | \\ Co \\ \\ NX \end{array}\right]^{n+} \quad (3)$$

$X^{n+} = Hg^{2+}$ or $H^+$

$$[RCo(CN)_5]^{3-} + H_3O^+ \underset{k_{-1}}{\overset{k_1}{\rightleftharpoons}} \underset{(\mathbf{5})}{[RCo(CN)_4(CNH)]^{2-}} \xrightarrow{k_2} [R{-}\underset{\underset{NH}{\|}}{C}{-}Co(CN)_4]^{2-} \quad (4)$$

Protonation of a cyanide ligand (probably *cis* to the alkyl group) of organopentacyanocobaltate(III) complexes is also rapid and reversible (equation 4), but the protonated complex **5** undergoes a subsequent intramolecular insertion of, or alkyl migration to, the HNC ligand. The rate law for the formation of the insertion product is shown in equation (5).

$$\text{Rate} = k_2(k_1/k_{-1})[\text{organocobalt}][H^+] = k_2[\text{protonated organocobalt}^+] \quad (5)$$

i.e. the rate increases linearly with increasing $[H^+]$ as long as the protonated intermediate is present in minor amounts, but becomes independent of the acid when all the complex is monoprotonated[13,14]. For $R = 4\text{-}CH_2pyH^+$ the value of $k_2$ is $1.4 \times 10^{-2}\ s^{-1}$ at 25 °C and $\mu = 3$ (ref. 14). These reactions are complicated by the reversible loss of a cyanide ligand in acidic solution, which can be suppressed by the addition of an excess of HCN.

Reversible loss of HCN also occurs, together with bridge formation and equatorial ligand protonation, in the reactions of alkylbis(dimethylglyoximato)cyanocobaltate(III) ions in aqueous acid (equation 6)[15]. Closely related to this reaction are

$$2[RCo(dmgH)_2CN]^- \underset{CN^-}{\overset{H_3O^+}{\rightleftharpoons}} [RCo(dmgH)_2CNCo(dmgH)_2R]^-$$

$$\overset{H_3O^+}{\rightleftharpoons} [RCo(dmgH)(dmgH_2)CNCo(dmgH)_2R]. \quad (6)$$

the Lewis acid-promoted carbonyl insertion (or methyl migration) reactions of organometal carbonyl complexes (equation 7)[16].

$$[MoMe(CO)_3(cp)] \xrightarrow{AlBr_3} \left[\begin{array}{c} Br_2Al{-}O \\ Br \quad C \\ Me{-}Mo{-}CO \\ Cp \quad CO \end{array}\right] \xrightarrow{CO} [Mo\{MeC({=}OAlBr_3)\}(CO)_3(Cp)] \quad (7)$$

## B. Irreversible Processes

A number of useful transformations of functional groups have been carried out on organic and other ligands of organotransition metal complexes by electrophilic reagents. These include (i) the acid hydrolysis of carboxylic esters and amides of organo-manganese, -cobalt and -iron complexes (e.g. equation 8)[17]; (ii) the halogenation[18] and deuteriation[19] of benzylmetal complexes (equations 9 and 10); (iii) the nitrosation of aminoalkylcobalt complexes (equation 10)[20]; (iv) the nitrosation of

$$[H_2NCOCH_2Mn(CO)_5] \xrightarrow{HCl} [HOOCCH_2Mn(CO)_5] \quad (8)$$

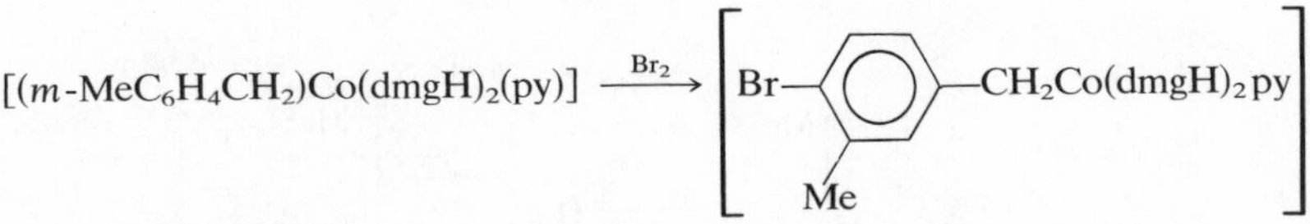

$$[(m\text{-}MeC_6H_4CH_2)Co(dmgH)_2(py)] \xrightarrow{Br_2} [\text{Br}-C_6H_3(\text{Me})-CH_2Co(dmgH)_2py] \quad (9)$$

$$[C_6H_5CH_2Fp] \underset{}{\overset{CF_3COOD}{\rightleftharpoons}} [\text{(D)(H)}C_6H_5^{+}CH_2Fp]\,CF_3COO^- \overset{CF_3COOH}{\rightleftharpoons} p\text{-}C_6H_4DCH_2Fp \quad (10)$$

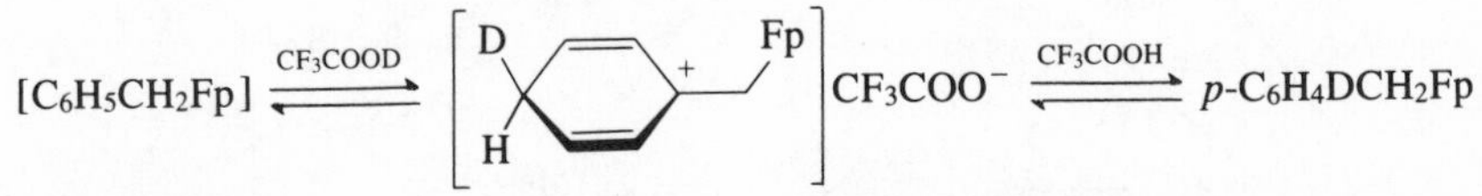

position 10 and the lactonization of side chains of the corrin ligand of coenzyme $B_{12}$ by nitrous acid[21]; and (v) the acid-catalysed hydrolysis of the ketal **6**, or of the corresponding phenylboronic ester, which is a key step in the chemical synthesis of coenzyme $B_{12}$ from cobalamin precursors (equation 12)[22]. Several of the above reactions take place in competition with reactions which do cause cleavage of the C—M bond.

$$[PhNHCH_2Co(dmgH)_2py \xrightarrow[HCl]{HNO_2} [PhN(NO)CH_2Co(dmgH)_2NO_2]^- \quad (11)$$

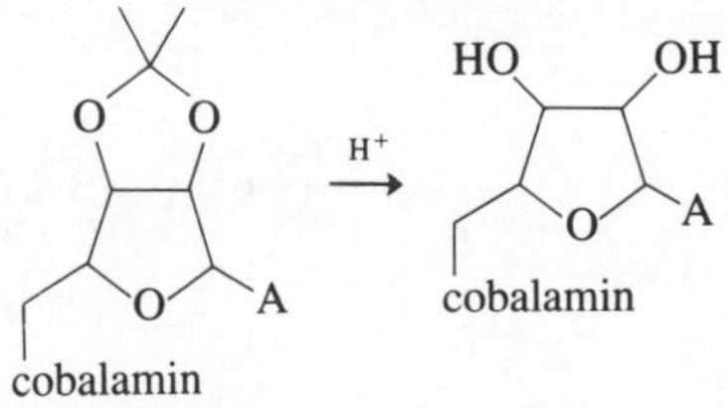

$$\mathbf{6} \xrightarrow{H^+} \text{(HO)(OH) diol, cobalamin, A} \quad (12)$$

cobalamin

A = adenine (**6**)

## III. ATTACK OF ELECTROPHILES AT THE ORGANIC LIGAND LEADING TO MODIFICATION OF THE CHARACTER OF THE CARBON—METAL BOND

There are a large number of reactions in which the electrophile directly attacks the organic ligand causing a change in the type of carbon—metal bond, e.g. from mono- to di-hapto or from substituted alkyl to carbene ligand. These frequently occur with alkenyl, alkynyl, or alkadienyl ligands having accessible, filled, olefinic or acetylenic $\pi$-orbitals, or with alkyl groups having strategically placed hydroxyl, alkoxyl, or acyloxy groups.

### A. Attact at the $\gamma$-Carbon of Alkenyl, Alkynyl, or Propadienyl Ligands

The earliest example of attack of an electrophile on the $\gamma$-carbon of an allyl ligand is the protonation reaction shown in equation 13a[23]. Rosenblum and coworkers subsequently developed this reaction to include a wide range of electrophiles, under non-nucleophilic conditions, which led to a useful series of cationic dihapto complexes[24] (equation 13b–f). In most cases the intermediate dihapto complex could be isolated and treated with base to a give a novel derivative of the substrate monohapto-allyliron complex. The reaction of similar intermediate dihapto complexes with carbanion is discussed in Section III.C. Similar reactions take place with

$$[\text{CH}_2{=}\text{CHCH}_2\text{Fp}] \xrightarrow{a:\ \text{dry HCl}} [(\text{MeCH}{=}\text{CH}_2)\text{Fp}]^+$$

$$[\text{CH}_2{=}\text{CHCH}_2\text{Fp}] \xrightarrow{b:\ \text{Me}_3\text{O}^+,\ \text{liq.SO}_2} [\text{MeSO}_2\text{CH}_2\text{CH}{=}\text{CH}_2 \to \text{Fp}]^+ \xrightarrow[0\,^\circ\text{C}]{\text{Et}_3\text{N}} [\text{MeSO}_2\text{CH}_2\text{CH}{=}\text{CHCH}_2\text{Fp}]\ \ 90\%$$

$$[\text{CH}_2{=}\text{CHCH}_2\text{Fp}] \xrightarrow{c:\ \text{Me}_3\text{O}^+,\ \text{CH}_2\text{Cl}_2} [\text{MeCH}_2\text{CH}_2\text{CH}{=}\text{CH}_2 \to \text{Fp}]^+ \xrightarrow{\text{Et}_3\text{N}} [\text{Me}{\sim}\text{CH}{=}\text{CHCH}_2\text{Fp}]$$

$$[\text{CH}_2{=}\text{CHCH}_2\text{Fp}] \xrightarrow{d:\ \text{CH}_3\text{COCl},\ \text{AgSbF}_6} [\text{CH}_3\text{COCH}_2\text{CH}{=}\text{CH}_2 \to \text{Fp}]^+ \xrightarrow[in\ situ]{\text{Et}_3\text{N}} [\text{CH}_3\text{C(O)CH}_2\text{CH}{=}\text{CHCH}_2\text{Fp}] \quad (13)$$

$$[\text{CH}_2{=}\text{CHCH}_2\text{Fp}] \xrightarrow{e:\ \text{2-R-1,3-dioxolan-2-ylium},\ \text{Ph}_3\text{C}^+\text{BF}_4^-} [\text{(2-R-1,3-dioxolan-2-yl)CH}_2\text{CH}{=}\text{CH}_2 \to \text{Fp}]^+ \xrightarrow[0\,^\circ\text{C}]{\text{Et}_3\text{N}} [\text{(2-R-1,3-dioxolan-2-yl)CH}_2\text{CH}{=}\text{CHCH}_2\text{Fp}]$$

$$[\text{CH}_2{=}\text{CHCH}_2\text{Fp}] \xrightarrow{f:\ (\text{MeO})_3\text{CH},\ \text{Ph}_3\text{C}^+\text{BF}_4^-} [(\text{MeO})_3\text{CCH}_2\text{CH}{=}\text{CH}_2 \to \text{Fp}]^+ \longrightarrow [\text{MeO}_2\text{CCH}_2\text{CH}{=}\text{CHCH}_2\text{Fp}]$$

propynyliron complexes (equation 14)[24] and with the nortricyclyl derivative **7** (equation 15)[25].

$$[\text{ArC}{\equiv}\text{CCH}_2\text{Fp}] + \text{HBF}_4 \longrightarrow [\text{MeC}{=}\text{C} \to \text{Fp}]^+ \quad (14)$$

$$\text{(7)} \xrightarrow{\text{H}^+} \text{norbornene} \to \text{Fp}^+ \quad (15)$$

However, if there is a suitable nucleophilic centre present within the complex, whether on the organic ligand or by virtue of the fact that the electrophile is an incipient zwitterion, intramolecular attack of that centre on the transient dihapto complex may lead directly to the formation of a new monohapto metal complex containing a cyclic ligand. Some idea of the scope of the reaction is shown in equation 16[26–29]. Some individual related examples are also shown in equations 17– M = Fp[29], $W(CO)_3(Cp)$[26], $Mo(CO)_3(Cp)$[26,27], $Co(dmgH)_2py$[26,28a], $Cr(NO)_2(Cp)$[26], etc.

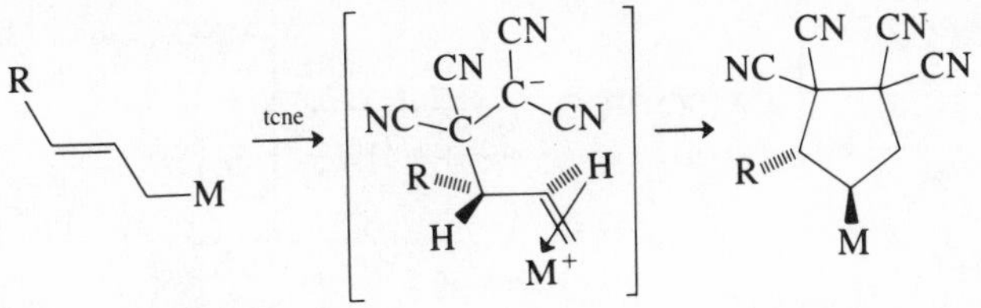

(16)

R = H, Me, Ph, etc.

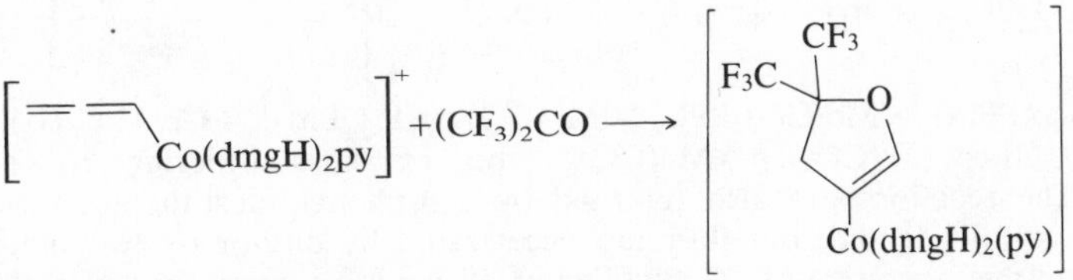

$$[CH_2{=}CH{-}CH_2{-}Co(dmgH)_2py]^+ + (CF_3)_2CO \longrightarrow [\text{cyclic } (CF_3)_2C{-}O \text{ ring}{-}Co(dmgH)_2(py)] \quad (17)^{28a}$$

23[30–32]. The crystal structure of the product of reaction 16 [R = Ph, M = $Co(dmgH)_2py$] has been determined and the *trans*-stereochemistry of phenyl and iron groups is consistent with an *antarafacial* $\eta^1$-$\eta^2$-$\eta^1$ migration of the Fe from the $\alpha$- to the $\beta$-carbon via a zwitterionic intermediate[33]. This is supported by the formation of both *cis* and *trans* isomers in reaction 19[30] and by the kinetics of reaction of a wide range of propargyl and allyl metal complexes with toluene sulphonylisocyanate (reaction 20)[31]. Similar reactions take place with a number of other isocyanates, such as chlorosulphonyl, trichloroacetyl, and methoxysulphonyl isocyanate[25] and with dichlorodicyanoquinone, dimethylmethylenemalonate, and the ketimine **8**[26,34,35].

The kinetic studies show a wide variation of reactivity of propynylmetal complexes towards toluenesulphonylisocyanate, the order of reactivity being

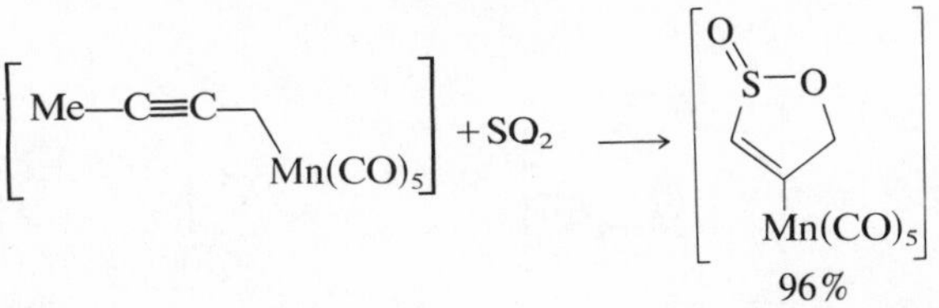

$$[Me{-}C{\equiv}C{-}CH_2{-}Mn(CO)_5] + SO_2 \longrightarrow [\text{cyclic } S(O){-}O \text{ ring}{-}Mn(CO)_5] \quad (18)^{28b}$$

96%

$$[PhC{\equiv}C{-}CH_2{-}Fp] + (F_3C)(NC)C{=}C(CF_3)(CN) \longrightarrow [\text{cyclopentene}, cis\text{-}CF_3,CF_3] + [\text{cyclopentene}, trans\text{-}CF_3,CN] \quad (19)^{30}$$

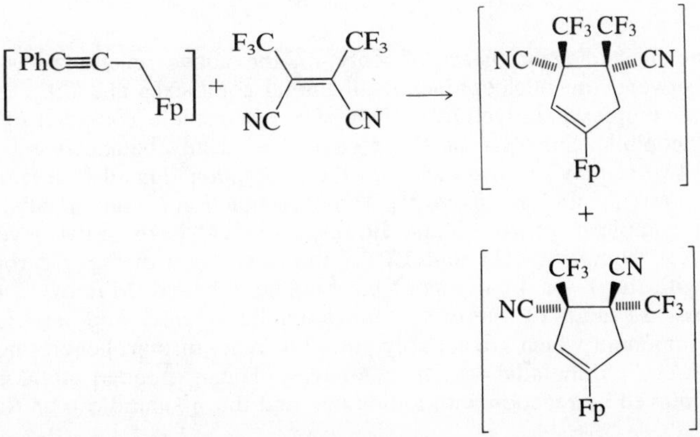

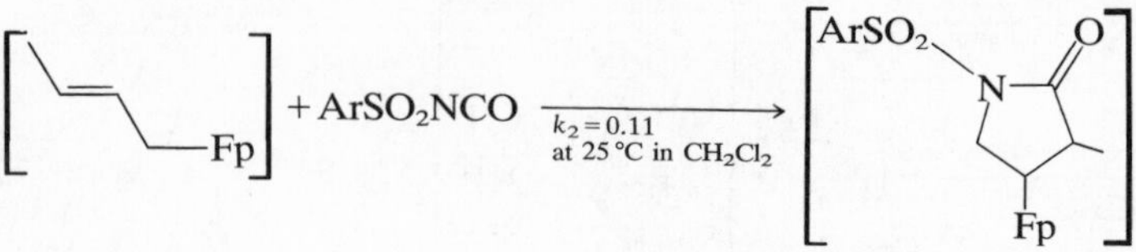

$$[\text{CH}_2=\text{CHCH(CH}_3)\text{Fp}] + \text{ArSO}_2\text{NCO} \xrightarrow[\text{at 25 °C in CH}_2\text{Cl}_2]{k_2=0.11} [\text{N-ArSO}_2\text{-pyrrolidinone, Fp}] \quad (20)^{31}$$

$$[\text{(cyclopropylmethyl)Fp}] + (\text{NC})_2\text{C}=\text{C(CN)}_2 \longrightarrow [\text{tetracyanocyclohexyl-Fp}] \quad (21)^{29}$$

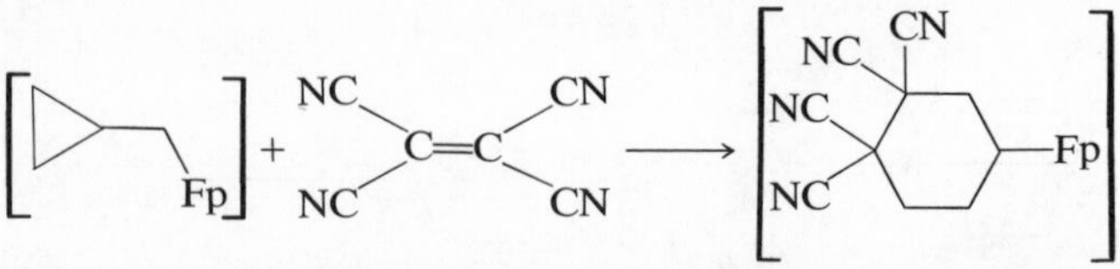

$Mo(CO)_2cp(PPh_3) > Mo(CO)_4PPh_3 > Mo(CO)_2cp[P(OPh)_3] > Cr(\eta^5\text{-}C_5H_5)(NO)_2 >$ $Fp > Mo(CO)_3cp > W(CO)_3cp > Mn(CO)_5$, the extremes differing by a factor of 500. The reactions are also retarded by a 3-phenyl substituent on either the propynyl or allyl ligand, but they are accelerated by one or by two methyl substituents on the 3-position of the allyl ligand. It should be noted, however, that many of these processes are accompanied by side reactions such as the insertion reactions discussed in Section VIII.

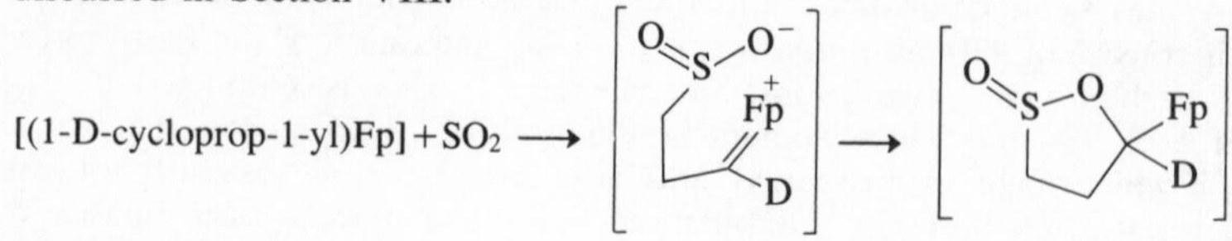

$$[\text{(1-D-cycloprop-1-yl)Fp}] + \text{SO}_2 \longrightarrow [\ldots] \longrightarrow [\ldots] \quad (22)^{32}$$

$$[\text{CH}_2=\text{CHCH}_2\text{Fp}] + \text{Me}_2\text{C}=\text{C}=\overset{+}{\text{N}}(\text{C}_5\text{H}_{10})\ \textbf{(8)} \longrightarrow [\ldots]^+ \longrightarrow [\ldots] \quad (23)^{35}$$

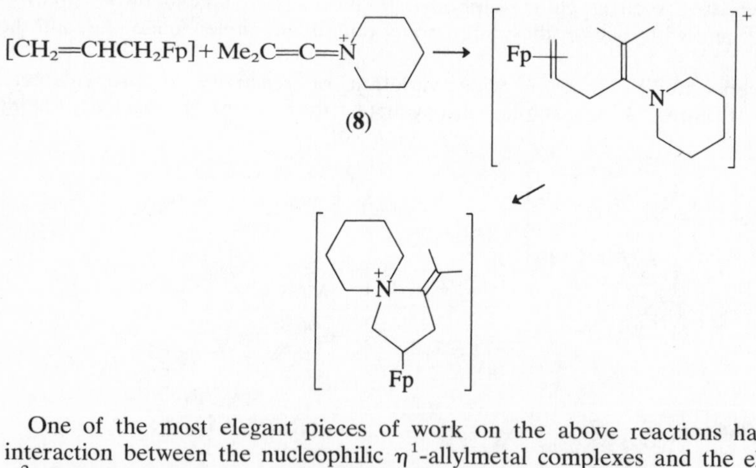

One of the most elegant pieces of work on the above reactions has used the interaction between the nucleophilic $\eta^1$-allylmetal complexes and the electrophilic $\eta^2$-alkeneiron complexes derived from them. The electrophilic character of the latter and the nucleophilic character of the former are ideally balanced and coupling between the two readily occurs, such that the $\eta^1$- and $\eta^2$-ligands are transformed, respectively, into $\eta^2$- and $\eta^1$-ligands[25]. Thus the reaction of the $\eta^1$-allyl with the $\eta^2$-butadiene complexes of iron (**9** and **10**, respectively)[36] leads to two seven-carbon dimetallic $\eta^1, \eta^2$-complexes (**11** and **12**) by the interaction of the $\gamma$-carbon of the allyl ligand with the 1- or 4-carbons of the butadiene ligand. Moreover, in each of the complexes **11** and **12** there is a nucleophilic $\eta^1$-allyl and an electrophile $\eta^2$-alkene component which are suitably placed to react further, generating two new seven-carbon cyclic dimetallic $\eta^1$-, $\eta^2$-complexes. The $\eta^2$-bonded metal cation can readily be removed by reaction with iodide ion, and the $\eta^1$-metal can be replaced by

Br, $CO_2Me$, etc., using appropriate electrophilic reagents as described in Sections V.A and IV.A, respectively. The potential of such reactions in organic synthesis is considerable.

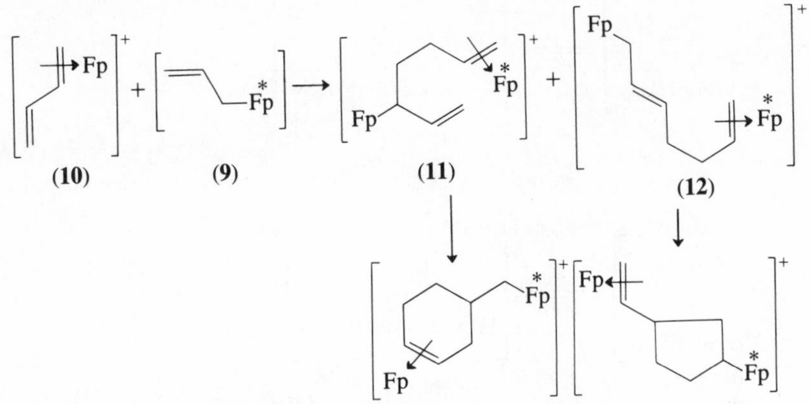

**(10)** + **(9)** → **(11)** + **(12)** (24)

## B. Attack at a γ-Oxygen Atom of the Organic Ligand

A cationic $\eta^2$-alkene complex has also been isolated from the reaction of the acylalkylcomplex **13**[37], but one of the best routes to such complexes is by protonation of the hydroxyl group of β-hydroxyethyl complexes. The latter can be made in good yield from halohydrins or epoxides, thus providing a good route to a wide range of $\eta^2$-olefin complexes (equation 26)[26]. However, with other metals the $\eta^2$-complexes may only be transient intermediates. For example, $\eta^2$-complexes of vinyl ethers and alcohols have been postulated as intermediates in the acid-catalysed hydrolysis of acetals of formylmethylcobalamins (equation 27). There reactions are reversible, as the same acetals can be made by reaction of appropriate vinyl ethers with cobaloxime(III) in alcoholic solution[38,39].

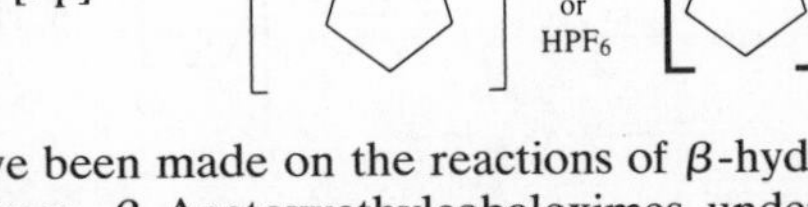

$$[RCOCH_2Fp] \xrightarrow{H^+} [R(OH)C{=}CH_2 \rightarrow Fp]^+ \quad (25)$$

**(13)**

$$\text{O(cyclopentene oxide)} + [Fp]^- \longrightarrow [\text{Fp-cyclopentyl-O}]^- \xrightarrow[\text{HPF}_6]{\text{HBF}_4 \text{ or}} [\text{Fp}(\eta^2\text{-cyclopentene})]^+ \quad (26)$$

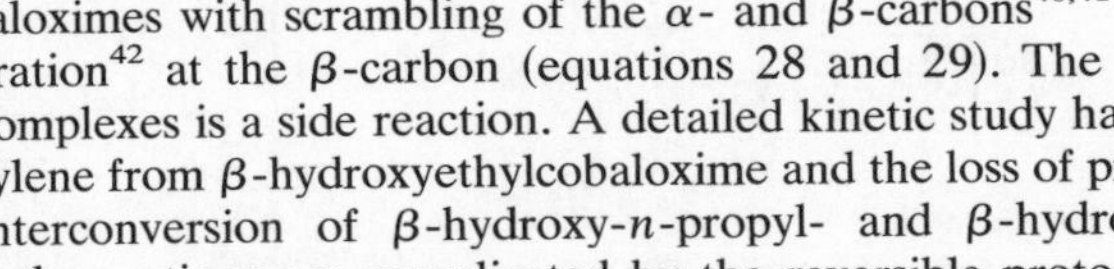

Extensive studies have been made on the reactions of β-hydroxy-, β-acetoxy, and β-alkoxy-ethylcobaloximes. β-Acetoxyethylcobaloximes undergo alcoholysis to β-alkoxycobaloximes with scrambling of the α- and β-carbons[40,41] and with retention of configuration[42] at the β-carbon (equations 28 and 29). The loss of olefin from these $\eta^2$-complexes is a side reaction. A detailed kinetic study has been made of the loss of ethylene from β-hydroxyethylcobaloxime and the loss of propene accompanying the interconversion of β-hydroxy-*n*-propyl- and β-hydroxyisopropyl-cobaloximes. Both reactions are complicated by the reversible protonation of the dioximato ligands (see Section II.A), but the overall reaction scheme for the β-hydroxypropyl complexes is as shown in equation 30, with $k_1 = 0.1$ and $k_2 = 0.7$–$10\ \text{mol l}^{-1}\ \text{s}^{-1}$ in aqueous solution at 25 °C and $\mu = 0.200\ \text{mol l}^{-1}$, $k_{-1}/k_3 \approx 0.2$. The loss of propylene is also accelerated by addition of chloride ion, probably through the formation of some of the hydroxypropylbis(dimethylglyoximato)chlorocobaltate(III)

$$\left[(RO)_2CHCH_2Co(dmgH)_2py\right] \underset{ROH}{\overset{H^+}{\rightleftharpoons}} \left[ROCH{=}CH_2 \rightarrow Co(dmgH)_2py\right]^+ \rightleftharpoons ROCH{=}CH_2 + [Co(dmgH)_2py]$$

$$\left[ROCH(HO)CH_2Co(dmgH)_2py\right] \underset{ROH}{\overset{H^+}{\rightleftharpoons}} \left[HOCH{=}CH_2 \rightarrow Co(dmgH)_2py\right]^+ \underset{H^+}{\rightleftharpoons} \left[OHC{-}CH_2Co(dmgH)_2py\right] \quad (27)$$

(with $H_2O$ / $H^+$ linking the two alkene complexes)

$$[AcOCH_2\overset{*}{C}H_2Co(dmgH)_2py] \overset{H^+}{\rightleftharpoons} \left[CH_2{=}CH_2 \rightarrow Co(dmgH)_2py\right]^+ \xrightarrow{EtOH} [EtOCH_2\overset{*}{C}H_2Co(dmgH)_2py] + [Et\overset{*}{O}CH_2CH_2Co(dmgH)_2py] \quad (28)$$

$\overset{*}{C}H_2=CD_2$ or $^{13}CH_2$

$$\left[AcO(Me)(H)C{-}CH_2Co(dmgH)_2py\right] \rightleftharpoons \left[CH_2{=}CH(Me)(H) \rightarrow Co(dmgH)_2py\right]^+ \overset{ROH}{\rightleftharpoons} \left[RO(Me)(H)C{-}CH_2Co(dmgH)_2py\right] \quad (29)$$

$$\left[HOCH_2CH(CH_3)Co(dmgH)_2aq.\right] \underset{H_2O,\ k_{-1}}{\overset{H^+,\ k_1}{\rightleftharpoons}} \left[CH_3CH{=}CH_2 \rightarrow Co(dmgH)_2aq.\right] \xrightarrow{k_3} C_3H_6 + [Co(dmgH)_2aq.]^+$$

$$\left[CH_3CH{=}CH_2 \rightarrow Co(dmgH)_2aq.\right] \underset{H^+,\ k_2}{\overset{H_2O,\ k_{-2}}{\rightleftharpoons}} [CH_3CH(OH)CH_2Co(dmgH)_2aq.] \quad (30)$$

ion[43]. The acid-catalysed decomposition of alkoxycarbonylethyl(1-hydroxy-2,2,3,3,7,7,8,8,12,12,13,13,17,17-hexadecamethyl-10, 20-diazaoctahydroporphimato)cobalt(III) has been used as a key step in the protection of carboxylic acids[44].

As with $\eta^2$-complexes from allylmetal complexes (Section III.A), the transient $\eta^2$-cobalt(III) complexes are susceptible to intramolecular attack by appropriately placed nucleophilic substituents. For example, the dihydroxyalkylcobaloxime (**14**) gives a moderate yield of the tetrahydrofuranylmethylcobaloxime during acidolysis[45].

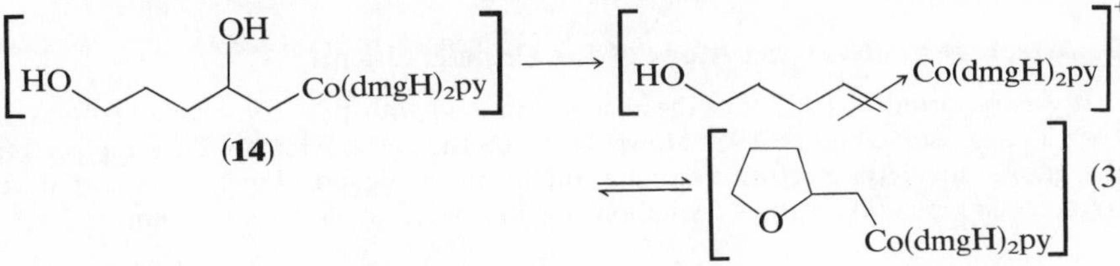

(31)

## C. Attack at a $\gamma$-Hydrogen Atom of the Organic Ligand

The abstraction of a hydride ion from the $\beta$-carbon of an alkylmetal complex is a much used route to $\eta^2$-olefin complexes. The reaction of the $\beta$-phenylethyl iron complex **15** with $Ph_3C^+BF_4^-$ gives two diastereoisomeric $\eta^2$-complexes which are cleaved by phosphine to the free styrenes. The deuterium distribution shows that the process takes place by a *trans anti periplanar* abstraction–migration mechanism (equation 32)[46]. The reaction is used for preparative purposes, a typical example being the formation of the diester **16** (equation 33)[47]. The $\eta^2$-bonded iron can be removed by treatment with iodide. An unusual complex is formed on abstraction of a hydride ion from the di-iron complex **17**. It has more of the characteristics of a di-$\eta^1$-complex than an $\eta^1$, $\eta^2$-complex, as if it cannot quite make up its mind which

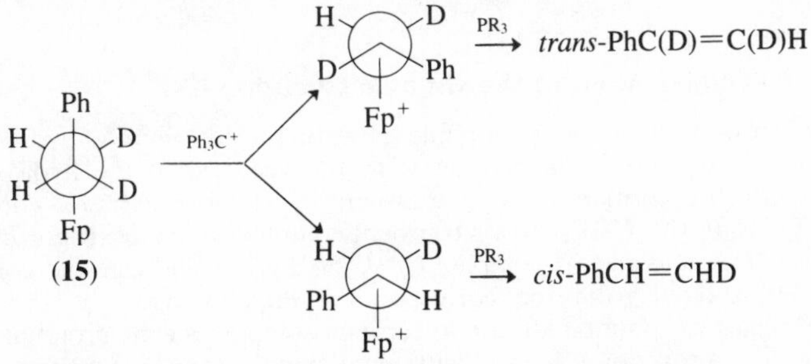

(32)

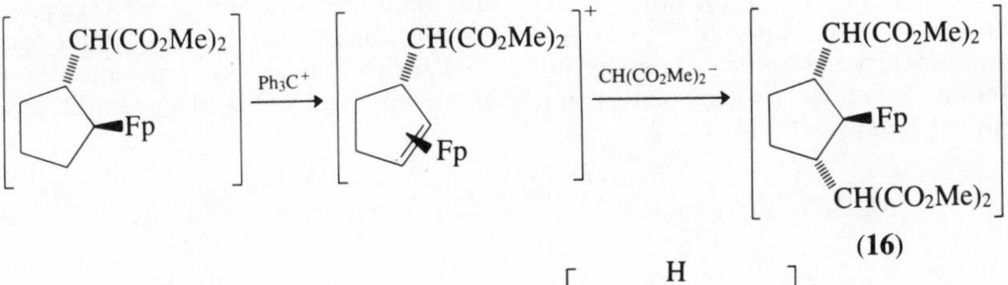

(33)

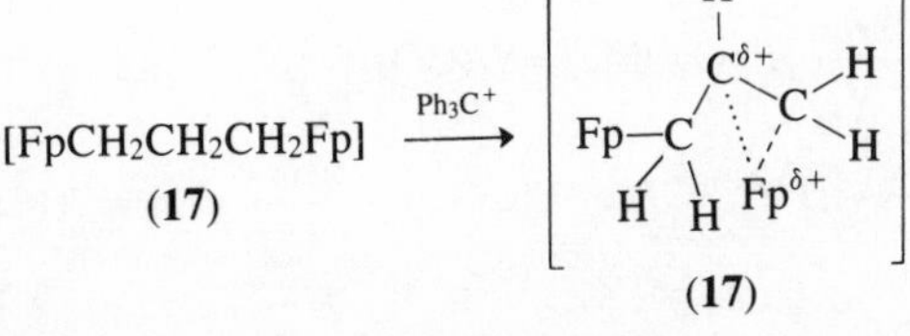

(34)

$\eta^2$-complex should form! In the crystal there is some asymmetry, the $C_2$—Fe distances being 2.59 and 2.72 Å and the $CH_2$—Fe distances are both greater than 2.08 Å which is the bond length in the neutral precursor complex. The asymmetry is also evident in solution at −20 °C for the $^1$H n.m.r. spectrum shows separate resonances for the two methylene groups at $\delta$ 2.25 and 2.92, the former resonance being a doublet of doublets with $J = 14$ and 3 Hz. The methine resonance is at $\delta$ 6.8[48].

## D. Attack at a $\gamma$-Nitrogen Atom of the Organic Ligand

It was reported in 1967 that the reaction of acid with the acetonitrile complex **18** gave an azaalene complex **19**[49]. However, when the carbonyl ligands are replaced by the more strongly electron-donating diphosphine ligand dppe, an unusual intramolecular reaction ensues (equation 36) to give a novel carbene complex[50].

$$[FpCH_2C{\equiv}N] \xrightarrow{H^+} [Fp(\eta^2\text{-}CH_2{=}C{=}NH]^+ \qquad (35)$$

(**18**) (**19**)

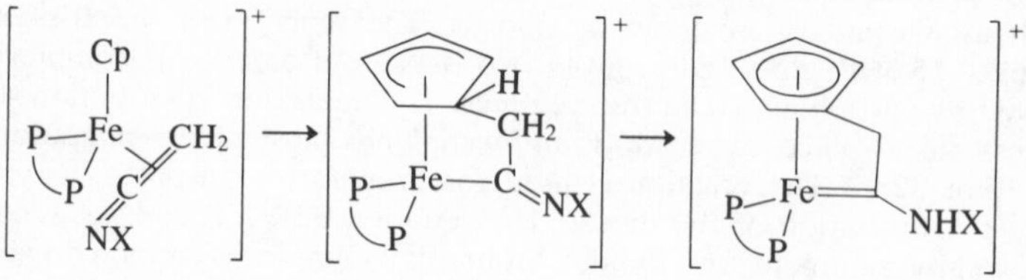

$$\left[(P\text{—}P)(Cp)Fe(\eta^2\text{-}CH_2{=}C{=}NX)\right]^+ \longrightarrow \left[(P\text{—}P)Fe(\eta^5\text{-}C_5H_4H\text{-}CH_2)\text{—}C{\equiv}NX\right]^+ \longrightarrow \left[(P\text{—}P)Fe{=}C(NHX)CH_2\text{-}C_5H_4\right]^+ \qquad (36)$$

X=H, Me

## E. Attack at a $\beta$-Oxygen Atom of the Organic Ligand

The attack of a proton or of an electrophilic alkylating agent on an $\alpha$-hydroxy- or $\alpha$-alkoxy-alkylmetal complex frequently leads to the removal of the hydroxyl or alkoxyl group with the formation of a transient or isolable carbene complex. Reaction of $MeOCH_2F$ with $HBF_4$ gives a transient complex which disproportionates even at −90 °C to $Fp^+$ and the $\eta^2$-complex $(\eta^2\text{-}C_2H_4)Fp^+$[51]. The carbene complex can, however, be isolated from the corresponding dppe complex as shown in equation 37[52]. At least as common a route to carbene complexes is by protonation of an $\alpha$-carbonyl group, a reaction observed with many metals, though most frequently in the Cr–W triad. Equation 38 shows an example which includes electrophilic attack both at $\alpha$-carbonyl and $\alpha$-alkonyl groups, in each case forming carbene complexes[53]. Equation 39 shows an unusual example of an intramolecular electrophilic attack catalysed by silver(I)[54].

$$[MeOCH_2Fe(dppe)(Cp)] \xrightarrow{HBF_4/Ac_2O} [CH_2Fe(dppe)(Cp)]^+BF_4^- \qquad (37)$$

$$[PhCOW(CO)_5] \xrightarrow{Me_2SO_4} [PhC(OMe){=}W(CO)_5]^+$$

$$\xrightarrow{PhLi} [Ph_2C(OMe)W(CO)_5] \xrightarrow[-78\,°C]{HCl} [Ph_2{=}W(CO)_5] \qquad (38)$$

$$\left[\text{cyclopentyl}(2\text{-Cl})\text{-O-C(=O)-Mn(CO)}_5\right] \xrightarrow{Ag^+} \left[\text{cyclopentane-fused dioxolanylidene}=\text{Mn(CO)}_5\right]^+ \quad (39)$$

## F. Attack at a β-Nitrogen Atom or a β-Sulphur Atom of the Organic Ligand

Directly analogous to the attack at an α-carbonyl group is the attack at an iminoacyl group (equation 40)[55]. The corresponding *N,N*-dimethylthioamide of platinum(II) reacts similarly.

$$trans\text{-}[(PhNH{=}CH)PtCl(PEt_3)_2] \xrightarrow{Me_2SO_4} \left[\begin{matrix}PhN(Me)\\ H\end{matrix}\!\!>C{=}PtCl(PEt_3)_2\right]^+ \quad (40)$$

## G. Attack at a β-Hydrogen Atom of the Organic Ligand

Only where γ-hydrogens are *not* present does one usually find abstraction of a β-hydride ion by an electrophile. The site must usually be activated by an electron-donating substituent as in equation 41[56]. A novel reaction is the removal of hydride ion from the benzacyclobutenyl complex **20**.

$$[(MeOCH_2)Fe(dppe)(Cp) \longrightarrow [MeOCH{=}Fe(dppe)(Cp)]^+ \quad (41)$$

$$\left[\text{benzocyclobutenyl-Fp}\right] \xrightarrow{Ph_3C^+} \left[\text{benzocyclobutenylidene}{=}\text{Fp}\right]^+ \quad (42)$$

**(20)**

## H. Attack at the β-Carbon of the Organic Ligand

Vinylidene carbene complexes are formed on alkylation of acetylides (equation 43)[57], although the dipolar electrophile hexafluoroacetone gives a cycloaddition product (equation 44) either by a concerted 1,2-addition or by an intramolecular attack of the internal nucleophilic centre on the electrophilic α-carbon of the initially formed adduct **21**[58]. See also Section VII for further examples.

$$[RC{\equiv}CFe(dppe)(Cp)] \xrightarrow{R'X} [RR'C{=}C{=}Fe(dppe)(Cp)^+X^- \quad (43)$$

$$[RC{\equiv}CFp^*]+(CF_3)_2CO \longrightarrow \left[\begin{matrix}(CF_3)_2C{-}O^-\\ |\\ R{-}C{=}C{=}Fp^*\end{matrix}\right]^+ \longrightarrow \textbf{(21)} \longrightarrow \left[\begin{matrix}(CF_3)_2C{-}O\\ R{-}C{=}C{-}Fp^*\end{matrix}\right] \quad (44)$$

## IV. ELECTRON TRANSFER REACTIONS

All of the above reactions have been described in terms of heterolytic processes in which there is full or partial transfer of an electron pair from substrate to electrophile. There is, however, evidence that some such reactions may proceed through an initial one-electron transfer process which is difficult to detect. As shown below, such one-electron processes are readily carried out electrochemically and may also be detected with certain electrophiles; it is the similarity between these reactions and some of the above processes which makes us suspect the oxidative character of some reactions formerly thought to be heterolytic.

For example, tetracyanoethylene forms charge-transfer complexes with a wide range of organic and organometallic substrates, such as aromatic hydrocarbons with donor substituents and organotin compounds. The energy of the charge-transfer bond in the electronic spectrum of such complexes is proportional, with some qualifications, to the energy required to transfer an electron from the substrate to the tetracyanoethylene. One can thus write alternative processes to the reaction shown in equation 6 in which a charge-transfer complex and the products of a subsequent complete electron transfer are on the reaction coordinate, the reaction being continued by coupling of the paramagnetic intermediates. Alternatively, charge-transfer complexes may be formed, but may not be on the reaction coordinate. In the case of the organotransition metal complexes described above, no direct evidence of such electron transfer has been forthcoming, but the need to consider such processes has been well described[59]. Attention to such processes should be paid where the organometallic complex has a particularly low ionization potential. To some extent, however, all electrophilic processes will be influenced by the ionization potentials of the substrate.

## A. One-electron Oxidative Processes

There are now a number of clearly defined one-electron processes which serve as models for the other, less readily detected processes. One feature of such processes is that the electrophilic moiety is frequently, but by no means always, absent from the reaction products derived from the substrate. Both inner- and outer-sphere oxidation can occur, but few investigations have so far been made in sufficient detail. Two-electron oxidation by direct attack of electrophiles on the metal is considered in a subsequent section.

Perhaps the clearest example of a one-electron oxidation process is in the reaction of one-electron oxidizing agents, such as cerium(IV) ammonium sulphate, with alkylcobaloxime(III) complexes (equation 45). When the reaction is carried out at

$$[RCo^{III}(dmgH)_2L] + Ce^{IV} \longrightarrow [RCo^{IV}(dmgH)_2L]^+ + Ce^{III} \qquad (45)$$

−50 °C in methanol, the product is an organocobaloxime(IV) cation which is sufficiently stable for its e.s.r. spectrum to be measured[60] and its subsequent reactions to be studied. The cobalt(IV) species has appreciable $d^5$ character and is susceptible to reaction with even mildly basic and nucleophilic species. Thus chloride ion reacts at the $\alpha$-carbon of the *sec*-octylcobaloxime(IV) cation with inversion of configuration to give optically active *sec*-octyl chloride; some elimination to give olefins also occurs (equation 46)[61]. In the absence of a nucleophile, the complex disproportionates to reform the organocobaloxime(III) precursor and give radical or olefinic products

$$Cl^- + \left[ \begin{array}{c} C_6H_{13} \\ H \cdots C-Co^{IV}(dmgH)_2L \\ H_3C \end{array} \right]^+ \longrightarrow Cl-C(C_6H_{13})(H)(CH_3) + [Co^{II}(dmgH)_2L] + \text{olefin} \qquad (46)$$

**(22)**

derived from the hypothetical organocobaloxime(V) complex[62]. The half-wave potentials for oxidation of a number of organo(aquo)cobaloxime(III) complexes vary from 0.85 to 0.90 V (*vs.* SCE) in 1 mol dm$^{-3}$ $HClO_4$[62]. A characteristic by-product of oxidation is the dimethylglyoxime derivative RON=C(Me)C(Me)=NOH[63].

$$2[RCo^{IV}(dmgH)_2aq]^+ \longrightarrow [RCo^{III}(dmgH)_{2aq.}] + [RCo^{V}(dmgH)_{2aq.}]^{2+} \longrightarrow \text{products} \quad (47)$$

Oxidation also takes place with hexachloroiridate(IV)[64] by an outer-sphere mechanism[65] and with halogens[66,67]. The latter reaction poses a number of very difficult problems, in part because a nucleophile (halide ion) is liberated from the reagent and may give rise to products indistinguishable from those formed by attack of the electrophile, in part because the initial rate-determining step may be followed by a whole sequence of rapid events.

In the reaction of substituted benzyl(pyridine)cobaloximes with iodide ion, the rate-determining step is the loss of the axial pyridine ligand, which is followed by a very rapid, probably inner-sphere, oxidation by iodine and attack of iodide ion on the benzylic carbon (equation 48)[68].

$$[(ArCH_2)Co(dmgH)_2aq] + X_2 \longrightarrow X_2^- + [(ArCH_2)Co(dmgH)_2aq.]^+ \xrightarrow{X^-} ArCH_2X \quad (48)$$

$$X_2^- \longrightarrow X^- + X\cdot$$

$$[(ArCH_2)Co(dmgH)_2aq.]^+ \xrightarrow{H_2O} ArCH_2OH$$

$$[(ArCH_2)Co(dmgH)_2aq.]^+ \longrightarrow ArCH_2ON{=}C(Me)C(Me){=}NOH$$

When the benzyl(aquo)cobaloximes are utilized, the rate of reaction is proportional to the iodine concentration, but also to both 1st *and* 2nd powers of the substrate concentration[69]. A pre-equilibrium between $I_2$ and one molecule of substrate is proposed. Complicating factors in these reactions are (a) the role(s) played by the iodine atom, and (b) the role(s) played by the cobaloxime(II) species which might capture the iodine atom but is also capable of reacting very rapidly with the bulk $I_2$ reagent to generate more iodine atoms. The complications of these reactions are illustrated by the fact that [RCo(salen)] and other complexes react with ICl to give mixtures of RI and RCl, the ratios of alkyl iodide to alkyl chloride depending on the nature of the alkyl groups and the chelate ligands[70,71]. Care has to be taken in interpreting such results because of the possible interconversion of alkyl iodides and alkyl chlorides *after* reaction. In the reaction of benzyl cobaloximes with bromine in aqueous sulphuric acid, very little benzyl bromide is found; the main products being benzyl alcohol and the *O*-benzyl derivative of dimethylglyoxime[72]. This is probably a reaction in which the solvent plays a key role in directing both rate determining and product-forming processes.

Most other oxidation processes proceed through intermediates that have not yet been characterized. A number of these, akin to equation 46, have been termed 'oxidatively induced nucleophilic substitution processes'. The earliest example was the observation by Johnson and Pearson[73] that optically active organometal—carbonyl complexes were cleaved by reaction with chlorine water, with retention of configuration at the $\alpha$-alkyl carbon, to give optically active carboxylic acids (equation 49)[73]. The same acids and esters are formed when an *acyl*—metal complex is oxidized in aqueous or alcoholic solvents[74]. Oxidation of organometalcarbonyl complexes has since been carried out with a variety of other oxidants. Using cerium(IV) or iridium(IV) as outer sphere oxidants, it seems that the initially formed oxidation product undergoes a rapid carbonyl insertion (or alkyl migration) reaction and then is attacked at the $\alpha$-acyl carbon by the nucleophilic solvent (water or an alcohol) to give the acid or ester and an, as yet, uncharacterized paramagnetic metal complex. However, if the solution contains sufficiently high a concentration of another nucleophile ($X^-$), the initially formed product can be intercepted (equation 50).

$$\text{D-(+)-Bu}^s\text{Br} + [\text{Fp}]^- \longrightarrow [(-)\text{-Bu}^s\text{Fp}] \xrightarrow{\text{Ph}_3\text{P}} [(-)\text{-Bu}^s\text{Fp*}] \xrightarrow[\text{H}_2\text{O}]{\text{Cl}_2} \text{L-(}-\text{)-Bu}^s\text{CO}_2\text{H} \quad (49)$$

$$[(\text{RCH}_2)\text{Mo(CO)}_3(\text{Cp})] \xrightarrow{\text{oxidant}} [(\text{RCH}_2)\text{Mo(CO)}_3(\text{Cp})]^+ \xrightarrow{X^-} \text{RCH}_2\text{X}$$

$$[(\text{RCH}_2\text{CO})\text{Mo(CO)}_3(\text{Cp})] \xrightarrow{\text{oxidant}} [(\text{RCH}_2\text{CO})\text{Mo(CO)}_2(\text{Cp})(\text{solvent})]^+ \xrightarrow{\text{R}'\text{OH}} \text{RCH}_2\text{CO}_2\text{R}' \quad (50)$$

Copper(II)[75,76] and halogens have also been used as oxidants, but each has its own special characteristics. For example, in the oxidation of organoironcarbonyl complexes with three equivalents of $CuX_2$ (X = Br or Cl) in $CH_2Cl_2$, ion pairs dominate the events following the electron transfer, and organic halides are usually the main products; consistent with this is the observation that *R*-(−)-PhCHDFp reacts with complete ($\geqslant$90%) inversion of configuration. Radical products are formed on oxidation of those substrates, such as neopentylmetal complexes, which are reluctant to undergo bimolecular nucleophilic substitution reactions because of steric hindrance of the alkyl ligand. Radical products are also formed more readily when phosphine ligands are present on the metal[75]. The formation of alkyl chlorides by nucleophilic attack on the $\alpha$-carbon of the oxidized complex occurs with inversion of configuration in the reaction of the diastereoisomeric palladium complex **23** with $CuCl_2$/LiCl (equation 51)[76], but radicals are formed in the reactions of many other organopalladium complexes[77]. An interesting exception to the normal course of oxidatively induced carbonyl insertion reactions occurs in the reaction of the iron complex **24** with cerium(IV) in alcoholic solutions (equation 52). The oxidized species is attacked by alkoxide ion both at the $\alpha$-carbon and at the $\beta$-hydrogen atom, the latter leading to the formation of a cyclic vinyl ether (equation 52)[78].

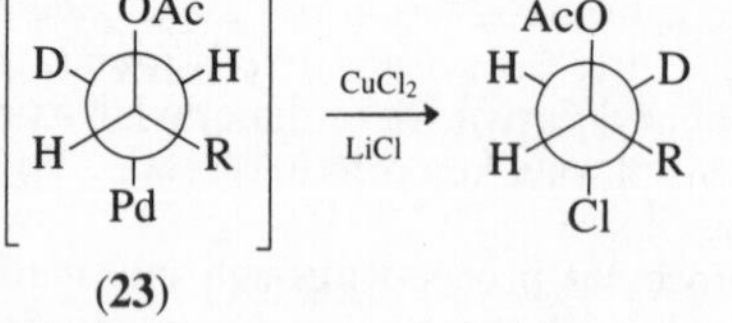

(51)

**(23)**

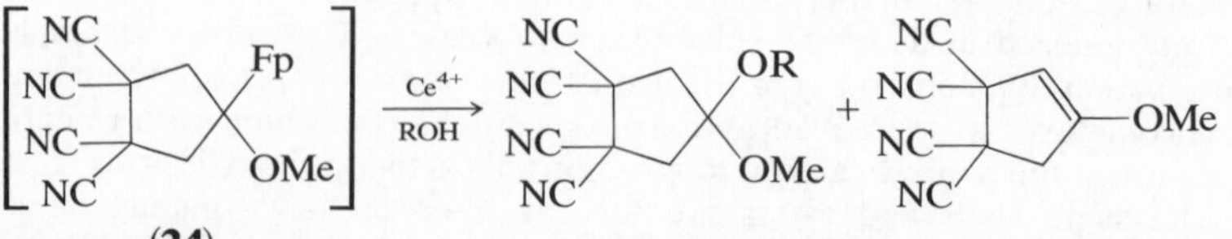

(52)

**(24)**

Oxidation of the anionic complexes $Fe[(RCO)(CO)_4]^-$ by $OCl^-$, oxygen, alkyl halides and halogens has been used for the synthesis of amides by carrying out the oxidation in the presence of amines[79]. The ease of oxidation appears to be in the order $[FeR(CO)_4]^- > [RFp] > [MoR(CO)_3(Cp)] > [WR(CO)_3(Cp)]$.

In many other cases, the evidence for an oxidation process has depended on the observation that the products of reaction neither contain the electrophile nor are the same as those formed in the unaccompanied thermal decomposition reaction. Examples include (i) the decomposition of diethylbis(bipyridyl)iron(II) in the presence of cobalt(III), cerium(IV), copper(II), or bromine, which gives mainly butane, not the mixture of ethylene and ethane formed on heating the substrate to 80 °C in the absence of the electrophile (equation 53)[80]. (ii) The oxidation of the tetramethylaurate(III) ion to tetramethylgold(IV), which subsequently undergoes successive reductive eliminations to give ethane and gold(0) via dimethylgold(II) (equation 54)[81]. The latter is also formed on oxidation of the dimethylaurate(I) ion by the same electrophiles. Reductive elimination also occurs on (iii) oxidation of the diarylnickel(II) complex **25** (equation 55)[82], but (iv) free-radical products are evident on oxidation of the dialkylplatinum(II) complexes $[PtR_2L_2]$, **26** (equation 56). The rate of oxidation

$$[Et_2Fe(bipy)_2] \xrightarrow{\text{oxidant}} [Et_2Fe(bipy)_2]^+ \longrightarrow C_4H_{10} + [Fe(bipy)_2]^+ \quad (53)$$

$$[Me_4Au]^- \xrightarrow[\text{slow}]{\text{oxidant}} [Me_4Au^{IV}] \longrightarrow C_2H_6 + [Me_2Au^{II}] \longrightarrow C_2H_6 + Au^0 \quad (54)$$

$$[Me_2Au]^- \xrightarrow{\text{oxidant}} [Me_2Au^{II}]$$

$$[Ph_2Ni(PEt_3)_2] \longrightarrow [Ph_2Ni^{III}(PEt_3)_2]^+ \longrightarrow PhPh + [Ni(PEt_3)_2]^+ \quad (55)$$

$$[R_2PtL_2] \xrightarrow{[IrCl_6]^{2-}} [R_2Pt^{III}L_2]^+ \longrightarrow [RPt^{II}L_2]^+ + R^{\bullet} \xrightarrow{[IrCl_6]^{2-}} RCl$$

$$[R_2Pt^{III}L_2]^+ \xrightarrow{[IrCl_6]^{2-}} [R_2PtClL_2]^+ \longrightarrow [R_2PtCl_2L_2] \quad (56)$$

$R = Me, Et; L = PPhMe_2, PPh_3$

of **26** by $[IrCl_6]^{2-}$ is faster for $R = Et$ than for $R = Me$ and much faster in each case for $L = PhMe_2P$ than for $Ph_3P$. Two decomposition modes have been detected for the platinum(III) intermediate; (a) homolysis followed by the abstraction of a chlorine atom from $[IrCl_6]^{2+}$ to give RCl, and (b) direct abstraction of a chlorine atom from $[IrCl_6]^{2-}$ by the coordinatively unsaturated paramagnetic oxidation product, followed by coordination of a chloride ion to give the diamagnetic six-coordinate dialkylplatinum(IV) complex $[PtR_2Cl_2L_2]$ (equation 56)[83]. (v) Free radical products, including those from homolytic displacement reactions, are also formed in the oxidation of allyliron complexes by silver(I) ion[84].

The decomposition of $\alpha$-hydroxyalkylpentaaquochromium(III) ions by copper(II) and iron(III) electrophiles takes place through electron transfer following an initial coordination of the electrophile to the hydroxyl oxygen (equation 57). Surprisingly, this oxidation generates chromium(II), which is subsequently rapidly oxidized by the copper(II) or iron(III) species (equation 58)[85]; the main victim of the primary oxidation process being the organic ligand which is converted into an aldehyde. The ease of oxidation of the hydroxyalkyl complex decreases markedly as the $\alpha$-carbon is substituted, and as the acid concentration is increased, through the retardation of complex formation at the hydroxyl oxygen (equation 57). It is also significant that the rate of the acidolysis which accompanies the oxidation reaction is not significantly different from those of other alkylpentaaquochromium(III) ions, which indicates that there is little or no tendency to carbene formation analogous to that shown in equation 37. A one-electron transfer is also apparent in the reaction of mercury(II) with the 1,2-dimethyl-1-hydroxymethylpentaaquochromium(III) ion (equation 60), though transient organomercurials are formed in the reactions of other hydroxyalkylchromium(III) ions (see Section V.A)[85]. In contrast, hydroxymethylcobalt(III) chelate complexes are reported to react with electrophiles by a *two*-electron oxidation process which is not available to the chromium(III) complexes[86] (equation 59).

$$Fe^{3+} + HOCHRCr^{2+} \longrightarrow H^+ + [FeOCHRCr]^{4+} \longrightarrow Fe^{2+} + RCHO + Cr^{2+} \tag{57}$$

$$Cr^{2+} + Fe^{3+} \longrightarrow Cr^{3+} + Fe^{2+} \tag{58}$$

$$[(HOCH_2)Co(N_4\ chelate)]^{2+} \longrightarrow H^+ + HCHO + [Co^I(N_4\ chelate)]^+ \tag{59}$$

## B. Reductively Induced Electrophilic Reactions

Whilst the reaction of an organometallic complex with a reducing agent does not fall directly into the orbit of electrophilic reactions, the transient organometallic complexes thus formed are usually much more susceptible to electrophilic attack than are their precursors, provided that dissociation to free radical products does not occur too rapidly. Thus, carbon dioxide is too weak an electrophile to react with organocobaloxime(III) complexes, but it does so in the presence of dithioerythritol (a highly reducing dithiol) giving the corresponding carboxylic acid (equation 61)[87]. This reaction has not only been used for the preparation of $^{13}C$-labelled amino acids, but is also probably of key importance in the biological formation of acetate from $N^5$-methyltetrahydrofolate via methylcobalamin. Similarly, the biological formation of methane from methylcobalamin, which also requires the presence of thiols, is likely to be due to the attack of even the very weakly electrophilic water molecules (as proton sources) on transient methylcobalamin(II) or methylcobalamin(I) complexes formed on reduction of the substrate by the thiol. This area of organometallic chemistry remains largely unexplored, in part because of the difficulties associated with having oxidizing electrophilic and reducing nucleophilic reagents in the same medium. It is significant, however, that whereas the direct attack of a proton at the $\alpha$-carbon of a stable diamagnetic alkylmetal complex is very rare (see Section VI.C), attack on the $\alpha$-carbon of unstable paramagnetic reduced complexes is very much more common.

$$Hg^{2+} + HOCMe_2Cr^{2+} \longrightarrow H^+ + [HgOCMe_2Cr]^{3+} \longrightarrow Hg^+ + Me_2CO + Cr^{2+} \tag{60}$$

$$\left[\text{phthalimido-}NCH_2Co(dmgH)_2py\right] \xrightarrow[H_2O]{^*CO_2,\ dte} \text{phthalimido-}NCH_2\overset{*}{C}O_2H \tag{61}$$

dte = dithioerythitol; $^*C = {}^{13}C$

# V. ATTACK AT THE METAL

For the majority of organotransition metal complexes with several filled non-bonding *d*-orbitals, the metal is liable to attack by the electrophile. Where the metal is already coordinatively saturated, such attack will usually lead, through a transition state of higher coordination number, either to an unstable intermediate or directly to substitution products. Where the metal is coordinatively unsaturated, the product of attack on the metal is more likely to be isolable. In principle, attack may also be on the carbon—metal $\sigma$-bonding orbital, but if the electrophile is to end up on the metal, this is likely to weaken the carbon—metal bond to such an extent as to cause the expulsion of a formally neutral organic ligand as a cation. Attack at the carbon—metal bonding orbital may result in the electrophile becoming attached to the $\alpha$-carbon, but this pathway is discussed later, in Section VI.

Attack at the metal without synchronous displacement of one of the other ligands is, in principle, a two-electron oxidation process and one may therefore expect it to occur most readily with low-valent complexes having an accessible oxidation states two units higher. The most obvious cases where attack at the metal may be suspected are the coordinatively unsaturated $d^8$ and $d^{10}$ complexes such as the alkyl and aryls of Pt(II), Pd(II), Au(I), Ir(I), and Rh(I); the least likely substrates are the $d^0$ and $d^3$ complexes such as the Ti(IV) and Cr(III) systems. Transient intermediates have, however, been inferred in the reactions of a great many other types of organotransition metal ion.

## A. Reactions of Coordinatively Unsaturated Alkyl and Aryl Complexes

The stability of adducts between organotransition metal complexes and electrophiles is very sensitive to the nature of the electrophile and its counter ion, where applicable, and to the nature of the organic and other ligands on the metal. Reductive elimination from these intermediates is the most likely mode of decomposition; for example, chlorine and thallium(III) chloride react with the $d^{10}$ gold(I) complexes, **27a–c**, (equation 62) to give the corresponding dihalogenogold(III) complexes[88], but other aryl and alkylgold(I) complexes apparently lead directly to the alkyl or aryl halide and the halogenogold(I) complex[89], probably, but not necessarily, by a reductive elimination of the organic halide from a transient gold(III) intermediate. By decreasing the reaction temperature oxidative addition products may be observed, which subsequently decompose when the temperature is raised[90].

$$[AuRL] \xrightarrow[\text{or } TlCl_3]{Cl_2} [R Au^{III} Cl_2 L] \xrightarrow{?} RCl + [AuClL] \longrightarrow [AuCl_3L] \qquad (62)$$

**(27a–c)**

$R = C_6F_5, C_6Cl_5; L = AsPh_3$

Similar problems arise with dialkyl- and diaryl-platinum(II) complexes (equation 63)[91], but the reaction of the bisgold(I) complex **28** with bromine takes a different course, giving the product, **29**, with a gold—gold bond, which can be subsequently cleaved by more bromine to give the expected bisgold(III) adduct (equation 64)[92]. In the halogen cleavage of alkylpalladium(II) complexes, free halide ion formed during the reaction can attack the $\alpha$-carbon of the alkyl group of the intermediate palladium(IV) complex with inversion of configuration at the carbon, thereby displacing a palladium(II) complex in a manner analogous to that shown in equation 51. This reaction therefore competes with the normal oxidative addition, reductive elimination pathway, both stages of which take place with retention of configuration at the metal[93,94].

$$cis\text{-}[PtR_2(PEt_3)_2] \xrightarrow{X_2} [Pt^{IV}X_2R_2(PEt_3)_2] \xrightarrow{?} RX + [PtXR(PEt_3)_2] \qquad (63)$$

$R = Me; X = I$ (direct pathway)

$R = Ph; X = Cl, I$ (via $[Pt^{IV}X_2R_2(PEt_3)_2]$)

$$\left[\begin{array}{c} CH_2AuCH_2 \\ Me_2P \quad\quad PMe_2 \\ CH_2AuCH_2 \end{array}\right] \xrightarrow{Br_2} \left[\begin{array}{c} Br \\ Au \\ Me_2P \quad\quad PMe_2 \\ Au \\ Br \end{array}\right] \xrightarrow{Br_2} \left[\begin{array}{c} Br \\ Au \\ Me_2P \quad Br \quad PMe_2 \\ Br \\ Au \\ Br \end{array}\right] \quad (64)$$

(**28**) (**29**) (**30**)

Alkyl halides, which are normally far too weakly electrophilic to attack an alkyl or an aryl ligand, also add to platinum(II) complexes, but here free radical processes may also obtrude[91]. In the reaction of alkyl halides with alkylgold(I) complexes, a *cis*-dialkylgold(III) adduct is first formed, but this reacts further with the monoalkylgold(I) substrate by a redistribution reaction which gives the trialkylgold(III) complex and a gold(I) halide (equations 65 and 66)[95]. Some dimerization products of the alkyl group are also formed either by a reductive elimination from the *cis*-dialkylgold(III) intermediate, or by free radical processes. As in the addition of halogens to gold(I) complexes, the observed course of reaction is dependent on both R and L[96]. Dialkylaurate(I) ions are among the most reactive species, giving the more stable trialkylgold(III) complexes, the stereochemistry of which depends on the method of preparation, as shown in equations 67 and 68[97].

$$[AuRL] + RI \longrightarrow cis\text{-}[AuIR_2L] \quad (65)$$

$$cis[[AuIR_2L] + [AuRL] \longrightarrow [AuR_3L] + AuIL \quad (66)$$

$$[Me_2Au]Li + EtI \xrightarrow{Ph_3P} trans\text{-}[Me_2(Et)Au(PPh_3)] \quad (67)$$

$$[AuMeR]Li + MeI \xrightarrow{Ph_3P} cis\text{-}[Me_2(Et)Au(PPh_3)] \quad (68)$$

The kinetics of acidolysis of several dialkyl-, diaryl-, and alkylaryl-platinum(II) complexes have been studied in some detail, but the exact mechanism still remains in some doubt. The formation of the monoalkyl or monoaryl complex, each of which is much less readily cleaved, takes place either through the formation of an unseen hydriodoplatinum(IV) intermediate which undergoes a reductive elimination of the alkane or arene, or by a direct attack of the proton at the $\alpha$-aromatic or $\alpha$-aliphatic carbon (equation 69). The *cis*- and *trans*-diphenyl complexes **31a** and **31b** react with

$$[PtR_2(PEt_3)_2] \overset{HX}{\rightleftharpoons} cis\text{-}[PtH(R)_2(S)(PEt_3)_2] \longrightarrow trans\text{-}[PtR(S)(PEt_3)_2] + RH$$

$$[PtR_2(PEt_3)_2] \overset{HCl}{\rightleftharpoons} cis\text{-}[PtClH(R)_2(PEt_3)_2] \longrightarrow cis\text{-}[PtCl(R)(PRt_3)_2] + RH \quad (69)$$

(**31a**) R = aryl, *cis*-isomer
(**31b**) R = aryl, *trans*-isomer
(**31c**) R = alkyl, *cis*-isomer
(**31d**) R = alkyl, *trans*-isomer
$X^-$ = non-coordinating anion

dry HCl in aprotic solvents to give the corresponding *cis*- and *trans*-monophenyl complexes, respectively but the *trans*-monophenyl complex is formed from both *cis*-

and *trans*-diphenyl complexes when the reaction is carried out in methanolic perchloric acid, unless an excess of chloride ion is present in solution, in which case the stereochemistry of the product is the same as that of the reagent[98–100]. The acidolysis of the diaryl complexes is accelerated by the presence of electron-donating groups in the aromatic rings, the Hammett $\rho$-constant being −4.6 for the diarylbis(triethylphosphine)platinum(II) complexes. The rate, which is first order in acid and first order in substrate, is unaffected by the presence of added chloride ion. In contrast, the rate law for the acidolysis of the *cis*-dimethylbis(triethylphosphine)platinum(II) complex contains two terms: the expected second-order term and a term which is also first order in chloride ion. The latter term has been ascribed to a nucleophilic acceleration of the acidolysis through coordination of chloride ion to the metal in the platinum(IV) intermediate (equation 69, path B), the implication being that, in the absence of chloride ion, the intermediate is a solvato complex.

The exact sequence of attack of electrophile and nucleophile in these reactions is unknown, but the chloride ion dependence is probably the best evidence in favour of the platinum(IV) intermediate, since it is difficult to comprehend how the chloride ion could significantly influence the acidolysis to the same extent in any other way. Further support for the formation of platinum(IV) intermediates comes from the observations that the alkyl group is cleaved more readily than the aryl group in a series of *trans*-monoalkylmonoarylbis(triethylphosphine)- and -bis(dimethylphenylphosphine) platinum(II) complexes[101,102]. In contrast, the aryl group is cleaved more readily than the alkyl group in the complex *cis*-[PtMe(*p*-tolyl)(COD)][102], which suggests that the balance between direct attack at carbon and oxidative addition to a platinum(IV) intermediate may be delicate and easily swayed by the nature of the other ligands on the metal, for it is reasonable to assume that direct electrophilic attack at the aromatic $sp^2$ carbon should be easier than direct electrophilic attack at an aliphatic $sp^3$ carbon. However, such a conclusion requires caution because the metal leaving group is not the same in the two cases (equation 70). The protonolysis of *trans*-[PtH($CH_2CN$)($Ph_3P$)$_2$] takes different courses in different media; in $HBF_4$–$Et_2O$ it gives $H_2$ and the Pt—C bond remains intact; in HCl–$Et_2O$–toluene it gives $CH_3CN$ and the Pt—H bond remains intact; whereas with an excess of HCl, both $H_2$ and $CH_3CN$ are formed[103]. This, and the fact that the rate law shows a chloride ion-dependent path, has been taken as evidence for an oxidative–addition, reductive–elimination pathway.

$$[PtR(Ar)(L)_2] \xrightarrow{XCl} RX + [PtClAr(PEt_3)_2]$$
$$[PtR(Ar)(L)_2] \xrightarrow{HCl} ArH + [PtClMe(COCl)] \qquad (70)$$

X = H or D; $L_2 = (PEt_3)_2$ or COCl

No hydridogold(III) complexes have been observed in the corresponding acidolysis of organogold(I) (equation 71) or of dialkylaurate(I) complexes[97], and it is the aryl group that is cleaved in the acidolysis of *cis*-[Au(Me)$_2$(Ph)($PPh_3$)][102], the rate of cleavage being dependent on the nature of the acid in the order $CF_3SO_3H > HNO_3 > CF_3COOH > AcOH$. In trialkylkylgold(III) complexes it is one of the mutually *trans*-alkyl groups which is cleaved. Both di- and mono-alkylpalladium(II) complexes are readily cleaved in aqueous acidic solution[105] and even by very weakly acidic reagents such as PhSH[104], EtOH, and ArC≡CH, in some cases by radical processes.

$$[AuR(PPh_3)] + HX \longrightarrow RH + [AuX(PPh_3)] \qquad (71)$$

R = aryl,alkyl; X = Cl, AcO, $PhCO_2$, $(CF_3)_2C(OH)O$

Similar ambiguities arise with the reactions of monoalkylgold(I) complexes with metal halides. Thus, methyl and ethyl triphenylphosphinegold(I) are cleaved by mercury(II) species, the rate law being first order in each component, the rate decreasing in the order $Hg(NO_3)_2 \gg Hg(OAc)_2 \gg HgI_2 > HgBr_2 > HgCl_2$. An $S_E2$ displacement at carbon was first suggested[106], but the alternative oxidative–addition, reductive–elimination mechanism is perhaps the more likely[95]. Synchronous attack of the mercury on carbon and of the halide ligand on gold is also possible with the mercury(II) halides, especially in aprotic solvents. The mechanisms of reactions of organogold complexes have also been reviewed[107].

$$[AuMe(PPh_3)] + HgCl_2 \longrightarrow MeHgCl + [AuCl(PPh_3)]. \tag{72}$$

## B. Reaction of Alkyl and Aryl Complexes of Coordinatively Saturated 18-Electron Complexes

The evidence for attack at the metal in 18-electron complexes is nearly all indirect and is therefore liable to reassessment in the light of future work. The main problems of interpretation lie first in the fact that the attack at the metal is a two-electron inner-sphere oxidation process, and this may have similar consequences to other electron-transfer processes between substrate and electrophile, and secondly in the similarity of products between oxidative–addition, reductive–elimination pathways and direct displacement at carbon.

The latter is clearly illustrated in the gas-phase reaction of $[MnMe(CO)_5]$ with proton donors (HA), for which two independent pathways have been detected: an attack at Mn leading to transient but identifiable $[MnMe(H)(CO)_5]^+$, which can subsequently lose carbon monoxide to give $[MnMe(H)(CO)_4]^+$ or eliminate methane, or an attack at the C—Mn bond, leading directly to methane. The choice between the paths is determined substantially by the basicity of the species A[108]. In solution, however, this dichotomy has not yet been clearly defined.

### 1. Halogenation

Probably the clearest example of an inferred attack at the metal is in the reaction of electrophiles with 2-phenylethyliron(II) complexes. The 2-phenylethyl ligand, as will be evident below, is unusual in that it accommodates reaction paths that are not available to most other 2-substituted ethyl complexes. For example, (a) the cleavage of the 1,1-dideuteriated-2-phenylethyliron complex **32** with bromine in aprotic solvents such as methylene chloride gives, as the main product, an equimolar mixture of the 1,1- and 2,2-dideuterio-2-phenylethyl bromide (equation 73), whereas

$$\underset{\textbf{(32)}}{[Fp(PhCH_2CD_2)]} \xrightarrow[CDCl_3]{Br_2} PhCH_2CD_2Br + PhCD_2CH_2Br \tag{73}$$

acidolysis of the same complex with trifluoroacetic acid or cleavage with mercury(II) chloride gives only 1,1-dideuteriated products (see Section VI.B)[109]. (b) Halogenation of the *threo*-2-phenyl-1,2-dideuterioethyliron complex **33** with chlorine, bromine, or iodine in a range of aprotic solvents (pentane, acetonitrile, chloroform, nitrobenzene, acetone, methylene chloride) gives the corresponding 2-phenyl-1,2-dideuterioethyl halide with substantial (minimum 63%, maximum 95+3%) retention of configuration. (c) In the reaction of **33** with bromine in the presence of an excess of chloride ion, 2-phenyl-1,2-dideuterioethyl chloride is formed with ≥78% retention of configuration, and in methanol as solvent 2-phenyl-1,2-dideuterioethylmethyl ether is formed with ⩾90% retention of configuration. (d) In the reaction of **33** with iodine chloride, *threo*-2-phenyl-1,2-dideuterioethyl chloride and iodide, predominantly the former, are also formed with substantial retention of configuration. Clearly, the product-forming reaction takes place predominantly by nucleophilic attack on some intermediate in which carbon-1 and carbon-2 are equivalent and shielded in such a way as to ensure retention of configuration. The special feature of the 2-phenylethyl ligand thus appears to be its ability to form spiro-phenonium species, such as the cation **34**, either detached from the metal or in some way still associated with the metal, which fulfils the above criteria[110] (equation 74).

(74)

**(33)** **(34)**

$XY = I_2, Br_2, Cl_2, ICl$

$Z^- = I^-, Br^-, Cl^-$

In contrast, the corresponding halogen cleavage of *cis*-[Mn(*threo*-PhCHDCHD)$(CO)_4(PEt_3)$] is distinctly less stereospecific with some degree of retention in all cases. The reaction with iodine chloride, however, gives mainly *erythro*-PhCHDCHDI compared with mainly the *threo*-chloride from the iron complex. Using $^{13}C$ labelling of the ethyl carbons, no scrambling of carbon-1 and carbon-2 could be detected[111] and it has been suggested that more than one mechanism, including free-radical mechanisms, may operate in this case.

The *erythro*-1,2-dideuterio-3,3-dimethylbutyliron complex **35**, which cannot give the corresponding spiro-cation, reacts with bromine in a variety of aprotic solvents, with iodine in carbon disulphide, and with chlorine in chloroform, to give the corresponding *threo*-halide, probably through an $S_N2$ displacement at carbon in the halogenated intermediate by halide ion with inversion of configuration (equation 75,

(75)

**(35)**

path A)[112,113]. When the same reactions are carried out in methanol, the ester $Me_3CCHDCHDCO_2Me$ is formed with retention of configuration at the $\alpha$-carbon, through nucleophilic attack of the solvent on an intermediate formed by oxidation (either a one-electron transfer as in Section IV.A, or a two-electron transfer through attack of the electrophile on the metal as in equation 75, path B) and alkyl migration.

That attack should be at the metal is not unreasonable in view of the fact that the photoelectron spectra of the corresponding methyl, $\eta^1$-allyl-, and $\eta^1$-cyclopentadienyliron complexes show that the three lowest ionization potentials (7.9–8.6 eV) are all associated with the loss of non-bonding $d$-electrons[114,115].

The stereochemistry of cleavage of the carbon—metal bond has also been elegantly determined for the iron centre. Fortunately, the chiral iron centre is configurationally stable in cyclopentadienyl, carbonyl, and phosphine alkyls and studies analogous to those at the saturated carbon can be made, i.e. using either enantiomeric or diastereoisomeric molecules, the former being preferable because of the absence of any possibility of asymmetric induction at the metal by any other chiral centre in the molecule. The earliest studies concerned the iodine cleavage of the diastereoisomerically pure complex **36**[116]. With a deficiency of iodine, both the product iodoiron complex **38** and the unreacted complex **36** were found to be only partially epimerized. It was suggested that, for the partial epimerization of **36**, a reversibly formed adduct such as **37** must be formed, which by square prism/trigonal bipyramid interconversion could racemize at a rate less than that at which it was converted, by attack of iodide at the $\alpha$-carbon, into products. The greater degree of epimerization of the product than of the reagent could be explained if some of the product was also formed by a reductive elimination from the intermediate **37** to a further intermediate of lower coordination number having only a limited degree of configurational stability.

In confirmation of these results, Flood *et al.*[117] established the absolute configuration of a number of complexes of the type (*S*)-**39**, where X = $CH_2O$-menthyl, $CH_2COO$-menthyl, alkyl (several), $PhCH_2$, and methyl, and they confirmed the observations relating to equation 76 with the methyliron complex. For example, a

$$\mathbf{(36)} + I_2 \underset{k_{-1}}{\overset{k_1}{\rightleftharpoons}} \mathbf{(37)} \xrightarrow{-MeI} \mathbf{(37a)}; \quad \mathbf{(37)} \xrightarrow[k_2]{{}^{*}I^-} \mathbf{(38)} + Me^{*}I; \quad \mathbf{(37a)} \xrightarrow{I^-} \mathbf{(38)} \qquad (76)$$

(36): [Ph–C₅H₃(Me)]Fe(CO)(PPh₃)Me; (37): [[Ph–C₅H₃(Me)]Fe(CO)(PPh₃)(I)Me]⁺; (37a): [[Ph–C₅H₃(Me)]Fe(CO)(PPh₃)]⁺; (38): [Ph–C₅H₃(Me)]Fe(CO)(PPh₃)I

very high enantiomeric excess of the iodoiron complex (94%) was obtained in the reaction of ($S$)-**39** (R = Me) with a two-fold excess of $I_2$ in the presence of a 10-fold excess of P($p$-tolyl)$_3$ at 15 °C, but the erratic nature of the results for the iodine cleavage suggests that there may be competition between heterolytic processes of high stereospecificity and non-stereospecific radical processes[117]. Similar studies on iron and other metal complexes have been carried out by Brünner and Wallner[118].

### 2. Acidolysis

In the corresponding acidolysis of ($S$)-**39** (R = Me) with trifluoroacetic acid, a moderate decrease of enantiomeric purity of the substrate accompanies the formation of alkane. The inorganic product also retains some enantiomeric excess, implying net retention or inversion at the metal. A chiral shift reagent was used to determine the enantiomeric purity in this case. However, it is clear that the alkane is not formed by nucleophilic attack at the $\alpha$-carbon of the cationic intermediate but by a reductive elimination to give an intermediate of lower coordination number (equation 77) similar to that postulated in equation 76, but with an undoubted

$$[\text{CpFe(CO)(PPh}_3\text{)R}]\ [(S)\text{-}\mathbf{39}] \underset{}{\overset{\text{CF}_3\text{COOH}}{\rightleftharpoons}} [\text{CpFe(CO)(H)(PPh}_3\text{)R}]^+\ \text{CF}_3\text{COO}^- \longrightarrow [\text{CpFe(CO)(PR}_3\text{)}]^+\ \text{CF}_3\text{COO}^- \longrightarrow [\text{CpFe(CO)(PPh}_3\text{)OCOCF}_3] \quad (77)$$

degree of configurational stability. It is clear, however, that whereas the capture of the intermediate **37a** by iodide ion may be influenced by the asymmetry of the cyclopentadienyl ligand, the capture of the corresponding intermediate in equation 77 by trifluoroacetate ion will be free from asymmetric induction.

However, detailed studies of the kinetics of reaction of a wide range of organoiron complexes with trifluoroacetic acid suggest that the simple view of the reaction of 1 mol of acid with 1 mol of substrate may be erroneous[119]. This is perhaps not surprising in view of the non-polar solvents used for most of these studies, and the likely role of ion pairs and higher agglomerates. The situation is further complicated by the fact that the reaction of [PhCH$_2$Fp] with $CF_3COOH$ and $CF_3COOD$ is dramatically influenced by oxygen, probably through oxidation of the transient hydroiron intermediate[120].

The ease of reaction of the complex [RFp] decreases in the order R = Ph, $Me_3SiCH_2$, Me > Bu > Et > $Me_3CCH_2$ > $PhCH_2CH_2$ > PhCHMe ≈ $PhCH_2$ ≫ $Me_2CH$, an order which is surprising and certainly not consistent with attack at the $\alpha$-carbon, but probably indicative of more than one mechanism over the whole reactivity range. In the reaction of the benzyliron complex with $CF_3COOD$, the cleavage is accompanied by ring deuteriation (Section I), but there is also a very high solvent/reagent deuterium kinetic isotope effect ($k_H/k_D \approx 14$), which has been considered to be a further indication of attack at the metal[120]. The cleavage of [arylFp] complexes by

$CF_3COOH$ shows a fairly large negative $\rho$ value, and the parallels with the cleavage of the diarylplatinum(II) complexes described in Section IV.A seem marked.

### *3. Cleavage by metal ions*

The problems associated with the reactions of metal halides and organotransition metal complexes are an amalgam of those found with halogenation and acidolysis. For example, the kinetics of reaction of several [RFp] complexes with mercury(II) chloride in aprotic solvents are first order in substrate, but of mixed order in the mercury(II) halide. Moreover, at least three distinct reaction paths have been detected (paths A–C; equation 78)[121]. The observation of high orders of reaction in non-aqueous solvents is not surprising because of the probable pre-association of reagents, but the choice of pathway is clearly a function of the character of the organic ligand; path A is followed where the group R is primary alkyl or aryl; path B is followed where R is a secondary alkyl or benzyl. It is possible that paths A and B proceed through a common intermediate formed by attack of the metal salt at the iron, so that reductive elimination leads to path A and a nucleophilic displacement by halide ions leads to path B. Path C, which is favoured when there are electron donating groups on the $\alpha$-carbon of the organic ligand, is clearly an oxidative pathway and may resemble the reactions shown in equation 56.

$$[RFp] \longrightarrow \begin{cases} \xrightarrow{A} RHgX + [FpX] \\ \xrightarrow{B} RX + [FpXHg] \\ \xrightarrow{C} Hg_2X_2 + \text{products} \end{cases} \qquad (78)$$

Studies of the stereochemistry of the formation of the alkylmercurial from $HgI_2$ and (*S*)-**39** (R = Me) in benzene indicate that a reaction analogous to equation 76 is applicable for those reactions leading to the organomercurial, some internal return to substrate being demonstrated by its loss of enantiomeric purity during reaction. However, as in the reaction of (*S*)-**39** with iodine, the halogeno product also loses enantiomeric purity in the presence of mercury(II) iodide[117]. The stereochemistry of the displacement at carbon depends on the nature of the complex. Thus, diastereoisomeric 2-phenyl-1,2-dideuterioethyl- and 1,2-dideuterio-3,3-dimethylbutyl-iron complexes both undergo cleavage by mercury(II) chloride in acetone or methylene chloride with retention of configuration at the $\alpha$-carbon (equation 79)[122,123]; *trans*-(*threo*-2-phenyl-1,2-dideuterioethyl)dicarbonyl(triethylphosphine)-cyclopentadienyltungsten also reacts with retention, but *cis*-(*threo*-2-phenyl-1,2-dideuterioethyl)tetracarbonylcyclopentadienylmanganese(I) reacts with substantial inversion of configuration at the $\alpha$-carbon. The latter reaction suggests that a fourth path, the direct attack at the $\alpha$-carbon, may also be possible (see Section VI).

[Newman projection: R; H, D; H, D; M] $\xrightarrow{HgCl_2}$ [Newman projection: R; H, D; H, D; HgCl] or [Newman projection: R; D, H; H, D; HgCl] (79)

M = Fe, W    M = Mn

R = Ph, $Bu^t$
M = Fp, $W(CO)_2(PEt_3)(Cp)$-*trans*
or $Mn(CO)_4(PEt_3)$-*cis*

No mercury—metal intermediates have been observed in the above reactions, but in the reaction of the molybdenum complex **40** with $Hg^{2+}$ in aqueous acid an intermediate is evident from the changes of electronic spectra during the course of reaction. Unfortunately, the kinetics of reaction do not allow us to distinguish between an intermediate on the reaction coordinate (equation 80, path A) and an addition complex formed reversibly but in competition with the reaction leading to the organomercurial (equation 80, path B), nor do they allow us to ascertain with certainty the nature of the intermediate[124].

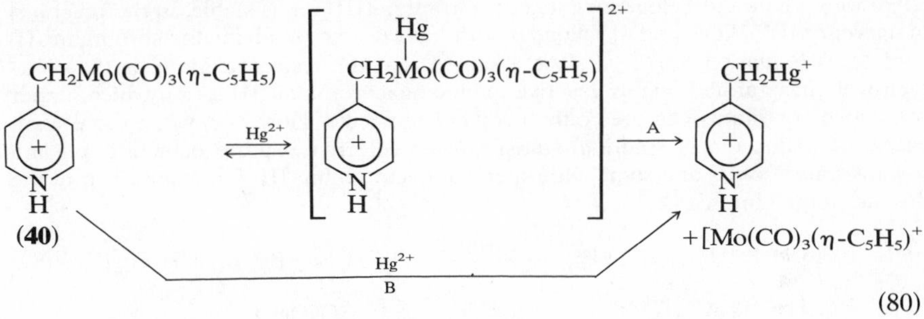


(80)

## VI. ATTACK AT SATURATED CARBON BY THE ELECTROPHILE

Electrophilic attack on carbon, i.e. at the filled carbon—metal bonding orbital, either from the rear, leading to inversion of configuration at carbon (equation 81, path A), or from the front, leading to retention of configuration (equation 81, path B), are both allowed processes in which the transition state contains a three-centre, two-electron orbital. A third type of attack in which there is synchronous attachment of the electrophile and its nucleophilic ligand on the $\alpha$-carbon and the metal, respectively, is also possible (equation 81, path C). A variety of symbols have been used to

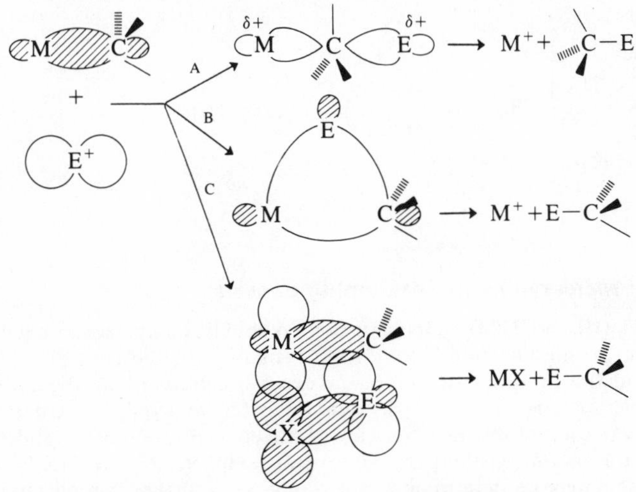


(81)

classify such reactions[110,125] but we prefer not to use them in this review, in part because of the difficulties in deciding between paths B and C in appropriate cases.

The cleanest examples of electrophilic attack at carbon occur with the organopentaaquachromium(III) ions which are (i) coordinatively saturated, (ii) substitution inert (with respect to exchange of the water ligands), (iii) not prone to oxidation, (iv) contain no fully occupied non-bonding *d*-orbitals, and (v) have a high formal positive charge on the metal. These complexes react in aqueous acidic solution (these cationic complexes are neither particularly soluble in other solvents nor stable in neutral aqueous solution) with a wide range of electrophilic reagents, including Tl(III) salts, Hg(II) salts, Hg(I) salts, $NO^+$, NOCl, $Cl_2$, $Br_2$, $I_2$, IBr, and $H_3O^+$. That the above reactions occur through attack at carbon by path A or B and not by path C is evident from the formation of only the hexaaquachromium(III) ion and halide ion, under conditions where the halogenopentaaquachromium(III) ion is stable, in the reactions of mercury(II) halides and of halogens with benzyl- and pyridylmethylchromium(III) complexes (equation 82)[126–130]. It is assumed that the displaced group is the coordinatively unsaturated, highly reactive, pentaaquachromium(III) cation which, under the above conditions captures only a solvent molecule. However, when there is an excess of halide ion in solution, statistical non-selective capture of halide ion and solvent water occurs and some chloropentaaquachromium(III) ion is observed in the product (equation 83)[131].

$$[(PhCH_2)Cr(H_2O)_5]^{2+} + Br_2 \xrightarrow{\text{aq. HClO}_4} PhCH_2Br + Br^- + [Cr(H_2O)_6]^{3+} \quad (82)$$

$$\left[ (3\text{-}C_5H_4NH^+)CH_2Cr(H_2O)_5 \right]^{2+} + HgCl_2 \xrightarrow{\text{aq. HCl}} (3\text{-}C_5H_4NH^+)CH_2HgCl + [Cr(H_2O)_5Cl]^{2+} + [Cr(H_2O)_6]^{3+} \quad (83)$$

The other good candidates for attack at the $\alpha$-carbon are the alkylzirconium(IV) complexes which, being $d^0$ are closest in analogy to the main group metal alkyls. Indeed, the 2-phenyl-1,2-dideuterioethylzirconium(IV) complex **41** reacts with bromine with retention of configuration in a similar manner to the tin(IV) alkyls (equation 84)[132].

$$\left[ Bu^tCHD\text{–}CHD\,ZrClCp_2 \right] \; (\mathbf{41}) + Br_2 \longrightarrow Bu^tCHD\text{–}CHDBr + [ZrBrClCp_2] \quad (84)$$

## A. Reactions with Mercury(II) and Thallium(III) Species

The reaction of Hg(II) and Tl(III) salts with organometallic complexes in aqueous solutions are subject to modification through the formation of the species $M^{n+}$ to $MX_4^{m-}$, especially in the presence of limited or excess amounts of the ligand $X^-$. Each of these species has its own reactivity (equation 85) and only by a detailed examination of the variation of rate as a function of the concentration of $X^-$ and with a knowledge of the formation constants for the several complexes $M^{n+}$ to $MX_4^{m-}$ can these individual reactivities be determined. The situation is further complicated by the partial hydrolysis of the complexes $M^{n+}$ in other than strongly acidic aqueous solution. For example, $Tl^{3+}_{aq.}$ is partially hydrolysed to $Tl(OH)^{2+}$, and $Tl(OH)_2^+$ even in 0.1 M $HClO_4$ and the reactivity of these species must also be taken into account. Clearly, such mechanistic subtleties need to be investigated and only very qualitative

$$RCH_2M + M'Cl_n^{m+} \longrightarrow RCH_2M'Cl_n^{(m-1)+} + M^+ \tag{85}$$
$$M' = Tl, Hg;\ n = 0\text{–}4;\ m = 3,2,1,0,-1,-2$$

(**42**) $M = Mn(CO)_5$; $R = Hpy^+$
(**43**) $M = Fp$; $R = Hpy^+$
(**44**) $M = Co(CN)_5^{3-}$; $R = PhCH_2$

$$\left[HN^{+}C_5H_4\text{–}CH_2Cr(H_2O)_5\right]^{2+} \xrightarrow{TlCl_2^+} HN^{+}C_5H_4\text{–}CH_2TlCl_2 + [Cr(H_2O)_6]^{3+}$$
$$\left[HN^{+}C_5H_4\text{–}CH_2Cr(H_2O)_5\right]^{2+} \xrightarrow{Hg_2^{2+}} HN^{+}C_5H_4\text{–}CH_2Hg^+ + Hg(0) + [Cr(H_2O)_6]^{3+} \tag{86}$$

conclusions can be drawn from both product and kinetic studies carried out under single sets of conditions, especially when it is noted that the rate of reaction of the manganese complex **42** with $Hg^{2+}$ is about seven orders of magnitude faster than with $[HgCl_4]^{2-}$ (see Figure 1).

The reaction of thallium(III) halides with the pyridylmethylchromium(III) ion (equation 86) provides one of the only known routes to monoalkylthallium(III) halides, but also, where the reagent is $Tl^{3+}$, it has a rather remarkable transition state of formal charge 6+. It is not surprising, therefore, that the rates of reaction of $Tl^{3+}$ and of $TlCl_4^-$ with this substrate are similar in magnitude[133]. The same substrate reacts directly with $Hg_2^{2+}{}_{aq.}$ to give the organomercury(II) complex, the hexaaquachromium(III) ion, and mercury(0). The mercury begins to precipitate during the reaction, but its electronic spectrum ($\lambda_{max.}$ 252 nm) can be observed in the early stages of the reaction[135]. However, mercury(I) will only react as an electrophile in its own right if its reactivity towards the substrate is more than *ca.* 1/65th of that of the corresponding $Hg^{2+}{}_{aq.}$ ion at ambient temperature. Therefore, only the most reactive organometallic complexes achieve this ratio and most other complexes, such as **42–44**, react through the small amount of mercury(II) that is always present in solutions of mercury(I) (equation 87). The kinetics of reaction of mercury(II) and

$$Hg_2^{2+} \xrightleftharpoons{\text{fast}} Hg^{2+} + Hg^0 \tag{87}$$

thallium(III) species with compounds **42–44** have also been studied in some detail in aqueous solution. Whilst no stereochemical studies have been carried out, the similarity of behaviour with that of the organochromium(III) ions, after due allowance for coulombic repulsions between the more highly charged species, indicates that these reactions take place by attack of the electrophile at the $\alpha$-carbon of each substrate. The variation of rate coefficients for some of these reactions as a function of the halide ion concentration is shown in Figure 1[136,137].

Mercury(0) is also formed in the reactions of several $\alpha$-hydroxyalkylchromium(III) ions with $Hg^{2+}$ [138]. Either the initially formed organomercury(II) product is unstable and collapses to the aldehyde and mercury(0) (equation 88) or a two-electron oxidation occurs giving the aldehyde and mercury(0) directly.

Reactions of organocobalt(chelate) complexes with mercury(II) species have been studied by a number of workers because of the probability that methylcobalamin plays a key role in the biological methylation of mercury(II) to the much more toxic methylmercurials in inland and coastal waters close to industrial estates in many

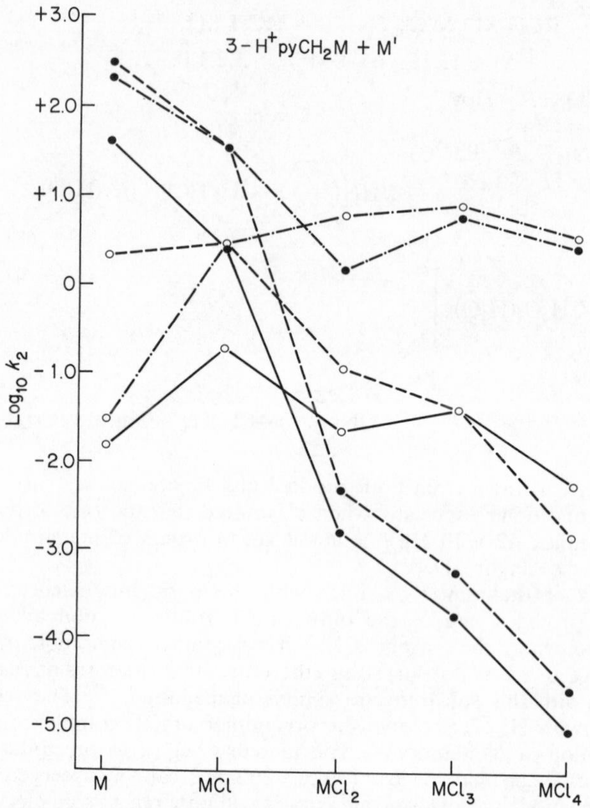


FIGURE 1. Calculated rate constants for the reaction of mercury(II) and thallium(III) species $M^{n+}$, $MCl^{(n-1)+}$, $MCl_2^{(n-2)+}$, $MCl_3^{(n-3)+}$, and $MCl_4^{(n-4)+}$ (charges omitted) with 3-pyridiomethylpentaaquachromium(III) ion (—·—·—), 3-pyridiomethyldicarbonyl-$\eta^5$-cyclopentadienyliron(II) (– – – –), and 3-pyridiomethylpentacarbonylmanganese(I) (———). ○, Tl(III) species; ●, Hg(II) species.

parts of the world[139]. The mercury wastes frequently come from chloralkali plants and the cobalamins from human wastes in the form of sewage.

$$[(H_2O)_5CrCHR^1OR^2]^{2+} + Hg^{2+} \xrightarrow{R^1=H;\ R^2=Me} [Cr(H_2O)_6]^{3+} + R^2OCHR^1Hg \rightarrow R^1CHO + R^2OH + H^+ + Hg^0$$

$$[(H_2O)_5CrCHR^1OR^2]^{2+} + Hg^{2+} \xrightarrow{R^1=Me;\ R^2=H,\ Me} [Cr(H_2O)_6]^{3+} + R^1CHO + R^2OH + Hg^0 \quad (88)$$

The reaction of methylcobalamin with $Hg^{2+}$ is complicated by the concurrent interaction of the latter and of acid with the sixth (axial) ligand (equation 3, Section I), but the process leading to methylmercury(II) is believed to be a bimolecular attack at the methyl carbon in the 'base on' form of the complex (equation 89)[140–150]. Studies of a range of other organocobalt(chelate) complexes suggest that the reactivity decreases as the organic ligand changes in the order Me ≈ $PhCH_2$ > Et > Pr, which

$$[\text{Me–Co–N}] + Hg^{2+} \longrightarrow [\text{Co}^+\text{–N}] + MeHg^+ \quad (89)$$

should be compared with the order shown in Section IV.B for attack at iron, and that a positive charge on the macrocyclic ligand has an adverse effect on the rate. In almost all cases, $CH_3Hg^+$ reacts at least two orders of magnitude more slowly than $Hg^{2+}$ [147–155]. The reactions of the chromium complex $[CrR([15]aneN_4)(H_2O)]^{2+}$ where R = Me, Et, $PhCH_2$, etc., with $Hg^{2+}$ and with $MeHg^+$ shows similar features[156]. The dimethylcobalt(III) chelate complexes tend to react much more readily than the monomethyl complexes, the first methyl group being displaced by $Hg^{2+}$, $MeHg^+$, or $PhHg^+$ much more readily than the second (equation 90)[142,146], indicative of a strong *trans*-influence of the methyl group in these octahedral complexes. The first methyl group is also displaced much more readily than subsequent methyl groups in the reactions of the trimethylgold(III) complexes with $HgCl_2$ (equation 91)[106]. Since this reaction gives exclusively the *cis*-dialkylgold(III) product, a *trans*-influence is also apparent in the square planar complexes.

$$[\text{R–M(N}_4\text{)–X}]^{n+} + M' \xrightarrow{H_2O} RM' + [H_2O\text{–M(N}_4\text{)–X}]^{(n+1)+} \xrightarrow[(X=R)]{M'} [H_2O\text{–M(N}_4\text{)–}OH_2]^{(n+2)+} \quad (90)$$

(**45**) M = Cr; $n$ = 2; X = $H_2O$; M′ = $MeHg^+$
(**46**) M = Co; $n$ = 1; X = R; M′ = $Hg^{2+}$

$$[(Me)_3Au(PPh_3)] + HgCl_2 \longrightarrow MeHgCl + cis\text{-}[(Me)_2AuCl(PPh_3)] \quad (91)$$

The displacement of cobalt(III) from organocobaloximes has been shown to take place with inversion of configuration (cf. retention in the corresponding organoiron complexes in Section IV.B) at the $\alpha$-carbon by the use of 2-phenyl-1,2-dideuterioethyl- and 1,2-dideuterio-3,3-dimethylbutylcobaloximes[157,158]. There are certain side reactions which take place with secondary alkylcobaloximes, but these are probably because electron-transfer processes arise when the displacement reaction is sterically hindered.

### B. Reaction With Other Metal Electrophiles

The transfer of alkyl, particularly methyl, groups from organocobalt(III) chelates to inorganic cobalt(III) complexes is probably also important in biological methylcobalamin chemistry. This reaction has also been carried out between synthetic methylcobalt(III) complexes and various cobalt(III) macrocyclic complexes (equation

92). One of the problems with the study of this reaction is that the corresponding homolytic transfer of methyl groups from methylcobalt(III) complexes to inorganic cobalt(II) complexes is very much faster, so that small traces of cobalt(II) in solution can make a vast difference to the rate of the transfer through interconversion between inorganic cobalt(III) and cobalt(II) complexes and methyl transfer to the latter[155,159]. Although rate studies of this reaction must be considered with caution, equilibrium studies should largely unaffected by this catalysis. From these equilibrium studies it is apparent that the order of receptor capability of the inorganic cobalt(III) complexes is $Co(dotnH)_{aq.}^{2+}$ > cobalamin(III) > Co(salophen) > Co(salen) > Co(7,7′-dimethylsalen) > Co(acacen) > $MeCo(dotnH)^+$, but this order can be influenced by the presence of axial ligands other than water. The above order applies in reverse to the ligand-donor capability of the methylcobalt(III) complexes; for example the dimethyl complex $[CoMe_2(dotnH)]$ is the most effective donor of methyl groups (equation 93), and the corresponding monomethyl complex $[CoMe(dotnH)_{aq.}^{+}]$ is the least effective. The factors controlling methyl transfer have been discussed in detail[160].

$$[MeCo(dmgH)_2L] + [Co^{III}(chelate)L]^+ \rightleftharpoons [MeCo(chelate)L] + [Co(dmgH)_2L]^+ \quad (92)$$

$$[(Me)_2Co(dotnH)] + [PhCo(dotnh)]^+ \rightleftharpoons [MeCo(Ph)(dotnH)] + [Co(dotnH)]^+ \quad (93)$$

The cleavage of the carbon—gold bond of $[AuMe(PPh_3)]$ by $[AuMe_2(I)(PPh_3)]$, to give $[AuMe_3(PPh_3)]$ and $[AuI(PPh_3)]$, takes place by a synchronous transfer of the methyl group and the iodine atom (equation 94). The transfer of methyl groups from

$$cis\text{-}[(Me)_2AuI(PPh_3)] + [(CD_3)Au(PPh_3)] \rightleftharpoons \left[ (Me)_2(PPh_3)Au \underset{\underset{D_3}{C}}{\overset{I}{\diamond}} Au(PPh_3) \right] \rightarrow \left[ Me\text{—}\underset{\underset{CD_3}{|}}{\overset{\overset{Me}{|}}{Au}}\text{—}PPh_3 \right] + [IAu(PPh_3)] \quad (94)^{161}$$

gold(I)[88e] and cobalt(III)[162] to palladium(II) has also been reported, but the mechanisms may not fall into this category. One problem with these reactions is that the organopalladium products are unstable and much of the information comes from kinetics and subsequent products only. The transfer of methyl groups from cobalt(III) to gold(I)[88] and to platinum(II)[88,163,164] are interesting because of the need for gold(III) or platinum(IV) catalysts, respectively. These may therefore be oxidatively induced nucleophilic displacements, or reductively induced electrophilic displacements, of the types described in Sections III.A and III.B, respectively. The transfer of organic ligands from cobalt(III) to palladium(II) and thence to organic substrates have been used in organic synthesis[165,166]. Transfer of alkyl groups from cobalt(III) to lead(IV)[167], cadmium(II)[168], lead(II)[168], and zinc(II)[168] and copper (II)[165] have also been reported as electrophilic displacements at the $\alpha$-carbon.

### C. Acidolysis

Attack of a proton at the $\alpha$-carbon does not appear to be a favoured process, except when the complex is activated by prior reduction (Section IV.B), and despite the high reactivity of many other electrophiles. The gas-phase reaction of $[MnMe(CO)_5]$ described earlier (Section VI.B) is one of the few unambiguous examples of attack at the $\alpha$-carbon; in solution chemistry much of the evidence is indirect. Thus, the acidolysis of several alkylpentaaquachromium(III) ions in aqueous acid follows the rate law[169] and it has been suggested that the second term refers to a

$$\text{Rate} = k_1[\text{substrate}] + k_2[\text{substrate}][H_3O^+]$$

proton attack at the $\alpha$-carbon with some assistance through hydrogen bonding to the coordinated water molecules. On the other hand, the acidolysis of the pyridylmethylmanganese complex **47** in 0.1–4 M perchloric acid takes place at a rate which is independent of the acid concentration (equation 95)[70]. In complete contrast, the pyridylmethylpentaaquachromium(III) ions shown in equations 83 and 86 do not undergo acidolysis in 4 M perchloric acid, but decompose through homolysis of the carbon—chromium bond. Only in more basic solutions do these complexes undergo an acidolysis, probably through a unimolecular dissociation of a species in which several of the water ligands have been replaced by hydroxide[71]. The mechanism of acidolysis of trialkylgold(III) complexes is not known, except that dialkylgold(III) compounds are formed with retention of configuration at the metal[96].

$$[4\text{-}(CH_2Mn(CO)_5)C_5H_4NH^+] \underset{\text{slow}}{\overset{H_2O}{\rightleftharpoons}} [Mn(CO)_5]^+ + CH_2{=}C_5H_4{=}NH \xrightarrow[\text{fast}]{H^+} [4\text{-}CH_3C_5H_4NH^+] \quad (95)$$

(**47**)

### D. Reactions with $NO^+$ and NOCl

Mildly acidic solutions of nitrous acid, with or without added halide ion, react with organochromium(III) and with some organopentacyanocobaltate(III) ions (equation 96) according to the rate law the first term of which is consistent with an attack of

$$\text{Rate} = k_3[\text{substrate}][H_3O^+][HNO_2] + k_4[\text{substrate}][H_3O^+][HNO_2][Cl^-]$$

$NO^+$, the second with attack of NOCl, on the substrate. The product of the reaction depends on the conditions, being either an oxime in more dilute solution or a dimer in more concentrated solution. Both products are, however, consistent with a direct attack of the nitrosating species at the $\alpha$-carbon with the formation of a monomeric nitrosoalkane (equation 96)[172].

$$[Co(ArCH_2)(CN)_5]^{3-} + NO^+ \xrightarrow{H_2O} [Co(CN)_5(H_2O)]^{2-} + ArCH_2NO \rightarrow ArCH{=}NOH \text{ or } \tfrac{1}{2}[ArCH_2NO]_2 \quad (96)$$

Ar = Ph, $Hpy^+$

## E. Other Electrophiles

An unusual reaction is the attack of the trityl cation, $Ph_3C^+$, on the trimethyltantalum complex **48**. The trityl cation is normally associated with hydride extraction, but in this case it becomes attached to carbon, either through a direct attack at the $\alpha$-carbon or through an oxidative–addition, reductive–elimination process. It is the central methyl group which is lost and the kinetic isotope ratio $k_H/k_D$ is 3.4[173].

$$Ph_3C^+ + \underset{(\mathbf{48})}{[Ta(Me)_3(Cp_2)]} \longrightarrow Ph_3CMe + [Ta(Me)_2Cp_2]^+ \xrightarrow{\text{base}} [Ta({=}CH_2)(Me)Cp_2] \quad (97)$$

# VII. ATTACK AT VINYLIC CARBON

Attack on vinyl complexes presents a number of mechanistic problems, not only because of the difficulties of distinguishing between attack on the metal and attack at the $\alpha$-carbon, but also because of the several consequences of attack at the relatively nucleophilic $\alpha$- and $\beta$-carbons. As shown in equation 98, attack of the electrophile at the $\alpha$-carbon can lead to a carbocation intermediate, **49**, which can either lose the

*cis*-RCH=CHX *trans*-RCH=CHX

$\uparrow -M^+$ $\uparrow -M^+$

*cis*-RCH=CHM $\xrightarrow{X^+}$ (**49**) R(H)C$^+$–C(M)(H)X $\rightleftharpoons$ (**50**) H(R)C$^+$–C(M)(H)X $\xrightarrow{-X^+}$ *trans*-RCH=CHM (98)

*cis*-RCH=CHM $\xrightarrow{X^+}$ (**51**) X(R)(H)C–C(H)=M$^+$ $\rightleftharpoons$ (**52**) X(H)(R)C–C(H)=M$^+$ $\xrightarrow{-X^+}$ *trans*-RCH=CHM

metal rapidly to give a vinyl compound with retention of configuration, or can undergo complete or partial rotation about the $C_{(1)}$—$C_{(2)}$ bond to give another carbocation, **50,** which can lose the metal to give the vinyl compound of opposite stereochemistry. The reversion of **50** to a vinyl metal complex thus provides a means of isomerization of the substrate. Alternatively, attack of the electrophile at the $\beta$-carbon can lead to a carbenoid intermediate **51** which can also undergo rotation about the $C_{(1)}$—$C_{(2)}$ bond to give a second carbenoid intermediate **52**. Reversion of **52** to a vinylmetal complex thus provides a further means of isomerization of the substrate. We may therefore expect a whole spectrum of properties within this framework, and the importance of observing the character of the substrate during the course of reactions in which products of mixed or different stereochemistry are formed cannot be overstressed.

Thus, in the reaction of halogens and mercury(II) acetate with both *cis*- and *trans*-styrylcobaloximes, very high yields of the corresponding *cis*- and *trans*-styryl products are formed without any apparent isomerization of the substrates. Since these substrates are unlikely to react through attack at the metal, a direct attack at the $\alpha$-carbon, with synchronous or near synchronous stereospecific expulsion of the cobaloxime group, is probable. Since the *cis*-styrylcobaloxime **53** is readily synthesized in good yield from phenylacetylene, equation 99 provides a useful synthesis of *cis*-styryl halides[174].

$$PhC{\equiv}CH + [Co(dmgH)_2py]^- \xrightarrow[\substack{ph>11\\75\%}]{MeOH} cis\text{-}[PhCH{=}CHCo(dmgH)_2py] \; (\mathbf{53}) \xrightarrow{95\%,\; X_2} cis\text{-}PhCH{=}CHX + [XCo(dmgH)_2py] \quad (99)$$

X = Cl, Br, I.

On the other hand, several other substituted vinylcobaloximes do not react stereospecifically with halogens and this may be indicative of long-lived carbocation intermediates or of the intrusion of free radical processes. Most of the other apparent displacements at the $\alpha$-carbon are of uncertain mechanism. These include the acidolysis of the vinylgold(I)gold(III) complex **54** (equation 100)[176] and the mercuration of styryltriphenylphosphinegold(I)[175], which proceed with retention of configuration at carbon, with preferential cleavage of the carbon—gold(I) bond.

$$[Me_3P(Me)_2Au(CF_3)C{=}C(CF_3)AuPMe_3] \; (\mathbf{54}) \xrightarrow{HCl} [Me_3P(Me)_2Au(CF_3)C{=}C(CF_3)H] + [ClAuPMe_3] \quad (100)$$

A number of these reactions are useful in organic synthesis, since vinyl metal complexes may readily be prepared from acetylenes. For example, the synthesis of $\alpha,\beta$-unsaturated ketones by the reaction of vinylzirconium complexes with aluminium chloride and then acetyl chloride (equation 101)[177], and the reaction of vinylcopper complexes with acyl chloride, $\alpha,\beta$-unsaturated ketones, and epoxides (e.g. equation 102)[178]. The stereospecific attack of a heterocyclic cation on the $\alpha$-carbon of styrylcobaloximes provides an unusual example (equation 103)[179].

$$[Pr^iCH{=}C(Me)ZrClCp_2(\eta\text{-}C_5H_5)_2] \xrightarrow{AlCl_3} [Pr^iCH{=}C(Me)AlCl_2]_n \xrightarrow{CH_3COCl} Pr^iCH{=}C(Me)COCH_3 \quad (101)$$

$$[R(Me)C{=}CH\,Cu(SMe_2)MgBr_2] \xrightarrow[hmpa]{oxirane} R(Me)C{=}CHCH_2CH_2O^- \quad (102)$$

$$[(cis\text{-}ArCH{=}CH)Co(dmgH)_2py] + \text{1,3-benzodithiolylium} \xrightarrow{96\%} cis\text{-}ArCH{=}CH\text{-}CH(\text{1,3-benzodithiol-2-yl}) + [Co(dmgH)_2py]^+ \quad (103)$$

Evidence for attack at the $\beta$-carbon has emerged more recently. For example, the vinyliron complex **55,** when treated with acid, gives mixtures of the $\eta^2$-ethene complex and of the novel cation **56** via the intermediacy of the carbene complex $[MeCHFp]^+$ (equation 104)[108]. The same mixture can also be obtained by reaction of the methoxymethyl-Fp complex **57** with the trityl cation (equation 104), an example of electrophilic attack at a $\beta$-oxygen atom (see Section III.E). Similarly, the niobium complex **58** reacts with methylfluorosulphonate at the $\beta$-carbon to give an intermediate carbene complex which then undergoes hydrogen transfer in the process of elimination of the olefin. The carbenoid intermediate is apparently sufficiently long-lived to allow rotation about $C_{(1)}—C_{(2)}$ (equation 105)[181]. Other examples of attack at the $\beta$-carbon of acetylides are given in Section III.H.

$$[CH_2{=}CHFp]\ (\mathbf{55}),\ [MeOCH(Me)Fp]\ (\mathbf{57}) \longrightarrow [CH_3CH{=}Fp]^+ + [Fp(\eta^2{-}C_2H_4)]^+ \longrightarrow [Fp\{\eta^2{-}CH_2{=}CHCH(Me)Fp\}]^+ \equiv [FpCH_2\overset{+}{C}HCH(Me)Fp]\ (\mathbf{56}) \quad (104)$$

RCH═C(R′) RCH(Me)c(r′)

$$\left[\begin{matrix} R & & R' \\ & C{=}C & \\ H & & Nb(CO)Cp_2 \end{matrix}\right] (\mathbf{58}) \xrightarrow{MeSO_3F} \left[Me{-}\underset{H}{\overset{R}{C}}{-}\overset{R'}{C}{=}Nb(CO)Cp_2\right]^+ \longrightarrow \underset{Me\quad H}{\overset{R\quad R'}{C{=}C}} + \underset{Me\quad R}{\overset{R\quad H}{C{=}C}} + [Nb(CO)Cp_2]^+ \quad (105)$$

$$[PhCH{=}CHCH_2Co(dmgH)_2py] \xrightarrow[-40\,°C,\ CDCl_3]{(SCN)_2} PhCH{=}CHCH_2SCN + PhCH(SCN)CH{=}CH_2 + [NCSCo(dmgH)_2py] \xrightarrow{-20\,°C} PhCH{=}CHCH_2NCS \quad (106)$$

$$ArSCl + [\text{cyclopentenyl-}CH_2CH_2Co(dmgH)_2py] \longrightarrow \text{ArS-spiro[2.4]heptane} + [CoCl(dmgH)_2py] \quad (107)$$

Ar = 2,4-dinitrophenyl

Attack at more remote unsaturated carbon frequently leads to modification of the C—M bond character (Section III), but can also lead directly to displacement of the metal. Attack at the $\gamma$-carbon of allylcobaloximes or at the $\delta$-carbon of but-3-enylcobaloximes, with electrophilic reagents such as halogens[182], 2,4-dinitrophenylsulphenyl chloride[183], or thiocyanogen[182], gives rearranged allyl and cyclopropylcarbinyl products, respectively (equations 106 and 107). Reaction of the 3-methylbut-3-enyl-iron[184] and -zirconium[185] complexes **59** and **60** with trifluoroacetic acid and *N*-bromosuccinimide, respectively, leads to cyclopropane derivatives (equations 108 and 109). The corresponding attack at the $\gamma$-oxygen atom of the acyl complex **61** leads to a vinyl ester (equation 110)[186] and attack at a $\delta$-oxygen to a cyclopropane derivative (equation 111)[187].

$$CF_3COOH + [\text{CH}_2{=}C(Me)CH_2CH_2Fp] \longrightarrow \text{(1,1-dimethylcyclopropane)} + [CF_3COOFp] \quad (108)$$

**(59)**

$$\text{CH}_2{=}CHCH_2CH(Me)ZrCl(\eta\text{-}C_5H_5)_2 \xrightarrow{\text{nbs}} \text{Br-cyclopropane product} + \text{CH}_2{=}C(Me)CH_2CH_2Br \quad (109)$$

**(60)**

$$[(MeCOCH_2Au(PPh_3)] + MeCOCl \longrightarrow MeCO_2C(Me){=}CH_2 + [ClAu(PPh_3)] \quad (110)$$

**(61)**

$$[\text{H, Co(CN)}_5, \text{O}]^{3-} \underset{}{\overset{H^+}{\rightleftharpoons}} [\text{H, Co(CN)}_5{}^{3-}, \text{OH}^+] \longrightarrow \text{H, OH} + [Co(CN)_5H_2O]^{2-} \quad (111)$$

## VIII. INSERTION REACTIONS

A large number of unsaturated electrophilic organic and inorganic molecules react with organotransition metal complexes by 'insertion' between the organic ligand and the metal. Sulphur dioxide, sulphur trioxide, tetracyanoethylene, hexafluorobut-2-yne, etc., react with a wide range of complexes (equation 112). Sulphur dioxide insertion reactions have been studied most extensively[188] but there are still a number of perplexing features of these reactions, probably because more than one mechanism can obtain and because evidence from one complex under one set of conditions has often been associated with that from other complexes and under different conditions where the mechanism need not be the same.

$$[RML_n] + XY \longrightarrow [R{-}X{-}Y{-}ML_n] \quad XY = CF_2CF_2,\ CF_3C{\equiv}CCF_3,\ (NC)_2C{=}C(CN)_2, \text{ etc.}$$

$$[RML_n] + XY \longrightarrow \left[R{-}\underset{\substack{\| \\ Y}}{X}{-}ML_n\right] \quad X = SO;\ Y = O \quad X = SO_2;\ Y = O \qquad (112)$$

The stereochemistry of the insertion at carbon has been studied in several substrates. Thus, diastereoisomeric [(PhCHDCHD)Fp][189], [(*t*-BuCHDCHD)Fp][113], $[Mn(PhCHDCHD)(CO)_4(PEt_3)]$[189], and $[W(PhCHDCHD)(CO)_2(PEt_3)(C_5H_5)]$[189] react with sulphur dioxide to give the corresponding insertion products with inversion of configuration at the $\alpha$-carbon, indicative of a rear-side attack of the sulphur at carbon, i.e. on the side away from the metal[190]. It has been suggested that this leads to an ion pair which collapses first to an *O*-metal sulphinate and subsequently, through rearrangement, perhaps by redissociation to the ion pair, to the thermodynamically favoured *S*-bonded sulphinate. The *O*-bonded sulphinates have been observed in solution[191] but have not been isolated. The insertion reaction also takes place with substantial retention of configuration at the iron centre, as shown by the formation of $[\{(RS, SR)\text{-}Me_3SiCH(Ph)(SO_2)\}Fe(CO)(Ph_3P)(\eta\text{-}C_5H_5)]$ from $[\{(RR, SS)\text{-}Me_3SiCH(Ph)\}Fe(CO)(Ph_3P)(\eta\text{-}C_5H_5)]$ with a rate constant of $12.3\times10^{-4}\ s^{-1}$ at 280 K in liquid sulphur dioxide[192], by the formation of the insertion product **63** from the diastereoisomeric substrate **62** (equation 113)[193], and other work[194]. When iodide is present during the insertion reaction, the iodoiron complex is formed, probably through interception of the ion pair.

[(Me, Ph-C₅H₃)Fe(CO)(PPh₃)Me] (**62**) —SO₂→ $[(Me,Ph\text{-}C_5H_3)Fe(CO)(PPh_3)]^+$ $CH_3SO_2^-$ ⇌ [(Me,Ph-C₅H₃)Fe(CO)(PPh₃)(O—S(=O)R)]

$[(Me,Ph\text{-}C_5H_3)Fe(CO)(PPh_3)]^+$ —I⁻→ [(Me,Ph-C₅H₃)Fe(CO)(PPh₃)I]

$[(Me,Ph\text{-}C_5H_3)Fe(CO)(PPh_3)]^+$ → [(Me,Ph-C₅H₃)Fe(CO)(PPh₃)(S(=O)₂R)] (**63**) (113)

The rates of insertion into a series of organoiron complexes[195] are not entirely characteristic of an electrophilic attack at the $\alpha$-carbon with inversion of configuration. Moreover, the variation in reactivity of different substrates is solvent dependent, a typical order being $Me_2CH > Me_3CCH_2 > Me > PhCH_2 > Me_3C$[194]. For a bimolecular attack at the $\alpha$-carbon, the *neo*pentyl group would be expected to be particularly unreactive and it seems likely, therefore, that at least two mechanisms operate in the above series. Even the order Me < Et, which obtains in some solvents, is unusual, being more consistent with an electron-transfer process than with a substitution reaction (see Section IV), and free radical processes have been detected in the sulphur dioxide insertion reactions of some benzylmetal complexes[196].

Support for the electrophilic character of some of the reactions comes from the observation that there is a 20 000-fold acceleration of the rate of insertion into organotungsten complexes by Lewis acids such as $BF_3$ and $AlBr_3$, and that the primary products are the complexes *O*-sulphinates, which, on addition of base, rearrange to the *S*-sulphinates (equation 114)[197].

$$[W(Me(CO)_3Cp] + SO_2 \longrightarrow [W\{OS(Me){=}OBF_3\}(CO)_3Cp] \xrightarrow{\text{base}} [W(SO_2Me)(CO)_3Cp] \quad (114)$$

The insertion of sulphur dioxide into organo-molybdenum, -ruthenium, -zirconium, -gold, and -platinum complexes has also been investigated in some detail[198–202]. The differences in mechanism are emphasized by the observation that the insertion of sulphur dioxide into the carbon—metal bond of complex **41** occurs with retention of configuration (equation 115)[201], and that the mixed alkylaryl-gold(III) complex **64** is subject to only a single insertion into the carbon—alkyl bond

$$\text{Newman projection: Bu}^t\text{, H, D, H, D, ZrClCp}_2 \ \textbf{(41)} \xrightarrow{SO_2} \text{Newman projection: Bu}^t\text{, H, D, H, D, SO}_2\text{ZrClCp}_2 \qquad (115)$$

rather than the carbon—aryl bond (equation 116)[202]. The latter is indicative of an attack at the metal followed by an intramolecular migration of the alkyl group. Transient intermediates have been detected in this reaction at low temperatures[203]. Even trialkylgold(III) complexes only undergo a single insertion reaction[203].

$$cis\text{-}[Au(PhMe_2)(PPh_3)] \ \textbf{(64)} \xrightarrow{SO_2} [AuMe(SO_2Me)(Ph)(PPh_3)] \qquad (116)$$

In contrast, certain organo-platinum and -iridium complexes are not prone to sulphur dioxide insertion, and may even be prepared by extrusion of sulphur dioxide from the preformed sulphinate complexes (equations 117 and 118)[205]. The formation of a five-coordinate sulphur dioxide complex from the platinum(II) sulphinate is, by the application of the principle of microscopic reversibility, good evidence that the insertion reaction would be via an attack on the metal.

$$[IrCl_2L_2(SO_2R)] \longrightarrow \text{Ir(R)(Cl)(L)(L)(SO}_2\text{)(Cl)} \qquad (117)$$

$$cis\text{-}[PtCl(SO_2Ph)(PPh_3)_2] \xrightarrow{300\,^\circ C} cis\text{-}[PtClPh(PPh_3)_2] + SO_2 \qquad (118)$$

The insertion of sulphur dioxide into allylmetal complexes frequently leads to rearranged products, frequently to unrearranged products (equation 119)[206], but the factors controlling the attack of sulphur on the $\alpha$- or $\gamma$-carbon of the allyl ligand are uncertain, since allylsulphinyl/metal ion pairs are likely to undergo a variety of intra- and inter-molecular rearrangements.

$$[MeCH{=}CHCH_2ML_n] \xrightarrow{SO_2} [MeCH{=}CHCH_2SO_2ML_n] \text{ and/or } [CH_2{=}CHCH(Me)SO_2ML_n] \qquad (119)$$

The insertion of tetracyanoethylene, hexafluoroacetone, etc., into gold—carbon and platinum—carbon bonds probably occurs through prior coordination of the electrophile to the metal. No intermediates have been observed in the reaction of tetrafluoroethylene with $[AuMe(PPh_3)]$[89,207] but the same substrate reacts with tetracyanoethylene to give an isolable $\eta^2$-complex[203], which slowly rearranges to the insertion product **65**[208]. Tetracyanoethylene insertion reactions are complicated by the formation of more than one type of product in several cases, and by the possible role of charge transfer complexes in the reactions (see Section IV)[209]. For example, the reaction of the alkylchromium complex **66** with tetracyanoethylene gives a mixture of the *C*-bonded and the *N*-bonded complexes **67** and **68**, which slowly interconvert, as well as the dinitrosyl complex **70** (equation 120)[210].

$$trans\text{-}[PtClMe(AsMe_2Ph)_2] + CF_3C{\equiv}CCF_3 \longrightarrow [PtClMe(AsMe_2Ph)_2[\eta^2\text{-}C_4F_6)] \quad (120)$$

21 days, ambient temp.

$$\left[ (CH_3)(CF_3)C{=}C(CF_3)PtCl(AsMe_2Ph)_2 \right]$$

**(65)**

$$[CrR(NO)_2Cp] + (NC)_2C{=}C(CN)_2 \longrightarrow [Cr\{C(CN)_2C(CN)_2R\}(NO)_2Cp]$$

**(66)** **(67)**

$$+[Cr\{N{=}C{=}C(CN)C(CN)_2R\}(NO)_2Cp] + [Cr(NO)_2(CN)Cp] \quad (121)$$

**(68)** **(69)**

## IX. REFERENCES


1. C. K. Ingold, *Structure and Mechanism in Organic Chemistry,* 3rd ed. Cornell University Press, Ithaca, 1969.
2. E.g. R. G. Wilkins, *The Study of Kinetics and Mechanism of Reactions of Transition Metal Complexes*, Allyn and Bacon, Boston, 1974, p. 255.
3. D. Forster, *Adv. Organomet. Chem.,* **17,** 255 (1979).
4. R. G. Pearson, *J. Chem. Educ.,* **45,** 581, 643 (1968).
5. E.g. G. Klopman, *Chemical Reactivity and Reaction Paths,* Wiley-Interscience, New York, 1974.
6. J. O. Edwards and R. G. Pearson, *J. Am. Chem. Soc.,* **84,** 16 (1962).
7. M. D. Johnson and N. Winterton, *J. Chem. Soc. A,* 507 (1970).
8. A. Bury and M. D. Johnson, unpublished work; A. L. Crumbliss and P. L. Gaus, *Inorg. Chem.,* **14,** 486 (1975); N. W. Alcock, M. P. Atkins, E. H. Curzon, B. T. Golding, and P. J. Sellars, *J. Chem. Soc., Chem. Commun.,* 1238 (1980).
9. H. B. Gjerde and J. H. Espenson, *Organometallics,* **1,** 435 (1982).
10. M. P. Atkins, B. T. Golding, A. Bury, M. D. Johnson, and P. J. Sellars, *J. Am. Chem. Soc.,* **102,** 3630 (1980).
11. A. Bazac and J. H. Espenson, *Inorg. Chem.,* **19,** 242 (1980).
12. S. M. Chemaly and J. M. Pratt, *J. Chem. Soc., Dalton Trans.,* 2267 (1980); D. Dolphin, A. W. Johnson, and R. Rodrigo, *Ann. N.Y. Acad. Sci.,* **112,** 590 (1964).
13. M. D. Johnson, M. L. Tobe, and L. Y. Wong, *J. Chem. Soc. A,* 491 (1967); 590 (1968).
14. M. D. Johnson, M. L. Tobe, and L. Y. Wong, *J. Chem. Soc. A,* 923 (1968).
15. D. Dodd and M. D. Johnson, *J. Chem. Soc., Dalton Trans.,* 1218 (1973); D. Dodd, M. D. Johnson, and C. W. Fong, *J. Chem. Soc., Dalton Trans.,* 58 (1974).
16. S. B. Butts, S. H. Strauss, E. M. Holt, R. E. Stimson, N. W. Alcock, and D. F. Shriver, *J. Am. Chem. Soc.,* **102,** 5093 (1980).
17. M. L. H. Green, J. K. P. Ariyaratne, A. M. Bierrum, M. Ishaq, and C. K. Prout, *J. Chem. Soc., Chem. Commun.,* 430 (1967); G. N. Schrauzer and R. J. Windgassen, *J. Am. Chem. Soc.,* **89,** 1999 (1967).
18. S. N. Anderson, D. H. Ballard, and M. D. Johnson, *J. Chem. Soc., Perkin Trans. 2,* 311 (1972).
19. S. N. Anderson, D. H. Ballard, and M. D. Johnson, *J. Chem. Soc., Chem. Comm.,* 779 (1979).
20. G. L. Blackmer, T. M. Vickerey, and J. N. Marx, *J. Organomet. Chem.,* **72,** 261 (1974).
21. F. Wagner, *Proc. Roy. Soc. A,* **288,** 344 (1965).
22. A. W. Johnson, L. Mervyn, N. Shaw, and E. L. Smith, *J. Chem. Soc.,* 4146 (1963); I. P. Rudakova, T. A. Pospelova, and A. M. Yurkevitch, *J. Gen. Chem. USSR,* **39,** 399 (1969); **40,** 2479 (1970).
23. M. L. H. Green and P. L. I. Nagy, *J. Chem. Soc.,* 189 (1963).
24. P. Lennon, M. Madhavarao, A. Rosan, and M. Rosenblum, *J. Organomet. Chem.,* **108,** 93 (1976).

25. M. Rosenblum, *Acc. Chem. res.*, **7,** 122 (1974).
26. A. Cutler, D. Ehntholt, W. P. Giering, P. Lennon, S. Raghu, A. Rosan, M. Rosenblum, J. Tancrede, and D. Wells, *J. Am. Chem. Soc.*, **98,** 3495 (1976).
27. S. R. Su and A. Wojcicki, *Inorg. Chim. Acta,* **8,** 55 (1974).
28. (a) C. Cooksey, D. Dodd, M. D. Johnson, and B. L. Lockman, *J. Chem. Soc., Dalton Trans.*, 1815 (1978); (b) J. E. Thomasson, P. W. Robinson, D. A. Ross, and A. Wojcicki, *Inorg. Chem.*, 2130 (1971); W. Bannister, B. L. Booth, R. N. Hazeldine, and P. Loader, *J. Chem. Soc. A,* 930 (1971); (c) D. W. Lichtenberg and A. Wojcicki, *Inorg. Chem.*, **14,** 1295 (1975).
29. W. P. Giering and M. Rosenblum, *J. Am. Chem. Soc.*, **93,** 5299 (1971).
30. J. P. Williams and A. Wojcicki, *Inorg. Chem.*, **20,** 1585 (1981).
31. P. B. Bell and A. Wojcicki, *Inorg. Chem.*, **20,** 1585 (1981).
32. A. Cutler, R. W. Fish, W. P. Giering, and M. Rosenblum, *J. Am. Chem. Soc.*, **94,** 4354 (1972).
33. D. Dodd, M. D. Johnson, I. P. Steeples, and F. D. McKenzie, *J. Am. Chem. Soc.*, **98,** 6399 (1976).
34. Y. Yamamoto and A. Wojcicki, *Inorg. Chem.*, **12,** 1779 (1973).
35. M. Rosenblum and P. S. Waterman, *J. Organomet. Chem.*, **187,** 267 (1980).
36. P. Lennon, A. Rosan, M. Rosenblum, J. Tancrede, and P. Waterman, *J. Am. Chem. Soc.*, **102,** 7033 (1980).
37. M. L. H. Green and J. K. P. Ariyaratne, *J. Chem. Soc., 1 (1964).*
38. R. B. Silverman and D. Dolphin, *J. Am. Chem. Soc.*, **98,** 4633 (1976).
39. T. Vickerey, R. Katz, and G. N. Schrauzer, *J. Am. Chem. Soc.*, **97,** 7248 (1975).
40. B. T. Golding, H. L. Holland, U. Horn, and S. Sakrikar, *Angew. Chem., Int. Ed. Engl.*, **9,** 959 (1970).
41. R. B. Silverman and D. Dolphin, *J. Am. Chem. Soc.*, **98,** 4626 (1976).
42. B. T. Golding and S. Sakrikar, *J. Chem. Soc., Chem. Commun.*, 1183 (1972).
43. J. H. Espenson and D. M. Wang, *Inorg. Chem.*, **18,** 2853 (1979).
44. R. Schaffold and E. Amble, *Angew. Chem., Int. Ed. Engl.*, **19,** 629 (1979).
45. T. G. Chervyakova, E. A. Parvenov, M. G. Edelev, and A. M. Yurkevitch, *J. Gen. Chem. USSR,* **44,** 449 (1974); E. A. Parvenov, T. G. Chervyakov, and M. G. Edelev, *J. Gen. Chem. USSR,* **43,** 2752 (1973).
46. D. Slack and M. C. Baird, *J. Chem. Soc., Chem. Commun.*, 701 (1974).
47. P. Lennon, A. M. Rosen, and M. Rosenblum, *J. Am. Chem. Soc.*, **99,** 8426 (1977).
48. M. Laing, J. R. Moss, and J. Johnson, *J. Chem. Soc., Chem. Commun., 656* (1977).
49. J. K. P. Ariyaratne and M. L. H. Green, *J. Chem. Soc.*, 2976 (1963).
50. P. M. Treichel, D. W. Firsich, and T. H. Lemmen, *J. Organomet. Chem.*, **202,** C77 (1980).
51. M. Brookhart and G. O. Nelson, *J. Am. Chem. Soc.*, **99,** 6099 (1977); M. Brookhart, J. R. Tucker, and G. R. Hush, *J. Organomet. Chem.*, **193,** C23 (1980).
52. M. Brookhart, J. R. Tucker, T. C. Flood, and J. Jensen, *J. Am. Chem. Soc.*, **102,** 1203 (1980).
53. C. P. Casey and T. J. Burkhardt, *J. Am. Chem. Soc.*, **95,** 5833 (1973); C. P. Casey, T. J. Burkhardt, C. A. Bunnell and J. C. Calabrese, *J. Am. Chem. Soc.*, **99,** 2127 (1977).
54. C. H. Game, M. Green, J. R. Moss, and F. G. A. Stone, *J. Chem. Soc. Dalton Trans.*, 351 (1974).
55. D. F. Christian, H. C. Clark, and R. C. Stepaniah, *J. Organomet. Chem.*, **112,** 227 (1976).
56. A. R. Cutler, *J. Am. Chem. Soc.*, **101,** 604 (1979).
57. R. D. Adams, A. Davison, and J. P. Selegue, *J. Am. Chem. Soc.*, **101,** 7232 (1979); R. A. Bell and M. H. Chisholm, *Inorg. Chem.*, **16,** 687 (1977).
58. A. Davison and J. P. Solar, *J. Organomet. Chem.*, **166,** C13 (1979).
59. J. K. Kochi, *Pure Appl. Chem.*, **52,** 571 (1980).
60. J. Halpern, J. Topich, and K. I. Zamaraev, *Inorg. Chim. Acta,* **20,** L21 (1976).
61. R. Magnusson, J. Halpern, I. V. Levitin, and M. E. Volpin, *J. Chem. Soc., Chem. Commun.*, 44 (1978).
62. J. Halpern, M. S. Chan, T. S. Roche, and G. M. Tom, *Acta Chem. Scand.*, **33A,** 141 (1979).
63. S. N. Anderson, D. H. Ballard, J. Z. Chrzastowski, D. Dodd, and M. D. Johnson, *J. Chem. Soc., Chem. Commun.*, 685 (1972).
64. P. Abley, E. R. Dockal, and J. Halpern, *J. Am. Chem. Soc.*, **94,** 659 (1972).
65. C. A. Chapman and M. D. Johnson, unpublished work.
66. B. D. Gupta and M. D. Johnson, unpublished work.

67. J. H. Espenson, unpublished work.
68. T. Okamoto, M. Goto, and S. Oka, *Inorg. Chem.*, **20,** 899 (1981).
69. R. Garlatti, G. Tauzher, and G. Costa, *J. Organomet. Chem.*, **139,** 179 (1977).
70. R. Dreos, G. Tauzher, N. Marisch, and G. Costa, *J. Organomet. Chem.*, **92,** 227 (1975).
71. J. P. Kitchin and D. A. Widdowson, *J. Chem. Soc., Perkin Trans. 1,* 1384 (1979).
72. S. N. Anderson and M. D. Johnson, unpublished work.
73. R. W. Johnson and R. G. Pearson, *J. Chem. Soc., Chem. Commun.*, 986 (1970).
74. S. N. Anderson, C. W. Fong, and M. D. Johnson, *J. Chem. Soc., Chem. Commun.*, 165 (1973); W. N. Rogers, J. A. Page, and M. C. Baird, *Inorg. Chem.*, **20,** 3521 (1981).
75. K. M. Nicholas and M. Rosenblum, *J. Am. Chem. Soc.,* **95,** 4449 (1973).
76. J. E. Backvall, *Tetrahedron Lett.*, 467 (1977).
77. R. A. Budnik and J. K. Kochi, *J. Organomet. Chem.*, **116,** C3 (1976).
78. T. S. Abram and R. Baker, *J. Chem. Soc., Chem. Commun.*, 267 (1979).
79. J. P. Collman, S. R. Winter, and R. G. Komoto, *J. Am. Chem. Soc.*, **95,** 249 (1973).
80. T. T. Tsou and J. K. Kochi, *J. Am. Chem. Soc.*, **100,** 1634 (1978).
81. S. Komiya, T. A. Albright, R. Hoffmann, and J. K. Kochi, *J. Am. Chem. Soc.*, **99,** 8440 (1977).
82. M. Almemark and B. Akermark, *J. Chem. Soc., Chem. Commun.*, 66 (1978).
83. J. Y. Chen and J. K. Kochi, *J. Am. Chem. Soc.*, **99,** 1450 (1977).
84. P. S. Waterman and W. P. Giering, *J. Organomet. Chem.*, **155,** C47 (1978).
85. A. Bakac and J. H. Espenson, *J. Am. Chem. Soc.*, **103,** 2721 (1981); J. H. Espenson and A. Bakac, *J. Am. Chem. Soc.*, **103,** 2728 (1981).
86. H. Elroi and D. Meyerstein, *J. Am. Chem. Soc.*, **100,** 5540 (1978).
87. G. L. Blackmer and C. W. Tsai, *J. Organomet. Chem.*, **155,** C17 (1978).
88. L. G. Vaughan and W. A. Sheppard, *J. Am. Chem. Soc.*, **91,** 6151 (1969); R. Uson, A. Laguna, and J. Vicente, *J. Organomet. Chem.*, **86,** 415 (1975); H. Schmidtbar and R. Franke, *Inorg. Chim. Acta,* **13,** 79 (1975).
89. E. G. Perevalova, T. V. Baukova, E. I. Goryunov, and K. I. Grandberg, *Izv. Akad. Nauk SSSR, Ser. Khim.*, 2148 (1970).
90. M. Aresta and G. Vasapollo, *J. Organomet. Chem.*, **50,** C51 (1973).
91. J. Chatt and B. L. Shaw, *J. Chem. Soc.*, 705, 4020 (1959).
92. H. Schmidbaur, *Acc. Chem. Res.*, **8,** 62 (1975); H. Schmidbaur and H. P. Scherm, *Chem. Ber.*, **110,** 1576 (1977).
93. D. R. Coulson, *J. Am. Chem. Soc.*, **91,** 200 (1969).
94. P. K. Wong and J. K. Stille, *J. Organomet. Chem.*, **70,** 121 (1974).
95. A. Johnson and R. J. Puddephat, *J. Organomet. Chem.*, **85,** 115 (1975); A. Tamaki and J. K. Kochi, *J. Organomet. Chem.*, **64,** 411 (1974); A. Shiotani and H. Schmidbaur, *J. Organomet. Chem.*, **37,** C24 (1972); R. J. Puddephat and C. E. E. Upton, *J. Organomet. Chem.*, **91,** C17 (1975).
96. A. Johnson and R. J. Puddephat, *J. Chem. Soc., Dalton Trans.*, 1360 (1976).
97. A. Tamaki and J. K. Kochi, *J. Chem., Soc., Dalton Trans.*, 2620 (1973).
98. R. Romeo, D. Minniti, S. Lanza, P. Uguagliati, and U. Belluco, *Inorg. Chim. Acta,* **19,** L55 (1976).
99. U. Belluco, U. Croatto, P. Uguagliati, and R. Pietropaolo, *Inorg. Chem.*, **6,** 718 (1967).
100. R. Romeo, D. Minniti, S. Lanza, P. Uguagliati, and U. Belluco, *Inorg. Chem.*, **17,** 2813 (1978).
101. R. Romeo, D. Minniti, and S. Lanza, *J. Organomet. Chem.*, **165,** C36 (1979).
102. J. K. Jaward and R. J. Puddephat, *J. Chem. Soc., Chem. Commun.*, 892 (1977).
103. P. Uguagliati, R. A. Michelin, and U. Belluco, *J. Organomet. Chem.*, **169,** 115 (1979).
104. R. J. Puddephat and P. J. Thompson, *J. Organomet. Chem.*, **117,** 395 (1976).
105. J. K. Stille and K. S. Y. Lau, *J. Am. Chem. Soc.*, **98,** 5841 (1976).
106. B. J. Gregory and C. K. Ingold, *J. Chem. Soc. B,* 276 (1969).
107. R. J. Puddephat, *The Chemistry of Gold,* Elsevier, Amsterdam, 1978, Chapter 9.
108. A. E. Stevens and J. L. Beauchamp, *J. Am. Chem. Soc.*, **101,** 245 (1979).
109. T. C. Flood and F. J. DiSanti, *J. Chem. Soc., Chem. Commun.*, 18 (1975).
110. D. Slack and M. C. Baird, *J. Am. Chem. Soc.*, **98,** 5539 (1976).
111. D. Dong and M. C. Baird, *J. Organomet. Chem.*, **172,** 467 (1979).
112. G. M. Whitesides and D. J. Boschetto, *J. Am. Chem. Soc.*, **93,** 1829 (1971).
113. P. L. Bock, D. J. Boschetto, J. R. Rassmussen, J. P. Demers, and G. M. Whitesides, *J. Am. Chem. Soc.*, **96,** 2814 (1974).
114. D. A. Symon and T. C. Waddington, *J. Chem. Soc., Dalton Trans.*, 2140 (1975).
115. B. D. Fabian, T. P. Fehlner, L.-S. J. Hwang, and J. A. Labinger, *J. Organomet. Chem.*, **191,** 409 (1980).

116. T. G. Attig and A. Wojcicki, *J. Am. Chem. Soc.*, **96,** 262 (1974).
117. T. C. Flood and D. C. Miles, *J. Organomet. Chem.*, **127,** 33 (1977).
118. H. Brünner and G. Wallner, *Chem. Ber.*, **109,** 1053 (1976).
119. N. De Luca and A. Wojcicki, *J. Organomet. Chem.*, **193,** 359 (1980).
120. S. N. Anderson, C. J. Cooksey, S. G. Holton, and M. D. Johnson, *J. Am. Chem. Soc.*, **102,** 2312 (1980).
121. L. J. Dizikes and A. Wojcicki, *J. Am. Chem. Soc.*, **99,** 5295 (1977).
122. D. Dong, D. A. Slack, and M. C. Baird, *Inorg. Chem.*, **18,** 188 (1979).
123. P. L. Bock and G. M. Whitesides, *J. Am. Chem. Soc.*, **96,** 2826 (1974).
124. J. Z. Chrzastowski and M. D. Johnson, *J. Chem. Soc., Dalton Trans.*, 2456 (1976).
125. M. H. Abraham, in *Comprehensive Chemical Kinetics* (Ed. C. H. Bamford and C. F. H. Tipper), Elsevier, Amsterdam 1973.
126. J. H. Espenson and D. A. Williams, *J. Am. Chem. Soc.*, **96,** 1008 (1974).
127. J. C. Chang and J. H. Espenson, *J. Chem. Soc., Chem. Commun.*, 233 (1974).
128. J. H. Espenson and G. J. Samuels, *J. Organomet. Chem.*, **113,** 143 (1976).
129. R. G. Coombes and M. D. Johnson, *J. Chem. Soc. A*, 1905 (1966).
130. W. Marty and J. H. Espenson, *Inorg. Chem.*, **18,** 1246 (1979).
131. J. P. Leslie and J. H. Espenson, *J. Am. Chem. Soc.*, **98,** 4839 (1976).
132. J. A. Labinger, D. W. Hart, W. E. Seibert, and J. A. Schwartz, *J. Am. Chem. Soc.*, **97,** 3851 (1975).
133. D. Dodd, M. D. Johnson, and D. Vamplew, *J. Chem. Soc. B*, 1841 (1971).
134. R. G. Coombes, M. D. Johnson, and D. Vamplew, *J. Chem. Soc. A*, 2297 (1968).
135. D. Dodd and M. D. Johnson, *J. Chem. Soc., Perkin Trans. 2*, 219 (1974).
136. D. Dodd, M. D. Johnson, and N. Winterton, *J. Chem. Soc. A*, 910 (1971).
137. D. Dodd and M. D. Johnson, *J. Chem. Soc. B*, 662 (1971).
138. J. H. Espenson and A. Bakac, *J. Am. Chem. Soc.*, **103,** 2728 (1981).
139. J. M. Wood, *Naturwissenschaften*, **62,** 357 (1975).
140. V. C. W. Chu and D. W. Gruenwedel, *Bioinorg. Chem.*, **7,** 169 (1977).
141. R. E. De Simone, M. W. Penley, L. Charbonneau, S. G. Smith, J. M. Wood, H. A. O. Hill, J. M. Pratt, S. Ridsdale, and R. J. P. Williams, *Biochim. Biophys. Acta*, **304,** 851 (1973).
142. J. S. Thayer, *Inorg. Chem.*, **18,** 1171 (1979).
143. P. J. Craig and S. F. Morton, *J. Organomet. Chem.*, **145,** 79 (1978).
144. M. Yakamoto, T. Yokoyama, J.-L. Chen and T. Kwan, *Bull. Chem. Soc. Jpn.* **48,** 844 (1975).
145. M. Imura, E. Sukegawa, S.-K. Pan, K. Nagao, J.-Y. Kim, T. Kwan, and T. Ukita, *Science*, **172,** 1248 (1971).
146. A. Dain and J. H. Espenson, *J. Chem. Soc., Chem. Commun.*, 653 (1971).
147. J. M. Wood, F. S. Kennedy, and C. G. Rosen, *Nature* (*London*), **220,** 170 (1968).
148. J. H. Espenson, W. R. Bushey, and M. E. Chmielewski, *Inorg. Chem.*, **14,** 1302 (1975).
149. J. H. Espenson and T. H. Chao, *Inorg. Chem.*, **16,** 2553 (1977).
150. R. J. Allen and C. A. Bunton, *Bioinorg. Chem.*, **5,** 311 (1976).
151. P. Abley, E. R. Dockal, and J. Halpern, *J. Am. Chem. Soc.*, **95,** 3166 (1973).
152. G. N. Schrauzer, J. H. Weber, T. M. Beckman, and R. K. Y. Ho, *Tetrahedron Lett.*, 275 (1971).
153. G. Tauzher, R. Dreos, G. Costa, and M. Green, *J. Organomet. Chem.*, **81,** 107 (1974).
154. V. E. Magnusson and J. H. Weber, *J. Organomet. Chem.*, **74,** 135 (1974).
155. J. H. Espenson, H. L. Fritz, R. A. Heckman, and C. Nicolini, *Inorg. Chem.*, **15,** 906 (1976).
156. G. J. Samuels and J. H. Espenson, *Inorg. Chem.*, **19,** 233 (1980).
157. H. L. Fritz, J. H. Espenson, D. Williams, and G. A. Molander, *J. Am. Chem. Soc.*, **96,** 2378 (1974).
158. H. Shinozaki, H. Ogawa, and M. Tada, *Bull. Chem. Soc. Jpn.*, **49,** 775 (1976).
159. G. Mestroni, C. Cocevar, and G. Costa, *Gazz. Chim. Ital.*, **103,** 273 (1973).
160. J. F. Endicott, K. P. Balakrishnan and C.-L. Wong, *J. Am. Chem. Soc.*, **102,** 5519 (1980).
161. G. W. Rice and R. S. Tobias, *J. Organomet. Chem.*, **86,** C37 (1975); cf. R. J. Puddephat, *The Chemistry of Gold,* Elsevier, Amsterdam, 1978, Chapter 9.
162. W. M. Scovell, *J. Am. Chem. Soc.*, **96,** 3451 (1974).
163. G. Agnes, S. Bendle, H. A. O. Hill, F. R. Williams, and R. J. P. Williams, *J. Chem. Soc., Chem. Commun.*, 850 (1971).
164. Y. T. Fanchiang, W. P. Ridley, and J. M. Wood, *J. Am. Chem. Soc.*, **101,** 1442 (1979).
165. J. Y. Kim, H. Yamamoto, and T. Kwan, *Chem. Pharm. Bull. Jpn.*, **23,** 1091 (1975); H. Yamamoto, T. Yokoyama, and T. Kwan, *Chem. Pharm. Bull.*, **23,** 2186 (1976).

166. M. E. Volpin, A. M. Yurkevitch, L. G. Volkova, E. G. Chauser, I. P. Rudakova, I. Y. Levitin, E. M. Tachkova, and T. M. Ushaleova, *J. Gen. Chem. USSR,* **45,** 150, 164 (1975).
167. R. T. Taylor and L. M. Hanna, *J. Environ. Sci. Health,* **A11,** 201 (1976).
168. M. W. Witman and J. H. Weber, *Inorg. Chem.,* **15,** 2375 (1976); **16,** 2512 (1977).
169. W. Schmidt, J. H. Swinehart, and H. Taube, *J. Am. Chem. Soc.,* **93,** 1117 (1971).
170. M. D. Johnson and N. Winterton, *J. Chem. Soc. A,* 511 (1970).
171. R. G. Coombes and M. D. Johnson, *J. Chem. Soc. A,* 177 (1966).
172. E. H. Bartlett and M. D. Johnson, *J. Chem. Soc. A,* 523 (1970).
173. R. R. Schrock and P. R. Sharp, *J. Am. Chem. Soc.,* **100,** 2389 (1978).
174. D. Dodd, M. D. Johnson, B. S. Meeks, D. M. Titchmarsh, K. N. V. Doung, and A. Gaudemer, *J. Chem. Soc., Perkin Trans. 2,* 1262 (1976).
175. A. N. Nesmeyanov, E. G. Perevalova, M. V. Ovchinnikov, and K. I. Grandberg, *Izv. Akad. Nauk SSSR,* 2282 (1975).
176. J. A. J. Jarvis, A. Johnson, and R. J. Puddenphatt, *J. Chem. Soc., Chem. Commun.,* 373 (1973).
177. D. B. Carr and J. Schwartz, *J. Am. Chem. Soc.,* **99,** 638 (1977).
178. P. R. McGuirk, A. Marfat, and P. Helquist, *Tetrahedron Lett.,* 1363, 2973 (1978).
179. K. Miura and M. Tada, *Chem. Lett.,* 1139 (1978).
180. T. Bodnar and A. R. Cutler, *J. Organomet. Chem.,* **213,** C31 (1981).
181. J. A. Labinger and J. Schwartz, *J. Am. Chem. Soc.,* **97,** 1596 (1975).
182. A. Bury and M. D. Johnson, unpublished work.
183. M. R. Ashcroft, B. D. Gupta, and M. D. Johnson, *J. Chem. Soc., Perkin Trans. 1,* 2021 (1980).
184. A. Bury, M. D. Johnson, and M. J. Stewart, *J. Chem. Soc., Chem. Commun.,* 622 (1980).
185. C. A. Bertelo and J. Schwartz, *J. Am. Chem. Soc.,* **98,** 262 (1976).
186. A. N. Nesmeyanov, K. I. Grandberg, E. I. Smyslova, and E. P. Perevalova, *Izv. Akad. Nauk SSSR, Ser. Khim.,* 2872 (1974).
187. J. Kwiatek, *Catal. Rev.,* **1,** 37 (1967).
188. A. Wojcicki, *Adv. Organomet. Chem.,* **12,** 31 (1974).
189. D. Dong, D. A. Slack, and M. C. Baird, *J. Organomet. Chem.,* **153,** 219 (1978).
190. S. Inayaki, H. Fujimoto, and K. Fukui, *J. Am. Chem. Soc.,* **98,** 4693 (1976).
191. S. E. Jacobson, P. Reich-Rohrwig, and A. Wojcicki, *Inorg. Chem.,* **12,** 717 (1973).
192. K. Stanley and M. C. Baird, *J. Am. Chem. Soc.,* **99,** 1808 (1977).
193. T. C. Flood, F. J. DiSanti, and D. C. Miles, *Inorg. Chem.,* **15,** 1910 (1976).
194. T. G. Attig and A. Wojcicki, *J. Am. Chem. Soc.,* **96,** 262 (1974); *J. Organomet. Chem.,* **82,** 397 (1974); S. L. Miles, D. L. Miles, R. Bau, and T. C. Flood, *J. Am. Chem. Soc.,* **100,** 7278 (1978).
195. D. Jacobson and A. Wojcicki, *J. Am. Chem. Soc.,* **95,** 6962 (1973).
196. A. E. Crease and M. D. Johnson, *J. Am. Chem. Soc.,* **100,** 8013 (1980).
197. R. G. Sverson and A. Wojcicki, *J. Am. Chem. Soc.,* **101,** 877 (1979).
198. D. W. Lichtenberg and A. Wojcicki, *Inorg. Chem.,* **14,** 1295 (1975).
199. J. O. Kroll and A. Wojcicki, *J. Organomet. Chem.,* **66,** 95 (1974).
200. S. E. Jacobson and A. Wojcicki, *J. Organomet. Chem.,* **72,** 113 (1974).
201. J. A. Labinger, D. W. Hart, W. E. Seibert, and J. Schwartz, *J. Am. Chem. Soc.,* **97,** 3851 (1975).
202. R. J. Puddephatt and M. A. Stalteri, *J. Organomet. Chem.,* **193,** C27 (1980).
203. A. Johnson and R. J. Puddephatt, *J. Chem. Soc., Dalton Trans.,* 384 (1977).
204. M. Kubota and B. M. Loeffler, *Inorg. Chem.,* **11,** 469 (1972).
205. C. D. Cook and G. S. Jauhal, *Can. J. Chem.,* **45,** 301 (1967).
206. F. A. Hartman and A. Wojcicki, *Inorg. Chim. Acta,* **2,** 289 (1963).
207. C. M. Mitchell and F. G. A. Stone, *J. Chem. Soc., Dalton Trans.,* 102 (1972).
208. H. C. Clark and R. J. Puddephatt, *J. Chem. Soc., Chem. Commun.,* 92 (1970).
209. S. R. Su and A. Wojcicki, *Inorg. Chem.,* **14,** 89 (1975). For other examples of related insertion reactions of unsaturated molecules, see also J. P. Williams and A. Wojcicki, *Inorg. Chem.,* **16,** 3116 (1977); Y. Yamamoto and H. Yamazaki, *Inorg. Chem.,* **16,** 3182 (1977); R. G. Sverson and A. Wojcicki, *J. Organomet. Chem.,* **149,** C66 (1978).
210. J. A. Hanna and A. Wojcicki, *Inorg. Chim. Acta,* **9,** 55 (1974).

The Chemistry of the Metal—Carbon Bond, Volume 2
Edited by F. R. Hartley and S. Patai


CHAPTER **8**

# Transition metal—carbon bond cleavage through β-hydrogen elimination

R. J. CROSS
*Chemistry Department, The University, Glasgow G12 8QQ, Scotland, UK*

# I. INTRODUCTION

Transfer of a hvdrogen atom from the $\beta$-carbon atom of an organotransition-metal compound, either to the metal or to another substituent, usually leads to metal—carbon bond scission. Such reactions are termed $\beta$-elimination reactions. They often proceed readily under mild conditions and examples are numerous. Indeed, $\beta$-elimination was the first well defined low-energy pathway leading to transition metal—carbon bond breaking. With the reverse of this reaction providing a route for hydrogenating olefins, and featuring in many catalytic cycles, the process is unquestionably important.

$$L_xM\overset{\alpha}{C}H_2\overset{\beta}{C}H_2\overset{\gamma}{C}H_2R$$

Probably owing to the ease of the $\beta$-elimination reaction, transfers of hydrogen from $\alpha$- or $\gamma$-carbon atoms were not recognized until later, and fewer examples are known. The three reactions are formally related, however, as shown in equations 1–3. The depiction of the olefin complex from the $\beta$-hydrogen transfer as a metallacyclopropane is a formalism, but serves to emphasize the similarity between the processes.

$$L_xMCH_2CH_2CH_2R \xrightarrow{\alpha\text{-elimination}} L_xHM{=}CHCH_2CH_2R \quad (1)$$

$$L_xMCH_2CH_2CH_2R \xrightarrow{\beta\text{-elimination}} L_x\overset{H}{M}\!-\!CHCH_2R \text{ (ring closed by } CH_2\text{)} \quad (2)$$

$$L_xMCH_2CH_2CH_2R \xrightarrow{\gamma\text{-elimination}} L_x\overset{H}{M}\!-\!CHR \text{ (ring closed by } CH_2{-}CH_2\text{)} \quad (3)$$

It is immediately apparent that none of the hydrogen transfers of equations 1–3 involves breaking a metal—carbon bond; indeed, new M—C bonds are formed. Subsequent or concurrent elimination reactions commonly (but not invariably) lead to the bond cleavages.

This chapter examines first the $\beta$-elimination process, reviewing progress in understanding its mechanisms and requirements and relating them to theoretical considerations, then surveying the scope and applications of the reaction. $\alpha$-Eliminations, which appear to be taking an increasingly important role in organometallic chemistry as new examples continue to emerge, are also examined and are discussed next. Lastly, although clear examples which lead to metal—carbon bond cleavage are still rare, $\gamma$-eliminations are described. Metallation reactions, which can produce four-, five-, or even six-membered rings, are related to the metallacyclobutane formation of equation 3, and since the factors which control these processes are relevant to $\gamma$-eliminations, they are briefly discussed with a few examples.

The chapter is devoted to transition metal compounds and most considerations are relevant only to these elements. Examples of the processes are sometimes found for main group compounds, however, and although the intimate details of the reactions might be different, a few examples are included for comparison.

## II. β-ELIMINATION REACTIONS

The reactions might proceed by either of the extremes depicted in equations 4 and 5, or by some intermediate route.

$$L_xM(X)(CH_2{-}CHRH) \longrightarrow L_xM\cdots CH_2\cdots CHR\cdots H \ (X) \longrightarrow L_xM(X)(H)(CH_2{=}CHR) \quad (4)$$

$$L_xM(X)(CH_2{-}CHRH) \longrightarrow L_xM\cdots CH_2\cdots CHR\cdots H\cdots X \longrightarrow L_xM{-}\|(CH_2{=}CHR) + HX \quad (5)$$

It is elimination of the coordinated olefin from the products of reaction 4 or 5 (spontaneously, or via replacement by solvent or another ligand) which results in M—C cleavage. HX elimination (a necessary consequence of reaction 5, and an option from the product of equation 4) also cleaves metal—carbon bonds if X is another carbon-bonded ligand. Some examples are shown in equations 6–8[1–3]. That

$$trans\text{-}[PtCl(Pr^n)(PEt_3)_2] \rightleftharpoons trans\text{-}[PtClH(PEt_3)_2] + CH_2{=}CHMe \quad (6)$$

$$[Cr(Bu^t)_3(thf)_3] \xrightarrow{thf} [Cr(Bu^t)(thf)_n] + CH_2{=}CMe_2 + CHMe_3 \quad (7)$$

$$cis\text{-}[Pt(octyl)_2(PPh_3)_2] \longrightarrow [Pt(PPh_3)_2] + \text{octane} + \text{oct-1-ene} \quad (8)$$

equation 4 more often approaches the true situation than equation 5 has been demonstrated by several selective deuteriation studies. Reaction 9 illustrates this[4].

$$(Bu_3P)CuCH_2CD_2Et \longrightarrow [CuD(Bu_3P)] + CH_2{=}CDEt$$

$$[CuD(Bu_3P)] \xrightarrow{H^+} HD; \qquad [CuD(Bu_3P)] \xrightarrow{[Cu(Bu_3P)(CH_2CD_2Et)]} CH_2DCD_2Et + 2Cu + 2PBu_3 \quad (9)$$

The position of the single deuterium label of the but-1-ene product confirmed that the atom transferred originated from the β-position. The copper deuteride was not isolated, but its formation was inferred from the products of its bimolecular reaction with unreacted starting material. Also, when the reaction was performed in the presence of acid, HD was produced from the intermediate.

At the other end of the transition series, elimination reactions are sometimes less selective. Deuteriation revealed, for example, that α-and β-hydrogen transfers were common amongst alkylchromium(III) compounds[2]. The power of the method is illustrated by Scheme 1, which involves transfer of deuterium *from* zirconium to both a β-carbon and (after CO insertion) an α-carbon, followed finally by a β-elimination to yield an isolable zirconium hydride[5]. Without deuterium labelling, the course of this reaction would surely have remained obscure.

In many cases, however, the information available from deuteriation studies is reduced owing to the operation of a rapid reverse reaction that leads to scrambling of the labelled sites. It is not possible yet to predict when such scrambling will occur. No complications due to H–D scrambling occur during the thermolysis by β-elimination

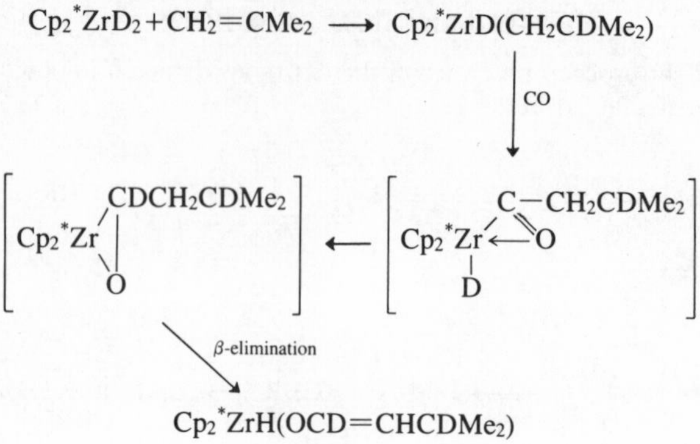

$$Cp_2^*ZrD_2 + CH_2{=}CMe_2 \longrightarrow Cp_2^*ZrD(CH_2CDMe_2)$$

$$\downarrow CO$$

$$\left[ Cp_2^*Zr(\text{-}CDCH_2CDMe_2\text{-}O\text{-}) \right] \longleftarrow \left[ Cp_2^*Zr(D)(\eta^2\text{-}C(CH_2CDMe_2){=}O) \right]$$

$$\xrightarrow{\beta\text{-elimination}} Cp_2^*ZrH(OCD{=}CHCDMe_2)$$

SCHEME 1

of the square-planar $d^8$ iridium(I) complex *trans*-$[Ir(CH_2CHDC_6H_{13})(CO)(PPh_3)_2]$[6]. On the other hand, easy reverse reactions at the related $d^8$ platinum complex *trans*-$[PtCl(CD_2CH_3)(PEt_3)_2]$ prevented confirmation of the site of hydrogen transfer (equation 10)[1].

$$[PtCl(CD_2CH_3)(PEt_3)_2] \longrightarrow trans\text{-}[PtHCl(PEt_3)_2] + trans\text{-}[PtDCl(PEt_3)_2] + \text{ethylene}(d_{1-3}) \quad (10)$$

Competition reactions coupled with selective deuteriation have revealed useful information, even in cases where H–D scrambling takes place. Thermolysis of a mixture of $[Pt(octyl)_2(PPh_3)_2]$ and $[Pt(CH_2CD_2Et)_2(PPh_3)_2]$ produced octane and octene (see equation 8) and deuteriated butane and butene[3].
No deuterium was incorporated in the $C_8$ products, so the eliminations were intramolecular. The $C_4$ products revealed extensive H–D scrambling, however, so although the $\beta$-transfer must be rapidly reversible in this case, once alkane or alkene is eliminated the process ceases to be reversible. No exchange between free and coordinated butene occurred, either.

One of the best proofs of the operation of equation 4 would be the isolation of a reaction product containing both hydride and olefin in the coordination sphere of the metal. Such compounds are rare, however, and the reason is now not hard to find. The ease of H–D scrambling indicates that the reverse of $\beta$-elimination (one step in olefin hydrogenation at metal atoms) is rapid. Thus if neither hydride nor olefin is lost from the metal after $\beta$-hydrogen transfer, the process will probably reverse. Equations 11–13 show three examples of detectable hydrido-olefin complexes[7–9].

$$[NbH(\eta^2\text{-}C_2H_4)Cp_2] \underset{-C_2H_4}{\overset{C_2H_4}{\rightleftharpoons}} [NbEt(\eta^2\text{-}C_2H_4)Cp_2] \quad (11)$$

$$[TaH(\eta^2\text{-}RCH{=}CH_2)Cp_2] \xrightarrow{CO} [Ta(CH_2CH_2R)(CO)Cp_2] \quad (12)$$

$$[Ni(\eta^3\text{-allyl})Et(PPh_3)] \rightleftharpoons [Ni(\eta^3\text{-allyl})H(C_2H_4)PPh_3]$$

$$\Updownarrow$$

$$Ni + PPh_3 + C_3H_6 \longleftarrow [Ni(\eta^3\text{-allyl})H(PPh_3)] + C_2H_4 \quad (13)$$

Some of the known hydride–olefin derivatives which do not participate in hydrogenation/$\beta$-elimination sequences may be prevented from doing so by an unfavourable geometry. Compound **2**, for example, has a *trans* arrangement ethylene and hydrogen, confirmed by X-ray crystallographic studies[10].

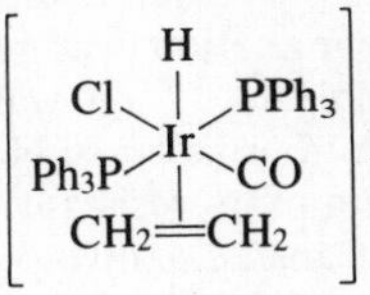

$$\left[\begin{array}{c} \text{H} \\ \text{Cl}\diagdown | \diagup\text{PPh}_3 \\ \text{Ir} \\ \text{Ph}_3\text{P}\diagup | \diagdown\text{CO} \\ \text{CH}_2{=}\text{CH}_2 \end{array}\right] \text{BF}_4$$

**(2)**

Before going on to discuss the chemical and geometrical requirements for β-elimination, it is worth pointing out one necessity, obvious from equations 4 and 5, namely that the α- and β-carbon atoms and one β-hydrogen atom need either to be within, or in close proximity to, the metal coordination sphere, or at least be able to approach sufficiently for a strong interaction at the transition state. There is plenty of independent evidence that this is indeed so. Both $^1H$ and $^{13}C$ n.m.r. spectra indicate strong coupling to $^{195}Pt$ of the β-$CH_2$ unit of the metallacyclobutane ring of $[(bipy)Cl_2\overline{PtCH_2CH_2CH_2}]$, and this probably results from a direct interaction[11]. The $\eta^1$-benzyl complex in equation 14 can be readily converted to an $\eta^3$-bonded moiety, suggesting ready availability of α, β, and γ-carbon atoms at the palladium atom[12].

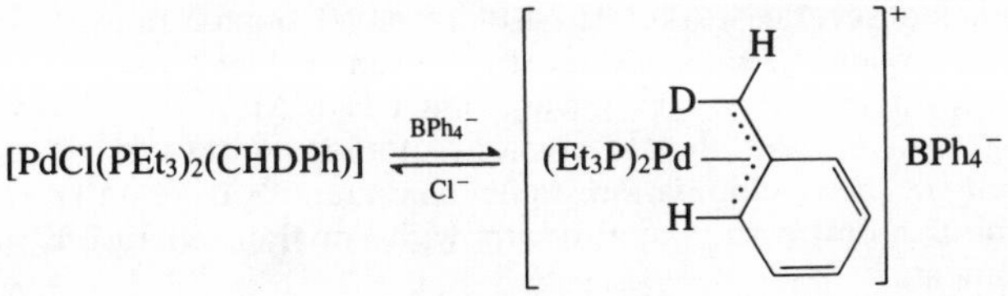

$$[PdCl(PEt_3)_2(CHDPh)] \underset{Cl^-}{\overset{BPh_4^-}{\rightleftharpoons}} \left[(Et_3P)_2Pd\text{—}(\eta^3\text{-CHD–C}_6\text{H}_5)\right]^+ BPh_4^- \qquad (14)$$

Also, the reduction of $[MoCp\{P(OMe)_3\}_2(PhC{\equiv}CPh)]^+$ by $[BH(Bu^s)_3]^-$ leads to a cyclic alkylidene compound, and it is likely that this comes about through the interaction of a β-carbon atom with Mo, despite an expected 120° angle at the α-carbon atom[13] (Scheme 2).

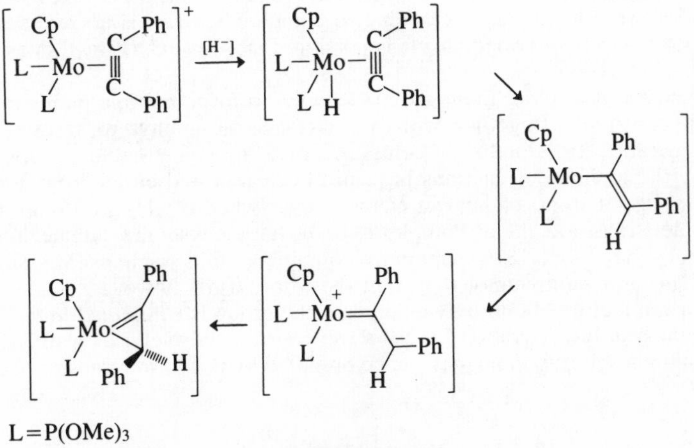

$$[MoCpL_2(PhC{\equiv}CPh)]^+ \xrightarrow{[H^-]} [MoCpL_2H(PhC{\equiv}CPh)] \rightarrow [MoCpL_2\text{–C(Ph)=C(Ph)H}] \rightarrow [MoCpL_2{=}\text{C(Ph)–CH(Ph)}]^+ \rightarrow [MoCpL_2(\text{cyclic C(Ph)–CH(Ph)})]$$

L=P(OMe)₃

SCHEME 2

### A. Elimination-stable Molecules: the Need for Transferable β-Substituents

Obviously a transferable group on the β-atom of the chain is necessary for β-elimination to proceed. Commonly this group is a hydrogen (or deuterium) atom. Long before the importance of β-elimination was recognized as a decomposition route of organotransition-metal complexes[14–16], it was realized that methyl derivatives, with no β-atoms, were more thermally stable than ethyl or propyl complexes[2,17,18] Aryl compounds also seem more resistant to thermolysis than long-chain alkyls[17,18]. Lack of a low-energy β-transfer pathway probably accounts for this, at least in part, although an aryltitanium complex has been shown to decompose via benzyne formation[19], and this could be by a β-hydrogen transfer (equation 15).

$$[TiCp_2(C_6D_5)_2] \longrightarrow [TiCp_2(C_6D_4)] + C_6D_6 \tag{15}$$

Benzyl compounds such as $[M(CH_2Ph)_4]$ (M = Ti, Zr, or Hf)[20] also have enhanced stability towards thermolysis, although it is unclear whether this results solely from a lack of transferable β-groups or from electronic or steric interactions with the metals[16]. The same considerations presumably apply to several pyridylmethyl complexes[21], which are more stable than their hydrocarbon-chain analogues.

The enhanced stability of these compounds can be taken as evidence that, in general, β-elimination is a lower energy process than other modes of thermal degradations of organotransition-metal complexes, including homolysis, reductive elimination, and binuclear elimination[16], as well as α- and γ-transfers. Recognition of this has enabled several workers to isolate novel organotransition-metal compounds by using organic groups specifically chosen to block β-elimination. These include $CH_2CMe_3$, $CH_2SiMe_3$, $CH_2SnMe_3$, and $CH_2P^+Me_3$[14,22–24] as well as variations involving bridging and chelate versions of these groups[24,25]. The success in isolating so many of these 'elimination stable' compounds suggests that transfer of H (or D) from the β-atom to the metal occurs with unparalleled ease compared with other β-substituents.

The ability of the transition metal to form a hydride, and the necessity to form a multiple bond between the α- and β-atoms, have occasionally been cited as requirements to enable β-elimination to operate, but this could be misleading. Thus the reluctance of silver to produce a silver hydride has been suggested[23] as a reason for the unimportance of β-elimination in the thermolysis of $[Ag(Bu^n)(PBu_3)]$[26]. This could, however, be due to orbital symmetry requirements (see below). Also, if β-elimination were to proceed according to equation 5, there is no requirement to form a hydrido-silver intermediate. It is possible that $[Ag(Bu^s)(PBu_3)]$ does decompose via β-elimination[26].

The reluctance of heavier main-group elements to form multiple bonds to carbon could well contribute (together with a reluctance of methyl to transfer to the transition metal) to the stability of complexes of 'elimination stable' groups such as $CH_2SiMe_3$. β-Elimination reactions have not been reported either from complexes where the α-ligand atom is a heavier element, even when β-$CH_2$ groups are present. For example, transition metal complexes of tertiary phosphines are legion[27], and many metal—metal bonded compounds containing units such as M—$SiMe_3$ are known[28]. This argument cannot represent the whole truth, however, as the requirement to form a multiple bond only emerges if the group is *eliminated from the metal* after β-atom transfer. Equation 2 shows the coordinated olefin as a metallacyclopropane, and phosphorus analogues can be prepared by β-hydride transfers (reactions 16 and 17)[29,30].

$$[Fe(PMe_3)_4] \rightleftharpoons \left[ Me_2P\begin{smallmatrix} CH_2 \\ | \\ FeH(PMe_3)_3 \end{smallmatrix} \right] \tag{16}$$

$$[RuCl_2(PMe_3)_4] + 2Na \longrightarrow 2NaCl + \left[ Me_2P\overset{CH_2}{\diagdown} RuH(PMe_3)_3 \right] \quad (17)$$

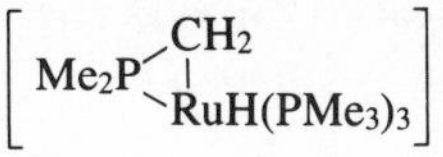

Although these $\beta$-transfers are intramolecular (no H–D exchange occurs when mixed with perdeuteriated compounds), a related bimolecular complex has been known for many years (**3**)[31]. Sulphur readily forms such cyclic compounds also (**4**)[32].

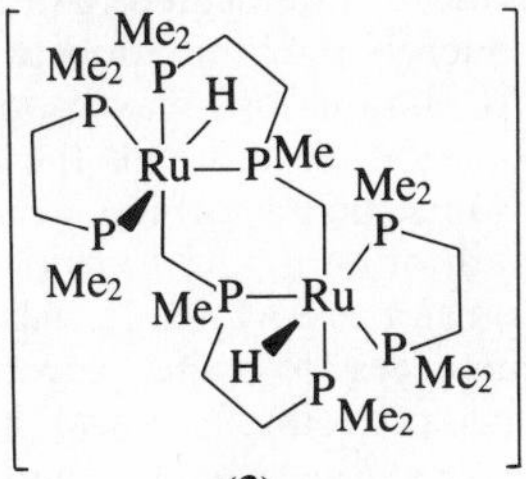

(**3**) 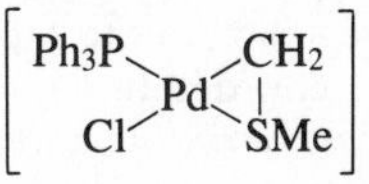 (**4**)

Thus, although the existence of many compounds with second-, third-, and fourth-row elements in an $\alpha$- or $\beta$-position to a transition metal, and which do *not* readily undergo $\beta$-substituent transfer reactions, is presumably a reflection that the activation energy for the process is higher than for compounds with $\alpha$- and $\beta$-atoms both carbon, the reason is not clear.

Equation 18 shows an interesting $\alpha$–$\beta$ rearrangement undergone by a family of

$$[Cp(CO)_2FeSiMe_2CH_2Cl] \xrightarrow{100\,°C} [Cp(CO)_2FeCH_2SiMe_2Cl] \quad (18)$$

iron—silicon compounds[33]. This reaction is unimolecular and could conceivably proceed via $\beta$-transfer of Cl to Fe, followed by $\alpha$-elimination of Si and Cl. This has not been proved, however, and a concerted Cl transfer via an intermediate such as **5**

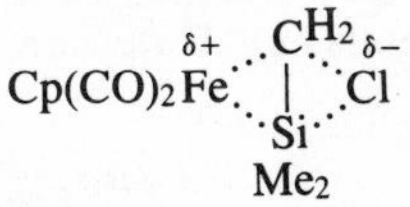

(**5**)

is plausible. Interestingly, the transfer of $\beta$-chloride substituents *to* silicon is known (equation 19)[34].

$$R_3SiCH_2CH_2Cl \longrightarrow R_3SiCl + C_2H_4 \quad (19)$$

## B. Availability of Coordination Sites

If the transfer of a $\beta$-atom substituent proceeds completely to the metal atom, as in equation 4, an extra coordination site is required on the metal whether or not any subsequent eliminations occur. The acknowledgement of this fact initiated a significant step forward in mechanistic organometallic chemistry[14a]. Thus, it was pointed out that the five-coordinate rhodium complexes $[RhR(CO)_2(PPh_3)_2]$ and $[RhR(CO)(PPh_3)_3]$, which readily lose CO or $PPh_3$ in solution, were labile towards $\beta$-elimination (equation 20)[35], whereas a variety of substitution-inert octahedral ions such as $[RhEt(NH_3)_5]^{2+}$ and $[Rh(C_2F_4H)(NH_3)_4(OH)_2]^{2+}$ resisted olefin elimination[36].

$$[\mathrm{RhEt(CO)(PPH_3)_3}] \rightleftharpoons [\mathrm{RhEt(CO)(PPh_3)_2}] + \mathrm{PPh_3}$$

$$\mathrm{C_2H_4} \Big\Uparrow\Big\Downarrow -\mathrm{C_2H_4} \qquad (20)$$

$$[\mathrm{RhH(CO)(PPh_3)_3}] \rightleftharpoons [\mathrm{RhH(CO)(PPh_3)_2}] + \mathrm{PPh_3}$$

It is now known that the true situation is more complicated than this. For example, the octahedral rhodium derivatives *do* undergo $\beta$-elimination at higher temperatures[36]. Even without a coordination site, $\beta$-eliminations may become possible via a geometry change or via a process like equation 5 (although the activation energy could be higher). Other geometrical constraints such as orbital symmetry restrictions might themselves dictate the thermolysis path of an organotransition metal complex, even if suitable coordination sites for $\beta$-transfer are available. Also, steric blocking from bulky *neighbouring* ligands might deny access to a site. Nevertheless, availability of coordination sites at the metal has proved of central importance to $\beta$-elimination, and we examine now several reactions where ligand eliminations are necessary prerequisites to $\beta$-eliminations.

### 1. *Cyclopentadienyl complexes*

Equation 21 shows the result of $\beta$-elimination from cyclopentadienyl iron complexes examined by Reger and Culbertson[37]. The reaction is first order in iron

$$[\mathrm{Fe(alkyl)Cp(CO)(PPh_3)}] \xrightarrow{\Delta} [\mathrm{FeHCp(CO)(PPh_3)}] + \text{alkene} \qquad (21)$$

complex when carried out in xylene solutions, but it is completely inhibited by addition of half an equivalent of $PPh_3$. Other donor species, such as thf or dioxane, also retard the reaction, indicating clearly that a critical step is loss of $PPh_3$ to make available a coordination site on iron. Thermolyses of both primary and secondary butyl complexes lead to the same product ratios of but-1-ene, *trans*-but-2-ene, and *cis*-but-2-ene, suggesting a fast isomerization by reversible $\beta$-hydrogen transfer prior to elimination. There is no significant deuterium kinetic isotope effect from either

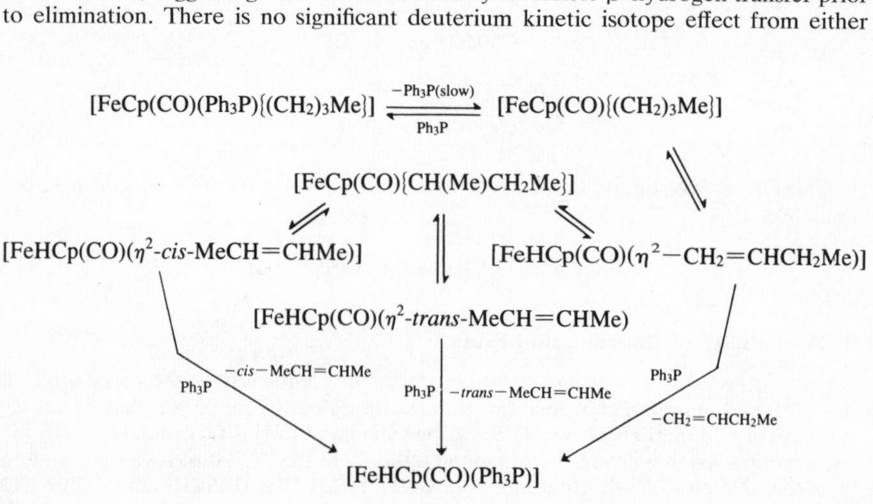

$$[\mathrm{FeCp(CO)(Ph_3P)\{(CH_2)_3Me\}}] \underset{\mathrm{Ph_3P}}{\overset{-\mathrm{Ph_3P(slow)}}{\rightleftharpoons}} [\mathrm{FeCp(CO)\{(CH_2)_3Me\}}]$$

$$[\mathrm{FeCp(CO)\{CH(Me)CH_2Me\}}]$$

$$[\mathrm{FeHCp(CO)}(\eta^2\text{-}cis\text{-}\mathrm{MeCH{=}CHMe})] \qquad [\mathrm{FeHCp(CO)}(\eta^2\text{-}\mathrm{CH_2{=}CHCH_2Me})]$$

$$[\mathrm{FeHCp(CO)}(\eta^2\text{-}trans\text{-}\mathrm{MeCH{=}CHMe})$$

$$\xrightarrow[\mathrm{Ph_3P}]{-cis-\mathrm{MeCH{=}CHMe}} \qquad \xrightarrow[\mathrm{Ph_3P}]{-trans-\mathrm{MeCH{=}CHMe}} \qquad \xrightarrow[\mathrm{Ph_3P}]{-\mathrm{CH_2{=}CHCH_2Me}}$$

$$[\mathrm{FeHCp(CO)(Ph_3P)}]$$

SCHEME 3

[Fe($CD_2CH_2$Et)Cp(CO)($PPh_3$)] or [Fe($CH_2CD_2$Et)Cp(CO)($PPh_3$)], indicating that H (or D) transfer is not rate determining. The deuterium labels are completely scrambled throughout the products *and* the alkyl groups of recovered starting material, however, indicating again reversible and rapid β-eliminations prior to final olefin loss. Reaction 21 is not reversible, however, so the final eliminations must be one-way steps. Scheme 3 accounts for all these features[37].

A very similar situation has been reported for [PdCp($PPh_3$)($CH_2CH_2Y$)] [Y = OMe or CH(COMe)$_2$] (equation 22), although failure to isolate or even detect hydride intermediates restricts the information accessible[38]. The β-elimination process in solution is retarded by added $PPh_3$, but dramatically accelerated if *m*-chloroperbenzoic acid (which oxidizes free tertiary phosphine) is added.

$$[PdCp(PPh_3)(CH_2CH_2Y)] \underset{Ph_3P}{\overset{-PPh_3}{\rightleftharpoons}} [PdCp(CH_2CH_2Y)]$$
$$\longrightarrow [PdHCp(\eta^2\text{-}CH_2{=}CHY)] \longrightarrow Pd + C_5H_6 + C_2H_3Y \quad (22)$$

Comparison with the iron complex previously described[37] suggests at first sight that the palladium compound should already have coordination sites available for β-transfer without further ligand loss. Such a transfer would lead to a 20-electron intermediate, however, and it is apparent that such constraints must be kept in mind in relation to coordination number.

Two further points emerge from this study of palladium compounds[38]. Firstly, it seems unlikely that the concerted elimination of the equation 5 type (which need not violate the 18-electron rule) can occur as a reasonably low-energy process for these compounds. Secondly, added phosphine produces some of the square-planar 16-electron complex [Pd($\eta^1$-$C_5H_5$)($PPh_3$)$_2$($CH_2CH_2Y$)]. Obviously β-elimination from *this* spcies also must be a process of higher energy than the route of equation 22.

The generality of this pattern is continued when cyclopentadienylcobalt(III) complexes are examined[39]. Loss of triphenylphosphine from [CoCpMe$_2$($PPh_3$)] allows ethylene to coordinate in its place, and this undergoes insertion into a cobalt—methyl bond to produce a propyl derivative. This propyl complex (**6** in Scheme 4) is

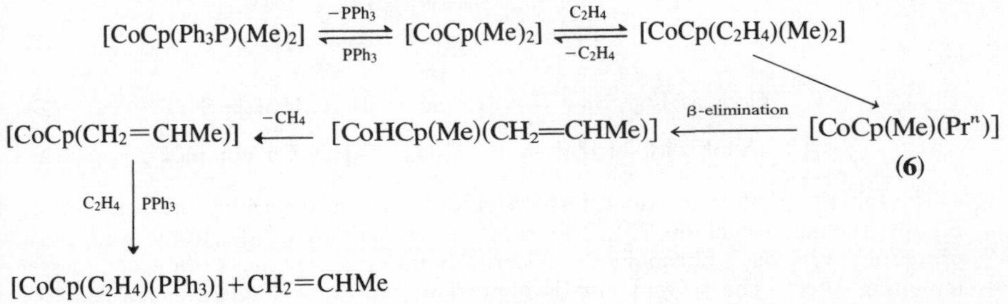

$$[CoCp(Ph_3P)(Me)_2] \underset{PPh_3}{\overset{-PPh_3}{\rightleftharpoons}} [CoCp(Me)_2] \underset{-C_2H_4}{\overset{C_2H_4}{\rightleftharpoons}} [CoCp(C_2H_4)(Me)_2] \longrightarrow [CoCp(Me)(Pr^n)]\ (\mathbf{6})$$

$$[CoCp(Me)(Pr^n)] \xrightarrow{\beta\text{-elimination}} [CoHCp(Me)(CH_2{=}CHMe)] \xrightarrow{-CH_4} [CoCp(CH_2{=}CHMe)]$$

$$[CoCp(CH_2{=}CHMe)] \xrightarrow{C_2H_4 \mid PPh_3} [CoCp(C_2H_4)(PPh_3)] + CH_2{=}CHMe$$

SCHEME 4

again coordinately unsaturated, and deuteriation studies proved β-hydrogen transfer from propyl to methyl, presumably via cobalt. Reversible loss of $PPh_3$ from [CoCpMe$_2$($PPh_3$)] and [CoCp($CD_3$)$_2$($PPh_3$)] can lead to intermolecular exchange of $CH_3$ and $CD_3$[40]. Presumably the same route is available for **6** (Scheme 4), so the absence of products derived from such exchanges indicates that the β-elimination process is faster.

An example where steric blocking prevents $\beta$-elimination, rather than actual occupation of a coordination site, is afforded by $[Zr(Bu^n)\{CH(SiMe_3)_2\}Cp_2]$. This complex is stable, whereas $[Zr(Bu^n)_2Cp_2]$, with a similar coordination number, but lacking the bulky bistrimethylsilylmethyl group, is thermally labile[41]. The unexpected stability of $[Cr(Bu^t)_4]$ might similarly be assigned to steric blocking[42].

Alkyluranium(IV) complexes, $[UR_4]$, prepared from RLi and $[UCl_4]$, decompose below room temperature to equal amounts of alkane and alkene, provided that R contains $\beta$-hydrogen. No free radicals are involved, and $\beta$-elimination probably operates[43]. The sterically crowded cyclopentadienyls $[URCp_3]$, on the other hand, do not decompose via $\beta$-hydrogen transfer to uranium[44]. Instead, the R group abstracts hydrogen from the $\pi$-bonded cyclopentadienyls. The restricted rotation when $R = Pr^i$ is evidence of the steric crowding, even at the large uranium ion.

The implications of these findings to metal—carbon bond scission via $\beta$-elimination are already clear. To proceed, the compound should be less than coordinately saturated and have an electron count lower than 18. This situation is most often reached by a ligand loss, so strongly bonding ligands should help prevent $\beta$-elimination. The process might still be prevented if the ligands are bulky enough to block transfer sites.

### 2. Octahedral complexes

A thermolysis study of the pseudo-octahedral cobalt complexes $[Co(acac)R_2L_2]$ (L = tertiary phosphine) found the activation energy for the reactions to decrease along the series $R = Me > Et > Pr^n > Bu^i$[45]. The compounds decomposed in either solid or solution, the methyl derivatives to ethane and the longer chain alkyls to 1:1 mixtures of alkane and alkene. The decompositions were first order, and $[Co(acac)(CH_2CD_3)_2(PMe_2Ph)_2]$ revealed a kinetic isotope effect of 2.3, indicating that $\beta$-hydrogen (deuterium) transfer was rate-determining[45]. N.m.r. spectroscopy showed that L was readily lost in solution. Also, added tertiary phosphine inhibited the thermolyses, so these compounds also fit the picture of ligand loss being necessary to generate a coordination site for $\beta$-elimination (equations 23–25), even though this step is not rate limiting.

$$[Co(acac)Et_2L_2] + \text{solvent (S)} \xrightleftharpoons{K_{23}} [Co(acac)Et_2LS] + L \tag{23}$$

$$[Co(acac)Et_2LS] \xrightarrow{\text{slow}} [Co(acac)EtH(C_2H_4)L] + S \tag{24}$$

$$[Co(acac)EtH(C_2H_4)L] \longrightarrow C_2H_6 + C_2H_4 + Co^I \text{ complex} \tag{25}$$

The inhibiting effect of added tertiary phosphine is in displacing the equilibrium, $K_{23}$, to the left (equation 23). The effect is transmitted by electronic *and* steric properties: very basic phosphines and sterically bulky phosphines both exert a lesser inhibiting effect. The solvent also is important: thermolysis rates decrease in the order acetone > toluene > dimethylformamide > pyridine. No proton abstraction from the solvent occurred.

When the deuterium-labelled material $[Co(acac)(CH_2CD_3)_2(PMe_2Ph)_2]$ was thermolysed rapidly as a crystalline solid (78 °C), or slowly in tolune (<5 °C), no H–D scrambling occurred and the volatile products of Co—C rupture were simply $CD_3CH_2D$ and $CH_2{=}CD_2$. In solution at ambient temperatures (28 °C), however, a complicated scrambling took place, presumably by a series of reversible $\beta$-transfers, and involving both ethyl groups. No exchange with free ethylene took place, however, and cleavage (by $H_2SO_4$) of unreacted starting material produced only

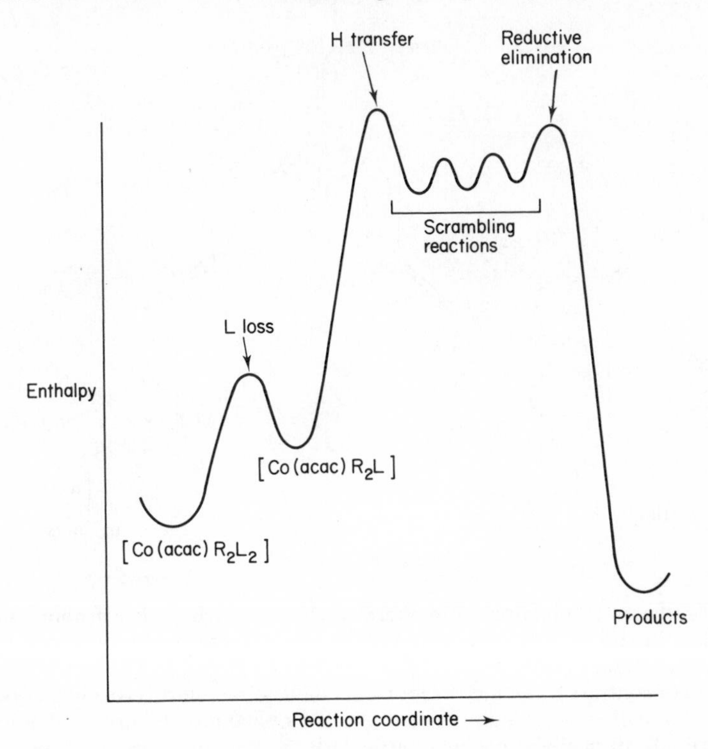


FIGURE 1. Reaction profile for the thermolysis of $[Co(acac)R_2L_2]$

$CH_3CD_3$, indicating that no starting material was regenerated after scrambling. Thus the rapid, reversible, $\beta$-transfers take place with lower energy barriers than for final elimination of product or regeneration of starting complex. This information, together with some n.m.r. derived enthalpies and activation energies, led the authors to propose the reaction profile shown in Figure 1.

Comparison of the $\beta$-elimination processes with the energetics of reductive elimination from the dimethyl derivative suggests that the $\beta$-hydrogen transfer took place after considerable 'thermal loosening of the metal—carbon bond'[45].

The iron complexes $[FeR_2(bipy)_2]$ with two chelating ligands, probably follow a similar thermolysis path to the cobalt compounds discussed, although less data are available for them[46]. When R is ethyl or $n$-propyl, $\beta$-elimination may well operate, although slightly more alkane than alkene was produced. The principal decomposition product of the dimethyl compound was methane, thus a proton abstraction of some sort can operate, and it might continue to operate at least as a minor path for the ethyl and propyl derivatives, accounting for the higher alkane yield. Creation of a coordination site again appears to be critical: kinetic studies indicate the importance of metal—ligand bond breaking, and excess bipyridyl inhibits the thermolyses. The decompositions are *accelerated* by the presence in solution of electronegative olefins, however, and solvent has a critical effect with the rate decreasing in the order acetonitrile > dmf > thf > furan > diethyl ether ≈ hexane. Scheme 5 was proposed to

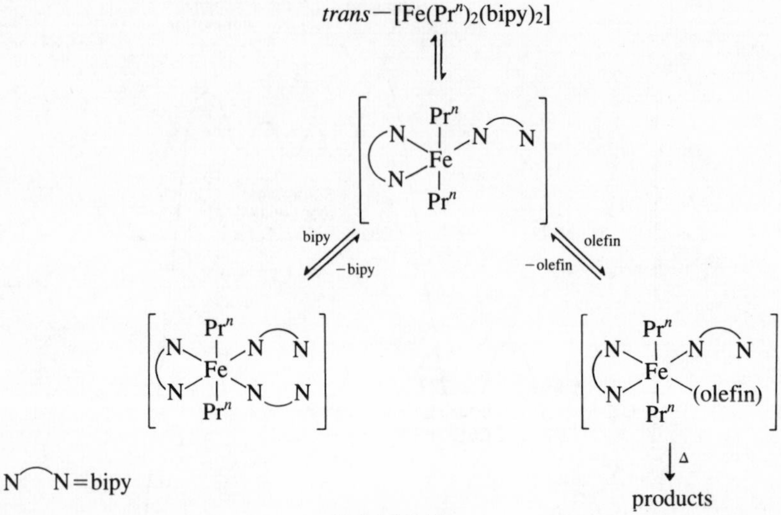

SCHEME 5

account for the $\beta$-eliminations, but there is clearly much to learn about metal—carbon bond breaking in this system, and its resemblance to the octahedral cobalt system may be superficial.

In complete contrast to all the reactions detailed so far, steric crowding appears to *initiate* $\beta$-elimination in pseudo-octahedral alkylcobalamins (Figure 2). These compounds are all sterically crowded, particularly if the alkyl group is other than primary. This crowding is partly relieved in acidic conditions when the axial 5,6-dimethylbenzimidazole base is protonated and the vacant site created allows the corrin ring system to bend away from R. Deprotonation allows the axial base to re-attack, with concurrent upwards movement of the corrin system (Scheme 6). This

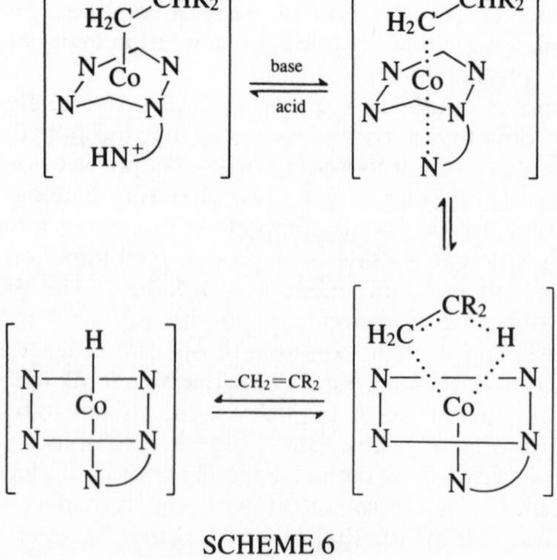

SCHEME 6

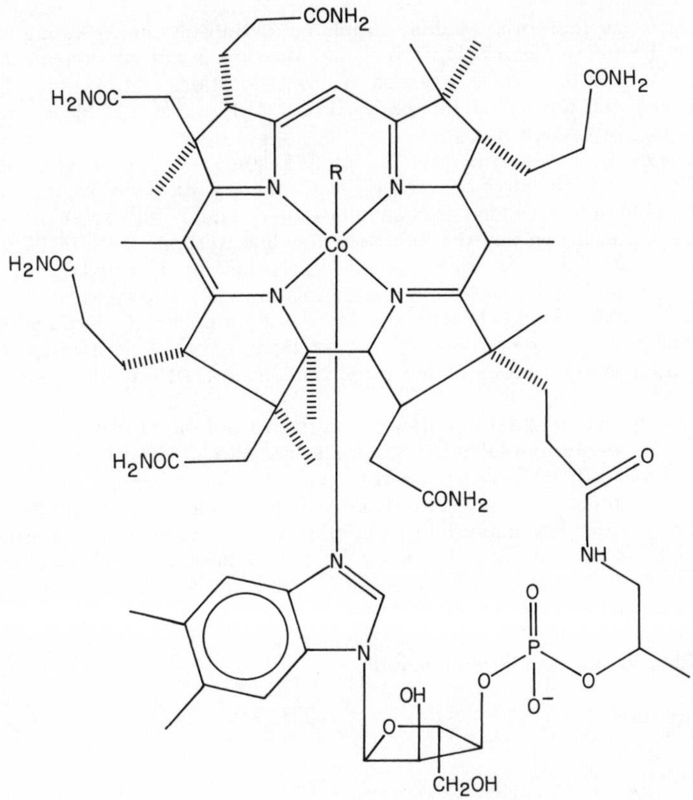

FIGURE 2. The structure of the alkylcobalamins

movement leads to rupture of the Co—C bond, by β-hydrogen transfer, moreover, if the β-carbon atom contains a hydrogen atom (otherwise homolytic fission results, for neopentyl, for example)[47]. This situation runs so counter to the requirements of coordination site and freedom from blocking ligands encountered so far that it is tempting to speculate that these β-transfers must be of a type where olefin loss occurs concurrently with Co—H formation (Scheme 6). Keeping in mind the proposal that thermal bond weakening was required to initiate β-hydrogen transfer in the compounds $[Co(acac)R_2L_2]$[45], it seems possible that a complete range of transition states may be plausible, from that depicted in equation 4, where the carbon metal bond(s) persists right to metal hydride formation, through to a situation like that for alkylcobalamins, where homolytic fission may be well advanced before hydrogen transfer begins.

### 3. Square-planar complexes

At first glance there would appear to be no problem here, as the sites above and below the square should be available for β-elimination. The compounds are 16-

electron species, moreover, so there should be no difficulty in expanding the electron count at the intermediates. However, the situation is not so simple, and there is controversy over the extent to which the axial locations can be utilized. We have already seen that the palladium species $[Pd(\eta^1\text{-}C_5H_5)(C_2H_4Y)(PPh_3)_2]$ are reluctant to undergo β-hydrogen transfer[38].

Whitesides *et al.*[3] carried out a detailed study on the thermolyses of *cis*-$[PtR_2(PPh_3)_2]$ (R = *n*-butyl or *n*-octyl; see equation 8). They found a first-order reaction, inhibited by added triphenylphosphine. N.m.r. and solubility studies led them to the conclusion that the inhibition was due to suppression of $PPh_3$ loss from Pt, rather than addition of excess ligand to an axial site. The chelate complex di-*n*-butyl[1,1′-bis(diphenylphosphino)ferrocene]platinum(II) decomposed (to *n*-butane and but-1-ene) more slowly than complexes of monodentate tertiary phosphines, presumably for the same reasons. It appears, then, that the lowest energy process for β-elimination must use one of the usual coordination sites of the platinum square plane.

Interesting additional information was derived from the preparation and thermolyses of the deuterium-labelled materials *cis*-$[Pt(CD_2CH_2Et)_2(PPh_3)_2]$ and *cis*-$[Pt(CH_2CD_2Et)_2(PPh_3)_2]$. There was no significant kinetic isotope effect from either, so the β-transfer step was not rate determining. Tertiary phosphine loss is probably rate limiting. Complete deuterium scrambling in the gaseous reaction products from both the 1,1-$d_2$ and the 2,2-$d_2$ complexes was apparent after thermolyses, but no exchange of free and coordinated butene was observed. Quenching unreacted starting material from thermolyses revealed no deuterium scrambling when no extra triphenylphosphine had been present, but significant H–D scrambling when $Ph_3P$ had been added to suppress β-elimination.

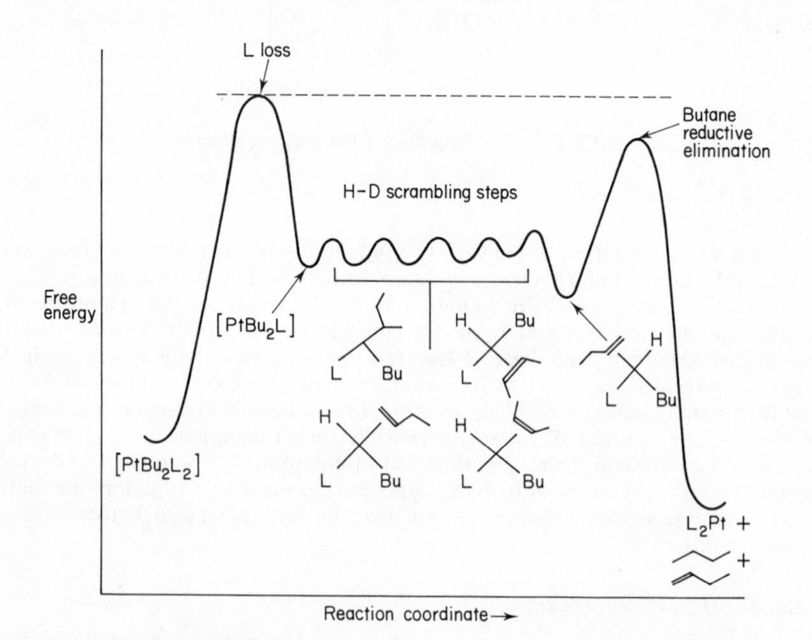


FIGURE 3. Thermolysis of $[PtBu_2(PPh_3)_2]$ in the absence of excess $PPh_3$(L)

Like the cobalt octahedral complex discussed previously[45], the H–D scrambling presumably occurs at facile reversible β-transfers after the initial ligand loss. But-1-ene must be thermodynamically favoured, since this is the only final olefin (equation 26). Reductive elimination of butane from the intermediate must be higher in energy than the H–D scrambling, in order for that to be complete before elimination. It must also be a higher energy process than butene elimination from the intermediate, or this process might lead to exchange with non-coordinated butene.

$$[PtH(Bu)(butene)(PPh_3)] \not\rightleftharpoons [PtH(Bu)(PPh_3)] + butene \quad (26)$$

Finally the energy barrier for the reverse reactions regenerating starting material from the intermediates must be higher than the barrier for butane elimination in the absence of excess ligand (hence no H–D scrambling in the starting complex), but lower than for butane elimination in the presence of excess phosphine. The two free energy diagrams in Figures 3 and 4 relate all these processes.

In a similar set of reactions, the more favourable reductive elimination step (compared with olefin elimination) has been exploited as a convenient preparation of zerovalent platinum olefin complexes[48a]. Only half the metal—carbon bonds are cleaved in these reactions (27 and 28), and the yields depend on both phosphine and alkyl groups.

$$cis\text{-}[PtEt_2(PEt_3)_2] \longrightarrow [Pt(C_2H_4)(PEt_3)_2] + C_2H_6 \quad (27)$$

$$cis\text{-}[Pt(C_5H_9)_2(PEt_3)_2] \longrightarrow [Pt(cyclopentene)(PEt_3)_2] + C_5H_{10} \quad (28)$$

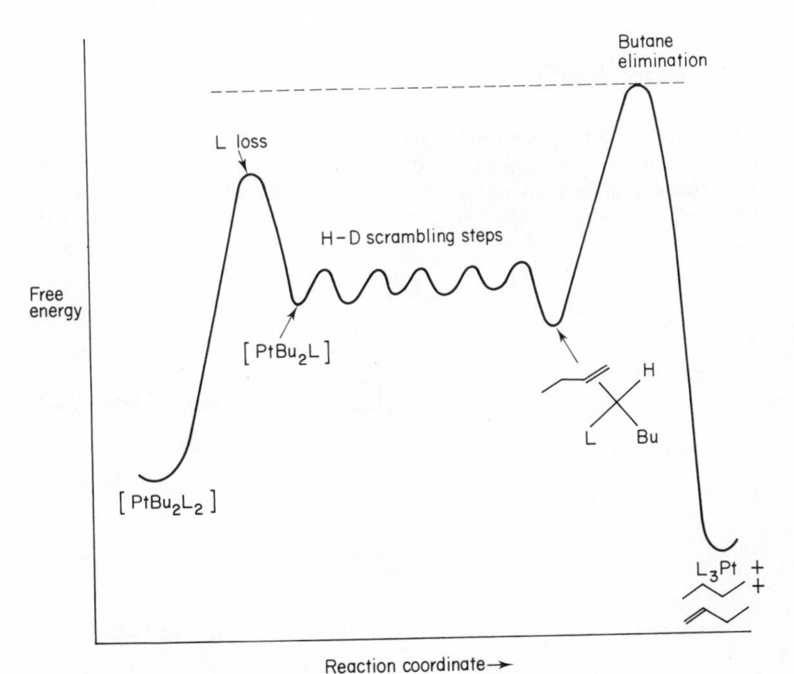


FIGURE 4. Thermolysis of $[PtBu_2(PPh_3)_2]$ in the presence of excess $PPh_3$(L)

Generally, trialkylphosphines and linear alkyls lead to better yields of $Pt^0$ complex (and thus less Pt—C bond cleavage) than triarylphosphines and cyclic alkyls. A similar route has led to syntheses of zerovalent palladium complexes[48b]. The mechanism of the reaction is probably similar to that for $[PtR_2(PPh_3)_2]$ above, but the details have not yet been elucidated. Since trialkylphosphines are less readily eliminated than triarylphosphines from Pt or Pd, the possibility cannot be dismissed that another route operates, possibly with a contribution from mechanisms such as equation 5.

The T-shaped three-coordinate intermediates essential to at least some of these $\beta$-eliminations were unexpected when first postulated, but evidence for many more such species featuring in a variety of processes has now accumulated[49], and the attainment of these 14-electron intermediates appears to be a relatively low-energy process in a number of cases, although their degree of solvation is unknown. Some closely related $\beta$-elimination/olefin hydrogenation sequences leading to Pt—C cleavage are shown in Scheme 7. These, too, require ligand loss from a square plane prior to the critical hydrogen migration, but in these cases a halide, $X^-$, is removed[50,51].

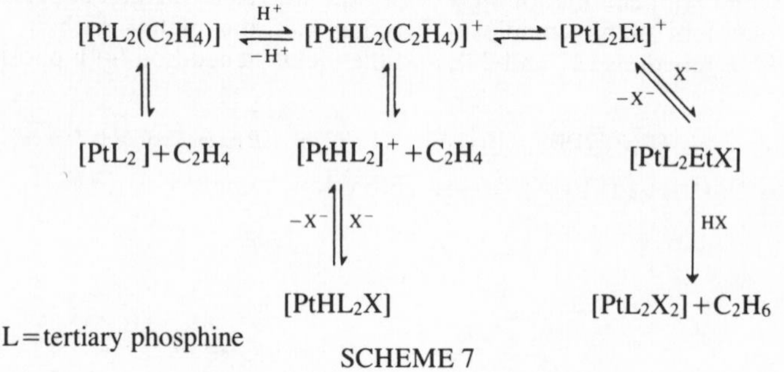

$$[PtL_2(C_2H_4)] \underset{-H^+}{\overset{H^+}{\rightleftharpoons}} [PtHL_2(C_2H_4)]^+ \rightleftharpoons [PtL_2Et]^+$$

$$[PtL_2(C_2H_4)] \rightleftharpoons [PtL_2] + C_2H_4$$

$$[PtHL_2(C_2H_4)]^+ \rightleftharpoons [PtHL_2]^+ + C_2H_4$$

$$[PtHL_2]^+ \underset{-X^-}{\overset{X^-}{\rightleftharpoons}} [PtHL_2X]$$

$$[PtL_2Et]^+ \underset{-X^-}{\overset{X^-}{\rightleftharpoons}} [PtL_2EtX] \xrightarrow{HX} [PtL_2X_2] + C_2H_6$$

L = tertiary phosphine

SCHEME 7

A further example of ligand loss from square-planar platinum(II) prior to Pt—C cleavage is afforded by the bipyridyl complex $[PtEt_2(bipy)]$[52]. Its decomposition is accelerated by olefins such as methyl acrylate, and the final products are ethylene and $[Pt\{CH(Me)CO_2Me\}_2(bipy)]$ only. Scheme 8 is supported by the first-order

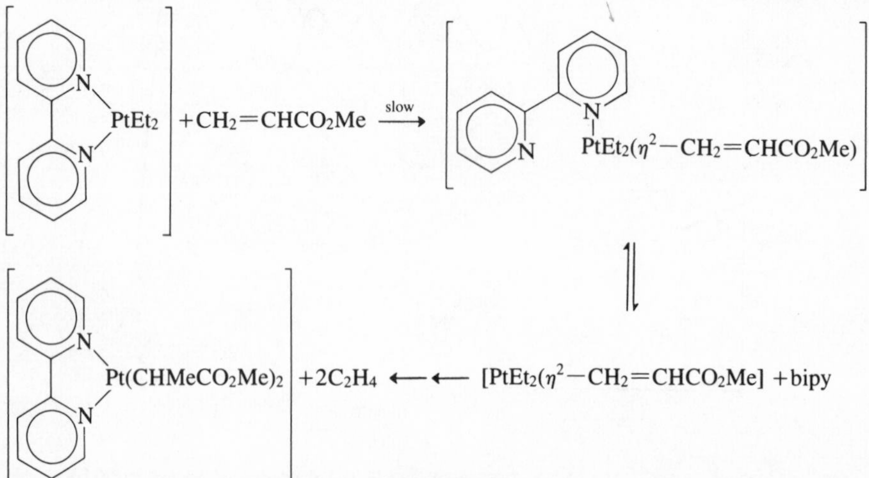

$$[PtEt_2(bipy)] + CH_2{=}CHCO_2Me \xrightarrow{slow} [PtEt_2(\eta^2\text{-}CH_2{=}CHCO_2Me)(\eta^1\text{-}bipy)]$$

$$\rightleftharpoons [PtEt_2(\eta^2\text{-}CH_2{=}CHCO_2Me)] + bipy \longrightarrow\longrightarrow [Pt(CHMeCO_2Me)_2(bipy)] + 2C_2H_4$$

SCHEME 8

dependence on the reactants and the strong retardation effect of added bipyridyl. The last steps, which could not be elucidated, involve β-hydrogen transfer from the ethyl groups, insertion of methyl acrylate into Pt—H, ethylene elimination, and bipyridyl recoordination[52].

The comparison of this olefin-accelerated decomposition with that of the octahedral complex $[FeR_2(bipy)_2]$ previously discussed[46] is interesting. Obviously the bipyridyl ligand can be readily displaced, at least from one site, and this limits its ability to stabilize organometallic compounds by blocking coordination positions[16]. A further interesting comparison is afforded by the nickel compounds $[NiR_2(bipy)]$[53]. The thermolysis of these compounds, too, is promoted by electron-withdrawing olefins such as methyl acrylate and tetracyanoethylene, and in some cases five-coordinate intermediates $[NiEt_2(olefin)(bipy)]$ can be isolated. Intermediates of this type probably also feature in the analogous platinum reactions[52], but subsequent decompositions of the nickel complexes[53] follow different courses. Some (such as the *n*-propyl complex) react by β-elimination (producing both propane and propene), but others (including the ethyl derivative) react by reductive elimination. The reason for the variations are not known.

Square-planar iridium(I) complexes, $[IrR(CO)(PPh_3)_2]$, undergo β-elimination as the rate-determining step in thermolyses reactions ($k_H/k_D = 2.2$)[6]. It is not known whether a rapid dissociation step precedes the β-atom transfer, however, because added triphenylphosphine coordinates with the primary product (reaction 29). This prevents its participation in a binuclear elimination of RH, thus ensuring a simpler reaction (30)[54].

$$[Ir(octyl)(CO)(PPh_3)_2] \xrightarrow{-octene} [IrH(CO)(PPh_3)_2] \xrightarrow{PPh_3} [IrH(CO)(PPh_3)_3] \quad (29)$$

$$[IrH(CO)(PPh_3)_2] + [Ir(octyl)(CO)(PPh_3)_2] \longrightarrow \text{octane} \quad (30)$$

Evidence against dissociation being a prerequisite to β-elimination from square-planar geometries has been provided by Yamamoto and coworkers[55]. Whereas *cis*-$[PdMe_2(PR_3)_2]$ and *cis*-$[PdEt_2(PR_3)_2]$ decompose by reductive elimination (which in these cases *does* require dissociation of $R_3P$), the *trans*-complexes $[PdR'_2(PR_3)_2]$ react cleanly by a β-hydrogen transfer process eliminating alkane and alkene. Unlike the related *cis*-triphenylphosphineplatinum complexes, excess ligand only slightly retards the process up to a limiting value, indicating that a direct elimination from the four-coordinate species predominates over the dissociative route (Scheme 9).

Deuteriation at the β-position shows no H–D scrambling in the products, and reveals a small but significant kinetic isotope effect ($k_H/k_D = 1.4$). The elimination

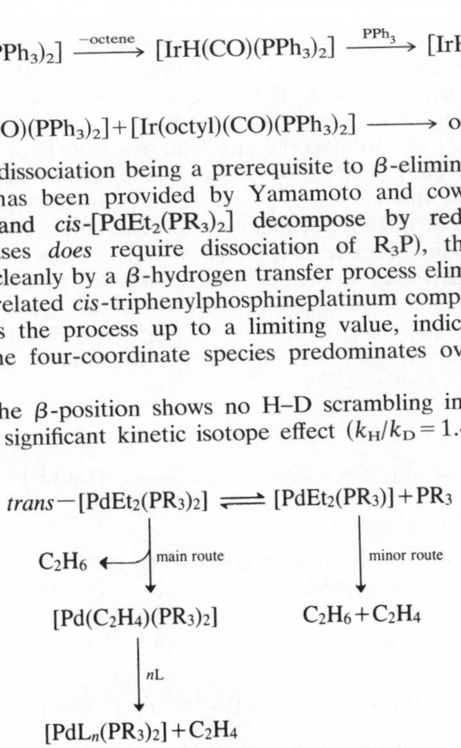

*trans*—$[PdEt_2(PR_3)_2] \rightleftharpoons [PdEt_2(PR_3)] + PR_3$

main route: $[PdEt_2(PR_3)_2] \rightarrow C_2H_6 + [Pd(C_2H_4)(PR_3)_2]$

minor route: $[PdEt_2(PR_3)] \rightarrow C_2H_6 + C_2H_4$

$[Pd(C_2H_4)(PR_3)_2] \xrightarrow{nL} [PdL_n(PR_3)_2] + C_2H_4$

SCHEME 9

reaction is accelerated by bulky tertiary phosphines, indicating elimination from a sterically crowded molecule (there is an electronic effect also, since small ligands such as pyridine can also accelerate the reactions). A positive entropy of activation (increasing with the bulk of $PR_3$) indicates a distortion to relieve strain prior to elimination. The small value of $k_H/k_D$ is in keeping with other processes as well as H (or D) transfer, being involved in the rate-determining step, and these are believed to involve both a lengthening of the Pd—C bond and distortion away from square planarity. The mechanism could involve activation by bond weakening, as postulated for the alkylcobalamin complexes (Scheme 6). It is also possible that reductive elimination could commence prior to complete transfer of hydrogen to metal. Scheme 10 outlines these possibilities.

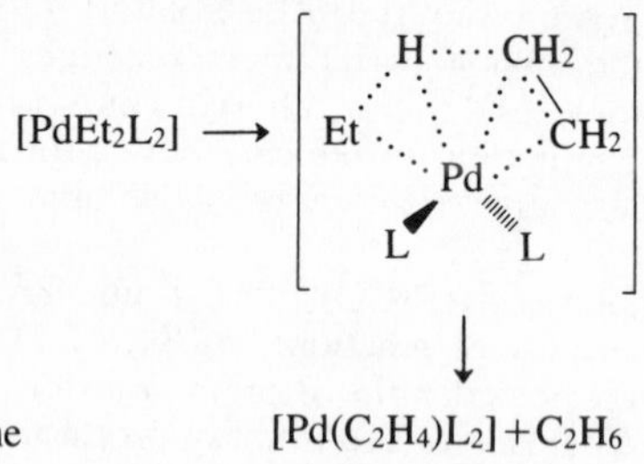

$$[PdEt_2L_2] \longrightarrow \left[ \text{Et} \cdots \text{Pd}(\text{H}\cdots\text{CH}_2\text{—CH}_2)\text{L}_2 \right] \longrightarrow [Pd(C_2H_4)L_2] + C_2H_6$$

L = tertiary phosphine

SCHEME 10

Further support for the importance of metal—carbon bond weakening prior to $\beta$-hydrogen transfer comes from a study of organocopper(I) compounds, $RCuL_n$ ($n = 1, 2,$ or 3)[56]. When R contains $\beta$-hydrogen atoms, the primary decomposition mode in both solid and solution is the $\beta$-elimination of olefins. The activation energy for elimination from $[CuEt(PPh_3)]$ is greater than that from $[CuEt(PPh_3)_2]$, indicating that ligand loss to create an extra coordination site is not of critical importance here. (Note, however, that these complexes are not square planar, and that different geometries may impose different symmetry requirements on the intermediates.) No loss of tertiary phosphine was detected cryoscopically in benzene. On the other hand, similar trends in activation enthalpies and entropies were noted for both $\beta$-eliminations from long-chain alkyls, *and* binuclear eliminations from methylcopper derivatives. The strength and ease of activation of Cu—C is obviously important in the latter cases, and is thus also likely to be important to the $\beta$-process. Equation 31 depicts the probable mechanism[56].

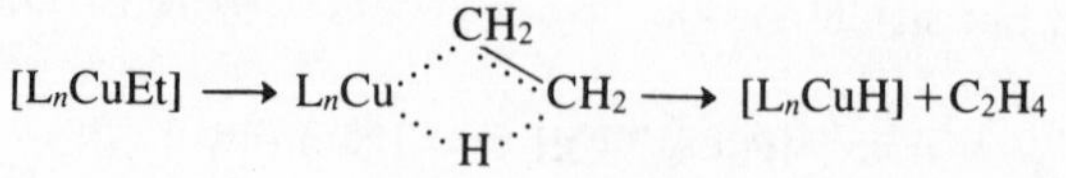

$$[L_nCuEt] \longrightarrow L_nCu\cdots(CH_2\text{—}CH_2\cdots H) \longrightarrow [L_nCuH] + C_2H_4 \qquad (31)$$

A clear case of a four-coordinate complex forming a five-coordinate species by $\beta$-elimination has been demonstrated for cobalt[57]. The five-coordinate cobalt(I) complex $[CoH(C_2H_4)(PMe_3)_3]$, itself a rare example of a compound containing both coordinated olefin and hydride, was isolated from reaction 32.

$$K[Co(C_2H_4)(PMe_3)_3] \xrightarrow{MeOH} [CoH(C_2H_4)(PMe_3)_3] + MeOK \qquad (32)$$

When the deuteriated solvent $CH_3OD$ was employed in the synthesis, a random distribution of the deuterium atom was found over the four olefinic sites *and* the cobalt site, indicating operation of a reversible β-elimination (reaction 33).

$$[CoEt(PMe_3)_3] \rightleftharpoons [CoH(C_2H_4)(PMe_3)_3] \quad (33)$$

Obviously no ligand loss is needed in this case.

In a similar vein, a five-coordinate hydridoplatinum olefin complex, generated photochemically, readily forms an ethyl complex by olefin insertion (reaction 34)[58].

$$[Pt(C_2H_4)(PPh_3)_2] \xrightarrow[254nm]{h\nu} [PtH(PPh_3)(PPh_2)(C_2H_4)] \longrightarrow [PtEt(PPh_3)(PPh_2)] \quad (34)$$

(in the two products, the PPh₃ and PPh₂ ligands are bridged by —R—) $R = 1,2\text{-}C_6H_4$

Although the geometry of the five-coordinate compound is unknown, this reaction represents the reverse step of reaction 33, and operates for platinum in a complex not unlike those found to proceed by ligand loss. It thus appears that for square-planar (and other coordinately unsaturated) species, use can be made of the spare coordination sites for β-atom migration leading to metal—carbon bond fission, but when ligand dissociation to create another site can occur readily, this affords a lower energy pathway for both β-elimination and its reverse, olefin hydrogenation.

Before moving on to discuss molecules with symmetry constraints, some further examples of β-elimination from square-planar molecules are worthy of mention at this stage. The gold(III) complex *trans*[$AuMe_2(Bu^t)(PPh_3)$] isomerizes readily to *trans*-[$AuMe_2(Bu^i)(PPh_3)$]. The reaction is first order in gold complex, and strongly retarded by added $PPh_3$, indicating again a dissociative process[59]. No exchange with free olefin occurs, suggesting a reaction profile similar to that proposed for *cis*-[$PtR_2(PPh_3)_2$][3]. Scheme 11 shows the likely mechanism.

$$trans\text{-}[AuMe_2(CMe_3)(PPh_3)] \rightleftharpoons [AuMe_2(CMe_3)] + PPh_3$$

$$\updownarrow$$

$$[AuMe_2(CH_2CHMe_2)] \longleftarrow [AuH(Me)_2(\eta^2\text{-}CH_2{=}CMe_2)]$$

$$\downarrow PPh_3$$

$$trans\text{-}[AuMe_2(CH_2CHMe_2)(PPh_3)]$$

SCHEME 11

Thermolyses of the mixed organoplatinum complexes *cis*-[$PtR^1R^2(PPh_3)_2$] generally produce products of β-elimination from both $R^1$ and $R^2$ (provided both have β hydrogen atoms) (reaction 35)[60]. In general, the amount of elimination from each R

$$cis\text{-}[PtEt(Pr^n)(PPh_3)_2] \xrightarrow{\Delta} \begin{cases} C_2H_4 + C_3H_8 \\ CH_2{=}CHCH_3 + C_2H_6 \end{cases} \quad (35)$$

group reflects the number of β-hydrogen atoms available. Thus, in the example above, the ratios of ethylene and propane to propylene and ethane are about 3:2.

Complications due to isomerizations arise with secondary alkyl derivatives, however, and once again the ready reversible isomerization, probably at the intermediate stages depicted in Scheme 11 and Figures 3 and 4, appear to operate.

Lastly, β-elimination from the organopalladium derivative shown in Scheme 12 is *promoted* by the addition of ligands such as $PPh_3$, *o*-phenanthroline, and iodide[61].

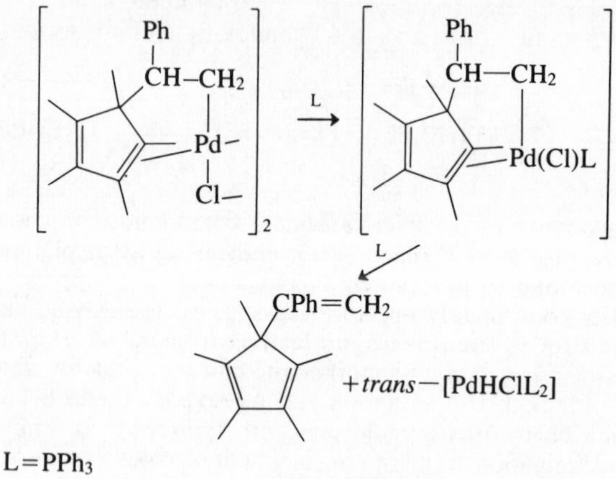

L=$PPh_3$

SCHEME 12

This does not necessarily mean that an increase in coordination number favours the β-transfer, as there is evidence that bridge cleavage precedes elimination. It does, however, serve as a reminder that the *nature* of the coordinating ligands is important as well as their number. The intimate mechanism is not known, and it is possible that the chelating nature of the organic ligand could be important, as discussed in the next section.

## C. Geometric Constraints and Metallacyclic Compounds

The constraints provided by some cyclic organic derivatives impose limitations on the β-hydrogen transfer process. They also introduce complications in the nature of competing reactions not encountered in acyclic molecules.

The square-planar complexes **7** and **8** containing five- and six-membered rings are

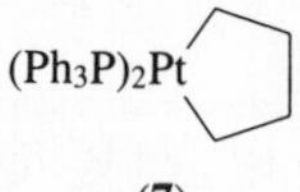

$(Ph_3P)_2Pt$ (**7**) $(Ph_3P)_2Pt$ (**8**)

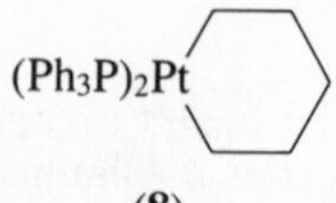

more thermally stable than their acyclic analogues[62]. Their thermolyses are a factor of $10^4$ slower than those of *cis*-$[PtEt_2(PPh_3)_2]$ or *cis*-$[Pt(Bu^n)_2(PPh_3)_2]$. The alkene products presumably arise via the usual β-transfer mechanism shown in equation 36

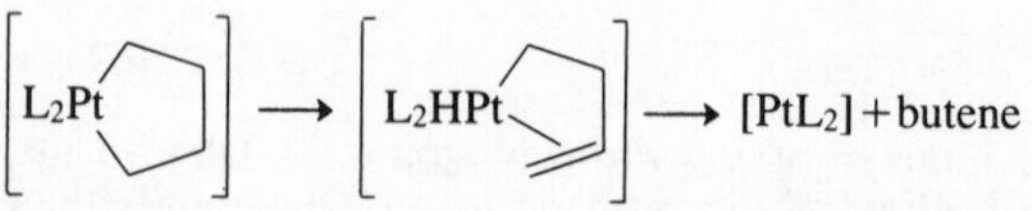

$$[L_2Pt] \rightarrow [L_2HPt] \rightarrow [PtL_2] + \text{butene} \quad (36)$$

(L = $PPh_3$). No detailed deuteriation studies to confirm this have been performed, however, so an $\alpha$-hydrogen transfer, which could lead to the same final products, cannot be ruled out (Scheme 13).

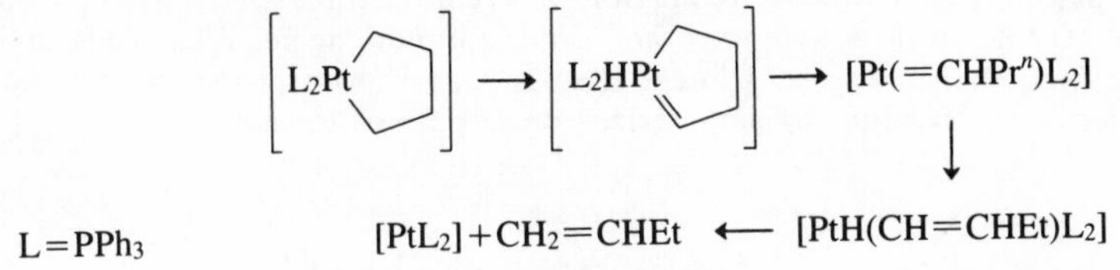

$$[L_2Pt(C_4H_8)] \longrightarrow [L_2HPt(C_4H_7)] \longrightarrow [Pt(=CHPr^n)L_2]$$

$$[PtL_2] + CH_2{=}CHEt \longleftarrow [PtH(CH{=}CHEt)L_2]$$

L = $PPh_3$

SCHEME 13

The additional stability of the cyclic compounds may well arise from the imposition of an unfavourable M—C—C—H dihedral. The optimum angle is presumably close to zero, allowing a close approach of hydrogen to the metal (structure **9**), but in five- or six-membered rings, angles near 90° are likely as in **10**. Seven-membered ring

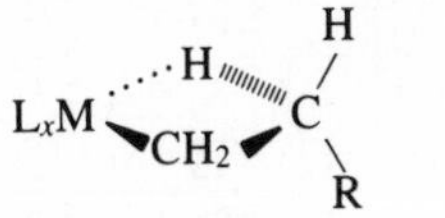

$L_xM$—$CH_2$—C(H)(H)(R) ··· H

(**9**)

$L_yM$—$CH_2$—C(H)(H), $CH_2$—$CH_2$ ··· H

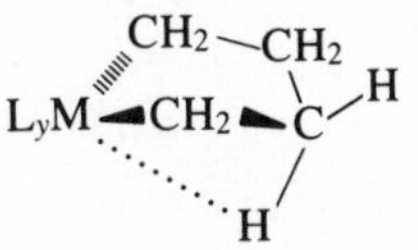

(**10**)

compounds, with greatly increased ring flexibility, more nearly resemble the acyclic complexes than **7** or **8**[62] in their propensity for $\beta$-elimination. The two $\alpha$-methyl-substituted derivatives **11** and **12** decompose at a similar rate to **7** and **8**, and

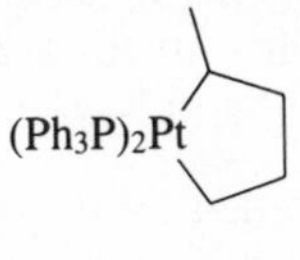

$(Ph_3P)_2Pt$

(**11**)

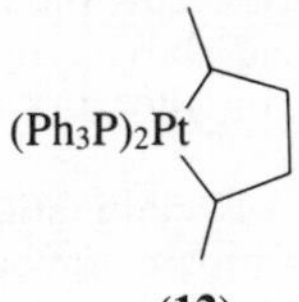

$(Ph_3P)_2Pt$

(**12**)

although the angular constraint of the ring would not affect the terminal methyl groups, $\beta$-hydrogen from the ring is transferred to Pt, and not from the side chains, for reasons unknown. It thus seems general that such cyclic compounds are less prone to M—C cleavage through $\beta$-elimination than linear alkyl derivatives.

Addition of tertiary phosphine to solutions of **7** or **8** *accelerates* their thermolyses, unlike acyclic analogues where an inhibiting effect is more common. The chelating diphosphine complex $[Pt(C_4H_8)(Ph_2PC_2H_4PPh_2)]$ is no more stable than its corresponding $PPh_3$ complexes. Equation 37 gives the rate law for thermolysis of **7**, with

$$-\frac{d[\mathbf{7}]}{dt} = (k_1 + k_2L)[\mathbf{7}] \qquad (37)$$

$k_1 = 0.54\,s^{-1}$ and $k_2 = 0.028\,s^{-1}\,l\,mol^{-1}$ at 120 °C. The evidence indicates that loss of tertiary phosphine to form a three-coordinate intermediate is *not* an important step in these thermolyses, but the geometry at platinum remains the same for both cyclic and acyclic compounds, so $Ph_3P$ loss presumably does take place in the cyclic derivatives also. Presumably then the $\beta$-hydrogen transfer at these three-coordinate cyclic intermediates must be unfavourable (even from side-chains in **11** and **12**). It thus appears that the lowest energy pathway for thermolysis by $\beta$-elimination from square planes occurs when it is possible to create a three-coordinate intermediate, and the organic group is able to rotate to the optimum angle. When these conditions are not met, then a five-coordinate species seems more prone to $\beta$-transfer than four-coordinate. Scheme 14 summarizes these options for compound **7**.

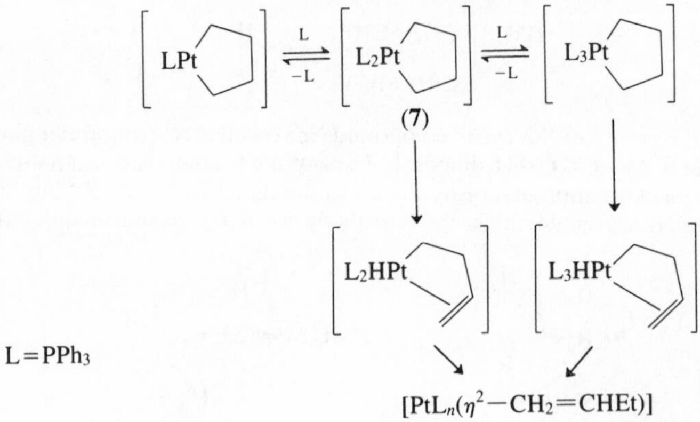

[LPt(ring)] ⇌ (L / −L) [$L_2Pt$(ring)] (**7**) ⇌ (L / −L) [$L_3Pt$(ring)]

[$L_2HPt$(butenyl)]    [$L_3HPt$(butenyl)]

L=$PPh_3$

[$PtL_n(\eta^2-CH_2=CHEt)$]

SCHEME 14

Thermolysis products from **7** are but-1- and -2-enes. No subsequent isomerization takes place after olefin elimination, so a situation resembling that for the thermolysis of *cis*-[$Pt(Bu^n)_2L_2$][3] must occur. Interestingly, those decompositions performed in the presence of added ligand show a significant shift towards but-1-ene. Practically no H–D scrambling between the ring and either $P(C_6D_5)_3$ ligand or $CD_2Cl_2$ solvent occurs[62].

The added barrier to $\beta$-elimination afforded by the cyclic compounds can allow other processes to compete favourably in some cases. Thus, whilst the *n*-butyltitanium(IV) complex [$TiCp_2(Bu^n)_2$] reacts unremarkably by $\beta$-transfer to yield butane and butenes, the cyclic compound [$TiCp_2(CH_2)_4$] reacts instead by carbon–carbon fission to produce mainly ethylene[63] and only a trace of but-1-ene (Scheme 15). The balance between $\beta$-elimination and C—C scission as metal—carbon cleav-

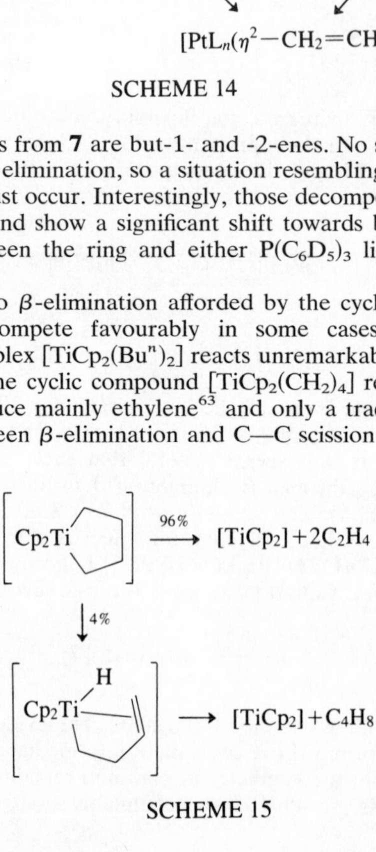

[$Cp_2Ti$(ring)] —96%→ [$TiCp_2$]+$2C_2H_4$

↓ 4%

[$Cp_2Ti$(H)(butenyl)] → [$TiCp_2$]+$C_4H_8$

SCHEME 15

age routes is subtle, as both compounds **13** and **14** react to give pentenes and only a trace of ethylene or propene.

$Cp_2Ti$ (13) $Cp_2Ti$ (14)

As with the platinum complexes discussed previously, the site of the hydrogen transferred has not been established, so although β-elimination is most likely, α-elimination cannot be ruled out[63]. (No C—C bond scission accompanied decomposition of the platinum complex **7**)[62].

A different type of competing reaction was found in the thermal decomposition of the platinacyclopentanes $[L_2PtC_4H_8]$ [$L_2 = (PBu^n_3)_2$, $(PEt_3)_2$, $Me_2PC_2H_4PMe_2$, or bipy][64]. Whilst the usual mixture of butenes arises from β-elimination processes, a molecule of $CH_2Cl_2$ solvent can add oxidatively to form new platinum(IV) species. These can decompose by reductive elimination of cyclobutane (equation 38), as well as by β-elimination and other pathways.

$$[L_2Pt(C_4H_8)] \xrightarrow{CH_2Cl_2} [L_2(CH_2Cl)ClPt(C_4H_8)] \longrightarrow \text{cyclobutane} + [PtCl(CH_2Cl)L_2] \quad (38)$$

The cyclobutane formed by reductive elimination from the tetravalent platinacyclopentanes was absent when chelating ligands were employed, and it has been pointed out that this is a necessary consequence of the symmetry requirements for these reactions[65]. To achieve a square-planar platinum(II) product after reductive elimination, monodentate ligands can simply move in the *xy* plane, an allowed process (equation 39).

$$[(Et_3P)_2Pt(CH_2Cl)(Cl)(C_4H_8)] \longrightarrow [(Et_3P)_2Pt(CH_2Cl)Cl] + \text{cyclobutane} \quad (39)$$

Chelate ligands are unable to do this, so the axial (*z*-axis) ligands would need to twist to *x* and *y* coordinates, a symmetry-forbidden sequence (equation 40).

$$[(Me_2PC_2H_4PMe_2)Pt(Cl)(CH_2Cl)(C_4H_8)] \not\longrightarrow [(Me_2PC_2H_4PMe_2)Pt(Cl)(CH_2Cl)] + \text{cyclobutane} \quad (40)$$

Thus, whilst the extra stability of the platinacycloalkanes can allow competing oxidative additions to interfere with β-eliminations, when reductive eliminations of cycloalkanes is disallowed the reactions of the platinum(IV) intermediates again include $\beta$-elimination[64]. The different thermolysis products from five- and six-membered platinacycloalkanes indicate that the compounds do not interconvert (equation 41), a possible consequence of reversible $\beta$-elimination.

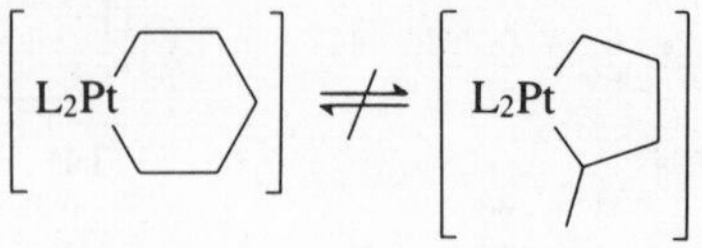

$$[L_2Pt(\text{cyclo-}C_5H_{10})] \not\rightleftharpoons [L_2Pt(\text{cyclo-}C_4H_7CH_3)] \quad (41)$$

The relationship of $\beta$-elimination to competing metal—carbon cleavage paths is even more complicated in cyclic nickel compounds[66]. $^{31}P$ n.m.r. and molecular weight studies have revealed both an easy loss and easy gain of $PPh_3$ by $[Ni(C_4H_8)(PPh_3)_2]$. The decomposition mode (Scheme 16) is entirely dependent on the coordination number.

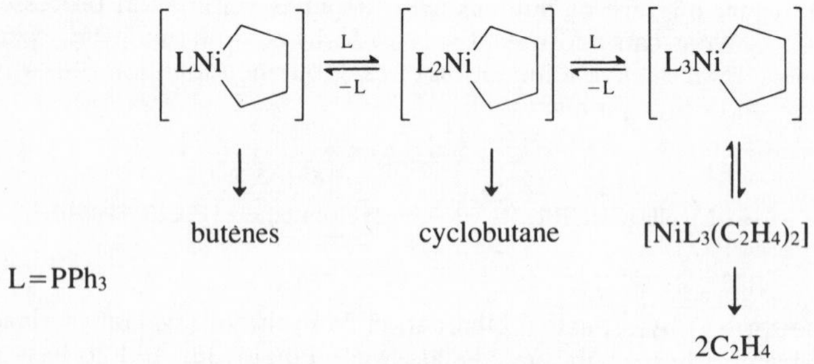

$$[LNi(C_4H_8)] \underset{-L}{\overset{L}{\rightleftharpoons}} [L_2Ni(C_4H_8)] \underset{-L}{\overset{L}{\rightleftharpoons}} [L_3Ni(C_4H_8)]$$

$[LNi(C_4H_8)] \rightarrow$ butenes

$[L_2Ni(C_4H_8)] \rightarrow$ cyclobutane

$[L_3Ni(C_4H_8)] \rightleftharpoons [NiL_3(C_2H_4)_2] \rightarrow 2C_2H_4$

$L = PPh_3$

SCHEME 16

The formation of cyclobutane from the four-coordinate material and of ethylene from the five-coordinate material have been shown to be, once again, consequences of orbital symmetry requirements[65]. The $\beta$-elimination from the three-coordinate species is surprising, however, as it is in direct conflict with the results of studies on cyclic platinum analogues[62]. The possibility remains, of course, that the products from either or both the nickel and platinum species do not in fact arise from genuine $\beta$-hydrogen transfers at all, but even so the differences between nickel and platinum are difficult to explain.

Yet another complication is introduced with the thermolysis of six-membered ring nickelaalkanes[66e]: and $\alpha$ C—C bond cleaves to form a carbene complex intermediate (reaction 42, $L = PPh_3$). The intermediate can be trapped by olefins (reaction 43). Of

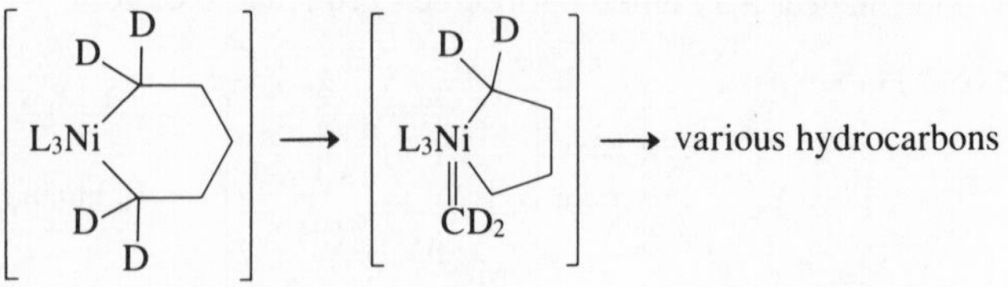

$$[L_3Ni(CD_2(CH_2)_3CD_2)] \longrightarrow [L_3Ni(=CD_2)(CD_2(CH_2)_3)] \longrightarrow \text{various hydrocarbons} \quad (42)$$

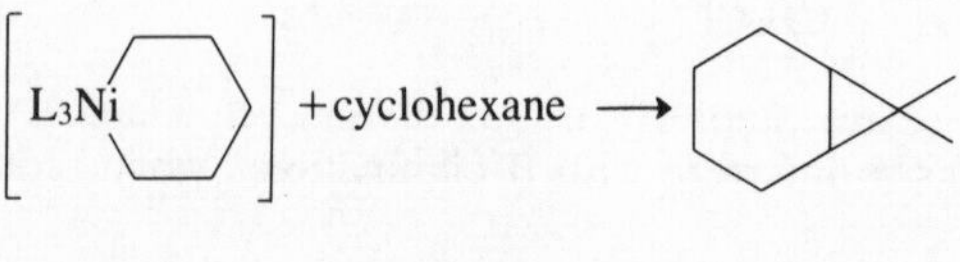

$$[L_3Ni(C_5H_{10})] + \text{cyclohexane} \longrightarrow \text{(bicyclic product)} \quad (43)$$

particular interest is the reaction with propene (reaction 44), as the nickelacyclobutanes formed decompose, among other routes, by $\beta$-elimination.

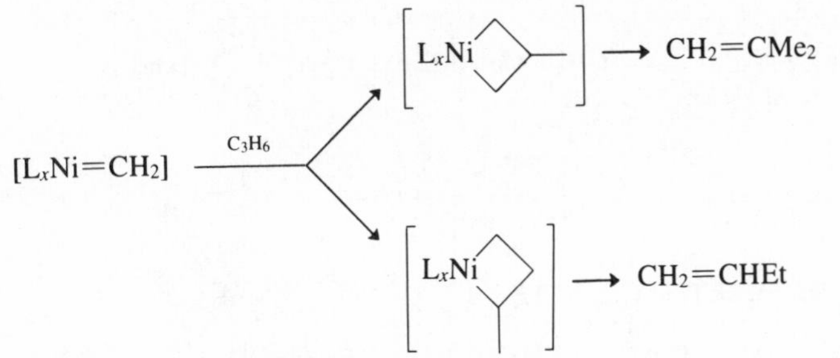

$$[L_xNi{=}CH_2] \xrightarrow{C_3H_6} \begin{cases} [L_xNi(\text{3-methylnickelacyclobutane})] \longrightarrow CH_2{=}CMe_2 \\ [L_xNi(\text{2-methylnickelacyclobutane})] \longrightarrow CH_2{=}CHEt \end{cases} \quad (44)$$

Among other four-membered metallacycles observed to undergo $\beta$-eliminations are some platinum(IV) derivatives. Other reactions, including isomerizations and reductive eliminations of cyclopropane compete[67], but $\beta$-eliminations have been proved in methyl-substituted compounds at least[68]. Selective deuteriation also showed that ring, rather than side-chain, groups were mainly involved (equations 45–47), as in the case of five-membered ring compounds[62].

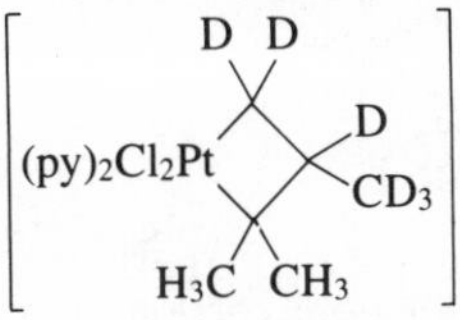

$$[(py)_2Cl_2Pt\text{-ring}(CD_2)(CD(CD_3))(C(CH_3)_2)] \longrightarrow D_3CC({=}CD_2)CDMe_2 + MeC({=}CH_2)CD(CD_3)(CD_2H) \quad (45)$$

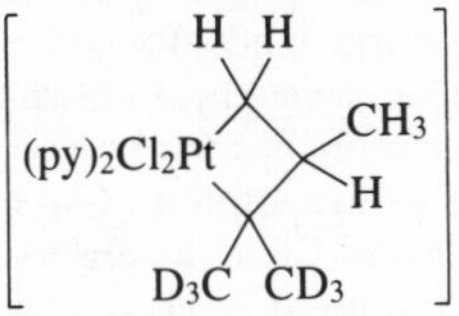

$$[(py)_2Cl_2Pt\text{-ring}(CH_2)(CH(CH_3))(C(CD_3)_2)] \longrightarrow D_3CC({=}CD_2)CHMe(CH_2D) + MeC({=}CH_2)CH(CD_3)_2 \quad (46)$$

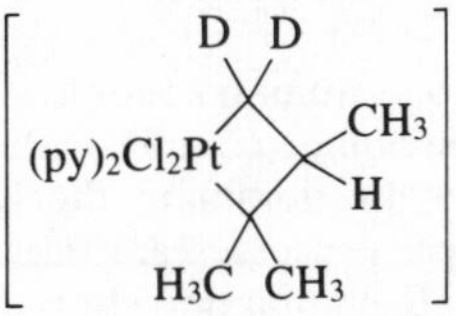

$$[(py)_2Cl_2Pt\text{-ring}(CD_2)(CH(CH_3))(C(CH_3)_2)] \longrightarrow MeC({=}CD_2)CHMe_2 + MeC({=}CH_2)CHMe(CD_2H) \quad (47)$$

Like the acyclic octahedral molecules discussed earlier, the process requires initial loss of a ligand to make a transfer site available. In this case pyridine is readily lost (reaction 48).

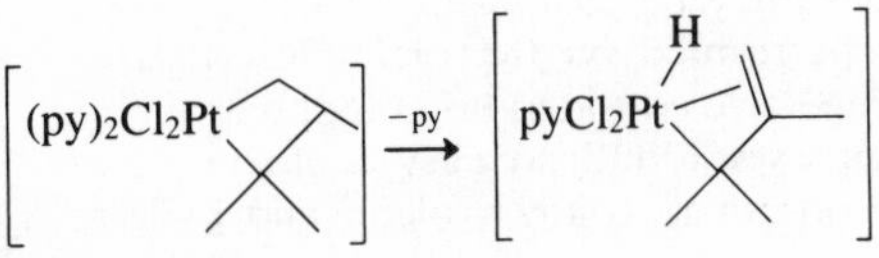

$$[(py)_2Cl_2Pt\text{-metallacycle}] \xrightarrow{-py} [pyCl_2Pt(H)(\text{alkene})] \xrightarrow{py} [PtCl_2(py)_2] + MeC({=}CH_2)CHMe_2 \quad (48)$$

Deuteriation studies on a binuclear titanium–aluminium metallacyclobutane also indicate the operation of the $\beta$-process (Scheme 17)[69].

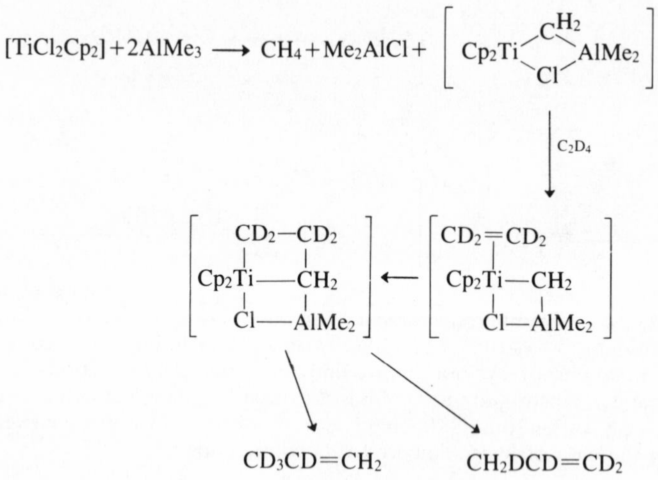

$$[TiCl_2Cp_2] + 2AlMe_3 \longrightarrow CH_4 + Me_2AlCl + \left[ Cp_2Ti \overset{CH_2}{\underset{Cl}{\diamond}} AlMe_2 \right]$$

$$\downarrow C_2D_4$$

$$\left[\begin{matrix} CD_2-CD_2 \\ | \quad\quad | \\ Cp_2Ti-CH_2 \\ | \quad\quad | \\ Cl-AlMe_2 \end{matrix}\right] \longleftarrow \left[\begin{matrix} CD_2=CD_2 \\ | \\ Cp_2Ti-CH_2 \\ | \quad\quad | \\ Cl-AlMe_2 \end{matrix}\right]$$

$$CD_3CD=CH_2 \qquad CH_2DCD=CD_2$$

SCHEME 17

## D. Theoretical Calculations

Before moving on to examples and applications of the $\beta$-elimination reaction, it is worth examining the results of some theoretical studies, as they can throw light on the processes involved. Considerations of orbital symmetry alone leads to some useful conclusions[70]. When the metal—ligand bonding is primarily through $d$-orbitals, it can be seen that the $\beta$-elimination/hydrogen migration is an allowed process (Figure 5). To proceed from left to right, the requirements are a filled M—H $\sigma$ orbital and empty M—ligand $\pi^*$ orbital. Electron transfer from filled to empty breaks (M—H) $\sigma$, (M—alkene) $\sigma$ and $\pi$, and (C—C) $\pi$, whilst forming (C—H) $\sigma$ and (M—C) $\sigma$. The process takes place in the plane of the paper, in keeping with the apparent experimental requirement of a zero dihedral[62].

With dominant $s$-orbital bonding from the metal, however, the symmetry requirements are not met (Figure 5). The orbital phase requirements cannot comply with both C—H and C—M bonding. This may be the reason why the dominant thermolysis modes of alkyls of $Ag^+$, $Hg^{2+}$, and $Tl^{3+}$ are free-radical processes and not $\beta$-eliminations. Copper(I) alkyls do, of course, decompose by $\beta$-elimination, but a possible rationalization may lie in the polymeric nature of many of these compounds. The eliminations could possibly be intermolecular[70]. $n$-Butyllithium, which is polymeric, decomposes to LiH and butene.

Finally, it can be seen from Figure 5c that dominant $p$-orbital bonding to the metal can comply with the requirements of $\beta$-elimination/hydrogen migration. Hydroboration, and $\beta$-elimination from aluminium alkyls, may be examples.

Molecular orbital calculations have been performed on the olefin insertion/$\beta$-elimination reactions at platinum(II)[71]. An exhaustive consideration of all reasonable geometries for phosphine/olefin/hydride complexes found no easy route via five-coordinate intermediates, nor a direct four-coordinate route if olefin and hydride

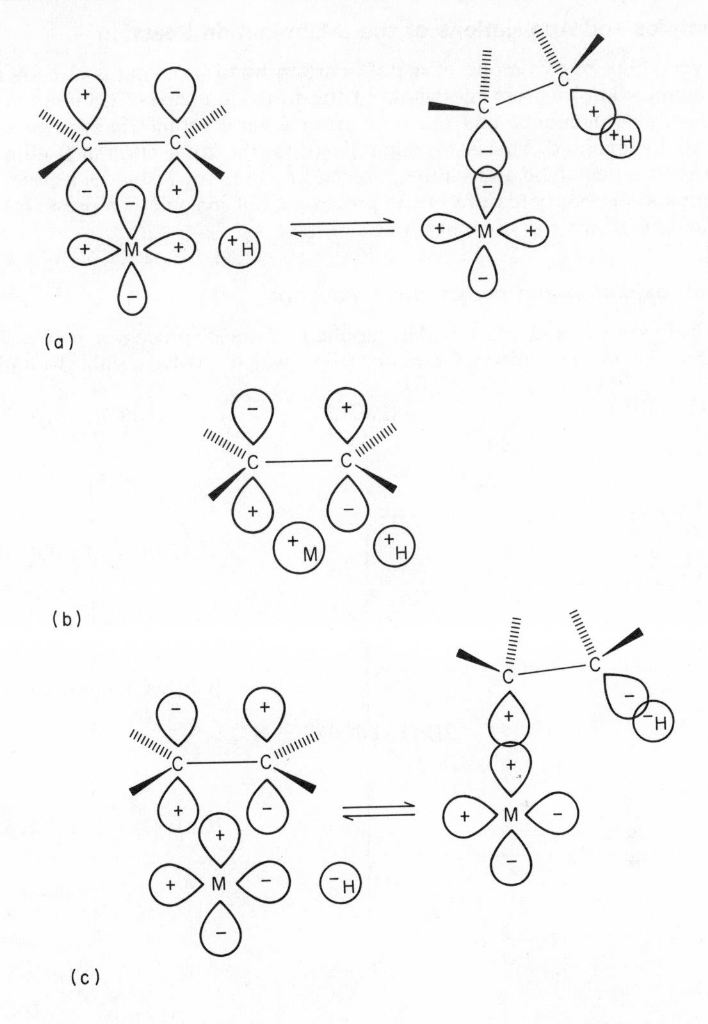


FIGURE 5. (a) $d$-orbital bonding; (b) $s$-orbital bonding; (c) $p$-orbital bonding

were *trans*. The lowest energy sequences involved three- and four-coordinate planar interactions, which fit well with the experimental evidence previously summarized.

It is worth noting that the nature of the metal—olefin bond produced by β-elimination may be flexible, so this may not represent a demanding requirement on the system. The traditional Chatt–Dewar–Duncanson model of synergic $d \rightarrow \pi^*$ and $\pi \rightarrow \sigma(M)$, dominated by the metal-to-olefin back-bonding[18], is probably largely correct. Theoretical and spectroscopic studies, however, reveal that the $\sigma$-donations from the olefin can be important, and even ionic forms can contribute significantly in some cases[72].

## E. Examples and Applications of the β-Elimination Reaction

The versatility of this mode of metal—carbon bond breaking is already apparent, with examples known from the whole of the transition series, including lanthanides and transuranic elements, and reactions from several geometric arrangements have been well documented. This last section illustrates the application of β-elimination in some important catalytic and synthetic reactions. In many instances its operation is a desirable and necessary feature of the processes, but in some it represents unwanted side-reactions leading to complications.

### 1. Olefin oxidation and substitution reactions

The best known and most widely applied of these processes is Wacker olefin oxidation. Scheme 18 outlines the major steps, which involve a vital β-elimination[73].

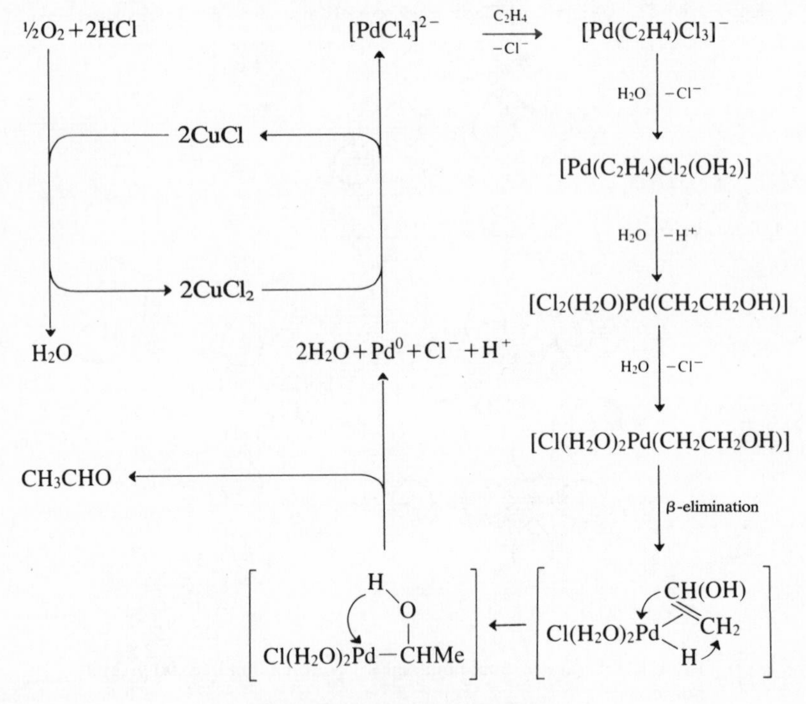


SCHEME 18 The Wacker process

These is now a lot of evidence that the attack of water on the olefin–palladium complex is *exo*, and not the reverse β-elimination type from the metal (although this might occur in certain situations[74]). *Exo*-attack was proved from the stereochemistry of the products from deuteriated olefins (reaction 49). The β-hydroxyethyl inter-

$$\left[ L_xPd-\| (CHD=CHD) \quad OH_2 \right] \xrightarrow{-H^+} \left[ L_xPd-CHD-CHD-OH \right] \qquad (49)$$

mediates were trapped by rapid CO insertion, and investigated as their ensuing lactones[75]. With substituted olefins, however, β-elimination was faster than CO insertion, and ketonic products resulted instead of lactones (Scheme 19).

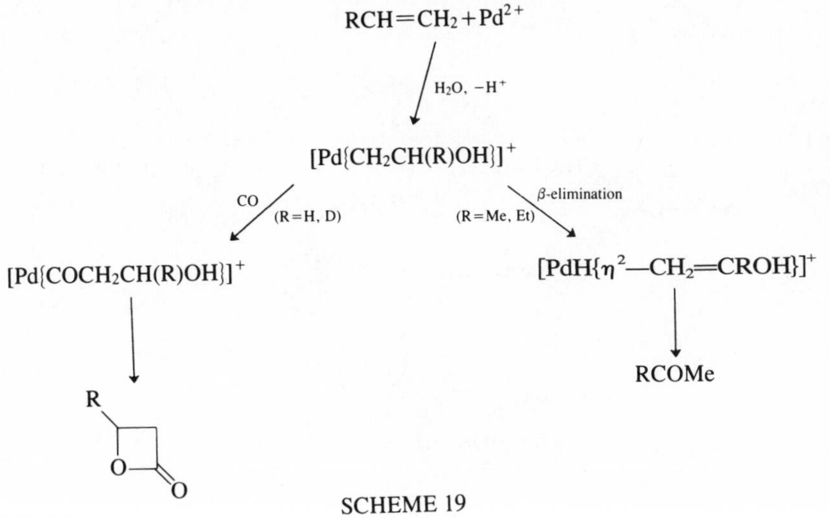

$$RCH{=}CH_2 + Pd^{2+} \xrightarrow{H_2O,\ -H^+} [Pd\{CH_2CH(R)OH\}]^+$$

$$[Pd\{CH_2CH(R)OH\}]^+ \xrightarrow[(R=H,\ D)]{CO} [Pd\{COCH_2CH(R)OH\}]^+ \longrightarrow \text{β-lactone (4-R-oxetan-2-one)}$$

$$[Pd\{CH_2CH(R)OH\}]^+ \xrightarrow[(R=Me,\ Et)]{\beta\text{-elimination}} [PdH\{\eta^2\text{—}CH_2{=}CROH\}]^+ \longrightarrow RCOMe$$

SCHEME 19

The rate-determining step of the Wacker process is that immediately prior to the β-elimination. In Scheme 18 this is shown as a ligand substitution, but it might well be a direct loss of $Cl^-$ to create a site for β-transfer[76] (equation 50).

$$[PdCl_2(H_2O)CH_2CH_2OH]^- \xrightarrow{-Cl^-} [PdCl(H_2O)(CH_2CH_2OH)] \longrightarrow [PdClH(H_2O)(\eta^2\text{-}CH_2{=}CHOH)] \quad (50)$$

In the related reaction of ethylene oxidation by acetic acid catalysed by $[Pd_2(OAc)_6]^{2-}$, there is kinetic evidence[77] that ligand loss is a necessary prerequisite to β-elimination (Scheme 20). β-elimination also features in another acetoxylation of olefins promoted by palladium[78]. This is shown in Scheme 21, and the process is made catalytic by a nitrocobalt(III) phenylporphyrin complex to oxidize palladium(0) byproduct. In a related system, however, the nitrocobalt complex plays a more direct role, and its interaction with the organic substrate is claimed to promote the β-elimination step itself[78] (Scheme 22).

It is obvious that the use of palladium(II) complexes in these processes is particularly well documented. Although this may be simply due to a concentration of effort on this element, certainly facile β-elimination to form the olefin products is a well proved feature of palladium chemistry. For example, a wide variety of organic groups can be transferred to palladium from mercury[79], and thence to coordinated olefins (*endo*-attack). β-Elimination then has the effect of liberating the substituted olefin[80] (equation 51), as in the previous examples.

$$[L_xPd^{II}] \xrightarrow{RHgX} [L_{x-1}PdR] \xrightarrow{\text{olefin}} [RCH_2CH_2PdL_y] \xrightarrow{\beta\text{-elimination}} RCH{=}CH_2 \quad (51)$$

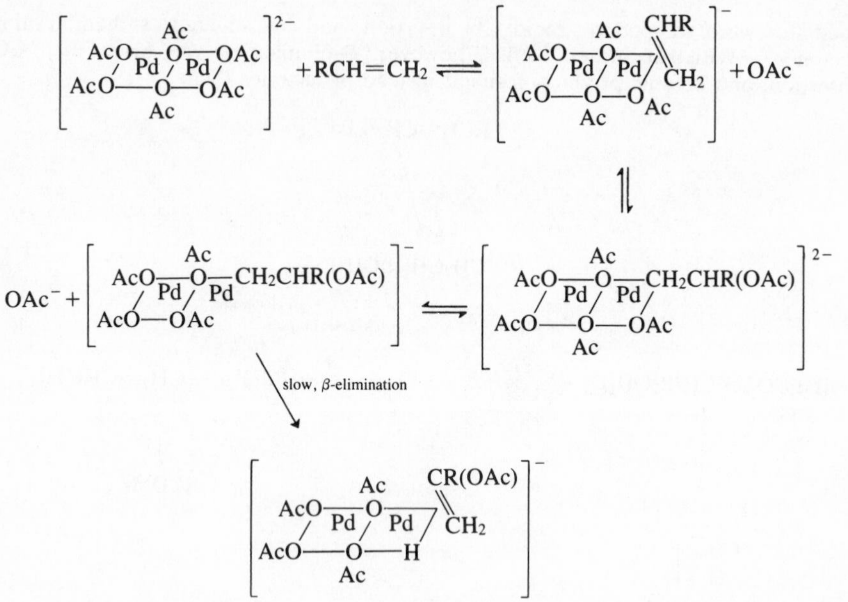

SCHEME 20

Recent useful examples include the acetoxylation of cyclooctadiene, accompanied by a photochemically induced rearrangement to a bicyclic olefin (Scheme 23)[81], the substitution by weakly basic carbanions of olefins (Scheme 24)[82], and an indole synthesis involving an *exo*-intramolecular attack (Scheme 25)[83]. On the debit side, $\beta$-elimination leads to unwanted olefin products in an arylation reaction designed to produce specifically labelled chlorides[84] (Scheme 26). The reversibility of this $\beta$-elimination also leads to H–D scrambling.

A ring closure of $\alpha$–$\omega$-diolefins is conveniently catalysed by $[(\eta^5\text{-}C_5Me_5)Cl_2Ta(C_2H_4)]$[85]. Scheme 27 outlines the proposed mechanism, which involves two $\beta$-eliminations at tantalum.

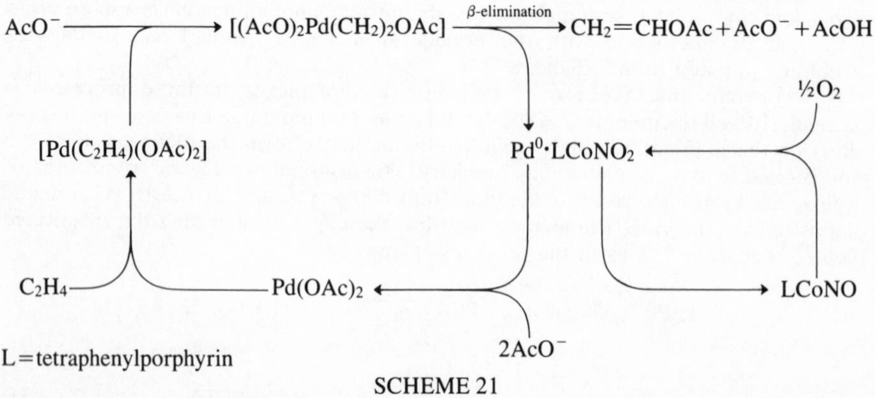

L = tetraphenylporphyrin

SCHEME 21

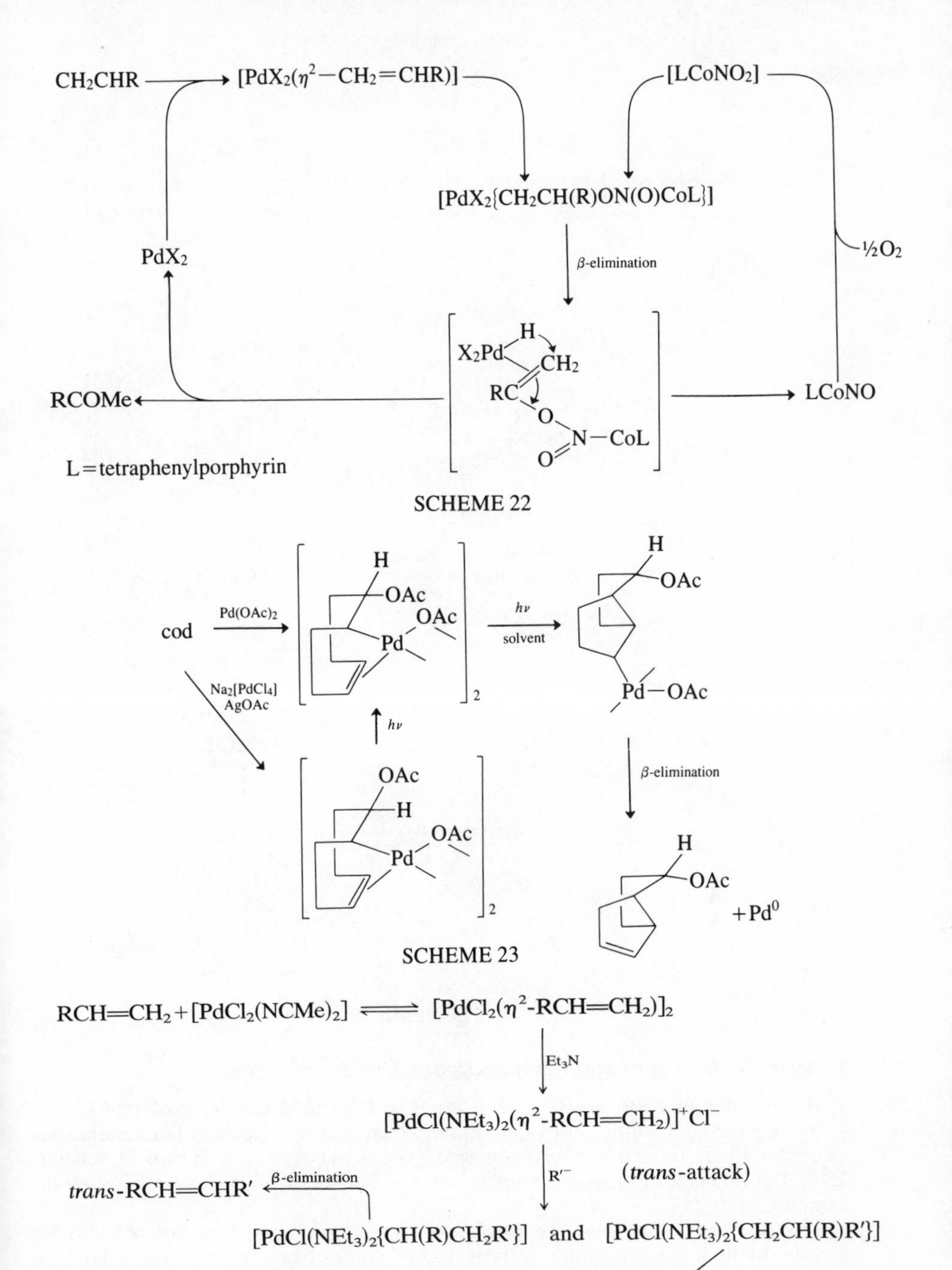


SCHEME 22



SCHEME 23

$$\text{RCH=CH}_2 + [\text{PdCl}_2(\text{NCMe})_2] \rightleftharpoons [\text{PdCl}_2(\eta^2\text{-RCH=CH}_2)]_2$$

$\downarrow \text{Et}_3\text{N}$

$$[\text{PdCl}(\text{NEt}_3)_2(\eta^2\text{-RCH=CH}_2)]^+\text{Cl}^-$$

$\downarrow \text{R}'^-$ (*trans*-attack)

*trans*-RCH=CHR′ ← β-elimination ← $[\text{PdCl}(\text{NEt}_3)_2\{\text{CH(R)CH}_2\text{R}'\}]$ and $[\text{PdCl}(\text{NEt}_3)_2\{\text{CH}_2\text{CH(R)R}'\}]$ → β-elimination → $\text{RR}'\text{C=CH}_2$

SCHEME 24

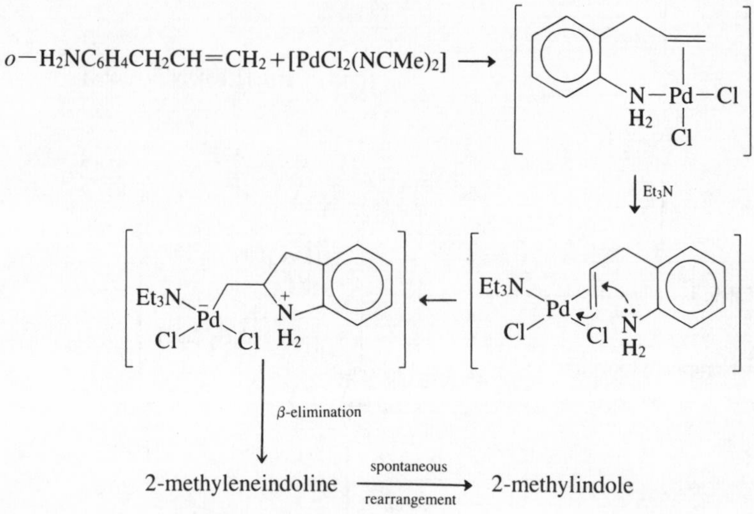

$o-H_2NC_6H_4CH_2CH=CH_2 + [PdCl_2(NCMe)_2] \longrightarrow$

Et3N

β-elimination

2-methyleneindoline $\xrightarrow[\text{rearrangement}]{\text{spontaneous}}$ 2-methylindole

SCHEME 25

$[PdClArL_x]$ + *trans*−DHC=CHD ⟶

ArPdClL$_y$

styrenes ← β-elimination

CuCl$_2$, LiCl

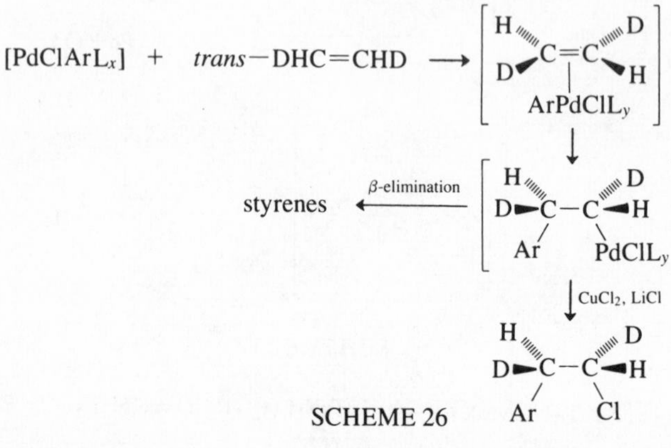

SCHEME 26

### 2. Metal hydride formation and hydrogen transfer reactions

The transfer of hydrogen from a $\beta$-atom to the metal can be exploited in the synthesis of metal hydrides, provided that these are stable to further eliminations. An example closely related to the olefin oxidation of Scheme 23 is shown in Scheme 28[86]. The detailed mechanism is unknown, but Pd—C breaking leaves the hydride complex.

This is a convenient place to emphasize that the $\alpha$-atom does not need to be carbon. Metal hydride formations from $\beta$-eliminations from, for example, ethoxides or formates are well known and have been used with $Pt^{II}$, $Ir^{III}$, $Os^{II}$ and many other elements (reactions 52 and 53)[87]. Any M—C cleavage in these cases is, of course, only secondary, but could result from reductive elimination of alkyl or aryl with the hydride formed.

$$[PtCl(OCH_2CH_3)(PEt_3)_2] \longrightarrow trans\text{-}[PtHCl(PEt_3)_2] + CH_3CHO \quad (52)$$

$$[Os(O_2CH)_2(CO)_2(PPh_3)_2] \longrightarrow [OsH_2(CO)_2(PPh_3)_2] + 2CO_2 \quad (53)$$

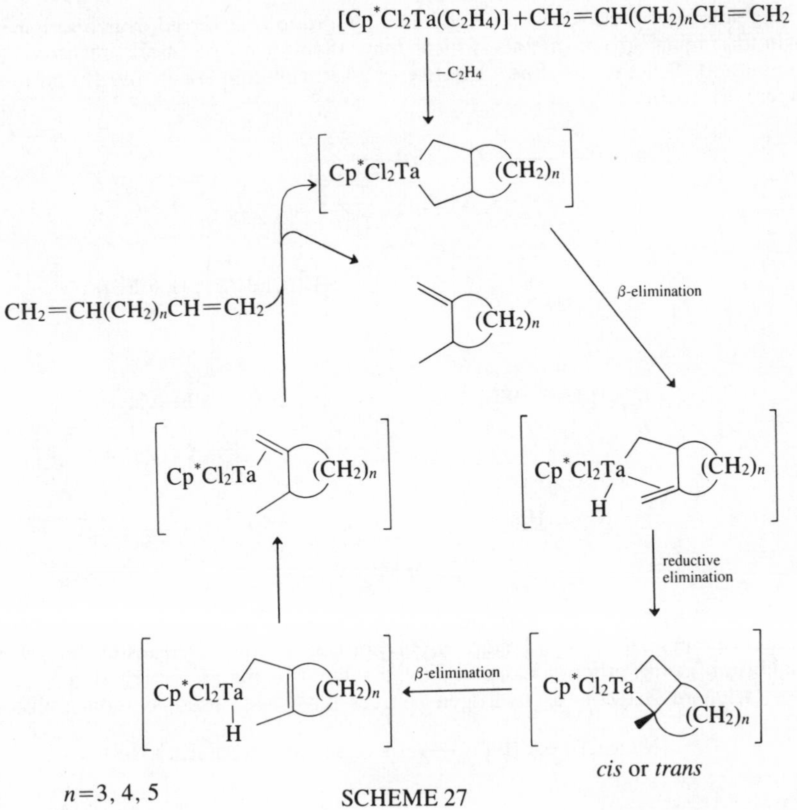

$[Cp^*Cl_2Ta(C_2H_4)] + CH_2{=}CH(CH_2)_nCH{=}CH_2$

$-C_2H_4$

$[Cp^*Cl_2Ta \; (CH_2)_n]$

$CH_2{=}CH(CH_2)_nCH{=}CH_2$

$(CH_2)_n$

β-elimination

$[Cp^*Cl_2Ta \; (CH_2)_n]$

$[Cp^*Cl_2Ta(H) \; (CH_2)_n]$

reductive elimination

$[Cp^*Cl_2Ta(H) \; (CH_2)_n]$

β-elimination

$[Cp^*Cl_2Ta \; (CH_2)_n]$

*cis* or *trans*

$n = 3, 4, 5$

SCHEME 27

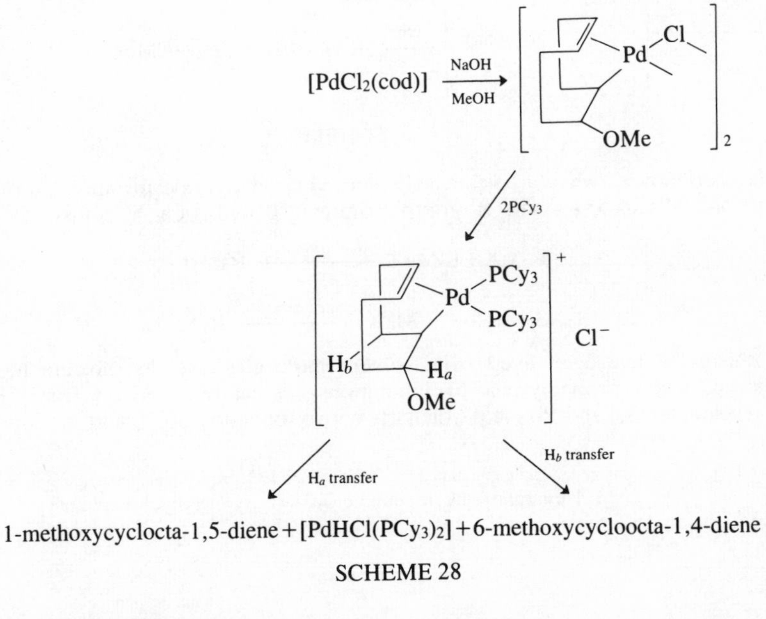

$[PdCl_2(cod)]$ $\xrightarrow[\text{MeOH}]{\text{NaOH}}$ $[\,Pd\,Cl\;\; OMe\,]_2$

$2PCy_3$

$[Pd(PCy_3)_2 \; H_b \; H_a \; OMe]^+ \; Cl^-$

$H_a$ transfer

$H_b$ transfer

1-methoxycyclocta-1,5-diene + $[PdHCl(PCy_3)_2]$ + 6-methoxycycloocta-1,4-diene

SCHEME 28

The source of the hydrogen can be an alkyl group transferred from another metal, including main group metals which may themselves be more resistant to $\beta$-elimination. Scheme 29 shows the use of triethylaluminium in the formation of a zirconium hydride[88].

$$[Cp_4Zr] \xrightarrow[-Et_2AlCp]{Et_3Al} [Cp_3ZrEt] \xrightarrow[-C_2H_6]{Et_3Al} [ZrCp_3(CH_2CH_2AlEt_2)]$$

$$[ZrCp_3(CH_2CH_2AlEt_2)] \longrightarrow [Cp_3HZr\cdots AlEt_2(vinyl)]$$

$$[Cp_3HZr\cdots AlEt_2(vinyl)] \xrightarrow[-Et_2Al(vinyl)]{Et_3Al} [Cp_3HZr\cdots AlEt_3] \longrightarrow [Cp_3ZrH]$$

SCHEME 29

The metal hydrides can often be used *in situ* as reducing agents to transfer H to another olefin. This second step is, of course, another $\beta$-transfer but with its equilibrium at the other side. An example related to the above reaction is Scheme 30[89]. Grignard reagents as hydrogen sources have been used to reduce silicon or

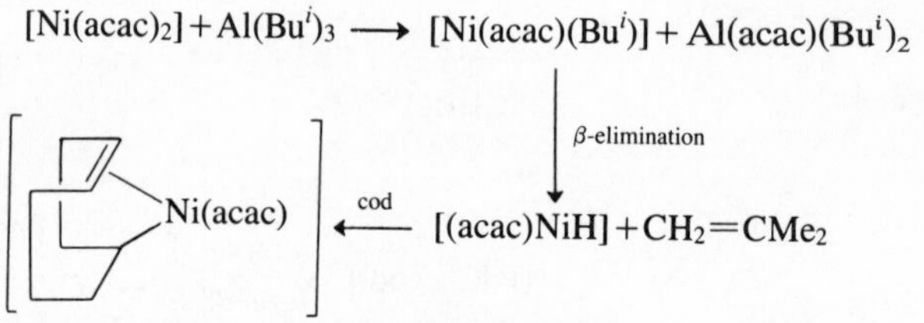

$$[Ni(acac)_2] + Al(Bu^i)_3 \longrightarrow [Ni(acac)(Bu^i)] + Al(acac)(Bu^i)_2$$

$$[Ni(acac)(Bu^i)] \xrightarrow{\beta\text{-elimination}} [(acac)NiH] + CH_2{=}CMe_2$$

$$[(acac)NiH] \xrightarrow{cod} [(C_8H_{13})Ni(acac)]$$

SCHEME 30

olefin derivatives, when a nickel salt is used as a hydride-forming intermediate (reactions 54 and 55)[90]. The R′ group requires a $\beta$ hydrogen, of course.

$$R_3SiX + R'MgX \xrightarrow{[NiCl_2(PPh_3)_2]} R_3SiH \tag{54}$$

$$R_2XSi(vinyl) + R'MgX \xrightarrow{[NiCl_2(PPh_3)_2]} R_2XSiEt \tag{55}$$

Transfer of hydrogen even from solvent molecules such as dioxane has been achieved, again probably via $\beta$-eliminations. A catalyst derived from $[Rh_2(\mu\text{-}Cl)_2(cyclooctene)_4]$ and $PR_3$ is particularly active (equation 56)[91], and the immediate

$$\text{1,4-dioxane} + \text{cyclopentene} \longrightarrow \text{1,4-dioxene} + \text{cyclopentane} \tag{56}$$

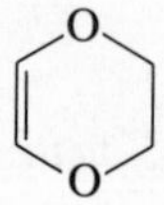

precursor to hydrogen transfer may be $[RhCl(PR_3)(C_4H_8O_2)]$. Another example of solvent-derived hydrogen transfer arises from molybdenum atom bombardment of thf[92]. The proposed mechanism is shown in Scheme 31.

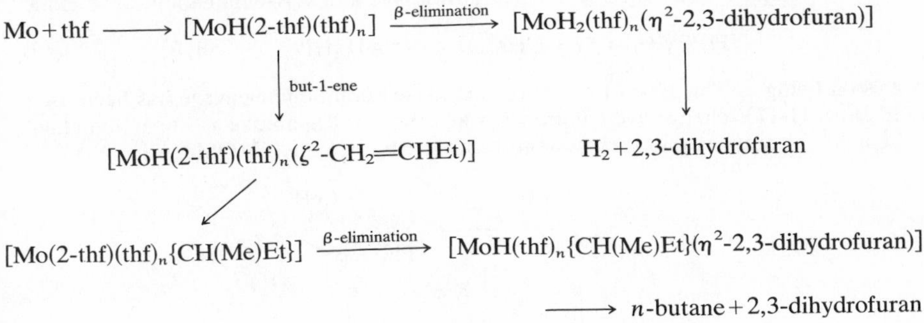

$$\mathrm{Mo + thf} \longrightarrow [\mathrm{MoH(2\text{-}thf)(thf)}_n] \xrightarrow{\beta\text{-elimination}} [\mathrm{MoH_2(thf)}_n(\eta^2\text{-2,3-dihydrofuran})] \longrightarrow \mathrm{H_2 + 2,3\text{-}dihydrofuran}$$

$$[\mathrm{MoH(2\text{-}thf)(thf)}_n] \xrightarrow{\text{but-1-ene}} [\mathrm{MoH(2\text{-}thf)(thf)}_n(\zeta^2\text{-}CH_2{=}CHEt)] \longrightarrow [\mathrm{Mo(2\text{-}thf)(thf)}_n\{\mathrm{CH(Me)Et}\}]$$

$$[\mathrm{Mo(2\text{-}thf)(thf)}_n\{\mathrm{CH(Me)Et}\}] \xrightarrow{\beta\text{-elimination}} [\mathrm{MoH(thf)}_n\{\mathrm{CH(Me)Et}\}(\eta^2\text{-2,3-dihydrofuran})] \longrightarrow n\text{-butane} + \text{2,3-dihydrofuran}$$

SCHEME 31

We end this section by noting that β-hydrogen transfer from O- or N-bonded species can often be of considerable synthetic or catalytic value. Scheme 32 is a

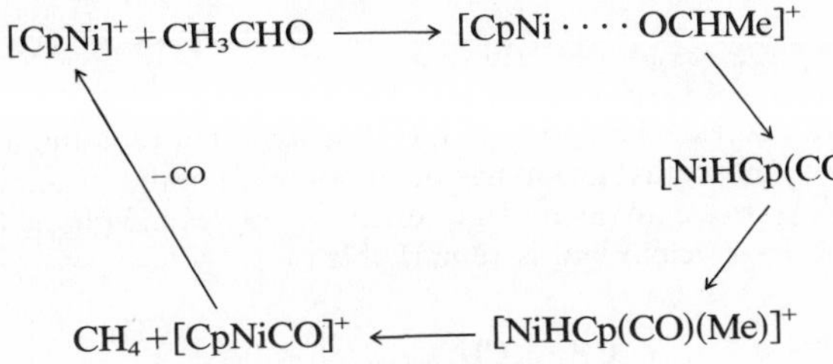

$$[\mathrm{CpNi}]^+ + \mathrm{CH_3CHO} \longrightarrow [\mathrm{CpNi}\cdots\mathrm{OCHMe}]^+ \longrightarrow [\mathrm{NiHCp(COMe)}]^+ \longrightarrow [\mathrm{NiHCp(CO)(Me)}]^+ \longrightarrow \mathrm{CH_4} + [\mathrm{CpNiCO}]^+ \xrightarrow{-\mathrm{CO}} [\mathrm{CpNi}]^+$$

SCHEME 32

decarbonylation route for aldehydes, which proceeds via β-elimination[93], and Scheme 33 is a method of catalysing the reaction of primary to secondary amines[94]; the catalyst is $[RuCl_2(PPh_3)_3]$. Many cases are known, of course, of the metal complex-catalysed reduction of ketones to alcohols, and in this context they are seen as simply the reverse of β-elimination from alcohols[95].

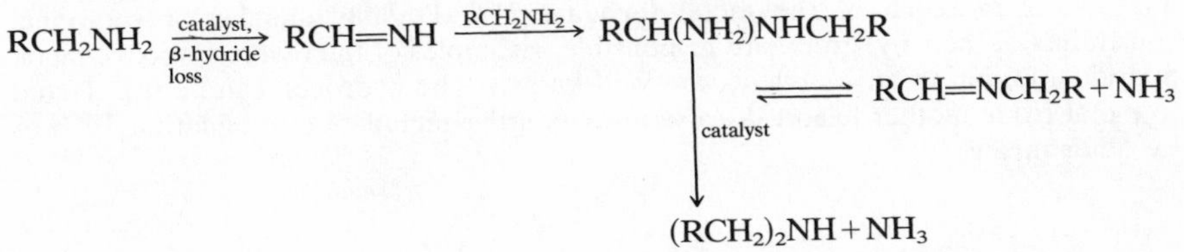

$$\mathrm{RCH_2NH_2} \xrightarrow[\beta\text{-hydride loss}]{\text{catalyst,}} \mathrm{RCH{=}NH} \xrightarrow{\mathrm{RCH_2NH_2}} \mathrm{RCH(NH_2)NHCH_2R} \rightleftharpoons \mathrm{RCH{=}NCH_2R + NH_3}$$

$$\mathrm{RCH(NH_2)NHCH_2R} \xrightarrow{\text{catalyst}} \mathrm{(RCH_2)_2NH + NH_3}$$

SCHEME 33

### 3. β-Elimination from unsaturated groups

The β-elimination creates unsaturation in the side-chain, whether or not it is eliminated cleaving the M—C bonds. A few examples are known where the organic substituent is already unsaturated, and the $\beta$-process increases the number of unsaturated bonds or the degree of unsaturation.

Hydrogen transfer from the methyl substituent of a coordinated olefin produces an $\eta^3$-allyl complex (equation 57)[96]. This transfer is strictly a $\beta$-elimination, the olefinic

$$[PtPCy_3(\eta^2\text{-}CH_2{=}CHMe)] \rightleftharpoons [PtHPCy_3(\eta^3\text{-allyl})] \quad (57)$$

carbon being $\alpha$. This type of reversible metal—carbon bond cleavage has been used to effect H–D exchange even at arenes, when they are capable of $\eta^2$- or $\eta^3$-bonding. Scheme 34 outlines one such system, catalysed by $[Co(allyl)\{P(OMe)_3\}_3]$[97].

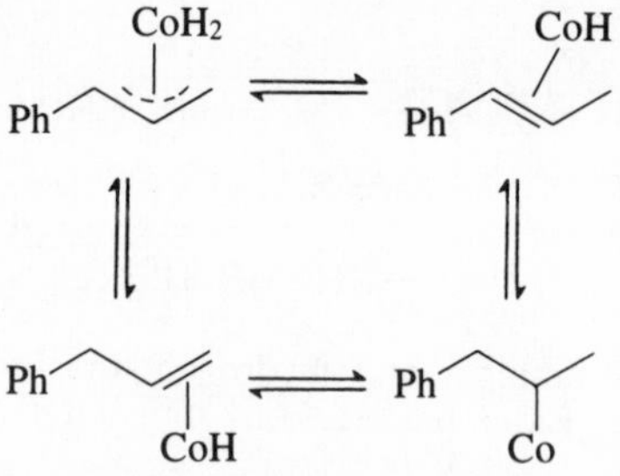

Ph, CoH2, CoH, Ph, Ph, CoH, Ph, Co

SCHEME 34

Room-temperature equivalence on the n.m.r. time scale between the hydrogens of a $\pi$-allyl and a substituent methyl group has been observed in the system in equation 58[98]. The ease of this process, involving an interaction between aliphatic C—H and a metal, and reversible Ir—C cleavage, is remarkable.

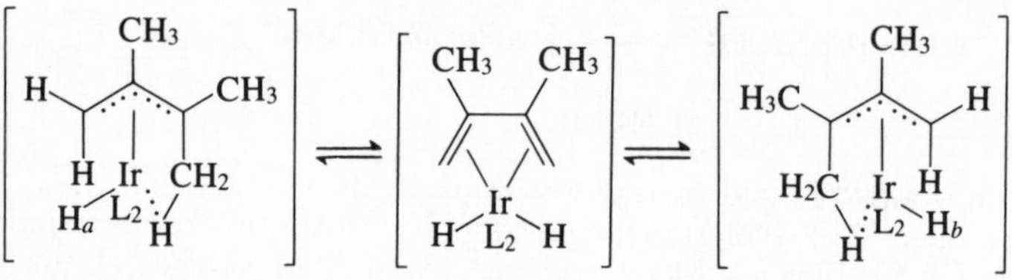

(58)

$\beta$-Hydrogen transfer from a vinyl group would produce a coordinated alkyne. Such transfers appear rare, however, although it does appear that the C—H in question is in reach of the metal (Scheme 2)[13]. Possibly angular or rotational constraints of the vinyl group are responsible. Examples of the reverse process, metal hydride addition to an acetylene, are well known. The hydrogen can be transferred to metal from another ligand (for example by *ortho*-metallation in equation 59[99]) or by other means[100].

$$[Pt(PPh_3)_2(MeO_2CC{\equiv}CCO_2Me)] \xrightarrow{130\ ^\circ C} [\text{Ph}_2\text{P}(C_6H_4)Pt(PPH_3)\{C(CO_2Me){=}CHCO_2Me\}] \quad (59)$$

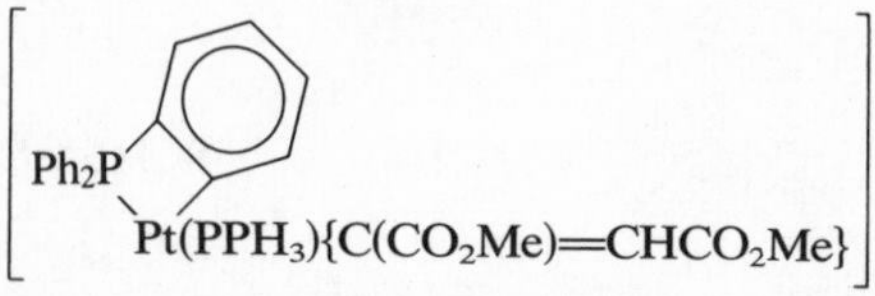

Two interesting hydrogen transfers from binuclear platinum complexes fall into the category of $\beta$-transfers involving unsaturated groups[101,102]. Both lead to severance of one end of a coordinated alkyne from the metal (reactions 60 and 61). The examples

$$[(cod)Pt(\mu\text{-}CF_3C{=}CCF_3)Pt(cod)] \xrightarrow{H^+} [(cod)Pt(\mu\text{-}H)(\mu\text{-}CF_3C{=}CCF_3)Pt(cod)]^+ \longrightarrow [(cod)Pt{-}Pt(cod)(CF_3C{=}CHCF_3)]^+ \quad (60)$$

$$[HPt(\mu\text{-}Ph_2PCH_2PPh_2)_2Pt(CF_3C{\equiv}CCF_3)]^+ \xrightarrow[C_2(CF_3)_2]{Cl^-} [(CF_3CH{=}C(CF_3))Pt(\mu\text{-}CF_3C{=}CCF_3)(\mu\text{-}Ph_2PCH_2PPh_2)_2PtCl]^+ \quad (61)$$

serve to introduce the special effects of metal atom clusters. Hydrogen transfers to and from metal clusters are numerous and generally facile[103]. There is every reason to believe that they proceed in a similar manner to $\beta$-elimination, leading to olefin and metal hydrides by M—C cleavage. Frequently the hydride and olefin groups produced can readily migrate around the cluster, so information on the stereochemistry of the processes is difficult to obtain, and the subject is too large to discuss here. It is worth keeping in mind, however, that lower energy barriers and/or additional reaction pathways allow a number of such processes to occur which might not proceed at mononuclear compounds.

We end this section with an example of unsaturation of a different kind. The insertion of $CO_2$ into molybdenum nitrogen bonds of $[Mo_2R_2(NMe_2)_4]$ forms new complexes $[Mo_2R_2(O_2CNMe_2)_4]$ with Mo$\equiv$Mo triple bonds[104]. When R is capable of $\beta$-hydrogen transfer, however, the molybdenum—carbon bonds are cleaved to eliminate alkene and alkane (the latter presumably via a binuclear process[16]), leading to increased unsaturation in the form of Mo$\equiv$Mo quadruple bonds (equation 62).

$$[Mo_2R_2(NMe_2)_4] + 4CO_2 \longrightarrow [Mo_2(O_2CNMe_2)_4] + RH + \text{alkene} \quad (62)$$

### 4. *Isomerization*

Reversible applications of the $\beta$-elimination process can lead to alkyl group isomerizations and/or olefin migrations and isomerizations. Examples of H–D scrambling have already been met. Olefin isomerizations can be catalysed either homogeneously or heterogeneously, and are well recognized[105]. Usually the thermodynamically favourable isomers finally predominate. Deuterium labelling established that a reversible anti-Markownikow addition of Pt—H across a double bond preceded olefin migration[106] (Scheme 35).

Isomerization of alkynes to dienes can also be brought about by the $\beta$-process (Scheme 36)[107].

Isomerization reactions often accompany other processes, when $\beta$-hydrogen atoms are present in alkyl substituents. They are often unwanted, as they lower the specificity of the reactions. For example, in the cross-coupling of chlorobenzene with isopropylmagnesium chloride catalysed by nickel complexes (Scheme 37), $\beta$-elimination at nickel at the intermediate stage leads to four products[108].

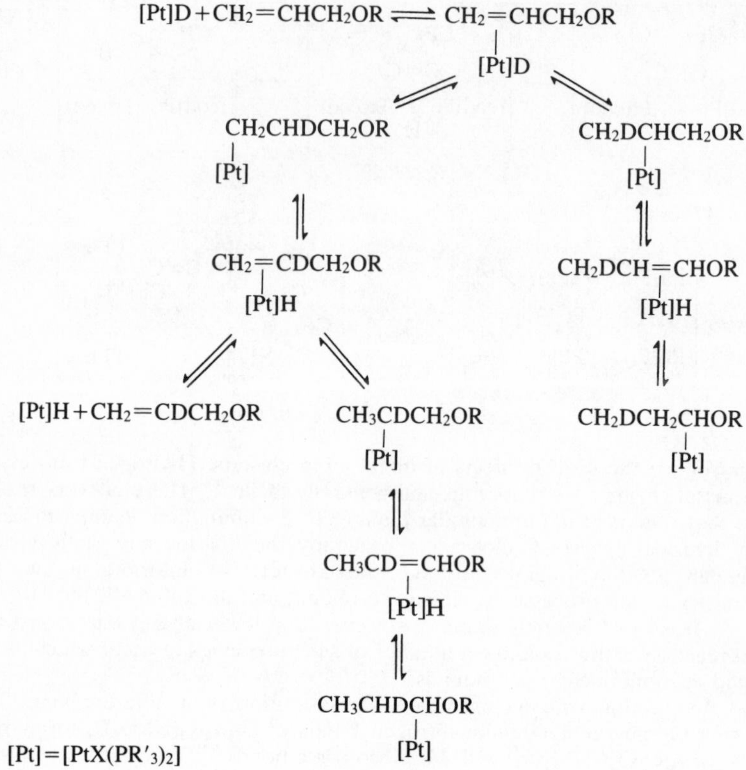

$$[Pt]D + CH_2{=}CHCH_2OR \rightleftharpoons \underset{[Pt]D}{CH_2{=}CHCH_2OR}$$

$$\underset{[Pt]}{CH_2CHDCH_2OR} \rightleftharpoons \underset{[Pt]D}{CH_2{=}CHCH_2OR} \rightleftharpoons \underset{[Pt]}{CH_2DCHCH_2OR}$$

$$\underset{[Pt]}{CH_2CHDCH_2OR} \rightleftharpoons \underset{[Pt]H}{CH_2{=}CDCH_2OR}$$

$$\underset{[Pt]}{CH_2DCHCH_2OR} \rightleftharpoons \underset{[Pt]H}{CH_2DCH{=}CHOR} \rightleftharpoons \underset{[Pt]}{CH_2DCH_2CHOR}$$

$$[Pt]H + CH_2{=}CDCH_2OR \rightleftharpoons \underset{[Pt]H}{CH_2{=}CDCH_2OR} \rightleftharpoons \underset{[Pt]}{CH_3CDCH_2OR}$$

$$\underset{[Pt]}{CH_3CDCH_2OR} \rightleftharpoons \underset{[Pt]H}{CH_3CD{=}CHOR} \rightleftharpoons \underset{[Pt]}{CH_3CHDCHOR}$$

$[Pt] = [PtX(PR'_3)_2]$

SCHEME 35

$$MeC{\equiv}CMe + [Ru]H \longrightarrow [Ru]C(Me){=}CHMe \rightleftharpoons CH_2{=}C{=}CHMe + [RuH]$$

$$CH_2{=}C{=}CHMe + [RuH] \rightleftharpoons CH_2{=}CHCH(Me)[Ru]$$

$$CH_2{=}CHCH{=}CH_2 + [Ru]H \rightleftharpoons CH_2{=}CHCH(Me)[Ru]$$

$[Ru] = [RuCl(PPh_3)_3]$

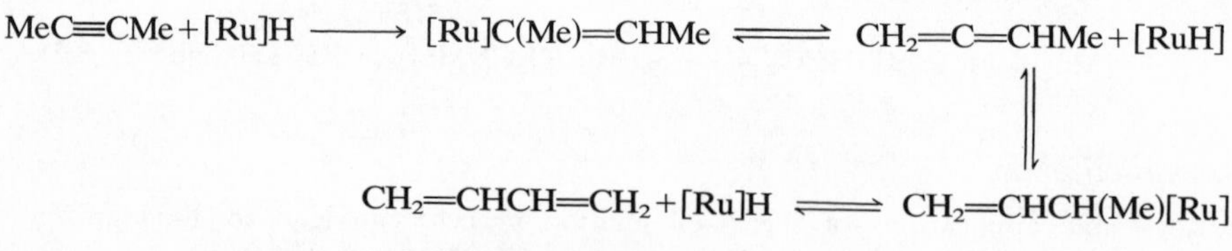

SCHEME 36

Reversible $\beta$-elimination reduces the value of a palladium-catalysed ketone synthesis (Scheme 38)[109], and a similar situation occurs with palladium-catalysed carboacylation of olefins[110]. When a primary $\sigma$-alkylpalladium intermediate forms, one product results and the process is of synthetic value. A secondary $\sigma$-alkyl complex with $\beta$-hydrogens rapidly rearranges, however, and a mixture is produced (Scheme 39). Similar migrations occur at metal cluster compounds[111].

### 5. Olefin hydrogenation and hydrosilation

As the reverse of $\beta$-elimination, olefin hydrogenation is not remarkable, and many examples have been cited already (see also ref. 105). An exciting recent development

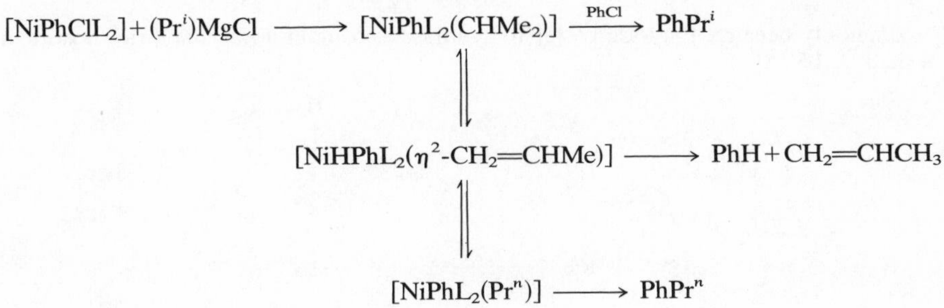

$$[NiPhClL_2] + (Pr^i)MgCl \longrightarrow [NiPhL_2(CHMe_2)] \xrightarrow{PhCl} PhPr^i$$

$$\Updownarrow$$

$$[NiHPhL_2(\eta^2\text{-}CH_2{=}CHMe)] \longrightarrow PhH + CH_2{=}CHCH_3$$

$$\Updownarrow$$

$$[NiPhL_2(Pr^n)] \longrightarrow PhPr^n$$

$L = PPh_3$

SCHEME 37

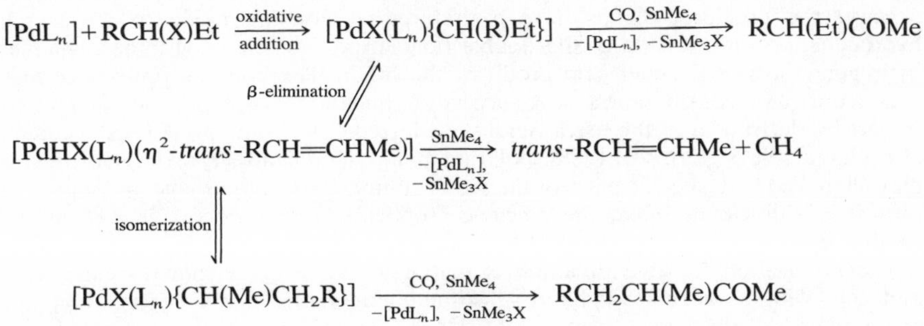

$$[PdL_n] + RCH(X)Et \xrightarrow[\text{addition}]{\text{oxidative}} [PdX(L_n)\{CH(R)Et\}] \xrightarrow[-[PdL_n],\ -SnMe_3X]{CO,\ SnMe_4} RCH(Et)COMe$$

$\beta$-elimination $\Updownarrow$

$$[PdHX(L_n)(\eta^2\text{-}trans\text{-}RCH{=}CHMe)] \xrightarrow[-[PdL_n],\ -SnMe_3X]{SnMe_4} trans\text{-}RCH{=}CHMe + CH_4$$

isomerization $\Updownarrow$

$$[PdX(L_n)\{CH(Me)CH_2R\}] \xrightarrow[-[PdL_n],\ -SnMe_3X]{CO,\ SnMe_4} RCH_2CH(Me)COMe$$

$L = AsPh_3$

SCHEME 38

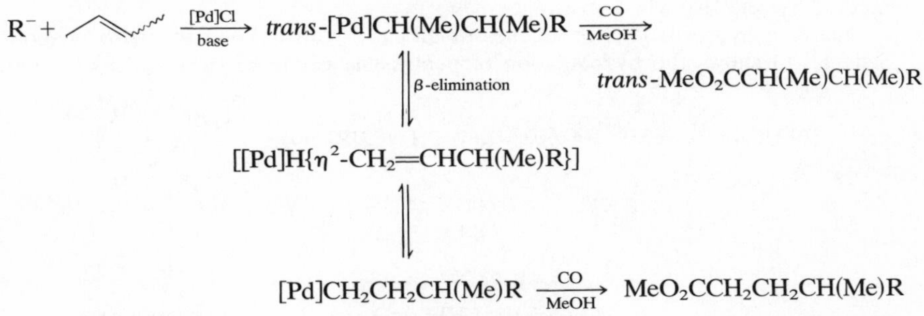

$$R^- + CH_2{=}CH\text{~~~} \xrightarrow[\text{base}]{[Pd]Cl} trans\text{-}[Pd]CH(Me)CH(Me)R \xrightarrow[MeOH]{CO} trans\text{-}MeO_2CCH(Me)CH(Me)R$$

$\Updownarrow$ $\beta$-elimination

$$[[Pd]H\{\eta^2\text{-}CH_2{=}CHCH(Me)R\}]$$

$$\Updownarrow$$

$$[Pd]CH_2CH_2CH(Me)R \xrightarrow[MeOH]{CO} MeO_2CCH_2CH_2CH(Me)R$$

$[Pd] = [PdCl(NEt_3)_2]$

SCHEME 39

is chiral or asymmetric hydrogenation: prochiral olefins are hydrogenated to form an excess of one enantiomer. There are implications for $\beta$-elimination from these experiments. Most of these reactions are catalysed by chiral rhodium phosphine complexes. The source of chirality is usually not the metal atom itself, but the phosphine ligands. These can be asymmetric at phosphorus, such as **15**[112], or, more commonly because they are easier to synthesize, contain an asymmetric substituent such as **16**[113].

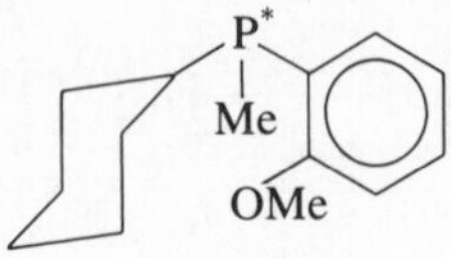

(**15**) (**16**) (diop)

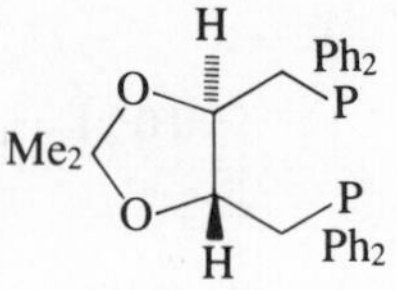

Many systems can produce optical purities of over 90%[105,114,115]. Recently, both metal-cluster compounds[116] and polymer-bound compounds[117] have been shown to behave similarly.

Although an understanding of the subject is still in its infancy, it appears to work because the prochiral olefins have a preferential bonding mode to the metal: hydrogenation on one side of the double bond produces one enantiomer, whereas hydrogenation at the other side produces the other. The bonding preference may arise from steric considerations: the success of chelating phosphines such as **16** may in part be derived from the extra rigidity conferred. Electronic properties no doubt also play a role in terms of olefin polarity: polar side-chains on olefin substituents may align also with specific parts of the metal complex. Thus the shape and nature of the olefin side-chains effect the efficiency (optical) of the process as well as the catalyst itself.

The implication for $\beta$-elimination is that if a source of asymmetry exists in a molecule with a racemic aliphatic substituent containing $\beta$-hydrogens, then one enantiomer of the substituent could be preferentially stabilized towards $\beta$-elimination, as the olefin so formed may not have preferential bonding. Similarly, a racemic metal centre (or a complex with a racemic ligand attached) and a chiral $\beta$-hydrogen-containing organic substituent might behave in the same way. Half the complex may undergo $\beta$-elimination more readily, affording a potential resolution method by selective M—C cleavage.

Catalysed hydrosilations of alkenes or alkynes[105] are related to hydrogenations. Scheme 40 shows the hydrosilation of pent-1-yne catalysed by a hydridorhodium

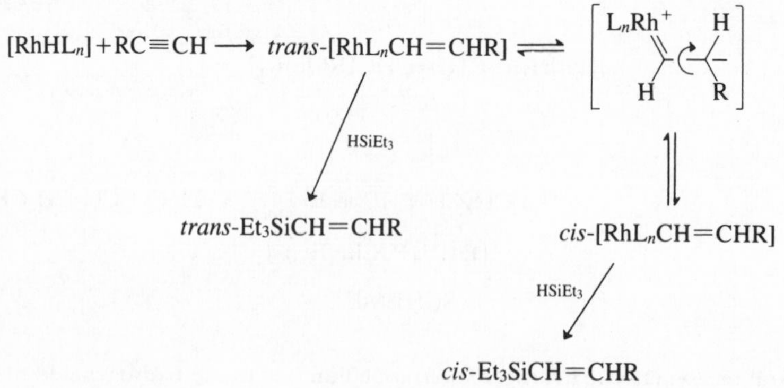

$$[RhHL_n] + RC{\equiv}CH \longrightarrow \textit{trans}\text{-}[RhL_nCH{=}CHR] \rightleftharpoons [L_nRh^+ \ldots]$$

$$\textit{trans}\text{-}[RhL_nCH{=}CHR] \xrightarrow{HSiEt_3} \textit{trans}\text{-}Et_3SiCH{=}CHR$$

$$\rightleftharpoons \textit{cis}\text{-}[RhL_nCH{=}CHR] \xrightarrow{HSiEt_3} \textit{cis}\text{-}Et_3SiCH{=}CHR$$

SCHEME 40

complex[118]. Binuclear complexes can also be active catalysts[119], and asymmetric hydrosilations are also known[120]. Dehydrogenative silylations (reaction 63) are less

$$RCH{=}CH_2 + R'_3SiH \longrightarrow \textit{trans}\text{-}RCH{=}CHSiR'_3 \tag{63}$$

common. A recent example uses the ruthenium cluster $[Ru_3(CO)_{12}]$ as catalyst[121]. Conventional β-elimination probably forms part of the cleavage mechanism in such reactions. In Scheme 41 silicon transfer takes place prior to β-eliminations[122].

$$[RhL_x(SiEt_3)[CH(Me)Bu^n\}\{\eta^2\text{-}CH_2{=}CHBu^n\}] \longrightarrow$$

$$[RhL_x\{CH(Me)Bu^n\}\{CH(CH_2SiEt_3)Bu^n\}]$$

$$\downarrow -L_xRh$$

$$C_6H_{14} + Et_3SiCH{=}CHBu^n + Et_3SiCH_2CH{=}CHPr^n$$

$L_x = \eta^5\text{-}C_5Me_5$ or $Cl(PPh_3)_2$

SCHEME 41

## III. α-ELIMINATION REACTIONS

These are less common than β-eliminations, and they are nearly always more energetic processes. There are two consequences of this. One is that α-eliminations are generally only observed when a competing β-transfer cannot proceed. This usually means that substituents with β-hydrogen should not be available: methyl, benzyl, neopentyl and trimethylsilylmethyl groups are among those prone to undergo α-eliminations*. The second consequence is that other reaction paths often compete under the same conditions. In some cases this has hampered clear elucidation of the actual reaction path followed by the α-process—transfer of hydrogen to the metal atom (reaction 64) or direct to a substituent (reaction 65).

$$[L_nXMCHR_2] \longrightarrow \left[L_nXM \overset{H}{\cdots} CR_2\right] \longrightarrow [L_nHXM{=}CR_2] \longrightarrow [L_nM{=}CR_2] + HX \tag{64}$$

$$[L_nXMCHR_2] \longrightarrow \left[\begin{matrix} L_nM \cdots CR_2 \\ \vdots \quad\ \vdots \\ X \cdots H \end{matrix}\right] \longrightarrow [L_nM{=}CR_2] + HX \tag{65}$$

Equations 64 and 65 illustrate an important difference between α- and β-eliminations. The carbenoid $CR_2$ group generated by α-elimination is generally not ejected from the metal or displaced by simple ligand-exchange reactions. Although they might react further by routes which may remove the $CR_2$ fragment from the metal, in general M—C fissions only occur when X is also a carbon-bonded moiety.

### A. Development of Metal—carbon Cleavage Reactions by α-Elimination

Although the process was first proposed as early as 1961 to account for the thermolysis products of $[TiCl_3CH_3]$[124], it was not until the 1970s that well established

* It is worth noting that main group organometallics, which are less likely to undergo β-elimination, often exhibit the α-process. This is demonstrated by the thermolysis of the lithium compounds $LiCCl_2CHR(OSiMe_3)$[123]. When R allows a conformation suitable for β-elimination, $LiOSiMe_3$ is generated, but when not, α-elimination of LiCl proceeds.

examples appeared. It was pointed out that the reactions should be favoured by complexes in high oxidation levels and with empty *d*-orbitals (the process is a reduction): steric crowding should also be favourable[16]. The first examples were detected in just such complexes. Unstable chromium alkyls with deuterium labels revealed $\alpha$-elimination (reaction 66)[2,125].

$$[Cr(CD_3)_3(thf)_3] \xrightarrow{20\,^\circ C} CD_3H + CD_2H_2 + [residue] \xrightarrow{H_2O} CD_3H + HD \quad (66)$$

Some hydrogen originated from the thf ligands, but the presence of HD after hydrolysis suggested an intermediate chromium deuteride from $\alpha$-migration (reaction 67). Generation of methane from $[CrCl_2(CH_3)(thf)_3]$ included one hydrogen

$$L_xCr{-}CD_3 \longrightarrow L_xDCr{=}CD_2 \quad (67)$$

from thf, but none from solvent, and evidence for a chromium carbene species was obtained[126]. Deuteriation studies on tetramethyltitanium supported an intramolecular hydrogen (deuterium) abstraction of one sort or another (equation 68)[127], and hexamethyltungsten appeared to decompose the same way (equation 69)[128]. The high coordination number for the latter material led to the proposal that a 1,2-elimination such as equation 65 operated in this case[25].

$$[TiMe_4] \longrightarrow CH_4 + [Me_2Ti{=}CH_2] \quad (68)$$

$$[WMe_6] \longrightarrow 3CH_4 + W + (CH_2)_n \quad (69)$$

Although a few examples of $\alpha$-elimination at elements from the other end of the transition series have been reported (usually minor reaction paths)[129,130], the titanium group continues to provide most examples. Neopentyl complexes of titanium, zirconium, and hafnium, $[M(CH_2CMe_3)_4]$, decompose by a first-order mechanism to eliminate 50% of their organic substituents as neopentane[131]. $[TiMe_2Cp_2]$ decomposes mainly to methane, with hydrogen being abstracted from both $CH_3$ and $C_5H_5$ groups, but not solvent. Deuteriation studies showed that no H–D scrambling occurred prior to elimination[132]. The large kinetic isotope effect (*ca.* 3) revealed that

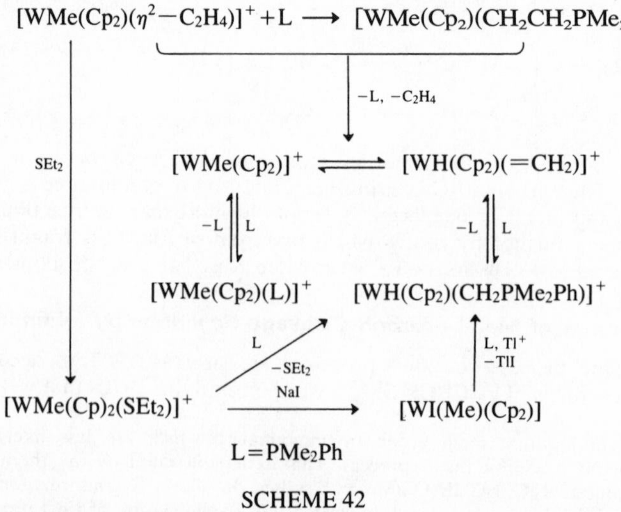

$$[WMe(Cp_2)(\eta^2{-}C_2H_4)]^+ + L \longrightarrow [WMe(Cp_2)(CH_2CH_2PMe_2Ph)]^+$$

$$\downarrow -L,\ -C_2H_4$$

$$[WMe(Cp_2)]^+ \rightleftharpoons [WH(Cp_2)({=}CH_2)]^+$$

$$-L \upharpoonleft\downharpoonright L \qquad\qquad -L \upharpoonleft\downharpoonright L$$

$$[WMe(Cp_2)(L)]^+ \qquad [WH(Cp_2)(CH_2PMe_2Ph)]^+$$

$$[WMe(Cp_2)(\eta^2{-}C_2H_4)]^+ \xrightarrow{SEt_2} [WMe(Cp)_2(SEt_2)]^+$$

$$[WMe(Cp)_2(SEt_2)]^+ \xrightarrow{L,\ -SEt_2} [WH(Cp_2)(CH_2PMe_2Ph)]^+$$

$$[WMe(Cp)_2(SEt_2)]^+ \xrightarrow{NaI} [WI(Me)(Cp_2)] \xrightarrow{L,\ Tl^+,\ -TlI} [WH(Cp_2)(CH_2PMe_2Ph)]^+$$

$$L = PMe_2Ph$$

SCHEME 42

hydrogen or deuterium transfer was involved in the rate-limiting step, but did not, of course, distinguish between routes 64 and 65. Exchange of ring and methyl hydrogen atoms is necessary to account for some products[133], and transfer of hydrogen to the metal at an intermediate has been postulated to account for this. Titanium(III) complexes [$TiR(C_5H_5)_2$] also decompose by hydrogen abstraction from a cyclopentadienyl ring[134]. Many competing reaction paths operate in all of these compounds, and much remains to be learned. Curiously, the stability order for the trivalent compounds is the reverse of the usual with ethyl>methyl, indicating that the β process is not available to these compounds.

Firm evidence for the transfer of an α hydrogen atom to the metal (equation 64) came from a study of cyclopentadienyltungsten compounds[135]. Treatment of the cation [$W(CD_3)Cp_2(C_2H_4)]^+$ with the tertiary phosphine $PMe_2Ph$ (L) produced first [$W(CD_3)Cp_2(CH_2CH_2P^+Me_2Ph)$], then [$WDCp_2(CD_2P^+Me_2Ph)$]. Prolonged heating of this latter cation yielded finally [$W(CD_3)Cp_2L]^+$, *containing the reconstituted $CD_3$ group.* This was interpreted as reversible H (or D) transfer between tungsten and α-carbon (Scheme 42). The rate of conversion of the phosphinemethylene derivatives to the methyltungsten phosphine complexes was found to be very dependent on the nature of the phosphines, as were the final equilibrium positions.

An orbital description of the α transfer is depicted in Figure 6[135]. After binding the two $C_5$ rings, three orbitals of *d* character are left on tungsten for attaching other ligands[136]. These are suitable for the transfer, provided the initial methyl compound is no more than a 16-electron complex. The tantalum compound [$TaMeCp_2(CH_2)$], isoelectronic with the proposed tungsten intermediates, has been isolated and examined[137].

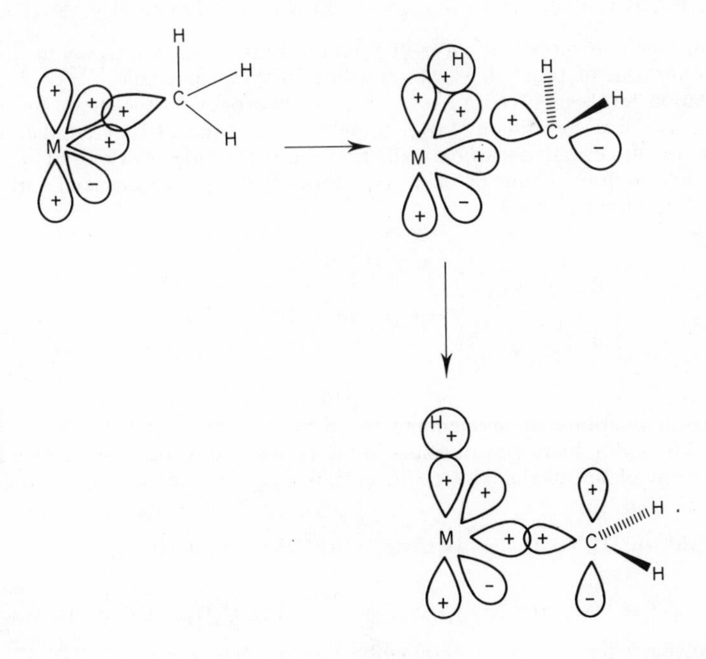


FIGURE 6. α-Hydrogen transfer

### B. Alkylidene and Alkylidyne Complexes of Niobium and Tantalum

The wealth of detail on $\alpha$-eliminations from compounds of these elements uncovered by Schrock[138] and his coworkers is exceptional. It is not yet clear if the principles discovered are general to other systems, so this and related work is treated separately.

The elimination of methane from $[NbMe_5]$ or $[TaMe_5]$[139] and of toluene from $[Ta(CH_2Ph)_5]$[139,140] are similar to those from high oxidation level compounds of titanium. The deuterium kinetic isotope effect indicates that C—H(D) bond breaking is rate determining. An important advance came with the elimination of toluene from $[TaCl_2(CH_2Ph)_3]$ induced by its reaction with cyclopentadienylthallium (equation 70)[141], as the benzylidene complex so produced could be isolated and characterized.

$$[TaCl_2(CH_2Ph)_3]+2CpTl \longrightarrow PhCH_3+[TaCp_2(CH_2Ph)(CHPh)] \quad (70)$$

A second elimination to produce a benzylidyne could also be induced[142]. Relief of steric strain appears to be an important factor: the bond angles at $\alpha$-carbon would formally increase (109° to 120° to 180°) at each stage. Tertiary phosphines and pentamethylcyclopentadienyl have been used to generate the strain required (equation 71)[142,143].

$$[TaCl(C_5Me_5)(CH_2Ph)(CHPh)]+2PMe_3 \longrightarrow PhCH_3+[TaCl(C_5Me_5)(CPh)(PMe_3)_2] \quad (71)$$

Not surprisingly perhaps, the bulky neopentyl group appears even more prone to $\alpha$-eliminations (reaction 72)[144].

$$[TaCl(CH_2CMe_3)_4] \longrightarrow CMe_4+[TaCl(CH_2CMe_3)_2(CHCMe_3)] \quad (72)$$

Deuterium labelling supported an $\alpha$-elimination route, with $k_H/k_D$ about 2.7.

The importance of relief of steric crowding in these molecules[141,142,145] and the high oxidation levels involved make transfer of hydrogen to the metal an unlikely process[144]. The five-coordinate $[MR_5]$ complexes are trigonal bipyramidal, and the hydrogens of equatorial $\alpha$-carbons will be relatively acidic, with axial $\alpha$-carbons relatively nucleophilic. Transfer of an equatorial hydrogen to an axial carbon was thus postulated (equation 73).

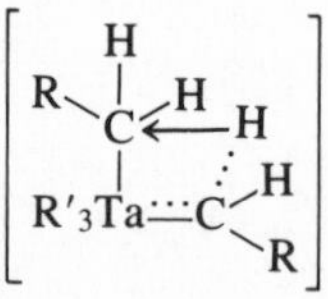

$$\left[\begin{array}{c} R\text{—}CH_2\text{—}H \cdots CHR \\ R'_3Ta \cdots \end{array}\right] \longrightarrow [R'_3Ta{=}CHR]+RCH_3 \quad (73)$$

Double $\alpha$-eliminations of neopentane to form dialkylidenes have been induced (reaction 74), and $\alpha$-hydrogen transfers in the opposite direction are involved in the rearrangement of an alkylalkylidyne to a dialkylidene (reaction 75)[146]. There is a

$$[M(CH_2CMe_3)_3(CHCMe_3)] \xrightarrow{2PMe_3} CMe_4+[M(CH_2CMe_3)(CHCMe_3)_2(PMe_3)_2] \quad (74)$$

$$[Ta(C_5R_5)(CH_2CMe_3)(CCMe_3)(PMe_3)_2] \xrightarrow{-PMe_3} [Ta(C_5R_5)(CHCMe_3)_2PMe_3] \quad (75)$$

marked solvent effect on the $\alpha$-elimination rate of RH from $[TaCpR_2X_2]$[147], those solvents which favour *cis* geometry favouring elimination. *Cis* ligand arrangements would be necessary to direct transfers of the type shown in equations 65 and 73.

X-ray diffraction studies on a number of compounds and neutron diffraction studies on $[TaCl_3(CHCMe_3)(PMe_3)]_2$ and $[Ta(C_5Me_5)(CHCMe_3)(C_2H_4)(PMe_3)]$[148] show that the α-hydrogens of alkylidenes closely approach Ta. At least part of the effect is steric: more bulky R groups interact with other ligands increasing ∠TaCR and decreasing ∠TaCH[147]. $^1J(^{13}C-^1H)$ decreases as the *p*-orbital character of CH increases. TaCH angles of less than 90° have been found, so the hydrogen might be described as semi-bridging (structure **17**). This effect is also partly electronic in

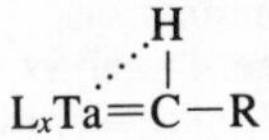

$$L_xTa{=}C(H){-}R$$

**(17)**

origin. Cases are known where R is held in a more crowded location than necessary by a Ta···H interaction. The same type of interaction appears to operate in the alkyl, as well as alkylidene, compounds.

The picture which emerges of α-elimination at these high oxidation level $d^0$ complexes is that the α-hydrogen is activated by attraction to the metal, an effect enhanced by steric crowding from other groups, and it is then transferred to other (*cis*) organic groups. $[TaCpBr_2(CH_2Ph)(CH_2CMe_3)]$ eliminates mainly toluene, in keeping with a more active α-hydrogen on neopentyl[147].

The role of tertiary phosphines in promoting α-eliminations is not simply steric, as trimethylphosphine is more active than arylmethylphosphines of greater bulk[138,149]. It has been suggested[135] that the mechanism may involve ligand attack at the α-carbon atom rather than attack at the metal. Scheme 43 outlines a possible route[135], although complete transfer of the α-hydrogen to metal may not be necessary prior to alkane elimination.

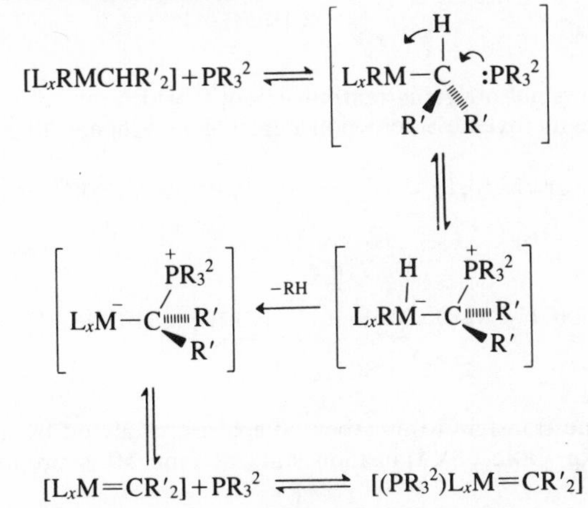

$$[L_xRMCHR'_2] + PR_3^2 \rightleftharpoons [L_xRM{-}C(H)(R')_2 \cdots :PR_3^2] \rightleftharpoons [L_xR\bar{M}(H){-}C(\overset{+}{P}R_3^2)R'_2]$$

$$[L_xR\bar{M}(H){-}C(\overset{+}{P}R_3^2)R'_2] \xrightarrow{-RH} [L_x\bar{M}{-}C(\overset{+}{P}R_3^2)R'_2] \rightleftharpoons [L_xM{=}CR'_2] + PR_3^2 \rightleftharpoons [(PR_3^2)L_xM{=}CR'_2]$$

SCHEME 43

Six-coordinate niobium and tantalum complexes $[MX_3(CHR)L_2]$ (L = tertiary phosphine) also undergo α-elimination to the alkylidynes when treated with tertiary phosphines[149]. The mechanism of Scheme 43 may operate, but seven-coordinate intermediates were postulated, as the methyl compound $[TaBr_3Me_2(PMe_3)_2]$ was isolated. Interestingly, a number of hydrido-alkylidene and hydrido-alkylidyne complexes of tantalum have been isolated[150]. Not only do these formally seven-coordinate species not undergo migration of the hydrogen to the $\alpha$ carbon atom, but also further $\alpha$-elimination (of $H_2$) can proceed (equation 76).

$$[TaHCl(C_5Me_5)(CHCMe_3)(PMe_3)] \xrightarrow[-H_2]{PMe_3} [TaCl(C_5Me_5)(CCMe_3)(PMe_3)_2] \quad (76)$$

Although it is not known to what extent hydrogen from an $\alpha$-carbon migrates to the metal in most of these $\alpha$-eliminations, there is evidence that other elements behave in related manner. Scheme 44 shows the formation of remarkable tungsten compounds with alkyl, alkylidene, and alkylidyne substituents[151].

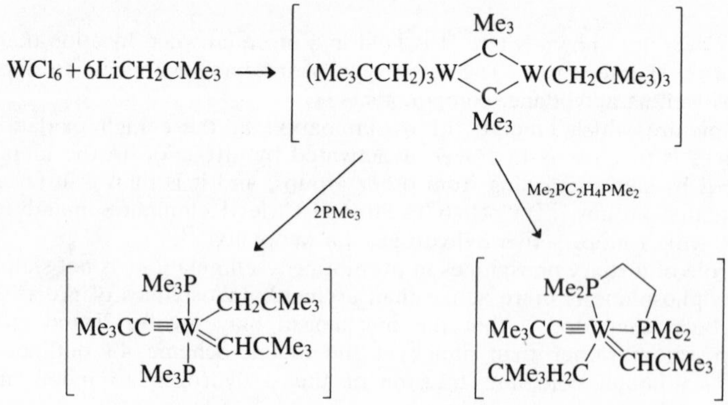

$$WCl_6 + 6LiCH_2CMe_3 \longrightarrow [(Me_3CCH_2)_3W(\mu\text{-CMe}_3)_2W(CH_2CMe_3)_3]$$

$\xrightarrow{2PMe_3}$ $[Me_3CC{\equiv}W(CH_2CMe_3)(=CHCMe_3)(PMe_3)_2]$

$\xrightarrow{Me_2PC_2H_4PMe_2}$ $[Me_3CC{\equiv}W(CH_2CMe_3)(=CHCMe_3)(Me_2PC_2H_4PMe_2)]$

SCHEME 44

Scheme 45 shows an intriguing rearrangement[152] which may represent the reverse of the phosphine–methylene interaction depicted in Scheme 43[135]. A coordinated

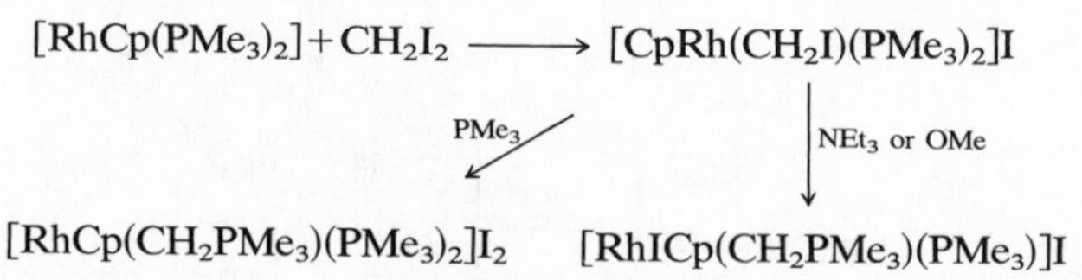

$$[RhCp(PMe_3)_2] + CH_2I_2 \longrightarrow [CpRh(CH_2I)(PMe_3)_2]I$$

$[CpRh(CH_2I)(PMe_3)_2]I \xrightarrow{PMe_3} [RhCp(CH_2PMe_3)(PMe_3)_2]I_2$

$[CpRh(CH_2I)(PMe_3)_2]I \xrightarrow{NEt_3 \text{ or } OMe} [RhICp(CH_2PMe_3)(PMe_3)]I$

SCHEME 45

trimethylphosphine transfers to a carbenoid species, catalysed by nucleophiles such as $PMe_3$, $NEt_3$, or $OMe^-$. A transition state of type **18** is proposed[152]. It seems

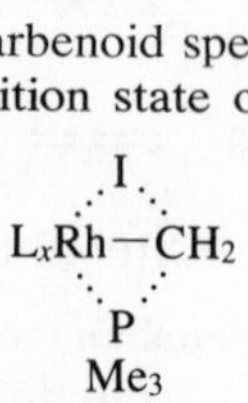

$$L_xRh{-}CH_2 \text{ (bridged by I and PMe}_3\text{, dotted bonds)}$$

**(18)**

probable that the transition metal ylide complexes[24] of Schmidbaur and others might undergo similar rearrangements, but none have been reported.

## C. Scope of the α-Elimination and Theoretical Considerations

α-Eliminations from trimethylsilylmethyl compounds of the lanthanides lutetium and erbium[153] extend the process to Group III and it is likely that most or all the transition elements of Groups III–VI, at least in their high oxidation levels, can undergo the process. Examples from the other end of the series are rare[129,130], but equation 77 is probably a genuine example[154].

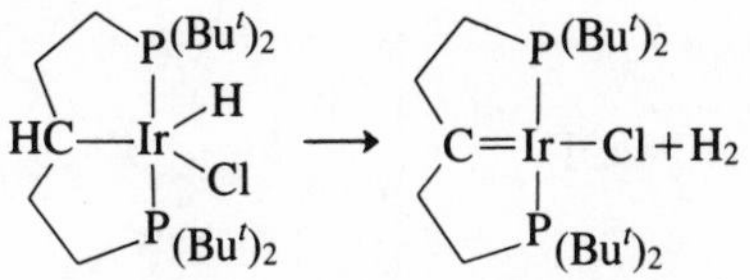

$$HC(CH_2CH_2P(Bu^t)_2)_2Ir(H)Cl \rightarrow C(CH_2CH_2P(Bu^t)_2)_2Ir{-}Cl + H_2 \qquad (77)$$

The analogous ruthenium compound undergoes a fast (on the n.m.r. scale) inversion of the α-CH, probably via a reversible α-elimination[155]. It too can eliminate $H_2$, but the olefin complex **19** is formed rather than a carbene compound[156], and β-elimination could be responsible.

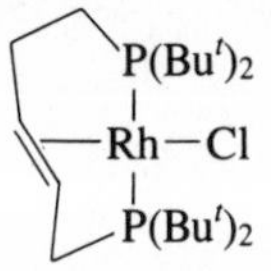

**(19)** 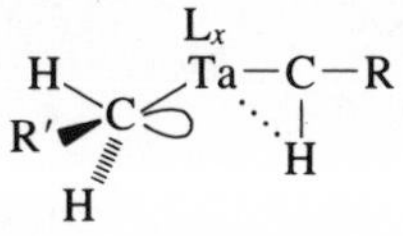 **(20)**

A detailed M.O. analysis addressed specifically to the large Ta—C—C angles of Schrock's carbene complexes produced results reassuringly in line with the deductions from experimental observation[157]. The process was described as a pivoting of the carbene, placing an α-hydrogen near tantalum, and an intramolecular electrophilic interaction of metal acceptor orbitals with carbene lone pair is seen as the cause. The main effect of large substituents on the carbene or other ligands is to protect the system from intermolecular reactions. In five-coordinate 14-electron species of these types, full transfer of hydrogen to the metal is a forbidden process, and this may be the most important factor in deciding whether the α-eliminations more closely resemble equation 64 or 65. The elimination was described as occurring via interaction of H (α) with a neighbouring R group, preferably after this has pivoted slightly towards it (**20**)[157].

## D. Multinuclear Compounds

Many reactions have been observed at metal surfaces which have (as yet) no counterpart at mononuclear metal complexes[158]. One reason for this may be that the arrangement of several metal atoms in close proximity allows additional reaction paths to operate.

A number of carbene or carbyne ligands have been observed bridging two or more metal atoms in binuclear or cluster compounds, and this structural modification could provide added impetus to the formation of such species by α-elimination. Curiously, the metals involved in these situations tend to be from the latter end of the transition series—the opposite to that for related mononuclear interactions. The mobility and versatility of these bridging carbenes are illustrated by equation 78[159] and Scheme

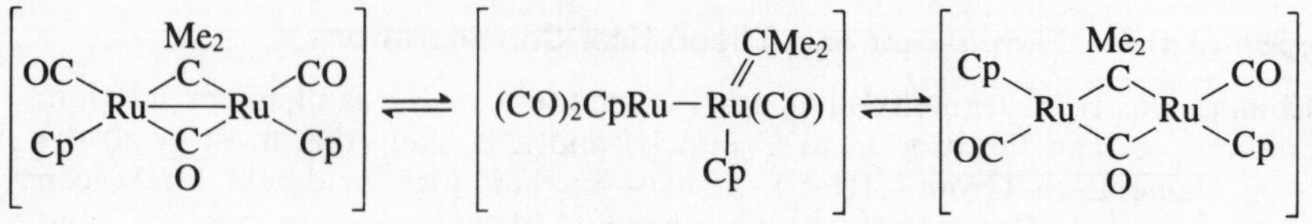

$$[Ru_2Cp_2(CO)_2(\mu\text{-}CO)(\mu\text{-}CMe_2)] \rightleftharpoons [(CO)_2CpRu{-}Ru(CO)Cp(=CMe_2)] \rightleftharpoons [Ru_2Cp_2(CO)_2(\mu\text{-}CO)(\mu\text{-}CMe_2)] \quad (78)$$

46[160]. Decompositions of compound **21** (Scheme 46) produces $Me_2C{=}CHMe$ as the

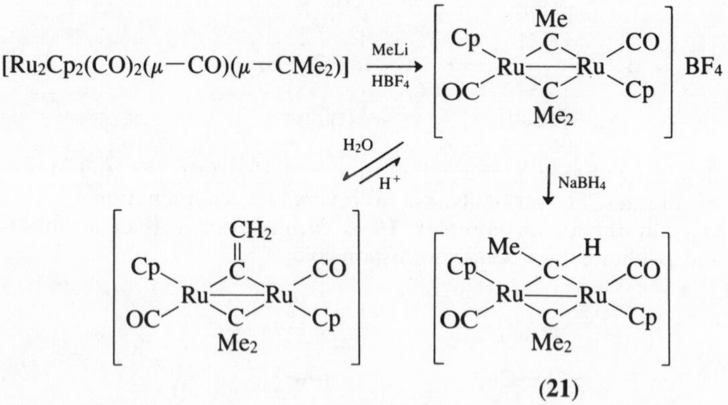

$[Ru_2Cp_2(CO)_2(\mu{-}CO)(\mu{-}CMe_2)]$ $\xrightarrow[HBF_4]{MeLi}$ $[Ru_2Cp_2(CO)_2(\mu\text{-}CMe)(\mu\text{-}CMe_2)]BF_4$

$H_2O$ / $H^+$: $[Ru_2Cp_2(CO)_2(\mu\text{-}C{=}CH_2)(\mu\text{-}CMe_2)]$

$NaBH_4$: $[Ru_2Cp_2(CO)_2(\mu\text{-}CHMe)(\mu\text{-}CMe_2)]$ (**21**)

SCHEME 46

major volatile product[160]. This represents a further route to M—C breaking after $\alpha$-elimination. Bridging and mobile hydrogen atoms in such clusters add more versatility to the processes. Reversible transfer of a hydrogen from a bridging methyl group to rhodium clusters has been elucidated (Scheme 47), but has no mononuclear

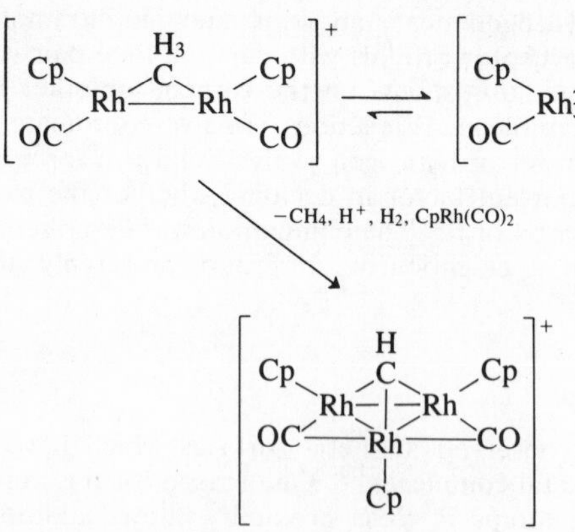

$[Rh_2Cp_2(CO)_2(\mu\text{-}CH_3)]^+ \rightleftharpoons [Rh_2Cp_2(CO)_2(\mu\text{-}CH_2)(\mu\text{-}H)]^+$

$-CH_4, H^+, H_2, CpRh(CO)_2$

$[Rh_3Cp_3(CO)_2(\mu_3\text{-}CH)]^+$

SCHEME 47

counterpart[161]. An apparently similar bridge-methyl to bridge-methylene conversion in the trinuclear osmium cluster $[Os_3(CO)_{10}(\mu\text{-}CH_3)(\mu\text{-}H)]$ reveals intriguing detail on X-ray crystallographic analysis[162,163]. The bridging methyl group is asymmetrically situated, with a C—H bond close to and edge on to one osmium atom (equation 79).

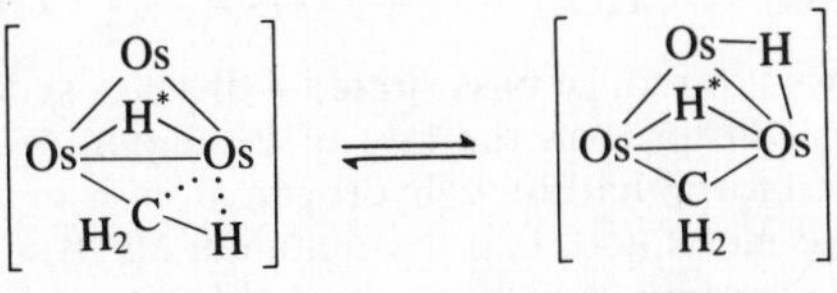

$$[\text{Os}_3(\mu\text{-H}^*)(\mu\text{-CH}_2\cdots\text{H})] \rightleftharpoons [\text{Os}_3(\text{H})(\mu\text{-H}^*)(\mu\text{-CH}_2)] \quad (79)$$

The situation resembles the intramolecular activation of alkyl and alkylidene hydrogens seen for tantalum(V) compounds and suggests that intermolecular processes may proceed by such a mechanism when the metal atoms can approach close enough.

The rhodium complexes of Scheme 48, elucidated by high-resolution n.m.r.

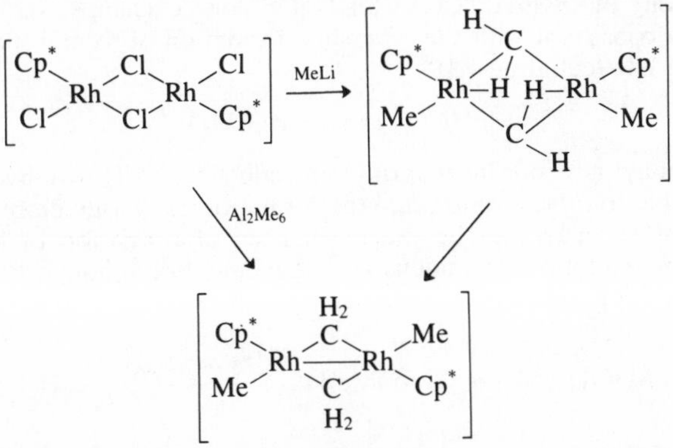

$$[\text{Cp}^*\text{RhCl}(\mu\text{-Cl})_2\text{RhCl}\text{Cp}^*] \xrightarrow{\text{MeLi}} [\text{Cp}^*(\text{Me})\text{Rh}(\mu\text{-CH}\cdots\text{H})_2\text{Rh}(\text{Me})\text{Cp}^*]$$

$$[\text{Cp}^*\text{RhCl}(\mu\text{-Cl})_2\text{RhCl}\text{Cp}^*] \xrightarrow{\text{Al}_2\text{Me}_6} [\text{Cp}^*(\text{Me})\text{Rh}(\mu\text{-CH}_2)_2\text{Rh}(\text{Me})\text{Cp}^*] \leftarrow [\text{Cp}^*(\text{Me})\text{Rh}(\mu\text{-CH}\cdots\text{H})_2\text{Rh}(\text{Me})\text{Cp}^*]$$

SCHEME 48

spectroscopy and X-ray crystallography, show yet another C–H interaction[164]. Pyrolysis gave some ethylene (presumably by coupling of bridging methylene to break the Rh—C bonds), but mainly methane and propene. The likely rearrangements to break these M—C bonds are outlined in equation 80, and amplify the versatility of multinuclear systems to α-eliminations.

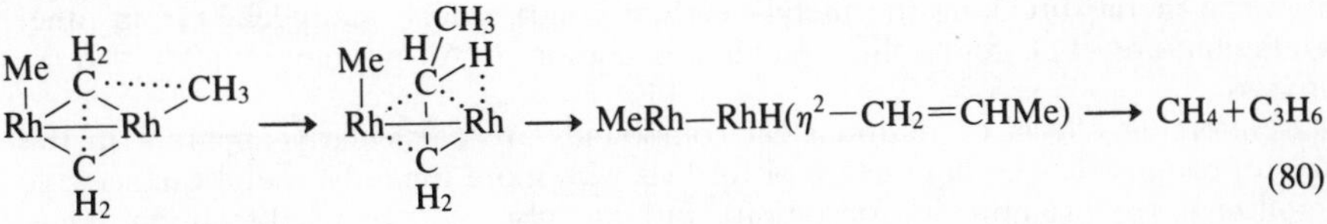

$$\text{Me}\text{Rh}(\mu\text{-CH}_2\cdots\text{CH}_3)(\mu\text{-CH}_2)\text{Rh} \longrightarrow \text{Me}\text{Rh}(\mu\text{-CH(H)CH}_3)(\mu\text{-CH}_2)\text{Rh} \longrightarrow \text{MeRh—RhH}(\eta^2\text{—CH}_2\text{=CHMe}) \longrightarrow \text{CH}_4 + \text{C}_3\text{H}_6 \quad (80)$$

## E. Formyl Complexes

Metal formyl complexes are still relatively rare, but interest in their formation and properties is intense, since they are likely intermediates in metal-catalysed reductions of carbon monoxide, Fischer–Tropsch synthesis, etc.[165]. Whilst the main interest is in their formation from metal carbonyls and metal hydrides, they are of interest here because all the mononuclear examples characterized decompose by α-elimination.

The formyl complex *trans*-$[\{(PhO)_3P\}(CO)_3Fe(CHO)]^-$ decomposed by complete transfer of hydrogen from the α-carbon of the formyl group to iron[168] (equation 81).

$$trans\text{-}[\{(PhO)_3P\}(CO)_3Fe(CHO)]^- \xrightarrow{67\,°C} (PhO)_3P + [(CO)_4FeH]^- \quad (81)$$

This hydrogen can also be used to reduce unsaturated molecules such as ketones, and the rate of such reductions is faster than the rate of formation of the hydrido-iron complex. It seems likely, therefore, that the α-hydrogen atom is in an activated state owing to its proximity to the metal and, like the tantalum alkyls and alkylidenes, it can be transferred directly to other substrates, probably coordinated. Unlike the tantalum examples, however, the hydrogen can also transfer to the metal.

Formyl iridium complexes, although fairly robust, appear to transfer hydrogen to the metal in the same way when heated (reaction 82)[167]. Kinetic measurements on

$$cis\text{-}[IrH(CHO)(PMe_3)_4]^+ \longrightarrow [Ir(CO)(PMe_3)_4] + H_2 \quad (82)$$

the isomerization of $[MnH(^{13}CO)(CO)_4]$ and on exchange of ligands with $[MnH(CO)_5]$ are consistent with the reversible formation of formyl, the equilibrium lying well to the left (equation 83)[168].

$$MnHCO \rightleftharpoons MnCHO \quad (83)$$

Rhenium carbonyl also produces formyl on reduction[169]. This α-hydrogen also is readily transferable to other groups, and the formation can easily be reversed. These α-hydrogen transfers may well be the mechanism of formation of hydrido-metal carbonyls from methanol or formaldehyde[170] (Scheme 49), although β-transfer from an oxygen-bonded formaldehyde is another possibility.

$$[RuHClL_3] + MeOH \longrightarrow [Ru(OMe)HL_3] \xrightarrow{\beta\text{-elimination}} [RuH_2L_3(CH_2O)]$$

$$\downarrow \text{solvent}$$

$$[RuH_2L_3(CO)] \xleftarrow{\alpha\text{-elimination}} [RuHL_3(S)(CHO)] \xleftarrow[-H_2]{CH_2O} [RuH_2L_3(S)]$$

$L = PPh_3$

SCHEME 49

These α-eliminations from formyls resemble most β-eliminations in two interesting aspects: the hydrogen *can* be transferred to the metal atom (equation 64), and the unsaturated substrate produced, carbon monoxide, can be readily eliminated from the metal (breaking the metal—carbon bond), unlike methylenes from other α-eliminations, but resembling olefin elimination or displacement after the β-process.

Reinforcing effects of multinuclear compounds are particularly apparent in this area of chemistry. The interaction of formyls with more than one metal can serve to stabilize it (or promote its formation), but can also lead to its disruption. Thus, although isonitriles can insert into Zr—H bonds[171] (the reverse of α-hydrogen transfer), the same does not happen with isoelectronic carbon monoxide, except via an intermolecular process (equation 84)[172]. Possibly the strength of the metal—

$$[WCp_2(CO)] + [ZrH_2Cp_2] \longrightarrow [Cp_2W{=}CH{-}O{-}ZrHCp_2] \quad (84)$$

oxygen bonds here is enough to change the equilibrium for hydrogen transfer from carbon-to-metal to metal-to-carbon. The reaction of $[ZrH_2Cp_2]$ with carbon monoxide involves such insertions, as well as α-hydrogen transfers and β-eliminations, leading finally to breaking of the Zr—C bonds and leaving only metal oxygen links (Scheme 50)[173].

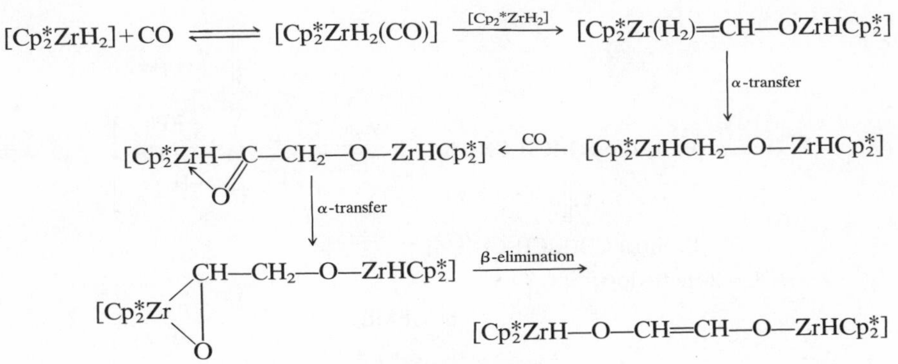

$$[Cp^*_2ZrH_2] + CO \rightleftharpoons [Cp^*_2ZrH_2(CO)] \xrightarrow{[Cp_2^*ZrH_2]} [Cp^*_2Zr(H_2){=}CH{-}OZrHCp^*_2]$$

$$\downarrow \alpha\text{-transfer}$$

$$[Cp^*_2ZrH{-}C(\text{–O–})H{-}CH_2{-}O{-}ZrHCp^*_2] \xleftarrow{CO} [Cp^*_2ZrHCH_2{-}O{-}ZrHCp^*_2]$$

$$\downarrow \alpha\text{-transfer}$$

$$[Cp^*_2Zr(\text{–O–})CH{-}CH_2{-}O{-}ZrHCp^*_2] \xrightarrow{\beta\text{-elimination}} [Cp^*_2ZrH{-}O{-}CH{=}CH{-}O{-}ZrHCp^*_2]$$

SCHEME 50

A very different structure arises from the related reaction of a tantalum hydride. Transfer of hydrogen *to* α-carbon proceeds, contrary to the norm for mononuclear compounds, but here the formyl is stabilized by symmetrically bridging two centres (equation 85)[174].

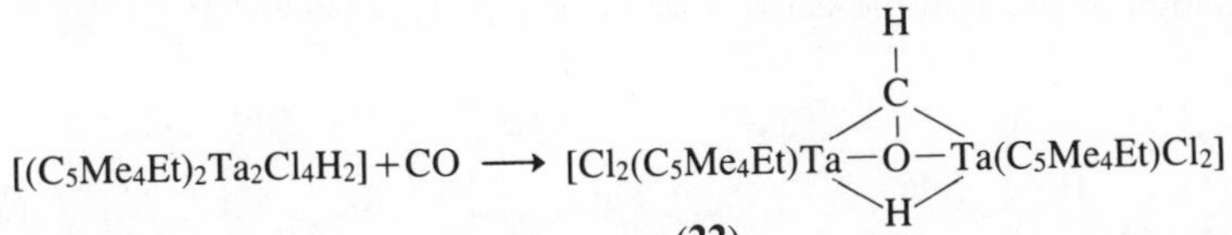

$$[(C_5Me_4Et)_2Ta_2Cl_4H_2] + CO \longrightarrow [Cl_2(C_5Me_4Et)Ta(\mu\text{-}CH{-}O)(\mu\text{-}H)Ta(C_5Me_4Et)Cl_2] \quad (85)$$

**(22)**

A further intriguing reaction occurs when this compound (**22**) is treated with $PMe_3$. The phosphine attacks the bridging formyl and brings about complete disruption of the C—O bond (reaction 86)[175]. The implications of this reaction for both CO reduction and nucleophile assistance of α-elimination are profound.

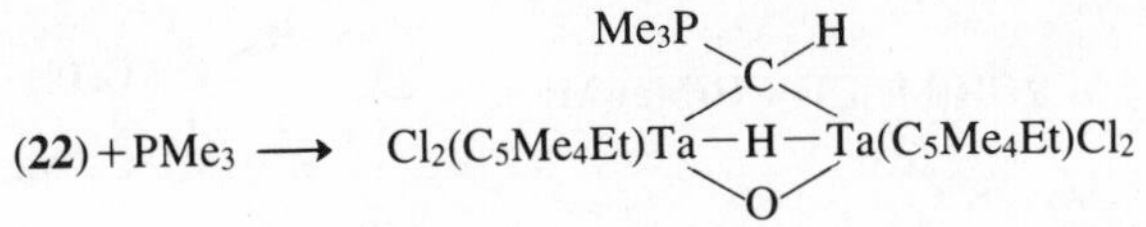

$$(\mathbf{22}) + PMe_3 \longrightarrow Cl_2(C_5Me_4Et)Ta(\mu\text{-}CH(PMe_3))(\mu\text{-}H)(\mu\text{-}O)Ta(C_5Me_4Et)Cl_2 \quad (86)$$

## F. Cyclic Complexes and Reactions of Generated Carbene Complexes

Steric restraints of cyclic organic derivatives helped give insight into the β-elimination process. The same does not appear to be the case with α-elimination reaction. Not only is the reaction less well understood, even for the acyclic compounds, operation of α-eliminations in metallacycles is rare and offers no additional information. Platinacyclobutanes have been observed to react with bulky nitrogen donors to yield carbene and olefin complexes. Both ylide formation and α-elimination, probably with total hydrogen transfer to metal, are probably involved (Scheme 51)[67,176].

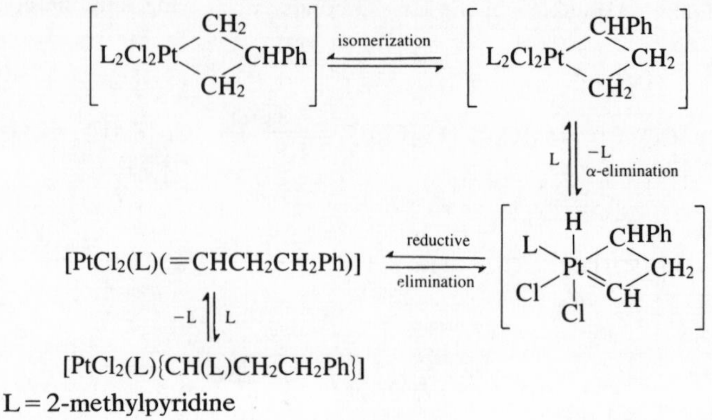

$$\left[L_2Cl_2Pt\begin{matrix}CH_2\\ \\ CH_2\end{matrix}CHPh\right] \underset{}{\overset{\text{isomerization}}{\rightleftharpoons}} \left[L_2Cl_2Pt\begin{matrix}CHPh\\ \\ CH_2\end{matrix}CH_2\right]$$

$$\left[L_2Cl_2Pt(CHPh)(CH_2)CH_2\right] \underset{L}{\overset{-L,\ \alpha\text{-elimination}}{\rightleftharpoons}} \left[PtHCl_2(L)(CHPh)(CH_2)(=CH)\right]$$

$$[PtCl_2(L)(=CHCH_2CH_2Ph)] \underset{\text{elimination}}{\overset{\text{reductive}}{\rightleftharpoons}} \left[PtHCl_2(L)(CHPh)(CH_2)(=CH)\right]$$

$$[PtCl_2(L)(=CHCH_2CH_2Ph)] \underset{-L}{\overset{L}{\rightleftharpoons}} [PtCl_2(L)\{CH(L)CH_2CH_2Ph\}]$$

L = 2-methylpyridine

SCHEME 51

Of much greater importance is the possibility of *formation* of metallacyclobutanes from olefins and metal carbene complexes, often generated by $\alpha$-elimination. It is postulated that Ziegler–Natta catalysis and olefin metathesis might both be catalysed by such complexes, successive carbon—metal bond making and breaking leading to the observed products. In a proposed mechanism for olefin polymerization, reversible $\alpha$-elimination operates in every cycle[177]. The initiating step is presumably also an $\alpha$-elimination to produce the active carbene complex (Scheme 52).

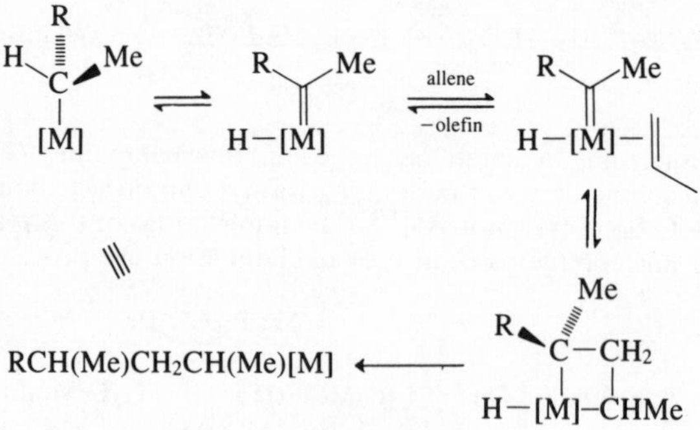

$$[M]-C(H)(R)(Me) \rightleftharpoons H-[M]=C(R)Me \underset{-\text{olefin}}{\overset{\text{allene}}{\rightleftharpoons}} H-[M](=C(R)Me)(\eta^2\text{-}CH_2=CHMe)$$

$$H-[M](=C(R)Me)(CH_2=CHMe) \rightleftharpoons H-[M]\begin{matrix}-C(R)(Me)-CH_2\\ -CHMe-\end{matrix}$$

$$H-[M]-C(R)(Me)-CH_2-CHMe \text{ (metallacycle)} \longrightarrow RCH(Me)CH_2CH(Me)[M]$$

SCHEME 52

Olefin metathesis via carbene complexes[178] operates according to equation 87.

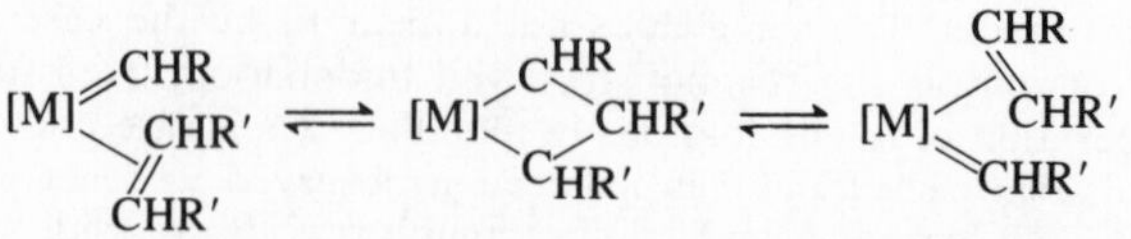

$$[M]=CHR + CHR'=CHR' \rightleftharpoons [M]\begin{matrix}CHR\\ \\ CHR'\end{matrix}CHR' \rightleftharpoons [M]=CHR' + CHR=CHR' \qquad (87)$$

Some of the metallacyclobutanes form olefins by $\beta$-elimination, however (Scheme 53)[179].

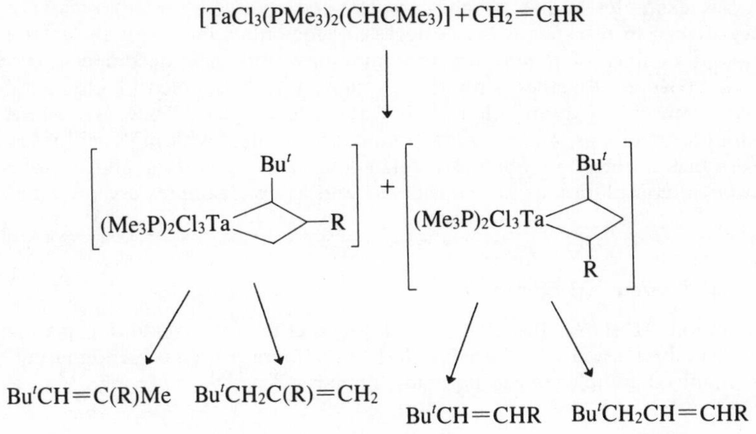

$$[TaCl_3(PMe_3)_2(CHCMe_3)] + CH_2{=}CHR$$

$$\downarrow$$

$$\left[(Me_3P)_2Cl_3Ta\text{-metallacyclobutane (Bu}^t\text{, R)}\right] + \left[(Me_3P)_2Cl_3Ta\text{-metallacyclobutane (Bu}^t\text{, R)}\right]$$

$$Bu^tCH{=}C(R)Me \quad Bu^tCH_2C(R){=}CH_2 \quad Bu^tCH{=}CHR \quad Bu^tCH_2CH{=}CHR$$

SCHEME 53

A common reaction of metal carbene complexes is to lead to the coupling of carbene fragments, presumably by an intermolecular process, to form olefins, and hence cause further metal—carbon bond scission. Other reactions of the generated carbene, such as addition to olefins to form cyclopropanes, are of course common. Precursors to these carbenes can often be formed by the α-elimination of halides, rather than hydrides, although the mechanism of the halogen transfer is equally obscure. Some examples include reactions 88–90[180–182].

$$[Fe(CH_2Cl)(Cp)(CO)_2] \xrightarrow{\text{cyclohexene}} [FeCl(Cp)(CO)_2] + 7,7\text{-dimethylbicyclo[4.1.0]heptane} \quad (88)$$

$$[Ir(CH_2Cl)(CO)(PPh_3)_2] \xrightarrow{\text{styrene}} [IrCl(CO)(PPh_3)_2] + \text{phenylcyclopropane} \quad (89)$$

$$2AgC(Cl){=}CPh_2 \longrightarrow 2AgCl + Ph_2C{=}C{=}C{=}CPh_2 \quad (90)$$

Finally, it should be noted that transfer of α-halide to main-group metal atoms has been used to generate carbenoid species with considerable success, perhaps most notably Seyferth's halomethylmercury compounds (equation 91)[183]. Transfer to silicon and other main group elements is also well known[184–188].

$$PhHgCCl_2Br + \text{cyclohexene} \longrightarrow PhHgBr + 7,7\text{-dichlorobicyclo[4.1.0]heptane} \quad (91)$$

## IV. γ-ELIMINATIONS

γ-Eliminations are the least known of the trio. Like α-eliminations, they appear to be more energetic than β-transfer reactions and are thus observed only when the β-process cannot operate. Whether the γ-eliminations are less favourable than α-eliminations also (this could be the case if the interaction of the γ-atom with the metal is significantly less than α and β), or whether it is simply coincidental that to date fewer have been documented, is uncertain. Like the α- and β-processes, it is the subsequent or accompanying reactions to the γ-elimination which lead to metal—carbon bond breaking.

A few examples of $\beta$-elimination were cited where the $\alpha$-atom was not carbon. For $\gamma$-hydrogen transfer, it is not necessary for either the $\alpha$- or the $\beta$-atoms to be carbon. Examples of $\beta$-heteroatom compounds and their decompositions are few, and are covered together with the $C_3$ materials. Examples of eliminations from ligands with the $\alpha$-atom other than carbon are legion. They are referred to as *metallation* reactions, and are far too numerous to deal with in detail in this chapter. Nevertheless, some are obviously related to the $\gamma$-transfers under discussion, so theoretical considerations are considered and a few examples are provided.

## A. $\gamma$-(and $\delta$-and $\varepsilon$-) Eliminations

Equation 92 shows the cleavage of tolylmethyl from platinum, presumably by reductive elimination after transfer of H to platinum from a $\delta$ carbon atom[189]. Cyclic ruthenium(II) complexes can be made similarly[190].

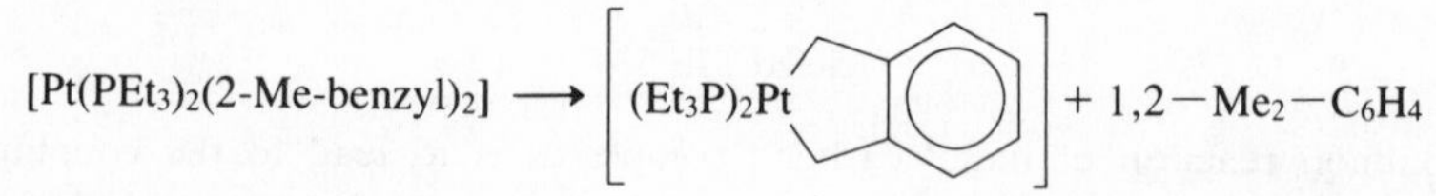

$$[Pt(PEt_3)_2(\text{2-Me-benzyl})_2] \longrightarrow \left[(Et_3P)_2Pt\text{(cyclic o-xylylene)}\right] + 1,2\text{-}Me_2\text{-}C_6H_4 \qquad (92)$$

Platinum(II) neopentyls decompose almost exclusively by $\gamma$-elimination (Scheme 54)[191], in contrast to the tantalum(V) complexes, where $\alpha$-elimination predominates.

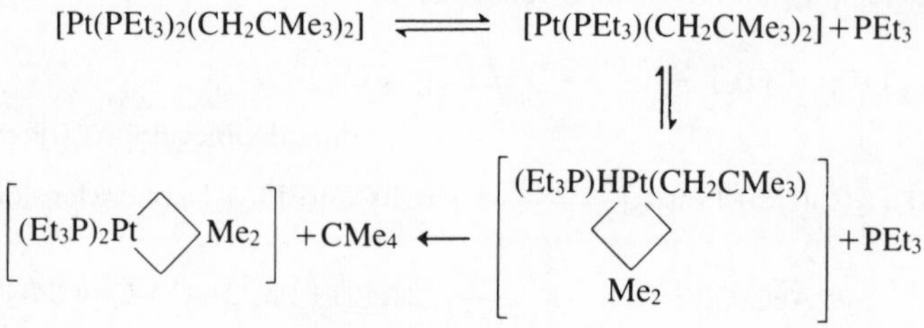

$$[Pt(PEt_3)_2(CH_2CMe_3)_2] \rightleftharpoons [Pt(PEt_3)(CH_2CMe_3)_2] + PEt_3$$

$$\Updownarrow$$

$$\left[(Et_3P)_2Pt\diamond Me_2\right] + CMe_4 \longleftarrow \left[(Et_3P)HPt(CH_2CMe_3)\diamond Me_2\right] + PEt_3$$

SCHEME 54

Although requiring loss of a ligand to initiate the process (presumably by making available a coordination site, a process closely resembling $\beta$-elimination from such square-planar complexes), $\gamma$-hydrogen transfer produces a five-coordinate intermediate prior to reductive elimination. Details of this process were confirmed by selective deuteriation studies, which also showed that any hydrogen abstraction from $PEt_3$ was less than 3% that from neopentyl, and any $\alpha$-hydrogen involvement must be less than 1% of $\gamma$-hydrogen transferred.

The difference between these platinum neopentyls, which clearly prefer $\gamma$-elimination, and neopentyls of niobium, tantalum, molybdenum, and tungsten, where $\alpha$-eliminations are common, is not easy to explain, but the high electron demand of the higher oxidation level compounds might well make the formation of carbenoid ligands desirable. A subsequent study by Whitesides and coworkers[192] indicated that the neopentylplatinum result is not unique. Products resulting from $\delta$- or $\varepsilon$-eliminations are found, as well as those of $\gamma$-transfer (equation 93 and 94).

$$[(Et_3P)_2Pt(\ldots)] \longrightarrow [(Et_3P)_2Pt(\ldots)] + [(Et_3P)_2Pt(\ldots)] \quad (93)$$

0.5% 99.5%

$$[(Et_3P)_2Pt(\ldots)] \longrightarrow [(Et_3P)_2Pt(\ldots)] + [(Et_3P)_2Pt(\ldots)] + [(Et_3P)_2Pt(\ldots)] \quad (94)$$

23% 68% 9%

All these thermolyses were inhibited by additional $PEt_3$ (indicating loss of a ligand as a key reaction step) and were first order in platinum complex. A deuterium kinetic isotope effect of *ca.* 3 was found for each process. The products and activation parameters did not vary according to the presence or absence of free ligand, and the rate-determining step was assigned as the reductive elimination of hydrocarbon[191,192]. An interesting and valuable result to emerge from this work is that the platinacyclobutane rings are not greatly strained, a significant finding in view of the importance of such four-membered rings in catalytic processes.

An interesting combination of α- and γ-eliminations from a mesityltantalum complex leads to a xylylmethylidene derivative[193]. The mesityl groups occupy the equatorial sites of trigonal bipyramidal tantalum(V), as may be expected for such bulky ligands. Treatment of a mesityl(neopentyl) tantalum derivative by $PMe_3$ leads to α-elimination and formation of mesitylene, a process with many analogues (equation 95).

$$[TaCl_3(\text{mesityl})(CH_2CMe_3)] \xrightarrow{PMe_3} [TaCl_3(PMe_3)_2(CHCMe_3)] + C_6H_3Me_3 \quad (95)$$

The mesityl (methyl) tantalum analogue behaves differently, however. γ-Elimination is involved in whichever route is followed to the product (Scheme 55). This time it is breaking of the tantalum—methyl bond which leads to elimination; the tantalum—mesityl bond does cleave, but a new one forms to a substituent methyl group. These results tend to confirm the reluctance of methyl groups to undergo α-elimination, and might also suggest that methyl substituents or aryls are particularly prone to interaction with metals.

Reactions of the ruthenium complex $[Ru_2(O_2CMe)_4Cl]$ with either $Mg(CH_2CMe_3)_2$ and $PMe_3$, or $Mg(CH_2SiMe_3)_2$ and $PMe_3$ lead to products from which one neopentyl or trimethylsilylmethyl has been cleaved, presumably by γ-elimination[194]. Rhodium

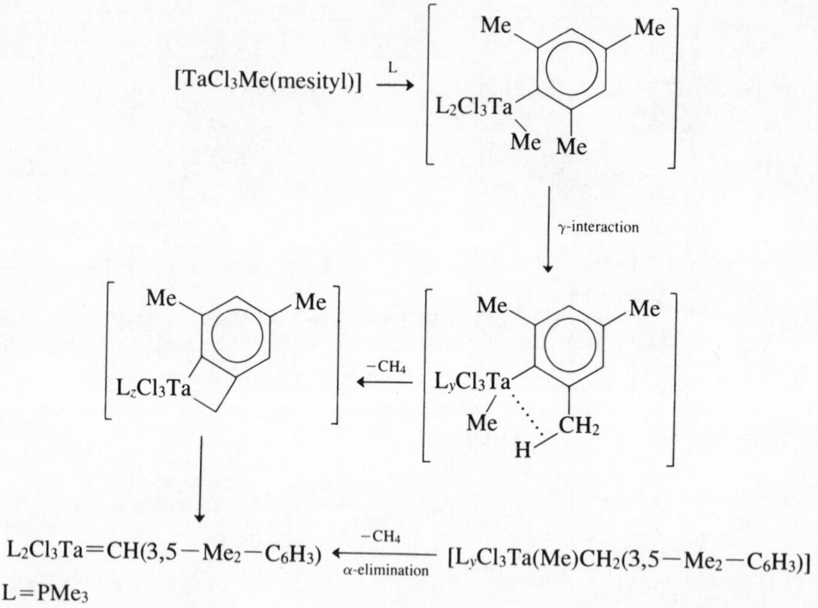

$$[TaCl_3Me(mesityl)] \xrightarrow{L} [L_2Cl_3Ta(2,4,6\text{-}Me_3C_6H_2)Me] \xrightarrow{\gamma\text{-interaction}} [L_yCl_3Ta(Me)\cdots H\text{-}CH_2(3,5\text{-}Me_2C_6H_2)] \xrightarrow{-CH_4} [L_zCl_3Ta(CH_2C_6H_2Me_2)]$$

$$L_2Cl_3Ta{=}CH(3,5{-}Me_2{-}C_6H_3) \xleftarrow[\alpha\text{-elimination}]{-CH_4} [L_yCl_3Ta(Me)CH_2(3,5{-}Me_2{-}C_6H_3)]$$

$L{=}PMe_3$

SCHEME 55

compounds can behave in a similar way (reaction 96). A related molybdenum

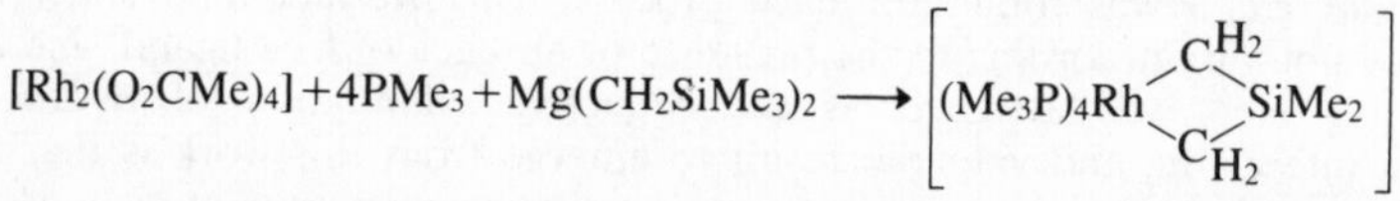

$$[Rh_2(O_2CMe)_4] + 4PMe_3 + Mg(CH_2SiMe_3)_2 \longrightarrow [(Me_3P)_4Rh(CH_2)_2SiMe_2] \quad (96)$$

compound behaves differently towards trimethylsilylmethylating agents, however, and two stages can be identified (Scheme 56).

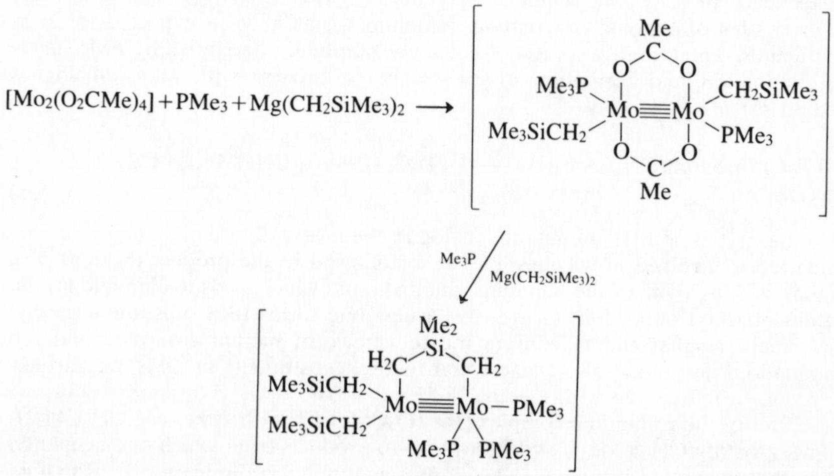

$$[Mo_2(O_2CMe)_4] + PMe_3 + Mg(CH_2SiMe_3)_2 \longrightarrow [(Me_3P)(Me_3SiCH_2)Mo(\mu\text{-}O_2CMe)_2Mo(CH_2SiMe_3)(PMe_3)]$$

$$\xrightarrow{Me_3P,\ Mg(CH_2SiMe_3)_2} [(Me_3SiCH_2)_2Mo(\mu\text{-}CH_2SiMe_2CH_2)Mo(PMe_3)_3]$$

SCHEME 56

The interaction of ylides with transition metal ions produces a class of β-elimination stabilized ylide complexes, and numerous examples are known[24]. These derivatives are analogous to the trimethylsilylmethyl complexes, with Si replaced with P. These ylide complexes appear to lose a proton from their methyl groups even more readily than do the silicon compounds, perhaps owing to the formal positive charge on phosphorus. The results of these γ-eliminations are either bridged or chelating dimethylenephosphorus compounds[24]. The byproducts are now tetramethylphosphonium salts. Although the reactions have been referred to as *trans*-ylidations[24], they are analogous to the γ-hydrogen processes we have examined. Some examples are shown in equations 97–99.

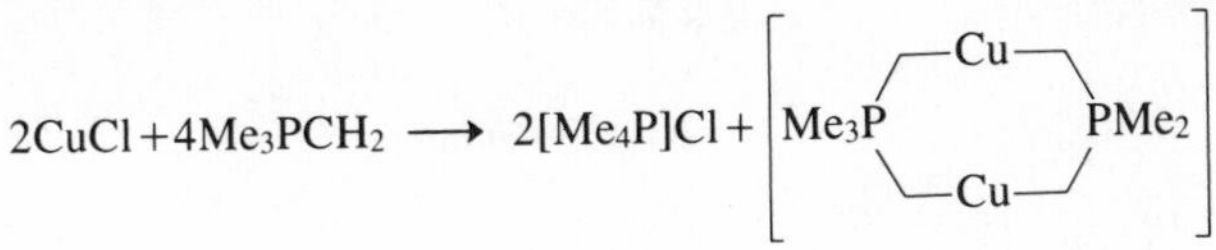

$$2CuCl + 4Me_3PCH_2 \longrightarrow 2[Me_4P]Cl + \left[Me_3P \begin{smallmatrix} -Cu- \\ \\ -Cu- \end{smallmatrix} PMe_2\right] \quad (97)$$

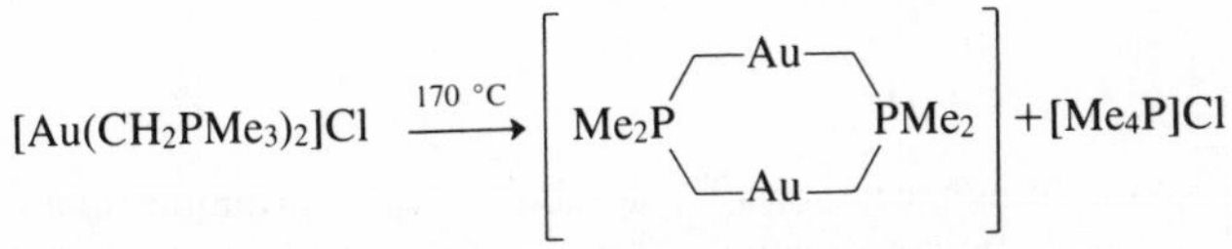

$$[Au(CH_2PMe_3)_2]Cl \xrightarrow{170\ ^\circ C} \left[Me_2P \begin{smallmatrix} -Au- \\ \\ -Au- \end{smallmatrix} PMe_2\right] + [Me_4P]Cl \quad (98)$$

$$[CoBrMe_2(PMe_3)_3] + 2Me_3PCH_2 \longrightarrow [Me_4P]Br + PMe_3 + \left[(Me_3P)_2Me_2Co\langle\rangle PMe_2\right] \quad (99)$$

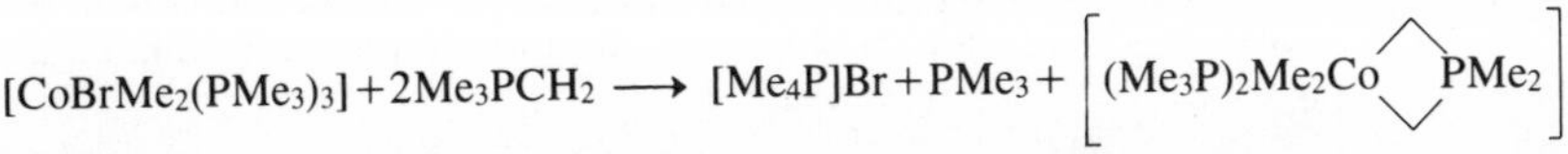

Scheme 57 shows a palladium catalysed formation of lactones from α,ω-haloalcohols and carbon monoxide[195]. Ring closure is accomplished by hydrogen transfer from an oxygen, often in the δ-position to the metal. These are rare examples of heteroatoms being responsible for ring closure.

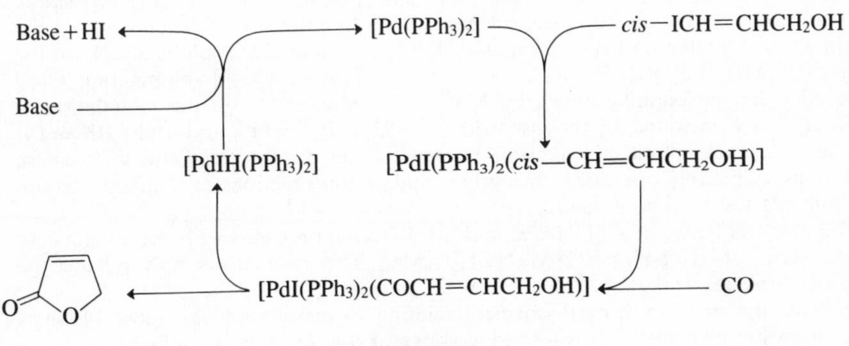

Base + HI ← → $[Pd(PPh_3)_2]$ — *cis*—ICH=CHCH$_2$OH

Base

$[PdIH(PPh_3)_2]$ $[PdI(PPh_3)_2(cis\text{—}CH{=}CHCH_2OH)]$

(furanone) ← $[PdI(PPh_3)_2(COCH{=}CHCH_2OH)]$ ← CO

SCHEME 57

## B. Metallation Reactions

When the α-atom is a heteroatom such as O, N, S, P, or As, elimination reactions are usually referred to as metallations, since a new bond is formed from part of the ligand to the metal. Metal—carbon bond cleavages resulting from these reactions are rare, simply because few of the complexes involved contain such bonds in the first instance. Equations 100–102 show some examples[170,196,197].

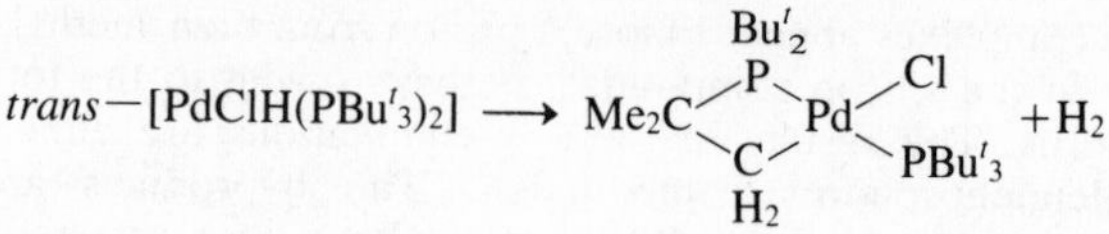

$$trans\text{—}[PdClH(PBu^t_3)_2] \longrightarrow Me_2C\langle{}^{P(Bu^t_2)}_{CH_2}\rangle Pd\langle{}^{Cl}_{PBu^t_3} + H_2 \qquad (100)$$

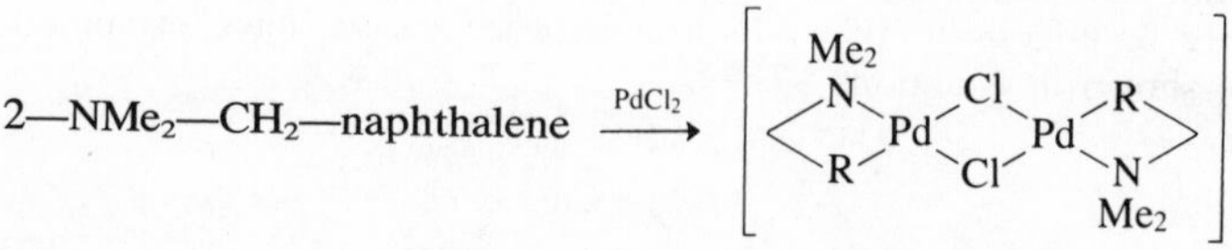

$$2\text{—}NMe_2\text{—}CH_2\text{—naphthalene} \xrightarrow{PdCl_2} \left[ \langle{}^{N(Me_2)}_{R}\rangle Pd\langle{}^{Cl}_{Cl}\rangle Pd\langle{}^{R}_{N(Me_2)}\rangle \right] \qquad (101)$$

R=2,3-naphthyl

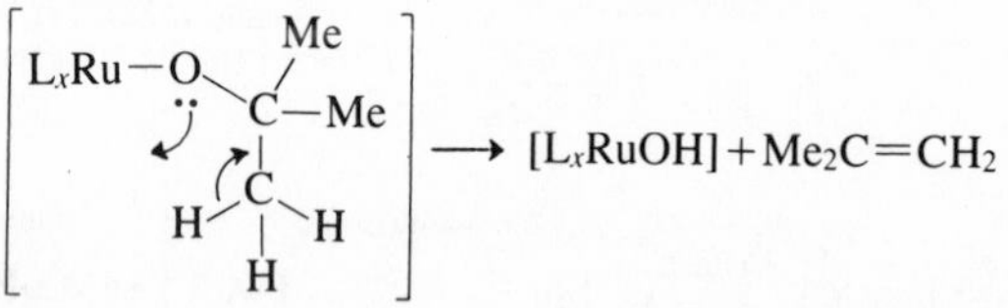

$$\left[ L_xRu\text{—}\ddot{O}\text{—}C(Me)_2\text{—}CH_3 \right] \longrightarrow [L_xRuOH] + Me_2C{=}CH_2 \qquad (102)$$

Recent reviews provide other examples[198]. Reactions are intramolecular: the ligands to be metallated must be coordinated. Thus 2-pyridylformate, **23**, is readily metallated by $[RhCl(PPh_3)_3]$[199], whereas **24**, with no nitrogen to coordinate, is

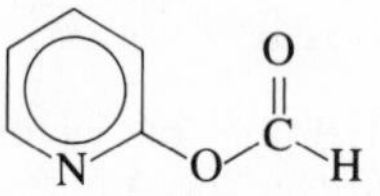

(23) 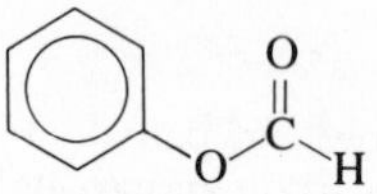 (24)

unaffected. It is also important that $\beta$-elimination should not be an available option in the compounds, or this would take precedence, being a lower energy process.

A study directed at elucidating electronic effects found a strong dependence on the metal atom. Metallating azobenzene with $PdCl_2$ revealed electrophilic attack on the ring by Pd(II), but metallating azobenzene by $[MeMn(CO)_5]$ (eliminating $CH_4$) proceeded by nucleophilic attack by Mn[200]. A similar study on the metallation of $R_2PCH_2C_6H_4F$-*m* found an activity series $Ir^I > Rh^I \gg Pd^{II} \approx Pt^{II}$, and whilst Rh and Ir displayed a nucleophilic mechanism, Pd reacted by electrophilic attack[201]. These variations will surely operate in the other elimination reactions, and might account for some of the variations found.

The reaction between azobenzene and $[MnR(CO)_5]$ proceeds with decreasing ease in the series $R = CH_2Ph > Et > Me > CH_2C_6Me_5$. This probably simply reflects the ease of elimination of RH[202].

Perhaps the most exciting discoveries relating to metallation are those of Shaw relating to steric effects[203]. It was readily apparent that more bulky substituents were instrumental in promoting metallations. For example, in reactions such as 103, the tendency to proceed to the right increases in the order $R = Me < Ph < Bu^t$. The steric effects are dominant over electronic factors.

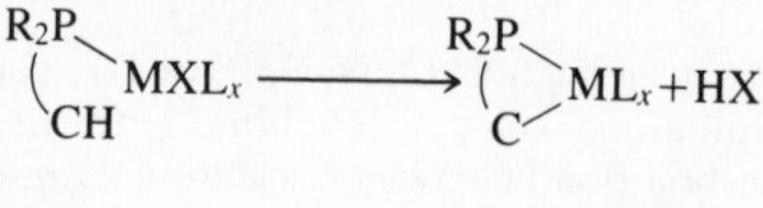

$$\left(\begin{matrix} R_2P \\ CH \end{matrix}\right) MXL_x \longrightarrow \left(\begin{matrix} R_2P \\ C \end{matrix}\right) ML_x + HX \qquad (103)$$

Shaw[203] pointed out that there was a close similarity between these steric effects and an organic effect known as the Thorpe–Ingold or *gem*-dialkyl effect. This relates to the formation of small rings, where replacement of a $CH_2$ by $CMe_2$ greatly increases the stability (and often the yield) of cyclic products. This effect hinges on two factors. An entropy effect operates: *gem*-dimethyls reduce the loss of internal rotational entropy which occurs on cyclization. An enthalpy effect also operates: the number of extra *gauche* interactions which are introduced on cyclization is reduced. The analogue for metallation is a '*gem*-di-*t*-butyl effect'. The bulky coordinated ligands will interact with neighbouring groups, producing a large energy barrier to rotation, and hence no loss of rotational freedom occurs on metallation. Since the metal atom becomes part of the ring, further steric bulk at the metal becomes just as important.

Other effects will enter also. The bulky *t*-butyl groups of $PBu^t[2,6\text{-}(OMe)_2C_6H_3]$ hold the methoxy group in the proximity of the metal when coordinated, making the best arrangement for an interaction of methyl and metal (**25**).

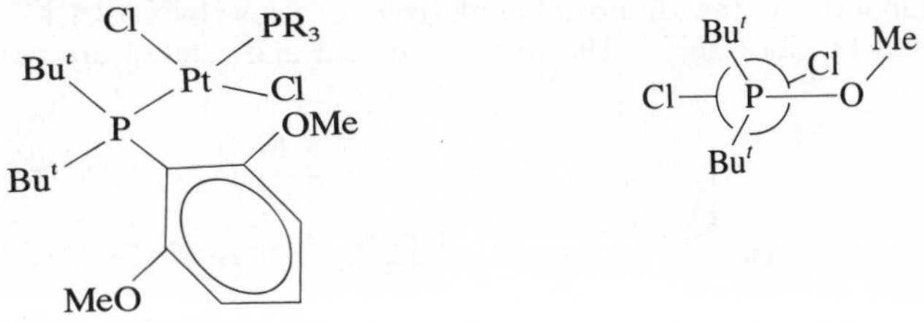

Cl, $PR_3$, $Bu^t$, Pt—Cl, P, OMe, $Bu^t$, MeO

$Bu^t$, Me, Cl, Cl—(P)—O, $Bu^t$

(**25**)

The implications of this steric effect for γ-eliminations are of enormous importance. Most of the authenticated examples occur in neopentyl or trimethylsilylmethyl compounds, where the *gem*-dimethyl effect itself will operate. This may provide an important boost to this reaction mode, enabling it to succeed instead of other reaction modes. The favourable energy of metallacyclobutanes found by Whitesides and coworkers may not be entirely general, since this contained *gem*-dimethyls also.

It has long been suspected that an interaction between the metal atom and hydrogen, or the C—H bond, is a necessary prerequisite to metallation, as indeed it appears to be for α-, β-, or γ-eliminations. Many interactions between hydrogen atoms of ligands and the metal atoms have been found[202], some extremely strong, but none of the compounds exhibiting these interactions undergo metallation! There are two limiting geometries for the approach of the C—H bond to the metal: linear and triangular.

C—H·····M
linear
(**26**)

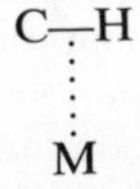

C—H
⋮
M
triangular
(**27**)

A comparison of the related metallations of 2-pyridyl formate and 8-quinolinyl formate at $[RhCl(PPh_3)_3]$ found very similar tendencies to undergo these reactions (and a deuterium isotope effect of 3.4, indicating rate-limiting C—H cleavage)[199]. Molecular models show that the C—Hs of both formates can approach by the triangular route, but only the pyridyl derivative is flexible enough to approach by the linear mode. The similarity of reaction rates indicates that the side-on route is favourable.

Evidence has been presented that metallation by palladium(II) can occur only in the square plane of the molecule[204]. This situation resembles the requirement of an in-plane coordination site for $\beta$- or $\gamma$-eliminations at platinum. It would be interesting to know if a ligand dissociation aided such metallations.

Two final features are worth noting with respect to these cyclization reactions, although both serve only to illustrate our ignorance on the subject. Metallation of *trans*-$[PtCl_2(PBu^t_3)_2]$ (reaction 104) is subject to a marked solvent effect, proceeding more readily in benzene than methylene chloride[205].

$$[(PhCN)_2PtCl_2] + 2PBu^t_3 \longrightarrow \left[ \text{cyclometallated } Bu^t_3P(Cl)Pt\text{–}CH_2CMe_2\text{–}PBu^t_2 \right] + [Bu^t_3PH][PtCl_4] \quad (104)$$

Metallation of the dithio ether derivative below by $[PdCl_4]^{2-}$ (reaction 105) is *catalysed* by silica gel[206]. The roles of solvent and catalyst are not understood.

$$\text{MeO-aryl(OCH}_2\text{SMe)(CH}_2\text{SMe)} \xrightarrow[\text{silica gel}]{Li_2[PdCl_4]} \left[ \text{MeO-aryl-Pd(Cl)(SMe)}_2 \text{ pincer complex} \right] \quad (105)$$

## V. CONCLUDING REMARKS

It is obvious from the preceding pages that while many trends and factors concerning $\alpha$-, $\beta$-, and $\gamma$-eliminations have emerged, allowing a certain amount of control to be exercised over them, much remains to be learnt.

One factor emerges clearly. The simple picture of the eliminations drawn in the Introduction (equations 1–3), showing transfer of hydrogen cleanly to the metal, are extremes which may never be achieved in reality. C—H bond making for subsequent alkane elimination, or M—C bond breaking for, for example, olefin ejection, may be well advanced before hydrogen transfer can approach completion. It appears that most (but not all) $\beta$-eliminations do manage in large part to transfer the $\beta$-hydrogen to the metal before eliminations of organic groups are far advanced. The reverse may be more common for $\alpha$-eliminations, but data here are limited to a narrow range of compounds and this could distort the impression given.

The situation may be likened to ligand substitution reactions. In the past they were often described as dissociative ($S_N1$) or associative ($S_N2$), but are now more accurately seen as the less extreme dissociatively controlled $I_d$ or associatively controlled $I_a$. The attempts of chemists to categorize the M—C bond breaking processes into simple steps ('$\beta$-hydrogen transfer followed by reductive elimination') as a means of understanding and classifying reactions may be too naive, and what are today described as separate processes may operate concomitantly. Rather than finding a common explanation to account for reactions, in the last analysis each molecule or molecular system may have to be treated as unique.

## VI. REFERENCES


1. J. Chatt, R. S. Coffey, A. Gough, and B. L. Shaw, *J. Chem. Soc. A,* 190 (1968).
2. (a) R. P. A. Sneedon and H. H. Zeiss, *J. Organomet. Chem.,* **22,** 713 (1970); (b) R. P. A. Sneedon and H. H. Zeiss, *J. Organomet. Chem.,* **26,** 101 (1971); (c) R. P. A. Sneedon and H. H. Zeiss, *Angew. Chem., Int. Ed. Engl.,* **7,** 951 (1968).
3. G. M. Whitesides, J. F. Gaasch, and R. R. Stedronsky, *J. Am. Chem. Soc.,* **94,** 5258 (1972).
4. G. M. Whitesides, E. R. Stedronsky, C. P. Casey, and J. San Fillipo, *J. Am. Chem. Soc.,* **92,** 1426 (1970).
5. J. M. Manriquez, D. R. McAlister, R. D. Sanner, and J. E. Bercaw, *J. Am. Chem. Soc.,* **100,** 2716 (1978).
6. J. Evans, J. Schwartz, and P. W. Urquhart, *J. Organomet. Chem.,* **81,** C37 (1974).
7. F. N. Tebbe and G. W. Parshall, *J. Am. Chem. Soc.,* **93,** 3793 (1971).
8. A. H. Klazinga and J. H. Teuben, *J. Organomet. Chem.,* **165,** 31 (1979).
9. H. Bönnemann, Ch. Grard, W. Kopp, and G. Wilke, *Proc. 23rd Int. Congr. Pure Appl. Chem.,* **6,** 265 (1971).
10. B. Olgemöller and W. Beck, *Angew. Chem., Int. Ed. Engl.,* **19,** 834 (1980).
11. R. J. Klinger, J. C. Huffman, and J. K. Kochi, *J. Organomet. Chem.,* **206,** C7 (1981).
12. Y. Becker and J. K. Stille, *J. Am. Chem. Soc.,* **100,** 845 (1978).
13. M. Green, N. C. Norman, and A. G. Orpen, *J. Am. Chem. Soc.,* **103,** 1267, 1269 (1981).
14. (a) G. Yagupsky, W. Mowat, A. Shortland and G. Wilkinson, *J. Chem. Soc., Chem. Commun.,* 1369 (1970); (b) M. R. Collier, M. F. Lappert, and M. M. Truelock, *J. Organomet. Chem.,* **25,** C36 (1970).
15. P. S. Braterman and R. J. Cross, *J. Chem. Soc., Dalton Trans.,* 657 (1972).
16. P. S. Braterman and R. J. Cross, *Chem. Soc. Rev.,* **2,** 271 (1973).
17. J. Chatt and B. L. Shaw, *J. Chem. Soc.,* 705, 4020 (1959).
18. M. L. H. Green, *Organometallic Compounds,* Vol. 2, Methuen, London, 1968.
19. (a) I. Dvorak, R. J. O'Brien, and W. Santo, *J. Chem. Soc., Chem. Commun.,* 411 (1970). (b) V. B. Shur, E. G. Berkovitch, L. B. Vasiljeva, R. V. Kudryautsev, and M. E. Vol'pin, *J. Organomet. Chem.,* **78,** 127 (1974); (c) C. P. Boekel, J. H. Teuben, and H. J. de Liefde Meijer, *J. Organomet. Chem.,* **81,** 371 (1974).
20. (a) U. Giannini and U. Zuchini, *Chem. Commun.,* 940 (1968); (b) I. W. Bassi, G. Allegra, R. Scordamaglia, and G. Chioccola, *J. Am. Chem. Soc.,* **93,** 3787 (1971).
21. M. D. Johnson and N. Winterton, *J. Chem. Soc. A,* 507 (1970).
22. P. J. Davidson, M. F. Lappert, and R. Pearce, *J. Organomet. Chem.,* **57,** 269 (1973).
23. P. J. Davidson, M. F. Lappert, and R. Pearce, *Chem. Rev.,* **76,** 219 (1976).
24. H. Schmidbaur, *Acc. Chem. Res.,* **8,** 62 (1975).
25. F. A. Cotton and G. Wilkinson, *Advanced Inorganic Chemistry,* 4th ed., Wiley–Interscience, New York, 1980.
26. G. M. Whitesides, D. E. Bergbreiter, and P. E. Kendall, *J. Am. Chem. Soc.,* **96,** 2806 (1974).
27. C. A. McAuliffe, *Transition Metal Complexes of Phosphorus, Arsenic and Antimony Ligands,* Macmillan, London, 1973.
28. E. H. Brooks and R. J. Cross, *Organomet. Chem. Rev. (A),* **6,** 227 (1970).
29. (a) J. W. Rathke and E. L. Muetterties, *J. Am. Chem. Soc.,* **97,** 3272 (1975); (b) H. H. Karsch, H.-F. Klein, and H. Schmidbaur, *Angew. Chem., Int. Ed. Eng.,* **14,** 637 (1975); (c) H. H. Karsch, H.-F. Klein, and H. Schmidbaur, *Chem. Ber.,* **110,** 2200, 2213, 2222 (1977); (d) T. V. Harris, J. W. Rathke, and E. L. Muetterties, *J. Am. Chem. Soc.,* **100,** 6966 (1978).
30. H. Werner and R. Werner, *J. Organomet. Chem.,* **209,** C60 (1981).
31. F. A. Cotton, B. A. Frenz, and D. L. Hunter, *J. Chem. Soc., Chem. Commun.,* 755 (1974); J. Chatt and J. N. Davidson, *J. Chem. Soc.,* 843 (1965).
32. G. Yoshida, H. Kurosawa, and R. Okawara, *J. Organomet. Chem.,* **113,** 85 (1976); *Chem. Lett.,* 1387 (1977).
33. C. Windus, S. Sujishi, and W. P. Giering, *J. Am. Chem. Soc.,* **96,** 1951 (1974); *J. Organomet. Chem.,* **101,** 279 (1975).
34. A. W. P. Jarvie, *Organomet. Chem. Rev. (A),* **6,** 153 (1970).

35. G. Yagupsky, C. K. Brown, and G. Wilkinson, *J. Chem. Soc. A,* 1392 (1970).
36. K. Thomas, J. A. Osborn, A. R. Powell, and G. Wilkinson, *J. Chem. Soc. A,* 1801 (1968).
37. D. L. Reger and E. C. Culbertson, *J. Am. Chem. Soc.,* **98,** 2789 (1976).
38. H. Kurosawa, T. Majima, and N. Asada, *J. Am. Chem. Soc.,* **102,** 6996 (1980).
39. E. R. Evitt and R. G. Bergman, *J. Am. Chem. Soc.,* **101,** 3973 (1979).
40. H. E. Bryndza, E. R. Evitt, and R. G. Bergman, *J. Am. Chem. Soc.,* **102,** 4948 (1980).
41. Ref. 23, p. 230.
42. W. Kruse, *J. Organomet. Chem.,* **42,** C39 (1972).
43. T. J. Marks, A. M. Seyam, and J. R. Kolb, *J. Am. Chem. Soc.,* **95,** 5529 (1973).
44. T. J. Marks and A. M. Seyam, *J. Organomet. Chem.,* **67,** 61 (1974).
45. T. Ikariya and A. Yamamoto, *J. Organomet. Chem.,* **120,** 257 (1976).
46. T. Yamamoto, A. Yamamoto, and S. Ikeda, *Bull. Chem. Soc. Jpn.,* **45,** 1104 (1972).
47. (a) J. H. Grate and G. N. Schrauzer, *J. Am. Chem. Soc.,* **101,** 4601 (1979); (b) G. N. Schrauzer and J. H. Grate, *J. Am. Chem. Soc.,* **103,** 541 (1981).
48. (a) R. G. Nuzzo, T. J. McCarthy, and G. M. Whitesides, *Inorg. Chem.,* **20,** 1312 (1981); (b) F. Ozawa, T. Ito, and A. Yamamoto, *J. Organomet. Chem.,* **168,** 375 (1979).
49. G. K. Anderson and R. J. Cross, *Chem. Soc. Rev.,* **9,** 185 (1980).
50. (a) H. C. Clark and C. R. Jablonski, *Inorg. Chem.,* **13,** 2213 (1974); (b) H. C. Clark and C. S. Wong, *J. Am. Chem. Soc.,* **96,** 7213 (1974).
51. P. Foley and G. M. Whitesides, *Inorg. Chem.,* **19,** 1402 (1980).
52. N. Chaudhury and R. J. Puddephatt, *J. Chem. Soc., Dalton Trans.,* 915 (1976).
53. T. Yamamoto, Y. Yamamoto, and S. Ikeda, *J. Am. Chem. Soc.,* **93,** 3350, 3360 (1971).
54. J. Schwartz and J. B. Cannon, *J. Am. Chem. Soc.,* **96,** 2276 (1974).
55. F. Ozawa, T. Ito, and A. Yamamoto, *J. Am. Chem. Soc.,* **102,** 6457 (1980).
56. A. Miyashita, T. Yamamoto, and A. Yamamoto, *Bull. Chem. Soc. Jpn.,* **50,** 1102, 1109 (1977).
57. H.-F. Klein, R. Hammer, J. Gross, and U. Schubert, *Angew. Chem., Int. Ed. Engl.,* **19,** 809 (1980).
58. S. Sostero, O. Traverso, M. Lenarda, and M. Graziani, *J. Organomet. Chem.,* **134,** 259 (1977).
59. A. Tamaki, S. A. Magennis, and J. K. Kochi, *J. Am. Chem. Soc.,* **96,** 6140 (1974).
60. S. Komiya, T. Yamamoto, and A. Yamamoto, *Chem. Lett.,* 1273 (1978).
61. T. Hosokawa and P. M. Maitlis, *J. Am. Chem. Soc.,* **95,** 4924 (1973).
62. (a) J. X. McDermott, J. F. White, and G. M. Whitesides, *J. Am. Chem. Soc.,* **95,** 4491 (1973); (b) J. X. McDermott, J. F. White, and G. M. Whitesides, *J. Am. Chem. Soc.,* **98,** 6521 (1976).
63. J. X. McDermott, M. E. Wilson, and G. M. Whitesides, *J. Am. Chem. Soc.,* **98,** 6529 (1976).
64. G. B. Young and G. M. Whitesides, *J. Am. Chem. Soc.,* **100,** 5808 (1978).
65. P. S. Braterman, *J. Chem. Soc., Chem. Commun.,* 70 (1979).
66. (a) R. H. Grubbs, A. Miyashita, M.-I. M. Liu, and P. L. Burk, *J. Am. Chem. Soc.,* **99,** 3863 (1977); **100,** 1300, 2418 (1978); (b) R. H. Grubbs and A. Miyashita, *J. Chem. Soc., Chem. Commun.,* 864 (1977); (c) R. H. Grubbs and A. Miyashita, *J. Am. Chem. Soc.,* **100,** 7416, 7418 (1978).
67. R. J. Puddephatt, *Coord. Chem. Rev.,* **33,** 149 (1980).
68. T. H. Johnson and S.-S. Cheng, *J. Am. Chem. Soc.,* **101,** 5277 (1979).
69. F. N. Tebbe, G. W. Parshall, and G. S. Reddy, *J. Am. Chem. Soc.,* **100,** 3611 (1978).
70. R. G. Pearson, *Chem. Br.,* **12,** 160 (1976).
71. D. L. Thom and R. Hoffmann, *J. Am. Chem. Soc.,* **100,** 2079 (1978).
72. (a) E. M. Bau, R. P. Hughes, and J. Powell, *Chem. Commun.,* 591 (1973); (b) T. Ziegler and A. Rank, *Inorg. Chem.,* **18,** 1558 (1979); (c) D. G. Cooper, G. K. Hamer, J. Powell, and W. F. Reynolds, *Chem. Commun.,* 249 (1973).
73. R. J. Cross, *Specialist Periodical Report, Catalysis,* 1982, Royal Society of Chemistry, London, Ch. 10.
74. N. Gragor and P. M. Henry, *J. Am. Chem. Soc.,* **103,** 681 (1981).
75. J. K. Stille and R. Divakaruni, *J. Organomet. Chem.,* **169,** 239 (1979).
76. J. E. Bäckvall, B. Åkermark, and S. O. Ljunggren, *J. Am. Chem. Soc.,* **101,** 2411 (1979).

77. S. Winstein, J. McCaskie, H.-B. Lee, and P. M. Henry, *J. Am. Chem. Soc.*, **98,** 6913 (1976).
78. B. S. Tourag, F. Mares, and S. E. Diamond, *J. Am. Chem. Soc.*, **102,** 6616 (1980).
79. R. J. Cross and R. Wardle, *J. Chem. Soc. A,* 840 (1970); R. J. Cross and N. H. Tennent, *J. Organomet. Chem.*, **72,** 21 (1974).
80. C. W. Bird, *Chem. Ind. (London),* 520 (1972).
81. M. Akbarzadeh and C. B. Anderson, *J. Organomet. Chem.*, **197,** C5 (1980).
82. L. S. Hegedus, R. E. Williams, M. A. McGuire, and T. Hayashi, *J. Am. Chem. Soc.*, **102,** 4973 (1980).
83. L. S. Hegedus, G. F. Allen, and E. L. Waterman, *J. Am. Chem. Soc.*, **98,** 2674 (1976).
84. J.-E. Bäckvall and R. E. Nordberg, *J. Am. Chem. Soc.*, **102,** 393 (1980).
85. G. Smith, S. J. McLain, and R. R. Schrock, *J. Organomet. Chem.*, **202,** 269 (1980).
86. A. B. Goel and S. Goel, *Inorg. Chim. Acta,* **45,** L85 (1980).
87. J. Chatt and B. L. Shaw, *J. Chem. Soc.*, 5075 (1962); J. Chatt, R. S. Coffey, and B. L. Shaw, *J. Chem. Soc.*, 7391 (1965); K. R. Laing and W. R. Roper, *J. Chem. Soc. A,* 1889 (1969); D. P. Arnold and M. A. Bennett, *J. Organomet. Chem.*, **199,** C17 (1980).
88. H. Sinn, W. Kaminsky, H.-J. Vollmer, and R. Woldt, *Angew. Chem., Int. Ed. Engl.*, **19,** 390 (1980).
89. K. Fischer, K. Jonas, P. Misbach, R. Stabba, and G. Wilke, *Angew. Chem., Int. Ed. Engl.*, **12,** 943 (1973).
90. R. J. P. Corriu and B. Meunier, *Chem. Commun.*, 164 (1973).
91. C. Masters, A. A. Kiffen, and J. P. Visser, *J. Am. Chem. Soc.*, **98,** 1357 (1976).
92. A. H. Reid, P. B. Shevlin, S. S. Yun, and T. R. Webb, *J. Am. Chem. Soc.*, **103,** 709 (1981).
93. R. R. Corderman and J. L. Beauchamps, *J. Am. Chem. Soc.*, **98,** 5700 (1976).
94. B. W. The-Khai, C. Concilio, and G. Porzi, *J. Organomet. Chem.*, **208,** 249 (1981).
95. R. A. Sanchez-Delgado and O. L. de Ochea, *J. Organomet. Chem.*, **202,** 427 (1980).
96. G. Carturan, A. Sirivanti, and F. Morandini, *Angew. Chem., Int. Ed. Engl.*, **20,** 112 (1981).
97. J. R. Bleeke and E. L. Muetterties, *J. Am. Chem. Soc.*, **103,** 556 (1981).
98. O. W. Howarth, C. H. McAteer, P. Moore, and G. E. Morris, *J. Chem. Soc., Chem. Commun.*, 506 (1981).
99. H. C. Clark and K. E. Hine, *J. Organomet. Chem.*, **105,** C32 (1976).
100. H. C. Clark and C. R. Milne, *J. Organomet. Chem.*, **161,** 51 (1978); H. C. Clark, A. B. Goel, and C. S. Wong, *J. Organomet. Chem.*, **190,** C101 (1980); H. C. Clark, C. Billard, and C. S. Wong, *J. Organomet. Chem.*, **190,** C105 (1980).
101. N. M. Boay, M. Green, and F. G. A. Stone, *J. Chem. Soc., Chem. Commun.*, 1281 (1980).
102. R. J. Puddephatt and M. A. Thomson, *Inorg. Chim. Acta,* **45,** 1281 (1980).
103. J. B. Keister and J. R. Shapley, *J. Am. Chem. Soc.*, **98,** 1057 (1976); J. Müller, B. Passon, and S. Schmitt, *J. Organomet. Chem.*, **195,** C21 (1980); P. Kalck, R. Poilblanc, R.-P. Martin, A. Rovera, and A. Gaset, *J. Organomet. Chem.*, **195,** C9 (1980).
104. M. H. Chisholm and D. A. Haitko, *J. Am. Chem. Soc.*, **101,** 6784 (1979).
105. G. W. Parshall, *Homogeneous Catalysis,* Wiley–Interscience, New York, 1980; C. Masters, *Homogeneous Transition-Metal Catalysis,* Chapman and Hall, London, 1981.
106. H. C. Clark and H. Kurosawa, *Inorg. Chem.*, **12,** 1566 (1973).
107. K. Hirai, H. Susuki, Y. Moro-oka, and T. Ikawa, *Tetrahedron Lett.*, **21,** 3413 (1980).
108. K. Tamao, Y. Kiso, K. Sumitani, and M. Kumada, *J. Am. Chem. Soc.*, **94,** 9268 (1972).
109. T. Kobayashi and M. Tanaka, *J. Organomet. Chem.*, **205,** C27 (1981).
110. L. S. Hegedus and W. H. Darlington, *J. Am. Chem. Soc.*, **102,** 4986 (1980).
111. C. U. Pittman, G. M. Wileman, W. D. Wilson, and R. C. Ryan, *Angew. Chem., Int. Ed. Engl.*, **19,** 478 (1980).
112. W. S. Knowles, M. J. Sabacky, and B. D. Vineyard, *J. Chem. Soc., Chem. Commun.*, 10 (1972).
113. H. B. Kagan and T. P. Dang, *J. Am. Chem. Soc.*, **94,** 6429 (1972).
114. H. Brunner, *Acc. Chem. Res.*, **12,** 250 (1979).
115. W. S. Knowles, M. J. Sabacky, B. D. Vineyard, and D. J. Weinkauff, *J. Am. Chem. Soc.*, **97,** 2567 (1975); M. Tanaka and I. Ogata, *J. Chem. Soc., Chem. Commun.*, 735

(1975); B. D. Vineyard, W. S. Knowles, M. J. Sabacky, G. L. Bachman, and D. J. Weinkauff, *J. Am. Chem. Soc.*, **99,** 5946 (1977); I. Ojima and T. Kogure, *J. Organomet. Chem.*, **195,** 239 (1980); J. M. Brown and P. A. Chaloner, *J. Am. Chem. Soc.*, **102,** 3040 (1980).
116. M. Bianchi, U. Matteoli, G. Menchi, P. Frediani, F. Piacenti, and C. Botteghi, *J. Organomet. Chem.*, **195,** 337 (1980).
117. N. Takaishi, H. Imai, C. A. Bortelo, and J. K. Stille, *J. Am. Chem. Soc.*, **100,** 264 (1978); T. Masuda and J. K. Stille, *J. Am. Chem. Soc.*, **100,** 268 (1978).
118. K. A. Brady and T. A. Nile, *J. Organomet. Chem.*, **206,** 299 (1981).
119. M. Green, J. L. Spencer, F. G. A. Stone, and C. A. Tsipis, *J. Chem. Soc., Dalton Trans.*, 1519 (1977).
120. J.-F. Peyronel, J.-C. Fiaud, and H. B. Kagan, *J. Chem. Res., (S)* 320; *(M)* 4057 (1980).
121. Y. Seki, K. Takeshita, K. Kawamoto, S. Murai, and N. Sonoda, *Angew. Chem., Int. Ed. Engl.*, **19,** 928 (1980).
122. A. Millan, E. Towns, and P. M. Maitlis, *J. Chem. Soc., Chem. Commun.*, 673 (1981).
123. J. Villiéras, C. Bacquet, and J.-F. Normant, *Bull. Chem. Soc. Fr.*, 1731 (1974).
124. H. de Vries, *Recl. Trav. Chim. Pays-Bas,* **80,** 866 (1961).
125. J. R. C. Light and H. H. Zeiss, *J. Organomet. Chem.*, **21,** 391 (1970).
126. K. Nishimura, H. Kuribayashi, A. Yamamoto, and S. Ikeda, *J. Organomet. Chem.*, **37,** 317 (1972); A. Yamamoto, Y. Kano, and T. Yamamoto, *J. Organomet. Chem.*, **102,** 57 (1975).
127. A. S. Khachaturov, L. S. Brester, and I. Yu. Paddubnyi, *J. Organomet. Chem.*, **42,** C18 (1972).
128. A. J. Shortland and G. Wilkinson, *J. Chem. Soc., Dalton Trans.*, 872 (1973).
129. S. Otsuka, A. Nakamura, T. Yoshida, M. Naruto, and K. Ataka, *J. Am. Chem. Soc.*, **95,** 3180 (1973).
130. T. Ikariya and A. Yamamoto, *J. Chem. Soc., Chem. Commun.*, 720 (1974).
131. P. J. Davidson, M. F. Lappert, and R. Pearce, *J. Organomet. Chem.*, **57,** 269 (1973).
132. G. J. Erskine, D. A. Wilson, and J. D. McCowan, *J. Organomet. Chem.*, **114,** 119 (1976); G. J. Erskine, J. Hartgerink, E. L. Weinberg, and J. D. McCowan, *J. Organomet. Chem.*, **170,** 51 (1979).
133. G. A. Razuvaev, V. P. Mar'in, and Yu. A. Andrianov, *J. Organomet. Chem.*, **174,** 67 (1979).
134. E. Klei, H. J. Telgen, and H. J. Teuben, *J. Organomet. Chem.*, **209,** 297 (1981).
135. (a) N. J. Cooper and M. L. H. Green, *J. Chem. Soc., Chem. Commun.*, 761 (1974); (b) N. J. Cooper and M. L. H. Green, *J. Chem. Soc., Dalton Trans.*, 1121 (1979); (c) M. Canestrani and M. L. H. Green, *J. Chem. Soc., Chem. Commun.*, 913 (1979).
136. J. C. Green, M. L. H. Green, and C. K. Prout, *J. Chem. Soc., Chem. Commun.*, 421 (1972); J. W. Lauher and R. Hoffmann, *J. Am. Chem. Soc.*, **98,** 1729 (1976).
137. R. R. Schrock, *J. Am. Chem. Soc.*, **97,** 6577 (1975); R. R. Schrock and L. J. Guggenberger, *J. Am. Chem. Soc.*, **97,** 6578 (1975); R. R. Schrock and P. R. Sharp, *J. Am. Chem. Soc.*, **100,** 2389 (1978).
138. R. R. Schrock, *Acc. Chem. Res.*, **12,** 98 (1979).
139. R. R. Schrock, *J. Organomet. Chem.*, **122,** 209 (1976).
140. V. Malatesta, K. U. Ingold, and R. R. Schrock, *J. Organomet. Chem.*, **152,** C53 (1978).
141. R. R. Schrock, L. W. Messerle, C. D. Woods, and L. J. Guggenberger, *J. Am. Chem. Soc.*, **100,** 3793 (1978).
142. S. J. McLain, C. D. Wood, L. W. Messerle, R. R. Schrock, F. J. Hollander, W. J. Youngs, and M. R. Churchill, *J. Am. Chem. Soc.*, **100,** 5962 (1978).
143. L. W. Messerle, P. Jennische, R. R. Schrock, and G. Stucky, *J. Am. Chem. Soc.*, **102,** 6744 (1980).
144. R. R. Schrock and J. D. Fellman, *J. Am. Chem. Soc.*, **100,** 3359 (1978).
145. M. R. Churchill, F. J. Hollander and R. R. Schrock, *J. Am. Chem. Soc.*, **100,** 647 (1978).
146. J. D. Fellmann, G. A. Rupprecht, C. D. Wood, and R. R. Schrock, *J. Am. Chem. Soc.*, **100,** 5964 (1978).
147. C. D. Wood, S. J. McLain, and R. R. Schrock, *J. Am. Chem. Soc.*, **101,** 3210 (1979).
148. A. J. Schultz, J. M. Williams, R. R. Schrock, G. A. Rupprecht, and J. D. Fellman, *J.*

Am. Chem. Soc., **101,** 1593 (1979); A. J. Schultz, R. K. Brown, J. M. Williams, and R. R. Schrock, *J. Am. Chem. Soc.*, **103,** 169 (1981).
149. G. A. Rupprecht, L. W. Messerle, J. D. Fellmann, and R. R. Schrock, *J. Am. Chem. Soc.*, **102,** 6236 (1980).
150. J. D. Fellmann, H. W. Turner, and R. R. Schrock, *J. Am. Chem. Soc.*, **102,** 6608 (1980).
151. D. N. Clark and R. R. Schrock, *J. Am. Chem. Soc.*, **100,** 6774 (1978).
152. R. Feser and H. Werner, *Angew. Chem., Int. Ed. Eng,* **19,** 940 (1980).
153. H. Schumann and J. Müller, *J. Organomet. Chem.*, **169,** C1 (1979).
154. H. D. Empsall, E. M. Hyde, R. Markham, W. S. McDonald, M. C. Norton, B. L. Shaw, and B. Weeks, *J. Chem. Soc., Chem. Commun.*, 589 (1977).
155. C. Crocker, R. J. Errington, W. S. McDonald, K. J. Odell, B. L. Shaw, and R. J. Goodfellow, *J. Chem. Soc., Chem. Commun.*, 498 (1979).
156. C. Crocker, R. J. Errington, R. Markham, C. J. Moulton, K. J. Odell, and B. L. Shaw, *J. Am. Chem. Soc.*, **102,** 4373 (1980).
157. R. J. Goddard, R. Hoffman, and E. D. Jemmis, *J. Am. Chem. Soc.*, **102,** 7667 (1980).
158. S. J. Thomson and G. Webb, *Heterogeneous Catalysis,* Oliver and Boyd, London, 1968.
159. A. F. Dyke, S. A. R. Knox, K. A. Mead, and P. Woodward, *J. Chem. Soc., Chem. Commun.*, 861 (1981).
160. M. Cooke, D. L. Davies, J. E. Guerchais, S. A. R. Knox, K. A. Mead, J. Roué, and P. Woodward, *J. Chem. Soc., Chem. Commun.*, 862 (1981).
161. W. A. Herrmann, J. Plank, D. Riedel, M. L. Ziegler, K. Weidenhammer, E. Guggotz, and B. Balbach, *J. Am. Chem. Soc.*, **103,** 63 (1981).
162. R. B. Calvert and J. R. Shapley, *J. Am. Chem. Soc.*, **99,** 5225 (1977).
163. A. J. Schultz, J. M. Williams, R. B. Calvert, J. R. Shapley, and G. D. Stuckey, *Inorg. Chem.*, **18,** 319 (1979); R. B. Calvert and J. R. Shapley, *J. Am. Chem. Soc.*, **100,** 7726, 7727 (1978).
164. K. Isobe, D. G. Andrews, B. E. Mann, and P. M. Maitlis, *J. Chem. Soc., Chem. Commun.*, 809 (1981).
165. E. L. Muetterties and J. Stein, *Chem. Rev.*, **79,** 479 (1979).
166. C. P. Casey and S. M. Newman, *J. Am. Chem. Soc.*, **100,** 2544 (1978).
167. D. L. Thorn, *J. Am. Chem. Soc.*, **102,** 7109 (1980).
168. B. H. Byers and T. L. Brown, *J. Organomet. Chem.*, **127,** 181 (1977).
169. J. A. Gladysz and W. Tam, *J. Am. Chem. Soc.*, **100,** 2545 (1978).
170. B. N. Chaudret, D. J. Cole-Hamilton, R. S. Nohr, and G. Wilkinson, *J. Chem. Soc., Dalton Trans.*, 1546, (1977).
171. P. T. Wolczanski and F. E. Bercaw, *J. Am. Chem. Soc.*, **101,** 6450 (1979).
172. P. T. Wolczanski, R. S. Threlkel, and J. E. Bercaw, *J. Am. Chem. Soc.*, **101,** 218 (1979).
173. P. T. Wolczanski and J. E. Bercaw, *J. Am. Chem. Soc.*, **101,** 6450 (1979).
174. (a) P. Belmonte, R. R. Schrock, M. R. Churchill, and W. J. Youngs, *J. Am. Chem. Soc.*, **102,** 2858 (1980); (b) M. R. Churchill and H. J. Wasserman, *J. Chem. Soc., Chem. Commun.*, 274 (1981).
175. M. R. Churchill, and W. J. Youngs, *Inorg. Chem.*, **20,** 382 (1981).
176. R. J. Al-Essa and R. J. Puddephatt, *J. Chem. Soc., Chem. Commun.*, 45 (1980).
177. K. J. Ivin, J. J. Rooney, C. D. Stewart, M. L. H. Green, and R. Mahtab, *J. Chem. Soc., Chem. Commun.*, 604 (1978); M. L. H. Green and A. Mahtab, *J. Chem. Soc., Dalton Trans.*, 262 (1979).
178. C. P. Casey and S. W. Polichnowski, *J. Am. Chem. Soc.*, **99,** 6097 (1977); C. P. Casey, S. W. Polichnowski, A. J. Shusterman, and C. R. Jones, *J. Am. Chem. Soc.*, **101,** 7282 (1979); M. Brookhart, M. B. Humphrey, H. J. Kratzer, and G. O. Nelson, *J. Am. Chem. Soc.*, **102,** 7802 (1980); T. R. Howard, J. B. Lee, and R. H. Grubbs, *J. Am. Chem. Soc.*, **102,** 6876 (1980); T. J. Katz and J. McGinnis, *J. Am. Chem. Soc.*, **99,** 1903 (1977).
179. S. M. Rocklage, J. D. Fellmann, G. A. Rupprecht, L. W. Messerle, and R. R. Schrock, *J. Am. Chem. Soc.*, **103,** 1440 (1981).
180. P. W. Jolly and R. Petit, *J. Am. Chem. Soc.*, **88,** 5044 (1966).
181. F. D. Mango and I. Dvoretzky, *J. Am. Chem. Soc.*, **88,** 1654 (1966).
182. G. Köbrich, H. Frölich, and W. Drischel, *J. Organomet. Chem.*, **6,** 194 (1966).
183. D. Seyferth, *Acc. Chem. Res.*, **5,** 65 (1972).

184. D. J. Cardin, B. Cetinkaya, M. J. Doyle, and M. F. Lappert, *Chem. Soc. Rev.*, **2,** 99 (1972).
185. A. G. Brook and P. J. Dillon, *Can. J. Chem.*, **47,** 4347 (1969).
186. W. I. Bevan, R. N. Haszeldine, and C. J. Young, *Chem. Ind. (London)*, 789 (1961).
187. W. I. Bevan and R. N. Haszeldine, *J. Chem. Soc., Dalton Trans.*, 2509 (1974).
188. W. I. Bevan, R. N. Haszeldine, J. Middleton, and A. E. Tipping, *J. Chem. Soc., Dalton Trans.*, 252, 260 (1975).
189. S. D. Chappell and D. J. Cole-Hamilton, *J. Chem. Soc., Chem. Commun.*, 238 (1980).
190. S. D. Chappell and D. J. Cole-Hamilton, *J. Chem. Soc., Chem. Commun.*, 319 (1981).
191. (a) P. Foley and G. M. Whitesides, *J. Am. Chem. Soc.*, **101,** 2732 (1979); (b) P. Foley, R. Di Cosimo, and G. M. Whitesides, *J. Am. Chem. Soc.*, **102,** 6713 (1980).
192. S. S. Moore, R. Di Cosimo, A. F. Sowinski, and G. M. Whitesides, *J. Am. Chem. Soc.*, **103,** 948 (1981).
193. P. R. Sharp, D. Astruce, and R. R. Schrock, *J. Organomet. Chem.*, **182,** 477 (1979).
194. R. A. Anderson, R. A. Jones, and G. Wilkinson, *J. Chem. Soc., Dalton Trans.*, 446 (1978).
195. A. Cowell and J. K. Stille, *J. Am. Chem. Soc.*, **102,** 4193 (1980).
196. H. C. Clark, A. B. Goel, and S. Goel, *J. Organomet. Chem.*, **166,** C29 (1979).
197. M. Julia, M. Duteil, and J. Y. Lallemand, *J. Organomet. Chem.*, **102,** 239 (1975).
198. J. Dehand and M. Pfieffer, *Coord. Chem. Rev.*, **15,** 327 (1976); M. I. Bruce *Angew. Chem., Int. Ed. Engl.*, **16,** 73 (1977).
199. J. W. Suggs and G. D. N. Pearson, *Tetrahedron Lett.*, **21,** 3853 (1980).
200. M. I. Bruce, B. L. Goodall, and F. G. A. Stone, *J. Chem. Soc., Dalton Trans.*, 687 (1978).
201. S. Hutkamp, D. J. Stufkens, and K. Vrieze, *J. Organomet. Chem.*, **168,** 351 (1979).
202. R. L. Bennett, M. I. Bruce, and F. G. A. Stone, *J. Organomet. Chem.*, **94,** 65 (1975).
203. B. L. Shaw, *J. Am. Chem. Soc.*, **97,** 3856 (1975); B. L. Shaw, *J. Organomet. Chem.*, **200,** 307 (1980).
204. A. J. Deeming and I. P. Rothwell, *J. Organomet. Chem.*, **205,** 117 (1981).
205. H. C. Clark, A. B. Goel, R. G. Goel, S. Goel, and W. O. Ogini, *Inorg. Chim. Acta*, **31,** L441 (1978).
206. R. A. Holton and R. V. Nelson, *J. Organomet. Chem.*, **201,** C35 (1980).

The Chemistry of the Metal—Carbon Bond, Volume 2
Edited by F. R. Hartley and S. Patai


CHAPTER 9

# Oxidative addition and reductive elimination

J. K. STILLE
*Department of Chemistry, Colorado State University, Fort Collins, Colorado 80523, USA*

## I. INTRODUCTION

In most organic reactions catalysed by homogeneous transition metal complexes, a carbon—metal $\sigma$-bond is formed and broken during the course of the catalytic cycle. In many such catalytic reactions, the carbon—metal $\sigma$-bond is generated by an oxidative addition reaction of an organic substrate to a low-valent transition metal complex, and ultimately the finished organic product is released from the transition metal by reductive elimination, regenerating the low-valent complex and completing the catalytic cycle.

In order for such a catalytic reaction to be efficient, the carbon—metal $\sigma$-bond cannot be particularly stable, since it is required to undergo some facile subsequent reaction leading ultimately to product with regeneration of the original catalytic species. Consequently, in many catalytic reactions, the product of oxidative addition is not readily isolated, thereby imposing restrictions on the study of this reaction. Similarly, the reductive elimination steps are often difficult to examine because the complexes bearing the partners to be eliminated are unstable.

Nevertheless, the oxidative addition reaction is one of the most available synthetic methods of preparing complexes containing carbon—transition metal $\sigma$-bonds. Further, there are a growing number of complexes containing a carbon—metal bond and a reductive elimination partner bonded to the same metal that either have been isolated or can by readily observed spectroscopically. However, our knowledge of the mechanisms of these reactions has expanded rapidly only in recent years. Consequently, the emphasis here is placed on the preparative and mechanistic aspects of these reactions without attempting to survey the literature exhaustively. Earlier chapters cover other methods of the synthesis of complexes containing metal—carbon bonds, particularly via a transmetallation reaction between a transition metal salt and an organometal such as organolithium and Grignard reagents (see Volume 1, Chapters 3–12). Finally, the catalytic cycles and mechanisms of some of the more important transition metal catalysed organic reactions are discussed in which oxidative addition and reductive elimination are key steps.

## II. OXIDATIVE ADDITION

Oxidative addition represents one of the most fundamental reactions in organometallic chemistry and a number of excellent reviews that cover various aspects of the numerous examples of this reaction have appeared[1–9]. Oxidative addition is the term given to a general class of reactions, regardless of mechanism, in which a metal is oxidized by the addition of a substrate XY. The increase in oxidation state is usually accompanied by an increase in coordination number. Low-spin transition metal complexes, particularly Group VIII $d^8$ and $d^{10}$ complexes, undergo two-electron oxidative addition reactions, although this type of oxidative addition is known for almost all even $d^n$ configurations. For example, coordinatively saturated (18-electron) Group VIII complexes of the $d^8$ configuration are five-coordinate complexes. The course of the oxidative addition reaction (equation 1) usually involves prior dissociation of ligands to a coordinatively unsaturated complex followed by the oxidative addition reaction. The two-electron oxidation produces a $d^6$ complex, which requires a coordination number of six in order to achieve an 18-electron configuration.

$$-\mathrm{M^I} \xrightarrow{-L} \mathrm{M^I} \xrightarrow{XY} \mathrm{M^{III}}(X)(Y) \text{ or } \mathrm{M}(X)(Y) \qquad (1)$$

$d^8$(18e) $d^8$(16e) $d^6$(18e)

It has been pointed out[1,10] that the tendency for $d^8$ complexes to undergo oxidative addition increases on descending a triad or in proceeding from right to left within Group VIII. The overall two-electron change may take place by concerted two-electron transformations or sequential one-electron changes. A family of mechanisms exist for this reaction that lie close enough to one another in energy that subtle changes in the substrate, the ligands, and the metal bring about changes in the mechanism. There are a smaller number of oxidative addition reactions that take place overall by one-electron changes.

$$\text{X—Y} + 2\text{M} \longrightarrow \text{X—M} + \text{Y—M} \quad \text{or} \quad \text{X—M—M—Y} \qquad (2)$$

As shown, either a *cis* or a *trans* octahedral complex can be obtained, for example, from the oxidative addition to a 16-electron $d^8$ square complex. In most reactions studied, it is not known whether the observed isomer is the kinetic product. There is ample evidence that an 18-electron $d^{10}$ square complex undergoes ligand dissociation to yield at least a 16-electron (coordinatively unsaturated) complex before a concerted two-electron oxidative addition takes place (dissociative mechanisms).

$$\mathrm{ML_4} \underset{+L}{\overset{-L}{\rightleftharpoons}} \mathrm{ML_3} \xrightarrow{XY} [\mathrm{X{-}ML_3}]^+ \mathrm{Y}^-$$

$$\mathrm{ML_3} \underset{+L}{\overset{-L}{\rightleftharpoons}} \mathrm{ML_2} \xrightarrow{XY} \text{X—M(L)(L)—Y or X—M(Y)(L)—L}; \quad [\mathrm{X{-}ML_3}]^+ \mathrm{Y}^- \xrightarrow{-L} \text{same products} \qquad (3)$$

In these reactions, a 16-electron (rather than an 18-electron) $d^8$ complex is the product. For example, the complexes $[M(PPh_3)_4]$ ($M = Ni^0$, $Pd^0$, $Pt^0$) having the $d^{10}$ configurations are coordinatively saturated and undergo the dissociation of phosphine ligands in solution to form three- and two-coordinate compounds[11–19] that are reactive toward oxidative addition. The phosphine ligands are $\sigma$-donors (lone-pair donation), which increases the electron density on the metal. This makes the metal a good nucleophile, favouring oxidative addition and, at the same time, increases the tendency for the phosphines to dissociate. Carbon monoxide,. on the other hand, is a $\pi$-acceptor ligand that decreases the electron density on the metal by delocalization of the unbonded $d$-electrons into the $\pi^*$ orbital of the carbonyl (back-bonding). Dicarbonylbis(triphenylphosphine)nickel(0), for example, is much less reactive than tetrakis(triphenylphosphine)nickel(0), primarily because the electron density on the metal is lower, and it does not undergo ligand dissociation readily because of the electronic balance of the $\sigma$-donor and $\pi$-acceptor ligands. There is good evidence, however, that steric effects are much more important than electronic effects in determining the dissociation of phosphine ligands from transition metal complexes. The greater the size of the ligand cone, the greater is the tendency for dissociation[16–19].

Of the Group VIII complexes having the $d^8$ configuration, the tendency for five-coordination increases in going from right to left and in ascending a triad. Thus, iron(0) tends to be five-coordinate (18-electron complex) whereas platinum(II) prefers to have four-coordination (16-electron complex).

If a $d^8$ complex is coordinatively saturated, usually only strong oxidizing agents add readily, and the reaction proceeds to give an intermediate octahedral cationic $d^6$ complex. Hence the coordination number of the metal (18-electron rule) and the type of ligand are important in determining the mechanism of the oxidative addition reaction.

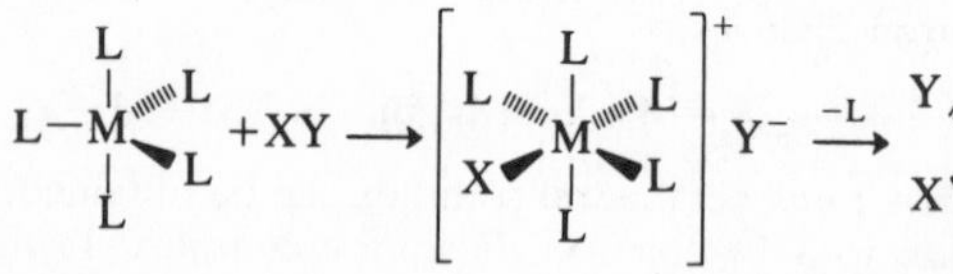

$$ML_5 + XY \longrightarrow [ML_5X]^+ Y^- \xrightarrow{-L} ML_4XY \qquad (4)$$

There are a wide variety of substrates that undergo oxidative addition reactions. Some of the reactions of substrates do not generate metal carbon $\sigma$-bonds[10], but represent key steps in catalytic cycles. For example, catalytic hydrogenation, oxidation, hydrosilylation, and hydrohalogenation reactions require oxidative addition reactions of $H_2$, $O_2$, SiH, and HX to generate key intermediates. These reactions will not be covered.

Of those substrates that do undergo oxidative addition to form metal—carbon $\sigma$-bonds, the classification into polar and non-polar oxidants can be made[9]. Polar substrates are susceptible to nucleophilic attack by the metal by virtue of their containing a good leaving group (in the sense of nucleophilic displacement). These substrates include acyl halides, alkyl halides, and, to a lesser extent, aryl and vinyl halides. Non-polar oxidants include hydrocarbons (ArH, RH, RC≡CH) and aldehydes that react by carbon—hydrogen bond breaking. The overall addition to the metal in this case is almost always stereospecifically *cis*, as might be expected from a concerted, three centred reaction. A special class of non-polar reactants are those which retain at least one bond of the reactant, XY, in the reaction product to give a metallocyclic product. Thus, acetylenes, alkenes, and small-ring (strained) hydrocarbons fall into this category.

The interaction of an alkene or alkyne with a transition metal can be viewed as simple coordination or oxidative addition to form a metallocycle. The relative contributions of $\pi$ *vs.* $\sigma$ (i.e. **1** and **3** *vs.* **2** and **4**) to the bonding are not easily

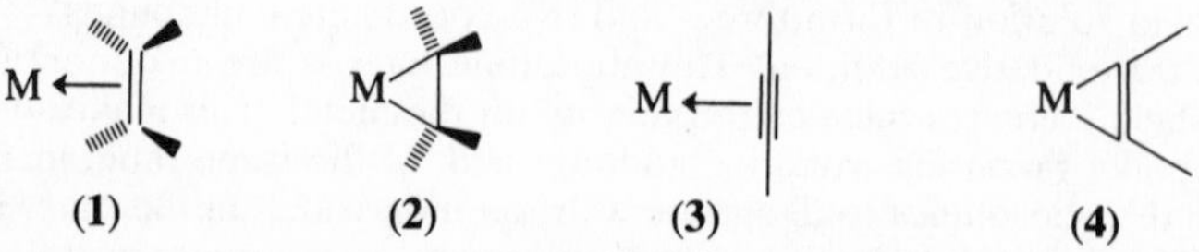

M← (**1**) M (**2**) M← (**3**) M (**4**)

determined, but there are, of course, the limiting cases for the extreme contributions of either[20]. The contribution of two bonding electrons in the forward $\sigma$-bond and two electrons in the $\pi$-back-bond is equivalent to two metal—carbon $\sigma$-bonds, and a metallocyclopropane structure.

A negligible contribution by the $\sigma$-forward bond and complete two-electron $\pi$-back-bonding is equivalent to the metal being oxidized; the olefin functions as a bidentate dicarbanion with two $\sigma$-bonds to the metal. This assesment of bonding can be made on the basis of X-ray structure, i.r. and n.m.r. spectra, and ESCA. If the binding energies of the electrons on the metal can be considered to be related to the degree of electron transfer from metal to ligand, then, on this basis, a degree of oxidation of the metal can be assigned. For $[PtL(Ph_3P)_2]$, where L=PhC≡CPh, $CH_2{=}CH_2$, and $O_2$, the oxidation states (degree of electron transfer) are 0.7, 0.8, and 1.8, respectively.

Hence the terminology of oxidative addition here is dependent on the molecular orbital picture or the selection of a number representing the degree of electron transfer (or some other parameter) that constitutes oxidative addition. These types of reaction will not be reviewed here.

These complexes are key, however, in subsequent organometallic reactions that generate metal—carbon $\sigma$-bonds, including cyclometallation reactions that produce larger metallocycles[21–23] and ultimately linear or cyclic oligomeric products. Cyclometallation reactions of this type are reviewed elsewhere.

M◁ + ‖ ⟶ M⬠ (5)

M◁(alkyne) ⟶ M⬠(diene) (6)

Some substrates that do form carbon—metal $\sigma$-bonds in oxidative addition reactions [RCN, $CS_2$, $(CF_3)_2CO$, $CH_2O$, RCONCO, RCSNCO, and certain carboranes] are not discussed here either, because at present there is little known about the mechanism of the reaction or because the reactions are less important in the synthesis of complexes and in catalytic processes.

Unlike the main group metals (particularly the alkali and alkaline earth metals) that undergo oxidative addition reactions readily in bulk, the transition metals react readily only in the 'atomic state'. This is accomplished usually by complexation of the metal with ligands to yield a mononuclear complex. 'Naked' transition metals also undergo facile oxidative addition reactions. The reduction of various transition metal salts produces reactive metal slurrys that undergo oxidative addition reactions[24]. Most transition metals give monatomic vapours (at high temperatures in a vacuum) that can be trapped on a cold surface and allowed to react with organic substrates via oxidative addition[25–28] (see Volume 1, Chapter 13).

## A. Organic Halides and Related Reactants

As indicated earlier, the mechanisms of the oxidative addition reactions are diverse. It could be expected that different mechanisms might be observed with different transition metals, electronic configurations or types of organic substrates (e.g. alkane C—H addition *vs.* organohalide C—X addition). Different mechanisms have been observed, however, even with the same metal, the same ligands and the same structure of the organic substrate, but with a change in halogen.

A variety of techniques have been used to determine the mechanisms of these reactions. The kinetics of a reaction do not necessarily distinguish between an $S_N2$

$$L_nM + RX \longrightarrow L_nMX + R^{\bullet} \tag{7}$$

mechanism and a radical process, for example, since the rate-determining step in a radical process may be the reaction of the metal with the organic halide to abstract a halogen atom. The relative rates of reactions of various organic halides are dependent on the halogen, but the order, $Cl < Br < I$, is the same for both radical and $S_N2$ reactions. Sulphonate esters such as *p*-toluenesulphonate esters undergo $S_N2$ reactions readily, while remaining relatively inert to radical reactions. Unfortunately, this test is not always definitive as to mechanistic pathway[29]. The effect of solvent on reaction rate and particularly the entropies of activation in reaction solvents of different polarity can provide an indication of the development or absence of change in the transition state.

Variation in the structure of the organic substrate and the observation of its effect and fate in the course of the reaction is probably the most definitive, unambiguous mechanistic probe. The order of reactivity for an $S_N2$-type reaction with alkyl halides, primary > secondary > tertiary, is reversed in a free-radical process. Rearrangement of neopentyl-type groups (reaction 7), particularly neophyl, provides evidence for an intermediate organic radical. The cyclization of the radical derived from 6-halohex-1-enes (reaction 8) not only provides evidence for a radical reaction,

$$PhC(Me)_2CH^{\bullet} \xrightarrow{\quad} Ph\dot{C}(Me)Et \tag{7a}$$

$$^{\bullet}CH_2(CH_2)_3CH{=}CH \xrightarrow{\quad} \text{cyclopentyl—}CH_2. \tag{8}$$

but can yield information concerning the rate of capture of the radical by the metal, since the rate constant for closure of the 5-hexenyl radical is known to be approximately $10^5\ s^{-1}$. The stereochemistry attending a particular organometallic reaction is a powerful mechanistic probe[30], inversion of configuration being taken as strong evidence for an $S_N2$-type reaction, while racemization is indicative of, but not conclusive for, a radical mechanism. Retention of geometry in reactions of vinyl halides is evidence for a non-radical process.

Enhanced adsorption or emission as a consequence of chemically induced dynamic nuclear polarization (CIDNP) allows the detection of a radical reaction, although the inability to observe CIDNP does not rule out such a reaction. The direct detection of a radical by e.s.r. or by using a spin trap such as $Bu^tNO$ to trap an organic radical $R^{\bullet}$ as a nitroxide, $Bu^tN(R)O^{\bullet}$, so that it may be observed by e.s.r. has been utilized to help elucidate the mechanisms of certain organometallic reactions[31]. The effect of radical initiators or inhibitors on the reaction and the addition of certain vinyl monomers that are known to polymerize by certain mechanisms have been used to deduce the mechanistic course of a reaction. In experiments of this type, where another reagent, such as a spin trap or a radical inhibitor, is added, control experiments must be carried out to establish that the added reagent does not alter the course of the reaction that is originally observed in the absence of such an addend.

Even though evidence such as e.s.r. may be obtained for the presence of radicals in a reaction, it is not necessarily true that a radical reaction is the main reaction pathway. The radical reaction could be a minor pathway to products, or even a side reaction, and still generate a high enough radical concentration to be detected.

In the following sections, there are many examples in which these techniques have been utilized in studies carried out to determine the mechanisms of oxidative addition reactions.

### 1. Group Ib

There are few examples of oxidative addition reactions of organic halides to copper(0) to generate $d^{10}$ copper complexes[32]. The addition of perfluoroalkyl iodides to copper in polar, aprotic solvents such as dmso or dmac at elevated temperatures give iodoperfluoroalkylcopper(II) complexes (**5**)[33]. Other functional groups, e.g. carboxyl, can be present in the perfluoroalkyl chain at a position remote from the iodo group. This reaction has been used in the cross coupling of various perfluoroalkyl halides with aryl and vinyl halides *in situ*. The homocoupling reaction of vinyl halides with copper(0) takes place with complete retention of geometry at the double bond[34].

$$R_fI + Cu \longrightarrow \underset{(\mathbf{5})}{R_fCuI} \tag{9}$$

$R_f = C_7F_{15}$, $X(CF_2)_nI$ $(X = H, Ph, CO_2H)$

$$MeCH{=}CHX + Cu \text{ or } CuPBu^n_3 \longrightarrow MeCH{=}CHCH{=}CHMe \tag{10}$$

The co-vapour deposition of gold, methyl iodide and dimethyl sulphide at −196 °C produce a gold(II) complex formulated as **6**, which on warming to 25 °C yields the iodo bridged dimer (**7**)[28].

$$\underset{\text{(atom)}}{Au} + MeI + Me_2S \xrightarrow{-196\,°C} \underset{\mathbf{6}}{[AuI(Me)(SMe_2)_2]} \xrightarrow{25\,°C} \underset{\mathbf{7}}{\left[Me_2Au(\mu\text{-}I)_2AuMe_2\right]} \tag{11}$$

The reaction of $d^{10}$ organocuprates with vinyl halides and alkyl halides probably takes place by a sequence of oxidative addition, reductive elimination reactions, although the oxidative addition product has not been isolated[35]. The cross-coupling reaction proceeds with predominant (84–92%) inversion of configuration at the alkyl halide carbon and retention of geometry in the vinyl halide[36,37]. This reaction requires that oxidative addition gives a $d^8$ square-planar *trans* Cu(III) complex (*trans* addition of the organic halide) followed by an exclusive reductive elimination of *cis* partners bound to copper. Retention of geometry in the vinyl halide suggests two retention reactions, in both the oxidative addition and the reductive elimination, whereas in the alkyl halide inversion of configuration in the oxidative addition reaction and retention in the reductive elimination are favored. Retention of configuration is also observed in R in the reductive elimination[36].

$$R_2Cu^ILi + (R')(H)C{=}C(X)(H) \xrightarrow{\text{rate}} \left[R{-}Cu^{III}(X)(CH{=}CHR'){-}R\right]Li^+ \xrightarrow{\text{fast}} (R')(H)C{=}C(R)(H) + RCu^I + LiX \tag{12}$$

$$R_2Cu^ILi + Me(R')(H)C{-}Br \xrightarrow{\text{rate}} \left[R{-}Cu(X)(CH(Me)R'){-}R\right]^- Li^+ \xrightarrow{\text{fast}} R{-}C(R')(H)Me + RCu^I + LiX \tag{13}$$

This mechanistic evidence is not inconsistent, however, with a direct displacement of bromide by an organo nucleophile (R) bound to copper.

$$R-\bar{Cu}-R \curvearrowright C-X \longrightarrow R-C + RCu^{I} + X^{-} \quad (14)$$

The reaction is overall second order, first order in copper reagent and first order in alkyl halide[38]. The order of reactivity $CH_3 >$ primary $>$ secondary $>$ tertiary, the greater reactivity of tosylate compared with halide, and the relative insensitivity of the reaction to free radical inhibitors, all support a rate-determining nucleophilic attack either to give product directly or to give the oxidative addition intermediate which undergoes reductive elimination in a fast step.

Silver atoms obtained by vapour deposition on a low-temperature surface react with perfluoroalkyl iodides to give oxidative addition products[39]. The $d^9$ bisperfluoroalkylsilver(II) complex (**8**) obtained from perfluoroisopropyl iodide is stable at 25 °C, whereas the products from perfluoromethyl and perfluoro-*n*-propyl iodides decompose above −100 °C.

$$\mathrm{Ag(atom)} + i\text{-}C_3F_7I \xrightarrow{-196\,^{\circ}C} \underset{\mathbf{8}}{[Ag\{CF(CF_3)_2\}_2]} + AgI \quad (15)$$

Dailkyl aurates undergo oxidative addition reactions with alkyl halides to give the $d^8$ trialkylgold(III) complexes (**9**)[40–42]. A predominance of the *trans* complex is

$$RX + (Me_2Au^{I}(PPh_3)^{-}Li^{+} \longrightarrow \underset{(\mathbf{9})}{trans\text{-}[AuR(Me)_2(PPh_3)]} + LiX \quad (16)$$

$X = Br, I;\ R = Me, CD_3, Et, Pr^n, Pr^i, Bu^n$

obtained in this reaction with alkyl halides, but these complexes do undergo *cis–trans* isomerization, particularly in the presence of phosphine. Rearrangement of the R group also may occur (see Section III). Generally, the sterically most demanding group occupies the position *trans* to the phosphine in competition with smaller alkyl groups (methyl). The oxidative addition reaction of iodobenzene, however, produces a *cis*-complex (**10**)[42]. The relative rates of oxidative addition, $Me > Et > Bu^n$ and $I > Br \gg Cl$[42] support an $S_N2$-type mechanism for this oxidative reaction.

$$PhI + [Au^{I}(Me)_2(PPh_3)]^{-}Li^{+} \longrightarrow \underset{\mathbf{10}}{cis\text{-}[AuPh(Me)_2(PPh_3)]} + LiI \quad (17)$$

The neutral $d^{10}$ methylgold(I) complex (**11**) reacts with methyl iodide to give ethane and iodotriphenylphosphinegold(I)[43]. The intermediate gold(III) complex (**12**) could not be detected in the reaction mixture, even though the spectrum of this stable, independently synthesized compound was available for comparison. In an independent experiment, however, **12** was shown to react rapidly with **11** to give the trimethylgold(III) complex **13**, which could be observed by n.m.r. and isolated at low temperature. Complex **13** also was shown to give ethane and methylgold(I) at moderate temperatures

$$\underset{\mathbf{11}}{[AuMe(PPh_3)]} + MeI \longrightarrow \underset{\mathbf{12}}{[AuI(Me)_2(PPh_3)]} \xrightarrow[\text{fast}]{} \underset{\mathbf{13}}{[Au(Me)_3(PPh_3)] + PPh_3]} \quad (18)$$

$$\mathbf{13} \longrightarrow C_2H_6 + [AuMe(PPh_3)]$$

### 2. *Group VIII*

a. *Oxidative additions to $d^{10}$ complexes of the nickel triad.* Oxidative addition reactions of $d^{10}$ complexes of the nickel triad have been carried out both with vapour-deposited (atomic) metal and metal(0) complexes bearing a variety of ligands, usually tertiary phosphines[44,45]. In a catalytic reaction involving a complex of the nickel triad, oxidative addition to produce a $d^8$ divalent complex may be followed by a second oxidative addition to give a $d^6$ metal(IV) complex. The $d^6$ tetravalent metal complexes are unstable, except in the case of platinum where a large number of platinum(IV) complexes have been isolated.

Nickel(0) complexes are stronger reducing agents than the analogous palladium(0) or platinum(0) complexes. For example, within the series $[M(PEt_3)_4]$, only the nickel complex reacts with benzonitrile, and reacts rapidly with pentafluorophenyl halide at ambient temperature, whereas the palladium and platinum complexes require several hours at 110 °C.

i. *Nickel*(0). The oxidative addition of a variety of vinyl halides to zerovalent nickel complexes takes place readily under mild conditions to give the complex containing the vinyl nickel(II) complex (Table 1). The vinyl fluorides are unreactive, compounds such as perfluoropropene, perfluorocyclobutene, and 1,1-difluoroethane yielding only the corresponding olefin complex[46]. The mechanism of the reaction probably involves prior coordination of the olefin to the complex, possibly forming subsequently an intermediate metallocyclopropane, particularly in reactions with electron-deficient olefins such as fluorinated olefins[21]. Indeed, such metallocyclopropenes can be isolated in the analagous reactions with platinum complexes (see

$$R^1(R^2)C{=}C(X)R^3 \longrightarrow \left[L_2Ni \leftarrow \|\,(C(X)R^3{=}C(R^1)R^2)\right] \longrightarrow \left[L_2Ni\langle\,C(X)R^3{-}C(R^1)R^2\,\rangle\right] \longrightarrow R^2(R^1)C{=}C(R^3){-}NiX(L)_2 \qquad (19)$$

below). *Trans* complexes are obtained. The reaction involves remarkable stereochemistry, retention of geometry being observed in all cases[47].

$$NiL_4 + (Z)\text{-}PhCH{=}CHBr \longrightarrow [NiBrL_2\{(Z)\text{-}CH{=}CHPh\}] + 2L \qquad (20)$$

$$NiL_4 + (E)\text{-}PhCH{=}CHBr \longrightarrow [NiBrL_2\{(E)\text{-}CH{=}CHPh\}] + 2L \qquad (20a)$$

$L = PEt_3, PPh_3$

The oxidative addition of aryl halides to different nickel(0) complexes takes place readily to generate an arylhalonickel(II) complex (Table 1)[50–53]. Nickel(0) complexes are more reactive than the analogous palladium or platinum complexes, and the order of reactivity for aryl halides is $I > Br > Cl > CN \gg F$[50,51].

The ligands on the tetrakisphosphine nickel(0) complex not only affect the rate of the reaction[51], but also the stability of the complex[52]. Although chlorobenzene reacts slowly with tetrakis(triphenylphosphine)nickel(0), an instantaneous reaction is observed with tetrakis(triethylphosphine)nickel(0)[51]. Benzonitrile reacts with tetrakis-(triethylphosphine)nickel(0), but not tetrakis(triphenylphosphine)nickel(0)[51].

$$ArX + [NiL_4] \longrightarrow \left[\begin{array}{c} L \\ | \\ Ar{-}Ni{-}X \\ | \\ L \end{array}\right] + 2L \qquad (21)$$

TABLE 1. Oxidative addition of organic halides to Ni(O) complexes

| Com-ounds | Ni(0) complex | L | L′ | Halide | Ni(II) complex | Ref. |
|---|---|---|---|---|---|---|
| Vinyl halides | $[NiL_2L']$ | $PEt_3$ | $CH_2{=}CH_2$ | $CH_2{=}CHBr$ | *trans*-$[NiBrL_2(CH{=}CH_2)]$ | 48 |
| | | $PEt_3$ | 1,5-COD | | | |
| | $[NiL_4]$ | $PEt_3$ | — | $CH_2{=}C(Cl)CH_2{=}CH_2$ | *trans*-$[NiClL_2\{C({=}CH_2)CH{=}CH_2\}]$ | 48 |
| | $[NiL_4]$ | $PPh_3$ | | (*Z*)-$PhCH{=}CHBr$ | *trans*-$[NiBrL_2\{(Z)\text{-}CH{=}CHPL\}]$ | 47 |
| | | $PEt_3$ | | (*E*)-$PhCH{=}CHBr$ | *trans*-$[NiBrL_2\{(E)\text{-}CH{=}CHPL\}]$ | 47 |
| | $[NiL_4]$ | $PhAsMe_2$ | | $CF_2{=}CFX$ ($X{=}Cl,Br$) | *trans*-$[NiXL_2(CF{=}CF_2)]$ | 49 |
| | | | | $CF_2{=}CCl_2$ | *trans*-$[NiClL_2(CCl{=}CF_2)]$ | |
| | | | | $CFCl{=}CFCl$ | *trans*-$[NiClL_2(CF{=}CFCl)]$ | |
| | $[NiL_2L']$ | $Ph_3P$ | $CH_2{=}CH_2$ | $CF_2{=}CFX$ ($X = Cl,Br$) | *trans*-$[NiXL_2(CF{=}CF_2)]$ | 46 |
| | | | | $CF_2{=}CCl_2$ | *trans*-$[NiClL_2(CCl{=}CF_2)]$ | |
| | | | | $CFCl{=}CFCl$ (*E,Z* mixture) | *trans*-$[NiClL_2(CF{=}CFCl)]$ | |
| Aryl halides | $[NiL_4]$ | $PPh_3$ | | $YC_6H_4X$ $X = Cl,Br$; $Y = H$, *o*-, *m*-, *p*-$CH_3$, *p*-Cl | *trans*-$[NiXL_2(C_6H_4Y)]$ | 50, 53,55 |
| | $[NiL_4]$ | $PPh_3$ | | $X = Cl,Br,I$; $Y = H$, *m*-, *p*-$CH_3$, Cl, CN, OPh, *p*-$OCH_3$, $COCH_3$, COPh, *m*-$CO_2CH_3$ | | 54 |
| | $[NiL_4]$ | $PEt_3$ | | *m*-, *p*-$FC_6H_4X$ $X = I$, Br, Cl, CN | *trans*-$[NiXL_2(C_6H_4F)]$ | 51 |
| | $[NiL_2]$ or | $PX_3$ (X = 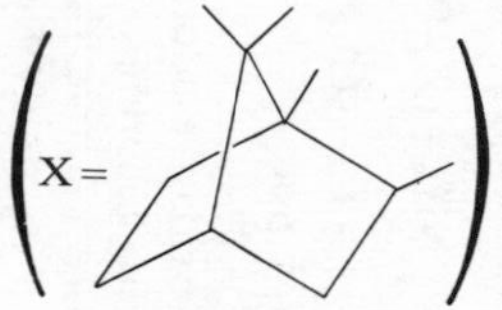 ) | | PhCl | *trans*-$[NiClL_2(Ph)]$ | 52 |
| | $[NiL_3]$ or | $PX_3$ | | | | |
| | $[NiL_2\mu L'N_1N_2]$ | $PX_3$ | 1,5-COD | | | |

| | | | | | | |
|---|---|---|---|---|---|---|
| | $[NiL_2L']$ or $[NiL_4]$ | $PEt_3$ | 1,5–COD or $C_2H_4$ | PhX<br>X = CL,Br | *trans*-$[NiXL_2(PL)]$ | 48 |
| | | $PEt_3$ | | *o*-, *m*-, *p*-$C_6H_4Cl_2$ | *trans*-$[NiClL_2(C_6H_4(Cl)]$ | |
| | | | | 1,2,4-$C_6H_3Cl_3$ | *trans*-$[NiClL_2(2,4\text{-}Cl_2C_6H_3)]$ | |
| | | | | $C_6F_5X$<br>X = Br, Cl, F | *trans*-$[NiXL_2(C_6F_5)]$ | |
| | | | | $C_6H_4ClBr$ | *trans*-$[NiBrL_2(C_6H_4Cl)]$ | |
| | $[Ni(acac)_2]$ + $Et_2AlOEt$ | $PEt_3$ | | $C_6F_5Br$ | *trans*-$[NiBrL_2(C_6F_5)]$ | 58 |
| | | $PPh_3$ | | *o*-$ClC_6H_4X$<br>X = Cl, Br | *trans*-$[NiXL_2(2\text{-}ClC_6H_4)]$ | |
| | | | | 1,2,5-$Cl_3C_6H_3$ | *trans*-$[NiClL_2(2,5\text{-}Cl_2C_6H_3)]$ | |
| | $[NiX_2'L_2]$ (cathode reduction) | $PPh_3$ | | PhX<br>X = Cl, Br, I | *trans*-$[NiXl_2(Ph)]$ | 59,60 |
| | $[NiX_2L]$ (cathode reduction) | $Ph_2PCH_2CH_2PPh_2$ | | | $[Ni(C_6H_5)X(Ph_2PCH_2CH_2PPh_2)]$ | |
| | $[NiI_2]$ (K reduction) | $PEt_3$<br>$PPh_3$ | | $C_6F_5Br$ | *trans*-$[NiBrL_2(C_6F_5)]$ | 61 |
| | Ni (vapour deposition) | $PEt_3$ | | $C_6F_5X$<br>X = Cl, Br | *trans*-$[NiBrL_2(C_6F_5)$ | 26,27, 28,62 |
| Allyl halides | $[NiL_4]$ | CO | | $CH_2{=}CHCH_2X$<br>X = Br, I | $[NiX(\eta^3\text{-allyl})]_2$ | 65–67 |
| | | | | $CH_2{=}C(R)CH_2Br$<br>R = $CH_3$, $CO_2Et$ | $[NiBr(\eta^3\text{-2-R-allyl})]_2$ | 67 |
| | | | | $RCH_2(Me)C{=}CHCH_2Br$<br>R = H, $(CH_3)_2C{=}CHCH_2CH_2$— | $[NiBr(\eta^3\text{-1-Me-1-}CH_2R\text{-allyl})]_2$ | 67 |
| | $[NiL_3'L_2]$ | $PPh_3$ | CO | $CH_2{=}CHCH_2Cl$ | $[NiClL(\eta^3\text{-allyl})]$ | 68 |
| | $[NiCp(CO)]^-$ | | | $CF_2{=}CFCF_2Cl$ | $[NiCp(CO)(CF_2CF{=}CF_2)]$ + $[NiCp(CO)(CF{=}CFCF_3)]$ | 88 |
| | | | | $CH_3CH{=}CHCH_2Cl$ | $[NiCp(\eta^3\text{-Me-allyl})]$ | 88 |

TABLE 1. ctd

| Compounds | Ni(0) complex | L | L′ | Halide | Ni(II) complex | Ref. |
|---|---|---|---|---|---|---|
| | $[NiL'_2]$ | | cod | $CH_2{=}CHCH_2OC(O)CH_{3-n}X_n$ | $[Ni(\eta^3\text{-allyl})(\eta^3\text{-}O_2CCH_{3-n}X_n)]$ | 69 |
| | | | | $n = 0$ | | |
| | | | | $n = 1, 2, 3; X = Cl$ | | |
| | | | | $n = 3; X = F$ | | |
| | $[NiL'_2]$ added $PR_3$ | | cod | $CH_2{=}CHCH_2OAc$ | $[Ni(\eta^3\text{-allyl})_2]$ | 70,71 |
| | | | cod | $CH_2{=}CHCH_2OAc$ | $[Ni(OAc)(PR_3)(\eta^3\text{-allyl})]$ | 71 |
| | | | | | $PR_3 = PPh_3$, $Ph_2PEt$, $Cy_3P$ | |
| | $[NiL'_2]$ added $PPh_3$ | | cod | $CH_2{=}CHCH_2OPh$ | $[Ni(OPh)(PPh_3)(\eta^3\text{-allyl})]$ | 71 |
| | Ni (atom) | | cod | $CH_2{=}CHCH_2X$ | $[NiX(\eta^3\text{-allyl})]_2$ | 27,28<br>73 |
| Alkyl halides | $[NiL_2L']$ | $L_2 = bipy$ | cod | $PhCH_2X$ | *cis*-$[NiXL_2(CH_2Ph)]$ | 78 |
| | | | | $X = Cl, Br$ | | |
| | | | | $o\text{-}BrC_6H_4CH_2Br$ | *cis*-$[NiL_2(o\text{-}CH_2C_6H_4Br)_2]$ | 78 |
| | $[NiL_2L']$ | $PPh_3$ | $C_2H_4$ | $PhCH_2X$ | *trans*-$[NiXL_2(CH_2Ph)]$ | 79,80 |
| | $[NiL_2L']$ | $Cy_3P$ | $C_2H_4$ | $p\text{-}C_6H_4(CH_2Cl)_2$ | $[ClNi(L_n)CH_2\text{-}p\text{-}C_6H_4\text{-}CH_2Ni(L_n)Cl]$ | 91 |
| | | $Ph_3P$ | $C_2H_4$ | | $(n = 1)$ | 91 |
| | | | | | $(n = 2)$ | |
| | $[NiL'_2]$ $+PR_3(L)$ | $Cy_3P$ | cod | $o\text{-}C_6H_4(CH_2Br)_2$ | $[BrNi(L)CH_2\text{-}o\text{-}C_6H_4CH_2Ni(L)Br]$ | |
| | | | | $o\text{-}C_{10}H_6(CH_2Br)_2$ | $[BrNi(L)CH_2\text{-}o\text{-}C_{10}H_6CH_2Ni(L)Br]$ | |
| | $[NiLL_2]$ | diphos | CO | $R_fI$ | $[NiIL(R_f)]$ | 81 |
| | | | | $R_f = C_2F_5$, $n\text{-}C_3F_7$ | | |
| | $[NiL_2L']$ | $Cy_3P$ | $N_2$ | RX | *trans*-$[NiXL_2(R)]$ | 81 |
| | | | | $R = Me, Et, CF_3$ | | |
| | Ni (atom) | $PMe_3$ | | $CF_3^{\bullet}$ | *rrans*-$[NiL_2(CF_3)_2]$ | 86b |

| | | | | | | |
|---|---|---|---|---|---|---|
| | $[NiL'_2]$ + bipy | | cod | $Br(CH_2)_4Br$ | [bipyNi(ring)] | 85 |
| | $[NiL_2]$ | $Ph_2P(CH_2)_nPPh_2$ $n = 2, 3, 4$ | | $(CN)_2$ | $[NiL_n(CN)_2]$ $n = 1, 1.5$ | 89 |
| | $[NiL_4]$ | $PR_3$ $P(OR)_3$ | | $(CN)_2$ | [L, L, L, CN, CN—Ni] | 90 |
| Acyl halides | $[NiL_2L']$ or $[NiL_4]$ | $PEt_3$ | $C_2H_4$ cod | ArCOCl Ar = $C_6F_5$-, o-$ClC_6H_4$ | *trans*-$[NiClL_2(COAr)]$ + *trans*-$[NiClL_2(Ar)]$ | 48 |
| | $[NiL_2]$ | $Ph_3P$ | $C_2H_4$ | $C_6F_5COCl$ | *trans*-$[NiClL_2(C_6F_5)]$ | 46 |
| | $[NiL_2L']$ | $Bu^tNC$ | | PhCOCl | [PhCO—Ni(L)(L)(L)—Cl] | 122 |
| | $[NiL_4]$ or $[NiL'L_2]$ | $PPh_3$ $PPh_3$ | $C_2H_4$ | PhCOCl | *trans*-$[NiClL_2(Ph)]$ | 92 |
| | | | | $ClCO_2R$ R = Me, Et | *trans*-$[NiClL_2(CO_2R)]$ | 92 |
| | $[NiL_2]$ | $Ph_2P(CH_2)_nPPh_2$ $n = 3, 4$ | | $ClC(=S)NMe_2$ | $[NiClL\{C(=S)NMe_2\}]$ | 94 |
| | $[NiL_4]$ | $PPh_3$ | | $ClC(=S)NMe_2$ | *trans*-$[NiClL_2\{C(=S)NMe_2\}]$ | 94 |
| | $[NiL'_2]$ + $PEt_3$ | | cod | $(RCO)_2O$ | *trans*-$[NiL_2(COR)(OCOR)]$ | 95 |

The electron density on nickel(0) and thus its ability to undergo oxidation could be expected to be determined by the ligands coordinated to it. The order of reactivity observed within a series of tricoordinate, 16-electron nickel(0) complexes is $[Ni(PR_3)_3] > [Ni(PAr_3)_3] > [(R_3P)_2Ni(olefin)] > [(Ar_3P)_2Ni(olefin)] > [Ni(COD)_2]$, which is the order of their anodic half-wave reduction potentials[45].

Two mechanisms have been proposed for this reaction, one analogous to nucleophilic aromatic substitution involving an intermediate Meisenheimer-type complex, and the other requiring an electron transfer from metal to aryl halide followed by collapse of the electron pair. The nucleophilic substitution mechanism was proposed as a result of a study of the linear free energy relationship for the reaction with substituted halobenzenes[54]. Electron-withdrawing groups ($+\sigma$) on the aryl

$$YC_6H_4X + [NiL_4] \rightleftharpoons \left[ \text{(X, } NiL_n \text{ Meisenheimer complex, Y)} \right] \longrightarrow trans\text{-}[NiXL_2(C_6H_4Y)] \quad (22)$$

$L = Ph_3P$

$Y = m, p\text{-}CH_3, Cl, CN, OC_6H_5, \; p\text{-}OCH_3, COCH_3, COC_6H_5, \; m\text{-}CO_2CH_3$

halide increase the reaction rates. A high value of $\rho$ (8.8) for those substituents with $\sigma > 0.23$ was observed. In addition to the oxidative addition product, nickel(I) halides are also obtained. Two competing mechanisms have been proposed[48] to account for the two types of products: a nucleophilic aromatic substitution (reaction 23a) for the oxidative addition product, and an electron transfer mechanism (reaction 23b) to account for the nickel(I) halide. When electron-withdrawing groups are present on the aromatic nucleus, the nucleophilic substitution mechanism (reaction 23a) is favored.

$$[NiL_3] + PhX \rightleftharpoons \left[ \text{(}NiL_3\text{, X Meisenheimer complex)} \right]^+ \longrightarrow trans\text{-}[NiXL_2(Ph)] \quad (23a)$$

$$[NiL_3] + PhX \longrightarrow \left[ {}^{\cdot +}NiXL_3 \; \bar{\cdot} C_6H_5 \right] \xrightarrow{\text{diffusion}} C_6H_5^{\cdot} + [Ni^IXL_3] \quad (23b)$$

(collapse of the radical pair leads to $trans\text{-}[NiXL_2(Ph)]$; $C_6H_5^{\cdot} \xrightarrow{SH} C_6H_6$)

The ratios of the two products vary with the solvent, the aryl group and the halide as well as the substitution on the aromatic ring[55]. When the nickel(I) halide is the major product, the aryl radical can abstract hydrogen from the solvent; the radical in

$$[NiL_4] \rightleftharpoons [NiL_3] + ArX \longrightarrow [NiXL_2(Ar)] + [Ni^IXL_3] \quad (24)$$

$L = PEt_3$

the case of 2,4,6-tri-$t$-butylbenzene can be observed by e.s.r. The yield of nickel(I) product is highest with aryl iodides and decreases in the order $ArI \gg ArBr > ArCl$. The reaction is second order, first order in each reactant, and shows an inverse dependence on phosphine concentration. Free radical inhibitors have no effect on the reaction. The effect of substituents on the aromatic ring ($+\rho$) also is consistent with the electron acceptor ability (reduction potential) of the aromatic substrate. One-electron acceptors such as tetracyanoethylene, chloranil, and 2,3-dichloro-5,6-dicyano-1,4-quinone are rapidly reduced by zerovalent nickel complexes in solution with the simultaneous generation of the corresponding radical anions[56]. These observations lend support to the electron transfer mechanism[55] in which both the oxidative addition product and the nickel(I) species evolve from a common intermediate. The aryl radical anion, $ArX^{-\bullet}$, is relatively stable, and thus the composition of the final product is essentially the same, regardless of the ligands on nickel(0). Coupling reactions of aryl halides by equivalent or higher amounts of bis(cycloocta-1,5-diene)nickel(0) in polar solvents such as DMF produce only insignificant amounts of product derived from aryl radicals[57].

$$[NiL_4] \underset{}{\overset{K_e}{\rightleftharpoons}} [NiL_3]+L \tag{25}$$

$$[NiL_3]+ArX \xrightarrow[\text{rate}]{k_2} [Ni^IL_3ArX^{-\bullet}] \tag{26}$$

$$[Ni^IL_3ArX^{-\bullet}] \longrightarrow \begin{cases} [NiXL_2(Ar)]+L \\ [Ni^IL_3]+X^-+Ar^\bullet \end{cases} \tag{27}$$

The nickel(0) phosphine complexes have been generated *in situ* from nickel(II) reagents by a variety of reducing agents. Nickel acetonylacetonate is reduced by diethylaluminium ethoxide, and when this is carried out in the presence of the aryl halide and phosphine, the oxidative addition product is obtained[58].

$$2L+Et_2AlOEt+C_2H_4+[Ni(acac)_2] \xrightarrow{25\,^\circ C} \xrightarrow{ArX} trans\text{-}[NiXL_2(Ar)] \tag{28}$$

Bisphosphinenickel(II) dihalides can be electrochemically reduced to give a coordinatively unsaturated nickel(0) complex that reacts rapidly with phenyl halides[59,60]. The reduction also can be effected with potassium to give a nickel(0) powder that reacts with pentafluorophenyl bromide in the presence of phosphine to give the oxidative addition product[61].

$$[NiX_2L_2] \longrightarrow L_2Ni \xrightarrow{PhX} [XL_2Ni(Ph)] \tag{29}$$

Vapor deposition of nickel on a cold (−196 °C) surface and codeposition of pentafluorophenyl halide gives the ligand-bare oxidative addition products[26–28,62]. These 12-electron complexes, which are stable at temperatures of −80 °C or below react with phosphines to yield the 16-electron complexes. In reactions with other aryl halides, the oxidative addition products are less stable, and nickel halide is produced in addition to coupling products.

$$Ni+C_6F_5Cl \longrightarrow [NiCl(C_6F_5)] \xrightarrow{PEt_3} trans\text{-}[NiCl(PEt_3)_2(C_6F_5)] \tag{30}$$

Reactions of 2-bromophenylchlorobis(triethylphosphine)nickel(II) (**14**) with lithium metal gives an unusual complex, the structure (**15**) of which has been suggested to contain two nickel(II) atoms in the ring[63].

$$trans\text{-}[NiClL_2(o\text{-}C_6H_4Br)] \xrightarrow{Li} [\,(o\text{-}C_6H_4)_2(NiL_2)_2\,] \rightleftharpoons [\,(o\text{-}C_6H_4)_2(NiL)_2\,]+2L \tag{31}$$

L = PEt3

(**14**) (**15**)

Allyl halides react rapidly with nickel(0) complexes to give high yields of the oxidative addition product[64]. The first reported reactions were carried out with nickel tetracarbonyl to give halogen bridged dimers (**16**)[65,66]. The reaction of $\pi$-allyl nickel

$$CH_2{=}CHCH_2X + [Ni(CO)_4] \longrightarrow \tfrac{1}{2}[NiX(\eta^3\text{-allyl})]_2 + 4CO \quad (32)$$

**16**

complexes of this general structure with reactive organic halides is a good method for effecting a cross-coupling reaction between an allyl halides and another organic halide[67]. The reaction of allyl chloride with triphenylphosphinetricarbonylnickel(0)

$$\tfrac{1}{2}[NiBr(\eta^3\text{-2-R-allyl})]_2 + R'X \longrightarrow R'CH_2C(R){=}OH_2 + [NiBrX] \quad (33)$$

gives a $\pi$-allyl monomeric complex[68]. Nickel(0) complexes also react with allyl acetates, but different oxidative addition products have been obtained by different

$$[(Ni(CO)_3(Ph_3P)] + CH_2{=}CHCH_2Cl \xrightarrow{40\,^\circ C} [NiCl(PPh_3)(\eta^3\text{-allyl})] + 3CO \quad (34)$$

groups under the same reaction conditions. Both the $\pi$-allylnickel acetate (**17**)[69] and the bisallylnickel(**18**)[70,71] complexes have been reported as the isolated products for the reaction of biscyclooctadienenickel(0) with allyl acetates. The reaction probably proceeds initially to give **17** which rapidly disproportionates. When phosphine is

$$[Ni(\eta^3\text{-allyl})(\eta^3\text{-}O_2CMe)] \; (\mathbf{17}) \longleftarrow CH_2{=}CHCH_2OAc + [Ni(cod)_2] \longrightarrow \tfrac{1}{2}[Ni(\eta^3\text{-allyl})_2] \; (\mathbf{18}) + \tfrac{1}{2}[Ni(OAc)_2$$

$$CH_2{=}CHCH_2OAc + [Ni(cod)_2] \xrightarrow{PR_3} [Ni(OAc)(PR_3)(\eta^3\text{-allyl})] \; (\mathbf{19}) \quad (35)$$

added to the reaction mixture, a monoallyl nickel complex (**19**) is obtained[71]. Allyl phenyl ether also is cleaved in the presence of phosphine to give a $\pi$-allylnickel complex containing phenoxide as the other valent ligand[71]. When phenyl esters are allowed to react with biscyclooctadienenickel(0) in the presence of triphenylphosphine, the ester undergoes acyl-oxygen cleavage in the oxidative addition instead of allyl-oxygen scission as in the reactions of allyl acetates and allyl phenyl ether[70,72].

When R is methyl, the oxidative addition product is stable, otherwise $\beta$-elimination is observed when R = ethyl.

$$RCO_2Ph \xrightarrow[55\,^\circ C,+L]{[Ni(cod)_2]} cis\text{-}[NiL_2(OPh)(COR)] \xrightarrow{-CO} cis\text{-}[NiL_2(OPh)(R)]$$

$$\xrightarrow{R(-H)} cis\text{-}[NiHL_2(OPh)] \xrightarrow[+CO,\ +L]{-PhOH} [Ni(CO)L_3] \quad (36)$$

$L = PPh_3$

The reaction of allyl halides with nickel atoms (vapour deposition) at −196 °C also gives the halogen-bridged dimer (**16**)[27,28,73].

The n.m.r. spectra of such $\pi$-allyl nickel complexes show a rapid, dynamic equilibrium between $\sigma$- and $\pi$-allyl complexes in solution[74–76] (equation 37). The

X-ray structures of allyl nickel complexes show that the allyl group lies in a plane orthogonal to the plane of the nickel and the other two ligands[76,77] (**20**).

$$\mathrm{Ni}(\eta^3\text{-alkyl}) \rightleftharpoons \text{(allyl)}\mathrm{Ni} \qquad (37)$$

(**20**)

The reaction of nickel(0) complexes with alkyl halides produced a stable oxidative addition product usually only when the alkyl group contains no $\beta$-hydrogen, since $\beta$-elimination of alkene occurs with the production of metal hydride. The oxidative addition reaction proceeds readily when the nickel(0) complex contains ligands that can dissociate readily, only two of the ligands being strongly coordinating ligands such as phosphine or 2,2′-bipyridine.

Thus, benzyl halides react with nickel complexes of the type $[NiL_2(olefin)]$ at low temperatures to give the oxidative addition products[78–80]. On standing,

$$[\mathrm{NiL_2(olefin)}] + \mathrm{PhCH_2X} \xrightarrow[-40\,^\circ\mathrm{C}]{-20\text{ to}} trans\text{-}[\mathrm{NiXL_2(CH_2Ph)}] + \text{olefin} \qquad (38)$$

$L = PPh_3$, $L_2 =$ bipy; X = Cl, Br; olefin = $CH_2{=}CH_2$, cod

the benzyl nickel complex gives bibenzyl and $[NiXL_2]$ or $[NiX_2L_2]$. Triphenylphosphine catalysed this decomposition[79,80]. As a result, when tetrakistriphenylphosphine or bistriphenylphosphinebipyridylnickel(0) are the reactants, 2 mol of phosphine are generated in the reaction and the halobistriphenylphosphinenickel(I) is reported to be the major product[79,80].

Perfluorohalides also oxidatively add to Ni(0) complexes of this type to give stable *cis*-nickel(II) complexes[46,81]. Both methyl iodide and ethyl bromide add to bis(tricyclohexylphosphine)nickel to give the *trans* 16-electron nickel(II) complex[82].

$$[\mathrm{Ni(Cy_3P)_2(CO)_2}] \xrightarrow{\mathrm{RX}} trans\text{-}[\mathrm{NiXR(PCy_3)_2}] \qquad (39)$$

$R = CH_3$, X = I
$R = CH_2CH_3$, X = Br

Facile $\beta$-elimination from this complex requires the dissociation of a phosphine (see below), but apparently dissociation does not take place readily in this complex containing the strong $\sigma$-donor phosphine ligands. Conflicting results have been reported[83,84] for the oxidative addition of a variety of alkyl halides to bistricyclohexylphosphinenickel(0); in no example was an alkyl nickel oxidative addition product obtained, even in the reaction with methyl halide. Alkyl halide bond scission was reported to give the nickel(I) halide[84]. The reaction with 1,4-dibromobutane gives the nickelacyclopentane and the nickel(II) dibromide[85]. Products of organic halide oxidative addition are known to undergo disproportionation so that in this case the intermediate oxidative addition product apparently undergoes an intramolecular disproportionation. The product is relatively stable since a low-energy $\beta$-hydride elimination pathway requires not only dissociation of the chelating ligand, but also a *cisoid* transition state for the metal and hydrogen for $\beta$-hydride elimination, a geometry not available in four and five-membered metallocycles (see below). The reaction of methyl iodide with a nickel complex containing four strong donor ligands has been reported to show different oxidative addition behaviour, yielding the dimethyl nickel(II) complex and nickel(I) iodide[86a].

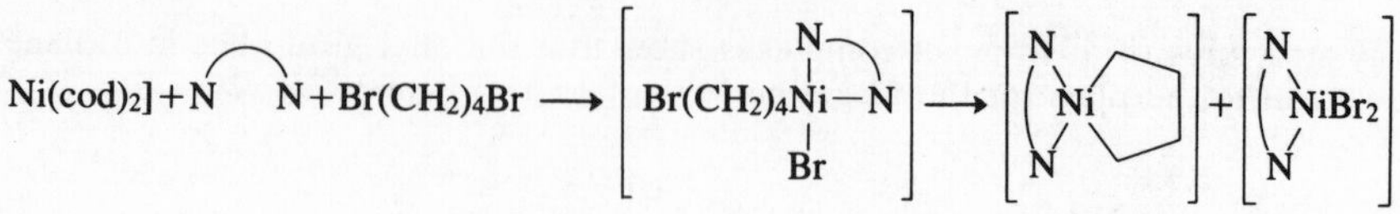

$$Ni(cod)_2 + N\frown N + Br(CH_2)_4Br \longrightarrow \left[ Br(CH_2)_4Ni(N\frown N)Br \right] \longrightarrow \left[ (N\frown N)Ni(CH_2)_4 \right] + \left[ (N\frown N)NiBr_2 \right] \quad (40)$$

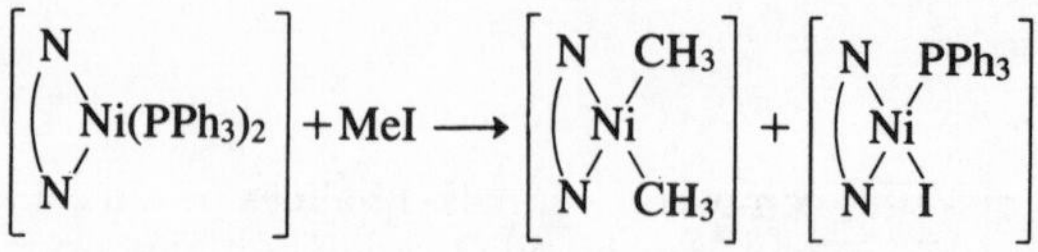

$$\left[ (N\frown N)Ni(PPh_3)_2 \right] + MeI \longrightarrow \left[ (N\frown N)Ni(CH_3)_2 \right] + \left[ (N\frown N)Ni(PPh_3)I \right] \quad (41)$$

$N\frown N$ = 2,2′-bipyridine

Trifluoromethyl radicals generated by the radiofrequency glow discharge homolysis of hexafluoroethane react with vapour-deposited nickel atoms at −196 °C, apparently to give the bis(trifluoromethyl)nickel(II) intermediate which reacts with trimethylphosphine at −78 °C to produce the *trans*-complex[86b].

$$Ni(0) + {}^{\bullet}CF_3 \xrightarrow{-196\,^{\circ}C} [Ni(CF_3)_2] \xrightarrow{PMe_3} trans\text{-}[Ni(CF_3)_2(PMe_3)_2] \quad (42)$$

Several mechanisms have been proposed for the reactions of alkyl halides not only with nickel(0) complexes, but also with $d^{10}$ palladium and platinum complexes. The three most cited mechanisms are an $S_N2$-type, a radical pair and a radical chain. The first two of mechanistic pathways are shown in Scheme 1, and can be used in the discussion of the mechanisms of organohalide reactions with the zerovalent nickel triad and also other Group VIII transition metals.

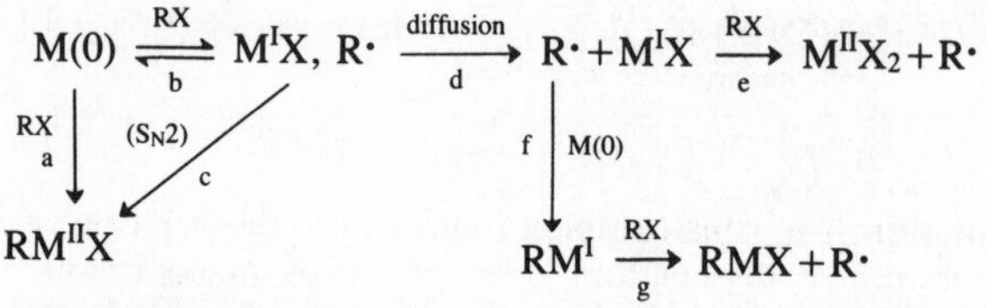

$$M(0) \underset{b}{\overset{RX}{\rightleftharpoons}} M^{I}X, R\cdot \xrightarrow[d]{\text{diffusion}} R\cdot + M^{I}X \xrightarrow[e]{RX} M^{II}X_2 + R\cdot$$

$$M(0) \xrightarrow[a]{RX} RM^{II}X \qquad M^{I}X, R\cdot \xrightarrow[c]{(S_N2)} RM^{II}X$$

$$R\cdot \xrightarrow[f]{M(0)} RM^{I} \xrightarrow[g]{RX} RMX + R\cdot$$

SCHEME 1. Mechanisms of oxidative addition reactions.

In an effort to determine whether the $S_N2$ mechanism was operative, optically active $\alpha$-deuteriobenzyl chloride was allowed to react with tetrakistriphenylphosphinenickel(0) in thf at −25 °C to give the oxidative addition product[a] [87]. This product was then treated with carbon monoxide to give the acyl nickel intermediate and then decomposed to the ester, which was essentially racemic. Since carbon monoxide ‘inserts’ into carbon—transition metal bonds stereospecifically with retention of configuration at carbon, the racemization occurred either in the oxidative addition step or the oxidative addition product racemized under the reaction

[a] Although not reported in ref. 87, the characteristic deep purple colour of the benzylnickel complex appeared immediately.

conditions. A small amount of dideuteriobibenzyl also was obtained from the reaction.

$$[NiL_4] + Ph\overset{*}{C}HClD \xrightarrow[-25°C]{thf} trans\text{-}[NiClL_2(CHDPh)]$$

$$\xrightarrow{CO,\ 1\ atm} trans\text{-}[NiClL_2(COCHDPh)]$$

$$\xrightarrow{MeOH} \underset{67\%}{PhCHDCO_2CH_3}(+10\%\ PhCHDCHDPh) \quad (43)$$

From these results and the products obtained from the various reactions of alkyl halides with nickel(0) complexes, it appears that an $S_N2$ pathway could be proceeding in some of the reactions to account for some of the oxidative addition products. Radical reactions also are taking place to account for other reaction products as well as the organonickel product.

The nickel(I) dimer $[NiCp(CO)]_2$ reacts with perfluoroalkyl halides to yield a nickel(II) oxidative addition product[88]. Cyanogen also is cleaved by nickel(0)[89,90].

$$[NiCp(CO)]_2 + R_fI \rightarrow [Ni(Cp(CO)(R_f)] + [NiICp(Co)] \quad (44)$$

$R_f = CF_3, C_2F_5, n\text{-}C_3F_7$

Acid chlorides rapidly undergo oxidative addition reaction with nickel(0) complexes, but decarbonylation of the resulting acyl complex usually takes place readily to give the alkyl or aryl nickel complex or mixtures of alkyl or aryl and acyl (Table 1)[46,48,92,93].

$$RCOCl + [NiL_n] \longrightarrow trans\text{-}[NiClL_2(COR)] \xrightarrow{-CO} trans\text{-}[NiClL_2(R)] \quad (45)$$

The tetrakis(*t*-butylisonitrile)nickel(0) complex gives a five-coordinate benzoyl complex that is stable in the solid state, but undergoes a rearrangement in solution at ambient temperature[122]. Chloroformates do not undergo decarbonylation as readily,

$$PhCOCl + [NiL_4] \longrightarrow \left[ Ph-\overset{O}{\overset{\|}{C}}-\underset{L}{\overset{L\ L}{Ni}}-Cl \right] \quad (46)$$

$L = Bu^tNC$

and this reaction provides a good synthetic route to nickel(II) carboxylates[92]. Chlorothioformamides react in a similar manner[94] as do acid anhydrides[95].

$$ClCO_2R + [NiL_4] \longrightarrow trans\text{-}[NiClL_2(CO_2R)] \quad (47)$$

$$ClC({=}S)NMe_2 + [NiL_4] \longrightarrow trans\text{-}[NiClL_2\{C({=}S)NMe_2\}] \quad (48)$$

$$(RCO)_2O + [NiL_4] \longrightarrow trans\text{-}[NiL_2(CO_2R)_2] \quad (49)$$

ii. *Palladium*(0). The oxidative addition of vinyl halides to palladium(0) complexes takes place rapidly, the ease of the reaction depending on the halide and the ligands on palladium[96–101] (Table 2). A variety of phosphine[96–99], arsine[100], and isonitrile[100] complexes react readily. Tris(triphenylphosphine)palladium carbonyl, having a lower electron density on palladium than the tetrakis(triphenylphosphine) complex, was reported[101] to react only very slowly with vinyl chloride, failing to react with tetrachloroethene and *cis*-1,2-dichloroethene. Its reaction product with vinyl chloride is the acryloyl palladium complex. The oxidative addition, as in the case of

TABLE 2. OXIDATIVE ADDITIONS OF ORGANIC HALIDES TO Pd(0) complexes

| Compounds | Pd(0) complex | L | L′ | Halide | Pd(II) complex | Ref. |
|---|---|---|---|---|---|---|
| Vinyl halides | $[PdL_4]$ | $PPh_3$, $Ph_2MeP$ | | $XC(Y)=C(Z)Cl$; X = Y = Z = Cl; X = H, Y = Z = Cl; X = Y = H, Z = Cl; X = Z = H, Y = Cl | $[Cl\text{-}PdL_2\text{-}C(X)=C(Y)Z]$ | 96, 98, 99 |
| | $[PdL_4]$ | $PPh_3$ | | $XCF{=}CF_2$; X = Br | *trans*-$[PdXL_2(CF{=}CF_2)]$ | 97 |
| | | $Ph_2MeP$ | | X = Cl, Br | | |
| | $[PdL_2]$ | $Bu^tNC$ | | $XCF{=}CF_2$; X = Cl, Br | *trans*-$[PdXL_2(CF{=}CF_2)]$ | 100 |
| | $[PdL_4]$ | $PhMe_2As$ | | | | |
| | $[PdL_3L']$ | $PPh_3$ | CO | $CH_2{=}CHCl$ | *trans*-$[PdClL_2(COCH{=}CH_2)]$ | 101 |
| Aryl halides | Pd (vapour deposition) | | | $C_6F_5Br$ | $[PdBr(C_6F_5)]$ | 26–28, 105–107 |
| | Pd (vapour deposition) | L | | $C_6F_5X$ | *trans*-$[PdXL_2(C_6F_5)]$ | 26–28 |
| | | | | | L = $PPh_3$ | 26, 28, 105–107 |
| | | | | | L = $PEt_3$, $PhMe_2P$ | |
| | | | | | L = $Me_2CO$, $Et_2O$, $SR_2$, $NHEt_2$, $NMe_3$, $NH_3$, Py, $AsR_3$ | 107 |
| | Pd (vapour deposition) | $PPh_3$ | | PhX; X = Cl, Br | *trans*-$[PdXPh(PPh_3)_2]$ | 26–28 |
| | Pd (powder reduction) | $PEt_3$ | | ArBr; Ar = $C_6H_5$, $C_6F_5$ | *trans*-$[PdBrAr(PEt_3)_2]$ | 61, 105 |

| | | | | | | |
|---|---|---|---|---|---|---|
| | $[PdL_4]$ | $PPh_3$ | | PhX<br>X = I, Br | *trans*-$[PdXPh(PPh_3)_2]$ | 102–104 |
| | $[PdL_4]$ | $PEt_3$ | | *m*-, *p*-$C_6H_4FX$<br>X = Cl, Br, I | *trans*-$[PdXL_2(m,p\text{-}C_6H_4F)]$ | 51 |
| | $[PdL_4]$ | $PPh_3$ | | *p*-$XC_6H_4Cl$<br>X = $NO_2$, CN, PhCO | *trans*-$[PdClL_2(p\text{-}XC_6H_4)]$ | 103 |
| | $[PdL_3L']$ | $PPh_3$ | CO | PhI | *trans*-$[PdIL_2(COPh)]$ | 101 |
| Allyl halides, acetates | Pd (sponge, powder) | | | $CH_2{=}CHCH_2Br$ | $[PdBr(\eta^3\text{-allyl})]_2$ | 109, 110 |
| | $[PdL_4]$ | $Ph_3P$ | | $CH_2{=}C(Me)CH_2Cl$ | *trans*-$[PdClL_2\{CH_2C(Me){=}CH_2\}]$ | 102 |
| | $[PdX_2L_2]$ (K reduction) | $Et_3P$ | | $CH_2{=}CHCH_2Br$ | $[PdBr(PEt_3)(\eta^3\text{-allyl})]$ | 105 |
| | $[Pd_2L_3]$ | dba | | $PhCH{=}C(R)CH{=}CHCH_2Cl$ | $[PdCl\{\eta^3\text{-1-(CR{=}CHPh)allyl}\}]$ | 114 |
| | | dba | | $RCH{=}C(R')CH_2X$<br>R, R′ = H, X = Cl, Br<br>R = H, R′ = $CH_3$, X = Cl<br>R = $CH_3$, R′ = H, X = Cl<br>R = Ph, R′ = H, X = Cl | $[PdX(\eta^3\text{-1-R-2-R'-allyl})]_2$ | 115 |
| | $[PdLL']$ | bipy | dba | $CH_2{=}C(Me)CH_2Cl$ | $[Pd(bipy)(\eta^3\text{-2-Me-allyl})]^+Cl^-$ | 115 |
| | $[PdL_3L']$ | $Ph_3P$ | CO | $CH_2{=}CHCH_2X$<br>X = Cl, Br | *trans*-$[PdXL_2(COCH_2CH{=}CH_2)]$ | 101 |
| | | | | $CH_2{=}C(Me)CH_2Cl$ | *trans*-$[PdClL_2\{COCH_2C(Me){=}CH_2\}]$ | 101 |
| | $[PdL_2]$ | $Cy_3P$ | | $CH_2{=}CHCH_2OAc$ | $[Pd(OAc)(PCy_3)(\eta^3\text{-allyl})]$ | 118 |
| Alkyl halides | Pd (vapour deposition) | $PMe_3$ | | $^{\bullet}CF_3$ | *trans*-$[PdL_2(CF_3)_2]$ | 86b |
| | Pd (vapour deposition) | $PEt_3$ | | $R_fX$<br>R = $CF_3$, $C_2F_5$, *n*-$C_3F_7$, $CF_2Br$, *i*-$C_3F_7$, $Cl_3C$ | *trans*-$[PdXL_2(R_f)]$ | 26–28, 62, 107, 119 |
| | | | | RCl<br>R = $PhCH_2$, $PhCH(CF_3)$ | *trans*-$[PdClL_2(R)]$ | 26–28, 120, 121 |
| | Pd (vapour deposition) | $PPh_3$, $PET_3$, $PPhMe_2$ | | $CF_3X$<br>X = Cl, Br, I | *trans*-$[PdXL_2(CF_3)]$ | 107 |

TABLE 2. ctd.

| Com-pounds | Pd(o) complex | L | L′ | Halide | Pd(II) Complex | Ref. |
|---|---|---|---|---|---|---|
| | Pd (vapour deposition) | | | $PhCH_2Cl$ | $CH_2$ PdCl $]_2$ | 120, 121 |
| | | | | $PhCH(CF_3)Cl$ | $CF_3$ CH PdCl $]_2$ | |
| | | | | 3,4-$Me_2$-$C_6H_3CH_2Cl$ | $CH_2$ PdCl Me Me $]_2$ | 121 |
| | | | | pyrid-2-yl-$CH_2Cl$ | Cl Pd $CH_2Cl$ N N $CH_2$ | 121 |
| | $[PdL_2]$ | $Bu^tNC$ | | MeI | *trans*-$[PdIL_2(Me)]$ | 122 |
| | $[PdL_2]$ | $Bu^tNC$ | | $R\overset{*}{C}H(Br)CO_2Et$ R = Me, Ph | *trans*-$[PdBrL_2\{CH(R)CO_2Et\}]$ | 92 |
| | $[PdL_4]$ | $PPh_3$ | | MeI | *trans*-$[PdIL_2(Me)]$ | 102 |
| | $[PdL_4]$ | $PPh_3$ | | $R_fI$ $R_f = CF, C_2F_5, C_3F_7$ | *trans*-$[PdIL_2(R_f)]$ | 123 |

| | | | | | |
|---|---|---|---|---|---|
| $[PdL_2]$ | $Ph_2PCH_2CH_2PPh_2$ | | $R_fI$<br>$R_f = CF_3, C_2F_5, C_3F_7$ | 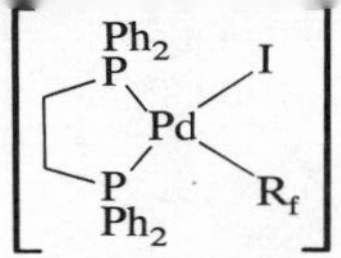 [structure: $Ph_2P$, $PPh_2$, Pd, I, $R_f$] | 123 |
| $[PdL_4]$ | $PPh_2Me$ | | $C_3F_7I$ | *trans*-$[PdIL_2(C_3F_7)]$ | 97 |
| $[PdL_4]$ | $PPh_3$ | | $PhCH_2Cl$ | *trans*-$[PdClL_2(CH_2Ph)]$ | 124 |
| $[PdL_4]$ | $PPh_3$ | | 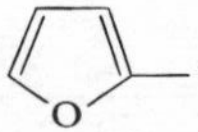 [furan]–$CH_2Cl$ (fur-2-yl-$CH_2Cl$) | *trans*-$[PdClL_2(CH_2\text{-fur-2-yl})]$ | 135 |
| $[PdL'L_3]$ | $PPh_3$ | CO | MeI | *trans*-$[PdIL_2(COMe)]$ | 101 |
| | | | $PhCH_2Br$ | *trans*-$[PdBrL_2(CH_2Ph)]$ | 101 |
| $[PdL_4]$<br>$[PdL_3]$ | $PPh_3$<br>$PEt_3$ | | PhCHDX<br>X = Cl, Br | *trans*-$[PdXL_2(CHDPh)]$ | 125–129 |
| $[PdL_3L']$ | $PPh_3$ | CO | PhCHDX | *trans*-$[PdXL_2(COCHDPh)]$ | 125–129, 133 |
| $[PdL_n]$<br>(added acac) | DBA | | $Ph_3CCl$ | 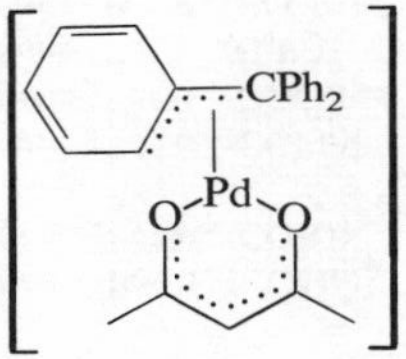 [structure: $CPh_2$, Pd, O, O] | 134 |
| $[PdL_3]$<br>(add $AgBF_4$, $NaBF_4$, or $NaBPh_4$) | $PEt_3$ | | $PhCH_2Cl$ | 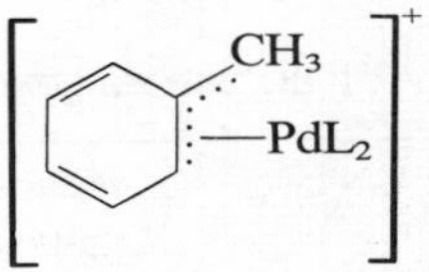 [structure: $CH_3$, $PdL_2$]$^+$ | 132, 133 |
| $[PdL_3]$<br>(add $AgBF_4$ or $NaBPh_4$) | $PEt_3$ | | PhCHDCl | 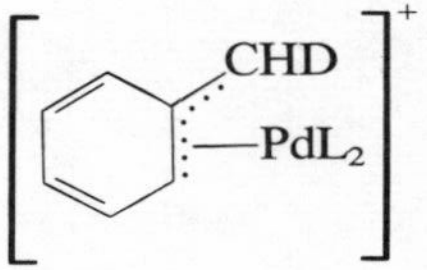 [structure: CHD, $PdL_2$]$^+$ | 132 |

TABLE 2. ctd.

| Compounds | Pd(0) complex | L | L′ | Halide | Pd(II) complex | Ref. |
|---|---|---|---|---|---|---|
| Acyl halides | Pd (vapour deposition) | $PEt_3$ | | $R_fCOCl$ $R_f = CF_3, C_3F_7$ | *trans*-$[PdClL_2(COR_f)]$ | 26–28, 62, 107 |
| | | | | $C_6F_5COCl$ | *trans*-$[PdClL_2(C_6F_5)]$ | 26–28, 107 |
| | $[PdL_4]$ | $PPh_3$ | | MeCOCl | *trans*-$[PdClL_2(COMe)]$ | 92, 102 |
| | $[PdL_2]$ | $Bu^tNC$ | | RCOCl R = Me, Ph | *trans*-$[PdClL_2(COR)]$ | 92 |
| | $[PdL_4]$ | $P(OPh)_3$ | | PhCOCl | *trans*-$[PdClL_2(COPh)]$ | 92 |
| | $[PdL_4]$ | $PPh_2Me$ | | $C_3F_7COCl$ | *trans*-$[PdClL_2(COC_3F_7)]$ | 97 |
| | | | | $C_6F_5COCl$ | *trans*-$[PdClL_2(C_6F_5)]$ | 97 |
| | $[PdL_4]$ | $PPh_3$ | | ClOCOR R = Me, Et | *trans*-$[PdClL_2(CO_2R)]$ | 92,102 |
| | | $P(OPh)_3$ | | ClOCOEt | *trans*-$[PdClL_2(CO_2Et)]$ | 92 |
| | $[PdL_2]$ | $Bu^tNC$ | | ClOCOR R = Me, Et, $CH_2Ph$ | *trans*-$[PdClL_2(CO_2R)]$ | 92 |
| | $[PdL_4]$ | $PPh_3$ | | $(CN)_2$ | *trans*-$[Pd(CN)_2(L)_2]$ | 136 |
| | $[PdL_2]$ | dba | | MeCOBr | $[PdBr(\eta^3\text{-1-Ph-3-OCOMe-3-CH=CHPh-allyl})]$ | 138 |

nickel(0), is stereospecific, retention of geometry being observed in the cases examined[96,98,99].

$$[L_nPd] + \underset{X}{\overset{Y}{}}\!\!>\!\!=\!\!<\!\!\underset{W}{\overset{Z}{}} \longrightarrow \left[ X{-}\underset{L}{\overset{L}{Pd}}{-}C(Y){=}C(Z)(W) \right] + (n-2)L \qquad (50)$$

When an olefin containing electron-withdrawing groups, but no chlorine, bromine, or iodine, is allowed to react with phosphine or arsine complexes of palladium, a product is formed that has appreciable palladacyclopropane character[96,100]. No olefin–palladacyclopropane has been isolated, however, when there is a reactive halogen (Cl, Br, I) on the alkene. Thus, by analogy with the reactions of platinum (see below) a reaction mechanism involving the rearrangement of the three-membered palladacycle has been suggested[100].

The oxidative addition reaction of palladium(0) complexes with aryl iodides or bromides is a clean reaction, giving the stable aryl palladium(II) complex in high yield[102]. Unlike tetrakis(triphenylphosphine)nickel(0), the analagous palladium complex does not react with chlorobenzene[103], even at elevated temperatures. Iodobenzene and bromobenzene react readily at 25 and 80 °C, respectively, but chlorobenzene does not react even at 135 °C[103]. Chlorobenzene substituted with electron-withdrawing groups ($+\sigma$) in the 4-position will undergo a reaction, the relative reactivities being $NO_2$ (86%, 80 °C), CN (97%, 100 °C), and PhCO (85%, 135 °C)[103]. With stronger donor ligands on palladium, for example triethylphosphine, the reaction with chlorobenzene takes place[51], whereas a palladium complex of lower electron donor ability, tris(triphenylphosphine)palladium carbonyl, reacts only with iodobenzene[101].

The order of reactivity is consistent with a mechanism similar to nucleophilic aromatic displacement in which aryl halide bond breaking is involved in the rate-determining step.

$$[PdL_2] + PhX \underset{k-1}{\overset{k}{\rightleftharpoons}} \left[ L_2Pd\diagdown\!\!\diagup X \text{ (cyclohexadienyl anion)} \right] \longrightarrow [PdXL(Ph)] \qquad (51)$$

The reaction is overall second order, first order each in palladium complex and aryl halide; the rate shows an inverse dependence on the phosphine concentration[104]. No palladium(I) species can be detected in this reaction (as is the case with nickel), accounting in part for a cleaner reaction. No free aryl radicals are generated in appreciable quantities, since aryl radicals would give biaryls and also scavenge hydrogen from the solvent. The reactive complex is bis(triphenylphosphine)-palladium(0), and $\rho$ for the reaction is +2, a value corresponding to the ease of electron transfer to the aromatic ring. Thus, either a nucleophilic displacement reaction or a one-electron transfer mechanism is consistent with the available data[103,104]. Collapse of the radical ions must occur nearly exclusively in this cage, without diffusion.

$$[PdL_2] + ArX \longrightarrow [Pd^{I}L_2ArI^{-\bullet}] \longrightarrow [PdIL_2(Ar)] \qquad (52)$$

Highly reactive transition metal powders obtained by the reduction of palladium chloride with potassium in the presence of phosphines has also been demonstrated[61,105] to produce an oxidative addition reaction with aryl halides. Palladium atoms, generated by metal vapour deposition[26–28,106–108], also react. The aryl palladium species (**21**) are more stable than the analogous nickel derivatives. The

$$ArX + Pd \longrightarrow \underset{\mathbf{21}}{[PdBrAr]} \xrightarrow{2L} [PdXL_2(Ar)] \quad (53)$$

perfluorophenylpalladium bromide is stable at 25 °C, undergoing decomposition only at 100 °C; phenylpalladium bromide is stable only below −100 °C. Reactions of these 12-electron complexes with a variety of phosphines produce the known palladium(II) complexes. Perfluorophenylpalladium bromide can be obtained as a stable powder[108] that forms 18-electron complexes with ligands such as acetone, diethyl ether, dienes, sulphides, and phosphines[106,107].

The oxidative addition of allyl bromide to palladium was reported initially by reactions of the halide with palladium sponge[109] or finely divided palladium[110]. The reaction also takes place with bis(triethylphosphine)palladium(0) generated *in situ* by the reduction of dichlorobis(triethylphosphine)palladium(II) with potassium metal[105].

$$CH_2{=}CHCH_2Br + Pd \longrightarrow [PdBr(\eta^3\text{-allyl})]_2 \quad (54)$$

The reaction of methallyl chloride with tetrakis(triphenylphosphine)palladium(0) was reported to give a rapidly equilibrating[74,75] σ-allyl complex[102]. It is important to tecognize that in reactions of allyl halides with tris- or tetrakis-phosphine ligated palladium complexes, the excess phosphine ligand can be scavenged by reaction with allyl halide, forming the allyl phosphonium salt. It is apparent, however, that when the chloro bridged dimer is allowed to react with phosphine, whether or not a

PPh3 / Pd / Cl / PPh3 ⇌ PPh3 / Pd / Cl / PPh3 (55)

cationic π-complex is generated depends on the polarity of the solvent, solvents such as acetone–water favouring the cationic complex[111] and thf favouring the σ-complex[112,113].

$$[PdCl(\eta^3\text{-alkyl})]_2 \xrightarrow{L} [PdClL(\eta^3\text{-alkyl})] \begin{cases} \xrightarrow[\text{polar solvent}]{L} [PdL_2(\eta^3\text{-alkyl})]^+Cl^- \\ \xrightarrow{L} trans\text{-}[PdClL_2(CH_2CH{=}CH_2)] \end{cases} \quad (56)$$

The tris(dibenzylideneacetone)palladium dimer also reacts with a variety of allyl halides to yield the π-allypalladium halo bridged dimers[114,115]. Allyl acetates undergo oxidative additions to a number of palladium(0) complexes. These allyl palladium complexes react with a variety of nucleophiles, and this reaction has found extensive application in organic synthesis. Catalytic alkylations by stabilized carbanions and aminations can be carried out[116,117]. The reaction of (*Z*)-3-acetoxy-5-carbomethoxycyclohexane with tetrakis(triphenylphosphine)palladium(0) in the presence of the sodium salt of methyl phenylsulphonylacetate gave a *cis*-alkylation product. This product is evidently the net result of two consecutive inversions at carbon, since conversion of the postulated intermediate $\pi$-allyl complex has been shown to occur with displacement of palladium by nucleophilic attack on the side of the ring opposite to the attachment of palladium. The initial displacement of acetate takes place, therefore, with inversion at the carbon bearing the acetate group.

$$\text{MeOCO-}C_6H_8\text{-OAc} \xrightarrow{[PdL_n]} \left[ (CO_2Me)C_6H_8\text{-}PdL_n \right] \xrightarrow{N^-} \text{MeOCO-}C_6H_8\text{-N} \quad (57)$$

Although the reactions of bis(tricyclohexylphosphine)palladium(0) with allyl acetate gives a $\pi$-allyl complex that can be isolated, tetrakis(triphenylphosphine)palladium(0) was reported to be unreactive in oxidative addition, only catalysing an allylic rearrangement of allyl acetate[118]. In view of the results obtained from the palladium-catalysed alkyation of allyl acetates by stabilized nucleophiles, there obviously are some questions concerning the mechanisms of these reactions that need to be resolved.

$$CH_2{=}CHCH_2OAc + [Pd(PCy_3)_2] \longrightarrow [Pd(OAc)(PCy_3)(\eta^3\text{-allyl})] + Cy_3\overset{+}{P}CH{=}CHCH_2OAc^- \quad (58)$$

$$CH_2{=}CHC(D_2)OAc \xrightleftharpoons{[Pd(PPh_3)_4]} AcOCH_2CH{=}CD_2 \quad (59)$$

Alkyl halides undergo oxidative addition reactions with palladium(0) complexes, but because $\beta$-elimination of the product takes place rapidly, the only isolated alkyl palladium(II) complexes are those which do not contain $\beta$-hydrogens. Methyl iodide rapidly reacts with palladium that has been vapour deposited[26–28,119], but the methyl-palladium halides are stable only below $-100\,°C$. The alkyl or perfluoroalkyl palladium halide obtained at $-96\,°C$ can be converted to stable phosphine complexes by the addition of phosphine[26–28,61–107,119–121] (Table 2). Generally, the 12-electron perfluoroalkylpalladium halides are more stable than the methyl or benzyl derivatives, the approximate order of stability being PhPdX (~100 °C) > CPhCH$_2$PdCl (~100 °C) > CF$_3$PdX (~90 °C) > C$_2$F$_5$PdI (>25 °C) > $n$-C$_3$F$_7$PdI (25 °C), Cl$_3$CPdBr (0 °C) > $i$-C$_3$F$_7$ (−78 °C) > PhPdBr (−116 °C) > MePdI ≈ EtPdI (−130 °C) > CF$_2$BrPdBr (−140 °C)[26,28,107]. The addition of benzyl chloride to atomic palladium at

$$RX + Pd(atom) \longrightarrow [PdXR] \xrightarrow{2L} [PdXL_2(R)] \quad (60)$$

$-196\,°C$ leads to the formation of the $\pi$-benzyl dimer which undergoes bridge cleavage in the presence of phosphine ligands to give the same benzylpalladium complexes obtainable from the oxidative addition of benzyl chloride to tetrakis(phosphine)palladium(0) complexes[26–28,120,121]. The reactions of ethyl and

$$PhCH_2Cl + Pd \longrightarrow \left[ (\eta^3\text{-}C_6H_5CH_2)PdCl \right]_2 \xrightarrow[PRt_3]{\text{excess}} trans\text{-}[PdCl(PEt_3)_2(CH_2Ph)] \quad (61)$$

$t$-butyl iodide with palladium (atoms) apparently yield the oxidative addition product at $-196\,°C$, but on warming give olefins resulting from $\beta$-elimination, followed by reductive elimination to give alkanes[119]. Neopentane undergoes an oxidative addition reaction and then decomposes to produce a complex mixture of products. Because added NO or toluene has no affect on the reaction, a radical chain reaction was discounted. Instead, a radical cage mechanism was favoured for the oxidative addition reaction. Trifluoromethyl radicals (glow discharge generated from hexafluoroethane) react with palladium atoms at $-196\,°C$ to give a bis(trifluoromethyl)-palladium intermediate that adds trimethylphosphine at $-78\,°C$ to yield the *trans* complex[86b].

Methyl iodide[101,102,122], perfluoroalkyl iodides[97,23] and benzyl halides[101,124] undergo oxidative addition to palladium(0) complexes bearing phosphine and isocyanide ligands. The *trans* complexes are obtained, except in reactions of complexes containing chelating ligands.

$$RX + [PdL_4] \longrightarrow [PdXL_2(R)] + 2L \tag{62}$$

The stereochemistry of the oxidative addition reaction of optically active benzyl halides to various phosphine palladium(0) complexes has been determined by a reaction sequence in which only the oxidative addition stereochemistry was unknown[125]. The remaining steps in the sequence either did not affect the active centre or had a known stereochemistry. The oxidative addition reaction was followed by carbonylation, in which the carbon monoxide 'insertion' into the carbon—palladium $\sigma$-bond is known to proceed with *retention* of configuration at carbon, and the decomposition to ester was completed by the solvent, a reaction that does not affect the asymmetric centre (Scheme 2). Thus, the stereochemistry of the oxidative addition reaction could be determined.

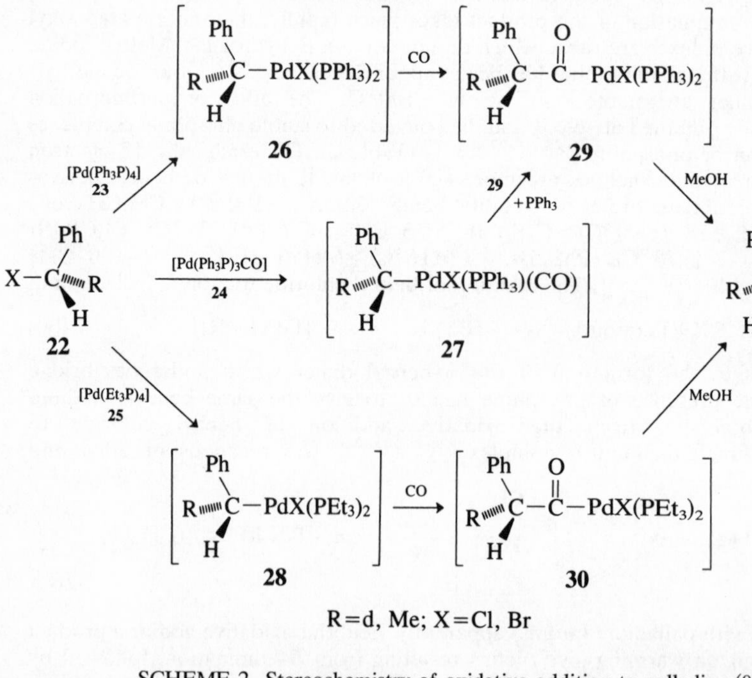

22: X—C(Ph)(H)R
[Pd(Ph3P)4] (23) → 26: [R(Ph)(H)C—PdX(PPh3)2] —CO→ 29: [R(Ph)(H)C—C(=O)—PdX(PPh3)2]
[Pd(Ph3P)3CO] (24) → 27: [R(Ph)(H)C—PdX(PPh3)(CO)] —+PPh3→ 29
[Pd(Et3P)4] (25) → 28: [R(Ph)(H)C—PdX(PEt3)2] —CO→ 30: [R(Ph)(H)C—C(=O)—PdX(PEt3)2]
29 —MeOH→ 31: R(Ph)(H)C—CO2Me
30 —MeOH→ 31

R=d, Me; X=Cl, Br

SCHEME 2. Stereochemistry of oxidative addition to palladium(0).

TABLE 3. Oxidative addition of benzyl halides to palladium(0)

| Pd(0) complex | No. | R | X | Net inversion (%) |
|---|---|---|---|---|
| $[Pd(Et_3P)_3]$ | **25** | D | Br | 30 |
| | | D | Cl | 72 |
| $[Pd(Ph_3P)_4]$ | **23** | D | Cl | 74 |
| $[Pd(Ph_3P)_4]$ | **23** | D | Cl | 100 (CO present) |
| $[Pd(Ph_3P)_3(CO)]$ | **24** | D | Cl | 100 |
| $[Pd(Ph_3P)_4]$ | **23** | $CH_3$ | Br | 90 |
| $[Pd(Ph_3P)_3(CO)]$ | **24** | $CH_3$ | Br | 90 |
| $[Pd(Ph_3P)_4]$ | **23** | $CF_3$ | Br | $<10$ |

The reaction proceeds with predominant *inversion* of configuration at carbon[126–132]. In some cases, 100% net inversion of configuration occurred, whereas in one example, only 30% enantiomeric excess was obtained (Table 3). Compounds **26** and **28** ($R=D$) could be isolated as stable complexes because they do not contain hydrogen $\beta$ to palladium, and optical rotations of the more soluble complexes **28** and **30** ($X=Cl$, Br) could be observed[131]. Because of the tendency for **26** ($R=CH_3$) to undergo $\beta$-elimination, it could not be isolated, but was converted directly to **29** either by use of carbonyl complex **27** or by the presence of carbon monoxide during oxidative addition.

It is apparent that tetrakis(triphenylphosphine)palladium(0) (**23**) and carbonyltris(triphenylphosphine)palladium(0) (**24**) behave differently than tris(triethylphosphine)palladium(0) (**25**). Although the net inversion of configuration in oxidative addition could be improved either by the use of the carbonyl complex (**24**) or by completing the reaction in the presence of carbon monoxide, the optical yields could not be similarly improved in the additions to **25** because this complex reacts rapidly with carbon monoxide to give inert carbonyltris(triethylphosphine)palladium(0) and other carbonylated palladium(0) complexes.

The loss of stereochemistry in the oxidative addition of **22** ($R=D$) to **23** can be accounted for, in part, by the partial racemization of **22** under the reaction conditions employed, because **22** recovered from an oxidative addition reaction suffered a 10% loss of its optical activity. Over longer periods of time, the reaction of excess **22** with **23** led to complete racemization of **22**. However, the optical activity of **28** was unchanged under the reaction conditions. In all of these reactions, no CIDNP could be observed, and in the reactions of **23** with **22** the presence of radicals could not be detected by chemical means. Thus, a nucleophilic exchange process that involves **26** ($R=D$) and **23** is a plausible explanation for racemization. Rapid transformation of alkyl complex **26** to acyl complex **29** would suppress racemization.

In the absence of carbon monoxide, complex **26** ($R=CH_3$) decomposes to styrene, ethylbenzene and bis(triphenylphosphine)palladium bromide. The reaction of 9-bromofluorene or ethyl $\alpha$-bromophenylacetate with **23** at 0 °C yields the coupled products **32** and **33**, respectively, and bis(triphenylphosphine)palladium bromide. In

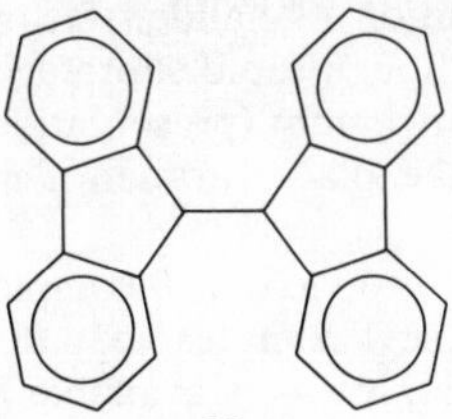

$PhCHCO_2Et$
|
$PhCHCO_2Et$

**32** **33**

the latter case a mixture of *erythro*- and *threo*-2,3-diphenyl succinate was obtained. When bis(*t*-butylisonitrile)palladium(0) reacts with ethyl $\alpha$-bromophenyl acetate or ethyl $\alpha$-bromopropionate, the oxidative addition product can be isolated[92]. However, chiral ester gave only racemic products. Although higher temperatures are usually required for the coupling reaction, the greater reactivity of 9-bromofluorene and ethyl $\alpha$-bromophenylacetate accounts for the lower reaction temperatures. For example **26** (R=H; X=Cl) undergoes a reaction with benzyl chloride at 80 °C to give bibenzyl[130]. In all of these reactions, no CIDNP could be observed, but chemical tests indicated the presence of radicals.

The oxidative addition product (**28**) is configurationally stable in solution over long periods of time, and its optical activity is unaffected by the presence of **22**, **25** or triethylbenzylphosphonium chloride, another product present when **22** (R=H; X=Cl) reacts with **25**. Thus, racemization in this reaction cannot be attributed to a nucleophilic exchange between **25** and **28**. In addition, the $\pi$-benzyl complex (**35**) maintains its configurational integrity in solution[132]. Even though CIDNP could not be observed from the reaction of **22** (R=H; X=Cl or Br), the possibility that racemization is occurring either by an oxidative addition reaction that involves a benzyl radical or by the dissociation of the oxidative product **28** into a radical pair, followed by rotation of the benzyl group, cannot be excluded. Only in the reaction of **28** (R=D; X=Br) with **25** is any coupling product, racemic dideuteriobibenzyl, obtained.

The inversion of configuration at carbon, the order of reactivity ($PhCH_2Br > PhCH_2Cl > PhCH(Me)Br > PhCH(Me)Cl > PhCH(CF_3)Cl$ and $[Pd(Et_3P)_3] > [(Pd(Ph_3P_4)] > [Pd(Ph_3P)_3(CO)]$, and the inability to detect free radicals during the

$$\left[ \text{H(D)(Ph)C}-Pd(PEt_3)_2 \right] \rightleftharpoons \left[ \pi\text{-(HDC)C}_6\text{H}_5 - Pd(PEt_3)_2 \right] \qquad (63)$$

**34** **35**

oxidative addition reaction of the optimally active benzyl halides clearly are in favour of a nucleophilic displacement mechanism. Radical species probably participate only during the decomposition of the oxidative addition product and, depending on the characteristics of the alkyl group of the organic halide, alternative pathways for the decomposition of the oxidative adduct are possible.

Benzylpalladium chloride complexes are easily converted to the corresponding $\pi$-benzyl complexes by removal of halide from palladium, usually by sodium tetraphenylborate or tetrafluoroborate or silver tetrafluoroborate. The $\sigma$-benzyl complex can be regenerated by readdition of halide. An optically active $\pi$-benzyl complex retains its activity because, although rapid suprafacial shifts of the $\pi$-benzyl groups take place, an antrafacial rearrangement (presentation of the opposite face of the phenyl ring to palladium) must take place through a $\sigma$-benzyl intermediate[132] (Scheme 3).

Acid chlorides react with palladium(0) to yield acyl complexes[92,102]. Depending on the ligands present on palladium, the acyl complex may decarbonylate to give the same palladium—carbon $\sigma$-bonded complex as is available from the direct addition of the appropriate organohalide to a palladium(0) species. Carbonylation of the

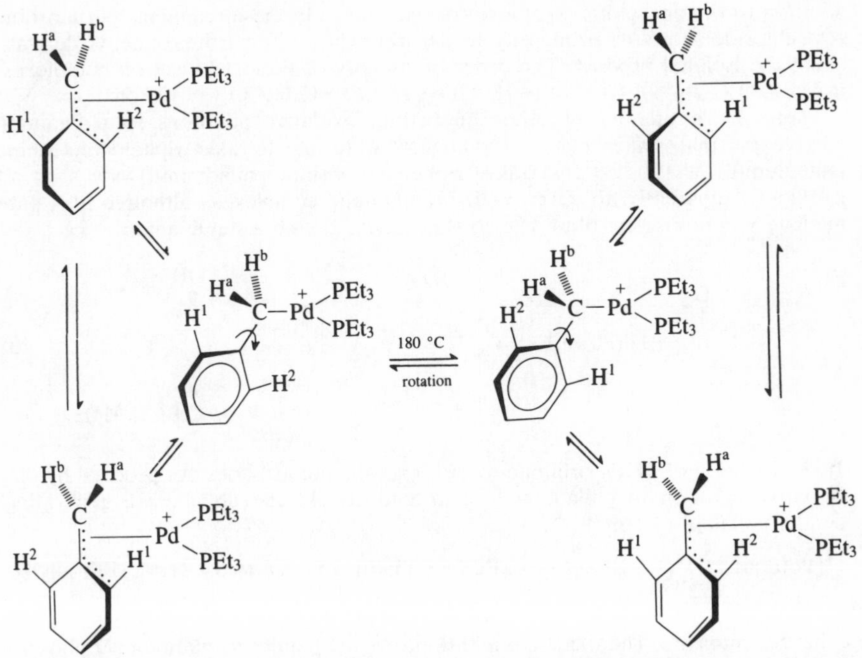

SCHEME 3. $\sigma \rightarrow \pi$ benzylpalladium rearrangement.

organopalladium(II) complex regenerates the acyl product. The ligands may be phosphines, phosphites and isonitriles. Perfluoro-allyl and -aryl acid chlorides undergo the oxidative addition reaction, but the acyl complex from perfluorobenzoyl chloride was not isolated, undergoing rapid decarbonylation to the perfluorophenyl product[97]. Chloroformates give palladium carboxylates on addition, an exception being the phenyl ester, which decomposes to phenyl benzoate[92].

$$\mathrm{RCOCl} + [\mathrm{PdL}_n] \longrightarrow trans\text{-}[\mathrm{PdClL_2(COR)}] \underset{+\mathrm{CO}}{\overset{-\mathrm{CO}}{\rightleftharpoons}} trans\text{-}[\mathrm{PdClL_2(R)}] \longleftarrow [\mathrm{PdL}_n] + \mathrm{RCl} \quad (64)$$

$$\mathrm{C_6F_5COCl} + [\mathrm{Pd(Ph_2MeP)}] \longrightarrow trans\text{-}[\mathrm{PdCl(PPh_2Me)_2(COC_6F_5)}] \xrightarrow{-\mathrm{CO}} trans\text{-}[\mathrm{PdCl(PPh_2Me)_2(C_6F_5)}] \quad (65)$$

$$[\mathrm{PdL}_n] + \mathrm{ClCO_2R} \longrightarrow trans\text{-}[\mathrm{PdClL_2(CO_2R)}] \underset{-\mathrm{CO}}{\xrightarrow{\mathrm{R=Ph}}} trans\text{-}[\mathrm{PdClL_2(OPh)}] \xrightarrow{\mathrm{ClCO_2Ph}} \mathrm{PhCO_2Ph} \quad (66)$$

$n = 4$: $\mathrm{L = PPh_3,\ P(OPh)_3}$
$n = 2$: $\mathrm{L = Bu^tNC}$

Acid chlorides also add to vapour deposited palladium at $-196\,°\mathrm{C}$[26–28,62,107]. At higher temperatures ($-78\,°\mathrm{C}$) the oxidative addition product reacts with triethylphosphine to give the same acyl palladium complexes as can be obtained by oxidative addition to the phosphine complexes of palladium. In these reactions, perfluorobenzoyl chloride shows its propensity to decarbonylate, the perfluorophenyl derivative being the isolated product. The order of stability of these 12-electron complexes is $n$-$C_3F_7CO$ (−78 °C), $CF_3CO$ (−78 °C) > $C_6F_5CO$ > $C_6H_5CO$ (−130 °C).

There are a number of other interesting oxidative additions to palladium(0) complexes that take place. Cyanogen adds to tetrakis(triphenylphosphine)-palladium(0) at 100 °C[136]. Tetrakis(triphenylphosphine)palladium(0) opens up allyl epoxides, apparently to give $\pi$-allyl palladium complexes, although this intermediate was not isolated but instead was alkylated with a stable anion[137].

$$\text{(cyclopentadiene monoepoxide)} + [Pd(Ph_3P)_4] \longrightarrow \left[ \text{(}O^-\text{-cyclopentenyl } \pi\text{-allyl } Pd^+L_2\text{)} \right] \xrightarrow{CH_2(CO_2Me)_2} \text{(OH, } CH(CO_2Me)_2 \text{ cyclopentene)} \quad (67)$$

The addition of acetyl bromide to bis(dba)palladium(0) does not proceed by direct oxidative addition to palladium, but instead attacks the dba ligand, generating a $\pi$-allyl palladium bromide[138].

$$[Pd(dba)_2] + MeCOBr \longrightarrow [PdBr(\eta^3\text{-1-Ph-3-OCOMe-3-CH=CHPh-allyl})]_n \quad (68)$$

iii. *Platinum*(0). The oxidative addition of vinyl halides to platinum(0) phosphine and arsine complexes ($Ph_3P$, $Ph_2MeP$, $Ph_3As$, or bisphosphines) take place less readily than the first two members of the nickel triad. Tetrakis(triphenylphosphine)-platinum(0), for example, will not undergo oxidative addition reactions with vinyl chloride[98,99], but substitution of a second halogen on the olefin usually produces a reaction. Oxidative addition to tetrakis(triphenylphosphine)platinum does proceed smoothly, however, with vinyl bromides or electronegatively substituted vinyl halides, particularly fluorovinyl halides (Table 4)[139–147].

$$[Pt(Ph_3P)_4] + \underset{B}{\overset{A}{}}C{=}C\underset{X}{\overset{C}{}} \longrightarrow \left[ X{-}\underset{PPh_3}{\overset{PPh_3}{Pt}}{-}C(C){=}C(A)(B) \right] \quad (69)$$

The reaction is stereospecific, retention of geometry being observed in reactions with both ($E$)- and ($Z$)-$\beta$-bromostyrene[147] as well as with other vinyl halides[98,99]. The stereochemistry of the ($E$)-$\beta$-bromostyrene oxidative addition product has been confirmed by an X-ray structure determination[147].

$$[PtL_3] \begin{cases} + (Z)\text{-PlCH=CHBr} \xrightarrow{-L} [PtBrL_2\{(Z)\text{-CH=CHPh}\}] \\ + (E)\text{-PhCH=CHBr} \xrightarrow{-L} [PtBrL_2\{(E)\text{-CH=CHPh}\}] \end{cases} \quad (70)$$

$L = PPh_3$

A number of different mechanisms have been suggested for this reaction. Retention of geometry at the trigonal carbon is consistent with the stereochemistry observed in nucleophilic substitution reactions of vinyl halides (addition–elimination) by the usual nucleophiles. However, when ordinary nucleophilic substitution reactions are compared with the oxidative addition reactions, there are inconsistencies in the relative reaction rates within a series of organic substrates and leaving groups. For nucleophilic substitution reactions with vinyl halides, the rates fall in the order F > Cl > Br, whereas the reverse order is observed in oxidative addition; I > Br > Cl (no reaction with F). Further, the reaction of nucleophiles is faster with methyl iodide than with $\beta$-bromostyrene, whereas the reverse is true in the oxidative addition reactions of zerovalent $d^{10}$ complexes[147].

The reaction probably proceeds by coordination of the olefin to the metal since the vinyl halide $d^{10}$ complexes can be converted to the oxidative addition product. Whether the key intermediate is a 'metallacyclopropane' in every case is uncertain. Because the reaction of certain perhaloethylenes with platinum(0) complexes affords products having considerable platinacylic character that rearrange to $\sigma$-vinyl complexes[139,140,142–146], the platinacylopropanes (and the nickel and palladium analogues) have been suggested as intermediates in the oxidative addition reactions of vinyl halides to platinum(0) complexes.

$$[L_nM] + (E)\text{-RCH=CHX} \rightleftharpoons \left[ L_2M\text{—}\| \right] \longleftrightarrow \left[ L_2M\text{(metallacyclopropane)} \right] \longrightarrow trans\text{-}[MXL_2((E)\text{-CH=CHR})] \quad (71)$$

A comparison of the rate constants for rearrangement of three platinacylopropane complexes in solvents of different polarity indicates two extreme mechanisms for the rearrangement reaction[98,99]. The rates of rearrangement plotted against Brownstein solvent parameters yield values of $R$ (slopes) that are consistent with an $S_N1$-type transition state (**36**) in the reaction of the trichloroethene adduct[148]. An intramolecular reaction, transition state **37**, takes over in the case of the arsine–bromotrifluoroethene complex.

$$L_2Pt(CXY\text{–}CZ_2) \longrightarrow \left[ L_2Pt(C^{\delta+}(Y)\cdots X^{\delta-})(CZ_2) \right]^{\ddagger}\ (\mathbf{36}) \quad \text{or} \quad \left[ L_2Pt\cdots X\cdots C(Y)\cdots CZ_2 \right]^{\ddagger}\ (\mathbf{37}) \quad (72)$$

| L | X | Y | Z | R |
|---|---|---|---|---|
| $PPh_3$ | Cl | Cl | Cl | $4.9 \pm 1.2$ |
| $PPh_3$ | Cl | H | Cl | $10.2 \pm 0.5$ |
| $AsPh_3$ | Br | F | F | 18.0 |

TABLE 4. Oxidative additions of organic halides to Pt(0) complexes

| Compounds | Pt(0) complex | L | L′ | Halide | Pt(II) complex | Ref. |
|---|---|---|---|---|---|---|
| Vinyl halides | $[PtL_4]$ | $PPh_3$ | | $CF_2{=}CFX$ $X = Cl, Br$ | *trans*-$[PtXL_2(CF{=}CF_2)]$ | 139, 140, 146 |
| | | | | $PhCH{=}CH, Br$ ($E, Z$) | *trans*-$[PtBrL_2(CH{=}CHPL)]$ | 141, 147 |
| | $[PtL_4]$ or | $PPh_3$ | | $Cl_2C{=}CCl_2$ | *trans*-$[PtClL_2(CCl{=}CCl_2)]$ | 142–144 |
| | $[PtL_2L']$ | $PPh_3$ | Olefin | $Cl_2C{=}CHCl$ | *trans*-$[PtClL_2(CH{=}CCl_2)]$ | 142 |
| | | | | $ClCH{=}CHCl$ ($E$) | *trans*-$[PtClL_2\{(E)\text{-}CH{=}CHCl\}]$ | 142 |
| | | $PPh_3$ | Olefin | $CFCl{=}CFCl$ ($E, Z$ mixture) | *trans*-$[PtClL_2(CF{=}CFCl)]$ | 145 |
| | $[PtL_4]$ | $PPh_2Me$ | | $CF_2{=}CFBr$ | *trans*-$[PtBrL_2(CF{=}CF_2)]$ | 140, 146 |
| | | | | $CCl_2{=}CCl_2$ | *trans*-$[PtClL_2(CCl{=}CCl_2)]$ | 98, 99 |
| | | | | $CHCl{=}CCl_2$ | *trans*-$[PtClL_2(CH{=}CCl_2)]$ | 98, 99 |
| | | | | $CHCL{=}CHCl$ ($E$) | *trans*-$[PtClL_2\{(E)\text{-}CH{=}CHCl\}]$ | 98, 99 |
| | | | | $CH_2{=}CHBr$ | *trans*-$[PtBrL_2(CH{=}CH_2)]$ | 98, 99 |
| | | | | $CHBr{=}CHBr$ ($E$) | *trans*-$[PtBrL_2\{(E)\text{-}CH{=}CHBr\}]$ | 98, 99 |
| | | | | $C(CF_3)CN{=}C(CF_3)CN$ ($E$ or $Z$) | *trans*-$[Pt(CN)L_2\{(E)\text{-}C(CF_3){=}C(CF_3)CN\}]$ | 145 |
| | $[PtL_4]$ | $AsPh_3$ | | $CF_2{=}CFBr$ | *trans*-$[PtBrL_2(CF{=}CF_2)]$ | 148 |
| | $[PtL_2L']$ | $PPh_3$ | tcne | tcne ($h\nu$) | *trans*-$[Pt(CN)(L)_2\{C(CN){=}C(CN)_2\}]$ | 153 |
| | $[PtL_4]$ | $PPh_3$ | | $IC{\equiv}CI$ | *trans*-$[PtIL_2(C{\equiv}CI)]$ | 141 |
| | | | | $PhC{\equiv}CBr$ | *trans*-$[PtBrL_2(C{\equiv}CPh)]$ | 141 |
| | $[PtL_2L']$ | $PPh_3$ | $NCC{\equiv}CCN$ | $NCC{\equiv}CCN$ | *cis*-$[Pt(CN)(L)_2(C{\equiv}CCN)]$ | 154, 155 |
| Aryl halides | $[PtL_3]$ | $PPh_3$ | | $PhBr$ ($h\nu$) | *trans*-$[PtBrL_2(Ph)]$ | 156 |
| | $[PtX_2L_2]$ (K reduction) | $PEt_3$ | | $ArX$ $Ar = Ph, C_6F_5$; $X = I, Br, Cl, CN$ | *trans*-$[PtXL_2(Ar)]$ | 105 |
| Allyl halides | $[PtL_4]$ | $PPh_3$ | | $CH_2{=}CHCH_2Cl$ | $[PdL_2(\eta^3\text{-allyl})]^+Cl^-$ | 157 |
| | $[PtL_4]$ | $PPh_3$ | | $RCH{=}CHCH_2X$ $R = H, CH_3$; $X = Cl, Br$ | *trans*-$[PtXL_2(CH_2CH{=}CH{\sim}R)]$ | 158 |

| | | | | | | |
|---|---|---|---|---|---|---|
| | Pt (vapour deposition) | $PPh_3$ | | $CH_2{=}CHCH_2Cl$ | $[PtClL(\eta^3\text{-allyl})]$ | 159 |
| | $[PtL_3]$ | $PPh_3$ | | $R^1CH{=}C(R^2)CH_2OAc$ | $[PtL(acac)\{CH_2C(R'){=}CHR^2\}]$ [with added $Tl(acac)_2$] $R^1 = R^2 = H$ $R^1 = Me, R^2 = H$ $R^1 = H, R^2 = Me$ | 160 |
| Alkyl halides | $[PtL_4]$ | $PPh_3$ | | MeI | *trans*-$[PtIL_2(Me)]$ | 156, 163 |
| | $[PtL_2L']$ | $PPh_3$ | $CH_2{=}CH_2$ | MeI | *trans*-$[PtIL_2(Me)]$ | 162 |
| | | | | $PhCH_2Br$ | *trans*-$[PtBrL_2(CH_2Ph)]$ | 162 |
| | | | | $CH_2Cl_2$ ($h\nu$) | *trans*-$[PtClL_2(CH_2Cl)]$ | 165 |
| | | | | CHRXY ($CH_2I_2$, $CH_2Br_2$, $CH_2BrCl$, $CH_2BrI$, $CH_2ClI$, $CHBr_3$) | *trans*-$[PtXL_2(CHRY)$ | 161 |
| | $[PtL_4]$ | $PEt_3$ | | $Bu^nBr$ | *trans*-$[PtBrL_2(Bu^n)]$ | 167 |
| | | | | $CH_2{=}CH(CH_2)_4Br$ | *trans*-$[PtBrL_2\{(CH_2)_4CH{=}CH_2\}]$ | 167 |
| | | | | | + *trans*-$[PtBrL_2(CH_2\text{-cyclopentyl})]$ | 167 |
| | | | | $MeCH(Cl)CO_2Et$ | *trans*-$[PtClL_2\{CH(Me)CO_2Et]$ | 168 |
| | $[PtL_3]$ | $PPh_3$ | | 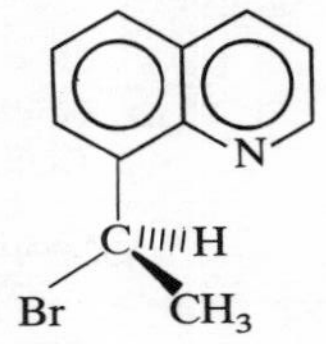 | 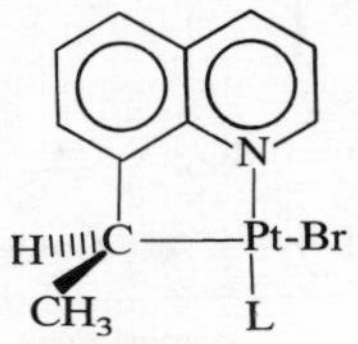 | 171 |
| | $[PtL_4]$ | $PPh_3$ | | $R_fI$ $R_f = CF_3, C_2F_5, C_3F_7$ | *trans*-$[PtIL_2(R_f)]$ | 123 |
| | $[PtL_2]$ | $Ph_2PCH_2CH_2PPh_2$ | | | 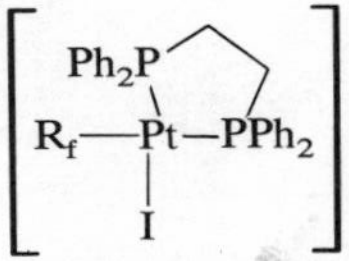 | 123 |
| | $[PtL_4]$ | $PPh_3$ | | $C_6F_5CH_2Cl$ | *trans*-$[PtClL_2(CH_2C_6F_5)]$ | 123 |

TABLE 4. ctd.

| Compounds | Pt(0) complex | L | L′ | Halide | Pt(II) complex | Ref. |
|---|---|---|---|---|---|---|
| Acid chlorides, acid anhydrides, and cyanides | $[PtL_4]$ | $PPh_3$ | | MeCOCl | *trans*-$[PtClL_2(COMe)]$ | 141 |
| | $[PtL_3]$ | $PPh_3$ | | | | |
| | $[PtL_2L']$ | $PPh_3$ | $C_2H_4$ | $[RCO)_2O$<br>$R = Me, CF_3, C_2F_5$ | *trans*-$[Pt(OCOR)(L)_2(COR)]$ | 172 |
| | | | | 3,3,4,4-tetrafluorosuccinic anhydride | 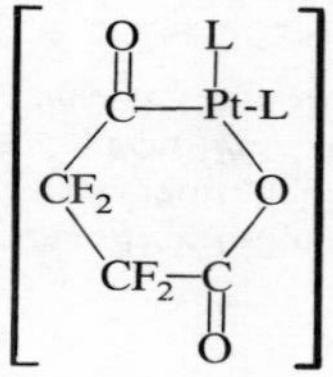 [ring: O=C—Pt(L)(L)—O—C(=O)—$CF_2$—$CF_2$—C] | 172 |
| | $[PtL_4]$ | $PPh_3$ | | $(CN)_2$ | $[PtL_2(CN)_2]$ | 136 |
| | $[PtL_4]$ | $PPh_3$ | | $CH_3C(CN)_3$ | *trans*-$[Pt(CN)(L)_2\{C(Me)(CN)_2\}]$ | 173 |
| | $[PtL_3]$ | $PEt_3$ | | PhCN | *trans*-$[Pt(CN)(L)_2(Ph)]$ | 174 |
| | $[PtL_4]$ | $AsPh_3$ | | 2,2,3,3-tetracyanooxirane | 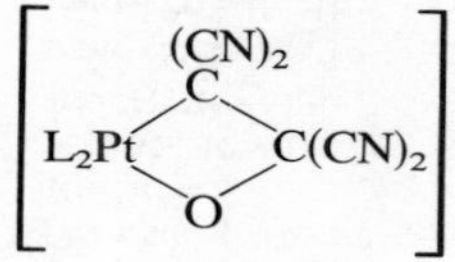 [ring: $L_2Pt$—$C(CN)_2$—$C(CN)_2$—O—] | 175 |
| | | $PPh_3$<br>$P(p\text{-}MeC_6H_4)$<br>$PEt_3$ | | | *cis*-$[Pt(CN)(L)_2\{OC(CN){=}C(CN)_2\}]$<br>($L = PEt_3$ only) | |

| | | | | | | |
|---|---|---|---|---|---|---|
| Organotins | $[PtL_2L'_2]$ | $PPh_3$ | $C_2H_4$ | $SnR_4$<br>R = Me, Ph | *cis*-$[PtL_2(R)(SnR_3)]$ | 176, 177, 179 |
| | | | | $ArSnMe_3$ | *cis*-$[PtL_2(Ar)(SnMe_3)]$ | 181, 182 |
| | | | | $R_3SnCl$<br>R = Me, Ph | *cis*-$[PtL_2(R)(SnClR_2)]$ | 177, 179, 180 |
| | | | | $Ph_2SnCl_2$ | *cis*-$[PtL_2(Ph)(SnPhCl_2)]$ | 177, 179 |
| | $[PtL_2L'_2]$ | $PMeBu^t_2$<br>$P(C_6H_{11})_3$, $PPh_3$ | $C_2H_4$ | $Me_3SnCH_2CH{=}CH_2$ | $[PtL(SnMe_3)(\eta^3\text{-allyl})]$ | 178 |
| Alkenes/alkynes and acids | $[PtL_2L']$ | $PPh_3$ | $CF_2{=}CF_2$ | $CF_3CO_2H$ | *cis*-$[PtL_2(CF_2CF_2H)(OCOCF_3)]$ | 183 |
| | $[PtL_2L']$ | $PPh_3$ | $CF_3C{\equiv}CCF_3$ | $CF_3CO_2H$ | $L_2Pt\{cis\text{-}C(CF_3){=}CHCF_3\}$ | 183 |
| | $[PtL_2L']$ | $PPh_3$ | $R{-}C{\equiv}C{-}R'$<br>R′ = Me, R = Ph<br>R′ = H, R = Me, Et, Ph, *p*-$MeC_6H_4$ | HX<br>X = Cl, $CF_3CO_2$ | *trans*-$[PtXL_2\{cis\text{-}CR{=}CR'H\}]$ | 184 |
| | | | cycloheptyne | HX<br>X = $CF_3CO_2$ | *trans*-$[PtXL_2$(cyclohept-1-enyl)] | 185 |
| | | | benzyne | HX<br>X = $CF_3CO_2$ | *trans*-$[PtXL_2$(cyclohex-1-enyl)] | 185 |
| | $[PtLL']$ | dppe | benzyne | HX<br>X = $OC_6H_4Me$-*p*, $CH_2NO_2$, $CH_2COCH_3$, $CH_2COPh$, CH(CN)Ph, OH, OMe | *trans*-$[PtXL_2$(cyclohex-1-enyl) | 186 |

Although the final rearrangement to the vinyl complex could be expected to retain the geometry at the vinyl halide site, this does not preclude a more direct, concerted reaction of the $\pi$-complexed vinyl halide with the metal centre in reactions of alkenes that are less electronegatively substituted and thereby have much less platinacyclopropane character. As indicated previously, a kinetic *cis* complex could be expected to rearrange to a thermodynamic *trans* complex.

$$\left[ L_2M \leftarrow \|(X,H;H,R) \right] \longrightarrow \left[ L_2M(X)\cdots C(H)\cdots C(H)(R) \right]^{\ddagger} \longrightarrow cis\text{-}[MXL_2\{(E)\text{-}CH{=}CHR\}] \downarrow trans\text{-}[MXL_2\{(E)\text{-}CH{=}CHR\}] \quad (73)$$

The reaction orders, the stereochemistry, and the fact that radical inhibitors have no effect on the rates of oxidative addition of vinyl halides to phosphine platinum(0) complexes does not support a radical mechanism[147]. The observed stereospecific retention of geometry at the $sp^2$ centre discounts the intervention of vinyl radicals which, if formed, would undergo configurational inversion[150–152].

Platinum(0) complexes also undergo reactions with electronegatively substituted olefins by cleaving a carbon—cyanide bond. (*E*)-Dicyanoethene forms a complex with tetrakis(diphenylmethylphosphine)platinum(0)[98,99], but does not undergo an oxidative addition reaction. Either (*E*)- or (*Z*)-2,3-dicyanoperfluorobut-2-ene reacts with a platinum(0) complex to yield the oxidative addition product[145]. The same (*E*)-oxidative addition product is obtained regardless of the geometry of the starting

$$[PtL_2(\text{stilbene})] + (Z)\text{-}CN(CF_3)C{=}C(CF_3)CN \longrightarrow [Pt(CN)(L)_2\{(E)\text{-}C(CF_3){=}C(CF_3)(CN\}] \quad (74)$$

$L = PPh_3$

alkene. Tetracyanoethylene undergoes oxidative addition only on irradiation[153]. A tcne radical anion, presumed to be an intermediate, was detected by e.s.r.[154]. Acetylenic halides[141] as well as dicyanoacetylene[154,155] add to platinum(0) complexes, but in the case of dicyanoacetylene irradiation is required.

$$[PtL_2(\text{tcne}) \xrightarrow{h\nu} [Pt(CN)(L)_2\{C(CN){=}C(CN)_2\}] \quad (75)$$

$L = PPh_3$

Aryl halides do not react as readily with platinum(0) complexes as they do with nickel(0) or palladium(0). The oxidative addition reaction of bromobenzene with tris(triphenylphosphine)platinum(0) proceeds slowly in the absence of a free radical initiator, while a 70% conversion or greater can be realized in the presence of AIBN[156]. Bis(triethylphosphine)platinum(0) generated *in situ* by the reduction of bis(triethylphosphine)platinum dihalide reacts with phenyl halides and perfluorophenyl halides to give the oxidative addition product[105]. In the reactions of aryl halides with platinum(0) species, apparently it is necessary to generate a bisphosphine complex.

$$PtL_2 + ArX \longrightarrow [PtXL_2(Ar)] \quad (76)$$

There are a few examples of $\pi$- or $\sigma$-allyl platinum complexes that have been prepared by oxidative addition to platinum(0). The preparation of allylbis(triphenylphosphine)platinum chloride is carried out by the oxidative addition of allyl chloride to tetrakis(triphenylphosphine)platinum(0)[157,158]. Both $\sigma$- and $\pi$-allyl structures, **38** and **39**, have been written for this product. Reaction of platinum(atoms) at −196 °C

$$[PtL_4] + CH_2{=}CHCH_2Cl \longrightarrow \underset{\mathbf{38}}{[PtL_2(\eta^3\text{-allyl})]^+Cl^-} \tag{77}$$

$L = Ph_3P$

$$\underset{\mathbf{39}}{[PtClL_2(CH_2CH{=}CH_2)]}$$

with allyl chloride gives a stable (160–170 °C) allylplatinum chloride tetramer that reacts with triphenylphosphine to yield the $\eta^3$-allyl complex[159].

$$Pt + CH_2{=}CHCH_2Cl \xrightarrow{-196\,^\circ C} [PtCl(C_3H_5)]_4 \xrightarrow{PPh_3} [PtCl(PPh_3)(\eta^3\text{-allyl})] \tag{78}$$

Allyl acetates react with platinum(0), and although the cationic $\eta^3$-allylplatinum acetate was not isolated, it was converted to an acetonylacetate derivative (containing $\sigma$-allyl bonding) that was isolated[160].

$$[PtL_3] + CH_2{=}CHCH_2OAc \longrightarrow [PtL_2(\eta^3\text{-allyl})]^+Cl^- \xrightarrow{acac} \left[(CH_2{=}CHCH_2){-}\overset{L}{\underset{O}{Pt}}{-}O \ldots\right] \tag{79}$$

The reaction of a variety of alkyl halides with platinum(0) complexes bearing phosphine ligands has been reported[123,156,161]. The reaction takes place readily with perfluoroalkyl halides[123] and methyl iodide[156] as well as di- and tri-halomethanes[161]. Both *cis*- and *trans*-complexes are obtained in the latter case, although *trans*-products are usually observed when two monodentate phosphines are present in the product.

$$RX + [PtL_n] \longrightarrow [PtXL_2(R)] + (n-2)L \tag{80}$$

The kinetics of the oxidative addition of methyl iodide either to ethylenebis-(triphenylphosphine)platinum(0)[162] or tetrakis(triphenylphosphine)platinum(0)[163] are second order, first order in platinum(0) complex and first order in methyl iodide. Benzyl bromide also shows second-order kinetics[162]. Measurement of the enthalpy of the reaction of methyl iodide with ethylenebis(triphenylphosphine)platinum(0) reveals that there is little difference in the bond dissociation energies of Pt—$CH_3$ and Pt—I in the product[164].

In the oxidative addition reaction of methyl iodide to tetrakis(triphenylphosphine)platinum(0), the concentration of this species can be neglected since dissociation into the trisphosphine platinum species is nearly complete[163]. The kinetic data are accommodated by the following reactions:

$$[PtL_3] \rightleftharpoons [PtL_2] + L \tag{81}$$

$$[PtL_3] + MeI \xrightarrow{k_2} [PtIL_2Me)] + L \tag{82}$$

$$[PtL_2] + MeI \xrightarrow{k_2'} trans\text{-}[PtIL_2(Me)] \tag{83}$$

Existing evidence supports a radical process for the reaction of platinum(0) complexes with most alkyl halides. Relatively unreactive halides such as methylene chloride will undergo oxidative addition in the presence of light or in the dark in the presence of the double ylide $(Me_3Si)_2N—P(NSiMe_3)_2$[165]. The reaction is inhibited by radical scavengers.

The exact mechanism of the reaction and even the products of the reactions are not without controversy. Initially, the oxidative addition of benzyl chloride and 2-halobutanes (bromo and iodo) to different platinum(0) complexes, $[PtL_n]$ ($n=3, 4$; $L=PPh_3$, $PPhMe_2$, $PEt_3$) was reported to generate the dihalobis(phosphine)-platinum(II) complexes as the only products[166]. Products of $\beta$-elimination were observed, particularly with organic halides such as $\alpha$-phenylethyl bromide.

The products of oxidative addition of alkyl halides containing $\beta$-hydrogens are obtained initially, but $\beta$-elimination does occur slowly[167,168]. It has been suggested that the course of the reaction depends heavily on the alkyl halide and its reactivity. The reaction of $n$-butyl bromide with tetrakis(triethylphosphine)platinum(0) yields predominately the butylpalladium halide (**40**) initially, but further reaction produces more dibromide, butane, and butene at the expense of **40**[167]. Duroquinone inhibits

$$Bu^nBr + [PtL_4] \longrightarrow \underset{\substack{\mathbf{40} \\ 95\%}}{trans\text{-}[PtBrL_2(Bu^n)]} + \underset{4\%}{trans\text{-}[PtBrHL_2]} + \underset{1\%}{[PtBr_2L_2]} + C_4H_{10} + C_4H_8 \qquad (84)$$

$L=PEt_3$

the reaction and radical abstraction from solvent (toluene) is observed when neopentyl bromide is the organic halide; benzylbromobis(triethylphosphine)platinum(II) is the product in this reaction. The oxidative addition reaction with 6-bromohex-1-ene gives a 3:1 ratio of open-chain to cyclized product. In addition, the oxidative addition to chiral ethyl $\alpha$-chloropropionate led to racemic products. In accord with these results a radical chain mechanism was written[167].

$$CH_2{=}CH(CH_2)_4Br + [PtL_4] \longrightarrow \underset{3}{trans\text{-}[PtBrL_2\{(CH_2)_4CH{=}CH_2\}]} + \underset{1}{trans\text{-}[PtBrL_2(CH_2cyclopentyl)]} \qquad (85)$$

More reactive halides ($\alpha$-bromo esters, benzyl bromides and secondary iodides) react too rapidly with tris(triethylphosphine)platinum(0) to be undergoing a radical chain reaction[168]. Thus, isopropyl iodide reacts with tris(triethylphosphine)platinum to give mainly hydridobis(triethylphosphine)platinum bromide and hydrocarbon products, with only a trace of isopropylplatinum iodide being formed. CIDNP is observed in this reaction. Benzyl bromide behaves similarly.

Thus, it has been suggested that very reactive halides are subject to a non-chain radical mechanism, platinum dihalides and organoplatinum halides being formed by reaction pathways b, d and e or f, respectively (Scheme 1). Diffusion from the cage (path d) is required for the generation of each of these products. For benzyl bromide, the radical path (b) was suggested for the formation of benzylplatinum bromide since solvent viscosity had no effect on the yield of the product. The benzyl chloride reaction was considered to proceed by the $S_N2$ path.

The reaction of various alkyl halides (MeI, $CD_3I$, EtI, $PhCH_2Br$ or $Ph_2CHBr$) with tris(triphenylphosphine)platinum(0) in the presence of $Bu^tNO$ gave e.s.r. spectra characteristic of the different alkyl moieties, trapped as $t$-butyl nitroso compounds, $Bu^tRNO$[31,169]. Similar experiments were carried out with other platinum(0) complexes ($[Pt(PPh_3)_n]$, $n=3, 4$; $[Pt(PPh_3)_2(C_2H_4)]$; $[Pt(PEt_3)_n]$, $n=3,4$) and with the nitrone trap, $PhCH{=}N(O)Bu^t$[170]. Control experiments were carried out to ensure that the spin trap was not altering the reaction course. When optically active 2-chloro- or-bromooctane was subjected to the reaction and then recovered before the reaction was complete, no loss in optical activity was observed.

The following non-chain radical mechanism was written, in accordance with these observations. It should be remembered, however, that the detection of radicals by e.s.r. does not require that the major reaction is proceeding via a radical mechanism involving the trapped or observed species.

$$[Pt(PPh_3)_3] \rightleftharpoons [Pt(Ph_3P)_2] + PPh_3 \tag{86}$$

$$[Pt(PPh_3)_2] + RX \overset{\text{slow}}{\rightleftharpoons} [Pt^{I}X(PPh_3)_2, R^{\cdot}] \tag{87}$$

$$[Pt^{I}X(PPh_3)_2, R^{\cdot}] \overset{\text{fast}}{\rightleftharpoons} [PtX(R)(PPh_3)_2] \tag{88}$$

On the other hand, the oxidative addition of a benzyl-type bromide, (α-bromoethyl)quinoline, to tris(triphenylphosphine)platinum(0) gives a stable, optically active oxidative addition product, the configuration of which was derived from Brewster's rules to have been formed with inversion of configuration at carbon[171]. In this case, the reaction could be greatly influenced by the chelating nitrogen, and even the expected[168] radical pair might maintain configurational identity until collapse to product.

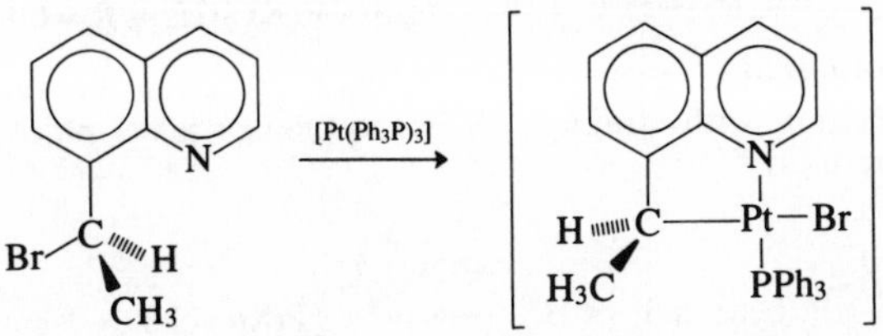

(89)

The reaction of tetrakis(triphenylphosphine)platinum(0) with acetyl chloride gives the acetyl oxidative addition product[141]. Acid anhydrides yield acyl products having the *trans* complex geometry; the use of cyclic anhydrides such as tetra-perfluorosuccinic anhydride gives the *cis*-cyclic product[172].

$$RCOX + [PtL_n] \longrightarrow \textit{trans}\text{-}[PtXL_2(COR)] \tag{90}$$

X = Cl, RCOO

There are a variety of other oxidative addition reactions that have been observed with platinum(0) complexes, some of thich undoubtedly proceed via a nucleophilic displacement reaction by platinum. Reaction of zerovalent phosphine complexes of platinum with cyanogen[136] or alkyl and aryl cyanides effects a cleavage of the

$$[PdL_4] + CMeC(CN)_3 \longrightarrow \textit{trans}\text{-}[Pt(CN)(L)_2\{C(Me)(CN)_2\}] \tag{91}$$

L = $PPh_3$

$$[PtL_3] + PhCN \longrightarrow \textit{trans}\text{-}[Pt(CN)(L)_2(Ph)] \tag{92}$$

L = $PEt_3$

carbon—carbon bond. The reaction of platinum(0) complexes with tetracyanoethylene oxide probably occurs by a nucleophilic displacement at carbon and the ultimate product depends on the ligands present[175]. With tetrakis(triphenylarsine)platinum, the platinocycle is isolated. With triethylphosphine as a ligand, the cyclic product is unstable and undergoes rearrangement.

$$[PtL_n] + \underset{(CN)_2}{\overset{(CN)_2}{C}}\!\!-\!O\!-\!C \longrightarrow [\overset{+}{P}tL_n\{C(CN)_2C(CN)_2O^-\}] \longrightarrow \left[ L_2Pt\underset{O}{\overset{C(CN)_2}{\diagdown\diagup}}C(CN)_2 \right] \longrightarrow [Pt(CN)(L)_2\{OC(CN){=}C(CN)_2\}] \quad (93)$$

Organotin compounds ($SnR_4$, $SnXR_3$ and $SnX_2R_2$) react with bis(ethylene)bis(phosphine)platinum(0) complexes by cleavage of the carbon—tin bond to yield an organoplatinum(II) complex, which has the *cis*-geometry in most cases[176–182]. In this reaction, aryl, alkyl, and allyl platinum(II) complexes can be obtained.

$$R_2SnXY + [Pt(PR'_3)(C_2H_4)_2] \longrightarrow cis\text{-}[Pt(PR'_3)_2(SnRXY)(R) \quad (94)$$

Vinyl platinum complexes can be prepared by the reaction of bisphosphine acetylene platinum(0) complexes with a strong acid such as hydrogen chloride or trifluoroacetic acid[183–186]. The reaction is stereospecific, protonation of the acetylene and carbon–platinum bond formation taking place by *syn* addition. With unsymmet-

$$[PtL_2(RC{\equiv}CR)] + HX \longrightarrow trans\text{-}[PtXL_2(cis\text{-}CR{=}CR'H)] \quad (95)$$

ric acetylenes, protonation occurs in a direction to yield the more stable vinyl cation[184]. The reaction with strained, cyclic acetylenes takes place so readily that weak acids can be used[186].

$$\left[ \left( \begin{matrix} L \\ L \end{matrix} \right) Pt(benzynyl) \right] + HX \longrightarrow \left( \begin{matrix} L \\ L \end{matrix} \right) PtX(cyclohex\text{-}1\text{-}enyl) \quad (96)$$

$$\widehat{L\quad L} = Ph_2PCH_2CH_2PPh_2$$

b. *Oxidative addition to $d^8$ complexes*

i. *Platinum(II)*. Although there is good evidence that oxidative addition of organic halides to $d^8$ complexes of nickel and palladium takes place, the resulting Ni(IV) and Pd(IV) complexes are unstable, and have not been isolated. Oxidative addition to platinum(II) complexes takes place readily and the resulting platinum(IV) products are stable, however (Table 5).

The reaction of methyl halides to both *cis*- and *trans*-bisphosphine methyl platinum halides gives the octahedral platinum(IV) complex[187–190]. The reaction generates the octahedral product in which the alkyl halide has added *trans*[189]. Arsine complexes behave similarly[190]. No reaction of methyl iodide or trifluoromethyl iodide occurs when the organo group on platinum is phenyl or trifluoromethyl[189].

$$trans\text{-}[PtXL_2(Me)] + RX' \longrightarrow \left[ \begin{matrix} & R & \\ Me & | & L \\ & Pt & \\ L & | & X \\ & X' & \end{matrix} \right] \quad (97)$$

Replacing the phosphine ligands with 3,5-lutidine or a chelating amine, tetramethylenediamine, apparently changes the geometry of the isolated oxidative addition product, giving complexes containing *trans* halogen, **41**[191].

[I, L, L, Me, Me, I around Pt]

**41**

Bisphosphinedialkylplatinum(II) complexes undergo the oxidative addition reactions with organohalides much more readily[189,192–196]. The reactions of bis(dimethylphenylphosphine)dimethylplatinum(II) have been studied the most extensively. Methyl iodide, acetyl chloride, and a variety of perfluoroalkyl halides give *trans* oxidative addition products, which may isomerize depending on the reaction conditions, particularly the solvent[189,193]. Changing the ligand also affects the

*cis*-$[PtL_2(Me)_2]$ + RX ⟶ [R, L, L, Me, Me, X around Pt] ⟶ [R, L, L, Me, Me, X around Pt] (98)

L = $Me_2PhP$ L = $Me_2PhAS$

stereochemistry of the final product. The analogous dimethylphenylarsine complex gives the product of *cis* oxidative addition while a bisnitrile complex gives *cis,trans* mixtures[189]. The stereochemistry of the oxidative addition reactions to *cis*-dimethylplatinum(II) complexes containing a chelating arsine ligand depends on the organic halide, methyl and acetyl halides generating the *trans* complex, while allylic halides give the *cis* oxidative addition product[197]. These complexes all can undergo

[(L L)Pt(Me)(Me)] + RX ⟶ [R, (L L), Me, Me, X around Pt] ⟶ [R, L, L, Me, X, Me around Pt] (99)

(L L) = $PhMeAs(CH_2)_2AsMePh$ R = $Me_3$, $CH_3CO$ R = $CH_2{=}CHCH_2$

isomerization. For example, from the oxidative addition product of perdeuteriomethyl iodide to *cis*-bis(dimethylphenylphosphine)dimethylplatinum(II), a *fac-trans* addition product is isolated which is stable at 33 °C but undergoes isomerization at 68 °C in deuteriochloroform to the *fac-trans*-isomer, producing a 1:2 ratio of isomers, respectively[196]. Generally, isomerization can be effected by removal of the halide (silver hexafluorophosphate), generating a solvated ionic complex[189].

[$CD_3$, L, L, Me, Me, I around Pt] ⇌ [Me, L, L, Me, $CD_3$, I around Pt] (100)

L = $Me_2PhP$

TABLE 5. Oxidative additions of organic halides to $d^8$ platinum(II) complexes

| Pt(II) | L | RX | Pt(IV) | Ref. |
|---|---|---|---|---|
| *trans*-[$PtIL_2(Me)$] | $PEt_3$ | RI | [R; Me, L; Pt; L, I, I] | 187 |
| | $P(Pr^n)_3$ | R = Me | | 187 |
| | $PMe_2Ph$ | R = Me<br>R = Me, $CF_3$ | | 188, 189 |
| | $AsR_3$ | MeI | [Me; Me, L; Pt; L, I, I] | 190 |
| *trans*-[$PtClL_2(Me)$] | $PMe_2Ph$<br>$P(CH_2{=}CHCH_2)_2Ph$<br>$PPh_2Me$ | MeI | [Me; Me, L; Pt; L, I, Cl] | 188 |
| *trans*-[$PtIL_2(Me)$] | 3,5-$Me_2$-pyridine | | [Me; Me, I; Pt; I, L, L] | 191 |
| [Me—Pt(L⌒L)—I] | $Me_2NCH_2CH_2NMe_2$ | | [Me; Me, I; Pt; I, L—L] | 191 |
| *cis*-[$PtL_2(Me)_2$] | $PMe_2Ph$ | RX<br>R = Me, X = I<br>R = $CD_3$, X = I | [R; Me, L; Pt; Me, X, L] | 188<br>193, 194, 196 |

| | | | | |
|---|---|---|---|---|
| | | R = $CF_3$, $CF_3CF_2$, $CF_3CF_2CF_2$, $CF_3(CF_2)_6$, X = I<br>R = $CF_3CF_2CF_2$, X = Br<br>R = $PhCH_2$, X = Br | | 189, 192 |
| | | | | 195 |
| *cis*-$[PtL_2(CD_3)_2]$ | $PMe_2Ph$ | MeI | $[Pt(CD_3)_2(CH_3)IL_2]$ | 196 |
| *cis*-$[PtL_2(Me)_2]$ | $P(CH_2{=}CHCH_2)_2Ph$<br>$PPh_2Me$ | MeI | $[PtMe_3IL_2]$ | 188 |
| *cis*-$[PtL_2(Ph)_2]$ | $PMe_2Ph$ | $CF_3I$ | $[Pt(CF_3)(Ph)_2IL_2]$ | 189 |
| *cis*-$[PtL_2(Me)_2]$ | $AsMe_3$ | RI<br>R = Me, $CF_3$ | $[PtRMe_2IL_2]$ | 189 |
| *cis*-$[PtL_2R_2]$<br>R = Me, Ph | 4-cyanotoluene | R′I<br>R′ = Me, $CF_3$ | $[PtR'R_2IL_2]$ | 189 |
| *cis*-$[PtL_2(Me)_2]$ | $AsMe_2Ph$ | $R_fCH_2I$<br>$R_f$ = $CF_3$, $CF_3CF_2$, $CF_3CF_2CF_2$ | $[PtR_fMe_2IL_2]$ | 192 |

TABLE 5. ctd.

| Pt(II) | L | RX | Pt(IV) | Ref. |
| --- | --- | --- | --- | --- |
| *cis*-$[PtL_2(C_6H_4Me)_2]$ ($o,p$-$C_6H_4Me$) | pyridine | RI R = Et, Pr | $[Pt(R)(MeC_6H_4)(I)(I)(Py)(Py)]$ | 198 |
| | | $CH_2{=}CHCH_2$ | $[Pt(C_3H_5)(C_3H_5)(I_3)(I_3)(py)(py)]$ | 198 |
| $[PtMe_2(L\frown L)]$ | 2,2′-bipyridine | MeI | $[Pt(Me)(Me)(Me)(I)(L\frown L)]$ | 201 |
| $[PtPh_2(L\frown L)]$ | 2,2′-bipyridine<br>1,10-phenanthroline | MeI | $[Pt(Ph)(Ph)(Me)(I)(L\frown L)]$ | 199, 200 |

| | | | | |
|---|---|---|---|---|
| 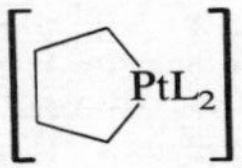 [platinacyclopentane $PtL_2$] | cod | MeX<br>X = Br, I | 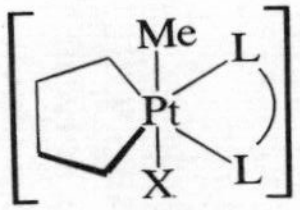 [Me, X, L–L chelate on Pt in platinacyclopentane] | 202 |
| | $PMe_2Ph$ | | 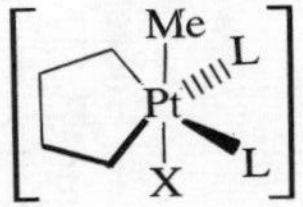 [Me, X, L, L on Pt in platinacyclopentane] | |
| *cis*-$[PtL_2(Me)_2]$ | 4-cyanotoluene | $CH_3COCl$ | 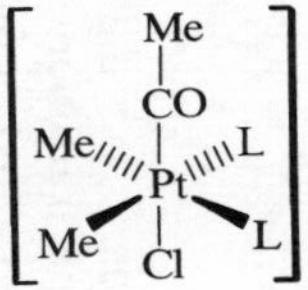 [Me–CO, Cl, Me, Me, L, L on Pt] | 189 |
| 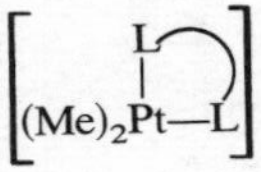 [$(Me)_2Pt$ with L–L chelate] | $PhMeAsCH_2CH_2AsMePh$ | RX<br>R=$CH_3$,<br>$CH_2$=$CHCH_2$, $CH_3CO$ | 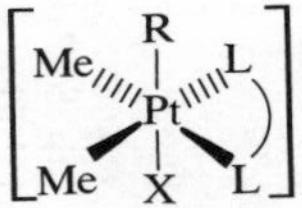 [R, X, Me, Me, L–L chelate on Pt] | 197 |

Changing the *cis* organic groups to phenyl gives an oxidative addition product with methyl iodide (*cis,trans* mixture) that is too unstable to isolate, and undergoes a reductive elimination of toluene[189]. The reaction of methyl iodide with *cis*-bis(4-methylbenzonitrile)dimethylplatinum(II) shows second-order kinetics and an activation energy of 8.6 kcal mol$^{-1}$ in chloroform[189].

The reactions of *cis*-dimethyl- or -diaryl-platinum(II) complexes, containing pyridine or bipyridine ligands, with a variety of organic iodides also gives products of *trans*-oxidative addition[198–201]. Alkyl, aryl, and allyl iodides react react readily[198].

$$[(N)_2PtR_2] + R'I \longrightarrow [(N)_2PtR_2(R')(I)] \qquad (101)$$

The reaction with methyl iodide is second order, and the rate increases with increasing solvent polarity, indicating an $S_N2$-type transition state[200]. The solvent effect on the rate of oxidative addition of methyl iodide to *cis*-diphenylbipyridylplatinum(II) is dominated by the entropy of activation term, which is typical of reactions proceeding through a highly polar transition state[201]. Diarylplatinum(II) complexes containing electron-donating groups on aryl react faster than those containing electron-withdrawing groups, a consequence of increase electron density on platinum[201].

A platinacyclopentane complex containing dimethylphenylphosphine or cyclooctadiene ligands also yields the *trans*-oxidative addition product on reaction with methyl halide[202].

$$[\text{C}_4\text{H}_8\text{PtL}_2] + \text{MeX} \longrightarrow [\text{C}_4\text{H}_8\text{Pt(Me)(I)L}_2] \qquad (102)$$

c. *Oxidative additions to complexes of the cobalt and iron triads.* Two-electron oxidative additions to $d^8$ complexes of the remaining Group VIII transition metals are common, and for a number of the transition metals the reactions of anionic complexes with organic halides are frequently used for the synthesis of metal alkyls. These reactions of anionic complexes are most prevalent with the first-row transition metals, including the earlier transition metal complexes of chromium and molybdenum. The reactions of anionic complexes of this type {$[MCp(CO)_3]^-$, M = Cr, Mo, W; $[MCp(CO)_2]^-$, M = Fe, Ru; $[Fe(CO)_4]^{2-}$; $[M(CO)_5]^-$, M = Mn, Re; $[Co(CO)_4]^-$; $[Co(PPh_3)(CO)_3]^-$} with organic halides have been reviewed[203–206]. Consequently, these will not be covered in detail here; only the more recent reactions of this type that provide new synthetic or mechanistic information will be discussed in detail. The relative nucleophilicities of some of these anion are $[FeCp(CO)_2]^- = 7 \times 10^7$, $[RuCp(CO)_2]^- = 7.5 \times 10^6$, $[NiCp(CO)]^- = 5.5 \times 10^6$, $[Re(CO)_5]^- = 2.5 \times 10^4$, $[WCp(CO)_3]^- = 5 \times 10^2$, $[Mn(CO)_5]^- = 77$, $[MoCp(CO)_3]^- = 67$, $[CrCp(CO)_3]^- = 4$, and $(Co(CO)_4]^- = 1$[206].

i. *Cobalt.* Cobalt(I) complexes containing the $d^{10}$ (anionic) and $d^8$ (neutral and anionic) configurations undergo two-electron oxidations, while cobalt(II) $d^7$ complexes undergo one-electron reactions to produce cobalt(III) complexes. The $d^{10}$ cobalt(I) tetracarbonyl anion was shown early on to react with the more reactive alkyl halides, allyl halides, and acid chlorides[207,208]. The acyl derivatives obtained from perfluoro acid chlorides are unstable above 0°C,

$$[Co(CO)_4]^- + RCH{=}CHCH_2X \longrightarrow [Co(CO)_4(\eta^3\text{-1-R-alkyl})] \qquad (103)$$

R = H, $CH_3$

and decarbonylate giving the perfluoro alkyl complex[208a]. The addition of methyl iodide in the presence of added ligand (triphenylphosphine) leads to an acyl complex. Cyclopropenyl bromide and the trimethylcyclopropenyl cation give allyl cobalt species in which carbon monoxide has 'inserted' into the three-membered ring[209]. The reaction of cobalt carbonyl with the triphenylcyclopropenyl cation, however, gives only the cyclopropenyl complex. Hydridocobaltcarbonyl or the corresponding

$$\text{Ph}_2\text{C}_3(\text{Ph})\text{Br} + [Co(CO)_4]^- \longrightarrow [R_3C_3CO\text{-}Co(CO)_3] \qquad (104)$$

$$R_3C_3^+ + [Co(CO)_4]^- \xrightarrow{R=Ph} [Ph_3C_3Co(CO)_2] \quad (R=CH_3 \text{ gives the } R_3C_3CO\text{-}Co(CO)_3 \text{ product})$$

anion reacts with oxiranes to yield the hydroxyethyl cobalt(I) product[210] (overall this cannot be considered an oxidative addition reaction). The reaction with cyclohexene oxide gives the *trans* product.

$$[CoH(CO)_4] \underset{+H^+}{\overset{-H^+}{\rightleftharpoons}} [Co(CO)_4]^- \xrightarrow{\text{oxirane}} [Co(CO)_4(CH_2CH_2OH)] \qquad (105)$$

Anionic complexes containing phosphine and phosphite ligands also react with acid derivatives[211,212], allyl iodides, benzyl iodide, and methyl iodide[213]. The reaction with ethyl chloride produces only ethylene and a hydridocobalt complex as a result of $\beta$-elimination[213].

$$[CoL_4]^- \quad L = P(OMe)_3 \qquad (106)$$

$$+PhN_2^+BF_4^- \longrightarrow [CoL_4(Ph)]$$

$$+RCH{=}CHCH_2I \xrightarrow{-L} [(\eta^3\text{-RC}_3H_4)CoL_3]$$

$$+PhCH_2I \longrightarrow [(\eta^3\text{-PhCH}_2)CoL_3]$$

Allyl and perfluoroalkyl halides add to neutral cobalt(I) $d^8$ complexes[214–217], particularly $\eta^5$-cyclopentadienylcobalt dicarbonyl[214–216]. The predominant cobalt(III) $d^6$ product of reactions with alkyl iodide is an 18-electron cationic complex[216]. The

$$[CoCp(CO)_2] + CH_2{=}CHCH_2I \longrightarrow \left[ (\eta^5\text{-}C_5H_5)Co(CO)(\eta^3\text{-}C_3H_5) \right]^+ I^- \quad (107)$$

reaction of alkyl halides with the analogous complex containing a phosphine ligand in place of one of the carbonyls gives the acyl product resulting from carbonyl insertion[218]. The reaction rates are greater for methyl than ethyl (400–1200 times),

$$[CoCp(CO)(L)] + RI \longrightarrow [CoCp(COR)(L)] \quad (108)$$

$L = PPh_3$, $PPhMe_2$, $PCy_3$; $R = CH_3$, $C_2H_5$

increasing with increasing solvent polarity (thf = 0.2, $Me_2CO = 0.9$, $CH_2Cl_2 = 1$, $CH_3CN = 2–3$) and decreasing in the order $L = PhPMe_2 > PPh_2Me = (10.7) > PPh_3 = (1) > P(C_6H_{11})_3$. These data plus a large negative entropy of activation ($\Delta S^{\ddagger} = -31$ e.u.) suggest an $S_N2$-type reaction[218].

Cobalt(I) complexes of the $d^8$ configuration containing nitrogen ligands, particularly ligands of the porphyrin and bisdimethylglyoximate type, have been studied extensively because they serve as models for vitamin $B_{12}$[219]. Anionic salen-type complexes[220–222] and bae-type complexes[223,224] undergo oxidative reactions with a variety of organic halides, including vinyl chloride[222].

The oxidative addition reactions of organic halides to the [bis(dimethylglyoximato)]cobalt(I) anion (cobaloxime) have been studied by a number of groups. Generally, oxidative addition takes place to give the neutral, $d^6$ cobalt(III) complex. Alkyl, benzyl, allyl, vinyl, and acyl cobalt(III) complexes as well as hydroxyethyl

$$[Co(dmgH)_2(L)]^- + RX \longrightarrow [CoR(dmgH)_2(L)] + X^- \quad (109)$$

$L = Py$, $H_2O$, $PBu^n_3$, $PhNH_2$

$R = Me$, Et, $Pr^n$, $Bu^n$, $C_5H_{11}{}^n$, $Pr^i$, $Bu^i$, $CH_2CN$, $CH_2OMe$, $CH_2CONH_2$, $CH_2{=}CHEt$

$CH_2{=}CHCHMe_2$, $PhCH_2$, $p\text{-}Bu^tC_6H_4CH_2$, $Ph_2CH$, PhCH(Me), $C_{10}H_6CH_2$,

$HO(CH_2)_n$, $HC{\equiv}CCH_2$, $R_1R_2C{=}C{=}CH(R_1 = R_2 = H, Me)$, $MeC{\equiv}CCH_2$,

1-ethynylcyclohexyl;

$X = Cl$, Br, I

derivatives have been obtained by oxidative addition of the appropriate halides (or epoxides)[225–233]. Although primary, secondary and tertiary alkyl halides react, secondary and tertiary halides, for example isopropyl or $t$-butyl, tend to give unstable alkyl cobaloximes, decomposition to give $\beta$-elimination products occurring. Since the nucleophilicity of cobalt in these complexes is dependent on the energy of the $3d_z{}^2$ orbital and the charge density on cobalt, the reaction is sensitive to the axial ligand structure, but relatively insensitive to the glyoxime ligand[233]. Propynyl halides give a product of rearrangement, an allenyl cobalt complex[231,232]. When the acetylenic group is not terminal, no such rearrangement takes place[232].

$$[Co(dmgH)_2py]^- \begin{cases} +HC{\equiv}CCH_2X \longrightarrow [CH_2{=}C{=}CH{-}Co(DMGH)_2py] \\ (\text{or } CH_2{=}C{=}CHCl) \\ +MeC{\equiv}CCH_2Cl \longrightarrow [MeC{\equiv}CCH_2{-}Co(dmgH)_2py \end{cases} \quad (110)$$

A porphyrin complex, similar to the cobaloximes but closer to the $B_{12}$ structure, also undergoes an oxidative addition reaction with alkyl and acyl halides (or anhydrides) to give the cobalt(III) products[234–236].

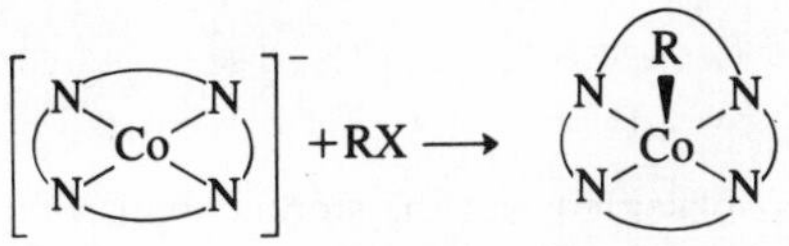

$$[\text{N}_4\text{Co}]^- + RX \longrightarrow \text{N}_4\text{Co–R} \quad (111)$$

$R{=}CH_3, C_2H_5, CH_3CO$

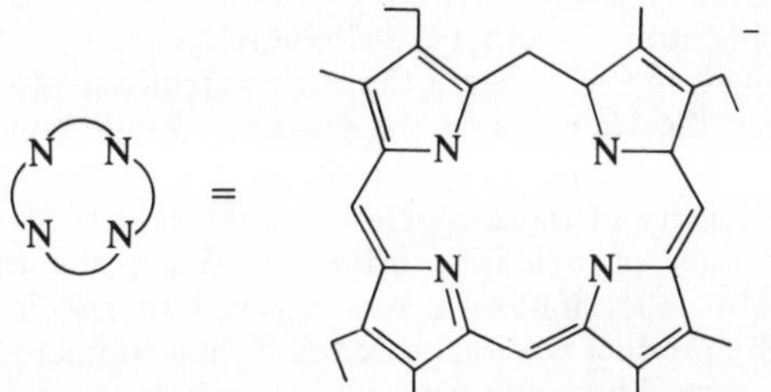

The reactions with the cobaloximes are second order, but this is a case in which the mechanism of the reaction ($S_N2$ or radical) may be dependent on the structure of the substrate. The rates have been reported[233] to decrease in the order I > Br > Cl, primary > secondary, and show large negative entropies of activation (−20 to −30 eu).

The rate of the reaction of *n*-butyl bromide with cobaloxime is half that of 1,4-dibromobutane[237]. The sensitivity of the reaction rates, and possibly the mechanism, to the structure of the organohalide is illustrated by the reaction of a cobalt(I) phthalocyanine anion[238]. The rates of alkylation by *n*-butyl halides decrease in the order I $(6\times10^3)$ > Br $(3\times10^2)$ > Cl (1). Introduction of a 1-phenyl or a 1-oxo substituent increases the rates by a factor of $10^4$–$10^5$, while branching in the 2-position decreases the rates. The relative rates for reactions of alkanes containing mono-, di- and tri-chloro-substituted carbons decrease in the order $RCHCl_2 > RCH_2Cl > RCCl_3$, suggesting a mechanistic change in the series.

The reaction is stereospecific with a number of substrates, however. The reactions of $[Co(dmgH)_2py]^-$ with *cis*- and *trans*-$\beta$-chloroacrylate[239], $\beta$-bromostyrene[239,240], and 1-bromooctene[241] give products of retention of geometry at the vinyl carbon. A reaction pathway proceeding via an elimination to give phenylacetylene followed by

$$[Co(dmgH)_2py]^- + cis\text{-}PhCH{=}CHBr \longrightarrow [cis\text{-}PhCH{=}CH{-}Co(dmgH)_2py] \quad (112)$$

a stereospecific addition of $[Co(dmg)_2py]^-$ does not take place, since the addition of this cobalt(I) complex to phenylacetylene is not regiospecific. A free-radical mechanism also is excluded, since the vinyl radical undergoes $sp^2$ inversion faster ($k_1 = 10^9 s^{-1}$) than radical pair coupling could be expected to take place. A nucleophilic addition–elimination mechanism as well as a concerted reaction (of the olefin–cobalt complex) similar to that proposed for $d^{10}$ complexes in the nickel triad (see above) are consistent with the observed stereochemistry.

There is sufficient evidence to suggest that the reactions of alkyl halides proceed by an $S_N2$-type mechanism[242]. Inversion of configuration at carbon in reactions with cyclohexyl bromides has been observed[243]. Stereospecific oxidative addition of chiral

$$[Co(dmgH)_2py]^- + Y\text{-}C_6H_{10}\text{-}X \longrightarrow [Y\text{-}C_6H_{10}\text{-}Co(dmgH)_2py] \quad (113)$$

X = Y = BR; X = OTs, Y = OH

2-bromooctane also was observed, but the stereochemistry of the reaction (inversion or retention) could not be determined[244]. Cleavage of the cobalt—2-octyl bond was carried out with halogen, which regenerated the starting material of the same absolute configuration, but the stereochemistry of the cleavage reaction is unknown. A similar reaction with chiral 1-methyl-2,2-diphenylcyclopropyl bromide gave an oxidative addition product, but here again the stereochemistry of the reaction could not be determined since the cleavage of the carbon—cobalt bond gave a racemic product, raising the question of the racemization step[245].

Loss of the stereospecificity of these reactions could be due also to racemization occurring with the product of oxidative addition. When the oxidative addition product of *trans*-1,4-dibromocyclohexane was allowed to age in the presence of $[Co(dmgH)_2py]^-$, and the product was reisolated and then subjected to cleavage with bromine, the resulting *trans*-dibromide was substantially racemized[246]. This racemization reaction has been attributed to a nucleophilic exchange process[247] on the basis of the stereochemical evidence and the absence of radical reactions as indicated by the usual tests. Although this evidence supports a typical $S_N2$ type mechanism for cyclohexyl halides, evidence for a 1-electron transfer reaction has been reported[248].

$$[py(dmgH)_2Co^*\text{—}CHRR'] + [Co(dmgH)py]^- \rightleftharpoons [py(dmgH)_2Co\cdots \overset{*}{C}(H)(R')(R)\cdots Co(dmgH)_2py]^- \rightleftharpoons [py(dmgH)_2Co^*]^- + [R'(H)(R)C\text{—}Co(dmgH)_2py] \quad (114)$$

Cobaloxime displaces the triflate anion from *threo*-2,3-dideuterio-3-*t*-butyl triflate with 95% inversion of configuration at carbon as determined by $^1H$ n.m.r.[249].

$$Bu^t(H)(D)C\text{—}C(H)(D)OTf + [Co(dmgH)_2py]^- \longrightarrow [Bu^t(D)(H)C\text{—}C(H)(D)Co(dmgH)_2py] \quad (115)$$

The reactions of cobalamine complexes with alkyl halides serve as models for vitamin $B_{12s}$ (reduced). The reactions of $B_{12s}$ with alkyl halides serve as a synthetic pathway to alkylated $B_{12}$. Alkyl, allyl, and acyl halides as well as ethylene oxide give remarkably stable products; bulky alkyl halides and phenyl bromide do not react. The literature on this subject is extensive and often misleading. A number of reviews have appeared[250–252].

The odd-electron ($d^7$) pentacyanocobalt(II) anion reacts by a one-electron charge to give $d^6$ alkylcobalt(III) complexes that are stable, provided that a $\beta$-hydrogen is not present in the alkyl group[4,253–258]. The reactions occurs with a wide variety of organic halides, including $\alpha$-haloacid derivatives and alkyl, allyl, benzyl, vinyl, and pyridyl halides[245,255]. The other reaction product is the pentacyano(halo)cobalt(III) anion. This second-order reaction shows reactivity patterns that are characteristic of the reaction of radicals with alkyl halides, $PhCH_2I > Bu^tI > Pr^iI > EtI > MeI$[257], and in general reveals an inverse rate dependence with bond strength[254,257]. The rate-determining step involves halogen atom abstraction from the organic halide by cobalt[4,257,258].

$$2[Co(CN)_5]^{3-} + RX \longrightarrow [RCo\text{–}CN)_5]^{3-} + [XCo(CN)_5]^{3-} \quad (116)$$

$$[Co(CN)_5]^{3-} + RX \xrightarrow[\text{determining}]{\text{rate}} [CoX(CN)_5]^{3-} + R^{\bullet} \quad (117)$$

$$R^{\bullet} + [Co(CN)_5]^{3-} \longrightarrow [CoR(CN)_5]^{3-} \quad (118)$$

The neutral bis(dimethylglyoximato)cobalt(II) complexes containing an axial ligand such as pyridine or triphenylphosphine react with benzyl halides by a similar radical mechanism[259]. The rates increase slightly with the basicity of the axial ligand; $\alpha$-phenylethyl bromides reacts faster than benzyl bromide. The reaction of bipyridyl(or phenanthroline)cobalt(II) with benzyl chloride in the presence of borohydride leads to the dibenzylcobalt(III) cation, **42**[260].

$$2[Co^{II}(dmgH)_2L] + PhCH_2Br \longrightarrow [Co(CH_2Ph)(dmgH)_2(L)] + [CoBr(dmgH)_2(L)] \quad (119)$$

$$\left[ (N\frown N)Co(CH_2Ph)_2 \right]^+ Cl^- \qquad N\frown N = \text{bipyridyl, phenanthroline}$$

**42**

Pentafluorobromobenzene reacts with cobalt ($d^9$) atoms (vapour deposition) at −196 °C to yield cobalt(II) bromide and bis(pentafluorophenyl)cobalt(II)[261]. Washing this complex with toluene generates the 17-electron $\eta^6$-arene complex.

$$2Co + 2C_6F_5Br \xrightarrow{-196\,^\circ C} [Co(C_6F_5)_2] + CoBr_2 \xrightarrow{PhMe} [Co(C_6F_5)_2(PhMe)] \quad (120)$$

ii. *Rhodium*. The reactions of chlorotris(triphenylphosphine)rhodium(I) (Wilkinson's complex) have been reviewed[262], as have the decarbonylation reactions of acid chlorides and aldehydes with this catalyst[263]. Homogeneous rhodium catalysts have gained importance industrially, not only owing to the ability of certain complexes to effect hydrogenation and hydroformylation reactions of olefins, but also because of their ability to carbonylate methanol to acetic acid.

Halocarbonylbis(phosphine)rhodium(I) complexes react only with the most reactive organic halides. Methyl iodide, benzyl halides, iodomethyl acetate, acyl halides, and allyl bromide give the octahedral rhodium(III) products[264,265].

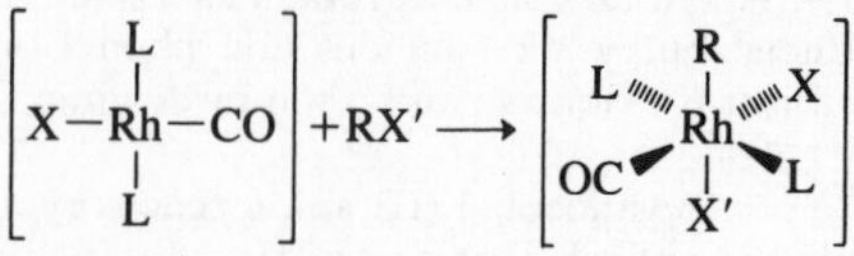

$$\left[X{-}Rh(L)_2{-}CO\right] + RX' \longrightarrow \left[Rh(R)(X)(L)_2(OC)(X')\right] \tag{121}$$

$L = Bu^n_3P$, $Me_2PhP$, $Me_2PhAs$;
$RX' = CH_3I$, $PhCH_2Cl$, $PhCH_2Br$, $ICH_2CO_2Me$, $CH_2{=}CHCH_2X$ ($X = Cl$, Br, I), RCOCl ($R = Me$, $CH_2{=}CH$, $CH_2{=}CHCH_2$, $Pr^i$)
$X = Cl$, Br, I, $N_3$

The reaction of methyl iodide with chlorocarbonyl(bisphosphine)rhodium is second order, with the better donor phosphines providing faster rates. The entropies of activation are large and negative: $L = P(p\text{-}MeOC_6H_4)$, −44 eu; $PPh_3$, −43 eu; $P(p\text{-}FC_6H_4)_3$, −29 eu; $AsPh_3$, −44 eu. The oxidative addition reactions occur only very slowly with methyl bromide but trifluoromethyl iodide, ethyl iodide, and butyl iodide do not react at 25 °C.

It has been shown that this reaction between methyl iodide and chlorocarbonyl-bis(trialkylphosphine)rhodium is not as straightforward as was originally thought. Not only is the *trans*-rhodium(III) product obtained, but the rhodium(III) acyl complex also is produced[267]. A mechanism was proposed in which methyl iodide reacts with dissociated phosphine to give a phosphonium iodide. This iodide then complexes with rhodium to give an anionic complex.

Replacement of halogen with vinyl gives a rhodium(I) species that undergoes the oxidative addition of methyl iodide to give a stable 18-electron diorganorhodium(III) complex containing *cis*-vinyl and -methyl groups[268].

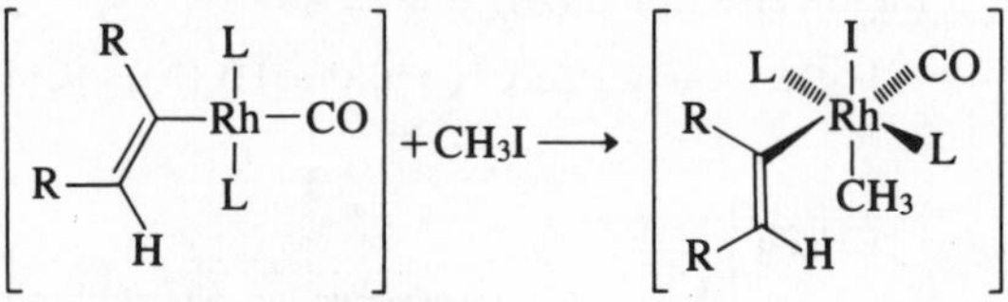

$$\left[(RHC{=}CR){-}Rh(L)_2{-}CO\right] + CH_3I \longrightarrow \left[(RHC{=}CR)Rh(I)(L)_2(CO)(CH_3)\right] \tag{122}$$

$L = Ph_3P$; $R = CO_2Me$

Monophosphinedicarbonylrhodium halide complexes also add reactive organic halides[269,270]. The exact geometry of the products has not been determined in most cases; in the reaction product shown, the carbonyls are *trans*. When R is methyl, rearrangement to the acyl complex takes place. The reaction with methyl iodide ($X = Cl$) is second order, the rates increasing with increasing solvent polarity[271]. The large negative entropy of activation ($-33 \pm 5$ eu) is characteristic of an $S_N2$-type reaction involving generation of charge in the transition state and ordering of solvent.

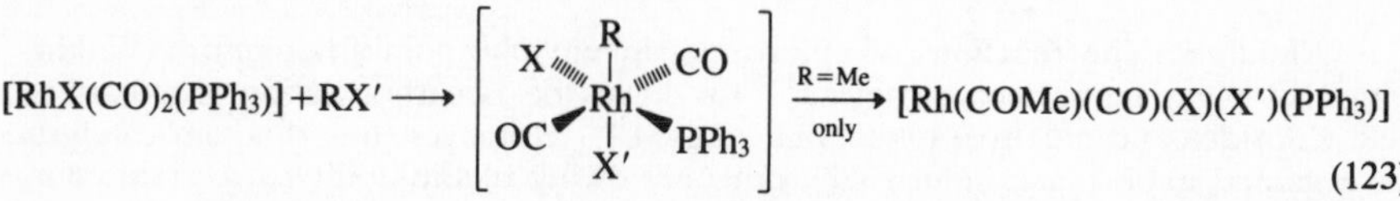

$$[RhX(CO)_2(PPh_3)] + RX' \longrightarrow \left[Rh(R)(X)(CO)(OC)(PPh_3)(X')\right] \xrightarrow[\text{only}]{R=Me} [Rh(COMe)(CO)(X)(X')(PPh_3)] \tag{123}$$

$X = Cl$, Br, I; $R = Me$, Ph, $CH_2CO_2Et$, $CH_3CO$

Perfluoropropyl iodide and pentafluorobenzoyl chloride react with acetylacetonatobis(diphenylmethylphosphine)rhodium(II) to give the perfluoro-alkyl and -acyl complexes, **43**[272].

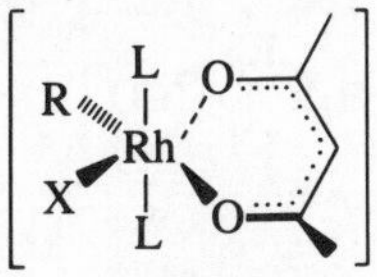

$R = CF_3CF_2CF_2$, X = I
$R = C_6F_5CO$, X = Cl
R = ($F_2$, $F_2$, Cl cyclobutenyl), X = Cl

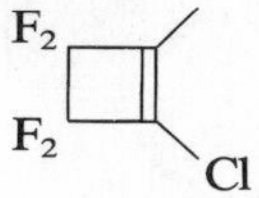

**43**

Oxidative additions to Wilkinson's complex take place only with the most reactive organic halides, methyl iodide, allyl halides, benzyl halides, and acid halides. Methyl iodide gives a product containing a molecule of coordinated methyl iodide[273,274], **44**, which is lost on recrystallization. The structure of this five-coordinate 16-electron

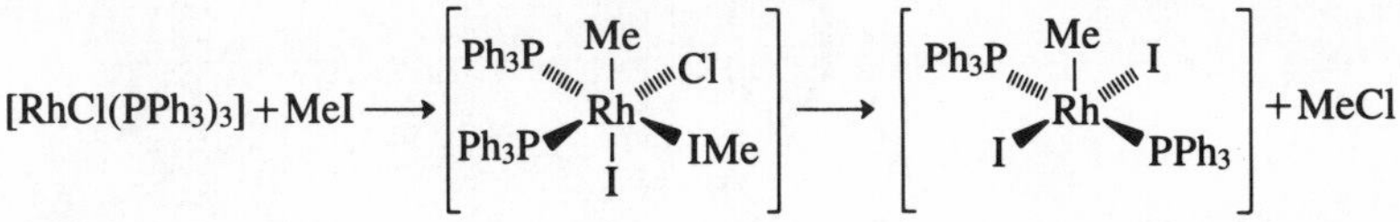

$$[RhCl(PPh_3)_3] + MeI \longrightarrow [RhCl(Me)(I)(IMe)(PPh_3)_2]\ (\mathbf{44}) \longrightarrow [RhMeI_2(PPh_3)_2] + MeCl \qquad (124)$$

complex has been determined by X-ray analysis[275]. Allyl chlorides give an initial $\sigma$-complex that rearranges to an 18-electron $\pi$-allyl complex[158,273,274]. Structures containing *cis*[274] and *trans*[158] halogens, **45** and **46**, have been proposed.

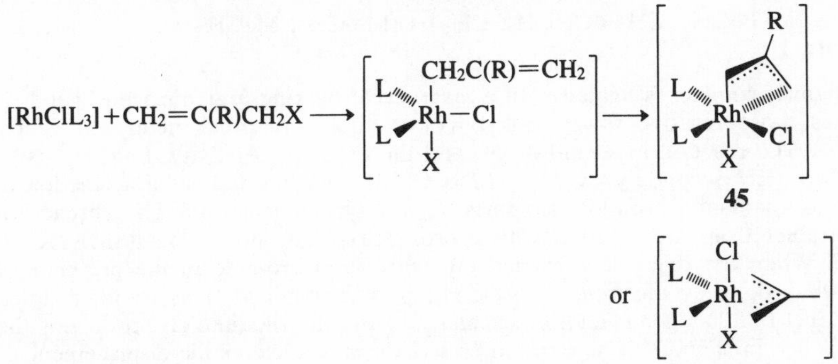

$$[RhClL_3] + CH_2{=}C(R)CH_2X \longrightarrow [L_2Rh(Cl)(X)(CH_2C(R){=}CH_2)] \longrightarrow \mathbf{45}\ \text{or}\ \mathbf{46} \qquad (125)$$

$L = PPh_3$; $R = H, CH_3$; X = Cl, Br

Benzyl chloride reacts to give a $\pi$-benzyl complex containing only one phosphine ligand, which gives the coordinatively saturated $\sigma$-benzyl complex on the addition of carbon monoxide or a coordinatively saturated $\pi$-benzyl complex on the addition of pyridine[276].

The $\eta^5$-cyclopentadienyl rhodium complexes also react only with the most reactive organic halides, the ultimate product of the reaction usually being that of carbon monoxide insertion[218,277,278]. Thus, methyl and ethyl iodide, benzyl halides, and allyl halides undergo this reaction, the first step giving the alkyl complex followed by conversion to the acyl complex. For allyl iodide, the first step is fast followed by a slow second step; thus the intermediate can be isolated[277]. The reaction with acid

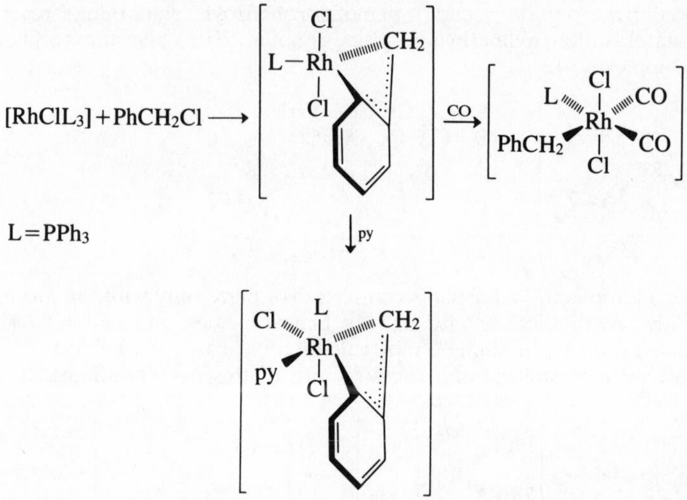

$$[RhClL_3] + PhCH_2Cl \longrightarrow [\text{L–Rh(Cl)}_2(\eta^3\text{-CH}_2C_6H_5)] \xrightarrow{CO} [\text{Rh(Cl)}_2(\text{L})(\text{CO})_2(\text{PhCH}_2)]$$

$L = PPh_3$

$$\downarrow py \qquad [\text{Rh(L)(Cl)}_2(\text{py})(\eta^3\text{-CH}_2C_6H_5)] \qquad (126)$$

halides ($CH_3COBr$, $CF_3COCl$) gives stable acyl complexes, $[RhCp(CO)(PPhMe_2)(COR)]^+X^-$ [278].

$$[RhCp(CO)(L)] + RX \longrightarrow [RhCp(Co)(L)(R)]^+X^- \longrightarrow [RhCl(Cp)(PPh_3)(COR)] \quad (127)$$

$L = PPh_3$; $R = PhCH_2$, $CH_2{=}CHCH_2$, $CH_2{=}C(Me)CH_2$, $MeCH{=}CHCH_2$, Me, Et; $X = Cl$, Br, I

Rhodium(I) complexes ligated with a macrocycle bearing four nitrogen ligands is $10^4$ times more reactive toward alkyl halides than any other neutral $d^8$ complex[279,280]. The reaction is second order, and the order of reactivity, I > OTs ≈ Br > Cl and Me > Et > secondary > $C_6H_{11}$ > $Me_3CCH_2Br$, and the failure of a reaction to occur with adamantyl bromide, supports an $S_N2$-type mechanism. The absence of cyclic product from the reaction with 6-bromohex-1-ene argues against a radical reaction. When the reaction is carried out with butyl bromide in the presence of lithium chloride, only the *trans* chloride complex is obtained. This result, coupled with the fact that halogen cannot be exchanged from the rhodium(II) product, is the best evidence in support of an ionic intermediate in a nucleophilic displacement.

Oxidative addition of methyl iodide to an anionic rhodium(I) complex (**47**) is a key step in the conversion of methanol to acetic acid by carbonylation[281] (Scheme 4). In the catalytic cycle shown, methyl iodide is generated by the reaction of methanol with hydrogen iodide. Thus, the reaction must be started with catalytic quantities of methyl iodide or hydrogen iodide.

The oxidative addition of acid chlorides to Wilkinson's complex is the first step in catalytic and stoichiometric decarbonylation of acid chlorides[262,263,282]. In almost every case, under the appropriate reaction conditions, the acyl complex can be isolated.

The reaction of acetyl chloride to tris(triphenylphosphine)rhodium chloride initially gives the *cis*-square-pyramidal complex in solution, which can be isolated as such[283]. In chloroform, at ambient temperature, isomerization to the *trans* complex takes place irreversibly. The acyl complex obtained from 3-phenylpropionyl chloride

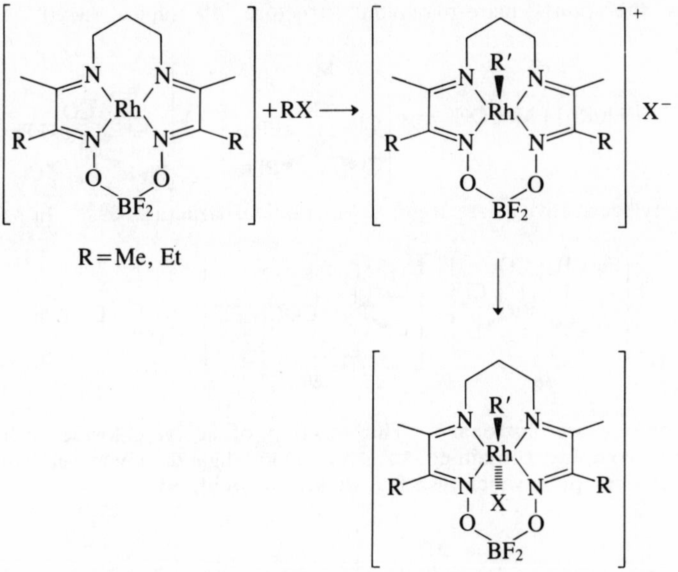

N N Rh R N N R O O $BF_2$ + RX ⟶ [ N R′ N Rh R N N R O O $BF_2$ ]$^+$ $X^-$

R = Me, Et

(128)

N R′ N Rh R N X N R O O $BF_2$

R′X = MeCOCl, PhCOCl, Me, EtI, $PhCH_2Cl$, $C_6H_{11}Br$, $Bu^nBr$

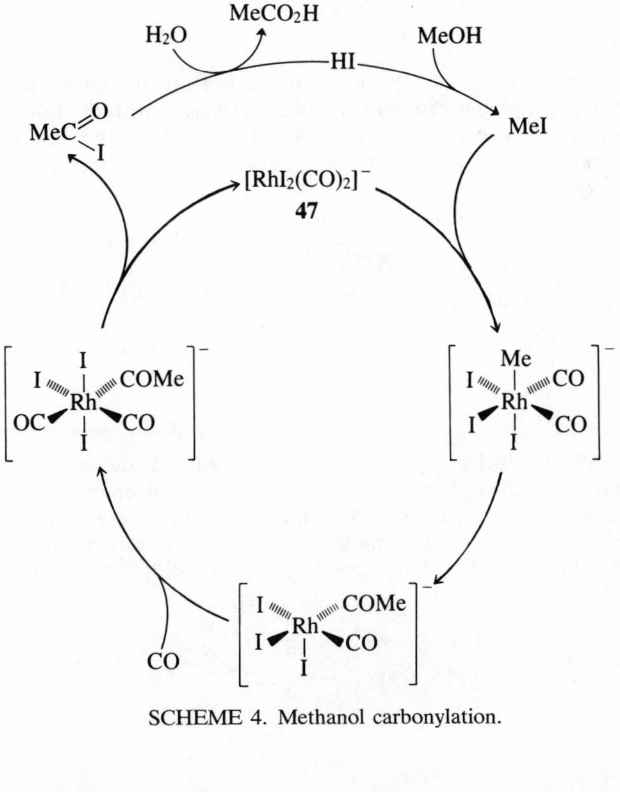

$MeCO_2H$

$H_2O$ MeOH

HI

MeC(=O)I MeI

$[RhI_2(CO)_2]^-$

**47**

$[RhI_3(COMe)(CO)_2]^-$: I, I, COMe, Rh, OC, CO, I

$[RhMeI_3(CO)_2]^-$: Me, I, CO, Rh, I, CO, I

$[RhI_3(COMe)(CO)]^-$: I, COMe, Rh, I, CO, I

CO

SCHEME 4. Methanol carbonylation.

also has the *trans*-square-pyramidal structure **48** (apical acyl)[283] while the

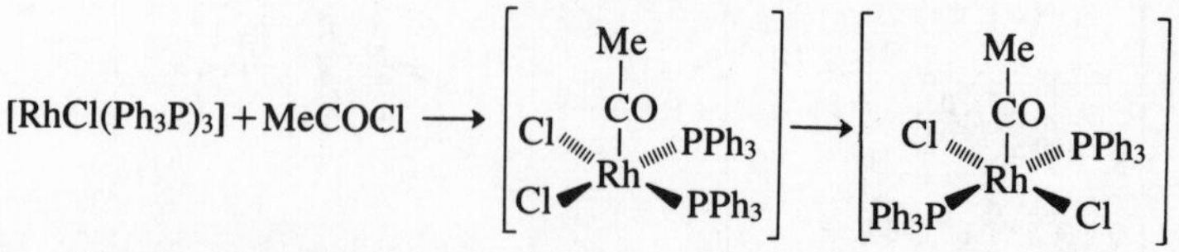

$$[RhCl(Ph_3P)_3] + MeCOCl \longrightarrow [RhCl_2(COMe)(PPh_3)_2] \longrightarrow [RhCl_2(COMe)(PPh_3)_2] \quad (129)$$

phenylacetyl derivative has a trigonal bipyramidal structure **49**[284]. In solution, the

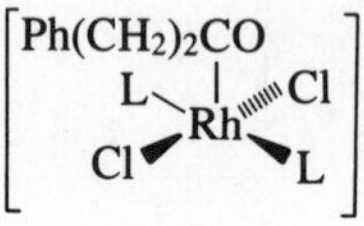

$$[Ph(CH_2)_2CO\,RhCl_2L_2] \quad (\mathbf{48}) \qquad [L_4Rh{-}COCH_2Ph] \quad (\mathbf{49}) \qquad L = PPh_3$$

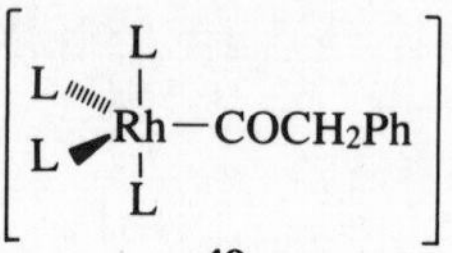

acyl complexes are fluxional[285]. The reaction of acetyl chloride with a similar rhodium(I) complex containing stronger donor ligands gives the cationic acyl rhodium(III) complex, which also is a square pyramid, **50**.

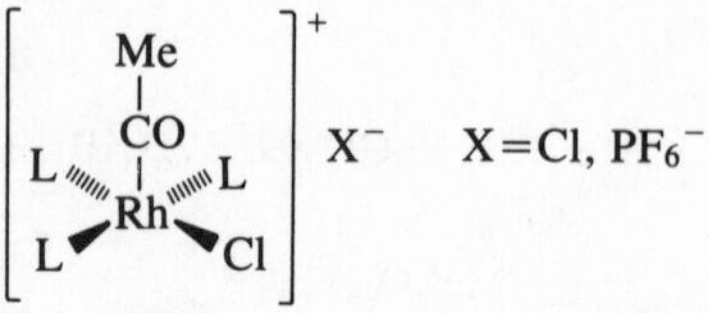

$$[MeCO\,RhL_3Cl]^+ X^- \quad X = Cl, PF_6^- \quad (\mathbf{50})$$

The rapid reaction rates have been attributed both to the ability of the filled $d_{z^2}$ orbital on the square-planar rhodium complex to overlap with the lowest unoccupied molecular orbital of the carbonyl carbon, **51**[263], and to the intermediacy of a zwitterionic complex, **52**[7].

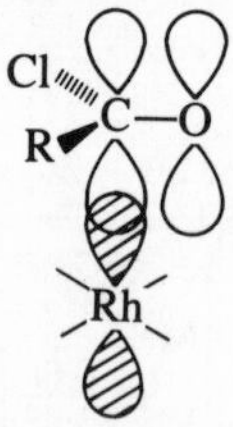

**51** 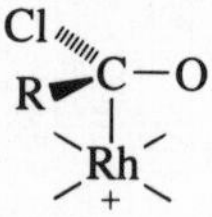 **52** 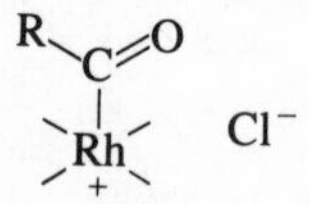 **53**

A wide variety of acyl complexes have been isolated and characterized (Table 6)[283–296]. Rearrangement to the rhodium carbonyl takes place in solution. When R is benzyl $k_1/k_{-1} \approx 7 \times 10^{-2}$ (reaction 130), but when R is aryl the rearrangement essentially is irreversible[287]. The most stable acyl complex is an acetylrhodium containing the chelating phosphine ligand $Ph_2P(CH_2)_3PPh_2$ [285].

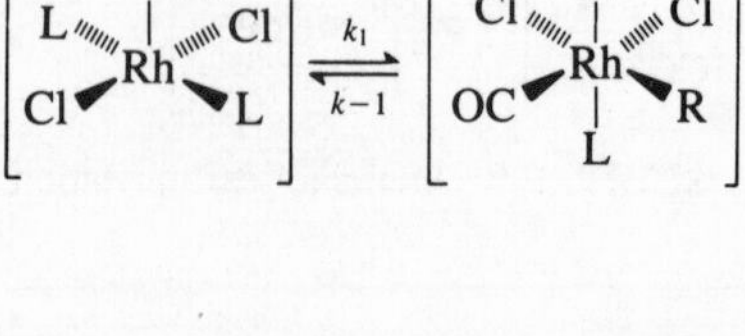

$$[RhCl_2(COR)L_2] \underset{k_{-1}}{\overset{k_1}{\rightleftharpoons}} [RhCl_2(CO)(R)L_2] \quad (130)$$

TABLE 6. Acylrhodium(III) complexes

$$RCOCl + [RhCl(PPh_3)_3] \longrightarrow [RhCl_2(COR)(L)_2]$$

| R | Ref. |
|---|---|
| $CH_3$ | 283, 288, 294 |
| $ClCH_2$ | 285 |
| $CH_3CH_2$ | 288 |
| $CH_3(CH_2)_2$ | 285, 288 |
| $CH_3(CH_2)_n$ ($n = 4, 5, 14, 16$) | 289 |
| cyclo-$C_8H_{15}$ | 289 |
| $PhCH_2$ | 284, 285, 287 |
| *p*-X-$C_6H_4CH_2$ (X = Cl, $NO_2$) | 287 |
| *erythro*- and *threo*-$CH_3CH(Ph)CH(Ph)$ | 291 |
| (*S*)$PhCH(CF_3)$ | 290 |
| $PhCH_2CH_2$ | 283, 289, 291 |
| $C_6D_5CD_2CH_2$ | 284, 291 |
| *threo*-PhCHDCHD | 292 |
| Ph | 287, 288, 293 |
| *p*-X-$C_6H_4$ (X = Cl, $NO_2$, OMe) | 287 |

| Acid halide | Complex | Ref. |
|---|---|---|
| MeCOCl | $(RhCl(COCH_3)(PMe_2Ph)_3]^+X^-$ <br> $X = Cl^-, PF_6^-$ | 286 |
| PhCOCl | $[RhCl(NO)(COPh)(Ph_3P)_2]$ | 296 |
| PhCOCl | $[RhCl_2(COPh)(Ph_2PCH_2CH_2PPh_2)]$ | 295 |
| $Me_2NCSCl$ | $[RhCl_2(CSNMe_2)(Ph_3P)_2]$ | 94 |

The mechanism for the decarbonylation of acid chlorides (Scheme 5) requires an oxidative addition of the acid chloride, acyl to alkyl rearrangement (**54** → **55**), and

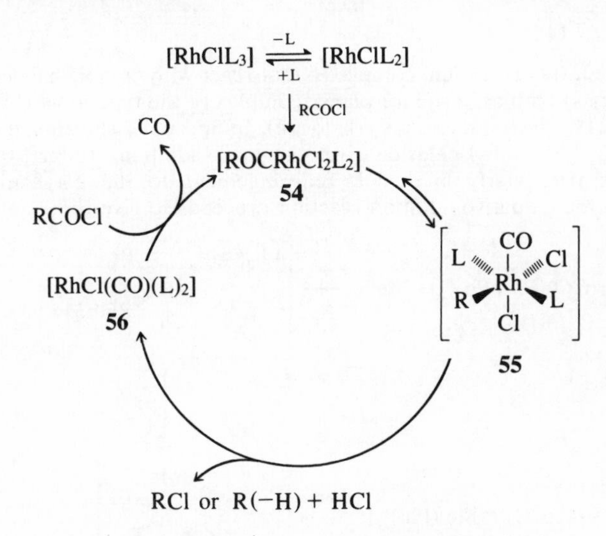


SCHEME 5. Acid chloride decarbonylation mechanism.

finally elimination of the organic group. Organic chlorides or alkenes and hydrogen chloride are eliminated, depending on the absence or presence of a $\beta$-hydrogen in the acid chloride and therefore $\beta$ to the carbon—rhodium $\sigma$-bond in the elimination step. The stoichiometric reaction to the chlorocarbonylbis(triphenylphosphine)rhodium stage (**56**), proceeds at relatively mild temperatures in most cases (*ca.* 80 °C), but higher temperatures (200 °C) are necessary to make the reaction catalytic. The reaction is not as simple as is written in many cases, however, the stoichiometric decarbonylation of aroyl halides taking place only above 150 °C. A number of other complexes are formed in equilibrium below 150 °C (including **56**) without the loss of the aryl chloride[297]. Furthermore, $\alpha,\beta$-unsaturated acid chlorides give vinyl phosphonium salts as an integral part of the decarbonylation reaction[298].

The oxidative addition reaction of alkyl halides and acid anhydrides occurs with an unusual complex containing two rhodium(I) atoms complexed to two different pairs of nitrogens in porphyrin complexes[299–303]. An oxidative reaction occurs when a chlorobis(triphenylphosphine)olefin rhodium(I) complex reacts with hydrogen chloride gas to yield a complex containing a rhodium—carbon $\sigma$-bond[304].

$$[\text{porphyrin}\{(CO)_2RhRh(CO)_2\}] \xrightarrow{R''X} [\text{porphyrin}Rh(R'')] \quad (131)$$

$R = R' = Et$; $R = Me$, $R' = Et$ $\qquad R'' = Me, Et, Pr^n, CH_3CO$

$$[RhCl(C_2F_3X(L)_2] \xrightarrow[CH_2Cl_2]{HCl} [RhCl_2(CF_2CFXH)(L)_2] \quad (132)$$

$X = F, Cl, H$; $L = PPh_3$

iii. *Iridium*. Of the $d^8$ iridium complexes that react with organic halides, the most studied is Vaska's complex, a square-planar complex of the type *trans*-$[IrX(CO)(L)_2]$ ($L = PPh_3$, $PMePh_2$, $PMe_2Ph$, etc.; $X$ = halogen). In non-polar solvents, the oxidative addition reaction of methyl chloride gives the *trans*-addition product, but in more polar solvents, particularly methanol, rearrangement to the *cis* complex takes place[305]. Thus, the oxidative addition reaction proceeds to give the kinetic product.

$$trans\text{-}[IrBr(CO)(PMePh_2)_2] + MeCl \longrightarrow [Ir(Me)(Cl)(Br)(CO)(PPh_2Me)(Ph_2MeP)] \xrightarrow{MeOH} \quad (133)$$

$$[IrCl(CO)\{P(Me)(Ph_2)\}_2] \xrightarrow{MeBr} [Ir(Me)(Br)(Cl)(I)(CO)(Ph_2MeP)]$$

Methyl halides have been shown to react with different facility, depending on the phosphine ligands and the halogen on the complex as well as that on methyl. The kinetic product is always that obtained from a *trans*-addition[305–307]. Methylfluorosulphonate also reacts to give the $d^6$, six-coordinate product[308], as do perfluoroalkyl iodides[305]. Allyl halides, however, gives the *cis*-addition product in non-polar solvents such as benzene, but isomerize in methanol to give the *trans*-

*trans*-$[IrX(CO)(L)_2]$ + RX′ ⟶ [Ir: R, L, X, OC, L, X′] (134)

product, which is obtained when methanol is the reaction solvent[309–311]. The reaction of either the *cis*- or the *trans*-complex with sodium tetraphenylborate gives an

*trans*-$[IrCl(CO)(L)_2]$ + R′CH=C(R)$CH_2X$ ⟶ [Ir: $CH_2C(R)$=CHR′, L, Cl, OC, X, L]

↙ (to π-allyl product) ; EtOH ↘ [Ir: $CH_2C(R)$=CHR′, L, Cl, OC, L, X]

[Ir: CO, L, L, Cl, π-allyl R′] ⟵ $NaBPh_4$ (135)

L=$PMePh_2$; X=Cl, Br; R=H, Cl, Me; R′=H, Me

18-electron π-allyl complex. Allenic or propargylic chlorides yield the same product, that arising from a formal oxidative addition of the allenic chloride[232]. There are a

$[IrCl(CO)(L)_2]$ + HC≡CC(R)(Me)Cl or CHCl=C=C(R)Me ⟶ [Ir: L, OC, CH=C=C(R)Me, Cl, Cl, L] (136)

R=H, Me

wide variety of other organic halides that undergo oxidative addition to Vaska's complex, including secondary alkyl, aryl, and acyl halides (Tables 7 and 8). The reaction apparently is more sensitive to the size of the phosphine ligand than to its donor ability. Although methyl iodide undergoes a reaction when the ligands are di(*t*-butyl)alkylphosphine (alkyl = Me, Et, Pr), allyl chloride and acetyl chloride will not[312]. For a series of $d^6$ octahedral complexes resulting from the oxidative addition reaction of various alkyl iodides to Vaska's complex (L = $PPh_3$, X = Cl), the bond dissociation energies for the iridium alkyl bond decrease in the order $H > CH_3 \approx I \approx CH_3CO > C_2H_5 > n\text{-}C_3H_7 > i\text{-}C_3H_7 > PhCH_2$[313,314].

The bulk of the evidence indicates that there are at least two mechanisms for the oxidative addition reactions, depending on the alkyl halide. Methyl halides and certain other primary halides, particularly benzyl bromide, allyl chloride, and chloromethyl methyl ether (see above), appear to undergo reaction by an $S_N2$-type

TABLE 7. Methyl iridium complexes

$$trans\text{-}[IrX(CO)(L)_2] + MeX' \longrightarrow [IrMe(CO)(L)_2XX']$$

| X′ | L | X | Ref. |
|---|---|---|---|
| I | $PPh_3$ | Cl | 315–318, 321 |
| I | $PPh_3$ | Br, I | 315 |
| $OSO_2F$ | $PPh_3$ | Cl, Br | 308 |
| I | $PPh_2Me$ | Cl | 305, 317 |
| Br, Cl | $PPh_2Me$ | Cl | 305 |
| Cl | $PPh_2Me$ | Br | 305 |
| I | $PMe_2Ph$ | Cl | 307, 317, 320 |
| Br | $PMe_2Ph$ | Cl | 307 |
| I | $PMe_2Ph$ | Br, I | 317 |
| I | $PEt_2Ph$ | Cl | 307, 321 |
| Br | $PEt_2Ph$ | Cl | 307 |
| Br | $PEt_2Ph$ | Br | 306 |
| I | $PEtPh_2$ | Cl | 321 |
| I | $P(p\text{-}CH_3C_6H_4)_3$ | Cl, Br, I | 317 |
| I | $P(p\text{-}ZC_6H_4)_3$ ($Z = Cl, F, CH_3, OMe$) | Cl | 319, 321 |
| I, Br | $PMe_3$ | Cl | 307 |
| I | $PEt_3$ | Cl | 317 |
| I | $P(OPh)_3$ | Cl | 317 |
| I | $PMe_2(o\text{-}MeOC_6H_4)$ | Cl | 320 |
| I | $PMe_2(p\text{-}MeOC_6H_4)$ | Cl | 320 |
| I | $PMe_2Bu^t$ | Cl | 312 |
| I | $AsPh_3$ | Cl, Br, I | 307 |
| I, Br | $AsMe_2Ph$ | Cl | 307 |

mechanism, while secondary alkyl halides and other primary halides undergo reaction by a radical chain mechanism.

The reaction of methyl iodide with halocarbonylbis(triphenylphosphine)iridium is second order, first order each in methyl iodide and iridium complex[315]. The rates of the reaction increase with increasing solvent polarity, a linear correlation being obtained from a plot of log $K$ *vs.* a solvent polarity parameter*, $q$[316].

The rates generally increase with the increasing basicity of the ligand [$PMe_2Ph > PEt_3 > PMePh_2 > P(p\text{-tolyl})_3 > PPh_3 > P(OPh)_3 > AsPh_3$][317]. the order $PMe_2Ph > PEt_3$ is probably a consequence of an overriding steric effect. Application of the Hammett $\sigma$-constant to substituents on the phosphine ligand yields a linear relationship with log $K$ with $\rho = 2.27$[318]. Replacement of carbonyl by other ligands such as nitrogen or phosphine increases the rate ($N_2 > PPh_3 > CO$), and replacement of chlorine by bromine or iodine decreases the rate ($Cl > Br > I$)[317]. These results, with the exception of the data for halogen replacement, can be attributed to an increased electron density on iridium producing a stronger nucleophile and higher reaction rates. It would be expected that because of the polarizability of the less electronegative iodine, iridium would acquire an even greater electron density and the iodo complex would afford the fastest rates. The inverse order observed is possibly a result of steric effects. The reaction rate also increases with increasing pressure on the reaction[316].

* $q = (D-1)/(2D+1)$, where $D$ = dielectric constant.

The entropy of activation is large and negative in non-polar solvents (−51 eu, $C_6H_6$), the magnitude decreasing in polar solvents[315]. The entropy of activation also is dependent on the phosphine ligand, being sensitive both to steric and electronic effects[319–321]. These data provide strong support for an $S_N2$-type mechanism involving a polar transition state and ionic products that collapse to a neutral complex.

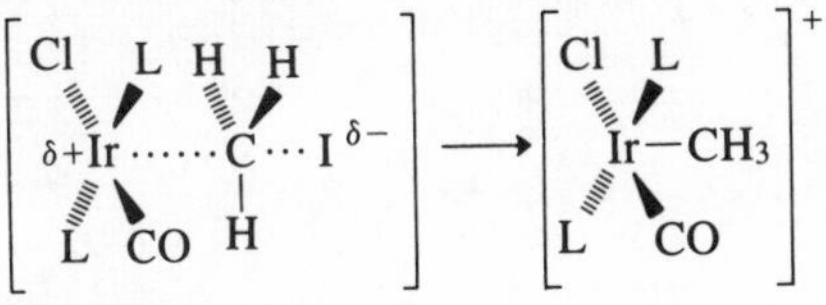

$$\left[ (Cl)(L)(L)(CO)\,^{\delta+}Ir \cdots\cdots CH_3 \cdots I^{\delta-} \right] \longrightarrow \left[ (Cl)(L)(L)(CO)Ir{-}CH_3 \right]^+ I^- \qquad (137)$$

The reaction of certain secondary alkyl halides proceeds by a radical chain process. In the absence of any free radical initiator under ultra-clean conditions (no traces of oxygen) the reaction does not take place[322]. An $S_N2$ reaction with secondary halides probably is excluded on steric grounds.

Reports that the reaction of Vaska's complex with chiral ethyl α-bromopropionate proceeds with retention of configuration at carbon[323] and that the reaction with *trans*-2-fluorocyclohexyl bromide proceeds with inversion of configuration[324] have now been shown to be incorrect[325–327]. The reaction is initiated by radicals, apparently including trace amounts of oxygen.

Loss of stereochemistry occurs in the reactions of primary alkyl bromides, *erythro*- and *threo*-deuterio-2-fluoro-2-phenylethyl bromide, and Vaska's complex as determined by $^1$H n.m.r.[325,327]. The reaction of a chiral secondary alkyl bromide, (+)-ethyl 2-bromo-3-fluorophenylacetate or *cis*- and *trans*-2-fluorocyclohexyl bromide, also gives products resulting from loss of stereochemistry. Identical 1:1 mixtures of *erythro*- and *threo*-oxidative addition products, **59** and **60**, result from the reaction of either geometrical isomer.

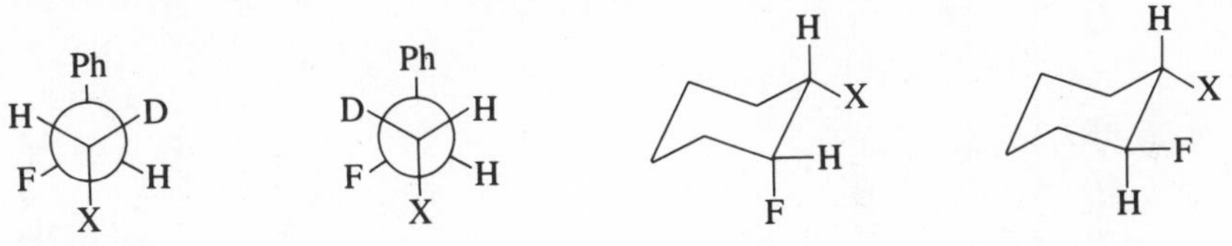

*erythro* **57**    *threo* **58**    *cis* **59**    *trans* **60**

X=Br or $IrClBr(CO)(L)_2$

The yield in these reactions are improved by free radical initiators, but decreased in the presence of inhibitors[325,328]. In some cases, in the presence of inhibitors, no reaction is observed. Traces of oxygen initiate the reaction, possibly via $[IrCl(CO)(L)_2(O)_2]$, which subsequently reacts to produce initiators. Ultraviolet irradiation accelerates the reaction. In competitive experiments, the order of reactivity of alkyl halides is tertiary > secondary > primary, an order which is inconsistent with an $S_N2$ process. Although galvinoxyl inhibits the reactions of ethyl iodide and ethyl α-bromopropionate, it does not inhibit the reactions of methyl iodide, benzyl bromide, allyl chloride, and chloromethyl methyl ether, the relative rates in this series being $CH_2{=}CHCH_2Br$ 12.7, $CH_2{=}CHCH_2Cl$ 5.2, MeI 1.0, $PhCH_2Br$ 0.93, $MeOCH_2Cl$ 0.39, and $PhCH_2Cl$ 0.02.

A radical chain mechanism is most consistent with those alkyl halides which show the characteristics of radical reactions[328].

TABLE 8. Alkyl and aryl iridium complexes

$$trans\text{-}[IrX(CO)(L)_2] + RX' \longrightarrow [IrR(X)(X')(L)_2]$$

| R | X' | L | X | Complex | Ref. |
|---|---|---|---|---|---|
| $CH_2{=}CHCH_2$ | Cl, Br | $PMePh_2$ | Cl | 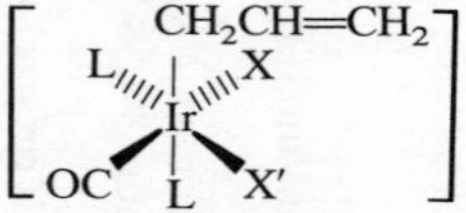 (A type) | 309 |
| | | | | 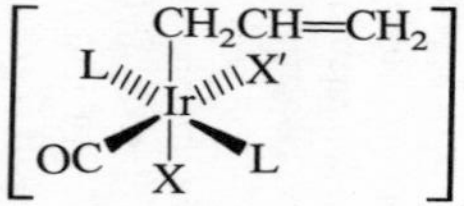 (B type) | 309 |
| | | | | 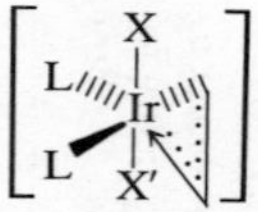 (C type) | 309 |
| $CH_2{=}CHCH_2$ | Cl, Br | $PMe_2Ph$ | Cl | A, B types | 311 |
| $CH_2{=}C(Me)CH_2$ | Cl, Br | | | A, B types | 310, 311 |
| $CH_2{=}C(Cl)CH_2$ | Cl, Br | | | A, B types | 310, 311 |
| $CH_2{=}C(Me)CH_2$ | Cl | $AsMe_2Ph$ | | A, B types | 310 |
| $CH_2{=}C(Me)CH_2$ | Cl | cod | Cl | $[IrCl_2(C_4H_7)(CO)(cod)]$ | 332 |
| $R_f$ ($R_f = CF_3, C_2F_5, n\text{-}C_3F_7$) | I | $PMePh_2$ | | 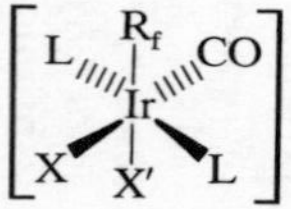 | 305 |

| | | | | | |
|---|---|---|---|---|---|
| (fluorocyclohexyl, axial F) or (fluorocyclohexyl, equatorial F) | Br | $PMe_3$ | | [(fluorocyclohexyl)–$IrClBr(L)_2(CO)$] | 324, 327, 328 |
| *erythro*- or *threo*-PhCFHCHD- | Br | $PMe_3$ | Cl | $[Ir(ClBr(L)_2(CO)(CHDCHFPh)]$ | 325, 327 |
| $CH_3CHCO_2Et$ | Br | $PPh_2Me$, $PMe_3$, $PMe_2Ph$ | Cl | $[IrClBr(L)_2(CO)\{CH(Me)CO_2Et\}]$ | 323, 326 |
| $PhCHFCHCO_2Et$ | Br | $PMe_3$ | Cl | $[IrClBr(L)_2(CO)\{CH(CHFPh)CO_2Et\}]$ | 326 |
| Ar | I, Br, Cl | $PPh_3$ | Cl | [Ir(Ar)(OC)(L)(L)(Cl)(X')] | 329, 330 |
| $R^1R^2C{=}C{=}CH$ | Cl | $PPh_3$ | Cl | [Ir(OC)(L)(L)(Cl)(Cl)(CH=CHCR$^1$R$^2$)] $R^1 = R^2 = H$; $R^1 = R^2 = Me$ | 232 |

$$[IrCp(CO)(L)] + RX \longrightarrow [IrCp(R)(CO)(L)]^+ X^-$$

| | | | | | |
|---|---|---|---|---|---|
| Me | I, Br, Cl | $PPh_3$, $PPhMe_2$, $PCy_3$ | | | 218, 334–336 |
| Et | | | | | 218 |
| $PhCH_2$ | | | | | 334 |
| $C_6H_{13}$ | | | | | 334 |
| $C_3F_7$ | | | | | 334 |

Initiation:

$$i^{\bullet} + Ir^{I} \longrightarrow iIr^{II} \tag{138}$$

$$iIr^{II} + RX \longrightarrow iIrX + R^{\bullet} \tag{139}$$

Propagation:

$$R^{\bullet} + Ir^{I} \longrightarrow RIr^{II} \tag{140}$$

$$RIr^{II} + RX \longrightarrow RIrX + R^{\bullet} \tag{141}$$

The data for the reaction of aryl halides with iridium(I) complexes are consistent either with an electron transfer mechanism or a nucleophilic displacement reaction proceeding through a Meisenheimer-type complex. Electron-withdrawing groups on the aryl halide increase the rate of the reaction[329] ($\rho = +0.4$–$0.6$[330] and I>Br>Cl[329]). The rate of the reaction is decreased by added phosphine[330].

$$ArX + trans\text{-}[IrCl(CO)(L)_2] \xrightarrow{140-180\ °C} \left[ \text{Ir}(Ar)(OC)(L)(L)(Cl)(X) \right] \tag{142}$$

A variety of other iridium complexes undergo reactions with alkyl halides. A nitrosotris(triphenylphosphine)iridium complex adds methyl iodide to give a 16-electron complex[331].

$$[IrL_3NO] + MeI \longrightarrow \left[ \text{Ir}(NO)(L)(I)(Me)(L) \right] \tag{143}$$

$$L = PPh_3$$

Methallyl chloride[332] and bromomethyl methyl ether[333] add to iridium(I) alkene complexes, in each case displacing an alkene ligand.

$$[IrCl(CO)(cod)_2]_2 + CH_2{=}C(Me)CH_2Cl \longrightarrow [IrCl_2(C_4H_7)(CO)(cod)] \tag{144}$$

$$[IrL_3(Me)(C_2H_4)] + BrCH_2OMe \longrightarrow \left[ \text{Ir}(Me)(Br)(L)(L)(CH_2OMe)(L) \right] \tag{145}$$

$$L = PMe_3$$

Carbon-$\eta^5$-cyclopentadienylphosphineiridium complexes add methyl and ethyl iodide[218,334,335]. The rates are faster than for either the cobalt or rhodium analogues, and the rates increase with increasing donor ability of the phosphine ligand[218,336]. The ionic products of these reactions are of the type expected initially from an $S_N2$ mechanism occurring with Vaska's complex, except that the ligands prevent collapse of the ion pair.

$$[IrCp(CO)(PPh_3)] + RI \longrightarrow \left[ \text{Ir}(Cp)(Ph_3P)(CO)(R) \right]^+ I^- \tag{146}$$

Acid chlorides react with Vaska's complex to give acyl derivatives, alkyl derivatives, or alkenes, depending on the structure of the acid chloride and the reaction conditions (Table 9)[306]. Although the dissociation energy for the acyl–iridium bond is relatively high[337] (greater than that of the alkyl—iridium bond), alkyl or aryl migration takes place readily. Once an alkyl iridium complex is obtained, $\beta$-elimination to give an alkene may occur.

At moderate temperatures, benzoyl and phenylacetyl chlorides give the *trans* acyl complexes, which rearrange to the phenyl and benzyl iridium complexes on warming[338–340]. The reaction is sensitive to the phosphine or arsine ligands, L, and the acid chloride. When the ligand is triphenylphosphine, benzoyl chloride will not react

*trans*-$[IrX(CO)(L)_2] + RCOCl \longrightarrow$ 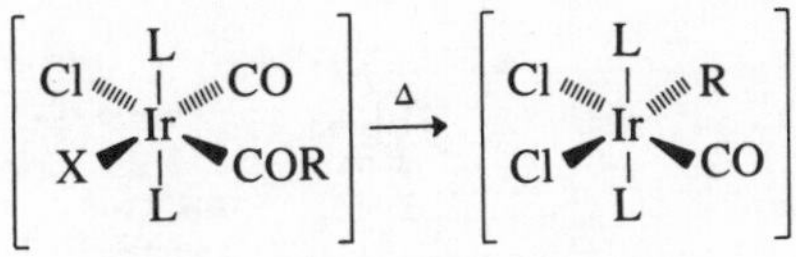 (147)

$R = Ar, ArCH_2$
$X = Cl, Br$

at a temperature below that at which rearrangement to the phenyl derivative is prevented[338,340]. With stronger ligand donors [$PPh_2Me$, $P(p\text{-}CH_3C_6H_4)_3$, $AsPh_3$] the acyl complex can be isolated. When the carbonyl ligand is replaced by nitrogen in Vaska's complex, a 16-electron acyl product is obtained, which rearranges rapidly to the 18-electron aryl or alkyl complex. The intermediate acyl complex cannot be isolated, however, in reactions with most aromatic acid chlorides[338].

*trans*-$[IrCl(N_2)(L)_2] + RCOCl \longrightarrow$ 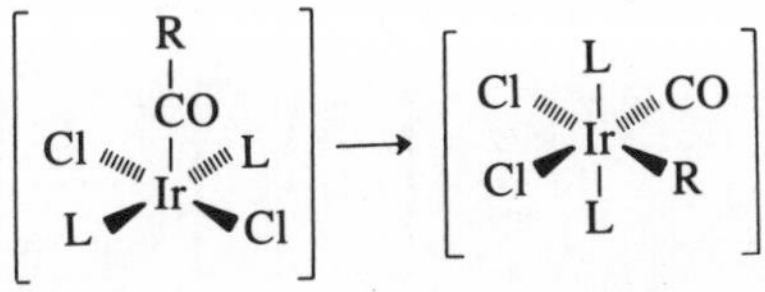 (148)

$R = Ar, ArCH_2 \quad X = Cl, Br$

In reactions with phenylacetyl chloride and its derivatives, the acyl complex is readily isolated, and the benzyl migration can be studied[339]. Electron-releasing substituents promote the migration, and thus for the series $ArCH_2$, $p$-$MeOC_6H_4CH_2 > p\text{-}CH_3C_6H_4CH_2 > C_6H_5CH_2 > p\text{-}NO_2C_6H_4CH_2 \gg C_6F_5CH_2$ ($\rho = 0.30$)

The reaction of aliphatic acid chlorides with Vaska's complex gives an acyl complex. Alkyl migration can occur, and if the alkyl group possesses a $\beta$-hydrogen, elimination to yield an olefin can take place[340]. Chlorotris(dimethylphenylphosphine)iridium adds aliphatic acid chlorides to give the acyl derivative[286]. Acyl to alkyl rearrangement takes place in solution giving an equilibrium mixture favouring the alkyl complex[341]. The alkyl complex obtained from branched-chain acid chlorides ultimately undergoes isomerization to a straight-chain isomer through a series of $\beta$-elimination, readdition reactions[342,343].

$[IrClL_3] + Me_2CHCOCl \longrightarrow \longrightarrow$ 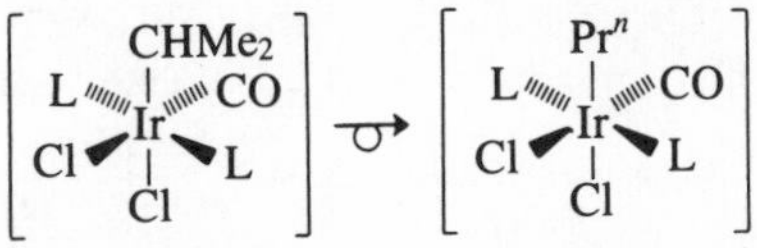 (149)

$L = PPh_3$

Methyl chloroformate[344], fluorinated acid chlorides[345], and acid anhydrides[172]

TABLE 9. Acyl iridium complexes

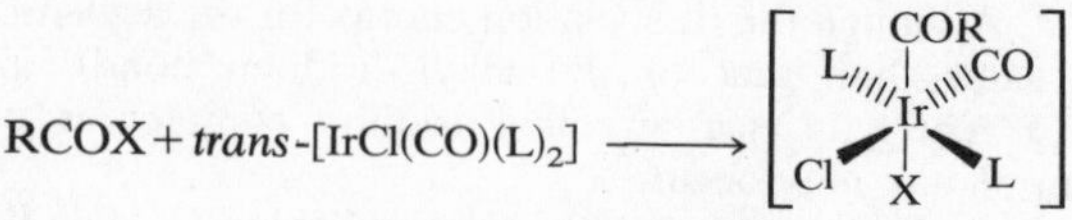

$$RCOX + trans\text{-}[IrCl(CO)(L)_2] \longrightarrow [IrCl(COR)(CO)(L)_2X]$$

| R | X | L | Ref. |
|---|---|---|---|
| Me | Br | $PEt_2Ph$ | 306, 307 |
| | Cl | | 306, 307 |
| | I | | 307 |
| | Cl | $PPh_3$ | 338 |
| | I, Br | $PMe_2Ph$, $PMe_3$, $AsMe_2Ph$ | 307 |
| Et, $Pr^n$, $Pr^i$, $n$-$C_5H_{11}$ | Cl, Br | $PEt_2Ph$ | 306 |
| cyclopropyl | Cl | $PMePh_2$ | 338 |
| cyclopropyl | Cl | $PPh_3$ | 347 |
| $CF_nH_{3-n}$ | Cl | $PMePh_2$ | 345 |
| $PhCH_2$ | Cl | $PMePh_2$ | 338 |
| $PhCH_2$ | Cl, Br | $PPh_3$ | 338 |
| Ph | Cl | $PMePh_2$, $P(p\text{-}CH_3C_6H_4)_3$, $AsPh_3$ | 338 |
| $YC_6H_4$ (Y = $p$-$OCH_3$, $p$-$NO_2$, $m$-Me, $o$-Me) | Cl | $PMePh_2$ | 338 |

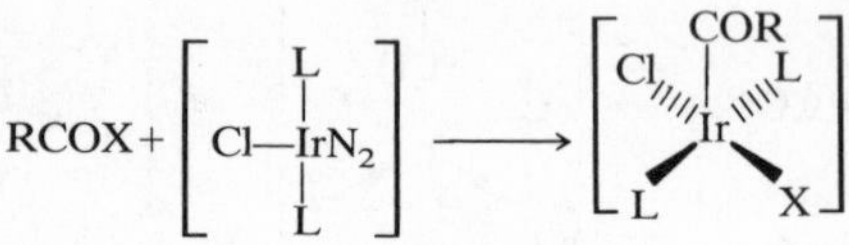

$$RCOX + [IrCl(N_2)(L)_2] \longrightarrow [IrCl(COR)(L)_2X]$$

| R | X | L | Ref. |
|---|---|---|---|
| Me, Et, $CMe(CH_2)_{10}$, cyclopropyl, cyclobutyl, $PhCH_2$, $CH_2{=}CH$ | Cl | $PPh_2Me$ | 338 |
| $CF_nH_{3-n}$ ($n$ = 1–3) | Cl | $PPh_2Me$ | 345 |
| $p$-$YC_6H_4CH_2$ (Y = $OCH_3$, $CH_3$, H, $NO_2$) | Cl | $PPh_3$ | 339 |
| $C_6F_5$ | | $PPh_3$ | 339 |
| Ph | | $PPh_2Me$ | 338 |
| Y-$C_6H_4$ (Y = $m$-Me, $p$-OMe, $p$-$NO_2$) | | $PPh_2Me$ | 338 |

| R | X | L | Complex | Ref. |
|---|---|---|---|---|
| Me, $Pr^n$, $Pr^i$ | Cl | $PMe_2Ph$ | $[IrCl_2(COR)(L)_3]$ | 286 |
| MeO, EtO, PhO | Cl | $PMe_3$, $AsMe_2Ph$ | 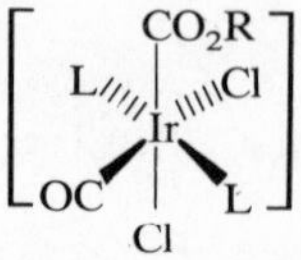 $[IrCl_2(CO_2R)(CO)(L)_2]$ | 344 |

TABLE 9. ctd.

| R | X | L | Complex | Ref. |
|---|---|---|---|---|
| RC(O)OC(O)R (R = Me, $CF_3$, $C_2F_5$) | Cl, Br | $PPh_3$, $PMePh_2$ | [Ir(COR)(L)(CO)(X)(L)(OCOR)] | |
| 2,2,3,3-tetrafluorosuccinic anhydride | | | [IrX(L)(L)(CO)(O–C(=O)–$CF_2$–$CF_2$–C(=O))] | |

oxidatively add to Vaska's complex, all giving the expected acyl-type 18-electron complex. Addition of acid chlorides to the iridium dimer $[IrCl(CO)(cod)_2]_2$ apparently produces the dimeric acyl complex, which cannot be isolated, but undergoes alkyl or aryl migration[346].

$$[IrCl(CO)(cod)_2]_2 + RCOCl \longrightarrow [(R)(CO)(OC)(Cl)Ir(\mu\text{-}Cl)_2Ir(Cl)(CO)(CO)(R)] \qquad (150)$$

R = Me, Et, $Pr^i$, Ph

iv. *Iron*. The zero-valent $d^8$-iron carbonyl, iron pentacarbonyl, and the $\eta^5$-cyclopentadienyldicarbonyl iron anion, as well as the $d^{10}$ iron tetracarbonyl dianion, and their derivatives all react with a variety of organic halides to give $\sigma$-carbon–iron complexes that are synthetically useful. These $d^8$ and $d^{10}$ complexes are all 18-electron complexes, yet they are reactive, the anionic complexes reacting without becoming unsaturated by loss of a ligand.

Iron pentacarbonyl reacts with perfluoroalkyl iodides to give the $d^8$ alkyl iron(II) complex, which loses carbon monoxide on heating to give the iodide-bridged dimer[348]. Alkyl halides containing $sp^3$-hybridized $\beta$-carbon atom bearing hydrogen tend to yield products of $\beta$-elimination. The reaction of allyl iodides with iron pentacarbonyl give the $\eta^3$-allyl complexes[349]. Neither allyl chloride nor bromide reacts, at least at temperatures below 40 °C. Irradiation is necessary to eject carbon monoxide, giving a coordinatively unsaturated complex which undergoes the oxidative addition reaction[350]. Thus, in many reactions with iron carbonyl, it is necessary to produce the tetracarbonyl before it will react.

$$R_fI + [Fe(CO)_5] \longrightarrow [FeI(R_f)(CO)_4] \xrightarrow{\Delta} \tfrac{1}{2}[FeI(R_f)(CO)_3]_2 + CO \qquad (151)$$

$$[Fe(CO)_5] + CH_2{=}CHCH_2X \longrightarrow [FeX(\eta^3\text{-allyl})(CO)_3] \qquad (152)$$

X = I; $h\nu$ necessary for X = Br, Cl

The reaction of 4-chlorobut-2-enol gives an acyl product resulting from an oxidative addition[350]. Loss of CO also is necessary for the reaction of iron carbonyl with vinyl epoxides[351,352].

$$HOCH_2CH{=}CHCH_2Cl + [Fe(CO)_5] \xrightarrow{h\nu} [\text{η}^3\text{-allyl}{-}Fe(CO)_3\text{ acyl complex}] + CO + HCl \quad (153)$$

$$[Fe(CO)_5] + \text{2-vinyloxirane} \xrightarrow{h\nu} [(CO)_3Fe{-}C(=O){-}O\text{ allyl complex}]$$

$$[Fe(CO)_5] + \text{vinyl cyclopentene oxide} \xrightarrow{h\nu} [\text{O}{-}C(=O)\text{, }Fe(CO)_3\text{ complex}] \quad (154)$$

(two isomers, *syn*- and *anti*-$\eta^3$-Fe)

Diironnonacarbonyl, however, reacts with allyl halides at 60 °C to give the $\eta^3$-complex[353], and with pentafluorobenzyl bromide to yield the dibenzyliron complex[354]. If benzyl bromide is used in place of the perfluorophenyl

$$R^1CH{=}C(R^2)CH_2X + [Fe_2(CO)_9] \xrightarrow[60\,°C]{C_6H_6} [FeX[\eta^3\text{-1-}R^1\text{-2-}R^2\text{-allyl})(CO)_3] + [Fe(CO)_5] + CO \quad (155)$$

| $R^1$ | $R^2$ | X |
|---|---|---|
| H | H | Cl, Br, I |
| Me | H | Cl, Br, I |
| $CO_2Me$ | H | Br |
| Br | H | Br |
| H | Br | Br |

analogue, the oxidative addition product is unstable, yielding dibenzyl ketone.

$$C_6F_5CH_2Br + [Fe_2(CO)_9] \longrightarrow [Fe(C_6H_5CH_2)_2(CO)_4] \quad (156)$$

The slow addition of iodine to iron pentacarbonyl in trifluoroacetic anhydride yields overall the product of oxidative addition of perfluoroacetyl iodide[355].

$$[Fe(CO)_5] + I_2 \xrightarrow{(CF_3CO)_2O} [FeI(CO)_4(COCF_3)] \quad (157)$$

Bis(trimethylphosphine)tricarbonyliron reacts with methyl iodide with loss of carbon monoxide to give the methyl derivative[356]. The reaction of methyl iodide with pentakis(trimethylphosphite) iron gives the methyl iron cationic complex, but the reaction with ethyl iodide yields only ethylene and the iron hydride[357]. Allyl halides react to give the $\eta^3$-allyl cation, while the reaction of benzyl iodide leads to bibenzyl and diiodotris(trimethylphosphite)iron(II)[357]. Tetrakis(trimethylphosphine)iron(0) gives the neutral oxidative addition product[358].

$$\left[\begin{array}{c} \text{CO} \\ | \\ \text{Me}_3\text{P}-\text{Fe}-\text{PMe}_3 \\ \text{CO}\ \ \text{CO} \end{array}\right] + \text{MeI} \longrightarrow [\text{FeI(Me)(CO)}_2(\text{PMe}_3)_2] + \text{CO} \qquad (158)$$

$$\text{FeL}_5 \begin{cases} +\text{MeI} \longrightarrow [\text{Fe(Me)(L)}_5]^+\text{I}^- & (159) \\ +\text{EtI} \longrightarrow [\text{Fe(Et)(L)}_5]^+\text{I}^- \longrightarrow [\text{FeH(L)}_5]^+\text{I}^- + \text{CH}_2=\text{CH}_2 & (160) \end{cases}$$

$L = P(OMe)_3$

$$[\text{Fe(PMe}_3)_4] + \text{MeX} \longrightarrow [\text{FeX(Me)(PMe}_3)_4] \qquad (161)$$

$X = Br, I$

The disodium tetracarbonyl ferrate 'supernucleophile'[359] is the most nucleophilic transition metal complex known. A comparison of the nucleophilicity of this complex with those of other complexes and nucleophiles by Pearson's nucleophilicity parameter[360] reveals that it is thirteen orders of magnitude more nucleophilic than the cobalttetracarbonyl anion (Table 10). The reactivity of the nucleophile is sensitive to the reaction solvent; solvents such as nmp, dmf and hmpa, which solvate the cations, yield solvent-separated ion pairs that produce a nucleophile of higher reactivity[361].

$$[\text{Fe(Na}^+:\text{S:)}_2(\text{CO})_4{}^{2-}] \rightleftharpoons \text{Na}^+ + \text{Fe[Na}^+:\text{S:)(CO)}_4{}^{2-}] \qquad (162)$$

This 'supernucleophile' reacts rapidly with organic halides to give the oxidative addition product[361]. Alkyl halides and tosylates can be coupled to give ketones[362,363] and converted to aldehydes[364], carboxylic esters, acids, or amides[365]. The oxidative addition reaction is limited to primary and secondary halides, since tertiary halides undergo elimination. Acid halides react to form the acyl product[362,366], but when the acid halide is a perfluoroacid halide, decarbonylation occurs as well as double acylation to give the bisperfluoroalkylirontetracarbonyl[348].

$$\left[\text{R}-\overset{\text{O}}{\overset{\|}{\text{C}}}-\text{Fe(CO)}_3\text{Cl}\right]^- \xleftarrow{\text{RCOCl}} \text{Na}_2[\text{Fe(CO)}_4] \xrightarrow{\text{RX}} \left[\text{R}-\text{Fe(CO)}_4\right]^- \qquad (163)$$

$$\text{Na}_2[\text{Fe(CO)}_4] \xrightarrow{\text{R}_f\text{COCl}} [\text{Fe(R}_f)_2(\text{CO})_4]$$

In the synthesis of ketones by successive additions of alkyl halides, the neutral dialkyl iron complex is not obtained, and the acyl–alkyl intermediate cannot be

TABLE 10. Nucleophilicity parameters for transition metal complexes

| Compound | $n = \log\left(\frac{k_n}{k_{\text{MeOH}}}\right)^a$ | Solvent |
|---|---|---|
| $Na_2[Fe(CO)_4]$ | 16.7 | nmp |
| $Na[FeCp(CO)_2]$ | 15 | glyme |
| Vitamin $B_{12(S)}$ | 14.4 | MeOH |
| $PhS^-$ | 9 | MeOH |
| $[Co(CO)_4]^-$ | 3.5 | Glyme |

[a] $k_n$ = rate constant for the reaction of the nucleophile with MeI; $k_{\text{MeOH}}$ = rate constant for the reaction of methanol with MeI[360].

detected[362]. The question of whether alkyl migration (CO insertion) takes place before a second oxidative addition occurs is unsettled. Note in this series of oxidative addition reactions, the $d^{10}$ supernucleophile is oxidized first to a $d^8$ iron(0) anion,

$$[Fe(CO)_4]^{2-} \xrightarrow{RX} \underset{d^8}{[FeR(CO)_4]^-} \xrightarrow{R'X} \left[Fe(R)(R')(CO)_4\right]_{d^6}$$

$$[FeR(CO)_4]^- \longrightarrow \underset{d^8}{[Fe(COR)(CO)_3]^-} \xrightarrow{R'X} \left[Fe(COR)(R')(X)(CO)_3\right]_{d^6} \longrightarrow RCOR' \quad (164)$$

which in turn is oxidized to a $d^6$ acyl iron(II) complex. All complexes except the 16-electron $d^8$ acyl iron(0) anion are coordinatively saturated. Added triphenylphosphine causes the alkyl rearrangement to the 18-electron complex, which also yields ketone on reaction with more alkyl halide.

$$[FeR(CO)_4]^- \xrightarrow{PPh_3} [Fe(COR)(CO)_3(PPh_3)]^- \quad (165)$$

An acyl–alkyl intermediate has been isolated, however. Reaction of 1,3-dibromopropane with disodium tetracarbonylferrate in the presence of triphenylphosphine yields a stable metallocycle[367]. When the organic halide is a primary

$$[Fe(CO)_4]^{2-} + Br(CH_2)_3Br \xrightarrow[thf]{PPh_3} \left[(CO)_3(PPh_3)Fe\text{—}CH_2CH_2CH_2C(=O)\text{—}(\text{ring})\right] \quad (166)$$

bromide containing an allene function, unsaturated ketones ultimately can be obtained[368].

$$[Fe(CO)_4]^{2-} + CH_2{=}C{=}CHCH_2CH_2Br \longrightarrow [Fe(CH_2CH_2CH{=}C{=}CH_2(CO)_4]^-$$

$$\downarrow \quad (167)$$

$$\left[(CO)_3Fe \text{ (2-methylenecyclopentanone complex) }\right]^- \longleftarrow \left[(CO)_3Fe\text{—}C(=O)\text{—}CH_2\text{—}CH_2\text{—}CH{=}C{=}CH_2\right]^-$$

The mechanism of the oxidative addition reaction of the iron dianion clearly involves a nucleophilic displacement reaction at carbon[362]. Reaction with (*S*)-2-octyl tosylate gave an alkyl derivative that underwent insertion of carbon monoxide with retention of configuration at carbon. A Baeyer–Villiger oxidation, also proceeding with retention at carbon, gave the acetate in 99% ee of the opposite configuration as the starting tosylate. Thus, inversion of configuration at carbon takes place in the oxidative addition.

$$\text{Me(TsO)C(H)C}_6\text{H}_{13} + \text{Na}_2[\text{Fe(CO)}_4] \longrightarrow [\text{Me(H)C(C}_6\text{H}_{13})\text{Fe(CO)}_4]^- \xrightarrow[\text{(2) MeI}]{\text{(1) CO}} \text{Me(H)C(C}_6\text{H}_{13})\text{COMe} \longrightarrow \text{Me(H)C(C}_6\text{H}_{18})\text{OAc} \quad (168)$$

The reaction overall is second order, first order each in halide and iron complex[369]. The reaction also shows a large negative entropy of activation ($-39 \pm 5$ eu) and a small enthalpy of activation, characteristic of an $S_N2$ reaction. The substrate reactivities, $Me > RCH_2 > (Et_2CHCH_2 > RR'CH > Me_3CCH_2 \gg$ adamantyl bromide (no reaction) and $I > Br > OTs \gg Cl$ also are characteristic of classic nucleophilic displacement reactions.

The oxidative addition reactions of other carbonyl anions, especially the $\eta^5$-cyclopentadienyldicarbonyl iron anion, $Fp^-$, have been reviewed[203–206,370], and the many organoiron derivatives have been tabulated[203,204]. Of the $\eta^5$-cyclopentadienyl metal carbonyl anions, this iron complex is the most nucleophilic, being about $2 \times 10^7$ more nucleophilic than $[CrCp(CO)_3]^-$[371]. Alkyl halides react readily[372–375], as do $\alpha,\omega$-dihalides[376]. Reaction with chloroiodomethane results in selection of the more reactive iodide, giving the chloromethyl derivative[377]. Although 2-adamantyl bromide reacts, 1-adamantyl bromide will not, as expected[378]. The reaction of propynyl bromide, originally reported to give the $\sigma$-acetylene complex, gives the $\sigma$-allene complex[231], and other bishalomethyl acetylenes are reported to give ordinary displacement products[376,379]. Benzyl-type halides also oxidatively add[380,382].

$$RX + Fp^- \longrightarrow FpR + X^- \quad (169)$$

R = Me, Et, $CH_2{=}CHCH(R')CH_2$ R' = H, Me), $CH_2Cl$, [2-adamantyl], $PhCH_2$

X = I, Cl, OTs I Br,Cl

(170)

$$Br(CH_2)_nBr + Fp^- \longrightarrow Fp(CH_2)_nFp$$

$n = 3, 4, 5, 6$

$$Fp^- + HC{\equiv}CCH_2Br \longrightarrow FpCH{=}C{=}CH_2 + Br^- \quad (171)$$

$$Fp^- + XCH_2C{\equiv}CCH_2X \longrightarrow FpCH_2C{\equiv}CCH_2Fp \quad (172)$$

X = Cl, $OSO_2Ph$

$$FpCH_2Ph \xleftarrow{PhCH_2Cl} Fp^- \xrightarrow{\text{1,2-dibromobenzocyclobutene}} \text{1,2-bis(Fp)benzocyclobutene} \quad (173)$$

A large number of other iron(II) alkyl derivatives have been synthesized by this reaction[383–388], and in a number of cases the alkyl groups contain other functionality. The reaction with *N,N*-dimethyl-2-chloroethylamine yields an acyl complex as a result of the nitrogen ligand inducing rearrangement (CO insertion)[386]. Bromoacetone dimethyl ketal gives the usual substitution product plus a vinyl ether[387], and ethyl-$\beta$-bromopyruvate diethylketal gives a stable iron complex in 63% yield[388].

$$FpCH_2C(OMe)_2Me + FpCH_2C(OMe){=}CH_2 \xleftarrow{BrCH_2C(OMe)_2Me} Fp^-$$

$$Fp^- \xrightarrow{BrCH_2C(OEt)_2CO_2Et} FpCH_2C(OEt)_2CO_2Et$$

$$Fp^- \xrightarrow{ClCH_2CH_2NMe_2} [FpCH_2CH_2NMe_2] \longrightarrow Cp(CO)Fe{\leftarrow}NMe_2 \text{ (chelate with } C(=O)CH_2CH_2 \text{ ring)}$$

Allyl halides give the $\sigma$-allyl complex[389–393] (Table 11); loss of carbon monoxide by photolysis yields the $\eta^3$-allyl complex[389]. The cyclic allyl chloride 2,3-dichlorocyclobutene (dccb) gives a ring-opened product instead of the usual substitution product[394–396].

$$Fp^- + CH_2{=}CHCH_2Cl \longrightarrow FpCH_2CH{=}CH_2 \xrightarrow{h\nu} [Cp(OC)Fe(\eta^3\text{-}C_3H_5)] \qquad (175)$$

$$dccb + Fp^- \longrightarrow FpCH{=}CHCH{=}CHFp \qquad (176)$$

Vinyl iodides[397], chlorides[398,99] and fluorides[400–403] give vinyl iron compounds. The reaction products show the characteristic of nucleophilic addition to the double bond followed by anion elimination. Tetrafluoroethylene[400], perfluorobutadiene[401], perfluorocyclopentene[402], and ethyl perfluoromethacrylate[403] react by replacement of a vinyl fluoride. An iron-bridged dimer is obtained in a reaction with 1,1-dicyano-2-chloroethene[404].

$$Fp^- + RCH{=}CHX \longrightarrow [FpCH(X)CHR]^- \longrightarrow FpCH{=}CHR \qquad (177)$$

R = MeCO, PhCO

$$Fp^- + (CN)_2C{=}CHCl \longrightarrow [Cp(OC)Fe(\mu\text{-}C{=}C(CN)_2)(\mu\text{-}CO)Fe(Cp)(CO)] \qquad (178)$$

Perfluoromethyl allenes undergo displacement of fluorine with an allylic-type rearrangement[405,406].

$$Fp^- + (CF_3)_2C{=}C(CF_3)_2 \xrightarrow{-70\,^{\circ}C} [FpC\{{=}C(CF_3)_2\}C({=}CF_2)CF_3] \qquad (179)$$

TABLE 11. σ-allyl-Fp complexes

| Structure | Substituents | | | | Ref. |
|---|---|---|---|---|---|
| | $R_1$ | $R_2$ | $R_3$ | $R_4$ | |
| $Fp-C(R^4)(R^3)-C(R^1)=C(R^2)(R^3)$ | H | H | H | H | 389 |
| | Me | H | H | H | 390–392 |
| | H | Me | H | H | 391 |
| | H | H | Me | H | 391, 392 |
| | H | Me | Me | H | 391 |
| | H | H | Ph | H | 391 |
| | H | H | H | Me | 391 |
| | H | D | D | H | 391 |
| | H | H | H | D | 391 |
| | Me | H | H | D | 391 |
| | Me | D | D | H | 391 |
| | H | H | OMe | H | 391 |
| | H | OMe | H | H | 391 |
| | H | $Me_3Si$ | H | H | 393 |
| Fp–cycloalkenyl, $(CH_2)_n$; $n$ = 1, 2, 3, 4, —CH=CH— | | | | | 391 |

Reaction of $[FeCp(CO)_2]^-$ with aryl halides gives low yields of product[372]. Hexafluorobenzene and other polyfluorinated aromatics give higher yields of aryl complexes[407,408]. The reaction with perfluorotoluene gives the *p*-trifluoromethyl product from reaction at an aryl fluorine rather than a benzyl.

$$Fp^- + ArX \longrightarrow FpAr \tag{180}$$

| Ar | X |
|---|---|
| $C_6H_5$ | I |
| $C_6F_5$ | F |
| $p$-$HC_6F_4$ | F |
| 2,3,6-$F_3C_6H_2$ | F |
| $p$-$BrC_6H_4$ | Br |
| $p$-$CF_3C_6F_4$ | F |

Acid chlorides give stable acyl products[347,374,378,409–414] (Table 12). The reaction with the 1-adamantyl acid chloride followed by decarbonylation is the only method of preparation of the 1-adamantyl iron derivative[378].

$$Fp^- + RCOCl \longrightarrow FpCOR \tag{181}$$

TABLE 12. Acyl derivatives

| Structure | R | Reference |
|---|---|---|
| FpCOR | Me | 314, 410, 411 |
| | Et | 409 |
| | $Pr^n$ | 410 |
| | $Cl(CH_2)_3$ | 411 |
| | $CF_3$ | 410[a] |
| | (−)PhCH(Me) | 412 |
| | Ph | 374 |
| | PhCH=CH | 374 |
| | cyclopropyl | 347 |
| | (adamantyl) | 378 |
| | 1,2-$Ph_2$-cyclopropen-3-yl | 413 |
| | CpFe–(C₅H₄) | 414 |
| FpC(O)RC(O)Fp | $(CH_2)_n$ ($n = 3,4$) | 411 |
| | $(CF_2)_3$ | 411 |

[a] From reaction with $(CF_3CO)_2O$.

The $[FeCp(CO)_2]^-$ nucleophile also couples with the cyclopropene cation and opens up epoxides stereospecifically.

$$Fp^- + [Ph_3C_3]^+ \longrightarrow \text{3-Fp-1,2,3-triphenylcyclopropene} \quad (182)$$

$$Fp^- + \text{cyclohexene oxide} \longrightarrow \textit{trans}\text{-2-Fp-cyclohexanol} \quad (183)$$

An 18-electron anionic $d^8$ iron(0) complex containing a $\sigma$-methyl group and phosphine ligands in place of the cyclopentadienyl ligand adds methyl iodide under mild conditions to give a complex whose structure has been shown by n.m.r to have *cis*-methyl and *trans*-phosphines[417]. The reaction of $\eta^5$-cyclopentadienyl (dppe) iron

$$Na^+[FeMe(CO)_2(PMe_3)_2]^- + MeI \longrightarrow [Fe(CO)_2(Me)_2(PMe_3)_2] \quad (184)$$

anion (as the magnesium bromide salt) with alkyl halides yields mostly alkane products and low yields of alkyl phosphines[418]. When the alkyl halide is 6-bromohex-1-ene only cyclized products, methyl cyclopentane, and the cyclopentylmethyl iron complex are obtained, indicating a radical mechanism.

The stereochemistry of the reaction of $[FeCp(CO)_2]^-$ with alkyl halides and sulphonates has been demonstrated in a number of cases to proceed with inversion of configuration at carbon. The reaction of *erthro*-1,2-dideuterioneohexyl brosylate produced the *threo* product with greater than 95% net inversion of configuration, as determined by the $^1$H n.m.r. coupling constants[419,420]. The same results were obtained when the *threo* isomer was employed or with the bromide instead of the brosylate[420]. Similarly, *threo*-1,2-dideuterio-2-phenylethyl tosylate alkylates

Newman projection (*erythro*; $Bu^t$, H, D, D, H, X) $+ Fp^-$ $\longrightarrow$ Newman projection (*threo*; $Bu^t$, D, H, D, H, Fp) $X = Br, -OSO_2C_6H_4Br$ (185)

*erythro* *threo*

$[FeCp(CO)_2]^-$ to give the *erythro* product[421,422]. Both *cis* and *trans*-4-methylcyclohexyl benzenesulphonate react with inversion, the stereochemical assignments being determined by the chemical shifts of the *cis*-and *trans*-methyl groups at low temperature[423]. Chiral 2-bromobutane also undergoes this reaction with *ca.* 75%

(cyclohexyl)$SO_2Ph$ $+ Fp^-$ $\longrightarrow$ (cyclohexyl)Fp (186)

net inversion of configuration at carbon[424]. The stereochemistry of the oxidative addition was determined from the 2-methylbutyric acid obtained via a carbon monoxide 'insertion' (retention)–oxidation sequence[424].

$$MeCH_2CH(Br)Me + Fp^- \longrightarrow [FpBu^s] \xrightarrow{PPh_3} [Fe(COBu^s)(CO)(PPh_3)(Cp)] \xrightarrow{Cl_2,\ H_2O} MeCH_2\overset{*}{C}(CO_2H)Me \quad (187)$$

The mechanistic course of this reaction is sensitive to the halide (bromide or iodide) and possibly to the structure of the organic group. The products of reaction of cyclopropyl carbinyl iodide and bromide are different. Whereas the bromide gave an unrearranged product of nucleophilic displacement, the iodide gave appreciable amounts of ring-opened compound, indicative of an electron transfer process and a cyclopropyl carbinyl radical[425]. In general, alkyl iodides and benzyl bromide give e.s.r. signals in reactions with $[FeCp(CO)_2]^-$, whereas alkyl bromides and chlorides do not. This information does not necessarily mean, however, that the major pathway leading to product is a radical.

$$\text{cyclopropyl-}CH_2X + Fp^- \longrightarrow \text{cyclopropyl-}CH_2Fp + CH_2{=}CHCH_2CH_2Fp \quad (188)$$

| | | |
|---|---|---|
| X = Br | >97% | — |
| | 70% | 30% |

A number of related reactions of various iron complexes have been reported. The cyclopentadienyldicarbonyliron dimer reacts with activated vinyl iodides to give the same products as would be obtained with the monomeric anion. These one-electron changes probably take place by a radical process[426,427]. Certain amine iron complexes containing four coordinating ligands undergo oxidative addition reactions with

$$[FeCp(CO)_2]_2 + RCH{=}CHI \longrightarrow FpCH{=}CHR \quad (189)$$

R = ArCO, R'OCO

various alkyl halides[428,429] and acid anhydrides[428] to give alkyl and acyl iron complexes. The oxidative addition of alkyl halides to the four-coordinate tetraazairon(II) complex gives a low-spin alkyl iron(III) product plus a high-spin iron(II) halide[429]. The reaction probably proceeds by a mechanism similar to that which occurs with pentacyanocobalt(II) (reactions 116–118[257]).

$$[Fe(C_{22}H_{22}N_4)] + RX \longrightarrow [Fe(C_{22}H_{22}N_4)X] + [Fe(C_{22}H_{22}N_4)R] \qquad (190)$$

$RX = MeI, EtI, PhCH_2Br$

v. *Ruthenium*. Most of the oxidative addition reactions of organic halides to ruthenium have been carried out with the ruthenium anion, $[RuCp(CO)_2]^-$. Bis(triphenylphosphine)tricarbonylruthenium(0) adds methyl iodide, however, to give the $d^6$ ruthenium(II) complex[430].

$$[Ru(CO)_3(PPh_3)_2] + MeI \longrightarrow [RuMe(I)(CO)_2(PPh_3)_2] \qquad (191)$$

The anion reacts with alkyl iodides and acid chlorides to give the alkyl and acyl complexes, respectively[203–206]. Methyl and ethyl ruthenium[373] as well as propionyl[409] and trifluoroacetyl[431] derivatives have been prepared. A number of perfluoro olefins and perfluoro aromatics also react[431].

$$[RuCp(CO)_2]^- \begin{cases} + RI \longrightarrow [RuCp(R)(CO)_2] \\ + RCOCl \longrightarrow [RuCp(COR)(CO)_2] \end{cases} \qquad (192)$$

A vinyl fluorine is displaced in the fluoro olefins. Displacement of fluorine in an aromatic ring occurs in a position activated towards nucleophilic displacement.

$$[RuCp(CO)_2^- \begin{cases} + CF_2{=}CFCF_3 \longrightarrow [RuCp(CO)_2\{(E)\text{—}CF{=}C(CH_3)F\}] \\ + C_6F_{10} \longrightarrow C_6F_9\text{—}Ru(CO)_2Cp \end{cases} \qquad (193)$$

$$[RuCp(CO)_2]^- \begin{cases} + C_6F_6 \longrightarrow [RuCp(CO)_2(C_6F_5)] \\ + C_5F_5N \longrightarrow 4\text{-}(C_5F_4N)\text{—}Ru(CO)_2Cp \\ + C_6F_5CN \longrightarrow [RuCp(CO)_2(4\text{-}CN\text{-}C_6F_4)] \\ + C_6F_5COCl \longrightarrow Cp(CO)_2Ru\text{—}C_6F_4\text{—}C(=O)\text{—}Ru(CO)_2Cp \end{cases} \qquad (194)$$

vi. *Osmium*. Two osmium carbonyl anions, $[Os(CO)_4]^{2-}$ ($d^{10}$) and $[HOs(CO)_4]^-$ ($d^8$), undergo oxidative addition reactions with alkyl compounds containing halogen or other leaving groups. Methyl bromide and iodide react with $[HOs(CO)_4]^-$ to give three products[432,433]. The *cis*-hydridomethyl osmium compound was not obtained pure, since it is unstable, eliminating methane at room temperature. The other two products are obtained from subsequent reactions of the hydride. A more satisfactory synthesis of the hydrido alkyl osmium complexes is through the displacement of fluorosulphonate groups[434,435]. In the displacement reaction of tosylates, deprotonation of the hydrido methyl product followed by reaction with more alkyl tosylate also

$$[OsH(CO)_4]^- + MeI \longrightarrow [cis\text{-}Os(CO)_4(Me)(H)] \xrightarrow{[HOs(CO)_4]^-} Os(Me)(CO)_4 \xrightarrow{MeI} [cis\text{-}Os(CO)_4(Me)_2]$$

$$[cis\text{-}Os(CO)_4(Me)(H)] \xrightarrow{MeI} [cis\text{-}Os(CO)_4(Me)I] + CH_4 \qquad (195)$$

gives the dimethyl osmium complex. Fluorosulphonates react rapidly, consuming all of the anion and preventing the formation of $[ROs(CO)_4]^-$. Disodium tetracarbonylosmium[434–437] is a more powerful nucleophile, reacting even with methyl

$$[OsH(CO)_4]^- + ROSO_2F \longrightarrow [cis\text{-}Os(CO)_4(H)(R)] \qquad (196)$$

$R = CD_3$, Me, $C_2F_2$

chloride[434,435]. Methyl and ethyl iodide[436] and tosylates[434,435] also yield the *cis*-dialkyl products.

$$[Os(CO)_4]^{2-} + RX \longrightarrow [cis\text{-}Os(CO)_4R_2] \qquad (197)$$

R = Me, Et
X = Cl, I, OTs

d. *Oxidative addition to Group VIB and VIIB complexes*

i. *Manganese and rhenium*. Most of the oxidative addition chemistry of the manganese triad is that of the $d^8$ anions, particularly the pentacarbonyl anions[203–206,370]. The pentacarbonyl anion of rhenium is more nucleophilic than that of manganese[371].

Primary alkyl halides react with the manganese pentacarbonyl anion to yield the $d^6$, alkylpentacarbonylmanganese(I), 18-electron complexes[373,438–440].

$$RX + [Mn(CO)_5]^- \longrightarrow [MnR(CO)_5] + I^- \qquad (198)$$

X = I; R = Me, Et, $Pr^n$
X = Cl; $R = PhCH_2$
$X = SO_4Me$; R = Me

$$XRX + 2[Mn(CO)_5]^- \longrightarrow [(CO)_5MnRMn(CO)_5] \quad (199)$$

$X = Br$; $R = (CH_2)_3$, $(CH_2)_2CH(Me)$

The same type of reaction takes place with the manganese anion in which a carbonyl has been replaced with phosphine[441,442]. Other functional groups in the organohalide can be tolerated[382,384–386], including acetylenic alcohols[443] and esters[444]. The presence of an amine group in the organic reactant can lead to rearrangement from the alkyl to acyl species[386].

$$[Mn(CO)_5]^- + ClCH_2C{\equiv}C(CH_2)_nC(OH)RR' \longrightarrow [Mn(CO)_5\{CH_2C{\equiv}C(CH_2)_nC(OH)RR'\}] \quad (200)$$

$R = H$, Me; $R' = H$, Me, Ph; $n = 0, 1$

$$[Mn(CO)_5]^- + ClCH_2O_2CBu^t \longrightarrow [Mn(CO)_5\{CH_2O_2CBu^t\}] \quad (201)$$

$$[Mn(CO)_5]^- + 2\text{-pyridyl-}CH_2Cl \longrightarrow \left[ \text{(OC)}_2(\text{CO})_2\text{Mn}(\text{C(=O)CH}_2\text{-2-pyridyl-}N) \right] \quad (202)$$

The reactions of perfluoroalkyl iodides generally produce the manganese iodide and the fluorocarbon portion decomposes by coupling or by $\beta$-elimination of fluoride[208,445]. Consequently, the perfluoroalkyl manganese derivatives usually are obtained by decarbonylation of the acyl derivatives (see below). The alkyl derivatives $[RMn(CO)_5]$, where R is a secondary or tertiary organic group, appear to de unstable, yielding hydrocarbons and dimanganese decacarbonyl[446]. Tropyllium bromide, for example, undergoes this coupling reaction.

$$[Mn(CO)_5]^- \begin{cases} + CF_3I \longrightarrow [MnI(CO)_5] + \frac{1}{2}CF_3CF_3 \\ + C_2F_5I \longrightarrow [MnI(CO)_5] + CF_2{=}CF_2 + F^- \end{cases} \quad (203)$$

Allyl halides give a $\sigma$-allyl derivative, although heating under vacuum induces loss of carbon monoxide and generates the $\pi$-allyl complex[209,447–449]. Perfluoroallyl chloride gives a $\sigma$-product[450] in which a rearrangement of the double bond has taken place to give the ($E$)-vinyl derivative[451].

Oxidative addition of vinyl halides occurs in a number of cases in which a stabilized intermediate anion from the addition to the double bond results[452,453]. In

$$[Mn(CO)_5]^- + R^2CH{=}C(R^1)CH_2Cl \longrightarrow [Mn(CO)_5\{CH_2C(R'){=}CH{\sim}R^2\}] \xrightarrow[\Delta]{-CO} [(\eta^3\text{-}CH_2C(R^1)CHR^2)Mn(CO)_4] \quad (204)$$

| $R^1$ | $R^2$ |
|---|---|
| H | H |
| H | Me |
| H | Cl |
| Me | H (20:80 $Z$:$E$) |

$$[Mn(CO)_5]^- \left\{ \begin{array}{l} + ClCX{=}C(CN)_2 \longrightarrow [Mn(CO)_5\{C(Cl)(X)C(CN)_2\}] \longrightarrow \\ \qquad \longrightarrow [(CO)_5MnCX{=}C(CN)_2] \\ + 1,2\text{-}Cl_2\text{-}3,3,4,4\text{-}F_4\text{-cyclobut-1-ene} \longrightarrow [(CO)_5Mn\text{-}C_4Cl_2F_4] \\ \qquad \longrightarrow (CO)_5Mn\text{-}C_4ClF_4 \end{array} \right. \quad (205)$$

these examples, the reaction appears to proceed by addition–elimination. On the other hand, $\alpha$-chloroenamines react to yield the acyl manganese complex[454].

$$[Mn(CO)_5]^- + Me_2NC(Cl){=}CMe_2 \longrightarrow [(CO)_4Mn\{C(O)C(=CMe_2)NMe_2\}] \quad (206)$$

Acid chlorides react to give an acyl derivative (Table 13) and, because decarbonylation occurs under relatively mild conditions, this is often the method of choice for the synthesis of the alkyl complexes. The ease with which decarbonylation occurs follows the order $[Co(OCR)(CO)_4] > [Mn(OCR)(CO)_5] > [(Re(COR)(CO)_5] \approx [Mo(COR)Cp(CO)_3] > [FeCp(COR)(CO)_2] > [WCp(COR)(CO)_3]$[206]. The acylation

TABLE 13. Acylation of $[Mn(CO)_5]^-$ and subsequent decarbonylation

| R | | |
|---|---|---|
| $[Mn(COR)(CO)_5]$ | $[MnR(CO)_5]$ | Ref. |
| Me | Me | 455 |
| Et | Et | 373, 455 |
| $Pr^i$ | $Pr^i$ | 455 |
| $ClCH_2CH_2$ | — | 456 |
| $C_5H_4M(CO)_3$ (M = Mn, Re) | — | 457 |
| $Cl(CH_2)_3$ | $(CH_2)_3$ [a] | 440 |
| $CH_3CO$ | — | 458 |
| Ph | Ph | 455 |
| EtO | — | 459 |
| $XCH_2$ (X = Cl, F) | | 460 |
| $H(CF_2)_2$ | $H(CF_2)_2$ | 445 |
| $R_f$ ($R_f = CF_3$, $i$-$C_3F_7$, $n$-$C_3F_7$, $Cl(CF_2)_4$, $H(CF_2)_4$, $CF_2{=}CFCF_2$) | $R_f$ | 208 |
| $(CF_2)_3$ | $(CF_2)_3$ [a] | 440 |
| Me [b] | — | 441 |
| $CF_3$ [b] | — | 441 |

[a] Complexes of the types MnCORCOMn or MnRMn are formed.
[b] Reaction with $[Mn(CO)_4\{P(C_6H_{11})_3\}]$

reaction proceeds readily at ambient temperature and decarbonylation can be achieved at 80–100 °C.

$$[Mn(CO)_5]^- + RCOCl \longrightarrow [Mn(COR)(CO)_5] \xrightarrow{\Delta} [RMn(CO)_5] \quad (207)$$

The reaction of varous alkyl halides with the manganese pentacarbonyl anion and its monophosphine substituted derivative follows a second-order rate law[461]. The order of reactivities for $[Mn(CO)_4L]$ is $L = PMe_2Ph > PPh_3 > P(OPh)_3 > CO$ (an order expected on the basis of nucleophilicity) and $PhCH_2Br > CH_2{=}CHCH_2Br > PhCH_2Cl > CH_2{=}CHCH_2Cl$. The rates are reduced significantly on the addition of certain complexing agents such as hmpa or crown ethers to the thf solvent. This reduction in rate has been attributed to a larger negative entropy of activation in the transition state.

The stereochemistry of the reaction proceeds with inversion of configuration at the carbon bearing the halide. Two different tests have been utilized. Chiral 1-phenylethyl bromide reacts with $[Mn(CO)_4]^-$ to give the corresponding derivatives containing a manganese—carbon $\sigma$-bond[424]. Treatment with triphenylphosphine causes an alkyl–acyl rearrangement, which is reasonably presumed to occur with retention of configuration at carbon. Subsequent cleavage to the acid gave a chiral acid in which overall inversion in the oxidative addition step had occurred. Also, reaction of $[Mn(CO)_4(PEt_3)]^-$ with *erythro*-1,2-dideuterio-1-phenylethyl tosylate gave the *threo*-alkyl manganese product[462]. Radical intermediates do not play an important role in these reactions[425].

$$Ph\overset{*}{C}H(Me)Br + [Mn(CO)_5]^- \xrightarrow{thf} [Mn(CO)_5\{CH(Me)Ph\}]$$

$$\xrightarrow{PPh_3} [Mn(CO)_4(PPh_3)\{COCH(Me)Ph\}] \xrightarrow[H_2O]{Br_2} Ph\overset{*}{C}H(Me)CO_2H \quad (208)$$

Ph; D, H; H, D; OTs $\xrightarrow{[Mn(CO)_4PEt_3]^-}$ [Ph; H, D; H, D; $Mn(CO)_4(PEt_3)$] (209)

The rheniumpentacarbonyl anion undergoes reactions with organic halides similar to those of the manganesepentacarbonyl anion. Alkyl halides[373,463] and the cycloheptatriene cation[464] react to give the $\sigma$-rhenium products.

$$RX + [Re(CO)_5]^- \longrightarrow [ReR(CO)_5] \quad (210)$$

| R | X |
|---|---|
| Me | I |
| Et | I |
| $PhCH_2$ | Cl |
| $C_7H_7^+$ | $BF_4^-$ |

The latter type of complex represents the first $\eta^1$-cycloheptatriene derivative of a transition metal.

The $d^6$ bis($\eta^5$-cyclopentadienyl)rhenium(I) anion reacts with alkyl halides in the presence of pentamethyldiethylene triamine to give the monoalkyl rhenium(III) $\sigma$-complex[465]. Alkylation also is achieved using methyl tosylate[466]. Addition of a second equivalent of methyl iodide followed by the addition of a soft anion yields the $d^2$ dimethylrhenium(V) cation[466].

$$[ReCp_2]^- Li^+ + RX \longrightarrow [ReRCp_2] \qquad (211)$$

X = Br; R = Et, $Pr^n$, CH=CHCH$_2$
X = Cl; R = Me
X = OTs; R = Me

$$[ReCp_2(Me)] \xrightarrow[(2)\ NH_4PF_6]{(1)\ MeI} [ReCp_2Me_2]^+PF_6^- \qquad (212)$$

The bis($\eta^5$-cyclopentadienyl)rhenium hydride also reacts with methyl iodide and allyl bromide to yield the $d^4$ methylrhenium(III) derivative and a rhenium(III) olefin complex, respectively[466].

$$[ReHCp_2] \begin{cases} +MeI \xrightarrow{-I^+} [ReHCp_2(Me)]^+ \xrightarrow{-H^+} Cp_2Re{-}Me \\ +CH_2{=}CHCH_2Br \xrightarrow{PF_6^-} [ReHCp_2(CH_2CH{=}CH_2)]^-PF_6^- \\ \qquad \longrightarrow [ReCp_2(\eta^2\text{-}CH_2{=}CHMe)]^+PF_6^- \end{cases} \qquad (213)$$

Rhenium pentacarbonyl anion reacts with perfluoro olefins, allenes, and acetylenes, some of the reactions occurring with the same patterns as exhibited by

$$[Re(CO)_5]^- \begin{cases} +CF_3C{\equiv}CCF_3 \longrightarrow [Re(CO)_5\{C(CF_3){=}C{=}CF_2\}] + \left[\text{bicyclic diene bearing } CF_3, Re(CO)_5, CF_3, Re(CO)_5\right] \\ +CF_2{=}CFCF{=}CF_2 \longrightarrow [Re(CO)_5(E\text{-}CF{=}CFCF{=}CF_2)] \\ +(CF_3)_2C{=}C{=}C(CF_3)_2 \longrightarrow [ReI(CO)_5\{C[{=}C(CF_3)_2]C(CF_3){=}CF_2\}] \end{cases} \qquad (214)$$

the manganese pentacarbonyl anion[401,406,467]. The reaction with an $\alpha$-chloroenamine yields an acyl complex, but with different bonding at the enamine end of the organic ligand[454].

$$[Re(CO)_5]^- + Me_2NC(Cl){=}CMe_2 \longrightarrow \left[(CO)_4Re\langle C({=}O)\text{-}CMe_2\text{-}C({=}NMe_2)\rangle\right] \qquad (215)$$

The reaction of rhenium pentacarbonyl anion with acid chlorides yields acyl derivatives that also can be decarbonylated by heat or irradiation (Table 14).

$$[Re(CO)_5]^- + RCOCl \longrightarrow [Re(COR)(CO)_5] \xrightarrow{-CO} [ReR(CO)_5] \qquad (216)$$

TABLE 14. Acylation of $[Re(CO)_5]^-$ and subsequent decarbonylation

| R | | |
|---|---|---|
| $[Re(COR)(CO)_5]$ | $[ReR(CO)_5]$ | Ref. |
| Me | Me | 463 |
| Et | — | 373 |
| $C_5H_4Mn(CO)_3$ | — | 466 |
| Ph | Ph | 463 |
| $C_7H_7$ | $C_7H_7$ | 464 |
| $R_f$ ($R_f = C_2F_5$, $C_3F_7$) | $R_f$ | 450 |

ii. *Chromium, molybdenum, and tungsten.* Within the series of anionic complexes of the general formula $[MCp(CO)_3]^-$, the order of nucleophilicity is W>Mo>Cr[203–206]; the anionic complexes of the type $[M(CO)_5CN]^-$ show very poor nucleophilicity[371]. Undoubtedly the difficulty or inability of the anions in this triad of the general formula $[MCp(CO)_3]^-$ to react with certain halides, particularly those which are reluctant to undergo nucleophilic substitution (vinyl and aryl, for example), is due to the poor nucleophilicity of these anions. The chromium complex displaces iodide from methyl and ethyl iodide giving a stable $\sigma$-complex[372,373]. Acetylide derivatives also have been obtained by this reaction[468].

$$[CrCp(CO)_3]^- + RX \longrightarrow [CrCp(CO)_3R] + X^- \qquad (217)$$

R = MeI, EtI, PhC≡CBr

An $\eta^3$-allyl chromium complex can be obtained from the reaction of a halotricarbonyl chromium anion containing a bidentate nitrogen ligand[469].

$$[Cr(CO)_3(\widehat{NN})X]^- + CH_2{=}CHCH_2Cl \longrightarrow [CrX(CO)_2(\eta^3C_3H_7)(\widehat{NN})] \qquad (218)$$

$\widehat{NN}$ = bipy or phen

Chromium(II) salts reduce organic halides to the corresponding hydrocarbon derivative; in some examples coupling products are obtained[470]. Benzyl chromium derivatives can be isolated from the reaction of chromium perchlorate and benzyl halides in aqueous medium[471,472], and these benzyl derivatives react to give the reduced product. The mechanism of this reaction has been demonstrated to proceed by a one-electron inner sphere reduction[473–475]. The relative rates of reduction of organic halides, tertiary>secondary>primary and I>Br>Cl, as well as the observation that cyclopropylcarbinyl chloride gives only but-1-ene and that 6-bromohex-1-ene produces a mixture of hex-1-ene and the cyclized product, methylcyclopentane, supports the mechanism shown[475].

$$PhCH_2X + Cr^{2+} \longrightarrow [Cr^{III}(CH_2Ph)(H_2O)_5]^{2+}ClO_4^- \qquad (219)$$

X = Cl, Br, I

$$RX + [Cr^{II}(en)_2]^{2+} \xrightarrow[\text{determining}]{\text{rate}} [Cr^{III}X(en)_2]^{2+} + R^\bullet \qquad (220)$$

$$R^\bullet + [Cr^{II}(en)_2]^{2+} \xrightarrow{\text{fast}} [Cr^{III}R(en)_2]^{2+} \qquad (221)$$

A complex of chromium(II) and a macrocyclic ligand containing four nitrogens yields alkyl chromium(III) products from reactions with alkyl halides in aqueous solvents[476].

$$[Cr(N_4\text{-macrocycle})]^{2+} + RX \longrightarrow [Cr(R)(H_2O)(N_4\text{-macrocycle})] \quad (222)$$

RX = MeI, EtBr, EtI, $Pr^iBr$, $Pr^iI$, cyclo-$C_5H_9Br$, cyclo-$C_6H_{11}Br$, $Bu^tBr$, 1-adamantyl-Br

A large number of neutral complexes containing a molybdenum—carbon bond, most of which are $\eta^3$, have been synthesized. Alkyl halides react with the $\eta^5$-cyclopentadienylmolybdenum(0) anion to give the molybdenum(II) $\sigma$-complexes in high yield[372,384–386], and methylene halides are metallated at only one of the two halogens[477]. This reaction with benzyl halides produces the $\sigma$-benzyl-type complex

$$[MoCp(CO)_3]^- + RI \longrightarrow [MoCp(CO)_3(R)] \quad (223)$$

R = Me, Et, $Pr^i$, $ClCH_2$, $ICH_2$

that goes to the $\pi$-benzyl species on irradiation[382,478–480]. The X-ray structure shows an alteration of the carbon—carbon bond distance in the benzene ring, characteristic of a non-aromatic alternating single–double bond structure with frozen suprafacial bonding[479]. In solution, the structure of this type of complex reveals that molybdenum has access to all four positions (edge and face combinations) for bonding to the benzyl group, demonstrating that both suprafacial and antrafacial shifts take place. The mechanism proposed[480] for the antarafacial shift included the generation of the short lived (undected) 16-electron $\sigma$-bonded benzyl intermediate that undergoes rapid 180° rotation of the phenyl ring to place molybdenum on the opposite face, then reverting to the $\pi$-benzyl structure capable of rapid suprafacial jumps.

$$[MoCp(CO)_2]^- + PhCH_2X \xrightarrow[(-CO)]{h\nu} \pi\text{-benzyl}(H_a, H_b)\text{–}M \rightleftharpoons H_aH_bC(C_6H_5)\text{–}M \rightleftharpoons \pi\text{-benzyl}(H_a, H_b)\text{–}M \quad (224)$$

M = $[MoCp(CO)_2]$

Allyl halides also give the $\sigma$-allyl product, which loses carbon monoxide *in vacuo* or

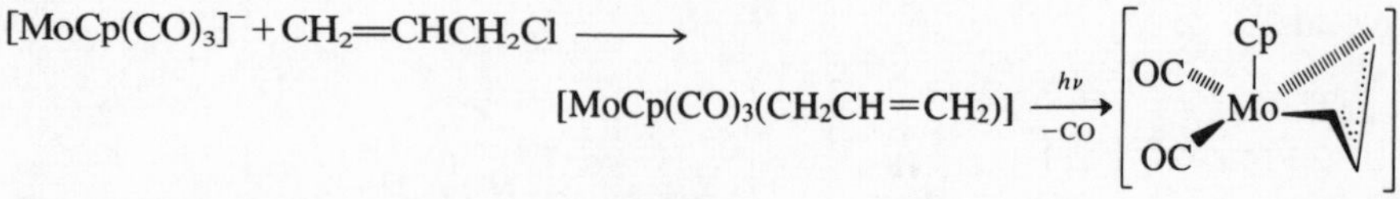

$$[MoCp(CO)_3]^- + CH_2{=}CHCH_2Cl \longrightarrow [MoCp(CO)_3(CH_2CH{=}CH_2)] \xrightarrow[-CO]{h\nu} [MoCp(CO)_2(\eta^3\text{-}C_3H_5)] \quad (225)$$

on irradiation to generate the $\pi$-allyl product[481,482]. This type of complex also is formed from the indenyl molybdenum analogue[483]. Other anionic molybdenum complexes containing ligands other than cyclopentadienyl ($\beta$-diketonate and halide) react rapidly with allyl halides under mild conditions[484,485].

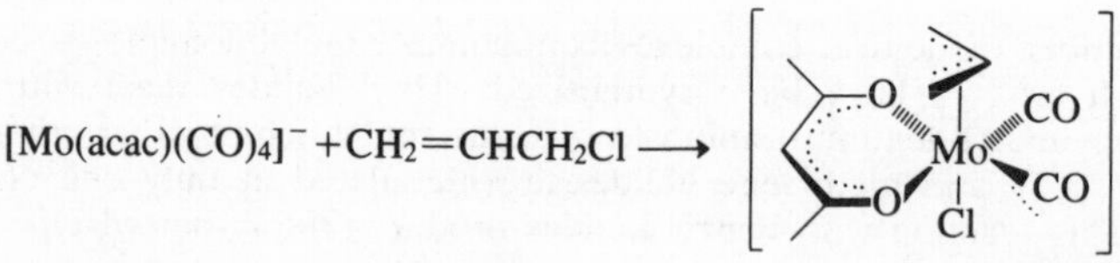

$$[Mo(acac)(CO)_4]^- + CH_2{=}CHCH_2Cl \longrightarrow [Mo(acac)(\eta^3\text{-}C_3H_5)(CO)_2Cl] \quad (226)$$

$$[(N\text{–}N)MoX(CO)_3]^- + R'CH{=}C(R)CH_2X \longrightarrow [(N\text{–}N)Mo(\eta^3\text{-}R'CHC(R)CH_2)(CO)_2X] \quad (227)$$

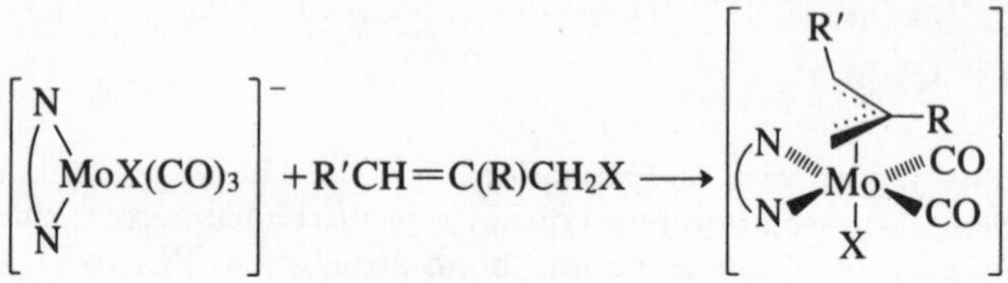

MN = bipy or phen;
X = Cl, Br, I

Certain neutral 18-electron molybdenum(o) complexes also are alkylated. The reaction of *cis*-$[Mo(CO)_2(dmpe)_2]$ with various alkyl halides that forms only *cis*-$[MoX(CO)_2(dmpe)_2]X$ and coupled product, R—R shows the characteristics of a radical chain mechanism[486,487]. Allyl halides, however, react to give the $\eta^3$-allyl complexes, a wide variety of these complexes being prepared[488,489].

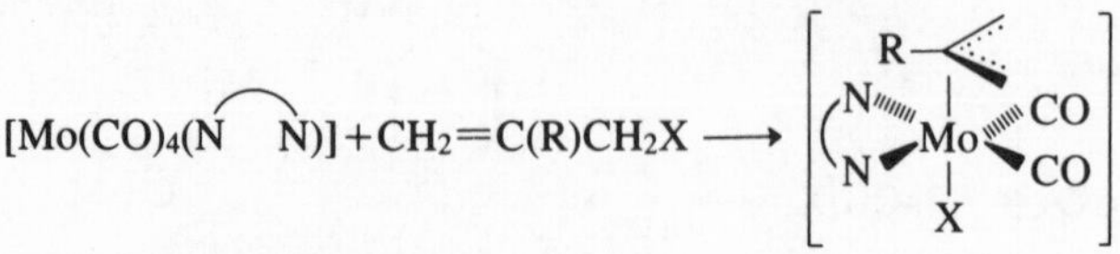

$$[Mo(CO)_4(N\frown N)] + CH_2{=}C(R)CH_2X \longrightarrow [(N\frown N)Mo(\eta^3\text{-}CH_2C(R)CH_2)(CO)_2X] \quad (228)$$

R = H, Me; X = Cl, Br, I;
$\widehat{NN}$ = bipy, phen di(2-pyridyl)amine

When the allyl halide is triphenylcyclopropenyl bromide, both the $\eta^3$-cyclobutenone and alkyl structures are obtained, probably by the reaction pathways shown[490].

Acetylenic[468], allenic[491], and aryl[492] halides all react with $\eta^5$-cyclopentadienylmolybdenumtricarbonyl anion. In the case of the allenic bromide, dimerization of the allene and carbon monoxide 'insertion' occur[491].

$$[Mo(CO)_4(N\frown N)] \longrightarrow \ldots \qquad (229)$$

$$[MoCp(CO)_3(C{\equiv}CPh)] \xleftarrow{PhC{\equiv}CBr} [MoCp(CO)_3]^- \xrightarrow{ArCl} [MoCp(CO)_3(Ar)]$$

$$[MoCp(CO)_3]^- \xrightarrow{Br(CH_2)_2C(R')=C=CHR''} [Cp(CO)_2Mo\text{—}C(=CH_2)CH_2CH_2C(R^1)=C=CHR^2] \qquad (230)$$

Ar = Ph, 1-naphthyl

| $R^1$ | $R^2$ |
|---|---|
| H | H |
| H | Me |
| Me | H |

Acyl derivatives, $[MoCp(CO)_3(COR)]$, are difficult to obtain since decarbonylation occurs readily and only small amounts of this type of product are obtained[409,410]. The acyl complex can be obtained from perfluoroacid chlorides, which are difficult to decarbonylate[450].

$$R_fCOCl + [MoCp(CO)_3]^- \longrightarrow [MoCp(CO)_3(COR_f)] \qquad (231)$$

The mechanism of the oxidative addition reaction of alkyl halides and their derivatives with $[MoCp(CO)_3]^-$ in most cases probably takes place by nucleophilic displacement[425]. The reaction is second order[493], and the reaction with *threo*-1,2-dideuterio-2-*t*-butylethyl triflate gives the *erythro* organometallic[249].

$$(Bu^t)(H)(H)C\text{—}C(D)(D)(OTf) + [MoCp(CO)_3]^- \longrightarrow [(Bu^t)(D)(H)C\text{—}C(H)(D)((CO)_3MoCp)] \qquad (232)$$

Oxidative addition reactions to $d^4$ molybdenum(II) and $d^2$ molybdenum complexes have been observed. Bis($\eta^5$-cyclopentadenyl)molybdenumcarbonyl yields the 18-electron molybdenum(II) cation[494], and bis $\eta^5$-cyclopentadienylmolybdenum dihydride reacts with trifluoromethyl fluorosulphonate to give 18-electron hexavalent $d^0$ molybdenum[495].

$$[MoCp_2(CO)] + RX \longrightarrow [MoCp_2(R)(CO)]^+ \quad (233)$$

R = Me, $CH_2{=}CHCH_2$, $MeOCH_2$, $CH_3CO$, $CH_2CO_2Me$, $CH_2CN$

$$[MoH_2Cp_2] + CF_3SO_3F \longrightarrow [MoH_2Cp_2(CF_3)]^+SO_3F^- \quad (234)$$

The tungsten anion, $[WCp(CO)_3]^-$, the most nucleophilic of the triad, oxidatively adds alkyl halides to give complexes containing metal—carbon $\sigma$-bonds that are more stable than those in the chromium or molybdenum derivatives[372,382,385,386,496]. The propensity to lose carbon monoxide on photolysis and convert the $\sigma$-bond to a $\pi$-benzyl-type complex[386] or to 'insert' carbon monoxide into the $\sigma$-bond[386] forming an acyl complex is considerably diminished in the tungsten complexes as compared with molybdenum.

$$RX + [WCp(CO)_3]^- \longrightarrow [WCp(CO)_3(R)] \quad (235)$$

RX = MeI, EtI, PhC≡CBr, $MeSCH_2Cl$, 2-pyridyl-$CH_2Cl$, 2-thienyl-$CH_2Br$

Reaction of an acetylenic bromide not only produces the expected organotungsten species, but also the tungsten dimer, $[WCp(CO)_3]_2$[468]. Reactive vinyl halides are metallated, probably by an addition to the double bond to yield a stabilized anion followed by halide elimination[398,497].

$$[WCp(CO)_3]^- + ROCCH{=}CHCl \longrightarrow \left[ R{-}\overset{O^-}{\overset{\|}{C}}{=}CH{-}\underset{Cl}{\underset{|}{C}}H_2{-}WCp(CO)_3 \right] \longrightarrow ROCCH{=}CHWCp(CO)_3 \quad (236)$$

R = Me, Ph, $p$-$BrC_6H_4$

Acid chlorides give acyl complexes which are generally stable towards decarbonylation[409,410].

$$[WCp(CO)_3]^- + RCOCl \longrightarrow [WCp(CO)_3(COR)] \quad (237)$$

In reactions with alkyl tosylates, inversion of configuration at carbon is observed; again, a nucleophilic displacement reaction appears to be the most likely mechanism[462].

$$[WCp(CO)_2(PEt_3)]^- + \text{(Newman projection: Ph; D, H; H, D; OTs)} \longrightarrow \text{(Newman projection: Ph; H, D; H, D; WCp(CO)}_2\text{(PEt}_3\text{))} \quad (238)$$

*erythro*

Allyl complexes of tungsten can be prepared by the oxidative addition reactions of allyl halides to a variety of anionic or neutral complexes. In a reaction with allyl iodide, the acetonylacetonate anion yields an $\eta^3$-allyl product[484].

$$Ph_4P^+[W(CO)_4(acac)]^- + CH_2{=}CHCH_2I \longrightarrow [(\eta^3\text{-allyl})W(acac)(CO)_2I] \quad (239)$$

Anionic tungsten(0) complexes containing amine bidentate ligands also give the $\eta^3$-allyl species in a reaction with allyl halides[485,489]. The neutral tungsten analogues, $(\widehat{NN})W(CO)_4$ ($\widehat{NN}$ = bipy or phen) do not react with allyl halides even in refluxing

$$[(\widehat{NN})WX(CO)_3]^- + R^2CH{=}C(R^1)CH_2X \longrightarrow [(\eta^3\text{-}R^2CHC(R^1)CH_2)W(\widehat{NN})(CO)_2X] \quad (240)$$

tetrahydrofuran. Replacement of one carbonyl with pyridine increases the nucleophilicity enough to promote reaction with allyl chloride, although little reaction takes place with metallyl chloride[488].

$$[(\widehat{N\ N})W(CO)_3(py)] + CH_2{=}CHCH_2Cl \longrightarrow [(\eta^3\text{-allyl})W(\widehat{NN})(CO)_2Cl] \quad (241)$$

A chloro-bridged anionic tungsten dimer reacts with allyl chloride in acetonitrile to yield a monomeric $\pi$-allyl complex[498]. Irradiation is required in order to promote the

$$[(CO)_3WCl_3W(CO)_3]^{3-} + CH_2{=}CHCH_2Cl \xrightarrow{MeCN} [(\eta^3\text{-allyl})W(MeCN)_2(CO)_2Cl] \quad (242)$$

reaction of tungsten hexacarbonyl with allyl halides[499]. Although allyl bromide and iodide give the expected $\pi$-allyl products, a dimer, $[W_2(CO)_6(C_3H_7)Cl_3]$, is obtained from the chloride.

$$[W(CO)_6] + CH_2{=}CHCH_2X \longrightarrow [WX(CO)_4(\eta^3\text{-allyl})] \quad (243)$$

X = Br, I

The reactions of bis($\eta^5$-cyclopentadienyl)molybdenum(II) (carbonyl or ethylene) complexes with reactive organic halides produces an oxidative addition product, a $d^2$, 18-electron tungsten(IV) cation, isolated as a hexafluorophosphate salt[494].

$$[WCp_2(L)] + RX \longrightarrow [WCp_2(L)(R)]^+PF_6^- \quad (244)$$

L = CO; R = Me, $CH_2CH{=}CH_2$, $CH_2OMe$, COMe, $CH_2CO_2Me$, $CH_2CN$

L = $C_2H_4$; R = $CH_2{-}CH{=}CH_2$, $CH_2CO_2Me$, $CH_2CN$

### 3. The early transition metals

Reports of oxidative addition reactions of the early transition metals with organic halides are scarce. Most of the organometallic complexes containing a metal—carbon bond are obtained by a reaction of a transition metal halide derivative with a Group I or II organometal, such as an organolithium or a Grignard reagent. The low oxidation state and, in many complexes, the high degree of coordinative unsaturation, make the complexes highly reactive even toward reaction solvents.

Vanadocene, a 15-electron paramagnetic complex, for example, reacts with organic halides (benzyl chloride, ethyl bromide, methyl iodide) to give vanadocene halides, a 16-electron vanadium(III) complex, and coupled product (e.g. bibenzyl)[500]. The anionic ($d^4$, 18-electron) vanadium(I) complex, $[VHCp(CO)_3]^-$, reacts with various alkyl halides and with cyclopropylcarbinyl halides (or tosylate) to yield $[VXCp(CO)_3]^-$ and, in the case of the cyclopropylcarbinyl compounds, either but-1-ene or methylcyclopropane, depending on the leaving group[501]. With the iodide, only but-1-ene is obtained, indicative of a radical chain reaction; oxidative addition of tosylate apparently takes place by a nucleophilic substitution, but rapid reductive elimination of alkyl and hydride does not allow the isolation of the alkyl vanadium species.

$$RCH_2OTs + [VHCp(CO)_3]^- \longrightarrow [VHCp(CO)_3(CH_2R)] \longrightarrow RMe + [VCp(CO)_3] \quad (245)$$

R = cyclopropyl

Nevertheless, certain anionic vanadium carbonyls or carbonyl derivatives give isolable oxidative addition products with reactive organic halides. Vanadium is a unique transition metal in that it forms a large number of neutral monomeric odd-electron carbonyls that do not tend to dimerize. Vanadium hexacarbonyl anion, an 18-electron complex, does not react with even the most reactive organic halides, including methyl iodide and perfluoroacetyl chloride[502]. Consequently, photolysis is necessary to promote the loss of carbon monoxide, giving a 16-electron anion that undergoes oxidative addition with allyl chloride[503,504]. Triphenylcyclopropenyl bromide also reacts, but evidently u.v. irradiation is not required[504].

$$Na^+[V(CO)_6]^- + R'CH{=}C(R)CH_2Cl \xrightarrow[20\,^\circ C]{h\nu} [V(CO)_5(\eta^3\text{-1-R'-2-R-allyl})] + CO + NaX \quad (246)$$

| R | R′ |
|---|---|
| H | H |
| Me | H |
| H | Me |
| H | Cl |
| Cl | H |

$$Na^+[V(CO)_6]^- + \text{(1-bromo-1,2,3-triphenylcyclopropene)} \longrightarrow [(\eta^3\text{-}C_3Ph_3)V(CO)_5] \quad (247)$$

Substitution of phosphines or arsines for carbonyl enhances the nucleophilicity at vanadium, and the oxidative addition of methyl iodide or allyl chloride does not require u.v. irradiation[502]. The reactions of a number of anionic vanadium carbonyls,

$$Et_4N^+[V(CO)_4(diars)]^- \begin{cases} + MeI \longrightarrow [V(CO)_4(diars)(Me)] \\ + CH_2{=}CHCH_2Cl \xrightarrow{thf} [V(CO)_3(diars)(\eta^3\text{-allyl})] \end{cases} \quad (248)$$

diars = (o-phenylene with two $AsMe_2$ groups)

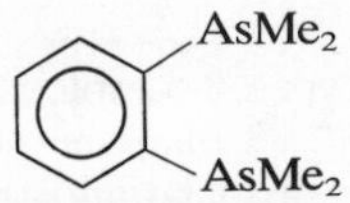

in which one or more carbonyls have been replaced by triphenylphosphine[504] or diphosphines[505,506] and diarsines[505], with a variety of allyl chlorides have been carried out and the crystal structures of two key complexes have been determined[507,508].

$$[V(CO)_5(PPh_3)]^- + R^1CH{=}C(R)CH_2Cl \longrightarrow [V(CO)_4(PPh_3)(\eta^3\text{-}R^1CHC(R)CH_2)] \quad (249)$$

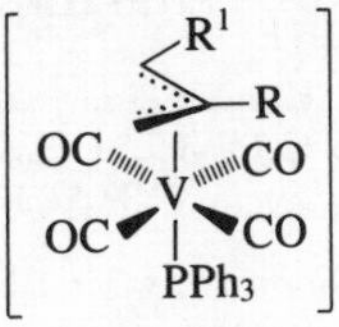

$$[V(CO)_4(L)_2]^- + R^2(R^3)C{=}C(R^1)CH_2X \longrightarrow [V(CO)_2(L_2)(\eta^3\text{-}R^2R^3CC(R^1)CH_2)] \quad (250)$$

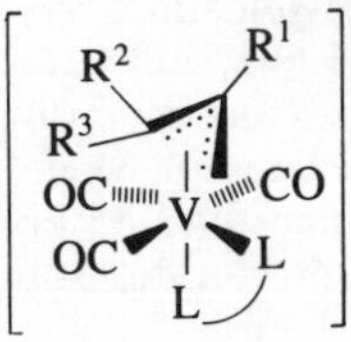

$L_2$ = diphos, diars, $Ph_2AsCH_2CH_2PPh_2$

| $R^1$ | $R^2$ | $R^3$ |
|---|---|---|
| H | H | H |
| Me | H | H |
| H | Me | H |
| H | Me | Me |
| H | Ph | H |

The reaction with the triphenylcyclopropenyl cation yields an acyl cyclopropyl derivative in which 1 mol of hydrogen has added to the cyclopropene double bond; the source of hydrogen is uncertain[509].

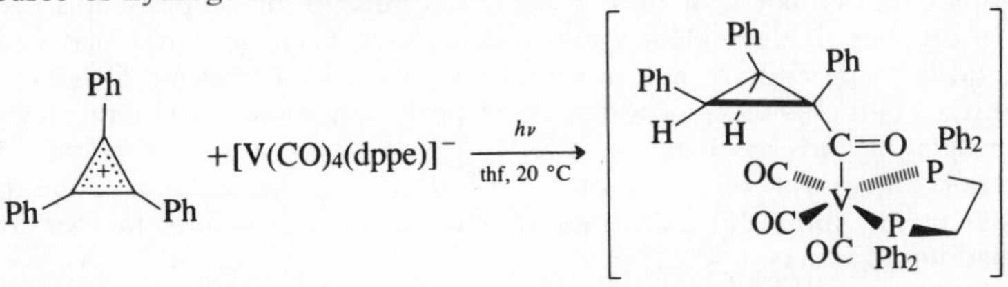

$$[C_3Ph_3]^+ + [V(CO)_4(dppe)]^- \xrightarrow[\text{thf, 20 °C}]{h\nu} [V(CO)_3(dppe)(C(=O)C_3H_2Ph_3)] \quad (251)$$

Although titanocene and its derivatives have received much attention, particularly with respect to nitrogen fixation and hydrocarbon activation, only a few oxidative addition reactions with organic halides are known[510]. Because titanocene is a 14-electron complex, it is extremely reactive, but its dicarbonyl 18-electron complex is much less so. Titanocenediarbonyl adds alkyl iodides with displacement of one carbonyl and 'insertion' of the other to give a 16-electron ($d^0$) titanium(IV) complex[511,512]. In the reaction of acid chlorides, both carbonyls are lost. The reaction with benzyl chloride yields only bibenzyl and titanocene dichloride[512].

$$[TiClCp_2(COR)] \xleftarrow{RCOCl} [TiCp_2(CO)_2] \xrightarrow{RI} [TiICp_2(COR)] \quad (252)$$

R = Me, Ph $\qquad$ R = Me, Et, $Pr^i$, $Bu^n$

The same type of oxidative addition reaction of methyl iodide takes place with zirconocene dicarbonyl at 100 °C[513]. The coordination of phosphine ligands in place

$$[ZrCp_2(CO)_2] + MeI \longrightarrow [ZrICp_2(Me)] \quad (253)$$

of carbonyls moderates the reactivity of zironocene, but this type of complex is still approximately ten times more reactive toward alkyl halides[514–516] than the neutral rhodium(I) dimethylglyoxime-type complexes[229,280]. Phosphine dissociation is necessary for oxidative addition. The rates of the reaction and the selectivity toward the alkyl zirconium product (*vs.* dihalide) fall in the order I>Br>Cl. The reaction with

$$[ZrCp_2(L)_2] + RX \longrightarrow [ZrXCp_2(R)] + [ZrX_2Cp_2] \quad (254)$$

L = $PPh_2Me$, $PPhMe_2$; $L_2$ = dppe, dmpe; RX = $Bu^nCl$, $Bu^nBr$, $Bu^nOTs$, *n*-$C_6H_{13}Br$, *n*-$C_8H_{17}I$, *n*-$C_8H_{17}Cl$, $CH_2{=}CH(CH_2)_4Cl$; *cyclo*-$C_6H_{11}Cl$, *cyclo*-$C_6H_{11}Br$, $Bu^sCl$, $Bu^sBr$, $Bu^sI$, *cyclo*-$C_6H_{11}CH_2Cl$, $Bu^tCl$, $Bu^tBr$

*erythro*-1,2-dideuterio-2-*t*-butylbromide leads to an *erythro*/*threo* product mixture, unreacted bromide remaining *erythro.* Cyclized product is obtained in the reaction with 6-chlorohex-1-ene, and CIDNP is observed during the reaction.

The amount of zirconocene dihalide is dependent on the structure of the organic halide. Tertiary halides in which the organic group is sterically bulky give higher yields of the dihalide than do primary, unhindered halides. Tertiary butyl chloride gives zirconocene dichloride, isobutane, and isobutene. These results are consistent with a radical mechanism.

$$[ZrCp_2(L)_2] \underset{+L}{\overset{-L}{\rightleftharpoons}} [ZrCp_2(L)] \xrightarrow{RX} [ZrXCp_2(L)] + R\cdot \xrightarrow{-L} [ZrXCp(R)]$$

$$\downarrow RX \quad (255)$$

$$[ZrX_2Cp_2] + R\cdot \longrightarrow \tfrac{1}{2}RH + \tfrac{1}{2}R(-H)$$

### 4. Oxidative addition to metal dimers

There are a number of metal clusters that undergo oxidative addition of organic halides, and hydrocarbons, usually with a gross rearrangement of the basic structure of the complex. A discussion of these reactions is beyond the scope of this review. Recently, a number of low-valent metal dimers have been prepared that contain metal—metal bonding and are held together with the aid of bridging ligands.

The synthesis and chemistry of a palladium dimer containing allyl and cyclopentadienyl bridging ligands has been reviewed[517]. In these complexes, each palladium has an average valence of +1 and a total of 32 electrons. The reaction of the dimer with methyl iodide splits the palladium—palladium bond resulting in monomeric organopalladium products.

$$\left[\begin{array}{c} Cp \\ L{-}Pd{-}Pd{-}L \\ R \end{array}\right] \xrightarrow{MeI} [PdCp(RC_3H_4)] + [Pd(MeI)(L)_2]$$

$$\downarrow \quad (256)$$

$$[PdCpMe_2] + [PdI(L)(\eta^3\text{-2-R-allyl})]$$

L = tertiary phosphine

Palladium and platinum dimers held together by bridging methylene phosphine and arsine ligands undergo oxidative addition reactions without the disruption of the dimeric complex. The average valence of each palladium in these complexes also is

+1 and the metal–metal distances in the palladium and the corresponding platinum dimer are 2.70 and 2.63–2.65 Å, respectively, comparable to, but slightly shorter than, those in the elemental metals (2.75 and 2.77 Å). Each metal can be considered to have 16 electrons. The platinum dimer adds diazomethane to yield the methylene-bridged platinum(II) complex, which no longer has platinum—platinum bonding[518,519]. The zerovalent palladium dimer $[Pd_2(dppm)_3]$ undergoes oxidative addition

$$[Cl-Pt(P\frown P)_2Pt-Cl] \xrightarrow{CH_2N_2} [Cl-Pt(P\frown P)_2(\mu\text{-}CH_2)Pt-Cl] \quad (257)$$

P⌒P = dppm

reactions with geminal dihalides, oxalyl chloride, methyl iodide, and *o*-diiodobenzene to give the dimeric palladium complexes in which palladium—palladium bonding is absent[520].

$$[Pd_2(dppm)_3] \xrightarrow{o-C_6H_4I_2} [Pd_2(\mu\text{-}C_6H_4)I_2(dppm)_2]; \quad [Pd_2(dppm)_3] \xrightarrow{RCHX_2} [X-Pd(dppm)_2(\mu\text{-}CHR)Pd-X];$$
$$[Pd_2(dppm)_3] \xrightarrow{MeX} [Me-Pd(\mu\text{-}X)_2(dppm)_2Pd-Me]; \quad [Pd_2(dppm)_3] \xrightarrow{ClCOCOCl} [Cl-Pd(dppm)_2(\mu\text{-}CO)Pd-Cl] \quad (258)$$

$$[CpCo(\mu\text{-}CO)_2CoCp]^{-\bullet} \xrightarrow{o-C_6H_4(CH_2Br)_2} [Cp-Co(\mu\text{-}CO)_2(o\text{-}C_6H_4(CH_2)_2)Co-Cp]$$
$$\xrightarrow{I(CH_2)_3I} [Cp-Co(\mu\text{-}CO)_2((CH_2)_3)Co-Cp]; \quad \xrightarrow{CH_2I_2} [Cp(OC)Co(\mu\text{-}CH_2)Co(CO)Cp]; \quad \xrightarrow{MeI} [Cp(Me)Co(\mu\text{-}CO)_2Co(Me)Cp] \quad (259)$$

Alkylation of bis($\eta^5$-cyclopentadienyl)di-$\mu$-carbonylcobaltate takes place with a variety of organic halides; dialkyl, bridging alkyl, and bridging benzyl-type complexes can be obtained[521–525]. The rhodium(I) dimer containing four bisisonitrile bridges undergoes a two-centred oxidative addition reaction with alkyl iodide and methyl tosylate, while maintaining the rhodium—rhodium bonding[526].

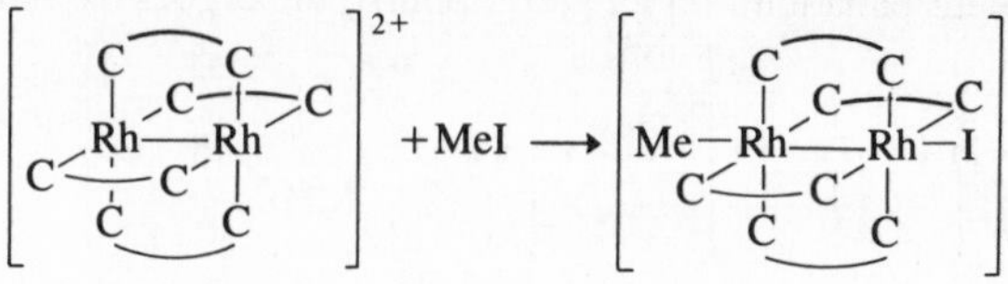

$$[Rh_2(C\frown C)_4]^{2+} + MeI \longrightarrow [Me{-}Rh(C\frown C)_4Rh{-}I] \quad (260)$$

$C\frown C = CN(CH_2)_3NC$

Diironoctacarbonyl dianion adds methylene iodide to give a bridged dimer that has been used as a homogeneous model for the Fischer–Tropsch reaction[527].

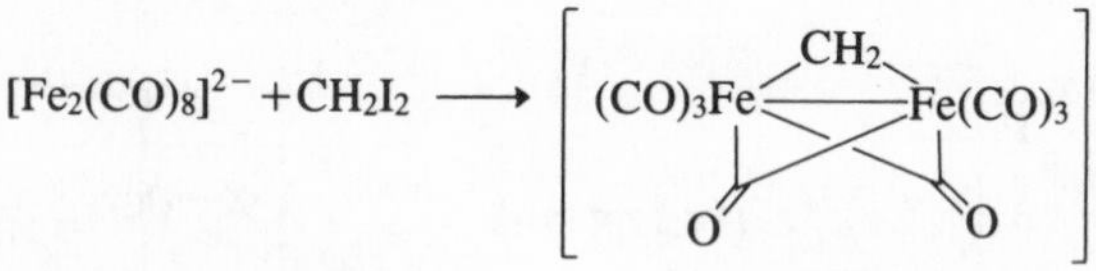

$$[Fe_2(CO)_8]^{2-} + CH_2I_2 \longrightarrow [(CO)_3Fe(\mu\text{-}CH_2)(\mu\text{-}CO)_2Fe(CO)_3] \quad (261)$$

## B. Carbon—Carbon Insertion

Transition metals catalyse a number of pericyclic reactions, including valence isomerization[528]. An early, key step in the sequence of reactions taking place in an isomerization is the oxidative addition of a hydrocarbon to a transition metal by breaking a carbon—carbon bond to form a metallocycle[529]. Generally this reaction occurs with small-ring, strained hydrocarbons, but there are a few examples in which carbon–carbon oxidative addition takes place in a strain-free hydrocarbon.

Although many transition metal complexes catalyse these rearrangements, the number of isolable compounds containing a carbon—metal—carbon bond formed from a carbon—carbon bond cleavage by a transition metal is surprisingly narrow. The subsequent reactions, particularly the reductive elimination step in most of the catalytic rearrangements, evidently have lower activation energies such that the oxidative addition intermediates are not isolated. Most of the isolated intermediates are those of platinum(II) and -(IV), rhodium(III), and iron(II). Thus, platinum(II) complexes, the zero valent $d^{10}$ platinum(0) phosphine complexes, $d^8$ iron carbonyls, and $d^8$ rhodium(I) complexes are active catalysts, but allow the isolation of certain metallocycles.

Cyclopropanes undergo ring-opening reactions with acids as a result of ring strain and the $\pi$-character of the carbon—carbon bonds. Thus, it is not surprising that cyclopropanes are opened by transition metals with a high electron affinity. Silver, which catalyses these rearrangements of cyclopropanes, has a high promotion energy indicative of a poor electron donor and a high electron affinity, showing that it is a good electron acceptor (Table 15). Platinum(II), however, is not only a good electron acceptor (high electron affinity) but utilizes sufficient back-bonding (low promotion energy). It is not unexpected, therefore, that platinum(II) reacts readily with the olefin-like cyclopropane bond. Certain highly electronegatively substituted cyclopropanes react with palladium and platinum zerovalent complexes, but are unreactive towards platinum(II) complexes[154].

TABLE 15. Donor/acceptor characteristics[530] of some transition metals

| Transition metal | Electronic configuration | Promotion energy (eV) | Electron affinity |
|---|---|---|---|
| Ni(0) | $d^{10}$ | 1.72 | 1.2 |
| Pd(0) | $d^{10}$ | 4.23 | 1.3 |
| Pt(0) | $d^{10}$ | 3.28 | 2.4 |
| Rh(I) | $d^8$ | 1.6 | 7.31 |
| Ir(I) | $d^8$ | 2.4 | 7.95 |
| Pd(II) | $d^8$ | 3.05 | 18.56 |
| Pt(II) | $d^8$ | 3.39 | 19.42 |
| Cu(I) | $d^{10}$ | 8.25 | 7.72 |
| Ag(I) | $d^{10}$ | 9.94 | 7.59 |

Bis(cyclooctadiene)nickel(0) reacts with the triphenylmethane dimer to give bis(triphenylmethyl)nickel(II)[531]. Nickel carbonyl reacts with spiro[2.4]hepta-4,6-diene with carbon—carbon bond scissions followed by carbon monoxide insertion to yield a $\sigma$-acyl-$\pi$-cyclopentadienyl complex as well as a dinuclear complex containing a nickel—nickel bond[532].

$[Ni(CO)_4]$ + (spiro[2.4]hepta-4,6-diene) →(70 °C, 1 h) [(C₅H₄CH₂CH₂C(=O))Ni(CO)] + [(CH₂)₂—C(=O)—(CH₂)₂ bridged bis(cyclopentadienyl)Ni—Ni(CO)₂] (262)

Palladium(0) and platinum(0) phosphine complexes yield metallacyclobutanes in reactions with highly electronegatively substituted cyclopropanes. Calculations of the electron densities, as provided by ESCA measurements on the carbon atoms in cyclopropanes, are instructive. Considering that the electron density at a cyclopropane carbon is −0.3. whereas the density at a cyano-bearing carbon on 1,1,2,2-tetracyanocyclopropane is 4.1, it is not surprising that the latter is susceptible to nucleophilic attack.

A recent review[533] on platinacyclobutane chemistry includes the ring-opening reactions of cyclopropanes and tabulations of these complexes. Consequently, only the essential features of the reactions with platinum(0) and platinum(II) complexes will be discussed. Phosphine and arsine complexes of palladium(0) and platinum(0) undergo oxidative addition of tetracyanocyclopropanes at the bond between the most electropositive carbons[154,534–537].

$[ML_n]$ + (cyclopropane with $(CN)_2$, $(CN)$, R′, R) → $[L_2M$ (metallacyclobutane with $(CN)_2$, $(CN)_2$, R′, R)$]$ (263)

$R = R' = H$, Me; $R = H$, $R' = Ph$; $R = Me$, $R' = Et$; R, $R' = (CH_2)_5$;

$M = Pt$, Pd; $L = PPh_3$, $PPh_2Me$, $AsPh_3$; $n = 3,4$

Bis(triphenylphosphine)ethyleneplatinum(0) also reacts with the appropriately substituted cyclopropane; retention of stereochemistry at carbon is observed[537].

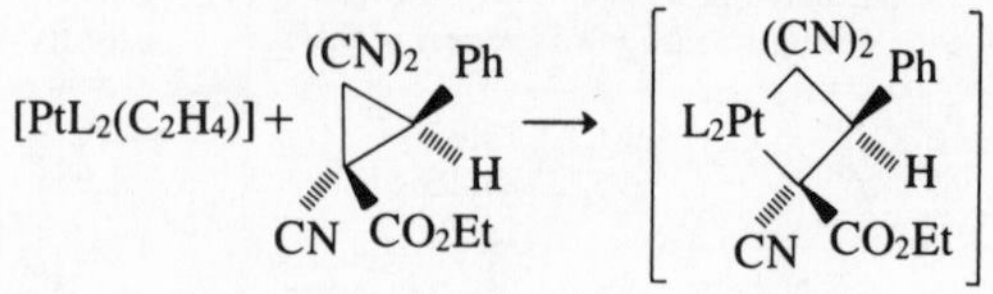

$$[PtL_2(C_2H_4)] + (CN)_2\text{-cyclopropane(Ph, H, CN, CO}_2\text{Et)} \longrightarrow [L_2Pt(CN)_2\text{-ring(Ph, H, CN, CO}_2\text{Et)}] \quad (264)$$

Platinacyclobutane complexes are produced from the reactions either with tetrakis(triphenylphosphine)platinum or the bis(triphenylphosphine)ethyleneplatinum[538,539]. With the ethylene complex, an intermediate cyclopropenoneplatinum $\pi$-olefin complex can be detected[539]. Substitution of a dicyanomethylene group for the carbonyl oxygen produces a similar reaction product[540].

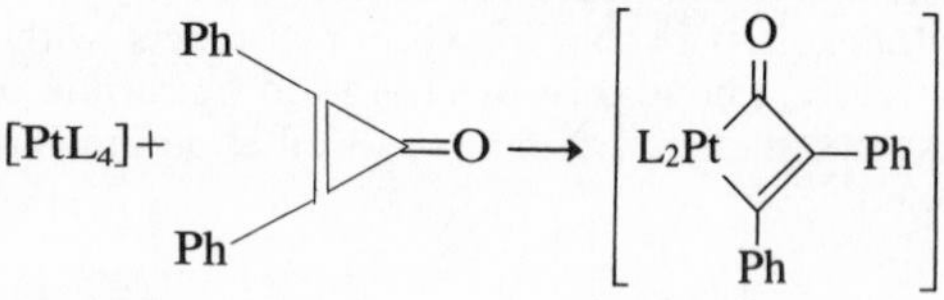

$$[PtL_4] + \text{Ph}_2\text{C}_3\text{=O} \longrightarrow [L_2Pt\text{-C(=O)-C(Ph)=C(Ph)}] \quad (265)$$

$$[PtL_2(C_2H_4)] + \text{Me}_2\text{C}_3\text{=O} \xrightarrow{-65\,°C} [L_2Pt \leftarrow \text{Me}_2\text{C}_3\text{=O}] \xrightarrow{-30\,°C} [L_2Pt\text{-C(Me)=C(Me)-C(=O)}]$$

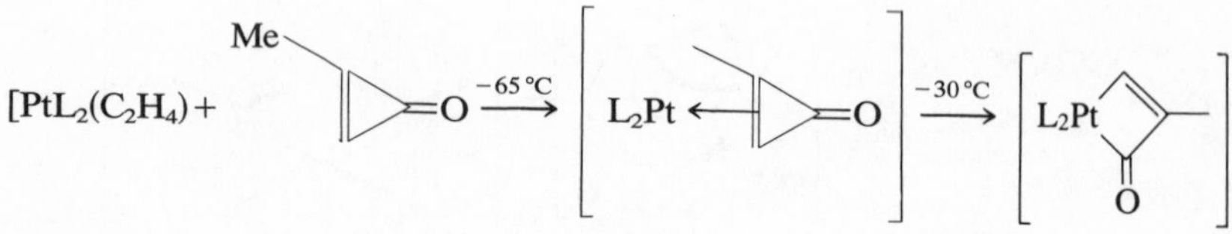

$L = PPh_3$ (266)

Platinum(II) complexes react with cyclopropane and substituted cyclopropanes on which the substituents are not electron withdrawing. The metallacyclobutanes obtained from Ziese's dimer are polymeric[541–554]. Electron-donating groups increase the reaction rate[541,543,544,555] whereas sterically hindered cyclopropanes fail to

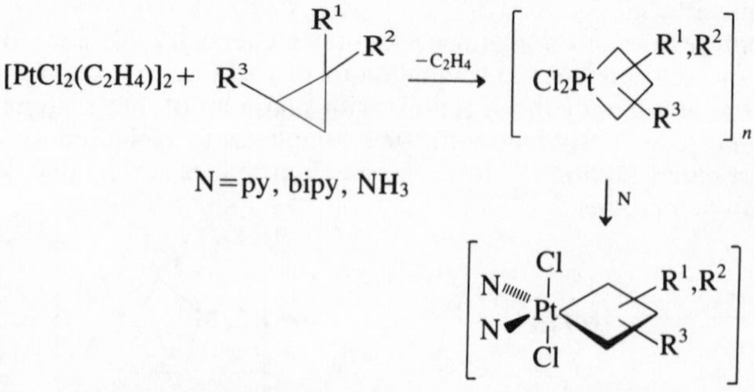

$$[PtCl_2(C_2H_4)]_2 + R^3,R^1,R^2\text{-cyclopropane} \xrightarrow{-C_2H_4} [Cl_2Pt\text{-ring}(R^1,R^2)(R^3)]_n$$

N = py, bipy, $NH_3$

$$\downarrow N \quad (267)$$

$$[N_2PtCl_2\text{-ring}(R^1,R^2)(R^3)]$$

react[547]. Addition of amine ligands to the polymeric platinacyclobutane yields a soluble monomeric complex[541,542,544,546–552,556,557] that has *trans*-chlorine atoms and *cis*-amines groups[558–560].

These complexes, containing hard bases *trans* to the carbon—platinum bond, are stable, but replacement with soft bases (cyanide, phosphine, arsine, carbon monoxide, olefins, etc.) usually results in elimination of the organic portion[541,545,561,562]. Platinacyclobutanes that are multiply substituted decompose even on addition of pyridine or acetonitrile by $\alpha$- or $\beta$-elimination, respectively, to yield ylide or olefin[563–565].

The reaction in tetrahydrofuran to form the platinacyclobutane is first order each in Zeise's dimer and cyclopropane[555]. The rate enhancement observed with electron-donating substituents[544,555] is possibly a result of the ability to generate a cyclopropane–platinum complex. The reaction is stereospecific, *trans*- and *cis*-1,2-diphenylcyclopropane yielding only the *trans- and cis*-diphenylplatinacyclobutanes, respectively[541,543,544,551,553]. Similar studies have been carried out with other substituted cyclopropanes[546], including deuteriated cyclopropanes[566,567]. This stereospecificity rules out ionic intermediates, and is more consistent with a concerted oxidative addition. However, since polar solvents enhance the rate of the reaction, the transition state may have some ionic character.

In many of the reactions the kinetic product formed initially rearranges to the thermodynamic isomer. Phenylcyclobutane reacts initially by insertion into the bond next to the phenyl group, but rearranges to the position remote from the platinum[549,556,568]. The rearrangement does not proceed by regeneration of the phenylcyclopropane and readdition to the platinum complex, $[PtCl_2py_2]$, but instead

$[PtCl_2(C_2H_4)]_2$ + Ph-cyclopropane ⟶ [$Cl_2Pt$(2-Ph-cyclobutane)] ⟶ [$PtCl_2py_2$ (2-Ph-platinacyclobutane)] ⟶ [$PtCl_2py_2$ (3-Ph-platinacyclobutane)] (268)

proceeds by an intramolecular rearrangement[569]. The rearrangement cannot proceed via a metal carbene, since this would lead to *cis–trans* isomerization in disubstituted cyclopropanes[566]. Similar rearrangements are observed with 1,2-diarylcyclopropanes. For example, *trans*-1,2-di(4-tolyl)cyclopropane initially gave a *trans*-2,4-di(4-tolyl)platinacyclopropane which isomerized to the *trans*-2,3-substituted product[546]. Hence the isolation of any product of insertion does not guarantee that it is the kinetic product.

Zeise's dimer also reacts at the central bond of bicyclo[1.1.0]butane at −45 °C to give the platinacycle which, on reaction with pyridine at −50 °C, yields a complex that can be isolated, and is relatively stable at ambient temperature[570]. In solution at −25 °C the complex decomposes to various $C_4$ products.

$$\text{bicyclobutane} + [PtCl_2(C_2H_4)]_2 \xrightarrow{-45\,^{\circ}C} [PtCl_2(C_4H_6)]_n \xrightarrow[-50\,^{\circ}C]{2py} [PtCl_2py_2(C_4H_6)] \quad (269)$$

The strained 1,2-diketone, benzocyclobutenedione, reacts with $d^8$ chlorotris(triphenylphosphine)cobalt(I) by undergoing cleavage between the acyl bonds[571]. Wilkinson's catalyst and iron tetracarbonyl also effect this cleavage. The chlorodicarbonylrhodium(I) dimer has been shown to cleave a number of small, strained ring

$$+[CoCl(Ph_3P)_3] \longrightarrow [\ CoCl(PPh_3)_2\ ] \quad (270)$$

compounds. In every case, carbon monoxide inserts into the carbon—rhodium bond that is formed as a result of the oxidative addition process. In ring-opening reactions of cyclopropanes, a chloro-bridged dimer is formed that can be converted to a monomeric rhodium(III) species by cleavage of the bridge with phosphine[572–574]. The bonds in phenylcyclopropane adjacent to phenyl, **61**, and in benzylcyclopropane

$$[RhCl(CO)_2]_2 + \text{cyclopropane} \longrightarrow [\ \text{CO, Rh, Cl}\ ]_2 \xrightarrow{PPh_3} [\ \text{CO, Rh, } PPh_3, PPh_3\ ] \quad (271)$$

remote from benzyl undergo insertion, **62**[573,574]. Carbon monoxide insertion in the unsymmetrical phenyl-substituted rhodacyclobutane intermediate takes place at the alkyl rather than the benzyl—rhodium bond **62**. Nortricyclene, which is isomerized **61**

**61** **62** **63**

to norbornadiene by rhodium catalysts, also reacts to yield an acyl metallacycle[575]. The reaction of cubane with the rhodium dimer gives a six-membered rhodacycle[576].

$$+[RhCl(CO)_2]_2 \longrightarrow [\ \text{Cl—Rh(CO)—C=O}\ ] \quad (272)$$

The isomerization reaction also is catalysed by chlororhodium diene dimers, $[RhCl(diene)]_2$, and $[RhCl(cod)(PPh_3)]$. The rearrangement reaction with a chlororhodium diene dimer is second order, first order each in rhodium and cubane. Wilkinson's catalyst and chlorocarbonylbis(triphenylphosphine)rhodium(I) are unactive.

$$+[RhCl(CO)_2]_2 \longrightarrow [\,Cl-Rh(CO)\,]\qquad (273)$$

Bicyclo[2.2.0]hexane and related hydrocarbons oxidatively add to the rhodium carbonyl dimer by cleavage of a central bond[577].

$$+[RhCl(CO)_2]_2 \xrightarrow{40\ ^\circ C} [\,OC-Rh(Cl)\,]\qquad (274)$$

$$+[RhCl(CO)_2]_2 \xrightarrow{50\ ^\circ C} [\,OC-Rh(Cl)\,]\qquad (275)$$

The rhodium octaethylporphyrin anion is alkylated by cyclopropyl methyl ketone by breaking the bond adjacent to the acyl group[578]. The source of hydrogen in this nucleophilic ring-opening reaction has not been identified.

$$[Rh(N_4)]^- + \text{cyclopropyl-COMe} \longrightarrow [Rh(N_4)CH_2CH_2CH_2COMe]\qquad (276)$$

(N N / N N) = octaethylporphyrin (Et, Et, Et, Et, Et, Et, Et, Et; N, N, N, N)

Bisacetylenic derivatives of ketones are decarbonylated by rhodium(I) compounds. One of the more striking oxidative addition reactions in which the product has been isolated involves the carbon—carbon bond cleavage of an acyl—acetylene bond by Wilkinson's catalyst[579].

$$+[RhClL_3] \longrightarrow \quad (277)$$

$L = PPh_3$; $R = Bu^t$, Ph

Iridium opens cyclopropanes, yielding either an $\eta^3$-iridium hydride[580] or a iridacyclobutane[581], depending on the iridium complex and the reaction conditions.

$$[IrCl(L)_2(N_2)] + \text{Ph-cyclopropane} \xrightarrow{-N_2} \quad (278)$$

$L = PPh_3$

Diironnonacarbonyl, a $d^8$ iron(0) dimer, undergoes oxidative addition reactions with strained ring hydrocarbons. Subsequently, carbon monoxide may be 'inserted' into a carbon—iron bond. Vinylcyclopropanes undergo ring-opening reactions readily, since breaking a bond adjacent to the vinyl substituent yields a $\pi$-allyl complex, in which case carbon monoxide insertion usually does not take place. A variety of these reactions are summarized in Table 16.

The reaction of a vinylcyclopropane derivative takes place by a disrotatory opening of the $C_{(3)}$—$C_{(4)}$ bond *cisoid* to the $C_{(1)}$—$C_{(2)}$ vinyl group to form an $\eta^3$ complex. The driving force for breaking two cyclopropane bonds in the strained $C_6H_6$ hydrocarbon

$$\longrightarrow \quad \underset{+CO}{\overset{-CO}{\rightleftharpoons}} \quad (279)$$

TABLE 16. Reactions of $[Fe_2(CO)_9]$ with strained ring hydrocarbons

| Substrate | Oxidative addition product | Ref. |
|---|---|---|
| 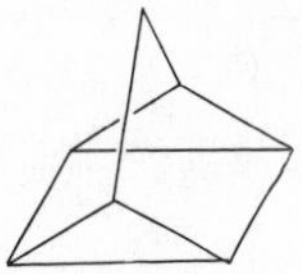 | 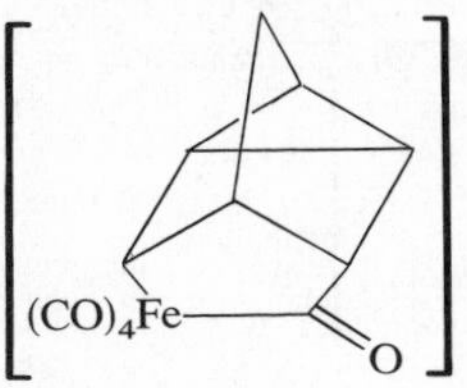 $(CO)_4Fe$; O | 582 |
| 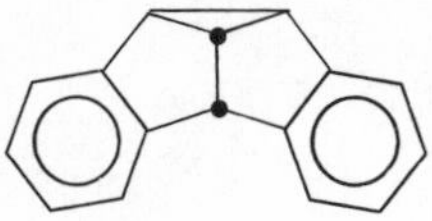 | 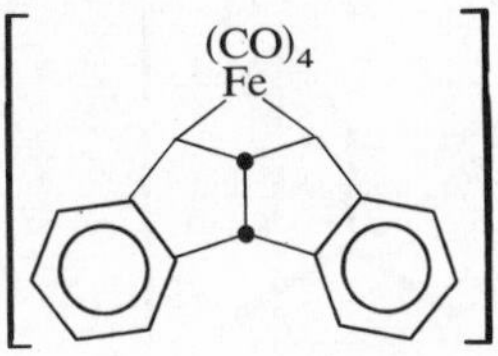 $(CO)_4$ Fe | 583–585 |
| | 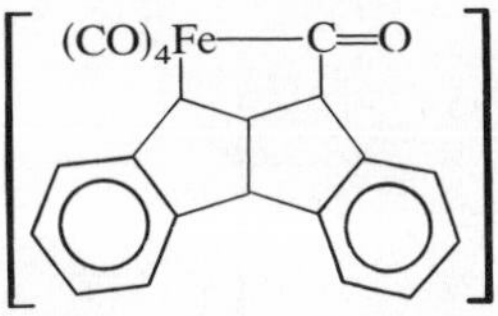 $(CO)_4Fe$—C=O | 585 |
| 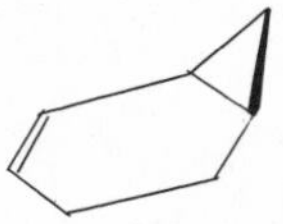 | 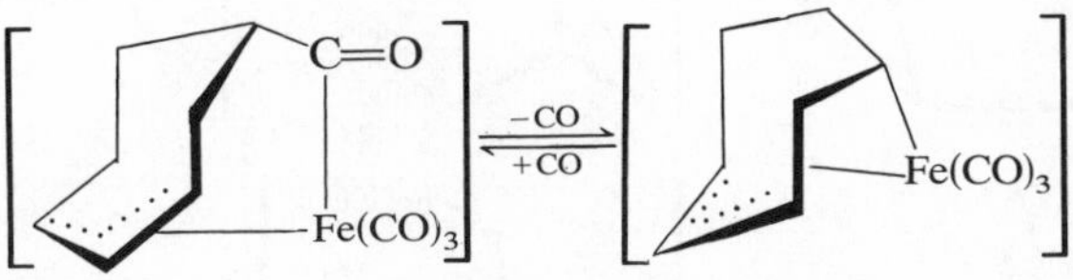 C=O; $Fe(CO)_3$; $\underset{+CO}{\overset{-CO}{\rightleftharpoons}}$; $Fe(CO)_3$ (under CO pressure) | 586 |
| 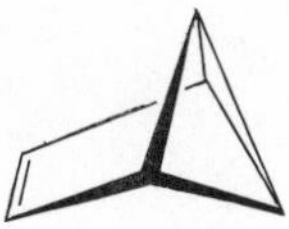 | 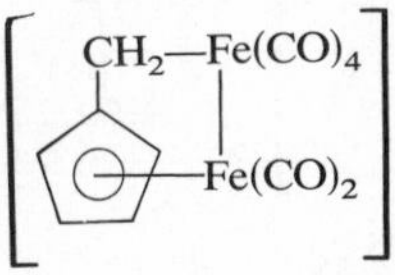 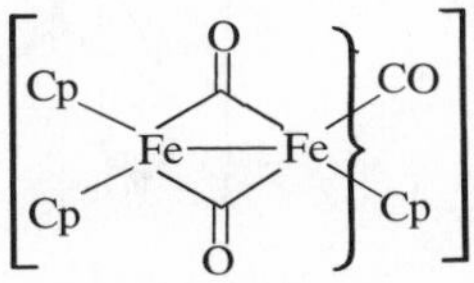 $CH_2$—$Fe(CO)_4$; $Fe(CO)_2$; Cp; O; CO; Fe—Fe; Cp; O; Cp | 587 |
| | 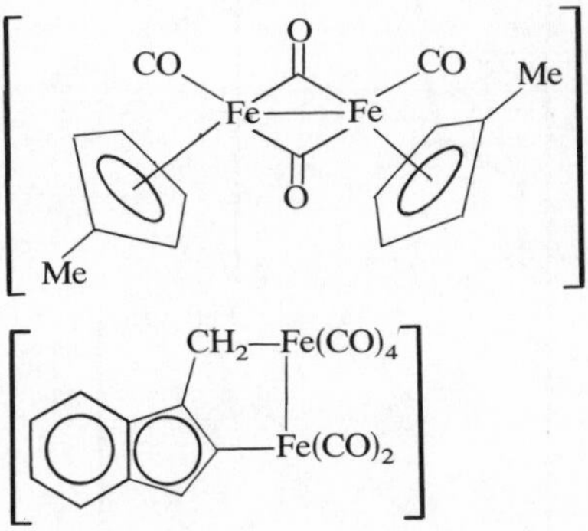 CO; O; CO; Me; Fe—Fe; O; Me | |
| | $CH_2$—$Fe(CO)_4$; $Fe(CO)_2$ | 587 |

TABLE 16. ctd.

| Substrate | Oxidative Addition Product | Ref. |
|---|---|---|
|  | H, $Fe(CO)_3$ | 588 |
| OMe, N | N, MeO, Fe $(CO)_3$ | 589 |
| O | O, $Fe(CO)_3$ | 590 |
|  | $Fe(CO)_3$ | 591 |
|  | $Fe(CO)_3$ | 591 |
|  | $Fe(CO)_3$ | 591 |

TABLE 16. ctd.

| Substrate | Oxidative addition product | Ref. |
|---|---|---|
| 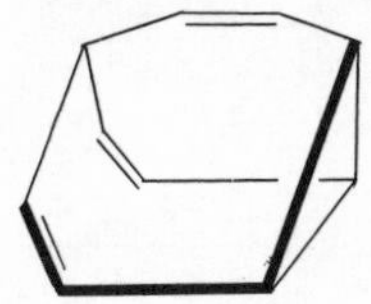 | 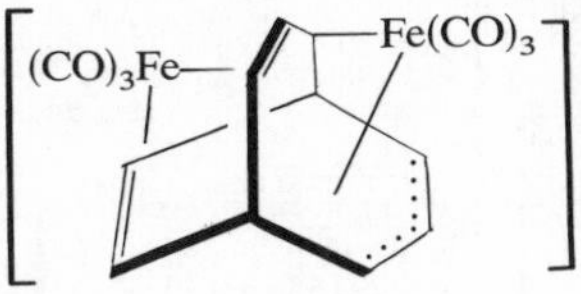 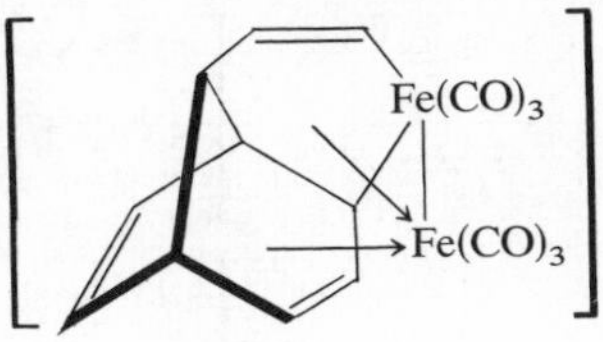 $[(CO)_3Fe \ldots Fe(CO)_3]$ $[Fe(CO)_3 \rightarrow Fe(CO)_3]$ | 592, 593 |
| 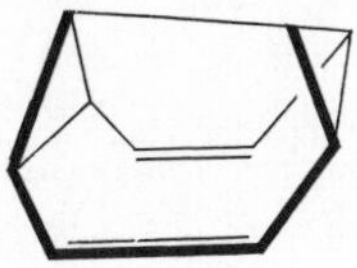 | 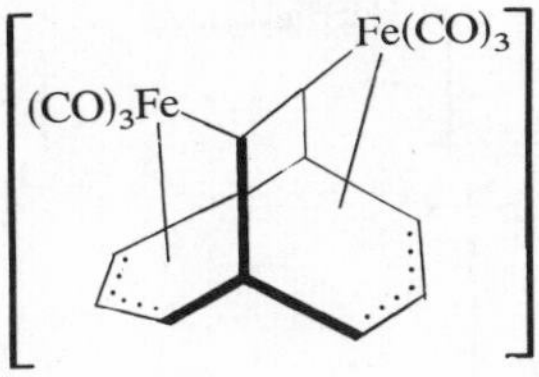 $[(CO)_3Fe \ldots Fe(CO)_3]$ | 594 |
| 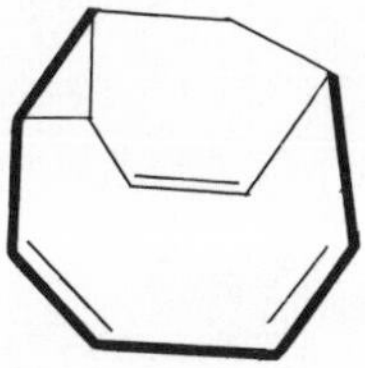 | 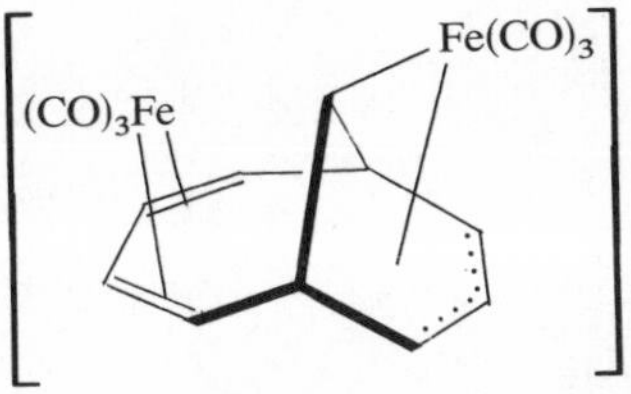 $[(CO)_3Fe \ldots Fe(CO)_3]$ | 595 |
| 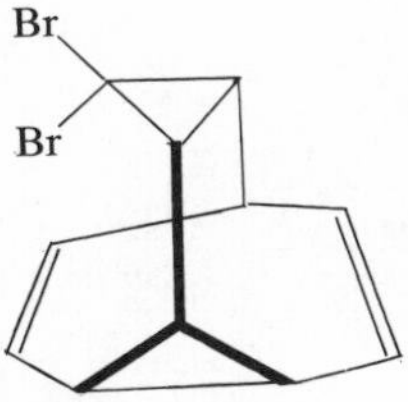 Br, Br | 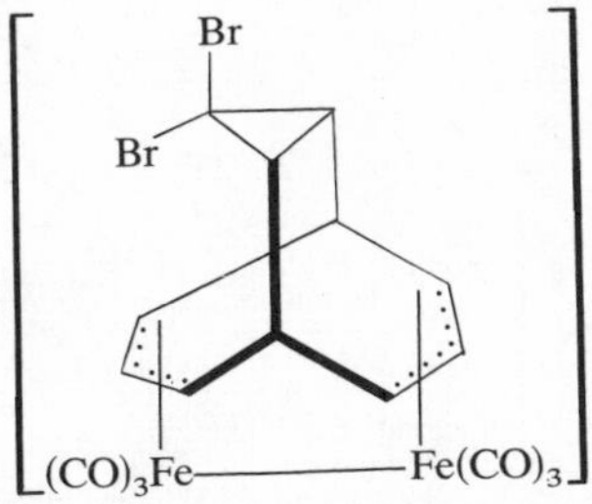 [Br, Br, $(CO)_3Fe$—$Fe(CO)_3$] | 596 |
| 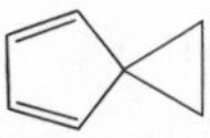 | 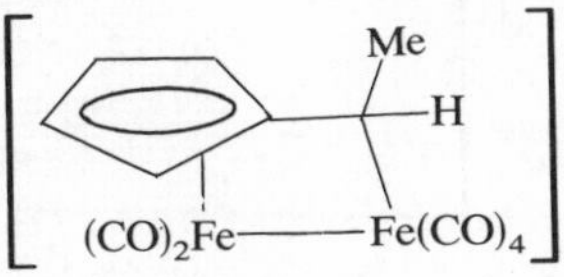 [Me, H, $(CO)_2Fe$—$Fe(CO)_4$] | 597 |
| 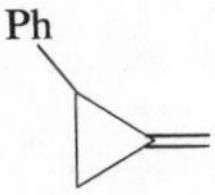 Ph | 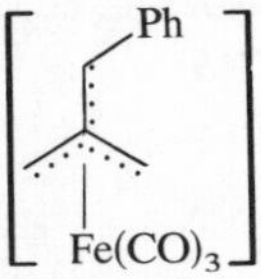 [Ph, $Fe(CO)_3$] | 598 |
| 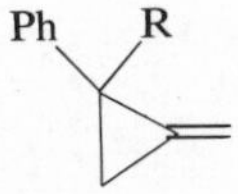 Ph, R; R = Ph, Me | 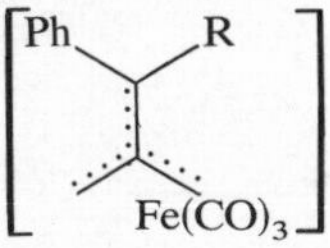 [Ph, R, $Fe(CO)_3$] | 598 |

TABLE 16. ctd.

| Substrate | Oxidative Addition Product | Ref. |
| --- | --- | --- |
|  $Me_2Si$ (cyclobutane) | 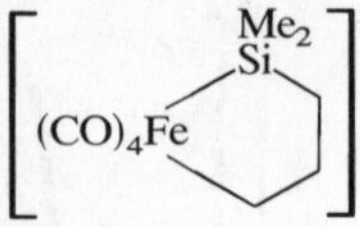 $[(CO)_4Fe$–$SiMe_2$ ring$]$ | 599 |
| 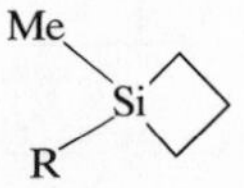 Me, R–Si (silacyclobutane) | 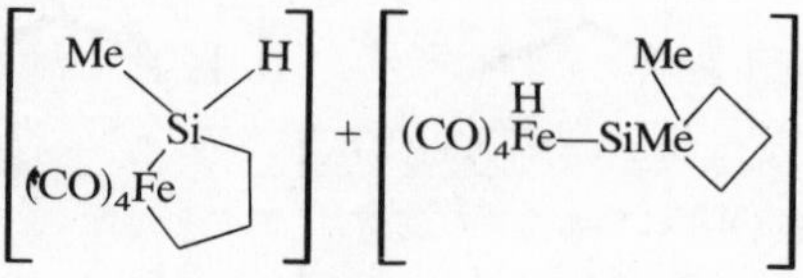 $[(CO)_4Fe$, Me, H, Si ring$]$ + $[(CO)_4Fe(H)$—SiMe (cyclobutyl)$]$ | 600 |
| 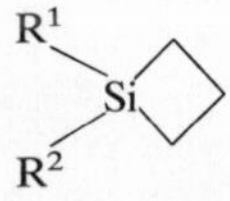 $R^1$, $R^2$–Si (silacyclobutane)<br>$R^1$ / $R^2$: Me / Me; Me / Cl; Me / OMe; Cl / Cl; OMe / OMe | 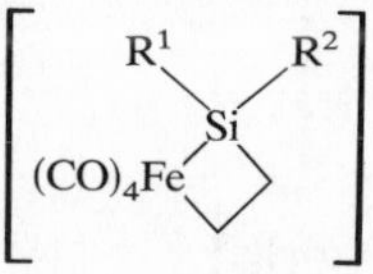 $[(CO)_4Fe$, $R^1$, $R^2$, Si$]$ | 601, 602 |
| 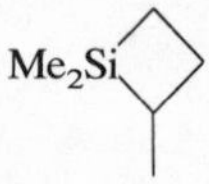 $Me_2Si$ (methyl cyclobutane) | 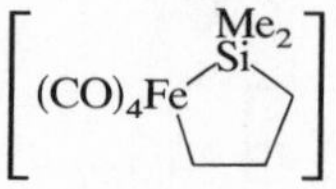 $[(CO)_4Fe$, $Me_2Si$ ring$]$ | 601, 602 |
| 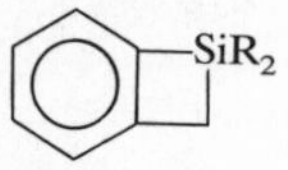 benzo–$SiR_2$<br>R = Me, Ph | 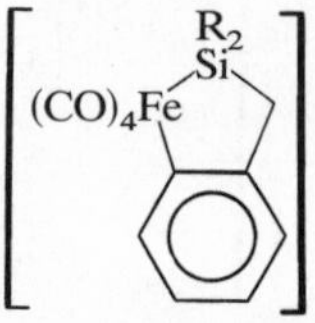 $[(CO)_4Fe$, $R_2Si$, benzo ring$]$ | 601, 602 |
| 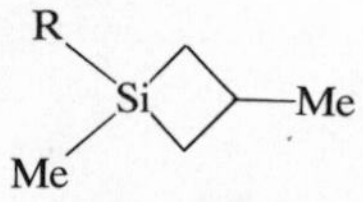 R, Me–Si (cyclobutane)–Me<br>R = $Bu^n$, OMe<br>$Z:E = 4:7$ | 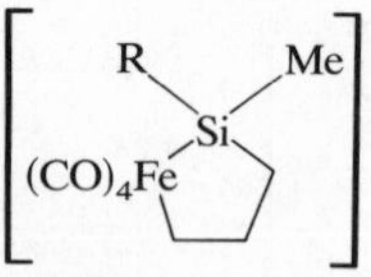 $[(CO)_4Fe$, R, Me, Si ring$]$<br>$Z:E = 4:7$ | 601, 602 |

is the generation of a $\eta^5$-cyclopentadienyl ligand (Table 16). This reaction has been proposed to proceed by the following sequence[587]. In the reaction of the spirocyclopentadiene with diironnonacarbonyl, a $\pi$-diironoctacarbonyl complex is formed

$$+[Fe_2(CO)_9] \longrightarrow [\,\cdots Fe(CO)_4\,] \longrightarrow [\,H\text{–}C(=Fe(CO)_4)\text{–H}\,] \downarrow [\,H_2C=\cdots\text{–}Fe(CO)_4\,] \qquad (280)$$

that slowly rearranges to the $\eta^5$-cyclopentadienyl iron dimer (Table 16)[597]. Opening of the silacyclopentane rings at the silicon—carbon bond also can be achieved with iron pentacarbonyl, but ultraviolet irradiation is needed to generate the coordinatively unsaturated iron tetracarbonyl[601].

Cyclopentadienylmanganesetricarbonyl also reacts with diphenylsilacyclobutane under an ultraviolet source to yield a manganacycle that decomposes at $-78\,°C$[603].

$$[MnCp(CO)_3] + Ph_2Si\diamond \longrightarrow [\,CpMn(CO)(CO)\,Si Ph_2\,] \qquad (281)$$

There is ample evidence that zirconium atoms (vapour deposited) at $-196\,°C$ react with neopentane by insertion into both carbon—hydrogen and carbon—carbon bonds, even though organometallic products were not isolated[604].

## C. Carbon—Hydrogen Insertion (Hydrocarbon Activation)

The activation of carbon—hydrogen bonds by homogeneous catalysts under mild reaction conditions is one of the most important but difficult problems in the area of homogeneous catalysis. Although the intramolecular oxidative addition of carbon—hydrogen bonds of ligands attached to a transition metal is frequently observed, the intermolecular oxidative addition reaction of hydrocarbon is less common. The transition metal alkyl hydrides resulting from such a reaction usually are not isolated, nor can these unstable intermediates generally be observed, even in the intramolecular reaction. Their existence has been implicated primarily in intramolecular reactions by an intramolecular metallation product, and in intermolecular reactions by hydrogen–deuterium exchange with solvent or by exchange experiments with molecular deuterium/hydrogen.

Hydrocarbons are stronger electron donors than dihydrogen and therefore should react more easily with electrophiles. The relative reactivity of alkanes and dihydrogen and the selectivity of attack at a carbon—hydrogen bond are the opposite for

$S_E2$ and $S_N2$ reactions. These trends are a reflection of the ionization potentials, electron and proton affinities, and kinetic acidities in the series.

$$S_E2{:}\quad H_2 < CH_4 < C_2H_6 < C_3H_8;\ \text{primary} < \text{secondary} < \text{tertiary H}$$

$$S_N2{:}\quad H_2 > CH_4 > C_2H_6 > C_3H_8;\ \text{primary} > \text{secondary} > \text{tertiary H}$$

Several reviews on hydrocarbon activation[605–610] and intramolecular or cyclometallation reactions[611–617] are available. However, in most cases, the oxidative addition products *per se* have not been isolated. Consequently, this section will give only brief reviews of cyclometallation and intermolecular oxidative addition with particular attention to reactions in which the oxidative addition products have been isolated.

### 1. *Intramolecular*[611–617]

Cyclometallation is recognized as a widely occurring reaction type in which a carbon—hydrogen bond of a ligand is cleaved to give a chelate ring containing a metal carbon bond. Many of these reactions that occur on an aromatic ring are electrophilic substitution reactions in which the metal is not oxidized at any stage of the reaction. Nucleophilic attack by the metal, however, results at least in an intermediate oxidative addition product, which in most cases is followed by reductive elimination. With the $d^8$, zerovalent complexes of the iron triad as well as rhodium(I) and iridium(I) complexes, the cyclometallated oxidative addition products are often isolated.

$$\text{Y–M(–Z⌒H–C)} \longrightarrow \text{Y–M(H)(–Z⌒C)} \longrightarrow \text{M(–Z⌒C)} + \text{HY} \qquad (282)$$

In reaction 282, the ligand Z is a Group V or VI donor and Y is usually an alkyl or halogen. Many of these reactions occur at elevated temperatures in an inert solvent. Most of cyclometallations occur with Group VI–VIII transition metals.

One of the first orthometallation reactions discovered was that between azobenzene and palladium(II) or platinum(II) chlorides ($K_2[MCl_4]$). When one of the azobenzene rings is substituted, the orthometallation reaction takes place on the ring which is more susceptible to electrophilic substitution[618]. The rate of orthometallation of $[IrClL_3]$ $[L = P(p\text{-}CX_6H_4)_3]$ increases as electron donors are placed on the aromatic ring[619]. Thus, electrophilic substitution mechanisms have been proposed[606], particularly for the orthometallations of palladium and platinum complexes containing nitrogen donor ligands. Since these orthometallations do not involve oxidative addition reactions, they will not be discussed.

Other Group VIII transition metals undergo oxidative addition reactions by insertion into an aromatic carbon—hydrogen bond. This type of reaction is favoured by groups on the aromatic ring that are electron withdrawing. Azobenzenes containing $+\sigma$ $m$-substituents, and thereby a substitution position which would most directly affect the position *ortho* to the azo group, react with methylmanganesepentacarbonyl in a manner consistent with nucleophilic attack[620]. If none of the ligands is lost (temporarily) before the oxidative addition step, then a 20-electron intermediate oxidative addition product is obtained, which loses methane. On the other hand, the reverse effect of substitution has been shown with phosphine complexes of methyl-

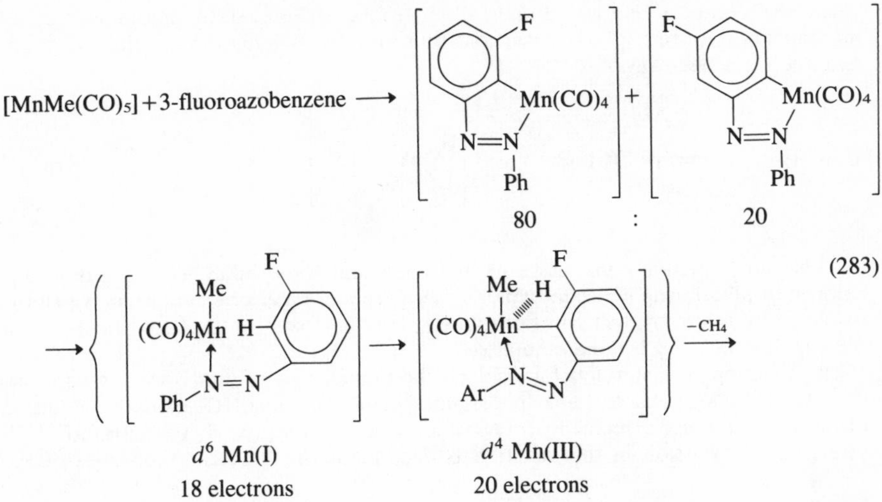

$[MnMe(CO)_5]$ + 3-fluoroazobenzene ⟶ [ortho-metallated Mn(CO)$_4$ complex, F adjacent] + [ortho-metallated Mn(CO)$_4$ complex, F remote]

80 : 20

(283)

⟶ { [(CO)$_4$Mn(Me) H–Ar, N=N–Ph] ⟶ [(CO)$_4$Mn(Me)(H), Ar–N=N] } $\xrightarrow{-CH_4}$

$d^6$ Mn(I) 18 electrons; $d^4$ Mn(III) 20 electrons

manganesetetracarbonyl[621]. Substitution of electron-withdrawing groups in the triphenylphosphine ligand retards the metallation reaction while electron donating groups increase the reaction rate.

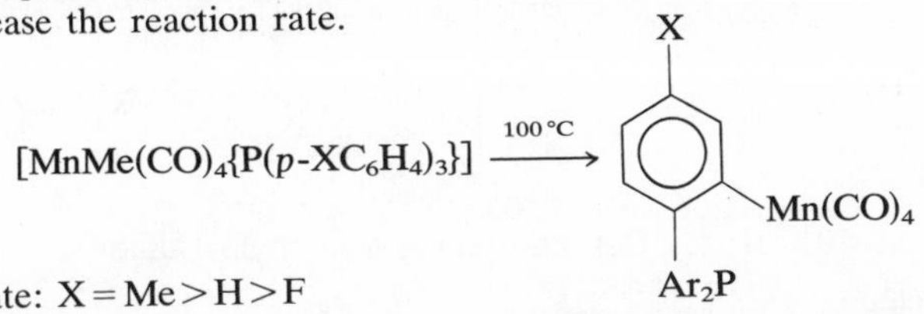

$$[MnMe(CO)_4\{P(p\text{-}XC_6H_4)_3\}] \xrightarrow{100\,°C} \text{(X-substituted ring bearing } Mn(CO)_4 \text{ and } Ar_2P) \quad (284)$$

relative rate: X = Me > H > F

Several other observations concerning the cyclometallation reactions have been made. With nitrogen donor ligands, five-membered rings are formed. In the case of phosphines, four-membered rings are often observed, particularly when large groups are substituted on phosphine. For a 'nucleophilic' oxidative addition reaction, the ease of orthometallation for $PhCH_2ZMe_2$ (Z = N, P, As) or for $PhCH_2XMe$ (X = O, S) is N > P > As and S > O.

Platinum and palladium complexes containing phosphine ligands that bear bulky aliphatic groups have been observed to undergo cyclometallation under mild conditions with the loss of hydrogen chloride to give complexes containing a carbon—metal σ-bond[622–625]. Intramolecular carbon—hydrogen insertion also takes place

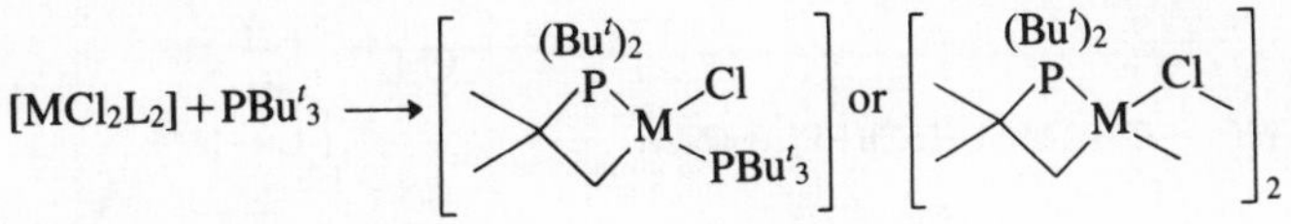

$$[MCl_2L_2] + PBu^t_3 \longrightarrow [(Bu^t)_2P\text{–}M(Cl)(PBu^t_3)\ \text{metallacycle}] \text{ or } [(Bu^t)_2P\text{–}M(Cl)\ \text{metallacycle}]_2 \quad (285)$$

L = nil, MeCN, PhCN; M = Pd, Pt

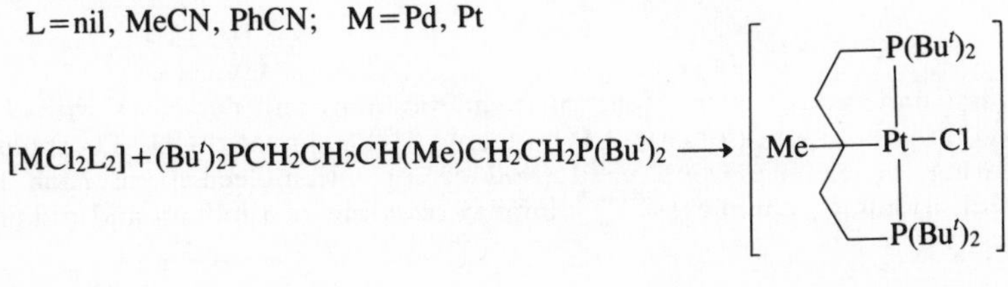

$$[MCl_2L_2] + (Bu^t)_2PCH_2CH_2CH(Me)CH_2CH_2P(Bu^t)_2 \longrightarrow [Me\text{–}C(CH_2CH_2P(Bu^t)_2)_2Pt\text{–}Cl] \quad (286)$$

when the complex contains bulky alkyl groups $\sigma$-bonded to platinum[626–628]. A mechanism requiring C—H insertion followed by elimination of the alkyl and hydrogen has been proposed[626,627].

$$[PtL_2(CH_2Bu^t)_2] \underset{}{\overset{-L}{\rightleftharpoons}} [PtL(CH_2Bu^t_2)] \longrightarrow \left[ \text{H(L)Pt(CH}_2\text{Bu}^t\text{)(cyclic)} \right] \longrightarrow \left[ L_2Pt(\text{cyclic}) \right] + Me_4C \quad (287)$$

Although reductive elimination of hydrogen and some other leaving group (e.g. halogen or alkyl) usually occurs rapidly, there are a number of examples in which the oxidative addition product can be isolated. This has been observed particularly for certain rhodium and iridium complexes.

The reaction of bis(cyclooctene)chloro-rhodium(I) or -iridium(I) dimers with azo ligands, Schiff's bases, or benzylphosphines yield the metal(III) oxidative addition products containing a metal hydride, which can be observed in the infrared[629–631]. The ease of insertion in this reaction is dependent on the basic properties of the

$$R_2PCH_2Ph + [MCl(coct)_2]_2 \xrightarrow{L} \left[ PhCH_2PR_2\text{—M(H)(L)(Cl)—PR}_2 \right] \quad (288)$$

$R = Ph, Bu^t$; $M = Rh, Ir$; $L = CO$, phosphine, 4-methylpyridine

coct = cyclooctene

$$PhX{=}NR + [MCl(coct)_2]_2 \xrightarrow{PR_3} \text{ or } [IrCl(PPh_3)_2(N_2)] \xrightarrow{-N_2} \left[ \text{C}_6\text{H}_4\text{—X{=}N(R)—M(H)(L)(L)Cl} \right] \quad (289)$$

$X = N, CH$

$$MeN{=}NC({=}CH_2)Me + [IrCl(PPh_3)_2(N_2)] \longrightarrow \left[ CH_3C{=}CH\text{—Ir(H)(L)(L)(Cl)—N(CH}_3\text{)—N=} \right] \quad (290)$$

metal (iridium undergoes a faster reaction than rhodium) and decreases with the hydrocarbon in the order aromatic CH > olefinic CH > aliphatic CH. Phosphine ligands bearing an olefinic group yield products of intramolecular insertion in reactions with iridium(I) complexes[632,633]. Similar reactions of rhodium and iridium

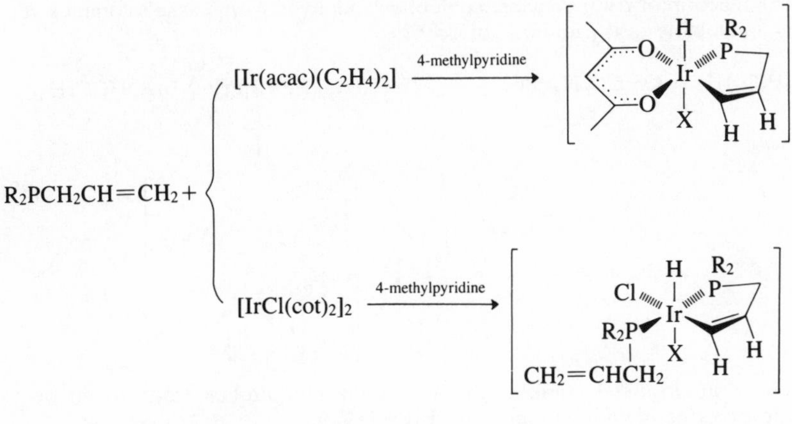

$$R_2PCH_2CH{=}CH_2 + \begin{cases} [Ir(acac)(C_2H_4)_2] \xrightarrow{\text{4-methylpyridine}} [\ldots] \\ [IrCl(cot)_2]_2 \xrightarrow{\text{4-methylpyridine}} [\ldots] \end{cases} \tag{291}$$

X = 4-methyl-1-pyridyl

have been observed with tricyclohexylphosphine ligands, phosphines containing aliphatic carbon—hydrogen bonds which undergo insertion[634]. In the presence of bulky chelating phosphines, rhodium chloride forms a complex in which C—H oxidative addition from the centre of the chelate bridge to rhodium has occurred[635].

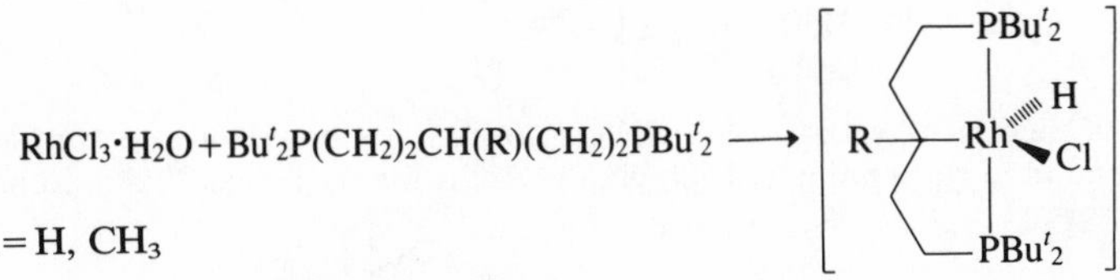

$$RhCl_3{\cdot}H_2O + Bu^t_2P(CH_2)_2CH(R)(CH_2)_2PBu^t_2 \longrightarrow [\ldots] \tag{292}$$

R = H, $CH_3$

Chlorotris(triphenylphosphine)iridium(I) undergoes orthometallation under mild conditions[619,636], particularly when the phosphine rings contain electron withdrawing substituents (see above)[619]. When iridium contains a methyl in place of chlorine, the

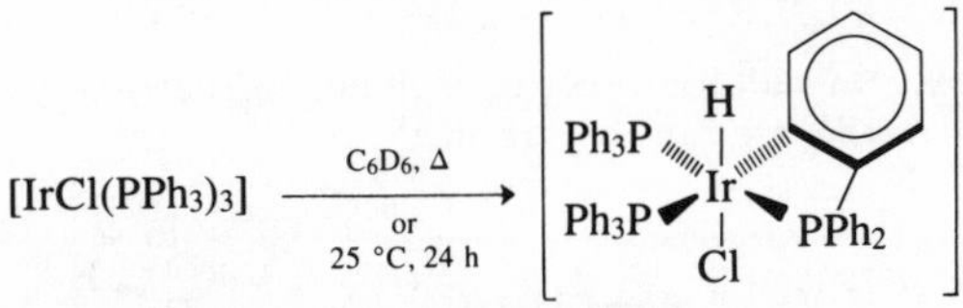

$$[IrCl(PPh_3)_3] \xrightarrow[\text{25 °C, 24 h}]{C_6D_6,\ \Delta \text{ or}} [\ldots] \tag{293}$$

intermediate oxidative addition product can be isolated, but rapidly loses methane at

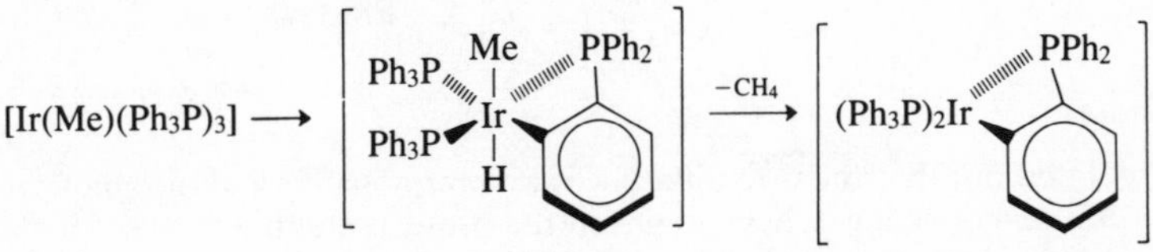

$$[Ir(Me)(Ph_3P)_3] \longrightarrow [\ldots] \xrightarrow{-CH_4} [(Ph_3P)_2Ir\ldots] \tag{294}$$

25 °C[637]. The reaction of *o*-diphenylphosphinobenzaldehyde with Vaska's complex at 25 °C gives an isolable acyl iridium hydride[638].

$$[IrCl(CO)(PPh_3)_2] + o\text{-}OHCC_6H_4PPh_2 \longrightarrow [IrCl(CO)(PPh_3)\{P(Ph)_2(o\text{-}OHCC_6H_4)\}] \quad (295)$$

Ph₂P, C=O, Cl, Ir, H, PPh₃, CO

Reduction of an iron(III) complex in the presence of diphos leads to an orthometallation product containing an iron—hydride bond[639].

$$[Fe(acac)_3] \xrightarrow[\text{diphos}]{EtOAlEt_2} [Fe(diphos)_2(\eta^2\text{-}C_2H_4)] \xrightarrow{h\nu} \quad (296)$$

Ph₂, Ph₂, P, P, Fe, P, P, Ph₂, Ph₂, ($\eta^2$—$C_2H_4$); Ph₂, Ph, P, P, Fe, P, P, Ph₂, Ph₂, H

The reduction of dichlorotetrakis(triphenylphosphine)ruthenium(II) in acetonitrile yields an orthometallated ruthenium hydride via the following sequence[640]. An arene

$$[RuCl_2L_4] \longrightarrow \longrightarrow [RuL'L_3] \longrightarrow \quad (297)$$

Ph₂P, L′, Ru, H, L, L

L = PPh$_3$, L′ = MeCN

osmium(0) complex, on addition of phosphine, oxidatively adds benzene intramolecularly to give the osmium(II) product[641].

$$[Os(PMe_3)(\eta^6\text{-}C_6H_6)(\eta^2\text{-}CH_2{=}CHR)] \xrightarrow{PMe_3} \quad (298)$$

L, L, H, Os, L, C₆H₅, L

L = PMe$_3$

## 2. Intermolecular

It has been pointed out that the rate differences in intermolcular and intramolecular C—H insertion reactions have their origins in the pre-exponential factors for the two reactions and the dependence of an intermolecular reaction on the hydrocarbon concentration (second-order reaction)[608]. With the assumption that the energies of activation for the oxidative addition of a carbon—hydrogen bond to a metal are the same regardless of the molecularity, knowing the pre-exponential factors for each ($A_{inter} = 2.5 \times 10^8$ l mol$^{-1}$ s$^{-1}$; $A_{intra} = 6 \times 10^{11}$ s$^{-1}$), and assuming a hydrocarbon concentration of 0.1–0.01 mol l$^{-1}$, then there will be a rate difference of $10^4$–$10^6$.

a. *Group VIII transition metals.* Cyclopentadiene reacts with nickel atoms (vapour deposition) to give a mixed $\eta^5$–$\eta^3$ complex which forms as a result of a formal hydrogen transfer from one cyclopentadienyl to the other[642]. The reaction probably takes place by oxidative addition followed by cyclopentadiene insertion into the metal hydride bond. Oxidative addition involving an allylic hydrogen is the usual mode of reaction of an alkene with a low-valent, coordinatively unsaturated transi-

Ni + cyclopentadiene ⟶ [($C_5H_5$)Ni···H] —cyclopentadiene→ [($C_5H_5$)Ni($C_5H_7$)] ⟶ [($\eta^5$-$C_5H_5$)Ni($\eta^3$-$C_5H_7$)] (299)

tion metal. Thus, the allyl nickel hydride can be prepared at low temperatures, but decomposes above −30 °C. At lower temperatures (−40 to −50 °C, L = $PF_3$) the $\pi$-olefin and the allyl hydride are in equilibrium[643].

$$[NiHL(\eta^3\text{-allyl})] \rightleftharpoons [NiL(\eta^2\text{-}CH_2{=}CHMe)] \qquad (300)$$

L = $PPh_3$, $PF_3$

Aryl platinum hydrides can be isolated from the oxidative addition of fluoroaromatics to coordinatively unsaturated zerovalent platinum complexes at 25 °C, while unfluorinated aromatics such as benzene, toluene, and naphthalene are unreactive[644]. Reaction with difluoroaromatics is sensitive to electronic effects (substitution isomers) since *o*-and *p*-difluorobenzene are unreactive. Certain acetylenes undergo oxidative addition reactions to platinum(0), but again the reaction is sensitive to the structure of the substrate[645].

$$[Pt(PCy_3)] + ArH \longrightarrow [PtHAr(PCy_3)_2] \qquad (301)$$

Ar = $C_6F_5$, 1,3,5-$F_3C_6H_2$, 1,3-$F_2C_6H_3$

1-ethynylcyclohexanol + [$PtL_n$] ⟶ [$PtH_2L_2(C{\equiv}CR)_2$] (302)

R = 1-cyclohexonal

Platinum(IV) oxidative addition products have been implicated in the deuterium exchange reactions at the benzyl positions of *p*-xylene, but metal hydride oxidative addition products have not been observed[646], Although platinum(II) also catalyses the exchange of deuterium on aromatic and aliphatic hydrocarbons, alkyl or aryl platinum hydrides have not been detected, even though $\sigma$-bonded platinum(IV) aryls have been synthesized from the reaction of the aromatic hydrocarbon with chloroplatinic acid in aqueous trifluoroacetic acid[647,648].

The decarbonylation of aldehydes to hydrocarbons by Wilkinson's complex[649,650] takes place under mild conditions to give a high yield of the corresponding alkane and in some cases minor amounts of alkene[262,263,282]. The mechanism requires a rate-determining oxidative addition of the aldehyde followed by alkyl migration and reductive elimination of alkyl and hydride (Scheme 6).

The reaction is stereospecific, taking place with retention of configuration[651,652] or geometry[653] at carbon. The reaction shows a primary deuterium isotope effect $k_H/k_D = 1.86$, consistent with the oxidative addition being the rate-determining step[654]. The stereochemistry is in accord with retention in the alkyl migration and reductive elimination steps.

The only oxidative addition intermediate isolated comes from an intramolecular C—H insertion reaction of quinoline-8-carbaldehyde which gives a product stable to its melting point, 175 °C[655]. In refluxing xylene, quinoline and chlorocarbonyl-

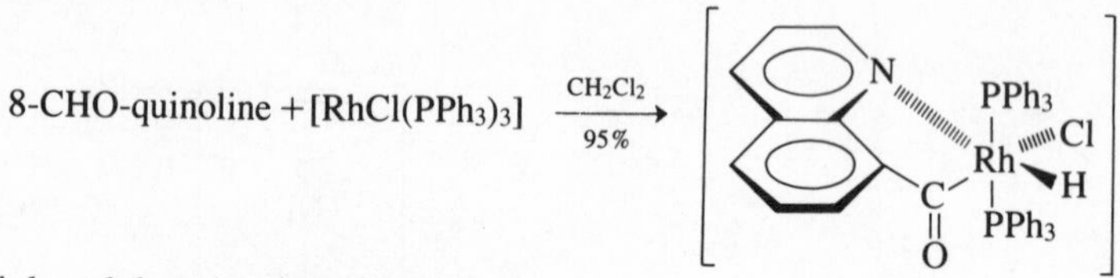

8-CHO-quinoline + $[RhCl(PPh_3)_3]$ $\xrightarrow[95\%]{CH_2Cl_2}$ [quinolinyl acyl Rh complex: N, PPh3, Cl, Rh, H, C=O, PPh3] (303)

bis(triphenylphosphine)rhodium(I) are obtained. Reactions of porphyrin-type rhodium dimers with aldehydes do not give a rhodium hydride but the acyl oxidative addition product can be isolated[656,657].

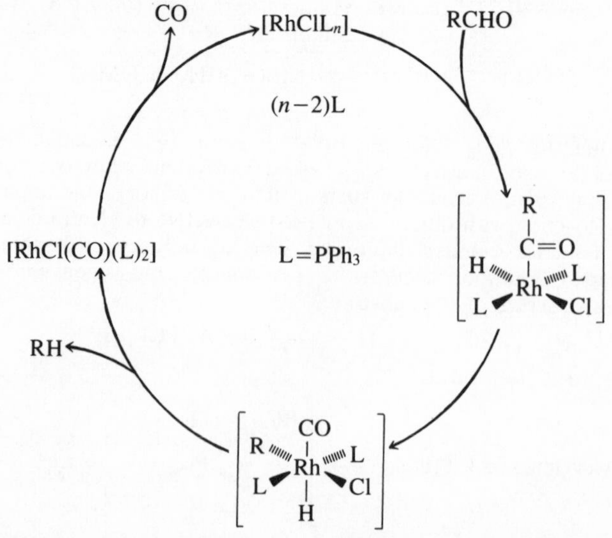

CO
$[RhClL_n]$
RCHO
$(n-2)L$
$[RhCl(CO)(L)_2]$
$L = PPh_3$
[R–C(=O)–Rh(H)(L)(L)(Cl)]
RH
[Rh(CO)(R)(L)(L)(Cl)(H)]

SCHEME 6. Decarbonylation of aldehydes by Wilkinson's catalysts.

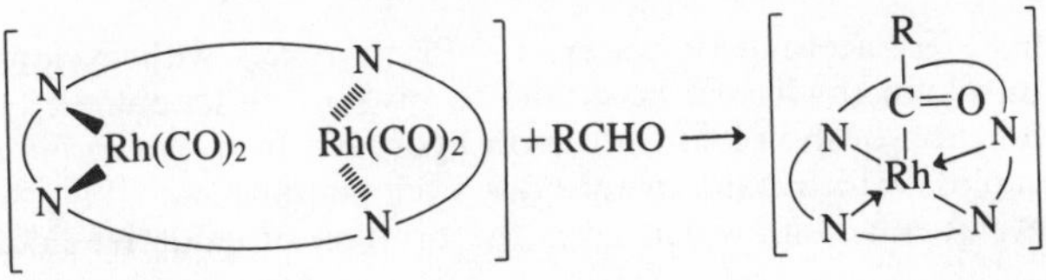

$$\left[ (N_4)Rh(CO)_2 \; Rh(CO)_2 \right] + RCHO \longrightarrow \left[ (N_4)Rh\text{—}C(=O)R \right] \quad (304)$$

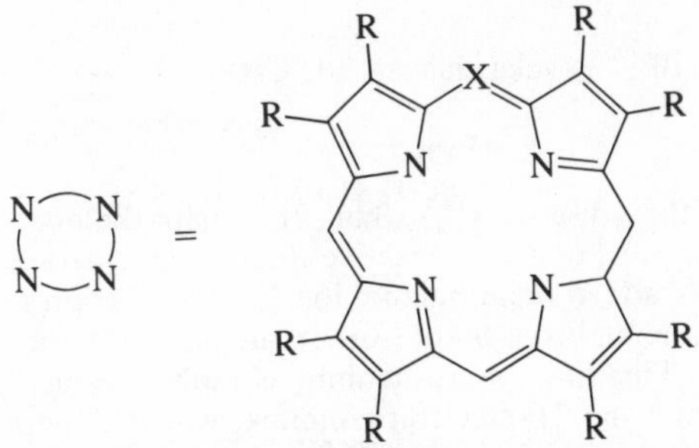

X = N, CH

Although propyne does not react with Vaska's complex[658], the more nucleophilic chlorotris(phosphine)iridium complexes undergo acetylenic C—H insertion to give stable $\sigma$-acetylenic iridium(II) hydrides[659]. When the phosphine is diphenylmethylphosphine and R = Ph, H, or $CO_2Me$, both isomers can be isolated. The iridium

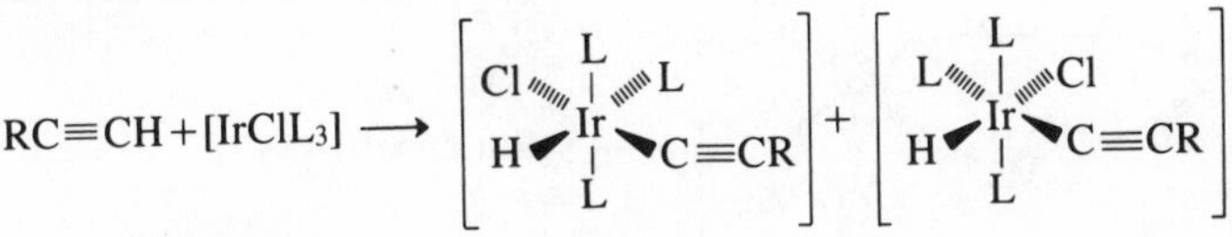

$$RC{\equiv}CH + [IrClL_3] \longrightarrow [IrClL_2H(C{\equiv}CR)] + [IrClL_2H(C{\equiv}CR)] \quad (305)$$

$L = PPh_3$; R = Ph, $CO_2Me$, H, $Pr^n$, $Bu^n$, $CH_2CH_2OH$

$L = PPh_2Me$; R = Ph

complex containing the chelating phosphine, $Et_2PCH_2CH_2PEt_2$, is sufficiently nucleophilic to 'activate' acetonitrile and yield a stable C—H insertion product[660]. Possibly the most important recent development in C—H activation is the discovery

$$[IrCl(Et_2PCH_2CH_2PEt_2)_2] + MeCN \longrightarrow [IrClH(CH_2CN)(Et_2PCH_2CH_2PEt_2)_2] \quad (306)$$

that the trimethylphosphine $\eta^5$-pentamethylcyclopentadienyliridium(I) complex reacts with benzene, cyclohexane, and neopentane by oxidative addition of a C—H bond to iridium[661]. The reactive 16-electron iridium(I) complex is generated by the photolysis of the corresponding iridium(III) hydride. The iridium complex

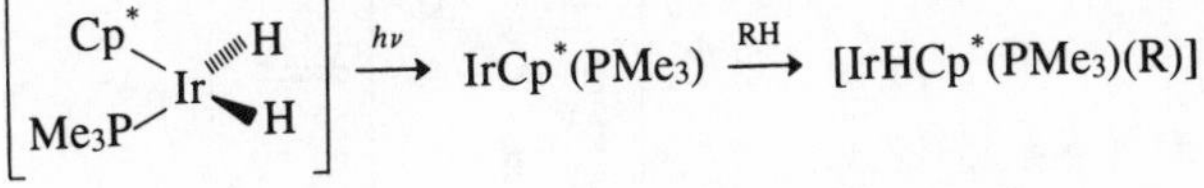

$$[Cp^*(Me_3P)IrH_2] \xrightarrow{h\nu} IrCp^*(PMe_3) \xrightarrow{RH} [IrHCp^*(PMe_3)(R)] \quad (307)$$

$Cp^*$ = $\eta^5$-$C_5Me_5$, $R = C_6H_5, C_6H_{11}, C_4H_{11}$

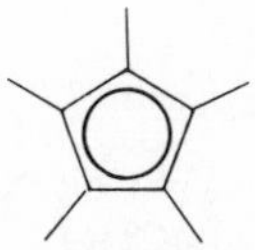

$[IrH_2S_2L_2]^+BF_4^-$ (S = acetone or water; L = $PPh_3$) reacts with cyclopentane in the presence of an olefin (hydrogen acceptor) to give $\eta^5$-cyclopentadienyliridium hydride[662]. The function of the olefin is to strip hydrogen from the iridium, yielding the iridium(I) complex, which oxidatively adds cyclopentane, and in a series of steps carries it to cyclopentadiene which gives the product of oxidative addition.

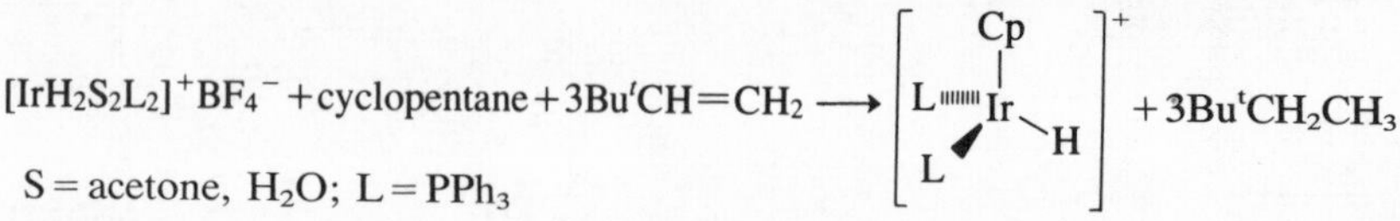

$$[IrH_2S_2L_2]^+BF_4^- + \text{cyclopentane} + 3Bu^tCH{=}CH_2 \longrightarrow [CpIrL_2H]^+ + 3Bu^tCH_2CH_3 \quad (308)$$

S = acetone, $H_2O$; L = $PPh_3$

Since the discovery[663] that the naphthalene radical anion would reduce $[RuCl_2(dmpe)_2]$ to give a reactive intermediate 16-electron ruthenium(0) species that oxidatively added naphthalene, the C—H insertion reactions of this complex and the analogous complexes of the other members of the iron triad have received much attention. Heating the ruthenium complex reductively eliminates naphthalene[663], regenerating the 16-electron complex, which dimererizes by C—H insertion of the methyl groups on the ligand[664,665]. In the solid state, the naphthalene ruthenium

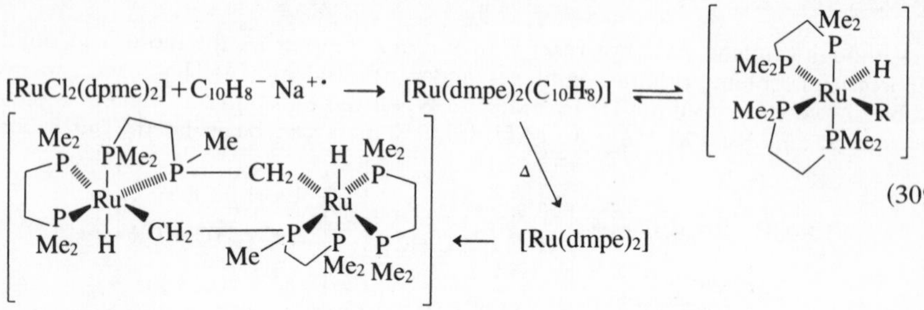

$$[RuCl_2(dpme)_2] + C_{10}H_8^-\ Na^{+\cdot} \longrightarrow [Ru(dmpe)_2(C_{10}H_8)] \rightleftharpoons [RuH(R)(dmpe)_2] \quad (309)$$

$$[Ru(dmpe)_2(C_{10}H_8)] \xrightarrow{\Delta} [Ru(dmpe)_2] \longrightarrow [\text{dimer}: (Me_2PCH_2CH_2PMe_2)Ru(H)(\mu\text{-}CH_2P(Me)CH_2CH_2PMe_2)_2Ru(H)(Me_2PCH_2CH_2PMe_2)]$$

R = 2-naphthyl

complex is as the hydride[666]. The *cis* complex in this triad (shown for ruthenium) is in equilibrium with a *trans* complex[667,668]. The iron naphthalene complex is the most reactive of the triad and therefore has received the most attention. Reactions of this series of complexes $[M(dmpe)_2(C_{10}H_8)]$ with aromatics, alkynes, alkenes, aldehydes, and compounds containing an activated $sp^3$ C—H bond take place by the reductive elimination of naphthalene followed by oxidative addition of the new substrate[668–679]. The reactivity sequence Fe ≫ Ru ≫ Os is a consequence of the decreasing rates of reductive elimination. Although the iron complex undergoes reaction with aromatics within a few minutes at ambient temperature, the ruthenium complex requires 8 h at 60 °C, and the osmium complex is unreactive. The position of insertion in aromatics is sterically controlled, yielding only *m*- and *p*-isomers of substituted benzenes.

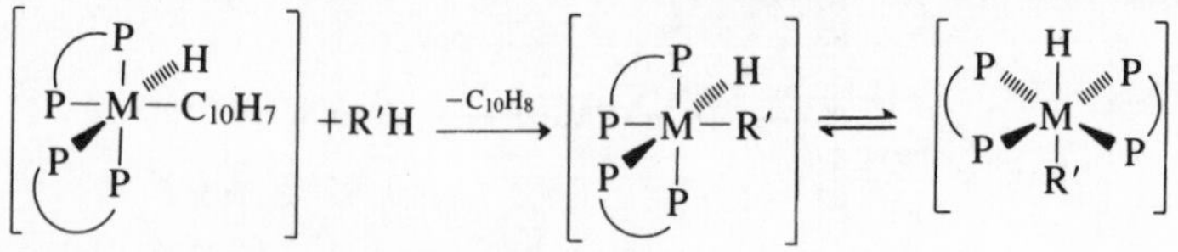

$$[cis\text{-}M(PP)_2(H)(C_{10}H_7)] + R'H \xrightarrow{-C_{10}H_8} [cis\text{-}M(PP)_2(H)(R')] \rightleftharpoons [trans\text{-}M(PP)_2(H)(R')] \quad (310)$$

PP = $R_2PCH_2CH_2PR_2$ (R = Me, Et); M = Fe, Ru:

R'H (in order of reactivity) = $(CF_3)_2C_6H_4$ > PhCN; $C_{10}H_8$ > $C_6H_6$ > PhMe > $PhNH_2$, MeCN, $Me_2CO$, $MeCO_2Et$, MeOR, $MeSO_2R$, $XCH_2Y$ (X = Y = CN; X = $CO_2Me$, Y = CN, etc.), HC≡CR″ (R″ = $Bu^t$, Ph, $CF_3$, $CO_2Et$, COMe), PhCHO

Atomic iron codeposited with cyclopentadiene by vapour deposition yields ferrocene with the evolution of hydrogen[642,675]. The first step in this reaction possibly is oxidative addition, which would be similar to that proposed for nickel (reaction 299), but instead of giving an $\eta^5$–$\eta^3$ complex the stable 18-electron complex is formed, possibly through a 20-electron oxidative addition intermediate.

$$Fe + C_5H_6 \longrightarrow [FeHCp] \xrightarrow{C_5H_6} [FeH_2Cp_2] \xrightarrow{-H_2} [FeCp_2] \quad (311)$$

Tetrakis(triphenylphosphine)ruthenium dihydride inserts into a vinyl C—H of methacrylates[676,677]. The acrylate acts as a hydrogen acceptor by an insertion–reductive elimination sequence to yield the 16-electron ruthenium(0) complex. Trirutheniumdodecacarbonyl and the analogous osmium cluster react with olefins,

$$[RuH_2L_4] + CH_2{=}C(Me)CO_2R \longrightarrow [RuHL_4\{C(Me)_2CO_2R\}] \longrightarrow [RuL_4] + Me_2CHCO_2R$$

$$\left[\text{Ru(H)(L)}_4\text{ chelate: Ru–O=C(OR)–C(Me)=C(H)–Ru}\right] \xleftarrow{CH_2{=}C(Me)CO_2R} \quad (312)$$

occasionally forming $\eta^3$-allyl complexes, and in some examples undergoing insertion into a vinyl C—H bond. The complex may remain as a cluster or may yield a monometallic complex. For example, bicyclo[3.2.1]octa-2,6-diene derivatives can give an $\eta^3$-ruthenium tricarbonyl[678] or a cluster complex derived from olefinic C—H insertion[679]. Cyclopentadiene yields the $\eta^5$-cyclopentadienyl hydride[680]. The intermediate diene complex, which could be isolated, reacts to give the oxidative addition product. Cyclooctene and cycloocta-1,5-diene react at the olefinic C—H bonds,

$$\tfrac{1}{3}[M_3(CO)_{12}] + C_5H_6 \xrightarrow{-CO} \left[(\eta^4\text{-}C_5H_6)M(CO)_3\right] \longrightarrow [MeCp(CO)_2H] \quad (313)$$

M = Ru, Os

maintaining the cluster[681–683], while cycloocta-1,3-diene generates a $\pi$-allyl ruthenium cluster as well[683]. Trienes, particularly cyclododeca-1,5,9-triene, afford $\eta^3$-allyl clusters[684–687]. In most of these reactions, bridging hydrides are obtained.

Triosmiumdodecacarbonyl shows similar behaviour. 1,3,5-Cyclooctatriene (cotr) gives a monomeric complex containing both $\eta^1$- and $\eta^3$-allyl bonding, hydrogen being lost upon the second oxidative addition[688]. The reaction with ethylene is one in which the

$$\tfrac{1}{3}[Os_3(CO)_{12}] + \text{cotr} \xrightarrow[-CO]{h\nu} \left[(\eta^1,\eta^3\text{-C}_8\text{H}_{10})Os(CO)_3\right] \quad (314)$$

cluster is maintained, but hydrogen is geminally stripped[689–692]. Vicinal $sp^2$ hydrogens undergo insertions when benzene or cyclopentene are the hydrocarbon reactants. A highly reactive (44-electron) triosmiumdecarbonyl can be obtained through the

$$[Os_3(CO)_{12}] + CH_2{=}CH_2 \xrightarrow[\text{reflux}]{n\text{-}C_8H_{18}} [(\mu\text{-H})_2Os_3(CO)_9(\mu_3\text{-}C{=}CH_2)] \quad (315)$$

reaction of the dihydride with an alkene[693,694]. The dodecacarbonyltriosmium(0) intermediate then reacts further with the alkene to give the alkenyl osmium hydride.

$$[Os_3H_2(CO)_{10}] + RCH{=}CHCO_2Et \longrightarrow [\text{intermediate}] \xrightarrow[-RCH_2CH_2CO_2Et]{25-50\ °C} [Os_3(CO)_{10}] \xrightarrow{CH_2{=}CHR'} [Os_3H(CO)_{10}(CH{=}CHR')] \quad (316)$$

R=H, $CO_2Et$ (*E*,*Z*);
R′=Ph, $Bu^n$

b. *Early transition metals.* Another important discovery, similar to the dehydrogenation of cyclopentane by iridium[662], is that the dehydrogenation of cyclopentane also takes place with $[ReH_7L_2]$ (L=$PPh_3$, $PEt_2Ph$) in the presence of a hydrogen acceptor[695]. Loss of hydrogen probably initiates the reaction, and the olefin further acts to strip hydrogen, leaving $[ReH_3L_2]$, a 14-electron complex. Vapour-deposited chromium metal also reacts with cyclopentadiene to yield chromacene[642,696].

$$[ReH_7L_2] + \text{cyclopentane} + Bu^tCH{=}CH_2 \longrightarrow [ReH_2L_2Cp] + Bu^tCH_2CH_3 \quad (317)$$

Both molybdenum and tungsten atoms (vapour deposition) react with cyclopentadiene to give the metallocene dihydride[697,698]. When the reaction is carried out with cycloheptatriene (cht), the intermediate monohydride undergoes olefin insertion instead of oxidative addition to form $\eta^7$, $\eta^5$ systems[698].

$$M + 2\ \text{cyclopentadiene} \longrightarrow [MH_2Cp_2] \quad (318)$$

$$M + \text{cht} \longrightarrow [MH(\eta^7\text{-cht})] \longrightarrow [\text{intermediate}] \longrightarrow [M(\eta^7\text{-cht})(\eta^5\text{-cht})] \quad (319)$$

M=Mo, W

Molybdenocene dihydride, on ultraviolet irradiation, loses hydrogen to give a molybdenum dimer via an intramolecular oxidative addition. The hydride dimer further loses hydrogen to yield a dimer containing a molybdenum—molybdenum bond[699]. Tungstenocene on irradiation in ether also yields the hydride dimer, among other products. The self-consuming reaction is a characteristic observed especially with the early transition metallocenes (see reactions 323 and 324). Tungstenocene also can be generated from the dihydride in the presence of a hydrogen acceptor[700,701]. The function of the hydrogen acceptor here again is to

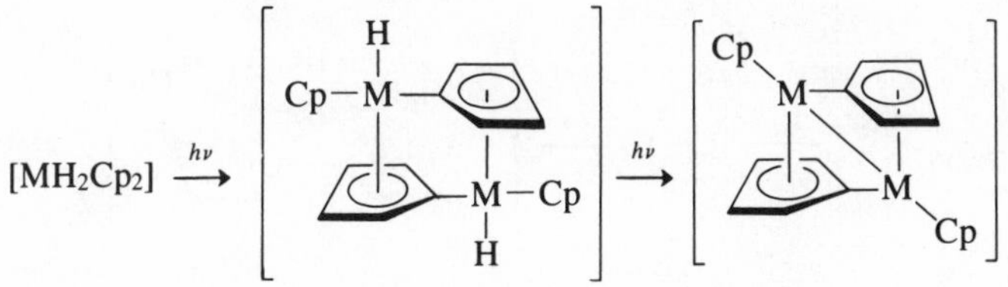

$$[MH_2Cp_2] \xrightarrow{h\nu} [\ldots] \xrightarrow{h\nu} [\ldots] \quad (320)$$

$$[Cp_2WH_2] \xrightarrow{-H_2^{h\nu} \text{ or } 120\,°C + CH_2{=}C(Me)CH{=}CH_2} [Cp_2W] + Me_2C{=}CHMe$$

$$[WHCp_2(ME)] \xrightarrow[\Delta]{-CH_4} [Cp_2W] + Me_2C{=}CHMe$$

$$[Cp_2W] \xrightarrow{ArMe} [WCp_2(CH_2Ar)_2] \qquad [Cp_2W] \xrightarrow{RH(\text{or } RD)} (WH(\text{or } D)Cp_2(R)] \quad (321)$$

Ar = $p$-$MeOC_6H_4$,3,5-$Me_2C_6H_3$, $p$-$CH_3C_6H_4$;
R = Ph,$C_6D_5$, $p$-$MeC_6H_4$, $p$-$MeOC_6H_4$, $MeOCOC_6H_4$(1 : 1 $m$-, $p$-), $C_6H_4F$(2 : 3 $m$-, $p$)

remove hydrogen by an insertion into the W—H bond followed by reductive elimination. The 16-electron tungstenocene oxidatively adds hydrocarbons[700–708]. Aromatic C—H insertion takes place with benzene and substituted aromatics, particularly those containing electron-withdrawing groups. Xylene, mesitylene, and 4-methylanisole undergo insertion at the benzyl C—H bond and ultimately yield the dialkyl tungsten derivative, apparently by the following process.

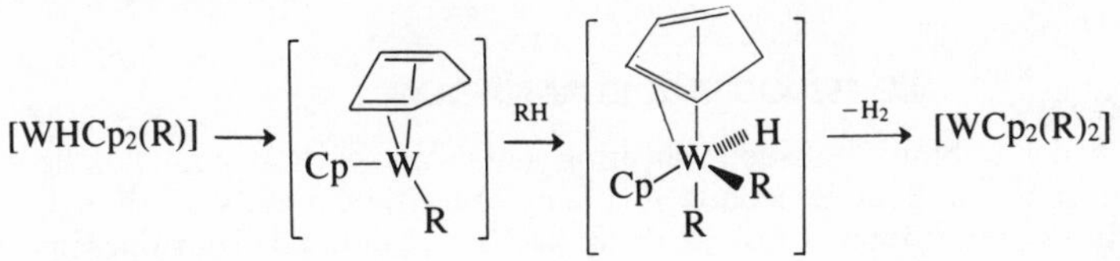

$$[WHCp_2(R)] \longrightarrow [\ldots] \xrightarrow{RH} [\ldots] \xrightarrow{-H_2} [WCp_2(R)_2] \quad (322)$$

Tantalocene and niobocene trihydrides lose hydrogen on thermolysis to yield the 14-electron metallocenes. In benzene, these reactive complexes undergo the characteristic intermolecular self-consumptive oxidative addition[709,710]. This reaction is

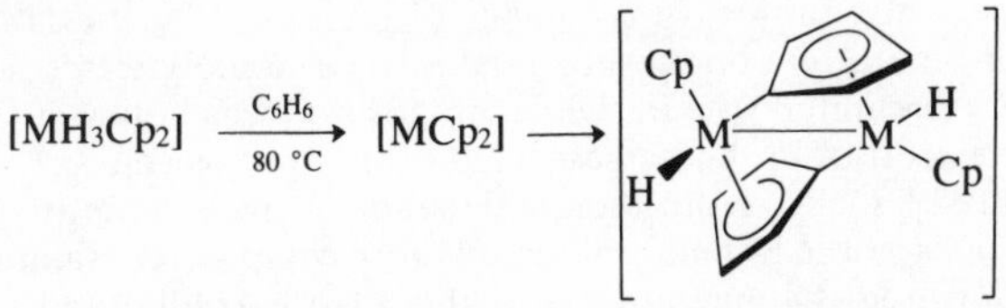

$$[MH_3Cp_2] \xrightarrow[80\,°C]{C_6H_6} [MCp_2] \longrightarrow [\ldots] \quad (323)$$

characteristic of titanocene and zirconocene, the chemistry of which has been reviewed[711]. Titanocene, generated by the reduction of titanocene dichloride, undergoes an intermolecular self-oxidative addition to yield a dimer, the structure of which depends on the conditions for the reduction and work-up. Pentamethyltitanocene undergoes an intramolecular C—H insertion, utilizing one of the methyl groups.

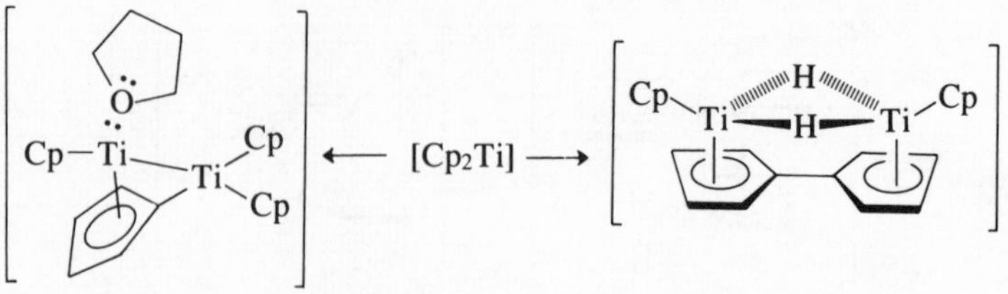

$$[Cp_2Ti] \qquad (324)$$

Zirconocene, prepared from the reduction of zirconocene dichloride by potassium–naphthalene at −80 °C, yields a naphthylhydridozirconium dimer, **64**, resulting from

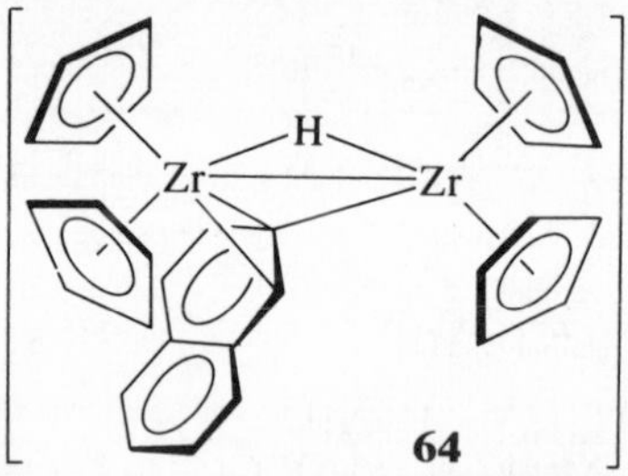

**64**

naphthalene C—H insertion[711,712]. Bisphosphine zirconocenes dissociate a ligand in solution and dimerize by C—H insertion followed by loss of hydrogen[713]. When the dissociation of a ligand takes place in the presence of toluene, the toluene oxidatively adds by aromatic C—H insertion to give a product that can be trapped[515].

$$[ZrCp_2L_2] \rightleftharpoons [ZrCp_2L] \xrightarrow{PhMe} [ZrHCp_2(C_6H_4Me)] \xrightarrow{Me_2CO} [ZrCp_2(C_6H_4Me)(OCHMe_2)] \quad (325)$$

## III. REDUCTIVE ELIMINATION

The ease with which $\sigma$-bonds between transition metals and carbon can be made and broken is central to catalytic reactions involving transition metals. Although the carbon—transition metal bond is not weak, its lability is derived from the variable oxidation states and coordination numbers of the transition metals that are not available to the main group analogues. The breaking of a transition metal—carbon $\sigma$-bond usually involves other coordination sites in addition to the $\sigma$-bond under consideration.

The stability of transition metal—carbon $\sigma$-bonds and the reactions by which they undergo scission has been the subject of a number of reviews[714–721]. Initial views on the stability of carbon—metal $\sigma$-bonds were related to a homolytic scission mode and ligand field theory. The initial step in the decomposition was believed to require the promotion of an electron from the highest filled orbital to a vacant $\sigma^*$ antibonding orbital. The stability of such a bond then is dependent on the energy difference between the orbitals, a larger $\Delta E$ being obtained in a complex containing strong $\sigma$-donor ligands that lowered the energy level of the filled $\sigma$-orbital and raised the energy level of the $\sigma^*$ orbital. Certainly the dissociation energy of such a bond is an important property of organometallic compounds, but the factors that influence the strength of such bonds and the overall stability of transition metal complexes containing $\sigma$-carbon bonds are silll not well understood, partly because there is scant reliable information on bond energies[722]. More recently, it has been recognized that one of the low-energy pathways for alkyl—transition metal complex decomposition is

$\beta$-elimination (see Chapter 8). If this pathway can be blocked, either by construction of the alkyl portion such that $\beta$-hydrogens are absent or $\beta$-hydrogen elimination is geometrically unfavourable, or by preventing the generation of an open coordination site, a requirement for $\beta$-elmination, then the transition metal $\sigma$-bonded complex may be fairly stable. The coupling reaction of organic compounds catalysed by transition metals is an important method of generating carbon—carbon bonds, the final step of which requires the elimination of the organic partners from the transition metal. The elimination can take one or more paths, categorized according to the mechanism (and products), including heterolytic as well as homolytic or concerted $\alpha$-elimination, $\beta$-elimination, 1,1-reductive elimination, and dinuclear elimination (Scheme 7).

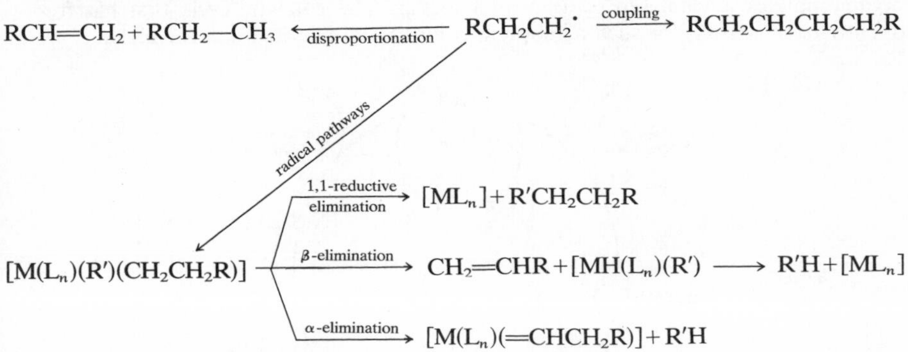

$$RCH{=}CH_2 + RCH_2{-}CH_3 \xleftarrow{\text{disproportionation}} RCH_2CH_2^{\bullet} \xrightarrow{\text{coupling}} RCH_2CH_2CH_2CH_2R$$

$$RCH_2CH_2^{\bullet} \xrightarrow{\text{radical pathways}} [M(L_n)(R')(CH_2CH_2R)]$$

$$[M(L_n)(R')(CH_2CH_2R)] \xrightarrow[\text{elimination}]{\text{1,1-reductive}} [ML_n] + R'CH_2CH_2R$$

$$[M(L_n)(R')(CH_2CH_2R)] \xrightarrow{\beta\text{-elimination}} CH_2{=}CHR + [MH(L_n)(R') \longrightarrow R'H + [ML_n]$$

$$[M(L_n)(R')(CH_2CH_2R)] \xrightarrow{\alpha\text{-elimination}} [M(L_n)({=}CHCH_2R)] + R'H$$

SCHEME 7. Elimination from transition metals.

It has been stated[715,719] that these reactions are usually concerted, proceeding by electron pair processes, and that no high-energy intermediates such as free radicals are involved, particularly in reductive elimination. This appears to be the case with most even-electron transition metal complexes, and, although these reactions appear to be concerted low-energy transformation, there is evidence for free radical pathways in a number of systems. In odd-electron complexes, decomposition by homolytic metal—carbon scission generally appears to be the mode of decomposition. Certainly, there is now substantial evidence for a radical process in the decomposition of alkyl cobalt(dmgH)$_2$ complexes[723]. In this reaction, there is a striking correlation between the base strength of the ligand *trans* to the cobalt alkyl bond and the dissociation energy or the enthalpy of activation, the more basic ligands increasing the dissociation energy.

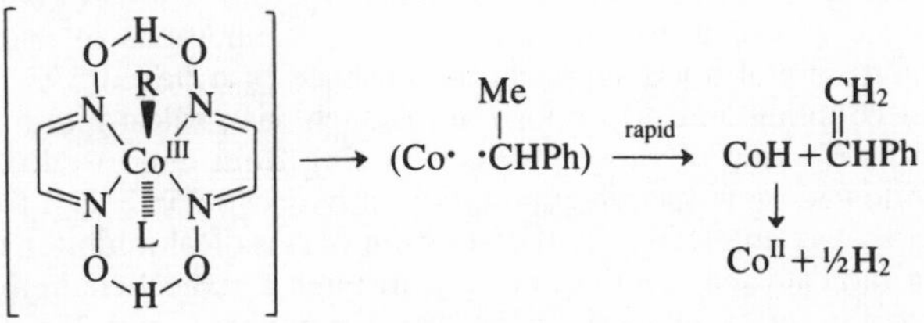

$$[R\text{-}Co^{III}(dmgH)_2L] \longrightarrow (Co^{\bullet}\ \ {}^{\bullet}CH(Me)Ph) \xrightarrow{\text{rapid}} CoH + CH_2{=}CHPh \qquad (326)$$

$$CoH \longrightarrow Co^{II} + \tfrac{1}{2}H_2$$

R = CH(Me)Ph

This section will be concerned, however, only with the 1,1-reductive elimination pathway for two organo-groups attached to the transition metal and the few examples in which isolated transition metal complexes containing an organo-group and a halide undergo elimination to form an organo-halide. In the 1,1-reductive elimination reaction, the formal oxidation state and the coordination number of the metal are reduced by two; bond breaking is accompanied by bond making. The reductive elimination reaction frequently follows an oxidative addition reaction, and this combination, oxidative addition–reductive elimination, is responsible for both stoichiometric and catalytic coupling reactions via transition metals, particularly those of the copper triad and Group VIII. The coupling reaction of organometallic reagents such as Grignard, organoaluminium, and organolithium reagents with organic halides is catalysed by transition metals. The catalytic cycle that has frequently been written for such a coupling is illustrated in Scheme 8 for nickel.

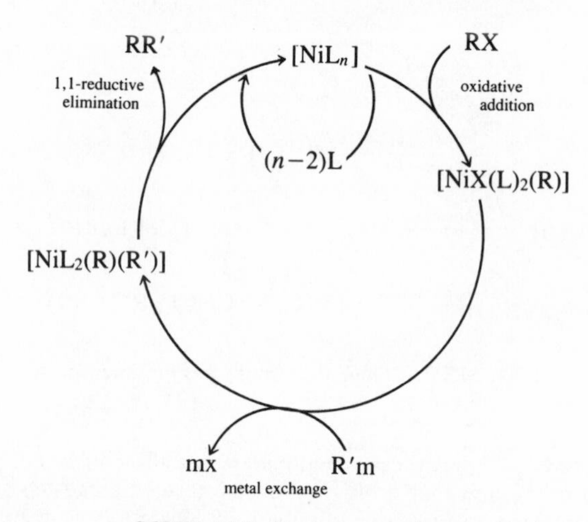


SCHEME 8. Catalytic coupling.

Not included in this section is the 1,1-reductive elimination of an organic group and hydrogen such as occurs subsequent to $\beta$-elimination (Scheme 7), the last step in homogeneous hydrogenation, and the last step in aldehyde decarbonylation. There are a few examples of elimination reactions involving univalent metals that couple via multinuclear elimination pathways.

A variety of different tests have been applied to the 1,1-reductive elimination reactions to confirm or disprove their non-radical nature: (1) retention of stereochemistry at an $sp^3$ carbon—metal bond or retention of geometry at a vinyl—metal centre during coupling is one of the most rigorous tests of a concerted process; (2) the absence of cyclic products in reductive eliminations of complexes containing a hex-5-enyl—metal bond supports the absence of a hexen-5-yl radical; (3) the distribution of coupling and disproportion products via radical intermediates is different than the distribution of those products arising from concerted coupling and $\beta$-elimination (followed by reductive elimination of hydrogen and alkyl from the metal); (4) a reaction that is unaffected by the addition of a radical inhibitor may be proceeding through a radical cage reaction, but not through a radical chain mechanism; (5) the introduction of a labeled chain transfer reagent and the lack of incorporation of the label into the product have the same implications; (6) the observation of an e.s.r. signal implicates a radical reaction, although radicals observed can be present in small concentrations, and may not be involved in the major reaction pathway; and (7) observation of CIDNP in the n.m.r. implicates radical intermediates, although failure to observe CIDNP does not exclude their presence.

## A. Organic Halide Elimination

The observation of reductive elimination of organic halide from an isolated transition metal complex containing a halogen and an organic group is rare. Platinum(IV) complexes containing at least two halogens and one or two methyl groups yield methyl halide on thermolysis (185–210 °C)[724]. In the dihalodimethylplatinum complexes, the groups undergoing elimination must occupy adjacent positions, and whether methyl halide or ethane is produced depends on the halogen and the geometry of the complex. The elimination of organic chloride is the final step in

$$\left[\begin{array}{ccc} & X & \\ L & | & Me \\ & Pt & \\ X & | & L \\ & X & \end{array}\right] \longrightarrow MeX + [PtCl_2L_2] \qquad (327)$$

X=Br, Cl

$$\left[\begin{array}{ccc} & Me & \\ L & | & Me \\ & Pt & \\ X & | & L \\ & X & \end{array}\right] \longrightarrow MeX + C_2H_6 \qquad (328)$$

X=Cl 75 25
X=Br 100

$$\left[\begin{array}{ccc} & Me & \\ X & | & Me \\ & Pt & \\ L & | & X \\ & L & \end{array}\right] \longrightarrow MeX + C_2H_6 \qquad (329)$$

X=Cl — 100
X=Br 10 90

$$\left[\begin{array}{ccc} & Me & \\ X & | & L \\ & Pt & \\ Me & | & L \\ & L & \end{array}\right] \longrightarrow MeX + C_2H_6 \qquad (330)$$

X=Cl 10 90
X=Br 60 40

the decarbonylation of acid chlorides in which there is no $\beta$-hydrogen (Scheme 5)[262,263]. The reaction is first order in complex. When R is methyl, the activation energy is 23 kcal mol$^{-1}$; free phosphile enhances the elimination by a nucleophilic

$$\left[\begin{array}{ccc} & R & \\ CO & | & PPh_3 \\ & Rh & \\ Ph_3P & | & Cl \\ & Cl & \end{array}\right] \longrightarrow RCl + [RhCl(CO)(PPh_3)_2] \qquad (331)$$

attack on the methyl, generating a phosphonium salt[725]. The rate of the reductive elimination of benzyl chlorides ($R = p\text{-}XC_6H_4CH_2$) is increased slightly by electron-withdrawing substituents ($\rho = 0.45$), consistent with a concerted process[287]. A small negative entropy of activation ($\Delta S^{\ddagger} = -3.6$) is observed. Although the analogous phenylrhodium(III) complex decomposes at relatively low temperatures[287], the elimination of chlorobenzene ($R = p\text{-}C_6H_5$) does not occur readily at ambient temperature[297]; temperatures of 150 °C are necessary for the elimination of chlorobenzene.

The reductive elimination takes place with retention of configuration at carbon. Although (*S*)-$\alpha$-trifluoromethylphenylacetyl chloride racemizes on decarbonylation[290], (*S*)-$\alpha$-deuteriophenylacetyl chloride decarbonylates to (*S*)-benzyl-$\alpha$-*d*-chloride with 20–27% net retention of configuration[284]. If the acyl–alkyl rearrangement does indeed proceed with retention of configuration at carbon, then the reductive elimination step also occurs with retention of configuration at carbon. Further support for this stereochemistry comes from the reaction of the (*S*)-$\alpha$-trifluoromethylbenzyl chlorosulphite with chlorocarbonlybis(diethylphenylphosphine)rhodium(I) to yield the optically active alkyl rhodium complex[290]. Reductive elimination produces (*R*)-$\alpha$-trifluoromethylbenzyl chloride, in which overall inversion of configuration at carbon takes place. Since there is precedence for the oxidative addition of chlorosulphite esters to take place with inversion of configuration at carbon, then the reductive elimination would necessarily take place with retention.

$$H(CF_3)(Ph)COSOCl + [RhCl(CO)(PEt_2Ph)_2] \xrightarrow{-SO_2} \left[ Ph(CF_3)(H)C{-}Rh(CO)(Cl)(Cl)(PEt_2Ph)_2 \right] \rightarrow Ph(CF_3)(H)CCl \quad (332)$$

The reductive elimination of akyl halide by the reaction of halogen with a *threo*-1,2-dideuterio-2-phenylethyliron(II) complex, postulated to proceed through an iron(IV) cation, results in cleavage of the iron—carbon bond with retention of configuration at carbon[726]. When the same cyclopentadienyl is subjected to oxidation

$$\left[ \text{Ph, H, H / D, D; } Fe(Cp)(CO)_2 \right] \rightarrow \left[ \text{Ph, H, H / D, D; } Cp{-}Fe(CO)_2{-}X \right]^{+} \rightarrow \text{Ph, H, H / D, D; } X \quad (333)$$

at an electrode in the presence of chloride ion, the alkyl chloride also is obtained with retention of configuration at carbon[727]. In this case, the mechanism of the reaction is believed to proceed by oxidation to an iron(III) intermediate.

The reaction of divalent allylmolybdenum or allyltungsten complexes with excess of phosphine results in the reductive elimination of allyl chlorides and the formation of the zerovalent metal complexes[728].

$$[MCl(CO)_2(\eta^3{-}RC_3H_4)(MeCN)_2] \xrightarrow{L} [MCl(CO)_2(\eta^3{-}RC_3H_4)L_2] \xrightarrow{L} CH_2{=}C(R)CH_2Cl + \left[ M(CO)_2(L)_3(NCMe) \right] \quad (334)$$

M = Mo, W; R = H, Me

L = PMePh$_2$

### B. 1,1-Reductive Elimination of Organic Partners

As discussed previously, when a β-hydrogen is available on an organic group bound to a transition metal, the β-elimination mode of decomposition usually is kinetically the most facile process. A vacant coordination site on the metal is necessary for this to occur. Thus, 1,1-reductive eliminations often are studied by using methyl, benzyl, neopentyl, and aryl groups. Even though a β-hydrogen is available, β-elimination may be prevented by preventing or retarding the release of a vacant coordination site. Often this is accomplished by charging excess of ligand to the reaction or using chelating ligands. If the position of a β-hydrogen is such that its elimination would produce a bridgehead double bond, this mode of decomposition can be prevented.

An important constraint for β-elimination is that the dihedral angle made by the metal, the α-and β-carbons, and the β-hydrogen in this concerted reaction be 0° or nearly so. In the smaller ring metallocycles, this coplanarity cannot be achieved easily and β-elimination is suppressed.

$$M{-}C_{\alpha}{-}C_{\beta}{-}H \longrightarrow [\,M\cdots C{=}C\cdots H\,] \longrightarrow C{=}C + M{-}H \qquad (335)$$

#### 1. *Copper, silver and gold*

Both copper(I) and copper(II) alkyls undergo decomposition to yield organic products, but copper(I) alkyls containing β-hydrogens yield alkenes[32,729]. It has been pointed out[729] that $d^9$ copper(I), silver(I), and gold(I) complexes, in which the metal cannot sustain a two-electron reduction, undergo the reductive elimination reactions by requiring at least two metal centres.

Aryl copper complexes decompose to give biaryls[730–732], the aromatic groups always being lost pairwise in the thermolysis of an octanuclear cluster[732]. Radicals

$$[Cu(2\text{-}C_6H_4CF_3)]_8 \xrightarrow[C_6H_6]{\Delta} 2,2'\text{-}(CF_3)_2\text{-biphenyl} + [Cu_8(2\text{-}C_6H_4CF_3)_6] \qquad (336)$$

are not involved in these reactions, even when the alkyl copper complexes containing a phosphine ligand undergo decomposition to give products of β-elimination[733]. In hexanuclear copper clusters containing mixed organic groups, only cross-coupling products are observed[734].

$$[Cu_6(Ar)_4(R)_2] \xrightarrow[C_6H_6]{80\,^{\circ}C} ArR + [Cu^{I}_4(Ar)_3RCu^{0}_2]$$

$R = C{\equiv}CR'$

Vinyl copper(I) species couple to give dienes, nearly complete retention of geometry being observed in all cases[735]. The decomposition reactions of neopentyl-

$$[(Z){-}CuCH{=}CHMe]_n \longrightarrow MeCH{=}CHCH{=}CHMe \quad 95\text{-}98\%\ Z,Z \qquad (337)$$

or

$$[\text{(Z)-propenyl}{-}CuPBu^n_3]_n$$

type copper(I) phosphine complexes, however, give products characteristic of a radical reaction[736]. Bis(triphenylphosphine)methylcopper(I) is moderately stable, but does yield products indicative of radical reactions[737].

$$[PhC(Me)_2CH_2CuPBu^n_3]_n \longrightarrow \underset{60\text{–}45\%}{PhCMe_3} + \underset{6\text{–}18\%}{PhCH_2CHMe_2}$$

$$+ \underset{4\text{–}14\%}{PhCH_2C(Me){=}CH_2} + \underset{0.3\text{–}1.7\%}{PhCH{=}CMe_2}$$

$$+ \underset{8\text{–}26\%}{[PhC(Me)_2CH_2]_2} + \underset{2\text{–}6.5\%}{PhC(Me)_2CH_2C(Me)_2CH_2Ph} + \underset{1\text{–}4.6\%}{[PhCH_2C(Me)_2]_2} \quad (338)$$

In contrast, neither dialkylcopperlithium nor divinylcopperlithium compounds in the presence of an electron acceptor couple via radical mechanisms, since double bond geometry is retained in the vinyl coupling and rearrangement reactions characteristic of alkyl radicals are not observed[738].

$$[CuR_2]Li \xrightarrow[\text{or } PhNO_2]{O_2} RR \quad (339)$$

$R = Bu^n$, $PhC(Me)_2CH_2$, $MeCH{=}CH$

The mechanism for the cross-coupling reaction of organic halides with organocuprates of the type $[CuR_2]Li$, as indicated in the oxidative addition section, is open to question[32,35]. However, a mechanism involving oxidative addition of the organic halide followed by reductive elimination finds support in the analogous chemistry of reductive elimination from trialkyl gold complexes. In the alkyl halide, inversion of configuration at carbon is observed, which is the expected result of an oxidative addition. Retention of configuration during reductive elimination also would be expected, consistent with the overall stereochemistry observed in coupling. Since cross coupling always is observed, and since the elimination of *cis* partners would be required, then the geometry of the intermediate copper(III) complex must place the two alkyls originally attached to copper trans from one another.

$$R'X + [CuR_2]Li \xrightarrow{-LiX} \left[ \begin{array}{c} R \\ | \\ R'\text{—}Cu \\ | \\ R \end{array} \right] \longrightarrow RR' + RCu \quad (340)$$

The thermolysis of an *n*-butyl silver phosphine complex results primarily in *n*-octane, with minor amounts of butene and butane[739]. The failure of radical scavengers to trap $C_4$ fragments also supports a concerted coupling process, at least through a dinuclear species. Aryl silver[730] and vinyl silver[735] complexes give coupling

$$[Ag(Bu^n)(PBu^n_3)] \longrightarrow \underset{93\%}{n\text{-}C_8H_{18}} + \underset{94\%}{PBu^n_3} + \underset{92\%}{Ag} \quad (341)$$

products, in the latter case the geometry at the $sp^2$ carbon bound to silver being retained. However, is with the copper analogues, phytyl silver phosphine complexes yield products characteristic of radical reactions[736].

Both gold(I) and gold(III) complexes undergo reductive elimination reactions[729,740], although in the former case a two-electron reduction requires at least two metal centres. Butane is obtained from ethyl(triphenylphosphine)gold(I). The rate of elimination of ethane from the methyl gold complex is first order and is retarded by added triphenylphosphine. Thus, a mechanism involving dissociation of phosphine followed by the reaction with additional methyl(triphenylphosphine)gold has been suggested[729,741].

$$[AuLR] \longrightarrow AuR + L \quad (342)$$

$$AuR + [AuLR] \longrightarrow RR + 2Au + L \quad (343)$$

The reductive elimination reaction from gold(III) complexes to yield coupled products was recognized early in the chemistry of organometallic complexes[742–744]. A number of different dialkylgold(III) species were shown to give coupled products, characteristic of a concerted 1,1-reductive elimination. The reductive elimination of ethane from dimethyl(triphenylphosphine)gold(III) complexes later was shown to be

$$[AuXR_2] \xrightarrow{-AuX} RR \xleftarrow{-AuBr} \left[ R_2Au\langle N\frown N \rangle \right]^+ Br^- \quad (344)$$

X=SCN, CN; $N\frown N$ =en or phen

intramolecular and first order in gold complex[475]. The reaction of methyl(triphenylphosphine)gold(I) with methyl iodide to yield ethane has been demonstrated to go through dimethylgold(III) iodide and ultimately through a trimethyl gold(III) complex[746,747].

$$[AuL(Me)] + MeI \longrightarrow [AuI(L)(Me)_2] \quad (345)$$

$$[AuI(L)(Me)_2] + [AuL(Me)] \xrightarrow{\text{fast}} [Au(L)(Me)_3] + [AuI(L)] \quad (346)$$

$$[Au(L)(Me)_3] \longrightarrow C_2H_6 + [Au(L)(Me)] \quad (347)$$

The 1,1-reductive elimination of coupled product from cationic bis(phosphine) dialkylgold(III) complexes is intramolecular[748,749]. The decrease in the rates of reductive elimination within a series of phosphine ligands ($Ph_3P > PMePh_2 > PMe_2Ph > PMe_3$), attributed to a decrease in cone angle[748,749], is probably due instead to an increase in the donor character of the ligand.

The recuctive elimination of ethane[750] or perfluorobiphenyl[751] from triorganogold complexes also has the characteristics of a concerted process. Trimethyl(triphenylphosphine)gold(III), for example, yields ethane and methyl(triphenylphosphine)gold(I). Detailed mechanistic studies[40,752,753] revealed that dissociation of phosphine takes place prior to reductive elimination, and that only *cis*-alkyl groups undergo coupling. Thus, while *trans*-ethyldimethylgold yields only propane, *cis*-ethyldimethylgold gives both ethane and propane. The reductive elimination is an intramolecular process, and although the *cis–trans* isomers do equilibrate in a polar solvent the *trans* isomer is favoured, particularly when the R group is large.

$$trans\text{-}[AuL(R)Me)_2] \xrightarrow{-L} trans\text{-}[AuR(Me)_2] \longrightarrow RCH_3 + [Au(L)(Me)] \quad (348)$$

$$cis\text{-}[AuL(R)(Me)_2] \xrightarrow{-L} cis\text{-}[AuR(Me)_2] \longrightarrow RMe + C_2H_6 + [Au(L)(Me)] + [Au(L)(R)] \quad (349)$$

R = $CD_3$, Et

Reductive elimination takes place from a three-coordinate intermediate, calculated to be Y- or T-shaped, but not through a complex of $C_3$ symmetry, which is a high energy species[753]. Isomerization from *cis*- to *trans*-T-shaped minima takes place

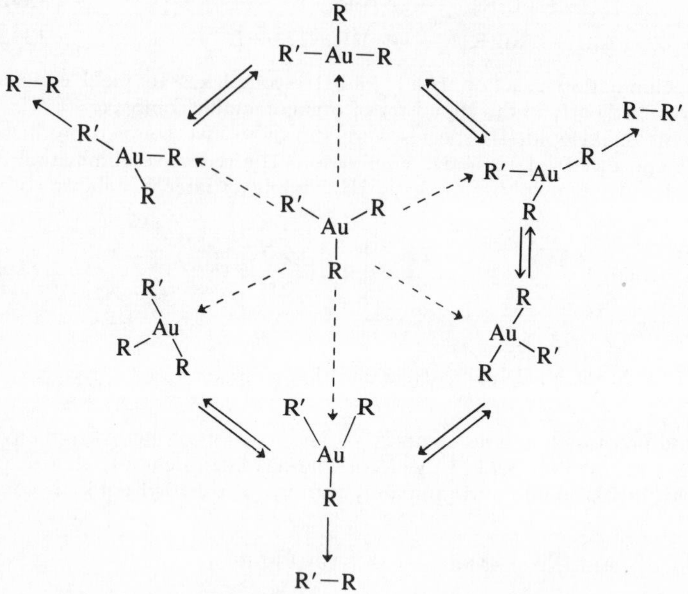

SCHEME 9. Intermediate gold(III) geometries.

through Y-shaped saddle points, which are channels for reductive elimination (Scheme 9). It is noteworthy that reductive elimination is taking place from a 14-electron complex. Thus, high orbital occupation is not necessary for reductive elimination[716], and in fact ligand dissociation to a complex of lower orbital occupation is required for reductive elimination to take place. The reaction of trimethylgold with methyl iodide yields ethane and dimethylgold iodide; a tetramethyl gold iodide complex has been postulated[754].

### 2. *Nickel, palladium, and platinum*

The concerted coupling of alkyl groups bound to transition metals of the nickel triad is an allowed process for adjacent partners[755], but may proceed by a variety of mechanistic pathways, depending on the metal, the ligands, and the organic groups. A summary of the various pathways (Scheme 10) shows, as will become apparent in this section, that both associative and dissociative pathways (18-, 16-, and 14-electron complexes) are available for nickel[45a]. A dissociative path is preferred for dialkylpalladium(II) complexes, while both platinum and palladium undergo reductive elimination through the tetravalent, 18-electron complex. Although stable platinum(IV) complexes can be isolated, and the reductive elimination reactions from these complexes can be studied, palladium(IV) complexes are not isolated and the coupling reactions of organic partners are only by implication, particularly in catalytic coupling reactions. Dialkylplatinum(II) complexes do not undergo 1,1-reductive elimination reactions readily.

Both bis- and tris-(phosphine)diorganonickel complexes undergo facile 1,1-reductive elimination[756,757]. The course of the elimination reactions of nickelocene phosphine complexes depends on the number of phosphines on nickel and therefore whether an 18-, 16-, or 14-electron complex is involved[758,762]. Although a $\beta$-

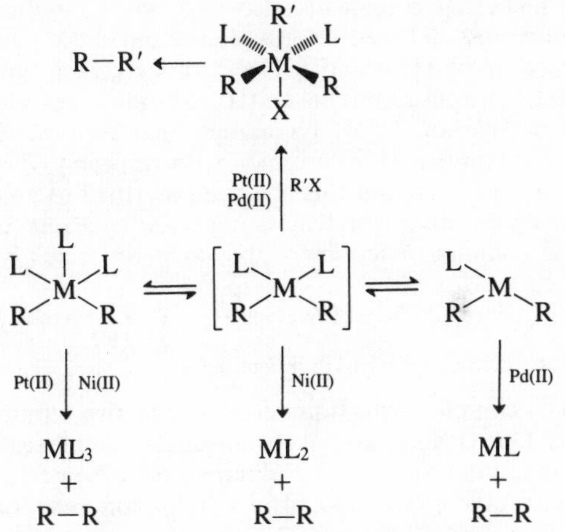

SCHEME 10. Elimination pathways for the nickel triad.

elimination reaction from nickelacyclopentane is suppressed because of the difficulty in achieving a coplanar transition state, this mode of decomposition is favoured for the 14-electron complex. The 1,1-reductive elimination takes place predominanantly from the 16-electron complex, while the five-coordinate nickel serves as an intermediate to a new four-coordinate geometry that yields ethylene. Bisphosphine coordinated nickelacyclohexanes give high yields of cyclopentane[763].

Extended Hückel calculations reveal that the square-planar geometry is favourable for the generation of cyclobutane (Scheme 11)[764]. The reverse cyclometallation reaction to give ethylene from this planar complex is forbidden, but is allowed from a geometry in which the ligands are orthogonal to the plane of nickelacyclopentane. Direct interconversion of these two four-coordinate geometries are symmetry forbidden, and must proceed through the 18-electron five-coordinate complex.

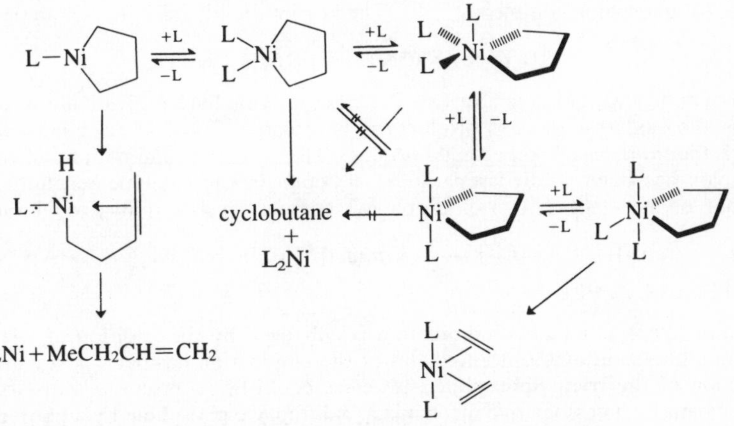

SCHEME 11. Elimination reactions of nickelacyclopentanes.

Vinyl and aryl nickel(II) complexes also undergo 1,1-reductive elimination, although there is much less detailed information about these coupling reactions. The cross-coupling reaction of (*E*)- and (*Z*)-$\beta$-bromostyrenes with phenylmagnesium bromide catalysed by bis(phosphine)nickel(II) chloride takes place with retention of geometry at the $sp^2$ carbon[765], which argues against a radical mechanism. The cross-coupling of phenylmagnesium bromide with (*E*)- and (*Z*)-1,2-dihaloethylenes, however, gives *E*, *Z*-mixtures, but this has been ascribed to an addition–elimination mechanism, primarily because acetylene is observed as a reaction product. The key intermediate in the coupling reactions of the styryl bromides is presumed to be the

$$(Z)\text{-PhCH}{=}\text{CHBr} + \text{PhMgBr} \xrightarrow{[\text{NiCl}_2\text{L}_2]} (Z)\text{-Ph}_2\text{PCH}{=}\text{CHPPh}_2 \qquad (350)$$

$L = PPh_3$; $L_2$ = dppe, dmpe, (*Z*)-$Ph_2PCH{=}CHPPh_2$

vinylphenylnickel(II) complex, which undergoes reductive elimination with retention of geometry. *trans*-Diaryl and -aryl(alkyl)nickel(II) complexes have been isolated, and these complexes undergo reductive elimination[51]. Isomerization of the *trans* to the *cis* isomer must take place first. The elimination reaction is first order and

$$[\text{NiBrL}_2(m,p\text{-FC}_6\text{H}_4)] \xrightarrow{\text{Rm}} [\text{NiRL}_2(\text{FC}_6\text{H}_4)] \longrightarrow \text{FC}_6\text{H}_4\text{R} \qquad (351)$$

$L = PEt_3$; Rm = PhMgBr, MeMgBr, MeLi

intramolecular. Because the elimination was retarded by added phosphines, prior dissociation to a 14-electron complex was postulated. Although these individual steps in the catalytic cross-coupling reactions can be separately documented, the overall catalytic process is faster than any of the individual steps. Further, added aryl halides were shown to give a six-fold rate increase in the reductive elimination. To explain these data, a mechanism involving an electron transfer reaction from the complex to the aryl halide, giving an unstable nickel(III) intermediate that undergoes rapid reductive elimination, was suggested[766].

The mechanism for the stoichiometric coupling of aryl halides by nickel(0) is much less clear[767]. The oxidative addition reaction is reasonable, but an exchange reaction of aryl for halogen to give a diaryl nickel complex does not take place. Electron transfer reactions to yield nickel(I) and nickel(III) intermediates appear to be central to the coupling[767].

The displacement reactions of halogen on aryl halides by cyanide is catalysed by nickel(0) phosphine complexes[768–772]. The key intermediates in the catalytic cycle can

$$\text{RC}_6\text{H}_4\text{X} + \text{NaCN} \xrightarrow{[\text{NiL}_3]} \text{RC}_6\text{H}_4\text{CN} + \text{NaX} \qquad (352)$$

be isolated, and their reactions to the next intermediate individually documented. Thus, the oxidative addition product can be isolated[769], and its reaction with cyanide yields the *trans*-arylnickel cyanide product. The reductive elimination reaction of this complex has been studied separately, although in the catalytic reaction it usually cannot be isolated. The *o*-chlorophenyl complex resists reductive elimination[768].

$$trans\text{-}[\text{NiX(Ar)(L)}_2] \xrightarrow{\text{CN}^-} trans\text{-}[\text{Ni(CN)(Ar)(L)}_2] \longrightarrow \text{ArCN} \qquad (353)$$

$L = PPh_3, PCy_3, PEt_3$

Because the 1,1-reductive elimination is enhanced by the addition of triethylphosphite, a five-coordinate intermediate for the elimination has been postulated[771]. The function of the triethylphosphite, however, could be to promote *trans* to *cis* rearrangement by an associative mechanism, and replace phosphine by a poorer $\sigma$-donor ligand, thereby enhancing reductive elimination.

The elimination reaction of dialkylbipyridinenickel(II) is first order and intramolecular[773–778]. Enhanced yields of coupling products are obtained when olefins such as acrylonitrile, containing electron-withdrawing groups, are added[773–775]. Oxidizing agents[779] (electron acceptors) such as aryl halides[777,778] enhance the rates of elimination, giving a bipyridine aryl nickel(III) chloride product. This provides a

$$[(N,N)Ni(R)_2] \xrightarrow{CH_2=CHX} [(N,N)Ni(R)_2(CH_2=CHX)] \longrightarrow RR + [(N,N)Ni \leftarrow CH_2=CHX] \quad (354)$$

$$[(N,N)Ni(R)_2] \xrightarrow{ArX} RR + [(N,N)Ni(X)(Ar)]$$

R = Me, Et, $Pr^n$, $Bu^t$; (N,N) = bipy

suitable explanation for the difference in the rates of the stoichiometric reductive elimination reaction and the catalytic reaction, since electron acceptors, such as aryl halides, and occasionally oxygen, are available in the catalytic reaction medium[766]. The enhanced lability of the organonickel complex has been attributed to a cation radical. This one-electron transfer lowers the activation energy for reductive elimination from 66 to 15.6 kcal $mol^{-1}$ [716]. It is also possible that in the 1,1-reductive elimination occurring in the presence of aryl halide, a nickel(IV) intermediate is the complex that generates the coupling product.

Palladium(0) catalyses the coupling of benzyl halides with organometals, such as Grignard reagents and organolithium compounds. In a number of studies the 1,1-reductive elimination of organic partners from bis(phosphine)diorganopalladium(II) complexes has been carried out as a model for that step in the catalytic coupling reaction[51,780]. For example, *trans*-bis(phosphine)methylphenylpalladium(II) complexes decompose thermally to give toluene. One of the problems to be examined in such a 1,1-reductive elimination reaction, therefore, is the mechanism by which the two *trans* organic partners eventually become coupled. In catalytic coupling reactions[780–782] proceeding by the oxidative addition–transmetallation sequence, the *trans* complex is obtained[51]. However, if isomerization to the *cis* complex were slow compared with reductive elimination, the transient *cis* complex might not be observed.

$$RX + [PdL_n] \longrightarrow trans\text{-}[PdX(R)(L)_2] + (n-2)L \quad (355)$$

$$trans\text{-}[PdX(R)(L)_2] + R'm \longrightarrow trans\text{-}[PdR(R')(L)_2] + mX \quad (356)$$

$$trans\text{-}[PdR(R')(L)_2] \longrightarrow RR' + [PdL_2] \quad (357)$$

In order for concerted thermal 1,1-reductive elimination to take place, it has been argued that the organic moieties must occupy adjacent positions in the complex[716,719–721]. Construction of an orbital correlation diagram for *cis* four-coordinate square-planar $d^8$ complexes reveals that the concerted elimination is symmetry allowed[755]. Although the thermal concerted elimination directly from the trigonal bipyramidal and the tetrahedral complexes is symmetry allowed, 1,1-reductive elimination from a trigonal three-coordinate species is symmetry forbidden[755].

There are a number of conceivable pathways by which the two organic groups in a *trans* complex could gain positions adjacent to one another prior to coupling: (1) oxidative addition of an organic halide to the palladium(II) complex; (2) prior dissociation of a phosphine to give a three-coordinate intermediate (dissociative mechanism); (3) prior association of a phosphine to give a five-coordinate complex (associative mechanism); (4) conversion of the complexes in (2) or (3) to the *cis* square-planar complex by recoordination or dissociation of phosphine (after rearrangement), respectively; and (5) distortion of the *trans* complex into a transient tetrahedral geometry.

The reductive elimination reactions of a number of dimethylpalladium complexes (Scheme 12) have been studied[783,784]. The *cis–trans*-isomerizations are rapid at

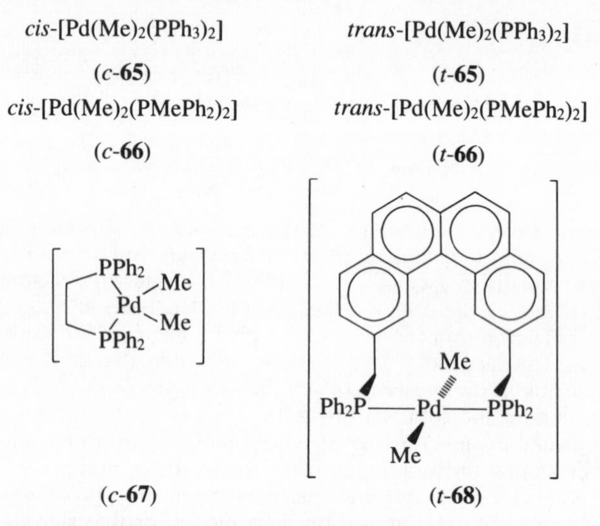


SCHEME 12. Dimethylpalladium(II) complexes.

moderate temperatures (45 °C) and no evolution of ethane takes place. The 1,1-reductive elimination takes place only from the *cis* isomers above 45–60 °C. At these temperatures, no reductive elimination takes place from the *trans* complexes. Although the dppe(dimethyl)palladium complex *c*-**67** undergoes 1,1-reductive elimination in dmso at 80 °C, the TRANSPHOS(dimethyl)palladium complex, *t*-**68**, fails to undergo reductive elimination of ethane.

Dmso solutions containing equimolar amounts of the *cis* isomers *c*-**65**, *c*-**66**, or *c*-**67** and their corresponding perdeuteriomethyl analogues undergo reductive elimination to give only ethane and $d_6$-ethane. No trideuterioethane could be detected, demonstraing that no exchange of methyl takes place between complexes, and that the reductive elimination is mononuclear and intramolecular.

The reductive elimination reactions from the *cis* complexes are first order, the relative rates being **65** > **66** > **67**. The addition of excess of diphenylmethylphosphine to a $d_6$-dmso solution of *c*-**66** significantly slowed the rate of reductive elimination of ethane. The addition of an equivalent of diphenylacetylene, however, gave reaction rates which were in agreement with the rates obtained for *c*-**66** without added acetylene. The palladium acetylene complex, **69**, could be isolated from the reaction solution.

$$[Pd(PhC{\equiv}CPh)(PMePh_2)_2]$$

(**69**)

There are certain requirements, therefore, for reductive elimination of ethane from bis(phosphine)dimethylpalladium(II) complexes to take place[783,784]. Firstly, only a *cis* complex will undergo the 1,1-reductive elimination. Secondly, polar/coordinating solvents enhance the dissociation of phosphine; initially 50% of the coordinated phosphine is displaced from complexes *c*-**65** and *c*-**66**. This dissociation does not occur readily in non-polar solvents. This *cis* chelating ligand dppe does not dissociate in a detectable amount from complex *c*-**67** even in the presence of strongly polar/coordinating solvents.

The rate constants for reductive elimination from the *cis* complexes ($1.04\times10^{-3}$, $6.5$–$6.9\times10^{-5}$ and $4.8\times0^{-7}\ s^{-1}$, respectively, at 54 °C) parallels the ability of the complex to dissociate phosphine; added phosphine retards the rate of elimination[783]. The *cis* chelating ligand dppe does not dissociate from *c*-**67** in a detectable amount,

$$cis\text{-}[Pd(Me)_2(PR_3)_2 + Solv \rightleftharpoons cis\text{-}[Pd(Me)_2(PR_3)(Solv)] + PR_3 \qquad (358)$$

$$cis\text{-}[Pd(Me)_2(PR_3)(Solv)] \longrightarrow \left[\begin{array}{c} R_3P \quad\quad CH_3 \\ Pd \\ (Solv) \quad\quad CH_3 \end{array}\right] \longrightarrow (R_3P)Pd\ Solv + C_2H_6 \qquad (359)$$

$$(R_3P)Pd(Solv) + PR_3 \longrightarrow (R_3P)_2Pd(Solv) \qquad (360)$$

accounting for a rate of reductive elimination which is 50–100 times slower than for *c*-**66**, a complex that is electronically and geometrically similar. It is not clear, however, whether the function of the polar/coordinating solvent is to aid in phosphine dissociation by solution or by occupying the coordination site vacated by phosphine. The $\sigma$-donating ability of the phosphines, which enhances oxidative addition, thus inhibits reductive elimination. Reductive elimination may occur either from the *cis* square-planar complex containing coordinated solvent or from a tricoordinate Y-shaped intermediate.

Further insight into the mechanism, and particularly the role of the solvent, was obtained by the examination of the rates of reductive elimination of ethane from *c*-**66** at different temperatures in solvents encompassing a wide range of polarity[785]. The rates of reductive elimination are faster by almost an order of magnitude in non-polar, aromatic solvents.

The energies of activation for the reductive eliminations in the non-polar aromatic solvents correspond to those calculated ($25\ kcal\ mol^{-1}$)[786] for the *cis*-dimethylphosphinepalladium(II) complex in either the T or Y geometry, **70** and **71**.

Me—Pd—$PH_3$ (with Me on Pd)

**(70)**

Me, Me — Pd — $PH_3$

**(71)**

The lower energies of activation and slower rates for those eliminations in polar/coordinating solvents are a reflection of the large negative entropies of activation. These large negative values are consistent with an elimination reaction that produces a coordinatively unsaturated palladium(0) complex, $L_2Pd(0)$, and a late transition state having some of the characteristics of the product. Thus, an ordering of coordinating solvent by palladium appears to be taking place as it proceeds to the zerovalent complex. Since this reaction apparently takes place by prior dissociation of phosphine[784], the observed activation energy is the sum of an endothermic ligand dissociation and the elimination of ethane. Coordination of solvent to the product to give $[Pd(S)_n(L)_2]$ (S = dmso, acetone, acetonitrile) lowers the energy of the $b_2$ orbital[786] in the product and thus the activation energy for the reductive elimination.

Theoretical calculations[786] for the reductive elimination of ethane from bis(phosphine)dimethylpalladium support this mechanism. Several important conclusions emerge from these calculations. Firstly, stronger donor ligands *trans* to the leaving groups increase the barrier to elimination. Secondly, the reductive elimination from the four-coordinate *cis*-bis(phosphine)dimethylpalladium is symmetry allowed, but is controlled by the energy of an antisymmetric $b_2$ oribtal. The energy barrier for elimination from this complex is substantial and is much higher than that of the corresponding nickel complex containing a lower energy $b_2$ orbital (Scheme 13). Thirdly, reductive elimination from a three-coordinate phosphinedimethylpalladium complex containing methyls in adjacent positions has a substantially lower activation energy than that of the four coordinate complex. Fourthly, the Y and T-shaped geometries of phosphinedimethylpalladium are more stable than the trigonal geometry; the trigonal geometry represents an energy hill (Scheme 14). Finally the barrier for isomerization from one T-shaped geometry to another is substantial, much higher than the energy barrier for reductive elimination from either a Y- or a T-shaped complex. There is a substantial energy barrier to transit around the Jahn–Teller wheel. A T-shaped *trans*-$[PdLR_2]$, which might be produced by liberating L from the *trans*-four-coordinate complex $[PdL_2R_2]$, will encounter a substantial energy barrier to rearrangement to *cis*-$[PdL_2R_2]$, which can undergo reductive elimination. Thus, *trans*-$[PdL_2R_2]$ must first isomerize to the *cis* isomer. Dissociation of phosphine from the *cis* isomer generates the T geometry directly (4 or 8 o'clock on the Jahn–Teller wheel). It is not clear, however, whether reductive elimination takes place from the T or the Y complex (6 o'clock), since the activation energies for each are nearly identical and the activation energy for isomerization from T to Y is low.

The fact that the groups undergoing coupling must occupy *cis*-positions is consistent with the observation that *cis*-diethyl complexes give *n*-butane, but the *trans* isomer yields ethene and ethane, as a result of $\beta$-elimination, followed by reductive elimination of hydride and ethyl[784]. Also, this required geometry is consistent with the observation that the carbonylation of a *cis*-diethylpalladium complex yields

$$\left[Et_2Pd(L)_2\right] \xrightarrow{-L} Et_2PdL \xrightarrow{CO} \left[Et_2PdL(CO)\right] \longrightarrow Et(EtC{=}O)PdL \xrightarrow{-CH_2{=}CH_2} H{-}Pd(L){-}C(=O)Et \longrightarrow EtC(=O)H \qquad (361)$$

$$\left[\textit{trans}\text{-}Et_2Pd(L)_2\right] \xrightarrow{-L} Et{-}Pd(L){-}Et \xrightarrow{CO} \left[Et{-}Pd(L)(CO){-}Et\right] \longrightarrow Et{-}Pd(L){-}C(=O)Et \longrightarrow Et{-}C(=O){-}Et \qquad (362)$$

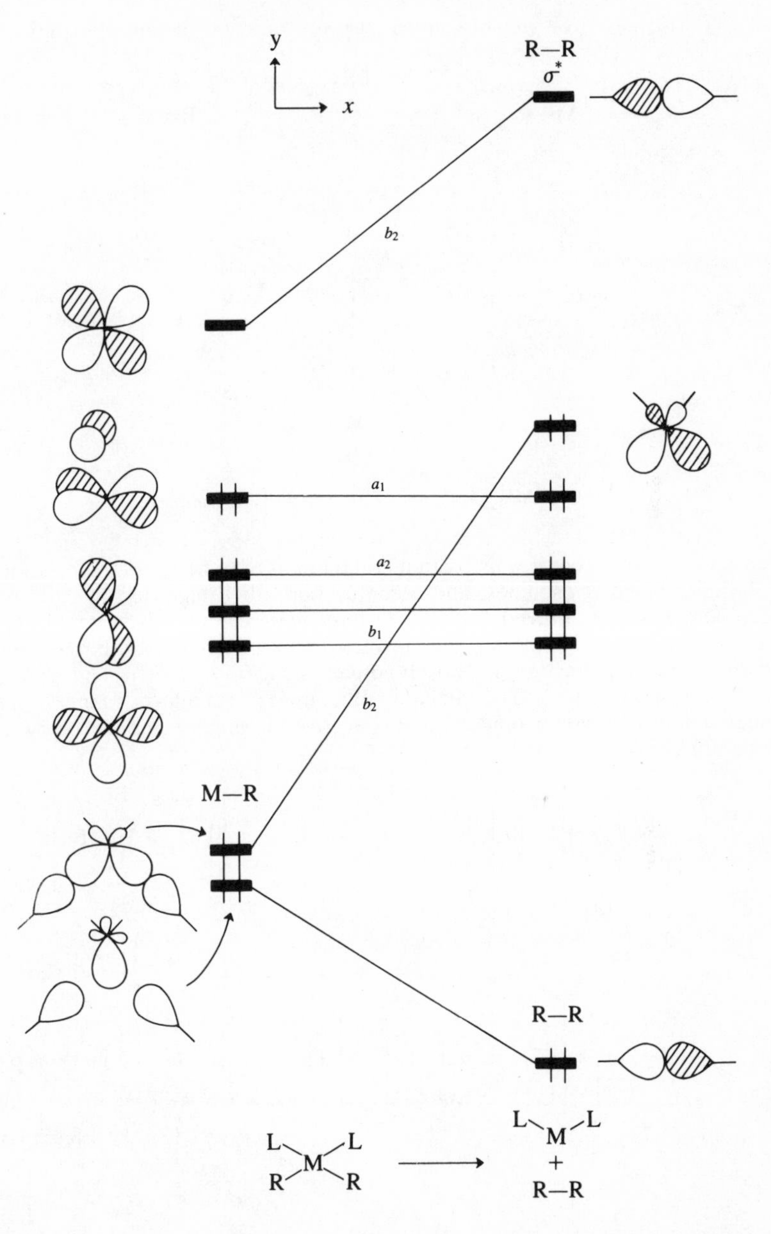


SCHEME 13. Schematic correlation diagram for the elimination of RR from a square-planar $d^8$ transition metal complex $[ML_2R_2]$. The reaction pathway maintains $C_{2v}$ symmetry.

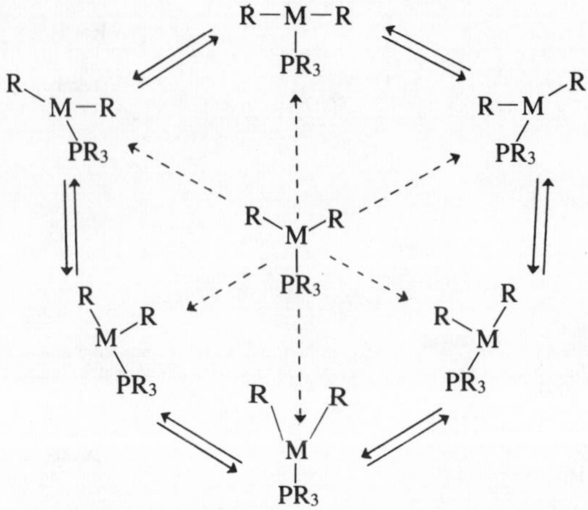

SCHEME 14. Isomerization of $PdR_2L$ complexes.

ethene and propanal, while the *trans* isomer produces diethyl ketone[787]. The sequence of reactions requires firstly that carbon monoxide occupies the position of one of the dissociated phosphines, and secondly that alkyl migration to carbonyl takes place. There is some precedent for this stereochemistry and these requirements.

The reductive elimination of propenylbenzene from both (*E*)-and (*Z*)-styrylmethylbis(diphenylmethylphosphine)palladium(II) give (*E*)- and (*Z*)-propenyl benzene, respectively[788]. The dialkyl palladium(II) complexes undergo *cis–trans* isomerization (complex geometry) and reductive elimination rapidly below ambient temperature.

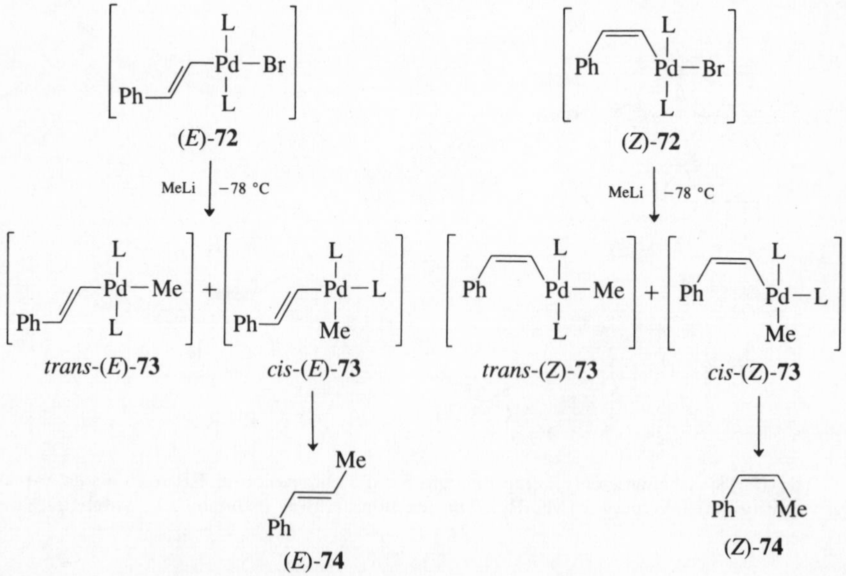

SCHEME 15. Synthesis, isomerization, and reductive elimination reactions of styrylmethylbis(diphenylmethylphosphine)palladium(II) complex

The oxidative addition reactions of (*E*)- and (*Z*)-$\beta$-bromostyrene with tetrakis(diphenylmethylphosphine)palladium(0) give the *trans* complexes (**72**) with 100% retention of geometry at the double bond (Scheme 15). Reaction of these complexes with methyllithium at ambient temperature produces (*E*)- and (*Z*)-propenylbenzenes [(*E*)-**74**, (*Z*)-**74**], respectively, uncontaminated with the other isomer. The styrylmethylpalladium(II) complexes are too unstable to be isolated at room temperature.

Both the (*E*)- and (*Z*)-styrylmethylbis(diphenylmethylphosphine)palladium(II) complexes [(*E*)-**73**, (*Z*)-**73**)] could be isolated from thf solutions by the reaction of (*E*)-**72** and (*Z*)-**72** with methyllithium at −78 °C. Mixtures of the *cis* and *trans* isomers of (*E*)-**73** and (*Z*)-**73** are obtained. Decomposition of these (*E*)- and (*Z*)-styrylmethylpalladium complexes ([*E*)-**73** and (*Z*)-**73**)] produced the corresponding (*E*)- and (*Z*)-propenylbenzenes [(*E*)-**74**, (*Z*)-**74**)], respectively.

The *cis*-(*E*)- and -(*Z*)-styrylmethylpalladium complexes undergo rapid decomposition, whereas the more thermally stable *trans* complexes isomerize to the *cis* geometry before reductive elimination can occur. Added diphenylmethylphosphine rapidly isomerizes the *trans* isomers to the *cis* isomers in either thf or toluene.

These results show that the same requirements are necessary for reductive elimination in these complexes as are demanded by the dimethylpalladium complex. The *trans* complex must isomerize first to place the alkyl groups in adjacent positions before reductive elimination can occur. The isomerization is catalysed by phosphine, but the *cis* isomers are preferred. In addition, the elimination takes place with complete retention of geometry at the double bond.

The elimination reaction also has been shown to involve the intermediate $\pi$-olefin complexes, (*E*)- and (*Z*)-(propenylbenzene)bis(diphenylmethylphosphine) palladium(0), which lose (*E*)- and (*Z*)-propenylbenzene, respectively.

$$(E)\text{-}\mathbf{73} \longrightarrow [^{\text{Ph}}\!\urcorner\!\llcorner - \text{Pd}(\text{PMePh}_2)_2] \longrightarrow (E)\text{-}\mathbf{74} \quad (363)$$

$$(Z)\text{-}\mathbf{73} \longrightarrow [^{\text{Ph}}\!\rceil\!\rfloor - \text{Pd}(\text{PMePh}_2)_2] \longrightarrow (Z)\text{-}\mathbf{74} \quad (364)$$

Two complexes, a *cis-trans* mixture of (*E*)-styrylmethylbis(diphenylmethylphosphine)palladium(II) and a *cis-trans* mixture of (*E*)-*p*-chlorostyryltrideuteriomethylbis(diphenylmethylphosphine)palladium(II) react to give only (*E*)-propenylbenzene and (*E*)-*p*-chloro-3,3,3-trideuteriopropenylbenzene, respectively. No crossover products are observed. Thus, exchange of neither the methyl nor the styryl groups takes place prior to reductive elimination. The elimination is intramolecular and does not take place from a dinuclear palladium complex. The reductive elimination of propenylbenzene from the styrylmethylpalladium complexes [(*E*)-**73**, (*Z*)-**73**] is first order. The rates of reductive elimination of ethane from *cis*-dimethylpalladium(II) complexes are much slower, however, than the elimination of propenylbenzene from the *cis*-styrylmethylpalladium complexes. Since excess of phosphine does not cause a rate change in the early stages of the reaction, coupling is probably taking place directly from the square-planar *cis* complex. This coupling of a *cis* $d^8[ML_2R_2]$ to a $d^{10}$ $ML_2$ and $R_2$ is symmetry allowed for a least motion $C_{2v}$ departure[786]. In the reductive elimination from the styrylmethylpalladium complexes, the product is a $d^{10}$ $[ML_2(\text{olefin})]$ complex. Thus. the energy of the $b_2$ orbital should be significantly lowered, allowing the pathway from the square-planar four-coordinate complex.

The 1,1-reductive elimination from the palladium(II) complexes generates palladium(0) complexes which, in a catalytic coupling reaction of an organohalide with an organometal, could undergo oxidative addition of the organic halide. Palladium(0) complexes catalyse the coupling of benzyl halides with Grignard reagents or tetraorganotin compounds. The reaction gives only cross-coupled products, provided that the benzyl halide is present during the reaction. As a result, the catalytic cycle often written for this coupling reaction involves Pd(0) $\rightleftharpoons$ Pd(II) transformations (Scheme 16). In a number of cases the rates of reductive elimination observed in the stoichiometric reactions are much slower than can be accommodated by a catalytic process.

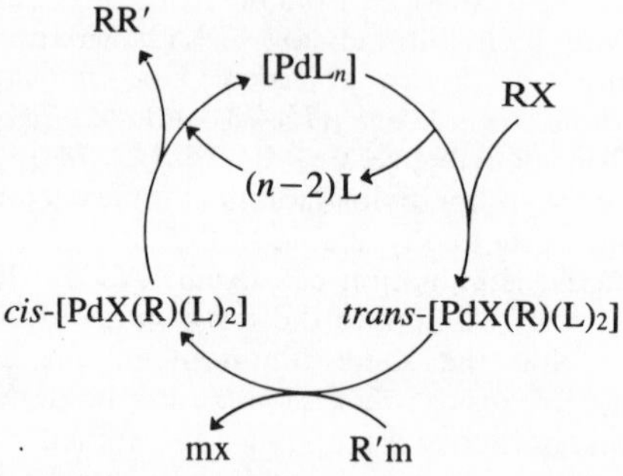

RR′

$[PdL_n]$

RX

$(n-2)L$

*cis*-$[PdX(R)(L)_2]$ *trans*-$[PdX(R)(L)_2]$

mx R′m

SCHEME 16

When a stoichiometric reaction is carried out by first isolating the oxidative addition product of benzyl bromide to palladium(0), and then allowing the transmetallation reaction to occur, in addition to coupling other reactions take place, and the rate of the reaction is much slower than when benzyl halide is present. When *cis*-dimethylbis(triphenylphosphine)palladium(II) (*c*-**65**) reacts with benzyl bromide, ethylbenzene is the only product, supporting a transient palladium(IV) intermediate in the reductive elimination and in the catalytic cycle[789].

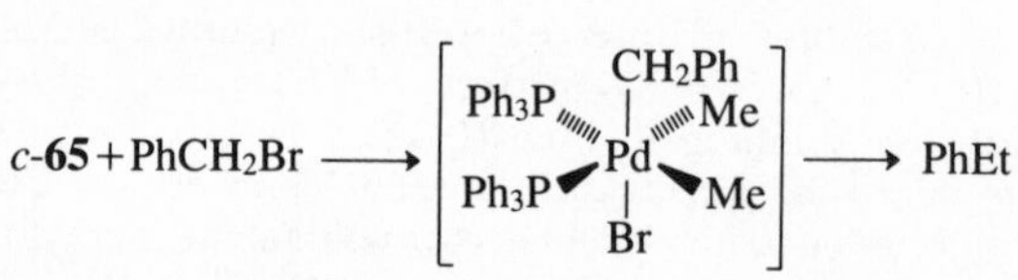

$$c\text{-}\mathbf{65} + PhCH_2Br \longrightarrow \left[ (Ph_3P)_2Pd(CH_2Ph)(Br)(Me)_2 \right] \longrightarrow PhEt \qquad (365)$$

Optically active chloro($\alpha$-deuteriobenzyl)bis(triphenylphosphine)palladium(II), on reaction with tetramethyltin (in the presence of a benzyl halide, e.g. *p*-nitrobenzyl bromide), yields optically active $\alpha$-deuterioethylbenzene in which retention of configuration at carbon has resulted in the reductive elimination (Scheme 17)[789,790]. Further, the reaction of $\alpha$-deuteriobenzyl bromide with *cis*-dimethylbis(triphenylphosphine)palladium(II) produces optically active $\alpha$-deuterioethylbenzene with inversion of configuration at carbon taking place. Since oxidative addition to low-valent palladium species takes place with inversion of configuration at carbon, then retention of configuration is taking place in the reductive elimination step. Thus, the mechanism of bond making and breaking in the 1,1-reductive elimination is a concerted or nearly concerted process from a palladium(IV) species.

Although complex *t*-**68**, which is held in a geometry such that the methyl groups are *trans*, will not undergo reductive elimination at 100 °C in $d_6$-dmso, the addition of methyl iodide to a solution of *t*-**68** at room temperature immediately produced

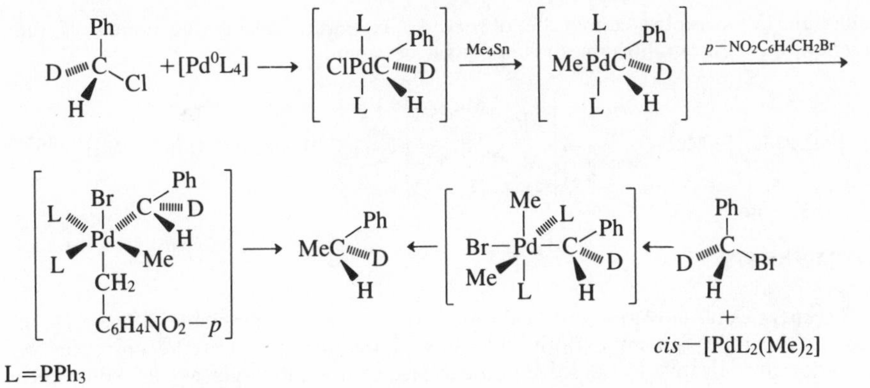

SCHEME 17. Stereochemistry of elimination from Pd(IV).

ethane. The addition of perdeuteriomethyl iodide produced only 1,1,1-trideuterioethane, which is compelling evidence for palladium(IV) intermediate[783].

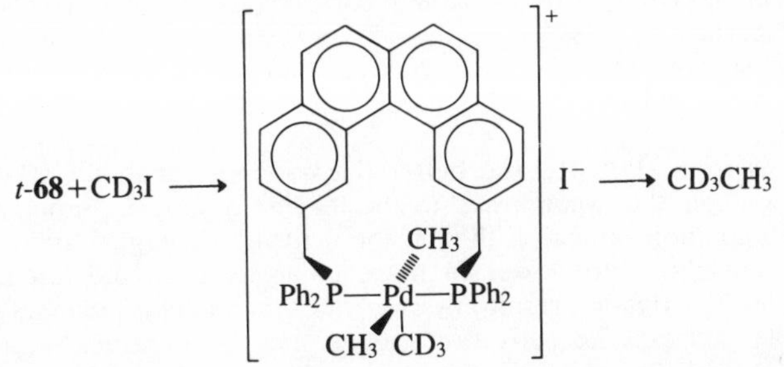

$$t\text{-}\mathbf{68} + CD_3I \longrightarrow [\ldots]^+ I^- \longrightarrow CD_3CH_3 \qquad (366)$$

Dimethyl(dppe)palladium(II) is known to yield ethane rapidly on addition of methyl iodide[780]. When the ligands are held *cis*, as they are in (dppe)dimethylpalladium, the oxidative addition of $CD_3I$ and the subsequent reductive elimination produces ethane in which the ratio of $CD_3CH_3$ to $CH_3CH_3$ is 2[785].

The reaction of methyl iodide with *cis*-bis(phosphine)dimethylpalladium (*c*-**65**, *c*-**66**) produces ethane rapidly, the intermediate trimethyliodopalladium(IV) species being too unstable to isolate. In a variety of solvents of different polarity, this reaction also gives the *trans*-bis(phosphine)iodomethylpalladium(II) products; the reaction is second order.

The rate constants are larger than those observed for the first-order reductive elimination of ethane from the *cis*-dimethylpalladium(II) complex *c*-**66** in the same solvents. The rates for the reaction of *c*-**66** with methyl iodide in the polar solvents are 400 times faster than the rates of reductive elimination of ethane from palladium(II). In non-polar solvents, the rate of reductive elimination of ethane from *c*-**66** is comparable to that of *c*-**66** with methyl iodide.

The second-order rate is consistent either with a rate-determining oxidative addition of methyl iodide to give the palladium(IV) intermediate followed by a fast 1,1-reductive elimination reaction, or a reversible oxidative addition reaction followed by a rate-determining reductive elimination reaction. Since the intermediate palladium(IV) complex cannot be observed ($^1$H n.m.r.) during the course of the reaction, any pre-equilibrium constant must be small.

$$cis\text{-}[PdL_2(Me)_2] + MeI \xrightarrow[\text{determining}]{\text{rate}} \left[ \begin{array}{c} \text{Me} \\ \text{Me}\quad | \quad \text{L} \\ \text{Pd} \\ \text{Me}\quad | \quad \text{L} \\ \text{I} \end{array} \right] \xrightarrow{\text{fast}} C_2H_6 + trans\text{-}[PdIL_2(Me)] \qquad (367)$$

*c*-**65**, *c*-**66**

L=$PMePh_2$, $PPh_3$

The energies of activation (16 kcal $mol^{-1}$) do not show appreciable differences, at least in the polar solvents. Both the rates of the reaction and the entropies of activation are sensitive to the solvent, much faster rates and larger negative entropies (−14 to −20 eu) being observed in polar media. This behaviour is characteristic of oxidative addition reactions of methyl iodide to other transition metals. In these reactions the solvent effect on the rate of oxidative addition is dominated by the entropy of activation term.

There are other examples in which the reductive elimination of *cis*-alkyl groups from palladium(II) complexes is promoted by the addition of organoiodides. The relatively stable dimethyl(bipyridyl)palladium complex, for example, yields ethane on the addition of perfluoropropyl iodide[791].

Thus, of the several possible 1,1-reductive elimination reaction pathways of organic groups from palladium, the reaction takes place only with the adjacent organic partners.

In eliminations from $[PdL_2R_2]$ the better the $\sigma$-donor capability of the leaving group and the weaker the donor *trans* to the leaving group, the more readily the 1,1-reductive elimination occurs. Although the activation energies for these reductive elimination reactions are lower in more polar solvents, the attendant large negative entropies of activation result in slower rates than in non-polar solvents. As a result, in a polar solvent the activation energy may be lowered because of the absence of the second phosphine ligand or its replacement by a polar solvent.

Strong donor ligands *trans* to the leaving group retard reductive elimination, yet at the same time enhance the rate of oxidative addition by increasing the electron density on palladium. Thus, a more facile pathway for reductive elimination, particularly in the catalytic coupling reactions of organometals and organohalides catalysed by palladium (where the organohalide is present in excess with respect to palladium throughout the course of the reaction) may be through a palladium(IV) intermediate. In many catalytic reactions the overall coupling rate and thus every individual step in the catalytic cycle are faster than the rate of the 1,1-reductive elimination from the diorganopalladium(II) complex. In polar solvents, the rates of the oxidative addition–reduction elimination sequence involving dimethylpalladium(II)→ iodotrimethylpalladium(IV)→iodomethylpalladium(II) is faster than the reductive elimination of ethane from dimethylpalladium(II), mostly because of the less negative entropies of activation for the oxidative addition step in the former sequence. In non-polar solvents, these two pathways have competitive rates since both the activation energies and entropies for the two pathways are similar.

Since the oxidative addition–transmetallation reaction gives a palladium(II) complex in which the organic partners are *trans*, this explains how a palladium complex containing adjacent organic groups can be realized. It also explains the presence of homo-coupling products, as well as cross-coupled material in many coupling reactions (Scheme 18).

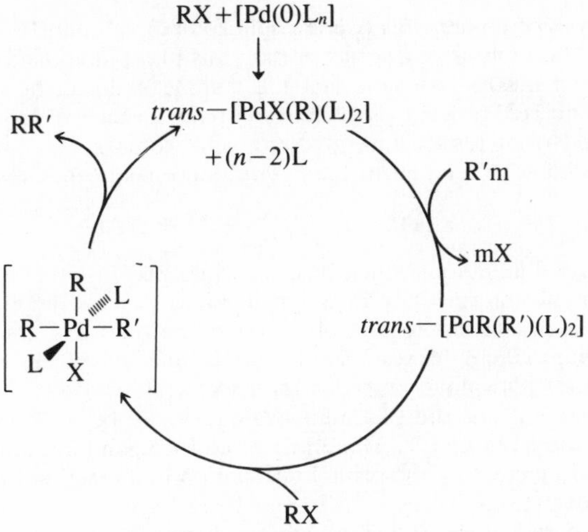

$$RX + [Pd(0)L_n] \rightarrow trans\text{-}[PdX(R)(L)_2] + (n-2)L \xrightarrow[-mX]{R'm} trans\text{-}[PdR(R')(L)_2] \xrightarrow{RX} [R\text{-}Pd(R')(L)_2(R)(X)] \rightarrow RR' + trans\text{-}[PdX(R)(L)_2]$$

SCHEME 18

Diorganobisphosphineplatinum(II) compounds that do not contain $\beta$-hydrogens on the organic group are relatively stable toward thermal decomposition. *cis*-Dimethylbis(triethylphosphine)platinum is stable above 85 °C[792]. Aryl complexes are more stable than benzyl derivatives[793]. In contrast to the nickel and palladium complexes containing *trans*-aryl and -methyl groups, the platinum analogue is stable[51].

When a $\beta$-hydrogen is present in a dialkyl platinum(II) complex, $\beta$-elimination is the preferred reaction pathway[794], even with platinacycles containing an unfavourable dihedral angle between the carbon platinum and $\beta$-carbon—hydrogen bonds. Cyclobutane can be the major product from the 1,1-reductive elimination of a platinacyclopentane bearing phosphines with bulky alkyl groups[795]. The generation of cyclobutane in high yields in dichloromethane has been shown to arise as a result

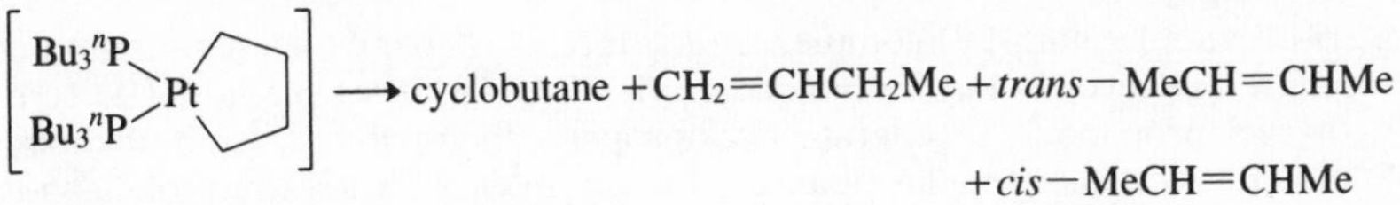

$$[(Bu^n_3P)_2Pt(CH_2)_4] \longrightarrow \text{cyclobutane} + CH_2{=}CHCH_2Me + trans\text{-}MeCH{=}CHMe + cis\text{-}MeCH{=}CHMe \quad (368)$$

of a 1,1-reductive elimination from a platinum(IV) complex generated by the oxidative addition of dichloromethane to the platinum(II) complex (see reaction 373).

Intramolecular C—H insertion of certain *cis*-dialkylplatinum(II) complexes ultimately can lead to a 1,1-reductive elimination of cyclopropanes via a platinacyclobutane[626,627].

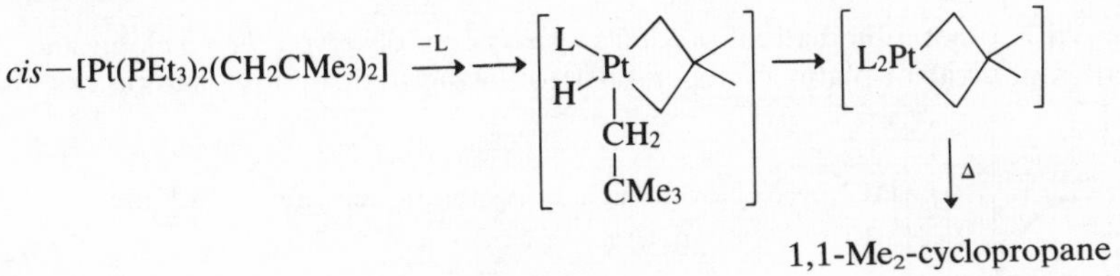

$$cis\text{-}[Pt(PEt_3)_2(CH_2CMe_3)_2] \xrightarrow{-L} \longrightarrow [L(H)Pt(CH_2CMe_2CH_2)(CH_2CMe_3)] \longrightarrow [L_2Pt(CH_2CMe_2CH_2)] \xrightarrow{\Delta} \text{1,1-Me}_2\text{-cyclopropane} \quad (369)$$

The thermolysis of diphenylbis(triphenylphosphine)platinum(II) yields biphenyl[796]. The solid-state thermolysis of a series of diarylbis(phosphine)platinum(II) complexes has demonstrated the requirement that the complexes have the *cis* geometry, and proceed by a concerted process, as evidenced by the absence of crossover products or products derived from radical intermediates and isomeric scrambling[797,798]. In this case, the presence of additional tertiary phosphine has been shown to enhance the

$$cis\text{-}[PtL_2Ar_2] \longrightarrow PtL_2 + ArAr \tag{370}$$

rate of reductive elimination, which has been interpreted either as prior formation of a five-coordinate intermediate or a synchronous process involving phosphine[799]. This behaviour is in contrast to that observed in the reductive elimination reactions of dialkylpalladium(II) complexes, in which phosphines inhibit the elimination and in fact dissociation of phosphine precedes reductive elimination[783,784].

The greater stability of the platinum analogues can be attributed to the higher energy level of the $b_2$ orbital[786]. The effect of added ligand in inducing coupling has been attributed to increasing the orbital occupation in a reaction in which the orbital occupation is decreased by two[716]. It has been argued that this effect is most noticeable at the end of the transition series where reducing the $d$-orbital energies and increasing energies for $nd \rightarrow (n+1)p$ promotion result in less energetically favorable 18-electron molecules. As a consequence of the lanthanide contraction, the progression for $nd \rightarrow (n+1)p$ promotional energies is not regular, and the trend to lower orbital occupation in the nickel triad is not regular; palladium shows the least tendency toward forming 18-electron molecules. Thus, the promotional energies are Ni 1.72 eV, Pd 4.23 eV, and Pt 3.28 eV, and this could reflect the trend to add a two-electron ligand.

Nevertheless, the reductive elimination of biaryls from platinum(II) is first order in platinum complex, and there is a pronounced negative entropy of activation ($\Delta S^{\ddagger} \approx -24$ eu)[800]. The rate of elimination from complexes containing chelating phosphines is slower than that from those having monophosphines. This is possibly a consequence of restricted transition-state geometry in the case of the chelating phosphine[721].

The loss of ethane from dimethylbipyridylplatinum(II) has been proposed to proceed by a radical process, however, based on the e.s.r. detection of a nitroso trapped intermediate[801].

Di- or tri-organoplatinum(IV) complexes undergo 1,1-reductive elimination much more readily. Platinacyclobutanes obtained from the reaction of platinum(II) complexes with cyclopropane[533] regenerate cyclopropane thermally[557,802], photochemically[803,804], or on treatment of the platinum(IV) complex with a variety of ligands, including cyanide[541,805,806], alkenes[562], phosphines[551,807], and arsines or stibines[561]. The thermal reductive elimination of bicyclobutane from a platinum(IV) complex,

$$[L_2X_2Pt(C_3H_6)] \longrightarrow \text{cyclopropane} + [Pt_2X_2L_2] \tag{371}$$

together with other elimination products, has been observed[570]. Cyclobutane is obtained from divalent platinacyclopentenes following the oxidative addition of the

$$[py_2Cl_2Pt(C_4H_6)] \longrightarrow [PtCl_2py_2] + \underset{20\text{-}30\%}{\text{bicyclobutane}} + \text{cyclobutene} + \text{buta-1,3-diene} \tag{372}$$

$$+\,\text{3-methylcyclopropene} + \text{methyclenecyclopropane}$$

solvent dichloromethane[808]. Other platinum(IV) metallocycles also yield cycloalkanes[808].

$$[L_2Pt(C_4H_8)] \xrightarrow{CH_2Cl_2} [PtCl(CH_2Cl)(L)_2(C_4H_8)] \longrightarrow \text{cyclobutane} + \text{butenes} + trans\text{-}[PtCl(CH_2Cl)(L)_2] \quad (373)$$

$$[PtF(CF_3)(L)_2(C_4H_8)] \xrightarrow{120\,°C} \underset{94\%}{\text{cyclobutane}} + \underset{8\%}{\text{but-1-ene}} + \underset{1\%}{\text{methylcyclopropane}} \quad (374)$$

$L = py$; $L_2 = bipy$

Thus, 1,1-reductive elimination of alkyl groups occurs readily from platinum(IV), but not platinum(II), $\beta$-elimination being preferred from the lower oxidation state. Whether or not $\beta$-elimination occurs or cycloalkanes are formed from the platinum(IV) complexes depends on the ligands and the third alkyl group on platinum, an electron-withdrawing alkyl group favouring cyclobutane formation.

The kinetic resolution of *trans*-1,2-dimethylcyclopropane by its enantioselective 1,1-reductive elimination from the platinacyclobutane with chiral phosphine has been achieved in 35% ee[809].

$$\underset{\text{racemic}}{\text{trans-1,2-dimethylcyclopropane}} + [PtCl_2(C_2H_4)]_2 \longrightarrow \underset{\text{racemic}}{[Cl_2Pt(C_3H_4Me_2)]} \xrightarrow{diop} \text{trans-1,2-dimethylcyclopropane} \quad (375)$$

Prolonged heating of dimethyldiiodobis(triethylphosphine)platinum generates ethane and the platinum(II) iodide[792]. Thermolysis (160–180 °C) of halotrimethyl- and tetramethyl-bis(phosphine)platinum(IV) complexes always yields ethane as a reductive elimination product[724]. Reductive elimination from the chloromethyl-platinum complexes generally produces both methyl halides and ethane (see above)[792].

$$[PtL_2(Me)_4] \longrightarrow C_2H_6 + cis\text{-}[PtL_2(Me)_2] \quad (376)$$

$$[PtX(L)_2(Me)_3] \longrightarrow C_2H_6 + [PtX(L)_2(Me)] \quad (377)$$

The elimination of ethane from trimethyliodobis(diphenylmethylphosphine) platinum(IV) containing one deuteriomethyl group shows a positive secondary isotope effect[193]. This type of complex, **75**, decomposes by an intramolecular

$$[PtI(L)_2(R_1)(R_2)_2] \qquad L = PPhMe_2$$

**75**

reductive elimination of ethane ($R_1 = R_2 = Me$), hexadeuterioethane ($R_1 = R_2 = CD_3$), a mixture of ethane and trideuterioethane ($R_1 = CD_3$, $R_2 = Me$), or a mixture of trideuterioethane and hexadeuterioethane ($R_1 = Me$, $R_2 = CD_3$)[194,196]. Kinetic studies indicate that ethane is eliminated from a five-coordinate intermediate generated by the dissociation of a phosphine ligand. When the ligand is diphosphine, however, elimination takes place without prior dissociation. The estimated activation energy (16.5 kcal mol$^{-1}$) is considerably lower than the methyl—platinum bond energy (35.8 kcal mol$^{-1}$) and lower than the activation energy for the elimination of ethane from $[PtCpMe_3]$[810]. Halobisphosphineplatinum(IV) complexes containing mutually *cis*-alkyl groups show a preference for reductive elimination partners, depending on the ligands *trans* to the alkyl groups and the electron-withdrawing or -donating ability of the alkyl group[189,192,195]. The leaving group order obtained from these reactions is $MeCO > Me > Ph \gg CF_3$[189].

$$[PtX(R_f)(Me)_2L_2] \longrightarrow C_2H_6 + trans\text{-}[PtI(R_f)(L)_2] \qquad (378)$$

$X = Br, I$; $R_f = CF_3, C_2F_5, C_3F_7. C_7F_{15}$; $L = PPhMe_2$

$$[PtBr(Me)_2(CH_2Ph)L_2] \longrightarrow \underset{58\%}{C_2H_6} + \underset{42\%}{PhEt} \qquad (379)$$

$L = PhMe_2P$

### 3. Other group VIII complexes

Reductive elimination of hydrocarbons from rhodium(III) in many cases is the final step in the isomerization of strained ring hydrocarbons[528,529]. The acylrhodium(III) intermediate ilsolated from the oxidative addition of the chlorodicarbonylrhodium(I) dimer undergoes reductive elimination on the addition of phosphine[576]. Similar

Cl—Rh(CO) acyl metallacycle $\xrightarrow{PPh_3}$ cage ketone (380)

reductive elimination reactions take place from other acylrhodium(III) metallacycles on the addition of carbon monoxide[811]. Reductive elimination of methyl and vinyl

[Cl(CO)Rh acyl metallacycle] $\xrightarrow[CO]{MeOH}$ bicyclic ketone with $CO_2R$ (381)

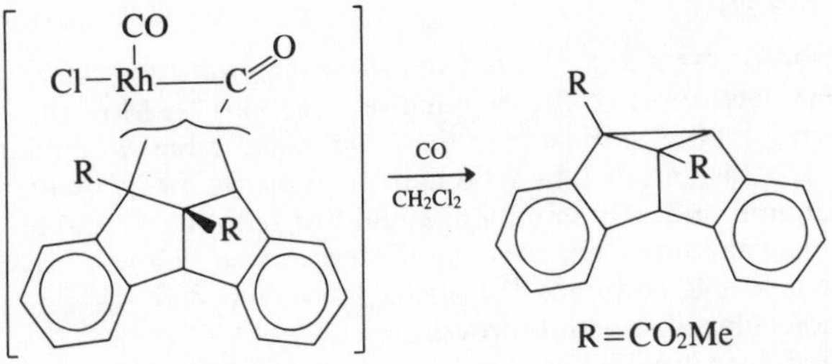

$$\xrightarrow[CH_2Cl_2]{CO} \qquad (382)$$

$R = CO_2Me$

groups from rhodium(III) takes place with retention of geometry at the $sp^2$ carbon, indicativke of a concerted reaction[268].

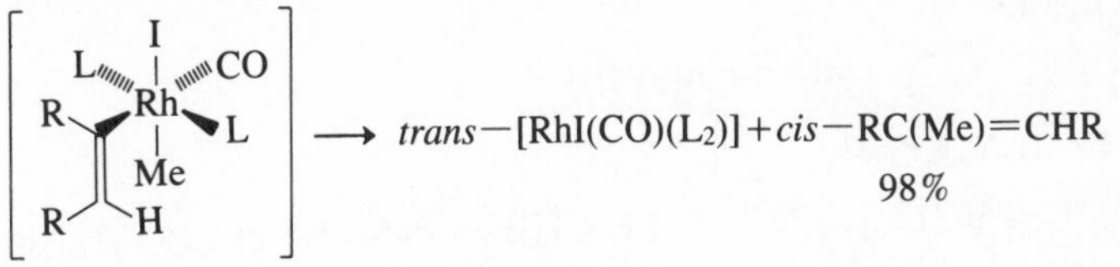

$$\longrightarrow \textit{trans}-[RhI(CO)(L_2)] + \textit{cis}-RC(Me){=}CHR \qquad (383)$$

98%

$R = CO_2Me$; $L = PPh_3$

The 1,1-reductive elimination of alkyl groups from $d^6$ iron(II) complexes does not give high yields of coupled products. Thermolysis of (dppe)dimethyliron(II) neat or in solution yields minor amounts of ethane. The majority of the elimination product is methane, which suggests an $\alpha$-elimination mechanism[812,813]. Diethyliron(II) complexes give primarily ethane and ethene, products resulting from an initial $\beta$-elimination[773].

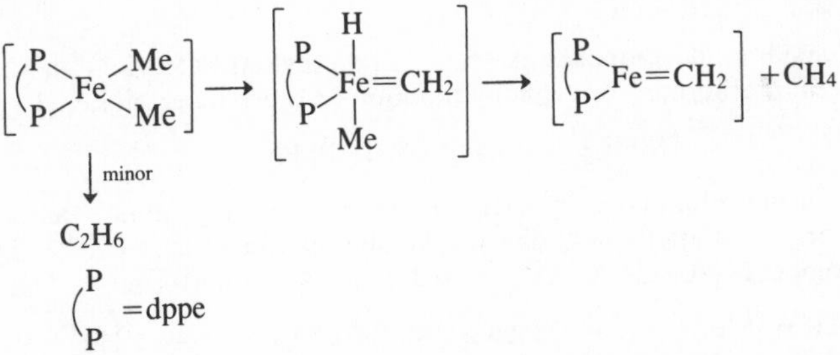

$$[(P)_2Fe(Me)_2] \longrightarrow [(P)_2Fe(H)(Me){=}CH_2] \longrightarrow [(P)_2Fe{=}CH_2] + CH_4 \qquad (384)$$

minor: $C_2H_6$

(P P) = dppe

The addition of an oxidizing agent or electrolytic oxidation gives high yields of 1,1-reductive elimination product[779,814,815]. The addition of an olefin containing electron-withdrawing groups also promotes this reductive elimination[816]. Olefins with

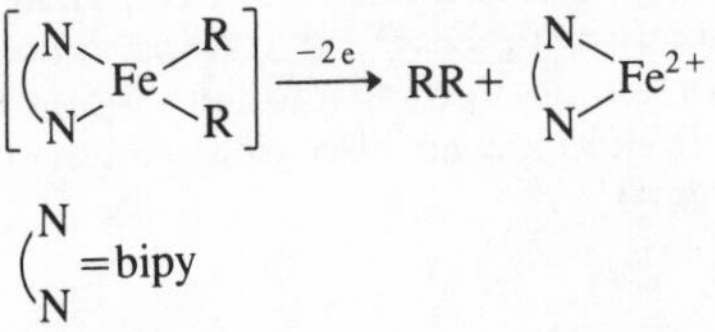

$$[(N)_2FeR_2] \xrightarrow{-2e} RR + (N)_2Fe^{2+} \qquad (385)$$

(N N) = bipy

*e* values of 1.30 or above (tcne, 2.25; maleic anhydride, 1.30; acrylamide, 1.30) effected the elimination; olefins with *e* values below 1.30 (e.g. acrylonitrile, 1.20) were polymerized. Similarly, dialkylcobalt(II) complexes undergo 1,1-reductive elimination of alkanes in the presence of oxidizing agents[779].

### 4. Early Transition Metals

Reductive elimination of organic partners from the early transition metals is rare, homolysis being observed in most of the elimination reactions[817]. Most of the early transition metals bearing methyl groups give methane, usually below ambient temperature. However, 1,2-(bisdimethylphenyl)ethane complexes of pentamethyltantalum, tetramethyltitanium, and pentamethylniobium are remarkably stable. Phenyl derivatives of titanium, vanadium, and zirconium decompose below ambient temperatrue to give biphenyl and benzene. Tetrabenzyl-titanium and -zirconium yield toluene by an intramolecular abstraction process.

Where $\beta$-elimination in tetraalkyl derivatives of the early transition metals is prevented, for example in certain 1-norbornyl derivatives, **76**, coupling products are observed[817]. In the coupling of tribenzylchromium, radical reactions best account for the coupling products[818].

$$[M(1\text{-norbornyl})_4]$$

**(76)**

M = Hf, Zr, Ti, V, Cr

$$[Cr(PhCH_2)_3] \xrightarrow{Et_2O} PhCH_2CH_2Ph + PhMe + o\text{-}PhCH_2C_6H_4Me \qquad (386)$$

The elimination of allyl chloride from $d^4$ $\eta^3$-allyl molybdenum and tungsten carbonyls, which takes place on the addition of phosphines, probably does not involve a radical mechanism[819].

$$[MCl(CO)_2(NCMe)_2(\eta^3\text{-2-R-allyl})] \xrightarrow{L} [MCl(CO)_2(L)_2(\eta^3\text{-2-R-allyl})]$$
$$\xrightarrow[MeCN]{L} CH_2{=}C(R)CH_2Cl + [M(CO)_2(NCMe)_2(L)_3 \qquad (387)$$

M = Mo; R = H, Me; L = phosphine
M = W; R = H

Tetraphenylvanadium decomposes at −50 °C in diethyl ether to yield biphenyl and a diphenylvanadium etherate. Tetraphenyltitanium also produces biphenyl.

$$[MPh_4] \longrightarrow [MPh_2] + PhPh \qquad (388)$$

Dimethyltitanocene eliminates mostly methane and only minor amounts of ethane[822], the origin of methane being an $\alpha$-elimination reaction[763,822]. Dibutyltitanocene decomposes at −55 °C to give $n$-butane and but-1-ene[823]. The metal-

$$[TiCp_2(Me)_2] \longrightarrow [TiHCp_2(Me)({=}CH_2)] \longrightarrow CH_4 \qquad (389)$$

locyclic titanocenes are more stable (to 0 °C), since the $\beta$-elimination pathway is not as available[823,824]. Titanacyclopentane and -hexane yield only minor amounts of cyclobutane[825] and cyclopentane, respectively[824]. The introduction of carbon monoxide into solutions containing titanacyclopentane, however, yields cyclopentanone via an 'insertion'–reductive elimination process[823–825]. Apparently the acylalkyl intermediate favours 1,1-reductive elimination over other decomposition pathways[826]. Although photolysis of dimethyltitanocene also yields methane, the photolysis of diphenyltitanocene gives biphenyl[827].

$$[Cp_2Ti(\text{titanacyclopentane})] \xrightarrow{CO} [Cp_2Ti(\text{acyl ring, C=O})] \longrightarrow \text{cyclopentanone} \qquad (390)$$

Thermolysis of tetrabenzylzirconium gives benzene, toluene, diphenylethane, and bibenzyl, products of homolysis of the benzylzirconium bond[828]. Tetraphenylzirconium decomposes at 0 °C in diethyl ether to yield biphenyl, benzene, and diphenylzirconium etherate[829].

## IV. REFERENCES


1. J. P. Collman and W. R. Roper, *Adv. Organomet. Chem.*, **7,** 53 (1968).
2. J. P. Collman, *Acc. Chem. Res.*, **1,** 136 (1968).
3. R. Ugo, *Coord. Chem. Rev.*, **3,** 319 (1968).
4. J. Halpern, *Acc. Chem. Res.*, **3,** 386 (1970).
5. R. D. W. Kemmitt and J. Burgess, *Inorg. React. Mech.*, **2,** 350 (1972).
6. J. Tsuji, *Fortschr. Chem. Forsch.*, **28,** 41 (1972).
7. J. K. Stille and K. S. Y. Lau, *Acc. Chem. Res.*, **10,** 434 (1977).
8. J. L. Davidson, *Inorg. React. Mech.*, **5,** 398 (1977).
9. J. P. Collman and L. S. Hegedus, *Principles and Applications of Organotransition Metal Chemistry,* University Science Books, Mill Valley, CA, 1980, Chapter 4.
10. L. Vaska, *Inorg. Chim. Acta,* **5,** 295 (1971).
11. A. Musco, W. Kuran, A. Silvani, and M. W. Anker, *J. Chem. Soc., Chem. Commun.*, 938 (1973).
12. A. Immirzi and A. Musco, *J. Chem. Soc., Chem. Commun.*, 400 (1974).
13. W. Kuran and A. Musco, *Inorg. Chim. Acta,* **12,** 187 (1975).
14. B. E. Mann and A. Musco, *J. Chem. Soc., Dalton Trans.*, 1673 (1975).
15. M. Matsumoto, H. Yoshioka, K. Nakatsu, T. Yoshida, and S. Otsuka, *J. Am. Chem. Soc.,* **96,** 3322 (1974).
16. C. A. Tolman, *J. Am. Chem. Soc.,* **92,** 2956 (1970).
17. C. A. Tolman, W. C. Seidel, and L. W. Gosser, *J. Am. Chem. Soc.,* **96,** 53 (1974).
18. L. E. Manzer and C. A. Tolman, *J. Am. Chem. Soc.,* **97,** 1955 (1975).
19. C. A. Tolman, *Chem. Rev.,* **77,** 313 (1977).
20. S. D. Ittel and J. A. Ibers, *Adv. Organomet. Chem.,* **14,** 33 (1976).
21. F. G. A. Stone, *Pure Appl. Chem.,* **30,** 551 (1972).
22. F. G. A. Stone, *Acc. Chem. Res.,* **14,** 318 (1981).
23. S. Otsuka and A. Nakamura, *Adv. Organomet. Chem.,* **14,** 245 (1976).
24. R. D. Rieke, *Top. Curr. Chem.,* **59,** 1 (1975).
25. D. Young and M. L. H. Green, *J. Appl. Chem. Biotechnol.,* **25,** 641 (1975).
26. K. J. Klabunde, *Angew. Chem., Int. Ed. Engl.,* **14,** 287 (1975).
27. K. J. Klabunde, *Acc. Chem. Res.,* **8,** 393 (1975).
28. P. L. Timms and T. W. Turney, *Adv. Organomet. Chem.,* **15,** 53 (1977).
29. R. G. Pearson and P. E. Figdore, *J. Am. Chem. Soc.,* **102,** 1541 (1980).
30. T. C. Flood, *Top. Stereochem.,* **12,** 37 (1981).
31. M. F. Lappert and P. W. Lednor, *Adv. Organomet. Chem.,* **14,** 345 (1976).
32. A. E. Jukes, *Adv. Organomet. Chem.,* **12,** 215 (1974).
33. V. C. R. McLoughlin and J. Thrower, *Tetrahedron,* **25,** 5921 (1969).
34. G. M. Whitesides and C. P. Casey, *J. Am. Chem. Soc.,* **88,** 4541 (1966).
35. G. H. Posner, *Org. React.,* **22,** 253 (1974).
36. G. M. Whitesides, W. F. Fischer, Jr., J. San Filippo, Jr., R. W. Bashe, and H. O. House, *J. Am. Chem. Soc.,* **91,** 4871 (1969).
37. C. R. Johnson and G. A. Dutra, *J. Am. Chem. Soc.,* **95,** 7777, 7783 (1973).
38. R. G. Pearson and C. D. Gregory, *J. Am. Chem. Soc.,* **98,** 4098 (1976).
39. K. Klabunde, *J. Fluorine Chem.,* **7,** 95 (1976).
40. A. Tamaki, S. A. Magennis, and J. K. Kochi, *J. Am. Chem. Soc.,* **95,** 6487 (1973).
41. A. Tamaki and J. K. Kochi, *J. Organomet. Chem.,* **51,** C39 (1973).
42. A. Tamaki and J. K. Kochi, *J. Chem. Soc., Dalton Trans.,* 2620 (1973).
43. A. Tamaki and J. K. Kochi, *J. Organomet. Chem.,* **64,** 411 (1974).
44. R. Ugo, *Coord. Chem. Rev.,* **3,** 319 (1968).

45a. D. R. Fahey, *Organomet. Chem. Rev., Sect. A,* **7,** 245 (1972).

45b. E. Uhlig and D. Walther, *Coord. Chem. Rev.,* **33,** 3 (1980).

46. J. Ashley-Smith, M. Green, and F. G. A. Stone, *J. Chem. Soc. A,* 3019 (1969).
47. L. Cassar and A. Giarrusso, *Gazz. Chim. Ital.,* **103,** 793 (1973).
48. D. R. Fahey and J. E. Mahan, *J. Am. Chem. Soc.,* **99,** 2501 (1977).
49. J. Browning, M. Green, and F. G. A. Stone, *J. Chem. Soc. A,* 453 (1971).
50. M. Hidai, T. Kashiwagi, T. Ikeuchi, and Y. Uchida, *J. Organomet. Chem.,* **30,** 279 (1971).
51. G. W. Parshall, *J. Am. Chem. Soc.,* **96,** 2360 (1974).
52. S. Otsuka, K. Tani, I. Kato, and O. Teranaka, *J. Chem. Soc., Dalton Trans.,* 2216 (1974).
53. L. Cassar, *J. Organomet. Chem.,* **93,** 253 (1975).
54. M. Foa and L. Cassar, *J. Chem. Soc., Dalton Trans.,* 2572 (1975).
55. T. T. Tsou and J. K. Kochi, *J. Am. Chem. Soc.,* **101,** 6319 (1979).
56. I. H. Elson, D. G. Morrell, and J. K. Kochi, *J. Organomet. Chem.,* **84,** C7 (1975).
57. M. F. Semmelhack, P. Helquist, L. D. Jones, L. Keller, L. Mendelson, L. S. Ryono, J. G. Smith, and R. D. Stauffer, *J. Am. Chem. Soc.,* **103,** 6460 (1981).
58. D. R. Fahey, *J. Am. Chem. Soc.,* **92,** 402 (1970).
59. M. Troupel, Y. Rollin, S. Sibille, J. F. Fauvarque, and J. Perichon, *J. Chem. Res.,* M137, S24, M173, S24 (1980).
60. S. Sibille, M. Troupel, J. F. Fauvarque, and J. Perichon, *J. Chem. Res.,* M2201, S147 (1980).
61. R. D. Rieke, W. J. Wolf, N. Kujunduić, and A. V. Kavaliunas, *J. Am. Chem. Soc.,* **99,** 4159 (1977).
62. K. J. Klabunde, J. Y. F. Low, and H. F. Efner, *J. Am. Chem. Soc.,* **96,** 1984 (1974).
63. J. E. Dobson, R. G. Miller, and J. P. Wiggen, *J. Am. Chem. Soc.,* **93,** 554 (1971).
64a. G. P. Chiusoli, *Proc. Int. Congr. Pure Appl. Chem., 23rd,* **6,** 169 (1971).
64b. R. Baker, *Chem. Rev.,* **73,** 487 (1973).
65. E. O. Fischer and G. Bürger, *Z. Naturforsch, Teil B,* **16,** 77 (1961).
66. E. O. Fischer and G. Bürger, *Chem. Ber.,* **94,** 2409 (1961).
67. E. J. Corey and M. F. Semmelhack, *J. Am. Chem. Soc.,* **89,** 2755 (1967).
68. R. F. Heck, J. Chien, and D. S. Breslow, *Chem. Ind.* (*London*), 986 (1961).
69. F. Dawans, J. C. Marechal, and Ph. Teyssié, *J. Organomet. Chem.,* **21,** 259 (1970).
70. J. Ishizu, T. Yamamoto, and A. Yamamoto, *Chem. Lett.,* 1091 (1976).
71. T. Yamamoto, J. Ishizu, and A. Yamamoto, *J. Am. Chem. Soc.,* **103,** 6863 (1981).
72. T. Yamamoto, J. Ishizu, T. Kohara, S. Komiya, and A. Yamamoto, *J. Am. Chem. Soc.,* **102,** 3758 (1980).
73. M. J. Piper and P. L. Timms, *Chem. Commun.,* 50 (1972).
74. B. Henc, P. W. Jolly, R. Salz, G. Wilke, R. Benn, E. G. Hoffmann, R. Mynott, G. Schroth, K. Seevogel, J. C. Sekutowski, and C. Kruger, *J. Organomet. Chem.,* **191,** 425 (1980).
75. R. Warin, M. Julémont, and Ph. Teyssié, *J. Organomet. Chem.,* **185,** 413 (1980).
76. P. W. Jolly and G. Wilke, *The Organic Chemistry of Nickel,* Vol. VI, Academic Press, New York, 1974.
77. M. R. Churchill and R. Mason, *Adv. Organomet. Chem.,* **5,** 93 (1967).
78. K. Jacob and R. Niebuhr, *Z. Chem.,* **15,** 32 (1975).
79. E. Bartsch, E. Dinjus, R. Fischer, and E. Uhlig, *Z. Anorg. Allg. Chem.,* **433,** 5 (1977).
80. E. Bartsch, E. Dinjus, and E. Uhlig, *Z. Chem.,* **15,** 317 (1975).
81. D. W. McBride, S. L. Stafford, and F. G. A. Stone, *J. Chem. Soc.,* 723 (1963).
82. P. W. Jolly, K. Jonas, C. Krüger, and Y. H. Tsay, *J. Organomet. Chem.,* **33,** 109 (1971).
83. G. Favero, A. Morvillo, and A. Turco, *Gazz. Chim. Ital.,* **109,** 27 (1979).
84. A. Morvillo, and A. Turco, *J. Organomet. Chem.,* **208,** 103 (1981).
85. S. Takanashi, Y. Suzuki, K. Sonogashira, and N. Hagihara, *J. Chem. Soc., Chem. Commun.,* 839 (1976).
86a. E. Uhlig, E. Dinjus, W. Poppitz, and R. Winter, *Z. Chem.,* **16,** 161 (1976).
86b. D. W. Firsich and R. J. Lagow, *Chem. Commun.,* 1283 (1981).
87. J. K. Stille and A. B. Cowell, *J. Organomet. Chem.,* **124,** 253 (1977).
88. D. W. McBride, E. Dudek, and F. G. A. Stone, *J. Chem. Soc.,* 1752 (1964).
89. M. Bressan, G. Favero, B. Corain, and A. Turco, *Inorg. Nucl. Chem. Lett.,* **7,** 203 (1971).

90. C. A. Tolman and E. J. Lukosius, *Inorg. Chem.*, **16,** 940 (1977).
91a. B. Hipler and E. Uhlig, *J. Organomet. Chem.*, **199,** C27 (1980).
91b. B. Hipler, E. Uhlig, and J. Vogel, *J. Organomet. Chem.*, **218,** C1 (1981).
92. S. Otsuka, A. Nakamura, T. Yoshida, M. Naruto, and K. Ataka, *J. Am. Chem. Soc.*, **95,** 3180 (1973).
93. B. Corain and G. Favero, *J. Chem. Soc., Dalton Trans.*, 283 (1975).
94. B. Corain and M. Martelli, *Inorg. Nucl. Chem. Lett.*, **8,** 39 (1972).
95. S. Komiya, A. Yamamoto, and T. Yamamoto, *Chem. Lett.*, 193 (1981).
96. P. Fitton and J. E. McKeon, *Chem. Commun.*, 4 (1968).
97. A. J. Mukhedkar, M. Green and F. G. A. Stone, *J. Chem. Soc. A,* 3023 (1969).
98. B. F. G. Johnson, J. Lewis, J. D. Jones, and K. A. Taylor, *J. Chem. Soc., Dalton Trans.*, 34 (1974).
99. J. Lewis, B. F. G. Johnson, K. A. Taylor, and J. D. Jones, *J. Organomet. Chem.*, **32,** C62 (1971).
100. H. D. Empsall, M. Green, S. K. Shakshooki, and F. G. A. Stone, *J. Chem. Soc. A,* 3472 (1971).
101. K. Kudo, M. Sato, M. Hidai, and Y. Uchida, *Bull. Chem. Soc. Jpn.*, **46,** 2820 (1973).
102. P. Fitton, M. P. Johnson, and J. E. McKeon, *Chem. Commun.*, 6 (1968).
103. P. Fitton and E. A. Rick, *J. Organomet. Chem.*, **28,** 287 (1971).
104. J. F. Fauvarque, F. Pflüger, and M. Troupel, *J. Organomet. Chem.*, **208,** 419 (1981).
105. R. D. Rieke and A. V. Kavaliunas, *J. Org. Chem.*, **44,** 3069 (1979).
106. K. J. Klabunde and J. Y. F. Low, *J. Organomet. Chem.*, **51,** C33 (1973).
107. K. J. Klabunde and J. Y. F. Low, *J. Am. Chem. Soc.*, **96,** 7674 (1974).
108. K. J. Klabunde, B. B. Anderson, and K. Neuenschwander, *Inorg. Chem.*, **19,** 3719 (1980).
109. J. Malatesta and M. Angoletta, *J. Chem. Soc.*, 1186 (1957).
110. E. A. Fischer and G. Bürger, *Z. Naturforsch., Teil B,* **16,** 702 (1961).
111. J. Powell and B. L. Shaw, *J. Chem. Soc. A,* 774 (1968).
112. F. A. Cotton, J. W. Faller, and A. Musco, *Inorg. Chem.*, **6,** 179 (1967).
113. B. Åkermark, G. Åkermark, L. S. Hegedus, and K. Zetterberg, *J. Am. Chem. Soc.*, **103,** 3037 (1981).
114. T. Ukai, H. Kawazura, Y. Ishi, J. J. Bonnet, and J. A. Ibers, *J. Organomet. Chem.*, **65,** 253 (1974).
115. T. Ito, S. Hasegawa, Y. Takahashi, and Y. Ishii, *J. Organomet. Chem.*, **73,** 401 (1974).
116. B. M. Trost, *Tetrahedron,* **33,** 2615 (1977).
117. B. M. Trost, *Acc. Chem. Res.*, **13,** 385 (1980).
118. T. Yamamoto, O. Saito, and A. Yamamoto, *J. Am. Chem. Soc.*, **103,** 5600 (1981).
119. K. J. Klabunde and J. S. Roberts, *J. Organomet. Chem.*, **137,** 113 (1977).
120. J. S. Roberts and K. J. Klabunde, *J. Organomet. Chem.*, **85,** C13 (1975).
121. J. S. Roberts and K. J. Klabunde, *J. Am. Chem. Soc.*, **99,** 2509 (1977).
122. S. Otsuka, A. Nakamura, and T. Yoshida, *J. Am. Chem. Soc.*, **91,** 7196 (1969).
123. D. T. Rosevear and F. G. A. Stone, *J. Chem. Soc. A,* 164 (1968).
124. D. Fitton, J. E. McKeon, and B. C. Ream, *J. Chem. Soc., Chem. Commun.*, 370 (1969).
125. J. K. Stille, *Ann. N.Y. Acad. Sci.*, **295,** 52 (1977).
126. K. S. T. Lau, R. W. Fries, and J. K. Stille, *J. Am. Chem. Soc.*, **96,** 4983 (1974).
127. P. K. Wong, K. S. Y. Lau, and J. K. Stille, *J. Am. Chem. Soc.*, **96,** 5956 (1974).
128. J. K. Stille, L. F. Hines, R. W. Fries, P. K. Wong, D. E. James, and K. S. Y. Lau, *Adv. Chem. Ser.*, No. 132, 90 (1974).
129. K. S. Y. Lau, P. K. Wong, and J. K. Stille, *J. Am. Chem. Soc.*, **98,** 5832 (1976).
130. J. K. Stille and K. S. Y. Lau, *J. Am. Chem. Soc.*, **98,** 5841 (1976).
131. Y. Becker and J. K. Stille, *J. Am. Chem. Soc.*, **100,** 838 (1978).
132. Y. Becker and J. K. Stille, *J. Am. Chem. Soc.*, **100,** 845 (1978).
133. R. R. Stevens and G. D. Shier, *J. Organomet. Chem.*, **21,** 495 (1970).
134. A. Sonoda, B. E. Mann, and P. M. Maitlis, *J. Chem. Soc., Chem. Commun.*, 108 (1975).
135. M. Onishi, T. Ito, and K. Hiraki, *J. Organomet. Chem.*, **209,** 123 (1981).
136. B. J. Argento, P. Fitton, J. E. McKeon, and E. A. Rick, *Chem. Commun.*, 1427 (1969).
137. B. M. Trost and G. A. Molander, *J. Am. Chem. Soc.*, **103,** 5969 (1981).

138. A. Sonoda, B. E. Mann, and P. M. Maitlis, *J. Organomet. Chem.,* **96,** C16 (1975).
139. M. Green, R. B. L. Osborn, A. J. Rest, and F. G. A. Stone, *Chem. Commun.,* 502 (1966).
140. M. Green, R. B. L. Osborn, A. J. Rest, and F. G. A. Stone, *J. Chem. Soc. A,* 2525 (1968).
141. C. D. Cook and G. S. Jauhal, *Can. J. Chem.,* **45,** 301 (1967).
142. W. J. Bland and R. D. W. Kemmitt, *J. Chem. Soc. A,* 1728 (1968).
143. W. J. Bland, J. Burgess, and R. D. W. Kemmitt, *J. Organomet. Chem.,* **14,** 201 (1968).
144. W. J. Bland, J. Burgess, and R. D. W. Kemmitt, *J. Organomet. Chem.,* **18,** 199 (1969).
145. J. Ashley-Smith, M. Green, and D. C. Wood, *J. Chem. Soc. A,* 1847 (1970).
146. A. J. Mukhedkar, M. Green, and F. G. A. Stone, *J. Chem. Soc. A,* 947 (1970).
147. J. Rajaram, R. Pearson, and J. Ibers, *J. Am. Chem. Soc.,* **96,** 2103 (1974).
148. J. Burgess, M. M. Hunt, and R. D. W. Kemmitt, *J. Organomet. Chem.,* **134,** 131 (1977).
149. R. G. Pearson, *Inorg. Chem.,* **12,** 712 (1973).
150. O. Simamura, K. Tokumaru, and H. Yui, *Tetrahedron Lett.,* 5141 (1966).
151. J. A. Kampmeier and R. M. Fantazier, *J. Am. Chem. Soc.,* **88,** 1959 (1965).
152. G. M. Whitesides, C. P. Casey, and J. K. Kreiger, *J. Am. Chem. Soc.,* **93,** 1379 (1971).
153. O. Traverso, V. Carassiti, M. Graziani, and U. Belluco, *J. Organomet. Chem.,* **57,** C22 (1973).
154. M. Graziani, M. Lenarda, R. Ros, and U. Belluco, *Coord. Chem. Rev.,* **16,** 35 (1975).
155. W. A. Baddley, C. Panottoni, G. Bandali, D. A. Clemente, and U. Belluco, *J. Am. Chem. Soc.,* **93,** 5590 (1971).
156. M. J. S. Gynane, M. F. Lappert, S. J. Miles, and P. P. Power, *J. Chem. Soc., Chem. Commun.,* 192 (1978).
157. A. J. Cohen, *Inorg. Synth.,* **6,** 209 (1960).
158. H. C. Volger and K. Vrieze, *J. Organomet. Chem.,* **9,** 527 (1968).
159. P. S. Skell and J. J. Havel, *J. Am. Chem. Soc.,* **93,** 6687 (1971).
160. H. Kurosawa, *J. Chem. Soc., Dalton Trans.,* 939 (1979).
161. N. J. Kermode, M. F. Lappert, B. W. Skelton, A. H. White, and J. Holton, *J. Chem. Soc., Chem. Commun.,* 698 (1981).
162. J. P. Birk, J. Halpern, and A. L. Pickard, *J. Am. Chem. Soc.,* **90,** 4491 (1968).
163. R. G. Pearson and J. Rajaram, *Inorg. Chem.,* **13,** 246 (1974).
164. C. T. Mortimer, M. P. Wilkinson, and R. J. Puddephatt, *J. Organomet. Chem.,* **165,** 265 (1979).
165. O. J. Scherer and H. Jungmann, *J. Organomet. Chem.,* **208,** 153 (1981).
166. R. G. Pearson, W. Louw, and J. Rajaram, *Inorg. Chim. Acta,* **9,** 251 (1974).
167. A. V. Kramer, J. A. Labinger, J. S. Bradley, and J. A. Osborn, *J. Am. Chem. Soc.,* **96,** 7145 (1974).
168. A. V. Kramer and J. A. Osborn, *J. Am. Chem. Soc.,* **96,** 7832 (1974).
169. M. F. Lappert and P. W. Lednor, *J. Chem. Soc., Chem. Commun.,* 948 (1973).
170. T. L. Hall, M. F. Lappert, and P. W. Lednor, *J. Chem. Soc., Dalton Trans.,* 1448 (1980).
171. V. I. Sokolov, *Inorg. Chim. Acta,* **18,** L9 (1976).
172. D. M. Blake, S. Shields, and L. Wyman, *Inorg. Chem.,* **13,** 1595 (1974).
173. J. L. Burmeister and L. M. Edwards, *J. Chem. Soc. A,* 1663 (1971).
174. D. H. Gerlach, A. R. Kane, G. W. Parshall, J. P. Jesson, and E. L. Muetterties, *J. Am. Chem. Soc.,* **93,** 3543 (1971).
175. R. Schlodder, J. A. Ibers, M. Lenarda, and M. Graziani, *J. Am. Chem. Soc.,* **96,** 6893 (1974).
176. C. Eaborn, A. Pidcock, and B. R. Steele, *J. Chem. Soc., Dalton Trans.,* 767 (1976).
177. G. Butler, C. Eaborn, and A. Pidcock, *J. Organomet. Chem.,* **144,** C23 (1978).
178. A. Christofides, M. Cirano, J. L. Spencer, and F. G. A. Stone, *J. Organomet. Chem.,* **178,** 273 (1979).
179. G. Butler, C. Eaborn, and A. Pidcock, *J. Organomet. Chem.,* **181,** 47 (1979).
180. G. Butler, C. Eaborn, and A. Pidcock, *J. Organomet. Chem.,* **185,** 367 (1980).
181. T. A. K. Al-Allaf, C. Eaborn, K. Kundu, and A. Pidcock, *J. Chem. Soc., Chem. Commun.,* 55 (1981).

182. C. Eaborn, K. Kundu, and A. Pidcock, *J. Chem. Soc., Dalton Trans.*, 1223 (1981).
183. D. M. Barlex, R. D. W. Kemmitt, and G. W. Littlecoat, *Chem. Commun.*, 613 (1969).
184. B. E. Mann, B. L. Shaw, and N. I. Tucker, *J. Chem. Soc. A,* 2667 (1971).
185. M. A. Bennett, G. B. Robertson, P. O. Whimp, and T. Yoshida, *J. Am. Chem. Soc.*, **93,** 3797 (1971).
186. M. A. Bennett, G. B. Robertson, P. O. Whimp, and T. Yoshida, *J. Am. Chem. Soc.*, **95,** 3028 (1973).
187. J. Chatt and B. L. Shaw, *J. Chem. Soc.*, 705 (1959).
188. J. D. Ruddick and B. L. Shaw, *J. Chem. Soc. A,* 2801 (1969).
189. T. G. Appleton, H. C. Clark, and L. E. Manzer, *J. Organomet. Chem.*, **65,** 275 (1974).
190. J. D. Ruddick and B. L. Shaw, *J. Chem. Soc. A,* 2969 (1969).
191. J. R. Hall and G. A. Swile, *J. Organomet. Chem.*, **76,** 257 (1974).
192. H. C. Clark, and J. D. Ruddick, *Inorg. Chem.*, **9,** 2556 (1970).
193. H. C. Clark and L. E. Manzer, *Inorg. Chem.*, **12,** 362 (1973).
194. M. P. Brown, R. J. Puddephatt, and C. E. E. Upton, *J. Organomet. Chem.*, **49,** C61 (1973).
195. M. P. Brown, R. J. Puddephatt, C. E. E. Upton, and S. W. Lavington, *J. Chem. Soc., Dalton Trans.*, 1613 (1974).
196. M. P. Brown, R. J. Puddephatt, and C. E. E. Upton, *J. Chem. Soc., Dalton Trans.*, 2457 (1974).
197. A. J. Cheney and B. L. Shaw, *J. Chem. Soc. A,* 3545 (1971).
198. C. R. Kistner, D. A. Drew, J. R. Doyle, and G. W. Rausch, *Inorg. Chem.*, **6,** 2036 (1967).
199. R. Usoń, J. Forniés, P. Espinet, and J. Gavin, *J. Organomet. Chem.*, **105,** C25 (1976).
200. J. K. Jawad and R. J. Puddephatt, *J. Organomet. Chem.*, **117,** 297 (1976).
201. J. K. Jawad and R. J. Puddephatt, *J. Chem. Soc., Dalton Trans.*, 1466 (1977).
202. M. P. Brown, A. Hollings, K. J. Houston, R. J. Puddephatt, and M. Rashisi, *J. Chem. Soc., Dalton Trans.*, 786 (1976).
203. R. B. King, *Adv. Organomet. Chem.*, **2,** 157 (1964).
204. P. M. Treichel and F. G. A. Stone, *Adv. Organomet. Chem.*, **1,** 143 (1964).
205. G. W. Parshall and J. J. Mrowca, *Adv. Organomet. Chem.*, **7,** 157 (1968).
206. R. B. King, *Acc. Chem. Res.*, **3,** 417 (1970).
207a. R. F. Heck and D. S. Breslow, *J. Am. Chem. Soc.*, **82,** 750 (1960).
207b. R. F. Heck and D. S. Breslow, *J. Am. Chem. Soc.*, **82,** 4438 (1960).
207c. R. F. Heck and D. S. Breslow, *J. Am. Chem. Soc.*, **83,** 1097 (1961).
208a. W. R. McClellan, *J. Am. Chem. Soc.*, **83,** 1598 (1961).
208b. W. R. McClellan, H. H. Hoehn, H. N. Cripps, E. L. Muetterties, and B. W. Howk, *J. Am. Chem. Soc.*, **83,** 1601 (1961).
209a. C. E. Coffey, *J. Am. Chem. Soc.*, **84,** 118 (1962).
209b. R. B. King and A. Efraty, *J. Organomet. Chem.*, **24,** 241 (1970).
209c. J. Potenza, R. Johnson, D. Mastropaolo, and A. Efraty, *J. Organomet. Chem.*, **64,** C13 (1974).
209d. T. Chaing, R. C. Kerber, S. D. Kimball, and J. W. Lauher, *Inorg. Chem.*, **18,** 1687 (1979).
210. R. F. Heck, *J. Am. Chem. Soc.*, **85,** 1460 (1963).
211. W. Hieber and H. Duchatsch, *Chem. Ber.*, **98,** 1744 (1965).
212. E. Lindner, H. Stich, K. Geibel, and H. Kranz, *Chem. Ber.*, **104,** 1524 (1971).
213. E. L. Muetterties and F. J. Hirsekorn, *J. Am. Chem. Soc.*, **96,** 7920 (1974).
214. E. O. Fischer and R. D. Fischer, *Z. Naturforsch., Teil B,* **16,** 475 (1961).
215. R. B. King, P. M. Treichel, and F. G. A. Stone, *J. Am. Chem. Soc.*, **83,** 3593 (1961).
216. R. F. Heck, *J. Org. Chem.*, **28,** 604 (1963).
217. R. B. King, *Inorg. Chem.*, **5,** 82 (1966).
218. A. J. Hart-Davis, and W. A. G. Graham, *Inorg. Chem.*, **9,** 2658 (1970).
219. J. M. Pratt and P. J. Craig, *Adv. Organomet. Chem.*, **11,** 331 (1973).
220. F. Calderazzo and C. Floriani, *Chem. Commun.*, 139 (1967).
221. G. Costa and G. Mestroni, *Tetrahedron Lett.*, 1783 (1967).
222. G. Costa, G. Mestroni, and G. Pellizer, *J. Organomet. Chem.*, **11,** 333 (1968).

223. G. Costa, G. Mestroni, and E. deSarorgnauia, *Inorg. Chim. Acta,* **3,** 323 (1969).
224. A. Bigotto, G. Costa, G. Mestroni, G. Pellizer, A. Puxeddu, E. Reisenhofer, L. Stefani, and G. Tauzher, *Inorg. Chim. Acta,* **4,** 41 (1970).
225. G. N. Schrauzer and J. Kohnle, *Chem. Ber.,* **97,** 3056 (1964).
226. G. N. Schrauzer, R. J. Windgassen, and J. Kuhule, *Chem. Ber.,* **98,** 3324 (1965).
227. G. N. Schrauzer and R. J. Windgassen, *J. Am. Chem. Soc.,* **88,** 3738 (1966).
228. G. N. Schrauzer and R. J. Windgassen, *J. Am. Chem. Soc.,* **89,** 143 (1967).
229. G. N. Schrauzer and R. J. Windgassen, *J. Am. Chem. Soc.,* **88,** 3738 (1967).
230. G. N. Schrauzer, E. Deutsch, and R. J. Windgassen, *J. Am. Chem. Soc.,* **90,** 2441 (1968).
231. M. D. Johnson and C. Mayle, *Chem. Commun.,* 192 (1969).
232. J. P. Collman, J. N. Cawse, and J. W. King, *Inorg. Chem.,* **8,** 2574 (1969).
233. G. N. Schrauzer and E. Deutsch, *J. Am. Chem. Soc.,* **91,** 3341 (1969).
234. D. Dolphin and A. W. Johnson, *Chem. Commun.,* 494 (1965).
235. D. A. Clarke, R. Grigg, A. W. Johnson, and H. Pinnock, *Chem. Commun.,* 309 (1967).
236. D. A. Clarke, D. Dolphin, R. Grigg, A. W. Johnson, and H. A. Pinnock, *J. Chem. Soc. C,* 881 (1968).
237. J. H. Espenson and T. H. Chao, *Inorg. Chem.,* **16,** 2553 (1977).
238. H. Eckert, I. Lagerlund, and I. Ugi, *Tetrahedron,* **33,** 2243 (1977).
239. K. N. V. Duong and G. Gaudemer, *J. Organomet. Chem.,* **22,** 473 (1970).
240. M. D. Johnson and B. S. Meeks, *J. Chem. Soc. B,* 185 (1971).
241. M. Tada, M. Kubota, and H. Shinozaki, *Bull. Chem. Soc. Jpn.,* **49,** 1097 (1976).
242. C. J. Cooksey, D. Dodd, C. Gatford, M. D. Johnson, G. J. Lewis, and D. M. Titchmarsh, *J. Chem. Soc., Perkin Trans. 2,* 655 (1972).
243. F. R. Jensen, V. Madan, and D. H. Buchanan, *J. Am. Chem. Soc.,* **92,** 1414 (1970).
244. D. Dodd and M. D. Johnson, *Chem. Commun.,* 571 (1971).
245. F. R. Jensen and D. H. Buchanan, *Chem. Commun.,* 153 (1973).
246. F. R. Jensen, V. Madan, and D. H. Buchanan, *J. Am. Chem. Soc.,* **93,** 5283 (1971).
247. J. Z. Chrzastowski, C. J. Cooksey, M. D. Johnson, B. L. Lockman, and P. N. Steggles, *J. Am. Chem. Soc.,* **97,** 932 (1975).
248. M. Okabe and M. Tada, *Chem. Lett.,* 831 (1980).
249. P. L. Bock and G. M. Whitesides, *J. Am. Chem. Soc.,* **96,** 2826 (1974).
250. R. Bonnett, *Chem. Rev.,* **63,** 573 (1963).
251. G. N. Schrauzer, *Acc. Chem. Res.,* **1,** 97 (1968).
252. D. G. Brown, *Prog. Inorg. Chem.,* **18,** 177 (1973).
253. J. Halpern and J. P. Maher, *J. Am. Chem. Soc.,* **86,** 2311 (1964).
254. J. Halpern and J. P. Maher, *J. Am. Chem. Soc.,* **87,** 5361 (1965).
255. J. Kwiatek and J. K. Seyler, *J. Organomet. Chem.,* **3,** 421 (1965).
256. P. B. Chock, R. B. K. Dewar, J. Halpern, and L. Y. Wong, *J. Am. Chem. Soc.,* **91,** 82 (1969).
257a. P. B. Chock and J. Halpern, *J. Am. Chem. Soc.,* **91,** 582 (1969).
257b. J. Halpern, *Pure Appl. Chem.,* **51,** 2171 (1979).
258. D. Dodd and M. D. Johnson, *J. Organomet. Chem.,* **52,** 1 (1973).
259. P. W. Schneider, P. F. Phelan, and J. Halpern, *J. Am. Chem. Soc,* **91,** 77 (1969).
260. G. Mestroni, A. Camus, and E. Mestroni, *J. Organomet. Chem.,* **24,** 775 (1970).
261. B. B. Anderson, C. L. Behrens, L. J. Radonovich, and J. K. Klabunde, *J. Am. Chem. Soc.,* **98,** 5390 (1976).
262. F. H. Jardine, *Prog. Inorg. Chem.,* **28,** 63 (1981).
263. M. C. Baird, in *The Chemistry of Acid Derivatives,* (Ed. S. Patai), Wiley, New York, 1979, Supplement B, Part 2, pp. 825.
264. R. F. Heck, *J. Am. Chem. Soc.,* **86,** 2796 (1964).
265a. J. Chatt and B. L. Shaw, *J. Chem. Soc. A,* 1437 (1966).
265b. A. J. Deeming and B. L. Shaw, *J. Chem. Soc. A,* 1437 (1966).
266. I. C. Douek and G. Wilkinson, *J. Chem. Soc. A,* 2604 (1969).
267. S. Franks, F. R. Hartley and J. R. Chipperfield, *Inorg. Chem.,* **20,** 3238 (1981).
268. J. Schwartz, D. W. Hart, and J. L. Holden, *J. Am. Chem. Soc.,* **94,** 9269 (1972).
269. A. J. Mukhedkar, M. Green, and F. G. A. Stone, *J. Chem. Soc. A,* 3023 (1969).
270. G. Deganello, P. Ugaugliati, B. Craciani, and U. Belluco, *J. Chem. Soc. A,* 2726 (1969).

271. P. Ugaugliati, A. Palazzi, G. Deganello, and U. Belluco, *Inorg. Chem.*, **9,** 724 (1970).
272. A. J. Mukhedkar, V. A. Mukhedkar, M. Green, and F. G. A. Stone, *J. Chem. Soc. A,* 3166 (1970).
273. M. C. Baird, D. N. Lawson, J. T. Mague, J. A. Osborn, and G. Wilkinson, *Chem. Commun.,* 129 (1966).
274. D. N. Lawson, J. A. Osborn, and G. Wilkinson, *J. Chem. Soc. A,* 1733 (1966).
275. P. G. H. Troughton and A. C. Skapski, *Chem. Commun.*, 575 (1968).
276. C. O'Connor, *J. Inorg. Nucl. Chem.,* **32,** 2299 (1970).
277. A. J. Hart-Davis and W. A. G. Graham, *Inorg. Chem.,* **10,** 1653 (1971).
278. A. J. Oliver and W. A. G. Graham, *Inorg. Chem.,* **9,** 243 (1970).
279. J. P. Collman, D. W. Murphy, and G. Dolcetti, *J. Am. Chem. Soc.,* **95,** 3687 (1973).
280. J. P. Collman and M. R. MacLaury, *J. Am. Chem. Soc.,* **96,** 3019 (1974).
281. D. Forester, *Adv. Organomet. Chem.,* **17,** 255 (1979).
282. J. Tsuji and K. Ohno, *Synthesis,* 157 (1969).
283. D. L. Egglestone, M. C. Baird, C. J. L. Lock, and G. Turner, *J. Chem. Soc., Dalton Trans.,* 1576 (1977).
284. K. S. Y. Lau, Y. Becker, F. Huang, N. Baenziger, and J. K. Stille, *J. Am. Chem. Soc.,* **99,** 5664 (1977).
285. D. A. Slack, D. L. Egglestone, and M. C. Baird, *J. Organomet. Chem.,* **146,** 71 (1978).
286. M. A. Bennett, J. C. Jeffery and G. B. Robertson, *Inorg. Chem.,* **20,** 323 (1981).
287. J. K. Stille and M. T. Regan, *J. Am. Chem. Soc.,* **96,** 1508 (1974).
288. M. C. Baird, J. T. Mague, J. A. Osborn, and G. Wilkinson, *J. Chem. Soc. A,* 1347 (1967).
289. K. Ohno and J. Tsuji, *J. Am. Chem. Soc.,* **90,** 99 (1968).
290. J. K. Stille and R. W. Fries, *J. Am. Chem. Soc.,* **96,** 1514 (1974).
291. J. K. Stille, F. Huang, and M. T. Regan, *J. Am. Chem. Soc.,* **96,** 1518 (1974).
292. N. A. Dunham and M. C. Baird, *J. Chem. Soc., Dalton Trans.,* 774 (1975).
293. J. Blum, E. Oppenheimer, and E. D. Bergman, *J. Am. Chem. Soc.,* **89,** 2338 (1967).
294a. D. Egglestone and M. C. Baird, *J. Organomet. Chem.,* **113,** C25 (1976).
294b. W. Strohmeier and P. Pföhler, *J. Organomet. Chem.,* **108,** 393 (1976).
295. M. F. McGuiggan, D. H. Doughty, and L. H. Pignolet, *J. Organomet. Chem.,* **185,** 241 (1980).
296. J. P. Collman, N. W. Hoffman, and D. E. Morris, *J. Am. Chem. Soc.,* **91,** 5659 (1969).
297. J. A. Kampmeier, R. M. Rodehorst, and J. B. Philip, Jr., *J. Am. Chem. Soc.,* **103,** 1847 (1981).
298. J. A. Kampmeier, S. H. Haris, and R. M. Rodehorst, *J. Am. Chem. Soc.,* **103,** 1478 (1981).
299. H. Ogoshi, T. Omura, and Z. Yoshida, *J. Am. Chem. Soc.,* **95,** 1666 (1973).
300. H. Ogoshi, J. I. Setsune, T. Omura, and Z. I. Yoshida, *J. Am. Chem. Soc.,* **97,** 6461 (1975).
301. R. Grigg, J. Trocha-Grimshaw, and V. Viswanatha, *Tetrahedron Lett.,* 289 (1976).
302. A. M. Abeysekera, R. Grigg, J. Trocha-Grimshaw, and V. Viswanatha, *J. Chem. Soc., Chem. Commun.,* 227 (1976).
303. A. M. Abeysekera, R. Grigg, J. Trocha-Grimshaw, and V. Viswanatha, *J. Chem. Soc., Perkin Trans. 1,* 36 (1977).
304. R. D. W. Kemmitt and D. I. Nichols, *Inorg. Nucl. Chem. Lett.,* **4,** 739 (1968).
305. J. P. Collman and C. T. Sears, Jr., *Inorg. Chem.,* **7,** 27 (1968).
306. J. Chatt, N. P. Johnson, and B. L. Shaw, *J. Chem. Soc. A,* 604 (1967).
307. A. J. Deeming and B. L. Shaw, *J. Chem. Soc. A,* 1128 (1969).
308. J. Burgess, M. J. Hacker, and R. D. W. Kemmitt, *J. Organomet. Chem.,* **72,** 121 (1974).
309. A. J. Deeming and B. L. Shaw, *Chem. Commun.,* 751 (1968).
310. A. J. Deeming and B. L. Shaw, *J. Chem. Soc. A,* 1562 (1969).
311. R. G. Pearson and A. T. Poulos, *Inorg. Chim. Acta,* **34,** 67 (1979).
312. B. L. Shaw and R. E. Stainbank, *J. Chem. Soc., Dalton Trans.,* 223 (1972).
313. G. Yoneda and D. M. Blake, *J. Organomet. Chem.,* **190,** C71 (1980).
314. G. Yoneda and D. M. Blake, *Inorg. Chem.,* **20,** 67 (1981).
315. P. B. Chock and J. Halpern, *J. Am. Chem. Soc.,* **88,** 3511 (1966).
316. H. Stieger and H. Kelm, *J. Phys. Chem.,* **77,** 290 (1973).

317. M. Kubota, G. W. Kiefer, R. M. Ishikawa, and K. E. Bencala, *Inorg. Chim. Acta,* **7,** 195 (1973).
318. W. H. Thompson and C. T. Sears, Jr., *Inorg. Chem.,* **16,** 769 (1977).
319. S. Carra and R. Ugo, *Inorg. Chim. Acta. Rev.,* **1,** 49 (1967).
320. E. M. Miller and B. L. Shaw, *J. Chem. Soc., Dalton Trans.,* 480 (1974).
321. E. Ugo, A. Pasini, A. Fusi, and S. Cenini, *J. Am. Chem. Soc.,* **94,** 7364 (1972).
322. F. R. Jensen and B. Knickel, *J. Am. Chem. Soc.,* **93,** 6339 (1971).
323. R. G. Pearson and W. R. Muir, *J. Am. Chem. Soc.,* **92,** 5519 (1970).
324. J. A. Labinger, R. J. Braus, D. Dolphin, and J. A. Osborn, *J. Chem. Soc., Chem. Commun.,* 612 (1970).
325. J. S. Bradley, D. E. Connor, D. Dolphin, J. A. Labinger, and J. A. Osborn, *J. Am. Chem. Soc.,* **94,** 4043 (1972).
326. J. A. Labinger, A. V. Kramer, and J. A. Osborn, *J. Am. Chem. Soc.,* **95,** 7908 (1973).
327. J. A. Labinger and J. A. Osborn, *Inorg. Chem.,* **19,** 3230 (1980).
328. J. A. Labinger, J. A. Osborn, and N. J. Coville, *Inorg. Chem.,* **19,** 3236 (1980).
329. J. Blum, M. Weitzberg, and R. J. Mureinik, *J. Organomet. Chem.,* **122,** 261 (1976).
330. R. J. Mureinik, M. Weitzberg, and J. Blum, *Inorg. Chem.,* **18,** 915 (1979).
331. C. A. Reed and W. R. Roper, *Chem. Commun.,* 155 (1969).
332. B. L. Shaw and E. Singleton, *J. Chem. Soc. A,* 1683 (1967).
333. D. L. Thorn and T. H. Tulip, *J. Am. Chem. Soc.,* **103,** 5984 (1981).
334. A. J. Oliver and W. A. G. Graham, *Inorg. Chem.,* **9,** 2653 (1970).
335. J. W. Kang and P. M. Maitlis, *J. Organomet. Chem.,* **26,** 393 (1971).
336. B. L. Shaw, *J. Organomet. Chem.,* **94,** 251 (1975).
337. G. Yoneda, S.-M. Lin, L.-P. Wang, and D. M. Blake, *J. Am. Chem. Soc.,* **103,** 5768 (1981).
338. M. Kubota and D. M. Blake, *J. Am. Chem. Soc.,* **93,** 1368 (1971).
339. M. Kubota, D. M. Blake, and S. A. Smith, *Inorg. Chem.,* **10,** 1430 (1971).
340. J. Blum, S. Kraus, and Y. Pickholtz, *J. Organomet. Chem.,* **33,** 227 (1971).
341. M. A. Bennett, and J. C. Jeffrey, *Inorg. Chem.,* **19,** 3763 (1980).
342. M. A. Bennett and R. Charles, *J. Am. Chem. Soc.,* **94,** 666 (1972).
343. M. A. Bennett, R. Charles, and T. R. B. Mitchell, *J. Am. Chem. Soc.,* **100,** 2737 (1978); M. A. Benett and R. Charles, *J. Am. Chem. Soc.,* **94,** 666 (1972).
344. A. J. Deeming and B. L. Shaw, *J. Chem. Soc. A,* 443 (1969).
345. D. M. Blake, A. Winkelman, and Y. L. Chung, *Inorg. Chem.,* **14,** 1326 (1975).
346. N. A. Bailey, C. J. Jones, B. L. Shaw, and E. Singleton, *J. Chem. Soc., Chem. Commun.,* 1051 (1967).
347. M. I. Bruce, M. Z. Iqbal, and F. G. A. Stone, *J. Organomet. Chem.,* **20,** 161 (1969).
348. R. B. King, S. L. Stafford, P. M. Treichel, and F. G. A. Stone, *J. Am. Chem. Soc.,* **83,** 3604 (1961).
349. R. A. Plowman and F. G. A. Stone, *Z. Naturforsch., Teil B,* **17,** 575 (1962).
350. R. F. Heck and C. R. Boss, *J. Am. Chem. Soc.,* **86,** 2580 (1964).
351. R. Aumann, H. Ring, C. Kruger, and R. Goddard, *Chem. Ber.,* **112,** 3644 (1979).
352. G. D. Annis, S. V. Ley, R. Sivaramakrishnan, A. M. Atkinson, D. Rogers, and D. J. Williams, *J. Organomet. Chem.,* **182,** C11 (1979).
353. H. D. Murdoch and E. Weiss, *Helv. Chim. Acta,* **45,** 1927 (1962).
354. A. N. Nesmeyanov, G. P. Zol'nikova, G. M. Babakhina, I. I. Kritskaya, and G. G. Yakobson, *Zh. Obshch. Khim.,* **43,** 2007 (1973).
355. H. G. Ang and D. Cheong, *J. Fluorine Chem.,* **9,** 247 (1977).
356. M. Pankowski and M. Bigorgne, *J. Organomet. Chem.,* **30,** 227 (1971).
357. E. L. Muetterties and J. W. Rathke, *J. Chem. Soc., Chem. Commun.,* 850 (1974).
358. H. H. Karsch, *Chem. Ber.,* **110,** 2699 (1977).
359. J. P. Collman, *Acc. Chem. Res.,* **8,** 342 (1975).
360. R. G. Pearson, H. Sobel, and J. Songstad, *J. Am. Chem. Soc.,* **90,** 319 (1968).
361. J. P. Collman, J. N. Cawse, and J. I. Brauman, *J. Am. Chem. Soc.,* **94,** 5905 (1972).
362. J. P. Collman, S. R. Winter, and D. R. Clark, *J. Am. Chem. Soc.,* **94,** 1788 (1972).
363. J. P. Collman and N. W. Hoffman, *J. Am. Chem. Soc.,* **95,** 2689 (1973).
364. M. P. Cook, Jr., *J. Am. Chem. Soc.,* **92,** 6080 (1970).
365. J. P. Collman, S. R. Winter, and R. G. Komoto, *J. Am. Chem. Soc.,* **95,** 249 (1973).

366. W. O. Siegl and J. P. Collman, *J. Am. Chem. Soc.*, **94,** 2516 (1972).
367. Y. Watanabe, T. Mitsudo, M. Yamashita, M. Tanaka, and Y. Takegami, *Chem. Lett.*, 475 (1973).
368. A. Guinot, P. Cadiot, and J. L. Roustan, *J. Organomet. Chem.*, **128,** C35 (1977).
369. J. P. Collman, R. G. Finke, J. N. Cawse, and J. I. Brauman, *J. Am. Chem. Soc.*, **99,** 2515 (1977).
370. J. E. Ellis, *J. Organomet. Chem.*, **86,** 1 (1975).
371. R. E. Dessy, R. L. Pohl, and R. B. King, *J. Am. Chem. Soc.*, **88,** 5121 (1966).
372. T. S. Piper and G. Wilkinson, *J. Inorg. Nucl. Chem.*, **3,** 104 (1956).
373. A. Davidson, J. A. McCleverty, and G. Wilkinson, *J. Chem. Soc.*, 1133 (1963).
374. J. A. Gladysz, G. M. Williams, W. Tam, D. L. Johnson, D. W. Parker, and J. C. Selover, *Inorg. Chem.*, **18,** 553 (1979).
375. M. L. H. Green and M. J. Smith, *J. Chem. Soc. A*, 3220 (1971).
376. R. B. King, *Inorg. Chem.*, **2,** 531 (1963).
377. R. B. King and D. M. Braitsch, *J. Organomet. Chem.*, **54,** 9 (1973).
378. S. Moorhouse and G. Wilkinson, *J. Organomet. Chem.*, **105,** 349 (1976).
379. T. E. Bauch and W. P. Giering, *J. Organomet. Chem.*, **114,** 165 (1976).
380. C. U. Pittman, Jr. and R. Felis, *J. Organomet. Chem.*, **72,** 399 (1974).
381. R. B. King, A. Efraty, and W. C. Zipperer, *J. Organomet. Chem.*, **38,** 121 (1972).
382. R. B. King and R. N. Kapoor, *Inorg. Chem.*, **8,** 2535 (1969).
383. K. H. Pannell, *Transition Met. Chem.*, **1,** 36 (1976).
384. R. B. King and M. B. Bisnette, *J. Am. Chem. Soc.*, **86,** 1267 (1964).
385. R. B. King and M. B. Bisnette, *Inorg. Chem.*, **4,** 486 (1965).
386. R. B. King and M. B. Bisnette, *Inorg. Chem.*, **5,** 293 (1966).
387. T. S. Abram and R. Baker, *Synth. React. Inorg. Met.-Org. Chem.*, **9,** 471 (1979).
388. T. C. T. Chang and M. Rosenblum, *J. Org. Chem.*, **46,** 4626 (1981).
389. M. L. Green and P. L. I. Nagy, *J. Chem. Soc.*, 189 (1963).
390. W. P. Giering and M. Rosenblum, *J. Organomet. Chem.*, **25,** C71 (1970).
391. A. Cutler, D. Ehntholt, W. P. Giering, P. Lennon, S. Raghu, A. Rosen, M. Rosenblum, J. Tancrede, and D. Wells, *J. Am. Chem. Soc.*, **98,** 3495 (1976).
392. J. Y. Merour and P. Cadiot, *C.R. Acad. Sci., Ser. C*, **271,** 83 (1970).
393. K. H. Pannell, M. F. Lappert, and K. Stanley, *J. Organomet. Chem.*, **112,** 37 (1976).
394. M. R. Churchill, J. Wormald, W. P. Giering, and G. F. Emerson, *Chem. Commun.*, 1217 (1968).
395. R. E. Davis, *Chem. Commun.*, 1218 (1968).
396. M. R. Churchill and J. Wormald, *Inorg. Chem.*, **8,** 1936 (1969).
397. A. N. Nesmeyanov, M. I. Rybinskaya, V. S. Kaganovich, T. V. Popova, and E. A. Petrovskaya, *Bull. Acad. Sci. USSR, Div. Chem. Sci.*, 2031 (1973).
398. A. N. Nesmeyanov, M. I. Rybinskaya, L. V. Rybin, V. S. Kaganovich, and P. V. Petrovskii, *J. Organomet. Chem.*, **31,** 257 (1971).
399. R. B. King and K. C. Hodges, *J. Am. Chem. Soc.*, **96,** 1263 (1974).
400. P. W. Jolly, M. I. Bruce, and F. G. A. Stone, *J. Chem. Soc.*, 5830 (1965).
401. M. Green, N. Mayne, and F. G. A. Stone, *Chem. Commun.*, 755 (1966).
402. R. E. Banks, R. N. Hazeldine, M. Lappin, and A. B. P. Lever, *J. Organomet. Chem.*, **29,** 427 (1971).
403. A. N. Nesmeyanov, I. B. Zlotkina, M. A. Khomutov, I. F. Lesheheva, N. R. Kolobova, and K. N. Anisimov, *Bull. Acad. Sci. USSR, Div. Chem. Sci.*, 6421 (1977).
404. R. B. King and M. S. Saran, *J. Am. Chem. Soc.*, **95,** 1811 (1973).
405. A. N. Nesmeyanov, N. E. Kolobova, G. K. Znobina, K. N. Anisimov, I. B. Zlotina, and M. D. Bargamova, *Bull. Acad. Sci. USSR , Div. Chem. Sci.*, 2127 (1973).
406. A. N. Nesmeyanov, N. E. Kolobova, I. B. Zlotina, B. V. Lokshin, I. F. Leshcheva, G. K. Znobina, and K. N. Anisimov, *J. Organomet. Chem.*, **110,** 339 (1976).
407. R. B. King and M. B. Bisnette, *J. Organomet. Chem.*, **2,** 38 (1964).
408. S. C. Cohen, *J. Chem. Soc., Dalton Trans.*, 553 (1973).
409. J. A. McCleverty and G. Wilkinson, *J. Chem. Soc.*, 4096 (1963).
410. R. B. King and M. B. Bisnette, *J. Organomet. Chem.*, **2,** 15 (1964).
411. R. B. King, *J. Am. Chem. Soc.*, **85,** 1918 (1963).
412. J. J. Alexander and A. Wojcicki, *Inorg. Chim. Acta*, **5,** 655 (1971).

413. C. E. Chidsey, W. A. Donaldson, R. P. Hughes, and P. F. Sherwin, *J. Am. Chem. Soc.*, **101,** 233 (1979).
414. A. N. Nesmeyanov, E. G. Perevalova, L. I. Leont'eva, S. A. Eremin, and O. V. Grigov'eva, *Bull. Acad. Sci. USSR, Div. Chem. Sci.*, 2558 (1974).
415. R. Gompper, E. Bartmann, and H. North, *Chem. Ber.*, **112,** 218 (1979).
416. W. P. Giering, M. Rosenblum, and J. Tancrede, *J. Am. Chem. Soc.*, **94,** 7170 (1972).
417. M. Pankowski, E. Samuel, and M. Bigorane, *J. Organomet. Chem.*, **97,** 105 (1975).
418. H. Felkin and B. Meunier, *Nouv. J. Chim.*, **1,** 281 (1977).
419a. G. M. Whitesides and D. J. Boschetto, *J. Am. Chem. Soc.*, **91,** 4313 (1969).
419b. G. M. Whitesides and D. J. Boschetto, *J. Am. Chem. Soc.*, **93,** 1529 (1971).
420. P. L. Bock, D. J. Boschetto, J. R. Rasmussen, J. P. Demers, and G. M. Whitesides, *J. Am. Chem. Soc.*, **96,** 2814 (1974).
421. D. A. Slack and M. C. Baird, *J. Chem. Soc., Chem. Commun.*, 701 (1974).
422. D. A. Slack and M. C. Baird, *J. Am. Chem. Soc.*, **98,** 5539 (1976).
423. K. M. Nicholas and M. Rosenblum, *J. Am. Chem. Soc.*, **95,** 4449 (1973).
424. R. W. Johnson and R. G. Pearson, *Chem. Commun.*, 986 (1970).
425. P. J. Krusic, P. J. Fagan, and J. San Filippo, *J. Am. Chem. Soc.*, **99,** 250 (1977).
426. A. N. Nesmeyanov, M. I. Rybinskaya, V. S. Kaganovich, T. V. Popova, and E. A. Petrovskaya, *Izv. Akad. Nauk SSSR, Ser. Khim.*, 2078 (1973).
427. A. N. Nesmeyanov, M. I. Rybinskaya, L. V. Rybin, E. A. Ptrovskaya, and V. A. Suova, *Bull. Acad. Sci. USSR, Div. Chem. Sci.*, 1511 (1976).
428. G. Huttner, S. Lange, and E. O. Fischer, *Angew. Chem., Int. Ed. Engl.*, **10,** 556 (1971).
429. V. L. Goedken and Y.-A. Park, *J. Chem. Soc., Chem. Commun.*, 214 (1975).
430. J. P. Collman and W. R. Roper, *J. Am. Chem. Soc.*, **87,** 4008 (1965).
431. T. Blackmore, M. I. Bruce, and F. G. A. Stone, *J. Chem. Soc. A*, 2158 (1968).
432. F. L'Eplattenier, *Chimia*, **23,** 144 (1969).
433. F. L'Eplattenier, *Inorg. Chem.*, **8,** 965 (1969).
434. J. Evans, S. J. Okrasinski, A. J. Pribula, and J. R. Norton, *J. Am. Chem. Soc.*, **98,** 4000 (1976).
435. W. J. Carter, J. W. Kelland, S. J. Okrasinski, K. E. Warner, and J. R. Norton, *Inorg. Chem.*, **21,** 3955 (1982).
436. F. L'Eplattenier and M. C. Pelichet, *Helv. Chim. Acta*, **53,** 1091 (1970).
437. R. D. George, S. A. R. Knox, and F. G. A. Stone, *J. Chem. Soc., Dalton Trans.*, 972 (1973).
438. W. Hieber and G. Wagner, *Justus Liebigs Ann. Chem.*, **618,** 24 (1958).
439. R. D. Clossen, J. Kozikowski, and T. H. Coffield, *J. Org. Chem.*, **22,** 598 (1957).
440. R. B. King, *J. Am. Chem. Soc.*, **85,** 1922 (1963).
441. W. Hieber, G. Faulhaber, and F. Theubert, *Z. Naturforsch., Teil B*, **15,** 326 (1960).
442. W. Hieber, G. Faulhaber, and F. Theubert, *Z. Anorg. Allg. Chem.*, **314,** 125 (1962).
443. J. Benaim and F. Giulieri, *J. Organomet. Chem.*, **165,** C28 (1979).
444. B. D. Dombek, *J. Am. Chem. Soc.*, **101,** 6466 (1979).
445. W. Beck, W. Hieber, and H. Tengler, *Chem. Ber.*, **94,** 862 (1961).
446. E. W. Abel, M. A. Bennett, R. Burton, and G. Wilkinson, *J. Chem. Soc.*, 4559 (1958).
447. H. D. Kaesz, R. B. King, and F. G. A. Stone, *Z. Naturforsch., Teil B*, **15,** 682 (1960).
448. V. A. Kormer, M. I. Lobach, N. N. Druz, V. I. Klepikova, and N. V. Kiseleva, *Dokl. Akad. Nauk SSSR*, **246,** 315 (1979).
449. M. K. Chaudhuri, *J. Organomet. Chem.*, **171,** 365 (1979).
450. H. D. Kaesz, R. B. King, and F. G. A. Stone, *Z. Naturforsch., Teil B*, **15,** 763 (1960).
451. E. Pitcher and F. G. A. Stone, *Spectrochim. Acta*, **17,** 1244 (1961).
452. R. B. King and A. Efraty, *J. Fluorine Chem.*, **1,** 283 (1972).
453. R. B. King and M. S. Saran, *J. Am. Chem. Soc.*, **95,** 1811 (1973).
454. R. B. King and K. C. Hodges, *J. Am. Chem. Soc.*, **97,** 2702 (1975).
455. T. H. Coffield, J. Kozikowski, and R. D. Closson, *J. Org. Chem.*, **22,** 598 (1957).
456. M. Dilgassa and M. D. Curtis, *J. Organomet. Chem.*, **172,** 177 (1979).
457. N. E. Kolobova, V. N. Khandozhko, V. F. Sizoi, Sh. Guseinov, O. S. Zhvanko, and Yu. S. Nekrasov, *Bull. Acad. Sci. USSR*, **28,** 573 (1979).
458. C. P. Casey, C. A. Bunnell, and J. C. Calabrese, *J. Am. Chem. Soc.*, **98,** 1166 (1976).

459. T. Kruck and M. Noack, *Chem. Ber.*, **97,** 1693 (1964).
460. W. Beck, W. Hieber, and H. Tengler, *Chem. Ber.*, **94,** 862 (1961).
461. M. Y. Darensbourg, D. J. Darensbourg, D. Burns, and D. A. Drew, *J. Am. Chem. Soc.*, **98,** 3127 (1976).
462. D. Dong, D. A. Slack, and M. C. Baird, *J. Organomet. Chem.*, **153,** 219 (1978).
463. W. Hieber, G. Braun, and W. Beck, *Chem. Ber.*, **93,** 901 (1960).
464. D. M. Heinekey and W. A. G. Graham, *J. Am. Chem. Soc.*, **101,** 6115 (1979).
465. R. I. Mink, J. J. Welter, P. R. Young, and G. D. Stucky, *J. Am. Chem. Soc.*, **101,** 6928 (1979).
466. D. Baudry and M. Ephritikhine, *J. Chem. Soc., Chem. Commun.*, 895 (1979).
467. A. N. Nesmeyanov, N. E. Kolobova, I. B. Zlotina, B. V. Lokshin, I. F. Leshcheva, G. K. Znobina, and K. N. Anisimov, *Bull. Acad. Sci. USSR,* **25,** 1093 (1976).
468. A. N. Nesmeyanov, L. G. Makarova, V. N. Vinogradova, V. N. Korneva, and N. A. Ustynyuk, *J. Organomet. Chem.*, **166,** 217 (1979).
469. B. J. Brisdon and G. F. Griffin, *J. Organomet. Chem.*, **76,** C47 (1974).
470. J. R. Hanson and E. Premuzic, *Angew. Chem., Int. Ed. Engl.*, **7,** 247 (1968).
471. F. A. L. Anet and E. Leblanc, *J. Am. Chem. Soc.*, **79,** 2649 (1957).
472. W. Marty and J. H. Espenson, *Inorg. Chem.*, **18,** 1246 (1979).
473. J. K. Kochi and D. D. Davis, *J. Am. Chem. Soc.*, **86,** 5264 (1964).
474. J. K. Kochi and D. Buchanan, *J. Am. Chem. Soc.*, **87,** 853 (1965).
475. J. K. Kochi and J. W. Powers, *J. Am. Chem. Soc.*, **92,** 137 (1970).
476. G. J. Samuels and J. H. Espenson, *Inorg. Chem.*, **18,** 2587 (1979).
477. R. B. King and D. M. Braitsch, *J. Organomet. Chem.*, **54,** 9 (1973).
478. R. B. King and A. Fronzaglia, *J. Am. Chem. Soc.*, **88,** 709 (1966).
479. F. A. Cotton and M. D. LaPrade, *J. Am. Chem. Soc.*, **90,** 5418 (1968).
480. F. A. Cotton and T. J. Marks, *J. Am. Chem. Soc.*, **91,** 1339 (1969).
481. M. Cousins and M. L. H. Green, *J. Chem. Soc.*, 889 (1963).
482. A. Davidson and W. C. Rode, *Inorg. Chem.*, **6,** 2124 (1967).
483. R. B. King and M. B. Bisnette, *Inorg. Chem.*, **4,** 475 (1965).
484. G. Doyle, *J. Organomet. Chem.*, **132,** 243 (1977).
485. B. J. Brisdon, D. A. Edwards, and J. W. White, *J. Organomet. Chem.*, **156,** 427 (1978).
486. J. A. Connor and P. I. Riley, *J. Chem. Soc., Chem. Commun.*, 149 (1976).
487. J. A. Connor and P. I. Riley, *J. Chem. Soc., Chem. Commun.*, 634 (1976).
488. C. G. Hull and M. H. B. Stiddard, *J. Organomet. Chem.*, *9,* 519 (1967).
489. B. J. Brisden and G. F. Griffen, *J. Chem. Soc., Dalton Trans.*, 1999 (1975).
490. M. G. B. Drew, B. J. Brisdon, and A. Day, *J. Chem. Soc., Dalton Trans.*, 1310 (1981).
491. J. Benain, J. Y. Merour, and J. L. Roustan, *Tetrahedron Lett.*, 983 (1971).
492. W. Seiel and W. Reichardt, *Z. Chem.*, **13,** 106 (1973).
493. M. Y. Darensbourg, P. Jimenez, and J. R. Sackett, *J. Organomet. Chem.*, **202,** C68 (1980).
494. M. L. H. Green and R. Mahtab, *J. Chem. Soc., Dalton Trans.*, 262 (1979).
495. D. Strope and D. F. Shriver, *J. Am. Chem. Soc.*, **95,** 8197 (1973).
496. R. B. King and M. B. Bisenette, *J. Organomet. Chem.*, **7,** 311 (1967).
497. A. N. Nesmeyanov, L. V. Rybin, M. I. Rybinskaya, and V. S. Kagnovich, *Izv. Akad. Nauk SSSR, Ser. Khim.*, 348 (1971).
498. F. Hohmann, *J. Organomet. Chem.*, **137,** 315 (1977).
499. C. E. Holloway, J. D. Kelly, and M. H. B. Stiddard, *J. Chem. Soc. A,* 931 (1969).
500. H. J. De Liefde Meijer, M. J. Hanosen, and G. J. M. Van Der Kerk, *Recl. Trav. Chim. Pays-Bas,* **80,** 831 (1961).
501. R. J. Kinney, W. D. Jones, and R. G. Bergman, *J. Am. Chem. Soc.*, **100,** 635 (1978).
502. J. E. Ellis and R. A. Faltynek, *J. Organomet. Chem.*, **93,** 205 (1975).
503. M. Schneider and E. Weiss, *J. Organomet. Chem.*, **73,** C7 (1974).
504. M. Schneider and E. Weiss, *J. Organomet. Chem.*, **121,** 345 (1976).
505. U. Franke and E. Weiss, *J. Organomet. Chem.*, **121,** 355 (1976).
506. U. Franke and E. Weiss, *J. Organomet. Chem.*, **153,** 39 (1978).
507. M. Schneider and E. Weiss, *J. Organomet. Chem.*, **121,** 189 (1976).
508. U. Franke and E. Weiss, *J. Organomet. Chem.*, **139,** 305 (1977).

509. U. Franke and E. Weiss, *J. Organomet. Chem.*, **165**, 329 (1979).
510. P. C. Wailes, R. S. P. Coutts, and H. Weigold, *Organometallic Chemistry of Titanium, Zirconium, and Hafnium,* Academic Press, New York, 1974.
511. C. Floriani and G. Fachinetti, *J. Chem. Soc., Chem. Commun.*, 790 (1972).
512. G. Fachinetti, C. Floriani, and H. Stoeckli-Evans, *J. Chem. Soc., Dalton Trans.*, 2297 (1977).
513. B. Demerseman, G. Bouquet, and M. Bigorne, *J. Organomet. Chem.*, **132,** 223 (1977).
514. J. Schwartz, *Pure Appl. Chem.*, **52,** 733 (1980).
515a. K. I. Gell and J. Schwartz, *J. Chem. Soc., Chem. Commun.*, 244 (1979).
515b. K. I. Gell and J. Schwartz, *J. Am. Chem. Soc.*, **103,** 2687 (1981).
516. G. M. Williams, K. I. Gell, and J. Schwartz, *J. Am. Chem. Soc.*, **102,** 3660 (1980).
517. H. Werner, *Adv. Organomet. Chem.*, **19,** 155 (1981).
518. M. P. Brown, J. R. Fischer, S. J. Franklin, and R. J. Puddephatt, *J. Chem. Soc., Chem. Commun.*, 749 (1978).
519. M. P. Brown, J. R. Fisher, R. J. Puddephatt, and K. R. Seddon, *Inorg. Chem.*, **18,** 2808 (1979).
520. A. L. Balch, C. T. Hunt, C.-L. Lee, M. M. Olmstead, and J. P. Farr, *J. Am. Chem. Soc.*, **103,** 3764 (1981).
521. N. E. Schore, C. Ilenda, and R. G. Bergman, *J. Am. Chem. Soc.*, **98,** 7436 (1976).
522. R. G. Bergman, *Acc. Chem. Res.*, **13,** 113 (1980).
523. K. H. Theopold and R. G. Bergman, *J. Am. Chem. Soc.*, **103,** 2489 (1981).
524. K. H. Theopold and R. G. Bergman, *J. Am. Chem. Soc.*, **102,** 5694 (1980).
525. W. H. Hersh and R. G. Bergman, *J. Am. Chem. Soc.*, **103,** 6992 (1981).
526. N. S. Lewis, K. R. Mann, J. G. Gordon, and H. B. Gray, *J. Am. Chem. Soc.*, **98,** 7461 (1976).
527. C. E. Sumner, Jr., P. E. Riley, R. E. Davis, and R. Pettit, *J. Am. Chem. Soc.*, **102,** 1752 (1980).
528. F. D. Mango, *Coord. Chem. Rev.*, **15,** 109 (1975).
529. K. C. Bishop, III, *Chem. Rev.*, **76,** 461 (1976).
530. R. S. Nyholm, *Proc. Chem. Soc.*, 273 (1961).
531. G. Wilke and H. Schott, *Angew. Chem., Int. Ed. Engl.*, **5,** 583 (1966).
532. P. Eilbracht, *Chem. Ber.*, **109,** 3136 (1976).
533. R. J. Puddephatt, *Coord. Chem. Rev.*, **33,** 149 (1980).
534. M. Lenarda, R. Ros, M. Graziani, and U. Belluco, *J. Organomet. Chem.*, **46,** 29C (1972).
535. M. Lenarda, R. Ros, M. Graziani, and U. Belluco, *J. Organomet. Chem.*, **65,** 407 (1974).
536. D. J. Yarrow, J. A. Ibers, M. Lenarda, and M. Graziani, *J. Organomet. Chem.*, **70,** 133 (1974).
537. J. Rajarm and J. A. Ibers, *J. Am. Chem. Soc.*, **100,** 829 (1978).
538. W. Wong, S. J. Singer, W. D. Pitts, S. F. Watkins, and W. H. Baddley, *J. Chem. Soc., Chem. Commun.*, 672 (1972).
539. J. P. Visser and J. E. Ramakers-Blom, *J. Organomet. Chem.*, **44,** C63 (1972).
540. M. Lenarda, N. B. Pahor, M. Calligaris, M. Graziani, and L. Randaccio, *Inorg. Chim. Acta,* **26,** L19 (1978).
541. W. J. Irwin and F. J. McQuillin, *Tetrahedron Lett.*, 1937 (1968).
542. G. W. Littlecott, F. J. McQuillin, and K. G. Powell, *Inorg. Synth.*, **16,** 113 (1976).
543. K. G. Powell and F. J. McQuillin, *Tetrahedron Lett.*, 3313 (1971).
544. F. J. McQuillin and K. G. Powell, *J. Chem. Soc., Dalton Trans.*, 2123 (1972).
545. D. B. Brown, *J. Organomet. Chem.*, **24,** 787 (1970).
546. R. J. Al-Essa, R. J. Puddephatt, C. F. H. Tipper, and P. J. Thompson, *J. Organomet. Chem.*, **157,** C40 (1978).
547. B. M. Cushman, S. E. Earnest, and D. B. Brown, *J. Organomet. Chem.*, **159,** 431 (1978).
548. B. M. Cushman and D. B. Brown, *J. Organomet. Chem.*, **152,** C42 (1978).
549. R. J. Puddephatt, M. A. Quyser, and C. F. H. Tipper, *J. Chem. Soc., Chem. Commun.*, 626 (1976).
550. R. J. Al-Essa, R. J. Puddephatt, M. A. Quyser, and C. F. H. Tipper, *J. Am. Chem. Soc.*, **101,** 364 (1979).

551. R. J. Al-Essa, R. J. Puddephatt, P. J. Thompson, and C. F. H. Tipper, *J. Am. Chem. Soc.,* **102,** 7546 (1980).
552. R. J. Al-Essa, R. J. Puddephatt, D. C. L. Perkins, M. C. Rendle, and C. H. F. Tipper, *J. Chem. Soc., Dalton Trans.,* 1738 (1981).
553. R. J. Al-Essa, R. J. Puddephatt, M. A. Quyser, and C. F. H. Tipper, *Inorg. Chim. Acta,* **34,** L187 (1979).
554. D. B. Brown and V. A. Viens, *J. Organomet. Chem.,* **142,** 117 (1977).
555. R. J. Al-Essa, R. J. Puddephatt, M. A. Quyer, and C. F. H. Tipper, *J. Organomet. Chem.,* **150,** 295 (1978).
556. P. W. Hall, R. J. Puddephatt, and C. F. H. Tipper, *J. Organomet. Chem.,* **71,** 145 (1974).
557. F. Iwanciw, M. A. Quyser, R. J. Puddephatt, and C. F. H. Tipper, *J. Organomet. Chem.,* **113,** 91 (1976).
558. J. A. McGinnety, *J. Organomet. Chem.,* **59,** 429 (1973).
559. N. A. Bailey, R. D. Gillard, M. Keeton, R. Mason, and R. R. Russell, *Chem. Commun.,* 396 (1966).
560. R. D. Gillard, M. Keeton, R. Mason, M. F. Pilbrow, and D. R. Russell, *J. Organomet. Chem.,* **33,** 247 (1971).
561. P. W. Hall, R. J. Puddephatt, and C. F. H. Tipper, *J. Organomet. Chem.,* **84,** 407 (1975).
562. R. J. Puddephatt, P. J. Thompson, and C. F. H. Tipper, *J. Organomet. Chem.,* **177,** 403 (1979).
563. B. M. Cushman and D. B. Brown, *J. Organomet. Chem.,* **152,** C42 (1978).
564. B. M. Cushman and D. B. Brown, *Inorg. Chem.,* **20,** 2490 (1981).
565. T. H. Johnson and S. S. Cheng, *J. Am. Chem. Soc.,* **101,** 5277 (1979).
566. C. P. Casey, D. M. Scheck, and A. J. Shusterman, *J. Am. Chem. Soc.,* **101,** 4233 (1979).
567. N. Dominelli and A. C. Oehlschlager, *Can. J. Chem.,* **55,** 364 (1977).
568. R. J. Al-Essa and R. J. Puddephatt, *J. Chem. Soc., Chem. Commun.,* 45 (1980).
569. T. H. Johnson, *J. Org. Chem.,* **44,** 1356 (1979).
570. A. Miyashita, M. Takahashi, and H. Takaya, *J. Am. Chem. Soc.,* **103,** 6257 (1981).
571. L. S. Liebeskind, S. L. Baysdon, M. S. South, and J. F. Blount, *J. Organomet. Chem.,* **202,** C73 (1980).
572. D. M. Roundhill, D. N. Lawson, and G. Wilkinson, *J. Chem. Soc. A,* 845 (1968).
573. K. G. Powell and F. J. McQuillin, *Chem. Commun.,* 931 (1971).
574. F. J. McQuillin and K. G. Powell, *J. Chem. Soc., Dalton Trans.,* 2129 (1972).
575. L. Cassar and J. Halpern, *J. Chem. Soc., Chem. Commun.,* 1082 (1970).
576. L. Cassar, P. E. Eaton, and J. Halpern, *J. Am. Chem. Soc.,* **92,** 3515 (1970).
577. M. Sohn, J. Blum, and J. Halpern, *J. Am. Chem. Soc.,* **101,** 2694 (1979).
578. H. Ogoshi, J.-I. Setsune, and Z.-I. Yoshida, *J. Chem. Soc., Chem. Commun.,* 572 (1975).
579. J. W. Suggs and S. D. Cox, *J. Organomet. Chem.,* **221,** 199 (1981).
580. T. H. Tulip and J. A. Ibers, *J. Am. Chem. Soc.,* **100,** 3252 (1978).
581. R. M. Tuggle and D. L. Weaver, *Inorg. Chem.,* **11,** 2237 (1972).
582. R. Aumann, *J. Organomet. Chem.,* **76,** C32 (1974).
583. R. M. Moriarty, K.-N. Chen, C.-L. Yeh, J. L. Flippen, and J. Karle, *J. Am. Chem. Soc.,* **94,** 8944 (1972).
584. J. L. Flippen, *Inorg. Chem.,* **13,** 1054 (1974).
585. S. W. Tam, *Tetrahedron Lett.,* 2385 (1974).
586. R. Aumann, *J. Organomet. Chem.,* **47,** C29 (1973).
587. R. M. Moriarty, K.-N. Chen, and J. L. Flippen, *J. Am. Chem. Soc.,* **95,** 6489 (1973).
588. R. M. Moriarty, C.-L. Yeh, and K. C. Ramey, *J. Am. Chem. Soc.,* **93,** 6709 (1971).
589. Y. Becker, A. Eisenstadt, and Y. Shvo, *J. Chem. Soc., Chem. Commun.,* 1156 (1972).
590. A. Eisenstadt, *Tetrahedron Lett.,* 2005 (1972).
591. R. M. Moriarty, C.-L. Yeh, K.-N. Chen, and R. Srinivasan, *Tetrahedron Lett.,* 5325 (1972).
592. R. Aumann, *Angew. Chem., Int. Ed. Engl.,* **10,** 188 (1971).
593. R. Aumann, *Angew. Chem.,* **84,** 583 (1972); *Angew. Chem., Int. Ed. Engl.,* **11,** 522 (1972).

594. R. Aumann, *Angew. Chem., Int. Ed. Engl.,* **10,** 189 (1971).
595. R. Aumann, *Angew. Chem., Int. Ed. Engl.,* **10,** 190 (1971).
596. R. Aumann and B. Lohmann, *J. Organomet. Chem.,* **44,** C51 (1972).
597. C. H. DePuy, V. M. Kobal, and D. H. Gibson, *J. Organomet. Chem.,* **13,** 266 (1968).
598. R. Noyori, T. Nishimura, and H. Takaya, *Chem. Commun.,* 89 (1969).
599. C. S. Cundy and M. F. Lappert, *J. Chem. Soc., Chem. Commun.,* 445 (1972).
600. C. S. Cundy and M. F. Lappert, *J. Organomet. Chem.,* **144,** 317 (1978).
601. C. S. Cundy, M. F. Lappert, J. Dubac, and P. Mazerolles, *J. Chem. Soc., Dalton Trans.,* 910 (1976).
602. C. S. Cundy and M. F. Lappert, *J. Chem. Soc., Dalton Trans.,* 665 (1978).
603. U. Schubert and A. Rengstl, *J. Organomet. Chem.,* **170,** C37 (1979).
604. R. J. Remick, T. A. Asunta, and P. S. Skell, *J. Am. Chem. Soc.,* **101,** 1320 (1979).
605. G. W. Parshall, *Chem. Tech.,* **4,** 445 (1974).
606. G. W. Parshall, *Acc. Chem. Res.,* **8,** 113 (1975).
607. G. W. Parshall, *Catalysis,* **1,** 335 (1977).
608. A. E. Shilov and A. A. Shteinman, *Coord. Chem. Rev.,* **24,** 97 (1977).
609. D. E. Webster, *Adv. Organomet. Chem.,* **15,** 147 (1977).
610. A. E. Shilov, *Pure Appl. Chem.,* **50,** 725 (1978).
611. M. I. Bruce and B. L. Goodall, in *The Chemistry of Hydrazo, Azo, and Azoxy Groups,* (Ed. S. Patai), Wiley, New York, 1975, pp. 259–311.
612. J. Dehand and M. Pfeffer, *Coord. Chem. Rev.,* **18,** 327 (1976).
613. H.-P. Abicht and K. Issleib, *Z. Chem.,* **17,** 1 (1977).
614. M. I. Bruce, *Angew. Chem., Int. Ed. Engl.,* **16,** 73 (1977).
615. I. Omae, *Coord. Chem. Rev.,* **28,** 97 (1979).
616. I. Omae, *Coord. Chem. Rev.,* **32,** 235 (1980).
617. A. J. Deeming and I. P. Rothwell, *Pure Appl. Chem.,* **52,** 649 (1980).
618. H. Takahashi and J. Tsuji, *J. Organomet. Chem.,* **10,** 511 (1967).
619. M. A. Bennett and D. L. Milner, *J. Am. Chem. Soc.,* **91,** 6983 (1969).
620. M. I. Bruce, B. L. Goodall, and F. G. A. Stone, *J. Chem. Soc., Chem. Commun.,* 558 (1973).
621. R. J. McKinney, R. Hoxmeier, and H. D. Kaesz, *J. Am. Chem. Soc.,* **97,** 3059 (1975).
622. R. Mason, M. Textor, N. Al-Salem and B. L. Shaw, *J. Chem. Soc., Chem. Commun.,* 292 (1976).
623. R. G. Goel and R. G. Montemayor, *Inorg. Chem.,* **16,** 2183 (1977).
624. H. Werner and H. J. Kraus, *J. Organomet. Chem.,* **204,** 415 (1980).
625. N. A. Al-Salem, W. S. McDonald, R. Markham, M. C. Norton, and B. L. Shaw, *J. Chem. Soc., Dalton Trans.,* 59 (1980).
626. P. Foley and G. M. Whitesides, *J. Am. Chem. Soc.,* **101,** 2732 (1979).
627. P. Foley, R. DiCosimo, and G. M. Whitesides, *J. Am. Chem. Soc.,* **102,** 6713 (1980).
628a. S. S. Moore, R. DiCosimo, A. F. Sowinski, and G. M. Whitesides, *J. Am. Chem. Soc.,* **103,** 948 (1981).
628b. R. DiCosimo, S. S. Moore, A. F. Sowinski, and G. M. Whitesides, *J. Am. Chem. Soc.,* **104,** 124 (1982).
629. J. F. van Baar, K. Vrieze, and D. J. Stufkens, *J. Organomet. Chem.,* **97,** 461 (1975).
630. S. Hietkamp, D. J. Stufkens, and K. Vrieze, *J. Organomet. Chem.,* **168,** 351 (1979).
631. J. F. van Baar, K. Vrieze, and D. J. Stufkens, *J. Organomet. Chem.,* **85,** 249 (1975).
632. S. Hietkamp, D. J. Stufkens, and K. Vrieze, *J. Organomet. Chem.,* **122,** 419 (1976).
633. S. Hietkamp, D. J. Stufkens, and K. Vrieze, *J. Organomet. Chem.,* **134,** 95 (1977).
634. S. Hietkamp, D. J. Stufkens, and K. Vrieze, *J. Organomet. Chem.,* **152,** 347 (1978).
635. C. Crocker, R. J. Errington, R. Markham, C. J. Moulton, K. J. Odell, and B. L. Shaw, *J. Am. Chem. Soc.,* **102,** 4373 (1980).
636. M. A. Bennett and D. L. Milner, *Chem. Commun.,* 581 (1967).
637. J. Schwartz and J. B. Cannon, *J. Am. Chem. Soc.,* **94,** 6226 (1972).
638. T. B. Rauchfuss, *J. Am. Chem. Soc.,* **101,** 1045 (1979).
639. G. Hata, H. Kondo, and A. Miyake, *J. Am. Chem. Soc.,* **90,** 2278 (1968).
640. D. J. Cole-Hamilton and G. Wilkinson, *J. Chem. Soc., Chem. Commun.,* 883 (1978).
641 .R. Werner and H. Werner, *Angew. Chem., Int. Ed. Engl.,* **20,** 793 (1981).
642. P. L. Timms, *Adv. Inorg. Radiochem.,* **14,** 121 (1972).

643. H. Bönnemann, *Angew. Chem., Int. Ed. Engl.,* **9,** 736 (1970).
644. J. Fornies, M. Green, J. L. Spencer, and F. G. A. Stone, *J. Chem. Soc., Dalton Trans.,* 1006 (1977).
645. J. H. Nelson, H. D. Jonassen, and D. M. Roundhill, *Inorg. Chem.,* **8,** 2591 (1969).
646. J. L. Garnett, *Catal. Rev.,* **5,** 229 (1972).
647. G. B. Shul'pin, L. P. Rozenberg, R. P. Shibaeva, and A. E. Shilov, *Kinet. Katal.,* **20,** 1570 (1979).
648. G. B. Shul'pin, A. E. Shilov, A. N. Kitaigorodskii, and J. V. Z. Krevor, *J. Organomet. Chem.,* **201,** 319 (1980).
649. J. A. Osborn, F. H. Jardine, J. F. Young, and G. Wilkinson, *J. Chem. Soc. A,* 1711 (1986).
650. M. C. Baird, C. J. Nyman, and G. Wilkinson, *J. Chem. Soc. A,* 348 (1968).
651. H. M. Walborsky and L. E. Allen, *Tetrahedron Lett.,* 823 (1970).
652. H. M. Walborsky and L. E. Allen, *J. Am. Chem. Soc.,* **93,** 5465 (1971).
653. J. Tsuji and K. Ohno, *Tetrahedron Lett.,* 2173 (1967).
654. J. A. Kampmeier, S. H. Harris, and D. K. Wedegaertner, *J. Org. Chem.,* **45,** 315 (1980).
655. J. W. Suggs, *J. Am. Chem. Soc.,* **100,** 640 (1978).
656. A. M. Abeysekera, R. Grigg, J. Trocha-Grimshaw, V. Viswanatha, and T. J. King, *Tetrahedron Lett.,* 3189 (1976).
657. A. M. Abeysekera, R. Grigg, J. Trocha-Grimshaw, and V. Viswanatha, *J. Chem. Soc., Perkin Trans. 1,* 1395 (1977).
658. C. K. Brown, D. Georgiou, and G. Wilkinson, *J. Chem. Soc. A,* 3120 (1971).
659. M. A. Bennett, R. Charles, and P. J. Fraser, *Aust. J. Chem.,* **30,** 1213 (1977).
660. A. D. English and T. Herskovitz, *J. Am. Chem. Soc.,* **99,** 1648 (1977).
661. A. H. Janowicz and R. G. Bergman, *J. Am. Chem. Soc.,* **104,** 352 (1982).
662. R. H. Crabtree, M. F. Mellea, J. M. Mihelcic, and J. B. Quirk, *J. Am. Chem. Soc.,* **104,** 107 (1982).
663. J. Chatt and J. M. Davidson, *J. Chem. Soc.,* 843 (1965).
664. F. A. Cotton, B. A. Frenz, and D. L. Hunter, *J. Chem. Soc., Chem. Commun.,* 755 (1974).
665. F. A. Cotton, D. L. Hunter, and B. A. Frenz, *Inorg. Chim. Acta,* **15,** 155 (1975).
666. U. A. Gregory, S. D. Ibekwe, B. T. Kilbourn, and D. R. Russell, *J. Chem. Soc. A,* 1118 (1971).
667. C. A. Tolman, S. D. Ittel, A. D. English, and J. P. Jesson, *J. Am. Chem. Soc.,* **100,** 4080 (1978).
668. U. Schubert and A. Rengstl, *J. Organomet. Chem.,* **166,** 323 (1979).
669. S. D. Ittel, C. A. Tolman, A. D. English, and J. P. Jesson, *J. Am. Chem. Soc.,* **98,** 6073 (1976).
670. S. D. Ittel, C. A. Tolman, A. D. English, and J. P. Jesson, *J. Am. Chem. Soc.,* **100,** 7577 (1978): **98,** 6073 (1976).
671. S. D. Ittel, C. A. Tolman, P. J. Krusic, A. D. English, and J. P. Jesson, *Inorg. Chem.,* **17,** 3432 (1978).
672. C. A. Tolman, S. D. Ittel, A. D. English, and J. P. Jesson, *J. Am. Chem. Soc.,* **101,** 1742 (1979).
673. S. D. Ittel, C. A. Tolman, A. D. English, and J. P. Jesson, *Adv. Chem. Ser.,* No. 173, 67 (1979).
674. G. W. Parshall, T. Herskovitz, F. N. Tebbe, A. D. English, and J. V. Zeile, in *Fundamental Research in Homogeneous Catalysis,* (Ed. M. Tsutsui), Plenum, New York, 1979, Vol. 3.
675. P. L. Timms, *J. Chem. Soc., Chem. Commun.,* 1033 (1969).
676. S. Komiya and A. Yamamoto, *Chem. Commun.,* 475 (1975).
677. S. Komiya, T. Ito, M. Cowie, A. Yamamoto, and J. A. Ibers, *J. Am. Chem. Soc.,* **98,** 3874 (1976).
678. A. J. P. Domingos, B. F. G. Johnson, and J. Lewis, *J. Chem. Soc., Dalton Trans.,* 145 (1974).
679. A. J. P. Domingos, B. F. G. Johnson, and J. Lewis, *J. Organomet. Chem.,* **36,** C43 (1972).

680. A. P. Humphries and S. A. R. Knox, *J. Chem. Soc., Dalton Trans.*, 1710 (1975).
681. A. J. Canty, B. F. G. Johnson, and J. Lewis, *J. Organomet. Chem.*, **43,** C35 (1972).
682. R. Mason and K. M. Thomas, *J. Organomet. Chem.*, **43,** C39 (1972).
683. A. J. Canty, A. J. P. Domingos, B. F. G. Johnson, and J. Lewis, *J. Chem. Soc., Dalton Trans.*, 2056 (1973).
684. M. I. Bruce, M. A. Cairns, A. Cox, M. Green, M. D. H. Smith, and P. Woodward, *J. Chem. Soc., Chem. Commun.*, 735 (1970).
685. A. Cox and P. Woodward, *J. Chem. Soc. A,* 3599 (1971).
686. M. I. Bruce, M. A. Cairns, and M. Green, *J. Chem. Soc., Dalton Trans.*, 1293 (1972).
687. M. Evans, M. Hursthouse, E. W. Randall, E. Rosenberg, L. Milone, and M. Valle, *J. Chem. Soc., Chem. Commun.*, 545 (1972).
688. M. I. Bruce, M. Cook, and M. Green, *Angew. Chem.*, **80,** 662 (1967); *Angew. Chem., Int. Ed. Engl.*, **7,** 639 (1968).
689. A. J. Deeming and M. Underhill, *J. Organomet. Chem.*, **42,** C60 (1972).
690. A. J. Deeming and M. Underhill, *J. Chem. Soc., Chem. Commun.*, 277 (1973).
691. A. J. Deeming and M. Underhill, *J. Chem. Soc., Dalton Trans.*, 1415 (1974).
692. A. J. Deeming, R. E. Kimber, and M. Underhill, *J. Chem. Soc., Dalton Trans.*, 2589 (1973).
693. J. B. Keister and J. R. Shapley, *J. Organomet. Chem.*, **85,** C29 (1975).
694. J. B. Keister and J. R. Shapley, *J. Am. Chem. Soc.*, **98,** 1056 (1976).
695. D. Baudry, M. Ephritikhine, and H. Felkin, *J. Chem. Soc., Chem. Commun.*, 1243 (1980).
696. P. S. Skell, D. L. Williams-Smith, and M. J. McGlinchey, *J. Am. Chem. Soc.*, **95,** 3337 (1973).
697. M. J. D'Aniello and E. K. Barefield, *J. Organomet. Chem.*, **76,** C50 (1974).
698. E. M. Van Dam, W. N. Brent, M. P. Silvon, and P. S. Skell, *J. Am. Chem. Soc.*, **97,** 465 (1975).
699. M. Berry, S. G. Davies, and M. L. H. Green, *J. Chem. Soc., Chem. Commun.*, 99 (1978).
700. M. L. H. Green and P. J. Knowles, *J. Chem. Soc. A,* 1508 (1971);
701. M. L. H. Green and P. J. Knowles, *J. Chem. Soc., Chem. Commun.*, 1677 (1970).
702. B. R. Francis, M. L. H. Green, and G. G. Roberts, *J. Chem. Soc., Chem. Commun.*, 1290 (1971).
703. C. Giannotti and M. L. H. Green, *J. Chem. Soc., Chem. Commun.*, 1114 (1972).
704. K. Elmitt, M. L. H. Green, R. A. Forder, I. Jefferson, and K. Prout, *J. Chem. Soc., Chem. Commun.*, 747 (1974).
705. M. L. H. Green, M. Berry, C. Couldwell, and K. Prout, *Nouv. J. Chim.* **1,** 187 (1977).
706. M. L. H. Green, *Pure Appl. Chem.*, **50,** 27 (1978).
707. N. J. Cooper, M. L. H. Green, and R. Mahtab, *J. Chem. Soc., Dalton Trans.*, 1557 (1979).
708. M. Berry, K. Elmitt, and M. L. H. Green, *J. Chem. Soc., Dalton Trans.*, 1950 (1979).
709. F. N. Tebbe and G. W. Parshall, *J. Am. Chem. Soc.*, **93,** 3793 (1971).
710. L. J. Gugsenberger and F. N. Tebbe, *J. Am. Chem. Soc.*, **93,** 5924 (1971).
711. G. P. Pez and J. N. Armor, *Adv. Organomet. Chem.*, **19,** 1 (1981).
712. G. P. Pez, C. F. Putnik, S. L. Suib, and G. D. Stucky, *J. Am. Chem. Soc.*, **101,** 6933 (1979).
713. T. V. Harris, K. I. Gell, and J. Schwartz, *Inorg. Chem.*, **20,** 481 (1981).
714. G. Wilkinson, *Pure Appl. Chem.*, **30,** 627 (1972).
715. P. S. Braterman and R. J. Cross, *J. Chem. Soc., Dalton Trans.*, 657 (1972).
716. P. S. Braterman and R. J. Cross, *Chem. Soc. Rev.*, **2,** 271 (1973).
717. M. C. Baird, *J. Organomet. Chem.*, **64,** 289 (1974).
718. R. R. Schrock and G. W. Parshall, *Chem. Rev.*, **76,** 243 (1976).
719. P. J. Davison, M. F. Lappert, and R. Pearce, *Acc. Chem. Res.*, **7,** 209 (1974).
720. P. J. Davidson, M. F. Lappert, and R. Pearce, *Chem. Rev.*, **76,** 219 (1976).
721. P. S. Braterman, *Topics in Current Chemistry* (*Fortschr. Chem. Forsch.*), **92,** 149 (1980), Springer Verlag, New York.
722. A. J. Connor, *Top. Curr. Chem.*, **71,** 71 (1977).
723. F. T. T. Ng, G. L. Rempel, and J. Halpern, *J. Am. Chem. Soc.*, **104,** 621 (1982).

724. J. D. Ruddick and B. L. Shaw, *J. Chem. Soc. A*, 2969 (1969).
725. E. L. Weinberg, and M. C. Baird, *J. Organomet. Chem.*, **179,** C61 (1979).
726. D. A. Slack and M. C. Baird, *J. Am. Chem. Soc.*, **98,** 5539 (1976).
727. W. Rogers, J. A. Page, and M. C. Baird, *J. Organomet. Chem.*, **156,** C37 (1978).
728. B. J. Brisdon, D. A. Edwards, and K. E. Paddick, *Transition Met. Chem.*, **6,** 83 (1981).
729. J. K. Kochi, *Acc. Chem. Res.*, **7,** 351 (1974).
730. H. Hashimoto and T. Nakano, *J. Org. Chem.*, **31,** 891 (1966).
731. A. Cairncross and W. A. Sheppard, *J. Am. Chem. Soc.*, **90,** 2186 (1968).
732. A. Cairncross and W. A. Sheppard, *J. Am. Chem. Soc.*, **93,** 247 (1971).
733. G. M. Whitesides, E. R. Stedronsky, C. P. Casey, and J. S. Filippo, Jr., *J. Am. Chem. Soc.*, **92,** 1426 (1970).
734. G. Van Koten and J. G. Noltes, *J. Chem. Soc., Chem. Commun.*, 575 (1974).
735. G. M. Whitesides, C. P. Casey, and J. K. Krieger, *J. Am. Chem. Soc.*, **93,** 1379 (1971).
736. G. M. Whitesides, E. J. Panek, and E. R. Stedronsky, *J. Am. Chem. Soc.*, **94,** 232 (1972).
737. A. Yamamoto, A. Miyashita, T. Yamamoto, and S. Ikeda, *Bull. Chem. Soc. Jpn.*, **34,** 1583 (1972).
738. G. M. Whitesides, J. S. Filippo, Jr., C. P. Casey, and E. J. Panek, *J. Am. Chem. Soc.*, **89,** 5302 (1967).
739. G. M. Whitesides, D. E. Bergbreiter, and P. E. Kendall, *J. Am. Chem. Soc.*, **96,** 2806 (1974).
740. B. Armer and H. Schmidbaur, *Angew. Chem., Int. Ed. Engl.*, **9,** 101 (1970).
741. A. Tamaki and J. K. Kochi, *J. Organomet. Chem.*, **61,** 441 (1973).
742. A. Burawoy, C. S. Gibson, and S. Holt, *J. Chem. Soc.*, 1024 (1935).
743. W. L. G. Gent and C. S. Gibson, *J. Chem. Soc.*, 1835 (1949).
744. M. E. Foss and C. S. Gibson, *J. Chem. Soc.*, 3063 (1949).
745. S. Komiya and J. K. Kochi, *J. Am. Chem. Soc.*, **98,** 7599 (1976).
746. A. Tamaki and J. K. Kochi, *J. Organomet. Chem.*, **40,** C81 (1972).
747. G. W. Rice and R. S. Tobias, *J. Organomet. Chem.*, **86,** C37 (1975).
748. C. F. Shaw, III, J. W. Lundeen, and R. S. Tobias, *J. Organomet. Chem.*, **51,** 365 (1973).
749. P. L. Kuch and R. S. Tobias, *J. Organomet. Chem.*, **122,** 429 (1976).
750. G. E. Coates, and C. Parkin, *J. Chem. Soc.*, 421 (1963).
751. L. G. Vaughan and W. A. Sheppard, *J. Organomet. Chem.*, **22,** 739 (1970).
752. A. Tamaki, S. A. Magennis, and J. K. Kochi, *J. Am. Chem. Soc.*, **96,** 6140 (1974).
753. S. Komiya, T. A. Albright, R. Hoffmann, and J. K. Kochi, *J. Am. Chem. Soc.*, **98,** 7255 (1976).
754. A. Shiotani and H. Schmidbaur, *J. Organomet. Chem.*, **37,** C24 (1972).
755. B. Åkermark and A. Ljungquist, *J. Organomet. Chem.*, **182,** 59 (1979).
756. J. Chatt and B. L. Shaw, *J. Chem. Soc.*, 1718 (1960).
757. H. F. Klein and H. H. Karsch, *Chem. Ber.*, **105,** 2628 (1972).
758. R. H. Grubbs, D. Carr, and P. Burk, *Org. Transition Met. Chem.*, 135 (1974).
759. R. H. Grubbs, A. Miyashita, M. I. M. Liu, and P. L. Burk, *J. Am. Chem. Soc.*, **99,** 3863 (1977).
760. R. H. Grubbs and A. Miyashita, *J. Am. Chem. Soc.*, **100,** 1300 (1978).
761. R. H. Grubbs, A. Miyashita, M. Liu, and P. Burk, *J. Am. Chem. Soc.*, **100,** 2418 (1978).
762. R. H. Grubbs and A. Miyashita, *J. Am. Chem. Soc.*, **100,** 7416 (1978).
763. R. H. Grubbs and A. Miyashita, *J. Am. Chem. Soc.*, **100,** 7418 (1978).
764. R. J. McKinney, D. L. Thorn, R. Hoffmann, and A. Stockis, *J. Am. Chem. Soc.*, **103,** 2595 (1981).
765. K. Tamao, M. Zemayashi, Y. Kso, and M. Kumada, *J. Organomet. Chem.*, **55,** C91 (1973).
766. D. G. Morrell and J. K. Kochi, *J. Am. Chem. Soc.*, **97,** 7262 (1975).
767. T. T. Tsou and J. K. Kochi, *J. Am. Chem. Soc.*, **101,** 7547 (1979).
768. L. Cassar, *J. Organomet. Chem.*, **54,** C57 (1973).
769. L. Cassar, S. Ferrara and H. Foá, *Adv. Chem. Ser.*, **132,** 252 (1974).
770. G. Favero, A. Morvillo, and A. Turco, *J. Organomet. Chem.*, **162,** 99 (1978).
771. G. Favero, M. Gaddi, A. Morvillo, and A. Turco, *J. Organomet. Chem.*, **149,** 395 (1978).

772. G. Favero, *J. Organomet. Chem.*, **202,** 225 (1980).
773. A. Yamamoto, K. Morifuji, S. Ikeda, T. Saito, Y. Uchida, and A. Misono, *J. Am. Chem. Soc.*, **87,** 4652 (1965).
774. T. Saito, Y. Uchida, A. Misono, A. Yamamoto, K. Morifuji, and S. Ikeda, *J. Am. Chem. Soc.*, **88,** 5198 (1966).
775a. T. Yamamoto, A. Yamamoto, and S. Ikeda, *J. Am. Chem. Soc.*, **93,** 3350 (1971).
775b. T. Yamamoto, A. Yamamoto, and S. Ikeda, *J. Am. Chem. Sic.*, **93,** 3360 (1971).
776. M. Uchino and S. Ikeda, *J. Organomet. Chem.*, **33,** C41 (1972).
777. M. Uchino, A. Yamamoto, and S. Ikeda, *J. Organomet. Chem.*, **24,** C63 (1970).
778. M. Uchino, K. Asagi, A. Yamamoto, and S. Ikeda, *J. Organomet. Chem.*, **84,** 93 (1975).
779. T. T. Tsou and J. K. Kochi, *J. Am. Chem. Soc.*, **100,** 1634 (1978).
780. T. Iso, H. Tsuchiya, and A. Yamamoto, *Bull. Chem. Soc. Jpn.*, **50,** 1319 (1977).
781. A. Sekiya and N. Ishikawa, *J. Organomet. Chem.*, **118,** 349 (1976).
782. A. Sekiya and N. Ishikawa, *J. Organomet. Chem.*, **125,** 281 (1977).
783. A. Gillie and J. K. Stille, *J. Am. Chem. Soc.*, **102,** 4933 (1980).
784. F. Ozawa, T. Ito, Y. Nakamura, and A. Yamamoto, *Bull. Chem. Soc. Jpn.*, **54,** 1868 (1981).
785. A. Moravskiy and J. K. Stille, *J. Am. Chem. Soc.*, **103,** 4182 (1981).
786. K. Tatsumi, R. Hoffmann, A. Yamamoto, and J. K. Stille, *Bull. Chem. Soc. Jpn.*, **54,** 1857 (1981).
787. F. Ozawas and A. Yamamoto, *Chem. Lett.*, 289 (1981).
788. M. K. Loar and J. K. Stille, *J. Am. Chem. Soc.*, **103,** 4174 (1981).
789. D. Milstein and J. K. Stille, *J. Am. Chem. Soc.*, **101,** 4981 (1979).
790. D. Milstein and J. K. Stille, *J. Am. Chem. Soc.*, **101,** 4992 (1979).
791. P. M. Maitlis and F. G. A. Stone, *Chem. Ind.* (*London*), 1865 (1962).
792. J. Chatt and B. L. Shaw, *J. Chem. Soc.*, 705 (1959).
793. J. Chatt and B. L. Shaw, *J. Chem. Soc.*, 4020 (1959).
794. J. X. McDermott, J. F. White, and G. M. Whitesides, *J. Am. Chem. Soc.*, **95,** 4451 (1973).
795. J. X. McDermott, J. F. White, and G. M. Whitesides, *J. Am. Chem. Soc.*, **98,** 6521 (1976).
796. F. Glockling, T. McBride, and R. J. I. Pollock, *J. Chem. Soc., Chem. Commun.*, 650 (1973).
797. P. S. Braterman, R. J. Cross, and G. B. Young, *J. Chem. Soc., Chem. Commun.*, 627 (1975).
798. P. S. Braterman, R. J. Cross, and G. B. Young, *J. Chem. Soc., Dalton Trans.*, 1306 (1976).
799. P. S. Braterman, R. J. Cross, and G. B. Young, *J. Chem. Soc., Dalton Trans.*, 1310 (1976).
800. P. S. Braterman, R. J. Cross, and G. B. Young, *J. Chem. Soc., Dalton Trans.*, 1892 (1977).
801. N. G. Hargreaves, R. J. Puddephatt, L. H. Sutcliffe, and P. J. Thompson, *J. Chem. Soc., Chem. Commun.*, 861 (1973).
802. P. W. Hall, R. J. Puddephatt, K. R. Seddon, and C. F. H. Tipper, *J. Organomet. Chem.*, **81,** 423 (1974).
803. G. Phillips, R. J. Puddephatt, and C. F. H. Tipper, *J. Organomet. Chem.*, **131,** 467 (1977).
804. D. C. L. Perkins, R. J. Puddephatt, and C. F. H. Tipper, *J. Organomet. Chem.*, **154,** C16 (1978).
805. C. F. H. Tipper, *J. Chem. Soc.*, 1045 (1955).
806. D. M. Adams, J. Chatt, R. G. Guy, and N. Sheppard, *Proc. Chem. Soc.* (*London*), 179 (1960).
807. D. C. L. Perkins, R. J. Puddephatt, M. C. Rendle, and C. F. H. Tipper, *J. Organomet. Chem.*, **195,** 105 (1980).
808. G. B. Young and G. M. Whitesides, *J. Am. Chem. Soc.*, **100,** 5808 (1978).
809. T. H. Johnson, T. F. Baldwin, and K. C. Klein, *Tetrahedron Lett.*, 1191 (1979).
810. K. W. Egger, *J. Organomet. Chem.*, **24,** 501 (1970).
811. B. F. G. Johnson, J. Lewis, and S. W. Tam, *Tetrahedron Lett.*, 3793 (1974).

812. T. Ikariya and A. Yamamoto, *J. Chem. Soc., Chem. Commun.*, 720 (1974).
813. T. Ikariya and A. Yamamoto, *J. Organomet. Chem.*, **118,** 65 (1976).
814. A. Yamamoto, K. Morifuji, S. Ikeda, T. Saito, Y. Uchida, and A. Misono, *J. Am. Chem. Soc.*, **87,** 4652 (1965).
815. A. Yamomoto, K. Morifuji, S. Ikeda, T. Saito, Y. Uchida, and A. Misono, *J. Am. Chem. Soc.*, **90,** 1878 (1968).
816. T. Yamamoto, A. Yamamoto, and S. Ikeda, *Bull. Chem. Soc. Jpn.*, **45,** 1104 (1972).
817. B. K. Bower and H. G. Tennent, *J. Am. Chem. Soc.*, **94,** 2512 (1972).
818. H. H. Zeiss and R. P. A. Sneeden, *Angew. Chem., Int. Ed. Engl.*, **6,** 435 (1967).
819. B. J. Brisdon, D. A. Edwards, and K. E. Paddick, *Transition Met. Chem.*, 83 (1981).
820. G. A. Razuvaev, V. N. Latyaeva, B. G. Zateev, and G. A. Kilyakora, *Proc. Acad. Sci. USSR,* **172,** 180 (1967).
821. V. N. Latyaeva, G. A. Ruzuvaev, A. V. Malishera, and G. A. Kiljakova, *J. Organomet. Chem.*, **2,** 388 (1964).
822. G. J. Erskine, D. A. Wilson, and J. D. McCowan, *J. Organomet. Chem.*, **114,** 119 (1976).
823. J. X. McDermott and G. M. Whitesides, *J. Am. Chem. Soc.*, **96,** 947 (1974).
824. J. X. McDermott, M. E. Wilson, and G. M. Whitesides, *J. Am. Chem. Soc.*, **98,** 6529 (1976).
825. R. H. Grubbs and A. Miyashita, *J. Chem. Soc., Chem. Commun.*, 864 (1977).
826. G. Fachinetti and C. Floriani, *J. Chem. Soc., Chem. Commun.*, 654 (1972).
827. M. D. Rausch, W. H. Boon, and H. G. Alt, *Ann. N.Y. Acad. Sci.*, **295,** 103 (1977).
828. K.-H. Thiele, E. Köhler, and B. Adler, *J. Organomet. Chem.*, **50,** 153 (1973).
829. G. A. Razuvaev, V. N. Latyaeva, L. I. Vishinskaya, and A. M. Rabinovitch, *J. Organomet. Chem.*, **49,** 441 (1973).

The Chemistry of the Metal—Carbon Bond, Volume 2
Edited by F. R. Hartley and S. Patai


CHAPTER **10**

# Structure and bonding of main group organometallic compounds

JOHN P. OLIVER
*Department of Chemistry, Wayne State University, Detroit, Michigan 48202, USA*



## I. INTRODUCTION

The structure and bonding of main group organometallic compounds are of major interest to chemists, not only because of the important role which they play in the reactivity of these species, but also because of the unusual and often surprising varieties of features observed. Among the earliest of these unexpected properties which was observed was the dimeric nature of trimethylaluminium[1]. This could not be readily accounted for by the bonding models in use at that time. For that reason it led to the recognition that this species, as well as many to be studied later, required a more complex model involving the use of multi-centred molecular orbitals. The concepts developed to treat the bonding in these systems now provide the basis for most of the bonding models used for main group organometallic compounds, and in addition have been extended to include many transition metal organometallics and mixed transition metal–main group metal species.

One of the simplest descriptions of multi-centred bonds was developed by Rundle[2] to account for the formation of the aluminium alkyl dimers. In this system the bonding results from the overlap of three $sp^3$ orbitals, one from each aluminium atom and the third from the bridging carbon atom to give rise to a set of molecular orbitals composed of a bond orbital occupied by the two electrons, a non-bonding orbital, and an antibonding orbital, **1**. This very simple description accounts for the

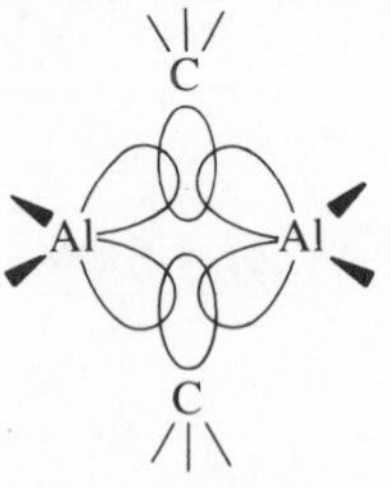

**1**

bonding in the aluminium alkyl dimers and serves to illustrate the approach taken in treating the bonding present in many of the organometallic systems to be dealt with in this chapter. Many of the descriptions in the literature are far more complex, involving more centres and/or orbitals. As yet a detailed and comprehensive treatment of the bonding in main group organometallic chemistry has not been developed, as has been the case for the boron hydrides[3], but significant steps are being made, as described below. The major features which are apparent in all of these treatments is that hydrogen, carbon, silicon, and possibly other elements may become a central atom in multicentre bonding, which by classical treatment yield bonds which are electron deficient.

The bridging atom has a higher coordination number than observed in simple covalently bound molecules, with values of 2 or 3 for hydrogen and 5 or 6 for carbon or silicon. Further, the bonds are all polarized to some extent and in the extreme cases have been described as 'totally ionic' with the units held together only by electrostatic interactions. These species may be represented by derivatives such as NaMe or KMe, which form essentially ionic lattices held together by electrostatic interactions with no apparent covalent character. At the other extreme, one may describe bonding in systems such as organomercury derivatives as essentially covalent two-electron $\sigma$ bonds with no unusual properties.

This review will include the pertinent information dealing with this subject and will rely heavily on the observed structural data obtained both from single-crystal X-ray structural determinations and from electron diffraction studies. Other pertinent physical measurements will also be discussed as they relate to the bonding in these systems. The focus will be placed almost exclusively on those systems which contain electron-deficient carbon bridge bonds with only occasional digression to consider other systems. It is not intended to be exhaustive and therefore references have been omitted, as have subjects which the author did not feel were essential for the major emphasis of this survey.

The attention of the readers is particularly directed to reviews by Oliver[4] and Haaland[5] on the structures of these molecules and to the monographs of Matteson[6] and Coates *et al.*[7], which deal with the mechanisms of reactions, and to the general field with extensive discussion of preparations, structures, and bonding, and to a survey by Armstrong and Perkins[8] covering the theoretical work on all of these systems. Other reviews dealing with specific metals or groups will be cited in the individual discussions.

## II. LITHIUM ALKYLS AND OTHER GROUP I DERIVATIVES

### A. Organolithium Derivatives

Organolithium compounds represent the most complex and least understood systems that will be discussed, despite the fact that they have been investigated very extensively. Much of the earlier work has been covered in reviews[9–11] and a

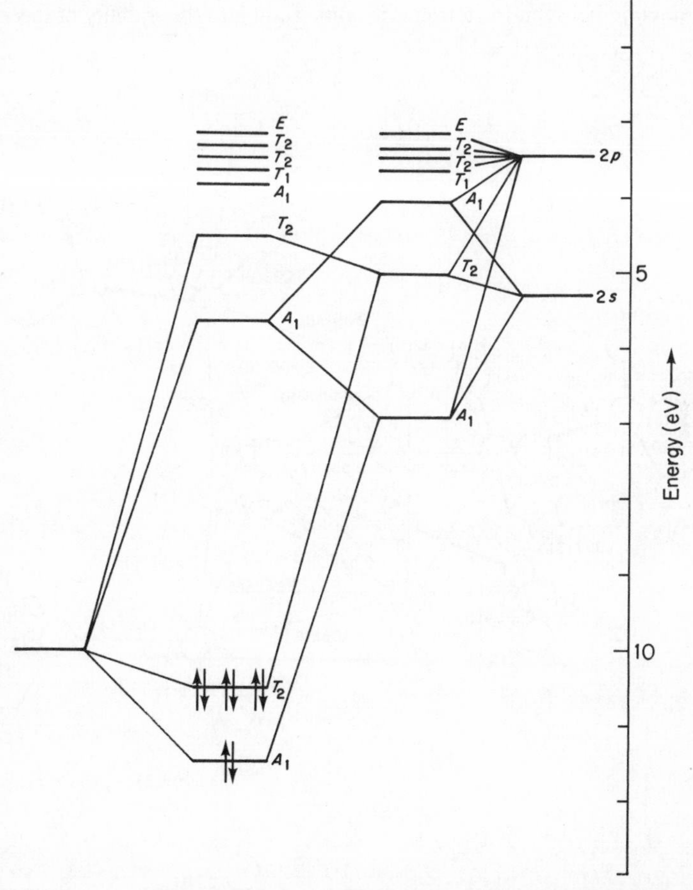


FIGURE 1. Energy level diagram for a localized four-centre molecular orbital in $[LiMe]_4$. Reproduced with permission from *J. Organomet. Chem.*, **2,** 197 (1964). The symmetry symbols refer to the pseudo-$C_{3v}$ symmetry of the localized MO.

monograph[12], with a more recent review dealing with their use as polymerization catalysts[13]. To develop our understanding of these derivatives we need to examine their structure in some detail. Several X-ray structural studies have been reported, including the determination of the structures of $[LiMe]_4$ [14,15], $[LiEt]_4$ [16,17], $[Li\text{-}cyclo\text{-}C_6H_{11}]_6$ [18], $[LiSiMe_3]_6$ [19,20], and $[LiCH_2SiMe_3]_6$ [21]. For the first structure determined, that of $[LiMe]_4$, Weiss and Lucken[14] proposed a bonding model based on the concept that the aggregate was held together by the overlap between an $sp^3$ orbital from the bridging carbon atom and three orbitals, one from each lithium atom, generating a four-centred bonding molecular orbital. A diagram of this model is shown in Figure 1 and is entirely consistent with the structure of both the methyl- and the ethyl-lithium tetramers. The structure of the ethyl derivatives is shown in Figure 2.

The recent redetermination of the structure of $[LiEt]_4$ at low temperature by Dietrich[17] shows this interaction with some indication of enhanced electron density occurring in the region in which bonding is anticipated, adding support to the initially proposed bonding model. The other feature which should be noted is the close proximity of the ethyl group to the adjacent tetrametric unit. This appears to result from interaction between the tetrameric units, adding to the stability of the solid, and

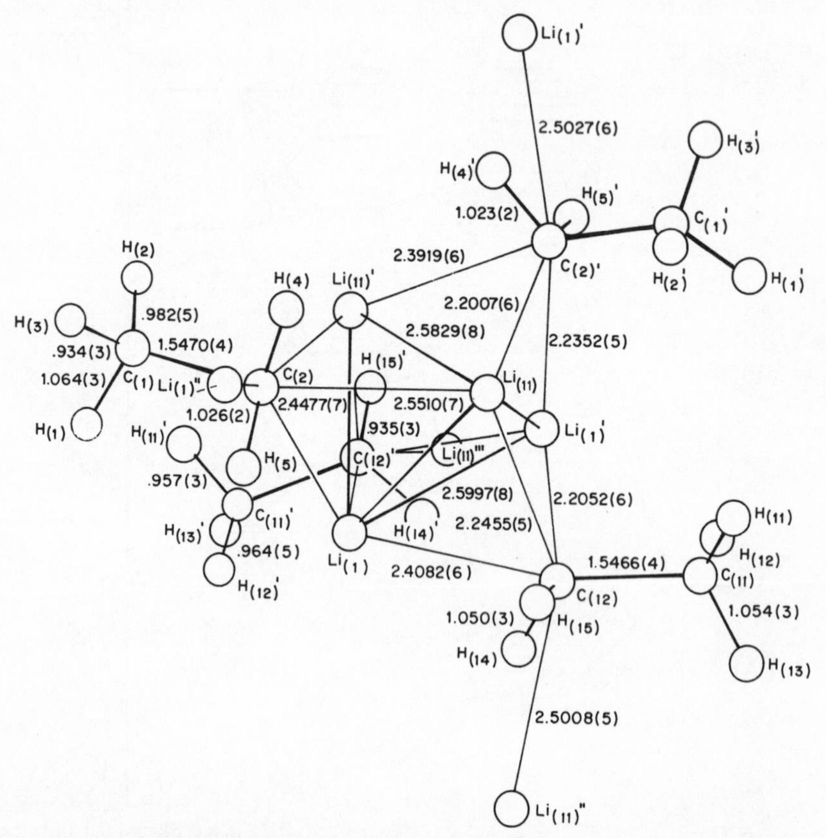


FIGURE 2. Low-temperature X-ray structure of $[LiEt]_4$. Reproduced with permission from *J. Organomet. Chem.*, **205,** 291 (1981).

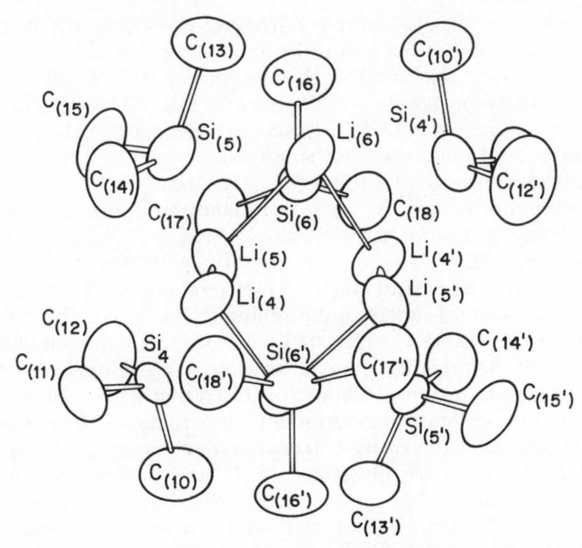


FIGURE 3. Perspective view of the $[LiSiMe_3]_6$ molecule. Reproduced with permission from *J. Am. Chem. Soc.*, **102,** 3769, (1980). Copyright 1980 American Chemical Society.

simultaneously leads to a far more complex bonding model when the secondary Li—C and/or Li—H bonding interations are taken into account.

These results should be compared with the structures observed for the hexameric systems which are illustrated by the structure of the $[LiSiMe_3]_6$ aggregate shown in Figure 3. There is no significant intermolecular interaction for the hexamers, with the major bonding interaction occurring between the $\alpha$-atom in the bridging moiety and the lithium framework, as indicated in Figure 4.

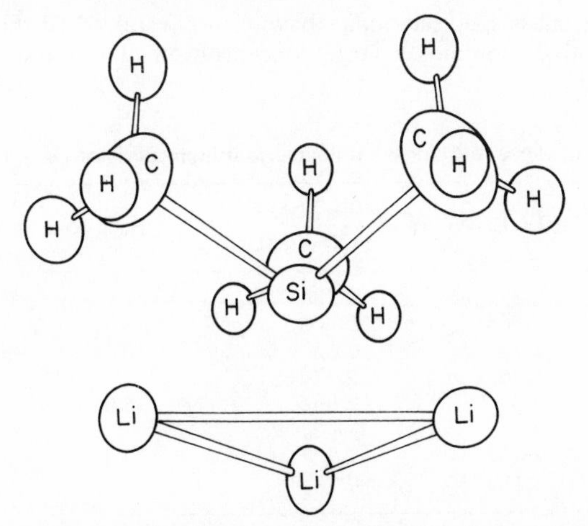


FIGURE 4. Partial view of the $[LiSiMe_3]_6$ molecule showing the location of the trimethylsilyl groups above the triangular face of the lithium atoms with calculated hydrogen positions included. Reproduced with permission from *J. Am. Chem. Soc.*, **102,** 3769 (1980). Copyright 1980 American Chemical Society.

An additional interaction between the β-hydrogen atoms of the bridging moiety and the lithium framework has been suggesteed by Zerger *et al.*[18] to account for the observed conformation of the cyclohexyl rings in $[Li\text{-cyclo-}C_6H_{11}]_6$ in a fashion similar to that originally proposed by Craubner[22]. This form of interaction seems unlikely in $[LiSiMe_3]_6$ where the Li–H distances are greater and the packing appears to result from steric repulsions between the methyl groups[20], and is completely ruled out in the recently determined structure of $[LiCH_2SiMe_3]_6$ hexamer, in which there are no β-hydrogen atoms, yet the basic atomic framework for both of these molecules is the same as that for the $[Li\text{-cyclo-}C_6H_{11}]_6$.

The relatively small differences in the overall geometry of the three known hexameric structures may be easily compared by examining the Li–Li, Li–C, or Li–Si distances and the seat-to-back angle in the lithium framework. These distances and angles are summarized in Table 1, which shows that there are only modest changes in geometry regardless of the bridging group or its geometry leaving the gross features of the structures unchanged. Further, it should be noted that the Li–H distances appear to be of only limited importance in determining the structures of these systems. This is shown from the fact that the overall structures have great similarity but marked difference in Li–H distances, as illustrated in Table 1, with little impact on the structure. It also might be noted that di-*t*-butylmethyleneaminolithium, $[LiN{=}CBu^t_2]_6$, forms a hexamer with a very similar structure in which the nitrogen atom serves as the bridging atom in the triangular lithium faces and in which neither α- nor β-hydrogen atoms are present[23].

The ability of these systems to retain their structure or at least to remain as aggregates in both solution and the gas phase has been clearly demonstrated. A number of molecular weight determinations have been carried out which show that the species typically have average molecular weights corresponding to tetramers or hexamers[9,24] in hydrocarbon solvents and have been extended to include both silyl-[19,20] and germyl-lithium[25] derivatives as well. With donor solvents, the degree of aggregation is reduced to four in the simple alkyllithium derivatives which have been studied[26]. Additional discussion of the interaction of donor molecules is given in Section II.B.

In the gas phase, mass spectral studies have clearly established that the vapour of alkyllithium derivatives contains a large concentration of tetramers and hexamers

TABLE 1. Selected distance and angles in hexmeric lithium derivatives

| Compound | Li–Li (Å) | Li–X(Å) (X = C, Si, N) | Back-to-seat angle (°) |
|---|---|---|---|
| $[Li\text{-cyclo-}C_6H_{11}]_6$ [a] | 2.397 | 2.184 | |
| | 2.968 | 2.300 | |
| $[LiSiMe_3]_6$ [b] | 2.73 | 2.65 | 70.5 |
| | 3.26 | 2.77 | |
| $[LiN{=}CBu^t_2]_6$ [c] | 2.35 | 2.06 | 85.0 |
| | 3.21 | | |
| $[LiCH_2SiMe_3]_6$ [d] | 2.45 | 2.20 | 79.5 |
| | 3.18 | 2.28 | |

[a] Ref. 18.
[b] Refs. 19 and 20.
[c] Ref. 23.
[d] Ref. 21.

which do not totally fragment on electron impact[27,28]. This suggests that these species have considerable thermodynamic stability towards dissociation.

## B. Theoretical Studies on Organolithium Aggregates

Having now provided a sound view of the observed structural features of these molecules, we should consider carefully the theoretical calculations which have appeared on the lithium aggregates to refine further our description of the bonding. These range from the relatively simple calculations initially performed by Weiss and Lucken[14] and others[29–32] which suggest that there is significant orbital overlap, to more recent calculations of a more sophisticated nature[33–36]. which in some cases have been interpreted as indicating no overlap with the bonding treated as resulting strictly from ionic interactions.

The simplest description provided for the tetrameric and hexameric units is that developed by Brown[9] in which the bonding interactions were suggested to occur through the overlap of three lithium orbitals and the bridging carbon orbitals as seen in **2**, giving rise to stable multi-centred molecular orbitals. Since these initial studies, a

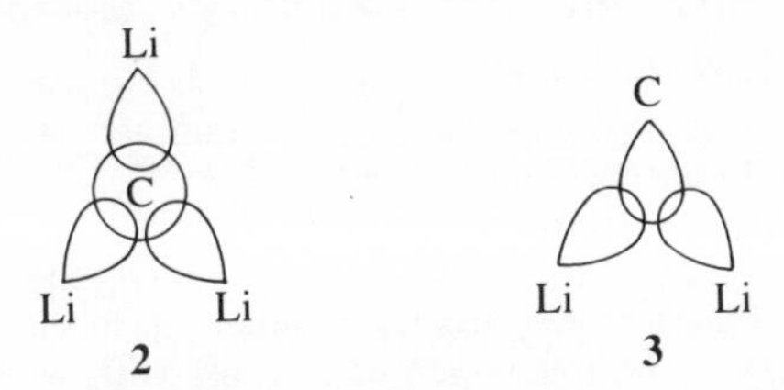


number of experimental and theoretical reports have appeared which suggest, modify, or attempt to refute this proposal. The most extreme view is that proposed by Streitwiser and co-workers[35,36] who suggested that the bonding in methyllithium monomer is totally ionic and extended this to suggest that bonding in all of the lithium aggregates is of this type. This also implies that there is no metal–metal interaction. Graham *et al.*[33], on the other hand, examined the structures of a variety of systems using different basis sets and concluded that there is significant covalent character associated with the bonding, on the basis of their calculations, and suggested that the bond has approximately 60% ionic character. In further studies, Graham *et al.*[34], using the PRDDO (partial retention of diatomic differential overlap) method with optimization of geometry, explored aggregates of methyllithium from $n = 1$ to 6 with two geometries for the higher aggregates, one based on cyclic species and the second on a closed form. Their results appear to be of particular interest since they treated a number of different aggregates in a consistent manner and, although they did not predict the observed geometry in all cases, they provided a reasonable model and suggested some interesting concepts which may aid the study of organolithium species.

The bonding model which they arrived at is essentially the type suggested earlier with two types of bonds: a three-centre bond of the form given in **3** and a four-centre bond similar to that in **2**. These are analogous to the bonding observed in $Al_2R_6$ and to the closed four-centre bonds observed in the boron hydrides, and seem to be reasonable models for the description of all of the aggregates observed. A further point should be made concerning the possibility of metal–metal interaction. These have both been proposed and excluded by numerous workers. Graham's *et al.* results[34] suggest these will be minimal, especially in the closed aggregates. These results again appear to be consistent with observed experimental results and particularly with spectroscopic studies on $[LiBu^t]_4$, which were interpreted as showing essentially no metal–metal interactions[37].

Several interesting suggestions have been made by Graham *et al.*[34]. One of these is that in the $D_{3d}$ hexamer, i.e. the distorted octahedral geometry which has now been observed in the three separate systems, may be described as two $D_{3h}$ trimer units held together by weaker interaction between the lithium and three-centred bridging carbon. This corresponds almost exactly to the suggestion made for the $[LiSiMe_3]_6$ species, which shows two relatively short Li—Si distances and one longer Li—Si distance[20]. The analogy does not follow as well for the two other hexamers, $[Li\text{-}cyclo\text{-}C_6H_{11}]_6$ [18] and $[LiCH_2SiMe_3]_6$ [21], which have all Li—C distances nearly equivalent.

The second suggestion which comes from these theoretical calculations is that the molecules are relatively floppy, with conversion from the closed aggregates to the open form requiring almost no energy. Simultaneously, the aggregates should have relatively high energies of dissociation to dimers or monomers.

## C. Exchange Phenomena Involving Alkyllithium Aggregates

The small differences in energy between different conformations of a given aggregate are entirely in keeping with a wide number of observed n.m.r. results, many of which have been previously reviewed[9–11,38,39], which show that both rapid inter- and intra-molecular exchange occur. Further, the small differences in aggregate energies, per LiR unit as $n$ varies, is completely consistent with recent n.m.r. results, which show that a much greater number of species exist in solution than previously considered, including larger aggregates with $n$ at least equal to 9[40–42]. They are not in full accord with some of the earlier n.m.r. studies which suggest that the mechanisms for exchange often involve dissociation[43,44] to monomers or dimers since, as previously noted, these dissociation processes appear to be highly unfavourable from the calculations.

Both the older n.m.r. studies and the more recent results show that a variety of mixed species and aggregates are formed. Further, they have been used to show that intermolecular exchange between aggregates usually occurs but that this may be slowed by lowering the temperatures or by use of bulkier groups. When the intermolecular exchange is slowed, then two classes of systems have been suggested based on n.m.r. results. One is said to involve slow intramolecular exchange on the n.m.r. time scale and leads to the postulate that the chemical shift of a given lithium nucleus is dependent only on the three nearest R groups. The second case assumes that very rapid intramolecular rearrangements occur, which lead both to averaged chemical shifts for a given aggregate and to averaged Li–C coupling constants where these are observed. There are data both from $^7Li$ and $^{13}C$ n.m.r. studies cited above and from the more recent $^6Li$ studies cited which support both cases, with differentiation between them arising as a result of steric interactions between the functional groups bound to lithium.

Several recent studies have appeared which show that $^6Li$ may be used very effectively to explore these systems in solution. The initial studies of Werhli[45–47] showed the possibility of using this technique and that the $^6Li$ quadrupole was not effective in causing relaxation; hence the $^6Li$ lines are sharp and give far better resolution than observed for $^7Li$. Fraenkel *et al.*[40,41] then made use of this technique to explore $LiPr^n$ in some detail, making use of the double labeled $^{13}C$–$^6Li$ systems. Their observations are of major importance since they have shown from these studies that more species occur in solution than previously thought. On the basis of the observed $^{6}Li$–$^{13}C$ coupling constants in each of the identifiable groups (both $^{6}Li$ and $^{13}C$), they concluded that there were five species present, namely hexamer, octamer and three different nonamers. The observation of five species has been independently confirmed[42] and extended to show that at least three species occur in $[LiBu^{n}]_n$, which have been proposed as hexamer, octamer, and nonamer. These studies have confirmed the formation of a number of mixed species previously proposed[9,10]. The results have also been interpreted as showing that the flexible nature of the molecules and their internal exchange are dependent on the alkyl substituents, with bulky groups stopping the rapid migration of groups within the aggregate in a fashion consistent with the 'local environment' postulate of Brown[9,10].

These results show two features which are of interest and need to be further explored, by both experimental and theoretical studies. These are the effect of the substituent in influence both intra- and inter-molecular exchange processes, and the ability, at least for some organolithium aggregates, to assume several different aggregation states differing by only small amounts of energy as shown by their temperature-dependent equilibria.

## D. Alkyllithium–Solvent Interactions

A point of major chemical interest is what occurs when simply alkyllithium aggregates interact with coordinating solvents. The importance of this type of interaction can be demonstrated very easily in two ways. Firstly, methyllithium and many of the aromatic lithium derivatives are insoluble in hydrocarbon solvents such as pentane or hexane, but soluble in ethers or amines. Secondly, in simple addition reactions, it has been shown that the stereochemistry changes with the solvent. For addition to a multiple bond in a hydrocarbon solvent, the process proceeds with *cis*

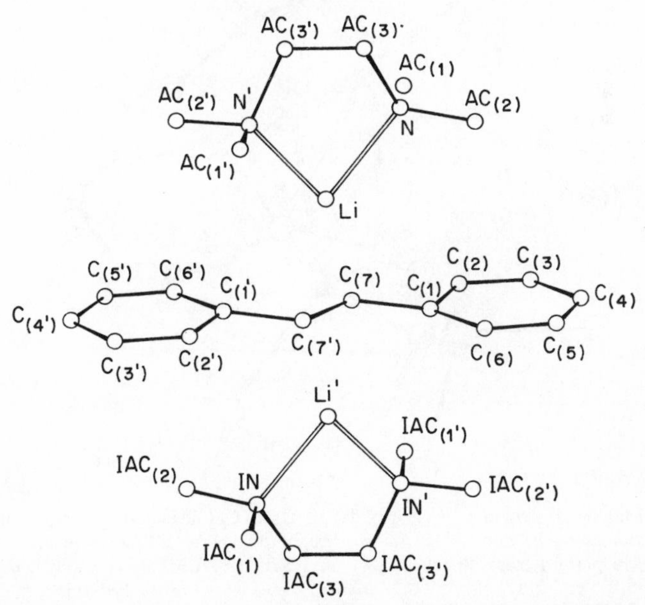


FIGURE 5. Molecular structure of stilbenebis(lithium tmeda). Reproduced with permission from *J. Am. Chem. Soc.*, **98,** 5531 (1976). Copyright 1976 American Chemical Society.

addition whereas in ethers the same reaction yields the *trans* product. In the earlier literature this was ascribed to the formation of 'reactive monomers' in the hydrocarbon which then added to the double bond in a 'concerted fashion'. In ethers the monomer units were reported to dissociate into ions with the reaction then accounted for in terms of the reaction of the simple 'carbanion' attacking the reactive substrate.

A large number of coordinated organolithium compounds have now been investigated by X-ray techniques. These studies have shown that for most aromatic systems, the description of ion-pair formation is reasonable. Stucky and co-workers[48–53] showed that this occurs and that the bonding in these systems can be described in terms of ion pairs. An example of one of these structures is shown in Figure 5. A similar situation also appears to be true for the sodium derivatives, but only one structure showing these features has been reported[54]. A structure containing two lithum atoms bound in a similar fashion has been reported by Lappert *et al.*[55] and is shown in Figure 6. This is of particular interest since it serves to some extent as a model for the dilithium species which have been investigated by theoretical calculations as described in Section II.F.

A bonding model for the simplest ion paired system, [LiCp], has been described by Alexandratos *et al.*[56], who concluded that the lithium atom was located over the centre of the ring in this system. When the organic moiety is less able to accept charge, however, other results are obtained. For the bicyclo[1.1.0]butan-1-yllithium–tmeda complex[57] and for the phenyllithium–tmeda complex[58], the species form stable

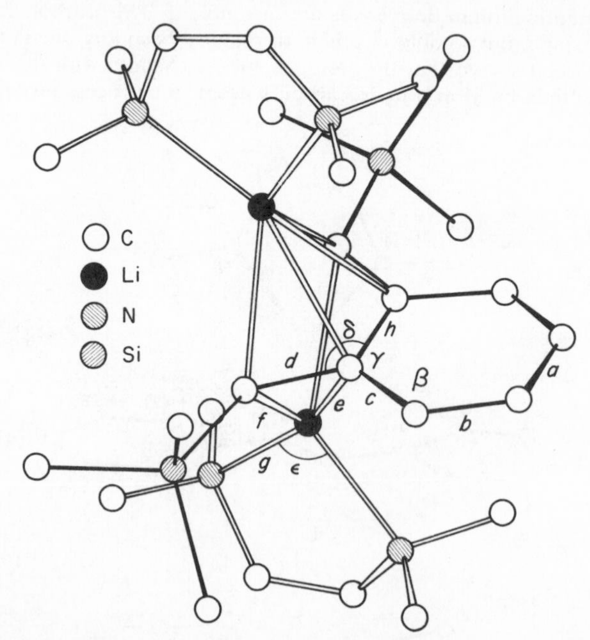


FIGURE 6. Molecular structure of [Li(tmeda)$_2${o-$C_6H_4(CHSiMe_3)_2$}]. Average values for the important distances (*a–h*) and angles (α–ε) designated are as follows (the e.s.d. of a single value is given in parentheses): distances (Å): *a*, 1.37(2); *b*, 1.36(2); *c*, 1.43(2); *d*, 1.42(1); *e*, 2.37(2); *f*, 2.38(2); g, 2.10(2); *h*, 1.45(1); angles(°): α, 119.6(11); β, 124.1(10); γ, 116.3(9); δ, 122.9(10); ε, 86.3(7). Reproduced with permission from *J. Chem. Soc., Chem. Commun.*, 14 (1982).

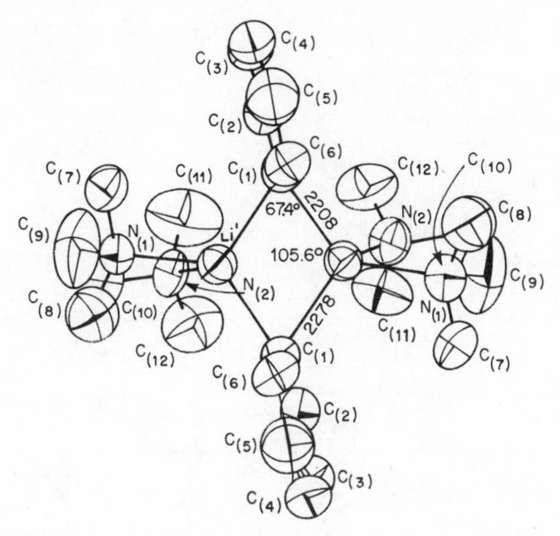


FIGURE 7. Structure of dimeric phenyllithium–tmeda complex. Reproduced with permission from *Chem. Ber.*, **111,** 3157 (1978).

dimers, with three-centre Li—C—Li bonds. The phenyl derivative is shown in Figure 7. For the simple alkyllithium amine complexes, all of the results so far obtained suggest that the aggregate size is reduced to 4 and that this framework then remains intact as indicated in $[LiMe]_4(tmeda)_2$, as shown in Figure 8, which consists of a tetramer in which bridging occurs between tetrameric units with tmeda serving as the bridge[59]. Dietrich and Rewicki[60] showed that a complex derivative internally coordinated to ether functional groups on the aryl substituent and to a bare oxygen atom located between two lithium tetrahedra can be formed[60].

A number of additional studies have been reported on species such as benzophenonedilithium adducts which have both oxygen and carbon bonded to the same lithium atom, which is also coordinated to base molecules[61], or to sulphur and carbon as in the 2-lithio-2-methyl- and 2-lithio-2-phenyl-1,3-dithiane derivatives reported by Amstutz *et al.*[62,63]. It has been shown that lithium can form a complex aggregate with ferrocene molecules which is stabilized by pentamethyldiethylenetriamine[64].

The only instance in which the aggregate has been shown to form a simple monomer without a highly delocalized aromatic ion present is in the silicon derivatives, $[LiSiMe_3]_2(tmeda)_3$, which is shown in Figure 9[65]. Preliminary data indicate that a few other simple organo and silyl derivatives may do this when sufficiently strong bases such as pmdeta are used, but no systematic studies have been reported to support this postulate.

The data obtained so far show that the description of the aggregates is difficult to predict in the solid state for alkyl derivatives and the limited solution studies which have appeared[66,67] seem to suggest stepwise addition to one of the higher aggregates, with these reverting to the tetrameric form when three or four base molecules/tetramer are present in solution.

An n.m.r. study[168] has suggested the formation of 1:1 and 2:1 tmeda–Li-sec-butyl which might be interpreted in terms of monomer or dimer formation, but no conclusive evidence has been provided and, until additional studies are carried out, the predominant evidence suggests that the alkyl lithium derivatives form solvated

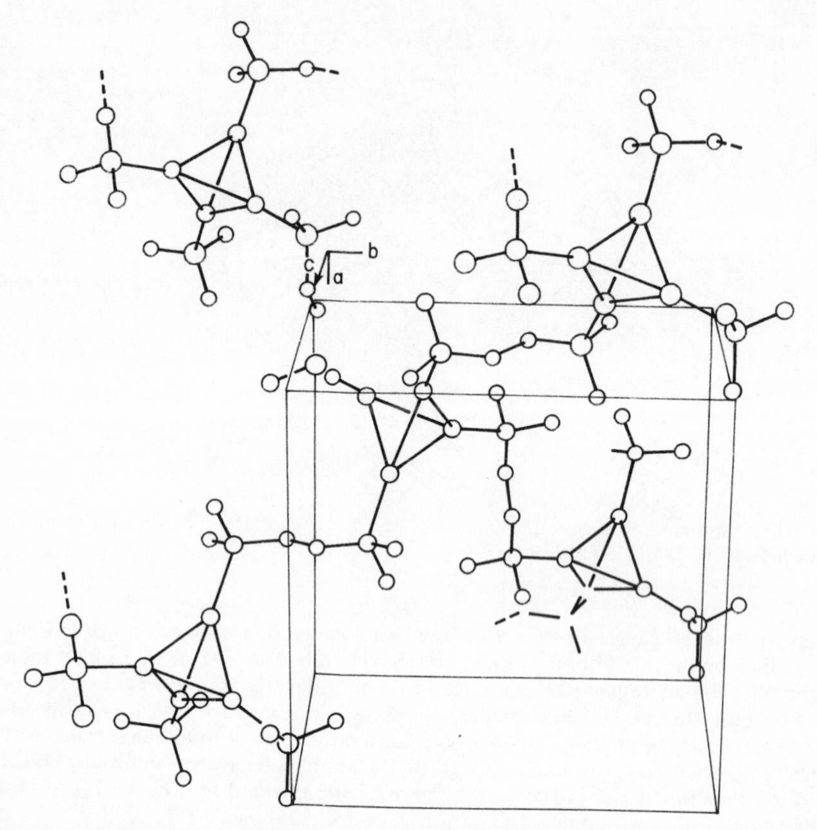


FIGURE 8. Structure of $[LiMe]_4(tmeda)_2$ showing the retention of the tetrameric unit. Reproduced with permission from *J. Organomet. Chem.*, **160,** 1 (1978).

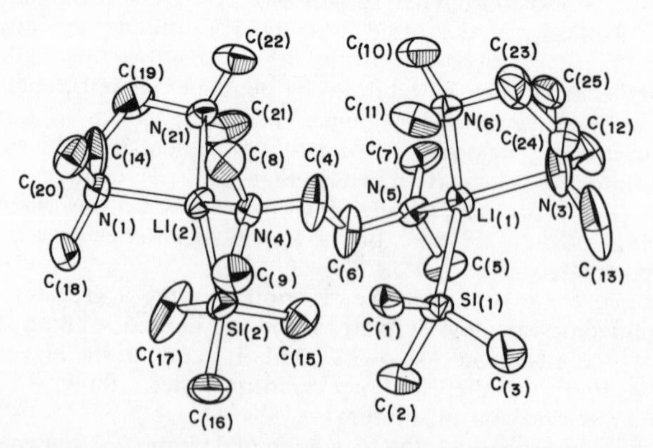


FIGURE 9. Diagram of the molecular unit of $[LiSiMe_3]_2(tmeda)_3$ with the atoms labelled. Positions 24 and 25 represent a disordered carbon atom, and in the refinement we assigned 50% occupancy factors. Reproduced with permission from *Organometallics*, **1,** 875 (1982). Copyright 1982 American Chemical Society.

tetramers while the aromatic species may form monomers or dimers in the presence of strongly coordinating molecules.

Recent attempts to quantify these interactions using CNDO/2 calculations have indicated that the stability of amine adducts increases in the order $NH_3 < NH_2Me < NHMe_2 < NMe_3$, but the asolute values obtained are clearly unreliable[69]. A recent report on the $^{13}C$ and $^{7}Li$ n.m.r. spectra of thf solutions of the halogenated derivatives, $LiCBr_3$ [70], and a more complex cyclopropylidene derivative[71], have been interpreted in terms of formation of solvated monomeric species as a result of the large Li–C coupling constant observed ($J_{^7Li-^{13}C} \approx 45$ Hz). Further, it was suggested that these derivatives may involve more complex equilibria, for example with dissociation of a bromine atom or with interaction between one of the bromine atoms and lithium as shown in **4** and **5**. Clark and Schleyer[72,73] have suggested

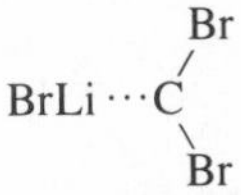

BrLi···C(Br)(Br)

**4**

Br···Li···C(Br)(Br)

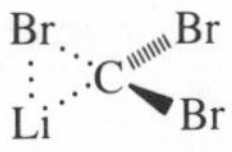

**5**

similar structures based on *ab initio* molecular orbital studies for halogenated organolithium compounds. These suggestions are intriguing but must await additional experimental support before they can be confirmed.

Most of the evidence suggests that aggregate formation greatly stabilizes the system and that dissociation of these aggregates require a significant amount of energy, a finding consistent with that of Graham *et al.*[34] but inconsistent with the proposal that the monomer units are the active species in solution. Before these proposals can be completely supported, however, additional studies must be carried out which will show better how the aggregates behave on interaction with coordinating solvents.

## E. Heavier Alkali Metal Derivatives

There have been far fewer studies carried out on the heavier Group I derivatives, but the limited X-ray studies which have appeared suggest that the heavier alkali metal derivatives form essentially ionic lattices. These include studies on both NaMe[74] and KMe[75], which crystallize in the NaCl lattice, and of RbMe and CsMe, which crystallize in the NiAS lattice type[76], and have also been extended to include $KSiH_3$ [77] and the alkali metal species $KSiMe_3$, $RbSiMe_3$, and $CsSiMe_3$ [78]. The result of these findings are consistent with the chemical and physical behaviour of these species. These observations include insolubility in hydrocarbon solvents and extreme reactivity towards species which can be attacked by a strongly basic or ionic species.

## F. Polylithium Derivatives

Priester *et al.*[79] and more recently Ship and Lagow[80] have described the preparation of a variety of polylithium organometallic derivatives. These species present a number of intriguing possibilities with regard to their structures and the bonding present in them. Unfortunately, these systems have so far defied quantitative experimental studies but have proved to be an interesting area for theoretical calculations.

Several groups have recently performed calculations on species such as $Li_2CH_2$, $Li_2C{=}CH_2$, $Li_2C_2H_4$, and $Li_2$biphenyl[81–87]. The results obtained suggested that unusual structures should be observed, with unique forms of bonding. For example, the *ab initio* calculations carried out on $Li_2C_2H_4$ suggest that there is only 1.9 kcal of energy difference between **6** and **7** and that the higher aggregates (from MNDO

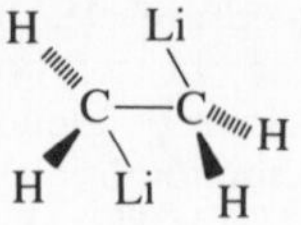

**6**  **7**

calculations) should be stable[82]. Similarly, they suggested that for the biphenyl derivative, **8**, the dibridged structure was 3.5 kcal $mol^{-1}$ more stable than the $\sigma$-

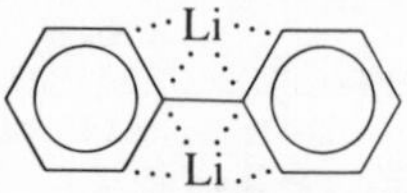

**8**

bonded species[83]. A number of calculations were carried out on other systems[84–87] which also indicated a variety of complex structures involving interaction between lithium atoms and several carbon centres. The crucial test of these reports will come when experimental evidence can be obtained on one or more systems which supports or indicates alterations in the proposed models.

## III. GROUP II DERIVATIVES

### A. Organo-beryllium and -Magnesium Compounds

The Group II derivatives behave in a significantly different fashion from their Group I counterparts as a result of having one less available vacant orbital. This leads to the formation of chains which are most readily described as metal centres tetrahedrally bound to ligands. These tetrahedra then share opposite edges to form infinite chains as shown in **9**. This chain structure has been clearly established for

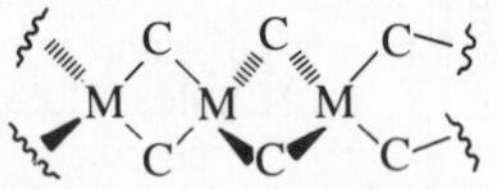

**9**

$BeMe_2$ [88], $MgMe_2$ [89], and $MgEt_2$ [90] in the solid state. The bonding in these derivatives is accounted for in terms of the simple model involving overlap of the metal orbitals with an orbital provided by the bridging group to yield two three-centred MOs holding the chains intact.

These chains may be broken down in one of two ways, by gas-phase dissociation to yield monomers, or by interaction with Lewis bases to give tetrahedral magnesium with one or two of the sites occupied by the donor atoms. The first of these is illustrated in the studies reported by Almenningen *et al.*[91,92] for dimethyl- and di-*t*-butylberyllium, which were shown to be linear monomeric molecules in the gas phase with Be—C bond distances of 1.698 Å. Infrared data also support these simple structures[93]. A similar gas-phase electron diffraction study has been carried out on di(neo-pentyl)magnesium[94], which has been found to be linear with an Mg—C bond distance of 2.126(6) Å. The molecule appears to be monomeric in solution as well[94a], and it seems probable that with sufficiently bulky substituents monomeric species may be observed in the solid state.

Studies of the interaction of bases with these systems have been carried on for some time and have provided a significant amount of very useful information. The simplest systems are those in which 2 mol of base are added, forming pseudo-tetrahedral four-coordinate metals, as seen in the quinuclidene adducts of dimethylberyllium[95] and -magnesium[96] or in the dimethylmagnesium–tmeda[97] and diphenylmagnesium–tmeda adducts[98]. In these systems, the C—M bonds are slightly longer than those observed for the monomer of beryllium or for the terminal groups in the magnesium chains which are broken down with base to small aggregate sizes. The aggregate size can be controlled by the amount of base added, as shown by studies on $MgEt_2$ with hmpa[99], on diethyl ether solutions of $MgEt_2$, as shown by the studies on their thermodynamic properties[100], or even better by the propynylberyllium species, which have been reported by two groups[101,102]. The results obtained suggest that the propynyl group is unique in its ability to bond to the metal, forming three-electron bridge bonds. Possibly a better way to describe this is a $\sigma$ bond to one metal and a $\pi$ bond to the second, as illustrated in **10**. This is discussed further in Section IV.

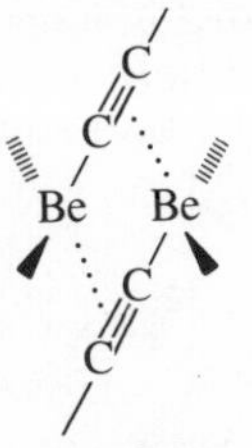

**10**

Several theoretical studies have appeared which treat the bonding in simple organo-beryllium and -magnesium compounds[32,103–106], and in the case of beryllium have been extended to include interactions with multiple bonds[107,108] and with methylene units[109]. All the calculations suggest that the bonding between the metals and the saturated alkyls involves primarily the $\sigma$-orbitals and can be described with a simple bonding model.

The calculations on the interaction of methylene[108] with beryllium indicate that the complex shown in **11** is stable with respect to the unbound element and $CH_2$ in

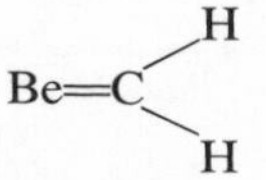

$$Be{=}CH_2$$

**11**

either the singlet or triplet state. The calculations on unsaturated systems suggest bonding between the beryllium and C═C or C≡C[107]. Calculations on cyclooctatetraeneberyllium[109] suggest that the beryllium should be displaced above the ring, which has a non-planar structure. The Saturn-like model, with the beryllium atom in the centre of the ring, did not represent a minimum. More discussion on bonding between beryllium and magnesium and unsaturated rings is presented in Section III.B.

N.m.r. studies have been reported which indicate that the polymeric $BeMe_2$ can be broken down into various sized chains by addition of controlled amounts of $SMe_2$ [110]. It is important to note this type of behaviour since it provides information concerning the relative strengths of the bridge bonds *vs.* the adduct bonds. In the Group II derivatives, it is apparent that the adduct bonds are always stronger than the bridge bonds. One may recall (Section II.D) that the opposite is often true for the lithium alkyls. A fascinating extension of this is to investigate systems which have the potential of internal coordination. This can be done by replacing one of the carbon atoms bound to the metal by a halogen, nitrogen, or an oxygen atom. The first is probably of most interest because of the wide use of Grignard reagents in organic synthesis. Much of the earlier literature suggests that these species are essentially solvent coordinated monomers of the form RMgX·2(solvent). Solid-state structural studies supports this formulation for some species such as $PhMgBr \cdot 2Et_2O$, which has a simple four-coordinate Mg atom as the central unit[111,112], or the similar structure observed for $EtMgBr \cdot 2Et_2O$[113] and $EtMgCl \cdot 2Et_2O$[114]. With stronger bases or different groups, either R or X, the structures become significantly more complex. For the MeMgBr·3thf, this distortion takes the form of a five-coordinate magnesium atom[115], while for $EtMgCl \cdot OEt_2$ it is suggested that the predominant species in solution is the chlorine-bridged dimer with an ether molecule bound to each magnesium atom[116]. In a subsequent paper[117], it was also claimed, based on i.r. and Raman data, that $MeMgI \cdot 2Et_2O$ and $MeMgBr \cdot 2Et_2O$ crystallize as two forms in the solid state, the predominant form being the disolvated monomer with tetracoordinate magnesium.

More complex species have been reported with halide[118] or amide[119] bridges. In these systems, the more negatively charged group always appears in the bridging site and leads to a dimer or high aggregate.

## B. Beryllocene and Its Derivatives

The chemistry of the cyclopentadienyl derivatives of beryllium has proved to be one of the more fascinating topics. These derivatives may bond either as $\eta^1$ or $\eta^5$ derivatives, with some evidence that other modes of bonding also are possible. For beryllium a wide number of studies, including both experimental and theoretical approaches, have been made. Almenningen *et al.*[120–122] carried out studies in the gas phase using electron diffraction techniques, and Schneider and Fisher[123] and Wong *et al.*[124] examined $[BeCp_2]$ in the solid state.

The initial structural reports in the gas phase and solid state differed significantly, as shown in **12** and **13**. The electron diffraction data best fit a model with $C_{5v}$

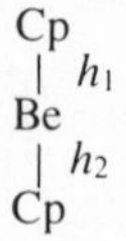

Cp
| $h_1$
Be
| $h_2$
Cp

$h_1 = 1.472$ Å; $h_2 = 1.903$ Å

**12**

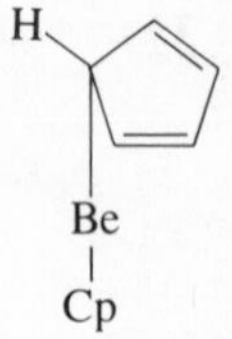

H
Be
Cp

**13**

symmetry with the beryllium atom offset from the centre of the molecule. The solid-state structure was described in terms of one $\eta^5$ ring with the second offset to the side in what was termed a 'slipped sandwich'[123,124]. These structural differences stimulated a number of theoretical calculations[125–131] in an effort to describe the system better. The results from these calculations all indicated that the $C_{5v}$ structure, suggested from the electron diffraction studies, was higher in energy than either a symmetrical structure with the beryllium atom located midway between the two rings or for a structure with one σ- and one π-bonded $C_5H_5$ ring. The latter was found to be the most stable from the calculations, and even this differs to some extent from the 'slipped sandwich' of the X-ray structure in that the two rings are not parallel.

As a result of these calculations, Almenningen *et al.*[132] redetermined the structure from electron diffraction data. They also coupled additional calculations with photoelectron spectroscopic studies to define the system better[133]. They again found that the $C_{5v}$ model fits the data well, but also found that a reasonable fit could be obtained with a 'slipped sandwich' and concluded that this model probably best describes the structure of beryllocene in the gas phase based on all available information. These results are consistent with Raman studies[134] in both the solid and liquid phases.

In contrast to the complexity and controversy surrounding the structure of beryllocene, which led to a number of papers, the magnesium compound, $[MgCp_2]$, has received only slight attention. The structure, both in the solid state[135] and in the gas phase[136], has been determined with the only significant differences being the observation of a staggered conformation in the solid state whereas the best fit for the gas-phase data is with the eclipsed form. Photoelectron spectra have been interpreted in terms of a bonding model which is largely ionic[137]; however, Haaland *et al.*[135] suggested that significant covalent interactions occur, based both on a structural study and on CNDO/2 calculations.

A number of studies[138–149] have been carried out on 'half-sandwich' derivatives of the form [BeXCp] and on the corresponding magnesium, [MgBrCp·1,2-tetramethylenediamine][149]. The results are less controversial than those obtained for the parent beryllocene with structure **14** being observed for all of the simple species. The

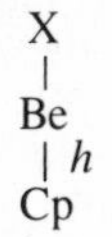

X
|
Be
| $h$
Cp

$h = 1.497$ Å (for $X = CH_3$)

**14**

bonding in these has been treated by several groups[150] and has been extended to include indenyl and fluorenyl complexes with the preferred bonding, suggested from the calculations, being $\eta^1$ in both cases rather than a π-complex[151].

In the $[BeCp(BH_4)]$ system, which has been extensively investigated in an effort to establish the precise structure, the calculations support the dihydrogen-bridged configuration shown in **15**, but the differences in energy between it and the tri-

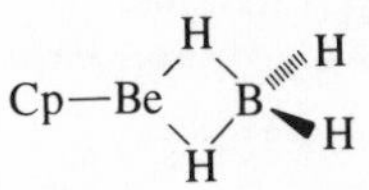

Cp—Be(μ-H)₂BH₂

**15**

hydrogen-bridged model are so small that one cannot rule out the latter model. The observation of the hydrogen bridge bonds leads into the area of metal hydrides and mixed metal hyrides.

A number of studies have been carried out on organohydrido derivatives, including species which contain both allyl and cyclopentadienyl organic groups. Structural

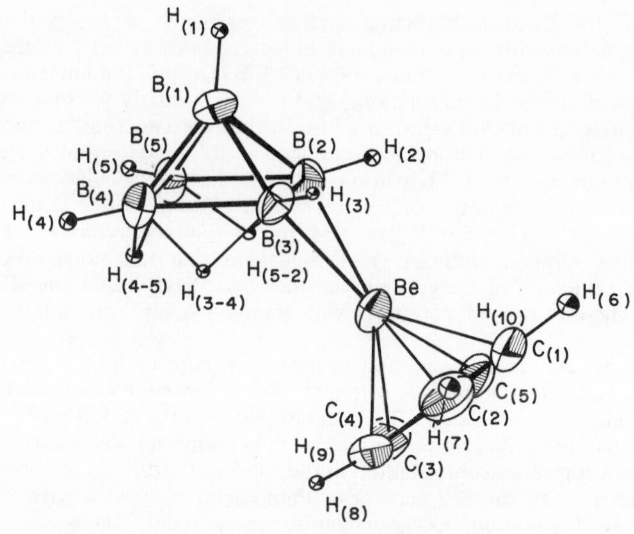


FIGURE 10. Molecular structure of $\mu$-[BeCp]$B_5H_8$. Reproduced with permission from *Inorg. Chem.*, **20,** 2185 (1981). Copyright 1981 American Chemical Society.

information has been reported for the simple borohydride, $Be(BH_4)_2$ [152], which shows that in this instance beryllium adopts the unusual (for beryllium) six-coordinate structure in the solid state. This is clearly an indication both of the ability of hydrogen to enter into bridge formation and of the small steric requirements for it. In the gas phase, electron diffraction studies have been reported which indicate a different structure with four-coordination about the beryllium atom[153]. These results have been interpreted in terms of a cyclic structure, B-$\mu$·H-Be-$\mu$-H-B($\mu$-H)$_2$, which may be thought of as the addition of a $BeH_2$ unit to two terminal hydrogen atoms on the $B_2H_6$ molecule. Several theoretical studies have appeared which deal with the bonding in $Be(BH_4)_2$ and for a variety of other simple hydride species[154–156], but these will not be discussed here. Gaines *et al.*[145] have shown that beryllium groups can be inserted directly into a borane cage with the structure shown in Figure 10, in which the beryllium atom is bound to the Cp ring in a fashion analogous to that observed in other Cp systems, while also being bound to $B_5H_8$ cage with Be—B bonds of *ca.* 2.05 Å and nearly equivalent bonding to all five Cp ring carbon atoms with Be—C distances of 1.84–1.89 Å, values which are comparable to those observed in other derivatives. Similar results were obtained by the same group in studies on the beryllaboranes, in which they showed that the beryllium atom can be incorporated into the cages of nidohexaborane(11) unit serving as a bridge between them, or can be incorporated into one boron cage and bound through hydrogen bridge bonds to a $BH_4$ unit[157–158].

## IV. GROUP III SPECIES

### A. Organoaluminium Derivatives

The organoaluminium compounds have been the subject of numerous investigations. Many of the results have been reviewed by Mole and Jeffrey[159], specific data on exchange[160,161], unsaturated derivatives[162], and structural work[4,5] being presented in other reviews. The alkyl derivatives provided the earliest and possibly the simplest system for the investigation of multi-centred bonds containing carbon. These investigations initially showed that trimethyl- and triethyl-aluminium consisted of equilibrium mixtures of monomers and dimers. Proposals concerning the nature of this bonding were put forward by Longuet-Higgins[163] and Lewis and Rundle[164], who established the solid-state structure from a classic X-ray investigation of the trimethylaluminium dimer. The structure is given in **16**. The details of this have been

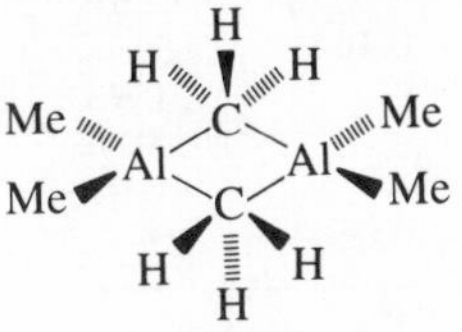

**16**

confirmed by more recent studies both in the solid state[165,166] and in the gas phase[167] and have been reviewed in detail elsewhere[4]. There have been suggestions that the bonding in this system is not best represented by the simple three-centred model, but that it involves a more complex pattern as shown in **7**[168]. The proposal, however, has been refuted on a variety of bases[169], including spectroscopic studies, which provide the most direct evidence against the involvement of the hydrogen atoms in bridge-bond formation[170,171]. The preferred model, therefore, is **18**, which shows the major

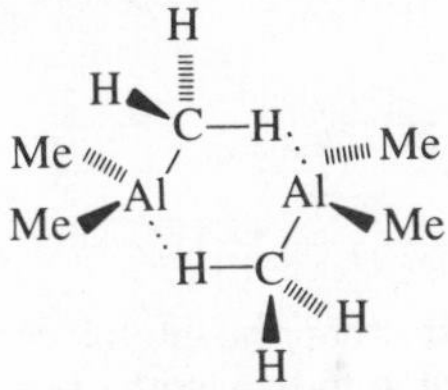

**17** 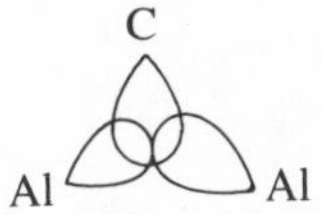 **18**

portion of the bonding resulting from the interaction of the carbon orbital with available orbitals on the two aluminium atoms. It is not clear from this model whether metal–metal interactions are involved. This has been proposed by Levison and Perkins[172–174] but the studies which have been carried out are insufficient to assess fully the extent, if any, of the interaction.

It should be noted that the strength of the bridge bond is of major importance both for the determination of the structure or aggregation state and, therefore, in the considerations of the reactivity of the compounds because the nature of the reactions is often dependent on the former[175,176]. Several studies have established that the degree of association is dependent on the chain length, with the longer alkyl derivatives giving rise to less association. A measure of this is shown in Table 2. If one extends these studies to include the alkyls of other Group III derivatives, it becomes immediately clear that only aluminium forms stable bridged dimers with simple alkyl groups in the electron-deficient site. The structure of $GaMe_3$ is unknown in the solid state and $InMe_3$ has been shown to form a weakly associated tetramer[177]. In solution both are monomeric[178] and in the gas phase they have been shown to have planar structures as shown in **19**[179,180]. A similar structure has been observed in the solid state for $In(CH_2SiMe_3)_3$ [181] and both triphenyl-indium and -gallium are

TABLE 2. Colligative properties and heats of dissociation for selected organo-aluminium compounds

| Compound | $\beta^a$ | $\bar{n}^b$ | $\Delta H_d$ (kcal mol$^{-1}$) |
|---|---|---|---|
| $Me_2Al_6$ [c,d] | 0.00029 | 1.97[e] | 19.40 |
| $Et_2Al_6$ [d] | 0.00501 | — | 16.94 |
| $(Pr^n)_6Al_2$ [f] | 0.0253 | — | 15.40 |
| $(Bu^n)_6Al_2$ [f] | 0.0376 | 1.96[e] | 15.02 |
| $(Bu^i)_6Al_2$ [f] | 0.164[g] | — | 8.1 |
| $[H_2C{=}CH(CH_2)_2]_3Al$ [e] | — | 0.98 | — |
| $[Me(CH_2)_4]_6Al_2$ [e] | — | 1.80 | — |
| $[H_2C{=}CH(CH_2)_3]_3Al$ [e] | — | 0.99 | — |
| $[Me(CH_2)_7]_6Al_2$ [f] | 0.0501 | — | 14.68 |

[a] The fraction dissociated at 20 °C in a 0.05 M solution in cyclohexane.
[b] Average degree of aggregation in freezing cyclohexane.
[c] Ref. 176.
[d] M. B. Smith, *J. Organomet. Chem.*, **70**, 13 (1974).
[e] T. W. Dolzine and J. P. Oliver, *J. Am. Chem. Soc.*, **96**, 1737 (1974).
[f] Ref. 175.
[g] 40 °C.

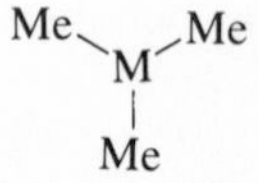

Me  Me
M
Me

M=Ga, In

**19**

monomeric in the solid state[182]. This suggests that the metal centre is simply not capable of becoming involved in this type of bonding with these weakly bridging units.

If we examine other organic groups, including cyclopropyl, vinyl, phenyl (when bound to aluminium), or ethynyl, we observe that bridge bonds appear to be more stable both kinetically and thermodynamically than those formed with alkyl groups. This has been clearly shown in a number of investigations, including n.m.r. studies which established that cyclopropyl[183–186], vinyl[187–190], phenyl[191,192], and phenylethynyl[193–195] groups are preferntially located in the bridging site when an equilibrium is established in the mixed systems $Me_3M$–$MR_3$.

The initial explanation put forward for this enhanced stability was that the $\pi$-electrons in the vinyl group were able to interact with the non-bonding orbital of the metals as indicated in **20**. This interaction, in addition to that involving the normal three-centred bonding, adds to the stability of the bridged system, leading to the formation of moderately stable $Ga_2(CH{=}CH_2)_6$ dimers[187,188]. A similar explanation was put forward to support the stability of the vinyl-bridged aluminium species with the structure of one of these derivatives now established from single-crystal X-ray studies as shown in Figure 11[196]. This enhanced stability is insufficient to keep the vinyl gallium from dissociating to the monomeric form in the gas phase. The structure, determined by electron diffraction, shows that it exists as a simple monomeric species with Ga—C bonds of 1.963(3) Å and no evidence of back-donation from the vinyl group into the vacant metal orbital[197]. No solid-state

**20**

structure has been determined to show if the dimer is present in the solid state. Similar studies have established that aryl groups typically form bridges with the same type of geometry, i.e. with the bridging ring perpendicular to the Al—Al axis[198,199]. For the heavier metals these groups, however, do not appear to be sufficiently good bridging units to result in dimer formation in the solid state. This is shown by the solid-state structures of triphenyl-gallium and -indium[182].

The other organic group which appears to stabilize bridge bonds by this type of interaction is the cyclopropyl moiety. This has been shown by n.m.r. studies to be preferentially located in the bridging site with preferred conformation *syn* at low temperatures[183,185,186], the same as observed in the solid state[6,7,184,200].

Simple CNDO calculations have been performed which suggest that the barrier to rotation is *ca.* 11 kcal $mol^{-1}$ for the bridging unit, assuming no bond dissociation[185]. These observations are in complete accord with variable temperature n.m.r. studies which yield on an activation energy for this process of 10.8 kcal/mol and are best interpreted in terms of a non-dissociative rotational process. Other studies[201] had

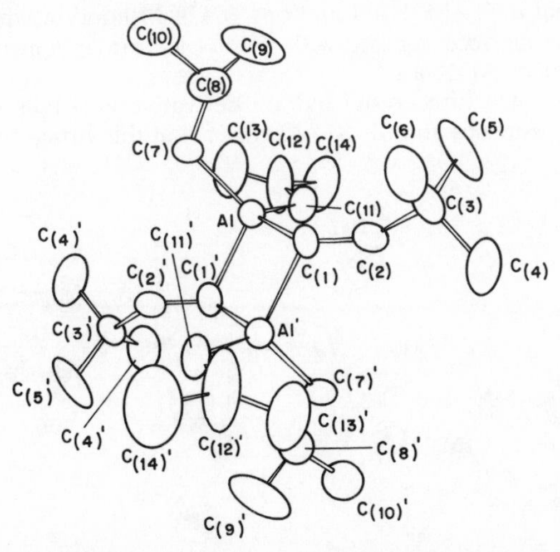


FIGURE 11. Structure of di($\mu$-*trans*-but-1-en-1-yl)(tetraisobutyl)dialuminium. Reproduced with permission from *J. Am. Chem. Soc.*, **98,** 3995 (1976). Copyright 1976 American Chemical Society.

suggested some alternative explanations for the behaviour of these systems in solution, but appear to be inconsistent both with the simple theoretical treatment and with the experimental data now available[185,186].

The heavier Group III derivatives, gallium and indium, are also bridged by cyclopropyl groups but these are bound less strongly, as indicated by the high degree of dissociation observed in solution[186]. No structural studies have been reported which will permit an assessment of the bond in the solid state or in the vapour phase.

We should now turn our attention to some of the other groups which have been shown to lead to stable or relatively stable bridged or chain structures in the solid state and to aggregate formation in solution. These are ethynyl, cyclopentadienyl, and most recently benzyl moieties. Ethynyl groups have now been shown to form stable dimeric derivatives with aluminium[202,203], gallium[204], and indium[205], all with approximately the same structures. Further, this structure appears to be maintained both in solution and in the gas phase, where electron diffraction studies[206] have shown almost identical parameters with those observed in the solid state for the aluminium derivatives. The structure for the gallium derivatives is given in Figure 12. The structures observed for aluminium and gallium derivatives both in the solid state and gas phase are identical in their important features with most features observed as normal. The exception occurs in the bridge bond, where two distinct M—C distances are observed, and the orientation of the bridging ethynyl unit precludes the type of *p*-orbital participation which was found for the vinyl, cyclopropyl, and aryl cases. It does, however, suggest that the bonding in these systems may be described as in **21**,

**21**

with one 'normal $\sigma$ C—M' bond and one $\pi$ C—M bond in which the vacant metal orbital serves as an acceptor site with the $\pi$-electron system of the ethynyl group serving as the electron donor.

The gas-phase structure of the indium derivative also falls into this group with different In—C distances and M—C—C angles, but this structure is distorted towards

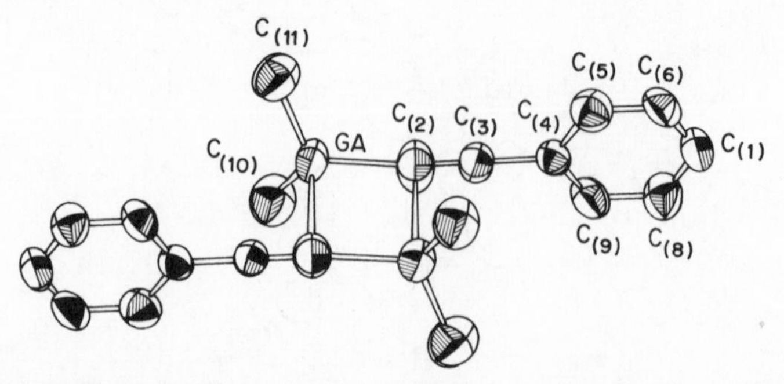


FIGURE 12. Molecular structure of $Ga_2(\mu\text{-}C{\equiv}CPh)_2Me_4$ with the atoms labelled. Reproduced with permission from *Inorg. Chem.*, **20,** 2335 (1981). Copyright 1981 American Chemical Society.

TABLE 3. Selected interatomic distances (Å) and angles (°) for ethynyl-bridged metal derivatives

| Compound | Bridging distance Short M | Long M | C≡C bridging) | M—C terminal) | ∠C≡C—C | ∠M—C—M (bridge) | M—C≡C | ∠M′—C≡C | ∠C—M—C terminal) |
|---|---|---|---|---|---|---|---|---|---|
| $Al_2(\mu$-C≡CMe$)_2Me_4$ [a] (gas phase) | 2.050 (15) | 2.15 (3) | 1.229 (4) | 1.956 (5) | 167.8 (1.6) | 92.0 (1) | 158.3 (1.9) | 108.7 (1.3) | 120.8 |
| $Al_2(\mu$-C≡CPh$)_2Ph_4$ [b] | 1.992 | 2.184 | 1.207 | 1.904 | | 91.73 | 171.6 | | |
| $Ga_2(\mu$-C≡CMe$)_2Me_4$ [c] | 2.004 (7) | 2.375 (7) | 1.183 (6) | 1.952 (8) | 180 (2) | 86.7 (3) | 172.8 (7) | 93.8 (5) | 126.86 |
| $Ga_2(\mu$-C≡CMe$)_2Me_4$ [d] (gas phase) | 2.02 (2) | 2.24 (3) | 1.22 (1) | 1.964 (6) | 174 (3) | | 169 (2) | | 120 (2) |
| $In_2(\mu$-C≡CMe$)_2Me_4$ [e] | 2.193 (14) | 2.933 (23) | 1.212 (20) | 2.185 (av) | 178.6 (1.3) | | 177 (1) | | 129.8 (4) |
| $In_2(\mu$-C≡CMe$)_2Me_4$ [d] | 2.19 (3) | 2.52 (4) | 1.23 (2) | 2.18 (2) | 170 (9) | | 156 (6) | | 123 (4) |
| $Be_2(\mu$-C≡CPh$)_2Me_2$ [f] | 1.85 | 1.89 | 1.17 | 1.75 | 178 | 77 | 147 | 1.36 | |
| $Cu_6(C_6H_4NMe_2$-2$)_4$ (C≡C$C_6H_4$Me-4$)_2$ [g] | 2.028 | 2.054 | 1.17 | | 177.9 | 75 | 148.1 | 137.2 | |
| HC≡CH [h] | | | 1.204 (2) | | | | | | |

[a] Ref. 206.
[b] Ref. 202.
[c] Ref. 204.
[d] Ref. 203.
[e] Ref. 205. Note that the In—C distance is 2.989 (24) Å, i.e. nearly equivalent to the In—$C_\alpha$ distances for the solid-state structure.
[f] Ref. 101.
[g] R. W. M. ten Hoedt, J. G. Noltes, G. van Koten, and A. Spek, *J. Chem. Soc., Dalton Trans.*, 1800 (1978).
[h] O. Kennard, D. G. Watson, F. H. Allen, N. W. Issacs, W. D. S. Motherwell, R. C. Petterson, and W. G. Gown (Eds.), *Molecular Structure and Dimensions*, Vol. Al, N.V.A. Ooshock, Utrecht, 1972, p. 52.

the three-centred electron-deficient bridge configuration[202]. In the solid state, additional interactions must be taking place which lead to the near equivalence of the In—$C_\alpha$ and In—$C_\beta$ distances and to the stacking of the molecules suggesting extended interactions through the crystal lattice. Several of these distances are summarized in Table 3, with comparisons with some other ethynyl metal compounds.

There are still a number of unresolved questions concerning the ethynyl-bridged species. Specifically, no-one has dealt with the bonding in a quantitative manner, and no studies have appeared which establish in a quantitative manner the stability of the bridge bond. Further, it seems plausible, especially in view of the dynamic properties of the cyclopropyl aluminium systems, that the bridging ethynyl groups may undergo some form of reorganization in solution. No suitable experiments to test this have been reported.

Turning now to more complex systems, i.e. those which contain the cyclopentadienyl moiety, we again find unusual behaviour. The initial studies on these derivatives were thought to show that association occurred with the postulated bridging taking place through the methyl substituent in [$AlCpMe_2$]. It has since been shown for aluminium[207], gallium[208], and indium[209] cyclopentadienyl derivatives that the mode of association in the solid state is through the Cp ring as indicated in Figure 13. In this system there are two distinct Al—C bonds between the bridging cyclopropyl group and the metal. These interactions are between $Al_{(1)}$ and $C_{(1)}$ and between $Al_{(1)'}$ and $C_{(3)}$. Thus, it is not a normal two-electron three-centred bond as found in trimethylaluminium, but a unique type of system involving the ring. A comparison of the atomic parameters for these systems is given in Table 4. The short metal—carbon bond distances are not significantly greater than those observed for metal—carbon—metal bridge bonds, at least in the aluminium system where there are good data. The observed distances are reasonable for the gallium and indium species compared with the metal—carbon distances observed in other derivatives. Another important comparison should be made with the gas-phase systems in which it has been suggested that the bonding is principally between the aluminium atom and $C_{(1)}$ and $C_{(2)}$ of the ring[210,211]. The Al—C distances here are 2.21 Å, again very near the value observed in bridged systems. It should be further noted that in these systems the metal (or ring) undergoes rapid migration so that all of the carbons interact with the metal as a function of time. This has been particularly well discussed by Haaland[5].

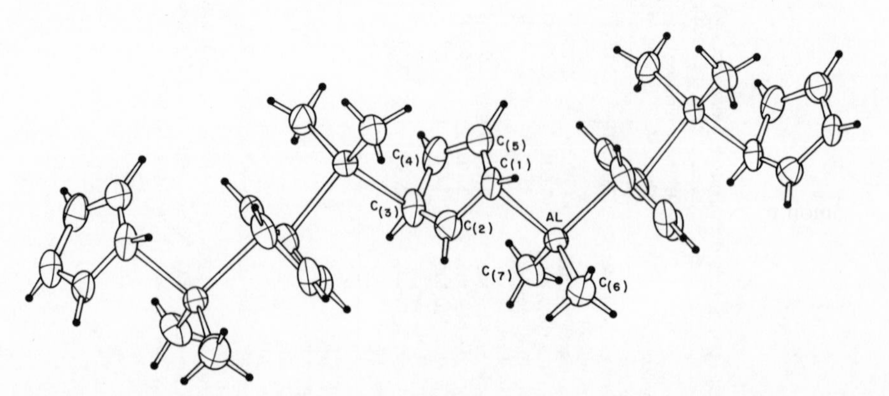


FIGURE 13. Perspective view of a portion of the $Al(\mu\text{-}C_3H_5)(CH_3)_2$ chain. Reproduced with permission from *Inorg. Chem.*, **21,** 458 (1982). Copyright 1982 American Chemical Society.

TABLE 4. Selected metal—carbon bond distances (Å) in Group III cyclopentadienyl-metal derivatives

| Compound | Metal—bridging carbon distances Ring A | Ring A′ | Other M—C distances |
|---|---|---|---|
| $[Al(\mu\text{-}C_5H_5)Me_2]$[a] | 2.203 (2) | 2.248 (2) | 1.959 (2) |
| | | | 1.947 (2) |
| $[Ga(\mu\text{-}C_5H_5)Me_2]$[b] | 2.215 (2) | 2.314 (2) | 1.972 (2) |
| | | | 1.962 (1) |
| $[In(\mu\text{-}C_5H_5)(Cp)_2]$[c] | 2.374 (7) | 2.466 (8) | |
| Ring B | | | 2.243 (9) |
| Ring C | | | 2.237 (9) |
| $[Al(Cp)Me_2]$[d] | $Al—C_{(1)} = Al—C_{(2)} = 2.21$ (2) | | |
| $[AlCl(C_5Me_5)(Me)]_2$[e] | | | |
| | $Al—C_{(1)}$ | | 2.100 (3) |
| | $Al—C_{(2)}$ | | 2.252 (3) |
| | $Al—C_{(3)}$ | | 2.282 (3) |
| $[AlCl(C_5Me_5)(Bu^i)]_2$[e] | $Al—C_{(1)}$ | | 2.096 (9) |
| | $Al—C_{(2)}$ | | 2.359 (8) |
| | $Al—C_{(3)}$ | | 2.243 (8) |

[a] Ref. 207.
[b] Ref. 208.
[c] Refs. 209 and 211; monomer, gas-phase $Al(C_5H_5)Me_2$, all C—C distances were assumed to be equivalent in the ring.
[e] Refs. 212 and 213.

A different modification of cyclopentadienylaluminium derivatives has been reported by Paine and co-workers[212,213a], who have shown that in the complexes $[AlCl(CpMe_5)(Me)]_2$ and $[AlClCp(CpMe_5)(Bu^t)]_2$, the pentamethylcyclopentadienyl rings occupy terminal positions. The bonding between the ring and the metal is described as trihapto with relatively short Al—C distances as seen in Table 4 comparable to the Al—C distances in simple bridged systems. Crude molecular orbital calculations were carried out on these systems which suggested significant $\sigma$-bonding contributions for the terminal $C_5$—Al interactions and, with the $\pi$-contributions added to this, led to the observed structures.

The final group of simple aluminium compounds which should be considered are those which contain allylic or benzylic functional groups. Only a few reports have appeared which are concerned with the allylaluminium species[214–217]. These reports show that there is an interaction of some type in solution and that there are temperature-dependent dynamic processes which occur. None of the derivatives have been explored sufficiently to allow definitive conclusions to be drawn about their structures or bonding, thus leaving this area open for speculation. The two most common proposals are that the aluminium systems take on the form indicated in **22**

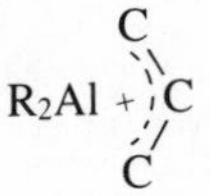

$R_2Al$ + C(—C—C) (allyl, dashed bonds)

**22**

with equivalent terminal positions or that they are $\sigma$-bonded with rapid reorientation. Benzylaluminium derivatives are now better documented. They again have been described as forming simple organoaluminium compounds, and from the limited n.m.r. studies it was suggested that the species is monomeric in solution[218]. There is no reason to refute this but in the solid state it has now been shown that the unusual chain structure shown in **23** is formed[219]. The bonding has not been dealt with

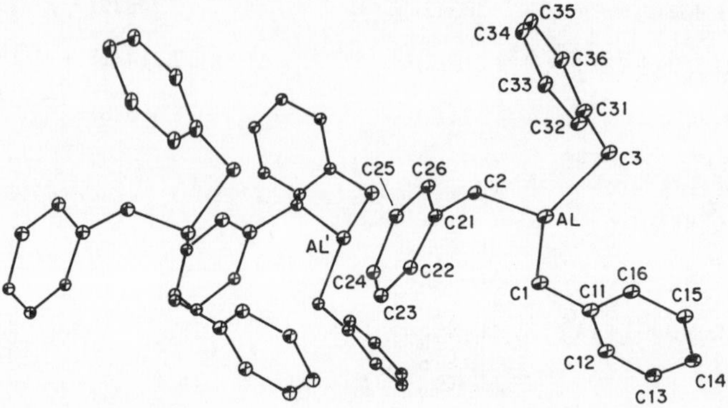

[Reproduced with permission from *Organometallics*, **1,** 881 (1982). Copyright 1982 American Chemical Society.]

quantitatively, but it is clear from the structural studies that the chain is formed by a strong interaction between one of the *ortho*-carbon atoms of a phenyl group and the aluminium atom in the adjacent $Al(CH_2Ph)_3$ unit. Further experimental and theoretical studies are clearly called for to provide some basis for understanding both in allylic and in the benzylic systems.

## V. COMPLEXES OF GROUP II, GROUP III, AND RELATED ORGANOMETALLIC DERIVATIVES

### A. Lithium and Group I Metallates

As indicated in Sections III and IV, the simple $MR_2$ and $MR_3$ derivatives such as $BeR_2$ and $AlR_3$ serve readily as acceptors for bases. This is especially true when the derivatives are reacted with $[LiR]_n$ in which the '$R^-$' unit serves as the basic moiety forming the well established 'ate' complexes. The beryllium derivative $Li_2BeMe_4$ has been characterized by X-ray studies[220], as have also the complex $[\{Li(tmeda)\}_2(MgMe_4)]$, which is shown in Figure 14[221], and the related complex $[(Li(tmeda)]_2[Ph_2MgPh_2MgPh_2]$[222]. A variety of Group I–Group III complexes also

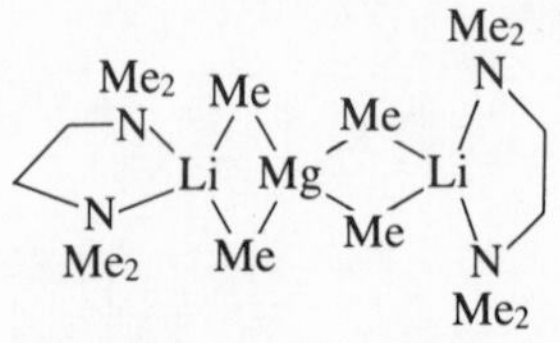

FIGURE 14. Structure of $[\{Li(tmeda)\}_2(MgMe_4)]$. Reproduced with permission from *Chem. Ber.*, **114,** 209 (1981).

have been characterized, including $LiBMe_4$ [223], the aluminium derivative $LiAlEt_4$ [224], and several other derivatives with both heavier Group I and Group III metals[225,226]. It is of interest to examine the cation–anion interaction in these systems. In $Li_2BeMe_4$, the beryllium atom is surrounded by a distorted tetrahedron of carbon atoms which are clearly bound to it. The Li—C distances are longer, but with the shortest Li—C distance of 2.52 Å there is a strong indication of Li—C interaction. In the $[\{Li(tmeda)\}_2(MgMe_4)]$ complex shown in Figure 14, the interactions yield Li—C distances of only 2.3 Å which are only slightly greater than those observed in the alkyllithium aggregates. In the Group I–Group III derivatives, this is even more pronounced for the boron derivatives, as shown in Figure 15 with essentially a linear Li—C—B system with very strong Li—C bonding suggested from the short (2.12 Å) Li—C distance. In the $LiAlEt_4$ derivative interaction is less pronounced but it still influences the structure, causing distortion around the aluminium atom and a short Li—C distance of 2.3 Å. When larger cations such as sodium or rubidium or larger central metals such as indium are used, the species appear to be best described in terms of ionic lattices[227].

In addition to the crystallographic studies described above, a number of infrared[228] and n.m.r. studies[229,230] have been carried out which support this general description and the trends suggested. These studies further show that in solution relatively strong ion pairing occurs, which can be reduced by making use of strongly coordinating solvents to surround the cation. One also may extend the ability of Group I derivatives to coordinate with other organometallic compounds such as the derivatives of zinc, cadmium, and mercury. There is less information available but it has been shown that $Li_2[ZnMe_4]$[231] has nearly tetrahedral zinc with the shortest Li—C distances being 2.52 Å, and that both $K_2[Zn(C{\equiv}CH)_4]$ and $K_2[Cd(C{\equiv}CH)_4]$[232] have

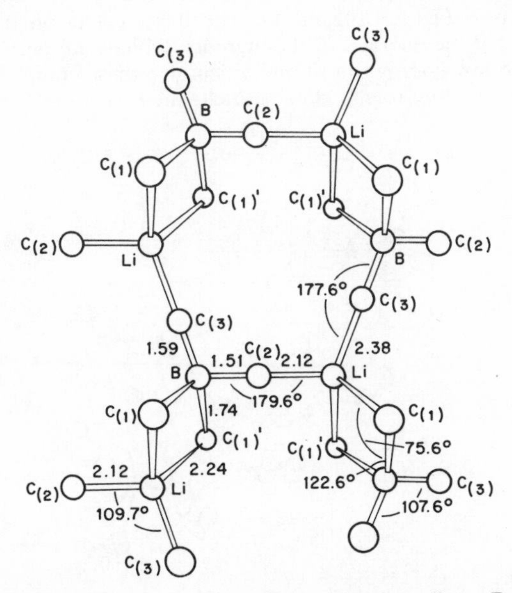


FIGURE 15. Molecular structure of $LiBMe_4$ showing the short, linear B—C—Li interaction. Reproduced with permission from *J. Am. Chem. Soc.*, **93,** 1553 (1971). Copyright 1971 American Chemical Society.

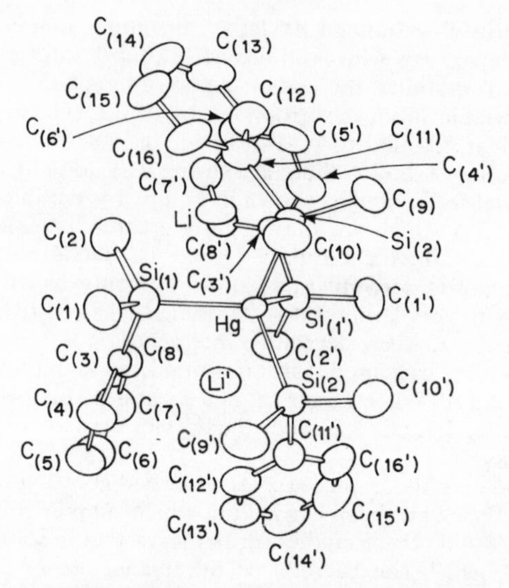


FIGURE 16. Structure of $Li_2 \cdot Hg(SiMe_2Ph)_4]$ showing encapsulation of the lithium atom. Reproduced with permission form *Inorg. Chem.*, **19,** 3577 (1980). Copyright 1980 American Chemical Society.

essentially tetrahedral zinc and cadmium with the potassium ions bound through electrostatic interactions. Complexes with silylmercury derivatives have been well documented with stable three- and four-coordinate[233,235,236] complexes being observed, as shown in Figures 16 and 17. The limited data on the three coordinate species suggest that the energies of the mercury orbitals are perturbed sufficiently to give rise to some low-energy transitions which give these complexes their distinctive colours[236]. Further, the n.m.r. data, which show a simple progression of the

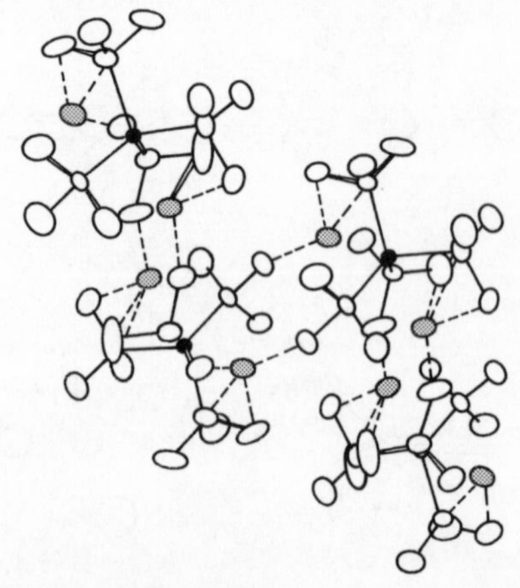

FIGURE 17. Structure of $Li_2[Hg(SiMe_3)_4]$ showing Li—C interactions between $[Hg(SiMe_3)_4]^{2-}$ units. Reproduced with permission from *Inorg. Chem.,,* **19,** 3577 (1980). Copyright 1980 American Chemical Society.

$^{199}Hg$-$^{1}H$ coupling with structure, suggests that the mercury *s*-orbital contribution to bonding decreases from 50 to 33 to 25% in going from the linear *sp* to tetrahedral $sp^3$ bound systems, with typical coupling constants of 40, 25, and 18 Hz, respectively[237]. Spectral data have also been reported for Hg—C complexes, but stable crystalline species have not been characterized[238].

### B. Group II–Group III Complexes

The Group II–Group III complexes have received little attention, but from these studies it has been readily established that these species form complexes as indicated in **24**[239]. In this derivative, the electron-deficient bridge may be described in the

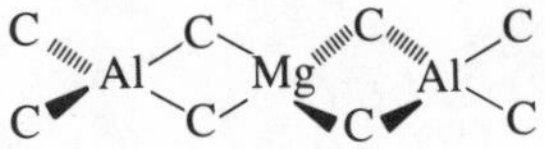

**24**

same way as for the simple aluminium dimers or $MgR_2$ chains, but with asymmetry introduced as a result of the differences in the two metal centres. Other systems of this type surely exist but have not been characterized.

### C. Mixed Metal Systems Contains Transition Elements

The Group I–Group III organometallic derivatives have the capacity to enter into bridging with a variety of other organometallic species to form unsymmetrically bridged species. These include the mixed copper—, silver—, or gold—lithium complexes[241–244]. This could further be extended to the copper, silver, and gold derivatives[245–247], which behave in a manner very similar to the main group compound with electron-deficient bridges, as shown in **25**. In a similar manner, one also

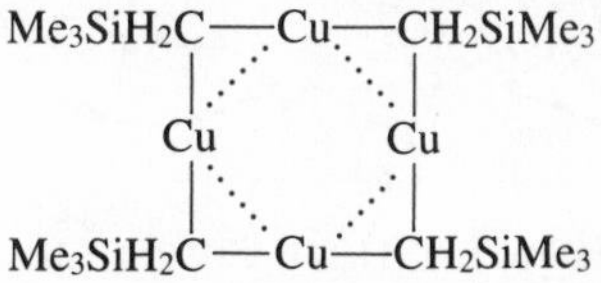

**25**

might include species such as $[LiMo(H)Cp_2]_4$, which contain lithium atoms bridging between two transition metal organometallic units[248], a feature which has so far been observed only for lithium, which appears to occupy a unique position with its ability to enter into many unusual structural types.

The other group of compounds which have been investigated to some extent and appear to be of major interest are those involving aluminium species. A few examples include the cyclopentadiene-bridged compound $Cp_2Ti$—$\mu$-H—$\mu$-$CpAlMe_3$, which contains both hydrogen and carbon bridging atoms between the two metal centres[249], the complex species $Cp_2Ti$—$\mu$-Cl—$\mu$-$CH_2AlMe_2$ with a proposed structure with halogen and methylene bridges[250], and may even be extended to include complexes such as $Cp_2Ti$—$\mu$-H—$\mu$-H—$AlR_2$—$\mu$-H)TiCp with a proposed structure containing six-membered rings with three metal and three hydrogen atoms, as shown in Figure 18[251]. There have even been some simple calculations in which the bonding in the mixed aluminium—lithium—carbon bridged systems are discussed[252].

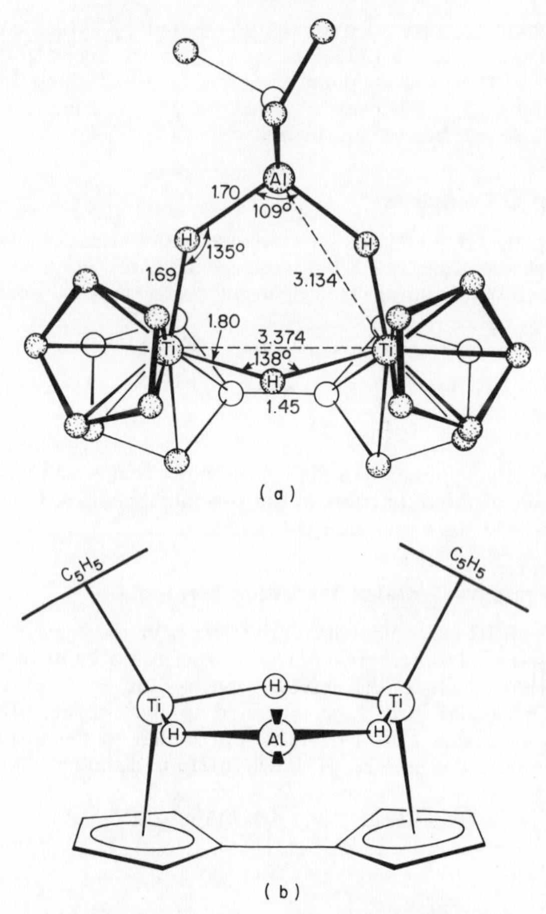


FIGURE 18. (a) Molecular structure of $[\{(C_5H_5)Ti\}_2Z(H)(H_2AlEt_2)(C_{10}H_8)]$ (hydrogen atoms omitted except for the hydride hydrogens). (b) Side view of the molecule. Reproduced with permission from *J. Am. Chem. Soc.*, **95,** 7870 (1973). Copyright 1973 American Chemical Society.

## D. Group III Addition Compounds

The relatively weak M—C—M bridge bonds observed in the aluminium derivatives and the availability of the vacant orbital in the heavier Group III derivatives makes the formation of Lewis basis addition complexes favourable, and many of these have been described. The structure of simple $R_3AlNR_3$ complexes have been determined in both the gas phase and the solid state. The structures of these in general have three-fold or pseudo-three-fold symmetry about the Al—N axis. A similar situation is observed for Al—O bonded species where the ether oxygen serves as the basic molecule. A similar situation presumably occurs for other Lewis bases such as $PPh_3$, but few data are available on these systems. The parameters of most interest in describing these systems are the M—base distance and the C—M—C or C—M—base angles, which will be sensitive to both the electronic effects (or base strength) and to the steric effects of the base and of the functional groups on the metal. The metal—carbon distances are also of interest, but their lack of sensitivity to the formation of the adducts and the relatively imprecise data make this parameter less useful. A number of the data related to these systems are collected in Table 5, which includes neutral Al—N compounds and Al—O bonded derivatives as well as heavier species involving gallium and indium, and show that these derivatives are readily formed with minimal distortion from the expected three-fold symmetry.

A second type of addition compound, which is directly related to the 'ate' derivatives discussed in Section V.A, are their complexes with anions such as $CN^-$. A number of these, including $K[Me_3AlCN]$[253], $K[Me_3AlN_3]$[254], the dimer $[(Me_3Al)_2N_3]$[255], the monomeric species $K[Me_3AlN_3]$[254] and the unusual hydride-bridged species $Na[Me_3Al—H—AlMe_3]$[256], which contains a linear Al—H—Al bridge, have been described. The last complex is of particular interest, since it precludes any significant metal—metal interaction, but yields a stable single hydrogen-bridged molecule. A related molecule, $[K\text{-dibenzo-18-crown-6}][Al_2Me_6O_2]\text{-}1{\cdot}5C_6H_6$, also has been described with an oxygen bridge between the two aluminium atoms, as shown in **26**[257].

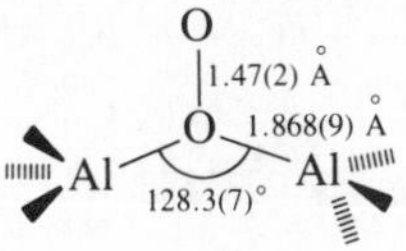

O
1.47(2) Å
O 1.868(9) Å
Al Al
128.3(7)°

**26**

TABLE 5. Carbon—metal bond distances and C—M—C bond angles for selected Group III addition compounds

| Compound | M—B distances (Å) | Angle (°) C—M—C | Ref. |
|---|---|---|---|
| Quinuclidine·$AlMe_3$ (solid) | 2.06 | 114 | *a* |
| $Me_3NAlMe_3$ (gas) | 2.099 | 114.8 | *b* |
| $Me_3NAlMe_2I$ (solid) | 2.02 | 120 | c |
| | | 110 (I—Al—C) | |
| $Me_3NAlCl_3$ (gas) | 1.945 | — | *d* |
| $Me_3NAlH_3$ (gas) | 2.063 | — | *e* |
| $MeCNAlMe_3$ (solid) | 2.02 | 116 | *f* |
| $Et_2OAl(o\text{-tol})_3$ (solid) | 1.928 (3) | 114.4 | g |
| $Me_2OAlMe_3$ | 2.014 | 117.8 | *h* |
| Dioxane $(AlMe_3)_2$ | 2.02 | 116.8 | *i* |
| $Me_3GaNMe_3$ | 2.20 | 116.7 | *j* |
| $Me_3GaPMe_3$ | 2.52 | 117.9 | *j* |

[a] C. D. Whitt, L. M. Parker, and J. L. Atwood, *J. Organomet. Chem.*, **32,** 291 (1971).
[b] G. A. Anderson, F. R. Forgaard, and A. Haaland, *Acta. Chem. Scand.*, **26,** 1947 (1972).
[c] J. L. Atwood and P. A. Milton, *J. Organomet. Chem.*, **36,** C-1 (1972); J. L. Atwood and P. A. Milton, *J. Organomet. Chem.*, **52,** 275 (1973).
[d] A. Almenningen, A. Haaland, T. Haugen, and D. P. Novak, *Acta. Chem. Scand.*, **27,** 1821 (1973).
[e] A. Almenningen, G. Gundersen, T. Haugen, and A. Haaland, *Acta. Chem. Scand.*, **26,** 3928 (1972).
[f] J. L. Atwood, S. K. Seale, and D. H. Roberts, *J. Organomet. Chem.*, **51,** 105 (1973).
[g] M. Barber, D. Liptak, and J. P. Oliver, *Organometallics*, **1,** 1307 (1982).
[h] A. Haaland, S. Samdal, O. Stokkeland, and J. Weidlein, *J. Organomet. Chem.*, **134,** 165 (1977).
[i] J. L. Atwood and G. D. Stucky, *J. Am. Chem. Soc.*, **89,** 5362 (1967).
[j] L. M. Golubinskaya, A. V. Golubinskii, V. S. Mastrgukov, L. V. Vilkov, and V. I. Bregadze, *J. Organomet. Chem.*, **117,** C-4 (1976).

### E. Heteroatom Bridged Complexes

Group III organometallic compounds tend to form a variety of bridged complexes with other atoms or groups in the bridging sites while retaining normal $\sigma$-bound carbon atoms in terminal positions. In some instances there are two types of bridging units in the same molecule. The best characterized of these is $Me_5Al_2NPh_2$, which has been studied by both X-ray crystallography[258,259] and n.m.r. spectroscopy[260]. It has been shown, from these studies, that interchange may occur between the bridging and terminal methyl groups without disruption of the Al—N—Al bridge. This clearly shows the difference in stability of the carbon and nitrogen bridges present in this system. This also appears to be true for the related complex $Me_5Al_2N(SiMe_3)_2$ [261] and for several other oxygen- and nitrogen-bridged systems which have been studied by means of n.m.r. spectroscopy but have not been further characterized.

The most common problem in dealing with the species containing mixed bridging units is the tendency to symmetrize to what appears to be the more stable species via equation 1. Many of these derivatives have been characterized and include the dimeric species such as bis($\mu$-isopropylamidodimethylaluminium) shown in **27**, which

$$2[R_2M(\mu\text{-}R)(\mu\text{-}R')MR_2] \longrightarrow [R_2M(\mu\text{-}R')_2MR_2 + [R_2M(\mu\text{-}R)_2MR_2] \qquad (1)$$

**27**

has been shown to crystallize in two forms with the iso-propyl groups in *cis* and *trans* positions[262]. Other species of this general form have been described and have the expected bridged structures with no unusual properties[263,264]. It has been shown that under appropriate conditions trimeric species are formed[265,266]. The conditions required appear to be limited steric interferences from the alkyl groups on the nitrogen. An alternative structural form has been observed for sulphur bridges in which infinite chains of $AlMe_2$ units are bound by SMe[267], whereas while in the case of $F^-$ bridging units it has been shown that a stable cyclic tetramer with alternating aluminium and fluorine atoms is formed[268,269]. Similar systems clearly exist with the heavier derivatives with both dimeric[270] and tetrameric[271] species established for gallium compounds. Although significantly less information is available and no mixed C—N bridged systems have yet been characterized.

Other combinations also are possible with two different metals present. The only one so far characterized is the complex $Mg[Al(\mu\text{-}OMe)_2Me_2]_2 \cdot 2C_4H_8O_2$ [272], which has magnesium surrounded by four bridging MeO groups and two oxygen atoms from dioxane molecules. The Al—C bonds are of normal length. The structure of this molecule, however, indicates the variety of possibilities available.

## VI. REFERENCES

1. A. W. Laubengayer and W. F. Gilliam, *J. Am. Chem. Soc.*, **63,** 477 (1941).
2. R. E. Rundel, *J. Am. Chem. Soc.*, **69,** 1327 (1947).
3. W. N. Lipscomb, Jr., *Boron Hydrides,* W. A. Benjamin, New York, 1963.

4. J. P. Oliver, *Adv. Organomet. Chem.*, **15,** 235 (1977).
5. A. Haaland, *Top. Curr. Chem.*, No. 53, 1 (1975).
6. D. S. Matteson, *Organometallic Reaction Mechanisms,* Academic Press, New York, 1974.
7. G. E. Coates, M. L. J. Green, and M. Wade *Organometallic Compounds, Vol. 1, The Main Group Elements,* Methuen, London, 1968.
8. D. R. Armstrong and P. G. Perkins, *Coord. Chem. Rev.*, **38,** 139 (1981).
9. T. L. Brown, *Adv. Organomet. Chem.* **3,** 365 (1965).
10. T. L. Brown, *Rev. Pure Appl. Chem.*, **23,** 447 (1970).
11. L. D. McKeever, in *Ions and Ion Pairs in Organic Reactions* (Ed. M. Szwarc), Vol. 1, Wiley, New York, 1972.
12. B. J. Wakefield, *Organolithium Compounds,* Pergamon Press, Oxford, 1974.
13. A. F. Halasa, D. N. Schulz, D. P. Tate, and V. D. Mochel, *Adv. Organomet. Chem.*, **18,** 55 (1980).
14. E. Weiss and E. A. C. Lucken, *J. Organomet. Chem.*, **2,** 197 (1964).
15. E. Weiss and G. Hencken, *J. Organomet. Chem.*, **21,** 265 (1970).
16. H. Dietrich, *Acta Crystallogr.*, **16,** 68 (1963).
17. H. Dietrich, *J. Organomet. Chem.*, **205,** 291 (1981).
18. R. Zerger, W. Rhine, and G. Stucky, *J. Am. Chem. Soc.*, **96,** 6048 (1974).
19. T. F. Schaaf, W. Butler, M. D. Glick, and J. P. Oliver, *J. Am. Chem. Soc.*, **96,** 7593 (1974).
20. W. H. Ilsley, T. F. Schaaf, M. D. Glick, and J. P. Oliver, *J. Am. Chem. Soc.*, **102,** 3769 (1980).
21. J. P. Oliver and B. Teclé, unpublished observations.
22. I. Craubner, *Z. Phys. Chem.*, **51,** 225 (1966).
23. H. M. M. Shearer, K. Wade, and G. Whitehead, *J. Chem. Soc., Chem. Commun.*, 943 (1979).
24. D. Margerison and J. D. Pont, *Trans. Faraday Soc.*, **67,** Pt. 2, 353 (1971).
25. V. I. Korsunsky, M. B. Taraban, T. V. Leshina, O. I. Margarskaya, and N. S. Vyazankin, *J. Organomet. Chem.*, **215,** 179 (1981).
26. P. West and R. West, *J. Am. Chem. Soc.*, **89,** 4395, (1967).
37. J. Berkowitz, D. A. Bafus, and T. L. Brown, *J. Phys. Chem.*, **65,** 1380 (1961).
28. W. McLean, J. A. Schultz, L. G. Pedersen, and R. C. Jarnagin, *J. Organomet. Chem.*, **175,** 1 (1979).
29. G. R. Peyton and W. H. Glaze, *Theor. Chim. Acta,* **13,** 259 (1969).
30. A. H. Cowley and W. D. White, *J. Am. Chem. Soc.*, **91,** 34 (1969).
31. M. F. Guest, I. H. Hillier, and V. R. Saunders, *J. Organomet. Chem.*, **44,** 59 (1972).
32. K. Ohkubo, H. Shimada, and M. Okada, *Bull. Chem. Soc. Jpn.*, **44,** 2025 (1971).
33. G. D. Graham, D. S. Merynick, and W. N. Lipscomb, *J. Am. Chem. Soc.*, **102,** 4572 (1980).
34. G. D. Graham, S. C. Richtsmeier, and D. A. Dixon, *J. Am. Chem. Soc.*, **102,** 5759 (1980).
35. A. Streitwieser, Jr., *J. Organomet. Chem.*, **145,** 1 (1978).
36. A. Streitweiser, Jr., J. E. Williams, Jr., S. Alexandratas, and J. M. McKelvey, *J. Am. Chem. Soc.*, **98,** 4778 (1976).
37. W. M. Scovell, B. Y. Kimura, and T. G. Spiro, *J. Coord. Chem.*, **1,** 107 (1971).
38. T. L. Brown, *Acc. Chem. Res.*, **1,** 23 (1968).
39. J. P. Oliver, *Adv. Organomet. Chem.*, **8,** 167 (1970).
40. G. Fraenkel, A. M. Fraenkel, M. J. Geckle, and F. Schloss, *J. Am. Chem. Soc.*, **101,** 4745 (1979).
41. G. Fraenkel, M. Henrichs, J. M. Hewitt, B. M. Su, and M. J. Geckle, *J. Am. Chem. Soc.*, **102,** 3345 (1980).
42. J. P. Oliver and R. D. Thomas, *Tenth International Conference on Organometallics, Toronto, Ontario, August 1981.*
43. L. M. Seitz and T. L. Brown, *J. Am. Chem. Soc.*, **88,** 2174 (1966).
44. M. Y. Darensbourg, B. Y. Kimura, G. E. Hartwell, and T. L. Brown, *J. Am. Chem. Soc.*, **92,** 1236 (1970).
45. F. W. Wehrli, *J. Magn. Reson.*, **23,** 527 (1976).

46. F. W. Wehrli, *Org. Magn. Reson.*, **11,** 106 (1978).
47. F. W. Wehrli, *J. Magn. Reson.*, **30,** 193 (1978).
48. S. P. Patterman, I. L. Karle, and G. D. Stucky, *J. Am. Chem. Soc.*, **92,** 1150 (1970).
49. J. J. Brooks and G. D. Stucky, *J. Am. Chem. Soc.*, **94,** 7333 (1972).
50. J. J. Brooks, W. Rhine, and G. D. Stucky, *J. Am. Chem. Soc.*, **94,** 7346, (1972).
51. J. J. Brooks, W. Rhine, and G. D. Stucky, *J. Am. Chem. Soc.*, **94,** 7339 (1972).
52. W. E. Rhine and G. D. Stucky, *J. Am. Chem. Soc.*, **97,** 737 (1975).
53. M. Walczak and G. D. Stucky, *J. Am. Chem. Soc.*, **98,** 5531 (1976).
54. H. Köster and E. Weiss, *J. Organomet. Chem.*, **168,** 273 (1979).
55. M. F. Lappert, C. L. Raston, D. W. Skelton, and A. H. White, *J. Chem. Soc., Chem. Commun.*, 14 (1982).
56. S. Alexandratos, A. Streitwieser, Jr., and H. F. Schaefer, III, *J. Am. Chem. Soc.*, **98,** 7959 (1976).
57. R. P. Zerger and G. D. Stucky, *J. Chem. Soc., Chem. Commun.*, 44 (1973).
58. D. Thoennes and E. Weiss, *Chem. Ber.*, **111,** 3157 (1978).
59. H. Köster, D. Thoennes, and E. Weiss, *J. Organomet. Chem.*, **160,** 1 (1978).
60. H. Dietrich and D. Rewicki, *J. Organomet. Chem.*, **205,** 281 (1981).
61. B. Bogdanovic, C. Krüger, and B. Wermeckers, *Angew. Chem., Int. Ed. Engl.*, **19,** 817 (1980).
62. R. Amstutz, D. Seebach, P. Seiler, B. Schweitzer, and J. D. Dunitz, *Angew. Chem., Int. Ed. Engl.*, **19,** 53 (1980).
63. R. Amstutz, J. P. Dunitz, and D. Seebach, *Angew. Chem., Int. Ed. Engl.*, **20,** 465 (1981).
64. M. Walczak, K. Walczak, R. Mink, M. D. Rausch, and G. Stucky, *J. Am. Chem. Soc.*, **100,** 6382 (1978).
65. B. Teclé, W. H. Ilsey, and J. P. Oliver, *Organometallics*, **1,** 875 (1982).
66. P. D. Bartlett, C. V. Goebel, and W. P. Weber, *J. Am. Chem. Soc.*, **91,** 7425 (1969).
67. H. L. Lewis and T. L. Brown, *J. Am. Chem. Soc.*, **92,** 4664 (1970).
68. J. M. Catala, G. Clouet, and J. Brossas, *J. Organomet. Chem.*, **219,** 139 (1981).
69. Z. Latajka, H. Ratajczak, K. Romanowska, and Z. Tomczak, *J. Organomet. Chem.*, **131,** 347 (1977).
70. H. Siegel, K. Hiltbrunner, and D. Seebach, *Angew. Chem., Int. Ed. Engl.*, **18,** 785 (1979).
71. D. Seebach, H. Siegel, K. Müllen, and K. Hiltbrunner, *Angew. Chem., Int. Ed. Engl.*, **18,** 784 (1979).
72. T. Clark and P. v. R. Schleyer, *Tetrahedron Lett.*, 4963 (1979).
73. T. Clark and P. v. R. Schleyer, *J. Am. Chem. Soc.*, **101,** 7747 (1979).
74. E. Weiss and G. Sauermann, *J. Organomet. Chem.*, **21,** 1 (1970).
75. E. Weiss and G. Sauermann, *Chem. Ber.*, **103,** 265 (1970).
76. E. Weiss and H. Köster, *Chem. Ber.*, **110,** 717 (1977).
77. M. A. Ring and D. M. Ritter, *J. Am. Chem. Soc.*, **83,** 802 (1961).
78. E. Weiss, G. Henchen, and H. Kuhr, *Chem. Ber.*, **103,** 2868 (1970).
79. W. Priester, R. West, and T. L. Chwang, *J. Am. Chem. Soc.*, **98,** 8413 (1976), and reference cited therein.
80. L. A. Ship and R. J. Lagow, *J. Org. Chem.*, **44,** 2311 (1979), and references cited therein.
81. S. Nagase and K. Morokuma, *J. Am. Chem. Soc.*, **100,** 1661 (1978).
82. A. J. Kos, E. D. Jemmis, P. R. Schleyer, R. Gleiter, U. Fishbach, and J. A. Pople, *J. Am. Chem. Soc.*, **103,** 4996 (1981).
83. A. J. Kos and P. v. R. Schleyer, *J. Am. Chem. Soc.*, **102,** 7928 (1980).
84. W. D. Laidig and H. F. Schaefer, III, *J. Am. Chem. Soc.*, **101,** 7184 (1979)
85. E. D. Jemmis, J. Chandrasekhar, and P. v. R. Schleyer, *J. Am. Chem. Soc.*, **101,** 2848 (1979).
86. J. Chandrasekhar and P. v. R. Schleyer, *J. Chem. Soc., Chem. Commun.*, 260 (1980).
87. G. Rauscher, T. Clark, D. Poppinger, and P. v. R. Schleyer, *Angew. Chem., Int. Ed. Engl.*, **17,** 276 (1978).
88. A. I. Snow and R. E. Rundle, *Acta Crystallogr.*, **4,** 348 (1951).
89. E. Weiss, J. *Organomet. Chem.*, **2,** 314 (1964).

90. E. Weiss, *J. Organomet. Chem.*, **4,** 101 (1965).
91. A. Almennigen, A. Haaland, and J. E. Nilsson, *Acta Chem. Scand.*, **22,** 972 (1968).
92. A. Almenningen, A. Haaland, and G. L. Morgan, *Acta Chem. Scand,* **23,** 2921 (1969).
93. J. Mounier, *J. Organomet. Chem.*, **38,** 7 (1972).
94. E. C. Ashby, L. Fernholt, A. Halland, R. Seip, and R. S. Smith, *Acta Chem. Scand., Ser. A,* **35,** 213 (1980).
94a. R. A. Anderson and G. Wilkinson, *J. Chem. Soc., Dalton Trans.*, 809 (1977).
95. C. D. Whitt and J. L. Atwood, *J. Organomet. Chem.*, **32,** 17 (1971).
96. J. Toney and G. D. Stucky, *J. Organomet. Chem.*, **22,** 241 (1970).
97. T. Greiser, J. Kopf, D. Thoennes, and E. Weiss, *J. Organomet. Chem.*, **191,** 1, (1980).
98. D. Thoennes and E. Weiss, *Chem. Ber.*, **111,** 3381 (1978).
99. J. Ducom, *C. R. Acad. Sci., Ser. C,* **268,** 1259 (1969).
100. J. Kess, *J. Organomet. Chem.*, **111,** 1 (1976).
101. B. Morosin and J. Howatson, *J. Organomet. Chem.*, **29,** 7 (1971).
102. N. A. Bell, I. W. Nowell, and H. M. M. Shearer, *J. Chem. Soc., Chem. Commun.*, 147 (1982).
103. D. S. Marynick, *J. Am. Chem. Soc.*, **103,** 1328 (1981).
104. M. A. Ratner, J. W. Moskowitz, and S. Topiol, *J. Am. Chem. Soc.*, **100,** 2329 (1978).
105. K. Ohkubo and F. Watanabe, *Bull. Chem. Soc. Jpn.*, **44,** 2867 (1971).
106. M. Astier and P. Millie, *J. Organomet. Chem.*, **31,** 139 (1971).
107. W. C. Swope and H. F. Schaefer, III, *J. Am. Chem. Soc.*, **98,** 7962 (1976).
108. J. S. Binkley, R. Seeger, and J. A. Pople, *Theor. Chim. Acta,* **45,** 69 (1977).
109. A. Streitwieser, Jr., and J. E. Williams, Jr., *J. Organomet. Chem.*, **156,** 33 (1978).
110. R. A. Kovar and G. L. Morgan, *J. Am. Chem. Soc.*, **91,** 7269 (1969).
111. G. D. Stucky and R. E. Rundle, *J. Am. Chem. Soc.*, **85,** 1002 (1963).
112. G. D. Stucky and R. E. Rundle, *J. Am. Chem. Soc.*, **86,** 5344 (1964).
113. L. J. Guggenberg and R. E. Rundle, *J. Am. Chem. Soc.*, **86,** 5344 (1964).
114. G. D. Stucky and J. D. Toney, *J. Organomet. Chem.*, **28,** 5 (1978).
115. M. Vallino, *J. Organomet. Chem.*, **20,** 1 (1969).
116. J. Kress and A. Novak, *J. Organomet. Chem.*, **99,** 23 (1975).
117. J. Kress and A. Novak, *J. Organomet. Chem.* **99,** 199 (1975).
118. J. Toney and G. D. Stucky, *J. Organomet. Chem.*, **28,** 5 (1971).
119. V. R. Magneson and G. D. Stucky, *Inorg. Chem.*, **8,** 1427 (1969).
120. A. Almenningen, O. Bastiansen, and A. Haaland, *J. Chem. Phys.*, **40,** 3434 (1964).
121. A. Haaland, *Acta Chem. Scand.*, **22,** 3030 (1968).
122. A. Almenningen, G. A. Anderson, F. R. Forgaard, and A. Haaland, *Acta Chem. Scand.*, **26,** 2315 (1972).
123. R. Schneider and E. O. Fischer, *Naturwissenschaften,* **50,** 349 (1963).
124. C. Wong, T. Y. Lee, T. W. Chang, and C. S. Liu, *Inorg. Nucl. Chem. Lett.*, **9,** 667, (1973).
125. M. J. S. Dewar and H. S. Rzepa, *J. Am. Chem. Soc.*, **100,** 777 (1978).
126. D. S. Marynick, *J. Am. Chem. Soc.*, **99,** 1436 (1977).
127. N. S. Chiu and L. Schafer, *J. Am. Chem. Soc.*, **100,** 2604 (1978).
128. E. D. Jemmis, S. Alexandratos, P. v. R. Schleyer, A. Streitwieser, Jr., and H. F. Schaefer, III, *J. Am. Chem. Soc.*, **100,** 5695 (1978).
129. C. Glidewell, *J. Organomet. Chem.*, **217,** 273 (1981).
130. J. R. Bews and C. Glidewell, *J. Organomet. Chem.*, **219,** 279 (1981).
131. O. P. Charkin, A. Veillard, J. Demuynck, and M. M. Rohmer, *Koord. Chim.*, **5,** 501 (1979).
132. A. Almenningen, A. Haaland, and J. Lusztyk, *J. Organomet. Chem.*, **170,** 271 (1979).
133. R. Gleiter, M. C. Böhn, A. Haaland, R. Johansen, and J. Lusztyk, *J. Organomet. Chem.*, **170,** 285 (1979).
134. W. Bünder and E. Weiss, *J. Organomet. Chem.*, **92,** 1 (1975).
135. A. Haaland, J. Lusztyk, J. Brunvoll, and K. B. Starowieyaki, *J. Organomet. Chem.*, **85,** 279 (1975).
136. S. Evans, M. L. H. Green, B. Jewitt, A. F. Orchard, and C. F. Pygall, *J. Chem. Soc., Faraday Trans. 2,* **68,** 1847 (1972).
137. J. Lusztyk and K. B. Starowieyaki, *J. Organomet. Chem.*, **170,** 293 (1979).

138. D. A. Drew and A. Haaland, *Chem. Commun.*, 1551 (1971).
139. D. A. Drew and A. Haaland, *Acta Chem. Scand.*, **26,** 3079 (1972).
140. D. A. Drew and A. Haaland, *Acta Chem. Scand.*, **26,** 3351 (1972).
141. D. A. Drew, G. Gundersen, and A. Haaland, *Acta Chem. Scand.*, **26,** 2147 (1972).
142. A. Haaland and D. P. Novak, *Acta Chem. Scand., Ser. A,* **28,** 153 (1974).
143. T. C. Bartke, A. Bjorseth, A. Haaland, K.-M. Marstokk, and H. Møllendal, *J. Organomet. Chem.*, **85,** 271 (1975).
144. M. C. Böhm, R. Gleiter, G. L. Morgan, J. Lusztyk, and K. B. Starowieyski, *J. Organomet. Chem.*, **194,** 257 (1980).
145. D. F. Gaines, K. M. Coleson, and J. C. Calabrese, *Inorg. Chem.*, **20,** 2185 (1981).
146. D. F. Gaines, K. M. Coleson, and J. C. Calabrese, *J. Am. Chem. Soc.*, **101,** 3979 (1979).
147. J. Bicerano and W. N. Lipscomb, *Inorg. Chem.*, **18,** 1565 (1979).
148. L. J. Allamandola and J. W. Nibler, *J. Am. Chem. Soc.*, **98,** 2096 (1976).
149. C. Johnson, T. Toney, and G. D. Stucky, *J. Organomet. Chem.*, **40,** C11 (1972).
150. D. A. Drew and A. Haaland, *Acta Chem. Scand.* **26,** 3074 (1972).
151. M. J. S. Dewar and H. S. Rzepa, *Inorg. Chem.*, **18,** 602 (1979).
152. D. S. Marynick and W. N. Lipscomb, *Inorg. Chem.*, **11,** 820 (1972).
153. A. Almenningen, A. Gunderson, and A. Haaland, *Acta Chem. Scand.*, **22,** 859 (1968).
154. D. S. Marynick, *J. Am. Chem. Soc.*, **101,** 6876 (1979).
155. Yu. B. Kirillov, A. I. Boldyrev, and N. M. Kliminko, *Koord. Khim.*, **6,** 1503 (1980).
156. C. Trindle and S. N. Datta, *Proc. Indian Acad. Sci., Sect. Chem. Sci.*, **89,** 175 (1980).
157. D. F. Gaines and J. L. Walsh, *Inorg. Chem.*, **17,** 1238 (1978).
158. D. F. Gaines, J. L. Walsh, and J. C. Calabrese, *Inorg. Chem.*, **17,** 1242 (1978).
159. T. Mole and E. A. Jeffery, *Organoaluminum Compounds,* Elsevier, Amsterdam, 1972.
160. J. P. Oliver, *Adv. Organomet. Chem.*, **8,** 167 (1970).
161. J. P. Oliver, *Adv. Organomet. Chem.*, **16,** 111 (1977).
162. J. P. Oliver and K. L. Henold, in *Organometallic Reactions,* (Eds. E. Becker and M. Tsutsui), Vol. 5, Wiley, New York, 1975, p. 387.
163. H. C. Longuet-Higgins, *J. Chem. Soc.*, 139 (1946).
164. P. H. Lewis and R. E. Rundle, *J. Chem. Phys.*, **21,** 986 (1953).
165. R. G. Vranka and E. L. Amma, *J. Am. Chem. Soc.*, **89,** 3121 (1967).
166. J. C. Hoffman and W. E. Streib, *J. Chem. Soc., D,* 911 (1971).
167. A. Almenningen, S. Halvorsen, and A. Haaland, *Acta Chem. Scand.*, **25,** 1937 (1971).
168. S. K. Byram, J. K. Fawcett, S. C. Nyberg, and R. J. O'Brien, *J. Chem. Soc., D,* 16 (1970).
169. F. A. Cotton, *Inorg. Chem.*, **9,** 2804 (1970).
170. M. J. S. Dewar and D. B. Patterson, *Chem. Commun.*, 544 (1970).
171. M. J. S. Dewar, D. B. Patterson, and W. T. Simpson, *J. Chem. Soc., Dalton Trans.*, 2381 (1973).
172. K. A. Levison and P. G. Perkins, *Discuss. Faradary Soc.*, No. 47, 183 (1969).
173. K. A. Levison and P. G. Perkins, *Theor. Chim. Acta,* **17,** 1 (1970).
174. K. A. Levison and P. G. Perkins, *Theor. Chim. Acta,* **17,** 15 (1970).
175. M. B. Smith, *J. Organomet. Chem.*, **22,** 273 (1970).
176. M. B. Smith, *J. Phys. Chem.*, **76,** 2933 (1972).
177. E. L. Amma and R. E. Rundle, *J. Am. Chem. Soc.*, **80,** 4141 (1958).
178. N. Muller and A. L. Otermat, *Inorg. Chem.*, **2,** 1075 (1963).
179. B. Beagley, D. G. Schmidling, and I. A. Steer, *J. Mol. Struct.*, **21,** 437 (1974).
180. T. Fieldberg, A. Haaland, R. Seip, Q. Shen, and J. Weidlein, *Acta Chem. Scand., Ser. A,* **36,** 495 (1982); see also G. G. Barbe, J. L. Hencher, A. Shen, and D. Tuck, *Can. J. Chem.*, **52,** 3936 (1974).
181. A. J. Carty, M. J. S. Gynane, M. F. Lappert, S. J. Miles, A. Singh, and N. J. Taylor, *Inorg. Chem.*, **19,** 3637 (1980).
182. W. S. McDonald and J. F. Malone, *J. Chem. Soc., A,* 3362 (1970).
183. D. A. Sanders and J. P. Oliver, *J. Am. Chem. Soc.*, **90,** 5910 (1968).
184. J. W. Moore, D. A. Sanders, P. A. Scherr, M. D. Glick, and J. P. Oliver, *J. Am. Chem. Soc.*, **93,** 1035 (1971).
185. D. A. Sanders, P. A. Scherr, and J. P. Oliver, *Inorg. Chem.*, **15,** 861 (1976).
186. R. D. Thomas and J. P. Oliver, *Organometallics,* **1,** 571 (1982).

187. J. P. Oliver and L. G. Stevens, *Inorg. Nucl. Chem.*, **24,** 953 (1962).
188. H. D. Visser and J. P. Oliver *J. Am. Chem. Soc.*, **90,** 3579 (1968).
189. H. D. Visser and J. P. Oliver, *J. Organomet. Chem.*, **49,** 7 (1972).
190. W. Fries, K. Sille, J. Weidlein, and A. Haaland, *Spectrochim. Acta, Part A,* **36,** 611 (1980).
191. E. A. Jeffrey, T. Mole, and J. K. Saunders, *Chem. Commun.*, 696, (1967).
192. E. A. Jeffery, T. Mole, and J. K. Saunders, *Aust. J. Chem.*, **21,** 137 (1968).
193. T. Mole and J. K. Surtees, *Chem. Ind. (London),* 1727 (1963).
194. G. Wilkie and W. Schneider, *Bull. Soc. Chim. Fr.*, 1462 (1963).
195. E. A. Jeffery and T. Mole, *J. Organomet. Chem.*, **11,** 393, (1968).
196. M. J. Albright, W. M. Butler, T. J. Anderson, M. D. Glick, and J. P. Oliver, *J. Am. Chem. Soc.*, **98,** 3995 (1976).
197. T. Fieldberg, A. Haaland, R. Seip, and J. Weidlain, *Acta Chem. Scand.*, 637 (1981).
198. J. F. Malone and W. S. McDonald, *J. Chem. Soc., Dalton* Trans., 2646 (1972).
199. J. F. Malone and W. S. McDonald, *J. Chem. Soc., Dalton Trans.*, 2649 (1972).
200. W. H. Ilsey, M. D. Glick, J. P. Oliver, and J. W. Moore, *Inorg. Chem.*, **19,** 3572 (1980).
201. G. A. Olah, G. K. S. Prakash, G. Liang, K. L. Henold, and G. B. Haigh, *Proc. Natl. Acad. Sci. USA,* **74,** 5217 (1977).
202. G. D. Stucky, A. M. McPherson, W. E. Rhine, J. J. Eisch, and J. L. Considine, *J. Am. Chem. Soc.,* **96,** 1941 (1974).
203. T. Fjeldberg, A. Haaland, R. Seip, and J. Weidlein, *Acta Chem. Scand., Ser. A,* **35,** 437 (1981).
204. B. Teclé, W. H. Ilsey, and J. P. Oliver, *Inorg. Chem.*, **20,** 2335, (1980).
205. W. Fries, W. Schwartz, H. D. Hausen, and J. Weidlein, *J. Organomet. Chem.*, **159,** 373 (1978).
206. A. Almenningen, L. Fernholt, and A. Haaland, *J. Organomet. Chem.*, **155,** 245 (1978).
207. B. Teclé, P. W. R. Corfield, and J. P. Oliver, *Inorg. Chem.*, **21,** 458 (1982).
208. K. Mentz, F. Zetter, H. D. Hausen, and J. Weidlein, *J. Organomet. Chem.*, **122,** 159 (1976).
209. F. W. B. Einstein, M. M. Gilbert, and D. G. Tuck *Inorg. Chem.*, **11,** 2832 (1972).
210. A. Haaland and J. Weidlein, *J. Organomet. Chem.*, **40,** 29 (1972).
211. D. A. Drew and A. Haaland, *Acta Chem. Scand.*, **27,** 3735 (1973).
212. P. R. Schonberg, R. T. Paine, C. F. Campana, and E. N. Duesler, *Organometallics,* **1,** 799 (1982).
213. P. R. Schonberg, R. T. Paine, and C. G. Campana, *J. Am. Chem. Soc.*, **101,** 7726 (1979).
214. H. Lehmkul and D. Reinehr, *J. Organomet. Chem.*, **23,** C25 (1970).
215. A. Stefani and P. Pino, *Helv. Chim. Acta,* **55,** 1110 (1972).
216. A. Stefani and P. Pino, *Helv. Chim. Acta,* **55,** 190 (1972).
217. A. Stefani, *Helv. Chim. Acta,* **56,** 1192 (1973).
218. J. J. Eisch and J.-M. Biedermann, *J. Organomet. Chem.*, **30,** 167, (1971).
219. A. F. M. Maqsudar Rahman, K. F. Siddiqui, and J. P. Oliver, *Organometallics,* **1,** 881 (1982).
220. E. Weiss and E. Wolfrum, *J. Organomet. Chem.*, **12,** 257 (1968).
221. T. Greiser, T. Kopf, D. Thoennes, and E. Weiss, *Chem. Ber.*, **114,** 209 (1981).
222. D. Thoennes and E. Weiss, *Chem. Ber.*, **111,** 3726 (1978).
223. D. Groves, W. Rhine, and G. D. Stucky, *J. Am. Chem. Soc.*, **93,** 1553 (1971).
224. R. L. Gerteis, R. E. Dickerson, and T. L. Brown, *Inorg. Chem.*, **3,** 872 (1964).
225. K. Hoffman and E. Weiss, *J. Organomet. Chem.*, **37,** 1 (1972).
226. J. L. Atwood and D. C. Hrncir, *J. Organomet. Chem.*, **61,** 43 (1973).
227. K. Mach, *J. Organomet. Chem.*, **2,** 410 (1964).
228. J. Yamamoto and C. A. Wilkie, *Inorg. Chem.*, **10,** 1129 (1971).
229. J. F. Ross and J. P. Oliver, *J. Organomet. Chem.*, **22,** 503 (1970).
230. T. D. Westmoreland, Jr., N. S. Bhacca, J. D. Wander, and M. C. Day, *J. Organomet. Chem.*, **38,** 1 (1972).
231. E. Weiss and R. Wolfrum, *Chem. Ber.*, **101,** 35 (1968).
232. E. Weiss and H. Plass, *J. Organomet. Chem.*, **14,** 21 (1968).
233. E. A. Sadurski, W. H. Ilsley, R. D. Thomas, M. D. Glick, and J. P. Oliver, *J. Am.*

*Chem. Soc.,* **100,** 7761 (1978).
234. M. J. Albright, T. F. Schaff, W. M. Butler, A. K. Hovland, M. D. Glick, and J. P. Oliver, *J. Am. Chem. Soc.,* **97,** 6261 (1975).
235. W. H. Ilsley, M. J. Albright, T. J. Anderson, M. D. Glick, and J. P. Oliver, *Inorg. Chem.,* **19,** 3577 (1980).
236. T. F. Schaaf, A. K. Hovland, W. H. Ilsey, and J. P. Oliver, *J. Organomet. Chem.,* **197,** 169 (1980).
237. T. F. Schaaf and J. P. Oliver, *J. Am. Chem. Soc.,* **91,** 4327 (1969).
238. L. M. Seitz and S. D. Hall, *J. Organomet. Chem.,* **15,** 7 (1968).
239. J. L. Atwood and G. D. Stucky, *J. Am. Chem. Soc.,* **91,** 2538 (1969).
240. D. B. Malpass and F. W. Fannin, *J. Organomet. Chem.,* **93,** 1 (1975).
241. G. Van Koten and J. G. Noltes, *J. Organomet. Chem.,* **174,** 367 (1979).
242. G. Van Koten and J. G. Noltes, *J. Am. Chem. Soc.,* **101,** 6593 (1979).
243. J. San Fillippo, Jr., *Inorg. Chem.,* **17,** 275 (1978).
244. J. Blenkers, H. K. Hofstee, J. Boersma, and G. J. M. VanderKerk, *J. Organomet. Chem.,* **168,** 251 (1969).
245. J. A. J. Jarvis, R. Pearce, and M. F. Lappert, *J. Chem. Soc., Dalton Trans.,* 999 (1977).
246. J. G. Noltes, R. W. M. ten Hoedt, G. van Koten, A. L. Spek, and J. C. Schoone, *J. Organomet. Chem.,* **225,** 365 (1982).
247. G. van Koten, C. A. Schaap, J. T. B. H. Jastrebekski, and J. G. Noltes *J. Organmet. Chem.,* **186,** 427 (1980).
248. R. F. W. S. Benfield, R. A. Forder, M. L. H. Green, G. A. Moser, and K. Prout, *J. Chem. Soc., Chem. Commun.,* 759 (1973).
249. F. N. Tebbe and L. J. Guggenberger, *J. Chem. Soc., Chem. Commun.,* 227 (1973).
250. F. N. Tebbe, G. W. Parshall, and G. S. Reddy, *J. Am. Chem. Soc.,* **100,** 3611 (1978).
251. L. J. Guggenberger and F. N. Tebbe *J. Am. Chem. Soc.,* **95,** 7870 (1973).
252. D. R. Armstrong, P. G. Perkins, and J. J. P. Stewart, *J. Chem. Soc., Dalton Trans.,* 1972 (1972).
253. J. L. Atwood and R. E. Cannon, *J. Organomet. Chem.,* **47,** 321 (1973).
254. J. L. Atwood and W. R. Newberry, III, *J. Organomet. Chem.,* **87,** 1 (1975).
255. J. L. Atwood and W. R. Newberry, III, *J. Organomet. Chem.,* **42,** C77 (1972).
256. J. L. Atwood, D. C. Hrncir, R. D. Rogers, and J. A. K. Howard, *J. Am. Chem. Soc.,* **103,** 6789 (1981).
257. D. C. Hrncir, R. D. Rogers, and J. L. Atwood, *J. Am. Chem. Soc.,* **103,** 4277 (1981).
258. V. R. Magneson and G. D. Stucky, *J. Am. Chem. Soc.,* **90,** 3269 (1968).
259. V. R. Magneson and G. D. Stucky, *J. Am. Chem. Soc.,* **91,** 2514 (1969).
260. J. E. Rie and J. P. Oliver, *J. Organomet. Chem.,* **133,** 147 (1977).
261. N. Wiberg and W. Baumeister, *J. Organomet. Chem.,* **36,** 277 (1972).
262. S. Amirkhalili, P. B. Hitchcock, A. D. Jenkins, J. Z. Nyathi, and J. D. Smith, *J. Chem. Soc., Dalton Trans.,* 377 (1981).
263. H. Hess, A. Hindener, and S. Steinhauser, *Z. Anorg. Allg. Chem.,* **377,** 1 (1970).
264. S. K. Seale and J. L. Atwood, *J. Organomet. Chem.,* **73,** 27 (1974).
265. J. L. Atwood and G. D. Stucky, *J. Am. Chem. Soc.,* **92,** 285 (1970).
266. G. M. McLaughlin, G. A. Sim, and J. D. Smith, *J. Chem. Soc., Dalton Trans.,* 2197 (1972).
267. D. J. Brauer and G. D. Stucky, *J. Am. Chem. Soc.,* **91,** 5462 (1969).
268. G. Gunderson, T. Haugen, and A. Haaland, *Chem. Commun.,* 708 (1972).
269. G. Andersen, T. Haugen, and A. Haaland, *J. Organomet. Chem.,* **54,** 77 (1973).
270. W. R. Nutt, R. E. Stimson, M. F. Leopold, and B. H. Rubin, *Inorg. Chem.,* **21,** 1909 (1982).
271. K. R. Breakell, D. F. Rendle, A. Storr, and J. Trotter, *J. Chem. Soc., Dalton Trans.,* 1584 (1975).
272. J. L. Atwood and G. D. Stucky, *J. Organomet. Chem.,* **13,** 53 (1968).

# Author Index

This author index is designed to enable the reader to locate an author's name and work with the aid of the reference numbers appearing in the text. The page numbers are printed in normal type in ascending numerical order, followed by the reference numbers in parentheses. The numbers in *italics* refer to the pages on which the references are actually listed.

# Subject index